Sustainable Intensification for Agroecosystem Services and Management

Manoj Kumar Jhariya • Arnab Banerjee •
Ram Swaroop Meena • Sandeep Kumar •
Abhishek Raj
Editors

Sustainable Intensification for Agroecosystem Services and Management

Springer

Editors
Manoj Kumar Jhariya
Department of Farm Forestry
Sant Gahira Guru Vishwavidyalaya
Ambikapur, Chhattisgarh, India

Arnab Banerjee
Department of Environmental Science
Sant Gahira Guru Vishwavidyalaya
Ambikapur, Chhattisgarh, India

Ram Swaroop Meena
Department of Agronomy
Banaras Hindu University
Varanasi, Uttar Pradesh, India

Sandeep Kumar
ICAR-Indian Institute of Pulses
Research, Regional Station
Phanda, Bhopal, India

Abhishek Raj
School of Agriculture
Lovely Professional University
Phagwara, Punjab, India

ISBN 978-981-16-3206-8 ISBN 978-981-16-3207-5 (eBook)
https://doi.org/10.1007/978-981-16-3207-5

This Springer imprint is published by the registered company Springer Nature Singapore Pte Ltd.
The registered company address is: 152 Beach Road, #21-01/04 Gateway East, Singapore 189721, Singapore

Preface

Agroecosystem services are important to perform various nature-based functions for humankind. Agroecosystem services act as a regulator for maintaining soil and water quality, gradual build-up of soil carbon stock and pool, and has a rich heritage of agricultural biodiversity. Secondarily, agroecosystem regulates the nutrient runoff, sedimentation process, maintenance of fauna, and other essential services for the benefit of human civilization. Above all, it is the basic unit of production of food, fodder, bioenergy, and various herbal products for the benefit of humans. In the modern world, food security is the biggest challenge of the twenty-first century. Further agriculture stands for approximately 40% of the global land area coverage. Therefore, proper sustainable intensification and management of agroecosystem along with maintaining ecological harmony is the future perspective of agroecosystem management and development. As per the latest projection across the globe, more than 800 million people remain hungry in various countries. Further, intensification of more than 70% in the food production sector would be required to fulfill the demand for food for more than 10 billion people till 2050. The sustainability of the agroecosystem is under severe stress due to the motto of overproduction and resource utilization. High input agriculture gives rise to a high energy footprint, agricultural pollution, resource depletion, loss of agro-biodiversity, and human health. In this connection, sustainable intensification helps to achieve advanced food security and sustainable approaches through eco-friendly technologies for the sustainability of the agroecosystem. Mismanagement of agroecosystem may lead to loss of agro-biodiversity, eutrophication, sedimentation, soil degradation, and pest infestation leading to a loss in yield and productivity.

More studies need to be done on the particular issue entitled "Sustainable Intensification for Agroecosystem Services and Management." Textbooks are available in the global market that addresses specific issues on agriculture, its production, environmental consequences. The present title would integrate all the concepts into a single dimension from which various scientists, research scholars, academicians, and policymakers can be benefitted from updated information. New insights are very much important in this particular aspect as our very existence depends on the sustainability of the agroecosystem. The present title consists of chapters addressing the issue of sustainable intensification, agroecosystem services, agroecosystem management, and overall sustainability of the agroecosystem. The present book consists of some specific research case studies considering some components of agroecosystem such as crop residue management, technological management of the rice–wheat system, anaerobic digestate, biochar, climatic influence over fruit quality and agroecosystem health, and watershed sustainability. These would provide new insights into the field of agroecosystem management. Some titles update the reader about the current scenario on the issue of food security, sustainable intensification, resource conservation, eco-designing, agroecosystem sustainability and services, and soil and crop management. Therefore, the present title would help to address current issues and their management holistically. The objectives that will be fulfilled by the present title are as follows: (1) present context of agroecosystem and its problem, (2) identify the key areas of research in the field of sustainable intensification, (3) identify the agroecosystem services and their potential role for ecosystem sustainability, (4) aware the globe in this context so that future policies can be framed from this for the betterment of human civilization, and (5) address sustainable intensification for agroecosystem management and services. It would help the academicians, researchers, ecologists, environmentalists, students, capacity builders, and overall the policymakers to have in-depth knowledge in the diverse field. Eminent academicians and scientists across the globe would be invited related to the theme of the book to share their scientific innovation, research outputs, views, and opinions, an experience that would enlighten the academic community. Each of the chapters has good scientific support in terms of scientific database, diagram, graph, image, picture, and flowchart as per the requirement with proper recent updated citation. All the chapters would be thoroughly reviewed by the respective individual of specific discipline which would enrich the chapter content from a future research perspective. The submission would be reviewed by the editorial team for further upgradation. It would set a roadmap for the preparation of sustainability in agroecosystems in the future.

Ambikapur, Chhattisgarh, India — Manoj Kumar Jhariya
Ambikapur, Chhattisgarh, India — Arnab Banerjee
Varanasi, Uttar Pradesh, India — Ram Swaroop Meena
Phanda, Bhopal, India — Sandeep Kumar
Phagwara, Punjab, India — Abhishek Raj

Acknowledgement

This book formulated through the financial support for faculty under the scheme of institution of eminence (IoE) scheme No. 6031, Banaras Hindu University, Varanasi (UP)-221005, India.

Contents

1 Sustainable Intensification for Agroecosystem Services and Management: An Overview 1
Manoj Kumar Jhariya, Arnab Banerjee, Ram Swaroop Meena, Sandeep Kumar, and Abhishek Raj

2 Food and Nutrition Security in India Through Agroecology: New Opportunities in Agriculture System 37
Jayati Chakraborti, Saikat Mondal, and Debnath Palit

3 Cost-Effective and Eco-Friendly Agricultural Technologies in Rice-Wheat Cropping Systems for Food and Environmental Security 69
Akbar Hossain, Rajan Bhatt, Sukamal Sarkar, Manashi Barman, Debjyoti Majumder, Saikat Saha, Mst. Tanjina Islam, Sagar Maitra, and Ram Swaroop Meena

4 Eco-Designing for Soil Health and Services 97
Abhishek Raj, Manoj Kumar Jhariya, Nahid Khan, Arnab Banerjee, Poonam, Ram Swaroop Meena, and Shish Ram Jakhar

5 Climate Change and Integrated Coastal and Agroecosystem Services 135
Zied Haj-Amor and Salem Bouri

6 Climate Change Vulnerability and Agroecosystem Services 163
Arnab Banerjee, Manoj Kumar Jhariya, Shailesh Kumar Yadav, Nahid Khan, Abhishek Raj, Ram Swaroop Meena, and Taher Mechergui

7 Intensification for Agroecosystem Services 197
Abhishek Raj, Manoj Kumar Jhariya, Nahid Khan, Arnab Banerjee, P. R. Paikra, Ram Swaroop Meena, and Sandeep Kumar

8 Crop Residue Management: A Novel Technique for Restoring Soil Health and Sustainable Intensification in India 229
Anita Kumawat, Ram Swaroop Meena, I. Rashmi, Ashok Kumar, S. D. Bamboriya, Devideen Yadav, Kuldeep Kumar, Dinesh Kumar, and Manoj Kumar Jhariya

9 Agroecology for Agricultural Soil Management 267
N. C. Temegne, A. F. Ngome, A. P. Agendia, and E. Youmbi

10 Agroecology Towards Environmental Sustainability 323
Shailesh Kumar Yadav, Arnab Banerjee, Manoj Kumar Jhariya, Abhishek Raj, Nahid Khan, Ram Swaroop Meena, and Sandeep Kumar

11 Agroforestry and Its Services for Soil Management and Sustainability 353
Nahid Khan, Manoj Kumar Jhariya, Abhishek Raj, Arnab Banerjee, Ram Swaroop Meena, Surendra Singh Bargali, Shailesh Kumar Yadav, and Anita Kumawat

12 Agroecosystem Service Management and Environmental Sustainability 379
Abhishek Raj, Manoj Kumar Jhariya, Arnab Banerjee, Nahid Khan, Ram Swaroop Meena, Prabhat Ranjan Oraon, and Shailesh Kumar Yadav

13 Ecological Intensification for Soil Management: Biochar – A Natural Solution for Soil from Agricultural Residues 403
Hana'a Burezq and Merlin K. Davidson

14 Resource Conservation for Sustainable Development 457
Abhishek Raj, Manoj Kumar Jhariya, Nahid Khan, Arnab Banerjee, Poonam, Ram Swaroop Meena, and Kavita Rani

15 Anaerobic Digestate: A Sustainable Source of Bio-fertilizer 493
Sedrah Akbar, Safia Ahmed, Samiullah Khan, and Malik Badshah

16 Soil Fertility Status and Sugarcane Growth Performance in the Mangrove Ecosystem of Nigeria 543
J. E. K. Aroh

17 Managing Natural Resources Through Ecological Intensification in Oil-Rich Niger Delta 615
Aroloye O. Numbere

18 Climatic Risks on Fruit Quality, Health, and Livelihoods: A Nigerian Case of Rural Women in Fruit Farming Business 657
Angela Oyilieze Akanwa and Uchechukwu B. Okoli

19 Allelopathic Effect of *Taraxacum officinale* L. on Germination and Physiology of Wheat 711
Syeda Naila, Zahoor ul Haq, Abdulllah, and Abdul Salam

20 Watershed Sustainability for Agricultural Intensification 743
Surendra Kumar Chandniha, Mitthan Lal Kansal, and Diwakar Naidu

21 Impact of Climate Change on Insects and their Sustainable Management 779
I. Merlin Kamala and I. Isaac Devanand

22 Ecointensification in Agriculture Under Changing Climate 817
Papiya Nandi Palit

23 Management of Agroecosystem for Food Security: An Overview 847
Pallavi Chattopadhyay, Pinaki Chattopadhyay, and Debnath Palit

Editors and Contributors

About the Editors

Manoj Kumar Jhariya is an Assistant Professor at the Department of Farm Forestry, Sant Gahira Guru Vishwavidyalaya (formerly Sarguja University), Ambikapur (Chhattisgarh), India, and the author or co-author of more than 70 scientific papers, 12 books, 50 book chapters, and several extension articles. He acquired his B.Sc. (agriculture) and M.Sc. and Ph.D. (both in forestry) degrees from Indira Gandhi Krishi Vishwavidyalaya, Raipur, Chhattisgarh, India. He was awarded a UGC-RGNF Fellowship, New Delhi, India, University Topper Gold Medal for Ph.D. and received the Chhattisgarh Young Scientist Award from Chhattisgarh Council of Science and Technology in 2013. He is an editorial board member for several journals and a life member of numerous societies.

Arnab Banerjee is an Assistant Professor, Department of Environmental Science, Sant Gahira Guru Vishwavidyalaya, Ambikapur, Chhattisgarh, India. He has completed M.Sc. and Ph.D. (environmental science) from Burdwan University and M.Phil. in environmental science from Kalyani University, West Bengal. He won the University Gold Medal for securing first class first position in M.Sc. examination. He has been awarded Young Scientist Award for best oral presentation at the international conference held at the University of Burdwan. He was a project fellow under UGC-sponsored major research project. To his credit, he has published 70 research papers in reputed national and international journals, 11 books, and 30 book chapters. He is life member of the Academy of Environmental Biology. He has supervised 25 postgraduate students and engaged in postgraduate teaching and research.

Ram Swaroop Meena is working as an Assistant Professor in the Department of Agronomy, I.Ag. Scs., BHU, Varanasi (UP). He has been awarded Raman Research Fellowship by the MHRD, GOI. He has completed his postdoctoral research on soil carbon sequestration under Padma Shari Prof. Rattan Lal, World Food Prize 2020 Laureate, Director, CMASC, Columbus, USA. He has supervised 20 PG and 6 PhD students and has 10 years of research and teaching experience. He is working on the three externally funded projects (DST, MHRD, and ICAR) with one patent. He has published more than 110 research and review papers and has H-index 42, as well as 4 published books at the national and 16 books (Springer, Elsevier, etc.) at the international levels and contributed in the books with 15 chapters at national and 50 at the international levels. He has worked as an expert for the NCERT, MHRD, GOI. He has contributed to several agricultural extension activities, trainings, meetings, workshops, etc.

Sandeep Kumar is currently working as a Scientist (agronomy) at the ICAR-Indian Institute of Pulses Research, Regional Station, Bhopal, India. He is born in a farmer family in the village Jaswantpura, Tehsil Rajgarh, District Churu, Rajasthan, India. He had his schooling in the same village and graduated in agriculture in 2013 from the College of Agriculture, Lalsot (SKRAU, Bikaner), Rajasthan. He obtained his master from the Banaras Hindu University, Varanasi and the doctorate degree in agronomy from Chaudhary Charan Singh Haryana Agricultural University, Hisar, and securing first division in all the classes with triple NET, JRF, and SRF from the Indian Council of Agricultural Research. His research interests have focused on enhancing resources use efficiencies for sustainable crop production. He has published 5 books and over 35 papers in leading national and international peer-reviewed journals and has also contributed to more than 20 book chapters. He has received numerous national awards, including Young Scientist, Best Research Scholar, and Best Master's Thesis awards.

Abhishek Raj is working as an Assistant Professor at the School of Agriculture, Lovely Professional University, Jalandhar (Punjab), India, and the author or co-author of more than 20 scientific papers, 2 books, 30 book chapters, and several extension articles with an H-index ~15. He acquired his B.Sc. (forestry) from SHUATS, Allahabad (UP) and M.Sc. and PhD. (both in forestry) degrees from Indira Gandhi Krishi Vishwavidyalaya, Raipur (Chhattisgarh) India. He has qualified ICAR-NET (national eligibility test) in agroforestry and received Young Scientist Award in 2016.

Contributors

Abdulllah Department of Plant Sciences, Quaid-i-Azam University, Islamabad, Pakistan

A. P. Agendia Department of Plant Biology, Faculty of Science, University of Yaounde I, Yaounde, Cameroon

Safia Ahmed Sustainable Bioenergy and Biorefinery Lab, Department of Microbiology, Quaid-i-Azam University, Islamabad, Pakistan

Angela Oyilieze Akanwa Department of Environmental Management, Faculty of Environmental Sciences, Chukwuemeka Odumegwu Ojukwu University, Uli, Anambra, Nigeria

Sedrah Akbar Sustainable Bioenergy and Biorefinery Lab, Department of Microbiology, Quaid-i-Azam University, Islamabad, Pakistan

J. E. K. Aroh Niger Delta University, Wilberforce Island, Bayelsa State, Nigeria

Malik Badshah Sustainable Bioenergy and Biorefinery Lab, Department of Microbiology, Quaid-i-Azam University, Islamabad, Pakistan

S. D. Bamboriya ICAR–Indian Institute of Maize Research, Ludhiana, Punjab, India

Arnab Banerjee Department of Environmental Science, Sant Gahira Guru Vishwavidyalaya, Sarguja, Ambikapur, Chhattisgarh, India

Surendra Singh Bargali Department of Botany, DSB Campus, Kumaun University, Nainital, Uttarakhand, India

Manashi Barman Department of Agronomy, Bidhan Chandra KrishiViswavidyalaya (BCKV), Nadia, West Bengal, India

Rajan Bhatt Regional Research Station, Kapurthala, Punjab Agricultural University, Ludhiana, Punjab, India

Hana'a Burezq Desert Agriculture and Ecosystems Program, Environment and Life Sciences Research Center, Kuwait Institute for Scientific Research, Safat, Kuwait

Surendra Kumar Chandniha Soil and Water Engineering, BRSM Collage of Agricultural Engineering and Technology and Research Station, IGKV, Mungeli, Chhattisgarh, India

Merlin K. Davidson Desert Agriculture and Ecosystems Program, Environment and Life Sciences Research Center, Kuwait Institute for Scientific Research, Safat, Kuwait

Zahoor ul Haq Department of Botany, Shaheed Benazir Bhutto University, Wari Campus, Pakistan

Akbar Hossain Department of Agronomy, Bangladesh Wheat and Maize Research Institute, Dinajpur, Bangladesh

Mst. Tanjina Islam Department of Agronomy, Hajee Mohammad Danesh Science and Technology University (HSTU), Dinajpur, Bangladesh

Shish Ram Jakhar Department of Soil Science and Agricultural Chemistry, Jawaharlal Nehru Krishi Vishwa Vidyalaya, Jabalpur, Madhya Pradesh, India

Manoj Kumar Jhariya Department of Farm Forestry, Sant Gahira Guru Vishwavidyalaya, Sarguja, Ambikapur, Chhattisgarh, India

Mitthan Lal Kansal Department of Water Resources Development and Management, Indian Institute of Technology, Roorkee, India

Nahid Khan Department of Farm Forestry, Sant Gahira Guru Vishwavidyalaya, Sarguja, Ambikapur, Chhattisgarh, India
Department of Farm Forestry, University Teaching Department, Sant Gahira Guru Vishwavidyalaya, Ambikapur, India

Samiullah Khan Sustainable Bioenergy and Biorefinery Lab, Department of Microbiology, Quaid-i-Azam University, Islamabad, Pakistan

Ashok Kumar ICAR–Indian Institute of Soil and Water Conservation, RC, Kota, Rajasthan, India

Dinesh Kumar ICAR–Indian Institute of Soil and Water Conservation, RC, Datia, Madhya Pradesh, India

Kuldeep Kumar ICAR–Indian Institute of Soil and Water Conservation, RC, Kota, Rajasthan, India

Sandeep Kumar ICAR-Indian Institute of Pulses Research, Regional Station, Phanda, Bhopal, India

Anita Kumawat ICAR–Indian Institute of Soil and Water Conservation, RC, Kota, Rajasthan, India

Sagar Maitra Centurion University of Technology and Management, Paralakhemundi, India

Debjyoti Majumder Uttar BangaKrishiViswavidyalaya (UBKV), Cooch Behar, West Bengal, India

Taher Mechergui Faculté des Sciences de Bizerte, Laboratoire des Ressources Sylvo-Pastorales de Tabarka, Tabarka, Tunisie

Ram Swaroop Meena Department of Agronomy, Institute of Agricultural Sciences, Banaras Hindu University, Varanasi, Uttar Pradesh, India

Diwakar Naidu Computer Science, BRSM Collage of Agricultural Engineering and Technology and Research Station, IGKV, Mungeli, Chhattisgarh, India

Syeda Naila Department of Botany, University of Peshawar, Peshawar, Pakistan

A. F. Ngome Institute of Agricultural Research for Development (IRAD), Yaounde, Cameroon

Aroloye O. Numbere Department of Animal and Environmental Biology, University of Port Harcourt, Port Harcourt, Nigeria

Uchechukwu B. Okoli Department of Environmental Management, Faculty of Environmental Sciences, Chukwuemeka Odumegwu Ojukwu University, Uli, Anambra, Nigeria

Prabhat Ranjan Oraon Department of Silviculture and Agroforestry, Faculty of Forestry, Birsa Agricultural University, Ranchi, Jharkhand, India

P. R. Paikra Krishi Vigyan Kendra Balrampur, IGKV, Balrampur, Chhattisgarh, India

Papiya Nandi Palit Department of Geography, Barjora College, Bankura, West Bengal, India

Poonam Krishi Vigyan Kendra, Navgaon (Alwar-I), Sri Karan Narendra Agriculture University (SKNAU), Jobner, Rajasthan, India

Abhishek Raj School of Agriculture, Lovely Professional University, Phagwara, Punjab, India

Kavita Rani Department of Microbiology, CCS Haryana Agricultural University, Hisar, Haryana, India

I. Rashmi ICAR–Indian Institute of Soil and Water Conservation, RC, Kota, Rajasthan, India

Saikat Saha Nadia Krishi Vigyan Kendra, Bidhan Chandra KrishiViswavidyalaya (BCKV), Nadia, West Bengal, India

Abdul Salam Department of Plant Sciences, Quaid-i-Azam University, Islamabad, Pakistan
Zhejiang Key Lab of Crop Germplasm, Department of Agronomy, College of Agriculture and Biotechnology, Zhejiang University, Hangzhou, China

Sukamal Sarkar Department of Agronomy, Bidhan Chandra KrishiViswavidyalaya (BCKV), Nadia, West Bengal, India

N. C. Temegne Department of Plant Biology, Faculty of Science, University of Yaounde I, Yaounde, Cameroon

Devideen Yadav ICAR–Indian Institute of Soil and Water Conservation, Dehradun, Uttarakhand, India

Shailesh Kumar Yadav Department of Environmental Science, Sant Gahira Guru Vishwavidyalaya, Sarguja, Ambikapur, Chhattisgarh, India
Department of Environmental Science, University Teaching Department, Sant Gahira Guru Vishwavidyalaya, Ambikapur, India

E. Youmbi Department of Plant Biology, Faculty of Science, University of Yaounde I, Yaounde, Cameroon

Sustainable Intensification for Agroecosystem Services and Management: An Overview

1

Manoj Kumar Jhariya, Arnab Banerjee, Ram Swaroop Meena, Sandeep Kumar, and Abhishek Raj

Abstract

Researches in agriculture are the essence of well-being for human civilization. It is a technology that boosts up the growth and development of the society considering the environmental aspect. Nowadays, the agriculture sector is suffering from multidimensional problem in terms of agro-pollution, resource depletion, climatic vulnerability, and reduction in productivity and yield followed by imbalance in the homeostatic of agroecosystem. Therefore, the major aim is to reduce the environmental consequences and achieve sustainable yield for the well-being of human society. Maintaining sustainable yield, pollution reduction, and eco-friendly approaches along with nutritive food production are some of the major agroecosystem services and part of management which needs to be addressed, scientifically, technically, and sustainably. Moving through this path, one needs to recognize the intensification practices followed by traditional

M. K. Jhariya (✉)
Department of Farm Forestry, Sant Gahira Guru Vishwavidyalaya, Sarguja, Ambikapur, Chhattisgarh, India

A. Banerjee
Department of Environmental Science, Sant Gahira Guru Vishwavidyalaya, Sarguja, Ambikapur, Chhattisgarh, India

R. S. Meena
Department of Agronomy, Institute of Agricultural Sciences, Banaras Hindu University, Varanasi, Uttar Pradesh, India
e-mail: meenars@bhu.ac.in

S. Kumar
ICAR-Indian Institute of Pulses Research, Regional Station, Phanda, Bhopal, India
e-mail: sandeep.kumar5@icar.gov.in

A. Raj
School of Agriculture, Lovely Professional University, Phagwara, Punjab, India

M. K. Jhariya et al. (eds.), *Sustainable Intensification for Agroecosystem Services and Management*, https://doi.org/10.1007/978-981-16-3207-5_1

culture that help to maintain ecosystem resiliency of the agroecosystem under the peril of climatic perturbations. Approach should be such that it should address the issue of poverty, food crisis and security, and gender sensitization followed by international collaborations. In such approach, farmers and local dwellers are the main actors who should play a key role in adopting the latest techniques and methodology to create a food-secure and ecologically sustainable world. This should be supported by the scientific community at both national and international levels with their technical expertise to achieve all-round sustainability and services of agroecosystem.

Keywords

Agroecosystem · Intensification · Resources · Sustainable management · Sustainability

Abbreviations

C	Carbon
FAO	Food and Agriculture Organization
NR	Natural resource
SD	Sustainable development

1.1 Introduction

Globally, agriculture shares a big contribution for providing food to millions of people across the nations. In Indian perspectives, agriculture has a significant contribution to the country's economy, therefore designating it as an agro-based country. The case is similar for various other developing nations where the only income source may be agriculture. Now, the unprecedented growth of the human population is putting pressure on agricultural resources all year round making agroecosystem mostly unsustainable (FAO 2020; Jhariya et al. 2021a, b). For example, for more yield, non-judicious use of agrochemicals has depleted the soil quality and biota, having its impact on human health through food chain migration of toxicants as well as reduction in the productivity of the soil (Meena et al. 2020a, b, c). Another big problem is the availability of land for cultivation purpose due to different land uses. As a consequence of that, altered land use practices have depleted various resources apart from agroecosystem leading to mass degradation of the environment (Banerjee et al. 2020; Raj et al. 2020; Jhariya et al. 2019a, b).

In agriculture, the 2030 agenda addresses the various issues of food security and crisis at global level and is attempting climate resiliency of agroecosystem. This in turn would help to conserve the natural resource (NR). Sustainable food production system is the central theme for sustenance of human society (Kumar et al. 2020;

Meena et al., 2018). It is evident that without proper nutrition and adequate food, the problem of poverty and hunger cannot be resolved. It is, therefore, the need of the hour that we transform our agriculture practice from synthetic approach toward climate-resilient approach which would be able to maintain the environmental quality and feed the growing population. The contribution of the rural stakeholders is no doubt very important for sustenance of the society. As per the Food and Agriculture Organization (FAO 2017), more than 2/3 of the food production comes from these rural stakeholders who actively participate in cultivation practices. Therefore, the policies and strategies need to be framed to promote agricultural research and create an open market to fulfill the 2030 agenda (Morales 2007; Jhariya et al. 2019c, d).

The sustainability of the agroecosystem is under challenge in that in which direction agriculture production should go and whether we go for more production in a limited land area or we adopt intensification practices for increasing the yield and productivity (Raj et al. 2021). Historical background in Indian agriculture reveals that the advancement of technology and process of agriculture under the banner of green revolution has cost much in terms of genetic erosion and loss of traditional variety followed by soil fertility. In the agricultural process, water requirement has increased significantly in order to provide proper irrigation to crop plants. Now, the problem starts with degraded and contaminated aquifer that causes shortage of water followed by irrigation through contaminated water. This in turn impacts the human health.

Now from world perspective, drastic transformation in the agricultural production process has taken place as manual process has been replaced by technological process. This has led to unprecedented growth in the agricultural sector. It was observed that 5 billion amount of production has increased in the global agroecosystem since 1900 through such technological progress. But the problem is that the development rate is not uniform across various nations in the globe. It is very interesting to note that the developed nations have shown a very systematic, steady growth in agriculture production and on the other hand the developing world has produced a different picture of food security and crisis (FAO 2020). This may be attributed toward unscientific planning and unprecedented growth of the human population. The condition has become even worse when people are fighting among themselves for food as revealed in African continent. Further, no proper steps have been taken to mitigate the problems. Other aspects include prior considerations need to be given to maintain the ecosystem stability and assimilative capacity of the agroecosystem. It is a common fact that depletion of soil resource would put our existence into question mark. The major reason behind the unsustainable approach of agriculture production lies within the concept of more production in less amount of land with short span of time. Commercialization and business-oriented thinking of the people have also drastically affected the agricultural sector.

Several examples of unsustainable approach of agricultural system have been observed in Asian subcontinents. With the development of the technology of production, the amount of food produced has increased quite a fold with the rising population (Pinstrup-Anderson 2009; Meena and Lal 2018). The problem of agroecosystem footprint has increased at significant level due to such unsustainable practices (Banerjee et al. 2021a, b). The major impact includes loss of NR base leading to decline in crop production. From the Indian perspective, it was observed that in the states of Punjab and Maharashtra, there is a significant rise in the cancer incidence as well as suicidal attempts of the farming community. This is due to the higher use of agrochemicals, and lack of money has become a common incidence which has affected the population very much. Further, the problem has aggravated due to rapid urbanization, infrastructure growth, as well as amenities. In Kerala, it was observed that higher population density has put pressure on the land resource which has become incapable to feed the entire population of Keralian people. In the development process, the involvement of illegal people has severe negative consequences that lead to altered land use as well as loss of soil quality.

From the world food production scenario, malnutrition and undernourishment have become big issues. For example, 17 million cases of undernourishment have been recorded across the world within a span of single year (FAO 2018). As per the data of the World Bank, more than 2/3 of the world's population reside under rural setup, and most of them belong to the farming community (IFPRI 2002). The food system is developed in such a way that it leads to social injustice and inequality in terms of distribution of food and optimum nutrition for the local people. All these aspects become much more problematic when it is considered in the context of amount of food waste generated, yield used for animal husbandry management, as well as bioenergy production which may have negative impact on the production process at various levels. Climatic irregularities pose adverse conditions that lead to decline in productivity under tropical to sub-tropical conditions (FAO 2018; Parfitt et al. 2010). Realizing the problems of the agricultural sector, the concept of agricultural intensification was integrated with the agroecological techniques for sustainable production of food and with minimum damage to the environment (Simon et al. 2020; Kumar et al. 2021).

Agricultural sustainability is an important aspect in the food production system as it should include the issue of environmental security, public health, and soil fertility. The thrust of agricultural production has increased these three issues leading to the total mismanagement of agroecosystem. Improper fertilizer input practices have led to nutrient loss followed by degeneration of soil carbon (C) pool; subsequently, the productivity also declines (Kumar et al. 2020a). The productivity has declined up to a significant level which may be attributed toward improper nutrient management of the soil. This matter needs to be considered with utmost care as the global population would be reaching up to nine billion till 2050 (FAO 2020). Therefore, restoring soil quality through proper nutrient management is the need of the hour. Strategy and policy formulation would be based upon the present nutrient status and the processes associated with it (Jhariya et al. 2018a, b).

To overcome this situation, strategy formulation and policy initiatives have been made across the globe starting from local to international levels. In this context, strategies for food assistance supported by the respective authority have already being attempted to solve the problems (Tranchant et al. 2019). However, beside these attempts, they do not address the bigger issue of poverty and food crisis and security. Therefore, modification of the green revolution concept through incorporation of intensification approach within the food production system may raise the productivity in the developing world with advance sustainable technology and management (Pretty 2018).

Agroecological principles can be suitable alternative as new food production system. It is highly beneficial as it is oriented toward agroecosystem and deals with various dimensions. It can be considered as a part of sustainable agriculture which includes various practices that allow the process of production in a sustainable way. For example, the biointensive model used in Nicaragua in African subcontinent has given fruitful results in terms of agricultural production. The industry-based agricultural practice in the form of synthetic fertilizer application and monoculture practice is being altered through agroecological principles under biointensive model (Banerjee et al. 2021c, d). The major advantages of this model include benefit sharing to the local people in terms of food for consumption, it increases the economic gain of poor rural families, and it can be practiced under limited resources. Such approaches would help to overcome the limitations of industrial agricultural and increase the scope of sustainability through enhanced food production (Wezel et al. 2009; Rosset and Altieri 2017; Giraldo and Rosset 2017; Sheoran et al. 2021).

The present chapter deals with the sustainable intensification perspective and its role toward maintaining sustainability of agroecosystem which has been attempted including the bigger issues of food security and crisis, lack of nourishment, and nutrient balance of soil ecosystem.

1.2 Sustainable Intensification in Agroecosystem

In present times, the entire globe is suffering from the problem of population explosion leading to higher stress on agricultural productivity. The modified land use pattern in terms of agricultural activity seems to be a global threat toward agrobiodiversity and other NRs. This therefore creates the question of judicious land use pattern in integrated form. In the present context, intensification process in the agricultural sector seems to have land conservation approach, but it mismatches with the real situation. In the developing countries, the problem of land holding seems to be the major factor that regulates the issue of food security at global context. The problem is further aggravated through wastage of food resources along with consumption pattern of the livestock population. The conventional systems of intensification often jeopardize the associated cost to the environment. The environment-friendly approaches toward intensification seem to benefit the ecosystem services as well as maintain the agrobiodiversity. It is therefore a common fact that improper management of the agroecosystem increases the various incidences of pest infestation and disease outbreaks. In this way, the functionality of the agrobiodiversity seems to be misinterpreted in terms of conventional intensification. Inter-linkage between intensification and biodiversity conservation demands proper policy and planning which is yet to be achieved properly

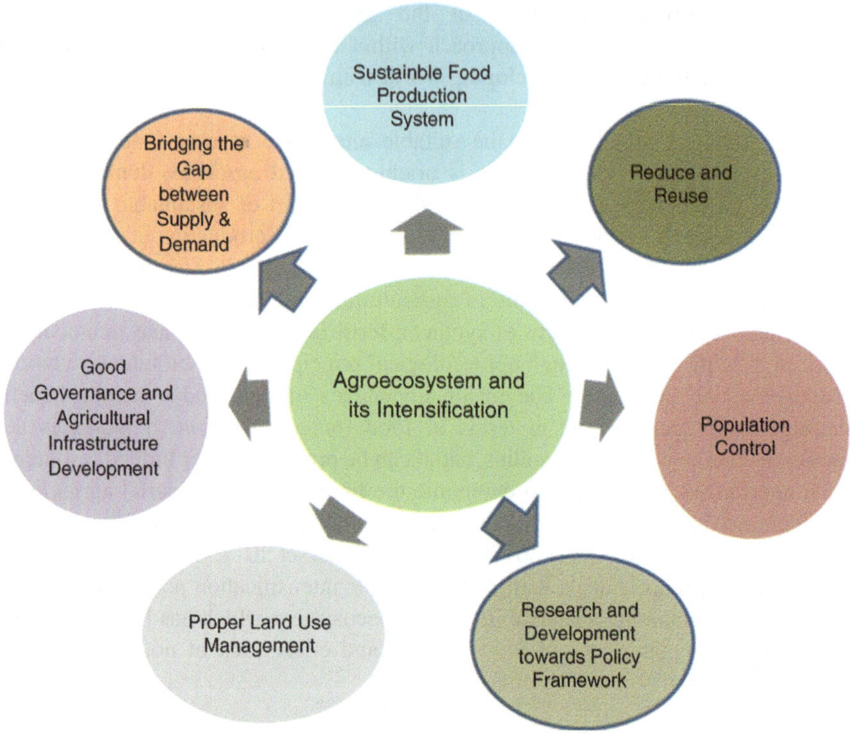

Fig. 1.1 Intensification approach in agroecosystem

(Mockshell and Kamanda 2017). Agroecosystem requires multidimensional approach to intensify its production which has been precisely described in Fig. 1.1.

The productivity of agroecosystem is very important for prosperity of life, sustainable use of NRs, as well as the economy and prevailing political system of a country (Table 1.1). In this perspective, sustainability issue is a crucial factor which has got various dimensions to be addressed as below.

1. Dimension from an Ecological Point of View.
   In this perspective, sustainable agricultural practices should be environmentally sound and aimed toward conservation of NRs along with proper functioning of the agroecosystem. It includes various issues in terms of land use, efficient water utilization, crop and soil management, and input management along with proper use of various agrochemicals (Hayati et al. 2010; Meena et al. 2020a, b, c). The main motto behind all these actions is to investigate the sustainable use of water resources, quality assurance and maintenance, conservation of germplasm of agrobiodiversity, and regulation of energy-intensive farming practices along with proper recycling of the NRs.
2. Dimension from an Economic Point of View.

Table 1.1 Approaches of intensification in agroecosystem

Factor/ indicators	Intensification through agroecological principles	Intensification through sustainable agriculture
Drivers	Various government and non-government, academic, and societal organization	Multinational companies, agro-business-based organization and sectors, R&D wing of government and non-government organizations
Principles	Conservative approaches for natural resources in agroecosystem Integrated approach for rural upliftment considering environmental and human dimension Ecology-based natural land husbandry practices	Improving production process with a conservative approach toward natural assets Improvement of resource utilization and intensification Sustainable approach to fulfill the future needs
Ideology	Poverty eradication, sustainability of environmental component, and suitable alternatives for modernized agricultural practices	Food security, sustainability of the environment and society and alternatives for industrial agriculture
Technology	Genetic alteration not permitted	Genetic alterations are permitted to some extent
Ecological perspective		
Land utilization	Unequal sharing of land between conservative and non-conservative approaches as well as in the case of intensive production process	Objective-oriented land distribution for production and biodiversity conservation purpose
Cropping land pattern	Integrated system with mixed culture and multipurpose practices of crop cultivation and livestock management	Sole cropping
Land area of broader scale	Climate- and stress-resilient agroecosystem with proper maintenance of agroecosystem services	Environment-friendly approach without altered land use
Cultivation approaches	Improving productivity through biological interactions	Good agronomic practices with high-yielding varieties
Economic approach		
Effectiveness	Resource use efficiency Production capability Input-output ratio	Adequate land availability Productivity of agroecosystem Prospect of agricultural production
Germplasm	Indigenous germplasm/seed application	High-yielding allochthonous germplasm
Input utilization	Lesser application of external inputs	Higher application of external inputs
Social context		
Awareness	Traditional knowledge Participatory management	Combination of knowledge, science, technology, and traditional practices
Farming community	Farmers with small land holding	Farmers with large land holding
Maintenance of livelihood	Rural livelihood maintenance at small scale	Livelihood maintenance at larger scale

Economy is a major aspect from the perspective of trade, business, and overall prosperity of human civilization. It also reflects the efficiency of the production unit based upon natural ecosystem. It includes various economic factors and above all the benefit-cost ratio to evaluate the overall yield and productivity (Hayati et al. 2010). However, the economic inputs to the agroecosystem are a crucial regulating factor that often influences the productive output of agroecosystem. This is a major challenge as the net economic gain is very important for the farming community as well as to maintain the farm productivity. It will also influence the well-being and enrichment of the lifestyle. On long-term basis, screening of the farm size along with off-farm economic development is the major target for the farming community all over the world. The economy of the production unit is also dependent upon the nature and type of farming practices. Apart from that, the type of crop used, size of the farm, and availability of the labor are some of the key factors which determine the sustainability of the agricultural practices. It was observed that the conventional practices provide benefits on short-term basis and have negative consequences on long-term basis. Therefore, considering the economic dimension, the sustainable agricultural approach should consider the diverse factors in order to bring sustainability on long-term basis.

3. Dimension from a Social Point of View.
   Societal development is an integral part of prosperity of human civilization. In this aspect, agriculture holds to be the key process to achieve the sustainability and social well-being of the human population. Some key factors need to be considered while considering sustainability from social dimension which include the recognition and maintenance of traditional knowledge as well as livelihood maintenance of the people. Sustainable production units also ensure good health of the community people which leads to a well-developed society. Creation of new job opportunities and better health provision along with proper benefits after retirement are some of the key factors which should be kept in mind while addressing social sustainability of agroecosystem (Hayati et al. 2010; Mockshell and Kamanda 2017; Meena et al. 2020a).

1.3 Sustainable Intensification Toward Agroecosystem Services

Ecosystem services from an agroecosystem point of view rely on proper nutrient management, pest control, and management of soil and water resource along with efficient biogeochemical cycling among the various components of agroecosystem (Power 2010). Apart from this, maintenance of agrobiodiversity is also another major ecosystem service obtained from agroecosystem. The current approach of sustainable intensification in agroecosystem may bring variable nature of output depending upon its functionality (Tscharntke et al. 2012). In this context, Bengtsson et al. (2005) have argued regarding higher species diversity and richness under organic farming. Research report revealed that absence of predators and pollinator species may increase the pest intensity and therefore hamper the yield and productivity. This therefore indicates that any form of malfunctional activity in terms of ecosystem services tends to hamper the productive capacity of the agroecosystem (Thies et al. 2011). Sustainable practices in the crop cultivation may lead to biological control of pests indigenously for various crops (Vandermeer et al. 2010). Researches have further revealed that diverse agroecosystems tend to retaliate the pest population autonomously and hence promote lesser use of agrochemicals (Letourneau et al. 2011; Meena et al. 2020a, b, c). This is an important ecosystem service as up to 40% of yield reduction is caused by pest outbreak (Oerke 2006). Pollination is another major event which is an important phase of plant life cycle. It was observed that more than 2/3 of the global production can be improved by suitable pollination by various species influencing up to 35% of the global food production (Garibaldi et al. 2011). Various researches have also revealed that diversity of pollinating species tends to increase the yield and productivity of the crops (Eilers et al. 2011). Maintenance of agrobiodiversity is another crucial agroecosystem service as it helps to reduce the agricultural pollution as well as maintains the ecological complexity within the agroecosystem (Brittain and Potts 2011). However, this aspect is nullified and yet to be understood and explored properly on an experimental basis across the globe (Letourneau et al. 2011; Tscharntke et al. 2012). Maintaining proper land use practices helps to restore the ecosystem quality along with high level of agrobiodiversity leading to sustainability in production system (Fig. 1.2) (Tscharntke et al. 2012).

1.4 Challenges for Ecointensification Toward Sustainability

Ecointensification is the concept that focuses on the increase in production considering the least damage to the environment. This can be considered as one mechanism for moving toward sustainability. Sustainable agriculture or sustainable food production system aims toward maintaining ecological integrity in the agroecosystem, without hampering the crop productivity and yield. Population explosion is the major challenge in front of the concept of ecointensification. As the population rises, the demand for food increases, and at the cost of more production, the environment of the agroecosystem gets degraded. Practices such as use of synthetic agrochemicals and energy-based agricultural practices followed by high-input agriculture support the aforesaid facts (Lichtfouse et al. 2009). This would lead to more crisis situation in the food production system where population rise is in common hike. To cope with this population increase, intensification practices have now become a must for the human population. This is evident from the functioning of the developed nations adopting agrochemical use for more production as well as involvement of the machines in the cultivation practices. As a consequence, soil resource depletion has initiated (Altieri 2005). Such mechanized agriculture practices would hamper the agroecosystem in terms of biodiversity loss and loss of ecosystem services. In the developing world, the situation is even worse due to population strength as well as faulty practice of cultivation making it a total

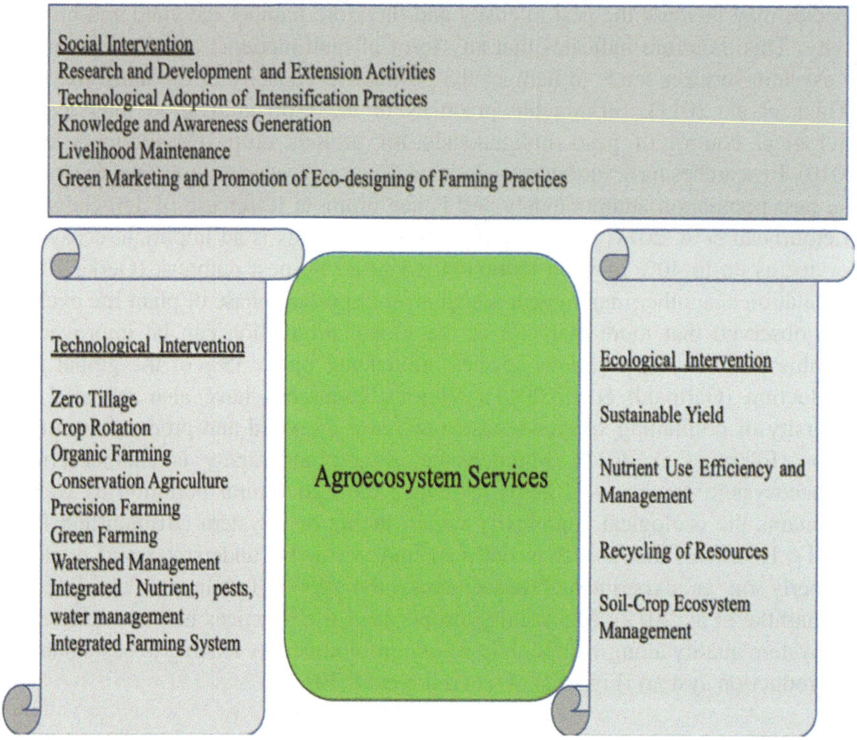

Fig. 1.2 Sustainability approaches in agroecosystem services

unsustainable system (Duru and Therond 2015). Such practices lead to various forms of environmental pollution, pest and disease outbreaks, and erosion of soil (Kremen and Miles 2012; Meena et al. 2020a). The problem is having long-term impact through reduction of productivity and soil health (van Ittersum et al. 2013). More interestingly, such events have taken place in areas where poverty is at its extreme (Chauvin et al. 2012). Sustainability has its own dimensions and integrated network. It is governed by societal aspects as well as policy and legal framework. Proper management of the environmental issues along with economic context needs to be addressed for sustainability (Fig. 1.3).

Maintenance of biodiversity and other associated ecological processes is the key for success in sustainable agriculture. Agriculture is such a sector that encompasses diverse ecological processes (Muller-Christ 2010; Gomiero 2016). The traditional practice in agriculture in the African subcontinent harbors diverse genetic pool and becomes a source of food for the farming community. They in turn help in germplasm conservation. From this, we can interpret that there is a strong relation between agroecosystem and agriculture practices along with associated agrobiodiversity (Johns et al. 2013).

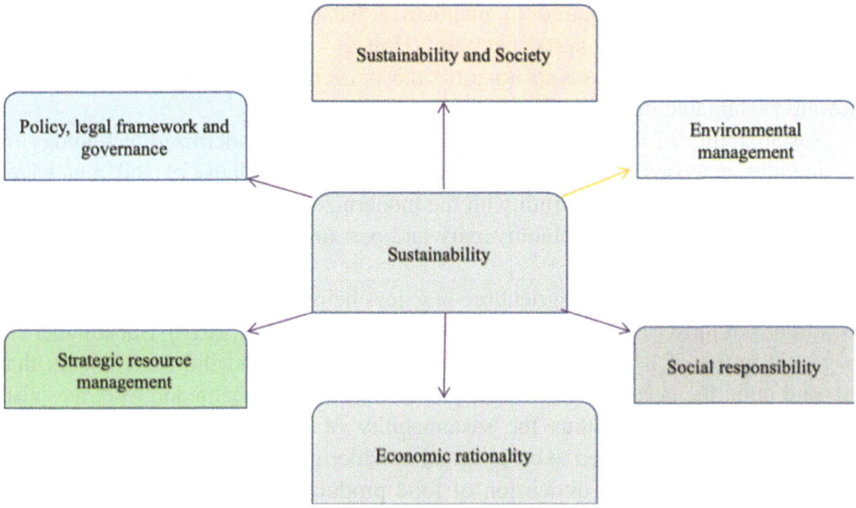

Fig. 1.3 Sphere of sustainability

Soil is an important component for both agroecosystem and the agriculture production. It encompasses indirectly the larger issues of food security and crisis, land quality, hunger, level of poverty, and other ecosystem services (Wood et al. 2000; Meena et al., 2020b). Soil can be considered as a production unit as well as heritage of agrobiodiversity. Various important ecological functions in the form of C balance, C stock, and nutrient balance take place within the soil system. Soil physical structure tends to hold water; soil biota helps in the organic matter decomposition process and also plays a significant role in nutrient exchange between plant and soil system.

Soil resources are severely affected through degeneration of soil quality and health across the globe. Population explosion followed by erosion of the top soils is the major problem across the globe at present moment (Ighodaro et al. 2013). As per research database, >25 t/ha^{-1} year^{-1} is getting depleted along with degeneration of crop land of about ten million hectare worldwide (Nyawade et al. 2018). Pimentel (2006) mentioned about the vulnerability of South America, Asian subcontinent, and Africa toward the erosive process of the soil. It approximates 30–40 t/ha in those areas. Changing land use pattern through anthropogenic factor leads to the degeneration of the land leading to unsustainable soil environment (Tesfahunegn et al. 2014).

Various issues such as the deforestation and loss of ecological services as well as the bioresources lead to the simplification of the agroecosystem. This is highly drastic in terms of ecosystem sustainability, and it further becomes problematic when combined with the event of changing climate. In order to keep sustainability in food production, we must address these issues separately (Landis 2017). Without hampering the crop production, proper care needs to be taken to maintain the biodiversity, ecological services, and above all soil health and productivity. Proper policy framework is required to maintain a balance between productivity and structural framework of agroecosystem (Holt et al. 2016). In this connection, agroecology has a step forward to solve the issue in a holistic manner and move toward sustainable development (SD).

Sustainability of agroecosystem is under pressure of modernized technology of agriculture practices (Tayleur et al. 2017). Various problems of soil and water resources are the inevitable truth with the modernized technologies. The secondary impact includes loss of agrobiodiversity and pest and disease outbreaks (Coll and Wajnberg 2017).

Apart from this, modern agriculture practices help to improve the socioeconomic conditions of most of the developing nations (Srivastava et al. 2016). But still there is a gap in terms of benefit sharing between the farmers and the stakeholders that depend upon the agriculture. The unequal distribution of wealth and resources also increases this gap, and thus the sustainability of agroecosystem is challenged. Therefore, approaches need to be designed considering the economic, technological, and socio-environmental dimension of food production and sustainability. In this perspective, sustainable agriculture plays a promising role (Allahyari 2012). However, the modernized technology in agriculture is not always fruitful to give adequate food for the growing population. Also, the consequences of environmental degradation come into forefront. It threatens the biota and various other abiotic components. As a consequence of that, the modern society of human civilization would be having issues such as food crisis, malnutrition, and many more in the coming future (OKP 2013). All these components are important under the consideration of sustainable agriculture, and therefore, integration of pollution prevention approach along with the harmony of the economic system needs to be considered on long timeframe.

The SD demands harmony between the production unit and the environmental component. At the present context, the production unit is attempting for more production leading to environmental degradation (Hope 2007). As per one projection, there may be cent-percent rise in the food production till 2050 in comparison to the level of 2005. Therefore, ecointensification practices based on extension system would be more applicable from future perspective to achieve SD (Brennan et al. 2016).

1.5 Agricultural Intensification and Environmental Sustainability

Nutrient exchange in the form of nutrient inflow and outflow within the agroecosystem makes it a different and unique system in comparison to other ecological system in nature (Sanchez 1994). There is a nice balance between nutrient inflow and outflow through the process of erosion and weathering in respective sequence. Therefore, moving toward sustainability requires the key element of nutrient balance to be explored properly in order to develop climate-resilient system (Fresco and Kroonenberg 1992). The two concepts of green revolution and sustainable agriculture deal with the nutrient issue in order to achieve sustainable yield.

Table 1.2 Review of agricultural sustainability practices in the world

Region coverage	Descriptions	References
Worldwide	Used life cycle tools and participatory methods for agricultural sustainability	De Luca et al. (2017)
	Carbon footprint analysis from energy crop cultivation for maintaining agricultural sustainability	Peter et al. (2017)
	Used life cycle tools for sustainable pig productions in agroecosystem	McAuliffe et al. (2016)
	Sustainable livestock's farming practices	Lebacq et al. (2013)
	Used life cycle assessment for sustainable beef productions and its impacts on environment	de Vries et al. (2015)
	Measurement of indicators for sustainable agricultural practices	Latruffe et al. (2016)
	Used life cycle assessment for sustainable milk productions	Baldini et al. (2017)
	Used indicator-based methods for assessing environment in sustainable agricultural systems	Acosta-Alba et al. (2011)
	Used Agri-environmental-based indicator for quantifying farming situations	Bockstaller et al. (2008)
	Assessing environmental impacts on sustainable farming practices	Payraudeau et al. (2005)
	Assessment of agricultural sustainability through varying indicators	Binder et al. (2010)
	Assessing environmental impacts on sustainable fruit productions	Cerutti et al. (2011)
Bangladesh	Assessment of agricultural sustainability through varying indicators	Roy et al. (2012)
Portugal	–	Morais et al. (2016)
Europe	Life cycle assessment for sustainable milk productions in European countries	Yan et al. (2011)

Proper management of nutrient is necessary in order to avoid the danger of nutrient erosion. So, meaningful approach for managing soil is required in order to achieve agricultural sustainability. Various approaches throughout the globe have been practiced to address the issue of agricultural intensification toward sustainability (Table 1.2).

The ecological balance of agroecosystem is reflected through prevailing agrobiodiversity as well as low level of pollution. Anthropogenic influence gradually reduces the demarcation between the agroecosystem and the natural environment (Khan et al. 2020a, b). Agroecosystem tends to have various inherent characteristics. The major aims include biological control of harmful insects and maintaining essential ecosystem services for improvement in productivity and yield (Cruz-Cárdenas et al. 2019). Further, gradual development in agrotechnology helps in the integration of various components along with involvement of farm-based

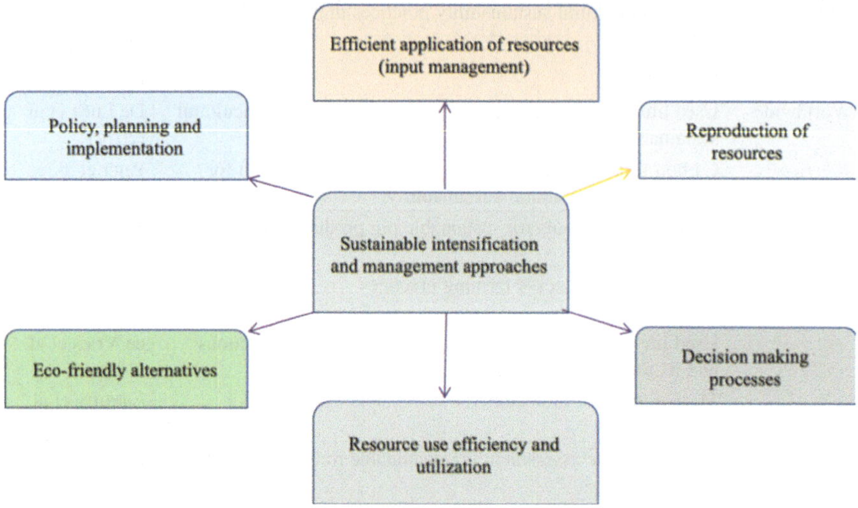

Fig. 1.4 Sustainable management of agroecosystem

technologies and its application in agriculture (Fig. 1.4) (Walter et al. 2017). The climatic perturbations may have some impact over the agroecosystem output as under natural condition, the influence of climatic element cannot be overruled. However, the impact may be positive or negative on case-to-case basis, and it depends upon the nature of crop cultivated (Neset et al. 2019). The best policy for mitigating climate change includes biodiversity enhancement, maintaining soil health, and sustainable use of water resources all together (Muller-Christ 2010; Altieri and Nicholls 2017).

1.6 Agroecosystem Management

Sustainable management of agroecosystem depends upon some key issues which include the soil C pool and its degeneration, erosion of soil, and nutrient imbalance. Such factors contribute significantly toward reduction in yield and productivity posing threat to global population. The triangle of productivity, degradation of the environment, and the level of poverty tends to have severe impacts over the people across the globe (Raman 2006). The trends have become worse by becoming static across the third world nations. The unequal availability of the resources has also created the problem of environmental degradation to some extent.

The issues of population explosion and growth of science and technology have led to the problem of food security and crisis. Particularly it is prevalent in the Africa and South Asian subcontinents which harbor half of the global population. As per the estimates, up to 800 million individuals across the globe are devoid of optimum food. However, attempt of addressing these issues with proper management has been initiated by the United States since 1948 by passing a bill on "The Right over Food."

From the Indian perspective, approximately 200 million people are devoid of food. Further, the problem is aggravated by major contribution to this group represented by women and children. Malnutrition has become a common disease in children for more than 170 million individuals affecting their physiological and psychological setup. Deficiencies in the nutrients make them weak and undernourished. The condition is such critical that it becomes the case of human rights (Raman 2006).

The NR distribution within the wealthier nations has developed the economy and technological progress in the developed world in comparison to the developing one. As a result, people of developed nations are enjoying the maximum benefits, and therefore only a meager portion of the population is being provided with adequate nutrient and food. Almost 95% of the people of the developing nations are suffering the worse situation of hunger, poverty, famine, and malnutrition. In order to address the issue, proper soil management is required to boost up the productivity and yield. Various revolutionary soil management approaches need to be implemented at the global level to combat the situation (Nair 2019).

From a global perspective, agriculture is a decision-making tool for both developing and developed nations. In the case of developed nations, the rapid growth and support of science and technology has helped them toward the increase in food production for their own consumption as well as maximum profit through net economic return. Data related to the issue reflects that up to 20% of the income at the family level is invested in food system, while a very small fraction of the population is engaged in agricultural activities. This therefore indicates that the agriculture does not share a good percentage in the national economy. On the other hand, the scenario is completely different from the developing countries' perspective. Here, most of the population is engaged in agricultural activities for income generation, fulfilling the demand of the growing population as well as maintaining livelihood. Therefore, agriculture has a big share in the economy of the developing world. Very interestingly, maximum of the populations resides under rural setup, and half of their income is invested in the food production system. Such process has taken place due to the introduction of industrialized agriculture followed by green revolution and the involvement of rich farmers having adequate irrigation facilities. However, the situation appears to be freak in the case of marginal farmers or farmers without land holding (Nair 2019).

The condition is made further problematic through the food production chain and bad practices prevailing in the country's economy system. From the Indian perspective, it was observed that the low price rate of the government as well as the subsidy did not reach to the people where it is needed. To reduce the population strength, policy has been framed, but its effective implementation is yet to be done. Further, beside the technological growth and green revolution, the amount of food that is required to be produced is not produced annually across the country making the food availability scarce. This in turn increases the price of the food and the poverty level of the developing countries creating the issue of food crisis and security. Even the maintenance of daily livelihood of the rural poor appears to be blink. Such conditions also make the people perplex regarding the land use. The policies of the government, in terms of extension activity, are also not well designed and misguide them to faulty land use practices. As a consequence, they invest maximum resource within a small area, and hence the productivity declines in due course of time. To address such issues, sustainable intensification in the area of yield management is very much necessary. Various techniques and approaches have been developed in this regard to orient a holistic approach of agricultural production toward agroecosystem management and environmental sustainability (Table 1.3).

1.6.1 Management of Crop Ecosystem

Management at the field level is the essential prerequisite for sustainable agroecosystem. Proper care should be taken toward maintaining ecological interactions between species as well as maintaining the harmony of the system. This would lead to long-term benefit of ecosystem services. It was observed that control of disease and pest population followed by maintenance of agrobiodiversity is very fruitful from sustainability perspectives. It was reported that agrobiodiversity is the internal regulator for the well-being of the agroecosystem (Altieri 1999). Most of the ecological services such as the climate regulation, mobility of nutrients, and C sequestration all involve the biotic components of agroecosystem. From this perspective, agrobiodiversity is the key for achieving sustainability in agroecosystem.

Maintenance of biodiversity helps to promote the positive interactions among the biotic components leading to effective functioning of the ecological process. Further, it increases efficiency in terms of resource use followed by biological control of harmful insects and other biotic populations under control. It also helps the marginal farmers in terms of reducing crop failure as well as acts as a subsidy during crop failure (Gliessman 1998). The most potential pathway in achieving this is the cultivation of traditional varieties and breeds.

Various measures in the form of agronomic practices as well as vegetation intervention are required for conserving the agroecosystem in sub-Saharan Africa from an economic point of view. Different techniques for managing crop ecosystem, soil and water conservation, nutrient and fertilizer management, and agriculture-cum-forestry practices need to be implemented as effective management processes. The technology of mulching is gaining rapid importance in the case of the arable land as it promotes both soil and water conservation. It was observed that adoption of suitable management strategies may help to bring the soil erosion below the tolerance range and improves soil health. Proper care should be taken in the sloppy areas through proper rotation practices to maintain the soil cover. Intercropping process has shown significant promise in the field of controlling soil erosion and improves the overall quality of soil. Mulching reduces the splashing action of raindrop, followed by decreasing runoff and improving the water infiltration rate in the soil. It also favors the earthworm activity along the profile which enhances the soil moisture level. In this process, combined application of hedgerows and mulching showed considerable promise in reducing erosion in Kenya (Kinama et al. 2007).

Table 1.3 Sustainable intensification approaches toward yield management of agroecosystem (Compiled: Branca et al. (2011), Ponniahh et al. (2008))

Techniques/ practices	Description	Impacts/output	Source
Modernized agronomic approach	Cover/shelter crops	Increase in yield due to reduced erosion and nutrient mobilization	Kaumbutho and Kienzle (2007)
	Cropping pattern/rotation	Increase in yield due to higher fertility of soils	Kwesiga et al. (2003)
	Improved varieties	Improves the production	Hine and Pretty (2008)
	Application of legume under crop rotation practice	Yield improvement due to high soil N content	
Nutrient management through integrated approach	Application of various forms of organic fertilizers	Improved productivity with enhanced N use efficiency resulting in higher soil fertility	
Crop residue management	Application of residue	Improved production due to higher soil fertility and water holding capacity	Lal (1981)
	Conservation agriculture	Increased yield on long-term basis with higher soil moisture value	Hine and Pretty (2008)
Management of water resource	Irrigation	Improved yield and intensification of land use	Khan et al. (2005)
	Ridge and furrow system bunds/zai	Improved yield under moisture stress condition	Lal (1981), Kasie et al. (2008)
	Contouring and terracing	Improved yield with reduction of erosive forces of wind and water and thus improved quality of soil	Dutilly-Diane et al. (2003), Posthumus (2005)
	Rain water collection and conservation	Improved production	Parrott and Marsden (2002)
Agriculture cum forestry practices	Fencing, shelter, and wind breaks	Improved yield and microclimate	Ellis-Jones and Mason (1999)
	Different approaches and models of agroforestry	Higher production, yield, reduction in erosive forces, proper management of water resource	Parrott and Marsden (2002), Verchot et al. (2007)
Technological extension and dissemination at farm level	Rapid rural appraisal (RRA); participatory rural appraisal and planning; participatory assessment and planning (PAP); participatory learning and action (PLA)	Evaluation and development of plan	Ponniah et al. (2008)

(continued)

Table 1.3 (continued)

Techniques/ practices	Description	Impacts/output	Source
	Participatory impact monitoring (PIM); participatory monitoring and evaluation (PME)	Evaluation and development of plan	
	Participatory farm management methods (PFM)	Evaluation and planning at domestic and farm level	
	Participatory rural communication appraisal (PRCA)	Dissemination of idea, facts, and knowledge	
	Rapid appraisal of agricultural knowledge systems (RAAKS)	Dissemination of idea, facts, and knowledge	
	Participatory technology development (PTD)	Monitoring, evaluation, and implementation	
	Participatory livelihood analysis	Predicting prospects and opportunities of people livelihood	
	Participatory poverty appraisal	Focus on poverty eradication	

1.6.2 Management of Soil Ecosystem

Soil is a key component in the agricultural production system. The ever-increasing population has put huge pressure on this particular NR, and as a result, it is getting readily depleted. The mechanism of depletion takes place through erosive forces of water and air and loss of soil fertility followed by unsustainable land use practices. As a consequence, proper management of soil resource is required.

Nutrient loss is the major aspect in terms of degeneration of soil quality. The situation is severe in the case of the east and central parts of Africa and other developing countries. According to an estimate, faulty cropping activities tend to cause loss of macro-nutrients up to 270 million tons on an annual basis in African subcontinents. Addressing these issues approaching organic farming in the form of application of farm yard manure (FYM) and other inorganic fertilizer may give good returns in terms of soil quality. This has been reported by various workers from different countries of Africa (Otieno et al. 2018; Okalebo et al. 2006; Pincus et al. 2016; Munyahali et al. 2017; Rodrigues et al. 2018). Recovering the productivity of soil does not include application of chemical fertilizer only but also demands proper cropping pattern, rotation, and other agronomic practices as required on a case-to-case basis (Nyawade et al. 2018).

Tillage is an important activity in agricultural production. It regulates the soil dynamics in terms of both physical and biological means. Appropriate application of the tillage practices helps to reduce the alteration in soil and increases soil water content and hence proper soil management. Various practices in the form of zero tillage, minimum tillage, and other forms of tillage activities serve the purpose of water conservation. Conservation tillage is an age-old practice which is rapidly gaining its popularity through its adoption in the dryer areas of African subcontinents. Such tillage practices are a bridge between various biological and physical methods that promotes soil and water conservation. It is an approach that helps to enhance the soil fertility followed by conservation of water. Under the conventional system, the major impact goes over soil physical structure altering its physical properties and therefore the total nature of the soil. As a consequence, the soil rhizospheric environment also degenerates due to improper practices. Through conventional system, the soil biota compositions alter significantly and are exposed to mega events of climate change. Conservation approach in tillage process leads to the maintenance of soil structure followed by optimum soil and water conservation.

New technology in the form of application of grass strips or vegetative buffers is considered to be an effective conservative measure of soil resource. Application of grass strips can be effectively done in areas of high rainfall and water channel along with fodder use. Usually the strips are planted on the contours at regular interval to maintain the spacing of the terrace. As a result, it acts as a suitable filter medium for runoff and thus reduces soil erosion. Spacing between the strips depends upon the nature of the land topography. However, trimming practices need to be done for the proper management of the strips in order to avoid any negative impact. It is also beneficial for the cattle hood that uses it as fodder.

Soil nutrient pool management is another large component under soil management practices. At present time, the abuse of chemical fertilizer is also changing the nature of soil resource causing a decline in the crop productivity. To combat such problem, proper management of fertilizer is the demand of the hour for sustainable crop growth. Improving the mechanism of fertilizer application along with fertilizer use efficiency of crops is an important aspect for agroecosystem management. It would lead to healthy growth and development of crop plants and thus reduce the erosive force of wind and water.

Reducing the runoff loss can be achieved through plantation at the slopes and on the contours. It was observed that up to 50% reduction of erosion can take place in moderate slope. This also helps to reduce the runoff rate and trap the sediment that is produced during erosion. The application of contour farming depends upon various factors in the form of land topography, climate, and land use practices. Contour bund reduces runoff and soil loss in the southern part of Africa (Thierfelder and Wall 2009). Contour bund helps to reduce the excess runoff from the steep slopes in the undulating topography. Integrated approach in the form of agroforestry is a suitable practice that integrates various components of agroecosystem under the aegis of sustainable agriculture (Kinama et al. 2007; Jhariya et al. 2015). Various agroforestry practices tend to improve the soil health status as well as the productivity of the crop ecosystem (Sjögren 2015; Raj et al. 2019a, b). Cultivation of legume and other species having higher growth rate can be a suitable alternative for crop production and soil management (Sanchez 1999).

1.7 Addressing Food Security Through Sustainable Agriculture

Growth in science and technology has improved the capacity of the human being for more production. This has led to the unsustainable exploitation of the NRs as well as diversified food production which is putting question mark in the quality of food that is produced (Godfray et al. 2010). Within a short span of time, the negative consequences of modernized agriculture have put the humankind in the backseat (Banerjee et al. 2021a, b, c, d; Jhariya et al. 2021a). Alteration in the land use in the form of agricultural conversion and pasture land acts as a crucial factor for loss of agrobiodiversity. Degenerating agrobiodiversity indirectly hampers our food resources. Therefore, food security tends to alter through critical approaches within the agroecosystem, thus reducing the capacity to produce quality food to address the issue of food security (Carolan 2013; Jhariya et al. 2021b). Population growth, biodiversity loss, ecological degradation, changing climate, and the global economy have made the issue of food security a central point of interest to which various nations across the globe are working on (Raj et al. 2018a, b; Khan et al. 2021a, b). The issue of food security has two major components which include the optimum production of foods to feed everyone and minimum ecological degradation (Godfray et al. 2010).

Developing industrial agricultural practices were approached in order to increase the rate of crop production. This is based upon technological interventions and altered land use along with more use of fertilizer. All these methods have severe negative consequences over the holistic integrity of agroecosystem and therefore produce unavoidable circumstances (Meena et al. 2020a, b, c). Increased use of fossil fuel and elevated level of greenhouse gas emission alter the climate of the globe which has some inherent impact over the agroecosystem and negative consequences on food security.

Beside the growth of science and technology and improvement in crop production, the poverty level still exists across the globe. The environmental dimension has created further problem to increase productivity and yield which may hamper the food security. Modernized agriculture practices have its overall impact on every component of agroecosystem leading to the decline in productivity and yield (Wood et al. 2000; Pretty 2018). Major policy framework to address the issue of food security includes involvement of the government bodies with scientific and technological intervention along with more investment in the agriculture sector (Fig. 1.5).

In the conventional practices, it was observed that various dimensions of food security are declining at a faster rate leading to its reduction in productivity and yield. Further, the modern approach in the agriculture practice leads to greater degradation of the environment and overshoots the productive capacity of the agroecosystem (Bennett et al. 2014). All the consequence is that we have to double our food production system till 2050 to feed the growing population of humankind in the upcoming times (FAO 2020).

Agro-intensification is an age-old practice which stresses upon more production on the same piece of land without giving proper care of the agroecosystem

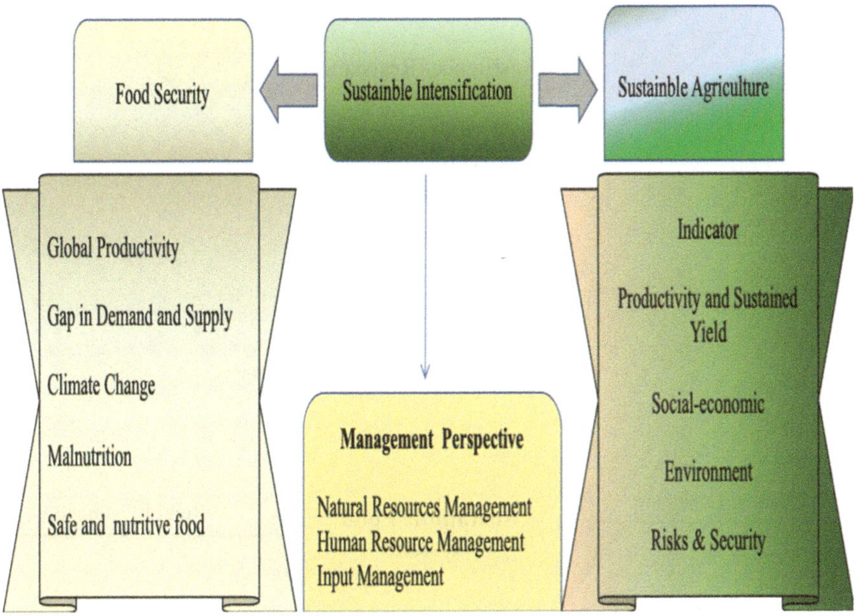

Fig. 1.5 Agroecosystem management through sustainable agriculture

environment (Peterson and Snappy 2015). It may be a suitable alternative for conventional system of production that may be deleterious for the environment. With the support of science and technology, the production has increased quite a fold than the population growth in the last 50 years but with more exploitation of NRs (FAO 2020). Sustainability is a holistic approach that addresses the issue of food security. The two components of agriculture and the security of food are found to be intensely correlated. Agriculture process acts as food production system and thus helps to reduce the hunger of human civilization across the globe. Secondarily, agriculture also creates self-employment opportunities for the community stakeholders through proper marketing mechanisms. This would enable to increase the socioeconomic level of rural livelihoods and help them in their well-being and prosperity (Fig. 1.6). However, it is regulated by the availability of the NRs and proper policy framework existing on an area basis (Peterson and Snappy 2015).

Considering the food security issue to achieve sustainability, one needs to produce enough food to feed the ever-increasing population in order to maintain the balance between demand and supply. The major purview under the approach of sustainable agriculture includes agricultural intensification through ecological management of the agroecosystem health. In this, challenge would be reduction of risks and integration of plan and policies with continuous monitoring of the ecosystem. It helps in the conservation of environment along with fulfilling the demand of the growing population. The two issues are very much interrelated to improvise the

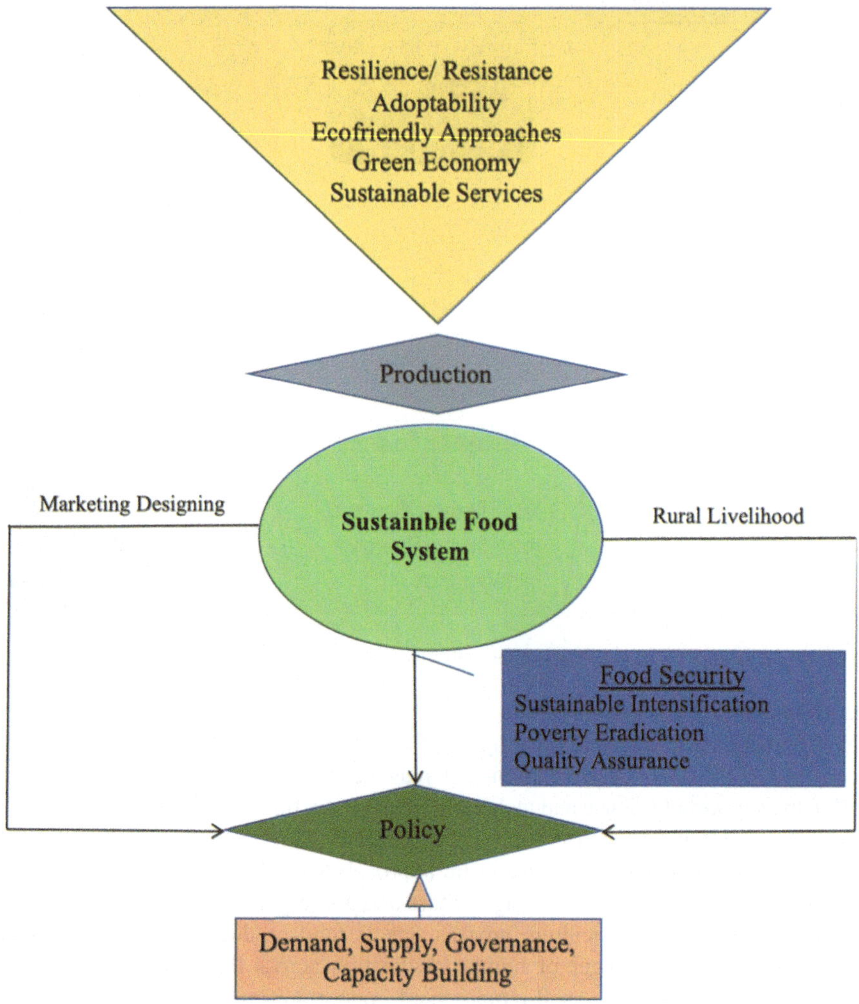

Fig. 1.6 Nexus between food security and sustainable food system

productivity as well as socioeconomic upliftment of the people along with least damage to the natural assets (Pretty 2018).

Sustainable agriculture aims toward developing eco-friendly practices in the agricultural system along with reduction of harmful effect on the environment (Jhariya et al. 2019a, 2021a). It is an integrated approach that encompasses the economic, social, and environmental aspect in relation to productivity and yield through various eco-friendly agricultural practices (Kumar et al. 2020). Sustainability in agriculture demands development of self-resiliency as well as developing adaptability of the agroecosystem under various external and internal changes. It involves practice that does not hamper the fertility status of soils as well as combating pest and disease outbreak problems (Bromilow 2013).

Broadly, agriculture is a concept of producing crop along with proper animal husbandry management. In order to achieve sustainability, all the practices need to be eco-friendly. In the modern world, most of the rural settlements are dependent upon the agriculture activities which govern the local economy. Therefore, proper priority needs to be given in this sector along with sustainable approaches. Under developing condition, the nation should consider agriculture activity in a sustainable way which is the need of the hour. Majority of the nation across the globe are still under the stress of poverty and lack of adequate food (FAO 2020).

The issue of sustainable agriculture in the developing world is very important for farmers having small land holding when compared to the consumer protection and environmental safety in the developed world. Considering these approaches, a new concept of sustainable intensification emerged in the forefront which emphasizes increase in yield and productivity followed by decline in environmental degradation (Foley et al. 2011). The fate of such approaches however remains still uncertain for large agricultural field.

The major target in the agricultural sector is to reduce the gap between demand and supply for agricultural crops. Further, in doing that, one needs to take care about the environmental perspective (Foley et al. 2011). This therefore indicates that the production unit of the food needs to be increased in terms of area. For example, gap reduction up to 95% for various food and cereal crops requires a production of additional 2.3 billion tons of crops which demands more land area (Foley et al. 2011). In perusing these specific approaches in the form of conservation agriculture, organic farming and integrated nutrient management could be the suitable policy to achieve the target (Foley et al. 2011).

At present moment, most of the cultivable land is under intense agriculture practice leaving very small area for its further expansion. Unscientific expansion and inappropriate practices may cause less yield and alteration of the land quality (FAO 2020). Therefore, waste reduction or wasteland formation can be reduced in order to achieve sustainability. It was observed that up to 40% of food material through agricultural practices is lost in the transport chain of production to consumption in the form of spoilage (Godfray et al. 2010). Further, the rising trend of the consumption due to increasing human population generates a significant amount of waste in the third world nations as well as the developed nation from a consumer point of view (Foley et al. 2011). This therefore indicates the challenge of proper management of waste which is huge in quantity, and therefore, urgent need is to convert them in useful materials (Foley et al. 2011). Consumption pattern depends upon the lifestyle in the form of eating habits. It was observed that negative outcomes appear depending upon the nature of consumption. For example, concentration of methane as a greenhouse gas automatically increases as because intensive animal husbandry practices and cattlehood maintenance for meat consumption purpose (Godfray et al. 2010). However, non-cultivable land can be effectively utilized by the third world nations to fulfill the protein requirement. On another aspect, livestock population becomes a major revenue for the poor people in the third world nations, and it has an inherent cultural significance to them. Under these circumstances, proper genetic manipulation and management may lead to enhanced production of animal husbandry products (Godfray et al. 2010).

Aquatic resources support approximately three billion people with 15% protein supply as food supplement (Godfray et al. 2010). Compared to agricultural system, aquaculture practice has a larger space for expansion to fulfill the objective of sustainable practices in agriculture. For the conservation of fishery resources, proper management and strategy planning need to be developed from fishery and coastal zone perspectives (Whitmarsh and Palmieri 2008). Development of aquaculture practices leads to capacity building of the farming community (Godfray et al. 2010). However, aquaculture has some inherent challenges in the form of water pollution and diseases which may hamper the objective of sustainable aquaculture (Godfray et al. 2010).

It was found that with global rising of human population, the availability of land and water resources is shrinking gradually and aggravating the food security issue (Godfray et al. 2010). In the past five decades, it was observed that there is a huge amount of agriculture expansion and more than 2/3 of freshwater resources are used for irrigation purpose (Foley et al. 2011). Further, researches on this particular area reflect that a huge amount of agricultural production comes from a very small piece of land that is irrigated. Therefore, one needs to improve the efficiency of irrigation in order to conserve the freshwater resource. Technological growth has led to dam construction which has enhanced the water availability for irrigation purpose for the upcoming times. This is required as most of the third world nations and the Middle East region would be facing shortage of water for food production. Researches on the agriculture sector reveal a 20% decline of the cereal crops due to lack of irrigation which demands more land for production (Foley et al. 2011). Further, the high nutrient and sediment content in the runoff also alters the productive capacity of the agricultural land in the downstream (Tilman et al. 2002). However, various strategies have been used to improve the irrigation facility, increase the water holding capacity of the soil, and develop crops that may withstand various forms of biotic and abiotic stress (Tilman et al. 2002).

Soil fertility is a crucial part to be given due consideration while approaching for sustainable agriculture. Since from past few decades, up to 17% of the lands are degraded through anthropogenic activities in the form of intense cultivation and changed cropping pattern (Tilman et al. 2002). Further, it was observed that various modernized practices such as use of chemical fertilizer, improper organic matter management, etc. have led to loss of soil fertility. For restoration of soil, various practices in the form of altered cropping pattern, increase in fallow period, and organic farming approach can be very much fruitful (Tilman et al. 2002). Soil fertility is the focus point of sustainable agriculture practice as crop productivity depends very much upon soil health. Improper management may lead to negative consequences. The water resources are also suffering problem of pollution and degradation across the globe (Godfray et al. 2010). Emission of N_2O acts as greenhouse gases and hence contributes significantly toward the changing climate. Further, overuse of nutrient causes increase in the demand of energy, and the requirement tends to be higher for artificial synthesis of chemical fertilizer. Across the globe, overuse and low availability of nutrients have negative output for the environment. It is a significant problem for agricultural productivity worldwide. The productivity and yield decline due to inappropriate nutrient supply (Cassman et al. 2002). Survey of agricultural practices across the globe reveals prevalence of more nutrient and nutrient-deficient region. As per the report, up to 10% of cropland globally approximates >30% N surplus and >35% P level. Such high level of nutrient may secondarily impact the water bodies through eutrophication (Cassman et al. 2002; Foley et al. 2011).

Proper management of environmental problems related to agricultural production can be mitigated through proper policy and strategy formulation (Foley et al. 2011). It was observed that proper pest control can improve the yield. Researches have revealed the importance of pesticide application, pest population, application of plant breeding for developing disease resistance varieties, origin of new crop varieties that results into diverse agroecosystem (Tilman et al. 2002). At the present century, the cultivable land covers 38% of the land area, and it is gradually expanding in the tropical region (Foley et al. 2011). It also reflected negative consequences in the ecosystem services. However, on an annual basis, up to ten million hectare forest areas are lost due to agricultural expansion (Foley et al. 2011). Therefore, proper maintenance of ecosystem services is necessary to address global SD. One typical example includes forest conservation which helps to prevent erosion, combat changing climate, as well as regulate the microclimate (Tilman et al. 2002).

While considering sustainable agriculture, one needs to consider the interrelationship between agriculture and climate change and how they affect each other. It was observed that up to 35% of greenhouse gas emission has been associated with agricultural production, deforestation, and animal husbandry practices (Foley et al. 2011). Various factors in terms of land use and land cover type, climatological features, and other ecological interactions significantly influence the event of climate change. On the other hand, the changing climate influences the agriculture through alteration in climatic elements and pest and disease outbreak leading to less production and loss of soil fertility.

1.8 Policy and Legal Perspectives

Policy framework and legal implementation of suitable strategies are very much essential to address the issue of food security and crisis and sustainability of agroecosystem. The production rate and the local economy will govern the fate of food production system. The component of conservation needs to be incorporated within the agroecosystem concept. New innovations and practices are required through research and developmental activities to fulfill the growing demand of food of the growing population. Ecological intensification is a suitable solution in this perspective which needs further future explorations. To do so, encouragement in the form of economic subsidy is also required.

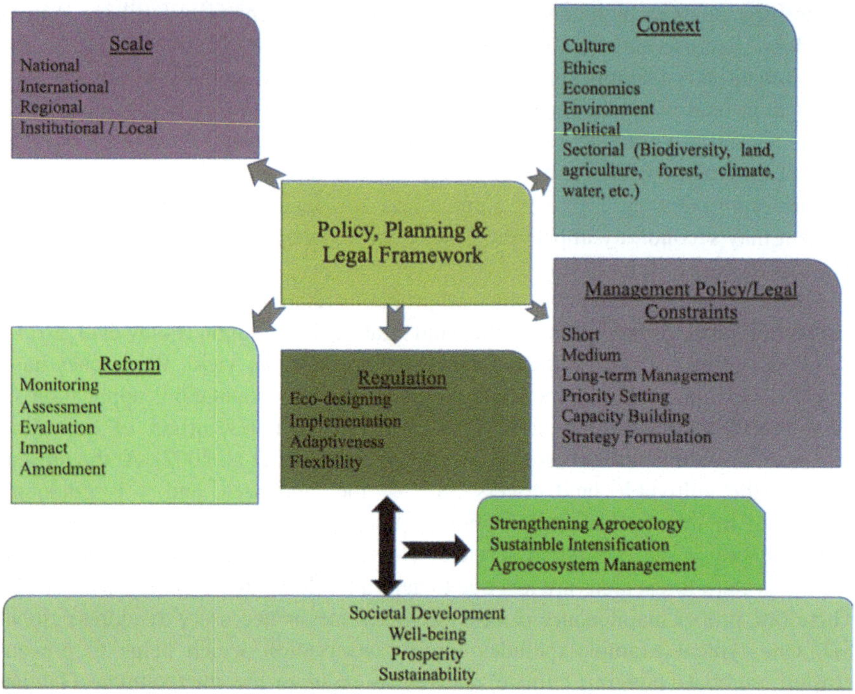

Fig. 1.7 Policy framework for sustainable intensification of agroecosystem

Further, research needs to be done focusing on increasing the efficiency of the firms as well as proper promotion of the technologies through extension activities (Kansiime et al. 2018). Transformations of the extension services in the agriculture sector may act as a suitable policy intervention (Aflakpui 2007). Further, this will empower poor rural farmers in a proper way. For strengthening the rural farmers toward adopting new farming techniques, women empowerment would be a significant step. Right to property, NR, and other assets needs to be properly recognized and scientifically blended with ecointensification practices. Level of education should be enhanced among the farming community to make them aware about the circumstances of adopting ecointensification and moving toward sustainability (Fig. 1.7). In this context, policy can be framed to develop social security for the farmers. More investment and more involvement would generate more income opportunities for the farmers as well as their socioeconomic upliftment. Social protection would help to reduce the issues of hunger, poverty, malnutrition, and food crisis and security. It would generate resiliency both from agroecosystem perspective and from the poor farming community (FAO 2020).

Crop insurance can be a suitable policy intervention for agriculture intensification. It would enhance more investment on the agriculture and thus involvement of more people in the food production system. Various workers have shown the importance of crop insurance as a policy to improve the economic security of the farming community under the era of climate change (Adiku et al. 2013).

1.9 Conclusions

At the present context, increased food production is the essential requirement for the sustenance of human civilization. Using synthetic chemicals for increasing food production and yield is doing no good as such approaches lead to the degradation of soil environment, water pollution, and reduction of yield and productivity on a long-term basis. In this perspective, ecointensification in sustainable manner would serve the purpose of reducing the environmental deterioration followed by sustainable yield and productivity. Proper policy and strategy should be formulated for capacity building of the rural farmers to adopt technologies such as organic farming, conservation agriculture, zero tillage, and eco-friendly practices at the farm level to ecologically intensify the crop production. However, there are some specific challenges in implementing agricultural intensification in a sustainable way which needs to be taken care of. Screening of suitable climate-resilient varieties along with promotion of traditional agriculture practices could be an effective way to promote sustainability in agroecosystem.

1.10 Future Perspectives

Future aspect of agricultural sustainability requires more research and development in the form of extension works in order to understand the complexity of agroecosystem and inclusion of agroecological principles and suitable policies for future development (Fig. 1.8). At the present context, the output from the agricultural sectors requires a multidimensional approach by not limiting only to yield but other aspects of agriculture and services. For the effective implementation of agroecological principles, one needs to undergo the cost-benefit approach of each of the strategies of agroecology. Incorporation of services of the ecosystem should be taken into account under productivity. The results of cost-benefit could be used to assess the subsidy requirement for a specific approach. It is very much important for third world nations to address the issue of food crisis under the context of population explosion. Land degradation is also an added problem to it. Future research should be attempted toward educating the farming community for adopting agroecological principles in their day-to-day practices along with proper design toward sustainability (Kremen and Miles 2012).

Such approaches in the form of organic farming, zero tillage, and conservation agriculture may help to achieve sustainable yields of some specific crops (Ponisio et al. 2015). Further, proper policy formulation and strategy implementation may help to address the issue of sustainability and ecointensification at the farm level.

The integrated approach should be a key component for agroecosystem management. It was observed that the health perspective of agroecosystem is much more

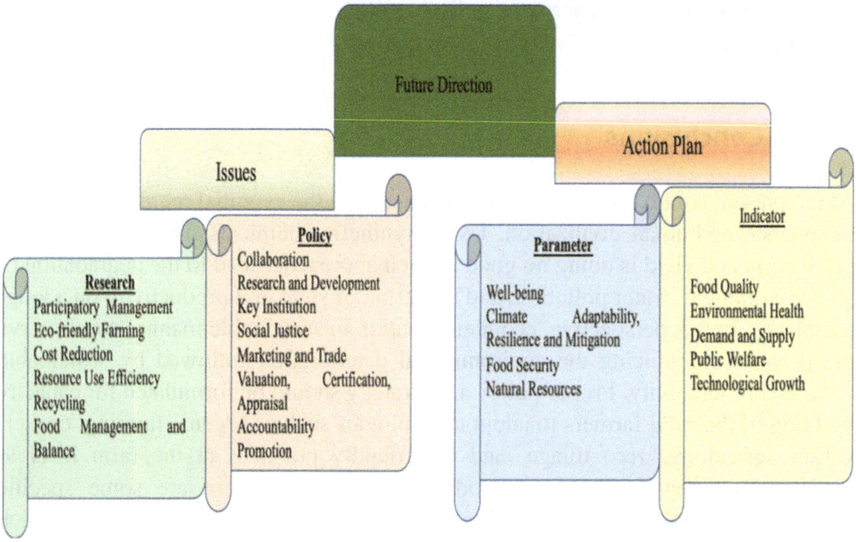

Fig. 1.8 Future perspective of sustainable intensification in agroecosystem services and management

important than the productivity of a specific crop or animal. The carrying capacity concept of the habitat needs to be incorporated in the food producing system (Silici 2014).

Popularizing the sustainable concept among the consumer level and broadly on the public level regarding public participation and promotion of ecointensification practices under the pressure of climate change is required. Such approaches would bring benefit to all stakeholders, and it would reduce the challenges of sustainability in agroecosystem. Involvement of scientists, researchers, and academic fraternities would come forward and involve the farming community for adopting ecointensification practices to achieve agricultural sustainability.

Screening of suitable varieties that are climate resilient through proper testing and cross-breeding is required to improve the productivity and yield. Testing of progeny would help to recognize the most suitable germplasm along with beneficial traits. The focus should be also on the requirement of the local stakeholders with gradual development of infrastructure. Community-based approaches and research work would be highly fruitful and improve the socioeconomic condition of local people making them self-sustainable. Involvement of advance technologies in the form of mobile app and web application such as the precision farming can be applied for the proper management of database and taking proper decision. This would work as the evaluation of each of the agroecological practices toward sustainability. Application of biotechnological principles would promote the crop breeding for sustainable yield. It would reduce the time and cost as well as help to screen suitable varieties for climate-smart agricultural practice.

References

Acosta-Alba I, van der Werf HMG (2011) The use of reference values in indicator-based methods for the environmental assessment of agricultural systems. Sustainability 3:424–442

Adiku SGK, Debrah E, Naab JB (2013) Towards the development of a micro-level weather index-based agricultural insurance to minimise the effect of climate change on crop production in the upper West region of Ghana. CCAFS report no. 1, Department of Soil Science, University of Ghana

Aflakpui GKS (2007) Present outlook and transformation in the delivery of agricultural extension services: implications for research–extension–farmer linkages. Outlook Agric 36:35–39

Allahyari MS (2012) Sustainable agricultural extension system. Islamic Azad University, Rasht

Altieri MA (1999) The ecological role of biodiversity in agroecosystems. In: Invertebrate biodiversity as bioindicators of sustainable landscapes. Elsevier, Amsterdam; New York, pp 19–31

Altieri MA (2005) Agroecology: principles and strategies for designing sustainable farming systems. https://www.biosafety-info.net/article.php?aid291

Altieri MA, Nicholls CI (2017) The adaptation and mitigation potential of traditional agriculture in a changing climate. Clim Change 140(1):33–45

Baldini C, Gardoni D, Guarino MA (2017) Critical review of the recent evolution of life cycle assessment applied to milk production. J Clean Prod 140:421–435

Banerjee A, Jhariya MK, Meena RS, Yadav DK (2021b) Ecological footprints in agroecosystem—an overview. In: Banerjee A, Meena RS, Jhariya MK, Yadav DK (eds) Agroecological footprints management for sustainable food system. Springer, Singapore, pp 1–23. https://doi.org/10.1007/978-981-15-9496-0_1

Banerjee A, Jhariya MK, Raj A, Yadav DK, Khan N, Meena RS (2021c) Land footprint management and policies. In: Banerjee A, Meena RS, Jhariya MK, Yadav DK (eds) Agroecological footprints management for sustainable food system. Springer, Singapore, pp 221–246. https://doi.org/10.1007/978-981-15-9496-0_7

Banerjee A, Jhariya MK, Raj A, Yadav DK, Khan N, Meena RS (2021d) Energy and climate footprint towards the environmental sustainability. In: Banerjee A, Meena RS, Jhariya MK, Yadav DK (eds) Agroecological footprints management for sustainable food system. Springer, Singapore, pp 415–443. https://doi.org/10.1007/978-981-15-9496-0_14

Banerjee A, Jhariya MK, Yadav DK, Raj A (2020) Environmental and sustainable development through forestry and other resources. Academic Press, New York, p 400. https://doi.org/10.1201/9780429276026

Banerjee A, Meena RS, Jhariya MK, Yadav DK (2021a) Agroecological footprints Management for sustainable food system. Springer, Singapore. https://doi.org/10.1007/978-981-15-9496-0

Bengtsson J, Ahnström J, Weibull AC (2005) The effects of organic agriculture on biodiversity and abundance: a meta-analysis. J Appl Ecol 42:261–269

Bennett E, Carpenter S, Gordon LJ, Ramankutty N, Balvanera P, Campbell B, Foley J, Folke C, Karlberg L, Liu J, Lotze-Campen H, Mueller ND, Peterson GD, Polasky S, Rockstrom J, Scholes RJ, Spierenburg M (2014) Toward a more resilient agriculture. Solutions 5(5):65–75

Binder CR, Feola G, Steinberger JK (2010) Considering the normative, systemic and procedural dimensions in indicator-based sustainability assessments in agriculture. Environ Impact Assess Rev 30:71–81

Bockstaller C, Guichard L, Makowski D, Aveline A, Girardins P, Plantureux S (2008) Review article agri-environmental indicators to assess cropping and farming systems—a review. Agron Sustain Dev 28:139–149

Branca G, McCarthy N, Lipper L, Jolejole MC (2011) Climate smart agriculture: a synthesis of empirical evidence of food security and mitigation benefits from improved cropland management. Mitigation of climate change in agriculture (MICCA) series. Food and Agriculture Organization of the United Nations, Rome, p 3

Brennan N, Ryan M, Hennessy T, Dillon EJ, Cullen P (2016) The role of extension in agricultural sustainability. FLINT Deliv 5:1

Brittain C, Potts SG (2011) The potential impacts of insecticides on the life-history traits of bees and the consequences of pollination. Basic Appl Ecol 12:321–331

Bromilow RH (2013) Paraquat and sustainable agriculture. Pest Manag Sci 60:340–349

Carolan M (2013) The food and human security index: rethinking food security and 'growth'. Int J Soc Agric Food 19:176–200

Cassman KG, Doermann A, Walters DT (2002) Agroecosystems, nitrogen-use efficiency, and nitrogen management. Ambio 31:132–140

Cerutti AK, Bruun S, Beccaro GL, Bounous G (2011) A review of studies applying environmental impact assessment methods on fruit production systems. J Environ Manage 92:2277–2286

Chauvin ND, Mulangu F, Porto G (2012) Food production and consumption trends in sub-Saharan Africa: prospects for the transformation of the agricultural sector. UNDP Regional Bureau for Africa, New York

Coll M, Wajnberg E (eds) (2017) Environmental Pest management: challenges for agronomists, ecologists, economists and policymakers. Wiley, New York

Cruz-Cárdenas CI, Cortés-Cruz M, Gardner CA, Costich DE (2019) Wild relatives of maize. In: North American crop wild relatives, vol 2. Springer, Cham, pp 3–39

De Luca AI, Iofrida N, Leskinen P, Stillitano T, Falcone G, Strano A, Gulisano G (2017) Life cycle tools combined with multi-criteria and participatory methods for agricultural sustainability: insights from a systematic and critical review. Sci Total Environ 595:352–370

De Vries M, van Middelaar CE, de Boer IJM (2015) Comparing environmental impacts of beef production systems: a review of life cycle assessments. Livest Sci 178:279–288

Duru M, Therond O (2015) Designing agroecological transitions: a review. Agron Sustain Dev 35:1237–1257

Dutilly-Diane C, Sadoulet E, de Janvry A (2003) How improved natural resource management in agriculture promotes the livestock economy in the sahel. J Afr Econ 12:343–370

Eilers EJ, Kremen C, Smith Greenleaf S, Garber AK, Klein AM (2011) Contribution of pollinator-mediated crops to nutrients in the human food supply. PLoS One 6(6):e21363

Ellis-Jones J, Mason T (1999) Livelihood strategies and assets of small farmers in the evaluation of soil and water management practices in the temperate inter-Andean valleys of Bolivia. Mt Res Dev 19:221–234

FAO (2017) Food and agriculture – driving action across the 2030 agenda for sustainable development. FAO, Rome. https://www.fao.org/3/a-i7454e.pdf

FAO (2018) The state of food security and nutrition in the world 2018: building climate resilience for food security and nutrition. FAO, Rome

FAO (2020) World's food systems rely on biodiversity. Food and Agriculture Organization, Rome. http://www.fao.org/news/story/en/item/1263301/icode/

Foley J, Ramankutty N, Brauman K, Cassidy ES, Gerber JS, Johnston M, Mueller ND, O'Connell C, Ray DK, West PC, Balzer C, Bennett EM, Carpenter SR, Hill J, Monfreda C, Polasky S, Rockstrom J, Sheehan J, Siebert S, Tilman D, Zaks DPM (2011) Solutions for a cultivated planet. Nature 478:337–342

Fresco LO, Kroonenberg SB (1992) Time and spatial scales in ecological sustainability. Land Use Policy 9:155–167

Garibaldi LA, Aizen MA, Klein AM, Cunningham SA, Harder LH (2011) Global growth and stability of agricultural yield decrease with pollinator dependence. Proc Natl Acad Sci U S A 14:5909–5914

Giraldo OF, Rosset PM (2017) Agroecology as a territory in dispute: between institutionality and social movements. J Peasant Stud 453:545–564

Gliessman SR (1998) Field and laboratory investigations in agroecology: a manual to accompany agroecology: ecological processes in sustainable agriculture. Ann Arbor Press, Chelsea

Godfray HCJ, Beddington JR, Crute IR, Haddad L, Lawrence D, Muir JF, Pretty J, Robinson S, Thomas SM, Toulmin C (2010) Food security: the challenge of feeding 9 billion people. Science 327(5967):812–818

Gomiero T (2016) Soil degradation, land scarcity and food security: reviewing a complex challenge. Sustainability 8:281

Hayati D, Ranjbar Z, Karami E (2010) Measuring agricultural sustainability. In: Lichtfouse E (ed) Biodiversity, biofuels, agroforestry and conservation agriculture. Springer, Cham, pp 73–100. https://doi.org/10.1007/978-90-481-9513-8_2

Hine R, Pretty J (2008) Organic agriculture and food security in Africa. United Nations Conference on Trade and Development and United Nations Environment Programme, Geneva

Holt AR, Alix A, Thompson A, Maltby L (2016) Food production, ecosystem services and biodiversity: we can't have it all everywhere. Sci Total Environ 573:1422–1429

Hope KR (2007) Poverty and environmental degradation in Africa: towards sustainable policy for reversing the spiral. Int J Environ Sustain Dev 6(4):451–472

IFPRI (International Food Policy Research Institute) (2002) Sustainable food security for all by 2020. International Food Policy Research Institute, Washington

Ighodaro ID, Lategan FS, Yusuf SFG (2013) The impact of soil erosion on agricultural potential and performance of Sheshegu community farmers in the eastern cape of South Africa. J Agric Sci 5:140–147

Jhariya MK, Banerjee A, Meena RS, Yadav DK (2019a) Sustainable agriculture, forest and environmental management. Springer, Singapore. https://doi.org/10.1007/978-981-13-6830-1

Jhariya MK, Banerjee A, Meena RS, Yadav DK (2019c) Agriculture, forestry and environmental sustainability—a way forward. In: Jhariya MK, Banerjee A, Meena RS, Yadav DK (eds) Sustainable agriculture, forest and environmental management. Springer, Singapore, pp 1–29

Jhariya MK, Banerjee A, Yadav DK, Raj A (2018b) Leguminous trees an innovative tool for soil sustainability. In: Meena RS, Das A, Yadav GS, Lal R (eds) Legumes for soil health and sustainable management. Springer, Singapore, pp 315–345. https://doi.org/10.1007/978-981-13-0253-4_10

Jhariya MK, Bargali SS, Raj A (2015) Possibilities and perspectives of agroforestry in Chhattisgarh. In: Zlatic M (ed) Precious forests-precious earth. InTech, Croatia, pp 237–257. https://doi.org/10.5772/60841

Jhariya MK, Meena RS, Banerjee A (2021a) Ecological intensification of natural resources for sustainable agriculture. Springer, Singapore. https://doi.org/10.1007/978-981-33-4203-3

Jhariya MK, Meena RS, Banerjee A (2021b) Ecological intensification of natural resources towards sustainable productive system. In: ecological intensification of natural resources for sustainable agriculture. Springer, Singapore. https://doi.org/10.1007/978-981-33-4203-3_1

Jhariya MK, Yadav DK, Banerjee A (2018a) Plant mediated transformation and habitat restoration: phytoremediation an eco-friendly approach. In: Gautam A, Pathak C (eds) Metallic contamination and its toxicity. Daya Publishing House, New Delhi, pp 231–247

Jhariya MK, Yadav DK, Banerjee A (2019b) Agroforestry and climate change: issues and challenges. Academic Press, New York, p 335. https://doi.org/10.1201/9780429057274

Jhariya MK, Yadav DK, Banerjee A, Raj A, Meena RS (2019d) Sustainable forestry under changing climate. In: Jhariya MK, Banerjee A, Meena RS, Yadav DK (eds) Sustainable agriculture, forest and environmental management. Springer, Singapore, pp 285–326

Johns T, Powell B, Maundu P, Eyzaguirre PB (2013) Agricultural biodiversity as a link between traditional food systems and contemporary development, social integrity and ecological health. J Sci Food Agric 93:3433–3442

Kasie MJ, Pender M, Yesuf G, Kohlin R, Buffstone P, Zikhali, Mulugeta E (2008) Sustainable land management practices improve agricultural productivity. Policy brief. Environment for Development (EFD), Addis Ababa

Kansiime MK, van Asten P, Sneyers K (2018) Farm diversity and resource use efficiency: targeting agricultural policy interventions in East Africa farming systems. NJAS-Wageningen J Life Sci 85:32–41

Kaumbutho P, Kienzle J (eds) (2007) Conservation agriculture as practiced in Kenya: two case studies. African Conservation Tillage Network and Food and Agriculture Organization of the United Nations, Nairobi. http://www.fao.org/ag/ca/doc/Kenya_casestudy.pdf

Khan M, Khan R, Wahab A, Rashid A (2005) Yield and yield components of wheat as influenced by intercropping of chickpea, lentil and rapeseed in different proportions. Pak J Sci 42(3–4):1–3

Khan N, Jhariya MK, Raj A, Banerjee A, Meena RS (2021a) Soil carbon stock and sequestration: implications for climate change adaptation and mitigation. In: Jhariya MK, Meena RS, Banerjee A (eds) Ecological intensification of natural resources for sustainable agriculture. Springer, Singapore. https://doi.org/10.1007/978-981-33-4203-3_13

Khan N, Jhariya MK, Raj A, Banerjee A, Meena RS (2021b) In: Jhariya MK, Meena RS, Banerjee A (eds) Eco-designing for sustainability. In: ecological intensification of natural resources for sustainable agriculture. Springer, Singapore. https://doi.org/10.1007/978-981-33-4203-3_16

Khan N, Jhariya MK, Yadav DK, Banerjee A (2020a) Herbaceous dynamics and CO_2 mitigation in an urban setup- a case study from Chhattisgarh, India. Environ Sci Pollut Res 27(3):2881–2897. https://doi.org/10.1007/s11356-019-07182-8

Khan N, Jhariya MK, Yadav DK, Banerjee A (2020b) Structure, diversity and ecological function of shrub species in an urban setup of Sarguja, Chhattisgarh, India. Environ Sci Pollut Res 27 (5):5418–5432. https://doi.org/10.1007/s11356-019-07172-w

Kinama JM, Stigter CJ, Ong CK, Ng'ang'a JK, Gichuki FN (2007) Contour hedgerows and grass strips in erosion and runoff control on sloping land in semiarid Kenya. Arid Land Res Manag 21 (1):1–19

Kremen C, Miles A (2012) Ecosystem services in biologically diversified versus conventional farming systems: benefits, externalities, and trade-offs. Ecol Soc 17:40

Kumar S, Meena RS, Datta R, Verma SK, Yadav GS, Pradhan GS, Molaei A, Rahman GKMM, Mashuk HA (2020a) Legumes for carbon and nitrogen cycling: an organic approach. carbon and nitrogen cycling in soil. Springer, Singapore. https://doi.org/10.1007/978-981-13-7264-3_10. 337-375

Kumar S, Meena RS, Jhariya MK (2020) Resources use efficiency in agriculture. Springer, Singapore. https://doi.org/10.1007/978-981-15-6953-1

Kumar S, Meena RS, Singh RK, Munir TM, Datta R, Danish S, Singh GS, Yadav, Kumar S (2021) Soil microbial and nutrient dynamics under different sowings environment of Indian mustard (*Brassica juncea* L.) in rice based cropping system. Sci Rep 11:5289. https://doi.org/10.1038/s41598-021-84742-4

Kwesiga F, Akinnifesi FK, Mafongoya PL, Mcdermott MH, Agumya A (2003) Agroforestry research and development in southern Africa during 1990s: review and challenges ahead. Agrofor Syst 53:173–186

Lal R (1981) Soil erosion problems on alfisols in western Nigeria. VI. Effects of erosion on experimental plots. Geoderma 25:215–230

Landis DA (2017) Designing agricultural landscapes for biodiversity-based ecosystem services. Basic Appl Ecol 18:1–12

Latruffe L, Diazabakana A, Bockstaller C, Desjeux Y, Finn J, Kelly E, Ryan M, Uthes S (2016) Measurement of sustainability in agriculture: a review of indicators. Stud Agric Econ 118:123–130

Lebacq T, Baret PV, Stilmant D (2013) Sustainability indicators for livestock farming- a review. Agron Sustain Dev 33:311–327

Letourneau DK, Armbrecht I, Rivera BS, Lerma JM, Carmona EJ, Daza MC, Escobar S, Galindo V, Gutiérrez C, López SD, Mejía JL, Rangel AMC, Rangel JH, Rivera L, Saavedra CA, Torres AM, Trujillo AR (2011) Does plant diversity benefit agroecosystems? A synthetic review. Ecol Appl 21:9–21

Lichtfouse E, Navarrete M, Debaeke P, Souchère V, Alberola C, Ménassieu J (2009) Agronomy for sustainable agriculture: a review. In: Sustainable agriculture. Springer, Dordrecht, pp 1–7

McAuliffe GA, Chapman DV, Sage CL (2016) A thematic review of life cycle assessment (LCA) applied to pig production. Environ Impact Assess Rev 56:12–22

Meena RS, Kumar S, Datta R, Lal R, Vijaykumar V, Brtnicky M, Sharma MP, Yadav GS, Jhariya MK, Jangir CK, Pathan SI, Dokulilova T, Pecina V, Marfo TD (2020a) Impact of agrochemicals

on soil microbiota and management: a review. Land (MDPI) 9(2):34. https://doi.org/10.3390/land9020034
Meena RS, Kumar V, Yadav GS, Mitran T (2018) Response and interaction of *Bradyrhizobium japonicum* and arbuscular mycorrhizal fungi in the soybean rhizosphere: a review. Plant Growth Regul 84:207–223. https://doi.org/10.1007/s10725-017-0334-8
Meena RS, Lal R (2018) Legumes for soil health and sustainable management. Springer, Singapore. https://doi.org/10.1007/978-981-13-0253-4_10
Meena RS, Lal R, Yadav GS (2020b) Long term impacts of topsoil depth and amendments on soil physical and hydrological properties of an Alfisol in Central Ohio, USA. Geoderma 363:1141164. https://doi.org/10.1016/j.geoderma.2019.114164
Meena RS, Lal R, Yadav GS (2020c) Long-term impact of topsoil depth and amendments on carbon and nitrogen budgets in the surface layer of an Alfisol in Central Ohio. Catena 2020194:104752. https://doi.org/10.1016/j.catena.2020.104752
Mockshell J, Kamanda J (2017) Beyond the agroecological and sustainable agricultural intensification debate: is blended sustainability the way forward? Discussion paper, GDI/DIE, ISBN 978-3-96021-039-9, p 34
Morais TG, Teixeira RFM, Domingos T (2016) Regionalization of Agri-food life cycle assessment: a review of studies in Portugal and recommendations for the future. Int J Life Cycle Assess 21:875–884
Morales FJ (2007) The mission and evolution of international agricultural research in developing countries. In: Loebenstein G, Thottappilly G (eds) Agricultural research management. Springer, Dordrecht, pp 9–36
Muller-Christ G (2010) Sustainable management. In: Coping with the dilemmas of resource-oriented management. Springer, Heidelberg. https://doi.org/10.1007/978-3-642-19165-7
Munyahali W, Pypers P, Swennen R, Walangululu J, Vanlauwe B, Merckx R (2017) Responses of cassava growth and yield to leaf harvesting frequency and NPK fertilizer in south Kivu, Democratic Republic of Congo. Field Crop Res 214:194–201
Nair KP (2019) Intelligent soil management for sustainable agriculture: the nutrient buffer power concept. Springer, Cham. https://doi.org/10.1007/978-3-030-15530-8.389
Neset TS, Wiréhn L, Klein N, Käyhkö J, Juhola S (2019) Maladaptation in Nordic agriculture. Clim Risk Manag 23:78–87
Nyawade SN, Karanja N, Gachene C, Schulte-Geldermann E, Parker M (2018) Susceptibility of soil organic matter fractions to soil erosion under potato-legume intercropping systems in Central Kenya. J Soil Water Conserv 73:568–577
Oerke EC (2006) Crop losses to pests. J Agric Sci 144:31–43
Okalebo JR, Othieno CO, Woomer PL, Karanja NK, Semoka JRM, Bekunda MA, Mugendi DN, Muasya RM, Bationo A, Mukhwana EJ (2006) Available technologies to replenish soil fertility in East Africa. Nutr Cycl Agroecosyst 76:153–170
OKP (2013) Republic of Turkey Ministry of Development Tenth Development Plan (2014–2018)/ Türkiye Cumhuriyeti Kalkınma Bakanlığı Onuncu Kalkınma Planı (2014–2018), Ankara
Otieno HM, Chemining'wa GN, Zingore S (2018) Effect of farmyard manure, lime and inorganic fertilizer applications on soil pH, nutrients uptake, growth and nodulation of soybean in acid soils of Western Kenya. J Agric Sci 10(4):199–208
Parfitt J, Barthel M, Macnaughton S (2010) Food waste within food supply chains: quantification and potential for change to 2050. Philos Trans R Soc B 365:3065–3081
Parrott N, Marsden T (eds) (2002) The real green revolution: organic and agroecological farming in the south. Greenpeace Environment Trust, London
Payraudeau S, Van Der Werf HMG (2005) Environmental impact assessment for a farming region: a review of methods. Agric Ecosyst Environ 107:1–19
Peter C, Helming K, Nendel C (2017) Do greenhouse gas emission calculations from energy crop cultivation reflect actual agricultural management practices?—a review of carbon footprint calculators. Renew Sustain Energy Rev 67:461–476

Peterson B, Snappy S (2015) What is sustainable intensification? Views from experts. Land Policy 46:1–10

Pimentel D (2006) Soil erosion: a food and environmental threat. Environ Dev Sustain 8:119–137

Pincus L, Margenot A, Six J, Scow K (2016) On-farm trial assessing combined organic and mineral fertilizer amendments on vegetable yields in Central Uganda. Agric Ecosyst Environ 225:62–71

Pinstrup-Andersen P (2009) Food security: definition and measurement. Food Sec 1:5–7

Ponisio LC, M'Gonigle LK, Mace KC, Palomino J, de Valpine P, Kremen C (2015) Diversification practices reduce organic to conventional yield gap. Proc R Soc B 282:1799. https://doi.org/10.1098/rspb.2014.1396

Ponniah A, Puskur R, Workneh S, Hoekstra D (2008) Concepts and practices in agricultural extension in developing countries: a source book. International Livestock Research Institute, Addis Ababa

Posthumus H (2005) Adoption of terraces in the Peruvian Andes. Doctoral Thesis, Wageningen University

Power AG (2010) Ecosystem services and agriculture: tradeoffs and synergies. Proc Roy Soc Lond B 365:2959–2971

Pretty J (2018) Intensification for redesigned and sustainable agricultural systems. Science 5:362

Raj A, Jhariya MK, Banerjee A, Yadav DK, Meena RS (2019b) Soil for sustainable environment and ecosystems management. In: Jhariya MK, Banerjee A, Meena RS, Yadav DK (eds) Sustainable agriculture, forest and environmental management. Springer, Singapore, pp 189–221

Raj A, Jhariya MK, Bargali SS (2018b) Climate smart agriculture and carbon sequestration. In: Pandey CB, Gaur MK, Goyal RK (eds) Climate change and agroforestry: adaptation mitigation and livelihood security. New India Publishing Agency (NIPA), New Delhi, pp 1–19

Raj A, Jhariya MK, Harne SS (2018a) Threats to biodiversity and conservation strategies. In: Sood KK, Mahajan V (eds) Forests, climate change and biodiversity. Kalyani Publisher, New York, pp 304–320

Raj A, Jhariya MK, Khan N, Banerjee A, Meena RS (2021) Ecological intensification for sustainable development. In: Ecological intensification of natural resources for sustainable agriculture. Springer, Singapore. https://doi.org/10.1007/978-981-33-4203-3_5

Raj A, Jhariya MK, Yadav DK, Banerjee A (2020) Climate change and agroforestry systems: adaptation and mitigation strategies. Academic Press, New York, pp 1–383. https://doi.org/10.1201/9780429286759

Raj A, Jhariya MK, Yadav DK, Banerjee A, Meena RS (2019a) Agroforestry: a holistic approach for agricultural sustainability. In: Jhariya MK, Banerjee A, Meena RS, Yadav DK (eds) Sustainable agriculture, forest and environmental management. Springer, Singapore, pp 101–131

Raman S (2006) Agricultural sustainability: principles, processes and prospects. The Hawoth Press, Birmingham, pp 13904–11580

Rodrigues MÂ, Ladeira LC, Arrobas M (2018) Azotobacter-enriched organic manures to increase nitrogen fixation and crop productivity. Eur J Agron 93:88–94

Rosset P, Altieri MA (2017) Agroecology: science and politics. Fernwood Publishing Company Limited, Winnipeg

Roy R, Chan NW (2012) An assessment of agricultural sustainability indicators in Bangladesh: review and synthesis. Environmentalist 32:99–110

Sanchez PA (1994) Tropical soil fertility research: towards the second paradigm. In: Proceeding XV ISSS symposium, Acapulco, Mexico, July 10–16, 1994, 1 pp 89–104

Sanchez PA (1999) Improved fallows come of age in the tropics. Agr Syst 47:3–12

Sheoran S, Kuma S, Kumar P, Meena RS, Rakshit S (2021) Nitrogen fixation in maize: breeding opportunities. Theor Appl Genet 134:1264. https://doi.org/10.1007/s00122-021-03791-5

Silici L (2014) Agroecology: what it is and what it has to offer. IIED issue paper. IIED, London

Simon X, Montero M, Bermudez O (2020) Advancing food security through Agroecological technologies: the implementation of the biointensive method in the dry corridor of Nicaragua. Sustainability 12:844. https://doi.org/10.3390/su12030844
Sjögren H (2015) Agroforestry systems with trees for biomass production in Western Kenya, PhD thesis, Department of Forest Ecology and Management Swedish University of Agricultural Sciences, Umeå, p 43
Srivastava P, Singh R, Tripathi S, Raghubanshi AS (2016) An urgent need for sustainable thinking in agriculture—an Indian scenario. Ecol Indic 67:611–622
Tayleur C, Balmford A, Buchanan GM, Butchart SHM, Ducharme H, Green RE, Milder JC, Sanderson FJ, Thomas DHL, Vickery J, Phalan B (2017) Global coverage of agricultural sustainability standards, and their role in conserving biodiversity. Conserv Lett 10 (5):610–618. https://doi.org/10.1111/conl.12314
Tesfahunegn GB, Tamene L, Vlek PLG (2014) Soil erosion prediction using Morgan-Morgan-Finney model in a GIS environment in northern Ethiopia catchment. Appl Environ Soil Sci. https://doi.org/10.1155/2014/468751
Thierfelder C, Wall PC (2009) Effects of conservation agriculture techniques on infiltration and soil water content in Zambia and Zimbabwe. Soil Tillage Res 105:217–227
Thies C, Haenke S, Scherber C, Bengtsson J, Bommarco R, Clement LW, Ceryngier P, Dennis C, Emmerson M, Gagic V, Hawro V, Liira J, Weisser WW, Winqvist C, Tscharntke T (2011) The relationship between agricultural intensification and biological control – experimental tests across Europe. Ecol Appl 21:2187–2196
Tilman DKG, Cassman PA, Matson RN, Polasky S (2002) Agricultural sustainability and intensive production practices. Nature 418:671–677
Tranchant JP, Gelli A, Bliznashka L, Sekou Diallo A, Sacko M, Assima A, Aurino E, Masset E (2019) The impact of food assistance on food insecure populations during conflict: evidence from a quasi-experiment in Mali. World Dev 119:185–202
Tscharntke T, Clough Y, Wanger TC, Jackson L, Motzke I, Perfecto I, Vandermeer J, Whitbread A (2012) Global food security, biodiversity conservation and the future of agricultural intensification. Biol Conserv 151:53–59. https://doi.org/10.1016/j.biocon.2012.01.068
van Ittersum MK, Cassman KG, Grassini P, Wolf J, Tittonell P, Hochman Z (2013) Yield gap analysis with local to global relevance – a review. Field Crop Res 143:4–17
Vandermeer J, Perfecto I, Philpott S (2010) Ecological complexity and pest control in organic coffee production: uncovering an autonomous ecosystem service. Bioscience 60:516–526
Verchot LV, Noordwijk MV, Kandji S, Tomich T, Ong C, Albrecht A, Mackensen J, Bantilan C, Anupama KV, Palm C (2007) Climate change: linking adaptation and mitigation through agroforestry. Mitig Adapt Strateg Glob Chang 12:901–918
Walter A, Finger R, Huber R, Buchmann N (2017) Opinion: smart farming is key to developing sustainable agriculture. Proc Natl Acad Sci 114(24):6148–6150
Wezel A, Bellon S, Doré T, Francis C, Vallod D, David C (2009) Agroecology as a science, a movement, and a practice. A review. Agron Sustain Dev 294:503–515
Whitmarsh D, Palmieri MG (2008) Aquaculture in the coastal zone: pressures, interactions and externalities. In: Holmer M, Black K, Duarte CM, Marba N, Karakassis I (eds) Aquaculture in the ecosystem. Springer, Berlin, pp 251–270
Wood S, Sebastien K, Scherr SJ (2000) Pilot analysis of global ecosystems. IFPRI and WRI, Washington
Yan MJ, Humphreys J, Holden NM (2011) An evaluation of life cycle assessment of European milk production. J Environ Manage 92:372–379

Food and Nutrition Security in India Through Agroecology: New Opportunities in Agriculture System

2

Jayati Chakraborti, Saikat Mondal, and Debnath Palit

Abstract

Food security of a country or community is the major condition to achieve the highest sustainable development. Climate change exerts enormous impact on global food security putting it to a risk most of the time. Thus, lack of food security is the most burning global problem emerging as a result of climate change. Depletion of environmental resources due to extreme weather conditions affects food production and instability. Countries and communities in the coastal areas, floodplains, high mountains, drylands, the Arctic region, and low-income people particularly in urban areas are most vulnerable to food insecurity. People's right to food can be secured by reducing the risk of hunger and undernutrition. Food security is the combination of food availability, food accessibility, food utilization, and food stability to the people. Thus, socioeconomic structure of a community is also dependent on food security. The increasing nature of social and biophysical vulnerabilities of a country determined the effect of the changing climate on food security. Food grain production in India has declined to 1.8% in 2015 than above 3% in the 1990s. India has become the second largest producer of rice and wheat and is leading in pulse production all over the world. To reduce the risk to food security from climate change, effective adaptive measures should be adopted. India scored 31.1 in the 2018 Global Hunger Index. Human impacts on the environment exacerbate changes in the climate which lead to the risk of food security. Global food security can be protected by acclimatization,

J. Chakraborti
Department of Zoology, Sreegopal Banerjee College, Magra, India

S. Mondal
Department of Zoology, Raghunathpur College, Purulia, India

D. Palit (✉)
Department of Botany, Durgapur Government College, Durgapur, India

M. K. Jhariya et al. (eds.), *Sustainable Intensification for Agroecosystem Services and Management*, https://doi.org/10.1007/978-981-16-3207-5_2

adaptation, and mitigation. Acclimatization and adaptation can be acquired through self-directed efforts whereas innovation policies by ecological managers. Development and adoption of new technologies are necessary to adopt with the changing climate and achieving total food security in the country. This chapter mainly focuses on the effects of climate change on food production and accessibility to healthy food irrespective of socioeconomic class and caste in India and also puts concern on the discussion regarding the recent policies, action plans, and strategies adopted worldwide to handle the risk of climate change.

Keywords

Climate change · Food access · Food security · Food stability · Mitigation

Abbreviations

ADB	Asian Development Bank
AFSIS	ASEAN Food Security Information System
AIFS	ASEAN Integrated Food Security
APTERR	ASEAN Plus Three Emergency Rice Reserve
ASEAN	Association of Southeast Asian Nations
BBC	British Broadcasting Corporation
BMI	Body mass index
Ca	Calcium
CFC	Chlorofluorocarbon
CO_2	Carbon dioxide
CSISA	Cereal Systems Initiative for South Asia
FAO	Food and Agricultural Organization of the United Nations
Fe	Iron
FIES	Food Insecurity Experience Scale
FSPWG	Food Secure Pacific Working Group
GEF	Global Environment Facility
GHG	Greenhouse gas
GHI	Global Hunger Index
HYV	High-yielding variety
IARI	Indian Agricultural Research Institute
IFPRI	International Food Policy Research Institute
IPCC	Intergovernmental Panel on Climate Change
IPFSP	Industry Partners for a Food Secure Pacific
IUCN	International Union for Conservation of Nature
K	Potassium
LPA	Long Period Average
N	Nitrogen
N_2O	Nitrous oxide
NVA	New Vision for Agriculture
O_3	Ozone

P	Phosphorus
PIF	Pacific Islands Forum
Ppm	Parts per million
SAARC	South Asian Association for Regional Cooperation
SPA-FS	Strategic Plan of Action on Food Security in the ASEAN Region
TSS	Total suspended solids
UN	United Nations
WB	World Bank
WWF	World Wildlife Fund
Zn	Zinc

2.1 Introduction

Nowadays, climate change is one of the most serious global issues and alarming challenges to the environment. Climate change refers to not only global warming but also a range of other extreme weather events or complex shifts that affect earth's weather systems and lead to wide long-term impacts like change in wildlife populations and habitat shifting, rising of sea water level, and shortage of food, water, and energy creating greater risk to livelihood, health, and life (Banerjee et al. 2020; Meena et al. 2018; Jhariya et al. 2019a, b; Raj et al. 2020). At present days, climate change is synonymous to anti-development. Greenhouse gas (GHG) emission and deforestation are mainly responsible for 3 degree centigrade rise in the earth's temperature by this century. The mean global temperature has raised by 0.74 °C and atmospheric carbon dioxide (CO_2) concentration from 280 to 368 ppm (part per million) in the year 2000 (Watson and McMichael 2001). The implication of climate change on the food security of a country or population is a matter of global concern because it has the potential to make life vulnerable (FAO 2008b).

India is a country of great diversity. The topography abruptly varies from region to region. It comprises mountains, coasts, forests, deltas, and deserts. This diverse geography is an intensive driver for climate change like rising temperatures, intense droughts, hot days, heat waves, sea water level rise due to melting ice and glaciers, erratic rainfall, heavy precipitation, dangerous flood storms, and differences in food habits of the people living here. The extreme weather patterns are the main causes behind loss of biodiversity and ecosystems which may result in the decline in agricultural productivity (Raj et al. 2018a, b). This may cause an increase of conditions like hunger and malnutrition and prevalence of diseases. India holds the second largest position in food production in the whole world. India is also the second highest in population among which only 35% is urban and the rest are rural based (Worldometers 2019). Nearly 195 million people are victims of food crisis (Food and Agricultural Organization of the United Nations, FAO 2007).

According to the record of the Government of India 2013, about 22% of its population lie under the poverty line. At the same time, many billionaires live in India. Rapid urbanization of rural India and rise in incomes are responsible for the change in food preferences and dietary intake from carbohydrate-rich cereals to expensive protein- and mineral-rich agricultural products such as fruits, fish, milk, meat, etc. (Law et al. 2019; Mittal 2006). This trend may lead to nutritional deficiency within a large section of population in India. Women and children are the most vulnerable section and worse victims of food crisis and malnutrition. Women constitute a major chunk of population whose right to food is affected to a large extent throughout the world, especially in developing countries, including India. A survey performed by the National Family Health in 2015–2016 reported that a maximum number of children under 5 years of age are underweight in the states of Bihar (43.9%), Madhya Pradesh (42.8%), and Andhra Pradesh (about 31.9%) in India (Chakrabarty 2016).

According to Mittal (2008), the demand for food in India is greatly increasing due to the rise in population and also income of the people, but food availability is constrained by low yield. She made predictions on demand and supply of food in India up to 2026 based on previous trends in her paper.

A latest report in a press conference by The Asian Development Bank and Postdam Institute for Climate Impact Research (2017) reflects on the long and hard struggle of the people of India for food security. According to the report, by the end of this century, southern parts of India will also observe an increase in temperature by approximately 6 °C as it has already been prevalent in the Ganga-Brahmaputra basin, i.e., northern part of the country. The International Union for Conservation of Nature (IUCN 2019) states that the rise in temperature will mostly affect the production of major crops or staples like rice, wheat, and maize. The increase in average temperature will lead to the rise in ocean temperature which will adversely influence the abundance and occurrence of marine species. Thus, this rise in temperature will hamper the income of farmers, fishermen, and many other agriculture associates.

2.2 India's Rank in Global Hunger Index 2018

The Global Hunger Index (GHI) is used to measure the hungriness and undernutrition annually all over the world (BBC News 2009; Withnall 2016). The results of GHI are calculated and published in October each year.

The 2018 GHI indicates that the level of hunger worldwide fell to 20.9 from 29.2 in 2000 but India scored 31.1 and got the rank 103rd out of the 119 qualifying countries (Pruthi 2018). Thus, India is standing at a serious level of hunger, and about 12 states are at an alarming situation of the GHI. In 1990, 210.1 million people in India were suffering from undernutrition. FAO reported that this number has declined to 194.6 million only in 2014 (FAO, IFAD and WFP 2014). In 2016, GHI score for India was 28.5 (von Grebmer et al. 2016), and the neighboring countries like Bangladesh, Nepal, Sri Lanka, and Myanmar got better GHI scores than India. India failed to reach the Millennium Development Goal. The goal attempted to lower down the number of undernourished people by 50 percent.

2.3 Climate Change

Climate change is the most significant and effective issue in the discussion of global development in every aspect. The change of climate refers to seasonal variations for a long period. The phenomenon is the outcomes of growing accumulation of greenhouse gases in the atmosphere. The Intergovernmental Panel on Climate Change (IPCC) defined climate change as "a change in the state of the climate that can be identified by the changes in the mean and/or the variability of its properties and that persists for an extended period, typically decades or longer" (IPCC 2013).

2.3.1 Drivers of Climate Change

There are many driving forces that induce climate change. They can be broadly divided into two categories: i) internal and ii) external. External processes are those that operate from outside the planet earth. Internal processes operate from inside the earth (Rehman et al. 2015) which include oceans and atmosphere. It is of two types, i.e., natural or climatic and artificial or human induced. Natural forces include changes in the composition of the atmosphere due to variation in ocean circulation, volcanic eruptions from earth's core, and earth's collisions with other comets or meteorites. Exploitation of natural resources, biodiversity, land use, etc. are influenced through anthropogenic activities. Alternations in the earth's atmospheric composition also occur due to human activity-induced pollution. These are actually responsible for recent climatic changes and will continue to exert their influence on climatic future (Rehman et al. 2015; Meena and Lal 2018).

2.3.1.1 Variation in Temperature

The temperature of the earth depends on the incoming radiation from the sun. Global mean temperatures vary as much as 5 °C due to change in the amount of solar radiation. There are three main factors that affect the amount of solar radiation; they are (1) the eccentricity of the earth's orbit, (2) the obliquity of the earth's axis (approximately 41,000 years), and (3) the variation of latitude which is a small deviation in the earth's axis of rotation relative to the solid earth. Huge amount of heat energy is necessary for a small rise of the average surface temperature of the planet. An average increase of only 0.07 °C (0.13 °F) in annual temperature of the earth per decade has been recorded since 1880 (Lindsey and Dahlman 2020).

2.3.1.2 Volcanic Eruptions

Volcanic eruptions are an important natural factor responsible for climatic changes. Over the last 100 years, volcanic eruptions may be the reason for global climate change. A weak volcanic activity might result in gas and particle emission in the troposphere (first layer of atmosphere), which consists of the larger portion of volcanic mass flux and GHGs. These materials emerging from volcanoes accumulate moderately in the atmosphere and form natural sources of pollution. Volcanic eruptions eject out large amount of sulfur-rich gases which are converted to dust

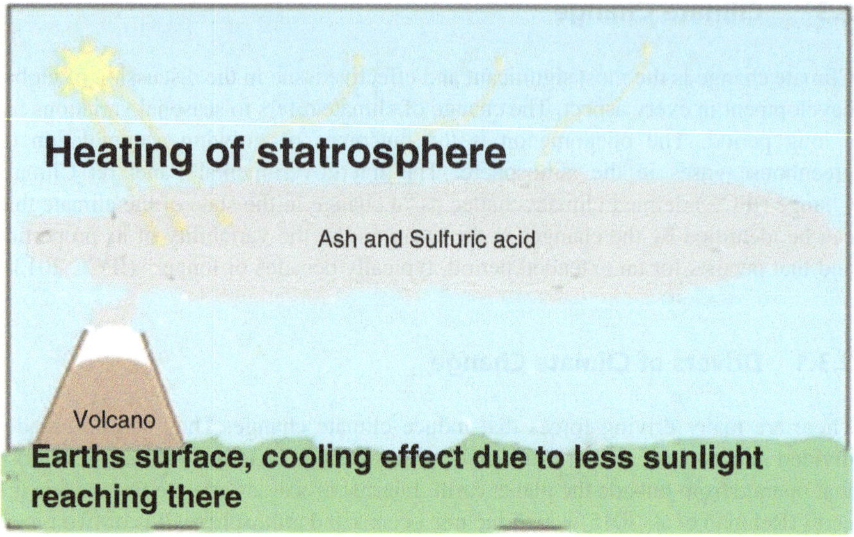

Fig. 2.1 Volcanic eruption and earth's climate change

particles and remain in the stratosphere layer for several years. Thereby, it affects the earth's climate. These sulfur-rich dust particles absorb enough solar radiation and prevent it from coming to the earth's surface. Thus, the stratosphere becomes hot and produces a cooling effect on the earth's surface resulting in an anomaly in average global temperature (Wolfe 2000) (Fig. 2.1).

2.3.1.3 Role of Greenhouse Gases

Since the twentieth century, GHGs are the most important driver of climate change. GHG absorbs some of the amount of radiant solar energy from the earth's surface, and thus the heat is retained in the lower atmosphere. Thus, increased concentrations of GHG in the atmosphere can produce a warming effect. From 1990 to 2015, the total warming effect of the earth's atmosphere produced from GHG due to anthropogenic causes has increased by 37% (Environmental Protection Agency 2017). The atmospheric concentration of GHG like CO_2 and certain other trace gases, including methane (CH_4), nitrous oxide (N_2O), chlorofluorocarbons (CFCs), and tropospheric ozone (O_3), has increased due to human activities. All of these gases absorb solar radiation resulting in the increase in atmospheric temperature.

2.3.2 Climate Change: World Scenario

Climate change is an immeasurable global threat today. There are so many reasons behind this problem. One of the major reasons is CO_2 emissions in the atmosphere. Climate change depends on the combined effect of GHG and aerosol emissions (IPCC 2013). Analysts suggest that the global temperature increase will lie between

0.5 °F and 1.3 °F (0.3–0.7 °C) in the next two decades (USGCRP 2016). This rise in temperature is very much sensitive to uncertainties in the emissions of GHG, black carbon, and other aerosols. If the concentrations of GHG were stabilized at existing level, it would ensure an increase in temperature of at least an additional 0.6 °C in this century in comparison to the last few decades. Deforestation is another major reason behind the O_3 layer depletion and climate change. In the last 15–20 years, the increase in global carbon emissions has been consistent. Increase in global temperature cannot be limited below 2 °C if the present trend of carbon emissions continues worldwide. At present day, the concentration of mean global atmospheric CO_2 concentration was assumed to occur about millions of years ago when this planet was unsuitable for living (USGCRP 2017). Climate change will seriously affect the coastal and marine ecosystems. This may also be an added stress for the fisheries and aquaculture sectors (Nema et al. 2012).

Various studies have put light on the fact that by 2095 world temperature and CO_2 concentration will increase by 3.4 °C and 1250 ppm, respectively (Pachauri and Reisinger 2007; Kumar et al. 2021). These changes have contributed to the melting of ice, glaciers, all over the world especially at the poles, West Antarctica and Greenland, and Arctic sea. The number of glaciers in Montana's Glacier National Park has decreased from 150 to less than 30. This event resulted in the global sea level rise by 0.13 inches (3.2 millimeters) every year which is increasing very fast and is expected to increase between 10 and 32 inches in recent years. It has challenged the survival of many species such as the Adélie penguin. Some species of foxes, polar bears, butterflies, and alpine plants have migrated toward further northern latitudes. Thus, rise in temperature is affecting wildlife population and their habitats resulting in the change in ecosystem (Nunez 2019).

The amount of average precipitation (rain and snowfall) has increased worldwide. But some regions are facing severe drought condition and increased risk of wildfires. Large parts of the world will face the risk of mega drought. Hurricanes and other such storms will occur frequently. There will be shortages of drinking water since glaciers are the source of three-fourths of the world's freshwater.

2.3.3 Climate Change: Indian Scenario

India is a land of diverse topography. So, climate change is a very critical issue in the context of the subcontinent. Irregular rainfall, drought, cyclonic storm, and drastic rise in temperature have become a common climatic trend in many parts of India. Over the past century, the amount of rainfall in monsoon season has decreased by 6–8% in the eastern, northeastern, and southern parts of the country where precipitation rate is known to be higher than other parts of the country (Lal et al. 2010). In West Bengal and Gujarat, storms have become more common (Khan and Hasan 2017). Sea level rise has taken place from 1.06 to 1.75 mm annually (IPCC 2007). Though monsoon days are reducing, the amount of rainfall will increase from 1 mm to 4 mm per day (Lal et al. 2010). Increase in rainfall in less rainfall areas will lead to flood and loss of fertile soil. Drought-prone areas like Rajasthan, Madhya Pradesh,

Gujarat, and Andhra Pradesh will receive more less rainfall and will experience severe droughts (Lal et al. 2010). Studies show that the average temperature of India will increase by 0.21 °C per 100 years and the level of groundwater will decrease by 1 to 3 meters per year (Khan and Hasan 2017). Surface temperature rise would result in the rapid meltdown of the Himalayan glaciers.

2.4 Food Security

Food security of a country or community is a very complex and multidimensional issue. Food security mainly refers to the regular crop production, sustainability, and thus its access to people at every level. A balance between the food availability, accessibility, utilization, and stabilization in a country controls the nation's food security (Kumar et al. 2020; Jhariya et al. 2021a, b). It has several dimensions and could not be restricted to a particular one. Food manufacturing, food division among various communities or regions, food grade, and food availability all together make up the total food security of a country (Kumar et al. 2020a). According to the World Food Summit in Rome (1996), food and nutrition security may be defined as a situation "when all people, at all times, have physical and economic access to sufficient, safe and nutritious food to meet their dietary demand according to food preferences for an healthy and active life" (FAO 2008a).

Food security is associated with producing healthy food for the whole population but not with protection of food. Food security of a country depends on the following basic steps: production of a large amount of cereal crops to make it available to the whole nation up to the limit necessary for survival. This includes:

- Pulses and cereals should be available sufficiently.
- It includes sufficient availability of milk and milk products.
- Considering non-vegetarians, availability of fish, meat, and egg should be secured.

2.4.1 How Can Food Security Be Ensured?

There are three dimensions of food security which are:

- Availability of enough food for all people according to preference.
- Absence of barrier on access to food.
- Ability to buy healthy food.

Food security and insecurity are of two types: transient and chronic. Transitory food security deals with raising policies related to price stability of food, crop insurance facility, and temporary employment facility for farmers who are vulnerable people during off-season, i.e., conditions like drought, flood, and inflation. In contrast, chronic food insecurity is associated with the problem of continuous non-availability or access to inadequate diet arising due to poverty or unequal socioeconomic status (viz., gender inequality) of the country.

According to Gopaldas (2006), hunger is of two types: (1) self-reported hunger in which there is a wish to satiate hunger and people themselves decide their own ability to fulfill their hunger and (2) chronic/endemic hunger in which the human body gets habituated in taking less amount food than that required for proper growth and development. The underlying causes of hunger in India are:

- The increase in population and gradual fall in crop yield per year.
- Scarcity of good-grade food grains due to export of healthy crops.
- Increase in socioeconomic inequality; the poors are forced to spend more money on medicines, education, transport, fuel, etc. from their low income, thus reducing their ability to spend on food.
- The price hike of cost of pulses, vegetables, and oils due to stagnant production and other commodities of basic requirement of every individual has reached to such an extent that the lower-income population is incapable of buying these food items, thus endangering their survival.
- Poor service of the government to fulfill the basic needs of people along with more accessibility for the low-income groups in order to afford proper nutrition and medical healthcare.
- Lack of employment in the last 10 years.
- Increase in regional discrepancies.

2.4.2 Effects of Food Crisis

Food insecurity or hunger may lead to the following consequences in the country or state:

- News of starvation, deaths, and farmers' suicides from many states.
- Acute anemia among more than 70% of the country's women population (Rammohan et al. 2011).
- Malnourishment and death in 46% of the nation's children from different states (Singh 2020).
- Decline in sex ratio of the country's population (Rao et al. 2017).

2.4.3 Food Security: World Perspective

Food security is a major global challenge. A committee of World Food Security in the United Nations defined food security as a condition in which sufficient amount of healthy and fresh food based on food preferences and dietary needs for an active and healthy life. It should be available and affordable throughout the year to every people, irrespective of caste, gender, and social status (FAO 2015). Everybody needs food, but the complexity of delivering sufficient food to the population for all countries, whether developing or developed, lies in the fact that it is not just about food and feeding people but also about all aspects of economy and society of the country (Breene 2016). According to FAO of the United Nations, almost a billion people around the world which includes 16% of the population in developing countries are found to be malnourished (Breene 2016).

Global food security is complex due to population growth across countries, their changing tastes, global climate change and water scarcity, rise in food prices, and several environmental stresses. All these have high impact on food security (Andersen and Lorch 1999). From 2007 to the first half of 2008, global food prices raised sharply worldwide from Bangladesh to Brazil and from Mexico to Mozambique leading to social unrest and riots in both developing and developed economies. War, poverty, and climate change are the reasons for food insecurity (Lawrence 2017). The Food Insecurity Experience Scale (FIES) contributes for the establishment of a food insecurity indicator for global monitoring. Global prevalence rate of moderate and severe food insecurity was developed by the survey data collected by FAO from 153 countries or territories nationally, once every year in 2014, 2015, and 2016 (Cafiero et al. 2018).

In 2009, the members and partners of the World Economic Forum took an initiative for a New Vision for Agriculture (NVA) to improve food security, environmental sustainability, and economic opportunities till 2050 (Breene 2016). Traditional, modernizing, and industrialized systems work simultaneously for producing and distributing food. These three major systems coexist in today's world. There are several social, economic, and political forces that enhance the transition or movements from traditional to modernized to industrialized system. Neoliberal globalization and financialization have profound role in reshaping the current global food system (Lawrence 2017).

Many strategies and policy should be undertaken urgently to generate adaptation responses. Those include maintenance of water distribution system, patterns of land use, postharvest processing of food, its trade, and market price. IFPRI works on global food security and has put emphasis on the implementation of sustainable agricultural technologies, analysis of cash transfers, building resilience to shocks, and balancing the nutritional value of food and costs of its production (Andersen and Lorch 1999). The World Economic Forum aimed to bring the farmers, government, society, and private sector together to facilitate partnerships and encourage exchange of knowledge at regional and country levels (Breene 2016).

2.4.4 Food Security: Indian Perspective

India could not make out a solution for the fulfillment of food and nutrition security of its population. The country has a poor performance in reducing hunger and malnutrition, and this has been considered as a matter of great concern. Many evidences collected from national surveys have reflected the fact that India is facing triple affliction of malnutrition presently (Singh 2019). According to the data collected from a survey made by the National Family Health in the year

Table 2.1 Position of India in world's agriculture (2018) (Source: FAO, World Food Situation, FAO Cereal Supply and Demand Brief 2020)

Item (million hectare)	India	World	India's position
Total population (crores)	135.26	759.43	Second
Total arable land	159.7	1407	Second
Irrigated land	667,000 sq. km	3,242,917 sq. km	Second (Date of information 2012)
Total cereal production (million tons)	277.49	2656	Third
Rice	107.8	514.6	Second
Wheat	98.0	732.4	Second
Pulses	20.26	42.33	First
Coarse cereal	43.06	1409.1	Second

Table 2.2 Comparison of average crop production between 2019/2020 and the last 5 years

Crop items (million tons)	Average crop production for the last 5 years	2019–2020
Total food grain	277.49	281.37
Rice	107.8	117.47
Wheat	94.61	106.21
Pulses	20.26	23.02
Coarse cereal	43.06	45.24
Oilseed	31.52	34.19
Sugarcane	349.78	353.84
Jute and jute-like fibers (million bales)	9.81	10.07
Cotton	28.04	34.89
Vegetables (tons/ha)	13.4	
Potatoes (tons/ha)	19.9	
Onion (tons/ha)	16.6	

2005–2006, in India, married women have a BMI (body mass index) less than normal.

According to the World Bank database in 2013, the cereal production of India is much lower than those of developed regions such as North America (6671 kg per ha), East Asia and the Pacific (5184 kg per ha), and the Euro area (5855.4 kg per ha) (Table 2.1).

From Table 2.2, it is clear that the total amount of food production has increased in 2018–2019 than the past 5 years according to government estimate. According to a press release in the Cooperation and Farmers Welfare today, the Department of Agriculture predicted that the production of major crops will be higher in the year 2019–2020 than the normal average production of the last few years due to 10% higher average monsoon rainfall in the last year than LPA (Table 2.3).

Table 2.3 Growth rate of yield per hectare (%) of food grains in India (Source: Reserve Bank of India 2015)

Year of production	Rice	Wheat	Pulses	Total food grains
1980 to 1990	2.7	3.4	2.0	3.0
1991 to 2001	0.9	1.7	−0.6	1.7
2002 to 2012	1.6	1.0	2.4	1.7
2010 to 2015	1.6	−1.0	1.9	1.8

Food security can be achieved by a nation only if it develops self-capabilities for harvesting crops without any assistance from foreign organizations neither financial nor technical. Specifically in the late 1960s, Indian farmers were motivated to cultivate HYV (high-yielding varieties) seeds for increased yield of food grains, particularly rice and wheat, evolved as a consequence of the Green Revolution. After the Green Revolution, India become technically sound enough to produce a large amount of food products and became capable of avoiding famines even during worse weather conditions. The major regions producing food grain in India are as follows (Abraham 2019):

The major rainfed rice-yielding states include West Bengal, Bihar, Assam, Eastern Uttar Pradesh, Andhra Pradesh, Tamil Nadu, Orissa, Karnataka, Kerala, and coastal areas of Maharashtra. Punjab and Haryana are also growing rice after adopting suitable modern irrigation techniques. These two states have become the country's top producer of rice contributing about 15% to India's rice production (Abraham 2019)

Wheat production is the major focus in the comparatively dry states where the annual rate of rainfall is less, viz., Punjab Haryana, Uttar Pradesh, Bihar, and parts of Rajasthan. Millets such as jowar, bajra, and ragi are also grown in India. Bajra is grown dominantly in Rajasthan as it provides the exact climatic condition for bajra harvesting. Ragi is a rainfed crop so it grows in huge amount in Tamil Nadu and Karnataka (TNAU Agritech Portal 2013). Pulses are the major sources of protein to the vegetarians in Bihar, Uttar Pradesh, Madhya Pradesh, Punjab, and Haryana. They can be grown as both *kharif* and *rabi* crops. The *rabi* season pulses are peas and masoor, and the *kharif* season pulses include urad, arhar, and moong. Gram seeds are mainly produced in Uttar Pradesh, Madhya Pradesh, and Rajasthan

2.4.5 Food Security and Climate Change

Stability in food security is the hardest global challenge among all other negative impacts of climate change. Research showed that changing climate does not exert equal effect on all countries of the world. But the impact is greatest on the tropical and equatorial countries of the planet (Ahmad et al. 2011). India could not ignore the bad impact of climate change for two reasons. First, as we have mentioned, in terms of geography, India has a unique identity. Plains, grasslands, mountains, deserts, glaciers, and rainforests constitute this subcontinent, and its geographical location is different from that of the northern countries. Second, it is surrounded by water bodies on its three sides: the Bay of Bengal in the east, Arabian Sea in the west, and Indian Ocean in the south which is the key regulator of country's rainfall. Crop production, water supply, biodiversity, and livelihoods are directly and indirectly linked to precipitation patterns (Chakrabarty 2016). Change in the pattern of rainfall in semi-arid and sub-humid regions constitute main cultivated lands of India results into decline in the production of rainfed crops. Many significant studies have been carried out to examine the effect of changing climate on the agriculture which shows that climate change may affect food system in several ways ranging from crop production to changes in markets, food price, and supply chain infrastructure. Temperature, rainfall, and humidity are the major climatic parameters that determine the quality and quantity of the crop yield. Any alteration in these factors may produce natural catastrophes which may cause huge crop loss, less and low-grade crop yield due to decay, and decline or erosion of cultivable lands (Chaudhry and Aggarwal 2007).

Climate change will affect all four dimensions of food security, namely, availability, affordability, utilization, and sustainability of food (Ranuzzi and Srivastava 2012), though most scholarly attention are focused on one dimension, i.e., food production. Accessibility and utilization of food largely depend on family or income and social rights. Climate can also determine the trends in price changes as well as the short-term variability of prices; thus, the stability of food system may be at risk under changing climate (Wheeler and von Braun 2013).

2.5 Climate Change and Food Production

Change in climate imparts huge challenge on the nation's food security. Risk of food security is becoming difficult to escape. Yields of both rainfed and irrigated crops are severely affected due to rise in temperature and decline in rainfall patterns (Lal et al. 2001). In India, harvesting of two main staple crops, i.e., wheat and rice, is found to be very much sensitive to temperature rise greater than 34 °C (Lobell et al. 2012). Reduction in the productivity of food is harming the livelihoods of the vulnerable class of the society. They are already on the verge of food insecurity (FAO 2008a). Global warming causes an increase in average surface temperature of 4 °C and a 10% increase in mean annual rainfall and a 15% increase in variation in year-to-year monsoon rainfall. The unchecked rise in sea levels leads to loss of land and infrastructure affecting availability and access dimensions of the food system. The states of Uttar Pradesh, Gujarat, Rajasthan, Madhya Pradesh, Maharashtra, and Karnataka are highly vulnerable to climate change. Table 2.4 shows season-wise variability of food grain production in India. Further, it indicates country's 3% increase in food grain production in the year 2017–2018 than 2016–2017. According to the Department of Agriculture and Cooperation, this increase in production is a result of normal rainfall during monsoon 2017 and also various effective policy initiatives taken by the government (Nadkar 2018).

Table 2.4 Season-wise production food grains in India from 2010 to 2018 (Source: Department of Agricultural and Farmers Welfare 2018)

Year	Kharif	Rabi	Total
2010–2011	121.0	124.0	245.0
2011–2012	131.0	128.0	259.0
2012–2013	128.1	129.2	257.3
2013–2014	128.7	136.5	265.2
2014–2015	128.1	124.0	252.1
2015–2016	125.2	126.5	251.7
2016–2017	138.0	137.0	275.0
2017–2018	141.0	140.0	281.0

2.6 Climate Change and Nutritional Deficiency of Crops

It is very clear from the definition of food security as established in the 1996 World Food Summit that food, nutrition, and health are interlinked with each other (Ramachandran 2013). Only sufficient production of food grain cannot stop reduction in undernutrition rates or nutrient deficiency among the population, but people need to intake adequate quantities of balanced diet to remain well nourished and healthy (Ramachandran 2013). Scientists have found that the long-term rise in temperature and decrease in rainfall patterns reduce soil quality, thus depleting the nutritional value of crops. Researches revealed that increased levels of atmospheric CO_2 have a harmful impact on the nutrient value of crops. Higher-level CO_2 decreases the zinc, iron, and protein levels in a crop. Different scientific studies on the impact of increased temperature and elevated atmospheric CO_2 on the nutritional level of cereals have predicted that by 2050, 148.4 million people globally and 53.4 million Indians may suffer from protein deficiency for which the staple food is rice (Sarkar 2018).

2.6.1 Effect of Temperature on Agriculture

In some part of India, variation in temperature and water availability results in reduced yields. In tropical regions, plants require optimum temperature conditions which are necessary for their growth; a slight increase in temperature may hamper its growth. For example, even 2 °C increase in temperature may hamper the expected yield of wheat crop (Hatfield and Prueger 2015). Anapalli et al. (2000) found that rice yield would decline about 6% for each one degree rise in temperature. Increase in temperature limits crop production mainly in two ways: (1) restricting vegetable growth and (2) adverse effects on fruiting. At high temperature, vegetable crops are subjected to very high transpiration losses (Rehman et al. 2015). At high temperature particularly above 40 °C, citrus fruit trees stop flowering, and growth is ceased even if soil nutrients and moisture are adequate. The best temperature for the growth of citrus fruits is 25–30 °C (Abobatta 2019). Recent global studies and IPCC reports have indicated that there is a probability of crop loss by 10–40% with increases in temperature in India by 2080–2100 (Ruchita and Rohit 2017). Lowering of temperature is also dangerous for plant growth as it is associated with freezing plant tissues which bring about certain histochemical changes in plant tissue and ultimately death (Rehman et al. 2015). Therefore, temperature limits the geographical areas suitable for harvesting various crops.

2.6.2 Effect of Rainfall and Drought on Agriculture

Quality and quantity of crops do not depend on temperature only. Prevalence of optimum rainfall is also an important determinant of cultivation. Temperature is again dependent on some other factors such as prevalence of wind and storm, water availability, the duration and intensity of sunlight, and various other factors. Different steps of farming like preparation of seedlings, transplantation of seedlings for proper plantation, and growth of *kharif* and *rabi* crops are fully dependent on regular and timely precipitation at threshold level and optimum temperature. Uncertainty and variation of rainfall in the growth season are expected to impact the pest-pathogen buildup and reduce rice yields. Increase in average global temperatures is increasing the evaporation rates which will lead toward increased risk of storms and untimely and heavy precipitation. Heavy rainfall may result in floods in some places, while droughts may occur in other areas receiving less rainfall (Gray 2019). Leaf structures may become changed and modified due to decreased photosynthetic and transpiration performance as an adaptive measure against drought condition (Fahad et al. 2017). Sufficient amount of water is required for the plants to absorb nutrients from the soil and its transport. Drought will limit the transport of nutrients in the plant body resulting in stunted growth and ultimately lowering the yield of crops (Fahad et al. 2017). It is clear from the above discussion that cultivation of crops or food production is very much climate dependent. Little variation in temperature, precipitation rate, soil salinity, etc. will affect the crop yield.

2.6.3 Effect of Carbon Dioxide Concentration on Agriculture

Le Quéré et al. (2018) reported that each year there is a regular increase in atmospheric CO_2 levels due to its emission from fossil fuel combustion, industry, and land use change and in 2018 atmospheric concentration of CO_2 reaches its highest level. Elevated CO_2 levels produce extreme heat, warming, and acidification of ocean water, resulting in disruptions in crop and livestock production in many ways: causing crop failure, reducing the abundance and range of fish varieties, and inundating coastal wetlands used for rice farming and aquaculture (Smith et al. 2018). The effect of rising CO_2 concentration is not the same on the macro- and micronutrient level for all crops. Concentration of protein and macronutrients like iron and zinc declined significantly in various major grain crops such as wheat, rice, and barley. Zinc and iron concentration are also likely decline in peas and soybeans, but experience little or no loss of protein (Smith et al. 2018). Rise in atmospheric

CO_2 level shifts regional precipitation patterns from high to low volume which would cause drought condition resulting in famine at various parts of the world. Plants respond to increased atmospheric CO_2 concentration by increasing its carbon assimilation rate and decrease the water loss by reducing stomatal conductance which may be considered useful under water-deficit situations. Elevated CO_2 level may also alter the protein synthesis of the plants (Taub 2010; Rehman et al. 2015; NASA 2016).

2.6.4 Impact of Salinity on Agriculture

Climate change induces rise in sea level, thus inundating coastal areas and bringing large area of fertile land under saline water. Salinization of soil reduces the soil fertility and leads to the loss of productivity. This phenomenon forces the residents of coastal area like Sundarbans region, the world's largest mangrove forest, to migrate in land areas of India and Bangladesh. Saline soil is unsuitable for cultivation because it has poor oxygen content and water is not present in liquid form in this type of soil; thus, the plant roots are unable to absorb water from the soil. Such type of soil is known as physiological dry soil, and the phenomenon such as the osmotic or water-deficit results due to salinity. This waterlogged anaerobic soil condition leads to nutrient deficiencies and ion toxicity in addition to osmotic and oxidative stress in the plants. Most of the vegetable crops grow in low salinity. The threshold value of salinity (ECt) ranges from about 1–3 dS/m approximately (Machado and Serralheiro 2017). The symptoms of salt stress in plants are stunted growth, wilting and necrosis of leaves, chlorosis, and burning of leaf tips (Machado and Serralheiro 2017). However, salt stress has some positive effects on some vegetable crops. It may increase the total soluble solids (TSS), dry matter content of fruit, acid content, carotenoid content, and antioxidant property of some fruits like melon, tomato, sweet pepper, and cucumber (De Pascale et al. 2015).

2.6.5 Climate Change and Food Accessibility

It is clear that agriculture is fully dependent on several physical parameters and highly vulnerable to the current changing climate situation. There has been a significant lowering in the production of crops in comparison to increasing demand. Thus, a hike in the prices of different food materials has been observed. Under such situation, the poor people are in big trouble. They had to sacrifice their total income and sometimes other assets also to meet their nutritional requirements for survival (Llyod et al. 2011). Lengthening of the crop growing season due to changes in the climatic parameters is adversely affecting the net income of farmers, agricultural laborers, and others depending solely on wages from agricultural fields. They cannot access good and healthy food due to shortage money (Schmidhuber and Tubiello 2007). In drought- and flood-prone regions, poor people, women, and children are the most vulnerable to food insecurity and inequality (Chakrabarty 2016).

2.6.6 Climate Change and Food Absorption

Food absorption is another major aspect of food security which gets hampered due to changing climate. It includes the utilization or consumption of food. While climate change reduces food production and affordability due to unsuitable weather and price hike, respectively; thus, proper utilization of food by the people cannot take place (FAO 2016). Climatic variation is also bringing about alternation rather reduction in the nutritional quality of foods. It will give rise to a neglected epidemic known as "hidden hunger" and nutrient deficiency among the common people of the country (Myers et al. 2014). Micronutrient deficiencies will decrease the immunity power and consequently increase the risk of infections and diseases (Phalkey 2015; Sheoran et al. 2021). Nowadays, in the eastern parts of India, the cases of diarrhea have increased by about 13% which may be one of the combined effects of food insecurity developed due to climate change (Moors et al. 2013). On the contrary, Ramachandran (2014) stated that climate change has been able to reduce diarrhea-related deaths in India and hence the argument continues.

2.6.7 Case Study

Let us now discuss about the case of Bihar. Agriculture forms the backbone of Bihar's economy, and it is one of the states of India which is vulnerable to climate change. According to a report, 60–70% of cultivation in Bihar is carried out by indigenous methods and thus fully depends on rainfall in terms of irrigation system which is inadequate (Bameti 2018). So any irregularity in rainfall will negatively influence the production of both *kharif* and *rabi* crops in Bihar (Bameti 2018). The trends of changes in seasonal patterns in the last few years are reduction in rainy days during monsoon, prolonged summer, and shorter winters, all affecting the cropping patterns. Thus, sowing of rice sapling gets delayed by over a month in the *kharif* (summer) season due to the late arrival of monsoon, and proper growth is hampered due to poor rainfall along with high temperatures (Khan 2018).

Sea level rise combined with land subsidence has salt-contaminated farm lands, and poor availability of high saline-tolerant crop varieties has made farming in low-lying islands unviable in the Indian Sundarbans. So farmers in the coastal land are cultivating just enough for their families' need, and they are shifting toward brackish water aquaculture production which will bring more job opportunities, reducing international migration. With the introduction of aquaculture industry, people who normally used to migrate abroad for jobs are engaging themselves in fish crop production to maintain their livelihood.

Risk of climate change is also prominent on the metro cities. Mumbai, Chennai, and Kolkata are especially affected by climate change (Pandve 2010; Dasgupta et al. 2012). Frequent floods have become a regular phenomenon in these cities and also along the banks of river Hooghly during monsoon season. The inhabitants in low-lying areas or wetlands get submerged and are affected every year (World Bank 2013).

Sinha and Swaminathan (1991) have said that malnutrition is a major problem in India though it produces an adequate amount of food and cereals. Thus, it is the time for a nutrition revolution in India. The country has not escaped the problem of nutrition insecurity. The government has to deliver multi-sectoral development involving agriculture, health, environment, and economy for rural development and to ensure total nutrition security in the country. In the Paris Climate Agreement held in the year 2015, almost all countries of the world had confirmed that in this twenty-first century, they will have to limit their average global temperature increase within 2 °C. The developed countries will extend appropriate financial and technical support to the developing and vulnerable countries to fulfill their ambition. The World Meteorological Organization has reported that the average temperature of earth has already increased by about 1.1 °C (Asrar 2019).

2.7 Ecological Footprint Under Changing Climate

Ecological footprint is the ratio of human consumption of available natural resources to the earth's capacity to regenerate them (Lim 2018). It is expressed in global hectares (gha) unit. It measures the biological productive area of the earth required to fulfill the raw material requirements, namely, fruits, vegetables, meat, fish, and wood, and fibers of the people, i.e., biocapacity (Global Footprint Network 2017; Meena et al. 2020; Banerjee et al. 2021a, b, c, d). The comparison between biocapacity and the ecological footprint determines the sustainability of our economy (Raj et al. 2021).

CO_2 emission is the major component of ecological footprint and known as carbon footprint (UN Climate Change News 2019). It is the measure of the area of forestland that is required to absorb all the carbon emissions from human activity and expressed in tons of CO_2 emitted. Two hundred years ago, the carbon footprint was essentially zero, and it has to be zero up to 2050 to reach the target of only 2 °C as decided in the Paris Agreement mentioned earlier (Global Footprint Network 2017).

According to a report published by the World Wildlife Fund (WWF), the total ecological footprint of our planet is 2.2 gha, while the biocapacity is only 1.8 gha. It has been calculated that our planet takes 18 months to regenerate its resources that people use in a 12-month period (Ingrid Beláková et al. 2017).

2.7.1 A Nexus Between Food Crisis, Food Security, and Ecological Footprint

A report by the IPCC has warned that agricultural activities directly account for about one-seventh of GHG emissions (Johnson 2010). Ecological footprint for food production or food-print can be measured by the GHG emissions produced during farming, rearing, processing, transporting, storing, cooking of food materials, and also disposing of the waste food (https://www.greeneatz.com/foods-carbon-footprint.html). Food production accounts for over a quarter (26%) of global GHG emissions (Poore and Nemecek 2018). The expansion of agriculture exerts greatest pressure on the environment and biodiversity increasing the impact on ecological footprint (Hannah and Max 2020). It would exploit natural resources and pollute ecosystems compromising the planet's biocapacity. If CO_2 emissions go unchecked, our future generation will experience fall in crop yields and animal growth rates and also decline in nutrition levels with continued CO_2 emissions (Leahy 2019).

2.7.2 Challenges of Climate Change on Food Security

It is clear from the above discussion that there are various challenges for the government regarding the management of food security in addition to climate change. It creates various stresses which have to be immediately solved to ensure food security. One such significant challenge is scarcity of both drinking and irrigation water. India depends mostly on monsoon rain and groundwater for irrigation. Availability of water for agriculture and also other activities in India is severely affected by climate change. Due to deviation of monsoon season and decrease in average monsoon precipitation pattern and increase in temperature and scorching heat, wetlands are drying, incidences of drought in India are increasing, and ecosystems are severely degrading (Cruz et al. 2007). Farmers have to rely fully on groundwater for irrigation. It is expensive to develop such irrigation facility. Punjab and Haryana are the two states in India that could be able to overcome extremely water-stressed condition by developing modern irrigation methods and growing the bulk of the country's agriculture crops. It needs huge sponsorship to adopt such modern techniques of irrigation and agriculture.

About 60% of the agricultural land in India is rainfed, and thus irrigation is mostly dependent on monsoon rain in those areas. Deficiency in rainfall is the main reason for the decline in its agriculture yield. Improved irrigation facility should be established to expand the cultivable area so that production can be increased to meet the growing food demand of the country. At present, approximately 69 million hectare of land in India is infertile, about 50% of which can be turned fertile after proper treatment and repeated farming. Almost the whole state of Rajasthan consists of barren land, followed by other semi-arid states like Gujarat, Uttar Pradesh, Madhya Pradesh, Maharashtra, Andhra Pradesh, and Karnataka (Ahmad et al. 2011).

The Indian Agricultural Research Institute (IARI) examined the impact of climate change on Indian agriculture by surveying the differences in crop production per hectare in different regions and states in the period from 2000 to 2007–2008 and has identified about 14% decline in rice production, i.e., from 17.96% in the 2000 to 3.45% in 2007–2008, and about 30% decrease in the production of wheat by the year 2007–2008 (Tripathi and Mishra 2017). Thus, if this situation continues, i.e., agricultural production cannot be raised up to optimum level, food security in India will be not only at risk but will collapse totally. There will be a serious shortage of food in the near future. There may be various reasons for this decline in crop production like scarcity of water, deviation of monsoon season due to global warming, less availability of well-irrigated fertile land, growing urbanization, and lastly shifting of job preferences.

The incomes of the farmers are highly affected due to less production and loss of profit. Indian farmers are poor and capable of small- and marginal-scale expenditures. So they had to rely on rainfed monocropping, which could provide food security for 3 to 4 months in a year (Ramachandran 2014). Farmers and landless agricultural laborers become jobless for the rest of the year. Severe shortage of food and money compelled the farmers to shift to different other jobs. Similar incidences happen with the fishermen and people earning livelihood from forest (Schmidhuber and Tubiello 2007).

In the present scenario, food security is not only a challenge to rural area, but urban food security is also a complex issue. People from rural areas are daily migrating to urban areas in search of apparently better job options and food security. They took shelter in urban slums and join urban informal sectors at a very minimum wages and start to live a very unhygienic, distressed, and insecure livelihood (Ramachandran 2014). Good food and health become a dream to them. More than 30% children in Bihar, Madhya Pradesh, and Karnataka are suffering from malnutrition and susceptible to disease. On the other hand, the proportion of children belonging to prosperous states like Karnataka and Maharashtra are suffering from stunted growth and obesity due to non-utilization of balanced diet.

The final component that can ensure food security is proper accessibility and utilization of food. Climate change in India generally decreases food production and increases loss which is expected to increase food prices. Utilization of food depends on food preferences which are driven by main three factors: financial ability which directs food affordability, general health and absence of disease that affects nutrient absorption or increases nutritional needs, and food availability in the market (Christian 2010). Absence of healthy and nutritious food increases incidence of obesity and non-communicable diseases.

Climate change and food insecurity increase socioeconomic pressures and worsen the stress and degradation of natural resources like forests, sea water, coastal zone groundwater, soil, biodiversity, etc. (Khan et al. 2020a, b). Thus, in a country like India with a large population whose economy and health are mostly dependent on weather-sensitive agricultural system, one must develop and implement effective strategies and government policies to solve the problem (Sathaye et al. 2006; Khan et al. 2021a, b).

2.8 Management Perspective

Climatic change is the greatest and fundamental threat to the survivability and sustainability of the human society. It is severely deteriorating the present global food security situation (Leisner 2020; Jiricka-Pürrer and Wachter 2019) by declining yields from agriculture, aquaculture, and fisheries (Kais and Islam 2018). Thus, the government should deal with food security threats and climate change issues at the international, regional, and national levels unanimously and must undertake mitigation strategies to ensure food security and economic growth (Thorpe and Figge 2018). But, it is only not sufficient to reduce climate change and hunger condition unless effective public policies are implemented (Nagoda and Nightingale 2017; Madumere 2017). There are following five main vital reasons for regional initiatives to protect food security (Islam and De Jesus 2012):

- Effective regulation of local markets and control over global food markets to stabilize import of vital food products to meet the demand.
- Food sellers should be aware of regional food preferences. Thus, they can ensure internal food security by supplying essential relevant food resources.
- The government and stakeholders should cooperate with each other.
- It increases food security by directing food production according to the food consumption patterns or public need and demand.
- Regional bodies must take initiatives to ensure food security by protecting the marginal communities.

Three major organizations, namely, the South Asian Association for Regional Cooperation (SAARC), the Association of Southeast Asian Nations (ASEAN), and the Pacific Islands Forum (PIF), undertook various strategies and preventive measures to cope effectively with the problems of food security which developed due to climate change prevalent from past decades (Islam and Kieu 2020).

Myanmar (Burma), Thailand, Indonesia, Singapore, Vietnam, Cambodia, Malaysia, the Philippines, Brunei, Laos, and three additional countries, China, Japan, and South Korea, are the members of ASEAN organization (Islam and Kieu 2020). All the countries were united by an agreement held by ASEAN Food Security Reserve in 1979 (ASEAN 2017a). ASEAN countries have developed food security mitigation strategies in the Vientiane Action Programme (2004–2010). The AIFS and SPA-FS are the two attempts that have been developed for the initial agreement to systematize an approach to food security (ASEAN 2017c). The AIFS framework works as a regional umbrella and looks after for the food security, stability, and improvement of farmer's livelihood in the ASEAN region. There are two other additional institutions under ASEAN, the ASEAN Food Security Information System (AFSIS) and ASEAN Plus Three Emergency Rice Reserve (APTERR), to provide support to manage food security mechanisms (ASEAN 2017b). APTERR is a reservoir for emergency food supply for the ASEAN region. It was established in 2008. APTERR banned export and increased import to improve the availability in the rice market. In contrast, AIFSIS undergoes several analyses to identify the areas likely to experience food insecurity (Desker et al. 2013; Meena et al. 2020a). These institutions help various countries to combat food insecurity in a combined way by sharing available resources and technical systems for the management and food policy programs.

Seven countries that took active role in the foundation of SAARC include India, Bangladesh, Pakistan, Nepal, Bhutan, Sri Lanka, and Maldives. Afghanistan joined as a member of SAARC later in 2007 (SAARC 2017). It also undertakes various declarations, statements, and commissions to mitigate the impacts of climate change on current food-insecure condition of the member countries (Islam and Kieu 2020). In 1987, the SAARC member states established the SAARC Food Security Reserve and CSISA to attain food security (CSISA 2017). It would act as an emergency food reserve for the SAARC countries during the time of food crisis and enhance regional inter-country partnership and integration to solve food crisis situations together (Mittal 2011).

A coordinated multi-sector approach is necessary to maintain food security. PIF invest in the research and development sector to find out adaptive measures against threats of climate change. The framework aims to develop a close inter-country partnership for strong cost-effective regulation of food security at the regional level. The agenda and plan of action set out by the framework during the 39th Pacific Islands Forum were unanimously accepted by all the members (Islam and Kieu 2020). Australia, the Cook Islands, the Federated States of Micronesian, Fiji, Kiribati, the Marshall Islands, Nauru, Niue, Palau, Papua New Guinea, Samoa, New Zealand, the Solomon Islands, Tonga, Tuvalu, and Vanuatu are the members of PIF. The forum developed a Food Secure Pacific Working Group (FSPWG) and Industry Partners for a Food Secure Pacific region for the implementation of actions on attaining food security regionally (Islam and Kieu 2020). FSPWG monitors the activities of the participants to approach good governance and accountability. IPFSP establishes collaboration between the food industry partners from some participating countries of PIF (Islam and Kieu 2020). PIF does not have any plan for storing food grains which could have been essential for emergency supply of essential food resources to the regional food market unlike ASEAN and SAARC, thus failing to maintain food accessibility, an important component of food security (Islam and Kieu 2020).

The action plan developed by all these regional organizations and forums aimed to (1) strengthen good governance by developing policies; (2) increase agriculture yield; (3) increase the supply of food resources to the regional market on the basis of public demand; (4) enhance food processing capacity by collaborating food industries from different countries to promote value-added food products, thus regulating food access; and (5) lastly build a strong food distribution network (Islam and Kieu 2020).

Apart from regional initiatives, there are many multilateral initiatives working under international organizations UN, FAO, WB, and the Global Environment Facility (GEF). Their main purpose is to extend financial support to different projects run by the South Asian countries and also other developing nations that have been undertaken worldwide (Mittal 2011). These organizations act through governments, non-governmental organizations, and local organizations to adapt global climate change and reduce the problem.

Management strategies are also to be adopted to lower the global ecological footprint. Is no viable technological solution to this problem? Only some easy and simple steps in our daily life transportation, housing, food, and goods can reduce it. About 20% of our total ecological footprint can be reduced by walking, cycling, and availing public transport. If you don't own and drive a car on average, you can reduce yours by as much as 20%. Flights release water vapor and N_2O at high altitude that may have two to four times more impact than CO_2 emissions. Thus, short-distance air trips are to be avoided to reduce ecological footprint. In our household, we must use energy-saving lamps and other appliances and biodegradable, non-toxic cleaning products to reduce our goods footprint (https://www.greeneatz.com/foods-carbon-footprint.html). Food footprint can be minimized by consuming organic fruits, vegetables, and food crops that are grown at minimal or no use of fertilizers and pesticides. Livestock industry contributes more GHG emissions globally than any other sector; thus, ecological footprint of vegetarians is estimated to be half than that of meat eaters. It is suggested to take at least one meatless meal in a week. Condition will also improve by planting drought-tolerant plants in your garden or yard (Hannah and Max 2020).

2.9 Future Research and Development

The twenty-first century aims to achieve global food security by 2030 putting back all the negativities associated with climate change (UNCTAD 2017). Implementation of new and modification of existing technologies are essential to combat biotic and abiotic stresses like improvement of seed quality, soil fertility, and water availability for potential increase in the amount of crop and livestock production. Innovations in relation to crop storage, refrigeration, transport, and processing are also needed to raise the dimension of food accessibility. New techniques should be introduced to produce nutrient-rich cereals and vegetables and also protein-rich legumes and dairy and husbandry products to eradicate malnutrition and improve food utilization (UNCTAD 2017).

Agricultural research development contributes a lot in the improvement of food production. Conventional cross-breeding techniques can improve the quality and quantity of crop. Genetic modification techniques are also useful for creating nutrient-fortified, drought- and heat-resistant, disease-resistant, herbicide- and pest-tolerant varieties (Buluswar et al. 2014). Conventional techniques limit plant improvements within the same family of crops. New scientific technique such as transgenic research involves the insertion of genetic material into the crop from unrelated organisms. So, transgenic varieties are improved in taste and appearance apart from those conferred by simple cross-breeding techniques (Buluswar et al. 2014; World Bank and FAO 2009). Well-known examples of transgenic crops are Bt cotton from India and China and Bt maize from Kenya. Biofortification is a new technique of incorporating micronutrients into staple food and its breeding. It is an effective approach to fight against malnutrition. The International Food Policy Research Institute has pioneered biofortification technique among the whole world. A variety of biofortified crops such as orange-fleshed sweet potatoes; iron- and zinc-fortified rice; vitamin A-enriched cassava, wheat, and pearl millet; and many more has been discovered (UNCTAD 2017).

Soil management is equally important to increase crop yields because fertile soil can yield healthy crop (Jhariya et al. 2015, 2018a, b). Soil is considered as a non-renewable resource (ITPS 2015). Fertilization overuse causes soil damage leading to economic wastes for smallholder farmers. Cultivation of nitrogen-fixing trees can improve the water-holding capacities of soil (Folberth et al. 2014; UN 2015). Thus, large-scale projects are undertaken by putting smallholder farmers for growing legume crops to focus on nitrogen fixation (Vanlauwe et al. 2019). The development of new technologies for the production of organic and biological fertilizers (composting, manure, or dung) is very much essential to replace the use of synthetic fertilizers. Biofertilizers are ecofriendly and more affordable to the smallholder farmers (Buluswar et al. 2014; Meena et al. 2020b). Water availability for the purpose of irrigation is another important factor to secure improved crop production. To solve such challenging problem, innovation of low-cost drills and methods for desalination is required for improved irrigation (UNCTAD 2017).

In spite of all good researches to increase yield, the vital aspect of food security is the people's accessibility to fresh food. To ensure freshness and biosafety of food, proper storage, refrigeration, handling, and transport of perishable items from farm to market are essential. Thus, scientist should develop new scientific post-harvest techniques which would be cost-effective so that food prices remain affordable to the poor people also (Buluswar et al. 2014).

Lastly, Internet-based data can be critically used to provide knowledge about new farming techniques, weather forecasting, and discussion with technologists, decision support, and insurance to farmers. Convergence or collective application of a number of new technologies such as genetic engineering to develop transgenic and biofortified crops, synthetic biology to invent bioflavoring, biological nitrogen fixation and other biofertilizers, and also artificial intelligence and robotics will produce profound impact on future food security worldwide (Raj et al. 2019a, b).

While science and technology has the key function in mitigating the dynamic problems of changing climate and maintaining environment and biosafety, a number of policies must be developed to arrange for sponsored research and development. FAO, IFAD and WFP (2015) have estimated that additional 267 billion dollar is necessary annually by the whole world for the purpose of research and development (UN 2015; FAO 2015).

2.10 Conclusion

In this twenty-first century, "right to food" is considered as a stepping stone to healthy livelihood. Thus, it is the exact time to make "Zero Hunger" a reality for all people belonging to all social status. Present changes in climatic conditions may be a boon to some temperate countries like Europe and North America where it is expected to improve both quality and quantity of crop yield, but it is a curse to the tropical countries.

India is a country of diversity in all aspects like geography, topography, climate, and socioeconomic structure. In the last two decades, India has been successful in improving the starved condition of its inhabitants a lot. India is one of the highest populated countries, so natural resources of the country are at the verge of extreme exploitation to meet the needs of the people of the subcontinent. This situation is further worsened due to variation in climate condition and its impact on agricultural production, availability of groundwater for agricultural growth, and resource conflict among its inhabitants. Availability of water to irrigate the vast arable land and the other resources that are required to sustain agricultural growth is under great stress due to climate change. Scarcity of resources and food crisis will increase the price of essential food commodities. It will hinder the country's progression toward food security and will indirectly lead to poverty and inequality.

References

Abobatta W (2019) Potential impact of global climate change on citrus. 4:308–312. https://doi.org/10.15406/mojes.2019.04.00168

Abraham N (2019) India must shift rice growing east from Punjab & Haryana to prevent desertification. India Spend. https://www.indiaspend.com/shifting-rice-cultivation-east-from-punjab-haryana-will-reduce-groundwater-stress/

Ahmad J, Alam D, Haseen S (2011) Impact of climate change on agriculture and food security in India. Int J Agric Environ Biotech 4(2):129–137

Anapalli S, Singh KK, Rathore L, Singh S, Sinha S (2000) Effects of climate change on rice production in the tropical humid climate of Kerala, India. Clim Chang 44(4):495–514. https://doi.org/10.1023/A:1005542414134

Andersen PP, Lorch RP (1999) Food security: a global perspective. In: Peters G, von Braun J (eds) Food security, diversification, and resource management: refocusing the role of agriculture, 1st edn. Ashgate Publishing, pp 51–76

ASEAN (25 Sept 2017a) Cooperation on environment ASEAN cooperation on climate change. http://environment.asean.org/asean-working-group-on-climate-change/

ASEAN (8 Dec 2017b) Integrated food security (AIFS) framework and strategic plan of action on food security in the ASEAN Region (SPA-FS) 2009–2013. https://www.asean.org/storage/images/archive/22338.pdf

ASEAN (9 Oct 2017c) Sustainable agrifood systems ASEAN integrated food security (AIFS) framework and strategic plan of action on food security in the ASEAN region (SPA_FS 2015–2020). http://asean-agrifood.org/?wpfb_dl=58

Asrar S (2019) How rise in Earth's average global temperature is affecting our planet. https://economictimes.indiatimes.com/news/environment/global-warming/how-rise-in-earths-average-global-temperature-is-affecting-our planet/articleshow/72039042.cms?utm_source=contentofinterest&utm_medium=text&utm_campaign=cppst

Bameti (2018) State profile, status of agriculture in Bihar. http://bameti.org/wp-content/uploads/2018/06/State-Profile.pdf

Banerjee A, Jhariya MK, Yadav DK, Raj A (2020) Environmental and sustainable development through forestry and other resources. Apple Academic Press Inc., CRC Press- a Taylor and Francis Group, US & Canada, 400 pp. https://doi.org/10.1201/9780429276026

Banerjee A, Meena RS, Jhariya MK, Yadav DK (2021a) Agroecological footprints management for sustainable food system. Springer, Singapore, 514 pp, p 514. https://doi.org/10.1007/978-981-15-9496-0

Banerjee A, Jhariya MK, Meena RS, Yadav DK (2021b) Ecological footprints in agroecosystem- an overview. In: Banerjee A, Meena RS, Jhariya MK, Yadav DK (eds) Agroecological footprints management for sustainable food system. Springer, Singapore, pp 1–23

Banerjee A, Jhariya MK, Raj A, Yadav DK, Khan N, Meena RS (2021c) Land footprint management and policies. In: Banerjee A, Meena RS, Jhariya MK, Yadav DK (eds) Agroecological footprints management for sustainable food system. Springer, Singapore, pp 221–246

Banerjee A, Jhariya MK, Raj A, Yadav DK, Khan N, Meena RS (2021d) Energy and climate footprint towards the environmental sustainability. In: Banerjee A, Meena RS, Jhariya MK, Yadav DK (eds) Agroecological footprints management for sustainable food system. Springer, Singapore, pp 415–443

BBC News (14 Oct 2009) Global hunger worsening, warns UN (Press release). http://news.bbc.co.uk/2/hi/europe/8306556.stm

Beláková I, Diviaková A, Belaňová E (2017) Ecological footprint in relation to climate change strategy in cities. IOP Conf Ser Mater Sci Eng 245:062021. https://doi.org/10.1088/1757-899X/245/6/062021

Breene K (18 Jan 2016) Food security and why it matters (Formative content). World Economic Forum

Buluswar S, Friedman Z, Mehta P, Mitra S, Sathre R (2014) 50 breakthroughs: critical scientific and technological advances needed for sustainable global development. In: Kumar UJ (ed) Food security and agricultural development report. Institute for Globally Transformative Technologies, Lawrence Berkeley National Lab, Berkeley, CA

Cafiero C, Viviani S, Nord M (2018) Food security measurement in a global context: the food insecurity experience scale. Measurement 116:146–152

Chakrabarty M (2016) Climate change and food security in India. ORF Issue Briefs 157:1–12

Chaudhry A, Aggarwal PK (2007) Climate changes and food security in India. Indian Agriculture Research Institute, New Delhi

Christian P (2010) Impact of the economic crisis and increase in food prices on child mortality: exploring nutritional pathways. J Nutr 140(1):177S–181S

Cruz RV, Harasawa H, Lal M, Wu S, Anokhin Y, Punsalmaa B, Honda Y, Jafari M, Li C, HuuNinh N (2007) Asia. Climate change 2007: impacts, adaptation and vulnerability. In: Parry ML, Canziani OF, Palutikof JP, van der Linden PJ, Hanson CE (eds) Contribution of working group II to the fourth assessment report of the intergovernmental panel on climate change. Cambridge University Press, Cambridge, pp 469–506

CSISA (9 Oct 2017) Cereal systems initiative for South Asia overview. http://csisa.org/about-csisa/overview/

Dasgupta S, Roy S, Sarraf M (2012) Urban flooding in a changing climate: case study of Kolkata, India. Asian-African J Econ Econom 12:135

De Pascale S, Maggio A, Orsini F, Stanghellini C, Heuvelink E (2015) Growth response and radiation use efficiency in tomato exposed to short-term and long-term salinized soils. Sci Hortic 189:139–149

Department of Agricultural and Farmers Welfare (28 Aug 2018) Fourth advance estimates of production of food grains and commercial crops (Press release). http://www.agricoop.nic.in/sites/default/files/4th_advest_E_0.pdf

Desker B, Caballero-Anthony M, Teng P (2013) Thoughts/issues on ASEAN food security: towards a more comprehensive framework, ERIA discussion paper series. ERIA, Senayan Jakarta Pusat, pp 1–43

Environmental Protection Agency (2017) Climate change indicators: greenhouse gases, 1200 Pennsylvania Ave. NW. Washington, DC 20460. https://www.epa.gov/climate-indicators/greenhouse-gases

Fahad S, Bajwa AA, Nazir U, Anjum SA, Farooq A, Zohaib A, Sadia S, Nasim W, Adkins S, Saud S, Ihsan MZ, Alharby H, Wu C, Wang D, Huang J (2017) Crop production under drought and heat stress: plant responses and management options. Front Plant Sci 8:1147

FAO (25 Jun 2007) The state of food and agriculture Food and Agriculture, UN. http://www.fao.org/3/a-a1200e.pdf

FAO (2008a) An introduction to the basic concepts of food security. In: EC - FAO Food Security (ed) Programme food security information for action practical guides. www.foodsec.org/docs/concepts_guide.pdf

FAO (2008b) Climate change and food security: a framework document. Rome. http://www.fao.org/3/k2595e/k2595e00.pdf

FAO (2015) The state of agricultural commodity markets, trade and food security: achieving a better balance between national priorities and the collective good. FAO, Rome

FAO (2016) Climate change and food security: risks and responses. http://www.fao.org/policy-support/tools-and-publications/resources-details/en/c/427091/

FAO (4 Jul 2020) World food situation: global cereal production, utilization, stocks and trade all set to rise to new records in 2020/21 (Press release). http://www.fao.org/worldfoodsituation/csdb/en/

FAO, IFAD and WFP (2014) The state of food insecurity in the world 2014. Strengthening the enabling environment for food security and nutrition, Rome. http://www.fao.org/3/a-i4030e.pdf

FAO, IFAD and WFP (2015) The state of food insecurity in the world. Meeting the 2015 international hunger targets: taking stock of uneven progress. FAO, Rome

Folberth C, Yang H, Gaiser T, Liu J, Wang X, Williams J, Schulin R (2014) Effects of ecological and conventional agricultural intensification practices on maize yields in sub-Saharan Africa under potential climate change. Environ Res Lett 9(4):1–12. https://doi.org/10.1088/1748-9326/9/4/044004

Global Footprint Network (2017) How ecological footprint accounting helps us recognize that engaging in meaningful climate action is critical for our own success. https://www.footprintnetwork.org/2017/11/09/ecological-footprint-climate-change/

Gopaldas T (2006) Hidden hunger: the problem and possible interventions. Econ Polit Wkly 41:3671–3674. https://doi.org/10.2307/4418615

Gray E (13 Jun 2019) Earth's freshwater future: extremes of flood and drought (feature article). Nasa Global Climate Change. https://climate.nasa.gov/news/2881/earths-freshwater-future-extremes-of-flood-and-drought/

Hannah R, Max R (2020) Environmental impacts of food production. Our World in Data. https://ourworldindata.org/environmental-impacts-of-food

Hatfield JL, Prueger JH (2015) Temperature extremes: effect on plant growth and development. Weather Clim Extrem 10(Part A):4–10

IPCC (2007) Climate change 2007 the physical science basis. Contribution of working group I to the fourth assessment report of IPCC. Cambridge University Press, Cambridge

IPCC (2013) Climate change 2013: the physical science basis. In: Parry ML, Canziani OF, Palutikof JP, van der Linden PJ, Hanson CE (eds) Contribution of working group I to the fifth assessment report of the intergovernmental panel on climate change. Cambridge University Press, Cambridge and New York, p 1535. http://www.climatechange2013.org/report

Islam MS, De Jesus IC (2012) Regional initiatives on food security. In: Rayfuse R, Weisfelt N (eds) The challenge of food security: international policy and regulatory frameworks, 1st edn. Edward Elgar Publishing Limited, Northampton, MA, pp 255–274

Islam MS, Kieu E (2020) Tackling, regional climate change impacts and food security issues: a critical analysis across ASEAN, PIF, and SAARC. Sustainability 12:883

ITPS (2015) Status of the world's soil resources (SWSR). FAO Intergovernmental Technical Panel on Soils, Rome

IUCN (2019) Issues-briefs. https://www.iucn.org/resources/issues-briefs/species-and-climate-change

Jhariya MK, Bargali SS, Raj A (2015) Possibilities and perspectives of agroforestry in Chhattisgarh. In: Zlatic M (ed) Precious forests-precious earth. InTech, Croatia, pp 237–257. https://doi.org/10.5772/60841

Jhariya MK, Yadav DK, Banerjee A (2018a) Plant mediated transformation and habitat restoration: phytoremediation an eco-friendly approach. In: Gautam A, Pathak C (eds) Metallic contamination and its toxicity. Daya Publishing House, A Division of Astral International, New Delhi, pp 231–247

Jhariya MK, Banerjee A, Yadav DK, Raj A (2018b) Leguminous trees an innovative tool for soil sustainability. In: Meena RS, Das A, Yadav GS, Lal R (eds) Legumes for soil health and sustainable management. Springer, pp 315–345. https://doi.org/10.1007/978-981-13-0253-4_10

Jhariya MK, Banerjee A, Meena RS, Yadav DK (2019a) Sustainable agriculture, forest and environmental management, 606 pp. Springer, Singapore. https://doi.org/10.1007/978-981-13-6830-1

Jhariya MK, Yadav DK, Banerjee A (2019b) Agroforestry and climate change: issues and challenges. Apple Academic Press Inc., CRC Press- a Taylor and Francis Group, US & Canada, 335 pp. https://doi.org/10.1201/9780429057274

Jhariya MK, Meena RS, Banerjee A (2021a) Ecological intensification of natural resources for sustainable agriculture. Springer, Singapore. https://doi.org/10.1007/978-981-33-4203-3

Jhariya MK, Meena RS, Banerjee A (2021b) Ecological intensification of natural resources towards sustainable productive system. In: Ecological intensification of natural resources for sustainable agriculture. Springer, Singapore. https://doi.org/10.1007/978-981-33-4203-3_1

Jiricka-Pürrer A, Wachter T (2019) Coping with climate change related conflicts-the first framework to identify and tackle these emerging topics. Environ Impact Assess Rev 79:106308

Johnson R (2010) Carbon footprints and food security. Food Business News

Kais SM, Islam MS (2018) Impacts of and resilience to climate change at the bottom of the shrimp commodity chain in Bangladesh: a preliminary investigation. Aquaculture 493:406–415

Khan MI (26 Oct 2018) Climate change hits Bihar farmers twice this year. https://indiaclimatedialogue.net/2018/10/26/bihar-farmers-face-sharp-end-of-climate-change/

Khan AA, Hasan A (2017) Climate change: concern for food security in India. IOSR J Humanit Soc Sci 22(10):52–57. https://doi.org/10.9790/0837-2210055257

Khan N, Jhariya MK, Yadav DK, Banerjee A (2020a) Herbaceous dynamics and CO_2 mitigation in an urban setup- a case study from Chhattisgarh, India. Environ Sci Pollut Res 27(3):2881–2897. https://doi.org/10.1007/s11356-019-07182-8

Khan N, Jhariya MK, Yadav DK, Banerjee A (2020b) Structure, diversity and ecological function of shrub species in an urban setup of Sarguja, Chhattisgarh, India. Environ Sci Pollut Res 27 (5):5418–5432. https://doi.org/10.1007/s11356-019-07172-w

Khan N, Jhariya MK, Raj A, Banerjee A, Meena RS (2021a) Soil carbon stock and sequestration: implications for climate change adaptation and mitigation. In: Jhariya MK, Meena RS, Banerjee A (eds) Ecological intensification of natural resources for sustainable agriculture. Springer, Singapore. https://doi.org/10.1007/978-981-33-4203-3_13

Khan N, Jhariya MK, Raj A, Banerjee A, Meena RS (2021b) Eco-designing for sustainability. In: Jhariya MK, Meena RS, Banerjee A (eds) Ecological intensification of natural resources for sustainable agriculture. Springer, Singapore. https://doi.org/10.1007/978-981-33-4203-3_16

Kumar S, Meena RS, Jhariya MK (2020) Resources use efficiency in agriculture. Springer, Singapore, 760 pp. https://doi.org/10.1007/978-981-15-6953-1

Kumar S, Meena RS, Datta R, Verma SK, Yadav GS, Pradhan GS, Molaei A, Rahman GKMM, Mashuk HA (2020a) Legumes for carbon and nitrogen cycling: an organic approach. In: Carbon and nitrogen cycling in soil. Springer, pp 337–375. https://doi.org/10.1007/978-981-13-7264-3_10

Kumar S, Meena RS, Singh RK, Munir TM, Datta R, Danish S, Yadav GS, Kumar S (2021) Soil microbial and nutrient dynamics under different sowings environment of Indian mustard (*Brassica juncea* L.) in rice based cropping system. Sci Rep 11:5289. https://doi.org/10.1038/s41598-021-84742-4

Lal M, Harasawwa H, Murdiyarso D (2001) Asia climate change 2001: impacts, adaptation and vulnerability. In: McCarthy JJ, Canziani OF, Leary NA, Doken DJ, White KS (eds.) The third assessment report of the intergovernmental panel on climate change. Cambridge University Press. Cambridge, UK, pp. 533–590

Lal R, Sivakumar MV, Faiz SMA, Rahman AM, Islam KR (2010) Climate change and food security in South Asia. Springer

Law C, Fraser I, Piracha M (2019) Nutrition transition and changing food preferences in India. J Agricult Econ 71(1):118–143

Lawrence G (2017) Re-evaluating food systems and food security: A global perspective. J Soc 53 (4):774–796

Le Quéré C, Andrew R, Friedlingstein P, Sitch S, Pongratz J, Manning A (2018) Global carbon budget 2017. Earth Syst Sci Data 10:405–448
Leahy S (2019) World food crisis looms if carbon emissions go unchecked, UN says. Environment, National geographic. https://www.nationalgeographic.com/environment/2019/08/ipcc-un-food-security/
Leisner CP (2020) Review: climate change impacts on food security-focus on perennial cropping systems and nutritional value. Plant Sci 293:1–7
Lim A (27 Dec 2018) What is ecological footprint? Definition and how to calculate it (press release). https://www.thoughtco.com/what-is-ecological-footprint-4580244
Lindsey R, Dahlman L (2020) Climate change: global temperature. NOAA Climate. https://www.climate.gov/news-features/understanding-climate/climate-change-global-temperature
Llyod SJ, Kovats RS, Chalabi Z (2011) Climate change, crop yields, and undernutrition: development of a model to quantify the impact of climate scenarios on child undernutrition. Environ Health Perspect 119:1817
Lobell DB, Sibley A, Monasterio JIO (2012) Extreme heat effects on wheat senescence in India. Nat Clim Change 2:186
Machado R, Serralheiro R (2017) Soil salinity: effect on vegetable crop growth management practices to prevent and mitigate soil salinization. Horticulture 3(30):1–13. https://doi.org/10.3390/horticulturae3020030
Madumere N (2017) Public enlightenment and participation-a major contribution in mitigating climate change. Int J Sustain Built Environ 6:9–15
Meena RS, Lal R (2018) Legumes for soil health and sustainable management. Springer, Singapore, 541 pp. https://doi.org/10.1007/978-981-13-0253-4_10
Meena RS, Kumar V, Yadav GS, Mitran T (2018) Response and interaction of *Bradyrhizobium japonicum* and arbuscular mycorrhizal fungi in the soybean rhizosphere: a review. Plant Growth Regul 84:207–223. https://doi.org/10.1007/s10725-017-0334-8
Meena RS, Lal R, Yadav GS (2020) Long term impacts of topsoil depth and amendments on soil physical and hydrological properties of an Alfisol in Central Ohio, USA. Geoderma 363:1141164. https://doi.org/10.1016/j.geoderma.2019.114164
Meena RS, Lal R, Yadav GS (2020a) Long-term impact of topsoil depth and amendments on carbon and nitrogen budgets in the surface layer of an Alfisol in Central Ohio. Catena 194:104752. https://doi.org/10.1016/j.catena.2020.104752
Meena RS, Kumar S, Datta R, Lal R, Vijaykumar V, Brtnicky M, Sharma MP, Yadav GS, Jhariya MK, Jangir CK, Pathan SI, Dokulilova T, Pecina V, Marfo TD (2020b) Impact of agrochemicals on soil microbiota and management: a review. Land (MDPI) 9(2):34. https://doi.org/10.3390/land9020034
Mittal S (2006) Structural shift in demand for food: projections for 2020 (working paper) 184. Indian Council for Research on International Economic Research, New Delhi
Mittal S (2008) Demand-supply trends and projections of food in India. Working paper no. 209. Indian Council of Research on International Economic Research, New Delhi
Mittal S (2011) Regional and international cooperation. In: Chatterjee B, Khadka M (eds) Climate change and food security in South Asia, 2nd edn. CUTS International, Jaipur, pp 87–106
Moors EJ, Singh T, Siderius C, Balakrishnan S, Mishra A (2013) Climate change and waterborne diarrhoea in northern India impacts and adaptation strategies. Sci Total Environ 468–469(1): S139–S151
Myers SS, Zanobetti A, Kloog I, Huybers P, Leakey ADB, Bloom AJ, Carlisle E, Dietterich LH, Fitzgerald G, Hasegawa T, Holbrook NM, Nelson RL, Ottman MJ, Raboy V, Sakai H, Sartor KA, Schwartz J, Seneweera S, Tausz M, Usui Y (2014) Rising CO_2 threatens human nutrition. Nature 510:139–142
Nadkar N (29 Aug 2018) Food grain output up 3.5% in 2017-18: Agri Dept (press release). Economic Outlook. https://www.cmie.com/kommon/bin/sr.php?kall=warticle&dt=2018-08-29%2010:24:33&msec=243

Nagoda S, Nightingale AJ (2017) Participation and power in climate change adaptation policies: vulnerability in food security programs in Nepal. World Dev 100:85–93

NASA (03 May 2016) Study: rising carbon dioxide levels will help and hurt crops (press release). https://www.nasa.gov/feature/goddard/2016

Nema P, Nema S, Roy P (2012) An overview of global climate changing in current scenario and mitigation action. Renew Sustain Energy Rev 16(4):2329–2336

Nunez C (13 May 2019) Carbon dioxide levels are at a record high. Here's what you need to know. National Geographic. https://www.nationalgeographic.com/environment/global-warming/greenhouse-gases/

Pachauri RK, Reisinger A (2007) Climate change 2007: synthesis report. Contribution of working group I, II and III to the fourth assessment report of the intergovernmental panel on climate change

Pandve HT (2010) Climate change and coastal mega cities of India. Ind J Occup Environ Med 14:22

Phalkey RK (2015) Systematic review of current efforts to quantify the impacts of climate change on under nutrition. Proc Natl Acad Sci U S A 112(33):E4522–E4529

Poore J, Nemecek T (2018) Reducing food's environmental impacts through producers and consumers. Science 360(6392):987–992

Pruthi R (2018) Global hunger index 2018: India ranks 103rd out of 119 countries. Current Affairs. https://www.jagranjosh.com/current-affairs/global-hunger-index-2018-india-ranks-103rd-out-of-119-countries-1539337074-1

Raj A, Jhariya MK, Harne SS (2018a) Threats to biodiversity and conservation strategies. In: Sood KK, Mahajan V (eds) Forests, climate change and biodiversity. Kalyani Publisher, New Delhi, pp 304–320

Raj A, Jhariya MK, Bargali SS (2018b) Climate smart agriculture and carbon sequestration. In: Pandey CB, Gaur MK, Goyal RK (eds) Climate change and agroforestry: adaptation mitigation and livelihood security. New India Publishing Agency (NIPA), New Delhi, pp 1–19

Raj A, Jhariya MK, Yadav DK, Banerjee A, Meena RS (2019a) Agroforestry: a holistic approach for agricultural sustainability. In: Jhariya MK, Banerjee A, Meena RS, Yadav DK (eds) Sustainable agriculture, forest and environmental management. Springer, Singapore, pp 101–131. https://doi.org/10.1007/978-981-13-6830-1

Raj A, Jhariya MK, Banerjee A, Yadav DK, Meena RS (2019b) Soil for sustainable environment and ecosystems management. In: Jhariya MK, Banerjee A, Meena RS, Yadav DK (eds) Sustainable agriculture, Forest and environmental management. Springer, Singapore, pp 189–221. https://doi.org/10.1007/978-981-13-6830-1

Raj A, Jhariya MK, Yadav DK, Banerjee A (2020) Climate change and agroforestry systems: adaptation and mitigation strategies. Apple Academic Press Inc., CRC Press- a Taylor and Francis Group, US & Canada, 383 pp. https://doi.org/10.1201/9780429286759

Raj A, Jhariya MK, Khan N, Banerjee A, Meena RS (2021) Ecological intensification for sustainable development. In: Ecological intensification of natural resources for sustainable agriculture. Springer, Singapore. https://doi.org/10.1007/978-981-33-4203-3_5

Ramachandran P (2013) Food & nutrition security: challenges in the new millennium. Ind J Med Res 138(3):373–382

Ramachandran N (2014) Persisting under nutrition in India, 1st edn. Springer, New Delhi, pp 3–240

Rammohan A, Awofeso N, Robitaille MC (2011) Addressing female Iron-deficiency anaemia in India: is vegetarianism the major obstacle? Int Sch Res Not 2012:8. https://doi.org/10.5402/2012/765476

Ranuzzi R, Srivastava R (16 May 2012) Impact of climate change and its impact on agriculture and food security. ICRIER Policy Series

Rao N, Pradhan M, Roy D (2017) Gender justice and food security in India. A review. IFPRI discussion paper 01600. http://www.indiaenvironmentportal.org.in/files/file/Gender%20Justice%20and%20Food%20Security%20in%20India.pdf

Rehman MU, Rather GH, Gull Y, Mir MR, Mir MM, Waida UI, Hakeem KR (2015) Effect of climate change on horticultural crops. In: Crop production and global environmental issues. Springer, Cham, pp 211–239. https://doi.org/10.1007/978-3-319-23162-4_9

Reserve Bank of India (16 Sept 2015) Handbook of statistics on indian economy 2014–15. https://www.rbi.org.in/Scripts/PublicationsView.aspx?id=16459

Ruchita S, Rohit S (2017) Effect of global warming on Indian agriculture. J Climatol Weather Forecast 5:195. https://doi.org/10.4172/2332-2594.1000195

SAARC (9 Oct 2017) A guide to the SAARC food bank. http://saarcsec.org/assets/responsive_filemanager/source/Files%20for%20Areas%20of%20Cooperation/ARD/06.a_A%20Guide%20to%20the%20SAARC%20Food%20Bank.pdf

Sarkar S (2018) There's more to the Indian diet than polished rice and wheat. Ecologic Agriculture, The Hindu. https://www.thehindu.com/sci-tech/agriculture/take-a-grain-check-theres-more-to-the-indian-diet-than-polished-rice-and-wheat/article22429984.ece

Sathaye J, Shukla P, Ravindranath N (2006) Climate change, sustainable development and India: global and national concerns. Curr Sci 90(3):314–325

Schmidhuber J, Tubiello FN (2007) Global food security under climate change. Proc Natl Acad Sci 104:19703

Sheoran S, Kuma S, Kumar P, Meena RS, Rakshit S (2021) Nitrogen fixation in maize: breeding opportunities. Theor Appl Genet. https://doi.org/10.1007/s00122-021-03791-5

Singh V (2019) Changing face of malnutrition, trends in the Indian context. ORF. https://www.orfonline.org/expert-speak/changing-face-of-malnutritiontrends-in-the-indian-context-57738/

Singh A (2020) Childhood malnutrition in India. In: Bhattacharya SK (ed) Perspective of recent advances in acute diarrhea. Intech Open. https://doi.org/10.5772/intechopen.89701

Sinha SK, Swaminathan MS (1991) Deforestation climate change and sustainable nutrients security. Clim Change 16:33–45

Smith RM, Thornton PK, Myers SS (2018) The impact of rising carbon dioxide levels on crop nutrients and human health. IFPRI GCAN Policy Note 10

Taub D (2010) Effects of rising atmospheric concentrations of carbon dioxide on plants. Nat Educ Knowl 3(10):21

The Asian Development Bank and Postdam Institute for Climate Impact Research (14 Jul 2017) A region at risk - the human dimensions of climate change in Asia and the Pacific [press release]. https://www.adb.org/news/events/publication-launch-region-risk-human-dimensions-climate-change-asia-and-pacific

Thorpe A, Figge F (2018) Climate change and globalisation as 'double exposure': implications for policy development. Environ Sci Pol 90:54–64

TNAU Agritech Portal (2013) ICAR, Expert system for finger millet. http://agritech.tnau.ac.in/expert_system/ragi/index.html

Tripathi A, Mishra A (2017) The wheat sector in India: production, policies and food security. In: Gomez y Paloma S et al (eds) The Eurasian wheat belt and food security: global and regional aspects. Springer, Cham, pp 275–296. https://doi.org/10.1007/978-3-319-33239-0_17

UN (4 Dec 2015) Agricultural technology for development: report of the secretary-general

UN Climate Change News (2019) Fossil fuels are biggest part of global ecological footprint. https://unfccc.int/news/fossil-fuels-are-biggest-part-of-global-ecological-footprint

UNCTAD (2017) The role of science, technology and innovation in ensuring food security By 2030. New York and Geneva, pp 1–55. https://unctad.org/en/PublicationsLibrary/dtlstict2017d5_en.pdf

USGCRP (2016) In: Crimmins A, Balbus J, Gamble JL, Beard CB, Bell JE, Dodgen D, Eisen RJ, Fann N, Hawkins MD, Herring SC, Jantarasami L, Mills DM, Saha S, Sarofim MC, Trtanj J, Ziska L (eds) The impacts of climate change on human health in the United States: a scientific assessment. U.S. Global Change Research Program, Washington, DC, 312 pp. https://doi.org/10.7930/J0R49NQX

USGCRP (2017) In: Wuebbles DJ, Fahey DW, Hibbard KA, Dokken DJ, Stewart BC, Maycock TK (eds) Climate science special report: fourth national climate assessment, vol I. U.S. Global Change Research Program, Washington, DC, 470 pp

Vanlauwe B, Hungria M, Kanampiu F, Giller KE (2019) The role of legumes in the sustainable intensification of African smallholder agriculture: lessons learnt and challenges for the future. Agric Ecosyst Environ 285:106601. https://doi.org/10.1016/j.agee.2019.106601

von Grebmer K, Bernstein J, Nabarro D, Prasai N, Amin S, Yohannes Y, Sonntag A, Patterson F, Towey O, Thompson J (2016) 2016 Global hunger index: getting to zero hunger. Welthungerhilfe, International Food Policy Research Institute, Concern Worldwide, Bonn, Washington, DC, Dublin. https://doi.org/10.2499/9780896292260

Watson R, McMichael A (2001) Global climate change — the latest assessment: does global warming warrant a health warning? Glob Change Hum Health 2:64–75. https://doi.org/10.1023/A:1011914326191

Wheeler T, von Braun J (2013) Climate change impacts on global food security. Sci NY 341:508–513. https://doi.org/10.1126/science.1239402

Withnall A (17 Oct 2016) Global hunger index: revealed - the worst countries in the world at feeding their own people (press release). The Independent. https://www.independent.co.uk/news/world/politics/2016-global-hunger-index-worst-countries-world-feeding-their-own-people-a7365761.html

Wolfe J (5 Sept 2000) Volcanoes and climate change (feature articles). https://cdn.earthdata.nasa.gov/conduit/upload/8684/NASA_SOP_2000_volcanoes_and_climate_change.pdf

World Bank (2013) Turn down the heat: climate extremes, regional impacts, and the case for resilience. http://www.worldbank.org/content/dam/Worldbank/document/Full_Report_Vol_2_Turn_Down_The_Heat_%20Climate_Extremes_Regional_Impacts_Case_for_Resilience_Print%20version_FINAL.pdf

World Bank and FAO (2009) Awakening Africa's sleeping giant. World Bank, Washington, DC

World Food Summit. Rome (1996) Declaration on world food security and world food summit plan of action. http://www.fao.org/docrep/003/w3613e/w3613e00.htm. Accessed 27 Jul 2015

Worldometers (2019) Countries in the world by population (2020). https://www.worldometers.info/world-population/population-by-country/

3 Cost-Effective and Eco-Friendly Agricultural Technologies in Rice-Wheat Cropping Systems for Food and Environmental Security

Akbar Hossain, Rajan Bhatt, Sukamal Sarkar, Manashi Barman, Debjyoti Majumder, Saikat Saha, Mst. Tanjina Islam, Sagar Maitra, and Ram Swaroop Meena

Abstract

Both rice and wheat in the "rice-wheat cropping systems" (RWCS) of South Asia and China feed more than 3.1 billion people. It is the most productive and vital agricultural systems worldwide to meet the food safety of the growing population. Although the RWCS have great concern for food security, however, one of the

A. Hossain (✉)
Department of Agronomy, Bangladesh Wheat and Maize Research Institute, Dinajpur, Bangladesh

R. Bhatt
Regional Research Station, Kapurthala, Punjab Agricultural University, Ludhiana, Punjab, India
e-mail: rajansoils@pau.edu

S. Sarkar · M. Barman
Department of Agronomy, Bidhan Chandra KrishiViswavidyalaya (BCKV), Nadia, West Bengal, India

D. Majumder
Uttar BangaKrishiViswavidyalaya (UBKV), Cooch Behar, West Bengal, India

S. Saha
Nadia Krishi Vigyan Kendra, Bidhan Chandra KrishiViswavidyalaya (BCKV), Nadia, West Bengal, India

M. T. Islam
Department of Agronomy, Hajee Mohammad Danesh Science and Technology University (HSTU), Dinajpur, Bangladesh

S. Maitra
Centurion University of Technology and Management, Paralakhemundi, India
e-mail: sagar.maitra@cutm.ac.in

R. S. Meena
Department of Agronomy, Institute of Agricultural Sciences, Banaras Hindu University (BHU), Varanasi, India
e-mail: meenars@bhu.ac.in

M. K. Jhariya et al. (eds.), *Sustainable Intensification for Agroecosystem Services and Management*, https://doi.org/10.1007/978-981-16-3207-5_3

foremost complications in the systems is that soils are puddled with repeated tillage by the traditional way for transplanting rice seedlings which lead to alteration in soil physical and chemical properties. Besides these, repeated tillages for puddling create a hard plow pan layer at the root zone of the rice plant that creates poor infiltration and waterlogging for the next dry season crop particularly wheat. Farmers in the systems generally use excessive synthetic fertilizers and pesticides for getting higher yield for both rice and wheat. As a result, repeated tillage and also imbalance application of inorganic fertilizers and pesticides increase the production cost as well as influence greenhouse gas (GHG) emission. Since the systems have several hostile effects on the environment due to traditional farming, it is already confirmed that the systems are the key source of food production for more than 3.1 billion people in the countries of South and Southeast Asia. Therefore, it will not be a wise decision to replace the system from the regions. In the meantime, researchers have recommended numerous advanced technologies in the RWCS for sustainable rice and wheat production. The chapter discusses cost-effective and ecological-friendly technologies for RWCS of South Asia for food and environmental security.

Keywords

Agroecology · Climate-smart technology · Rice-wheat systems · South Asia · Sequential technology

Abbreviations

CA	Conservation agriculture
CH_4	Methane
CO_2	Carbon dioxide
CPF	Carbon footprint
CRM	Crop residue management
CRs	Crop residues
CSA	Climate-smart agriculture
DSS	Decision support system
FFP	Farmers' fertilizer practice
GHGs	Greenhouse gases
GIS	Geographical information system
GPS	Global positioning system
IGP	Indo-Gangetic Plains
LCC	Leaf color chart
N_2O	Nitrous oxide
PF	Precision farming
RCTs	Resource conservation tillages
RS	Remote sensing
RWCS	Rice-wheat cropping system

SA	South Asia
SSNM	Site-specific nutrient management
VRT	Variable rate technology
WPF	Water footprint
ZT	Zero tillage

3.1 Introduction

The "rice-wheat cropping system" (RWCS) of South Asia (SA) is most vital for the food security of the growing population, and it is considered as the highest productive cropping systems in the world (Ladha et al. 2003; Nawaz et al. 2019). The intensive RWCS that include rice, wheat, and recently added maize crop are prevalent throughout SA of the Indo-Gangetic Plains (IGP) fertile land under the four countries (Fig. 3.1), including the western region of Pakistan; the northern, north-western, and eastern parts of India; the western and north-western regions of Bangladesh; and a portion of mid-hills of the Himalayan and the Terai plains of

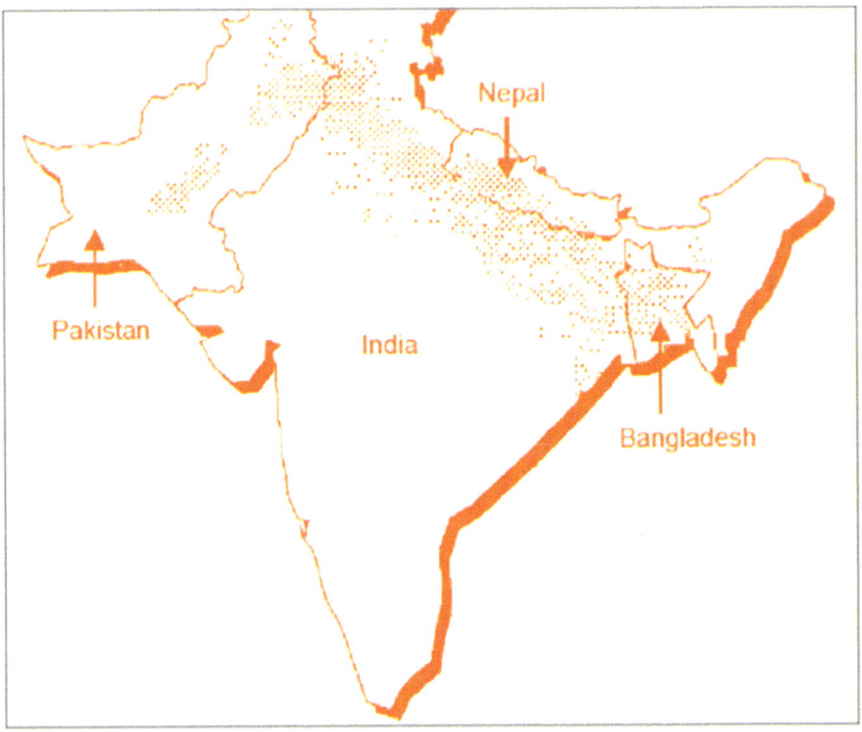

Fig. 3.1 Schematic map of the IGP showing the rice-wheat growing regions (shaded areas) of Bangladesh, India, Nepal, and Pakistan

Nepal (Kataki et al. 2001). These systems establish the chief cost-effective activity in various rural regions of SA and deliver the main food for million people. In the system, generally, rice is cultivated during the monsoon (under rainfed condition), and then wheat grows as a rabi crop (winter).

More recently, a significant reduction in the productivity of the RWCS is reflected due to the changing climate. Therefore, the declining tendencies of cereal production in RWCS of SA create great concern for the growing population. Concurrently, questions of natural resource destruction, labor unavailability, and climate variability are also encountered for millions who depend on this system for their livings (Srivastava and Mukhopadhyay 1997; CSISA 2019). One of the main difficulties of this system is that soils of rice and wheat are managed differently. For example, rice seedlings are transplanted into puddled soils traditionally (soils plowed with repeated tillage with wet basis) (Jat et al. 2011). The frequent cycles of wet puddling for transplanted rice for several years have promoted the deterioration of the soil physical and chemical properties, creation of a hardpan at a surface depth, deprived infiltration, and waterlogging leading to poor rooting (Nawaz et al. 2019; Islam et al. 2019). This unnecessary plowing generates poor soil structure which ultimately creates difficulties in irrigation, for example, the poor infiltration creates oxygen stress in wheat plants particularly after irrigation, ultimately turning the plants yellowish and leading to the stunted growth of the affected plants. Besides these problems, macro- and micronutrients are becoming limited in the RWCS (Duxbury et al. 2000; Nawaz et al. 2019; Meena et al. 2018). In many areas in IGP of SA, water layers are deteriorating quickly as more water is uptaken than recharged. In other areas, particularly in the coastal regions of SA countries, water tables are mounting, which leads to waterlogging and also salinity or sodicity. Under constant cultivation, the spread of the single weed such as *Phalaris minor* takes place due to its resistance against herbicide isoproturon.

Recently, scientists suggested climate-smart (CSA) and next-generation sequential technologies for resolving these difficulties; among them, the zero tillage (ZT) system is one of the best ones (Erenstein 2009; Jat et al. 2014; Raj et al. 2018), since it has numerous friendly welfares comprising of fewer C-emission and less global warming, fewer usage of synthetic fertilizers and pesticides, and higher water productivity (Sapkota et al. 2015; Meena and Lal 2018). In the cropping systems, generally, wheat is established into the rice residues with least trouble. Without puddling, soil physical and chemical belongings are enhanced, and the succeeding growth and yield of the wheat crop are improved (Gupta and Sayre 2007). The ZT system does not disturb soil surface and reduces weed germination, which leads to the control of the weeds. Besides the ZT, usage of bed planting system can also reduce weed infestation, ultimately limiting the more use of herbicides.

Since RWCS are a principal source of cereals particularly rice and wheat grains in the region, therefore, it is hard to find out a substitute system to meet the cereal demand of the growing population. Farmers find very few alternatives for rice and wheat that provide similar low risk and profit. Researchers have to search for options for supporting this system and making it more efficient and profitable. The chapter highlights the major consequences in the RWCS of SA under the changing climates. The overview of the chapter also suggested several next-generation sequential technologies to encounter the nutrition and dietary safety of an increasing population who are depending on RWCS.

3.2 Rice-Wheat Ecosystems of South Asia

The RWCS of SA are inhabiting about 12.3, 0.5, 2.2, and 0.8 million ha agricultural land in India, Nepal, Pakistan, and Bangladesh, where around 85% of this area falls in the productive IGP regions (Bhatt et al. 2016). Eastern India, Nepal, and Bangladesh were having higher rice-growing areas followed by little under wheat, while the reverse is true for the N-W India and Pakistan up to the early 1960s. Afterward, a significant portion of land during the *kharif* season shifted to the rice due to government policies, assured minimum support price, and the need of the time. Further, semi-dwarf short-duration cultivars followed by the assured irrigation and fertilization facilities improved the yields to manifolds as per the time requirements and led to the "Green Revolution" (Bhatt et al. 2016). Earlier, rice and wheat cultivars were of longer duration which are susceptible to insect pest and disease attack and finally require a huge quantum of irrigation water. Further, flood irrigation in rice and puddling was responsible for the lowering of the underground water table and soil health and arising micronutrient deficiencies and finally resulted in the decline in the availability of water, groundwater depletion, and increased pest and disease susceptibility (Aggarwal et al. 2004; Bhatt et al. 2016, 2019; Meena et al. 2017).

Further, earlier practices led to the degradation of the land and declining soil health (Paul et al. 2014; Dass et al. 2016a, b; Kumar et al. 2021), while the adoption of the short-duration crop cultivars followed by the advanced methods of crop establishment results in improved productivities along with soil health and livelihoods. Further, assured irrigations (Yadav et al. 1998) and timely transplantation of rice (Jalota et al. 2007) cut off the unnecessary water loss through the evaporation which further partition higher fraction of the crop ET to the productive transpiration. Further, this leads to a higher intake of the nutrient in the plants through the roots which ultimately results in lesser water footprints of the rice-wheat cropping system in general and of rice in particular. Further, among the major problems of RWCS is the production of the huge crop residue biomass whose fate is still a big question. Among rice and wheat residues, later used in animal husbandry while rice straw not preferred because of higher silica content. Thereby, farmers because of the shorter window period and a huge volume of the residues prefer to burn it in the open which further results in the production of large volumes of GHGs causing climate change having its consequences on agriculture because of higher temperature and CO_2 levels (Jhanavi and Bhatt 2020). In SA, more than 20 cropping systems are practiced; among them, RWCS are the dominant systems (Yadav et al. 1998). Consequently, it could be established that intensively cultivated conventional RWCS in SA are responsible for declining of water table, soil health, productivity of land and water, and overall the livelihoods of the farmers. This needs to be replaced with the resource conservation technologies for sequestering higher C back in the soil, improving water levels which further led to sustainable agriculture in the region.

3.3 Challenges for Sustainable Cropping Systems Under the Changing Climate

The estimation conducted by IPCC (2001a) revealed that the atmospheric constitutes, regional and global temperature, and precipitation are changing across the globe particularly due to anthropologic activities, and it is anticipated that these fluctuations will likely to be continued over the upcoming era (Bindi and Howden 2004). The enhancement of the higher GHGs (i.e., CH_4, CO_2, N_2O, etc.) in the atmosphere limits the absorption of long-wave radiation from the surface of the earth, leading to warmth by both the surface of the earth and the lower atmosphere (Petit et al. 2000). Among the GHGs, CO_2 concentration and climatic factors are considered as the major reasons for affecting agricultural production. Therefore, it is not unexpected that fluctuations in these environmental elements are varied widely, depending upon the cropping system and regions. Although the outlook of global warming may bring new prospects for increasing agricultural productivity in many temperate areas through increasing temperature and rainfalls, in warmer and tropical areas, the impressions may be substantial and harmful by enhancing the limitation of water stress, uneven precipitation, and difficulties in association with the fluctuation of temperature. It is expected that the temperature across the globe near to the surface of the earth has increased by 0.6 °C since 1850, when the estimations were made, and is now greater than at any time during at least the past 2000 years (IPCC 2001b; Mann and Jones 2003). Future climate changes are extremely unreliable, since a climate model projected that the mean temperature across the globe will further rise between 2 and 6 °C for the twenty-first century (IPCC 2001a).

The increased level of atmosphere GHG emission has already affected the biophysical process of agroecosystems (Bindi and Howden 2004). For example, most cereals and pulses need specific day-night temperature for their growth and development, while the fluctuation of temperature as a result of global warming is shortening the length of life span which ultimately reduces the crops' yield (Tubiello et al. 2000; vanIttersum et al. 2003). In the case of root and tuber crops, the increasing CO_2 may increase the ground sinks as a result of available carbon and apoplastic loading of phloem (Bindi and Howden 2004; Sheoran et al. 2021), whereas due to global warming, the increasing temperature may shorten the growing season, which leads to a decline in the yield (Wolf 2002). Horticultural crops including most of the field-grown vegetables and high-value crops generally need sufficient agro-climatic condition, ample water, and nutrient supply, while temperature and CO_2 fluctuations may reduce the growth and yield of the crop (Wurr et al. 1998).

3.4 Importance of the Ecosystem Services for Food and Environmental Security

The ecosystems, such as agro-ecosystem, forest ecosystem, and aquatic ecosystem, denoted the numerous forms of welfares to humans through the natural environment (MEA 2005) and subsequently pay attention to food and environmental security (Jhariya et al. 2019a, b, 2021). Generally, the ecosystem includes four diverse components, namely, (1) provisioning, (2) regulating, (3) cultural, and (4) supporting services. The first three facilities straightly affect the people, but the inevitability of supportive services is to uphold other facilities (La Notte et al. 2017). The major categories of ecosystem services can be subdivided into 17 subcategories (Costanza and Kubiszewski 2012) (Fig. 3.2). Researchers estimated that the values of these global ecosystem services for 16 biomes are US$33 trillion per year (de Groot et al. 2012).

Food and nutritious safety can be defined as people always have social, physical, and economic entree to healthy food. Food safekeeping in comprehensive aspect shelters the backgrounds of convenience, accessibility, and consumption of food stuffs but does not spell out the nutritional dimension (Fig. 3.3). Whereas nutritional security attains the issues related to the nutritional dimension along with quality and quantity of food materials to encounter the demands of a balanced diet of the people, supported by the environment (Capone et al. 2014), environmental security (ES) attributes the intimidations that happen through ecological measures and propensities to individuals, societies, or nations. Besides, ES is the possibility of the surroundings to sustain life system across the world with its diverse mechanisms (Mathews 1989). The ES may cover an extensive erraticism of anxieties linked to the global environs of the current trends of climate change. The impression of human struggle and worldwide associations among diverse nations on ecological concerns also may be a theme for the center of magnetism (Brown 1977; Westing 1986).

Food production across the globe has improved over the past periods, but food uncertainty over a mass population has been rising day by day (FAO 2009). The FAO clearly indicates the number of malnourished populations across the globe is getting higher in the last decades. The condition is worst in developing nations, and statistical evidence pointed out that half of the population across the globe suffer from malnutrition and out of total deaths count 40% are due to environmental degradation (Godfray et al. 2010; Meena et al. 2020; Pimentel et al. 2007). On the other hand, an unexpected discouragement of essential natural resources upkeeps the boosting of food improvement (Daily et al. 1998). Soil degradation and the sinking of groundwater, deforestation, and frequent incidence of water-borne diseases are some of the clear evidence (Pimentel et al. 1997). This key cause of unrestrained reduction of natural means hampers not only upcoming food production but also the direction of a countless hazard to the environment (Khan et al. 2021a, b). However, the restoration of natural resources should be a better practice to enjoy these resources sustainably for food grain production in the future (Richardson 2010; Banerjee et al. 2020; Kumar et al. 2020b; Raj et al. 2020). The policy in different

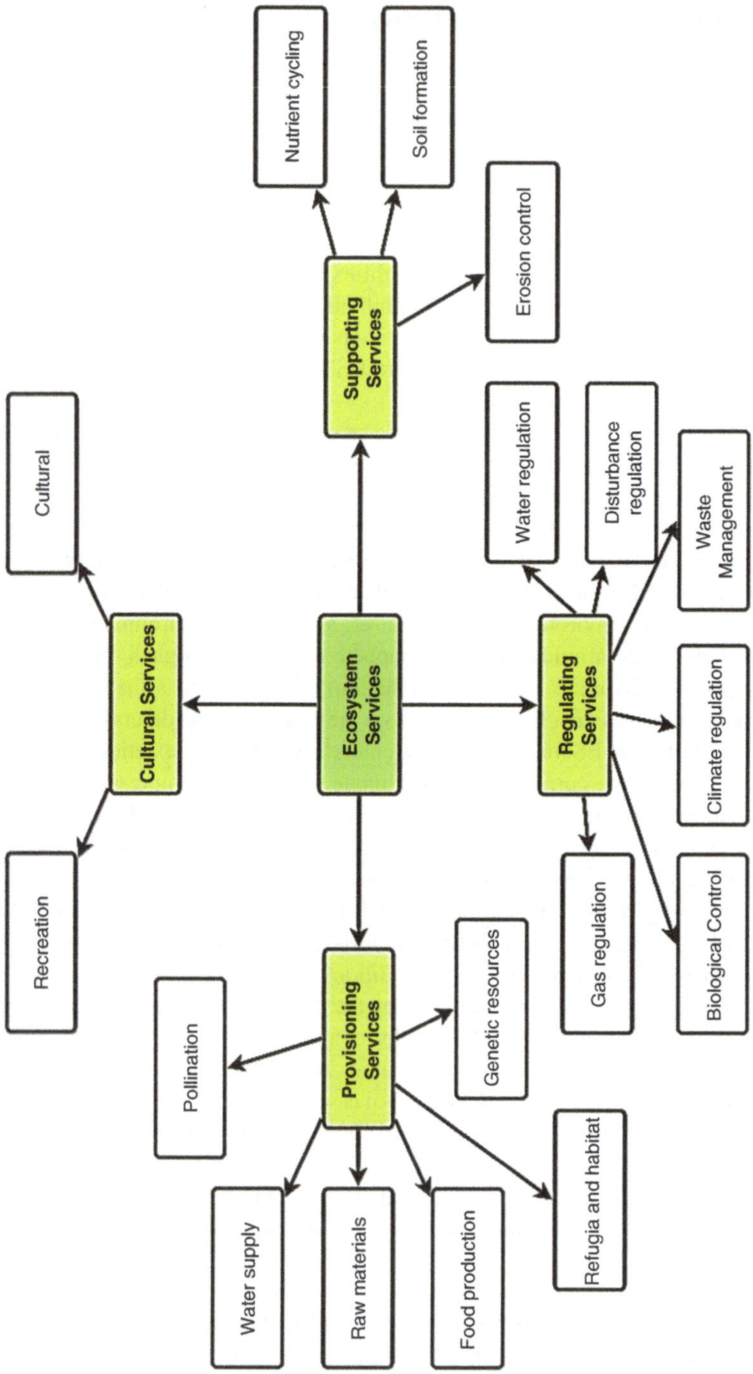

Fig. 3.2 Different ecosystem services (Adapted from Richardson 2010)

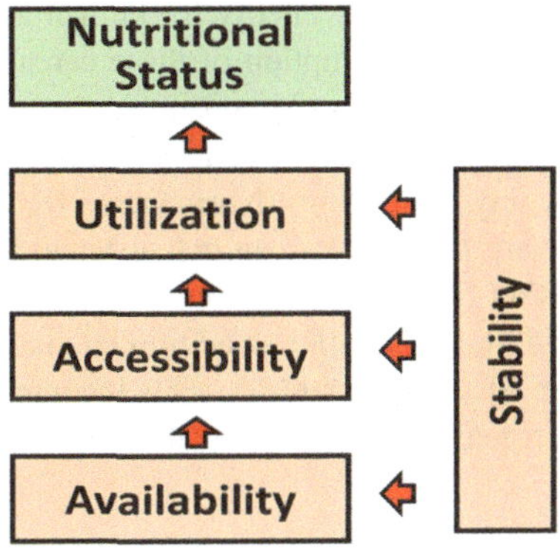

Fig. 3.3 Food and nutritional security dimensions

literatures also clearly indicates the importance of ecosystem services for food as well as environmental security.

3.5 Ecosystem Services in Rice-Wheat Cropping System

Among the cropping system in the world, the RWCS comprises the largest share for food and nutritional safety (Timsina and Connor 2001), while the traditional RWCS has also delivered ecosystem amenities since ancient eras. This traditional system fails to provide food to a mass population, since the past decades. The ever-growing population on earth sets a growing burden for more food production in a declining land area. It is necessary to produce more on decreasing land area through RWCS and to become a more input-intensive system across the globe especially in SA (Chivenge et al. 2020). These two directions indicate that optimal land utilization and efficient use of inputs are the major opportunities to intensify the production for ecosystem services in future decades (Kumar et al. 2020a; Banerjee et al. 2021).

RWCS provides plenty of ecosystem services and provides food to a large number of people across the globe to serve a better livelihood status for the farming community (Chivenge et al. 2020). As discussed in the earlier part, these ecosystem services also depend on numerous factors. The ecosystem services are well-defined as benefits derived by human beings from a healthy ecosystem, and these can be categorized into four distinct groups such as (1) provisioning, (2) regulating, (3) cultural, and (4) supporting which are discussed as follows.

3.5.1 Provisioning Services

The most important part of under-provisioning service in RWCS is food provisioning for the people of lower- and middle-income groups in SA and Southeast (SE) Asian countries (Norman and Kebe 2006). Besides, rice and wheat are the most popular cereals across the globe and consumed by a large number of people.

Supplementary energy for human and nutritional security can be achieved through regular consumption of these cereal crops (FAO 2016; GRiSP 2013). Worldwide, millions of people consume rice and wheat as a staple food (>70% of the total world's population), and RWCS provides livelihood as well as food and nutritional safety for millions (Norman and Kebe 2006). On the other side, most of the poor and marginal farmers can get sustenance and develop their livelihood security through RWCS in SE Asian countries (GRiSP 2013). Postproduction operations and value addition also offer an additional income for women with little or no formal education in SE Asian countries. This income certainly helps in sustaining their livelihood in remote villages (Demuyakor et al. 2013; Akpokodje and Erenstein 2001).

3.5.2 Regulating and Maintenance Services

Rice-wheat ecosystems considerably contribute to energy balancing, nutrient flow, and water balancing in agricultural production systems (Burkhard et al. 2015). However, as we know that, soils in paddy fields consist of hefty water storing capability and can easily govern floods, particularly in the wet season also. Groundwater recharge through the percolation of irrigation water from rice fields was also reported in many earlier findings (Chivenge et al. 2020). Besides, the buildup of soil organic carbon due to the decomposition of organic matter gets a favorable environment in soil submergence (Shirato and Yokozawa 2005). Apart from this, crop diversification and crop rotation in RWCS help in the proper nutrient cycling (Chivenge et al. 2020). Rice is a semi-aquatic plant species in nature and proposing shelter, food, breeding, and nesting habitat for diverse natural organisms. Therefore, RWCS provide several ecosystem services, i.e., organic matter decomposition/ nutrient recycling and pest regulation. Food provisioning via wildlife harvesting is also a significant ecosystem service in this cropping system (Schoenly et al. 1996).

3.5.3 Cultural Services

Rice is life, and the cultivation of rice has created many riverbank cultures worldwide. A great countable amount of art, song, rituals, and social habits related to rice cultivation is properly documented in different studies throughout the world (Chivenge et al. 2020). Rice, which is native to SE Asia, gradually becomes an integral part of many festivities across the globe. Some of the food products and beverages (alcoholic and non-alcoholic) are made from a variety of cereal crops, i.e., rice, wheat, barley, etc. (Juliano and Hicks 1996; Cicero and Gaddi 2001). Worldwide, a huge number of places on cultural identity, landscape aesthetics, and especially cultural heritage cite are related to rice and wheat farming (Settele et al. 2015).

3.5.4 Environmental Security

Most disservice action toward dreadful environmental conditions we can found in RWCS is stable burning in many countries. After the harvesting of rice, stable burning releases a huge amount of particulate matter which is associated with respiratory troubles (Chivenge et al. 2020). Though harvesting with manual labors will ultimately increase the cost of cultivation but side by side help in reducing the environmental pollution level. Besides, CH_4 and N_2O produced from rice fields are a major source of GHGs and global warming potential. Many findings also cited that prolonged flooding followed by a dry cycle may be a cause of releasing these obnoxious GHGs (FAO 2016). Indiscriminate use of agrochemicals (chemical pesticides and fertilizers) signals a major threat in RWCS (Power 2010). In rice fields, the use of excess nitrogenous fertilizers leads to ammonia volatilization, which ultimately resulted in acid rains. Production of nitrous oxide is more in rice fields due to denitrification (Prasad and Nagarajan 2004). Crop residue management (use of green manuring or grown dual-purpose summer legumes), crop diversification, and conservation agriculture are some of the ways to cope with the adverse application of chemical fertilizers (Chauhan et al. 2012).

The productivity of the rice-wheat ecosystem has been increasing significantly in the last few eras after the Green Revolution. Intensification in the usage of better-quality good crop varieties coupled with chemical pesticides and fertilizers was also evident at the time of the Green Revolution. While intensive agriculture has boosted food grain production, it has contributed toward malnutrition and poverty, predominantly in emerging nations. This was connected with the indiscriminate application of chemical plant nourishments initiating ecological contamination and GHGs. Undoubtedly, RWCS suggest a variety of environmental facilities for food provision and other issues but possess a threat in the case of environmental security. The GHGs remain the foremost challenge under these RWCS. Therefore, it is essential to find out alternatives to mitigate both ecological contamination and GHGs.

3.6 Intensive Rice-Wheat System and Dark Revolution

Conventional RWCS helps in fighting against hunger but still has many dark sides/limitations as it finally results in many sustainability issues. Among them, the declining underground water table is the main issue which creates huge hue and cry nowadays, and several technologies are recommended for reducing huge water intake of different crops in general and of rice in particular which are popularly known as resource-conserving. India has the annual share of water more than one fourth (230 km^3) globally (Tyagi et al. 2012). In spite of consuming a huge quantum of water in agriculture for irrigation, scarcity exists due to the competitive demand from the other sectors, viz., industry, etc. (Singh et al. 2010; Meena et al. 2020a). Further, RWCS is also responsible for the higher levels of underground water pollution because of excessive use of fertilizers particularly of N followed by flood irrigation in rice; results seem to be more severe in the coarse-textured soils.

The use of this polluted water for irrigation or drinking in the dairy sector results as end products consumed by human beings. Sustainable and need-based use of fertilizers entirely depends on the soil analysis reports or LCC and need-based irrigation as per the soil matric potentials through tensiometers, and both are the latest technologies recommended for the region (Bhatt 2013; Bhatt et al. 2019; Bhatt and Meena 2020). Further, intensive cultivation of RWCS results in the evolution of some hardy weed which competes with the crop plants for resources and thereby reduces the grain yields. Further, insect pest attack under the lavish environment is another sustainability issue that attacks the main plants and significantly cuts down the crop grain yields and thus water productivities.

Degraded soil structure is the outcome of the conventional puddling operations which is required for rice establishments and to reduce the drainage losses for creating the anaerobic environment. In the long run, puddling operation reduces the aeration under the effect of which next upcoming aerobic crop, viz., wheat, got adversely affected due to poor root growth. This is why most of the time wheat followed by puddled transplanted rice faces aeration stress and becomes yellow. However, the negative effects of structural degradation on upland crops, viz., wheat, etc., were reported by many workers (Kukal and Aggarwal 2003; Bhatt and Kukal 2018). Deficiency of nutrients, both macro and micro, is also reported at a higher rate in different locations, followed by RWCS intensively (Biswas and Tewatia 1991), and this needs to be addressed as soon as possible, as it affects the quality of the ultimate product. Under these conditions, biofortification of these micronutrients, viz., of Zn, is of utmost importance in improving human nutrition (Bhatt et al. 2020). Conventional tillage operation results in the primary particles which earlier bids with each other in the shape of large aggregates. After the disintegration of the bigger aggregates, soil organic matter earlier hidden in the aggregate is now exposed to the microorganisms which oxidize it to CO_2 as a GHG. Thereby, adopting the ZT, which prohibits the breakage of the larger aggregates into the smaller ones on one side sequester this C there on the other side also reduces the cost of cultivation by reducing the fuel consumption. Hence, conventional practices of the establishment of rice-wheat crops and their irrigation practices all are intensive as far as water, energy, and other inputs are involved, which finally break down the bigger aggregates and declined the underground water table. Finally, RWCS results in the production of huge crop residues on to the field after combined harvesting whose management in itself is a great challenge. In fact, out of the two crop residues, rice residue management is not an easy job though some methods for its sustainable use are suggested in the region (Yadwinder and Sidhu 2014).

Labor shortage, particularly during the peak seasons of paddy transplantation, is another dark side of RWCS which was experienced particularly from the last few years, and for that, many government schemes such as MGNREGA (GOI 2015) are responsible which provides them assured working days. Several researchers already reported the beneficial effects of the application in the RWCS (Tiwari et al. 1989; Katyal 2003), but the quantified recommendation of "S" in texturally divergent soils still needs to be worked out. In coarse-textured soils, wheat grown after puddled transplanted rice suffered from manganese deficiency and aeration stress due to the formation of a plow pan at a depth of 7–10 cm. Further, the deterioration of wheat productivity due to B insufficiency was observed in soils of West Bengal, India, in the intensively cultivated rice and wheat under RWCS (Chatterjee et al. 1987). Therefore, RWCS, even claimed to be time-, labor-, energy-, and capital-intensive, helps in the filling of empty stomachs (whenever required), but ultimately we and our future generations have to pay much for it for non-judicious exploitation of natural and non-renewable resources, and this is the ultimate dark side reported with this system.

3.7 Rice-Wheat Ecosystem and Footprints Under the Changing Climate

The RWCS refers to the annually growing of these crops in a sequence, through conventional practices of the intensive tillage in wheat, while puddling operation in rice (Sarkar 2015). Despite being the most potential cropping system of SA, a number of serious problems are now coming. The total factor productivity of the RWCS is now gradually decreasing (Chauhan et al. 2012). The farmers of this region are mostly marginal, and they are less exposed to modern agricultural practices (Malik et al. 2016). The present days of rice that came to boost up production in Asian countries; however, these traditional cultivars of rice have already reached a yield stagnation situation in recent years (Prasad 2005), and the grain production associated with the Green Revolution has now ceased.

In the upcoming 50 years, the Asian population will increase by about 1.5 billion (Sarkar et al. 2016), and the climate change will cause extreme variations in rainfall, wind speed, sunshine hours, etc. This will adversely affect the production as a whole causing erratic growth, and the water scarcity will come forward as a burning problem. This would lead to a loss in the irrigated cultivation system and also an increase in the demand for fossil fuel and non-renewable energy resources (Sarkar et al. 2016). The capability of this RWCS to sustain productivity is now endangered by many factors (Chauhan et al. 2012; Asseng et al. 2015; Balwinder-Singh et al. 2015). The chief factors that are liable for stalling output of this sequence are presented in Fig. 3.4.

Resource-intensive RWCS ecosystem generated a number of environmental-related issues. Among the different quantitative indicators, the carbon footprint (CPF) and water footprint (WPF) have gained a wider acceptance and are applied in agriculture for evaluating the performance of different RCTs (Tjandra et al. 2014). Since the beginning of the twentieth century, the global energy budget has been increased by ten times (Tandon and Singh 2010) due to RCTs which further reduces the evolution of the GHGs (Tubiello et al. 2015). Besides, the sustainability of any production system depends on the increased and efficient use of C-based inputs, viz., organic manures, etc. (Lal 2004). In SA countries, the primary production stage contributes to the major share in agricultural GHG emissions. Extensive use of inputs might be in terms of fertilizers, pesticides, or irrigation water which ultimately results in the decreased sustainability (Vetter et al. 2017; Meena et al. 2020b).

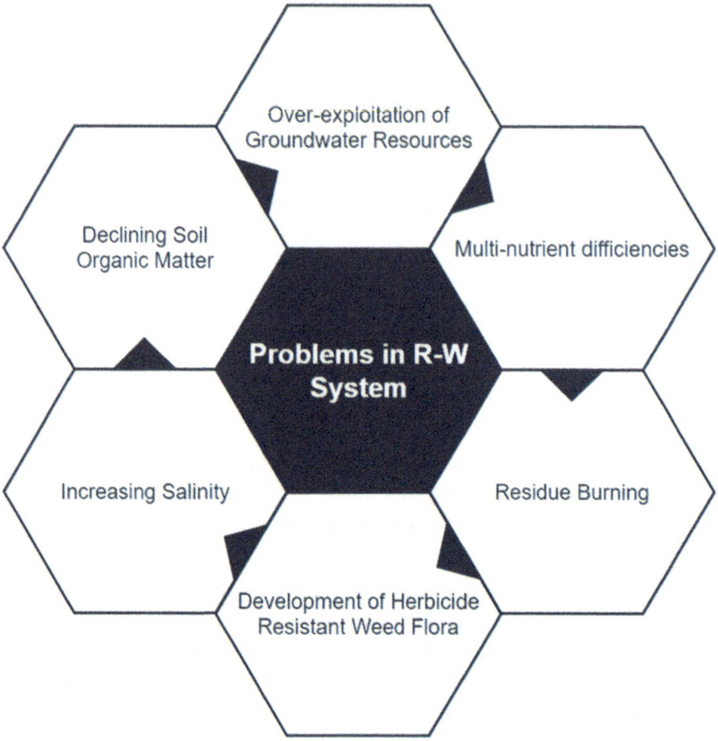

Fig. 3.4 Major problems in RWCS in South Asian regions (Source: Chauhan et al. 2012)

Emissions from different rice-wheat-based ecosystems vary significantly due to management factors like input use, labor employment, etc. and the cultivars. The RWCS contributed to the GHGs in the atmosphere in SA countries. In India, puddled rice fields contributed around 3.37 Mt of CH_4 that accounts for 24% of the total agricultural GHG emission of India (Mittal et al. 2018). On the other hand, CH_4 emission from the rice field has remained constant over the years, but N_2O emission has increased from 169 to 217 thousand tons during the last four decades (Mittal et al. 2018). It is pointed out in different scientific literatures that CH_4 emitted in the aerobic soils, viz., direct-seeded rice fields, and an excess amount use of fertilizers particularly of N led to N_2O evolution from the soils into the atmosphere (Gupta et al. 2016). Earlier study also reported that emissions of CH_4 RWCS higher by 172% particularly when compared with split application of the N fertilizers compared to the single split (Balasubramanian et al. 2017b). The nitrogen fertilization results up to 50% of total emissions, while strategies are recommended for the reduction of C footprints to a significant extent (Dhaliwal et al. 2020). Residue burning in the RWCS ecosystem is also a serious threat which finally adds to the CH_4 and N_2O to the tune of 0.25 and 0.007 million tons in India, respectively

(Balasubramanian et al. 2017a). As per Pathak et al. (2002), rice-based food products have elevated CPF than wheat-based products (Pathak et al. 2002).

The WFP per ton of crop produced vary appreciably among crops, soil textural classes, and agro-climatic conditions. The average WFP for breakfast cereal, wheat, and rice varied as per 1827, 1644, and 1673 m^3ton^{-1}, respectively (Hoekstra and Mekonnen 2010). It is reported that the total water required for paddy cultivation from seed to seed was just around 1500 mm, which further varied as per different soil textural classes, agro-climatic factors, and crop establishment methods (Balasubramanian et al. 2017a). Balasubramanian et al. (2017a) reported that the average WFP of rice produced in India is about 3150 Lkg^{-1} under conventional rice establishments which could be easily reduced to 1953 Lkg^{-1}using SRI technologies. Sustainable water resource management including land leveling, efficient irrigation techniques, etc. must be adopted to reduce the increasing trend of WFP for the RWCS ecosystem.

The current scenario of climate change with higher intensity of rains, etc. adversely affected the land and water productivity of the RW ecosystem of SA countries (Asseng et al. 2015; Balasubramanian et al. 2017a; Jat et al. 2018; Xiao et al. 2018). According to the IPCC report, temperatures in the Indian subcontinent (mainly in the lower IGP) are supposed to hike by approximately 4 °C by 2080–2099 (IPCC 2020). Further, yearly minimum temperature also hiked top around 18–21% higher than through emissions. The frequency of tropical cyclones in the Bay of Bengal may increase, and rainfall may upsurge by 20–30% (Caesar et al. 2015). It is reported that the land productivity of the rice reduced due to climate change due to a reduction in different growth parameters (Balasubramanian et al. 2017a). The CFP and GHG emission from the puddled rice may significantly increase due to increasing temperature and rising atmospheric CO_2 concentration (Independent 2020). The researchers also predicted that by the end of the twenty-first century, the rising of atmospheric CO_2 and warming will increase GHG emission by two times in rice production (Van Groenigen et al. 2013). Thus, the adoption of resource-efficient management strategies is essential to optimize the production of the crops under the RW ecosystem and to cope with the ill impacts of climate change.

3.8 Next-Generation Technologies for Sustainable Rice-Wheat System

In developing countries, the indigenous and conventional practices involve the higher application of fertilizers. In general, different soil properties varied across the field, and the application is not the best and effective practice (Gaston et al. 2001). Utilization of inputs in the traditional way has led to low productivity, soil health degradation, low input use efficiency, and non-judicious use of natural resources. The RWCS across the SA countries has donated extremely to the food security of the regions, but now this system productivity reached an equilibrium rather starts declining. Among the different reasons identified for this, the decline in the soil organic matter and deterioration of soil structure are the main which further have adverse effects on the other properties, viz., soil aeration, soil infiltration, etc. A decline in soil fertility, particularly of organic C and N, deterioration in soil physical characteristics, a delay in sowing of wheat, and declining water availability are some of the causes of this slowdown in productivity. Water productivity (g kg^{-1}) can be defined as the quantity of irrigation water used to produce per unit of the grains. However, during the twenty-first century, decreasing water productivity is a major cause of concern as reported by many workers throughout the region under different agro-climatic conditions (Humphreys et al. 2010; Bhatt 2015). Therefore, a sustainable paradigm shift is an immediate need for enhancing the system's productivity and sustainability. Future agriculture thus needs to focus on non-destructive, resource-efficient, environmentally friendly practices with the reduced cost of the production system. In this respect, precision farming solves the purpose which focuses on site-specific crop management with spatial variability and without environmental intervention.

3.8.1 Future Strategies for Rice-Wheat Cropping System

By reducing GHG emissions, the CSA practices help to enhance the overall land and water productivity in a climate-smart way. Climate-resilient agriculture is an integrated approach that guides the actions needed for transforming and restructuring agriculture under changing climate scenarios. CSA aims to tackle three main objectives: (1) improving the livelihoods of the farmers, (2) building up new climate-smart effective technologies, and (3) reducing the evaluation of GHGs (FAO 2010).

There is a need to sustainably reduce the C, water, and energy footprints in the RWCS which is possible only through improvement in input use efficiency of these parameters by adopting new recommended RCTs. Adoption of strategies involving judicious use of natural resources and harnessing solar energy in the best possible way may lead to achieving our target. The following strategies are an urgent need to achieve the target.

3.8.1.1 Water Smart

Interventional strategies to enhance water use efficacy through improved RCTs, viz., DSR, GPS-mounted laser land leveling, use of tensiometer for need-based irrigation, micro-irrigation like drip and sprinkler, rainwater harvesting, raised bed planting, and medium-range weather forecast-based irrigation application.

3.8.1.2 Nutrient Smart

By adopting site-specific nutrient management for enhanced nutrient use efficiency through the use of LCC, SPAD, and polymer-coated urea (PCU) as slow-release fertilizers. PCU reported having higher use efficiency of applied nutrients (Bhatt and Singh 2020). The purpose is to apply the fertilizers as per the need of the plant.

3.8.1.3 Carbon Smart

Here the objective is to enhance the C-sequestration rate through different interventions, viz., ZT, legume intercropping, and using the techniques of residue retention in soil and incorporation.

3.8.1.4 Weather Smart

The purpose here is to smartly deal with the climatic variables by one or different approaches such as growing of short-duration plant cultivars, timely paddy seedling transplanting, and applied crop residues as mulch on the soil surface for finally harvesting better grain yields.

3.8.1.5 Energy Smart

The purpose is to invent new crop establishment technologies which ultimately result in reduced energy consumptions, viz., happy seeder, laser leveler, and use of crop residues as mulch harnessing solar energy for irrigation.

3.8.1.6 Knowledge Smart

Indigenous technical knowledge (ITK) is received from the earlier generations which need to be improved with scientific interventions for overall reducing the livelihoods and to reduce the C, water, pesticide, and energy footprints.

3.8.2 Precision Farming (PF) as a New Age Tool

The increasing global population with limited or rather decreasing resources which are being used for agricultural production results in a great challenge to provide food security to the entire world. To cope with these challenges or constraints in the region, farmers must adopt some improved RCTs which help them to enhance their yields and, hence, livelihoods on a sustainable basis. The concept of precision farming helps in resorting to both spatial and temporal variabilities with reduced environmental pollution and depends on the integrated approach of the new interventions. Precision land leveling, precision planting, SSNM by using GreenSeeker, LCC, and soil moisture assessment-based irrigation management have tremendous potential for increasing crop productivity and input use efficiency under field conditions while reducing the cost of production and deleterious impacts on the environment. In developing countries mostly in Southeast Asia, there is a wide scope for the application and operations in a farmer's field for practicing a part of PF technologies under a rice-wheat cropping system.

For enhancing the efficiency of farm input use efficiency, increasing productivity and returns of crop production and reducing potential environmental pollution one should ensure to apply farm inputs (1) at right time (2) right dose and, (3) right place. Precision farming helps in bringing overall sustainability by taking care of other factors, reducing the emission of GHGs, improving soil health and improving the soil physic-chemical properties in one or another way. Sharma et al. (2005) showed that several technological interventions have been made during the early twentieth century on an experimental basis.

3.8.2.1 Site-Specific Seed Planting

The seeds of crops have become one of the costly inputs, and hence planting it at a desirable depth and in the right amount to prevent excess loss needs special surveillance. Manual sowing/planting is one of the most common practices in countries like India. Low-cost seed planter devices have served the purpose of the farmers. Development of new farmer-friendly machinery, viz., happy seeder, zero tillage, mechanical transplanter, etc., is the major player in bringing overall sustainability in the region. Many scientists believe that the performance of these machines is due to the saving in terms of the water-saving which further reported with better water productivity in the region. Further, Chandra et al. (2007) in Haryana revealed that both irrigation and gross water productivity of wheat were significantly increased under zero tillage with 15–60 L ha^{-1} of fuel savings (Hobbs and Gupta 2003; Laxmi et al. 2003; Malik et al. 2004). On-field research trials conducted across the IGP reported significant increases (3–73%) in wheat yields under zero tillage practice (Dhiman et al. 2003; Prasad et al. 2002). Further, many researchers also revealed to have better yield benefits and hence higher profits even with lesser use of the wheat seeds, which also reduced the overall costs involved.

3.8.2.2 Site-Specific Nutrient Management (SSNM)

Under normal conditions, farmers used to apply more and more fertilizers to their crop for having higher land and water productivity. As these recommendations are made for the broader area, they are always erroneous and vary from field to field within small areas. Site-specific nutrient management (SSNM) which is a general concept for improving the availability and demand of nutrients based on spatial and temporal variability is being tried in most of the countries across the globe for achieving the higher or potential yield targets. It takes into consideration the plant nutrient requirements at each growth stage and the soil's ability to supply those nutrients and incorporates the information to areas in the field that require different agronomic management practices based on field average. SSNM permits to fine-tuning of crop management systems based upon the 4R nutrient relationship – the right product, right rate, right time, and place of nutrient use.

Right Product

Appropriate and affordable fertilizer should be recommended based on the availability and accessibility of the farmers according to crop needs and soil type characteristics to safeguard a well-adjusted supply of nutrients.

Right Rate

The optimum dose of fertilizer should be applied as per the crop needs, taking into account the current supply of nutrients within the soil. Imbalance fertilizer leads to a detrimental effect on the environment, including runoff, leaching and emissions of

GHGs, as well as monetary losses, whereas too little fertilizer will exhaust soils, leading to the deterioration of soil health and poor stand establishment of crops.

Right Time

Availability of nutrients is essential during crop requirement is important which may be done by assessing crop nutrient dynamics. Thus, split uses of inorganic fertilizers in combination with organic nutrient deliver the slow releasing of nutrients, which may be the right pathway of sustainable crop productions.

Right Place

Placement of nutrients at the optimal distance from the crop rhizosphere and placing them in the right soil depth so that crops can use them with ease is key to minimizing nutrient losses. Generally, incorporating nutrients into the soil is recommended rather than applying them to the surface due to enhanced loss due to exposure to the open environment. The ideal method for placement of nutrient fertilizers depends on the characteristics of the soil, crop, tillage regime and type of fertilizer.

Yield, Profitability, and NUE Under SSNM

Indigenous practice of farmers apply higher doses of fertilizer that resulted in 7% yields while 12% loss of profits than SSNM (Dobermann et al. 2002). A study on RWCS with SSNM and conventional practices in North-West India showed that yield and profit were increased 12 and 14% for rice and 17 and 13% for wheat under the practice with SSNM (Bhatt et al. 2016). Another study conducted by Pasuquin et al. (2010) in Indonesia, the Philippines, and Vietnam showed that maize yield was increased (16%) significantly under SSNM than indigenous practice; for example, under SSNM, grain yield of maize was increased by 0.89, 1.16, and 1.25 t ha^{-1} in Indonesia, the Philippines, and Vietnam. Similarly, Abdulrachman et al. (2002) conducted 45 field trials on irrigated rice in different countries of South Asia and found that the amount of N fertilizer rate was reduced significantly under SSNM; for example, N fertilizer rate was reduced by 10–20% in China, Vietnam, and Indonesia. In India, research conducted by Dass et al. (2012) at semi-arid regions of Northern India on chlorophyll meter-based N application (30 kg basal + 30 kg N ha^{-1} at SPAD value 37.5) saved 30 kg N ha^{-1} and increased *kharif* maize grain yield by almost 10% as compared to soil test-based N application.

3.8.3 Remote Sensing and GIS as a Tool of Sustainable Crop Production

Remote sensing (RS) is an integral and potential component of the precision farming system. RS and GIS provide better alternatives option for precision agriculture with timely assessment and monitoring, fast, reliable, low-cost techniques with high accuracy. RS and GIS offer the solution of monitoring the spectral and spatial and temporal changes at a higher resolution to capture the spatial variability which is helpful to detect over a period which provides us with an understanding of the associated variability in crop and crop field. GIS and RS are multi-layer based systems that provide the user with the flexibility of overlaying the various real-world layers and finding the best possible model for précising the agriculture practices that need to be undertaken. The technology has prevalent solicitation but on a larger scale, which is a prerequisite for accuracy in agriculture. In precision farming depending upon the user criterion, it may require different spatial resolutions, spectral coverage and frequencies. For example, measurement of the crop stress monitoring in Rice or Wheat crops or incidence of any pest or diseases will require higher resolutions than what required for crop growth monitoring or yield mapping. National-level datasets are being generated at the micro-level scale. Random soil sampling is done at the micro-level and analyzed for developing repository data banks. Digital soil maps also help in the judicious use of the fertilizers as it also evaluates the site fertility from where samples were not collected, thus proved to be an accurate tool for the climate-smart and sustainable use of the fertilizers in the region which on one side improved the yields while on other mitigate the adverse effects of the climate change.

3.8.4 Role of Crop Residue Management

Conservation agriculture (CA) involves the least disturbance to the soils along with maximum retention of the crop residues onto the bare soil surface with an idea to improve the soil organic matter status, reduce the production of GHGs, and improve the livelihoods sustainably (Singh and Sidhu 2014). CA-based ZT-wheat in the RWCS of north-western India showed an optimistic impression on the yield, profitability, and resource use efficiency (Erenstein and Laxmi 2008; Ladha et al. 2009). The introduction of HYV crop cultivars gives a higher grain yield and also huge quantities of crop residues (CRs) under RWCS north-western India, while due to the huge amount of CRs, farmers generally burn CRs, particularly rice CRs, causing nutrient losses, and havoc air pollution, leads to threat for environment and human health safety. Mulch is one of the best alternative options for rice residue management during wheat crop cultivation, especially under ZT conditions. Mulch can increase yield, water use efficiency, and profitability while reducing weed infestations. Surplus residue from the previous wheat crop can also be incorporated into the paddy fields with no adverse effect on rice yield. Long-term studies of residue recycling have proved improvements in soil physical, chemical, and biological properties. Due to high nutrient contents, the crop residues tend to improve the soil health by one or other mode, which further helps to reduce the fertilizer doses as per plant requirements as we have to feed the plant but not the soil. Since CRs contain significant quantities of plant nutrients, their long-term application will have a significant positive effect on fertilizer management in the RW system. Further, biochar could also be a viable option for managing the paddy residues which further can be used for improving soil health for increasing nutrient use efficiency and to minimize air pollution (Bijay-Singh et al. 2008).

3.9 Summary and Conclusion

Based on all the above discussion, it may be revealed that RWCS of SA is the best productive agricultural system but suffers from some sustainability issues though it has a wide potential for meeting the food and nutritional security for an increasing population. Scientists are busy evaluating the technologies which help in improving the input use efficiency of the applied inputs. However, due to the changing climate, remarkable declining tendencies of the production systems are observed in recent decades which create great concern for the rising population. In the systems, for rice cultivations, repeated plowing for puddling of transplanted rice are worsening soil properties, creation of a hard plow pan layer at a surface-depth which generates an unfortunate infiltration, leads to water-logging and poor-rooting for wheat, as a result of less recharge than more uptake of groundwater tables are declining which leads to arsenic problem in some areas, in same land and seasons are repeatedly same crops that increase the infestation of pests-diseases and also imbalance application of inorganic fertilizers and pesticides and traditional cultivations systems rise the GHG emission. The chapter also revealed that if the systems have several negative effects on the agroecology of these regions, it is difficult to replace the systems, since the systems are a key source of staple food production for the growing population in the region. Therefore, researchers are trying to find alternative ways to sustain these systems and to make it more efficient and profitable by minimizing the adverse effect on agroecological conditions. Recently, researchers recommended several climate-smart and next-generation sequential technologies for undertaking these difficulties, but all are site- and situation-specific. Therefore, an integrated approach of these RCTs might help the farmers to enhance their livelihoods through improving the land and water productivity in a climate-smart mode for bringing long-term sustainability in the region.

References

Abdulrachman S, Gines HC, Nagarajan R, Satawathananont S, Son TT, Wang GH (2002) Variation in the performance of site-specific nutrient management among different environments with irrigated rice in Asia. Better Crops Int 16(2):18–23

Aggarwal PK, Joshi PK, Ingram JSI, Gupta RK (2004) Adapting food systems of the Indo Gangetic plains to global environmental change: key information needs to improve policy formulation. Environ Sci Pol 7:487–498

Akpokodje GLF, Erenstein O (2001) Nigeria's rice economy: state of the art, the Nigerian rice economy in a competitive world; constraints opportunities and strategic choices. Final report presented to West Africa Rice Development Association (WARDA) Bouake, Cote d'Ivore, p 55

Asseng S, Ewert F, Martre P, Rötter RP, Lobell DB, Cammarano D, Kimball BA, Ottman MJ, Wall GW, White JW, Reynolds MP (2015) Rising temperatures reduce global wheat production. Nat Clim Chang 5(2):143. https://doi.org/10.1038/nclimate2470

Balasubramanian R, Saravanakumar V, Boomiraj K (2017a) Ecological footprints of and climate change impact on rice production in India. Elsevier

Balasubramanian S, Nelson A, Koloutsou-Vakakis S, Lin J, Rood MJ, Myles L, Bernacchi C (2017b) Evaluation of DeNitrification DeComposition model for estimating ammonia fluxes

from chemical fertilizer application. Agric For Meteorol 23:123–134. https://doi.org/10.1016/j.agrformet.2017.02.006
Balwinder-Singh, Humphreys E, Gaydon DS, Sudhir-Yadav (2015) Options for increasing the productivity of the rice-wheat system of north West India while reducing groundwater depletion. Part 2. Is conservation agriculture the answer? For Crop Res 173:81–94. https://doi.org/10.1016/j.fcr.2014.11.019
Banerjee A, Jhariya MK, Yadav DK, Raj A (2020) Environmental and sustainable development through forestry and other resources. Apple Academic Press Inc., CRC Press- a Taylor and Francis Group, US & Canada, 400 pp. https://doi.org/10.1201/9780429276026
Banerjee A, Meena RS, Jhariya MK, Yadav DK (2021) Agroecological footprints management for sustainable food system. Springer, Singapore, 514 pp. https://doi.org/10.1007/978-981-15-9496-0
Bhatt R (2013) Soil test based fertilization to improve production of oil seed crops in Kapurthala district of Punjab. Int J Sci Environ Technol 2(3):521–526
Bhatt R (2015) Soil moisture dynamics during intervening period in rice–wheat sequence as affected by different tillage methods at Ludhiana, Punjab, India. Soil Environ 34(1):82–88
Bhatt R, Kukal SS (2018) Delineation of soil water balance in wheat-dry direct seeded rice system under conventional and zero-till conditions in semi-arid tropics. Oryza 55(4):574–589
Bhatt R, Meena RS (2020) Delineation of soil moisture potentials and balance components. In: Soil Moisture Importance. InTechOpen. https://doi.org/10.5772/intechopen.92587
Bhatt R, Kukal SS, Busari MA, Arora S, Yadav M (2016) Sustainability issues on rice-wheat cropping system. Int Soil Water Conserv Res 4:68–83. https://doi.org/10.1016/j.iswcr.2015.12.001
Bhatt R, Kaur R, Gosh A (2019) Strategies to practice climate smart agriculture to improve the livelihoods under rice-wheat systems in South Asia. In: Sustainable soil and environmental management. Springer, pp 29–72. https://doi.org/10.1007/978-981-13-8832-3_2
Bhatt R, Hossain A, Kumar P (2020) Zinc biofortification as an innovative technology to alleviate the zinc deficiency in human health: a perceptive of paddy-wheat rotation of South Asia. Open Agric 5:1–12. https://doi.org/10.1515/opag-2020-0018
Bijay-Singh, Shan YH, Johnson-beeebout SE, Yadvinder-Singh, Buresh RJ (2008) Crop residue management for lowland rice-based cropping systems in Asia. Adv Agron 98:118–199
Bindi M, Howden M (2004) Challenges and opportunities for cropping systems in a changing climate. In: Proceedings of the 4th international crop science congress, 26 Sep – 1 Oct 2004, Brisbane, Australia. Published on CDROM, pp 52–54. www.cropscience.org.au
Biswas BC, Tewatia RK (1991) Nutrient balance in agro-climatic regions of India- an overview. Fertil News 36:13–17
Brown L (1977) Redefining security. World watch paper 14. World Watch Institute, Washington, D.C.
Burkhard B, Muller A, Muller F, Grescho V, Anh Q, Arida G, Bustamante JVJ, Van Chien H, Heong K, Escalada M, Marquez L, Truong DT, Villareal S, Settele J (2015) Land cover-based ecosystem service assessment of irrigated rice cropping systems in Southeast Asia: an explorative study. Ecosyst Serv 14:76–87
Caesar J, Janes T, Lindsay A, Bhaskaran B (2015) Temperature and precipitation projections over Bangladesh and the upstream Ganges, Brahmaputra and Meghna systems. Environ Sci Process Impacts 17:1047–1056. https://doi.org/10.1039/c4em00650j
Capone R, Bilali HE, Debs P, Cardone G, Driouech N (2014) Food system sustainability and food security: connecting the dots. J Food Secur 2(1):13–22
Chatterjee C, Sinha P, Nautical S, Aggarwal C, Sharma CP (1987) Metabolic changes associated with boron-calcium interaction in Maize. Soil Sci Plant Nutr 33:607–617
Chauhan BS, Mahajan G, Sardana V, Timsina J, Jat ML (2012) Productivity and sustainability of the rice–wheat cropping system in the Indo-Gangetic Plains of the Indian subcontinent: problems, opportunities, and strategies. In: Advances in agronomy, vol 117. Academic Press, pp 315–369. https://doi.org/10.1016/B978-0-12-394278-4.00006-4

Chivenge P, Angeles O, Hadi B, Acuin C, Connor M, Stuart A, Johnson-Beebout S (2020) Ecosystem services in paddy rice systems. In: The role of ecosystem services in sustainable food systems. Academic Press, pp 181–201

Cicero A, Gaddi A (2001) Rice bran oil and g-oryzanol in the treatment of hyper lipoproteinaemias and other conditions. Phytother Res 15:277–289

Costanza R, Kubiszewski I (2012) The authorship structure of "ecosystem services" as a transdisciplinary field of scholarship. Ecosyst Serv 1:16–25. https://doi.org/10.1016/j.ecoser.2012.06.002

CSISA (2019) Cereal systems initiative for South Asia: phase III - building resilience in South Asia's cereal systems. Semi-annual report September 30, 2018 to March 31, 2019. https://csisa.org/wp-content/uploads/sites/2/2019/06/Cereal-systems-initiative-for-South-Asia-CSISA-Phase-III-Semi-annual-report-March-2019.pdf. Accessed June 2020

Daily G, Dasgupta P, Bolin B, Crosson P, Du Guerny J, Ehrlich P, Folke C, Jansson AM, Jansson BO, Kautsky N, Kinzig A (1998) Food production, population growth, and the environment. Science 281:1291–1292

Dass A, Singh DK, Dhar S (2012) Precise supply of nitrogen and irrigation to hybrid maize using plant sensors In: Proceedings of the international agronomy congress: agriculture, diversification, climate change management and livelihoods; Nov 26–30; New Delhi, India, pp 534–535

Dass A, Chandra S, Choudhary AK, Singh G, Sudhishri S (2016a) Influence of field re-ponding pattern and plant spacing on rice root–shoot characteristics, yield, and water productivity of two modern cultivars under SRI management in Indian Mollisols. Paddy Water Environ 14 (1):45–59

Dass A, Shekhawat K, Choudhary AK, Sepat S, Rathore SS, Mahajan G, Chauhan BS (2016b) Weed management in rice using crop-competition – a review. Crop Prot 95:45–52

De Groot R, Brander L, Van Der Ploeg S, Costanza R, Bernard F, Braat L, Christie M, Crossman N, Ghermandi A, Hein L, Hussain S (2012) Global estimates of the value of ecosystems and their services in monetary units. Ecosyst Serv 1:50–61. https://doi.org/10.1016/j.ecoser.2012.07.005

Demuyakor B, Dogbe W, Owusu R (2013) Parboiling of paddy rice, the science and perceptions of it as practiced in Northern Ghana. Int J Sci Technol Res 2:13–18

Dhaliwal SS, Naresh RK, Gupta RK, Panwar AS, Mahajan N, Singh R, Mandal A (2020) Effect of tillage and straw return on carbon footprints, soil organic carbon fractions and soil microbial community in different textured soils under rice–wheat rotation: a review. Rev Environ Sci Biotechnol 19:103–115

Dhiman SD, Kumar S, Om H (2003) Shallow tillage and drill technology for wheat. Indian Farm 53:10–13

Dobermann A, Witt C, Dawe D (2002) Increasing the productivity of intensive rice systems through site-specific nutrient management. Science Publishers and International Rice Research Institute (IRRI), New Delhi and Makati City

Duxbury JM, Abrol IP, Gupta RK, Bronson KF (2000) Analysis of soil fertility experiments with rice-wheat rotations in South Asia. In: Abrol IP, Bronson KF, Duxbury JM, Gupta RK (eds) Long term soil fertility experiments in rice-wheat cropping systems. Rice-wheat consortium paper series number 6. RWC for IGP, New Delhi

Erenstein O (2009) Adoption and impact of conservation agriculture based resource conserving technologies in South Asia. In: Proceedings 4th world congress on conservation agriculture, February 4–7, 2009, New Delhi, India 2009 Feb 4, pp 439–444

FAO (2010) Climate-smart agriculture, policies, practices and financing for food security, adaptation and mitigation. The Hague conference on agriculture, food security and climate change, Rome, Italy

FAO (2016). Available from: http://www.fao.org/faostat/en/-data/QC.2016

Gaston LA, Locke MA, Zablotowicz RM, Reddy KN (2001) Spatial variability of soil properties and weed populations in the Mississippi delta. Soil Sci Soc Am J 65:449–459

Godfray HCJ, Beddington JR, Crute IR, Haddad L, Lawrence D, Muir JF, Pretty J, Robinson S, Thomas SM, Toulmin C (2010) Food security: the challenge of feeding 9 billion people. Science 327:812–818

GOI (2015) State of Indian agriculture. Ministry of Agriculture and Farmers Welfare department of Agriculture, Cooperation and Farmers Welfare Directorate of Economics and Statistics New Delhi

GRiSP (2013) Rice Almanac, 4th edn. International Rice Research Institute, Los Baños, p 283

Gupta R, Sayre KD (2007) Conservation agriculture in South Asia. J Agric Sci 145(3):207–214

Gupta DK, Bhatia A, Kumar A, Das TK, Jain N, Tomer R, Malyan SK, Fagodiya RK, Dubey R, Pathak H (2016) Mitigation of greenhouse gas emission from rice-wheat system of the Indo-Gangetic plains: through tillage, irrigation and fertilizer management. Agric Ecosyst Environ 230:1–9. https://doi.org/10.1016/j.agee.2016.05.023

Hobbs PR, Gupta RK (2003) Resource-conserving technologies for wheat in the rice-wheat system. In: Ladha JK, Hill JE, Duxbury JM, Gupta RK, Buresh RJ (eds) Improving the productivity and sustainability of rice wheat systems: issues and impacts, ASA special publication number 65. ASA-CSSA-SSSA, Madison, WI, pp 149–172

Hoekstra AY, Mekonnen MM (2010) The green, blue and grey water footprint of crops and derived crop products. Volume 1: main report. Value Water Res Rep Ser No 471:80

Humphreys E, Kukal SS, Christen EW, Hira GS, Sharma RK (2010) Halting the groundwater decline in north-west India—which crop technologies will be winners? In: Advances in agronomy, vol 109. Academic Press, pp 155–217

Independent (2020) Rice farming up to twice as bad for climate change as previously thought, study reveals. The Independent. https://www.independent.co.uk/environment/rice-farming-climate-change-global-warming-india-nitrous-oxide-methane-a8531401.html. Accessed 1 Apr 2020

IPCC (2001a) Climate change 2001: the scientific basis. In: Houghton JT, Ding Griggs DJ, Noguer M, van der Linden PJ, Xiaosu D (eds) Contribution of working group I to the third assessment report of the intergovernmental panel on climate change (IPCC). Cambridge University Press, Cambridge, p 944

IPCC (2001b) Climate change 2001: impacts, adaptation and vulnerability. In: McCarthy JJ, Canziani OF, Leary NA, Dokken DJ, White KS (eds) Contribution of working group II to the third assessment report of the intergovernmental panel on climate change (IPCC). Cambridge University Press, Cambridge, p 1032

IPCC (2020) Global climate data and information. In: World Bank. https://climateknowledgeportal.worldbank.org/country/india/climate-data-projections. Accessed 21 Jan 2020

Islam S, Gathala MK, Tiwari TP, Timsina J, Laing AM, Maharjan S, Chowdhury AK, Bhattacharya PM, Dhar T, Mitra B, Kumar S, Srivastwa PK, Dutta SK, Shrestha R, Manandhar S, Sherestha SR, Paneru P, Siddquie NEA, Hossain A, Islam R, Ghosh AK, Rahman MA, Kumar U, Rao KK, Gérard B (2019) Conservation agriculture based sustainable intensification: increasing yields and water productivity for smallholders of the eastern Gangetic Plains. Field Crops Res 238:1–17. https://doi.org/10.1016/j.fcr.2019.04.005

Jalota SK, Khera R, Arora VK, Beri V (2007) Benefits of straw mulching in crop production: a review. J Res Punjab Agric Univ 44:104–107

Jat ML, Saharawat YS, Gupta R (2011) Conservation agriculture in cereal systems of South Asia: nutrient management perspectives. Karnataka J Agric Sci 24(1):100–105

Jat RK, Sapkota TB, Singh RG, Jat ML, Kumar M, Gupta RK (2014) Seven years of conservation agriculture in a rice–wheat rotation of eastern Gangetic Plains of South Asia: yield trends and economic profitability. Field Crops Res 164:199–210

Jat ML, Stirling CM, Jat HS, Tetarwal JP, Jat RK, Singh R, Lopez-Ridaura S, Shirsath PB (2018) Soil processes and wheat cropping under emerging climate change scenarios in South Asia. In: Advances in agronomy, vol 148. Academic Press, pp 111–171

Jhariya MK, Banerjee A, Meena RS, Yadav DK (2019a) Sustainable agriculture, forest and environmental management. Springer, Singapore, 606 pp. https://doi.org/10.1007/978-981-13-6830-1

Jhariya MK, Yadav DK, Banerjee A (2019b) Agroforestry and climate change: issues and challenges. Apple Academic Press Inc., CRC Press- a Taylor and Francis Group, US & Canada, 335 pp. https://doi.org/10.1201/9780429057274

Jhariya MK, Meena RS, Banerjee A (2021) Ecological intensification of natural resources for sustainable agriculture. Springer, Singapore, 655 pp. https://doi.org/10.1007/978-981-33-4203-3

Juliano BO, Hicks PA (1996) Rice functional properties and rice food products. Food Rev Int 12:71–103

Kataki PK, Hobbs P, Adhikary B (2001) The rice-wheat cropping system of South Asia. J Crop Prod 2:1–26. https://doi.org/10.1300/J144v03n02_01

Katyal JC (2003) Soil fertility management—a key to prevent desertification. Ind Soc Soil Sci 51:378–387

Khan N, Jhariya MK, Raj A, Banerjee A, Meena RS (2021a) Soil carbon stock and sequestration: implications for climate change adaptation and mitigation. In: Jhariya MK, Meena RS, Banerjee A (eds) Ecological intensification of natural resources for sustainable agriculture. Springer, Singapore, pp 461–489. https://doi.org/10.1007/978-981-33-4203-3_13

Khan N, Jhariya MK, Raj A, Banerjee A, Meena RS (2021b) Eco-designing for sustainability. In: Jhariya MK, Meena RS, Banerjee A (eds) Ecological intensification of natural resources for sustainable agriculture. Springer, Singapore, pp 565–595. https://doi.org/10.1007/978-981-33-4203-3_16

Kukal SS, Aggarwal GC (2003) Puddling depth and intensity effects in rice– wheat system on a sandy loam soil. I. Development of subsurface compaction. Soil Till Res 72:1–8

Kumar S, Meena RS, Jhariya MK (2020a) Resources use efficiency in agriculture. Springer, Singapore, 760 pp. https://doi.org/10.1007/978-981-15-6953-1

Kumar S, Meena RS, Datta R, Verma SK, Yadav GS, Pradhan GS, Molaei A, Rahman GKMM, Mashuk HA (2020b) Legumes for carbon and nitrogen cycling: an organic approach. In: Carbon and nitrogen cycling in soil. Springer, pp 337–375. https://doi.org/10.1007/978-981-13-7264-3_10

Kumar S, Meena RS, Singh RK, Munir TM, Datta R, Danish S, Yadav GS, Kumar S (2021) Soil microbial and nutrient dynamics under different sowings environment of Indian mustard (*Brassica juncea* L.) in rice based cropping system. Sci Rep 11:5289. https://doi.org/10.1038/s41598-021-84742-4

La Notte A, D'Amato D, Mäkinen H, Paracchini ML, Liquete C, Egoh B, Geneletti D, Crossman ND (2017) Ecosystem services classification: a systems ecology perspective of the cascade framework. Ecol Indic 74:392–402

Ladha JK, Pathak H, Tirol-Padre A, Dawe D, Gupta RK (2003) Productivity trends in intensive rice–wheat cropping systems in Asia. In: Improving the productivity and sustainability of rice–wheat systems: issues and impacts, vol 65, pp 45–76

Ladha JK, Kumar V, Alam MM, Sharma S, Gathala M, Chandana P, Saharawat YS, Balasubramanian V (2009) Integrating crop and resource management technologies for enhanced productivity, profitability, and sustainability of the rice-wheat system in South Asia. In: Integrated crop and resource management in the rice–wheat system of South Asia. International Rice Research Institute, Los Banos, pp 69–108

Lal R (2004) Carbon emission from farm operations. Environ Int 30:981–990

Laxmi V, Gupta RK, Swarnalatha A, Perwaz S (2003) Environmental impact of improved technology - farm level survey and farmers' perception on zero tillage (Case Study). Presented at roles of agriculture workshop 20–22 October, 2003 - Rome, Italy. Indira Gandhi Institute of Development Research, Mumbai

Malik RK, Gupta RK, Singh CM, Brar SS, Singh SS, Sinha RK, Singh AK, Singh R, Naresh RK, Singh KP, Thakur TC (2004) Accelerating the adoption of resource conservation technologies for farm level impact on sustainability of rice–wheat system of the indo-Gangetic Plains. NATP Prog Rep 2004:165–174

Malik AI, Ali MO, Zaman MS, Flower K, Rahman MM, Erskine W (2016) Relay sowing of lentil (Lens culinaris subsp. culinaris) to intensify rice-based cropping. J Agric Sci 154:850–857. https://doi.org/10.1017/S0021859614001324

Mann ME, Jones PD (2003) Global surface temperatures over the past two millennia. Geophys Res Lett 30(15):1820

Mathews JT (1989) Redefining security. Foreign Aff 68(2): 162–177

MEA (2005) Ecosystems and human well-being: synthesis. Island Press, Washington, DC

Meena RS, Lal R (2018) Legumes for soil health and sustainable management. Springer, Singapore. https://doi.org/10.1007/978-981-13-0253-4_10

Meena RS, Gogaoi N, Kumar S (2017) Alarming issues on agricultural crop production and environmental stresses. J Clean Prod 142:3357–3359

Meena RS, Kumar V, Yadav GS, Mitran T (2018) Response and interaction of *Bradyrhizobium japonicum* and Arbuscular mycorrhizal fungi in the soybean rhizosphere: a review. Plant Growth Regul 84:207–223. https://doi.org/10.1007/s10725-017-0334-8

Meena RS, Lal R, Yadav GS (2020) Long term impacts of topsoil depth and amendments on soil physical and hydrological properties of an Alfisol in Central Ohio, USA. Geoderma 363:1141164. https://doi.org/10.1016/j.geoderma.2019.114164

Meena RS, Lal R, Yadav GS (2020a) Long-term impact of topsoil depth and amendments on carbon and nitrogen budgets in the surface layer of an Alfisol in Central Ohio. Catena 194:104752. https://doi.org/10.1016/j.catena.2020.104752

Meena RS, Kumar S, Datta R, Lal R, Vijaykumar V, Brtnicky M, Sharma MP, Yadav GS, Jhariya MK, Jangir CK, Pathan SI, Dokulilova T, Pecina V, Marfo TD (2020b) Impact of agrochemicals on soil microbiota and management: a review. Land (MDPI) 9(2):34. https://doi.org/10.3390/land9020034

Mittal R, Chakrabarti B, Jindal T, Tripathi A, Mina U, Dhupper R, Chakraborty D, Jatav RS, Harit RC (2018) Carbon footprint is an indicator of sustainability in rice-wheat cropping system: a review. Chem Sci Rev Lett 7:774–784

Nawaz A, Farooq M, Nadeem F, Siddique KH, Lal R (2019) Rice–wheat cropping systems in South Asia: issues, options and opportunities. Crop Pasture Sci 70(5):395–427

Norman J, Kebe B (2006) African smallholder farmers: rice production and sustainable livelihoods. Int Rice Comm Newslett 55:33–44

Pasuquin JM, Witt C, Pampolino M (2010) A new site-specific nutrient management approach for maize in the favourable tropical environments of Southeast Asia. In: Proceedings of the 19th world congress of soil science, soil solutions for a changing world; 2010 1–6 Aug; Brisbane, Australia

Pathak H, Bhatia A, Prasad S, Singh S, Kumar S, Jain MC, Kumar U (2002) Emission of nitrous oxide from rice-wheat systems of Indo-Gangetic plains of India. Environ Monit Assess 77:163–178. https://doi.org/10.1023/a:1015823919405

Paul J, Choudhary AK, Suri VK, Sharma AK, Kumar V, Shobhna (2014) Bio–resource nutrient recycling and its relationship with bio–fertility indicators of soil health and nutrient dynamics in rice wheat cropping system. Commun Soil Sci Plant Anal 45:912–924

Petit JR, Jouzel J, Raynaud D, Barkov NI, Barnola JM (2000) Climate and atmospheric history of the past 420000 years from the Vostok ice core, Antarctica. Nature 399:429–436

Pimentel D, McNair M, Buck L, Pimentel M, Kamil J (1997) The value of forests to world food security. Hum Ecol 25:91–120

Pimentel D, Cooperstein S, Randell H, Filiberto D, Sorrentino S, Kaye B, Nicklin C, Yagi J, Brian J, O'hern J, Habas A (2007) Ecology of increasing diseases: population growth and environmental degradation. Hum Ecol 35:653–668

Power AG (2010) Ecosystem services and agriculture: tradeoffs and synergies. Philos Trans Royal Soc B Biol Sci 365(1554):2959–2971

Prasad R (2005) Rice-wheat cropping systems. Adv Agron 86:255–339. https://doi.org/10.1016/S0065-2113(05)86006-7

Prasad R, Nagarajan S (2004) Rice-wheat cropping system-food security and sustainability. Curr Sci 87:1334–1335

Prasad B, Sinha RK, Singh AK (2002) Studies on the effect of nitrogen management on yield of wheat in rice-wheat cropping system under zero tillage. In: Malik RK, Balyan RS, Yadav A, Pahwa SK (eds) Herbicide resistance management and zero tillage in rice-wheat cropping system. CCSHAU, Hisar, pp 120–122

Raj A, Jhariya MK, Bargali SS (2018) Climate smart agriculture and carbon sequestration. In: Pandey CB, Gaur MK, Goyal RK (eds) Climate change and agroforestry: adaptation mitigation and livelihood security. New India Publishing Agency (NIPA), New Delhi, pp 1–19

Raj A, Jhariya MK, Yadav DK, Banerjee A (2020) Climate change and agroforestry systems: adaptation and mitigation strategies. Apple Academic Press Inc., CRC Press- a Taylor and Francis Group, US & Canada, 383 pp. https://doi.org/10.1201/9780429286759

Richardson RB (2010) Ecosystem services and food security: economic perspectives on environmental sustainability. Sustainability 2(11):3520–3548

Sapkota TB, Jat ML, Aryal JP, Jat RK, Khatri-Chhetri A (2015) Climate change adaptation, greenhouse gas mitigation and economic profitability of conservation agriculture: some examples from cereal systems of Indo-Gangetic Plains. J Integr Agric 22:1917

Sarkar S (2015) Management practices for enhancing fertilizer use efficiency under rice-wheat cropping system in the Indo-Gangetic Plains. Innovare J Agric Sci 3(2):5–10

Sarkar S, Sengupta K, Karim MJ (2016) Better photosynthesis in rice (*Oryza sativa* L.) by introduction of the C 4 pathway: an evolutionary approach towards a sustainable system. Int J Bio-resour Stress Manag 7:1186–1193. https://doi.org/10.5958/0976-4038.2016.00182.2

Schoenly K, Cohen JE, Heong KL, Litsinger JA, Aquino GB, Barrion AT, Arida G (1996) Food web dynamics of irrigated rice fields at five elevations in Luzon, Philippines. Bull Entom Res 86:451–466

Settele J, Spangenberg JH, Heong KL, Burkhard B, Bustamante JV, Cabbigat J, Van Chien H, Escalada M, Grescho V, Harpke A, Horgan FG (2015) Agricultural landscapes and ecosystem services in South-East Asia: the LEGATO-project. Basic Appl Ecol 8:661–664

Sharma SK, Jat ML, Biswas C (2005) Components of precision farming. Indian J Fertil 1:13–26

Sheoran S, Kuma S, Kumar P, Meena RS, Rakshit S (2021) Nitrogen fixation in maize: breeding opportunities. Theor Appl Genet. https://doi.org/10.1007/s00122-021-03791-5

Shirato Y, Yokozawa M (2005) Applying the Rothamsted carbon model for long-term experiments on Japanese paddy soils and modifying it by simple tuning of the decomposition rate. Soil Sci Plant Nutr 51:405–415

Singh Y, Sidhu HS (2014) Management of cereal crop residues for sustainable rice-wheat production system in the indo-Gangetic plains of India. Proc Indian Natl Sci Acad 80(1):95–114

Singh R, Kundu DK, Bandyopadhyay KK (2010) Enhancing agricultural productivity through enhanced water use efficiency. J Agric Phys 10:1–15

Srivastava JP, Mukhopadhyay M (Mar 1997) Sustainable intensification of rice-wheat cropping systems in India. A report of the World Bank, Washington, DC. http://documents.worldbank.org/curated/en/640421468749987153/pdf/multi0page.pdf

Tandon SK, Singh S (2010) Energy balance in conservation agriculture and conventional farming: a comparison. In: Joshi PK, Challa J, Virmani SM (eds) Conservation agriculture innovations for improving efficiency equity and environment. National Academy of Agricultural Sciences (NAAS), New Delhi, pp 259–280

Timsina J, Connor DJ (2001) Productivity and management of rice–wheat cropping systems, issues and challenges. Field Crop Res 69:93–132

Tiwari KN, Sharma DN, Tripathi SK (1989) Salt affected soils of Uttar Pradesh, their reclamation and management. CS Azad University of Agriculture and Technology, Kanpur, pp 1–34

Tjandra TB, Ng R, Yeo Z, Song B (2014) Framework and methods to quantify carbon footprint based on an office environment in Singapore. J Clean Prod 112:4183–4195. https://doi.org/10.1016/j.jclepro.2015.06.067

Tubiello FN, Donatelli M, Rosenzweig C, Stockle CO (2000) Effects of climate change and elevated CO_2 on cropping systems: model predictions at two Italian locations. Eur J Agron 13:179–189

Tubiello FN, Cóndor-Golec RD, Salvatore M, Piersante A, Federici S, Ferrara A, Rossi S, Flammini A, Cardenas P, Biancalani R, Jacobs H (2015) Estimating greenhouse gas emissions in agriculture: a manual to address data requirements for developing countries. In: Tubiello FN, Cóndor-Golec RD, Salvatore M, Piersante A, Federici S, Ferrara A, Rossi S, Flammini A, Cardenas P, Biancalani R, Jacobs H (eds) Estimating greenhouse gas emissions in agriculture: a manual to address data requirements for developing countries. Food and Agriculture Organization of the United Nations, Rome, 193 pp. https://www.cabdirect.org/cabdirect/abstract/20163174327

Tyagi SK, Datta PS, Singh R (2012) Need for proper water management for food security. Curr Sci 105:690–695

Van Groenigen KJ, Van Kessel C, Hungate BA (2013) Increased greenhouse-gas intensity of rice production under future atmospheric conditions. Nat Clim Chang 3:288–291. https://doi.org/10.1038/nclimate1712

vanIttersum MK, Howden SM, Asseng S (2003) Sensitivity of productivity and deep drainage of wheat cropping systems in a Mediterranean environment to changes in CO_2, temperature and precipitation. Agric Ecosyst Environ 97:255–273

Vetter SH, Sapkota TB, Hillier J, Stirling CM, Macdiarmid JI, Aleksandrowicz L, Green R, Joy EJ, Dangour AD, Smith P (2017) Greenhouse gas emissions from agricultural food production to supply Indian diets: implications for climate change mitigation. Agric Ecosyst Environ 237:234–241. https://doi.org/10.1016/j.agee.2016.12.024

Westing AH (1986) An expanded concept of international security. In: Global resources and international conflict. Oxford University Press, Oxford

Wolf J (2002) Comparison of two potato simulation models under climate change. II. Application, of climate change scenarios. Clim Res 21:187–198

Wurr DCE, Hand DW, Edmondson RN, Fellows JR, Hannah MA, Cribb DM (1998) Climate change: a response surface study of the effects of CO_2 and temperature on the growth of beetroot, carrots and onions. J Agric Sci 131:125–133

Xiao D, Bai H, Liu D (2018) Impact of future climate change on wheat production: a simulated case for China's wheat system. Sustainability 10:1277. https://doi.org/10.3390/su10041277

Yadav RL, Yadav DS, Singh RM, Kumar A (1998) Long term effects of inorganic fertilizer input and crop productivity in rice–wheat cropping system. Nutr Cycl Agroecosyst 51:193–200

Eco-Designing for Soil Health and Services

4

Abhishek Raj, Manoj Kumar Jhariya, Nahid Khan, Arnab Banerjee, Poonam, Ram Swaroop Meena, and Shish Ram Jakhar

Abstract

Soil health and quality are key aspects upon which various ecosystem processes depend. Ongoing series of land degradations, deforestation, intensive agricultural practices, etc. affects the soil health. These deleterious unsustainable practices deprive soil fertility and affect overall ecosystem services (ES). Depleting nature of soil affects tree-crop productivity that is not fruitful for satisfying global hunger populations. Healthy soil promises food-income-climate security and maintains overall environmental sustainability and ecological stability. Human and livestock's health are entirely dependent upon soil quality. Therefore, the query "how does soil maintain plant-human-animal health and productivity?"

A. Raj
School of Agriculture, Lovely Professional University, Phagwara, Punjab, India

M. K. Jhariya (✉) · N. Khan
Department of Farm Forestry, Sant Gahira Guru Vishwavidyalaya, Sarguja, Ambikapur, Chhattisgarh, India

A. Banerjee
Department of Environmental Science, Sant Gahira Guru Vishwavidyalaya, Sarguja, Ambikapur, Chhattisgarh, India

Poonam
Krishi Vigyan Kendra, Navgaon (Alwar-I), Sri Karan Narendra Agriculture University (SKNAU), Jobner, Rajasthan, India

R. S. Meena
Department of Agronomy, Institute of Agricultural Sciences, Banaras Hindu University, Varanasi, Uttar Pradesh, India
e-mail: meenars@bhu.ac.in

S. R. Jakhar
Department of Soil Science and Agricultural Chemistry, Jawaharlal Nehru Krishi Vishwa Vidyalaya, Jabalpur, Madhya Pradesh, India

M. K. Jhariya et al. (eds.), *Sustainable Intensification for Agroecosystem Services and Management*, https://doi.org/10.1007/978-981-16-3207-5_4

arises. This indicates toward synergistic concept between soil and living organisms. However, adopting eco-model in varying land use (agriculture, forestry, agroforestry, and other farming practices) helps to minimize the soil degradation and ensures higher productivity. But the main problem is that "how does eco-designing of varying land use systems ensure healthy and quality soil?". Climate-smart agriculture, conservation agriculture, zero-tillage practices, use of cover crop, mulching, and soil water conservation practices are intrinsic parts of eco-designing or eco-models. These practices ensure healthy and productive ecosystem that makes a pathway for sustainable development (SD). Eco-designing for sustainable soil management practices promotes the storage and sequestration of carbon (C) as soil organic C pools which leads to C balance. Above- and belowground biomass productions, rhizosphere biology, microbial populations, earthworm and other organisms, etc. modify soil health and productivity. Higher nutrient use efficiency, C cycling, water regulation and purification, erosion control, higher biomass and C stocks, food and nutritional security, and higher economy of farmers can be ensured through healthy eco-models. Therefore, eco-designing of different land use systems ensures a healthy ecosystem and environment. Eco-modeling modifies ES in more sustainable ways without disturbing our environment. Thus, adopting eco-designing models in soils promises higher productivity and profitability and ensures SD of the world. In this context, a government and public policy will strengthen the ecosystem health by adopting a sustainable soil-based eco-model. A scientific-based research and design add another effort to drive these eco-design practices in more efficient and productive way to ensure the global SD.

Keywords

Carbon sequestration · Eco-designing · Ecology · Land use system · Soil health · Sustainability

Abbreviations

AF	Agroforestry
AFs	Agroforestry system
C	Carbon
ES	Ecosystem services
GHGs	Greenhouse gases
Mha	Million hectare
N	Nitrogen
SD	Sustainable development
SOC	Soil organic carbon

4.1 Introduction

Soil supports the life of living organisms and is regarded as the largest natural resources on the earth. Healthy soil ensures healthy ecosystem processes and related services to organisms. Plants, animals, and humans are variable elements of the biodiversity which rely upon soils for their lives. Soils adhere to all these organisms in an intrinsic bond that are essential for ecosystem functioning and healthy biomes (Lal 2014; Banerjee et al. 2020; Khan et al. 2021a, b). From a time immemorial, soils have been nourishing forest, agriculture, agroforestry (AF), and other ecosystem components (Raj et al. 2020; Meena et al. 2018). The totality of health and productivity depends on healthy and quality soil. In this context, one can ask that "how does soil maintain ecosystem processes?". The overall ecosystem processes are possible through healthy living components (human, plant, and animals) that can be ensured by healthy soil. Thus, somehow a healthy soil synergizes with healthy ecosystem.

A dramatic series of land degradation, deforestation, and intensive farming practices decline the quality of soils. The global loss of forest areas was reported up to 7.80 million ha year^{-1} in the decade of 1990 than today's value of 4.70 Mha (million hectare) year^{-1} (FAO and UNEP 2020). Around 98.0 Mha of forest areas was lost due to frequent fires in the year 2015. Similarly, as per one estimate, around 7% (301 Mha) of naturally regenerating forest areas have been lost during the past 30 years (FAO 2020). Unsustainable land use practices lead to poor soil fertility and productivity. Deforestation, illicit felling of timber, overexploitation of forest resources, etc. affect the topmost organic layer of soil by promoting erosion and organic soil losses (Oraon et al. 2018). As per FAO and ITPS (2015), approx. 1500 Pg carbon (C) has been reported in 1 m depth of soils. Both tropical and permafrost world comprising high amount of soil organic carbon (SOC) pools in wetlands and peatland regions (Gougoulias et al. 2014; Köchy et al. 2015). Around 0.47–1.30 Pg C per year has been lost from tropical forest (Mackey et al. 2020). Using high synthetic and chemical fertilizers, heavy mechanization, etc. deprive the quality of soil under intensive farming practices (Meena et al. 2020a). No doubt the food production is higher in intensive agriculture and agroforestry system (AFs) but at the cost of human, animal, and environmental health. It is tough to digest these consequences through intensive farming practices. All these deleterious and unsustainable practices release greenhouse gases (GHGs) to the atmosphere apart from soil health and quality loss that leads to climate change phenomenon. Therefore, it would be interesting to note "how do unsustainable land use practices lead to poor soil ecosystem services (ES)?". Unscientific and unsustainable land use practices deprive the quality of soil by affecting fertility, nutrient availability, and resource use efficiency that affects the overall health and quality of soil (Kumar et al. 2020). This will entirely affect overall plant productivity, ecosystem health, and environmental sustainability (Raj et al. 2021).

However, soil orders, types, and management practices modify the extent and value of organic C stocks in different land use practices. For example, as per order-wise, the maximum (316 billion tons) C pools have been observed in Gelisols followed by 190 billion tons in Inceptisols and 20 billion tons in Andisols, respectively (Eswaran et al. 2000). SOC reflects fertility and productivity that can be governed by herbivores, invertebrates, and microbial and fungal populations. These organisms entirely affect overall soil fertility index, rhizosphere biology, nutrient availability and its use efficiency, and overall soil health (Zdruli et al. 2017).

Applying eco-designing model in land use practices ensures healthy and quality soil on which ecosystem health and productivity depends (Khan et al. 2021a, b). Sustainable-based intensification in different land use practices such as agriculture, forestry, and AF is based on ecological concept. These practices are further modified by applying no-tillage, mulching, and conservation agriculture that maximize soil fertility and health. Eco-designing-based models in land use system enhance biodiversity which intensify ES. Provisioning, cultural, regulatory, and social or aesthetic services are different wings of ES. Therefore, it would be interesting to note "how do eco-designing models promote the soil ES in sustainable ways?". This indicates toward adoption of an effective model/tool based on ecological concept that will ensure healthy soil which in turn provides soil-based ES. A healthy soil reflects higher soil fertility and productivity. An eco-model in any farming system minimizes nutrient leaching losses and makes these nutrients available to plant for proper growth and development. Healthy soil ensures higher biomass and efficient C sequestration for climate change mitigation (Raj et al. 2018a, b). High soil fertility, maximum SOC pools, erosion control, watershed management, healthy rhizosphere biology, efficient nutrient cycling, C balance, and climate change mitigation are important soil-based ES (Lal 2009; Brevik 2013; Pimentel and Burgess 2013).

Thus, overall soil-food-climate security can be possible through the adoption of ecological design-based farming models in any land use systems (Jhariya et al. 2021a, b). This practice provides a regulatory framework for any further improvement in any research and design in agroecosystem. It will not only maximize productivity and profitability but also check both C and environmental footprints (Banerjee et al. 2021a, b, c, d). In lieu of the above, this chapter discusses the eco-designing-based land use practices for ensuring healthy and quality soil. A rigorous discussion is also made on C sequestration, food security, and climate resiliency through ecological-based sustainable soil management practices.

4.2 Soil: A Friend or Foe

It is a very common fact regarding soil as "soil is a friend or foe." Uncountable and multifarious benefits from soil ecosystem confirm a friendly nature with our environment. Soil performs significant functions, regulates ecosystem processes, and ensures environmental sustainability and ecological stability. Soil holds aboveground and belowground organisms as biomass and sustains their life. It nurtures different life forms as tree, herbs, shrubs, climbers, fern, and various flora and fauna. Belowground populations of various microorganisms, protozoans, fungi, earthworms, and other invertebrates are regulated by soil ecosystem. That's why we say "soil is soul of infinite life" because the whole life and productivity of ecosystem depend on soils. Soil makes an intrinsic connection with our soul and society. Soil regulates life by providing several ES in terms of tangible and intangible ways. Therefore, soil health and quality play an important role in the provision of benefits that can nurture a variety of organisms. It is interesting to know that this dirt holds billions of peoples and animals and a variety of unseen organisms. A single spoon of soil adheres to billions of bacteria and invisible organisms. Thus, a healthy soil is a perfect indicator of our healthy environment that regulates healthy and proper ecosystem processes in sustainable ways (Raj et al. 2019a, b).

4.3 Soil Health and Quality: An Ecological Perspective

Assessment of soil health and quality depends on topography, altitudes, prevailing climatic situations, physicochemical properties, and biotic interference in the tropics. A different management practice including intensive and sustainable practices also affects soil quality. This will further initiate an interest toward in-depth study of edaphology and pedology. Land use practices entirely affect the soil quality that may be further modified by different practices and models. Intensified agricultural practices destroy organic soil layers which entirely affects soil fertility, productivity, and overall health. Thus, the physical, biological, and chemical properties of soils affect overall health and quality. An adverse impact on soils affects nutrient content and microbial populations which results in soil pollution. An unscientific farming practice will deprive the soil fertility resulting in lesser SOC pools (Kumar et al. 2020a). This will not only affect overall productivity and profitability but also influcnces human-animal-environmental health in the long term. Thus, healthy soils perform enormous ES in terms of greater productivity and maintaining soil-food-climate security.

As per one quote, "a healthy soil regulates healthy ecological services." No doubt, healthy soils perform multiple ES. For example, it supports different life forms of flora and fauna and supplies various essential nutrients and resources for terrestrial primary productions. Water purification, erosion control, and climate regulations are significant regulatory services from soil resource (Laishram et al. 2012). A scientific study and its relevancy are much needed to see the varying benefits and services from healthy and quality soil. Several scientific indicators exist for healthy and quality soils. For example, indicator relevancy for soil health and quality depends on soil management variations and its related sensitiveness (Parisi et al. 2005). Land manger compatibility with soil management practices also affects soil quality. Similarly, organic matter content, macro-fauna, microbial populations, soil morphology, physical and chemical quality, etc. are sub-indicators identified by Velasquez et al. (2007) (Meena and Lal 2018).

4.4 Soil in the Tropics: A Global Perspective

The type, nature, and extent of soil resources are diversely distributed throughout the tropical world. Aspect, altitude, and topography affect soil types and variability which is further modified by climatic situations. As per pedological views, the term tropical soil indicates soil of the tropical climate which is poorly developed, i.e., pedogenically young excluding low-latitude desert soils. The tropical soil has gone through various stages of biochemical weathering processes for a long period of time. Thus, the definition of tropical soils depends on types of parent material, topography, genetic developmental stages, and prevailing climatic conditions (Chesworth et al. 2008).

Soils of the biomes provide uncountable ES and provide food and fiber to nurture billions of peoples and other organisms. As per one estimate, only 0.20 ha of total arable land (1.35 billion ha) are available on individual basis which is also not distributed properly. Only 46% of arable lands are shared by Asian and African continents having almost 71% world populations with low-quality, poorly weathered, and unproductive soils. However, most developed countries such as Europe and N. America comprise mostly fertile and high-quality land resources along with more land per capita basis. Soil and other resource conservation measures are also prominent in these developed regions of the world. Land conservation is very essential for productions and sustainable development (SD) (Anderson 2010). The global population was recorded as 6.80 billion having 1.35 billion ha of arable land in 2009 (World Fact Book 2009). This value represents only 0.20 ha of arable land that would be available per person. The Asian continent contributed the highest arable land areas, that is, 32%, followed by 17% in N. America and 14% in African continent, respectively. Soil degradation and erosion problems accelerate arable land unsuitability (0.3–0.8%) for agricultural productions (den Biggelaar et al. 2004a, b). As per Lal (2007), 45% of arable soils are rendered unsuitable due to varying forms of degradations globally. As per one estimate, nearly 99% of the total human-consumed food items come from the land/soil (Pimentel and Pimentel 2000; Kumar et al. 2021).

Vertisols and Mollisols types of soils are reported in grassland ecosystem. Of these, 6.9% of the world land areas are shared by Mollisols which sustains approximately 6.7% of global populations. This soil is highly productive and fertile particularly when it is managed by adopting conservation tillage practices in the tropics (Blum Winfried and Eswaran 2004). Tropical peatlands are more fertile and prominently distributed in S-E Asia. However, it used particularly 1.5 Mha for palm oil productions in the Indonesian countries (Anderson 2010).

4.5 Eco-Friendly Practices in Different Land Use Systems for Soil Health and Management

Today, an unsustainable and unscientific land use practice creates many environmental problems. Deforestation and intensified farming practices deprive quality of soils. Deforestation leads to soil water erosion and loss of topmost organic soil. This will lead to huge loss of soil and C that affects ecosystem health. High synthetic inputs and heavy mechanizations in farming systems promote soil degradation that overall affects food-health-climate security. Similarly, continuous and overgrazing system beyond land carrying capacity declines soil health and quality along with poor forage productions. This practice also affects overall health and productivity of livestock. In this context, adopting eco-friendly and eco-designing models in varying land use systems ensures soil-food-climate security. Adopting various sustainable management practices in land use systems can enhance biodiversity that intensify ES. Adopting ecologically sustainable model is the smart choice in terms of maintaining soil health and quality. These varying models ensure healthy and productive soils which are the pillar of SD. Several models have been developed to check soil erosion and quality. For example, Renard et al. (1991) have used C factor commonly known as cover-management factor within RUSLE (Revised Universal Soil Loss Equation) as an indicator of soil protection under different land use systems. However, it is a highly significant tool for checking erosion and other soil losses (Panagos et al. 2015). These practices not only maintain soil quality but also reduce global hunger problem by bumper and quality food grain productions. Thus, eco-designing of land use practices ensures environmental sustainability and ecological stability.

4.5.1 Agriculture

Soil is one of the important natural resources that sustains all agricultural components. Soil stores billions of microorganisms that decompose agricultural residues and enhance productivity and health of agroecosystem. Healthy and quality soils ensure health and productivity of crops. Soil provides essential micro- and macronutrients to plants for proper growth and development. The unsustainable and unmanaged farm practices destroy soil quality which is entirely connected with healthy environment. High synthetic inputs as chemical and inorganic fertilizers lead to poor soil fertility. Heavy mechanizations in farm disturb original top organic soils that affect overall growth of plants. An intensive agricultural practice not only deprives soil quality but also affects overall health and productivity of crops. Agricultural production is higher in intensive practices but at the cost of human-animal-environmental health. Intense agricultural practices release GHGs into the atmosphere that leads to global warming and changing climate. Thus, adopting eco-designing will ensure greater soil-crop productivity with promising environmental health. Varying practices of crop rotations and its impacts on SOC pools in different regions of the world are depicted in Table 4.1.

Table 4.1 Varying practices of crop rotations and its impacts on soil organic carbon pools in different regions of the world

Regions	Soil types and annual precipitation	Description of crop rotation and durations	Ranges of soil organic carbon	References
Wooster region of the USA	Well-drained silt loam soil types experienced an annual precipitation of 1000 mm	Corn and corn (R1) and corn and soybean (R2) types of rotations were observed for the duration of 49 years	Maximum value (16.2 g/kg) was observed in R1 in the depth of 0–10 cm followed by R2 having 15.7 g/kg	Kumar et al. (2012)
Iowa region in the USA	Clay loam types of soil experienced an annual precipitation of 711 mm	Corn and soybean (R1) and corn+oats +meadow+meadow (R2) types of rotation were involved for 42 years of duration in this region	Maximum value (32.3 g/kg) was observed in R2 than R1 having 28.2 g/kg	Robinson et al. (1996)
Cruz Alta region of Brazil	Clay soil having 1774 mm of annual precipitations	Wheat and soybean (R1), wheat + soybean + black oats (R2), and black oats + soybean + black oat + maize + oilseed radish (R3) types of crop rotations for 26 years of duration were observed	Maximum value (6.2 mg/ha) was observed in R3 in the depth of 0–5 cm followed by R2 as 5.12 mg/ha and least value in R1 as 4.03 mg/ha, respectively	Campos et al. (2011)
The region of North Dakota in the USA	Silt clay loam soil	Two types of rotations were observed as R1 (oat + sorghum + soybean_corn) and R2 (corn + soybean) for the duration of 20 years	Higher value was observed in R1 as 15.9 mg/ha in compared to R1 having 13.9 mg/ha	Varvel (2006)
Sicily region of Italy	Clay type of soil having 481 mm of annual precipitations	Wheat and wheat (R1) and wheat and faba bean (R2) types of crop rotations were observed for 19 years	Higher value was observed under R2 as 21.2 g/kg compared to R1 having 18.2 g/kg	Barbera et al. (2012)
Illinois provenance of the USA	Silt clay loam types of soil experienced 978 mm of annual precipitations	Three types of rotations, viz., R1 (corn and corn), R2 (soybean and soybean), and R3 (corn, soybean, and wheat), were observed for 19 years of duration	Maximum value was reported under R3 as 47.1 mg/ha in the depth of 0–20 cm followed by 44.6 mg/ha in R1 and 41.2 mg/ha in R2, respectively	Zuber et al. (2015)
The region of Saskatchewan in CA	Silt loam of semiarid climate	Two types of rotations as R1 (fallow-canola-wheat) and R2 (wheat-oilseed-pulse) were observed for 17 years of duration	Higher value was reported under R2 as 30.8 mg/ha followed by 29.3 mg/ha in R1, respectively	Shrestha et al. (2013)

The healthy and quality soils are the foundation of sustainable agriculture production. In this context, replacing older models by adopting new scientific-based eco-designing in agricultural system ensures higher production and is economically profitable for poor farmers. Therefore, the query "how do eco-based agricultural models maintain soil-crop-environmental health?" arises. Soil-crop system must be designed on ecological principles that maintains overall agricultural services in sustainable ways. Eco-designing models will ensure soil fertility, crop productivity, and less GHG emissions, enhance biodiversity, and maintain overall soil-crop-climate security. Models must be developed to see nitrogen (N) cycling and soil water dynamics in soil-plant systems which is highly complex due to the variability in soil physicochemical properties and prevailing climatic conditions. In this context, various models have been used to explore these complex systems in both spatial and temporal ways. Also, these models were utilized for better understanding of water and N simulation, crop growth, and organic matter turnover. In this context, models such as WOFOST, EPIC, HYDRUS-1D, WNMM, DNDC, SPACSYS, RZWQM, DAISY, HERMES, DSSAT, and APSIM were already used (Penning de Vries et al. 1989; Ahuja et al. 2000; Li et al. 2007; Šimůnek et al. 2008). Thus, these soil-crop models are ecologically designed and justify specific questions and simulate different processes to acquire their relevant objectives (Kersebaum et al. 2015).

Likewise, a model toolkit, namely, HYDRUS-1D, has been used to simulate salt, nutrient, and water movement along with various contaminants of pathogens, pesticides, and heavy metals. This model explores one-dimensional movement of water, heat flow, and multiple solutes through a series of differential and kinetic equations. However, this model is not suitable for understanding soil-crop interactions and crop growth. Similarly, understanding of suitable fertilizer doses and its recommendation are not covered by this model. This is a major drawback of HYDRUS-1D model in agricultural system. But recently, integrating this model with WOFOST crop model is used to create an innovative and more efficient model which can overcome these issues and help to study the varying irrigation management in semiarid regions (Zhou et al. 2012).

DNDC is another ecologically designed model which simulates soil N and C dynamics in agricultural system. This model is also helpful in tracing various gas emissions into the atmosphere. In this context, this model helps in monitoring a level of GHG emission, and accordingly we can modify manure and chemical fertilizer applications in the farm. Therefore, this model certainly helps in recognizing the emission level through high synthetic inputs through intensive practices which can be helpful for further improvement in soil-climate health maintenance (Li et al. 2010). Similarly, a WHCNS model was also used to evaluate N and water management in agricultural system of North China. This model is highly significant in assessing the soil C value, soil temperature, water flow, soil N value, and crop growth pattern in intensive cropping system (Liang et al. 2016; Sheoran et al. 2021).

Thus, adopting eco-designing models in agricultural system gives an idea about system simulation and related impacts on health and productivity of any farming systems on the environment. These models of course intensify productivity and profitability and maintain soil-food-climate security in sustainable ways.

4.5.2 Forestry

Forest covers almost 4 billion ha land (30% of total earth land areas) which holds a huge amount of C under tropical and boreal regions of S-E Asia and peatlands (Pan et al. 2013). But, approximately 25% of total SOC pools are lost due to anthropogenic and deforestation activity (FAO and ITPS 2015). These databases raise the question "how does forest hold tremendous C and make C stability, storage, and flux in the ecosystem?". However, sustainable forest management (SFM) practice-based eco-designing model helps in overcoming these situations. Creating ecology-based conservative model in forestry land use practices ensures healthy ES with greater soil productivity. These models act as toolkits which provide a framework to study forest productivity and health regulation in an ever-changing climate. Ensuring soil health, productivity, economic profitability of farmers, water purification, greater C sequestration capacity, and soil-food-climate security are possible through eco-modeling-based SFM practices. Different forest management practices, its impacts on SOC pools, existing knowledge gaps, and accordingly research priorities are depicted in Table 4.2 (Mayer et al. 2020; Meena et al. 2020a).

Forest is reported as world air conditioner that regulates temperature and precipitation worldwide. Forest types vary as per varying soil and climatic conditions. The maximum C stock (553 Pg C) has been observed in tropical forest followed by 395 Pg C in boreal and 292 Pg C in temperate forest biomes. Of these, soil C stocks were reported maximum in boreal and temperate forest as 318 and 153 Pg C than vegetational C, while tropical forests are diversified and have a higher vegetational C (340 Pg C) as compared to soil C pools (213 Pg C). These dynamics and variations of C in soils and vegetation vary as per varying forest types and species compositions (Prentice 2001). Similarly, C stocks also vary as per varying latitudes. As per increasing latitudes from low, middle, and peak, the C content in both vegetations and soils varied as 37, 14, and 49% of total 1240 Pg C in forest. Likewise, boreal forest shared 85% SOC of the total terrestrial C value. This value is further followed by 60 and 50% of total terrestrial C in both temperate and tropical rainforest. The highest value of SOC pools was reported from taiga, tundra, and pre-tundra regions. In young to wetland regions, the SOC pools varied from 0 to 50%, respectively (Lal 2005).

These databases reflect a wide variation in the value of SOC which entirely affects overall soil health and forest productivity. These SOC values indicate forest as a biggest source of C sink that plays atmospheric C balance and climate change mitigation. Efficient nutrient cycling, SOC pools, water regulation, soil fertility enhancement, and microbial populations in soils are affected by litter addition and its decompositions in forest. However, eco-designing-based forestry models ensure healthy soils that regulate efficient ecological processes. Adoption of SFM is typically based on ecological principle that certainly ensures a better health and productivity of soils and forests. These practices ensure forest- and soil-based ES that ensures environmental sustainability and ecological stability. Controlling soil erosion, maintenance of top organic soil layers, soil infiltrations, higher soil resource use efficiency, and regulation of soil water system can be possible by adopting

Table 4.2 Different forest management practices, its impacts on SOC pools, existing knowledge gaps, and accordingly research priorities (Compiled: Mayer et al. 2020)

Different management practice in forest ecosystem	Impact on SOC pools	Knowledge gaps	Needs of research priorities
Afforestation practices	Positive impacts on soil organic carbon were observed when applied on a land that was previously a cultivable land	Knowledge gap of various processes, stability, and flux of carbon in soil ecosystem	Research study must be focused on a long-term study on organic soils' and mineral' chronosequences
Management pertaining to primary forest conversion into plantations and secondary forest land areas	Negative impacts on soil carbon were observed due to land conversion that promotes heavy carbon losses	Knowledge gap in understanding the clear definition and benchmarking of primary forest	Research study must be focused on a long-term study on research sites
Harvesting practices in forests	Negative impacts on soil carbon were observed due to harvesting of forest timber and other produces	Lack of knowledge on varying soil types and related organic carbon pools	Research study must be focused on a long-term study on research sites including varieties of soil types in forest ecosystem
Collection of harvested residues	Negative impacts on soil carbon pools were observed due to the removal of harvested residues	Lack of knowledge on long-term effects of residue removal on forest health and deep organic carbon pools	Research study must be focused on a long-term study on nutrient dynamics and carbon pools in large research sites of forest ecosystems
Management including site preparation	Promotes soil disturbance that leads to negative impacts on SOC pools	Lack of knowledge regarding the duration of impacts on soil disturbance; lack of substitution between intended and unintended impacts on soil carbon and health	Research is intended to quantify soil carbon changes and losses due to disturbance causes by site preparations
Addition of nitrogen	Positive impacts as promoting nitrogen availability and organic carbon values	Knowledge gap regarding nitrogen addition impacts on stabilized SOC pools	A need of research studies on linking the concept between composition and abundance of mycorrhizal fungi along with their interaction with other nutrients' (P and K) inputs in soil across the forest sites

(continued)

Table 4.2 (continued)

Different management practice in forest ecosystem	Impact on SOC pools	Knowledge gaps	Needs of research priorities
Species selection including leguminous and N_2-fixing associates as management practices	Positive impacts were observed on soil organic carbon pools	Knowledge gap on tree growth enhancement due to nitrogen availability, carbon allocation pattern in soil, and losses of nitrogen as nitrous oxide emissions	Balance and extensive researches are needed to see carbon sequestration process for SOC pools and losses of nitrogen as N_2O emissions
Selection of tree species across the forest sites as management practices	Positive impacts are differentiated into coniferous and broadleaved species in which more carbon pools were observed in forest floor under coniferous species than SOC in soil mineral under broadleaved species	Interaction understanding of tree species with changing site quality and climate	Extensive researches are needed to explore the size of soil carbon pools and its stability and fluxes in common garden experiments
Management pertaining to tree species diversity in forests	Varied from positive to null impacts on SOC pools	Limited knowledge was available on the interactive roles of species and functional diversity	A research and design must be made on long-term experiments in varying levels of species and functional diversity across the forest sites
Tree density and thinning management practices	Neutral impact on SOC was observed	Lack of knowledge on long-term impact of thinning intensities on soil carbon pools	Required a research on tree density and related thinning intensities across the forest sites
Regulation of herbivores in forests	Positive impacts observed	Lack of better understanding on herbivores' impacts on plant-soil interactions in tropical forests	Extensive experiments are needed to see sustainable population densities of herbivores along the gradient of soil in forests
Practices involving biomass and litter removal for fuel and fodder	Exert a negative impact on soil carbon stocks	Limited information on litter decaying and decomposition rate and belowground fine/coarse root productions. Biomass removal impacts on	Long-term researches are needed in the context of fodder, fuelwood, and timber production in sustainable ways

(continued)

Table 4.2 (continued)

Different management practice in forest ecosystem	Impact on SOC pools	Knowledge gaps	Needs of research priorities
		carbon pools are also scare	
Management of forest fire	Positive or negative impacts were observed in the case of prescribed burning and method of fire suppression	A limited understanding on soil heating levels and related changing soil physicochemical properties and carbon pools across the forest sites in tropical world	Comprehensive experiments are needed to see the long-term impacts of intense fires on microbial carbon and labile and non-labile carbon pools

eco-designing forest models. SFM stands better for regulating all these services in efficient and sustainable ways.

There is an intrinsic relationship between forest and soil. The practices of sustainable soil management provide anchorage system to higher plant extensive root system. Water regulation and nutrient availability to higher trees provide significant services under healthy soil. In turn, forest adds litter, and its decomposition releases nutrient into the soil. Increase in SOC pools, erosion control, less leaching loss, etc. enhance soil health by creating eco-friendly model of forest ecosystem. Creating eco-friendly models of course influences both soil and forest health and productivity. Designing a model must be in accordance with ecological concept. This eco-model enhances biodiversity which intensify ES for both forest and soils. In this context, SFM practices stand and regulate every soil parameter by promoting soil health and quality for the long term. Soil water conservation, efficient nutrient cycling, and regulating soil physicochemical properties are significant services provided by SFM. Thus, adopting a scientific management of soil ensures forest health and productivity. Adopting eco-based model promises soil-forest-climate security that makes a pathway for achieving the SD at global scale under climate change. An eco-designing model for sustainable forest and soil management is depicted in Fig. 4.1 (Jhariya et al. 2019a, b; Rodrigues et al. 2020; Mayer et al. 2020).

Soil erosion is one of the land degradations which is widely prevalent in the world in which soil transforms into sediments (Lal 2014; Weil and Brady 2017; Meena et al. 2020a). The physical structure and organic layers are destroyed in the erosion process that is highly significant for plant root development. Moreover, destruction of surface layers leads to water and nutrient losses which decrease productivity and increase pollution of surface waterways. However, erosion results in poor soil ES that affects overall food requirement and raw material availability (Hurni et al. 2008). Water-mediated soil erosions are entirely linked with desertification processes. It exerts a pressure on the ecosystem due to water imbalance that calls for further management strategies for soil protections (IPCC 2014; Seidl et al. 2016). In

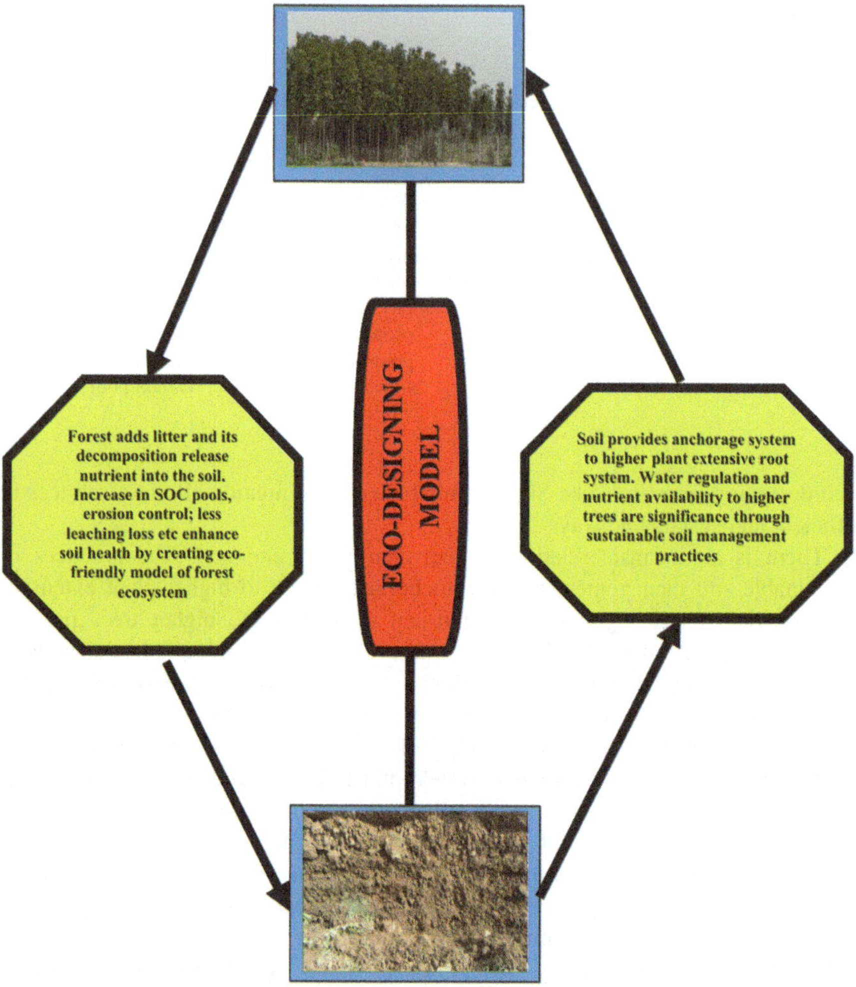

Fig. 4.1 Eco-designing model for sustainable forest and soil management (Compiled: Jhariya et al. 2019a, b; Rodrigues et al. 2020; Mayer et al. 2020)

this context, tree works as a boon which maintains overall hydrological cycle to maintain water status and quality (Carvalho-Santos et al. 2014). Similarly, vegetation covers protect water-mediated soil erosions and maintain soil water balance in the ecosystem. Eco-designing model-based forest management along with effective afforestation techniques regulates water dynamics and soil productivity for healthy ecosystem processes (Ganasri and Ramesh 2016). Water purification, regulations, filtration, proper infiltration, soil protections, erosion controls, etc. are ecological services possible through eco-designing models of forest management practices (Ellison et al. 2017).

4.5.3 Agroforestry

Modeling of AF is of utmost importance for proper structure, productivity, and functioning. Tree, crop, and animals are key components, and their intrinsic relationship maintains AF structure. A variety of AF models exists which are generally location specific in nature. Nature, types, number of components, and tree-crop interactions differentiate one AF model from another in any agroecological zones (Jhariya et al. 2015; Singh and Jhariya 2016). A tree-crop combination in different AFs in the world is depicted in Table 4.3. As per one estimate, a total of one billion ha land area are under AFs which is verified for higher production and has become

Table 4.3 Tree-crop combinations in different agroforestry systems in the world

Regions	Agroforestry systems (AFs)	Tree-crop combinations	References
France	Agrisilviculture system	Tree as *Juglans nigra* (walnut) and herbaceous crops as cereals	Battie-Laclau et al. (2020)
	Agrisilviculture system	Tree as *Populus deltoides* × *P. nigra* (*Populus* sp.), *Alnus glutinosa* (black alder), and wheat as herbaceous crops	Clivot et al. (2020)
Global	Diverse types	Diversified combinations of tree-crop systems	Bayala and Prieto (2020); Isaac and Borden et al. (2020); Marsden et al. (2020); Zhu et al. (2020)
Ghana region in Africa	Shaded perennial tree-crop system	Combinations of shade trees for cacao plants (*Theobroma cacao*)	Borden et al. (2020)
China	Agrisilviculture system	Combination of *Ziziphus jujuba* (ber plant), canola, and daylily plant (*Hemerocallis* species)	Huo et al. (2020); Ling et al. (2020)
	Shaded perennial tree-crop system	Combinations of shade trees for coffee plant (*Coffea arabica*)	Rigal et al. (2020)
Ethiopia and Rwanda (Africa)	Parkland system	Mixture of verities of species such as *Faidherbia albida* (ana tree), *Acacia tortilis* (umbrella thorn), *Grevillea robusta* (silver oak tree) × cereals as herbaceous crop	Sida et al. (2020)
Ethiopia (Africa)	Fallows system	System comprising multiple tree species and cereals as herbaceous crop	Terefe and Kim (2020)
Sulawesi (Indonesia)	Shaded perennial tree-crop system	Combinations of shade trees for cacao plants (*Theobroma cacao*)	Wartenberg et al. (2020)

the potential source for the income of the farming community (Zomer et al. 2016; Abdulai et al. 2018). Tree components modify numerous functions of soils and make better tree-crop-soil interactions (van Noordwijk et al. 2019). Tree-crop system also promotes belowground productions and related ES (Bayala et al. 2015).

AF structure and productions are entirely dependent upon soil health and quality. Soil is an important natural resource that sustains every element of AFs. Soil provides physical supports to both tree-crop root systems and provides essential nutrients to them. Extensive root systems of woody perennial trees and herbaceous crops withdraw soil nutrients for their proper growth and development. Integrating N_2-fixing leguminous tree species makes soil productive and fertile for proper health and productivity of AFs. In this context, the leguminous tree *Acacia nilotica* (babul) plays a key role in soil enrichment along with gum production which is a source of income (Raj 2015; Raj and Singh 2017; Jhariya et al. 2018a). Additions of leaf litter and other residues and its decomposition will add organic matter into the soils. These organic residues are decomposed by various microorganisms and release essential nutrients and promote organic C contents into the soils. Thus, additions of organic C and nitrogen and availability of essential nutrients enrich soil fertility and productivity (Jhariya 2017a, b; Jhariya et al. 2018b).

A great link exists between healthy soil and AF sustainability. Healthy soils support AFs by intensifying crop diversity and productivity, promoting nutrient availability, and anchoring plant roots to create more sustainable eco-designing model. A better designed AF obviously influences soil health and productivity. Promoting efficient nutrient cycling, improving soil fertility by decomposition, adding tree-crop residues and litters, etc. lead to increased SOC pools. Thus, a better designed AF model promotes compatible tree-crop interactions that enhance overall health and productivity. Water regulations, soil microbial populations, healthy rhizosphere, and better nutrient use efficiency are the various ES through AF soil-based models. Climate-resilient eco-designing models ensure C balance and promise soil-climate security with sustainable AF production (Jhariya et al. 2019c). This model synergizes soil health and AF sustainability. In this context, an eco-designing model for soil sustainability in AFs is depicted in Fig. 4.2 (Raj et al. 2019a, b; van Noordwijk et al. 2020).

One question revolves in this context, i.e., "how do eco-designing AF models alleviate nutrient losses and maintain soil health?". A close and efficient type of nutrient cycling has been observed in AFs rather than open type in sole cropping system. Therefore, the chance of nutrient losses through leaching is minimized due to regular capture of nutrients by extensive root systems of woody perennial trees. This process will certainly help nutrient loss and promotes proper capture of essential nutrients and its mobility in tree-crop systems. Thus, AF is characterized by close nutrient cycling that promotes soil health and quality along with higher AF productivity (Burgess and Rosati 2018). Recent research topics were based on soil fertility in AFs which was further extended in the context of soil quality and health (Barrios et al. 2012; Muchane et al. 2020). Higher value of SOC was reported in AFs as compared to sole cropping system (Cardinael et al. 2018). It was due to higher litter inputs and plant residues that increase nutrient inputs into the soils.

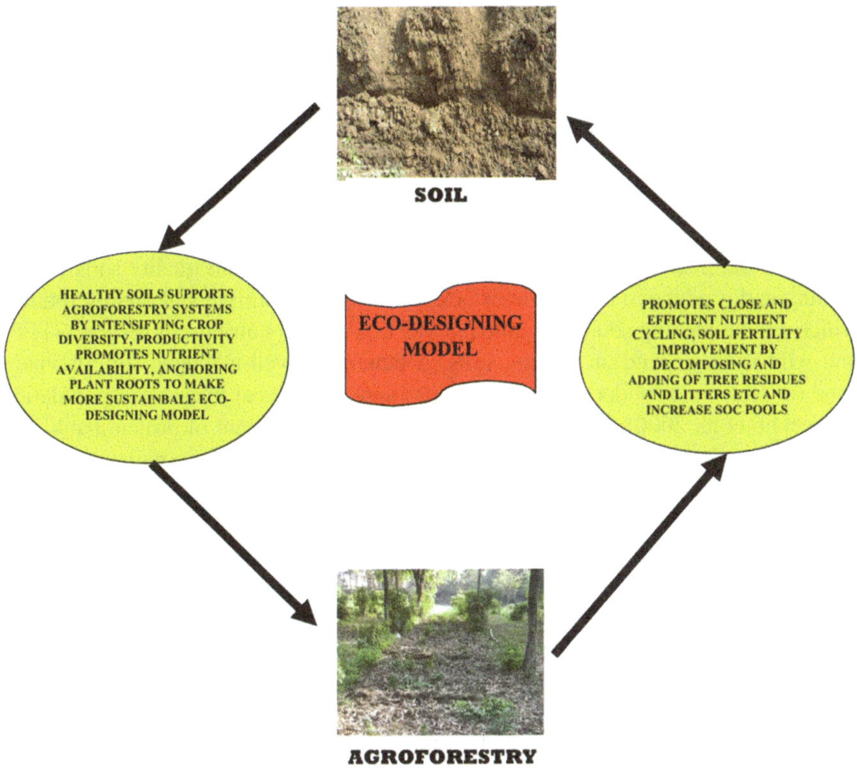

Fig. 4.2 Eco-designing model for soil sustainability in agroforestry system (Compiled: Raj et al. 2019a, b; van Noordwijk et al. 2020)

Well-designed and ecology-based AF models ensure a variety of ES. Timber, fuelwood, fodder, nutritious fruits, and NTFPs are important tangible products derived through AFs. Efficient nutrient cycling, SOC pools, soil fertility enhancement, soil water conservation, climate change mitigation, etc. are intangible services under well-managed AF models. Enhancing soil health and quality is another key function of eco-designing AF models. Proper microbial populations, healthy rhizosphere biology, soil enrichment, efficient nutrient use efficiency, erosion controls, etc. are possible through well-designed AFs. AFs involve better water cycling, infiltrations, retention, and run-off reductions as compared to intensive sole cropping system (Pavlidis and Tsihrintzis 2018). Meylan et al. (2013) developed a model for identifying various constraints and related trade-offs in four types of coffee-based AFs. This model helps in assessing climatic and management factors involved in tree-crop productions, soil erosion losses, hydrological processes, and economic gains. Similarly, Hi-sAFe model is a key tool used for understanding scientific processes in the adoption of AFs in ecological regions. This model identifies ecological interaction between tree and crops in AFs. This model further couples with STICS-based crop model to develop new tree model to assess the competition for natural resources among tree-tree and tree-crop systems (Dupraz et al. 2019).

4.5.4 Livestock

Soil also supports livestock and animals either alone or for any part of farming system, rangeland, forest, agriculture, and AFs. Soil nurtures wild animals in forests and domestic livestock through farming practices. Healthy and quality soils ensure healthy and productive animals/livestock. Soil ensures nutritive and quality fodders which are the feeding source of animals. The quality and nutrition of several NTFPs and wild fruits depend on healthy soils. A better and well-managed soil alleviates food (as fruits and fodder) and nutrition problems for global livestock's population (Muyekho et al. 2000). For example, an efficient management of protein bank AFs delivers protein-rich palatable fodder and grasses for livestock population. These products provide feeds to animals under integrated farming system. In turn, healthy animals provide quality products in the form of meat, milk, egg, wool, and other tangible products for peoples. Thus, a healthy soil manages the whole ecosystem and ecological processes in sustainable ways without affecting our environment. A better management practice of livestock helps to minimize GHG emission into the atmosphere and maintain climate security. Soils also sustain grassland agroecosystem by nurturing cattle populations. Livestock and cattle population improve soil health and fertility through excreta which is a good source of nutrient (N, phosphorus, C, etc.) (Sheldrick et al. 2002). Therefore, livestock enrich soil fertility on which overall health and quality depends upon. An eco-friendly designing of livestock-based farming system for soil sustainability is depicted in Fig. 4.3 (Weishaupt et al. 2020).

These all ecological processes are effectively possible by adopting eco-designing models in any land use systems. Eco-designing models in land use practices ensure healthy soil which is entirely linked with healthy and productive livestock populations. Integrated farming models work more efficiently and provide various ES. Similarly, AFs must be oriented in more ecological concept to provide better productive soil and livestock. The quality fodder productions in AFs are dependent on soil management practices. Integrating leguminous fodder plant gives extra benefits in terms of healthy soil that ensure soil-animal-climate security. Thus, there is a great nexus between healthy soil and productive livestock which can be possible by adopting eco-designing models. Livestock grazing systems entirely affect the status of SOC in both positive and negative ways. However, various models have been developed to assess SOC pools under the influence of livestock and wild animals. These models are based on ecological feature of grazing types, durations, and potential. For example, SNAP model was used recently to assess primary production system due to episode grazing in land use system. This model is an ecology-based dynamic model which is used extensively for improving the production in a sustainable way.

Likewise, SNAPGRAZE is a combined model which was used to quantify grazing management impacts on SOC pools in varying climatic situations. This

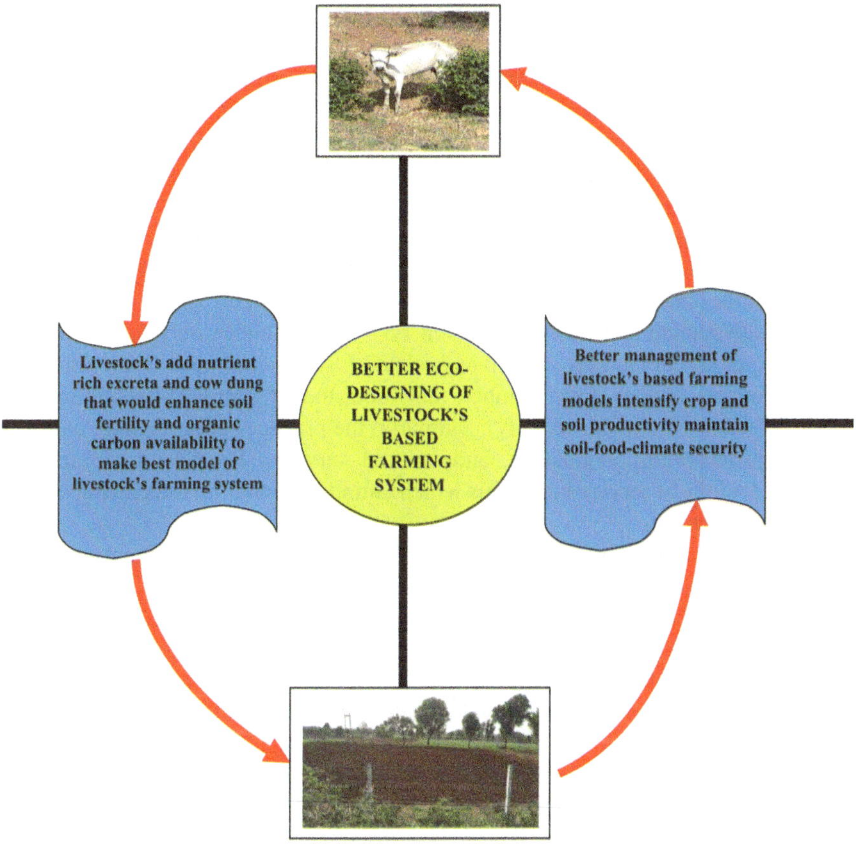

Fig. 4.3 Eco-friendly designing of livestock-based farming system for soil sustainability (Compiled: Weishaupt et al. 2020)

model predicts the impact of continuous grazing of high stocking densities on forage productions and SOC pools. These predictions were further modified by prevailing climatic variability, i.e., temperature and precipitation (Ritchie 2020). GRASIM model is used for updating grazing impact on forage productions along with C and N cycling on a daily basis (Mohtar et al. 1997). However, testing of soil C dynamic models ensures effective results of grazing impacts on forage biomass and production along with types of nutrient cycling in managed agroecosystem (Ritchie 2014). Thus, applying eco-designing models in land use systems helps in assessing grazing potential, forage productions, and livestock's health. These models give an in-depth idea about soil-livestock interrelationship which is quite important for a healthy and sustainable ecosystem.

4.6 Eco-Designing for Soil Ecosystem Services

ES have direct and indirect benefits through various land use practices. Provisioning, regulatory, cultural, and aesthetic services are possible through better management of natural resources (forest, agriculture, AF, horticulture, soil, and rangeland) (MEA 2005). Different land use practices ensure a variety of ES. Various tools are used for modeling and evaluating the ES (Table 4.4). Unsustainable land use practices such as intensive farming system leads to poor services and benefits. Sustainable land use system, based on ecological concept, helps to improve biodiversity which intensifies the ES. Further, the queries "how do different land use systems synergize with ES?" and "do eco-designing models ensure soil ES?" arise. This entirely focus on types and management of land use system and related ES to environmental sustainability. Forest provides various direct (timber, fuelwood, fodder, and NTFPs) and indirect services (soil-food-climate security). Similarly, the practices of agriculture and AFs provide provisioning, regulatory, and cultural services. Rangeland provides an essential ecological service to animals which entirely depends on the grazing system and its management.

Of all these resources and land use systems, soil is a very important terrestrial resource that supports other natural resources. A healthy and quality soil enhances biodiversity in forestry and other farming practices that intensify ES in sustainable ways. Poor and degraded soil affects overall food-nutrition-climate security. Therefore, adopting eco-designing models in various land use system ensures soil fertility, controls erosion, regulates water cycling and C balance, and promotes soil water conservation. Efficient nutrient cycling, nutrient use efficiency, better soil fertility, higher SOC pools, etc. are important ES possible by adopting eco-designing model of sustainable soil management practices (Prado et al. 2016). A healthy soil promises food-health-climate security. Soil provides essential nutrients to the plants for better and quality food and fruit production. Nutritive foods ensure human and animal health system that maintains overall environmental health and sustainability. Therefore, a great attention must be drawn on soil health which is the pillar of environmental health and SD. Thus, an uncountable and multifarious ES is a good indicator that reflects quality of soils (Vogel et al. 2018; van Leeuwen et al. 2019). Belowground soil processes and delivery of ES are depicted in Table 4.5. Adopting a healthier and scientific-based ecological model in different land use practices promises healthier soil-based ES. This would make a great nexus among soil-food-climate security which maintains overall environmental sustainability and ecological stability. Moreover, the goal of SD can be achieved through eco-designing-based land use practices that ensure healthy and quality soils.

4.7 Eco-Designing for Soil Carbon Sequestration

Carbon is the essential component that prevails in our environment in various forms as atmospheric C, form of living biomass in flora and fauna, and SOC. C is an integral component of our life in the form of biomass and sustains lives on the earth.

Table 4.4 Tools for modeling and evaluating the ecosystem services

Tools	Description	References
Tools for modeling and evaluating site-based ecosystem services (TESSA)	Low-cost and site-specific tools for measuring ecosystem services. It helps users by identifying, measuring, and assessing types of ecosystem services at site scale	Peh et al. (2013)
InVEST toolkit which is commonly known as "integrated valuation of ecosystem services and tradeoffs"	Low-cost software and production function-based model which is used to map and assess the ecosystem services from nature. Terrestrial-, marine-, and freshwater-based models are used under this tool	Natural Capital Project (2015)
ARIES toolkit commonly known as "artificial intelligence for ecosystem services"	This tool is build up as a software- or web-based application which assesses the value and importance of ecosystem services in an environmental decision-making process	Villa et al. (2014)
MIMES toolkit commonly known as "multiscale integrated models of ecosystem services"	Tool comprising model that evaluates land use pattern and related ecosystem services ranging from local to global scale. Assess material transfers within ecosystem and related services	Boumans et al. (2015)
SPASMO tool commonly known as "soil-plant-atmosphere system model"	Assess soil processes, formations, and related ecosystem services. It also investigates impacts of management practices on plant-soil health and productivity from regional to global scale	Green et al. (2003)
MOSES toolkit also known as "modelling soil ecosystem services"	This tool comprises process-based soil model that gathers an information soil profile and related ecosystem services such as SOC pools, soil carbon sequestration processes, water regulation and buffering, biomass productivity, etc.	Aitkenhead et al. (2011)
NVE toolkit commonly known as "the nature value explorer"	It is basically a web-based software/application which is used to assess a comparison of ecosystem services from two different groups such as regulatory and cultural services. Assessing the impacts of land use change on ecosystem services is also quantified	Broekx et al. (2013)

C balance in the atmosphere is very important for the regulation and proper functioning of the ecosystem. However, a slight change in C value breaks the ecosystem processes and ecological integrity. Increase in atmospheric C due to various anthropogenic and natural factors ensures warming effects. Unsuitable land use practices, deforestation, and intensive farming practices lead to land degradation followed by higher emission of C into the atmosphere. The developmental projects, mining, illicit

Table 4.5 Belowground soil processes and delivery of ecosystem services (ES)

Belowground soil processes	Trigger mechanisms	Management options	Delivery of ecosystem services (ES)	References
Run-off or water infiltration	Soil-inhabiting organisms and extensive root systems triggered belowground soil processes	Managed through deep-rooted higher plant and soil tillage and mulching activities	Regulation of water cycling, flood, and water purification	Bayala and Prieto (2020); Huo et al. (2020); Ling et al. (2020); Zhu et al. (2020)
The process of hydraulic lift	Plant root systems influence the processes	Integrating deep- and extensive-rooted trees	Delivery of provisioning services as a production benefit	Bayala and Prieto (2020); Huo et al. (2020)
Interaction processes as niche complementarities, competition and facilitation, etc.	Network of mycorrhizal biology and plant root systems affects overall belowground soil processes	Managing tree-crop population, density, root pruning activities, root architecture, etc. soil fertilization and tillage practices also for the proper management of belowground activities	Delivery of provisioning services as a production benefit and efficient nutrient cycling process	Battie-Laclau et al. (2020); Bayala and Prieto (2020); Borden et al. (2020); Isaac and Borden 2020; Rigal et al. (2020); Sida et al. (2020)
Litter addition and decomposition in soils	Existing soil organisms and microbial populations triggered the process of belowground litter decompositions	Litter types and quality due to nature of plant species; soil tillage practices and fertilization also helpful in belowground soil management	Regulation of prevailing climate and efficient nutrient cycling are important ecosystem services	Marsden et al. (2020)
Storage and sequestration of organic carbon into the soil (SOC pools)	Soil organisms, mycorrhizal networks, and extensive root systems affect overall sequestration processes	Presence of deep- and extensive-rooted tree-crop system; use of cover crops and better soil tillage practices are management options	Climate regulation is an important ecosystem service	Rigal et al. (2020); Terefe and Kim (2020)

(continued)

Table 4.5 (continued)

Belowground soil processes	Trigger mechanisms	Management options	Delivery of ecosystem services (ES)	References
Belowground symbiotic N_2 fixation	Overall bacterial populations triggered these processes	Integration of N-fixing plant and *Rhizobium* inoculation are management options	Efficient nutrient cycling process	Isaac and Borden 2020
Mineral weathering, movement, and deposition processes	Both mycorrhizal network and extensive root systems influence the processes	Deep-rooted plants, types, and interaction nature of tree-crop system involved as management options	Soil formation process	Isaac and Borden 2020
Nutrient loss and leaching	Nature and types of root system triggered overall processes	Integrating deep- and extensive-rooted tree-crop system and applying soil amendment and fertilizations are essential management options	Water regulation and purification and efficient nutrient cycling are key ecosystem services	Bayala and Prieto (2020); Zhu et al. (2020)
Soil aggregate stabilization	Soil faunal populations, mycorrhizal network, and extensive root systems	Plant types and density, use of soil covers, tillage practices, and amendments applications are varying management options	Water and climate regulations are key ecosystem services	Marsden et al. (2020); Wartenberg et al. (2020); Zhu et al. (2020)
Soil porosity formation and its proper maintenance	Soil faunal populations and plant root systems influence soil porosity and maintenance	Deep- and extensive-rooted tree-crop system and varying tillage practices are management options	Flood and climate regulations, water purification, etc. are key ecosystem services	Ling et al. (2020); Marsden et al. (2020); Zhu et al. (2020)

timber cutting, intensive grazing and rangeland practices, and high synthetic inputs in the agriculture system have led to C emissions. These deleterious and unscientific land use practices have negative impacts on our environment and ecosystem processes. In this context, sustainable-based land management practice includes various eco-designing models that promote C retention into the resources in the form of either vegetational biomass or soil C stocks. Soil plays a key role in the storage and sequestration of atmospheric C. A healthy and quality soil restores organic C and maintains SOC pools. As per Lal (2004a), around 1550 and 950 Pg of organic and inorganic C pools up to 1 m soil depth represent soils as the third largest C sinks in the world.

An ecologically designed AF model enhances the soil C status by C additions into the soil through litter and other residue inputs. Addition of litter and plant residues promotes decomposing microbial biomass that releases C with other essential nutrients for plant growth and development. This maintains the health status of soil. Similarly, around 0.4–1.2 Gt of C were found to be sequestered annually by soils in agriculture land use system (Lal 2004a). Therefore, eco-designing-based agriculture practices can ensure a potential sequestration of 8–10 Gt of C annually as SOC pools (Hansen et al. 2013). Thus, better agricultural practices are of utmost importance for efficient soil C sequestration. This ensures higher biomass productivity along with proper C balance in the atmosphere. Similarly, adopting conservation agriculture, no-tillage practices, mulching, and use of cover crops maintains soil C status with minimal soil disturbance (Lal 2004b; Baker et al. 2007; Meena et al. 2020c). Moreover, an optimal fertilizer dose, applying FYM and green manure, litter, and plant residue input can potentially enhance SOC stocks (Chaudhury et al. 2016). Better rangeland practices with healthy and sustainable grazing systems ensure a better SOC beside significant forage productions. Thus, adopting eco-designing models in these land use systems enhances soil C sequestration and regulates C balance in the ecosystem (Khan et al. 2020a, b).

4.8 Eco-Designing for Soil Ensuring Food Security

Food unavailability, shortage, and insecurity are the major problems of today's world due to the unexpected weather and changing climate (IPCC 2007; FAO 2015). However, a rigorous discussion has been made on climate change issues at various national and international platforms. Climate change is the curse of the nature that is continuing at an alarming rate due to intensive farming practices, deforestation activity, and various unsustainable land use practices. Of all known curses, poor food quality and its availability are the major concerns today globally. Extreme weather resulted in poor soil quality which induced food and nutrition insecurity. Also, unsustainable land use practices deprive the quality and quantity of foods due to land and soil degradations. Moreover, anthropogenic activity, deforestation, natural calamities such as landslides, and soil water erosion activity deprive soil C stocks and quality resulting in poor food production (Oldeman 1998). Intensive farming practices ensure higher food productions but at the cost of quality food grains. Further, heavy use of chemical and inorganic fertilizers affects soil nutrient availability, and these chemicals enter into the food chain system to affect overall crop productions. This resulted in higher production without compromising human-animal-ecosystem health (Painkra et al. 2016; Meena et al. 2020b).

Soil health and quality are directly connected with food and nutritional security. However, ongoing global hunger necessitates agricultural land expansion and higher application of synthetic inputs. This activity employs heavy fertilizer application for higher global production. A good example is the "Green Revolution." Thanks to this initiative for higher and bumper production but land degradation and declining soil fertility were seen in parallel. Declining trend of human-animal-environmental health was observed in past and present era globally. That's why focusing on bumper crop productions in parallel with maintaining food quality and environmental health is a smart choice of today. In this context, for promoting soil quality better eco-designing-based farming models is encouraged worldwide. Sustainable land use practices increase soil quality which in turn maintains food and climate security.

Healthy soil provides essential nutrients that can be captured by extensive root systems of tree-crop systems. Soil nutrient availability and its mobility in plants decide the production and quality level of food grains. Therefore, ensuring healthy soil will promise higher productions with quality maintenance. Thus, focusing on soil and food quality along with higher production is an urgent need of today. In this context, adopting sustainable-based farming practices promotes soil fertility and nutrient availability for higher crop productions. Higher SOC pools enhance other essential nutrient availability which can be captured by plants. No doubt, SOC is a good indicator of soil fertility, and good soil fertility reflects better soil quality that ensures higher agricultural productions. Thus, we must focus on better farming management practices which are quite linked with healthy soil that ensures higher and quality food production for satisfying global populations.

SOC pools become a boon for African farmers in terms of higher agricultural productions. This country suffered irrigation problems and experienced shortage of water and poor availability of fertilizers. The country has adopted ecology-oriented farming model that enhances SOC stocks which in turn ensure higher productions with quality food (Lal 2004a). As per De Moraes Sá et al. (2017), eco-designed agricultural practices ensure higher SOC stocks that can potentially enhance food productions up to 17.60 Mt. annually. This made a revolutionary era that linked SOC stocks with food production and quality. This indicates that a great nexus and synergy exist among eco-designing farming models with healthy soil and quality food productions. Thus, we can say "eco-designing land use systems ensure food and nutritional security." One more benefit of eco-designing land use practices is that it minimizes GHG emission. This will promote climate security along with maintaining food and nutritional security. In a nutshell, healthy and quality soils promise healthy environment through sustainable production. This will help in maintaining human-animal health which is a prerequisite for nation development and sustainability.

4.9 Eco-Designing for Soil Ensuring Ecological Stability

Greater productions, aggregate soil structure and formation, soil water infiltration and storage, soil fertility enhancement, and climate change mitigation are possible by adopting efficient eco-based model for healthy soil. Ecological stability is of utmost importance for environmental sustainability and better ecosystem processes. Soil-mediated efficient biogeochemical cycle (including C, N, and water cycle) regulates ecosystem structure and functions. However, higher plants' extensive root system, meso-fauna, microbial populations, and healthy rhizosphere biology promote a better soil health and quality. Thus, ecological stability can be possible by ensuring an area for eco-based model of sustainable soil management in land use practices. In this context, Fig. 4.4 depicted an eco-designing model for healthy soil-mediated ecological stability (Prokopov et al. 2019; Thakur et al. 2020). This figure justifies one question "how does eco-designing of soil ensure ecological stability in land use systems?"

4.10 Research and Design

Enhancing land use productiveness, quality, and soil-food-climate health are the major challenges for today. Forestry and agricultural productions along with quality maintenance are very concerning topics. Maintaining soil health and quality is the most important issue for better production and ecosystem processes. All these services can be achieved by adopting various ecology-oriented researches and designs in land use systems. In this context, promoting SFM and sustainable soil management practices is based on this principle. These practices can intensify various ecological and environmental services. Better research and design are prerequisites for better soil-based ES. Research must be linked in accordance to soil types, natures, tree-crop interactions, species nature, topography, and biophysical attributes. An efficient eco-designing model should be used for the proper functioning of ecosystem. An eco-based climate-resilient farming model should be adopted to improve land and plant productiveness under changing climate. Unsustainable land use management declines soil fertility and health. Intensive practices such as high synthetic inputs and heavy mechanization destroy soil health and quality. These intensive practices obviously release GHGs into the atmosphere resulting in global warming. In this context, climate-resilient eco-designing model that makes healthy and sustainable soil along with pollution-free environment. Farming system must be designed in an ecology-based model that can be developed by adopting mulching, no-tillage practices, use of cover crop and litter inputs, etc. These practices minimize GHG emissions, minimize climate change, and develop climate-resilient healthy farming systems. In this context, a climate-resilient eco-designing model has been developed for ensuring soil sustainability (Fig. 4.5) (El Chami et al. 2020; Michler et al. 2019).

Various ecological indicators should be applied in the farm level to check soil quality and productiveness. Similarly, the design needs to be framed to observe

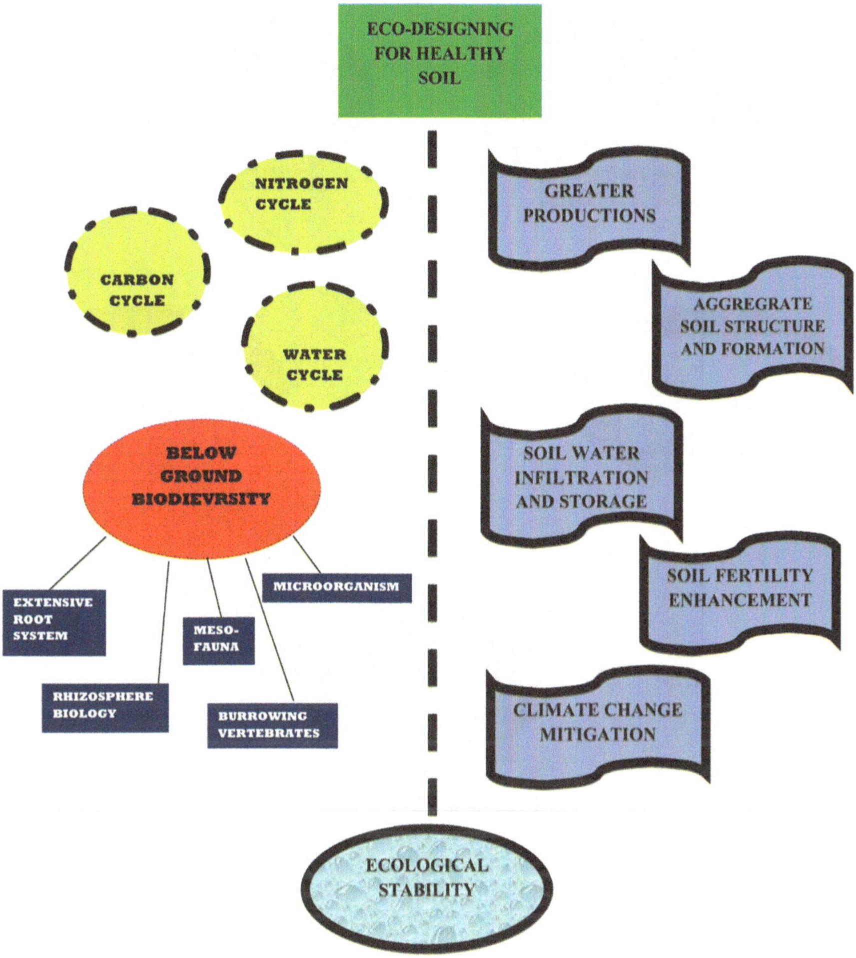

Fig. 4.4 Eco-designing model for healthy soil-mediated ecological stability (Compiled: Prokopov et al. 2019; Thakur et al. 2020)

tree-crop interaction and related ES in AF models through proper research. A well-designed model is needed to understand tree-crop interaction and its impacts on soil health and fertility. This eco-design not only raises plant-soil productivity but also maintains the health and economy of poor farmers. Thus, a design must be applied to confirm social-economic-climate integrity which makes a pathway for achieving the goal of SD (Bradshaw and Sykes 2014).

A land use-based research must be performed to draw the attention of researchers, academicians, policy makers, and stakeholders. Research and design need to be performed to maintain a balance between ecologist and economist. This will become a pillar for developing sustainable world (Chaudhary et al. 2015). A better research and ecological design would be helpful in understanding soil's complex processes

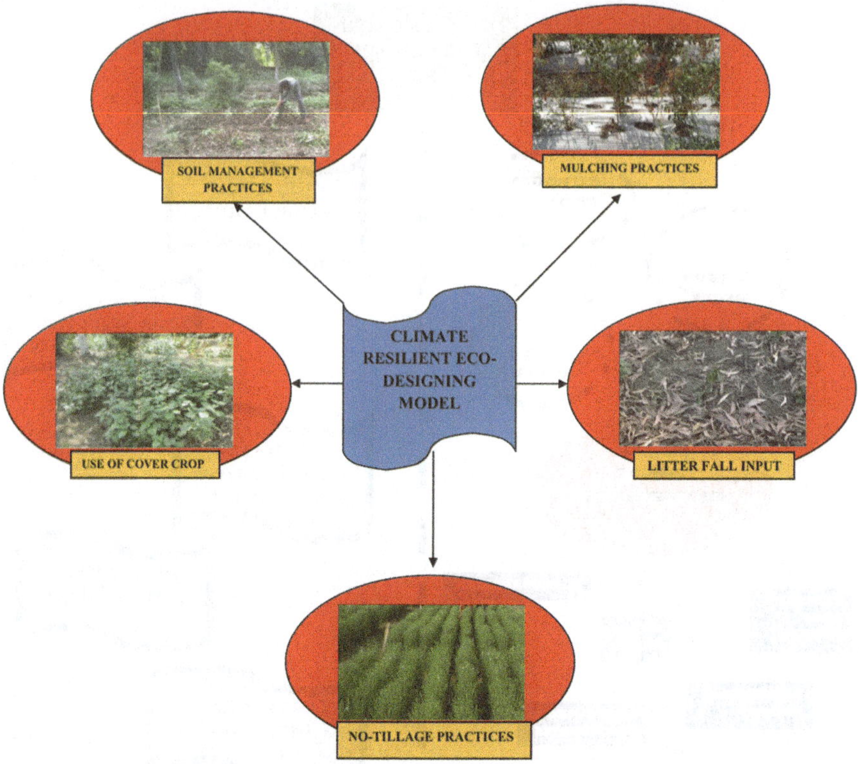

Fig. 4.5 Climate-resilient eco-designing model for soil sustainability (Compiled: El Chami et al. 2020; Michler et al. 2019)

that explore soil quality and land suitability for any farming systems. Enhancing the processes of soil C sequestration and SOC pools is another interesting topic of the ecologist. A well-designed ecological model in land use practices ensures soil health and climate change mitigation through better soil C sequestration. This activity also builds biomass and makes C balance in the ecosystem. A better soil quality assessment in any land use system is of utmost importance for understanding a variety of ES on local to global scale (Rutgers et al. 2012).

4.11 Governance and Policy

The topic of soil-based ES is placed in many national and international platforms. Maintaining the integrity between scientists and decision-makers is attributed by various global initiatives. For example, GSBI (Global Soil Biodiversity Initiative), IPCC (Intergovernmental Panel on Climate Change), GEO BON (GEO's Biodiversity Observation Network), FAO (Food and Agriculture Organization), ESP (Ecosystem Services Partnership), and MEA (Millennium Ecosystem Assessment) are involved in assessing soil ES (Prado et al. 2016). A soil conservation strategy must be developed under public and government sector for improving ES. Similarly, a climate-resilient and regulation policy should be reframed in various land use systems (agriculture, forestry, AF, and livestock). An effective policy for sustainable land practices needs to be employed from regional to global level. An effective policy and good governance shall be employed to achieve sustainable food system for feeding world population (FAO 2020). A policy must be framed for raising awareness among farmers to adopt eco-designing-based farming models in agroecological regions. A policy for good and scientific training should be employed among farmers in rural areas for raising awareness of soil's importance in land use practices. This will enhance people's understanding in soil quality maintenance and its role in food-health-climate security. Around 2.05 billion ha which is 54% of the global forest areas requires forest management plan for the long term. The areas of national parks, game reserves, and conservation areas are designated legally under protected areas of 700 million hectare which is 18% of the world forest areas (FAO 2020). Various global initiatives have certain policy toward proper growth, restoration, and conservation of one trillion trees throughout the world (WEF 2020). In this context, SFM practice ensures proper growth and conservation of forest vegetations for a long time along with overall ecosystem management.

4.12 Conclusions

Land degradation mediated poor soil quality due to anthropogenic and natural factors are important concern today across the globe. These deleterious and unscientific land use practices not only destroy land productivity but also reduce soil-people-environmental health. Soil-mediated poor ES was observed due to deforestation and intensive agricultural and overgrazing practices. Soil holds billions of organisms and supports a variety of flora and fauna. Soil C sequestration is a good strategy to minimize GHG emission and mitigate climate change. Sustainable and eco-designing land use practices ensure higher biodiversity that intensify ES. Sustainable land use practices promise healthy soil which in turn maintains food-health-climate security. Thus, an eco-design-based farming model is a boon for healthy soil that ensures food and nutritional security. These practices ensure environmental sustainability and ecological stability that are pillars of SD.

4.13 Future Roadmap

The present and future of our lives entirely rely on soil resource. Soil health is a central concept that determines ecosystem health and environmental quality. Nowadays, ongoing series of land degradation decline soil productiveness and related ES. Intensive land use management ensures poor soil fertility, less nutrient availability, declining soil C sequestration capacity, and other important ES. Soil works from food production to economical profit. The demand of food is rising continuously due to burgeoning populations. This will induce a pressure on land use system. These will necessitate land conversion into agricultural practices. Deforestation and intensive farming practices destroy the health and quality of soil. An organized manner of farm model is necessary for minimizing the negative impacts on the environment and ecosystem.

A future roadmap must be framed toward modeling and designing land use system based on ecological concept for healthy and productive soil. Soil health assessment is quite necessary for eco-designing of any farming systems in any regions of the world. A model must be developed to check erosion problems and promotes soil water conservation in any arid and degraded zones. Effective eco-designing models must be employed for the reclamation of desertification and high saline and alkaline soils that entirely affect overall plant productions and quality. Soil fertility is a good indicator of soil health and quality. Thus, there should be a soil health assessment prior to the adoption of any AF model which is further designed based on ecological principle. These processes ensure higher plant and soil productivity and maintain food-income-climate security. A climate-resilient farming model must be framed to mitigate changing climate through higher soil climate sequestration. This will not only enhance SOC pools but also improve soil fertility, resulting into healthy microbial populations, and maintain the C balance in the environment. Thus, a roadmap for soil-food-climate maintenance should depend on better eco-designing-based land use practices that ensure our future goal of sustainable world.

References

Abdulai I, Jassogne L, Graefe S, Asare R, Van Asten P, Läderach P, Vaast P (2018) Characterization of cocoa production, income diversification and shade tree management along a climate gradient in Ghana. PLoS One 13:1–17. https://doi.org/10.1371/journal.pone.0195777

Ahuja LR, Ma LW, Terry A (2000) The root zone water quality model. Water Resources Publications LLC, Littleton, CO

Aitkenhead MJ, Albanito F, Jones MB, Black HIJ (2011) Development and testing of a process-based model (MOSES) for simulating soil processes, functions and ecosystem services. Ecol Model 222:3795–3810. https://doi.org/10.1016/j.ecolmodel.2011.09.014

Anderson D (2010) Global perspective of arable soils and major soil associations. Prairie Soils Crops J 3:1–8

Baker JM, Ochsner TE, Venterea RT, Griffis TJ (2007) Tillage and soil carbon sequestration- what do we really know? Agric Ecosyst Environ 118:1–5

Banerjee A, Jhariya MK, Yadav DK, Raj A (2020) Environmental and sustainable development through forestry and other resources. Apple Academic Press Inc., CRC Press- a Taylor and Francis Group, US & Canada, 400 pp. https://doi.org/10.1201/9780429276026

Banerjee A, Meena RS, Jhariya MK, Yadav DK (2021a) Agroecological footprints management for sustainable food system. Springer, Singapore, 514 pp. https://doi.org/10.1007/978-981-15-9496-0

Banerjee A, Jhariya MK, Meena RS, Yadav DK (2021b) Ecological footprints in agroecosystem- an overview. In: Banerjee A, Meena RS, Jhariya MK, Yadav DK (eds) Agroecological footprints management for sustainable food system. Springer, Singapore, pp 1–23. https://doi.org/10.1007/978-981-15-9496-0_1

Banerjee A, Jhariya MK, Raj A, Yadav DK, Khan N, Meena RS (2021c) Land footprint management and policies. In: Banerjee A, Meena RS, Jhariya MK, Yadav DK (eds) Agroecological footprints management for sustainable food system. Springer, Singapore, pp 221–246. https://doi.org/10.1007/978-981-15-9496-0_7

Banerjee A, Jhariya MK, Raj A, Yadav DK, Khan N, Meena RS (2021d) Energy and climate footprint towards the environmental sustainability. In: Banerjee A, Meena RS, Jhariya MK, Yadav DK (eds) Agroecological footprints management for sustainable food system. Springer, Singapore, pp 415–443. https://doi.org/10.1007/978-981-15-9496-0_14

Barbera V, Poma I, Gristina L, Novara A, Egli M (2012) Long-term cropping systems and tillage management effects on soil organic carbon stock and steady state level of C sequestration rates in a semiarid environment. Land Degrad Dev 23:82–91

Barrios E, Sileshi GW, Shepherd K, Sinclair F (2012) Agroforestry and soil health: linking trees, soil biota, and ecosystem services. In: Wall DH, Bardgett RD, Behan-Pelletier V et al (eds) Soil ecology and ecosystem services. Oxford University Press, Oxford, pp 315–330

Battie-Laclau P, Taschen E, Plassard C, Dezette D, Abadie J, Arnal D, Benezech P, Duthoit M, Pablo A-L, Jourdan C, Laclau J-P, Detrand I, Taudiere A, Hinsinger P (2020) Role of trees and herbaceous vegetation beneath trees in maintaining arbuscular mycorrhizal communities in temperate alley cropping systems. Plant Soil 453:153–171. https://doi.org/10.1007/s11104-019-04181-z

Bayala J, Prieto I (2020) Water acquisition, sharing and redistribution by roots: applications to agroforestry systems. Plant Soil. https://doi.org/10.1007/s11104-019-04173-z

Bayala J, Sanou J, Teklehaimanot Z, Ouedraogo SJ, Kalinganire A, Coe R, Noordwijk MV (2015) Advances in knowledge of processes in soil–tree–crop interactions in parkland systems in the West African Sahel: a review. Agric Ecosyst Environ 205:25–35. https://doi.org/10.1016/j.agee.2015.02.018

Blum Winfried EH, Eswaran H (2004) Soils for sustaining global food production. J Food Sci 69:38–42

Borden KA, Thomas SC, Isaac ME (2020) Variation in fine root traits reveals nutrient-specific acquisition strategies in agroforestry systems. Plant Soil. https://doi.org/10.1007/s11104-019-04003-2

Boumans R, Roman J, Altman I, Kaufman L (2015) The multiscale integrated model of ecosystem services (MIMES): simulating the interactions of coupled human and natural systems. Ecosyst Serv 12:30–41. https://doi.org/10.1016/j.ecoser.2015.01.004

Bradshaw RHW, Sykes MT (2014) Ecosystem dynamics: from the past to the future. Wiley Blackwell, Chichester, 336 p

Brevik EC (2013) The potential impact of climate change on soil properties and processes and corresponding influence on food security. Agriculture 3:398–417

Broekx S, Liekens I, Peelaerts W, De Nocker L, Landuyt D, Staes J, Meire P, Schaafsma M, Van Reeth W, Van den Kerckhove O, Cerulus T (2013) A web application to support the quantification and valuation of ecosystem services. Environ Impact Assess 40:65–74. https://doi.org/10.1016/j.eiar.2013.01.003

Burgess PJ, Rosati A (2018) Advances in European agroforestry: results from the AGFORWARD project. Agrofor Syst 92(4):801–810. https://doi.org/10.1007/s10457-018-0261-3

Campos BH, Amado TJC, Bayer C, Nicoloso RDS, Fiorin JE (2011) Carbon stock and its compartments in a subtropical oxisol under long-term tillage and crop rotation systems. Rev Bras Ciênc Solo 35:805–817

Cardinael R, Guenet B, Chevallier T, Dupraz C, Cozzi T, Chenu C (2018) High organic inputs explain shallow and deep SOC storage in a long-term agroforestry system - combining experimental and modeling approaches. Biogeosciences 15:297–317. https://doi.org/10.5194/bg-2017-125

Carvalho-Santos C, Honrado JP, Hein L (2014) Hydrological services and the role of forests: conceptualization and indicator-based analysis with an illustration at a regional scale. Ecol Complex 20:69–80. https://doi.org/10.1016/j.ecocom.2014.09.001

Chaudhary S, Mcgregor A, Houston D, Chettri N (2015) The evolution of ecosystem services: a time series and discourse centered analysis. Environ Sci Pol 54:25–34. https://doi.org/10.1016/j.envsci.2015.04.025

Chaudhury S, Bhattacharyya T, Wani SP, Pal DK, Sahrawat KL, Nimje A, Chandran P, Venugopalan MV, Telpande B (2016) Land use and cropping effects on carbon in black soils of semi-arid tropical India. Curr Sci 110(9):1692–1698

Chesworth W, Spaargaren O, Hadas A, Groenevelt PH, Otero XL, Ferreira TO, Vidal P, Macías F, Chesworth W, Finkl CW Jr, Doran JW, Mielke LN, Kirkham MB, Groenevelt PH, Finkl CW (2008) Tropical soils. In: Chesworth W (ed) Encyclopedia of soil science. Encyclopedia of earth sciences series. Springer, Dordrecht. https://doi.org/10.1007/978-1-4020-3995-9_607

Clivot H, Petitjean C, Marron N, Dalle E, Genestier J, Blaszczyk N, Santenoise P, Laflotte A, Piutti S (2020) Early effects of temperate agroforestry practices on soil organic matter and microbial enzyme activity. Plant Soil 453:189–207. https://doi.org/10.1007/s11104-019-04320-6

De Moraes Sá JC, Lal R, Cerri CC, Lorenz K, Hundria M, Cesar P, Carvalho F (2017) Low-carbon agriculture in South America to mitigate global climate change and advance food security. Environ Int 98:102–112

den Biggelaar C, Lal R, Wiebe K, Breneman V (2004a) The global impact of soil erosion on productivity I: absolute and relative erosion-induced yield losses. Adv Agron 81:1–48

den Biggelaar C, Lal R, Wiebe K, Breneman V, Reich P (2004b) The global impact of soil erosion on productivity II: effects on crop yields and production over time. Adv Agron 81:49–95

Dupraz C, Wolz KJ, Lecomte I, Talbot G, Vincent G, Mulia R, Bussière F, Ozier-Lafontaine H, Andrianarisoa S, Jackson N, Lawson G, Dones N, Sinoquet H, Lusiana B, Harja D, Domenicano S, Reyes F, Gosme M, van Noordwijk M (2019) Hi-sAFe: a 3D agroforestry model for integrating dynamic tree–crop interactions. Sustainability 11:2293

El Chami D, Daccache A, El Moujabber M (2020) How can sustainable agriculture increase climate resilience? A systematic review. Sustainability 12(8):3119. https://doi.org/10.3390/su12083119

Ellison D, Morris CE, Locatelli B, Sheil D, Cohen J, Murdiyarso D, Gutierrez V, van Noordwijk M, Creed IF, Pokorny J, Gaveau D, Spracklen DV, Tobella AB, Ilstedt U, Teuling AJ, Gebrehiwot SG, Sands DC, Muys B, Verbist B, Springgay E, Sugandi Y, Sullivan CA (2017) Trees, forests and water: cool insights for a hot world. Glob Environ Chang 43:51–61. https://doi.org/10.1016/j.gloenvcha.2017.01.002

Eswaran H, Reich FP, Kimble JM, Beinroth FH, Padamnabhan E, Moncharoen P (2000) Global carbon stocks. In: Lal R, Kimble JM, Eswaran H, Stewart BA (eds) Global climate change and pedogenic carbonates. CRC/Lewis, Boca Raton, FL

FAO (2015) Learning tool on nationally appropriate mitigation actions (NAMAs) in the agriculture, forestry and other land use (AFOLU) sector. FAO, Rome

FAO (2020) Global forest resources assessment 2020 – main report. FAO, Rome

FAO and ITPS (2015) Status of the world's soil resources. FAO, Rome

FAO and UNEP (2020) The state of the world's forests 2020. Forests, biodiversity and people. Rome. https://doi.org/10.4060/ca8642en

Ganasri BP, Ramesh H (2016) Assessment of soil erosion by RUSLE model using remote sensing and GIS - a case study of Nethravathi Basin. Geosci Front 7:953–961. https://doi.org/10.1016/j.gsf.2015.10.007

Gougoulias C, Clark JM, Shaw LJ (2014) The role of soil microbes in the global carbon cycle: tracking the below-ground microbial processing of plant-derived carbon for manipulating carbon dynamics in agricultural systems. J Sci Food Agric 94:2362–2371

Green SR, Dijssel C, van den Snow V, Clothier BE, Webb T, Russell J, Ironside N, Davidson P (2003) Spasmo: a risk assessment model of water, nutrient and chemical fate under agricultural lands. In: Currie LD, Hanly JA (eds) Tools for nutrients and pollutant management: applications to agriculture and environmental quality. Fertilizer and Lime Research Centre, Massey University, Palmerston North, pp 321–335

Hansen J, Kharecha P, Sato M, Masson-Delmotte V, Ackerman F, Beerling DJ, Hearty PJ, Hoegh-Guldberg O, Hsu SL, Parmesan C, Rockstrom J, Rohling EJ, Sachs J, Smith P, Steffen K, Van Susteren L, Von Schuckmann K, Zachos JC (2013) Assessing "dangerous climate change": required reduction of carbon emissions to protect young people, future generations and nature. PLoS One 8(12):e81648

Huo G, Zhao X, Gao X, Wang S (2020) Seasonal effects of intercropping on tree water use strategies in semiarid plantations: evidence from natural and labelling stable isotopes. Plant Soil. https://doi.org/10.1007/s11104-020-04477-5

Hurni H, Herweg K, Portner B, Liniger H (2008) Soil erosion and conservation in global agriculture. In: Braimoh AK, Vlek PLG (eds) Land use and soil resources. Springer, Dordrecht, pp 41–71. https://doi.org/10.1007/978-1-4020-6778-5_4

IPCC (2007) Climate change 2007: mitigation. Contribution of working group III to the fourth assessment report of the intergovernmental panel on climate change. Cambridge University Press, Cambridge, UK and New York

IPCC (2014) Climate change 2014: synthesis report. Contribution of working groups I, II and II to the fifth assessment report of the intergovernmental panel on climate change. IPCC, Geneva

Isaac ME, Borden KA (2020) Nutrient acquisition strategies in agroforestry systems. Plant Soil. https://doi.org/10.1007/s11104-019-04232-5

Jhariya MK (2017a) Vegetation ecology and carbon sequestration potential of shrubs in tropics of Chhattisgarh, India. Environ Monit Assess 189(10):518. https://doi.org/10.1007/s10661-017-6246-2

Jhariya MK (2017b) Influences of forest fire on forest floor and litterfall dynamics in Bhoramdeo Wildlife Sanctuary (C.G.), India. J For Environ Sci 33(4):330–341. https://doi.org/10.7747/JFES.2017.33.4.330

Jhariya MK, Bargali SS, Raj A (2015) Possibilities and perspectives of agroforestry in Chhattisgarh. In: Zlatic M (ed) Precious forests-precious earth. InTech, Croatia, pp 237–257. https://doi.org/10.5772/60841

Jhariya MK, Banerjee A, Yadav DK, Raj A (2018a) Leguminous trees an innovative tool for soil sustainability. In: Meena RS, Das A, Yadav GS, Lal R (eds) Legumes for soil health and sustainable management. Springer, Singapore, pp 315–345. https://doi.org/10.1007/978-981-13-0253-4_10

Jhariya MK, Yadav DK, Banerjee A (2018b) Plant mediated transformation and habitat restoration: phytoremediation an eco-friendly approach. In: Gautam A, Pathak C (eds) Metallic contamination and its toxicity. Daya Publishing House, A Division of Astral International, New Delhi, pp 231–247

Jhariya MK, Yadav DK, Banerjee A, Raj A, Meena RS (2019a) Sustainable forestry under changing climate. In: Jhariya MK et al (eds) Sustainable agriculture, forest and environmental management. Springer, Singapore, pp 285–325. https://doi.org/10.1007/978-981-13-6830-1_9

Jhariya MK, Yadav DK, Banerjee A (2019b) Agroforestry and climate change: issues and challenges. Apple Academic Press Inc., CRC Press- a Taylor and Francis Group, US & Canada, 335 pp. https://doi.org/10.1201/9780429057274

Jhariya MK, Banerjee A, Meena RS, Yadav DK (2019c) Sustainable agriculture, forest and environmental management. Springer, Singapore, 606 pp. https://doi.org/10.1007/978-981-13-6830-1

Jhariya MK, Meena RS, Banerjee A (2021a) Ecological intensification of natural resources for sustainable agriculture. Springer, Singapore. https://doi.org/10.1007/978-981-33-4203-3

Jhariya MK, Meena RS, Banerjee A (2021b) Ecological intensification of natural resources towards sustainable productive system. In: Ecological intensification of natural resources for sustainable agriculture. Springer, Singapore. https://doi.org/10.1007/978-981-33-4203-3_1

Kersebaum KC, Boote KJ, Jorgenson JS, Nendel C, Bindi M, Frühauf C, Gaiser T, Hoogenboom G, Kollas C, Olesen JE, Rötter RP, Ruget F, Thorburn PJ, Trnka M, Wegehenkel M (2015) Analysis and classification of data sets for calibration and validation of agro-ecosystem models. Environ Model Softw 72:402–417. https://doi.org/10.1016/j.envsoft.2015.05.009

Khan N, Jhariya MK, Yadav DK, Banerjee A (2020a) Herbaceous dynamics and CO_2 mitigation in an urban setup- a case study from Chhattisgarh, India. Environ Sci Pollut Res 27(3):2881–2897. https://doi.org/10.1007/s11356-019-07182-8

Khan N, Jhariya MK, Yadav DK, Banerjee A (2020b) Structure, diversity and ecological function of shrub species in an urban setup of Sarguja, Chhattisgarh, India. Environ Sci Pollut Res 27 (5):5418–5432. https://doi.org/10.1007/s11356-019-07172-w

Khan N, Jhariya MK, Raj A, Banerjee A, Meena RS (2021a) Soil carbon stock and sequestration: implications for climate change adaptation and mitigation. In: Jhariya MK, Meena RS, Banerjee A (eds) Ecological intensification of natural resources for sustainable agriculture. Springer, Singapore. https://doi.org/10.1007/978-981-33-4203-3_13

Khan N, Jhariya MK, Raj A, Banerjee A, Meena RS (2021b) Eco-designing for sustainability. In: Jhariya MK, Meena RS, Banerjee A (eds) Ecological intensification of natural resources for sustainable agriculture. Springer, Singapore. https://doi.org/10.1007/978-981-33-4203-3_16

Köchy M, Hiederer R, Freibauer A (2015) Global distribution of soil organic carbon - part 1: masses and frequency distributions of SOC stocks for the tropics, permafrost regions, wetlands, and the world. Soil 1:351–365

Kumar S, Kadono A, Lal R, Dick W (2012) Long-term no-till impacts on organic carbon and properties of two contrasting soils and corn yields in Ohio. Soil Sci Soc Am J 76:1798–1809

Kumar S, Meena RS, Jhariya MK (2020) Resources use efficiency in agriculture. Springer, Singapore, p 760. https://doi.org/10.1007/978-981-15-6953-1

Kumar S, Meena RS, Datta R, Verma SK, Yadav GS, Pradhan GS, Molaei A, Rahman GKMM, Mashuk HA (2020a) Legumes for carbon and nitrogen cycling: an organic approach. In: Carbon and nitrogen cycling in soil. Springer, New York, pp 337–375. https://doi.org/10.1007/978-981-13-7264-3_10

Kumar S, Meena RS, Singh RK, Munir TM, Datta R, Danish S, Yadav GS, Kumar S (2021) Soil microbial and nutrient dynamics under different sowings environment of Indian mustard (*Brassica juncea* L.) in rice based cropping system. Sci Rep 11:5289. https://doi.org/10.1038/s41598-021-84742-4

Laishram J, Saxena KG, Maikhuri RK, Rao KS (2012) Soil quality and soil health: a review. Int J Ecol Environ Sci 38(1):19–37

Lal R (2004a) Soil carbon sequestration impacts on global climate change and food security. Science 304(5677):1623–1627. https://doi.org/10.1126/science.1097396

Lal R (2004b) Soil carbon sequestration to mitigate climate change. Geoderma 123:1–22

Lal R (2005) Forest soils and carbon sequestration. For Ecol Manag 220:242–258

Lal R (2007) Anthropogenic influences on world soils and implications to global food security. Adv Agronom 93:69–93

Lal R (2009) Soil degradation as a reason for inadequate human nutrition. Food Secur 1:45–57

Lal R (2014) Soil conservation and ecosystem services. Int Soil Water Conserv Res 2:36–47. https://doi.org/10.1016/S2095-6339(15)30021-6

Li Y, White R, Chen D, Zhang J, Li B, Zhang Y, Huang Y, Edis R (2007) A spatially referenced water and nitrogen management model (WNMM) for (irrigated) intensive cropping systems in the North China plain. Ecol Model 203(3–4):395–423

Li H, Qiu J, Wang L, Tang H, Li C, Van Ranst E (2010) Modelling impacts of alternative farming management practices on greenhouse gas emissions from a winter wheat–maize rotation system in China. Agric Ecosyst Environ 135:24–33

Liang H, Hu K, Batchelor WD, Qi Z, Li B (2016) An integrated soil-crop system model for water and nitrogen management in North China. Sci Rep 6:5755. https://doi.org/10.1038/srep25755

Ling Q, Zhao X, Wu P, Gao X, Sun W (2020) Effect of the fodder species canola (*Brassica napus* L.) and daylily (Hemerocallis fulva L.) on soil physical properties and soil water content in a rainfed orchard on the semiarid Loess Plateau, China. Plant Soil 453:209–228. https://doi.org/10.1007/s11104-019-04318-0

Mackey B, Kormos CF, Keith H, Moomaw WR, Houghton RA, Mittermeier RA, Hole D, Hugh S (2020) Understanding the importance of primary tropical forest protection as a mitigation strategy. Mitig Adapt Strateg Glob Change. https://doi.org/10.1007/s11027-019-09891-4

Marsden C, Martin-Chave A, Cortet J, Hedde M, Capowiez Y (2020) How agroforestry systems influence soil fauna and their functions—a review. Plant Soil. https://doi.org/10.1007/s11104-019-04322-4

Mayer M, Prescott C, Abaker WEA, Augusto L, Cécillon L, Ferreira GWD, James J, Jandl R, Katzensteiner K, Laclau JP, Laganière J, Nouvellon Y, Paré D, Stanturf JA, Vanguelova EI, Vesterdal L (2020) Influence of forest management activities on soil organic carbon stocks: a knowledge synthesis. For Ecol Manag 466:1–25

MEA (2005) Millennium ecosystem assessment: ecosystems and human well-being. Island Press, Washington, DC

Meena RS, Lal R (2018) Legumes for soil health and sustainable management. Springer, Singapore, p 541. https://doi.org/10.1007/978-981-13-0253-4_10

Meena RS, Kumar V, Yadav GS, Mitran T (2018) Response and interaction of *Bradyrhizobium japonicum* and arbuscular mycorrhizal fungi in the soybean rhizosphere: a review. Plant Growth Regul 84:207–223. https://doi.org/10.1007/s10725-017-0334-8

Meena RS, Kumar S, Datta R, Lal R, Vijaykumar V, Brtnicky M, Sharma MP, Yadav GS, Jhariya MK, Jangir CK, Pathan SI, Dokulilova T, Pecina V, Marfo TD (2020a) Impact of agrochemicals on soil microbiota and management: a review. Land (MDPI) 9(2):34. https://doi.org/10.3390/land9020034

Meena RS, Lal R, Yadav GS (2020b) Long term impacts of topsoil depth and amendments on soil physical and hydrological properties of an Alfisol in Central Ohio, USA. Geoderma 363:1141164. https://doi.org/10.1016/j.geoderma.2019.114164

Meena RS, Lal R, Yadav GS (2020c) Long-term impact of topsoil depth and amendments on carbon and nitrogen budgets in the surface layer of an Alfisol in Central Ohio. Catena 194:104752. https://doi.org/10.1016/j.catena.2020.104752

Meylan L, Merot A, Gary C, Rapidel B (2013) Combining a typology and a conceptual model of cropping system to explore the diversity of relationships between ecosystem services: the case of erosion control in coffee-based agroforestry systems in Costa Rica. Agric Syst 118:52–64. https://doi.org/10.1016/j.agsy.2013.02.002

Michler JD, Baylis K, Arends-Kuenning M, Mazvimavi K (2019) Conservation agriculture and climate resilience. J Environ Econ Manag 93:148–169

Mohtar RH, Jabro JD, Buckmaster DR (1997) A grazing simulation model: GRASIM B: field testing. Trans Am Soc Agric Eng 40:1495–1500

Muchane MN, Sileshi GW, Gripenberg S, Jonsson M, Pumarino L, Barrios E (2020) Agroforestry boosts soil health in the humid and sub-humid tropics: a meta-analysis. Agric Ecosyst Environ 295:106899. https://doi.org/10.1016/j.agee.2020.106899

Muyekho FN, Cheruiyot DT, Kamidi M, Wanyonyi M, Akuno F, Gitahi F, Mwania N (2000) Participatory evaluation of forages for increased herbage dry matter yields in smallholder farms in Trans Nzoia District: Preliminary experiences. In: Mureithi JG, Gachene CKK, Muyekho FN, Onyango M, Mose L, Magenya O (eds) Participatory technology development for soil management by smallholders in Kenya. Proceedings of the 2nd Scientific Conference of the Soil Management and Legume Research Network Projects, June 2000 Mombasa, Kenya, pp 321–326

Natural Capital Project (2015) Integrated valuation of ecosystem services and tradeoffs. http://www.naturalcapitalproject.org/InVEST.html

Oldeman LR (1998) Soil degradation: a threat to food security. International Soil Reference and Information Center, Wageningen

Oraon PR, Singh L, Jhariya MK (2018) Forest floor biomass, litterfall and physico-chemical properties of soil along the anthropogenic disturbance regimes in tropics of Chhattisgarh, India. J For Environ Sci 34(5):359–375. https://doi.org/10.7747/JFES.2018.34.5.359

Painkra GP, Bhagat PK, Jhariya MK, Yadav DK (2016) Beekeeping for poverty alleviation and livelihood security in Chhattisgarh, India. In: Narain S, Rawat SK (eds) Innovative technology for sustainable agriculture development. Biotech Books, New Delhi, pp 429–453

Pan Y, Birdsey RA, Phillips OL, Jackson RB (2013) The structure, distribution, and biomass of the world's forests. Ann Rev Ecol Evol Syst 44:593–622

Panagos P, Borrelli P, Meusburger K, Alewell C, Lugato E, Montanarella L (2015) Estimating the soil erosion cover-management factor at the European scale. Land Use Policy 48:38–50. https://doi.org/10.1016/j.landusepol.2015.05.021

Parisi V, Menta C, Gardi C, Jacomini C, Mozzanica E (2005) Microarthropod communities as a tool to assess soil quality and biodiversity: a new approach in Italy. Agric Ecosyst Environ 105:323–333

Pavlidis G, Tsihrintzis VA (2018) Environmental benefits and control of pollution to surface water and groundwater by agroforestry systems: a review. Water Resour Manag. https://doi.org/10.1007/s11269-017-1805-4

Peh KSH, Balmfor A, Bradbury RB, Brown C, Butchart SHM, Hughes FMR, Stattersfield A, Thomas DHL, Walpole M, Bayliss J, Gowing D, Jones JPG, Lewis SL, Mulligan M, Pandeya B, Stratford C, Thompson JR, Turner K, Vira B, Willcock S, Birch JC (2013) Tessa: a toolkit for rapid assessment of ecosystem services at sites of biodiversity conservation importance. Ecosyst Serv 5:51–57. https://doi.org/10.1016/j.ecoser.2013.06.003

Penning de Vries FWT, Jansen DM, ten Berge HF, Bakema A (1989) Simulation of ecophysiological processes of growth in several annual crops. Wageningen University, Wageningen

Pimentel D, Burgess M (2013) Soil erosion threatens food production. Agriculture 3:443–463

Pimentel D, Pimentel M (2000) Feeding the world's population. Bioscience 50:387

Prado RB, Fidalgo ECC, Monteiro JMG, Schuler AE, Vezzani FM, Garcia JR, Oliveira APD, Viana JHM, Pedreira BCCG, Mendes IC, Reatto A, Parron LM, Clemente EP, Donagemma GK, Turetta APD, Simões M (2016) Current overview and potential applications of the soil ecosystem services approach in Brazil. Pesq Agrop Brasileira 51(9):1021–1038. https://doi.org/10.1590/s0100-204x2016000900002

Prentice IC (2001) The carbon cycle and atmospheric carbon dioxide. Climate change 2001: the scientific basis IPCC. Cambridge University Press, Cambridge, pp 183–237

Prokopov M, Salvati L, Egidi G, Cudlín O, Vˇcelakova R, Plch R, Cudlín P (2019) Envisioning present and future land-use change under varying ecological regimes and their influence on landscape stability. Sustainability 11:4654

Raj A (2015) Evaluation of gummosis potential using various concentration of ethephon. M.Sc. thesis, I.G.K.V., Raipur (C.G.), pp 89

Raj A, Singh L (2017) Effects of girth class, injury and seasons on Ethephon induced gum exudation in *Acacia nilotica* in Chhattisgarh. Indian J Agrofor 19(1):36–41

Raj A, Jhariya MK, Harne SS (2018a) Threats to biodiversity and conservation strategies. In: Sood KK, Mahajan V (eds) Forests, climate change and biodiversity. Kalyani Publisher, New Delhi, pp 304–320

Raj A, Jhariya MK, Bargali SS (2018b) Climate smart agriculture and carbon sequestration. In: Pandey CB, Gaur MK, Goyal RK (eds) Climate change and agroforestry: adaptation mitigation and livelihood security. New India Publishing Agency (NIPA), New Delhi, pp 1–19

Raj A, Jhariya MK, Yadav DK, Banerjee A, Meena RS (2019a) Soil for sustainable environment and ecosystems management. In: Jhariya MK et al (eds) Sustainable agriculture, forest and environmental management. Springer, Singapore, pp 189–221. https://doi.org/10.1007/978-981-13-6830-1_6

Raj A, Jhariya MK, Yadav DK, Banerjee A, Meena RS (2019b) Agroforestry: a holistic approach for agricultural sustainability. In: Jhariya MK, Banerjee A, Meena RS, Yadav DK (eds)

Sustainable agriculture, forest and environmental management. Springer, Singapore, pp 101–131

Raj A, Jhariya MK, Yadav DK, Banerjee A (2020) Climate change and agroforestry systems: adaptation and mitigation strategies. Apple Academic Press Inc., CRC Press- a Taylor and Francis Group, US & Canada, 383 pp. https://doi.org/10.1201/9780429286759

Raj A, Jhariya MK, Khan N, Banerjee A, Meena RS (2021) Ecological intensification for sustainable development. In: Ecological intensification of natural resources for sustainable agriculture. Springer, Singapore. https://doi.org/10.1007/978-981-33-4203-3_5

Renard KG, Foster GR, Weesies GA, Porter JP (1991) RUSLE: revised universal soil loss equation. Soil Water Conserv 46:30–33. https://doi.org/10.1002/9781444328455.ch8

Rigal C, Xu J, Vaast P (2020) Young shade trees improve soil quality in intensively managed coffee systems recently converted to agroforestry in Yunnan Province. China Plant Soil. https://doi.org/10.1007/s11104-019-04004-1

Ritchie ME (2014) Plant compensation to grazing and soil carbon dynamics in a tropical grassland. PeerJ 2:e233

Ritchie ME (2020) Grazing management, forage production and soil carbon dynamics. Resources 9:49. https://doi.org/10.3390/resources9040049

Robinson C, Cruse R, Ghaffarzadeh M (1996) Cropping system and nitrogen effects on Mollisol organic carbon. Soil Sci Soc Am J 60:264–269

Rodrigues AR, Botequim B, Tavares C, Pecurto P, Borges JG (2020) Addressing soil protection concerns in forest ecosystem management under climate change. For Ecosyst 7:34. https://doi.org/10.1186/s40663-020-00247-y

Rutgers M, Wijnen HJ, van Schouten AJ, Milder C, Kuiten AMP, Brussaard L, Breure AM (2012) A method to assess ecosystem services developed from soil attributes with stakeholders and data of four arable farms. Sci Total Environ 415:39–48. https://doi.org/10.1016/j.scitotenv.2011.04.041

Seidl R, Spies TA, Peterson DL, Stephens SL, Hicke JA (2016) Searching for resilience: addressing the impacts of changing disturbance regimes on forest ecosystem services. J Appl Ecol 53:120–129. https://doi.org/10.1111/1365-2664.12511

Sheldrick WF, Syers JK, Lingard J (2002) A conceptual model for conducting nutrient audits at national, regional, and global scales. Nutr Cycl Agroecosyst 62:61–72

Sheoran S, Kuma S, Kumar P, Meena RS, Rakshit S (2021) Nitrogen fixation in maize: breeding opportunities. Theor Appl Genet. https://doi.org/10.1007/s00122-021-03791-5

Shrestha BM, McConkey BG, Smith WN, Desjardins RL, Campbell CA, Grant BB, Miller PR (2013) Effects of crop rotation, crop type and tillage on soil organic carbon in a semiarid climate. Can J Soil Sci 93:137–146. https://doi.org/10.4141/cjss2012-078

Sida TS, Baudron F, Ndoli A, Tirfessa D, Giller KE (2020) Should fertilizer recommendations be adapted to parkland agroforestry systems? Case studies from Ethiopia and Rwanda. Plant Soil 453:73–188. https://doi.org/10.1007/s11104-019-04271-y

Šimůnek J, Saito H, Sakai M, Genuchten TM (2008) The HYDRUS-1D software package for simulating the one-dimensional movement of water, heat, and multiple solutes in variably-saturated media. University of California

Singh NR, Jhariya MK (2016) Agroforestry and agrihorticulture for higher income and resource conservation. In: Narain S, Rawat SK (eds) Innovative technology for sustainable agriculture development. Biotech Books, New Delhi, pp 125–145

Terefe B, Kim DG (2020) Shifting cultivation maintains but its conversion to mono-cropping decreases soil carbon and nitrogen stocks compared to natural forest in Western Ethiopia. Plant Soil. https://doi.org/10.1007/s11104-019-03942-0

Thakur MP, Phillips HRP, Brose U, De Vries FT, Lavelle P, Loreau M, Mathieu J, Mulder C, Van der Putten WH, Rillig MC, Wardle DA, Bach EM, Bartz MLC, Bennett JM, Briones MJI, Brown G, Decaëns T, Eisenhauer N, Ferlian O, Guerra CA, König-Ries B, Orgiazzi A, Ramirez K, Russell DJ, Rutgers M, Wall SH, Cameron EK (2020) Towards an integrative understanding of soil biodiversity. Biol Rev 95:350–364 van Leeuwen JP, Creamer RE, Cluzeau D, Debeljak M, Gatti F, Henriksen CB, Kuzmanovski V, Menta C, Pérès G, Picaud C, Saby NPA, Trajanov A, Trinsoutrot-Gattin I, Visioli G, Rutgers M (2019) Modeling of soil functions for assessing soil quality: soil biodiversity and habitat provisioning. Front Environ Sci 7:113. https://doi.org/10.3389/fenvs.2019.00113

van Noordwijk M, Barrios E, Shepherd K, Bayala J, Oborn I (2019) Soil science as part of agroforestry. In: Sustainable development through trees on farms: agroforestry in its fifth decade. World Agroforestry (ICRAF) Southeast Asia Regional Program, Bogor, pp 63–92

van Noordwijk M, Speelman E, Hofstede GJ, Farida A, Abdurrahim AY, Miccolis A, Hakim AL, Wamucii CN, Lagneaux E, Andreotti F, Kimbowa G, Assogba GGC, Best L, Tanika L, Githinji M, Rosero P, Sari RR, Satnarain U, Adiwibowo S, Ligtenberg A, Muthuri C, Peña-Claros M, Purwanto E, Oel PV, Rozendaal D, Suprayogo D, Teuling AJ (2020) Sustainable agroforestry management: changing the game. Land 9:243

Varvel GE (2006) Soil organic carbon changes in diversified rotations of the western Corn Belt. Soil Sci Soc Am J 70:426–433

Velasquez E, Lavelle P, Andrade M (2007) GISQ, a multifunctional indicator of soil quality. Soil Biol Biochem 39:3066–3080

Villa F, Bagstad KJ, Voigt B, Johnson GW, Portela R, Honzak M, Batker D (2014) A methodology for adaptable and robust ecosystem services assessment. PLoS One 9:1–18. https://doi.org/10.1371/journal.pone.0091001

Vogel HJ, Bartke S, Daedlow K, Helming K, Kögel-Knabner I, Lang B, Rabot E, Russell D, Stößel B, Weller U, Wiesmeier M, Wollschlager U (2018) A systemic approach for modeling soil functions. Soil 4:83–92. https://doi.org/10.5194/soil-4-83-2018

Wartenberg AC, Blaser WJ, Rosshetko JM, Noordwijk MV, Six J (2020) Soil fertility and Theobroma cacao growth and productivity under commonly intercropped shade-tree species in Sulawesi, Indonesia. Plant Soil 453:87–104. https://doi.org/10.1007/s11104-018-03921-x

WEF (2020) One trillion trees – World Economic Forum launches plan to help nature and the climate. In: World economic forum [online]. Geneva. https://www.weforum.org/agenda/2020/01/one-trillion-trees-world-economic-forum-launches-plan-tohelp-nature-and-the-climate/

Weil RR, Brady NC (2017) The nature and properties of soils, 15th edn. Pearson Press, Upper Saddle River, NJ

Weishaupt A, Ekardt F, Garske B, Stubenrauch J, Wieding J (2020) Land use, livestock, quantity governance, and economic instruments—sustainability beyond big livestock herds and fossil fuels. Sustainability 12:2053

World Fact Book (2009) The world factbook. https://www.cia.gov/library/publications/theworld factbook/geos/xx.html

Zdruli P, Lal R, Cherlet M, Kapur S (2017) New world atlas of desertification and issues of carbon sequestration, organic carbon stocks, nutrient depletion and implications for food security. In: Carbon management, technologies, and trends in Mediterranean ecosystems. Springer, Berlin, pp 13–25

Zhou J, Cheng GD, Li X, Bill X, Wang GX (2012) Numerical modelling of wheat irrigation using coupled HYDRUS and WOFOST models. Soil Sci Soc Am J 76:648–662

Zhu X, Liu W, Chen J, Bruijnzeel LA, Mao Z, Yang X, Cardinael R, Meng FR, Sidle RC, Seitz S, Nair VD, Nanko K, Zou X, Chen C, Jiang XJ (2020) Reductions in water, soil and nutrient losses and pesticide pollution in agroforestry practices: a review of evidence and processes. Plant Soil 453:45–86. https://doi.org/10.1007/s11104-019-04377-3

Zomer RJ, Neufeldt H, Xu J, Ahrends A, Bossio D, Trabucco A, van Noordwijk M, Wang M (2016) Global tree cover and biomass carbon on agricultural land: the contribution of agroforestry to global and national carbon budgets. Sci Rep 6:1–12. https://doi.org/10.1038/srep29987

Zuber SM, Behnke GD, Nafziger ED, Villamil MB (2015) Crop rotation and tillage effects on soil physical and chemical properties in Illinois. Agron J 107:971–978

Climate Change and Integrated Coastal and Agroecosystem Services

5

Zied Haj-Amor and Salem Bouri

Abstract

In the coastal areas, under some critical conditions such as coastal erosion (which costs roughly $500 million per year for coastal property loss) and coastal flooding (floods comprise 46% of all disaster events between 1998 and 2012), there is a huge need of integrated coastal agroecosystem management in order to ensure a satisfied agricultural productivity with a significant decrease in environmental effects. Climate change is having widespread effects on coastal areas with significant damage to ecosystems. After a deep presentation of coastal agroecosystems (roles, global distribution, and potential threats), the present chapter summarized the aspects of climate change and their effects on coastal agroecosystems. Then, the chapter describes the key policies and measures required for ensuring integrated coastal agroecosystem management. All these issues (i.e., climate change aspects, climate change effects, and integrated coastal agroecosystem management) are clearly discussed. Based on the findings of this chapter, coastal agroecosystems are highly affected by climate change, and these effects (e.g., decrease in agricultural productivity) will continue to affect them in a variety of aspects (i.e., shoreline erosion, coastal storms, flooding, etc.). These effects, therefore, require a more strategic plan to coastal risk management in order to restore the sediment balance, decrease damage of coastal erosion and floods, and achieve a better development of coastal agroecosystems.

Keywords

Climate change · Coastal agroecosystems · Management · Measures · Policies

Z. Haj-Amor (✉) · S. Bouri
Sfax University, Sfax, Tunisia

M. K. Jhariya et al. (eds.), *Sustainable Intensification for Agroecosystem Services and Management*, https://doi.org/10.1007/978-981-16-3207-5_5

Abbreviations

FAO Food and Agriculture Organization
IPCC Intergovernmental Panel on Climate Change
NR Natural resources
SWR Soil and water resources

5.1 Introduction

Coastal regions are frontiers with 356,000 km^2 globally (Central Intelligence Agency 2016). They cover 20% of the land on earth and contain an extensive tapestry of natural ecosystems. They are home to 50% of the global population and make an important contribution (>60%) to international agricultural production due to their favorable biophysical and climatic conditions (Burke et al. 2000). As reported in Martinez et al. (2007), these coastal regions provide numerous environmental benefits. For example, the physical features of coastal ecosystems (e.g., reefs of mangrove) are vital for natural functions such as land accretion and help to control land erosion and other damage arising from wind and wave action. Furthermore, the coastal regions are used extensively and increasingly for several activities such as agriculture, trade, industry, and amenity (Varallyay 2007).

Worldwide, the coastal agroecosystems represent 40% of the earth's land (IPCC 2000). Due to this important area, several goods and functions are ensured from these systems such as agricultural production, materials (e.g., for construction), climate regulation, water purification, soil retention, and climate change mitigation. Several management actions (e.g., soil and water conservation actions) should be practiced in order to ensure the sustainability of these goods and functions. As per the Intergovernmental Panel on Climate Change (IPCC 2019), in recent years, the sustainability of the coastal agroecosystems has been threatened by several serious problems such as land degradation, pollution of soil and water resources (SWR), urbanization, and intensification of industrial activities. These in turn affect human welfare through their effects on productivity, health, and amenity. Land degradation is considered as the most critical problem among the mentioned problems (Soomere et al. 2011; Kumar et al. 2020b). The Food and Agriculture Organization (FAO 2010) has mentioned that climate change, unsuitable agricultural water practices, and soil mismanagement are important contributors to land degradation in the coastal agroecosystems.

Due to their localization (i.e., transition between land and sea) and climate change, coastal agroecosystems are negatively impacted by changes in rainfall and temperature patterns (IPCC 2013). Climate change can affect the coastal agroecosystems through direct and indirect ways. Direct effects include the impact of sea level rise. Indirect effects occur through events such as river floods, pulses, and quality of runoff that originate off-site but that affect the coasts. Also, climate change may pose the greatest threat to agriculture by increasing the demand for water and available water supply, soil salinization and fertility, and crop yield (Karmakar et al. 2016; Banerjee et al. 2020; Raj et al. 2020; Meena et al. 2018).

It is important that greater attention should be paid to understand the contribution of climatic factors in coastal agroecosystems degradation (Singh et al. 2011). Responding to the challenges of climate change impacts on coastal agroecosystems requires urgent adaptation strategies at the local, regional, national, and global levels. Costal countries are urged to improve and consolidate their coastal agroecosystem management and to develop and implement practical measures (i.e., integrated coastal agroecosystem management), which have positive development outcomes that are resilient to climate change (IPCC 2019). First of all, this chapter discusses the properties of coastal agroecosystems (such as roles, threats, etc.). Secondly, in-depth analysis of the climate change effects on coastal agroecosystems is presented. Then, key adaptation options (i.e., integrated coastal agroecosystem management) are presented for minimizing the effects of climate change on coastal agroecosystems. Lastly, the chapter outlines policy framework and future perspectives for integrated coastal agroecosystem management. Finally, the chapter concludes by outlining priorities for adapting strategies to mediate the effects of climate change on coastal agroecosystems.

5.2 Coastal Ecosystems and Coastal Agroecosystems

5.2.1 Definition

The coastal ecosystems (e.g., estuaries, salt marshes, mangroves, etc.) represent the intersection areas between land and water. This intersection creates an exceptional environment with specific properties such as high biodiversity, distinct structure, and specific climate condition (Burke et al. 2001). The coastal ecosystem areas provide the setting for several human activities such as fishing, agriculture, and tourism. Based on 2018 statistics, 2.996 million people live within the coastal areas (49% of world total population) (IPCC 2019). The shape of each coastal ecosystem depends on land processes (e.g., land erosion), climate conditions, and water dynamic (waves, currents, etc.) (Green and Bruckner 2000).

Based on the definition proposed by Loeuille et al. (2013), the coastal agroecosystems are systems of remarkable biological and agricultural productivity and significant interactions between ecological processes and agricultural activities that take place at coastal areas. Accordantly, several goods and services (e.g., foods, construction materials, animal goods, environment protection, etc.) are ensured from these systems. The interactions between land, sea, climate conditions, and human activities (Fig. 5.1) highly influence the properties of each coastal agroecosystem. Furthermore, a coastal agroecosystem is referred to the entire system of production, distribution, and consumption of food resources in the coastal areas, in all its components (i.e., agricultural, agronomic, economic, environmental, and social components) (Hoorweg and Muthiga 2009).

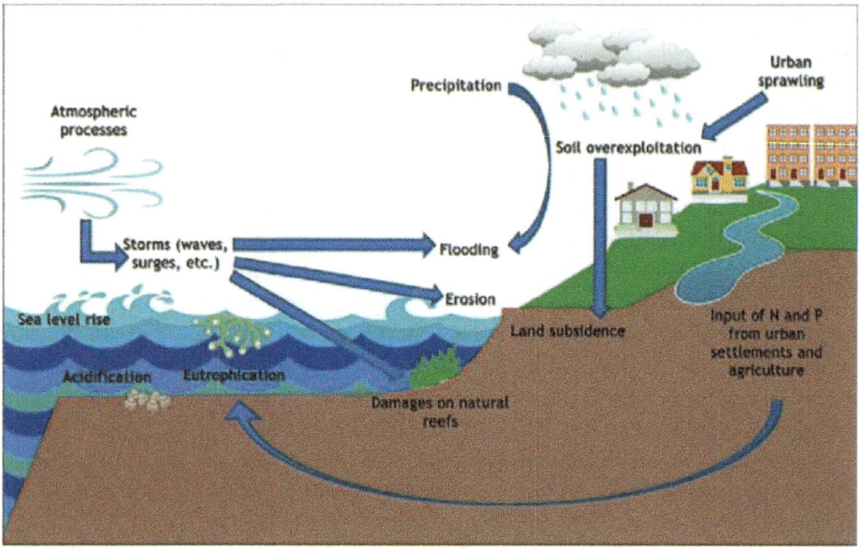

Fig. 5.1 Interactions between land, sea, climate conditions, and human activities (Source: Vitelletti and Bonaldo (2020))

As reported in Malézieux (2012), an agroecosystem is mainly composed of living (e.g., animals) and non-living components (e.g., soil, air, and water). Both components constitute an important unit of agricultural activity. The strong interaction between these components depends on many abiotic (e.g., temperature, soil moisture, water chemistry, etc.) and biotic (e.g., crops, herbivores, detritivores, etc.) factors (Fig. 5.2). This composition of agroecosystems plays a key role in agroecosystem services provision (Jhariya et al. 2019a, b; Meena and Lal 2018). Some management actions must be practiced in order to enhance agricultural productivity within the agroecosystems and produce more sustainability with insignificant negative environmental effects and fewer external inputs.

Over the past years, it was observed that intensive agriculture may contribute to several ecological issues such as water resources' degradation, biodiversity loss, soil fertility loss, and soil properties' deterioration (Hoorweg and Muthiga 2009; Jhariya et al. 2021a, b). For example, in many coastal areas, about 40% of biodiversity loss has already been revealed (Banerjee et al. 2018). Recently, an agroecosystem is defined as an approach that can help to avoid (or reduce) several ecological issues through responsible, resilient, and sustainable agricultural production (Mertz et al. 2009; Banerjee et al. 2021a, b, c, d). In this approach, the agricultural production is highly considered as ecosystem manipulation (Weiner 2003). As a systemic approach, agroecology relies on the enhancement of agricultural production as a whole, with the constant objective of ameliorating environmental performance (Wezel et al. 2009; Raj et al. 2021). In the coastal lands, various agroecosystems are observed, and each system is characterized by some specific properties such as

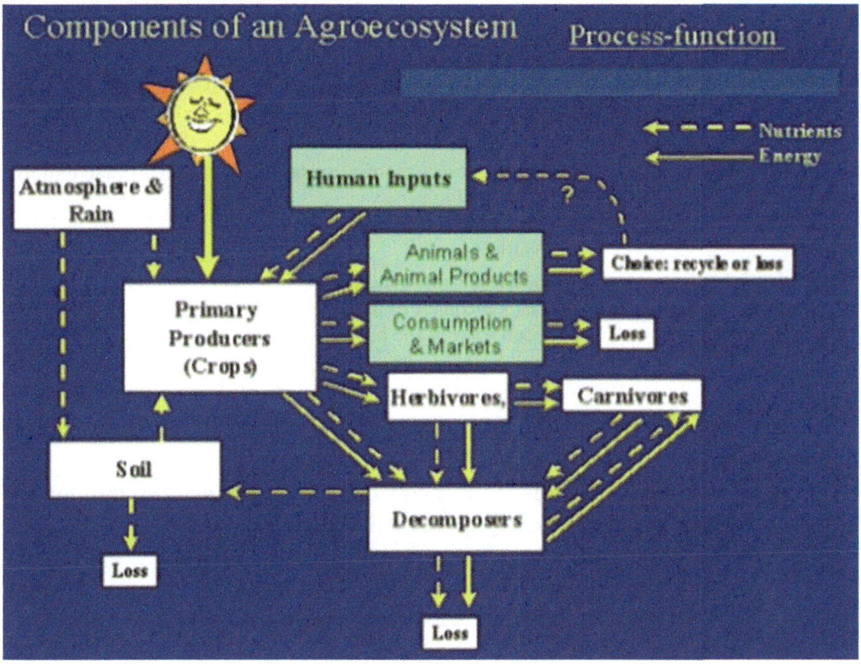

Fig. 5.2 Components of an agroecosystem. (Source: https://www.slideshare.net/shahjee1522/agro-ecology-concept)

farming method (e.g., organic farming), field size, natural resources (NR) management, etc. (Hoorweg and Muthiga 2009; Banerjee et al. 2018).

5.2.2 Roles

A lot of roles are associated with coastal agroecosystems. Generally, four roles can be distinguished: fundamental, supporting, regulating, and environmental roles (Millennium Ecosystem Assessment 2003). Usually, coastal agroecosystems have been considered only as a major source of agricultural productivity (Power 2010). But, they also provide with lot of further roles (Table 5.1), such as biodiversity preservation (Zhang et al. 2007), shoreline stabilization (Dominati et al. 2014), soil resources' conservation (Huang et al. 2015), microclimate regulation, and climate change mitigation (Smith et al. 2008). All these roles are very important due to the frequent exposition of these vulnerable ecosystems to aberrant climate conditions and global warming (Villanueva et al. 2015). A lot of international studies (e.g., Arriaza et al. 2008; Lobell and Gourdji 2012; Senapati and Gupta 2014) expected that changes in air temperature, rainfall, wind velocity, and sea level rise would have significant effect on coastal agroecosystem roles. For example, Lobell and Gourdji

Table 5.1 Roles of agroecosystems (Millennium Ecosystem Assessment 2003)

Roles	Actions (examples)
Fundamental roles	Detoxification of polluted waters a waste disposal Shoreline stabilization Flood control
Supporting roles	Soil properties' enhancement (e.g., soil structure and fertility) Genetic biodiversity Water provision Nutrient cycling
Regulating roles	Soil resources' conservation (e.g., soil retention) Natural control of plant pests Food sources and habitat for beneficial insects
Environmental roles	Regulation of microclimate Climate change mitigation Regulation of environmental processes

(2012) evaluated the effect of global warming on crop growth and revealed that the crop and livestock productivity may decrease due to warmer climate.

5.2.3 Global Distribution of Coastal Ecosystems

Due to the large length of coastline (worldwidely), it is noted that the coast is home to different ecosystems (Allan 2004). Biodiversity is the main characteristic of each costal ecosystem. Usually, four major costal ecosystem types can be distinguished: coral reefs, mangroves, salt marshes, and sand dunes. Some details about these four costal ecosystems are presented in Fig. 5.3. The coral reefs and mangroves have tropical distributions, whereas the salt marshes and sand dunes are common globally (Abson and Termansen 2011). Currently, the most significant problem facing these ecosystems is runoff from industrial, agricultural, and municipal areas. Usually, this runoff may contribute to higher nutrient and pollutant levels in coastal waters (i.e., pollution of coastal and ocean waters). Such critical levels may have dangerous effect on both humans and marine life. Furthermore, current climate change aspects (e.g., sea level rise) urge the decision-makers to protect these ecosystems (Alongi 2008).

In addition to the cultivation of common crops (e.g., rice, cucurbits, tomato, beet, betel vine, etc.), an important number of specific crops such as mangroves, seaweed, and seagrasses are also observed in many coastal agroecosystems (Bandyopadhyay et al. 2003). Saline condition, especially saline water, is highly required for ensuring the growth of these crops (Hoorweg and Muthiga 2009). For example, in India, the mangrove agroecosystems are distributed along the coastal areas of West Bengal ($87^{0}25'$–890 E latitude; $21^{0}30'$ N–$23^{0}15'$ longitude). Furthermore, the forests are a major component of the coastal agroecosystems due to their significant ecological resources (Islam 2003). Worldwide, about 90% of the coastal agroecosystem land is used for various agricultural activities within the forests (FAO 2010). The sustainability of agricultural productivity in these systems, therefore, requires a

a) Coral reefs:

Coral reefs are a unique marine ecosystem. They are built up entirely of living organisms. They are huge deposits of calcium carbonate made up mainly of corals. Coral reefs with the highest biodiversity occur in South-east Asia. The Great Barrier Reef is in Australia and is renowned not just for its great biodiversity, but also its extent and excellent condition.

b) Mangroves:

Mangroves are most common in South-East Asia. They literally live on the coastline – they have one 'foot' on the land and the other in the sea. Because they grow in the intertidal zone, they live in a constantly changing environment. They are regularly flooded by the sea. At low tide, especially during periods of high rainfall, there may be floods of fresh water.

c) Salt marshes:

Salt marshes occupy a midway location between mudflats that are permanently submerged by water and land vegetation lying above the high-tide mark. Coastal salt marshes develop in locations sheltered from the open sea, namely at the heads of bays and in estuaries. Since estuaries are where a river meets the sea, the water is brackish (partly salty and partly fresh). In bays, the water is salty.

d) Coastal sand dunes:

Coastal sand dunes are accumulations of sand shaped into mounds and ridges by the wind. They develop best where: there is a wide beach and large quantities of sand, the prevailing wind is onshore (from the sea to the shore), and there are suitable locations for the sand to accumulate.

Fig. 5.3 Four major costal ecosystem types: coral reefs, mangroves, salt marshes, and sand dunes (Abson and Termansen 2011) more strategic approach to coastal risk management in order to mitigate climate change and achieve a better agricultural development of coastal agroecosystems.

5.2.4 Potential Threats

Even though a significant increase in agricultural productivity was noted over the recent decades, however, several current coastal agroecosystems are inefficient and environmentally unsound (Bennett et al. 2005). The unsuitable management of coastal agroecosystems may be the cause of many threats such as land degradation and erosion, nutrient runoff, water resources' degradation, and food production decline (Jackson et al. 2013). The different aspects of land degradation such as wind erosion, water erosion, soil properties' degradation (e.g., salinity, acidification, fertility decline, etc.), nutrient depletion, soil pollution, and sealing are considered as major causes of hampering the growth in agricultural productivity (Haj-Amor and Bouri 2020; Raj et al. 2019a, b). For example, the effect of land degradation on yields in coastal agroecosystems of China was evaluated as a decrease in food production capacity on the current arable land area from 482 Mt. in 2005 to about 400 Mt. by 2050 (Bennett et al. 2009; Bindraban et al. 2012).

Recently, land degradation has significantly increased in several coastal agroecosystems as a consequence of unsuitable SWR management such as poor control of soil erosion and poor management of fertilizers (Bossio et al. 2008). In Africa, about 39% of the coastal agroecosystems are threatened by land degradation (Vlek et al. 2010). Also, some human activities such as entire crop removal (so loss of organic matter) may highly decrease nutrient availability, may decrease water infiltration rates and porosity, and consequently may affect the resilience of coastal agroecosystems (Bouma et al. 2011). It is important to note that large land degradation may result in substantial yield losses and contributes to downstream sedimentation, which may deteriorate water resources such as costal aquifers and fill up water storage reservoirs (Vlek et al. 2010). Accordantly, there is a close relationship between land degradation occurrence, low water productivity, and impaired coastal agroecosystem roles (Bossio et al. 2008). This relationship is usually associated with high population pressure (Muchena et al. 2005). This latter can in itself trigger investments in labor-intensive conservation actions and SWR use (Nelson 2005).

Recent investigations have revealed that about 50% of all coastal agroecosystems has more than 10% tree cover. This indicates that trees are a mainstream element of coastal agroecosystems and may ensure some forest services (Zomer et al. 2009). Tree cover in coastal agroecosystems may have an important effect on the water infiltration and consequently on catchment hydrology (Carroll et al. 2004). Also, when tree cover is changed, other coastal agroecosystem roles may also be changed (e.g., carbon storage, agricultural productivity, etc.) (Harvey et al. 2006).

In addition to the above threats, recently, several coastal agroecosystems are under threat mainly due to climate change. Climate change is one of the major threats to sustainable development of coastal agroecosystems due to its negative impacts on coastal ecosystems (IPCC 2019). Food security in coastal agroecosystems, threatened by climate change, is the most significant challenge over the recent years due to the continuous increasing population in many coastal countries (Molden 2007). Conservation agriculture is usually claimed to provide important services to cope with land degradation and enhance livelihood such as soil fertility enhancement, soil erosion decrease, carbon accumulation, energy conservation, and biodiversity increase (FAO 2002; Reicosky and Saxton 2006; Kumar et al. 2020a, 2021) assured from decreased tillage, soil cover, and crop rotation (Kassam et al. 2009). Various effects of climate change on land degradation must be continuously evaluated in order to ensure these good services (Khan et al. 2020a, b, 2021a, b).

5.3 Climate Change and Its Effect on Coastal Ecosystems and Coastal Agroecosystems

Currently, a lot of coastal ecosystems and coastal agroecosystems are under climate change threat. Climate variability is also considered as a threat for these ecosystems. To avoid or at least minimize the negative effects of these threats, it was noted over the past years that several coastal agroecosystems have coped with climate change (Mertz et al. 2009). Climate variability and climate change may have critical effects on ecosystems (IPCC 2007). Due to the continuous population increase in the coastal countries, food security and imbalance in the coastal agroecosystems are considered as the major challenges (Lal 2005). Accordantly, great attention has been recently allocated to the effects of climate change on food production (Dickson et al. 2007). Even though some coastal agroecosystems may easily adapt to climate change due to some specific environmental conditions and some successful adaptation strategies, however, some challenges associated with climate change (food security, soil properties' degradation, etc.) pose a threat to many coastal agroecosystems (Cronk and Fennessy 2001). This section intends to illustrate climate change aspects and effects in coastal ecosystems and coastal agroecosystems and to discuss climate change mitigation.

5.3.1 Climate Change Aspects

The most significant aspects of climate change influencing food security and coastal agroecosystem are elevated atmospheric CO_2 concentration, sea level rise, increasing air temperature, and rainfall pattern change (Fuhrer 2003). All these aspects of climate change may have significant effect on crop growth, SWR, and agroecosystem roles (Amthor 2001). Over the recent years, these aspects already affect many coastal agroecosystems, especially in arid areas such North Africa, China, India, etc. (Earman and Dettinger 2011).

5.3.1.1 Elevated Atmospheric CO_2 Concentration

Worldwide, the increase in atmospheric CO_2 concentration is always considered as the most significant aspect in the ongoing issue of climate change (Jacobson 2005). Based on the recent assessments, it is estimated that the current atmospheric CO_2 concentration is about 412 ppm (global value for the year 2018) (IPCC 2019). Compared to a concentration of 370 ppm in 2000, this value represents about 12% increase. This rapid increase in atmospheric CO_2 concentration is mainly attributed to human activities, especially the combustion of fossil fuels and deforestation activity (Haj-Amor and Bouri 2020). Under this situation, a significant increase in air temperature was also revealed over the recent decades. The rapid increase in atmospheric CO_2 concentration and its potential effect on the climate may have several effects on coastal agroecosystems (Caldeira and Wickett 2003). For example, the absorption of atmospheric CO_2 by the surface water bodies (e.g., sea, lakes, etc.) of coastal agroecosystems contributes usually to a considerable acidification of water resources (i.e., a decrease in water pH) which may in turn have negative effects on several marine organisms that build shells and other structures out of calcium carbonate (Feely et al. 2004).

Furthermore, the increase in atmospheric CO_2 concentration may have negative effect on the structure of dunes and beaches of many coastal ecosystems. For example, a study conducted by Università Ca' Foscari Venezia (2018) has revealed that the dunes and beaches of Sardinia will soon be changed due to the increase in CO_2 emissions. Due to this change, some adaptation strategies are urgently required such as a switch from fossil fuels to clean renewable energies and more effective actions for energy conservation. This is more needed over the next years due to the expected increase in the atmospheric CO_2 concentration to 1000 ppm by 2100 (IPCC 2007).

5.3.1.2 Increasing Air Temperature

The global average temperature is increasing as a direct consequence of the greenhouse impact caused mainly by the rapid increase in atmospheric CO_2 concentration (Clarke and Smethurst 2010). Based on the recent investigations, it was revealed that from 1951–1978 period to 2010–2019 period, the global air temperature has increased by 0.2–4 °C (Fig. 5.4).

The responses of lands to the rising temperature are not uniform across the earth (Klein and Nicholls 1999). Usually, coastal lands are more sensitive to this rising more than other lands. The coastal lands may suffer severe effects from sea level rise (i.e., a direct consequence of rising temperature). Also, northern high latitudes change more quickly than the tropics (Willows and Connell 2002).

5.3.1.3 Rainfall Pattern Change

As revealed by several authors (Ge et al. 2010; Al Charaabi and Al-Yahyai 2013; Putnam and Broecker 2017), climate change is altering rainfall patterns in several many coastal agroecosystems, especially arid ecosystem in North Africa, China, India, Australia, etc. Generally, wet ecosystems get more rainfall, and dry ecosystems become warmer (Held and Soden 2006). Elevated air temperature has

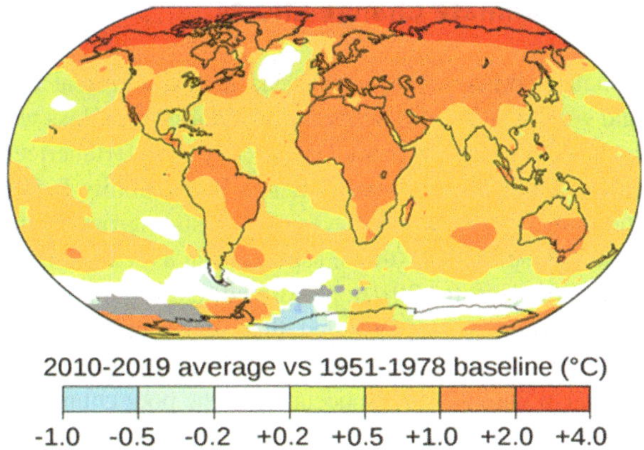

Fig. 5.4 Global air temperature change in the last 50 years (Source: NASA)

contributed to more moisture evaporation from land and water into the atmosphere. Based on the recent investigations (IPCC 2019), it was revealed that for each 1 °F increase in temperature, the atmosphere may hold 4% more water vapor. Due to this evaporation process, more rainfall and heavy downpours were prevalent. But, this extra rainfall is not evenly spread worldwide, and some coastal agroecosystems, especially the arid ecosystems, might actually get less rainfall, because climate change causes shifts in air and ocean currents, which may highly modify rainfall patterns. The current climate investigations coupled with future climate projections (based on performant climate models) have revealed that the arid coastal agroecosystems would get drier. Therefore, the effect of rainfall patterns change on water resources will be more critical in these ecosystems (Leung et al. 2004). In the arid coastal agroecosystems, the implications of climate change on water sector can be one of the major risks for improving uninterrupted water supply for several purposes, such as agricultural purpose.

5.3.1.4 Sea Level Rise

This aspect of climate change is the result of the expansion of water due to the rising temperature and the addition of water to the oceans from the melting of land ice (Gregory et al. 2001). For example, based on the recent investigations, it was revealed that loss from the ice sheets of Greenland and Antarctica was a major contributor to sea level rise during the 1993–2003 periods (IPCC 2007). Over 1961–2003 period, the global average sea level rose at an average rate of 1.8 mm $year^{-1}$ (IPCC 2007). Based on climate projections for the year 2090, the global sea level will rise 0.44 mm $year^{-1}$ above 1990 levels (Leatherman et al. 2000). Due to this potential rise, it is expected that the frequency of extreme high water levels and storm waves will heavily rise (IPCC 2019). Therefore, it is expected that these storm waves will have critical effects on many coastal agroecosystems.

5.3.2 Climate Change Effects

Usually, climate change may lead to many critical impacts on coastal ecosystems and coastal agroecosystems such as biodiversity loss which may be up to about 40% by 2100 in several coastal agroecosystems (Banerjee et al. 2018; Sheoran et al. 2021). The CCVI (Climate Change Vulnerability Index) is usually used to evaluate the vulnerability status of various coastal agroecosystems to some climate change aspects (especially extreme aspects such as sea level rise, storms, extreme droughts, etc.). Maplecroft (2010) performed CCVI analyses for more than 160 coastal agroecosystems and revealed that several coastal agroecosystems in Africa and Asia are at "extreme risk" from the effects of climate change. Often, the most critical effects of climate change occurred in the poor countries with limited NR (e.g., water resources), unsuitable infrastructure, and high population density (McGranahan et al. 2007). Therefore, the coastal agroecosystems in African and Asian countries such as Somalia, Haiti, the Philippines, and Indonesia are considered as the most vulnerable ecosystems for extreme climate change events.

Coastal erosion, sediment deficit, and land loss are considered as the major impacts of climate change on coastal agroecosystems. Usually, coastal erosion is the consequence of natural processes and human activities. The sea level rise is a major contributor to coastal erosion process (especially in sandy shorelines). About 70% of the world's coastal agroecosystems have been eroded during the past 100 years. For example, it has been estimated that 25% of the Italian coastal agroecosystems already show erosion process (IPCC 2019). Also, sea level rise may affect quality and availability of water resources in many coastal agroecosystems (Haj-Amor and Bouri 2020).

On the other hand, the coastal agroecosystems are very rich in species diversity (flora and fauna) such as mangrove forests and aquaculture (Dasgupta et al. 2007). Recently, the current aspects of climate change have highly contributed to significant modification in the ecological structure and roles of many coasts. As a consequence, an important biodiversity loss was reported (Guiteras 2007). Under this condition, maintaining high biodiversity in coastal agroecosystems makes agricultural production more sustainable and economically viable.

In addition to the negative impact of climate change on biodiversity, it is important to note that climate change aspects may also have critical impacts on agriculture (Debnath 2013). For example, increase in air temperature will decrease crop yields as crops become intolerant to high temperatures. Because of the intrusion of seawater in the fertile land due to cyclone or water surge, the soil fertility may be destroyed (Fig. 5.5). Also, warmer temperatures in the arid coastal areas will decrease soil moisture; thus, more irrigation will be required (Haj-Amor and Bouri 2020). Some past studies (e.g., Lobell et al. 2011) have revealed that, at global level, the climate change that occurred over 1981 to 2002 has contributed to a decrease of 40 million tons per year for maize, wheat, and other major crops. Furthermore, Jones and Thornton (2003) performed agricultural productivity projections under different climate change scenarios of 2055 for maize and revealed that a 10% decrease in

Fig. 5.5 Seawater intrusion in an Indian coastal area (Banerjee et al. 2018)

Table 5.2 Impacts of climate change on agriculture (Compiled: Mahato 2014)

Climate change aspect	Aspect in 2050	Impacts on agriculture
CO_2 concentration increase	CO_2 concentration up to 600 ppm	Enhance photosynthesis Decrease irrigation application
Sea level rise	Sea level rise up to 15 cm	Water resources' salinity Increase water use
Temperature increase	Increase by 1–2 °C	Earlier growing season Increase crop water requirement
Rainfall pattern change	Seasonal change by 10%	Many crop and soil problems Increase water use

maize production for Africa and Latin America will occur by 2055 which may lead to a loss of US$2 billion per year.

In many coastal agroecosystems, crop yields and agricultural production are highly impacted by climate change (Mendelsohn 2008). Several studies (e.g., Hassan 2010; Mahato 2014) revealed that climate change aspects (e.g., sea level rise, warmer climate) will have various effects on agriculture (Mahato 2014). Table 5.2 summarizes some of the effects.

5.3.3 Climate Change Mitigation

The management of coastal ecosystems has been considered as a key element in international and national climate change mitigation actions (Hadwen et al. 2011). In

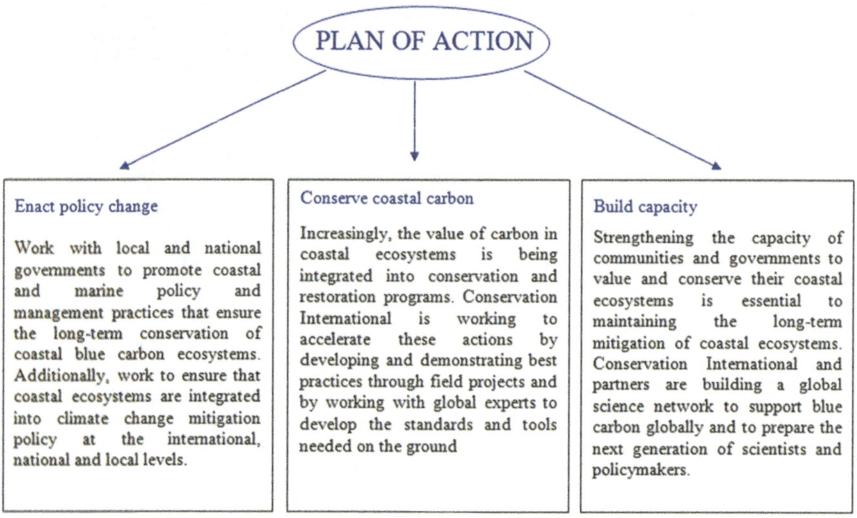

Fig. 5.6 Plan of action adapted for climate change mitigation in the coastal agroecosystems

order to mitigate the negative effects of climate change on these fragile ecosystems (i.e., coastal ecosystems), the United Nations Framework Convention on Climate Change (UNFCCC) encourages the decrease of greenhouse gases and conservation of all available coastal resources such as SWR. UNFCCC has reported useful actions that can help to ensure suitable land use leading to carbon sequestration or decreased emissions from coastal ecosystems (Wilby and Keenan 2012). Furthermore, the UNFCCC urges the decision-makers to focus on climate change mitigation by the adaptation of the following plan of action (Fig. 5.6):

5.4 Integrated Coastal Agroecosystem Management

Based on the information reported in Vallega (2002), the "integrated coastal agroecosystem management" term can be defined as a dynamic management strategy in which all policies, sectors (social, cultural, economic, and environmental), authorities (local, regional, and national), and dimensions (i.e., temporal and spatial scales) are properly taken into account for achieving sustainable use of coastal agroecosystems. The main objective of this strategy is to conserve coastal resources (e.g., SWR), their ecological roles, and ultimately their values by applying suitable land use planning within a social, institutional, and economic context (Cicin-Sain et al. 2000; Meena et al. 2020a). In this section, we discuss needs and planning process for integrated coastal agroecosystem management and measures adopted for this management.

5.4.1 Needs for Integrated Coastal Agroecosystem Management

A coastal agroecosystem is formed by dynamic interactions between living (e.g., crops, insects, microbes, etc.) and non-living elements (e.g., climate conditions) in a coastal region. Each element has its specific properties and functions. Therefore, there is a high need to take into account these specific properties and functions for minimizing coastal risks and achieving sustainable use of coastal agroecosystems (Pokhrel 2018). Avoiding or minimizing coastal risks and ensuring suitable usages for the coastal agroecosystems are not immediately compatible objectives. Indeed, various human activities in these ecosystems have led to a serious deterioration of the coastal areas (Martí et al. 2007). In this context, the coastal agroecosystems are considered as difficult ecosystems to manage due to several causes such as temporal issues (e.g., current and tides), geographical issues (e.g., inshore and shoreline), and needs of various interventions (OECD 2019). For a given costal ecosystem, usually, several different local, national, and regional government agencies are responsible for conserving various services (e.g., fisheries, environment, agriculture, etc.), by applying suitable land management (Wilby and Keenan 2012).

Coastal agroecosystems are highly productive but significantly threatened ecosystems in the world (Magadza 2000). A lot of roles and services are ensured by the coastal agroecosystems. However, these roles have been damaged by several threats. The major threats can be summarized as follows: (1) the coastal agroecosystems are highly affected by human activities (e.g., industry) and some population pressures (e.g., population increasing) in the coastal regions (McGranahan et al. 2007); (2) issues associated with climate change aspects such as infrastructure damage, loss of some agricultural activities, degradation of SWR, loss of some aquaculture, etc. (Nicholls and Lowe 2004); and (3) increase in urbanization near the coast with various environmental effects (based on recent updates, about 700 million people are living in coastal areas). Researchers across the world revealed that these major threats may bring significant physical, biological, ecological, and biogeochemical changes to coastal ecosystems. Accordantly, there is a great need to ensure integrated management of coastal agroecosystems (Tompkins and Adger 2005).

5.4.2 Planning Process for Integrated Coastal Agroecosystems

Intensive agriculture, developed in the coastal areas throughout the twentieth century, has contributed to a massive food production. However, this production was associated with a lot of ecological issues such as soil deterioration, loss of biodiversity, and homogenization of soil crops (Harman et al. 2015). Recently, the scientific community agrees that these issues are direct consequence of intensive agriculture and climate change (Spalding et al. 2014). Accordantly, the scientific community is advocating for the development of a clear planning process for integrated coastal agroecosystems in order to produce a more resilient farming system. The scientific community tries to define this planning process because it aims to produce a more sustainable and ecological farming system (Martí et al. 2007). Based on the outcomes of many studies (e.g., Gladstone et al. 2003; Tompkins and Adger 2005; Wilby and Keenan 2012; Meena et al. 2020b), the general process for integrated coastal agroecosystems must consider the following concepts:

- Resource assessment: This concept refers to a comprehensive inventory of coastal natural and human resources (e.g., physical and biological data, resource uses such as soil and water uses, cultural heritage, traditional land uses and activities). Also, it includes long-term in-depth bio-complexity research studies.
- Impact assessment: The main objective of this concept is to assess the coastal agroecosystems' vulnerability to various activity impacts. Based on the best available knowledge and acknowledging uncertainties, it is a good tool to help in making decisions and evaluate options for the mitigation and environmental sound management (e.g., spatial and use conflict analysis, GIS models).
- Policy and regulatory framework: It is a basic tool for training and education and for local community participation in the decision-making process. Based on the analysis of existing institutional and legal mechanisms, the main objective of this concept is to develop comprehensive policy framework to address coastal agroecosystem issues.

The above three concepts constitute the dynamic planning process. This process whereby values, ecological processes, environmental risks, and actions are developed and evaluated should be considered as a dynamic and continuous planning process.

5.4.3 Measures Adopted for Integrated Coastal Agroecosystem Management

The overall aim of integrated coastal agroecosystem management is to maintain ecological processes of these ecosystems and meet human needs for goods and services. Therefore, management measures need to be developed on the basis of available science resources regarding ecological processes of coastal agroecosystems and a comprehensive understanding of human needs and expectations, which are both tangible and intangible (Gladstone et al. 2003). Planning and organization of management measures should be based on a clear definition of general and specific objectives, followed by a series of management actions, the most important of which are summarized in Fig. 5.7. These management actions must take into consideration the potential effects of climate change on agroecosystems. Indeed, as reported in many studies (Skelly and Weinstein 2003; Diaz 2007; Costello et al. 2009), it is expected that the climate change will affect the health of several coastal agroecosystems (Meena et al. 2020c). The effects of climate change, therefore, require a more strategic plan to coastal risk management in order to restore the sediment balance, decrease damage of coastal erosion and floods, and achieve a better development of coastal agroecosystems (McInnes et al. 2000).

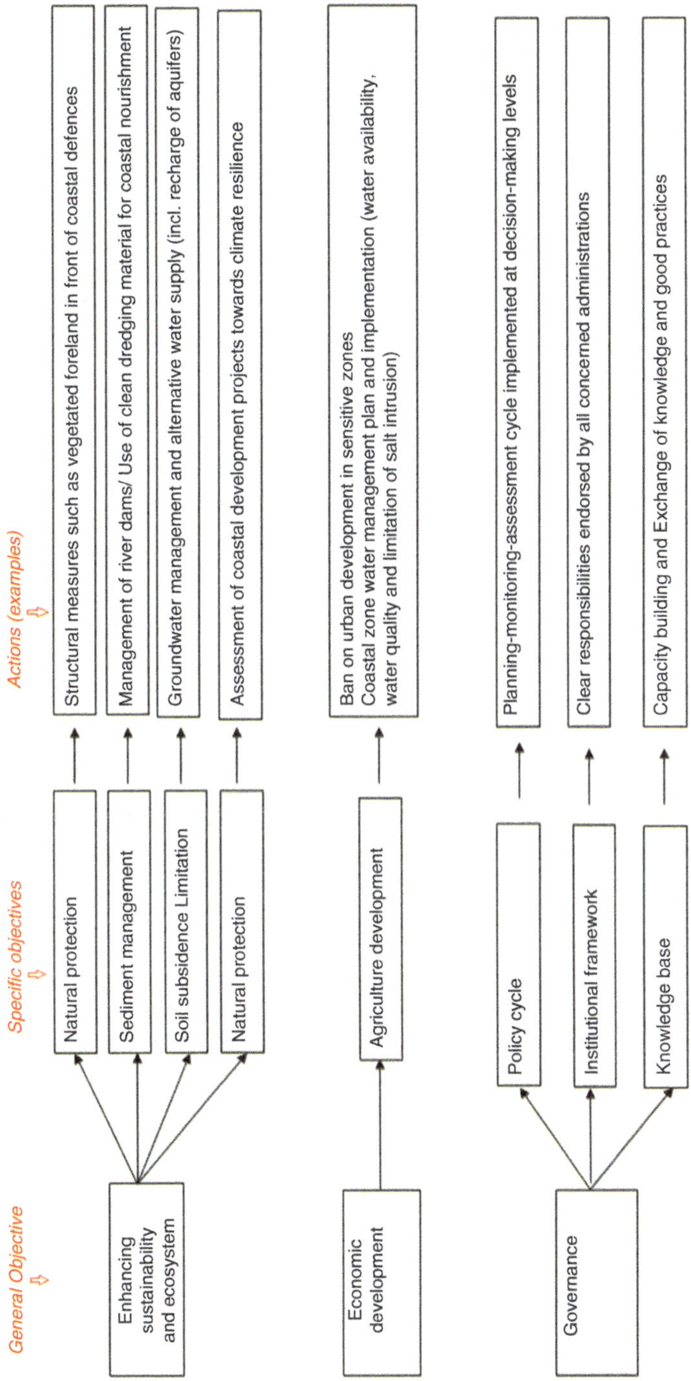

Fig. 5.7 General and specific objectives of measures adopted for integrated coastal agroecosystem management

Consideration of coastal erosion and flooding risks into long-term plans can highly help the decision-makers to decrease risks in agroecosystems (Brunsden and Lee 2000).

Furthermore, some helpful soil management actions at the local level (i.e., within the coastal agroecosystems) can highly reduce the effects of climate change. For example, land leveling, subsurface drains, and perennial crop land use systems are so useful to conserve the physical, chemical, and biological properties (ISDR 2002). These practices tend to reduce crop damage from flooding (after a heavy rainfall event) and to control runoff without causing soil erosion. This leads to perfect crop growth and sustainable agricultural productivity (Haj-Amor and Bouri 2020). Finally, it is important to note that in addition to the technical parameters, the success of these soil management actions will depend on the participation of local people with their traditional knowledge.

5.5 Research and Development in Coastal Agroecosystem Management

Many international studies (e.g., Campbell 2006; Billgren and Holmén 2008; McFadden et al. 2009) have focused on the measures that can enhance agricultural productivity and ensure integrated management of coastal agroecosystems. Based on the findings of these studies, we can summarize the major measures such as follows: (1) adequate coastal risk management in order to restore the sediment balance, decrease damage of coastal erosion and floods, and achieve a better development of coastal agroecosystems; (2) suitable management of NR such as SWR; (3) suitable control of crop properties such as crop growth; (4) continuous control of some natural factors and climate change conditions; (5) apply technologies (e.g., water-saving technologies) within the coastal agroecosystems; and (6) decrease the environmental effect of energy use.

Furthermore, Banerjee et al. (2018) summarized all the climate change aspects that may have negative effects on coastal agroecosystems. These major aspects include elevated atmospheric CO_2 concentration, air temperature increase, change in rainfall pattern, and sea level rise. Under these climate change aspects, the IPCC (2019) has proposed an integrated management of all NR in the coastal agroecosystems. For example, an integrated water resource management includes efficient irrigation, use of treated wastewater, and suitable irrigation practices when saline water is used for irrigation. Also, an integrated crop management includes enhancing crop genetics, good selection of crop types (e.g., select suitable crops under saline conditions), and continuous control of crop growth, especially under some harsh climate conditions (Barba de la Rosa et al. 2009). All these measures could be so helpful to ensure sustainable agricultural productivity and reduce environmental issues in the coastal agroecosystems.

Over the recent years, a great research effort was performed in order to select the minor crops' advanced measures that have a high potential for facilitating the integrated management of coastal agroecosystems. Based on the findings of these recent investigations (Sabuncu et al. 2016; Almonacid-Caballer et al. 2016), it was observed that a lot of advanced measures and technologies could be so helpful to ensure sustainable agricultural productivity and reduce environmental issues (e.g., erosion, flooding, etc.) in the coastal agroecosystems. For example, satellite images can be powerful tools for scientists to get a lot of useful data such as water temperature, sea level, shoreline modification, soil properties, and potential threats to the coastal agroecosystems. The integration of these images within a geographic information system (GIS) is required to predict critical events that may have serious effects on the coastal agroecosystems (Liu et al. 2013).

5.6 Policy Framework for Integrated Coastal Agroecosystem Management

Development of adequate policy framework is so helpful for ensuring integrated management of coastal agroecosystems and increasing food production in each agroecosystem (Gladstone et al. 2003). The management measures should be developed on the basis of the best science available about ecological processes of coastal agroecosystems and a comprehensive understanding of human needs and expectations, which are both tangible and intangible (Costello et al. 2009). The FAO (2010) reported that the policies of integrated management of coastal agroecosystems need to focus on integrated production systems with close collaboration with the smallholders (e.g., capacity building and exchange of knowledge and good practices). Over the past few years, some successful environmental services of the policies of integrated coastal agroecosystem management were noted (reported in IPCC 2019). However, further policy options should be developed in order to respond to many environmental and agricultural challenges (e.g., climate change issue, food production decrease, etc.). Integrated coastal agroecosystem management could only ensure if a lot of key policy options and technologies are adapted. The following are some examples of these key policy options (Table 5.3).

Despite the above recommended key policy options and their current positive outcomes, there is still a lot of work which needs to be done to ensure increase in

Table 5.3 Policy options for integrated coastal agroecosystem management

Policy option	Actions (examples)
Maximize agricultural productivity	Mechanization Suitable management of irrigation water
Support techniques and technologies	Satellite images for coast control Water-saving technologies
Reduce environmental effects	Regulation of environmental processes Suitable use of energy with each agroecosystem
Governance practices	Continuous planning and assessment (by the decision-makers) Exchange of knowledge and good practices

food production in the coastal agroecosystems without or with minimal negative impacts on ecosystems.

5.7 Conclusion

Integrated management of coastal agroecosystems is an urgent and fundamental process in the coastal areas. It is highly required for coastal areas that are complex and risky (i.e., climate change risks). Furthermore, it is also a useful and management process when decision-makers are faced with major environmental issues such as climate change. In coastal areas, human activities (e.g., agriculture and industry) and climate change and their impacts are often highly complex. Integrated coastal agroecosystem management provides a context for considering all of these complexities and then deciding what is important to be done in the coastal agroecosystems. As discussed in the present chapter, the measures adopted for integrated coastal agroecosystems may allow, and even force, the political process to allocate resources to the most significant issues. In essence, the process allows the decision-makers to strike a balance between maintaining ecological processes of coastal agroecosystems and meeting human needs for goods and services (especially agricultural productivity).

5.8 Future Perspectives

In the present chapter, the major future perspectives can be summarized as follows:

- To achieve sustainable agroecosystems in the coastal areas, it is very much important to decrease greenhouse gas emissions. Therefore, more future research works on the possible ways that can help to achieve this decrease are required.
- More future research works on the development of various policies that can help to achieve efficient use of NR (e.g., water and soil resources) in the coastal agroecosystems are required.
- The integrated coastal agroecosystem management measures reported in this chapter are considered as being helpful activities to maintain and improve the productive capacity of the NR. These measures must be implemented according to the respective local conditions; i.e., the strategy is adapted at the local level. Therefore, more future research works on some adaptation ways are required in this perspective.
- In addition to the technical parameters, the success of measures adopted for integrated coastal agroecosystem management will depend on the participation of local people with their traditional knowledge. Therefore, more field research works on these direction need to be done.

References

Abson DJ, Termansen M (2011) Valuing ecosystems in terms of ecological risks and returns. Conserv Biol 25:205–258

Al Charaabi Y, Al-Yahyai S (2013) Projection of future changes in rainfall and temperature patterns in Oman. J Earth Sci Clim Change 4:154

Allan JD (2004) Landscapes and riverscapes: the influence of land use on stream ecosystems. Annu Rev Ecol Evol Syst 35:257–284

Almonacid-Caballer J, Sánchez-García E, Pardo Pascual JE, Balaguer-Beser AA, Palomar-Vázquez J (2016) Evaluation of annual mean shoreline position deduced from Landsat imagery as a mid-term coastal evolution indicator. Mar Geol 372:79–88

Alongi DM (2008) Mangrove forests: resilience, protection from tsunamis, and responses to global climate change. Estuar Coast Mar Sci 76:1–13

Amthor JS (2001) Effects of atmospheric CO_2 on wheat yield: review of results from experiments using various approaches to control CO_2 concentration. Field Crop Res 73:1–34

Arriaza M, Gomez-Limon JA, Kallas Z, Nekhay O (2008) Demand for non-commodity outputs from mountain olive groves? Agric Econ Rev 9(1):5–23

Bandyopadhyay BK, Maji B, Sen HS, Tyagi NK (2003) Coastal soils of West Bengal – their nature, distribution and characteristics, Bulletin No. 1/2003. Central Soil Salinity Research Institute, Regional Research Station, Canning Town, West Bengal, p 62

Banerjee S, Samanta S, Chakraborti PK (2018) Impact of climate change on coastal agro-ecosystems. In: Lichtfouse E (ed) Sustainable agriculture reviews, vol 33. Springer, Berlin

Banerjee A, Jhariya MK, Yadav DK, Raj A (2020) Environmental and sustainable development through forestry and other resources. Apple Academic Press Inc., CRC Press- a Taylor and Francis Group, US & Canada, 400 pp. https://doi.org/10.1201/9780429276026

Banerjee A, Meena RS, Jhariya MK, Yadav DK (2021a) Agroecological footprints management for sustainable food system. Springer, Singapore, 514 pp. https://doi.org/10.1007/978-981-15-9496-0

Banerjee A, Jhariya MK, Meena RS, Yadav DK (2021b) Ecological footprints in agroecosystem- an overview. In: Banerjee A, Meena RS, Jhariya MK, Yadav DK (eds) Agroecological footprints management for sustainable food system. Springer, Singapore, pp 1–23

Banerjee A, Jhariya MK, Raj A, Yadav DK, Khan N, Meena RS (2021c) Land footprint management and policies. In: Banerjee A, Meena RS, Jhariya MK, Yadav DK (eds) Agroecological footprints management for sustainable food system. Springer, Singapore, pp 221–246

Banerjee A, Jhariya MK, Raj A, Yadav DK, Khan N, Meena RS (2021d) Energy and climate footprint towards the environmental sustainability. In: Banerjee A, Meena RS, Jhariya MK, Yadav DK (eds) Agroecological footprints management for sustainable food system. Springer, Singapore, pp 415–443

Barba de la Rosa AP, Fomsgaard IS, Laursen B, Mortensen AG, Olvera-Martinez L, Silva-Sanchez C, Mendoza-Herrera A, Gonzalez-Castaneda J, De Leon-Rodrıguez A (2009) Amaranth (*Amaranthus hypochondriacus*) as an alternative crop for sustainable food production. Phenolic acids and flavonoids with potential impact on its nutraceutical quality. J Cereal Sci 49:117–121

Bennett EM, Peterson GD, Levitt EA (2005) Looking to the future of ecosystem services. Ecosystems 8:125–132

Bennett EM, Peterson GD, Gordon LJ (2009) Understanding relationships among multiple ecosystem services. Ecol Lett 12:1394–1404

Billgren C, Holmén H (2008) Approaching reality: comparing stakeholder analysis and cultural theory in the context of natural resource management. Land Use Policy 25(4):550–562

Bindraban PS, van der Velde M, Ye L, van den Berg M, Materechera S, Kiba DI, Tamene L, Ragnarsdottir KV, Jongschaap REE, Hoogmoed M, Hoogmoed WB, Beek CL, van Lynden GWJ (2012) Assessing the impact of soil degradation on food production. Curr Opin Environ Sustain 4:478–488

Bossio D, Noble A, Molden D, Nangia V (2008) Land degradation and water productivity in agricultural landscapes. In: Bossio D, Geheb K (eds) Conserving land, protecting water. Comprehensive assessment of water management in agriculture series 6. CAB International, Wallingford, UK in association with CGIAR Challenge Program on Water and Food, Colombo and International Water Management Institute (IWMI), Colombo, pp 20–32

Bouma J, Droogers P, Sonneveld MPW, Ritsema CJ, Hunink JE, Immerzeel WW, Kauffman S (2011) Hydropedological insights when considering catchment classification. Hydrol Earth Syst Sci 15:1909–1919

Brunsden D, Lee EM (2000) Understanding the behaviour of coastal landslide systems - an inter-disciplinary view. In: Landslides - research, theory and practice. Thomas Telford, London

Burke LK, Yumiko K, Kassem K, Revenga C, Spalding M, McAllister M (2000) Pilot analysis of global ecosystems: coastal ecosystems. World Resources Institute, Washington, DC, 94 pp

Burke L, Kura Y, Kasem K, Revenga C, Spalding M, McAllister D (2001) Coastal ecosystems. World Resources Institute, Washington, DC, 93 pp

Caldeira K, Wickett ME (2003) Anthropogenic carbon and ocean pH. Nature 425(6956):365

Campbell A (2006) The Australian natural resource management knowledge system. Land and Water Australia, Canberra

Carroll ZL, Bird SB, Emmett BA, Reynolds B, Sinclair FL (2004) Can tree shelterbelts on agricultural land reduce flood risk? Soil Use Manag 20:357–359

Central Intelligence Agency (2016) The World Factbook 2016–17: geography: world: geographic overview: coastline. Central Intelligence Agency, Washington, DC

Cicin-Sain B, Knecht RW, Vallega A, Harakunarak A (2000) Education and training in integrated coastal management: lessons from the international arena. Ocean Coast Manag 43:291–330

Clarke D, Smethurst J (2010) Effects of climate change on cycles of wetting and drying in engineered clay slopes in England. Q J Eng Geol Hydrogeol 43(4):473–486

Costello A, Abbas M, Allen A, Bell S, Bellamy R, Friel S, Groce N, Johnson A, Kett M, Lee M, Levy C, Maslin M, McCoy D, McGuire B, Montgomery H, Napier D, Pagel C, Patel J, Oliveira JAPD, Redclift N, Rees H, Rogger D, Scott J, Stephenson J, Twigg J, Wolff J, Patterson C (2009) Managing the health effects of climate change. Lancet 373:1693–1734

Cronk JK, Fennessy MS (2001) Wetland plants: biology and ecology. Lewis Publishers, Washington, DC

Dasgupta S, Laplante B, Meisner C, Wheeler D, Yan J (2007) The impact of sea level rise on developing countries: a comparative analysis. World Bank policy research working paper 4136

Debnath A (2013) Condition of agricultural productivity of Gosaba C.D. Block, South24 Parganas, West Bengal, India after severe cyclone Aila. Int J Sci Res Publ 3(7):1–4

Diaz JH (2007) The influence of global warming on natural disasters and their public health outcomes. Am J Disaster Med 2(1):33–42

Dickson ME, Walkden MJ, Hall JW (2007) Systemic impacts of climate change on an eroding coastal region over the twenty-first century. Clim Chang 84(2):141–166

Dominati E, Mackay A, Green S, Patterson M (2014) A soil change-based methodology for the quantification and valuation of ecosystem services from agro-ecosystems: a case study of pastoral agricultural in New Zealand. Ecol Econ 100:119–129

Earman E, Dettinger M (2011) Potential impacts of climate change on groundwater resources – a global review. J Water Clim Change 2(4):213–229

FAO (2002) Conservation agriculture: case studies in Latin America and Africa. FAO soils bulletin 2002/78. FAO, Rome, 70 pp

FAO (2010) Forest resources assessment 2010. UN Food and Agriculture Organization, Rome

Feely RA, Sabine CL, Lee K, Berelson W, Kleypas J, Fabry VJ, Millero FJ (2004) Impact of anthropogenic CO_2 on the $CaCO_3$ system in the oceans. Science 305(5682):362–366

Fuhrer J (2003) Agroecosystem response to combinations of elevated CO2, ozone, and global climate change. Agric Ecosyst Environ 97:1–20

Ge X, Li T, Zhang S, Peng M (2010) What causes the extremely heavy rainfall in Taiwan during Typhoon Morakot (2009)? Atmos Sci Lett 11(1):46–50

Gladstone W, Krupp F, Younis M (2003) Development and management of a network of marine protected areas in the Red Sea and Gulf of Aden region. Ocean Coast Manag 46:741–761
Green EP, Bruckner AW (2000) The significance of coral disease Epizootiology for coral reef conservation. Biol Conserv 96(3):347–361
Gregory JM, Church JA, Dixon KW, Flato GM, Jackett DR, Lowe JA, Oberhuber JM, O'Farrell SP, Stouffer RJ (2001) Comparison of results from several AOGCMs on global and regional sea level change 1900-2100. Clim Dyn 18(3/4):225–240
Guiteras R (2007) The impact of climate change on Indian agriculture. Working paper, Department of Economics, MIT
Hadwen WL, Capon SJ, Kobashi D, Poloczanska ES, Rochester W, Martin TG, Bay LK, Pratchett MS, Green J, Cook BD, Berry A, Lalonde A, Hall A, Fahey S (2011) Climate change responses and adaptation pathways in Australian coastal ecosystems: synthesis report. National Climate Change Adaptation Research Facility, Gold Coast, 359 pp
Haj-Amor Z, Bouri S (2020) Climate change impacts on coastal soil and water management. CRC Press, Boca Raton. https://doi.org/10.1201/9780429356667
Harman BP, Heyenga S, Taylor BM, Fletcher CS (2015) Global lessons for adapting coastal communities to protect against storm surge inundation. https://doi.org/10.2112/JCOASTRES-D-13-00095.1
Harvey CA, Medina A, Sanchez DM, Vilchez S, Hernandez B, Saenz JC, Maes JM, Casanoves F, Sinclair FL (2006) Patterns of animal diversity in different forms of tree cover in agricultural landscapes. Ecol Appl 16:1986–1999
Hassan RM (2010) Implications of climate change for agricultural sector performance in Africa: policy challenges and research agenda (dagger). J Afr Econ 19(Supplement 2):77–105
Held IM, Soden BJ (2006) Robust responses of the hydrological cycle to global warming. J Clim 19 (21):5686–5699
Hoorweg J, Muthiga N (2009) Advances in coastal ecology people, processes and ecosystems in Kenya. African studies collection (20). African Studies Centre, Leiden
Huang J, Tichit M, Poulot M, Darly S, Li S, Petit C, Aubry C (2015) Comparative review of multifunctionality and ecosystem services in sustainable agriculture. J Environ Manag 149:138–147
IPCC (2000) In: Watson RT, Noble IR, Bolin B, Ravindranath NH, Verardo DJ, Dokken DJ (eds) Intergovernmental panel on climate change, land use, land use change, and forestry. Cambridge University Press, Cambridge. http://www.ipcc.ch/pub/SPM_SRLULUCF.pdf
IPCC (2007) Climate change 2007: impacts, adaptation and vulnerability, contribution of working group II to the fourth assessment report of the intergovernmental panel on climate change. Cambridge University Press, Cambridge
IPCC (2013) Summary for policymakers. In: Stocker TF, Qin D, Plattner GK, Tignor M, Allen SK, Boschung J, Nauels A, Xia Y, Bex V, Midgley PM (eds) Climate change 2013: the physical science basis. Contribution of working group to the fifth assessment report of the intergovernmental panel on climate change. Cambridge University Press, Cambridge and New York
IPCC (2019) Climate Change and Land: an IPCC special report on climate change, desertification, land degradation, sustainable land management, food security, and greenhouse gas fluxes in terrestrial ecosystems - summary for policymakers
ISDR (2002) Living with risk: a global review of disaster reduction initiatives. Switzerland, UN
Islam S (2003) Sustainable eco-tourism as a practical site management policy. AH Development Publishing House, Dhaka
Jackson B, Pagella T, Sinclair F, Orellana B, Henshaw A, Reynolds B, Mcintyre N, Wheater H, Eycott A (2013) Polyscape: a GIS mapping framework providing efficient and spatially explicit landscape scale valuation of multiple ecosystem services. Landsc Urban Plan 112:74–88
Jacobson MZ (2005) Studying ocean acidification with conservative, stable numerical schemes for nonequilibrium air-ocean exchange and ocean equilibrium chemistry. J Geophys Res Atmos 110:D07302

Jhariya MK, Banerjee A, Meena RS, Yadav DK (2019a) Sustainable agriculture, forest and environmental management. Springer, Singapore, p 606. https://doi.org/10.1007/978-981-13-6830-1

Jhariya MK, Yadav DK, Banerjee A (2019b) Agroforestry and climate change: issues and challenges. Apple Academic Press Inc., CRC Press- a Taylor and Francis Group, US & Canada, 335 pp. https://doi.org/10.1201/9780429057274

Jhariya MK, Meena RS, Banerjee A (2021a) Ecological intensification of natural resources for sustainable agriculture. Springer, Singapore. https://doi.org/10.1007/978-981-33-4203-3

Jhariya MK, Meena RS, Banerjee A (2021b) Ecological intensification of natural resources towards sustainable productive system. In: Ecological intensification of natural resources for sustainable agriculture. Springer, Singapore. https://doi.org/10.1007/978-981-33-4203-3_1

Jones PG, Thornton PK (2003) The potential impacts of climate change on maize production in Africa and Latin America in 2055. Glob Environ Chang 13:51–59

Karmakar R, Das I, Dutta D, Rakshit A (2016) Potential effects of climate changes on soil properties: a review. Sci Int 4(2):51–73

Kassam AH, Friedrich T, Shaxson F, Pretty J (2009) The spread of conservation agriculture: justification, sustainability and uptake. Int J Agric Sustain 7:292–320

Khan N, Jhariya MK, Yadav DK, Banerjee A (2020a) Herbaceous dynamics and CO_2 mitigation in an urban setup- a case study from Chhattisgarh, India. Environ Sci Pollut Res 27(3):2881–2897. https://doi.org/10.1007/s11356-019-07182-8

Khan N, Jhariya MK, Yadav DK, Banerjee A (2020b) Structure, diversity and ecological function of shrub species in an urban setup of Sarguja, Chhattisgarh, India. Environ Sci Pollut Res 27 (5):5418–5432. https://doi.org/10.1007/s11356-019-07172-w

Khan N, Jhariya MK, Raj A, Banerjee A, Meena RS (2021a) Soil carbon stock and sequestration: implications for climate change adaptation and mitigation. In: Jhariya MK, Meena RS, Banerjee A (eds) Ecological intensification of natural resources for sustainable agriculture. Springer, Singapore. https://doi.org/10.1007/978-981-33-4203-3_13

Khan N, Jhariya MK, Raj A, Banerjee A, Meena RS (2021b) Eco-designing for sustainability. In: Jhariya MK, Meena RS, Banerjee A (eds) Ecological intensification of natural resources for sustainable agriculture. Springer, Singapore. https://doi.org/10.1007/978-981-33-4203-3_16

Klein RJT, Nicholls RJ (1999) Assessment of coastal vulnerability to climate change. Ambio 28 (2):182–187

Kumar S, Meena RS, Jhariya MK (2020aa) Resources use efficiency in agriculture. Springer, Singapore, 760 pp. https://doi.org/10.1007/978-981-15-6953-1

Kumar S, Meena RS, Datta R, Verma SK, Yadav GS, Pradhan GS, Molaei A, Rahman GKMM, Mashuk HA (2020bb) Legumes for carbon and nitrogen cycling: an organic approach. In: Carbon and nitrogen cycling in soil. Springer, Berlin, pp 337–375. https://doi.org/10.1007/978-981-13-7264-3_10

Kumar S, Meena RS, Singh RK, Munir TM, Datta R, Danish S, Yadav GS, Kumar S (2021) Soil microbial and nutrient dynamics under different sowings environment of Indian mustard (*Brassica juncea* L.) in rice based cropping system. Sci Rep 11:5289. https://doi.org/10.1038/s41598-021-84742-4

Lal R (2005) Climate change, soil carbon dynamics, and global food security. In: Lal R, Stewart B, Uphoff N (eds) Climate change and global food security. CRC Press, Boca Raton, pp 113–143

Leatherman SP, Zhang K, Douglas BC (2000) Sea level rise shown to drive coastal erosion. EOS Trans Am Geophys Union 81(6):55–57

Leung YK, Yeung KH, Ginn EWL, Leung WM (2004) Global climate change: cause of climate change by human activities. Hong Kong Observatory N° 107

Liu Y, Huang H, Qiu Z, Fan J (2013) Detecting coastline change from satellite images based on beach slope estimation in a tidal flat. Int J Appl Earth Obs Geoinf 23(1):165–176

Lobell DB, Gourdji SM (2012) The influence of climate change on global crop productivity. Plant Physiol 160:1686–1697

Lobell DB, Schlenker W, Costa-Roberts J (2011) Climate trends and global crop production since 1980. Science 333:616–620

Loeuille N, Barot S, Georgelin E, Kylafis G, Lavigne C (2013) Eco-evolutionary dynamics of agricultural networks: implications for sustainable management. Adv Ecol Res 49:339–435

Magadza CHD (2000) Climate change impacts and human settlements in Africa: prospects for adaptation. Environ Monit Assess 61(1):193–205

Mahato A (2014) Climate change and its impact on agriculture. Int J Sci Res Publ 4(4):1–11

Malézieux E (2012) Designing cropping systems from nature. Agron Sustain Dev 32:15–29

Maplecroft (2010) Climate change risk atlas 2010. https://maplecroft.com/about/news/climate_change_risk_list_highlights_vulnerable_nations_and_safe_havens_05.html

Martí X, Lescrauwaet A-K, Borg M, Valls M (2007) Indicators Guidelines to adopt an indicators-based approach to evaluate coastal sustainable development. Deduce project, Department of the Environment and Housing, Government of Catalonia. https://climate-adapt.eea.europa.eu/metadata/tools/deduce-indicators-guidelines-to-adopt-an-indicators-based-approach-to-evaluate-coastal-sustainable-development

Martinez ML, Intralawan A, Vazquez G, Perez-Maqueo O, Sutton P, Landgrave R (2007) The coasts of our world: ecological, economic and social importance. Ecol Econ 63(2–3):254–272

McFadden L, Tapsell S, Penning-Rowsell E (2009) Strategic coastal flood risk management in practice: actors' perspectives on the integration in flood risk management process in London and the Thames estuary. Ocean Coast Manag 52(12):636–645

McGranahan G, Balk D, Anderson B (2007) The rising tide: assessing the risks of climate change and human settlements in low-elevation coastal zones. Environ Urban 19(1):17–37

McInnes RG, Tomalin D, Jakeways J (2000) Coastal change, climate and instability. Final report of the EU LIFE Environment project. Ventnor

Meena RS, Lal R (2018) Legumes for soil health and sustainable management. Springer, Singapore, p 541. https://doi.org/10.1007/978-981-13-0253-4_10

Meena RS, Kumar V, Yadav GS, Mitran T (2018) Response and interaction of *Bradyrhizobium japonicum* and arbuscular mycorrhizal fungi in the soybean rhizosphere: a review. Plant Growth Regul 84:207–223. https://doi.org/10.1007/s10725-017-0334-8

Meena RS, Lal R, Yadav GS (2020a) Long term impacts of topsoil depth and amendments on soil physical and hydrological properties of an Alfisol in Central Ohio, USA. Geoderma 363:1141164. https://doi.org/10.1016/j.geoderma.2019.114164

Meena RS, Lal R, Yadav GS (2020b) Long-term impact of topsoil depth and amendments on carbon and nitrogen budgets in the surface layer of an Alfisol in Central Ohio. Catena 194:104752. https://doi.org/10.1016/j.catena.2020.104752

Meena RS, Kumar S, Datta R, Lal R, Vijaykumar V, Brtnicky M, Sharma MP, Yadav GS, Jhariya MK, Jangir CK, Pathan SI, Dokulilova T, Pecina V, Marfo TD (2020c) Impact of agrochemicals on soil microbiota and management: a review. Land (MDPI) 9(2):34. https://doi.org/10.3390/land9020034

Mendelsohn R (2008) The impact of climate change on agriculture in developing countries. J Nat Resour Policy Res 1(1):5–19

Mertz O, Halsnæs K, Olesen JE, Rasmussen K (2009) Adaptation to climate change in developing countries. Environ Manag 43:743–752

Millennium Ecosystem Assessment (2003) Ecosystem and human well-being: framework for assessment. Island Press, Washington DC

Molden D (ed) (2007) Water for food, water for life: comprehensive assessment of water management in agriculture. Earthscan, London, in association with International Water Management Institute (IWMI), Colombo

Muchena FN, Onduru DD, Gachin GN, de Jager A (2005) Turning the tides of soil degradation in Africa: capturing the reality and exploring opportunities. Land Use Policy 22:23–31. https://doi.org/10.1016/j.landusepol.2003.07.001

Nelson GC (2005) Drivers of ecosystem change: summary chapter. In: Hassan R, Scholes R, Ash N (eds) Ecosystems and human well-being: current state and trends, volume 1. Findings of the condition and trends working group of the millennium ecosystem assessment. World Resources Institute and Island Press, Washington, DC, pp 73–76

Nicholls RJ, Lowe JA (2004) Benefits of mitigation of climate change for coastal areas. Glob Environ Chang 14:229–244

OECD (2019) Responding to rising seas: OECD country approaches to tackling coastal risks. OECD Publishing, Paris. http://www.oecd.org/environment/cc/policy-highlights-responding-to-rising-seas.pdf

Pokhrel S (2018) Effect of climate change on agro-ecosystem services and the adaptation measures in Nepal. J Environ Sci 4:92–104

Power AG (2010) Ecosystem services and agriculture: tradeoffs and synergies. Philos Trans R Soc B 365:2959–2971

Putnam AE, Broecker WS (2017) Human-induced changes in the distribution of rainfall. Sci Adv 3: e1600871

Raj A, Jhariya MK, Yadav DK, Banerjee A, Meena RS (2019a) Agroforestry: a holistic approach for agricultural sustainability. In: Jhariya MK, Banerjee A, Meena RS, Yadav DK (eds) Sustainable agriculture, forest and environmental management. Springer, Singapore, pp 101–131. https://doi.org/10.1007/978-981-13-6830-1

Raj A, Jhariya MK, Banerjee A, Yadav DK, Meena RS (2019b) Soil for sustainable environment and ecosystems management. In: Jhariya MK, Banerjee A, Meena RS, Yadav DK (eds) Sustainable agriculture, forest and environmental management. Springer, Singapore, 606 pp, pp 189–221. https://doi.org/10.1007/978-981-13-6830-1

Raj A, Jhariya MK, Yadav DK, Banerjee A (2020) Climate change and agroforestry systems: adaptation and mitigation strategies. Apple Academic Press Inc., CRC Press- a Taylor and Francis Group, Burlington, ON, p 383. https://doi.org/10.1201/9780429286759

Raj A, Jhariya MK, Khan N, Banerjee A, Meena RS (2021) Ecological intensification for sustainable development. In: Ecological intensification of natural resources for sustainable agriculture. Springer, Singapore. https://doi.org/10.1007/978-981-33-4203-3_5

Reicosky DC, Saxton KE (2006) The benefits of no-tillage. In: Baker CJ, Saxton KE, Ritchie WR, Chamen WCT, Reicosky DC, Ribeiro MFS, Justice SE, Hobbs PR (eds) No-tillage seeding in conservation agriculture. CABI, Wallingford, pp 11–20

Sabuncu A, Dogru A, Ozener H, Turgut B (2016) Detection of coastline deformation using remote sensing and geodetic surveys. In: The international archives of the photogrammetry, remote sensing and spatial information sciences - ISPRS archives. Prague, pp 1169–1174

Senapati S, Gupta V (2014) Climate change and coastal ecosystem in India: issues in perspectives. Int J Environ Sci 5(3):530–543

Sheoran S, Kuma S, Kumar P, Meena RS, Rakshit S (2021) Nitrogen fixation in maize: breeding opportunities. Theor Appl Genet. https://doi.org/10.1007/s00122-021-03791-5

Singh BP, Cowie AL, Chan KY (2011) Soil health and climate change, Soil biology. Springer, Berlin, pp 1–414

Skelly C, Weinstein P (2003) Pathogen survival trajectories: an eco-environmental approach to the modelling of human campylobacteriosis ecology. Environ Health Perspect 111(1):19–28

Smith P, Martino D, Cai Z, Gwary D, Janzen H, Kumar P, McCarl B, Ogle S, O'Mara F, Rice C, Scholes B, Sirotenko O, Howden M, McAllister T, Pan GR, Omanenkov V, Schneider U, Towprayoon S, Watternbach M, Smith J (2008) Greenhouse gas mitigation in agriculture. Philos Trans B 363(1492):789–813

Soomere T, Viška M, Lapinskis J, Räämet A (2011) Linking wave loads with the intensity of erosion along the coasts of Latvia. Est J Eng 17:359–374

Spalding MD, Ruffo S, Lacambra C, Meliane I, Hale LZ, Shepard CC, Beck MW (2014) The role of ecosystems in coastal protection: adapting to climate change and coastal hazards. https://doi.org/10.1016/J.OCECOAMAN.2013.09.007

Tompkins EL, Adger WN (2005) Defining a response capacity for climate change. Environ Sci Pol 8:562–571

Università Ca' Foscari Venezia (2018) Beaches at risk due to the increase in atmospheric CO_2. Science Daily. www.sciencedaily.com/releases/2018/11/181109101435.htm. Accessed 5 Sep 2020

Vallega A (2002) Coastal management: the integration principle. http://marinespecies.org/introduced/wiki/Some_definitions_of_Integrated_Coastal_Zone_Management_(ICZM)

Varallyay G (2007) Potential impacts of climate change on agro-ecosystems. Agric Conspec Sci 72:1–8

Villanueva AJ, Gómez-Limón JA, Arriaza M, Rodríguez-Entrena M (2015) The design of agri-environmental schemes: farmers' preferences in Southern Spain. Land Use Policy 46:142–154

Vitelletti ML, Bonaldo D (2020) Response to climate change in coastal and marine protected areas: threats and opportunities. In book: Governing future challenges in Mediterranean protected areas (pp 61). CNR Edizioni

Vlek PLG, Le QB, Tamene L (2010) Assessment of land degradation, its possible causes and threat to food security in sub-Saharan Africa. In: Lal R, Stewart BA (eds) Food security and soil quality. Advances in soil science series. CRC Press, Boca Raton, FL, pp 57–86

Weiner J (2003) Ecology – the science of agriculture in the 21st century. J Agric Sci 141:371–377

Wezel A, Bellon S, Dore T, Francis C, Vallod D, David C (2009) Agroecology as a science, a movement or a practice. A review. Agron Sustain Dev 29(4):503–515

Wilby RL, Keenan R (2012) Adapting to flood risk under climate change. https://doi.org/10.1177/0309133312438908

Willows R, Connell R (eds) (2002) Climate adaptation: risk, uncertainty and decision-making. UKCIP Technical Report. www.ukcip.org.uk

Zhang W, Ricketts T, Kremen C, Carney K, Swinton S (2007) Ecosystem services and dis-services to agriculture. Ecol Econ 64(2):253–260

Zomer RJ, Trabucco A, Coe R, Place F (2009) Trees on farm: analysis of global extent and geographical patterns of agroforestry. ICRAF working paper no. 89, World Agroforestry Centre (ICRAF), Nairobi

Climate Change Vulnerability and Agroecosystem Services

6

Arnab Banerjee, Manoj Kumar Jhariya, Shailesh Kumar Yadav, Nahid Khan, Abhishek Raj, Ram Swaroop Meena, and Taher Mechergui

Abstract

The mega event of climatic perturbations has its severe impact on human health and also on the well-being of the global ecosystem. The major issue of changing climate has affected various ecosystems globally in terms of acidification of oceans followed by elevated level of carbon dioxide. It has its severe impacts in various forms of habitat degeneration leading to huge loss of biodiversity. Therefore, there is an urgent need to inventory the climatic risks and its vulnerability issues and their subsequent management for developing ecosystem resiliency toward climate change. Mitigating the changes in the climate solution based upon natural systems needs to be scientifically explored. The present chapter is an attempt to understand the climatic risks and vulnerabilities of ecosystem along with suitable strategies for the effective management of ecosystem change. The chapter concludes by finding the challenging opportunities and research

A. Banerjee (✉) · S. K. Yadav
Department of Environmental Science, Sant Gahira Guru Vishwavidyalaya, Sarguja, Ambikapur, Chhattisgarh, India

M. K. Jhariya · N. Khan
Department of Farm Forestry, Sant Gahira Guru Vishwavidyalaya, Sarguja, Ambikapur, Chhattisgarh, India

A. Raj
School of Agriculture, Lovely Professional University, Phagwara, Punjab, India

R. S. Meena
Department of Agronomy, Institute of Agricultural Sciences, Banaras Hindu University, Varanasi, Uttar Pradesh, India
e-mail: meenars@bhu.ac.in

T. Mechergui
Faculté des Sciences de Bizerte, Laboratoire des RessourcesSylvo-Pastorales de Tabarka, Tabarka, Tunisie

M. K. Jhariya et al. (eds.), *Sustainable Intensification for Agroecosystem Services and Management*, https://doi.org/10.1007/978-981-16-3207-5_6

initiatives toward the issue of nexus between climatic changes and ecosystem vulnerability and risks.

Keywords

Climatic alterations · Risks · Vulnerabilities · Ecosystem change

Abbreviations

AC	Adaptive capacity
C	Carbon
CBD	Convention on biological diversity
COP	Conference of the parties
FEMA	Federal Emergency Management Administration
GHGs	Greenhouse gases
ICZM	Integrated coastal zone management
IPCC	Intergovernmental Panel on Climate Change
IWRM	Integrated water resource management
NPP	Net primary productivity
UNFCCC	United Nations Framework Convention on Climate Change
USA	United States of America

6.1 Introduction

Since the past century, the earth surface mean temperature level has risen with altered precipitation pattern followed by frequent occurrence of extreme events. The level of changes is variable in nature depending upon a case-to-case basis (IPCC 2007). Such changes have posed global threat upon the scenario of ecological process and important ecosystem services (Jhariya et al. 2021a, b; Meena et al. 2018). The most important effect was observed at the species level within the ecosystem. The occurrence, distribution, and pattern of species distribution changed. There is rapid species migration toward the pole, and there is altitudinal migration. Population density and distribution of various species also changed considerably. Alteration in the phenological events has taken place through climatic alterations. Avifaunal and arthropod populations have reflected altered mode of life cycle.

From an ecosystem perspective, there is expansion of ecosystems as well as changes in species composition within the forest ecosystem. However, the impact is moderate in level where less changes have taken place in the pattern of climatic element (Raj et al. 2020; Banerjee et al. 2020). The future projection is much more than the present level. For the terrestrial ecosystem, the reports of the Intergovernmental Panel on Climate Change (IPCC) reflect that up to one tenth of species at the global level is under severe threat of extinction with 1 °C elevation of global temperature. Various aquatic systems and polar and high-altitude biome ecosystems would be specifically vulnerable under climatic perturbations. It was observed that species under montane forest biome are highly vulnerable due to their narrow distribution and lack of long-range dispersal ability as well as other abiotic and biotic factors. The interrelationship of the genetic diversity with the changing climate is yet to be explored properly although it is assumed that it would help to develop climate resiliency of species and ecosystem (Jhariya et al. 2019a, b). Various modeling studies have reflected individual response of species toward climatic alteration which tend to show their impact over the composition of the ecosystems. The proper functioning of the ecosystems would also change. It was reported that in the northern part of Europe, there is an increase in NPP (net primary productivity) followed by a decrease under water stress condition as revealed from the modeling studies. The productivity of the ecosystem would also alter in association with the changing climate due to the alteration in the fall of litter. Changing climate would also have its impact on various ecological services making the ecosystem much more vulnerable (Fig. 6.1).

Ecosystem change associated with climatic perturbations would also have negative consequences over the socioeconomic condition of the community stakeholders. The effect may be drastic enough causing irreversible change of the ecosystem whose function would be abruptly inhibited. Assessment of vulnerability of ecosystem is a big issue to deal with on a global perspective. Although vulnerability assessment focuses on socioeconomic and natural hazard perspective, it can be used to assess the impact of changing climate on ecosystem dynamics. Various modeling approaches were used to assess the climatic variability on the species as well as on the ecosystem. Therefore, vulnerability of the ecosystem under the threat of climate change needs to be assessed from risk, adaptability, as well as tenure of exposure perspective.

6.2 Concept of Vulnerability Assessment

As per IPCC (2001), the assessment of vulnerability to ecosystems depends upon three principal factors which include the exposure of the ecosystem to particular climatic extremes, sensitivity assessment of the ecosystem toward changing climate, and ecosystem resiliency through adaptability with the changing climate (Fig. 6.2).

6.2.1 Ecosystem Exposure to Climatic Perturbations

It refers to the alteration in the climatic elements to which a particular ecosystem is exposed. Dawson et al. (2011) argued that the consideration of indices for the suitability of habitats should be the major component of vulnerability assessment. Alterations in the climatic elements are assessed and summarized under the exposure assessment. Similar exposure assessment studies were done by Lal et al. (2002) in the case of island states. The concept of exposure assessment was broadened from

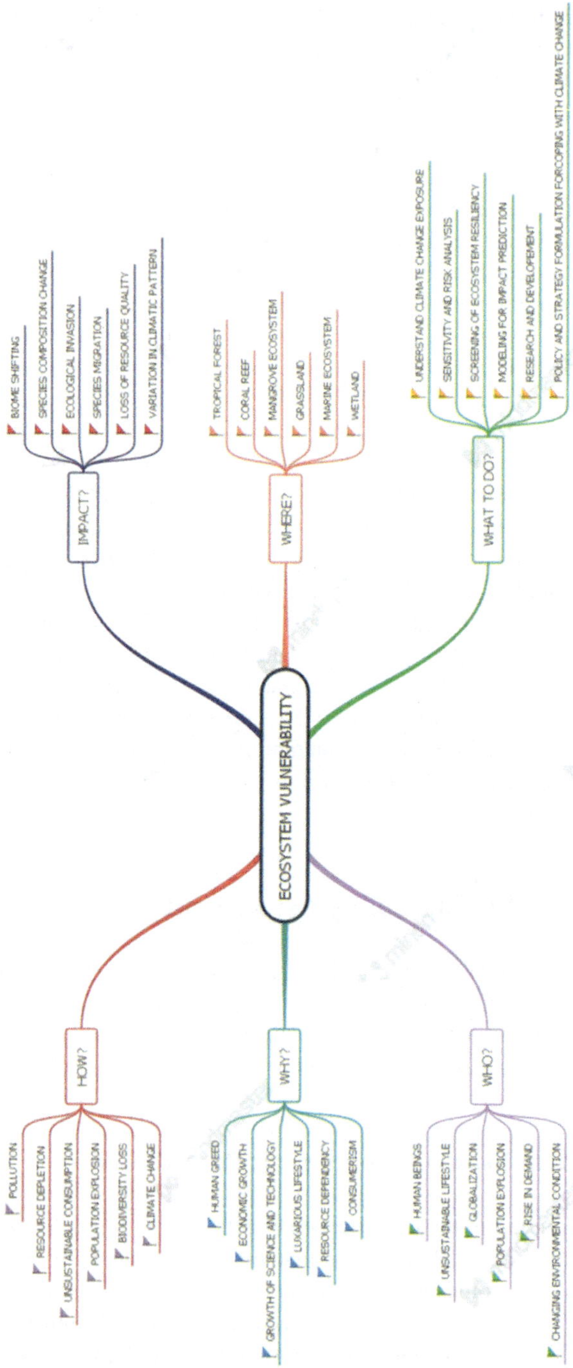

Fig. 6.1 Ecosystem vulnerability toward climate change

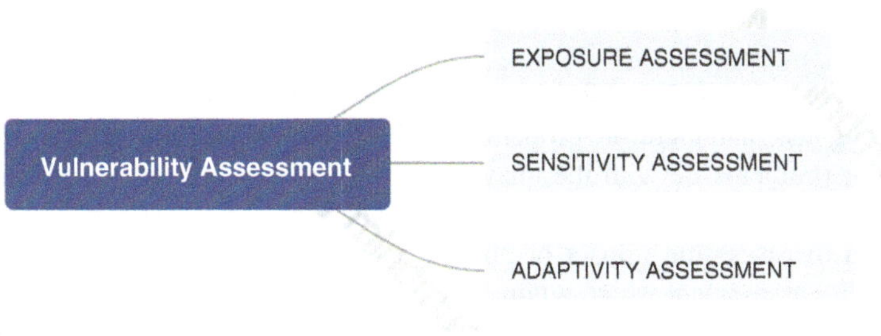

Fig. 6.2 Basic concept of vulnerability assessment of ecosystem

the perspectives of vulnerability of ecosystems as it includes climatic changes as a stress factor. Under such circumstances, various indicators were used, apart from climatic changes. Thus, various modeling approaches were used and framed for exposure assessment.

6.2.2 Analysis of Ecosystem Sensitivity

Sensitivity of ecosystem toward climate change is usually measured through the changes in the environment with response to the alteration in climatic elements. For example, rise in sea level is associated with rise in temperature. Sensitivity tends to have a higher variation in comparison to exposure. Further, the concept of exposure and sensitivity has been combined in the form of potential impact for better interpretation of ecosystem sensitivity (Metzger et al. 2005). Various forms of modeling approach are very much useful for the analysis of climatic sensitivity of ecosystems. Interaction between ecosystem components needs to be analyzed properly for finding out better outcomes.

6.2.3 Adaptive Capacity of the Ecosystem Toward Climate Change

In the process of vulnerability assessment, determining the adaptive capacity (AC) is the most important factor under the impact of climate change. Basically AC is the combination of all various factors that determine the ecosystem ability to develop the resiliency in the climatic changes. These very changes would be dependent upon the variation of the local environment. For determining AC at the species level and at the ecosystem level, it requires a complex database of species interactions, its life cycle, and their functional role within the ecosystem (Khan et al. 2020a, b; Meena and Lal 2018). The major factors used for determining the AC include species response to climate change, migration of species, and existing climatic perturbations in the surrounding environment. Czucz et al. (2011) mentioned about the interrelationship between habitat fragmentation and the connectiveness as the key for estimating the AC for natural ecosystems.

6.2.4 Ecosystem Vulnerability

Under the context of vulnerability assessment, assessing ecosystem vulnerability is a challenging task, and therefore proper methodology for its assessment is yet to come. The major problems lie with the unavailability of proper indicators for ecosystem vulnerability assessment. Considering the human factor as an external issue for the ecosystem to assess the impact or change associated with ecosystem is used for vulnerability assessment which ignores the inherent change in the ecosystem. Sometimes on the basis of species extinction from a particular ecosystem, the vulnerability assessment is usually done. However, different workers have adopted different approaches for vulnerability assessment in various times which have been represented in Table 6.1.

However, vulnerability assessment of ecosystems relies mostly on assessing the various form of ecosystem services. Such assessments can be done through applying modeling approach. This may not give fruitful results as it is regulated through the dynamics of ecosystem which cannot be demarcated properly.

6.3 Climate Change and Ecosystem Degradation

Ecosystem change in the form of degradation is a major issue that alters the ecosystem homeostasis at the global level. Extreme climatic conditions have resulted in increased frequency of the natural hazards (Banerjee et al. 2021a, b). This causes lesser resiliency of the ecosystem followed by a higher risk of natural disaster. The degradation of the ecosystem attribute lies with the transformation of the ecosystem from C (carbon) sink to C source (Khan et al. 2021a, b; Kumar et al. 2021).

Climatic perturbation tends to influence the well-being of the society as well as the lives of the people through extreme climatic events. Further, this shrinks the scope of resiliency of the ecosystem toward climatic vulnerability. As per the reports of IPCC (2007), the incidents of natural hazards have increased significantly since 1990. The problem is that the increase of frequency of natural hazards has risen along with increasing temperature in the past few decades but this trend would continuously rise even if reduction in the GHGs (greenhouse gases) was undertaken by the society. This would make the entire human society much more vulnerable under the climatic risks.

As per the IPCC Fourth Assessment Report, the future projection reflects the ecological overshoot condition as well as overcoming the resilient capacity of the ecosystem by the end of this present century through the mega event of climate change. Most of the ecosystems of the tundra region, coastal region, and Mediterranean climatic ecosystems are mostly vulnerable in front of the climate change event. Structural and functional alteration has taken place from an ecosystem perspective. It was assumed that 2 to 3 °C rise in surface temperature would cause one third proportion of species loss. Further, increase of temperature up to 4 °C would lead to the transformation of more than one fourth part of the ecosystem triggering a mass extinction event.

Table 6.1 Approaches on ecosystem vulnerability assessment in relation to climate change

Approaches	Details	References
Case study-based approach	Various types of ecosystems	Buytaert et al. (2011); Thomas (2008); Lasco et al. (2008)
	Assessment of forest vulnerability in Canada	Ogden and Innes (2007)
	Changing climate and forest adaptation in Australia for protection to bioresources in Australia	Williams (2000)
Statistical interpretation-based approaches	Species and ecosystem distribution modeling for assessing climatic requirements	Holdridge (1947)
	Maximum entropy modeling	Phillips et al. (2006)
	Habitat prediction for future bullfrogs in South America	Nori et al. (2011)
	Assessment of parkland agroforestry systems in west African Sahel region	Luedeling and Neufeldt (2012)
Indicator-based approaches	ATEAM project for vulnerability assessment through proxy indicators	Rounsevell and Metzger (2010)
	Use of socioeconomic indicators for vulnerability assessment of states of India	Malone and Brenkert (2008)
	Hydrological indicators for predicting suitable habitat condition for British salmon fishes in river ecosystems	Walsh and Kilsby (2007)
	Vulnerability assessment of Nyala antelopes of national park in Malawi by using ecological indicators	Mkanda (1996)
Mechanistic modeling	Assessment of coastal retreat by modeling coastline elevation and rise in sea levels	Chu-Agor et al. (2011); Boateng (2012)
	Modeling of NPP, ozone level, and climate change in the republic of China	Ren et al. (2011); Imhoff and Bounoua (2006); Moldenhauer and Lüdeke (2002)
	CropSyst, InfoCrop, and CERES-maize models for assessing crop vulnerability	Aggarwal et al. (2006); Tingem and Rivington (2009); Tingem et al. (2009); Makadho (1996)
	IBIS model for assessing forest vulnerability toward climate change in India, PICUS model for Austrian forests, and LPJGUESS model for global assessment of dynamics of vegetation	Gopalakrishnan et al. (2011); Seidl et al. (2011a, b); Gritti et al. (2006); Seaquist et al. (2009); Scholze et al. (2006)
	WETSIM and WETLANDSCAPE for wetlands and savanna ecosystems	Johnson et al. (2005, 2010); Christensen et al. (2004)
	Integration of modeling approach by combining CENTURY crop-ecosystem model and econometric model, for studying impact of climate change on basin area in California	Antle et al. (2004); Quinn et al. (2001)

(continued)

Table 6.1 (continued)

Approaches	Details	References
	Conceptual model of the factors leading to recent re-greening in some parts of the Sahel agent-based modeling for predicting impact of changing climate and policy formulation toward watershed management in Oregon	Sendzimir et al. (2011); Nolin (2012)

Moreover, the degradation of the ecosystems under the influence of climatic extremes would become more severe reducing the buffering action of the ecosystems followed by the frequent occurrence of natural calamities (Banerjee et al. 2021c, d). Further, loss of biodiversity and gene pool would make the ecosystem more climate vulnerable by the alteration in the trophic level, food web, and food chain, and the ecological invasion would predominate (Raj et al. 2018a, b). Thus, the vulnerability of the natural ecosystem tends to increase under climatic extremes.

6.4 Climate Change and Ecosystem Vulnerability

Alteration in the climatic elements leads to alteration in the structure and species composition (Winkler et al. 2012). It is a well-known fact that the ecosystem is a collection of various flora, fauna, and other living organisms which may respond to climatic changes individually leading to total alteration in the ecosystem. Changes within the ecosystem type will depend upon the local condition of the area followed by the existing environmental conditions. For assessing the long-term trends of the ecosystem change, simulation of the responses of the ecosystem under projected climatic changes can be used. MC1 dynamic model was used by Lenihan et al. (2008) who reported the associated changes in the classes of vegetation due to changing climate and fire impact. Kunkel (2011) reported loss of boreal forest and its gradual change into deciduous forest of temperate region in the Midwest region. The boreal forest ecosystem will be replaced by a further type of grassland ecosystems under the bane of climate change. Various researches on the Midwest zone have revealed the expansion of savanna ecosystem followed by its conversion into various forms of grassland ecosystem (Lenihan et al. 2008; Frelich and Reich 2010). In the process of ecosystem conversion, species tend to migrate from one place to another leading to the formation of newer communities (Davis et al. 2005).

It was observed that species respond specifically against the climatic perturbations which cause an increase in the density of dominant species at the cost of depletion of non-dominating species (Morin et al. 2008). Prasad et al. (2007) reported the migration of various tree species due to changing climate in the eastern part of the USA. Other workers measured various ecological processes associated with the migration of tree species (Morin et al. 2008). The prevailing local climatic conditions and different forms of habitat degradation will govern and regulate the

Table 6.2 Methodology for ecosystem vulnerability assessment toward climate change

S. no.	Methodology	References
1.	Variation of net primary productivity (NPP) on time scale	Wu et al. (2007); Zhao and Wu (2014)
2.	Track the changes in vegetation	Gonzalez et al. (2010)
3.	Assessment of vegetation intactness and prediction of future trend under the forefront of climate change	Watson et al. (2013); Segan et al. (2016)
4.	Relative species richness	Eigenbrod et al. (2015)
5.	Shifting of biome	Gonzalez et al. (2010)
6.	Stability of vegetation	Xu et al. (2020)
7.	Productivity of vegetation	Beck et al. (2011)

occurrence and distribution of species. Various research works have been undertaken by several research workers on variable basis which has been mentioned in Table 6.2.

6.4.1 Vulnerability of Terrestrial Ecosystems Toward Climate Change

Terrestrial ecosystems include the land-based ecosystems with various forms of vegetal cover in the form of agroecosystem, forest, urban ecosystem, etc. The level of vulnerability would be different for diverse ecosystems depending upon the ecosystem dynamics and influence of various biotic and abiotic factors. The activities of human beings would put pressure on the various terrestrial ecosystems through altered land use practices (Raj et al. 2019a, b). This would cause drastic loss of biodiversity in the form of patchy distribution of species followed by reduced migration rate under the stress of climate change. The day-to-day event of climatology has little impact on the terrestrial ecosystem. However, the issue becomes complex when the frequency and duration of climatic event increase which makes the ecosystem more vulnerable to climatic vagaries. With more frequent occurrence of climatic extremes, the system gets lesser time for adjustment with the changing climate. The occurrence of multiple event tends to cause total destruction of the adaptive capacity of the ecosystem leading to the formation of new ecosystem with different species composition in comparison to the past. According to one case study of Yellowstone National Park, the increased forest fire frequency would severely affect the vegetal component with very few remnants of seeds of the indigenous species. Further, due to low seed content, the regeneration of indigenous species is difficult after post-burning time period (Turner et al. 2019). The problem becomes further worse as the climatic change of the local climate has made the regeneration process unfavorable. The present conditions under which the forest will regenerate also differ from their previous state because of climate change. The migration of species across various directions is also very evident under the mega event of climate change (Fei et al. 2017).

According to a report, globally 11% of area is crop land followed by 22% pastureland (Ramankutty et al. 2008). At present times, larger production within a small area has taken place with gradual growth and development of agricultural technology. This is creating the nuisance of agricultural pollution and land degradation, and therefore the current agroecosystem is running deficit to feed the growing human population. So, conversion of land uses is taking place very frequently (Molotoks et al. 2018).

The nexus between the agroecosystem and climate change at the global level is making the issue much worse. Agricultural practices are acting as the drivers of climate change. As per the research reports, agriculture along with other types of land uses accounts for one fourth of the global GHG emission (Molotoks et al. 2018; Smith et al. 2014; Sheoran et al. 2021) which clearly reflects that these activities are promoting climate change although agroecosystem is affected by such events. However, agroecosystem has the capability of combating climate change through CO_2 uptake (Kumar et al. 2020).

6.4.1.1 Climatic Vulnerability and Urban Forests

The major perspective includes the magnitude of higher impacts over urban forests. Urban forests are basically the managed forest ecosystem that occurs in and around the cities. In the Midwest region of the USA (United States of America), 33.2% of tree cover is present (Nowak and Crane 2002) which stands to be the second for most of the region in the USA. It was observed that forests in the urban setup tend to occupy as an admixture of ornamental and ground vegetation (Woodall et al. 2010). The biomass content is much lower than the normal forest area. Urban forests tend to regulate the climate of cities and towns and also serve the aesthetic and amenity values for the urban people (Younger et al. 2008). Changing climate would have both direct and indirect impact over the urban forest (Khan et al. 2020a, b). Climate change will act as a stress factor over these urban ecosystems leading to its further degeneration (Roloff et al. 2009). Major consequences include pest and disease prevalence followed by the inhibition of pollination with higher occurrence of heat waves. Additional stresses include ambient air pollution, conversion of urban areas into heat island, altered hydrology, etc. The ecological range of tree species has shifted from their original place. However, such shifting mechanism helps to act the urban forest as refugia and paves the pathway of future dispersal range (Woodall et al. 2010).

6.4.2 Vulnerability of Freshwater and Marine Ecosystems Under Climate Change

From the global perspective, only a small portion of water is available in the usable form. This small fraction of freshwater tends to support 10% of the fauna of global aquatic ecosystems. Gradual and irreversible changes in the aquatic ecosystem are the inevitable truth of the climatic extremes. The distribution and ethology of animals of the aquatic system have changed in the aquatic system. Changing climate influences the climatic elements and leads to occurrences of events such as flood and drought which also hamper the integrity of the ecosystem. Under the influence of climatic perturbations, the freshwater ecosystems also get various forms of problematic issues in the form of water pollution, incidence of ecological invasion, anthropogenic activities such as the dam construction, and fishing on the freshwater habitat. According to a research database, the problem of water scarcity has engulfed four billion people across the globe (Castello and Macedo 2016; Mekonnen and Hoekstra 2016; Meena et al. 2020).

6.4.2.1 Marine Ecosystems

Climatic alterations have also impacted the oceanic ecosystems through frequent occurrence of heat waves leading to the warming of the oceanic water temperature. This has led to heat stress for the aquatic organisms (Oliver et al. 2018). The warming of the water surface at the deep ocean and nutrient pollution on the shoreline have depleted the oxygen creating inhospitable condition for the aquatic organisms (Breitburg et al. 2018). Absorption of atmospheric CO_2 by the oceanic surface is also leading to ocean acidification which hampers the shell formation process. Sea level rise due to melting of polar glaciers under rising temperature level has caused severe damage to all the coastal ecosystems and its surrounding communities. Coral reef ecosystems have become extremely vulnerable under the event of ocean acidification (NASEM 2018). The temperature extremes have promoted coral bleaching which is actually the death of the corals. Once the process has initiated for a particular coral reef system, it is subjected to further degradation at a faster rate (Castello and Macedo 2016; Hughes et al. 2018). Various coastal ecosystems have become mostly vulnerable under the climatic changes followed by depletion of almost half of the organisms of marine population (WWF 2015). Habitat loss has become frequent in the form of various coastal ecosystems. Various anthropogenic uses of the marine ecosystems in the form of aquaculture practices, waste disposal ground, transportation, and tourism activity have become a potential threat for marine ecosystem (Danovaro et al. 2017).

6.4.2.2 Coral Reef Ecosystem

Coral reefs are the unique ecosystem which tend to remain in and around the continental shelf region of the ocean ecosystem. These ecosystems occupy a very small geographic area of the oceanic ecosystem ($>1\%$) representing a higher level of oceanic diversity of marine species across the globe (Reaka-Kudla 1996). Coral reefs perform various important ecosystem services in the form of protection from water erosion and act as a breeding ground for various marine fishes. The vagaries of climate change in the form of elevated level of temperature and rising CO_2 concentration followed by sea level rise alter the environment of coral reef ecosystem leading to coral bleaching and death of the coral species. As per the scientific reports, increase in water temperature acts as a heat stress for them which ultimately leads to the death of the corals (Hoegh-Guldberg 1999). Thus, warming of the water has elevated the level of bleaching of corals in the past few decades. During 1997–1998, a temperature rise of up to 6 °C has caused bleaching up to 90% level. It is much more prevalent in the years of El Niño (Wilkinson et al. 1999). Hoegh-Guldberg (1999) mentioned the elevated level bleaching frequency due to temperature increase. Future prediction of the rise of bleaching frequency has been reported by various workers. This trend is so higher that it overshoots the regeneration capacity from bleaching of corals (Hoegh-Guldberg 1999). Reduction of the calcification process of corals was reported by various workers due to the rise in atmospheric CO_2 level (Kleypas et al. 1999). Further, a 30% decrease of calcification process has been predicted till 2100. Research reports further reveal that up to 20% decrease in calcification process hampers the distribution of corals in higher latitudes (Kleypas et al. 1999).

6.4.2.3 Mangrove Ecosystems

Mangrove ecosystems reflect a smaller distribution pattern. They have their unique biodiversity pattern and help in storm protection of the coastal areas. Climate change has a drastic impact on the mangrove ecosystem in terms of salinity level and inundation of mangrove ecosystem through sea level rise. According to one prediction, a 45 cm rise in sea level would inundate two thirds of the Sundarbans area of the Gangetic delta. Further prediction says that a 1 m rise would completely engulf the Sundarbans area. Habitat loss by inundation may inhibit the species migration pattern due to the anthropogenic encroachment of the natural habitat. Higher level of loss of mangrove forest has taken place during the twentieth century due to anthropogenic encroachment. In the African continent, similar loss of mangrove forest was reported from Jamaica. However, to some extent, mangrove can combat the climatic changes. The sediment flux of the coastal zone determines the fate of mangrove ecosystem in front of climatic perturbations. As a consequence of that, there is no uniformity of the fate of mangrove under changing climate. However, migration of mangrove species toward the land surface was recorded in some cases.

The small islands are particularly vulnerable under the threat of climate change. Rise in sea level leads to the total inundation of the small islands. Frequent coastal and sea storms may also impact the small island surfaces. Most of the economic and anthropogenic activities in these coastal region take place in the coastal area making the situation more worse (Hay and Kaluwin 1993). Various natural hazards and coastal processes are very much problematic for these small island ecosystems. Such processes would have its impact on the economy of the islands, and their existence would be under question. Plant species of these island are mostly endemic in nature, and the avifaunal species belong to the threatened category (Nurse et al. 1998). Climatic changes would lead to change in species composition of these island ecosystems.

6.5 Climate Change and Species Vulnerability

In the upcoming decades, changing climate will have its severe impact on the biodiversity at the global level, and there would be a high rate of species extinction (Loarie et al. 2009). Response from the biotic components of the ecosystem takes place on a spatiotemporal scale across the various habitats (Staudinger et al. 2012). The phenology and the life cycle events of organisms within the ecosystem also change under the influence of changing climate. This, on the other hand, would have its impact over spatial distribution and existence of the living community. The effect is further aggravated at the trophic level, community shifting, and inhibiting beneficial species interactions (Miller-Rushing et al. 2010; Meena et al. 2020a). Overall the impact is observed on the rate of productivity and reducing the product output for human uses (Mace et al. 2014).

The type of exposure to climatic perturbations, the sensitivity of the organisms toward the climatic extremes, and the level of adaptability of the ecosystem toward changing climate determine the level of vulnerability at the species and ecosystem level (Glick et al. 2011). Therefore, proper identification of the traits that makes the species or ecosystem vulnerable to the changing climate needs to be done at utmost priority. Apart from climate change, the various ecosystems on the earth surface are highly vulnerable to various processes such as altered land uses, ecological invasion, as well as environmental pollution (Staudt et al. 2013).

To properly assess the species vulnerability, historical account of the species distribution is required. From these databases, it would be clear to us about their past experience or response toward climate change. Further, this information would be helpful to assess the level of vulnerability of the species, habitat, and ecosystem toward changing climate. The interrelationship between the various environmental factors followed by altered land use pattern, habitat disturbances, and various ecological processes with the changing climate needs to be known properly in order to assess the vulnerability properly. The direct impact of changing climate on specific species which are vulnerable, endangered, or in the verge of extinction needs to be studied. Further, the nexus between biodiversity, ecosystem, and services needs to explored properly. The two major aspects of impact identification followed by resiliency and sustainability of the ecosystem need to be explored properly (Raj et al. 2021). Predictive models can be used for the screening of climate-resilient ecosystems, species, and habitat. Link models can be used for spatiotemporal alteration in the flora and fauna.

From the past experiences of climatic changes, it is revealed that temperature alteration inhibits the migration of tree species (Davis et al. 2005). Modeling approaches have predicted the low rate of colonization in the eastern part of the USA from their normal range of distribution for the upcoming 100 years (Iverson et al. 2004). Loss of habitat and habitat fragmentation inhibit the rate of tree migration.

Iverson et al. (2004) made one projection that 15% of habitat would be occupied by five tree species in the upcoming 100 years (Kunkel 2011; Scheller and Mladenoff 2008). The impact of changing climate has also led to range contractions of species (Zhu et al. 2011). There would be a shift of about 400–600 km regarding the species distribution within a century which would be not enough to combat the impact of climate change (Prasad et al. 2007).

The mega event of changing climate is imposing a significant level of influence over the world's biodiversity and various ecosystem services. The response of the ecosystem and species tends to vary site-wise. From an animal perspective, warming of temperature has caused changes in the behavioral attitude in the form of alteration in feeding behavior, altering the biological clock, or their geographical distribution. The speed of adaptation of species reflecting variable response toward climatic perturbations tends to be very slow.

6.6 Climate Change and Ecosystem Services

Ecosystem as a unit performs various essential services for the humankind. Climate change would have its impact across various dimensions over these ecosystem services. It was observed that the supportive services for food production tend to alter under the pressure of the changing climate. Any climate-induced changes in the agriculture system of the USA would have its impact over the world in terms of price hikes and reduce the standard of living (Parry et al. 2007). As per the predictions given by Nelson et al. (2009), the decrease in the productivity of the agroecosystem would cause major price hikes for the staple food crop till 2050. The warmer climate and increased temperature would reduce the yield of the crops significantly (Schlenker and Roberts 2009). Further, inadequate irrigation facility followed by irregular pattern of rainfall and rise of temperature would hamper the production process (Alston et al. 2010). The cost of some of the crops would be high enough for cultivation purposes under the vagaries of climate change. The required environmental condition for the cultivation of specific crops would become difficult under the changing climate (Luedeling et al. 2011). Various workers have reported about the degeneration of the quality of wine grape in the USA under the pressure of climatic perturbations (Hannah et al. 2013; Lobell and Field 2011).

From a management perspective, coping with climate change requires improvement of the cropland fertility status and health (Cong et al. 2014). Simultaneously, it would also help toward reducing the negative impacts of modern synthetic agriculture practices (Bossio et al. 2010). In the case of mismatch between water demand and supply in the agriculture sector, it would reduce the yield and productivity of the agroecosystem and necessitate more requirement of water. However, this problem could be addressed by capturing the rainwater during the gap of no cultivation and then use the same during the cultivation practices (Baker et al. 2012). The storage area of water may be the wetlands and the ponds. This type of approach would provide secondary ecological services in terms of improving the wetland ecology and increasing the wildlife diversity.

Wildfire is a mega event that is destroying the forest area with massive deforestation throughout the world. There is sufficient economic loss due to forest wildfire (Bowman et al. 2013). Climatic perturbation has increased the frequency and intensity of forest fire and has gradually brought more area under the adverse impact of forest fire (Westerling et al. 2006; Jhariya 2017a, b; Jhariya and Singh 2020, 2021). According to one case study from Arizona, the thinning practice has reduced the impact of wildfire and other ecosystem services (McRae et al. 2001).

Marine fishery is an important ecosystem service that provides earning for millions of people across the globe. Most of the marine fishery practice usually takes place in colder temperature. Under the era of climate change, the warming of the ocean surface water makes the oceanic species migrate toward the pole (Ruckelshaus et al. 2013). Use of various advanced technologies would help in the proper management of the fishery practices and also to understand the climate-oriented risks of the stocks of fishery resources (Link et al. 2011). Suitable policies could be framed for the better management of the marine habitat and juveniles of the species and develop resistance against the climatic perturbations (Perry et al. 2010). The livelihood of the local people is under severe threat as they usually sustain through the income coming from fishery practices (Nye et al. 2009). Further, the problem is aggravated through poleward migration of species due to ocean warming. Species having low dispersal ability or unable to migrate comes under the verge of extinction due to changing climatic condition (Cheung et al. 2010; Shanks 2009). The coral reef ecosystem and mollusk population are under severe threat of extinction due to climatic extremes. However, biomass contribution by the aquatic vegetation would help in the survival of fish species (Branch et al. 2013).

The climatic changes would alter the local hydrological process of an area (Staudinger et al. 2012; Karl et al. 2009). As a consequence, the problem of water stress is being felt all over the world. As water is an essential commodity for the people, these changes would have significant impact over the humankind. Various parts of the USA would feel the water stress condition due to climatic alterations (Walker et al. 2011). It was estimated that the storage stock of snow water would show a declining trend till 2040 in the USA (Adam et al. 2009). Other associated factors such as the population pressure and various forms of human use of water may aggravate the problem of water stress under changing climatic condition. The USA would be under higher sustainability risk in terms of water availability till 2050. There may be up to 14-fold increase in the sustainability risk with respect to water resource till 2050 (Roy et al. 2012). Sustainable water management, development of water-efficient technology, and framing of suitable policies toward the conservation of water in the form of water markets are the timely requirements (Gober et al. 2013). Changing climate will alter the functions of riparian ecosystems as well as deplete other natural resources.

On a broad scale, climate change would severely impact the recreation and amenity values of the ecosystem. Due to the rising temperature and sea level, recreation value of sea beaches would also decline along with less fishing activity (Pendleton et al. 2011). The amenity value of North Carolina sea beach reduced up to $1 billion within a span of 74 years (2006 to 2080) (Whitehead et al. 2009). Loss of various indigenous fish species would take place under the climatic extremes, and various exotic species would replace them under the changing environmental condition (Moyle et al. 2013).

Impact of changing climate has reflected significant level of variation on a regional basis. The impact is important from producer perspectives as they only are capable of producing foods and form the base of ecosystem food web. Researches on coastal systems have revealed that the rising of water temperature of areas under tepid sea inhibits the oceanic circulation of nutrients which hampers the productivity of the marine ecosystem. On the other hand, under cold region, warming of temperatures of surface water tends to enhance the productivity of the particular ecosystem. The variable response of species toward changing climate would change the structure and function of a particular ecosystem. The pattern of climatic extremes is also imposing severe impacts over the ecosystems. Alteration in the level of primary production may aid in maintaining the sustainable aquaculture practices.

6.6.1 Water Resources

Forest watersheds tend to supply safe water for the biota of forest ecosystems. Changing climate would adversely impact the forested watersheds and disable its clean water supply facility. Forests play various vital ecosystem services in terms of maintaining hydrological cycle (Furniss Michael et al. 2010). Management of forested watershed would provide safe and clean water. Through research, it was found that forested watershed has a key potential toward supply of drinking water and thus helps to reduce the cost for treatment of polluted water. It was observed that decontamination of water sources is much more fruitful than the various treatment facilities (Ernst et al. 2004).

The nexus between the changing climate, habitat loss and fragmentation, and loss of forest cover would cause decline in the species diversity through rapid and severe extinction mechanism. Further, if the migration in those areas is inhibited, there would be no proper successful colonization of the migratory species. The capacity of the forested watersheds in water purification process would be governed by the influence of changing climate. This would regulate the quality of municipal water supplies at different regions of the globe. Rainfall pattern of a particular area will determine the fate of the water quality from these forest ecosystems. Rising rainfall in some areas would create problem for the forested areas to provide clean water as they would be incapable of absorbing the excess amount of rainfall. Proper distribution of rainfall and maintenance of the hydrological cycle are key ecosystem services which require regular monitoring and successful evaluation for the upcoming times.

6.6.2 Carbon Storage

Forest acts as C sink (Flickinger 2010; Raeker et al. 2010). C stock and storage in forest is a valuable function toward combating the climatic perturbations. Variation in terms of C stock and storage under the climatic perturbations is prevalent which is a key ecosystem service for the well-being of the humankind. There would be a mixed impact of increase and decrease of C stock in various areas under the changing climate. As per the research reports, it was observed in the USA, on an annual basis, the amount of GHG concentration would reduce up to 10% of emission from the fossil fuel (McKinley et al. 2011). Further, it was observed that alteration in the land use would hamper the aboveground C storage as well as alter the distribution pattern of the C pool (Rhemtulla et al. 2009).

6.6.3 Episodic Disturbances

Various forms of climatic perturbances as well as other biotic effects result in significant loss of C (Lenihan et al. 2008). Suppression of fire leads to the reduction in C loss up to 6%. Deforestation and forest fire tend to reduce C storage and productivity of the forest ecosystem which is further aggravated by the issue of changing climate (Gough et al. 2008). However, the results may vary on a case-to-case basis (Scheller and Mladenoff 2005). Insect and disease outbreak may hamper the C stock and storage development of forest ecosystem (Hicke et al. 2011). Recovery from this disturbed state would take much longer time.

6.6.4 Ecosystem Productivity

Climatic perturbations have a significant level of influence over the productivity of the forest ecosystems. Researches on open forest stands have revealed greater response of the forest ecosystem under the elevated level of CO_2 which may increase the net primary production (Norby and Zak 2011). A 25% increase in biomass production in different forest types of northern Wisconsin was reported by Ainsworth and Long (2005). Higher NPP were recorded from the northern part of Wisconsin as reflected from the higher growth rate of cherry and oak species (Chiang et al. 2008).

6.6.5 Recreational Opportunities

Forest becomes a significant spot of recreation in terms of biodiversity and wildlife. Climate change would hamper these ecosystem services through the alteration in the timing for recreation purposes. It was observed that recreation in forest areas is solely season based. From the climate change perspective, it was observed that a significant level of change has taken place over the past century, and from future perspective, climate change would generate summer of longer duration followed by short-duration winters. So, there would be a huge shift in the seasonal pattern of the world (Andresen et al. 2012). This may influence the recreation and tourism activities in the forest area during the winter season. However, it may bring some benefits in the form of recreation which is nature based (Dawson and Scott 2010). Various winter-based recreation activities would be reduced due to climatic perturbations (Notaro et al. 2011). Climatic extremes and variation would reduce the quality of recreation facility through lesser visits in the recreation site.

6.6.6 Cultural Values

Human beings are intricately associated with the nature in terms of culture. Climate change may impose significant influence on the cultural aspect and may be modified. Therefore, the cultural aspect would be under severe risk due to climate change.

6.6.7 Climate Change, Ecosystem Service, and Corporate Sector

In various locations of the USA, flooding hazards were significantly mapped by FEMA (Federal Emergency Management Administration), and they are properly assessing proper use of funds in eco-restoration practices (King 2013). Recognition of interrelationship between the changing climate and its impact over the various ecosystem services is now being widely acknowledged in the corporate sector also. Corporate institutions are recognizing the importance of maintaining the ecosystem services in front of the challenge of the climate change in relation to their profit which very much depends upon the ecosystem services without climatic perturbations. Therefore, they are developing climate-resilient practices and adopting the strategies for mitigating the climate change (Lydenberg et al. 2010; Meena et al. 2020b). Companies such as Coca-Cola are now relying on the nature's inherent ability of water purification before using them in the bottling plant as ecosystem service. The company is trying to capitalize the ecosystem service of water purification through natural process in terms of reducing their costs in maintaining water quality. As a major importer of the maize and other allied crops for the production of sugar, the company is monitoring the impact of changing climate over the cultivation practices of this particular crop along with their economic potential in the market mechanism (BSR 2013). Similar assessment of risk associated with flood is usually being monitored by the Dow Chemical Company of Texas. Both these companies have taken the help of reputed NGOs regarding the preparation of balance sheets for them for the estimation of climate risks associated with their production potential. This type of corporate mechanism has helped in cost reduction in terms of fuel load based on weather condition and forecasts as revealed by Qantas Airways Limited of Australia. Similar trend was observed in the cotton industry of Australia (Anaman et al. 1997).

6.7 Challenges of Ecosystem in Climate Change

Changing climate has a diverse impact over the terrestrial and aquatic ecosystem along with interaction through diverse climatic elements and different types of ecosystems. The major aspect includes to determine the appropriate climate-induced changes within the ecosystems and determine the stress threshold for irreversible change along with trophic interactions among the diverse organisms involved in the ecosystem. It was observed that the alteration in the ecosystem is taking place in different ecosystems under the influence of climatic elements and other natural

Table 6.3 Researches on vulnerability assessment of ecosystem on climate change

Research area	Reference
Impact of climate on mountainous region of Rwenzori between Congo and Uganda border	Eggermont et al. (2010)
Landslide hazard estimation of Mt. Elgon region	Claessens et al. (2007)
Assessment of impact of previous storm experiences and future prediction of tropical storm over forests of Taiwan	Lin et al. (2011)
Vulnerability assessment of hurricane at Mexico	Alayón-Gamboa and Ku-Vera (2011)
Assessment of interrelationship between grazing and climate change	Pyke and Marty (2005)
Climatic factor manipulation in Mediterranean Macchia ecosystem	Ripullone et al. (2009)

events. The response of the ecosystem is the cumulative outcome of the biotic and abiotic interactions that takes place within an ecosystem. Various research works have been reported in relation to vulnerability assessment of ecosystems under the changing climate (Table 6.3).

Assessment of changing climate on various properties and functions of ecosystem along with its impact on various ecological communities is a herculean task. Turner et al. (2020) tried to establish the interrelationship between the climatic extremes and the associated irreversible changes within the ecosystem. Change detection is a big issue as the climatic extremes are beyond the predictable limit. Further, some components of the ecosystem tend to be highly susceptible under changing climate. Further, the components that frame the ecological system tend to make them resilient toward the climate (Bardgett and Caruso 2020).

The life history and food web pattern of the soil biota tend to develop resiliency in terms of ecological functioning. For example, bacteria are much more efficient in nutrient cycling in comparison to fungi and hence help to develop resiliency under changing condition. On the other hand, fungi develop resistance to change. Therefore, such different modes of operations can help for the ecological restoration of habitat from climatic alterations or may destabilize the system. Therefore, climatic perturbations that have an impact over soil biota on a long-term basis are yet to be understood properly. Further, response toward climatic extremes and potential for sudden change remain as a big knowledge gap in this aspect. Ancient studies related to paleo-ecology revealed the influence of fire event to the alteration in the species composition in the north and south hemispheres across the world (Iglesias and Whitlock 2020). However, there was involvement of various differential factors. The changing climate significantly influences the loss of biodiversity as well as C storage potential through altered land use practices. Molotoks et al. (2020) used modeling approach to evaluate the degeneration of various ecosystem services due to increase in agriculture production in Latin American countries.

6.8 Management and Policy Implications Toward Reducing Ecosystem Vulnerability and Combating Climate Change

Climatic alterations that have led to severe changes in global climate ecosystem services appear to be very critical in adapting and reducing risks toward vulnerability. Various natural and ecological processes often create a challenge for the effective management of the scenario of changing climate. Various ecological services form the central point of ecosystem management through adaptation toward changing climate. Proper management at the ecosystem level helps to develop the resiliency among the various ecosystems under the changing climatic condition. Management options in terms of maintaining ecosystem services would have significant impact toward mitigating climate change. This would help to promote healthy ecosystems in terms of development of ecosystem resiliency and adaptability toward changing climate.

Adaptation strategies based on the ecosystem are very much important across various sectors. Examples include the development of coastal forest, mangrove vegetation for the protection against flood and coastal hazards, and maintenance of genetic resource in the agriculture sector. Various eco-friendly practices in terms of the development of shelterbelt, eco-restoration of mangroves, and climate-resilient agriculture practices would do world good toward reducing the vulnerability of ecosystem toward climate change. Apart from their benefits, these approaches are very much limited in their application. Integrated approaches in terms of IWRM (integrated water resource management) and ICZM (integrated coastal zone management) for managing the water resource and rising sea level need to be employed for the effective management of ecosystem and reduction of vulnerability toward climate change.

Proper management at the ecosystem level also brings additional benefits in terms of mitigating climate change. Proper land use practices may aid in the reduction of habitat loss as well as increased C stock in various ecosystems. C sequestration is a significant approach to reduce the climatic vulnerability of the ecosystem and is a good mitigatory strategy for climate change. Emission reduction through sustainable approaches would help to reduce the GHG emission in the atmosphere.

Development of physical barriers through ecosystem management helps to reduce the vagaries of climatic extremes. For example, development of vegetal layer at the downstream area helps to reduce climate-induced natural hazards. Coral reef ecosystem tends to develop resiliency against the climatic perturbations of surges and oceanic storms. Mangroves tend to reduce the impact of coastal surges up to 90% (UNEP-WCMC 2006).

From the perspective of ecosystem vulnerability toward climatic risk, disaster management should be an important policy for the effective management of the ecosystem. One needs to take proper preventive measure toward ecosystem vulnerability reduction under disasters as well as climatic extremes. Proper evaluation of the effects of changing climate helps in the process of climatic adaptation and thus helps in vulnerability reduction. Reduction in the risk of occurrence of natural hazards develops adaptability of the ecosystem to mitigate climate change. Various

Table 6.4 Ecosystem resiliency toward climate change

Climate-resilient ecosystem	Functional role	Reference
Tropical forest and coral reef ecosystems	Network system for the protection of coral reefs and other important areas would develop resiliency Grazing and seed dispersal through various biota would help in developing resiliency in forest ecosystem against changing climate	França et al. (2020)
Mangroves and salt marshes	Proper conservative measures help in optimum C sequestration leading to the development of enhanced resiliency toward changing climate	Roberts et al. (2020)
Soft sediment of the benthic zone of marine ecosystem	The benthic organism performs key ecological processes and thus helps to develop climate-resilient ecosystem	Solan et al. (2020)
Marine ecosystems	The icthyofauna and marine mammal population help in the regulation of nutrient cycle and other processes that helps the marine ecosystem to combat against the changing climate	Roberts et al. (2020)

novel approaches in different countries for developing ecosystem resiliency have been shown in Table 6.4.

It is a clearly stated fact that proper management of ecosystem helps in reducing the risk of disaster followed by proper climate change adaptation. This has become a worldwide agenda under the United Nations Framework Convention on Climate Change (UNFCCC). Thus, ground-level work and intervention are required to improve the situation and work effectively toward ecosystem sustainability. The Convention on Biological Diversity (CBD) is now addressing the issue of reducing climatic vulnerability of ecosystem and other valuable ecosystem services. However, integrated approaches interlinking the changing climate and ecosystem degradation are yet to be done properly. Regional and limited approaches are available across the globe for the effective management of ecosystem to reduce the climatic vulnerability. However, this matter is gaining importance as reflected from the 14th meeting in 2008 organized by COP (Conference of the Parties) under the aegis of UNFCCC. Post-meet of COP15 in 2009, an international level of agreement came out to give importance toward proper ecosystem management. Political support is lacking in this aspect which needs to be promoted significantly. Proper economic, scientific, and technological know-how needs to be formulated toward reduction of climatic risk and for the effective management of ecosystem. This would also come under the dimension of disaster risk reduction. Further, awareness generation, capacity building, and suitable planning need to be incorporated in the national policy which would help to reduce the climatic risk in the developing countries.

Appropriate funding and technical support should be there in time for the developing countries. In this connection, ecosystem-based approaches and suitable disaster risk reduction strategies would be highly fruitful. Collaboration between the policy makers and scientific community and academicians would help to fight against the multidimensional factors of changing climate, natural disaster, and degradation of the ecosystem.

6.9 Conclusion

The time scale factor is a crucial one for assessing the climatic vulnerability. It is very difficult to assess the future changes as the anticipated response is very much unpredictable. Past experiences have been given major importance in assessing the vulnerability. But, the trend may not be similar for future perspective. Across the globe, the work on vulnerability and climatic risk assessment is improper or not completed, and therefore the quantitative evaluation is very problematic. Various statistical predictions used for vulnerability assessment are also not very much fruitful. Climate change has its own impact over the ecosystem and its services along with individual species. Therefore, proper assessment of the vulnerability in relation to climate change is very much essential and is also a big challenge. Exposure, sensitivity, and adaptivity of the ecosystem should be the key element of vulnerability assessment of the ecosystem at the frontier of changing climate. Proper evaluation and identification of the impact of changing climate on various natural resources, on ecosystem services, and on various ecosystems need to be done very precisely. Proper ecosystem management needs to be adopted for mitigating the effect of changing climate. Such approaches would help to develop climate-smart practices as well as climate-resilient ecosystems. This would help in reducing the vulnerability of ecosystems under the face of climate change.

6.10 Research and Development Activities Toward Reducing Ecosystem Vulnerability and Future Perspectives

Ecosystem vulnerability toward changing climate is a big issue which requires a comprehensive learning and exploration for properly predicting the future changes. Such studies would help to formulate the policies and strategies about the mitigatory measures against climatic alterations. Proper understanding of the ecological process that helps in the proper functioning of the ecosystem is very much essential. It also helps to acquire knowledge in relation to changes that take place within the ecosystem and provide an insight over the ways and approaches for ecological restoration of the ecosystems. Future research should be aimed toward developing proper communication of scientific knowledge toward the decision-making system that would help to develop political will of the policy makers to adopt mitigatory measures against climate change. Future research should be aimed toward exploring the knowledge gap in relation to ecosystem vulnerability toward changing climate. Identifying the components that make the ecosystem more resilient and more adaptive toward the changing climate is very much essential at the present context. The time frame of ecosystem vulnerability requires monitoring and assessment of

Table 6.5 Researches on vulnerability assessment of ecosystem on climate change

Research area	Reference
Impact of climate on mountainous region of Rwenzori between Congo and Uganda border	Eggermont et al. (2010)
Landslide hazard estimation of Mt. Elgon region	Claessens et al. (2007)
Assessment of impact of previous storm experiences and future prediction of tropical storm over forests of Taiwan	Lin et al. (2011)
Vulnerability assessment of hurricane at Mexico	Alayón-Gamboa and Ku-Vera (2011)
Assessment of interrelationship between grazing and climate change	Pyke and Marty (2005)
Climatic factor manipulation in Mediterranean Macchia ecosystem	Ripullone et al. (2009)

the ecosystem on a long-term basis along with the effectivity of the each of the management interventions.

From future perspective, the diversity, trophic relationship, and heterogeneity of the habitat are the major attributes that would produce climate-resilient ecosystems and would help to reduce the adverse impact. In this perspective, ecological restoration with an aim toward reducing fossil fuel emission would do world good toward combating climate change (Jhariya et al. 2018a, b). Researches should be aimed toward understanding the connectivity between the ecosystems to reduce the climatic vulnerability and get well adapted under the changing climate. Table 6.5 represents the various research approaches that have already been done in the field of ecosystem vulnerability assessment toward climate change. Such types of studies also aim toward future prediction of various changes that would take place with respect to changing climate. In this vulnerability assessment process, a better knowledge regarding various components of the ecosystem is required.

Such studies would help us to formulate the baseline for climate-resilient ecosystems, identify the changes in ecosystem species composition, identify the ecosystem threshold toward changing climate, as well as identify the response of ecosystems, species, populations, and communities toward the changing climate. This would add further knowledge to identify the successful species against climatic extremes as well as climate-vulnerable species (Stein and Rebecca Shaw 2013).

References

Adam JC, Hamlet AF, Lettenmaier DP (2009) Implications of global climate change for snowmelt hydrology in the twenty-first century. Hydrol Process 23(7):962–972

Aggarwal PK, Kalra N, Chander S, Pathak H (2006) Info crop: a dynamic simulation model for the assessment of crop yields, losses due to pests, and environmental impact of agro-ecosystems in tropical environments. I. Model description. Agric Syst 89:1–25

Ainsworth EA, Long SP (2005) What have we learned from 15 years of free-air CO_2 enrichment (FACE)? A meta-analytic review of the responses of photosynthesis, canopy properties and plant production to rising CO_2. New Phytol 165(2):351–372

Alayón-Gamboa JA, Ku-Vera JC (2011) Vulnerability of smallholder agriculture in Calakmul, Campeche, Mexico. Indian J Tradit Knowl 10(1):125–132

Alston J, Anderson M, James J, Pardey P (2010) Persistence pays: US agricultural productivity growth and the benefits from R&D spending. Springer, New York

Anaman KA, Lellyett SC, Drake L, Leigh R, Henderson-Sellers A, Noar PF, Sullivan PJ, Thampapillai DJ (1997) Economic and social benefits of meteorological services produced by the Australian Bureau of Meteorology: final project report. Graduate School of the Environment and the Climatic Impacts Centre, Macquarie University, Sydney, Australia, 51 pp (unpublished)

Andresen J, Hilberg S, Kunkel K (2012) Historical climate and climate trends in the Midwestern USA. In: Winkler J, Andresen J, Hatfield J, Bidwell D, Brown D (coordinators) U.S. national climate assessment Midwest technical input report. Available from the Great Lakes Integrated Sciences and Assessment (GLISA) Center. http://glisa.umich.edu/docs/NCA/MTIT_Forestry.pdf

Antle JM, Capalbo SM, Elliott ET, Paustian KH (2004) Adaptation, spatial heterogeneity, and the vulnerability of agricultural systems to climate change and CO_2 fertilization: an integrated assessment approach. Clim Chang 64:289–315

Baker JM, Griffis TJ, Ochsner TE (2012) Coupling landscape water storage and supplemental irrigation to increase productivity and improve environmental stewardship in the US Midwest. Water Resour Res 48:W05301

Banerjee A, Jhariya MK, Yadav DK, Raj A (2020) Environmental and sustainable development through forestry and other resources. Apple Academic Press Inc., CRC Press- a Taylor and Francis Group, US & Canada, 400 pp. https://doi.org/10.1201/9780429276026

Banerjee A, Meena RS, Jhariya MK, Yadav DK (2021a) Agroecological footprints management for sustainable food system. Springer, Singapore, 514 pp. https://doi.org/10.1007/978-981-15-9496-0

Banerjee A, Jhariya MK, Meena RS, Yadav DK (2021b) Ecological footprints in agroecosystem- an overview. In: Banerjee A, Meena RS, Jhariya MK, Yadav DK (eds) Agroecological footprints management for sustainable food system. Springer, Singapore, pp 1–23. https://doi.org/10.1007/978-981-15-9496-0_1

Banerjee A, Jhariya MK, Raj A, Yadav DK, Khan N, Meena RS (2021c) Land footprint management and policies. In: Banerjee A, Meena RS, Jhariya MK, Yadav DK (eds) Agroecological footprints management for sustainable food system. Springer, Singapore, pp 221–246. https://doi.org/10.1007/978-981-15-9496-0_7

Banerjee A, Jhariya MK, Raj A, Yadav DK, Khan N, Meena RS (2021d) Energy and climate footprint towards the environmental sustainability. In: Banerjee A, Meena RS, Jhariya MK, Yadav DK (eds) Agroecological footprints management for sustainable food system. Springer, Singapore, pp 415–443. https://doi.org/10.1007/978-981-15-9496-0_14

Bardgett RD, Caruso T (2020) Soil microbial community responses to climate extremes: resistance, resilience and transitions to alternative states. Philos Trans R Soc Lond B Biol Sci 375 (1794):20190112

Beck PSA, Juday GP, Alix C, Barber VA, Winslow SE, Sousa EE, Heiser P, Herriges JD, Goetz SJ (2011) Changes in forest productivity across Alaska consistent with biome shift: changes in forest productivity across Alaska. Ecol Lett 14(4):373–379. https://doi.org/10.1111/j.1461-0248.2011.01598.x

Boateng I (2012) GIS assessment of coastal vulnerability to climate change and coastal adaption planning in Vietnam. J Coast Conserv 16:25–36

Bossio D, Geheb K, Critchley W (2010) Managing water by managing land: addressing land degradation to improve water productivity and rural livelihoods. Agric Water Manag 97:536–542

Bowman DMJS, Murphy BP, Boer MM, Bradstock RA, Cary GJ (2013) Forest fire management, climate change, and the risk of catastrophic carbon losses. Front Ecol Environ 11:66–67

Branch TA, De Joseph BM, Ray LJ, Wagner C (2013) Impacts of ocean acidification on marine seafood. Trends Ecol Evol 28:178–186

Breitburg D, Levin LA, Oschlies A, Grégoire M, Chavez FP, Conley DJ, Garçon V, Gilbert D, Gutiérrez D, Isensee K, Jacinto GS, Limburg KE, Montes I, Naqvi SWA, Pitcher GC, Rabalais NN, Roman MR, Rose KA, Seibel BA, Telszewski M, Yasuhara M, Zhang J (2018) Declining oxygen in the global ocean and coastal waters. Science 359(6371):7240. https://doi.org/10.1126/science.aam7240

BSR (2013) Private sector uptake of ecosystem services concepts and frameworks: the current state of play. www.bsr.org/en/ourinsights/report-view/private-sector-uptake-of-ecosystem-services-concepts-and-frameworks. Accessed 26 Sep 2013

Buytaert W, Cuesta-Camacho F, Tobon C (2011) Potential impacts of climate change on the environmental services of humid tropical alpine regions. Glob Ecol Biogeogr 20:19–33

Castello L, Macedo MN (2016) Large-scale degradation of Amazonian freshwater ecosystems. Glob Change Biol 22(3):990–1007

Cheung WWL, Lam VWY, Sarmiento JL, Kearney K (2010) Large-scale redistribution of maximum fisheries catch potential in the global ocean under climate change. Glob Change Biol 16:24–35

Chiang JM, Iverson LR, Prasad A, Brown KJ (2008) Effects of climate change and shifts in forest composition on forest net primary production. J Integr Plant Biol 50(11):1426–1439

Christensen L, Coughenour MB, Ellis JE, Chen ZZ (2004) Vulnerability of the Asian typical steppe to grazing and climate change. Clim Chang 63:351–368

Chu-Agor ML, Munoz-Carpena R, Kiker G, Emanuelsson A, Linkov I (2011) Exploring vulnerability of coastal habitats to sea level rise through global sensitivity and uncertainty analyses. Environ Model Softw 26:593–604

Claessens L, Knapen A, Kitutu MG, Poesen J (2007) Modelling landslide hazard, soil redistribution and sediment yield of landslides on the Ugandan footslopes of Mount Elgon. Geomorphology 90(1):23–35. https://doi.org/10.1016/j.geomorph.2007.01.007

Cong R, Hedlund K, Brady M (2014) Managing soil natural capital: an effective strategy for mitigating future agricultural risks? Agric Syst 129. https://doi.org/10.1016/j.agsy.2014.05.003

Czucz B, Csecserits A, Botta-Dukat Z, Kroel-Dulay G, Szabo R, Horvath F, Molnar Z (2011) An indicator framework for the climatic adaptive capacity of natural ecosystems. J Veg Sci 22:711–725

Danovaro R, Aguzzi J, Fanelli E, Billett D, Gjerde K, Jamieson A, Ramirez-Llodra E, Smith CR, Snelgrove PV, Thomsen L, Dover CL (2017) An ecosystem-based deep-ocean strategy. Science 355(6324):452–454

Davis MB, Shaw RG, Etterson JR (2005) Evolutionary responses to changing climate. Ecology 86 (7):1704–1714

Dawson J, Scott D (2010) Climate change and tourism in the great lakes region: a summary of risks and opportunities. Tour Marine Environ 6(2–3):119–132

Dawson TP, Jackson ST, House JI, Prentice IC, Mace GM (2011) Beyond predictions: biodiversity conservation in a changing climate. Science 332:53–58

Eggermont H, Verschuren D, Audenaert L, Lens L (2010) Limnological and ecological sensitivity of Rwenzori mountain lakes to climate warming. Hydrobiologia 648(1):123–142. https://doi.org/10.1007/s10750-010-0140-z

Eigenbrod F, Gonzalez P, Dash J, Steyl I (2015) Vulnerability of ecosystems to climate change moderated by habitat intactness. Glob Chang Biol 21(1):275–286. https://doi.org/10.1111/GCB.12669

Ernst C, Hopper K, Summers D (2004) Protecting the source: land conservation and the future of America's drinking water. Trust for Public Land

Fei S, Desprez JM, Potter KM, Jo I, Knott JA, Oswalt CM (2017) Divergence of species responses to climate change. Sci Adv 3(5):e1603055. https://doi.org/10.1126/sciadv.1603055

Flickinger A (2010) Iowa's forests today: an assessment of the issues and strategies for conserving and maintaining Iowa's forests. Iowa Department of Natural Resources

França FM, Cassandra E, Benkwitt CE, Guadalupe P, Robinson JPW, Graham NAJ, Tylianakis JM, Berenguer E, Lees AC, Ferreira J, Barlow J (2020) Climatic and local stress or interactions threaten tropical forests and coral reefs. Phil Trans R Soc B 375:20190116. https://doi.org/10.1098/rstb.2019.0116

Frelich LE, Reich PB (2010) Will environmental changes reinforce the impact of global warming on the prairie forest border of Central North America? Front Ecol Environ 8(7):371–378

Furniss Michael J, Staab Brian P, Sherry H, Clifton Cathrine F, Roby Kenneth B, Ilhadrt Bonnie L, Larry Elizabeth B, Todd Albert H, Reid Leslie M, Hines Sarah J, Bennett Karen A, Luce Charles H, Edwards Pamela J (2010) Water, climate change, and forests: watershed stewardship for a changing climate. Gen. Tech. Rep. PNW-GTR-812. U.S. Department of Agriculture, U.S. Forest Service, Pacific Northwest Research Station, Portland, OR

Glick P, Stein BA, Edelson NA (2011) Scanning the conservation horizon: a guide to climate change vulnerability assessment. National Wildlife Federation, Washington, DC

Gober P, Larson KL, Quay R, Polsky C, Chang H, Shandas V (2013) Why land planners and water managers don't talk to one another and why they should. Soc Nat Resour 26:356–364

Gonzalez P, Neilson RP, Lenihan JM, Drapek RJ (2010) Global patterns in the vulnerability of ecosystems to vegetation shifts due to climate change: global vulnerability to climate change. Glob Ecol Biogeogr 19(6):755–768. https://doi.org/10.1111/J.1466-8238.2010.00558.X

Gopalakrishnan R, Jayaraman M, Bala G, Ravindranath NH (2011) Climate change and Indian forests. Curr Sci 101:348–355

Gough CM, Vogel CS, Schmid HP, Curtis PS (2008) Controls on annual forest carbon storage: lessons from the past and predictions for the future. Bioscience 58(7):609–622

Gritti ES, Smith B, Sykes MT (2006) Vulnerability of Mediterranean Basin ecosystems to climate change and invasion by exotic plant species. J Biogeogr 33:145–157

Hannah L, Roehrdanz PR, Ikegami M, Shepard AV, Shaw MR, Tabor G, Zhi L, Marquet PA, Hijmans RJ (2013) Climate change, wine, and conservation. P Natl Acad Sci USA 110:6907–6912

Hay JE, Kaluwin C (1993) Climate change and sea level rise in the south pacific region. In: Proceedings of the second SPREP meeting, Nomea New Caledonia, 6–10 April, 1993. South Pacific Regional Environment Programme, Apria, West Samoa, 238 pp

Hicke JA, Allen CD, Desai AR, Dietze MC, Hall RJ, Hog EH, Kashian DM, Moore D, Raffa KF, Sturrock RN, Vogelmann J (2011) Effects of biotic disturbances on forest carbon cycling in the United States and Canada. Glob Change Biol 18:7–34

Hoegh-Guldberg O (1999) Climate change, coral bleaching and the future of the world's coral reefs. CSIRO Marine Freshw Res 50(8):839–866

Holdridge LR (1947) Determination of world plant formations from simple climatic data. Science 105:367–368

Hughes TP, Anderson KD, Connolly SR, Heron SF, Kerry JT, Lough JM, Baird AH, Baum JK, Berumen ML, Bridge TC, Claar DC, Eakin CM, Gilmour JP, Graham NAJ, Harrison H, Hobbs JPA, Hoey AS, Hoogenboom M, Lowe RJ, McCulloch MT, Pandolfi JM, Pratchett M, Schoepf V, Torda G, Wilson SK (2018) Spatial and temporal patterns of mass bleaching of corals in the Anthropocene. Science 359(6371):80–83. https://doi.org/10.1126/science.aan8048

Iglesias V, Whitlock C (2020) If the trees burn, is the forest lost? Past dynamics in temperate forests help inform management strategies. Phil Trans R Soc B 375:20190115. https://doi.org/10.1098/rstb.2019.0115

Imhoff ML, Bounoua L (2006) Exploring global patterns of net primary production carbon supply and demand using satellite observations and statistical data. J Geophys Res-Atmos 111:D22

IPCC (2001) IPCC third assessment report. In: Climate change 2001: working group II: impacts, Adaptation and vulnerability

IPCC (2007) Climate change 2007: the physical basis. Summary for policy-makers. Contribution of working group I to the fourth assessment report of the Intergovernmental Panel on Climate Change. Cambridge University Press, Cambridge

Iverson LR, Schwartz MW, Prasad AM (2004) How fast and far might tree species migrate in the eastern United States due to climate change? Glob Ecol Biogeogr 13(3):209219

Jhariya MK (2017a) Vegetation ecology and carbon sequestration potential of shrubs in tropics of Chhattisgarh, India. Environ Monit Assess 189(10):518. https://doi.org/10.1007/s10661-017-6246-2

Jhariya MK (2017b) Influences of forest fire on forest floor and litterfall dynamics in Bhoramdeo Wildlife Sanctuary (C.G.), India. J For Environ Sci 33(4):330–341. https://doi.org/10.7747/JFES.2017.33.4.330

Jhariya MK, Singh L (2020) Herbaceous diversity and biomass under different fire regimes in a seasonally dry forest ecosystem. Environ Dev Sustain 22:1–19. https://doi.org/10.1007/s10668-020-00892-x

Jhariya MK, Singh L (2021) Effect of fire severity on soil properties in a seasonally dry forest ecosystem of Central India. Int J Environ Sci Technol:1–12. https://doi.org/10.1007/s13762-020-03062-8

Jhariya MK, Yadav DK, Banerjee A (2018a) Plant mediated transformation and habitat restoration: phytoremediation an eco-friendly approach. In: Gautam A, Pathak C (eds) Metallic contamination and its toxicity. Daya Publishing House, A Division of Astral International, New Delhi, pp 231–247

Jhariya MK, Banerjee A, Yadav DK, Raj A (2018b) Leguminous trees an innovative tool for soil sustainability. In: Meena RS, Das A, Yadav GS, Lal R (eds) Legumes for soil health and sustainable management. Springer, Singapore, pp 315–345. https://doi.org/10.1007/978-981-13-0253-4_10

Jhariya MK, Banerjee A, Meena RS, Yadav DK (2019a) Sustainable agriculture, forest and environmental management. Springer, Singapore, p 606. https://doi.org/10.1007/978-981-13-6830-1

Jhariya MK, Yadav DK, Banerjee A (2019b) Agroforestry and climate change: issues and challenges. Apple Academic Press Inc., CRC Press- a Taylor and Francis Group, US & Canada, p 335. https://doi.org/10.1201/9780429057274

Jhariya MK, Meena RS, Banerjee A (2021a) Ecological intensification of natural resources for sustainable agriculture. Springer, Singapore. https://doi.org/10.1007/978-981-33-4203-3

Jhariya MK, Meena RS, Banerjee A (2021b) Ecological intensification of natural resources towards sustainable productive system. In: Ecological intensification of natural resources for sustainable agriculture. Springer, Singapore. https://doi.org/10.1007/978-981-33-4203-3_1

Johnson WC, Millett BV, Gilmanov T, Voldseth RA, Guntenspergen GR, Naugle DE (2005) Vulnerability of northern prairie wetlands to climate change. Bioscience 55:863–872

Johnson WC, Werner B, Guntenspergen GR, Voldseth RA, Millett B, Naugle DE, Tulbure M, Carroll RWH, Tracy J, Olawsky C (2010) Prairie wetland complexes as landscape functional units in a changing climate. Bioscience 60:128–140

Karl TR, Melillo JM, Peterson TC (eds) (2009) Global climate change impacts in the United States. Cambridge University Press, New York

Khan N, Jhariya MK, Yadav DK, Banerjee A (2020a) Herbaceous dynamics and CO_2 mitigation in an urban setup- a case study from Chhattisgarh, India. Environ Sci Pollut Res 27(3):2881–2897. https://doi.org/10.1007/s11356-019-07182-8

Khan N, Jhariya MK, Yadav DK, Banerjee A (2020b) Structure, diversity and ecological function of shrub species in an urban setup of Sarguja, Chhattisgarh, India. Environ Sci Pollut Res 27 (5):5418–5432. https://doi.org/10.1007/s11356-019-07172-w

Khan N, Jhariya MK, Raj A, Banerjee A, Meena RS (2021a) Soil carbon stock and sequestration: implications for climate change adaptation and mitigation. In: Jhariya MK, Meena RS, Banerjee A (eds) Ecological intensification of natural resources for sustainable agriculture. Springer, Singapore. https://doi.org/10.1007/978-981-33-4203-3_13

Khan N, Jhariya MK, Raj A, Banerjee A, Meena RS (2021b) Eco-designing for sustainability. In: Jhariya MK, Meena RS, Banerjee A (eds) Ecological intensification of natural resources for sustainable agriculture. Springer, Singapore. https://doi.org/10.1007/978-981-33-4203-3_16

King RO (2013) The National Flood Insurance Program: status and remaining issues for congress. R42850. Congressional Research Service, Washington, DC

Kleypas JA, Buddemeier RW, Archer D, Gattuso JP, Langdon C, Opdyke BN (1999) Geochemical consequences of increased atmospheric carbon dioxide on coral reefs. Science 284:118–120

Kumar S, Meena RS, Datta R, Verma SK, Yadav GS, Pradhan GS, Molaei A, Rahman GKMM, Mashuk HA (2020) Legumes for carbon and nitrogen cycling: an organic approach. In: Carbon and nitrogen cycling in soil. Springer, pp 337–375. https://doi.org/10.1007/978-981-13-7264-3_10

Kumar S, Meena RS, Singh RK, Munir TM, Datta R, Danish S, Singh GS, Yadav, Kumar S (2021) Soil microbial and nutrient dynamics under different sowings environment of Indian mustard (*Brassica juncea* L.) in rice based cropping system. Sci Rep 11:5289. https://doi.org/10.1038/s41598-021-84742-4

Kunkel K (2011) Midwest region climate outlooks (unpublished draft). National Climatic Data Center, Asheville, NC, 29. Accessed 5 Oct 2011

Lal M, Harasawa H, Takahashi K (2002) Future climate change and its impacts over small island states. Clim Res 19:179–192

Lasco RD, Pulhin FB, Sanchez PAJ, Villamor GB, Villegas KAL (2008) Climate change and Forest ecosystems in the Philippines: vulnerability, adaptation and mitigation. J Environ Sci Manag 11:1–14

Lenihan JM, Bachelet D, Neilson RP, Drapek R (2008) Simulated response of conterminous United States ecosystems to climate change at different levels of fire suppression, CO_2 emission rate, and growth response to CO_2. Glob Planet Change 64(1–2):16–25

Lin TC, Hamburg SP, Lin KC, Wang LJ, Chang CT, Hsia YJ, Vadeboncoeur MA (2011) Typhoon disturbance and forest dynamics: lessons from a northwest Pacific subtropical forest. Ecosystems 14:127–143

Link JS, Nye J, Hare J (2011) Guidelines for incorporating fish distribution shifts into a fisheries management context. Fish 12:461–469

Loarie S, Duffy P, Hamilton H, Asner GP (2009) The velocity of climate change. Nature 462 (7276):1052–1055. https://doi.org/10.1038/nature08649

Lobell DB, Field CB (2011) California perennial crops in a changing climate. Clim Chang 109:317–333

Luedeling E, Neufeldt H (2012) Carbon sequestration potential of parkland agroforestry in the Sahel. Clim Chang 115:443–461

Luedeling E, Girvetz EH, Semenov MA, Brown PH (2011) Climate change affects winter chill for temperate fruit and nut trees. PLoS One 6:e20155

Lydenberg S, Rogers J, Wood D (2010) From transparency to performance: industry-based sustainability reporting on key issues. Harvard University and the Initiative for Responsible Investment, Cambridge, MA

Mace GM, Reyers B, Alkemade R, Biggs R, Chapin FS III, Cornell SE, Diaz S, Jennings S, Leadley P, Mumby PJ, Purvis A, Scholes RJ, Seddon AWR, Solan M, Steffen W, Woodward G (2014) Approaches to defining a planetary boundary for biodiversity. Glob Environ Chang 28:289–297

Makadho JM (1996) Potential effects of climate change on corn production in Zimbabwe. Clim Res 6:147–151

Malone E, Brenkert A (2008) Uncertainty in resilience to climate change in India and Indian states. Clim Chang 91:451–476

McKinley D, Ryan M, Birdsey R, Giardina C, Harmon M, Heath L, Houghton R, Jackson R, Morrison J, Murray B (2011) A synthesis of current knowledge on forests and carbon storage in the United States. Ecol Appl 21(6):1902–1924

McRae DJ, Duchesne LC, Freedman B, Lynham TJ (2001) Comparisons between wildfire and forest harvesting and their implications in forest management. Environ Rev 9:223–260

Meena RS, Lal R (2018) Legumes for soil health and sustainable management. Springer, Singapore, p 541. https://doi.org/10.1007/978-981-13-0253-4_10

Meena RS, Kumar V, Yadav GS, Mitran T (2018) Response and interaction of *Bradyrhizobium japonicum* and Arbuscular mycorrhizal fungi in the soybean rhizosphere: a review. Plant Growth Regul 84:207–223. https://doi.org/10.1007/s10725-017-0334-8

Meena RS, Lal R, Yadav GS (2020) Long term impacts of topsoil depth and amendments on soil physical and hydrological properties of an Alfisol in Central Ohio, USA. Geoderma 363:1141164. https://doi.org/10.1016/j.geoderma.2019.114164

Meena RS, Lal R, Yadav GS (2020a) Long-term impact of topsoil depth and amendments on carbon and nitrogen budgets in the surface layer of an Alfisol in Central Ohio. Catena 194:104752. https://doi.org/10.1016/j.catena.2020.104752

Meena RS, Kumar S, Datta R, Lal R, Vijaykumar V, Brtnicky M, Sharma MP, Yadav GS, Jhariya MK, Jangir CK, Pathan SI, Dokulilova T, Pecina V, Marfo TD (2020b) Impact of agrochemicals on soil microbiota and management: a review. Land (MDPI) 9(2):34. https://doi.org/10.3390/land9020034

Mekonnen MM, Hoekstra AY (2016) Four billion people facing severe water scarcity. Sci Adv 2 (2):e1500323

Metzger MJ, Leemans R, Schröter D (2005) A multidisciplinary multi-scale framework for assessing vulnerabilities to global change. Int J Appl Earth Observ Geoinf 7:253–267

Miller-Rushing AJ, Høye TT, Inouye DW, Post E (2010) The effects of phenological mismatches on demography. Philos Trans R Soc Lond Ser B Biol Sci 365(1555):3177–3186. https://doi.org/10.1098/rstb.2010.0148

Mkanda FX (1996) Potential impacts of future climate change on Nyala *Tragelaphus angasii* in Lengwe National Park, Malawi. Clim Res 6(2):157–164

Moldenhauer O, Lüdeke MKB (2002) Climate sensitivity of global terrestrial net primary production (NPP) calculated using the reduced-form model NNN. Clim Res 21:43–57

Molotoks A, Stehfest E, Doelman J, Albanito F, Fitton N, Dawson TP, Smith P (2018) Global projections of future cropland expansion to 2050 and direct impacts on biodiversity and carbon storage. Glob Change Biol 24(12):5895–5908

Molotoks A, Henry R, Stehfest E, Doelman J, Havlik P, Krisztin T, Alexander P, Dawson TP, Smith P (2020) Comparing the impact of future cropland expansion on global biodiversity and carbon storage across models and scenarios. Phil Trans R Soc B 375:20190189. https://doi.org/10.1098/rstb.2019.0189

Morin X, Viner D, Chuine I (2008) Tree species range shifts at a continental scale: new predictive insights from a process-based model. J Ecol 96(4):784–794

Moyle PB, Kiernan JD, Crain PK, Quiñones RM (2013) Climate change vulnerability of native and alien freshwater fishes of California: a systematic assessment approach. PLoS One 8:e63883

NASEM (National Academies of Sciences, Engineering, and Medicine) (2018) A research review of interventions to increase the persistence and resilience of coral reefs. The National Academies Press, Washington, DC. https://doi.org/10.17226/25279

Nelson GC, Rosegrant MW, Koo J, Robertson R, Sulser T, Zhu T et al (2009) Climate change: impact on agriculture and costs of adaptation. International Food Policy Research Institute, Washington, DC

Nolin AW (2012) Perspectives on climate change, mountain hydrology, and water resources in the Oregon Cascades, USA. Mt Res Dev 32:S35–S46

Norby RJ, Zak DR (2011) Ecological lessons from free-air CO_2 enrichment (FACE) experiments. Ann Rev Ecol Evol Syst 42:181–203

Nori J, Urbina-Cardona JN, Loyola RD, Lescano JN, Leynaud GC (2011) Climate change and American bullfrog invasion: what could we expect in South America? PLoS One 6(10):e25718. https://doi.org/10.1371/journal.pone.0025718

Notaro M, Lorenz DJ, Vimont D, Vavrus S, Kucharik C, Franz K (2011) 21st century Wisconsin snow projections based on an operational snow model driven by statistically downscaled climate data. Int J Climatol 31:1615–1633. https://doi.org/10.1002/joc.2179

Nowak DJ, Crane DE (2002) Carbon storage and sequestration by urban trees in the USA. Environ Pollut 116(3):381389

Nurse L, McLean RF, Suarez AG (1998) Small island states. In: Watson RT, Zinyowera MC, Moss RH (eds) The regional impacts of climate change: an assessment of vulnerability. Special report of IPCC working group II. Intergovernmental Panel on Climate Change, Cambridge University Press, Cambridge and New York, pp 331–354

Nye JA, Link JS, Hare JA, Overholtz WJ (2009) Changing spatial distribution of fish stocks in relation to climate and population size on the Northeast United States continental shelf. Mar Ecol-Prog Ser 393:111–129

Ogden AE, Innes JL (2007) Perspectives of forest practitioners on climate change adaptation in the Yukon and Northwest Territories of Canada. For Chron 83:557–569

Oliver ECJ, Donat MG, Burrows MT, Moore PJ, Smale DA, Alexander LV, Benthuysen JA, Feng M, Gupta AS, Hobday AJ, Holbrook NJ, Perkins-Kirkpatrick SE, Scannell HA, Straub SC, Wernberg T (2018) Longer and more frequent marine heatwaves over the past century. Nat Commun 9:1324

Parry ML, Canziani OF, de Alcaraz E, Allali A, Kajfež-Bogataj L, Love G, Stone J, van Ypersele JP, Palutikof JP (2007) Fourth assessment report: impacts, adaptation and vulnerability. IPCC working group II. Cambridge University Press, New York

Pendleton L, King P, Mohn C, Webster DG, Vaughn R, Adams PN (2011) Estimating the potential economic impacts of climate change on Southern California beaches. Clim Chang 90:475–492

Perry RI, Cury P, Brander K, Jennings S, Mollmann C, Planque B (2010) Sensitivity of marine systems to climate and fishing: concepts, issues and management responses. J Marine Syst 79:427–435

Phillips SJ, Anderson RP, Schapire RE (2006) Maximum entropy modelling of species distributions. Ecol Model 190:231–259

Prasad AM, Iverson LR, Matthews SN, Peters M (2007) A climate change atlas for 134 forest tree species of the eastern United States [database]. Northern Research Station, USDA Forest Service, Delaware, OH

Pyke CR, Marty JT (2005) Cattle grazing mediates climate change impacts on ephemeral wetlands. Conserv Biol 19(5):1619–1625. https://doi.org/10.1111/j.1523-1739.2005.00233.x

Quinn NWT, Miller NL, Dracup JA, Brekke L, Grober LF (2001) An integrated modelling system for environmental impact analysis of climate variability and extreme weather events in the San Joaquin Basin, California. Adv Environ Res 5:309–317

Raeker G, Fleming J, Morris M, Moser K, Treiman T (2010) Missouri's forest resource assessment and strategy. Missouri Department of Conservation and USDA Forest Service, Northern Research Station–Forest Inventory and Analysis Program

Raj A, Jhariya MK, Harne SS (2018a) Threats to biodiversity and conservation strategies. In: Sood KK, Mahajan V (eds) Forests, climate change and biodiversity. Kalyani Publisher, New Delhi, pp 304–320

Raj A, Jhariya MK, Bargali SS (2018b) Climate smart agriculture and carbon sequestration. In: Pandey CB, Gaur MK, Goyal RK (eds) Climate change and agroforestry: adaptation mitigation and livelihood security. New India Publishing Agency (NIPA), New Delhi, pp 1–19

Raj A, Jhariya MK, Yadav DK, Banerjee A, Meena RS (2019a) Agroforestry: a holistic approach for agricultural sustainability. In: Jhariya MK, Banerjee A, Meena RS, Yadav DK (eds) Sustainable agriculture, forest and environmental management. Springer, Singapore, pp 101–131

Raj A, Jhariya MK, Banerjee A, Yadav DK, Meena RS (2019b) Soil for sustainable environment and ecosystems management. In: Jhariya MK, Banerjee A, Meena RS, Yadav DK (eds) Sustainable Agriculture, Forest and Environmental Management. Springer, Singapore, pp 189–221

Raj A, Jhariya MK, Yadav DK, Banerjee A (2020) Climate change and agroforestry systems: adaptation and mitigation strategies. Apple Academic Press Inc., CRC Press- a Taylor and Francis Group, US & Canada, 383 pp. https://doi.org/10.1201/9780429286759

Raj A, Jhariya MK, Khan N, Banerjee A, Meena RS (2021) Ecological intensification for sustainable development. In: Ecological intensification of natural resources for sustainable agriculture. Springer, Singapore. https://doi.org/10.1007/978-981-33-4203-3_5

Ramankutty N, Evan AT, Monfreda C, Foley JA (2008) Farming the planet: 1. Geographicdistribution of global agricultural lands in the year 2000. Glob Biogeochem Cycles 22(1):1–19. https://doi.org/10.1029/2007gb0029

Reaka-Kudla ML (1996) The global biodiversity of coral reefs: a comparison with rainforests. In: Reaka-Kudla ML, Wilson DE, Wilson EO (eds) Biodiversity II: understanding and protecting our natural resources. Joseph Henry and National Academy Press, Washington, DC, pp 83–108
Ren W, Tian HQ, Tao B, Chappelka A, Sun G, Lu CQ, Liu ML, Chen GS, Xu XF (2011) Impacts of tropospheric ozone and climate change on net primary productivity and net carbon exchange of China's forest ecosystems. Glob Ecol Biogeogr 20:391–406
Rhemtulla JM, Mladenoff DJ, Clayton MK (2009) Historical forest baselines reveal potential for continued carbon sequestration. Proc Natl Acad Sci 106(15):6082–6087
Ripullone F, Borghetti M, Raddi S, Vicinelli E (2009) Physiological and structural changes in response to altered precipitation regimes in Mediterranean macchia ecosystem. Trees 23 (4):823–834
Roberts CM, O' Leary BC, Hawkins JP (2020) Climate change mitigation and nature conservation both require higher protected area targets. Phil Trans R Soc B 375:20190121. https://doi.org/10.1098/rstb.2019.0121
Roloff A, Korn S, Gillner S (2009) The climate-species-matrix to select tree species for urban habitats considering climate change. Urban For Urban Green 8(4):295–308
Rounsevell MDA, Metzger MJ (2010) Developing qualitative scenario storylines for environmental change assessment. Wiley Interdiscip Rev Clim Change 1:606–619
Roy SB, Chen L, Girvetz EH, Maurer EP, Mills WB, Grieb TM (2012) Projecting water withdrawal and supply for future decades in the US under climate change scenarios. Environ Sci Technol 46:2545–2556
Ruckelshaus M, Doney S, Galindo HM, Barry JP, Chan F, Duffy JE, English CA, Gaines SD, Grebmeier JM, Hollowed AB, Knowlton N, Polovina J, Rabalais NN, Sydeman WJ, Talley LD (2013) Securing ocean benefits for society in the face of climate change. Mar Policy 40:154–159
Scheller RM, Mladenoff DJ (2005) A spatially interactive simulation of climate change, harvesting, wind, and tree species migration and projected changes to forest composition and biomass in northern Wisconsin, USA. Glob Change Biol 11(2):307–321
Scheller RM, Mladenoff DJ (2008) Simulated effects of climate change, fragmentation, and inter-specific competition on tree species migration in northern Wisconsin, USA. Clim Res 36 (3):191–202
Schlenker W, Roberts MJ (2009) Nonlinear temperature effects indicate severe damages to US crop yields under climate change. Proc Natl Acad Sci U S A 106:15594–11598
Scholze M, Knorr W, Arnell NW, Prentice IC (2006) A climate-change risk analysis for world ecosystems. Proc Natl Acad Sci U S A 103:13116–13120
Seaquist JW, Hickler T, Eklundh L, Ardo J, Heumann BW (2009) Disentangling the effects of climate and people on Sahel vegetation dynamics. Biogeosciences 6:469–477
Segan DB, Murray KA, Watson JEM (2016) A global assessment of current and future biodiversity vulnerability to habitat loss–climate change interactions. Global Ecol Conserv 5:12–21. https://doi.org/10.1016/J.GECCO.2015.11.002
Seidl R, Rammer W, Lexer MJ (2011a) Climate change vulnerability of sustainable forest management in the Eastern Alps. Clim Chang 106:225–254
Seidl R, Rammer W, Lexer MJ (2011b) Adaptation options to reduce climate change vulnerability of sustainable forest management in the Austrian Alps. Can J For Res-Revue Canadienne De Recherche Forestiere 41:694–706
Sendzimir J, Reij CP, Magnuszewski P (2011) Rebuilding resilience in the Sahel: Regreening in the Maradi and Zinder Regions of Niger. Ecol Soc 16(3):1
Shanks AL (2009) Pelagic larval duration and dispersal distance revisited. Biol Bull 216:373–385
Sheoran S, Kuma S, Kumar P, Meena RS, Rakshit S (2021) Nitrogen fixation in maize: breeding opportunities. Theor Appl Genet. https://doi.org/10.1007/s00122-021-03791-5
Smith P, Bustamante M, Ahammad H, Clark H, Dong H, Elsiddig EA, Haberl H, Harper R, House J, Jafari M, Masera O, Mbow C, Ravindranath NH, Rice CW, Robledo Abad C, Romanovskaya A, Sperling F, Tubiello F (2014) Agriculture, forestry and other land use (AFOLU). In: Edenhofer O, Pichs-Madruga R, Sokona Y, Farahani E, Kadner S, Seyboth K,

Adler A, Baum I, Brunner S, Eickemeier P, Kriemann B, Savolainen J, Schlömer S, von Stechow C, Zwickel T, Minx JC (eds) Climate change 2014: mitigation of climate change. Contribution of working group III to the fifth assessment report of the intergovernmental panel on climate change. Cambridge University Press, Cambridge and New York

Solan M, Bennett EM, Mumby PJ, Leyland J, Godbold JA (2020) Benthic-based contributions to climate change mitigation and adaptation. Phil Trans R Soc B 375:20190107. https://doi.org/10.1098/rstb.2019.0107

Staudinger M, Grimm NB, Staudt A, Carter SL, Stuart FS III, Kareiva P, Ruckelshaus M, Stein BA (2012) Impacts of climate change on biodiversity, ecosystems, and ecosystem services: technical input to the 2013 national climate assessment. Cooperative Report to the 2013 National Climate Assessment, Washington, DC

Staudt A, Leidner A, Howard J, Brauman K (2013) The added complications of climate change: understanding and managing biodiversity and ecosystems. Front Ecol Environ 11(9):494–501. https://doi.org/10.1890/120275

Stein BA, Rebecca Shaw M (2013) Biodiversity conservation for a climate-altered future. In: Moser S, Boykoff M (eds) Successful adaptation: linking science and practice in managing climate change impacts. Rutledge Press, New York, pp 50–66

Thomas RJ (2008) Opportunities to reduce the vulnerability of dryland farmers in central and West Asia and North Africa to climate change. Agric Ecosyst Environ 126:36–45

Tingem M, Rivington M (2009) Adaptation for crop agriculture to climate change in Cameroon: turning on the heat. Mitig Adapt Strateg Glob Change 14:153–168

Tingem M, Rivington M, Bellocchi G (2009) Adaptation assessments for crop production in response to climate change in Cameroon. Agronom Sustain Dev 29:247–256

Turner MG, Braziunas KH, Hansen WD, Harvey BJ (2019) Short-interval severe fire erodes the resilience of subalpine lodge pole pine forests. Proc Natl Acad Sci U S A 116(23):11319–11328. https://doi.org/10.1073/pnas.1902841116

Turner MG, Calder WJ, Cumming GS, Hughes TP, Jentsch A, LaDeau SL, Timothy M, Lenton TM, Shuman BN, Turetsky MR, Ratajczak Z, Williams JW, Williams AP, Carpente SR (2020) Climate change, ecosystems and abrupt change: science priorities. Phil Trans R Soc B 375:20190105. https://doi.org/10.1098/rstb.2019.0105

UNEP-WCMC (2006) Spatial data layer of Mangrove distribution derived through Landsat image classification, UNEPWCMC, Cambridge, UK. Data analysis, July 2006. Cambridge, UK

Walker JF, Hay LE, Markstrom SL, Dettinger MD (2011) Characterizing climate-change impacts on the 1.5-yr flood flow in selected basins across the United States: a probabilistic approach. Earth Interact 15:1–16

Walsh CL, Kilsby CG (2007) Implications of climate change on flow regime affecting Atlantic salmon. Hydrol Earth Syst Sci 11:1125–1141

Watson JEM, Iwamura T, Butt N (2013) Mapping vulnerability and conservation adaptation strategies under climate change. Nat Clim Chang 3(11):989–994. https://doi.org/10.1038/NCLIMATE2007

Westerling AL, Hidalgo HG, Cayan DR, Swetnam TW (2006) Warming and earlier spring increase western US forest wildfire activity. Sci 313:940–943

Whitehead JC, Poulter B, Dumas CF, Bin O (2009) Measuring the economic effects of sea level rise on shorefishing. Mitig Adapt Strat Glob Change 14:777–792

Wilkinson C, Linden R, Cesar H, Hodgson G, Rubens J, Strong AE (1999) Ecological and socioeconomic impacts of 1998 coral mortality in the Indian Ocean: an ENSO impact and a warning of future change? Ambio 28:188–196

Williams JE (2000) The biodiversity crisis and adaptation to climate change: a case study from Australia's forests. Environ Monit Assess 61:65–74

Winkler JA, Arritt R, Pryor S (2012) Climate projections for the Midwest: availability, interpretation and synthesis. In: Winkler J, Andresen J, Hatfield J, Bidwell D, Brown D (coordinators) U.S. National Climate Assessment Midwest technical input report. Available from the Great

Lakes Integrated Sciences and Assessment (GLISA) Center. http://glisa.umich.edu/docs/NCA/MTIT_Forestry.pdf

Woodall CW, Nowak DJ, Liknes GC, Westfall JA (2010) Assessing the potential for urban trees to facilitate forest tree migration in the eastern United States. For Ecol Manag 259(8):1447–1454

Wu S, Dai E, Huang M, Shao X, Li S, Tao B (2007) Ecosystem vulnerability of China under B2 climate scenario in the 21st century. Chin Sci Bull 52(10):1379–1386. https://doi.org/10.1007/S11434-007-0197-X

WWF (2015) Living blue planet report: species, habitats and human well-being. WWF, Gland

Xu K, Wang X, Jiang C, Sun OJ (2020) Assessing the vulnerability of ecosystems to climate change based on climate exposure, vegetation stability and productivity. For Ecosyst 7:23

Younger M, Morrow-Almeida HR, Vindigni SM, Dannenberg AL (2008) The built environment, climate change, and health: opportunities for co-benefits. Am J Prev Med 35(5):517–526

Zhao D, Wu S (2014) Vulnerability of natural ecosystem in China under regional climate scenarios: an analysis based on eco-geographical regions. J Geogr Sci 24(2):237–248. https://doi.org/10.1007/S11442-014-1085-3

Zhu K, Woodall CW, Clark JS (2011) Failure to migrate: lack of tree range expansion in response to climate change. Glob Change Biol 18(3):1042–1052

Intensification for Agroecosystem Services 7

Abhishek Raj, Manoj Kumar Jhariya, Nahid Khan, Arnab Banerjee, P. R. Paikra, Ram Swaroop Meena, and Sandeep Kumar

Abstract

Agroecosystem itself represents a managed ecosystem in agricultural land by human-managed crops and livestock's integration that are highly productive, profitable, and ecologically sustainable. Growing populations and related food demands necessitate intensive practices in agriculture systems. Deforestation and other anthropogenic factors promote forest land conversion into arable lands. Intensive agroecosystem ensures higher crop productions but at the cost of ecosystem and environmental health. High intensive inputs of chemical fertilizers and heavy mechanizations resulted in land degradation and poor soil health. Intensive agroecosystem practices further destroy soil and environmental quality

A. Raj
School of Agriculture, Lovely Professional University, Phagwara, Punjab, India

M. K. Jhariya (✉) · N. Khan
Department of Farm Forestry, Sant Gahira Guru Vishwavidyalaya, Sarguja, Ambikapur, Chhattisgarh, India

A. Banerjee
Department of Environmental Science, Sant Gahira Guru Vishwavidyalaya, Sarguja, Ambikapur, Chhattisgarh, India

P. R. Paikra
Krishi Vigyan Kendra Balrampur, IGKV, Balrampur, Chhattisgarh, India

R. S. Meena
Department of Agronomy, Institute of Agricultural Sciences, Banaras Hindu University, Varanasi, Uttar Pradesh, India
e-mail: meenars@bhu.ac.in

S. Kumar
ICAR-Indian Institute of Pulses Research, Regional Station, Phanda, Bhopal, India
e-mail: sandeep.kumar5@icar.gov.in

M. K. Jhariya et al. (eds.), *Sustainable Intensification for Agroecosystem Services and Management*, https://doi.org/10.1007/978-981-16-3207-5_7

along with poor ecosystem services. In this context, applying sustainable practices in agroecosystem is based on ecological concept that enhances crop-soil productivity in sustainable ways without destroying our environment. Sustainable intensification in agroecosystem enhances biodiversity that intensifies ecosystem services in both tangible (direct) and intangible (indirect) ways. Production services (tangible) include the timber biomass, fuelwood, food products, and several non-wood forest products that are delivered directly from the agroecosystem. Climate change mitigation, soil fertility improvement, watershed management, pest disease control, water regulation, food and nutritional security, etc. come under the protection services. Sustainable intensification-based agroecosystem enhances climate-resilient and soil health management. Climate-resilient agroecosystem ensures less emission of greenhouse gases (GHGs) and makes sustainable ecosystem. Conservation agriculture, use of cover crops, and no-tillage practices are key drivers that promote sustainable agroecosystem. An effective policy for scientific research and design must be included to promote sustainable agroecosystem practices that promise food-soil-climate security at global scale. This chapter discusses about ecosystem services through sustainable-based agroecosystem rather than intensive practices. A rigorous discussion is also made on theoretical models of agroecosystem, significance of sustainable agroecosystem, and drivers for sustainable intensification in agroecosystem. Climate- and soil-resilient agroecosystem makes this chapter more comprehensive and informative for academicians, policy makers, and researchers worldwide.

Keywords

Agroecosystem · Biodiversity · Climate change · Ecology · Ecosystem services · Sustainable intensification

Abbreviations

C	Carbon
CA	Conservation agriculture
CO_2	Carbon dioxide
GHGs	Greenhouse gases
NPP	Net primary productivity
SOC	Soil organic carbon

7.1 Introduction

Agroecosystem covered approximately 30% of the earth's surface area (Altieri 1999). It is defined as "the human exerted ecosystem in agriculture land which harmonized a synergy between crops and livestock's to create more natural ecosystem for maintaining environmental sustainability and ecological stability" (Swift et al. 1996). Agroecosystem provides uncountable ecosystem services which are influenced by varying management practices in any agroecological regions (Tilman et al. 2002; Jhariya et al. 2019a, b). The services are in tangible (timber, food, fruits, fuelwood, fodder, NTFPs, etc.) and intangible (soil health, microclimate ameliorations, climate security, etc.) ways (Raj et al. 2020; Banerjee et al. 2020; Meena et al. 2018).

Agroecosystem is disturbed nowadays due to practices of intensifications that affect overall health and productivity and reduce biodiversity (Gliessman 2015). Increasing global population necessitates food and shelter requirement that leads to deforestation and conversion of forest land into cultivable land. Further, the intensifying practices in agricultural land by high synthetic inputs and heavy mechanizations of course boom food production but at the cost of environmental health and sustainability (Meena et al. 2020a, b, c; Kumar et al. 2020). These are the major concern today that affects overall environmental health and sustainability at global scale. As per Krishna (2010), a chemical- and mechanical-based intensive agriculture covered 50% of global arable lands that require high synthetic inputs.

There are two schools of thoughts: The first is intensification of course intensify food productions but at the cost of environmental health. Various questions revolve in my mind: (a) *How do intensifications affect our natural environment?* (b) *Does any practice exist that harmonizes food production along with the formation of sustainable and vibrant environment*? The first questions have lots of story, and you might be aware about climate change. Land expansions and intensifications in agroecosystem release various GHGs (greenhouse gases) into the environment which leads to global warming. Deforestation, illicit felling of timbers, mining, and other developmental projects destroy forest health and productivity (Raj et al. 2018a, b). Moreover, conversion of forest and other land into agricultural systems promotes land degradation and affects overall soil and climate security (Raj et al. 2019a, b). However, intensification makes a bad picture among all researchers, scientists, academicians, and policy makers which affects overall soil-food-climate security. These problems can be resolved by adopting some ecological- and sustainable-based intensification practices in agroecosystem (Jhariya et al. 2021a, b; Meena and Lal 2018).

Ecological-oriented and sustainable-based intensifications are good strategies that make a balance between food productions, soil health maintenance, and environmental health in sustainable ways (Raj et al. 2021). The practices involve less application of synthetic inputs which release less GHG into the environment. The principle behind sustainable intensification is "How do we make a vibrant environment with sustainable food productions?". Sustainable intensification is a good alternative to chemical and mechanical intensification which is based on the principle of ecological intensification that ensures a sustainable environment (Gaba et al. 2014). Enhancing food productions, biodiversity maintenance, soil enrichment, climate security, etc. are the important significance achieved by practices of sustainable agroecosystem. However, sustainable intensification practices include climate-smart practices, conservation agriculture (CA), mulching practices, no-tillage practices, cover crop system, and diversified crop rotation systems. These practices make agroecosystem in more sustainable ways by enhancing soil organic carbon (SOC), minimizing pest and diseases, improving system productivity, and intensifying overall ecosystem services (Kumar et al. 2020a). In the context of soil, unsustainable land management practices affect SOC value and increase erosion continuously in agroecosystem. However, diversification in agroecosystem makes more productive and protective systems which paves the way towards achieving the goal of sustainability for a better world (Gliessman 2015).

Protecting the environment through climate change mitigation and adaptation is another goal that can be achieved by sustainable-based practices of agroecosystem. Diversified agroecosystem and its scientific-based management practices ensure carbon (C) absorption from the atmosphere and its fixation into plants and soils. Thus, sustainable and ecological intensification reduces GHG emission and maintains SOC pools and vegetational biomass in agroecosystem to make a more vibrant ecosystem (Emmerson et al. 2016). However, sustainable intensification can beat intensification practices by promising food balance for the burgeoning nine billion global populations without affecting soil-climate security (Lennon 2015). Thus, a question arises "How can we achieve food-soil-climate security through sustainable agroecosystem practices?"

Soil health and quality maintenance are the other important aspects which can be studied under sustainable agroecosystem services (Jhariya et al. 2018a, b). Soil enrichment, fertility enhancement, efficient nutrient cycling, healthy rhizosphere biology, nutrient use efficiency, etc. are important services provided by sustainable-based agroecosystem practices. CA, use of cover crops, mulching, better tree-crop combinations, and integration of leguminous multipurpose tree species in agroecosystem are becoming boon for soil health management. However, integration of perennial grasses and legumes can fix nitrogen and add more C into the soil SOC pools and increase nutrient availability and its retentions into the soils (Congreves et al. 2015). Perennial grasses and legumes enhance soil physicochemical properties which reduce erosions, minimize nutrient loss, and control acidifications and salinization to make a more productive and climate-resilient agroecosystem (Lal et al. 2011). Thus, better and efficient soil ecosystem services are possible through the application of sustainable-and ecological-based intensification practices in agroecosystem. An intimate balance between soil degradation and conservation/reclamation practices ensures agroecosystem health. Failure of agroecosystem in making balances between the two aspects is a major hurdle behind the success of soil health maintenance and sustainability.

This chapter contains a comprehensive discussion about the intensification in agroecosystem and related impacts on the environment. A rigorous discussion on ecological and sustainable agroecosystem services is also made in this chapter. Thus, sustainable agroecosystem is based on ecological concept which intensifies ecosystem services that ensure food-soil-climate security along with environmental sustainability and ecological stability at global scale.

7.2 Concepts of Agroecosystem

Nowadays, agroecosystem concepts are very clear among academicians, researchers, scientists, and policy makers which are discussed enormously in national and international platforms. It is a system of crops and livestock which is designed for sustainable agricultural productions and nurtured by human wisdom. Natural resource management and its conservation are gaining prime importance while practicing agroecosystem in any ecological regions. A subtle modification of natural ecosystem into more productive and protective agroecosystem delivers multifarious ecosystem services (Singh and Jhariya 2016; Jhariya et al. 2015). Application of ecological principle in sustainable-based agroecosystem makes more productive and profitable system that ensures a sustainable environment (Zhang et al. 2007). A sustainable intensification in agroecosystem involves climate-smart practices, CA, no-tillage practices, cover crops and mulching practices, and integrated and organic farming systems, respectively. These practices are more viable which makes more productive and sustainable agroecosystem in contrary to intensive and conventional agriculture (Felipe-Lucia et al. 2014). Ensuring sustainable intensification can enhance biodiversity that intensifies agroecosystem services which are the pillar of sustainable development. However, agroecosystem practices have been originated from various dimensions of ecosystem (environment), economic, and human level that must be explored to analyze agroecosystem complexity in both spatial and temporal scales (Bernués et al. 2014). Thus, the study of agroecosystem depends on the interactive components among human, crops, animals, and environment that needs to explore viability of agroecosystem sustainability.

7.3 Theoretical Model of Agroecosystem for a Sustainable Future

The practices of agroecosystem are based on various goals and principles having multifarious and uncountable significance. In this context, a theoretical model of agroecosystem for a sustainable future is depicted in Fig. 7.1. As per the figure, stage 1 comprises goals, principles, and significance of agroecosystem practices which include (a) proper blending of natural resources and its conservation, (b) increased land and water availability for high productive agroecosystem, (c) based on climate-resilient agro-farming principles, (d) principles of better ecosystem services that maintain soil-food-climate security, and (e) better ecological stability and environmental sustainability, while stage 2 adds some points such as ecological intensification, social and gender equality, effective agricultural policy, and scaling of better technology that must be considered for strengthening the practices of agroecosystem

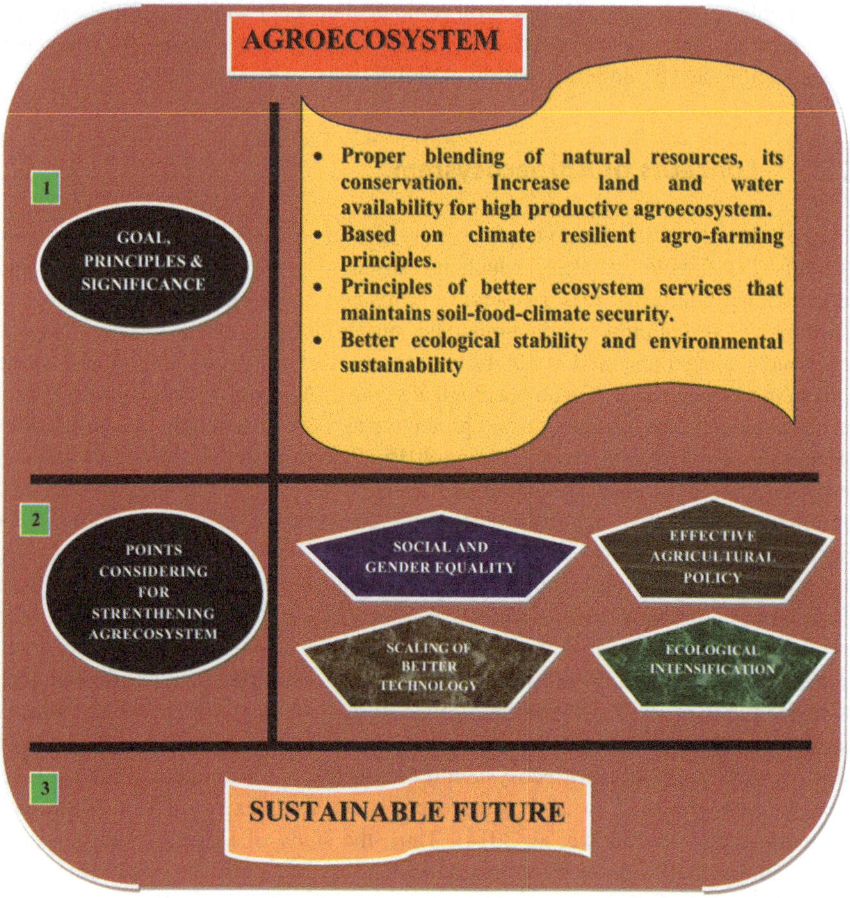

Fig. 7.1 Theoretical model of agroecosystem toward sustainability (Compiled: Peterson et al. 2018; Córdoba et al. 2020)

in any ecological zones of the regions. Therefore, the integration of stage 2 would promise and ensures a sustainable future of the mother earth (Peterson et al. 2018; Córdoba et al. 2020).

Thus, economic and ecological modeling approaches are two pillars that make a vibrant environment by promoting sustainable agroecosystem. However, both are anti-parallel, i.e., economical growth can be achieved but at the cost of ecological health. For that, a combined model can be possible in the form of theoretical model that promotes sustainable agroecosystem based on ecologically intensified principle. This model can be adaptable in diversified agroecological regions and their proper management through scientific research which promote crop and soil production in sustainable ways. A modeling approach must be placed in practical ground to see "How variable are these sustainable agroecosystem models in economical, social, and ecological context?" and "Does any theoretical model exist for sustainable agroecosystem services?". These questions merely depend on proper practicing of sustainable intensification in agroecosystem.

7.4 Ecosystem Services of Agroecosystem

Ecosystem services are an abstract concept that delivers uncountable benefits in terms of tangible and intangible ways to mankind. These services are delivering through either managed or unmanaged ecosystems, categorized into provisional, cultural, regulatory, and supporting services (Khan et al. 2020a, b). Timber, fuelwood, food, fruits, fodder (for livestock), non-wood products, etc. are delivered by provisioning services. These services of course are tangible and providing direct benefits as economic and market values. Pollution reductions, clean water and air, pest and disease control, erosion control, microclimate regulation, climate change mitigation, etc. are designed under regulatory services. These services are intangible benefits to our environment that maintains ecosystem processes. Likewise, aesthetic and ecotourism values are recognized under cultural services that are non-material benefits to humankind. The process of soil formation and efficient nutrient cycling are delivered by supporting services that are required to produce all other services for better ecosystem (MEA 2005).

As per one estimate, approx. 75% of earth's glacier-free land areas are covered by human-managed ecosystems (Ellis and Ramankutty 2008; Kumar et al. 2021). Agroecosystem practices are one of the human-managed practices that deliver multifarious and uncountable ecosystem services. A well-managed and ecology-oriented agroecosystem delivers both tangible (direct) and intangible (indirect) services. Food productions for sustaining populations, fuelwood and firewood productions, fodder/grass productions for animals, timber from multipurpose trees, and several harvestable goods are tangible agroecosystem services, whereas ecosystem maintenance, microclimate amelioration, soil enrichment, watershed management, erosion control, pest disease management, water regulations, and climate security are covered under intangible agroecosystem services. However, it is necessary to evaluate the ecosystem services from agroecosystem landscape for better understanding of on-farm and off-farm benefits to humans. In this context, Table 7.1 represents the key description of ecosystem services in agroecosystem and related on-farm and off-farm benefits (Garbach et al. 2014). As per Foley et al. (2005), approx. 40% of total earth's land areas are contributed by cropland- and pasture-based ecosystems that deliver valuable ecosystem services. However, average values (USD/ha/year) of ecosystem services from non-cropland terrestrial biomes are depicted in Fig. 7.2 (Costanza et al. 1997; Porter et al. 2009).

Among the indirect agroecosystem services, a pollination mechanism plays a key role in sexual reproduction in various plants, plantation crops, woody perennial trees, fruits, vegetables, seeds, and nuts (Klein et al. 2007; Painkra et al. 2016). Humans get micronutrients and calories by consumption of these wild edible plants (Sundriyal and Sundriyal 2004; Sheoran et al. 2021). As per Klein et al. (2007), around 60–90% of all plant species are regulated by pollination, while 35% of world

Table 7.1 Key description of ecosystem services in agroecosystem and related on-farm and off-farm benefits (Compiled: Garbach et al. 2014)

Types of ecosystem service in agroecosystem	Key description	On-farm benefits	Off-farm benefit
Delivery of tangible products such as food, fiber, fuelwood, and fodders (for livestock)	Goods harvested from agroecosystem	Consumption and sale of foods and other agricultural products in the same farm	Agricultural goods are utilized in off-farm markets
Regulation of water flow as hydrological services	Control and buffering of hydrological cycle which includes an efficient soil infiltration and runoff moderation	Better ground water recharge, water availability, and use efficiency for plant growth	Water flows and recharges to main stream and benefits for aquifer body. Reducing the flooding and stream stabilized
Ecosystem services such as soil fertility improvement along with better structure and texture	Maintenance of soil structure, nutrient availability, proper mobility, efficient nutrient cycling, and organic matter additions as SOC pools	Promotes plant growth and development that reduce fertilizer inputs	Off-farm manufacture of chemical/inorganic fertilizer that reduces the chance of mining
Pest disease controlling services	Insect pest controlling through natural enemies such as releasing of some pathogens, predators, and parasites	Minimize competition among crops and several harmful damages	Of course, it will definitely reduce and limit the requirement of harmful pesticides that threaten human health and environment
Carbon storage and sequestration	Process of absorbing and fixing atmospheric carbons into the plant (as biomass) and soil (as SOC pools) ecosystems	Biomass in terms of products as timber, food, and fuels	Mitigating GHGs and carbon balancing effects
Weed controlling services	Controlling populations and minimizing negative impacts of undesirable plants as IAS (Indian alien weeds), etc.	On-farm weed controls and reducing competitions with desirable plants in agroecosystem	It may definitely reduce and limit the requirement of harmful herbicides that threaten human health and environment
Aesthetic and cultural services	Maintain landscape values for social and cultural importance, heritage	Promotes socio-cultural, aesthetic, and religious value of important plants	Cultural, heritage, aesthetic, and ecotourism values

(continued)

Table 7.1 (continued)

Types of ecosystem service in agroecosystem	Key description	On-farm benefits	Off-farm benefit
	maintenances, religious values of plants, recreations, and ecotourism	and promotes ecotourism through on-farm services	through off-farm services

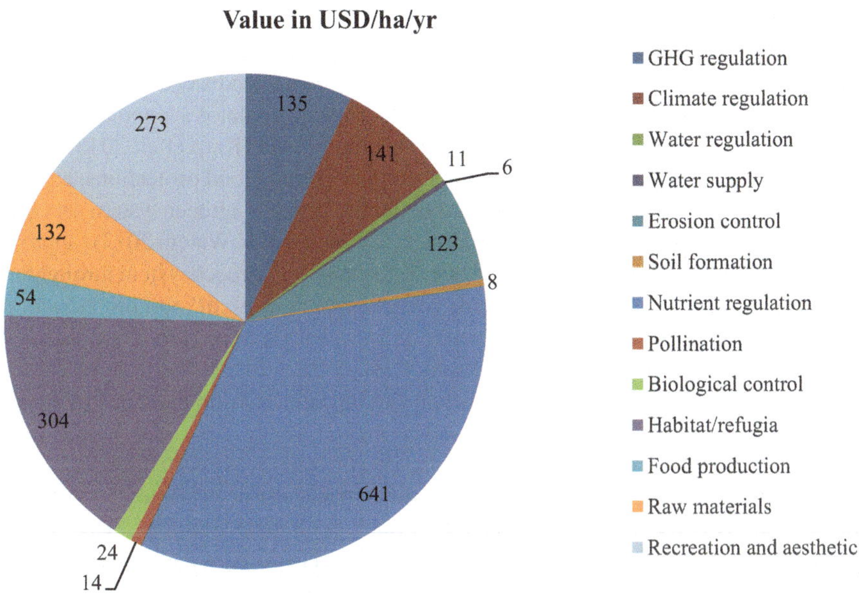

Fig. 7.2 Average value (USD/ha/year) of ecosystem services from non-cropland terrestrial biomes (Compiled: Costanza et al. 1997; Porter et al. 2009)

crop productions are covered by animal pollinators. Crops like sugarcane, cereals, and many important palatable grasses are pollinated through wind and other means. Thus, pollination-based ecosystem services help in balancing food production systems and agricultural economy.

The process of soil formation, structure maintenance, fertility enhancement, efficient nutrient cycling and nutrient use efficiency, soil microbial biomass, etc. is soil-based agroecosystem services. A scientific-based management and conservation of soils, etc. are important practices that intensify ecosystem services from agroecosystem (Zhang et al. 2007; Khan et al. 2021a, b). Healthy and quality soils indicate higher soil fertility and SOC pools that ensure healthy microbial populations and climate security. Sustainable soil management practices promise healthy ecosystem and maintain food and climate security. In this context, questions appear "How does sustainable agroecosystem ensure a better soil health and quality?" and

"Is there any synergy between soil health and sustainable agroecosystem services?". Study reveals that indeed a link exists between healthy soil and sustainable agroecosystem services. Applying sustainable intensification in agroecosystem based on ecological concept will ensure healthy and quality soils. In turn, well-maintained, healthy, and productive soils promise food and climate security that paves a way for achieving the goal of sustainable development.

Pest and disease management is another very important service provided by agroecosystem services. As per Oerke (2005), the global production of some important crops like maize, rice, wheat, cotton, and soybean is reduced by 8–15% due to animal pest attacks and resulted in 30 USD loss in the USA (Pimentel et al. 2005). Understanding pest behavior and predator ecology and their relationship ensures a good strategy for pest control services in agroecosystem (Landis et al. 2000). In this context, pest control services enhanced by increasing predator's population have been observed in farms through conserving natural habitat (Karp et al. 2013).

Similarly, water supply and regulations, air purification, flood protections, etc. are water-related ecosystem services that are delivered through agroecosystem services (Brauman et al. 2007; Meena et al. 2020a, b, c). As per UN Water (2013), approx. 70% of global freshwater consumption are contributed by agroecosystem throughout the world. Irrigated agricultural systems support approx. 40% of crop production in the world. Similarly, 90% of total water withdrawals in fast-growing economies are consumed by agriculture land use systems (USDA ERS 2013). Thus, these figures and estimates reflect water footprints which can be regulated and managed by better agroecosystem services.

In agroecosystem, low and poor biodiversity performs ecosystem disservices and poor services that need some better scientific management practices to promote higher biodiversity for better ecosystem and ecological stability. Diversified and prominent agroecosystem has inevitable potentials in ecosystem management and services to biomes. Many authors have reported biodiversity management and ecosystem services through better agroecosystem management. Production of food grains and quality fruits is required to maintain food and nutrition security through agricultural services. Production (timber, fuelwood, fodders, and various non-wood products) and protection benefits were obtained in the form of forest services, income security by diversified agroecosystem under economic services, and ameliorating microclimate by diversified systems. This ensures better climate services and C storage and sequestration to mitigate climate change issues and promotes environmental sustainability under environmental services. Further, diversified agroecosystem promotes healthy and nutritious food, fruits, and fodder for maintaining health status of human and animal. Moreover, it supports toward soil enrichment, efficient nutrient cycling, water and nutrient availability, pest disease control, and better pollination under regulatory and ecological services that are recognized in various agroecological regions of the world (Fig. 7.3) (Duru et al. 2015; Garbach et al. 2014).

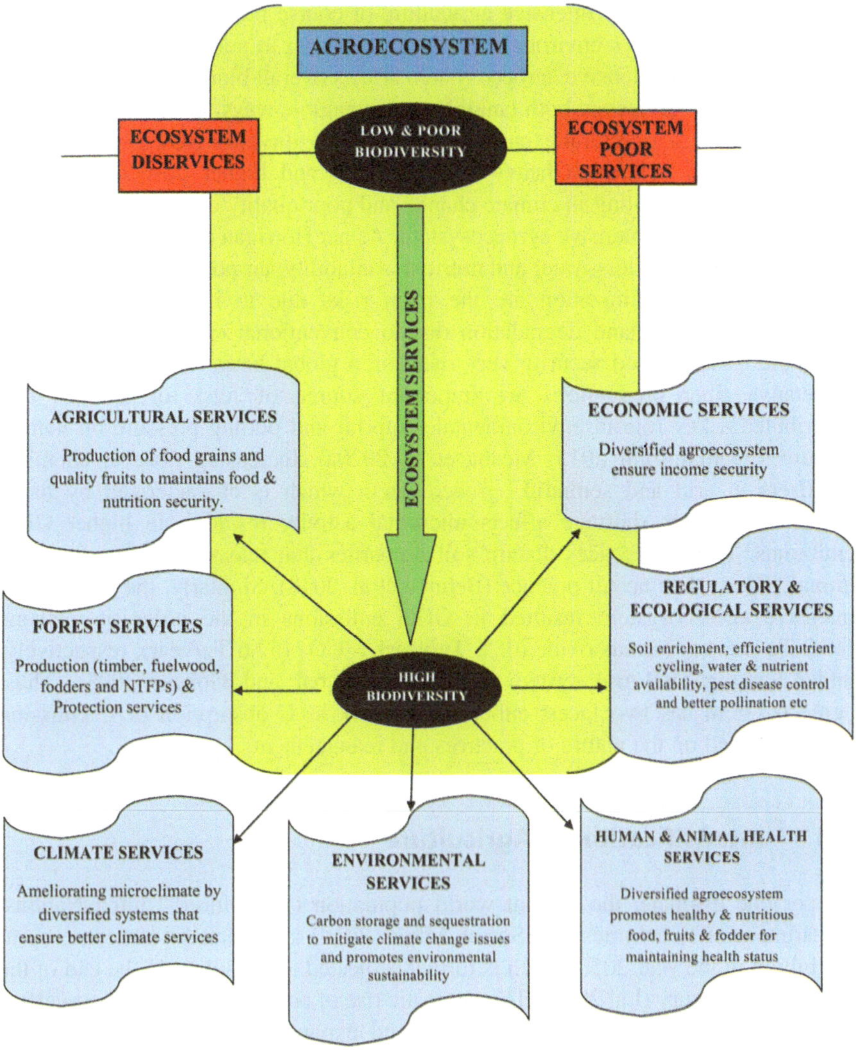

Fig. 7.3 Ecosystem services through agroecosystem (Compiled: Duru et al. 2015; Garbach et al. 2014)

7.5 Intensification in Agroecosystem

Growing populations necessitate food and shelters that lead to agricultural land expansion and intensifying practices in agroecosystem. Conventional cropping systems increase the overall productivity but at the cost of environment and human health. Intensive use of high synthetic inputs as chemical fertilizers and heavy mechanizations resulted in the degradation of land and environment

(Alexander et al. 2015). Intensive agriculture of course booms the agroecosystem production but destroys environmental health resulting in poor ecosystem services (Duru et al. 2015). Intensive agroecosystem affects overall biodiversity that leads to poor ecosystem services in both tangible and intangible ways. Deforestation due to land conversion, agricultural expansions, poor soil fertility, less SOC pools, and soil health degradation through heavy mechanization and higher dose of fertilizers increases GHGs resulting in climate change and poor quality of food which are the negative impacts of intensive agroecosystem. As per Horrigan et al. (2002), soil and water contamination, less water and nutrient availability, air pollution due to GHGs, erosion, and soil salinization are the main risks due to intensive agricultural practices. However, land degradation due to conventional cropping systems has made the issue of food security very risky on a global basis (Hurni et al. 2015). Vegetative fiber productions are important source of feed for animals that contributed a key role in environmental footprint and putting pressure on natural resources (Jankielsohn 2015; Meena et al. 2020a). Increasing dose of chemical fertilizers in arid and semiarid agroecosystem which is characterized by harsh climate and water shortage affects microbial activity resulting in higher GHG emissions. Tillage practices disturb soil properties that release 21% more carbon dioxide (CO_2) than no-till practice (Behnke et al. 2018). Similarly, the practices of intensive agroecosystem resulted in GHG emissions in the order of methane (4.13 Tg/year) > nitrous oxide (0.26 Tg/year) > CO_2 (52.6 Tg/year), respectively, under varying land use system in Pakistan (Iqbal and Goheer 2008). Thus, agroecosystem has two faces; either it can work as C absorption or C emission which is based on the nature of practices and management.

7.6 Intensification in Agriculture

As per one estimate, the current world population (7.7 billion, United Nations, Department of Economics and Social Affairs 2019) is expected to increase up to 9.5 billion in the year 2050 which is further projected to 12 billion in the end of the twenty-first century (Lal 2015). This dramatic rise of population obviously escalates food demands and promotes agriculture land expansion and intensive practices which affects the environment negatively. However, conversion of natural land to arable land and higher inputs of chemical and inorganic fertilizers destroy agriculture sustainability in the long term. Thanks to the Green Revolution for increasing higher production by intensive agricultural practices, it leads to land and environment degradation due to GHG emissions (Ajmal et al. 2018). Therefore, a query appears "how and to what extent do the intensive agriculture practices affect natural vegetations in the tropics?". This can be justified by one report of Ordway et al. (2017), and according to them, approx. 70% of natural vegetation is converted for agriculture in the tropics. However, decreasing biodiversity of wild flora and fauna are other deleterious impacts that have been observed due to intensive agricultural practices. This results in the decline in natural predators that control the pest damage which is further modified by increasing chemical inputs.

The practices of intercropping can mitigate the impact of intensive agriculture (Mkenda et al. 2019). Further, field margin management rather than field manipulation might be an effective strategy that enhances biodiversity of the farmland (Wiggers et al. 2016). In this context, it would be interesting to note "how does managing field margin ensure pest control in agroecosystem?". Greater management of field margins ensures food provisioning for predator and parasitoids that work as pest control services in well-managed agroecosystem (Ramsden et al. 2015).

7.7 Examples of Sustainable Intensification

The agriculture sector is the backbone for the world's economy of 80% population (FAO 2003). Nearly 80% of cropping areas comes under mixed cultivation in Western Africa (Steiner 1982). Some crops, such as *Triticum aestivum* (wheat), *Gossypium hirsutum* (cotton), *Oryza sativa* (rice), *Zea mays* (corn), *Coffea arabica* (coffee), *Arachis hypogaea* (ground nut), and *Theobroma cacao* (cocoa), hold major positions in world agroecosystem. While looking on the negative impacts of chemical and mechanical inputs, these practices are reduced up to a certain extent in many regions of the world. Instead of that, a sustainable intensification based on ecological concept needs to be practiced in agroecosystem.

Many sustainable-based practices are used for better health and productivity of agroecosystem in the tropics. For example, conservation practices, use of cover crops, and no-tillage practices are standard practices that maintain food-soil-climate sustainability in agroecosystem. Among all, no-tillage activity is widely adopted under sustainable practices that involves less soil disturbance and maintains soil and crop productivity. Promoting crop diversification, use of cover crops, and integrated farming practices can help in enhancing the effectiveness of no-tillage in agroecosystem. However, in certain extent, a negative impact of tillage practices is also observed in the region of North America where poor soil health occurs due to less soil productivity and erosion (Carr et al. 2012). Thus, no-tillage practices are becoming widely adopted farming techniques that spread throughout the world. For example, no-tillage-based farming systems covered 35.5% of total cropland areas in the USA (Horowitz et al. 2010). In this regard, queries appear (a) "how do no-tillage practices make agroecosystem sustainability?" and (b) "does any synergy exist between no-tillage practices and sustainable agroecosystem services?". These questions are based on one hypothesis as minimum soil disturbance and high fertility are observed under no-tillage practices. This enhances soil microbial biomass, nutrient availability, less leaching losses, efficient nutrient cycling, etc. that are enough to reflect better soil health and quality that ensure healthy and sustainable agroecosystem services. A combination of no-tillage with diverse crop rotation practices gives remarkable results in terms of SOC pools and related nutrient reserve in the overall agroecosystem. As compared to conventional and intensive practices, a sustainable-based agroecosystem promises higher SOC pools that ensure a better crop performance in the long term (Ibrahim et al. 2015).

Furthermore, use of cover crops, conservation practices, and climate-smart practices improves SOC pools, enhances efficient nutrient cycling, and minimizes leaching such as nitrogen losses that maintain environmental health and sustainability (Lemaire et al. 2014). Thus, these practices are pillars of a sustainable agroecosystem that makes higher productivity and profitability and maintains overall food-soil-climate security for a long-term sustainable basis (Reese et al. 2014).

7.8 Sustainable Intensification in Agroecosystem: Need and Potential

Intensive practices such as high synthetic inputs (chemical fertilizers, herbicides, pesticides, etc.) and heavy mechanizations obviously disturb the productivity and health of our environment and ecosystem. These practices intensify plant production but release some GHGs into the atmosphere. Higher application of inorganic fertilizers affects human and animal health by entering into the food chain and food web. The residual impacts of chemicals are undeniable that disturb the overall ecosystem health and processes. These deleterious impacts of intensive practices necessitate the application of sustainable intensification in agroecosystem based on ecological concept. Thus, sustainable-based agroecosystem not only minimizes negative impacts but also promotes food-soil-climate security that paves a way to achieve the goal of sustainable development (Whitfield et al. 2018).

The potential of sustainable intensification should not be underestimated due to its multifarious and uncountable significance. Ecological- and sustainable-based intensification maintains overall crop and soil productivity and profitability in a sustainable basis in the long term without affecting our environment (Garnett et al. 2013). Thus, sustainable intensification is a process and goal that makes a balance between economic and ecological performance (Gadanakis et al. 2015). Thus, sustainable-based agroecosystem promotes food productions, enhances biodiversity, intensifies ecosystem services, and mitigates climate change issue (Bernard and Lux 2017). Sustainable intensification-based agroecosystem and related ecosystem services are depicted in Table 7.2. Upscaling performance through sustainable intensification of agroecosystem is depicted in Fig. 7.4 (Andres and Bhullar 2016).

7.9 Sustainable Intensification for Climate-Resilient Agroecosystem

Extreme weather and uncertain rainfall promote unexpected monsoon that makes a blurred image on farmers. Climate change leads to the emergence of insect pest and infectious disease that destroy overall health and productivity of agroecosystem. Moreover, species-species interactions, vegetational shifting, and crops' morphological and phonological changes are induced by extreme weather conditions. Further, intensification increases the highly synthetic inputs in agroecosystem that increase GHGs into the atmosphere. In this context, the practices of ecological and

Table 7.2 Sustainable intensification-based agroecosystem and related ecosystem services

Type of ecosystem services	Benefits	Keywords	References
Maintenance, regulatory, and ecological services	Minimizing deforestation and biodiversity conservation in agroecosystem	Habitat and germplasm protections, forest health and productivity maintenance, sustainable forest management practices, etc.	Clough et al. (2011); Tscharntke et al. (2012); Lal et al. (2015)
	Soil fertility improvement, nutrient cycling, carbon sequestration, and watershed management	Soil health maintenance and climate security	Barrios et al. (2012); Mbow et al. (2014); Lal et al. (2015); Andres and Bhullar (2016)
	Controlling harmful pest and diseases through applying biological predators	Pest and disease control	Smith Dumont et al. (2014); Bieng et al. (2013); Lal et al. (2015); Gavinelli et al. (2020)
Provisioning services	Sustainable and diversified production of food, fuel, and timber, improved pollinations, enhanced lifespan of fruits and perennial crops	Food and nutritional security, biomass accumulations, and stable production system	Bisseleua et al. (2013); Lal et al. (2015); Duru et al. (2015)
Aesthetic, social, and cultural services	People's positive interactions with farming ecosystem, promotes moral and social values of plants in agroecosystem	Ecotourism and religious belief	Lal et al. (2015); Andres and Bhullar (2016)

sustainable intensification in agroecosystem can help in minimizing the climate change phenomenon without affecting global development. Sustainable intensification makes viable agroecosystem in terms of climate-resilient system and relates to climate-smart agriculture which is highly discussed by various international organizations (CGIAR, World Bank, and FAO) (Campbell et al. 2020). Thus, sustainable intensification-induced climate-resilient agroecosystem is based on the ecological principle of climate-smart agriculture that ensures triple-win situations as high productivity and reduces GHGs and resilience to climate change phenomenon (FAO 2019).

Climate-smart agroecosystem practices are based on ecological principles that imply sustainable productions of crops and soil productivity along with climate security by minimizing GHG emissions into the atmosphere. In this context, it is important to think "how can climate-smart agroecosystem practices achieve the goal of sustainable development?". Sustainable intensification-induced climate-smart

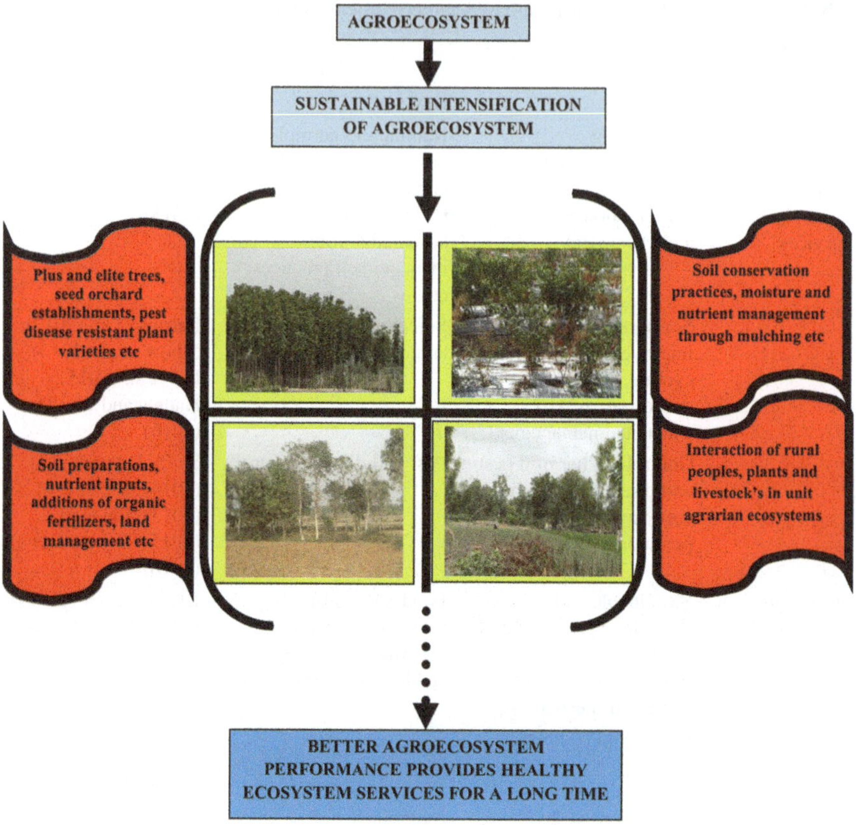

Fig. 7.4 Upscaling performance through sustainable intensification of agroecosystem (Compiled: Andres and Bhullar 2016)

agroecosystem has positive impacts on the health and productivity of humans, animals, soil, and climate that of course make a pathway for achieving the goal of sustainable development on a long-term basis (Clay and Zimmerer 2020). Enteric fermentation, extensive rice cultivation, tillage practices, biomass removal, and burning in agroecosystem emit GHGs into the atmosphere. These emissions destroy our environment health and break sustainability chain for nation development. Sustainable intensifying practices impair all health issue and productivity of agroecosystem which directly or indirectly affects our environment and ecology in positive ways. Thus, climate-smart agroecosystem protects our climate by minimizing the negative impacts of climate change and maintaining environmental sustainability and ecological stability at global scale.

7.10 Sustainable Intensification in Agroecosystem for Soil Health

Soil is the largest terrestrial natural resources that nurture diversified flora and fauna and inhabiting microbes. The existence of life on earth is meaningless in the absence of soils. Soil delivers multifarious and uncountable ecosystem services that help in achieving the goal of sustainability. Soil stores abundant organic C as 1500 gigaton which is quite more than vegetational and atmospheric C (Lal 2004; Tarnocai et al. 2009). This figure reflects the importance of soils in SOC pools which indirectly maintain C balance in the atmosphere and play a key role in climate change mitigation. But due to unsustainable land use practices, intensive agroecosystem, tillage practices, high synthetic inputs, and heavy mechanizations disturb the soil structure, health, and quality. As per Janssens et al. (2005), approx. 10% declining in SOC was observed due to unscientific and intensive farming practices in European country. Similarly, intensive and conventional farming practices reduce total SOC pools by 30–50% in the regions of the USA (Reicosky 2003). These figures are enough to express the negative impacts of intensive agroecosystem practices that make a fear for unsustainability. In this context, sustainable intensification practices in agroecosystem overcome these problems by enhancing both productivity and health status of soils.

Minimum practices of tillage and addition of optimum organic matter and its proper decomposition could be potentially viable for healthy and quality soils (Lal 2003). Less disturbance of soil due to reduced tillage practices can enhance soil formation process, better soil structure, enhance soil fertility and nutrient availability, and lessen erosion problems in sustainable intensifying agroecosystem (Stavi et al. 2016). The impact of tillage management on SOC in agroecosystem is depicted in Table 7.3.

Adoption of CA, mulching, and use of cover crops are important sustainable-based intensification that impairs soil health and quality in agroecosystem for a long-term basis. Mulching and use of cover crops have significant effects on soil through moisture conservation and evaporation reduction. Therefore, these practices enhance C sequestration capacity of soil that maintains the overall SOC status along with soil structure and quality improvement. As per Basch et al. (2012), well-managed ecologically and sustainable-based agroecosystem practices can store 0.2–-0.7 t ha^{-1} $year^{-1}$ of organic C pools into the soils. Intensive and conventional monocropping system has declined the soil quality by leaching and loss of essential nutrients that affects the overall productivity and farmer's profitability in the Indo-Gangetic Plain region of India. Sustainable intensification comprising CA played a viable role in restoring and managing the quality of soils (Choudhary et al. 2018). In this context, practices of managed agroecosystem are based on ecological principles which work as boon for healthy, productive soil and environment.

Table 7.3 Tillage management impacts on soil organic carbon in agroecosystem

Regions	Soil and climate	Types of tillage and durations	Impacts on SOC	References
Ohio state (USA)	Silt loam type of soil characterized by 1000 mm of annual precipitation	No-tillage, conventional tillage, and plow tillage for 49 years of duration	Higher value was observed in 0–20 cm depth under no-tillage practices	Kumar et al. (2012)
Colorado region in the USA	Silt loam type of soil characterized by 418 mm of annual precipitation	No-tillage, conventional tillage, and rotary tillage for 39 years of duration	Higher value (32.0 mg/ha) was observed in 0–30 cm depth under no-tillage practices followed by 31.6 and 26.5 mg/ha in rotary tillage and conventional tillage practices	Mikha et al. (2013)
Minnesota region in the USA	Silt loam type of soil is prevalent in this region	No-tillage, moldboard plow, and chisel plow for 23 years of duration	Higher value (30%) of SOC was observed in the depth of 0–20 cm under no-tillage practices than moldboard plow and chisel plow	Dolan et al. (2006)
Texas region in the USA	Silt clay loam type of soil characterized by 980 mm of annual precipitation	No-tillage and conventional tillage for 23 years of duration	Higher value (72%) was observed in the depth of 0–5 cm under conventional tillage than no-tillage	Wright and Hons (2005)
Lleida region in Spain	Silt loam type of soil characterized by 430 mm of annual precipitation	No-tillage, conventional tillage, rotary tillage, and subsoil tillage for 21 years of duration	Higher value (12.8 mg/ha) was observed in the depth of 0–5 cm under no-tillage practices followed by 9.1 mg/ha in rotary tillage, 7.7 mg/ha in subsoil tillage, and 5.6 mg/ha in conventional tillage, respectively	Álvaro-Fuentes et al. (2008)
Harare region in Zimbabwe	Clay type of soil characterized by 800–1000 mm of annual precipitation	Conventional tillage, clean ripping, and tied ridging for 19 years of duration	Higher value (20.4 mg/g) was observed under tied ridging followed by 16.8 mg/g in clean ripping and 14.9 mg/g in conventional tillage, respectively	Chivenge et al. (2007)

(continued)

Table 7.3 (continued)

Regions	Soil and climate	Types of tillage and durations	Impacts on SOC	References
Hebei region in China provenance	Silt loam type of soil characterized by 480.7 mm of annual precipitation	No-tillage, plow tillage, and rotary tillage for 14 years of duration	Higher value (12%) was observed in 0–5 cm depth under no-tillage followed by 5.6% in 10–20 cm depth under plow tillage	Zhao et al. (2015)

7.11 Drivers for Agroecosystem Sustainability

The word "sustainable agroecosystem" implies a highly productive and profitable agricultural ecosystem regulated by sustainable intensification and ecological concept without destroying environmental health. CA, use of cover crops, and no-tillage practices are important drivers that promote agroecosystem sustainability with healthy environment. CA is indeed a good driver based on sustainable intensification that improves productivity and profitability, maintains food security, and promotes resource conservations and environmental sustainability. CA was practiced over 157 Mha by 2013. However, CA is based on three principles such as minimum soil disturbance, crop diversifications, and maintenance of SOC pools in any agroecosystem (FAO 2015). Thus, drivers like CA, non- or zero-tillage (NT), and use of cover crops improve soil physicochemical properties and regulate water infiltration and overall food-soil-climate security through a sustainable agroecosystem.

Therefore, it can be discussed "is sustainable intensification possible in agroecosystem?". If it is possible, then "how can we relate and synergize intensification to sustainability for agroecosystem?". Further, "what are the drivers that promote sustainable agroecosystem?". Four relations exist while we talk about the synergy between intensification and sustainability in agroecosystem management which is depicted in Fig. 7.5. Substitute and complement relations appeared under the study of intrinsic relationship between intensification and sustainability in agroecosystem. This will signify our understanding while exploring the agroecosystem practices and management in any agroecological regions of the world. For example, low intensification will promote high sustainability of agroecosystem and vice versa. This is due to the low addition of external inputs (chemical fertilizers, using heavy machinery, etc.) that will exert less pressure on land that leads to high sustainability in productions and protection of agroecosystem. Similarly, this hypothesis will ensure healthy agroecosystem in terms of higher productivity, soil enrichment, fertility enhancement, maximizing microbial population, better nutrient use efficiency, healthy rhizosphere, food and nutritional security, and overall soil-food-climate security at global scale. However, sustainable land management practices, soil fertility improvement practices, organic fertilizer inputs,

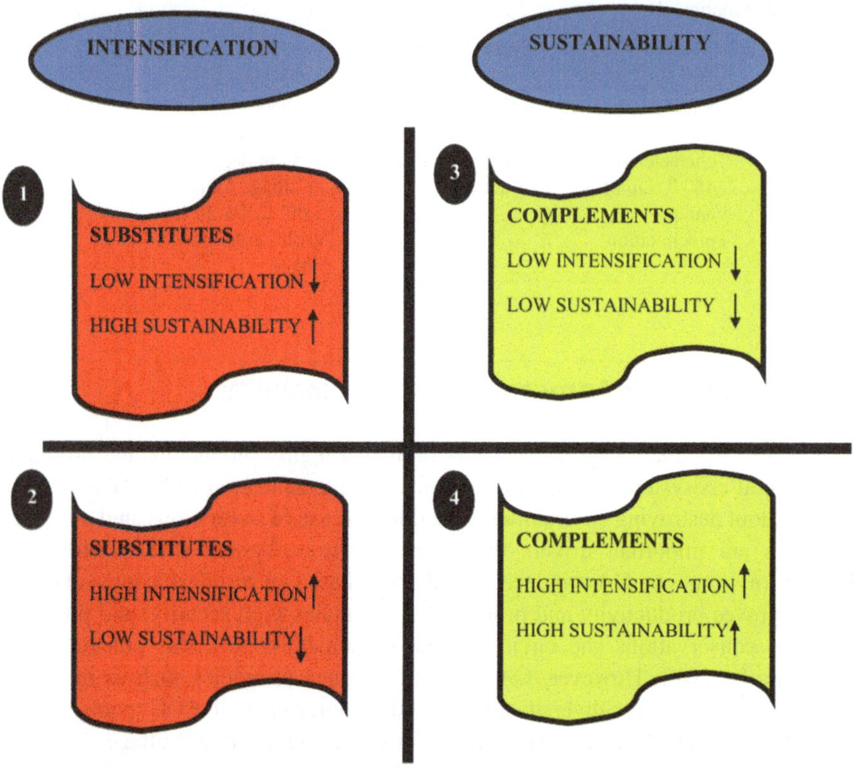

Fig. 7.5 Synergy between intensification and sustainability in agroecosystem (Compiled from Mutyasira et al. 2018)

mulching, soil conservation practices, and litter inputs from woody perennial trees and crops are key drivers that accelerate healthy and quality agroecosystem that promotes environmental sustainability and ecological stability (Mutyasira et al. 2018; Meena et al. 2020b).

7.12 Net Primary Productivity and Carbon Footprint in Agroecosystem

Lowering C footprints in parallel to increasing net primary productivity (NPP) in agroecosystem needs to be connected with sustainable land management practices. NPP represents net biomass production of plants per year in any geographical locations. As per Janzen (2004), total NPP productions were 6.80 and 5.30 Pg C/year in cropland and pastureland areas. Regional climate affects overall NPP values which are further modified by water, soil, and biological processes. In turn, agriculture systems and world NPP play a key role in the global C cycle. As per Lal (2004), increasing CO_2 level due to anthropogenic factors affects the overall food and climate security. As per IPCC (2013), around 848.0 Pg C have been surpassed by earth's atmosphere, and this value is increasing at the rate of 4.0 Pg C/year Around 11% (1.43 billion ha) and 27% (3.51 billion ha) of the total world land area are shared by global croplands and pasturelands which store around 765 Pg C (FAOSTAT 2015). Soil C budget relates to NPP which reflects the potential of agroecosystem. Altered C dynamics and varying land use systems affect global C balance and footprints (Banerjee et al. 2021a, b). As per Houghton et al. (1999), the value of NPP is controlled by varying climatic situations in regional scale rather than agroecosystem impacts on NPP at local scales. However, Monfreda et al. (2008) have reported agroecosystem impacts on NPP at global scales. In this context, the major question is "how does sustainable agroecosystem affect the overall NPP and carbon footprint?"

NPP estimations are necessary to see the structure and functions of ecosystem which helps in the exploration of productivity and sequestration of C (Deal 2011). Cultivation practices, site quality, soil nutrient load, soil physicochemical properties, and climatic variability are the major factors that affect overall NPP. However, both grazing and removal of biomass affect the overall NPP. Removal of plant residues after harvesting has deleterious impacts on water conservation, overall SOC, and plant yields (Blanco-Canqui and Lal 2007). As per the World Bank (2012), 30% of world GHG emissions are contributed by intensive agricultural practices. Globally, 70% and 40% of nitrous oxide and methane are emitted from intensive agricultural practices (World Bank 2015a, b). The practice of sustainable agroecosystem can potentially sequester the atmospheric CO_2 and promotes NPP. Intensive agroecosystem practice comprises fertilizer inputs and heavy mechanization directly or indirectly affecting overall NPP and might result in negative C footprints by influencing overall soil health and related ecosystem services (Banerjee et al. 2021c, d).

7.13 Challenges Toward Sustainable Intensification in Agroecosystem

Climate is a key factor that determines the overall health and productivity of agroecosystem. Extreme weather and changing climate are the major challenges today that affect the overall sustainability of agroecosystem. For example, extreme temperature and moisture variability have created difficulties in achieving the goal of agroecosystem sustainability in semiarid region (Stewart et al. 1991). Rising temperature in parallel with decreasing rainfall simultaneously affects the overall sustainable agroecosystem in the arid tropics. This climatic variability affects crop and soil productivity and health of sustainable agroecosystem. Declining soil health and quality are other challenges in agroecosystem that can be overcome by applying sustainable-based agroecosystem approaches. However, frequent changing soil types and varying nutrient loads affect the applicability of sustainable intensification in agroecosystem.

7.14 Research and Design for Sustainable Agroecosystem

The researchon agricultural production is now shifted to study on environmental footprints (Walbridge and Shafer 2011). Traditionally, research and design of agroecosystem are prevailed into small plots but it should be extended into large areas up to temporal and spatial scale (Robertson et al. 2008).However, a research on smaller scale of agroecosystem could not reflecta better understanding of agroecosystem functions at global scale (Carpenter et al. 2006). Therefore, these large scales of long term agroecosystem studies reflect sustainability of the overall systems. Furthermore, a network of LTAR (Long-Term Agroecosystem Research) has been initiated by USDA which has proposed around 18 sites in this network (Karlen et al. 2014). A variety of information such as soil management, watershed development, cropping and grazing systems etc are provided by LTAR based sites (Sadler et al. 2015).

A world of sustainable agroecosystem can be framed by doing effective research and development. Various research and development have been prioritized, and these are (a) germplasm and genetic resources, (b) land and water productivity enhancement, (c) effective policies for enhancing value chains, (d) resilience buildup, and (e) research for climate-resilient practices. These practices reform the agroecosystem in a more sustainable way by collecting diverse types of heat-, cold-, and disease-resistant plant varieties undermining the genetic resources by proper blending of natural resources and its conservation that enhance the land and water productivity in irrigated, rainfed, and agrosilvopastoral systems and by creating some effective policies for enhancing value chains of diversified agricultural products and its value additions that will ensure income security for poor farmers. Similarly, an integration of plant-animal farming systems promises socioeconomic and environmental security along with the adoption of research for climate-resilient practices, i.e., conventional, breeding, and molecular research ensures resilience of crops to climate change and makes a more sustainable agroecosystem at global scale. In lieu of the above, research and development priorities for a sustainable agroecosystem world are depicted in Fig. 7.6 (Andres and Bhullar 2016; Smith et al. 2017).

Sustainability in agroecosystem is utmost important for proper functioning of ecosystem. This sustainability can be achieved through well managed and ecology oriented agroecosystem practices. Proper understanding of tree-crop interaction, crop combinations, soil management practices, etc. based research and design are considered for sustainable agroecosystem. An efficient research approach and related design can intensify crop and soil productivity of agroecosystem in sustainable way. Therefore, design should be in favor of high crop diversification, maximizing crop productivity, improving soil health and fertility, enhancing C sequestration, etc. in both vegetation and soils. Thus, a proper design and research in agroecosystem management can ensure social, economic and sustainability goals for long-term.

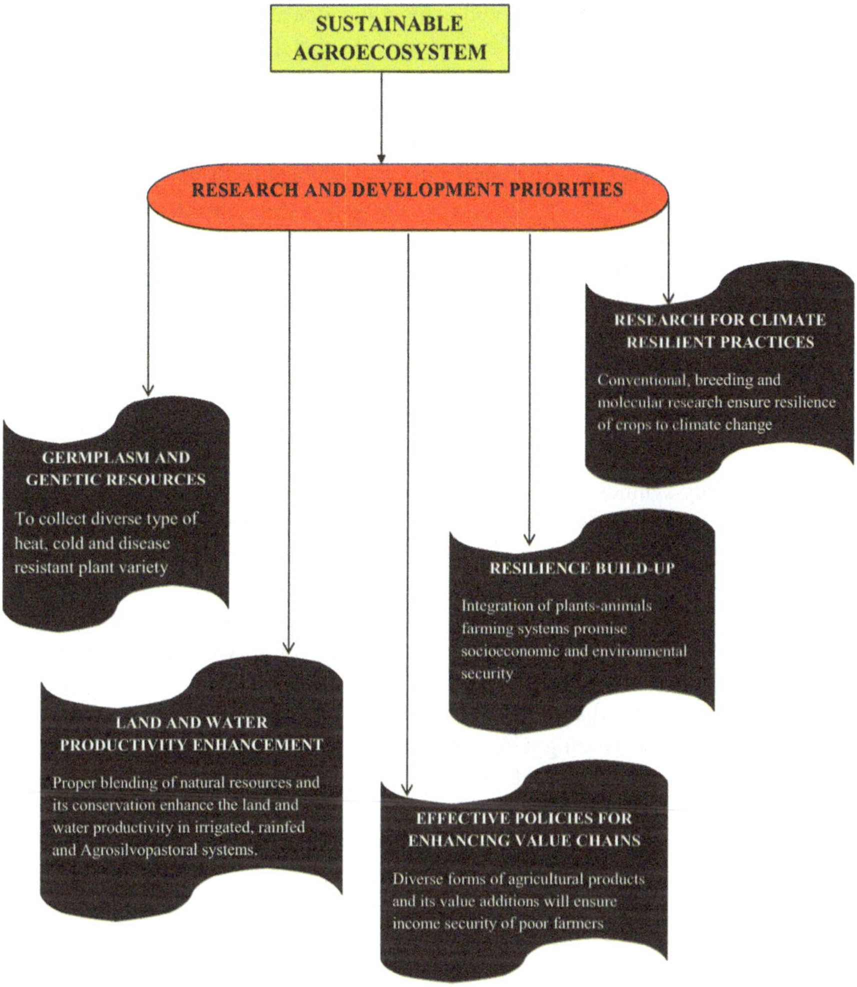

Fig. 7.6 Research and development priorities for a sustainable agroecosystem world (Modified: Andres and Bhullar 2016; Smith et al. 2017)

7.15 Policy and Regulatory Frameworks

The policy and regulatory frameworks must be practiced in agroecosystem ground for creating sustainable models that enhance crop-soil productivity and maintain ecosystem processes. Applying sustainable intensification, improved agroecosystem practices, proper rice management in agroecosystem, proper land use practices, and integrated nutrient management are thrust areas of research that should be regulated by effective policy frameworks. Policy must be adopted to promote climate-resilient agroecosystem that minimizes GHG emissions (Pattey et al. 2005). Integrating leguminous tree in agroecosystem will explore the idea about the nature of tree-crop interaction, N_2 fixation ability, soil enrichment, etc. that also fulfill the forest and tree covers given by the National Forest Policy. Sustainable management of tree outside forestry (TOF) as social and farm forestry promotes agroecosystem structure and services. Policy must be framed in accordance to better tree-crop combination, integration of plus trees in agroecosystem, and establishment of seed orchard for highly productive and disease-free agroecosystem. Thus, an effective and rigid policy is required for building up of sustainable agroecosystem that maintains overall food-soil-climate security at global scale.

7.16 Conclusions

Intensive agroecosystem destroys the natural resources, declining the environmental health and quality. Poor ecosystem services are observed in intensified agricultural practices. Highly synthetic inputs and heavy mechanization destroy soil quality and release GHGs leading to global warming. Applying sustainable intensification based on ecological concept is boon for healthy agroecosystem. Ecosystem services through sustainable agroecosystem are highly significant in terms of productivity and profitability. Soil enrichment, sustainable food production, food and nutritional security, climate security, etc. are verified ecosystem services possible through sustainable agroecosystem. Practicing sustainable intensification promotes climate-resilient agroecosystem that minimizes the GHG emissions and mitigates changing climate. However, practicing CA, zero-tillage, use of cover crops, etc. reduce emissions and enhance crop and soil productivity for a long time. Great attention must be drawn by policy makers for promoting sustainable agroecosystem globally that could help in creating a sustainable world.

7.17 Future Roadmap

Integration of leguminous N_2-fixing tree, multipurpose tree, underutilized crops, medicinal and aromatic plants, disease-free and highly productive plants, etc. in agroecosystem makes a future roadmap plan for research and development. A roadmap must be created to see a sustainable world through intensive free agroecosystem. Less use of chemical and fertilizer in agroecosystem promotes a pollution-free world. A roadmap should be framed on location-specific agroecosystem models and related varying productivity and profitability. This, of course, gives a better understanding of agroecosystem potential in ecosystem services for creating a sustainable world. Transforming food system in a sustainable form and maintaining nutritional and health security are the major concerns today (Gliessman 2016). This can be achieved by applying sustainable intensification approaches in agroecosystem. Thus, making a broader roadmap for a highly productive agroecosystem not only satisfies global food needs but also makes a healthier and sustainable world. A paradigm shift in agroecosystem to explore soil health, crop productivity, timber quality, livestock's health, and environmental health is necessary for achieving the goal of sustainable development. Roadmap for climate-resilient and productive agroecosystem is an important concern today that could be achieved by applying sustainable intensification. This could not only help in maintaining food-soil-climate security but also maintains environmental sustainability and ecological stability.

References

Ajmal M, Ali HI, Saeed R, Akhtar A, Tahir M, Mehboob MZ, Ayub A (2018) Biofertilizer as an alternative for chemical fertilizers. Res Rev J Agric Allied Sci 7(1):1–7

Alexander P, Rounsevell MDA, Dislich C, Dodson JR, Engstrom K, Moran D (2015) Drivers for global agricultural land use change: the nexus of diet, population, yield and bioenergy. Glob Environ Chang 35:138–147. https://doi.org/10.1016/j.gloenvcha.2015.08.011

Altieri MA (1999) The ecological role of biodiversity in agroecosystems. Agric Ecosyst Environ 74:19–31

Álvaro-Fuentes J, López M, Cantero-Martinez C, Arrúe J (2008) Tillage effects on soil organic carbon fractions in Mediterranean dryland agroecosystems. Soil Sci Soc Am J 72:541–547

Andres C, Bhullar GS (2016) Sustainable intensification of tropical agro-ecosystems: need and potentials. Front Environ Sci 4:5. https://doi.org/10.3389/fenvs.2016.00005

Banerjee A, Jhariya MK, Yadav DK, Raj A (2020) Environmental and sustainable development through forestry and other resources. Apple Academic Press Inc., CRC Press- a Taylor and Francis Group, US & Canada, 400 pp. https://doi.org/10.1201/9780429276026

Banerjee A, Meena RS, Jhariya MK, Yadav DK (2021a) Agroecological footprints management for sustainable food system. Springer, Singapore, 514 pp. https://doi.org/10.1007/978-981-15-9496-0

Banerjee A, Jhariya MK, Meena RS, Yadav DK (2021b) Ecological footprints in agroecosystem- an overview. In: Banerjee A, Meena RS, Jhariya MK, Yadav DK (eds) Agroecological footprints management for sustainable food system. Springer, Singapore, pp 1–23. https://doi.org/10.1007/978-981-15-9496-0_1

Banerjee A, Jhariya MK, Raj A, Yadav DK, Khan N, Meena RS (2021c) Land footprint management and policies. In: Banerjee A, Meena RS, Jhariya MK, Yadav DK (eds) Agroecological footprints management for sustainable food system. Springer, Singapore, pp 221–246. https://doi.org/10.1007/978-981-15-9496-0_7

Banerjee A, Jhariya MK, Raj A, Yadav DK, Khan N, Meena RS (2021d) Energy and climate footprint towards the environmental sustainability. In: Banerjee A, Meena RS, Jhariya MK, Yadav DK (eds) Agroecological footprints management for sustainable food system. Springer, Singapore, pp 415–443. https://doi.org/10.1007/978-981-15-9496-0_14

Barrios E, Sileshi GW, Shepherd K, Sinclair F (2012) Agroforestry and soil health: linking trees, soil biota and ecosystem services. In: Wall DH, Bardgett RD, Behan-Pelletier V, Herrick JE, Jones TH, Ritz K, Six J, Strong DR, van der Putten WH (eds) Soil ecology and ecosystem services. Oxford University Press, Oxford, pp 315–330

Basch G, Kassam A, Gonzalez-Sanchez EJ, Streit B (2012) Making sustainable agriculture real. CAP 2020. Life+Agricarbon project. http://www.ecaf.org/docs/ecaf/ca%20and%20cap%202020.pdf

Behnke GD, Zuber SM, Pittelkow CM, Nafziger ED, Villamil MB (2018) Long-term crop rotation and tillage effects on soil greenhouse gas emissions and crop production in Illinois, USA. Agric Ecosyst Environ 261:62–70

Bernard B, Lux A (2017) How to feed the world sustainably: an overview of the discourse on agroecology and sustainable intensification. Reg Environ Chang 17:1279–1290. https://doi.org/10.1007/s10113-016-1027-y

Bernués A, Rodríguez-Ortega T, Ripoll-Bosch R, Alfnes F (2014) Socio-cultural and economic valuation of ecosystem services provided by Mediterranean mountain agroecosystems. PLoS One 9(7):e102479

Bieng MAN, Gidoin C, Avelino J, Cilas C, Deheuvels O, Wery J (2013) Diversity and spatial clustering of shade trees affect cacao yield and pathogen pressure in Costa Rican agroforests. Basic Appl Ecol 14:329–336. https://doi.org/10.1016/j.baae.2013.03.003

Bisseleua HBD, Fotio D, Yede Missoup AD, Vidal S (2013) Shade tree diversity, cocoa pest damage, yield compensating inputs and farmers' net returns in West Africa. PLoS One 8: e56115. https://doi.org/10.1371/journal.pone.0056115

Blanco-Canqui H, Lal R (2007) Soil and crop response to harvesting corn residues for biofuel production. Geoderma 141:355–362

Brauman KA, Daily GC, Duarte TK, Mooney HA (2007) The nature and value of ecosystem services: an overview highlighting hydrologic services. Ann Rev Environ Resour 32:67–98

Campbell BM, Thornton P, Zougmoré R, van Asten P, Lipper L (2020) Sustainable intensification: what is its role in climate smart agriculture? Curr Opin Environ Sustain 8:39–34

Carpenter SR, De Fries R, Dietz T, Mooney HA, Polasky S, Reid WV (2006) Millennium ecosystem assessment: research needs. Science 314:257–258

Carr PM, Mäder P, Creamer NG, Beeby JS (2012) Editorial: overview and comparison of conservation tillage practices and organic farming in Europe and North America. Renew Agric Food Syst 27:2–6

Chivenge P, Murwira H, Giller K, Mapfumo P, Six J (2007) Long-term impact of reduced tillage and residue management on soil carbon stabilization: implications for conservation agriculture on contrasting soils. Soil Tillage Res 94:328–337

Choudhary M, Jat HS, Datta A, Yadav AK, Sapkota TB, Mondal S, Meena RP, Sharma PC, Jat ML (2018) Sustainable intensification influences soil quality, biota, and productivity in cereal-based agroecosystems. Appl Soil Ecol 126:189–198

Clay N, Zimmerer KS (2020) Who is resilient in Africa's green revolution? Sustainable intensification and climate smart agriculture in Rwanda. Land Use Policy 97:104558. https://doi.org/10.1016/j.landusepol.2020.104558

Clough Y, Barkmann J, Juhrbandt J, Kessler M, Wanger TC, Anshary A, Buchori D, Cicuzza D, Darras K, Putra DD, Erasmi S, Pitopang R, Schmidt C, Schulze CH, Seidel D, Steffan-Dewenter I, Stenchly K, Vidal S, Weist M, Wielgoss AC, Tscharntke T (2011) Combining high biodiversity with high yields in tropical agroforests. Proc Natl Acad Sci U S A 108:8311–8316. https://doi.org/10.1073/pnas.1016799108

Congreves KA, Hayes A, Verhallen EA, Van Eerd LL (2015) Long-term impact of tillage and crop rotation on soil health at four temperate agroecosystems. Soil Till Res 152:17–28

Córdoba C, Triviño C, Toro Calderón J (2020) Agroecosystem resilience. A conceptual and methodological framework for evaluation. PLoS One 15(4):e0220349. https://doi.org/10.1371/journal.pone.0220349

Costanza R, d'Arge R, De Groot RS, Farber S, Grasso M, Hannon B, Limburg K, Naeem S, O'Neill RV, Paruel J, Raskin RG, Sutton P, Van den Belt M (1997) The value of the world's ecosystem service and natural capital. Nature 387:253–260

Deal MW (2011) Net primary production in three bioenergy crop systems following land conversion. Thesis and dissertation. The University of Toledo Digital Repository, The University of Toledo, Toledo, OH

Dolan M, Clapp C, Allmaras R, Baker J, Molina J (2006) Soil organic carbon and nitrogen in a Minnesota soil as related to tillage, residue and nitrogen management. Soil Tillage Res 89:221–231

Duru M, Therond O, Martin G, Martin-Clouaire R, Magne MA, Justes E, Journet EP, Aubertot JN, Savary S, Bergez JE, Sarthou JP (2015) How to implement biodiversity-based agriculture to enhance ecosystem services: a review. Agron Sustain Dev 35:1259–1281. https://doi.org/10.1007/s13593-015-0306-1

Ellis EC, Ramankutty N (2008) Putting people in the map: anthropogenic biomes of the world. Front Ecol Environ 6:439–447
Emmerson M, Morales MB, Oñate JJ, Batáry P, Berendse F, Liira J, Aavik T, Guerrero I, Bommarco R, Eggers S, Part T, Tscharntke T, Weisser W, Clement L, Bengtsson J (2016) How agricultural intensification affects biodiversity and ecosystem services. Adv Ecol Res 55:43–97
FAO (2003) Synthesis report on the role of agriculture on environmental externalities. Food and Agriculture Organization. http://ftp.fao.org/es/esa/roa/pdf/2.../Environment_MaliNA.pdf
FAO (2015) Conservation agriculture. Food and Agriculture Organization of the United Nations, Rome. http://www.fao.org/ag/ca/1a.html
FAO (Food and Agriculture Organization of the United Nations) (2019) Climate-smart agriculture and the sustainable development goals: mapping interlinkages, synergies and trade-offs and guidelines for integrated implementation, 2019. FAO, Rome
FAOSTAT (2015) Food and Agriculture Organization of the United Nations Statistics Division. FAO, Rome
Felipe-Lucia MR, Comín FA, Bennett EM (2014) Interactions among ecosystem services across land uses in a floodplain agroecosystem. Ecol Soc 19(1):20
Foley JA, DeFries R, Asner GP, Barford C, Bonan G, Carpenter SR, Chapin FS, Coe MT, Daily GC, Gibbs HK, Helkowski JH, Holloway T, Howard EA, Kucharik CJ, Monfreda C, Patz JA, Prentice IC, Ramankutty N, Snyder PK (2005) Global consequences of land use. Science 309 (5734):570–574. https://doi.org/10.1126/science.1111772
Gaba S, Bretagnolle F, Rigaud T, Philippot L (2014) Managing biotic interactions for ecological intensification of agroecosystems. Front Ecol Evol 2:29. https://doi.org/10.3389/fevo.2014.00029
Gadanakis Y, Bennett R, Park J, Areal FJ (2015) Evaluating the sustainable intensification of arable farms. J Environ Manag 150:288–298. https://doi.org/10.1016/j.jenvman.2014.10.005
Garbach K, Milder JC, Montenegro M, Karp DS, DeClerck FAJ (2014) Biodiversity and ecosystem services in agroecosystems. Encycl Agric Food Syst. https://doi.org/10.1016/B978-0-444-52512-3.00013-9
Garnett T, Appleby MC, Balmford A, Bateman IJ, Benton TG, Bloomer P, Burlingame B, Dawkins M, Dolan L, Fraser D, Herrero M, Hofmann I, Smith P, Thornton PK, Toulmin C, Vermeulen SJ, Godfray HCJ (2013) Sustainable intensification in agriculture: premises and policies. Science 341:33–34. https://doi.org/10.1126/science.1234485
Gavinelli F, Fusaro S, Ivan D, Ragusa S, Paoletti MG (2020) Unpredicted ecological and ecosystem services of biodiversity. Spontaneous vegetation, hedgerows, and maple trees as useful landscape components to increase predatory mite population in agroecosystems. Appl Soil Ecol 154:103584. https://doi.org/10.1016/j.apsoil.2020.103584
Gliessman SR (2015) Agroecology: the ecology of sustainable food systems, 3rd edn. CRC Press, Boca Raton, FL
Gliessman S (2016) Transforming food systems with agroecology. Agroecol Sustain Food Syst 40 (3):187–189. https://doi.org/10.1080/21683565.2015.1130765
Horowitz J, Ebel R, Ueda K (2010) "No-till" farming is a growing practice. Economic information bulletin no. 70. USDA Economic Research Service, Washington, DC
Horrigan L, Lawrence RS, Walker P (2002) How sustainable agriculture can address the environmental and human health harms of industrial agriculture. Environ Health Perspect 110:445–456
Houghton RA, Hackler JL, Lawrence KT (1999) The U.S. carbon budget: contributions from land-use change. Science 285:574–578
Hurni H, Giger M, Liniger H, Studer RM, Messerli P, Portner B, Schwilch G, Wolfgramm B, Breu T (2015) Soils, agriculture and food security: the interplay between ecosystem functioning and human well-being. Curr Opin Environ Sustain 15:25–34. https://doi.org/10.1016/j.cosust.2015.07.009
Ibrahim M, Alhameid A, Kumar S, Chintala R, Sexton P (2015) Long-term tillage and crop rotation impacts on a northern great Plainsmollisol. Adv Crop Sci Tech 3:2

IPCC (2013) Summary for policymakers. In: Stocker TF, Qin D, Plattner GK, Tignor M, Allen SK, Boschung J (eds) Climate change 2013: the physical science basis. Working group I contribution to the fifth assessment report of the intergovernmental panel on climate change. Cambridge University Press, Cambridge, UK and New York, pp 1–27

Iqbal MM, Goheer MA (2008) Greenhouse gas emissions from agro-ecosystems and their contribution to environmental change in the Indus Basin of Pakistan. Adv Atmos Sci 25:1043

Jankielsohn A (2015) The hidden cost of eating meat in South Africa: what every responsible consumer should know. J Agric Environ Ethics 28:1145–1157. https://doi.org/10.1007/s10806-015-9581-8

Janssens IA, Freibauer A, Schlamadinger B, Ceulemans R, Ciais P, Dolman AJ, Heimann M, Nabuurs GJ, Smith P, Valentini R, Schulze ED (2005) The carbon budget of terrestrial ecosystems at country scale – a European case study. Biogeosciences 2:15–26

Janzen HH (2004) Carbon-cycling in earth systems – a soil science perspective. Agric Ecosyst Environ 104:399–417

Jhariya MK, Bargali SS, Raj A (2015) Possibilities and perspectives of agroforestry in Chhattisgarh. In: Zlatic M (ed) Precious forests-precious earth. InTech, Croatia, pp 237–257. https://doi.org/10.5772/60841

Jhariya MK, Yadav DK, Banerjee A (2018a) Plant mediated transformation and habitat restoration: phytoremediation an eco-friendly approach. In: Gautam A, Pathak C (eds) Metallic contamination and its toxicity. Daya Publishing House, A Division of Astral International, New Delhi, pp 231–247

Jhariya MK, Banerjee A, Yadav DK, Raj A (2018b) Leguminous trees an innovative tool for soil sustainability. In: Meena RS, Das A, Yadav GS, Lal R (eds) Legumes for soil health and sustainable management. Springer, Singapore, pp 315–345. https://doi.org/10.1007/978-981-13-0253-4_10

Jhariya MK, Banerjee A, Meena RS, Yadav DK (2019a) Sustainable agriculture, forest and environmental management. Springer, Singapore, 606 pp. https://doi.org/10.1007/978-981-13-6830-1

Jhariya MK, Yadav DK, Banerjee A (2019b) Agroforestry and climate change: issues and challenges. Apple Academic Press Inc., CRC Press- a Taylor and Francis Group, US & Canada, p 335. https://doi.org/10.1201/9780429057274

Jhariya MK, Meena RS, Banerjee A (2021a) Ecological intensification of natural resources for sustainable agriculture. Springer, Singapore. https://doi.org/10.1007/978-981-33-4203-3

Jhariya MK, Meena RS, Banerjee A (2021b) Ecological intensification of natural resources towards sustainable productive system. In: Ecological intensification of natural resources for sustainable agriculture. Springer, Singapore. https://doi.org/10.1007/978-981-33-4203-3_1

Karlen DL, Peterson GA, Westfall DG (2014) Soil and water conservation: our history and future challenges. Soil Sci Soc Am J 78:1493–1499

Karp DS, Mendenhall CD, Sandi RF, Chaumont N, Ehrlich PR, Hadly EA, Daily GC (2013) Forest bolsters bird abundance, pest control, and coffee yield. Ecol Lett 16(11):1339–1347. https://doi.org/10.1111/ele.12173

Khan N, Jhariya MK, Yadav DK, Banerjee A (2020a) Herbaceous dynamics and CO_2 mitigation in an urban setup- a case study from Chhattisgarh, India. Environ Sci Pollut Res 27(3):2881–2897. https://doi.org/10.1007/s11356-019-07182-8

Khan N, Jhariya MK, Yadav DK, Banerjee A (2020b) Structure, diversity and ecological function of shrub species in an urban setup of Sarguja, Chhattisgarh, India. Environ Sci Pollut Res 27 (5):5418–5432. https://doi.org/10.1007/s11356-019-07172-w

Khan N, Jhariya MK, Raj A, Banerjee A, Meena RS (2021a) Soil carbon stock and sequestration: implications for climate change adaptation and mitigation. In: Jhariya MK, Meena RS, Banerjee A (eds) Ecological intensification of natural resources for sustainable agriculture. Springer, Singapore. https://doi.org/10.1007/978-981-33-4203-3_13

Khan N, Jhariya MK, Raj A, Banerjee A, Meena RS (2021b) Eco-designing for sustainability. In: Jhariya MK, Meena RS, Banerjee A (eds) Ecological intensification of natural resources for sustainable agriculture. Springer, Singapore. https://doi.org/10.1007/978-981-33-4203-3_16

Klein AM, Vaissiere BE, Cane JH, Steffan-Dewenter I, Cunningham SA, Kremen C, Tscharntke T (2007) Importance of pollinators in changing landscapes for world crops. Proc R Soc B: Biol Sci 274:303–313. https://doi.org/10.1098/rspb.2006.3721

Krishna KR (2010) Agroecosystems of South India: nutrient dynamics, ecology and productivity. Brown Walker Press, p 551. https://www.universal-publishers.com/m/book.php?method=ISBN&book=1599425335

Kumar S, Kadono A, Lal R, Dick W (2012) Long-term no-till impacts on organic carbon and properties of two contrasting soils and corn yields in Ohio. Soil Sci Soc Am J 76:1798–1809

Kumar S, Meena RS, Jhariya MK (2020) Resources use efficiency in agriculture. Springer, Singapore, p 760. https://doi.org/10.1007/978-981-15-6953-1

Kumar S, Meena RS, Datta R, Verma SK, Yadav GS, Pradhan GS, Molaei A, Rahman GKMM, Mashuk HA (2020a) Legumes for carbon and nitrogen cycling: an organic approach. In: Carbon and nitrogen cycling in soil, pp 337–375. https://doi.org/10.1007/978-981-13-7264-3_10

Kumar S, Meena RS, Singh RK, Munir TM, Datta R, Danish S, Singh GS, Yadav, Kumar S (2021) Soil microbial and nutrient dynamics under different sowings environment of Indian mustard (*Brassica juncea* L.) in rice based cropping system. Sci Rep 11:5289. https://doi.org/10.1038/s41598-021-84742-4

Lal R (2003) Global potential of soil carbon sequestration to mitigate the greenhouse effect. Crit Rev Plant Sci 22:157–184

Lal R (2004) Soil carbon sequestration impacts on global climate change and food security. Science 304:1623–1627

Lal R (2015) Restoring soil quality to mitigate soil degradation. Sustainability 7(5):5875–5895. https://doi.org/10.3390/su7055875

Lal R, Delgado JA, Groffman PM, Millar N, Dell C, Rotz A (2011) Management to mitigate and adapt to climate change. J. Soil Water Conserv 66:276–285

Lal R, Singh BR, Mwaseba DL, Kraybill D, Hansen DO, Eik LO (2015) Sustainable intensification to advance food security and enhance climate resilience in Africa. Springer International Publishing, Cham

Landis DA, Wratten SD, Gurr GM (2000) Habitat management to conserve natural enemies of arthropod pests in agriculture. Ann Rev Entomology 45:175–201

Lemaire G, Franzluebbers A, de Faccio Carvalho PC, Dedieu B (2014) Integrated crop–livestock systems: strategies to achieve synergy between agricultural production and environmental quality. Agric Ecosyst Environ 190:4–8

Lennon JJ (2015) Potential impacts of climate change on agriculture and food safety within the island of Ireland. Trends Food Sci Technol 44:1–10

Mbow C, Smith P, Skole D, Duguma L, Bustamante M (2014) Achieving mitigation and adaptation to climate change through sustainable agroforestry practices in Africa. Curr Opin Environ Sustain 6:8–14. https://doi.org/10.1016/j.cosust.2013.09.002

Meena RS, Lal R (2018) Legumes for soil health and sustainable management. Springer, Singapore, p 541. https://doi.org/10.1007/978-981-13-0253-4_10

Meena RS, Kumar V, Yadav GS, Mitran T (2018) Response and interaction of *Bradyrhizobium japonicum* and Arbuscular mycorrhizal fungi in the soybean rhizosphere: a review. Plant Growth Regul 84:207–223. https://doi.org/10.1007/s10725-017-0334-8

Meena RS, Kumar S, Datta R, Lal R, Vijaykumar V, Brtnicky M, Sharma MP, Yadav GS, Jhariya MK, Jangir CK, Pathan SI, Dokulilova T, Pecina V, Marfo TD (2020a) Impact of agrochemicals on soil microbiota and management: a review. Land (MDPI) 9(2):34. https://doi.org/10.3390/land9020034

Meena RS, Lal R, Yadav GS (2020b) Long term impacts of topsoil depth and amendments on soil physical and hydrological properties of an Alfisol in Central Ohio, USA. Geoderma 363:1141164. https://doi.org/10.1016/j.geoderma.2019.114164

Meena RS, Lal R, Yadav GS (2020c) Long-term impact of topsoil depth and amendments on carbon and nitrogen budgets in the surface layer of an Alfisol in Central Ohio. Catena 194:104752. https://doi.org/10.1016/j.catena.2020.104752

Mikha MM, Vigil MF, Benjamin JG (2013) Long-term tillage impacts on soil aggregation and carbon dynamics under wheat-fallow in the central Great Plains. Soil Sci Soc Am J 77:594–605

Millennium Ecosystem Assessment (MEA) (2005) Millennium ecosystem assessment, ecosystems and human well-being. World Resources Institute, Washington, DC

Mkenda PA, Ndakidemi PA, Mbega E, Stevenson PC, Arnold SEJ, Gurr GM, Belmain SR (2019) Multiple ecosystem services from field margin vegetation for ecological sustainability in agriculture: scientific evidence and knowledge gaps. PeerJ 7:e8091. https://doi.org/10.7717/peerj.8091

Monfreda C, Ramankutty N, Foley JA (2008) Farming the planet: geographic distribution of crop areas, yields, physiological types, and net primary production in the year 2000. Global Biogeochem Cycles 22(1):GB1022

Mutyasira V, Hoag D, Pendell D, Manning D (2018) Is sustainable intensification possible? Evidence from Ethiopia. Sustainability 10(11):4174. https://doi.org/10.3390/SU10114174

Oerke EC (2005) Crop losses to pests. J Agric Sci 144:31–43

Ordway EM, Asner GP, Lambin EF (2017) Deforestation risk due to commodity crop expansion in sub-Saharan Africa. Environ Res Lett 12(4):044015. https://doi.org/10.1088/1748-9326/aa6509

Painkra GP, Bhagat PK, Jhariya MK, Yadav DK (2016) Beekeeping for poverty alleviation and livelihood security in Chhattisgarh, India. In: Narain S, Rawat SK (eds) Innovative Technology for Sustainable Agriculture Development. Biotech Books, New Delhi, pp 429–453

Pattey E, Trzcinski MK, Desjardins RL (2005) Quantifying the reduction of greenhouse gas emissions as a result of composting dairy and beef cattle manure. Nutr Cycl Agroecosyst 72:173–187

Peterson CA, Eviner VT, Gaudin ACM (2018) Ways forward for resilience research in agroecosystems. Agric Syst 162:19–27

Pimentel D, Zuniga R, Morrison D (2005) Update on the environmental and economic costs associated with alien-invasive species in the United States. Ecol Econ 52:273–288

Porter J, Costanza R, Sandhu H, Sigsgaard L, Wratten S (2009) The value of producing food, energy, and ecosystem services within an agro-ecosystem. Ambio 38:186–193

Raj A, Jhariya MK, Harne SS (2018a) Threats to biodiversity and conservation strategies. In: Sood KK, Mahajan V (eds) Forests, climate change and biodiversity. Kalyani Publisher, New Delhi, pp 304–320

Raj A, Jhariya MK, Bargali SS (2018b) Climate smart agriculture and carbon sequestration. In: Pandey CB, Gaur MK, Goyal RK (eds) Climate change and agroforestry: adaptation mitigation and livelihood security. New India Publishing Agency (NIPA), New Delhi, pp 1–19

Raj A, Jhariya MK, Yadav DK, Banerjee A, Meena RS (2019a) Agroforestry: a holistic approach for agricultural sustainability. In: Jhariya MK, Banerjee A, Meena RS, Yadav DK (eds) Sustainable agriculture, forest and environmental management. Springer, Singapore, pp 101–131

Raj A, Jhariya MK, Banerjee A, Yadav DK, Meena RS (2019b) Soil for sustainable environment and ecosystems management. In: Jhariya MK, Banerjee A, Meena RS, Yadav DK (eds) Sustainable Agriculture, Forest and Environmental Management. Springer, Singapore, pp 189–221

Raj A, Jhariya MK, Yadav DK, Banerjee A (2020) Climate change and agroforestry systems: adaptation and mitigation strategies. Apple Academic Press Inc., CRC Press- a Taylor and Francis Group, Burlington, ON, p 383. https://doi.org/10.1201/9780429286759

Raj A, Jhariya MK, Khan N, Banerjee A, Meena RS (2021) Ecological intensification for sustainable development. In: Ecological intensification of natural resources for sustainable agriculture. Springer, Singapore. https://doi.org/10.1007/978-981-33-4203-3_5

Ramsden MW, Menéndez R, Leather SR, Wäckers F (2015) Optimizing field margins for biocontrol services: the relative role of aphid abundance, annual floral resources, and overwinter habitat in enhancing aphid natural enemies. Agric Ecosyst Environ 199:94–104. https://doi.org/10.1016/j.agee.2014.08.024

Reese CL, Clay DE, Clay SA, Bich AD, Kennedy AC, Hansen SA, Moriles J (2014) Winter cover crops impact on corn production in semiarid regions. Agron J 106(4):1479–1488. https://doi.org/10.2134/agronj13.0540

Reicosky DC (2003) Tillage-induced CO_2 emissions and carbon sequestration: effect of secondary tillage and compaction. In: Garcia-Torres L, Benites J, Martinez-Vilela A, Holgado-Cabrera A (eds) Conservation agriculture. Kluwer Academic Publishers, Dordrecht, pp 291–300

Robertson GP, Allen VG, Boody G, Boose ER, Creamer NG, Drinkwater LE, Gosz JR, Lynch L, Havlin JL, Jackson LE, Pickett STA, Pitelka L, Randall A, Reed AS, Seastedt TR, Waide RB, Wall DH (2008) Long-term agricultural research: a research, education, and extension imperative. Bioscience 58(7):640–645

Sadler EJ, Lerch RN, Kitchen NR, Anderson SH, Baffaut C, Sudduth KA, Prato AA, Kremer RJ, Vories ED, Myers DB, Broz R, Miles RJ, Young FJ (2015) Longterm agroecosystem research in the Central Mississippi River Basin: introduction, establishment, and overview. J Environ Qual 44(1):3–12

Sheoran S, Kuma S, Kumar P, Meena RS, Rakshit S (2021) Nitrogen fixation in maize: breeding opportunities. Theor Appl Genet. https://doi.org/10.1007/s00122-021-03791-5

Singh NR, Jhariya MK (2016) Agroforestry and agrihorticulture for higher income and resource conservation. In: Narain S, Rawat SK (eds) Innovative technology for sustainable agriculture development. Biotech Books, New Delhi, pp 125–145

Smith Dumont E, Gnahoua GM, Ohouo L, Sinclair FL, Vaast P (2014) Farmers in Côte d'Ivoire value integrating tree diversity in cocoa for the provision of ecosystem services. Agrofor Syst 88:1047–1066. https://doi.org/10.1007/s10457-014-9679-4

Smith A, Snapp S, Chikowo R, Thorne P, Bekunda M, Glover J (2017) Measuring sustainable intensification in smallholder agroecosystems: a review. Glob Food Secur 12:127–138. https://doi.org/10.1016/j.gfs.2016.11.002

Stavi I, Bel G, Zaady E (2016) Soil functions and ecosystem services in conventional, conservation, and integrated agricultural systems: a review. Agron Sustain Dev 36:1–12

Steiner KG (1982) Intercropping in tropical smallholder agriculture with special reference to West Africa, Schriftenreihe der Deutsches Gesellschaft fürTechnischeZusammenarbeit (137)

Stewart BA, Lal R, El-Swaify SA (1991) Sustaining the resource base of an expanding world agriculture. In: Lal R, Pierce FJ (eds) Soil Management for Sustainability. Soil and Water Conservation Society, Ankeny, IA, pp 125–144

Sundriyal M, Sundriyal R (2004) Wild edible plants of the Sikkim Himalaya: nutritive values of selected species. Econ Bot 55:286–299

Swift M, Vandermeer J, Ramakrishnan P, Anderson J, Ong C, Hawkins B (1996) Biodiversity and agroecosystem function. In: Mooney H, Cushman J, Medina E, Sala O (eds) Functional roles of biodiversity: a global perspective. Wiley, New York, pp 261–298

Tarnocai C, Canadell JG, Schuur EAG, Kuhry P, Mazhitova G, Zimov S (2009) Soil organic carbon pools in the northern circumpolar permafrost region. Glob Biogeochem Cycles 23:1029–1040

Tilman D, Fargione J, Wolff B, D'Antonio C, Dobson A, Howarth R (2002) Forecasting agriculturally driven global environmental change. Science 292:281–284

Tscharntke T, Clough Y, Wanger TC, Jackson L, Motzke I, Perfecto I, Vandermeer JJ, Whitbread AM (2012) Global food security, biodiversity conservation and the future of agricultural intensification. Biol Conserv 151:53–59. https://doi.org/10.1016/j.biocon.2012.01.068

UN Water (United Nations Water) (2013) Agriculture and food security. United Nations, Rome. www.unwater.org/statistics

United Nations, Department of Economics and Social Affairs (2019) World population prospects. United Nations, San Francisco

USDA ERS (U.S. Department of Agriculture Economic Research Service) (2013) Farm practices, management, and irrigation water use. Washington, DC. www.ers.usda.gov/topics/farm-practices-management/irrigation-water-use

Walbridge MR, Shafer SR (2011) A long-term agro-ecosystem research (LTAR) network for agriculture. In: Medley CN, Patterson G, Parker MJ (eds) Observing, studying, and managing change: proceedings of the fourth interagency conference in the watersheds. US Geological Survey Scientific Investigations Report # 2011–5169, pp 26–30

Whitfield S, Challinor AJ, Rees RM (2018) Frontiers in climate smart food systems: outlining the research space. Front Sustain Food Syst. https://doi.org/10.3389/fsufs.2018.00002

Wiggers JMRH, Van Ruijven J, Berendse F, De Snoo GR (2016) Effects of grass field margin management on food availability for black-tailed godwit chicks. J Nat Conserv 29:45–50. https://doi.org/10.1016/j.jnc.2015.11.001

World Bank (2012) Carbon sequestration in agricultural soils. Economic and Sector Work. Report number: 67395- GLB. ftp://ftp.fao.org/ag/agp/ca/CA_CoP_May12/ARD_ESW12_CarbonSeq_web[1].pdf

World Bank (2015a) Agricultural nitrous oxide emissions. http://data.worldbank.org/indicator/EN.ATM.NOXE.AG.ZS/countries/1W?display=graph

World Bank (2015b) Agricultural methane emissions. http://data.worldbank.org/indicator/EN.ATM.METH.AG.ZS/countries/1W?display=graph

Wright AL, Hons FM (2005) Soil carbon and nitrogen storage in aggregates from different tillage and crop regimes. Soil Sci Soc Am J 69:141–147

Zhang W, Ricketts TH, Kremen C, Carney K, Swinton SM (2007) Ecosystem services and dis-services to agriculture. Ecol Econ 64:253–260

Zhao X, Xue JF, Zhang XQ, Kong FL, Chen F, Lal R, Zhang HL (2015) Stratification and storage of soil organic carbon and nitrogen as affected by tillage practices in the North China plain. PLoS One 10(6):e0128873

8

Crop Residue Management: A Novel Technique for Restoring Soil Health and Sustainable Intensification in India

Anita Kumawat, Ram Swaroop Meena, I. Rashmi, Ashok Kumar, S. D. Bamboriya, Devideen Yadav, Kuldeep Kumar, Dinesh Kumar, and Manoj Kumar Jhariya

Abstract

India has achieved a record food grain production of ~300 million tonnes in 2019–2020. Simultaneous production of a large volume of crop residues (CRs) is natural. It is documented that ~700 million tonnes (Mt) of CRs are generated every year in India. But the proper disposal of CRs is of serious concern, especially in the irrigated and mechanized cropping systems of India. Hence, a huge quantity of CRs is burnt on-farm to clear the field for timely sowing of the next crop. The burning of CRs causes environmental pollution and loss of soil

A. Kumawat · I. Rashmi · A. Kumar · K. Kumar
ICAR–Indian Institute of Soil and Water Conservation, RC, Kota, Rajasthan, India
e-mail: anita.kumawat@icar.gov.in; rashmi.i@icar.gov.in; ashok.kumar26@icar.gov.in; kuldeep.kumar1@icar.gov.in

R. S. Meena (✉)
Department of Agronomy, Institute of Agricultural Sciences, Banaras Hindu University, Varanasi, Uttar Pradesh, India
e-mail: meenars@bhu.ac.in

S. D. Bamboriya
ICAR–Indian Institute of Maize Research, Ludhiana, Punjab, India
e-mail: shanti.bamboriya@icar.gov.in

D. Yadav
ICAR–Indian Institute of Soil and Water Conservation, Dehradun, Uttarakhand, India
e-mail: devideen.yadav@icar.gov.in

D. Kumar
ICAR–Indian Institute of Soil and Water Conservation, RC, Datia, Madhya Pradesh, India
e-mail: dinesh.kumar12@icar.gov.in

M. K. Jhariya
Department of Farm Forestry, Sant Gahira Guru Vishwavidyalaya, Sarguja, Ambikapur, Chhattisgarh, India

M. K. Jhariya et al. (eds.), *Sustainable Intensification for Agroecosystem Services and Management*, https://doi.org/10.1007/978-981-16-3207-5_8

organic carbon (SOC) and nutrients, reduces microbial activities, and enhances soil erodibility. The continuous burning/removal of CRs leads to higher input costs in the short term and a decline in soil productivity in the long term. The burning of 1 ton of paddy straw release 1460 kg carbon dioxide (CO_2), 60 kg carbon monoxide (CO), 3 kg particulate matter (PM), 200 kg ash, and 2 kg sulfur dioxide (SO_2). Moreover, CR burning results in the loss of entire carbon (C), 80% of nitrogen (N), 25% of phosphorus (P), 20% potassium (K), and 50% sulfur (S). The inappropriate management of CRs will further lead to continuous depletion of soil fertility and deterioration of atmospheric quality. Hence, there is a need to develop efficient crop residue management (CRM) strategies to prevent the wastage of this valuable natural resource. The recycling of CRs offers a sustainable and ecologically sound option for restoring soil health and agricultural intensification. It can play an important role in C sequestration at 0.2×10^{15} g year^{-1} to improve the soil organic carbon (SOC) pool. This book chapter explains all the efficient CRM practices with respect to eco-intensification. Retaining CRs as mulch on the soil surface, in situ incorporation, and producing compost and biochar are the most effective approaches to improve soil, air, and water quality. Hence, the aim of this chapter is to explore the feasibility of different CRM options for replenishing and sustaining soil health and environmental security. This chapter is focused on the possible alternatives for efficient recycling of surplus CRs to improve soil and environmental security and sustainable crop production in cereal-centric intensive cropping systems of India. It will help producers, researchers, academicians, and policymakers to achieve the "Sustainable Development Goals" in India.

Keywords

Biochar · Carbon sequestration · Composting · Crop residues · Environment · Mulching · Soil health

Abbreviations

C	Carbon
CA	Conservation agriculture
CH_4	Methane
CO	Carbon mono-oxide
CO_2	Carbon dioxide
CRM	Crop residue management
CRR	Crop residue retention
CRs	Crop residues
CT	Conventional tillage
DAP	Diammonium phosphate
GHGs	Greenhouse gases
IGPs	Indo-Gangetic plains

K	Potassium
MBC	Microbial biomass carbon
MBN	Microbial biomass nitrogen
Mt	Million tonnes
N	Nitrogen
N_2O	Nitrous oxide
P	Phosphorus
PM	Particulate matter
S	Sulphur
SO_2	Sulphur dioxide
SOC	Soil organic carbon
SOM	Soil organic matter
VOCs	Volatile organic compounds
ZT	Zero tillage

8.1 Introduction

India being an agrarian economy, different types of crops are grown in its varied agroecosystems. With a production ~273 Mt (million tonnes) of cereals, 358 Mt of sugarcane, 23 Mt of pulses, and 33.5 Mt of oilseeds in 2019–2020, a vast amount of CRs (crop residues) has also been generated (Yadav and Arora 2018; Meena et al. 2018; MOA&FW 2020). Punjab generates about 23 and 17 Mt of rice and wheat residues per annum, respectively. Of this, 18.4 Mt of rice residues (>80%) and 8.5 Mt of wheat residues (50%) are burnt on-farm (Kumar et al. 2015a, 2019a). Generally, the CRs are being used as livestock feed, cooking fuel, packaging material, electricity and energy generation, composting, thatching, etc. (Prasad et al. 2012, 2020). However, a significant quantity of CR remains unutilized in the fields and their proper disposal is of serious concern (Naresh et al. 2017). Hence, a huge volume of CRs is burnt in situ in the fields for carrying out well-timed field operations and sowing of succeeding crops (Fig. 8.1). The short period (10–20 days) between the harvesting of the preceding crop and sowing of the next crop is a major reason for the on-farm burning of CRs in modern mechanized and intensive cropping systems (Jain et al. 2014). In situ burning is a fast and easy approach to handle a large amount of CRs (Naresh et al. 2017). Further, the labor shortage, high cost of removing CRs through traditional methods, and the use of combined harvesters are also accelerating the on-farm burning of CRs (Pathak et al. 2012; Meena and Lal 2018).

The burning of CRs produce a wide range and amount of toxic elements and gases such as PM, aerosols, soot particles, CO, CO_2, nitrous oxide (N_2O), ammonia (NH_3), SO_2, volatile organic compounds (VOCs), etc. (Singh et al. 2020a; Nagar et al. 2020). These pollutants badly affect human health by causing asthma, eye irritation, respiratory diseases, lung cancer, etc. (Chen et al. 2017). The release of

Fig. 8.1 On-farm burning of crop residues (source: thehindubusinessline.com/2018)

greenhouse gases (GHGs) because of CR burning causes global warming and climate change and also contributes 40% of CO_2 and 30% of ozone (O_3) in the troposphere (Bhuvaneshwari et al. 2019; Prasad et al. 2020). The CR burning not only pollutes the atmosphere but also degrades soil quality through the loss of organic matter (OM) and nutrients. It also results in the loss of whole C, 80–90% of N, 25% of P, 21% of K, and 4–60% of S (Singh et al. 2020a; Nagar et al. 2020; Kumar et al. 2021). Paddy straw burning in Punjab alone leads to a loss of 3.9 Mt of C, 59 thousand tons of N, 20 thousand tons of P, and 34 thousand tons of K (Jat 2016). Besides the loss of primary nutrients, CR burning also leads to the loss of secondary and micronutrients, SOM, and microbial biomass (Bisen and Rahangdale 2017). Burning increases soil temperature which causes reduced microbial diversity and their activities and increased evaporation rate (Mehta et al. 2013; Jhariya and Singh 2020, 2021). Burning or removal of the protective cover of CRs from soil surface enhances the water flow and soil susceptibility to erosion (Sanscartier et al. 2014; Turmel et al. 2015). Thus, continuous burning or removal of CRs result in net nutrient loss, which ultimately results in higher nutrient requirement and input cost in the short term and a decline in soil productivity in the long term (Singh and Sidhu 2014).

The ecosystem functions/services (i.e., provisioning, regulating, supporting, and recreational services) are also influenced by the intensive burning of CRs (Kumar et al. 2019a). The adverse effect of CR burning on regulating services includes the release of soot particles and PM, emission of GHGs, and global warming (Milham et al. 2014). The loss of supporting ecosystem services is related to soil ecological health which includes a decline in soil fertility, increased soil compaction, accelerated soil erosion, loss of nutrients, SOC, and soil biodiversity (Huang et al. 2012; Sharma et al. 2017; Sheoran et al. 2021). Further, CR burning also influences pollination by affecting the population of pollinators and their diversity (Carbone and Aguilar 2017). The negative impact of CR burning on provisioning services involves loss of productivity, fodder, employment opportunities, raw material for small industries, and higher cost of inputs (Kumar et al. 2019a). The loss of recreational services includes health risks to living beings (human, animal, and birds) and distress due to delay or cessation of road and air transportation services (Sawlani et al. 2019; Tripathi et al. 2019). Moreover, the burning of CRs also enhances medical expenses along with the loss of working days. Approximately >one million dollars is incurred in human health welfare every year in Punjab, India (Kumar and Kumar 2016).

CRs offer a sustainable and ecologically sound option for restoring soil and environmental health (Jat et al. 2014). Multiple advantages of crop residue retention (CRR) or in situ incorporation have been reported by several researchers in sustaining the health of agricultural lands (Wilhelm et al. 2007; Turmel et al. 2015; Meena et al. 2020). CR adds OM, improves soil fertility and microbial diversity, and conserves soil moisture (Singh et al. 2020b). Residue mulching also provides physical protection to the soil against wind and water erosion (Bertol et al. 2007; Kumawat et al. 2020a). CRs sustain C and N balance and improve the labile C pool in the soil (Thakuria et al. 2009). Besides an organic source of nutrients, CRs also help to improve the quality of soil, water, and air (Yadav and Arora 2018). Thus, CRs are of incredible value for the stakeholders. However, the faulty management of leftover CRs has resulted in deterioration of soil and atmospheric quality and also impaired the health of living beings. In addition, no market for selling, declined demand, low and unpredictable price of straw, etc. compels farmers to on-farm burning (Erenstein 2002). Recycling of CRs can enhance SOC, nutrient content, and also soil and environmental quality (Das et al. 2013). CRR on the soil surface or incorporation into the soil for a long period replenishes soil fertility and improves ecosystem functions (Lehtinen et al. 2014; Sidhu et al. 2015). CRR also facilitates improved soil bio-physicochemical properties, viz., soil structure, soil reaction, microbial activity, their diversity, etc. (Gathala et al. 2011; Lohan et al. 2018). Moreover, residue mulch creates a better environment for microorganisms, which facilitates improved biological N fixation in legume crops and increased dehydrogenase and phosphatase enzymatic activities. The population of bacteria and fungi increased by 5–10 and 1.5–11 times, respectively, under residue retention conditions (Chauhan et al. 2012). Residue mulching on soil surface also leads to significant improvement in C sustainability index (Jat et al. 2011). The C sequestration rate of 0.2×10^{15} g year^{-1} can be achieved if 15% of C reserved in the CRs gets transformed into inert SOC fraction (Lal 1997). About 7–10 tyear^{-1} residues are generated from the rice-wheat system which exhausts 730 kg primary nutrients from the soil (Gupta et al. 2002). This amount of nutrients can be used to meet the nutrient requirement of crops through proper management or in situ recycling (Jat et al. 2015). Thus, CR recycling may be beneficial in reducing the gap between the recommended dose of nutrients and farmers' practice in key production systems of India (Singh et al. 2017, 2020a).

CRR is although a key component to enhance SOC; however, its impact may be affected by the type of soils and climate and management strategies (Govaerts et al. 2009; Kumar et al. 2020). CA is an imperative crop production technology for efficient CRM, for restoration of degraded lands, and for increasing C sequestration into the soil (Lal 1997). SOC sequestration would further help to lower the emission of GHGs, boost productivity, and decrease ecological harm due to CR burning or removal (Jat 2013; Purakayastha et al. 2015; Meena et al. 2020a). Zero tillage (ZT) with residue mulching under CA-based cropping systems is also considered as a good agronomic practice to maintain a continuous organic cover on the soil surface (Fuentes et al. 2009; Kumawat et al. 2020a). Composting is another technique of managing the large quantity of leftover CRs on the field. Paddy straw from a 1 ha area can produce nearly 3.2 t of enriched compost containing 1.78, 0.95, and 2.06% of N, P, and K, respectively. About 206 kg of N, P, and K can be obtained from the compost of paddy straw of a 1 ha area, which costs around 150 USD (Goswami et al. 2020). The application of OM-rich compost sustains soil physical, chemical, and biological quality which can replace or supplement the fertilizer requirement of crops (Bhuvaneshwari et al. 2019). Similarly, biochar produced from CRs can be a sustainable option to mitigate climate change through C stabilization or sequestration while reducing wastage of CRs (Rogovska et al. 2011; Fang et al. 2014; Athira et al. 2019). It can hold C into the soil for up to 100 years and also helps to decrease C footprints by 38–49% from the paddy production system (Mohammadi et al. 2016). The porous nature of biochar facilitates higher water and nutrient retention and improves aggregate stability, beneficial microbial mass, and soil biological activities (Palansooriya et al. 2019). Besides, the adoption of Turbo Happy Seeder is also a profitable and efficient technology to cope up with the CRs load in the fields without burning. A profit of about INR 22,254 (+44%) can be gained by the average farmer through switching from burning to use of the Happy Seeder invention. Further, Happy Seeder has huge potential to replenish soil quality and to cut down the air pollution as it can reduce nearly 78% of GHG emissions per ha over CR burning (Pal et al. 2019). Thus, residue mulching, in situ incorporation, composting, biochar production, mechanization, etc. are some efficient and sustainable approaches that can curb the CR burning and recycle the nutrients into the soil itself. However, the effectiveness of these technologies depends on application methods of CRs and their degradation rate and type of soil, climate, crops, cropping system, and tillage practices. The CRs can be economically viable, agronomically productive, and socially amicable if managed properly. Therefore, the identification and development of location-specific interventions are crucial for the large-scale adoption of recommended technologies and for sustainable soil health and environmental security (Banerjee et al. 2020, 2021; Meena et al. 2020b). Implementation of government policies is essential to curb residue burning and to prevent air pollution. Further, the adoption of new inventions and approaches is needed to mitigate the adverse impact of air pollution and for sustainable utilization of CRs especially in the agriculture, energy, and industrial sector. This chapter discusses various sustainable alternatives of CRM to curtail the negative impact of CR burning on soil, environment, and human health. There is also a need to encourage the farmers for adoption of various approaches of CRM through workshops, trainings, and public meetings at local and regional levels. Further, farmers should also make aware of the long-term soil health, environmental, and monetary benefits of CRs. Thus, the aim of this chapter is to highlight the adverse impact of CR burning on ecosystem services and various strategies for effective CRM. This chapter is focused on efficient technologies for sustainable agricultural intensification in India where land degradation is a serious threat to the sustainability of crop production.

8.2 Availability of Crop Residues

In India, the quantity of annual CR production has been estimated by different sources. According to the Ministry of New and Renewable Energy, government of India, about 501.8 Mt per annum of CRs is produced, out of which 140.8 Mt are surplus residues (MNRE 2009). Pathak (2004) estimated 523 Mt total annual CR production, of which 127 Mt was surplus in India. Moreover, annual CR production of 686 Mt (generated by 26 crops) has been estimated by Hilodhari et al. (2014). The state-wise volume of CR generation, surplus residues, and in situ burnt CRs (Mt) per annum is given in Table 8.1 (Yadav et al. 2019). Out of the total CR burnt (92.8 Mt), more than 50% of CRs are burnt in Punjab, Uttar Pradesh, and Haryana.

The quantity of residue generated by different crops is the function of crop production, residue to crop ratio, and dry matter fraction (Devi et al. 2017). Cereal crops have different residues to economic yield ratios such as 1.5 for paddy, maize, and millets, 1.5–1.7 for wheat, and 3.0 for rapeseed and mustard and cotton (Table 8.2) (Hilodhari et al. 2014; Jain et al. 2014). However, a wide gap exists between the quantity of CRs produced and their actual utilization due to variability in types of crop cultivated, production system, crop intensity, and agronomic yields under variable agroecological parts of the country (Singh and Sidhu 2014).

Of the total residue generated in India, 352 Mt is produced by cereal crops, 66 Mt by fiber crops, 29 Mt by oilseeds, 13 Mt by pulses, and 12 Mt by sugarcane (MNRE 2009) (Fig. 8.2). At the individual crop level, rice alone contributes 34% of total residue generated from cereals followed by wheat (29%) (Pathak et al. 2012; Ravindra et al. 2019).

The quantity of surplus CRs (gross residues generated minus residues used for a range of purposes) generated in India is estimated 84–141 Mt/year, which are typically burnt on-farm (MNRE 2009). Of the total available surplus residues, 58% is produced by cereal crops, 23% by fiber crops, and the remaining 19% by other crops (Fig. 8.3). Around 80% of surplus cotton residues are also burnt in fields (Pathak et al. 2012).

Regarding the surplus residue production potential of India, Uttar Pradesh generates 40 Mt, Maharashtra 31 Mt, and Punjab 28 Mt (Hilodhari et al. 2014). The surplus residue production potential of different states of India under different crop groups is presented in Fig. 8.4.

Table 8.1 State-wise quantity of residue generation, surplus, and burnt crop residues in India (data source: Yadav et al. 2019)

State	Total crop residue generation (Mt year^{-1})	Surplus residue generation (Mt year^{-1})	Crop residue burnt (Mt year^{-1})
Andhra Pradesh	43.9	7.0	2.7
Arunachal Pradesh	0.4	0.1	0.0
Assam	11.4	2.3	0.7
Bihar	25.3	5.1	3.2
Chhattisgarh	11.3	2.1	0.8
Goa	0.6	0.1	0.0
Gujarat	28.7	8.9	3.8
Haryana	27.8	11.2	9.1
Himachal Pradesh	2.9	1.0	0.4
Jammu and Kashmir	1.6	0.3	0.9
Jharkhand	3.6	0.9	1.1
Karnataka	33.9	9.0	5.7
Kerala	9.7	5.1	0.2
Madhya Pradesh	33.2	10.2	1.9
Maharashtra	46.5	14.7	7.4
Manipur	0.9	0.1	0.1
Meghalaya	0.5	0.1	0.1
Mizoram	0.1	0.0	0.0
Nagaland	0.5	0.1	0.1
Odisha	20.1	3.7	1.3
Punjab	50.8	24.8	19.6
Rajasthan	29.3	8.5	1.8
Sikkim	0.6	0.0	0.0
Tamil Nadu	19.9	7.1	4.1
Tripura	0.0	0.0	0.1
Uttarakhand	2.9	0.6	0.8
Uttar Pradesh	60.0	13.5	21.9
West Bengal	35.9	4.3	5.0
India	501.8	140.8	92.8

8.3 Impact of Crop Residue Burning

CR burning is a common phenomenon of intensive rice-wheat rotation in the Indo-Gangetic Plains (IGPs) of India. It is an important contributor to poor air quality in the region and nearby states (Singh and Sidhu 2014). Moreover, the burning of CRs

Table 8.2 Residue generations by major crops in India (data source: Jain et al. 2014)

Crop	Residue to crop ratio	Dry matter fraction	Residue generation (Mt)
Paddy	1.5	0.9	197.8
Wheat	1.7	0.9	120.7
Maize	1.5	0.9	26.8
Millets	1.5	0.9	21.6
Sugarcane	0.4	0.9	107.5
Rapeseed and mustard	3.0	0.8	17.3
Groundnut	2.0	0.8	11.4
Cotton	3.0	0.8	90.9
Jute	2.2	0.8	31.5

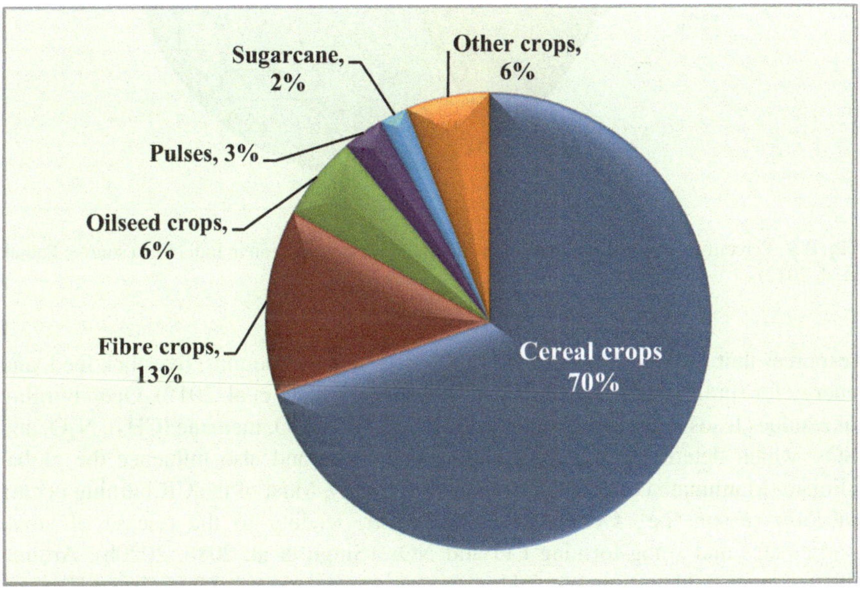

Fig. 8.2 Percent share of different crops in total residue generation in India (data source: Pathak et al. 2012)

aggravates health risks like respiratory problems, cardiovascular disease, asthma, eye irritation, chronic bronchitis, and skin diseases. The emissions of toxic gases also exert health hazards to animals, birds, and other living entities of the region (Khatun et al. 2015; Rao 2020). A case study carried out in rural Punjab (three villages), India, revealed that the major health problems associated with the residue burning were itching in the eyes and congestion (Table 8.3). About 20–94% of peoples in the selected villages of Punjab were suffering from these problems only. Similarly, about 50% of villagers felt aggravation of health problems during the straw burning (Kumar et al. 2015a). Moreover, the burning of CRs is a loss of imperative natural

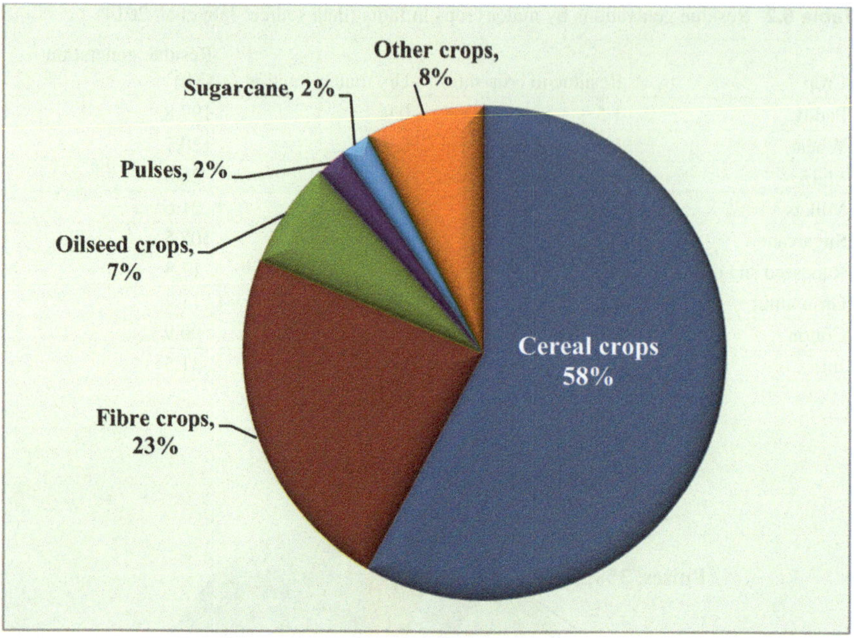

Fig. 8.3 Percent share of different crops in surplus residue generation in India (data source: Pathak et al. 2012)

resources that are valuable input for SOC, bioactive compounds, livestock feed, and energy for rural households and small industries (Pathak et al. 2012). Open burning of residues leads to the emission of GHGs, viz., CO_2, CO, methane (CH_4), N_2O, and SO_2, which deteriorate the surrounding air quality and also influence the global climate (Manjunatha et al. 2015; Lohan et al. 2018). Most of the CR burning occurs in *Rabi* season, i.e., October and November, leading to the release of small particulates and smog-forming CO and NOx (Singh et al. 2018, 2020b). Around 48% of open-field burning of total burning occurs in the state of Punjab and Haryana alone, and that is thought to have contributed about 20% to Delhi's pollution due to heavy pollutant transportation in the atmosphere (Bhuvaneshwari et al. 2019). Complete burning of cereal CRs (especially rice and wheat) leads to 100%, 20.1–22.2%, 19.8–21.8%, and 75.0–80.2% loss of N, P, K, and S (Dutt 2018). Further, the burning of dry crop biomass emits a large amount of CO_2 and other pollutants into the atmosphere, which can reduce addable C amount to the soil (Nath et al. 2017).

Further, the burning of the CRs adversely affects the beneficial soil microbiome along with the loss of essential plant nutrients and polluting the surrounding environment (Fig. 8.5) (Trivedi et al. 2017; Lohan et al. 2018). Although the farmers know the negative impact of residue burning, still they prefer the burning of CRs due to several advantages, viz., timely conducting of field operations, lower cost of

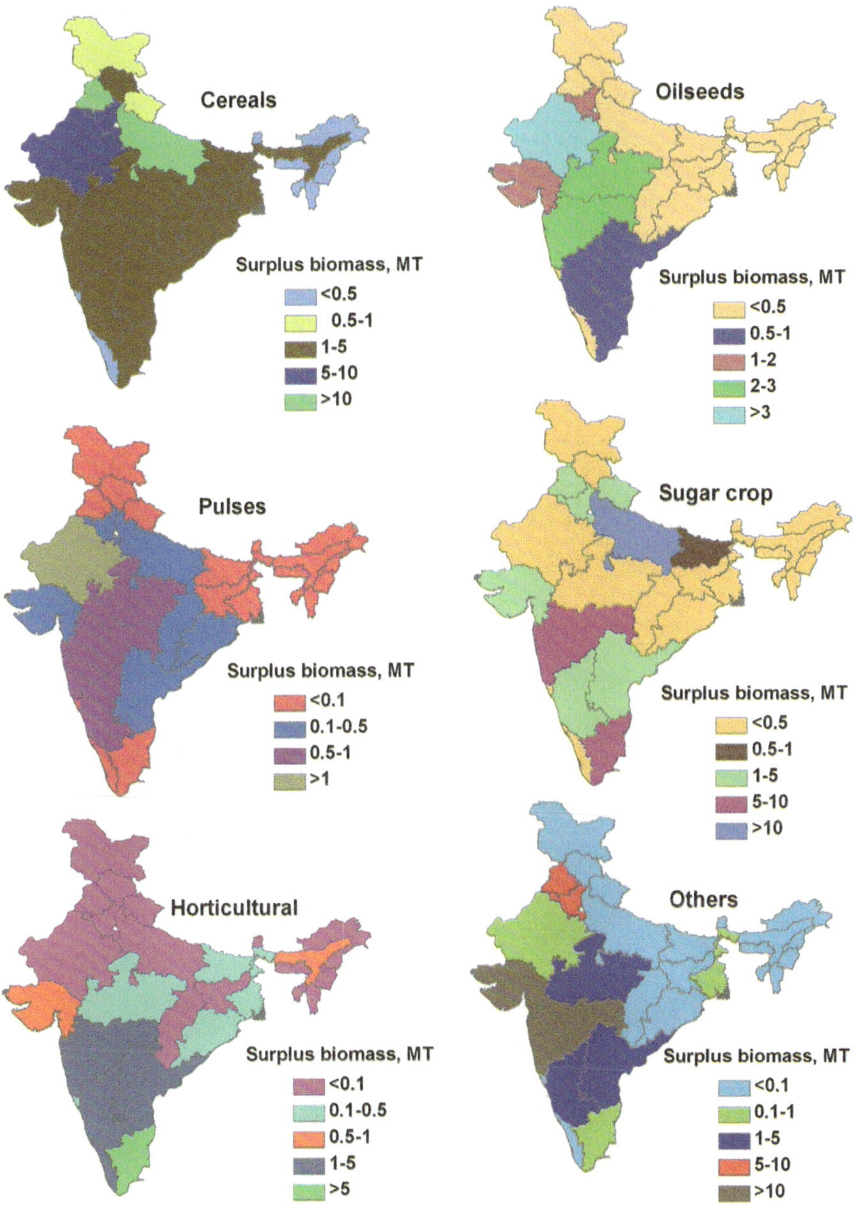

Fig. 8.4 Surplus crop residue generation in different parts of India (source: Hilodhari et al. 2014)

Table 8.3 Percentage of human beings faced different health risks due to residue burning (data source: Kumar et al. 2015a)

Health problems	Human beings (%)
Chronic bronchitis (infection due to straw burning cause swelling in lungs)	1.6–2.0
Eye itching or irritation (burning sensation in eyes)	73.5–93.7
Congestion in the chest (coughing problem)	20.4–44.4
The feeling of sore throat and nasal problems (smoke cause itching in nose and throat)	0.7–4.3
Asthma (difficulty in breathing, coughing in the chest)	3.1–23.8
Emphysema (release of harmful gases, particulate matter, chemical, etc. causes infection in the lungs)	0.0
Respiratory diseases (release of dust particles, smoke, toxic pollutants, etc. affect the immune system which causes fever or illness)	7.8–15.9
Other respiratory and cardiac diseases	0.0
Diseases not related to residue burning	1.6–8.2
Aggravation of health risks due to straw burning	42.6–52.0

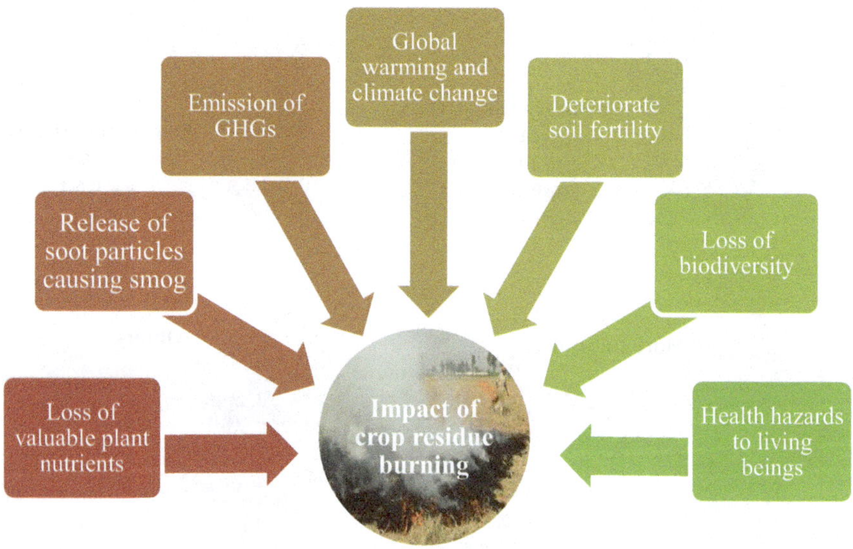

Fig. 8.5 Impact of crop residue burning of soil and environmental quality (modified by Lohan et al. 2018)

residue management practices, higher agronomic yields, and better control of insect-pest, diseases, and weeds (Chen et al. 2005).

8.3.1 Loss of Essential Plant Nutrients

It has been reported that the burning of 1 ton of CRs results in the release of 1515 kg CO_2, 92 kg CO, 3.83 kg oxides of N (NOx), 0.4 kg SO_2, 2.7 kg CH_4, and 15.7 kg non-methane volatile compound (Andreae and Merlet 2001). Similarly, the burning of 1 ton of paddy straw results in the loss of about 6 kg N, 2 kg P, 25 kg K, and 1 kg S. In general, dead material obtained from different crops contains 80% of N, 25% of P, 20% of K, and 50% of S (MOA&FW 2019; Nagar et al. 2020). Besides the loss of primary and secondary nutrients, micronutrients are also lost due to the residue burning (Kumar et al. 2019a). Thus, CRs are a valuable organic source of plant nutrients, and when these residues are incorporated or used as in-situ mulch, they will enhance SOC and essential plant nutrients (Kushwaha et al. 2000; Kumawat et al. 2020a).

8.3.2 Impact on Soil Properties

The heat produced from residue burning increases the soil temperature which leads to the death of beneficial soil microorganisms. Although the effect is temporary, frequent residue burning causes loss of soil microbial population permanently (Mehta et al. 2013). In addition, the CR burning enhances the $NH_4 \pm N$ and HCO^{3-}-P concentration, but the enrichment of soil fertility remains negligible (Jain et al. 2014). The continuous burning over a long period reduces total N, C, and potentially mineralizable N in the upper soil layer (Pathak et al. 2012).

8.3.3 Impact on Environment

CR burning releases a wide range of pollution-causing particles, viz., toxic gasses, VOCs, PM, carcinogenic hydrocarbons, and other tiny pollutants (Fig. 8.6) (Awasthi et al. 2011, 2017). Residue burning emits a large quantity of GHGs; aerosols, viz., CH_4, CO, N_2O, NO_X, etc.; and other volatile harmful elements and compounds that cause global warming and climate change (Manjunatha et al. 2015; Ravindra et al. 2019). It is estimated that C present in CRs emit as CO_2 (70% of C present), CO (7%), and CH_4 (0.66%), and 2.09% of N is emitted as N_2O from straw burning (Sahu et al. 2015). The burning of 1 ton of rice residues releases around 3 kg of PM, 60 kg CO, 1460 kg CO_2, 199 kg ash, and 2 kg SO_2 (Gadi et al. 2003; Gupta et al. 2004). Moreover, biomass burning also releases a large number of minute particulates and smoke which degrades the quality of the surrounding environment and augment the chances of road accidents by decreasing visibility (Mandal et al. 2014; Yadav and Devi 2018). A large number of pollution-causing particles present in the residue smoke are recognized as carcinogens and may cause several breath issues and lung diseases (Chakrabarti et al. 2019; Singh et al. 2020b).

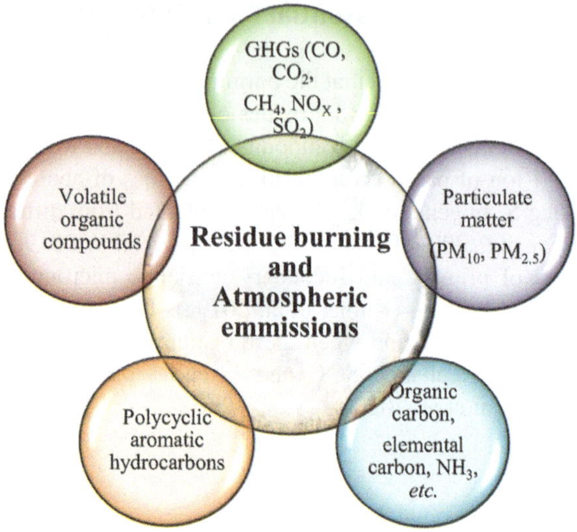

Fig. 8.6 Atmospheric emissions of various compounds through CR burning (modified by Ravindra et al. 2019)

8.4 Factors Responsible for Residue Burning

The deteriorative impact of in situ CR burning is well-known by the stakeholders. However, to clear the field and shortage of available period between the harvesting of former crop and the sowing of the subsequent crop, farmers opt for in situ residue burning. Further, the necessity of timely agronomic practices for higher crop productivity also leads to on-farm residue burning. The major reasons for increased residue burning are as follows:

- Increased mechanization, especially the use of combine harvesters (combine left 80% residues after harvesting in the fields is often subjected to on-farm burning).
- Residue burning is economical and inexpensive as more labor is required for the manual removal of residues.
- Decreasing numbers of livestock and declining interest in animal husbandry.
- Long duration required for compost preparation from CRs (4–6 months).
- Nonavailability of alternate economically viable options for residue burning.
- Labor scarcity, high wages of field workers during harvesting season, the uncertainty of the weather, and ease of harvesting and threshing operations, etc. increased the use of combine harvesters.
- Burning facilitates timely field operations, viz., land preparation and sowing.
- The easy way of controlling weeds, insect-pest, and diseases under conventional tillage (CT).
- Availability of short period (10–15 days) for the sowing of succeeding crop.
- Low crop productivity and economic returns under CA.
- The complexity of nutrient management under ZT with CRs mulching.
- Negative attitudes or perceptions towards CA and institutional constraints.

Table 8.4 Primary nutrient concentrations in residues of different crops

Crop	N (%)	P (%)	K (%)	Reference
Rice	0.5–0.8	0.16–0.27	1.4–2.0	Dobermann and Fairhurst (2002)
Wheat	0.61–0.65	0.085–0.09	1.67–1.69	Yadav et al. (2019)
Barley	0.52–0.68	0.08–0.10	1.25–1.79	Lal (2009); Behera (2018)
Oats	0.68	0.14	0.45	Lal (2009)
Maize	0.84–0.90	0.10–0.14	1.09–1.19	Logah (2011)
Pearl millet	0.45	0.07	0.95	Behera (2018)
Sorghum	0.52–0.78	0.12	1.01–1.21	Lal (2009); Behera (2018)
Mung bean	1.32	0.33	1.78	Singh et al. (2008a)
Soybean	2.83–3.07	0.62–0.64	1.0–1.23	Almaz et al. (2017)
Sugarcane	0.45	0.08	0.95	Behera (2018)
Cotton	0.98	–	–	Lal (2009)
Potato	0.52	0.09	0.85	Behera (2018)

8.5 Crop Residue as Source of Plant Nutrients

CRs are the principal source of OM and also a rich source of essential plant nutrients (Table 8.4) (Lal 1997). Dry biomass contains around 40% C, and when this biomass is applied to the soil, it stabilizes and ensures the sustainability of agricultural ecosystems (Singh and Sidhu 2014; Raj et al. 2019a, b, 2020, 2021). From 141.2 Mt of straw, about 7.0 lakh tons of N, 8.4 lakh tons P, and 21 lakh tons K are supplied to the soil in India (Dutt 2018). Around 25% of N, 25% P, 50% of S, and 75% K uptake by cereals are reserved in residues themselves that makes them a vital source of nutrients (Byous et al. 2004). Likewise, paddy reserves around 40% of N, 30–35% of P, 80–85% of K, and 40–50% of S absorption in its biomass at the harvesting stage. Equally, almost 25–30% of N and P, 35–40% S, and 70–75% K uptake remain reserved in the wheat residues (Sharma 2016; Singh et al. 2018). Usually, 1 ton of rice straw (dry-weight basis) contains 5–8 kg N, 0.7–1.2 kg P, 12–17 kg K, 0.5–1 kg S, 3–4 kg calcium (Ca), 1–3 kg magnesium (Mg), and 40–70 kg silicon (Si) at harvest stage (Prasanthkumar and Kannan 2018). Likewise, 1 ton of wheat straw reserves 4–5 kg N, 0.7–0.9 kg P, and 9–11 kg K. Based on several studies, an average amount of around 5.5 kg N, 2.3 kg P, 25 kg K, and 1.2 kg S per tonne of straw has been estimated (Kocher et al. 2017; Kaushal and Prashar 2020). In addition to macronutrients, 1 ton of rice and wheat straw also contains about 9–11 kg S, 100 g zinc (Zn), 777 g iron (Fe), and 745 g manganese (Mn) (Singh et al. 2018). Further, the amount of nutrients in CRs depends on soil type, crop production techniques, type of crop, variety, etc. Around 25 kg higher K per tonne of rice straw has been reported in the northwest IGP region of India than other regions of India (Singh et al. 2014). Hence, CRs restrain a considerable quantity of essential plant nutrients that can be used in different ways for the recycling of OM to restore soil health and improve agronomic productivity under different land-use systems.

Fig. 8.7 Turbo Happy Seeder for the sowing of crops in residue retained fields under conservation agriculture

8.6 Management of Crop Residues

CRs are primarily used as livestock feed and bedding material. Moreover, these are also used as fuel, composting, thatching, packaging, power generation, biofuel production, and energy in industries (Thorat et al. 2015; Singh et al. 2020b). However, the straw of paddy is not preferred by the farmers of Punjab, Haryana, and neighboring states because of greater silica concentration (12–16%) in its straw (Singh et al. 2014). The CRs remained unused are mostly subjected to in situ burning (Yadav and Arora 2018). CR burning is a quick, labor-saving practice, but it deteriorates atmosphere quality and harms ecological services (Kaur et al. 2019; Rao 2020). Thus, the maintenance of CRs is a serious challenge due to the nonavailability of the right residue recycling technology and a very short turnaround period between harvesting and sowing of two crops. Singh et al. (2008b) argued that the evaluation of CRM practices should be based on productivity potential, ecological impact and sustainability aspect of cropping systems. Farmers have multiple approaches to maintain and restore the soil health and environmental quality such as in situ residue incorporation, complete or partial CRR on the soil surface, composting, biochar formation, and adoption of conservation tillage. Further, the use of Turbo Happy Seeder machinery (Fig. 8.7) for sowing of crops in CRs retained fields without soil disturbance under CA. It helps in reducing labor requirement, saves irrigation water and energy inputs, lowers the cost of production, and sustains the soil health while facilitating timely sowing of crops with more or less similar crop yields (Sidhu et al. 2015).

8.6.1 Incorporation of Crop Residues into the Soil

In situ turning of CRs into the soil is although vital for the recycling of essential plant nutrients and the buildup of OM (Kumari et al. 2014; Goswami et al. 2020). However, the CRs with a higher C/N ratio lead to temporary microbial immobilization of N, which results in N deficiency in the crops sown immediately after residue incorporation (Malhi et al. 2001; Bisen and Rahangdale 2017). The problem of N immobilization can be avoided through the application of additional N fertilizers during residue incorporation (Singh et al. 2005a). The phase of net N immobilization and net supply from cereal CRs to the following crop is governed by the decomposition phase prior to sowing of subsequent crops, residue quality, and ecological state of soil (Singh et al. 2005a, 2014). The sowing of subsequent crops immediately after the incorporation of cereal CRs can lead to decreased crop yields as compared to residue removal or burning (Singh and Sidhu 2014). The N deficiency due to N immobilization in the wheat crop after paddy can be avoided by allowing adequate time (2–3 weeks) between paddy straw incorporation and sowing of succeeding wheat crop (Singh et al. 2009; Kumar et al. 2015b). However, the incorporation of rice residues before wheat sowing is challenging for the farmers due to the short window period between rice harvesting and wheat sowing. This rice residue incorporation preceding wheat sowing can also delay planting by 10–15 days (Kumar et al. 2015a). Mandal et al. (2004) suggested that the turning of residues of paddy and wheat into the soil have no harmful impact on the performance of crops in the system. Farmers of IGPs collect wheat straw for animal feeding after harvesting through a combined harvester, and the remaining 20–30% (1.5–2.0 t ha^{-1}) of straw left in the field can be incorporated before the rice planting (Singh and Sidhu 2014). A study conducted at PAU, Ludhiana, reported no adverse effects of wheat residue incorporation (2–3 t ha^{-1}) on yields of succeeding paddy crop (Kumar et al. 2019b). Moreover, a window period of ~60–65 days amid rice harvesting and wheat sowing can be utilized for the sowing of green manure crops or dual-purpose pulses, viz., green gram and black gram for grain and in situ green manuring (Singh et al. 1991).

8.6.2 In Situ Mulching of Crop Residue

Conservation of natural resources especially soil and water is imperative for sustainable agronomic production under arid and semiarid conditions, having a high rate of evapotranspiration (Jat et al. 2014). Except for the alteration in the sowing period, mulching is the only practice to reduce the evaporative loss of water (Mupangwa et al. 2012). Mulching increases productivity (4–29%), reduces irrigation numbers, and improves utilization efficiency of nutrients and water by altering the hydrothermal properties of soil (Jat et al. 2012; Lu 2020). However, the benefits of mulching depend on temperature, precipitation, water management, soil texture, type of mulching material, and quantity of mulch applied (Mechergui et al. 2020). The response of residue mulching is superior in the regions receiving low rainfall, high temperature, and having coarse-textured soils. Further, high moisture in the root zone of soil profile under residue mulching improves the germination process and seedling vigor (Singh et al. 2014). Nandan et al. (2019) reported increased grain yields of maize, wheat, and rice by 7–10, 5–11, and 3–8%, respectively, under ZT with residue mulching over CT without residue mulching. ZT and residue mulching enhanced productivity and water-use efficiency due to reduced mechanical resistance, the deeper and dense rooting pattern of crops, regulated soil temperature, and improved moisture conservation (Jat et al. 2012; Lohan et al. 2018; Kumawat et al. 2020b).

8.6.2.1 Effect of Residue Mulching on Crop and Water Productivity

CRs and organic manures are the key sources of C supply to the soil, which have a positive impact on soil physical, chemical, and biological properties (Gathala et al. 2011; Yadav et al. 2019). CRs modify the physicochemical condition of soil which subsequently affects the microbial activities and their diversity and consequently nutrient transformation (Kumawat et al. 2020a). Soil enzyme activities are a good indicator of soil health as they make plant nutrients in available form through nutrient cycling which results in improved crop growth and productivity. The on-farm trial conducted at different locations of Punjab reported 3.24% higher wheat yield under happy seeder planting over CT during 2007–2010 (Sidhu et al. 2015). Likewise, in a medium-term study (1995–2004), Nangia et al. (2010) noted that ZT with straw mulching significantly increased the crop productivity by up to 36% and water productivity by up to 28% and decreased the runoff by up to 93% in Yellow River Basin of China as compared with CT. Irrigation water can be saved up to 20% by the sowing of wheat without pre-sowing irrigation on residual soil moisture, which eventually saved 80 kWh of electricity and decreased the release of CO_2 by up to160 kg (Singh et al. 2014). It has been also reported that higher soil water storage and lower evapotranspiration losses under ZT with residue mulching significantly increased the grain water productivity of wheat, soybean, and maize by 3.3, 1.9, and 3.2 kg $ha^{-1}mm^{-1}$as compared to CT without residue retention (Yang et al. 2018).

8.6.2.2 Effect of Residue Mulching on Soil Erosion

Soil erosion is a global concern, which impairs soil quality through the loss of SOC, nutrients, and biodiversity (Gregg and Izaurralde 2010; Pimentel 2006). Hence, erosion control is vital for C sequestration into the soil which provides multiple advantages including improving soil fertility and water and nutrient holding capacity and consequently enhances agronomic productivity (Stockmann et al. 2013). CR maintained overground surface as mulch reduces erosion processes, runoff, soil, and nutrient losses through the buffering action of raindrops (Table 8.5). Besides, the

Table 8.5 Effect of residue mulching on soil erosion, runoff, soil, and C and nutrient loss

Parameters	Findings	References
Soil erosion	Residue mulch creates a physical barrier against the erosive impact of raindrops and minimizes further erosion processes	Park et al. (2020)
	Retaining 2–4 t ha^{-1} CRs on soil surface reduces soil erosion by 80% over the bare soil	Ranaivoson et al. (2017)
	Soil erosion can be reduced up to 36% under high rainfall intensity by applying 4.5 t ha^{-1}residue mulch, while under low rainfall intensity, the same amount of biomass can reduce soil erosion up to 16%	Scopel et al. (2005)
Runoff	Retaining 1.5–4.5 t ha^{-1}CRs as mulch decrease runoff up to 50%	Ranaivoson et al. (2017)
	Surface application of 4.5 t ha^{-1}CR decreased surface water flow by 20–40% in soil with a slope of 3–7%	Scopel et al. (2005)
	Residue mulch can decrease water runoff by >90% by reducing the velocity of flowing water and increased infiltration	Cerda et al. (2016); Prosdocimi et al. (2016)
	Application of 1.5 and 4.5 t ha^{-1}CRs as mulch decreased the runoff by 7.7–56.3% and by 46.2–75.0%, respectively	Scopel et al. (2005)
	Runoff depth was reduced by 81% under heavy rainfall due to straw mulching	Dai et al. (2018)
Soil and sediment loss	Application of 1.5–4.5 t ha^{-1}dry biomass of maize, rye, and rice reduced the soil loss by 50%, while application of 2.0 t ha^{-1}residues of wheat is lowered by up to 90%	Woyessa and Bennie (2004)
	Covering 60% ground surface with CRs reduced soil loss by 80%	Ranaivoson et al. (2017)
	Residue mulching reduced soil loss by 96.5%, while the burning of CRs increased the soil loss by 192%	Rasoulzadeh et al. (2019)
	Straw mulching decreased the soil loss and sediment concentration by 81.5 and 39.3%, respectively, over without mulching	Grum et al. (2017)
	Straw mulching reduced sediment yield by 97% under heavy rainfall	Dai et al. (2018)
C and nutrients loss	Surface cover with rice straw reduced the loss of total and sediment-bound C, N, and P by up to 82%	Park et al. (2020)
	The annual loss of C, N, P, K, Ca, and Mg in eroded soil was 200–1000, 39–176, 0.6–5.0, 1.4–14, 18–90, and 1–13 $kgha^{-1}$, respectively, from non-mulched treatments	Lal (1976)
	Wheat straw mulching reduced the loss of total N and P by 57.5 and 63.3%, respectively	Grum et al. (2017)

residues of small grain crops are more efficient in reducing erosion than residues of course grain (i.e., sorghum and maize) crops (Woodruff and Siddoway 1973). Graham et al. (2007) suggested that soil erosion can remain below 0.5 tha^{-1} through the removal of only 28% of maize residues under existing production techniques. Izaurralde et al. (2007) noted reduced soil erosion and loss of SOC and nutrients under ZT than CT. Likewise, Guy and Lauver (2007) stated that retention of CR as mulch reduced soil erosion and improved water retention capacity of the soil. The stubble mulching was initiated in the late 1930s or early 1940s for effective control of soil erosion caused by either intensive tillage practices or complete removal of CRs from the soil surface (Lal 2002). Operations, such as ZT with residue retention, stubble mulching, chiseling, etc., create rough and cloddy soil surface, which helps in reducing the direct impact of rainfall on soil and eventually reduced soil erosion (Mehra et al. 2018; Kumawat et al. 2020b). Small grains planters, viz., deep furrow and hoe drills, are efficient in creating about 5–13 cm height of erosion-resistant surfaces through the sowing of crops in the presence of CRs (Woodruff and Siddoway 1973).

8.6.2.3 Effect of Residue Mulching on Soil Moisture and Temperature

Conservation tillage with residue mulching has several advantages, viz., soil and water conservation, improves soil health and hydrothermal properties, and reduces the emission of GHGs (Mehra et al. 2018). CRs left on the soil surface improve fallow efficiencies and protect the soil surface from the erosive impact of raindrops, reduce runoff and increase infiltration, and thereby improved moisture content in the soil profile under rainfed agroecologies (Mulumba and Lal 2008; Kumar et al. 2019b). CRR also reduces evaporation loss of water, regulates soil temperature, and buffers soil moisture as well as canopy temperature (Jat et al. 2009), which ultimately helps in adapting to terminal heat stress in crops (Akter and Islam 2017; Kajla et al. 2015). Retaining CRs on the soil surface moderate the soil temperature and can decrease soil temperature up to 10 °C at a depth of 5.0 cm during summer. Besides, residue retention also helps to mitigate heat stress in wheat at the reproductive stage by lowering the canopy temperature up to 2.9 °C (Jat et al. 2009; Dadhich et al. 2015). Kumawat et al. (2020a) found that retaining CRs on the soil surface stored about 10.5–26.8% and 10.2–19.3% higher gravimetric and volumetric moisture content, respectively, at 0–10 cm soil depth during the chickpea-growing period in the Vertisol of Central India. CRs facilitate an increased rate of infiltration and create a physical barrier against evaporative loss of water surface, which leads to increased water availability to plants (Mulumba and Lal 2008; Yadav et al. 2021).

8.6.2.4 Effect of Residue Mulching on Nutrient Management

Optimization of N fertilizer under surface mulch condition (i.e., CA) is crucial as CRs supply about 35–50 kg N ha^{-1} (Ibewiro et al. 2000; Singh et al. 2014). During initial years of residue incorporation, 20–40 kg ha^{-1} additional N is recommended to compensate for the various losses of N fertilizer, and subsequently prescribed dose of fertilizers can be applied to attain higher crop productivity (Singh et al. 2005b). Reduced yield under ZT with residue mulching is associated with low N availability to plants during the early growth period due to increased immobilization and volatilization of N and reduced N mineralization over CT (Malhi et al. 2001; Kachroo et al. 2006; Srinivas et al. 2006). Direct drilling of N fertilizer below the soil surface minimizes the contact of N fertilizer with straw mulch and thus reduces immobilization and volatilization (Kushwaha et al. 2000; Singh and Timsina 2005).

The response of the wheat crop to 120 kg N ha^{-1} under rice straw mulching was found similar to wheat crop under CT (Rahman et al. 2005; Singh et al. 2015a, b). The continuous residue incorporation minimizes the amount of N fertilizer due to increased SOM and available N into the soil without any reduction in crop productivity (Mishra et al. 2017). In this context, alteration in timing and rate of N fertilizer is needed to synchronize N supply to crops under residue retained conditions (Roozbeh and Rajaie 2020). Applying nitrogenous fertilizers on CR retaining soil surface under CA is not appropriate because of immobilization and volatilization loss of N fertilizers (Saurabh et al. 2018). Therefore, N fertilizers should be placed in the furrow beneath the OM-enriched soil surface under CA. Application of diammonium phosphates (DAP) at a rate of 130 kg ha^{-1} and 56 kg ha^{-1} urea in seed row showed no adverse effect on agronomic yields. However, the application of a higher quantity of urea and DAP can adversely affect seed germination and yield of the crop (Jat et al. 2014). Hence, placing N at a rate of 120 kg ha^{-1} and recommended dose of P and K during sowing of wheat in between two rows of crops significantly increased the productivity as compared to topdressing in two equal splits (Singh et al. 2015a, b; Meena et al. 2015).

8.6.2.5 Effect of Residue Mulching on Soil Microbiome

Soil microorganisms are the living component of SOM, and they are considered as a key player in CR decomposition and nutrient cycling (Franzluebbers et al. 1999; Gonzalez-Quiñones et al. 2011; Larsen et al. 2014). Therefore, the presence of diverse and active soil microbial life is very important to sustain soil health and agricultural production (Hendgen et al. 2018). Management practices especially, residue returns to the soil and conservation tillage, are the major factors that alter microbial diversity and their activities inside the soil (Fig. 8.8) (Hao et al. 2019). Labile fractions of CRs supply C, which acts as a regular source of energy for soil microbiomes (Wang et al. 2003). Moreover, surface application of CR along with conservation tillage creates a favorable soil environment for microbial growth by moderating soil moisture and temperature (Jiang et al. 2018). It has been reported that surface retention of CRs under ZT significantly increased the microbial biomass carbon and N (MBC and MBN) by 42 and 79%, respectively, at 0–15 cm soil depth as compared to the CT. Further, the increased organic biomass and improved soil environmental conditions under CA-based practices led to increased alkaline phosphatase, dehydrogenase, and β-glucosidase activity by 58, 14, and 13%, respectively, over CT (Jat et al. 2019, 2020). Surface retention of residues increased the SOC, microbial biomass carbon (MBC), and total soil bioactivity by 15%, 14%, and 37%, respectively. Govaerts et al. (2007) also observed 1.2- and 1.3-fold increased MBC and microbial biomass N (MBN), respectively, under residue mulching over the residue removal. Besides, a combination of ZT and straw mulching also enhances microorganism metabolic activity, Shannon index, Simpson index (Xiao et al. 2019), and soil fungal richness (Yadav et al. 2021).

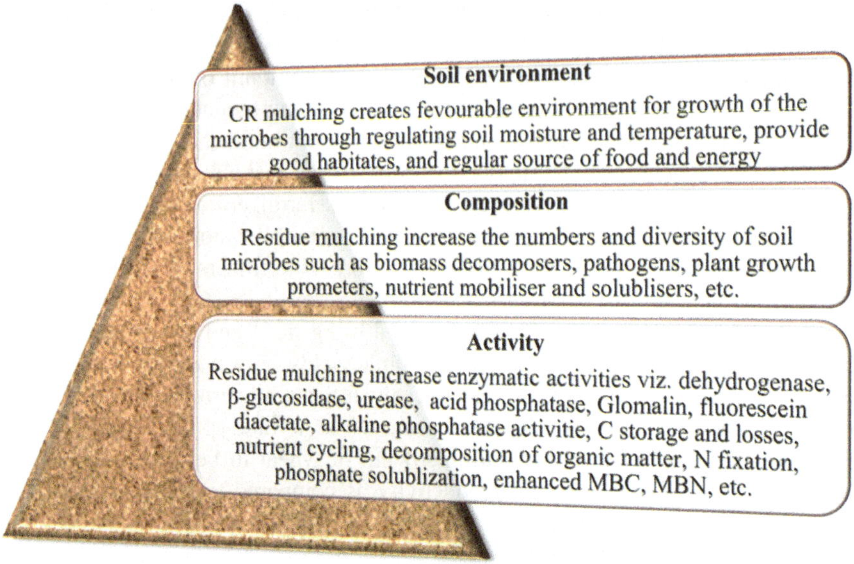

Fig. 8.8 Effect of CR mulching on soil environment and microbial diversity and activity

8.6.3 Compost Making from Crop Residues

Converting unused CRs into compost is a good option for efficient utilization of residues along with various environmental and economic benefits (Mor et al. 2016; Bindu and Manan 2018). The organic manure prepared from composting of CRs and farmyard manure helps in improving soil moisture, nutrient content, microbial diversity, their activities, etc. (Lohan et al. 2018). However, the composition and C/N ratio of CRs, pH, moisture, temperature, and aeration may affect the process of composting (Bhuvaneshwari et al. 2019). A study conducted at ICAR-IARI, New Delhi, observed that a microorganism consortium accelerates the rate of decomposition and takes ~75–90 days in compost preparation. PAU, Ludhiana, used raw rock phosphate along with rice straw to make nutrient-rich compost having 1.5% N, 2.3% P_2O_5, and 2.5% K_2O (Sidhu and Beri 2005). About 3 tons of nutrient-rich compost can be made from the paddy straw of 1 ha through the composting process (Sidhu and Beri 2008; Bindu and Manan 2018). Around 50% of N and P requirements can be reduced by using CR compost in paddy-wheat rotation (Jat et al. 2014). Moreover, each kilogram of CR used in an animal shed contains around 2–3 kg urine which is a rich source of N (Sidhu and Beri 2008).

8.6.4 Crop Residues as Biochar for Carbon Stabilization

Biochar is obtained through the pyrolysis process by burning of plant biomass at 300–600 °C under the limited or absence of oxygen (Jat et al. 2014). It offers a significant opportunity to convert large amounts of agricultural waste into valuable assets (Purakayastha et al. 2015). If CRs are transformed into biochar, 45–55% of biomass C can be recovered initially and 10–20% after 5–10 years through decomposition over open burning (Purakayastha et al. 2015). Being highly stable and resistant to microbial decomposition, it offers a promising strategy for enduring C stabilization (Lehmann et al. 2011; Joseph et al. 2020). Further, biochar has the potential to convert up to 50% of C in a chemically altered form (C sequestration), and it can remain active in the physical and chemical state in the soil for centuries. Joseph et al. (2020) documented that biochar application improved the aggregation and SOM stabilization which increased the C sequestration. The recalcitrant nature of biochar also contributes to SOC sequestration. Biochar provides physical protection to SOC in aggregates which facilitate long-term C stabilization (Six et al. 2002). Further, partly carbonized and highly degradable CRs requiring low temperature also lead to C stabilization (Zimmerman 2010). Weng et al. (2017) reported a 10% increase in SOC stabilization in occluded particulate OM and mineral-bound SOM fractions in biochar-amended soil. Similarly, improved aggregation and C stabilization have been reported in Vertisols due to straw-derived biochar application (Rahman et al. 2018). Biochar improves soil bio-physicochemical properties and consequently crop yields (Glaser et al. 2002; Lehmann et al. 2006; Yamato et al. 2006). Biochar produced by using paddy straw was found advantageous for rice-based production systems, while the real impact on soil health, crop yields, and SOC content is governed by location-specific crop production techniques (Singh et al. 2018; Joseph et al. 2020). Moreover, biochar stimulates native microbial activities, enhances moisture and nutrient supply through increased vesicular-arbuscular mycorrhizae fungal colonization, and also promotes biological N fixation in leguminous plants. Purakayastha et al. (2015) observed that biochar made from corn residues was rich in N and P and is alkaline, which can be used as an amendment for acidic soils. Further, paddy straw biochar enhanced microbial activities that can be used for the restoration of biological fertility of the soil, and wheat straw biochar was rich in K with greater potential of C sequestration over several years. Although, there are few studies available about the differential impact of biochar under varied crop production systems. Hence, long-term studies are required to understand the behavior of biochar under different cropping systems and soil types.

8.7 Soil and Environmental Impact of Residue Management

Generally, Indian soils are low in OC content and are being continuously exhausted by intensive tillage practices and cropping systems (Jat et al. 2014; Yadav and Arora 2018). The complete removal of CRs from the field leads to an extreme level of K mining as 80–85% of K uptake is retained in the straw of rice and wheat crops (Singh et al. 2004). Continuous retention of CR buildup SOM enriches the soil with essential plant elements and improves microbial dynamics and their activities (Balota et al. 2003; Saha et al. 2010; Busari et al. 2015). The SOC pool can be increased through an increased rate of OM in the form of CRs. However, the extent of the increase in the SOC pool is influenced by management practices in combination with CRs (Singh et al. 2000). Sharma et al. (2019) documented that ZT with residue mulching in wheat increased the total SOC content by about 3–13% and oxidizable C by 8.2–8.5%, respectively, at various surfaces of up to 60 cm depth over CT. Likewise, total SOC and oxidizable C content were enhanced by 6.5–12.5% and 4.2–28.2%, respectively, under ZT-based direct-seeded rice as compared to conventional practice. SOC has a beneficial impact on a range of soil physical, chemical, and biological health parameters (Smith et al. 2000; Jhariya et al. 2018, 2019, 2021). These soil qualities have strong interactions and crucial for defining the productive potential of a soil (Blanco and Lal 2010; Mehra et al. 2018).

The increased numbers of bio-pores under residue retention condition lead to higher hydraulic conductivity and infiltration rate over CT (He et al. 2009; Busari et al. 2015; Choudhary et al. 2019). CRR also reduces the negative impact of hardpan in subsequent crops developed during the puddling process of rice cultivation (Singh et al. 2005c). CRs also help in enhancing the di-nitrogen fixation by enhancing the activity of a symbiotic bacteria, i.e. *Azotobacter chroococcum* and *A. agilis*, into the soil (Singh et al. 2014). Moreover, a huge quantity of C can be stored in the soil, although the pace of C sequestration is affected by soil characteristics and climate (West and Post 2002). Based on the alteration in SOC and application of C, about 12–25% of applied C in the form of rice straw was sequestrated in the soil over 7 years of residue application (Singh et al. 2009). A critical level of 2.47 t C ha^{-1} $year^{-1}$ has been calculated to retain an antecedent quantity of SOC in rice-based crop rotations (Srinivasarao et al. 2013).

The SOC sequestration would help in mitigating GHG emissions and increasing soil production efficiency (Wang et al. 2010; Frank et al. 2017; Khan et al. 2021a, b). It protects atmospheric quality which is spoiled through residue burning and inappropriate intensive tillage operations. Besides, restoration of soil health and improvement in environmental quality and residue retention also enhance different ecosystem services which lead to higher and sustained agronomic yields (Palm et al. 2014; Lal 2009) (Fig. 8.9).

The enduring sustainability of different cropping systems is primarily governed by the amount of C input, output, and C-use efficiency (Jat et al. 2011). Gupta et al. (2007) stated that the incorporation of CRs increases total P fractions, reduces P adsorption, and increases P availability in the soil. Moreover, residue retention can also supplement a considerable amount of K fertilizers as paddy residue contains more than 80% of total plant K at harvest (Singh et al. 2008c). Jat et al. (2020) reported that the incorporation of CRs improves OC and available N, P, and K and essential microelements in the soil (Nandan et al. 2019).

The availability of micronutrients is also affected by CR recycling. About 50–80% of micronutrients absorbed by paddy and wheat can be transformed into available form through residue incorporation or mulching (Singh and Sidhu 2014). Singh et al. (2014) stated that use of Zn at a rate of 2.5 kg ha^{-1} in the preceding crop, and incorporating 50% of CRs of every crop was more efficient over 10 kg Zn ha^{-1}alone, indicating that residue incorporation can save a considerable amount of Zn fertilizer (Jat et al. 2014). Further, soil salinity and alkalinity can also be reduced

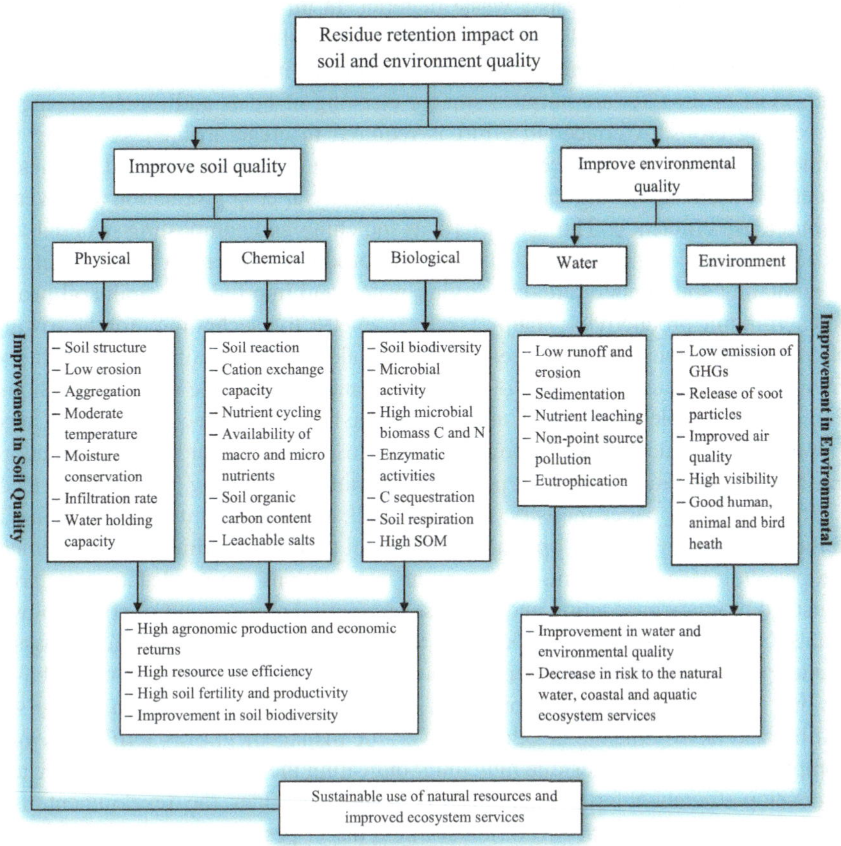

Fig. 8.9 Positive roles of residue retention on soil and environmental quality and ecosystem processes (adopted from Lal 2009)

by CR mulching (Basak et al. 2020; Soni et al. 2020). In intensive cropping systems of IGPs and other regions of the country, ZT with residue mulching is gaining attention because of minimum production cost, saving of energy and irrigation water, enhanced agronomic production, economic returns, and better quality of soil and environment (Kumawat et al. 2020a; Jat et al. 2020). The CRs with higher C/N ratios release hydroxyls, especially during the decomposition process which significantly helps in correcting soil acidity (Singh et al. 2014; Neina 2019).

8.8 Conclusion and Future Prospective

CRs are the universally available, largest, and cheapest source of lignocellulosic biomass that has a large number of applications. Since ancient times CRs are utilized as livestock feed, cooking fuel, thatching material, etc., and the leftover CRs are burnt by considering them as a waste material. In this context, CR recycling is a sustainable and ecologically effective approach for fulfilling crop nutrient requirement, restoring and sustaining soil health, and enhancing atmospheric value. However, most of the CRs are burnt in situ for cleaning of fields and timely sowing of succeeding crops. A concerted effort is required to create awareness among the farmers about the significance and role of CRs in nutrient recycling and soil health benefits towards achieving the goal of sustainability in crop production and overall resource use efficiency. The use of CRs as mulching and composting material and incorporation of it into the soil and in biochar production has been found beneficial in regulating soil temperature, conserving water, and improving overall soil health and agronomic productivity. Happy Seeder is an effective technology for the sowing of crops under residue retained field conditions along with sustained soil health and environmental security, low cost of production, and reduced time of field operations, limits gradual depletion of SOM, and avoids crop burning. Moreover, there is a need to aware and promote farmers about alternative uses of straw for biogas production, mushroom cultivation, soil incorporation/retention, composting, gasification, electricity generation, paper industry, etc. Due to its renewable nature and local availability, CR may be the best option for bioethanol production which can supplement the petrol/diesel requirement. Government interventions are needed to popularize densified biomass production and their use as a partial replacement of coal in power plants. Cost-effective ethanol production technology from lignocellulosic biomass is needed to be developed. Furthermore, awareness about the benefits of CA-based residue management practices for in situ residue recycling is needed to restore soil fertility and environmental health. However, the management of N fertilizer is very complex under ZT with residue mulching, indicating the necessity of more research to develop efficient N fertilizer application methods. Besides, the development of an efficient package of practices (nutrients, water, insect-pest and disease management, etc.) is required for CRs based crop production systems. Moreover, the development of appropriate simulation models to make available integrated and location-specific CRM technologies to attain the multiple goals of enhancing nutrient and input use efficiency, profitability, and soil productivity. There is also a need to support in situ CRM practices through supplying machinery on subsidized rates and promoting custom hiring systems to avoid the burning of CRs and the spread of CRM practices in large as well as small and marginal fields. Appropriate institutional and policy support is required for large-scale spreading and adoption of CRM practices.

References

Akter N, Islam MR (2017) Heat stress effects and management in wheat. A review. Agron Sustain Dev 37(5):1–7

Almaz MG, Halim RA, Yusoff MM, Wahid SA (2017) Effect of incorporation of crop residue and inorganic fertilizer on yield and grain quality of maize. Indian J Agric Res 51(6):574–579

Andreae MO, Merlet P (2001) Emission of trace gases and aerosols from biomass burning. Global Biogeochem Cycles 15:955–966

Athira G, Bahurudeen A, Appari S (2019) Sustainable alternatives to carbon intensive paddy field burning in India: a framework for cleaner production in agriculture, energy, and construction industries. J Clean Prod 236:117598

Awasthi A, Agarwal R, Mittal SK, Singh N, Singh K, Gupta PK (2011) Study of size and mass distribution of particulate matter due to crop residue burning with seasonal variation in rural area of Punjab, India. J Environ Monit 13(4):1073–1081

Awasthi A, Hothi N, Kaur P, Singh N, Chakraborty M, Bansal S (2017) Elucidative analysis and sequencing of two respiratory health monitoring methods to study the impact of varying atmospheric composition on human health. Atmos Environ 171:32–37

Balota EL, Colozzi-Filho A, Andrade DS, Dick RP (2003) Microbial biomass in soils under different tillage and crop rotation systems. Biol Fertil Soils 38:15–20

Banerjee A, Jhariya MK, Yadav DK, Raj A (2020) Environmental and Sustainable Development through Forestry and Other Resources. Apple Academic Press Inc., CRC Press- a Tayler and Francis Group, US & Canada, 400 pp. https://doi.org/10.1201/9780429276026

Banerjee A, Meena RS, Jhariya MK, Yadav DK (2021) Agroecological footprints management for sustainable food system. Springer, Singapore. https://doi.org/10.1007/978-981-15-9496-0

Basak N, Barman A, Sundha P, Rai AK (2020) Recent trends in soil salinity appraisal and management. In: Rakshit A, Ghosh S, Chakraborty S, Philip V, Datta A (eds) Soil analysis: recent trends and applications. Springer, Singapore, pp 143–161

Behera B (2018) Recycling of crop residues for improved soil nutrient status and farm income, department of agronomy CA. Odisha University of Agriculture & Technology, Bhubaneswar

Bertol OJ, Rizzi NE, Bertol I, Roloff G (2007) Soil and water loss and quality of surface runoff associated with interrill erosion in no- tillage area treated with chemical and organic fertilizers. Rev Bras Cienc Solo 31:781–792

Bhuvaneshwari S, Hettiarachchi H, Meegoda JN (2019) Crop residue burning in India: policy challenges and potential solutions. Int J Environ Res Public Health 16(5):832

Bindu SM, Manan J (2018) Ways and means of paddy straw management: a review. J Rev Agric Allied Fields 1(1):1–12

Bisen N, Rahangdale CP (2017) Crop residues management option for sustainable soil health in rice-wheat system: a review. Int J Chem Stud 5(4):1038–1042

Blanco H, Lal R (2010) Principles of soil conservation and management. Springer, Dordrecht, New York

Busari MA, Kukal SS, Kaur A, Bhatt R, Dulazi AA (2015) Conservation tillage impacts on soil, crop and the environment. Int Soil Water Conserv Res 3(2):119–129

Byous EW, Williams JE, Jonesa GE, Horwath W, Kessel C (2004) Nutrient requirements of rice with alternative straw management. Better Crops 36:6–11

Carbone LM, Aguilar R (2017) Fire frequency effects on soil and pollinators: what shapes sexual plant reproduction? Plant Ecol 218(11–12):1283–1297

Cerda A, Gonzalez-Pelayo O, Gimenez-Morera A, Jordan A, Pereira P, Novara A, Brevik EC, Prosdocimi M, Mahmoodabadi M, Keesstra S, Orenes FG, Ritsema CJ (2016) Use of barely straw residue to avoid high erosion and runoff rates on persimmon plantations in Eastern Spain under low frequency-high magnitude simulated rainfall events. Soil Res 54:154–165. https://doi.org/10.1071/SR15092

Chakrabarti S, Khan MT, Kishore A, Roy D, Scott SP (2019) Risk of acute respiratory infection from crop burning in India: estimating disease burden and economic welfare from satellite and national health survey data for 250 000 persons. Int J Epidemiol 48(4):1113–1124

Chauhan BS, Mahajan G, Sardana V, Timsina J, Jat ML (2012) Productivity and sustainability of the rice–wheat cropping system in the Indo-Gangetic Plains of the Indian subcontinent: problems, opportunities, and strategies. Adv Agron 117:315–369

Chen Y, Tessier S, Cavers C, Xu X, Monero F (2005) A survey of crop residue burning practices in Manitoba. Appl Eng Agric 21:317–323

Chen J, Li C, Ristovski Z, Milic A, Gu Y, Islam MS, Wang S, Hao J, Zhang H, He C, Guo H (2017) A review of biomass burning: emissions and impacts on air quality, health and climate in China. Sci Total Environ 579:1000–1034

Choudhary M, Rana KS, Meena MC, Bana RS, Jakhar P, Ghasal PC, Verma RK (2019) Changes in physico-chemical and biological properties of soil under conservation agriculture based pearl millet–mustard cropping system in rainfed semi-arid region. Arch Agron Soil Sci 65 (7):911–927

Dadhich RK, Meena RS, Reager ML, Kansotia BC (2015) Response of bio-regulators to yield and quality of Indian mustard (*Brassica juncea* L. Czernj. and Cosson) under different irrigation environments. J Appl Nat Sci 7(1):52–57

Dai C, Liu Y, Wang T, Li Z, Zhou Y (2018) Exploring optimal measures to reduce soil erosion and nutrient losses in southern China. Agric Water Manag 210:41–48

Das TK, Bhattacharyya R, Sharma AR, Das S, Saad AA, Pathak H (2013) Impacts of conservation agriculture on total soil organic carbon retention potential under an irrigated agro-ecosystem of the western Indo-Gangetic Plains. Eur J Agron 51:34–42

Devi S, Gupta C, Jat SL, Parmar MS (2017) Crop residue recycling for economic and environmental sustainability: the case of India. Open Agric 2(1):486–494

Dobermann A, Fairhurst TH (2002) Rice straw management. Better crops Int 16:7–11

Dutt S (2018) Solving straw burning problem. https://www.thehillstimes.in/featured/solving straw-burning-problem/

Erenstein O (2002) Crop residue mulching in tropical and semi-tropical countries: an evaluation of residue availability and other technological implications. Soil Till Res 67:115–133

Fang Y, Singh B, Singh BP, Krull E (2014) Biochar carbon stability in four contrasting soils. Eur J Soil Sci 65(1):60–71

Frank S, Havlík P, Soussana JF, Levesque A, Valin H, Wollenberg E, Kleinwechter U, Fricko O, Gusti M, Herrero M, Smith P (2017) Reducing greenhouse gas emissions in agriculture without compromising food security? Environ Res Lett 12(10):105004

Franzluebbers AJ, Nazih N, Stuedemann JA, Fuhrmann JJ, Schomberg HH, Hartel PG (1999) Soil carbon and nitrogen pools under low- and high-endophyte-infected tall fescue. Soil Sci Soc Am J 63:1687–1694

Fuentes M, Govaerts B, De León F, Hidalgo C, Dendooven L, Sayre KD, Etchevers J (2009) Fourteen years of applying zero and conventional tillage, crop rotation and residue management systems and its effect on physical and chemical soil quality. Eur J Agron 30:228–237

Gadi R, Kulshrestha UC, Sarkar AK, Garg SC, Parashar DC (2003) Emissions of SO_2 and NO_X from biofuels in India. Tellus B 55(3):787–795

Gathala M, Ladha JK, Balyan V, Saharawat YS, Kumar V, Sharma PK (2011) Effect of tillage and crop establishment methods on physical properties of a medium-textured soil under 7-year rice–wheat rotation. Soil Sci Soc Am J 75:1–12

Glaser B, Lehmann J, Zech W (2002) Ameliorating physical and chemical properties of highly weathered soils in the tropics with charcoal–a review. Biol Fertil Soils 35:219–230

Gonzalez-Quiñones V, Stockdale EA, Banning NC, Hoyle FC, Sawada Y, Wherrett AD, Jones DL, Murphy DV (2011) Soil microbial biomass– interpretation and consideration for soil monitoring. Soil Res 49:287–304

Goswami SB, Mondal R, Mandi SK (2020) Crop residue management options in rice–rice system: a review. Arch Agron Soil Sci 66(9):1218–1234

Govaerts B, Mezzalama M, Unno Y, Sayre KD, Luna-Guido M, Vanherck K, Dendooven L, Deckers J (2007) Influence of tillage, residue management, and crop rotation on soil microbial biomass and catabolic diversity. Appl Soil Ecol 37(1–2):18–30

Govaerts B, Verhulst N, Castellanos-Navarrete A, Sayre KD, Dixon J, Dendooven L (2009) Conservation agriculture and soil carbon sequestration: between myth and farmer reality. Crit Rev Plant Sci 28:97–122

Graham RL, Nelson R, Sheehan J, Perlack R, Wright LL (2007) Current and potential US corn stover supplies. Agron J 99(1):11

Gregg JS, Izaurralde RC (2010) Effect of crop residue harvest on long-term crop yield, soil erosion and nutrient balance: trade-offs for a sustainable bioenergy feedstock. Biofuels 1:69–83

Grum B, Assefa D, Hessel R, Woldearegay K, Kessler A, Ritsema C, Geissen V (2017) Effect of in situ water harvesting techniques on soil and nutrient losses in semi-arid Northern Ethiopia. Land Degrad Dev 28(3):1016–1027

Gupta RK, Shukla AK, Ashraf M, Ahmed ZU, Sinha RKP, Hobbs PR (2002) Options for establishment of rice and issues constraining its productivity and sustainability in eastern Gangetic plains of Bihar, Nepal and Bangladesh Rice-Wheat Consortium Travelling Seminar Report Series 4. Rice-Wheat Consortium for the Indo-Gangetic Plains, New Delhi, p 36

Gupta PK, Sahai S, Singh N, Dixit CK, Singh DP, Sharma C, Tiwari MK, Gupta RK, Garg SC (2004) Residue burning in rice–wheat cropping system: causes and implications. Curr Sci 87:1713–1717

Gupta RK, Yadvinder-Singh LJK, Bijay-Singh SJ, Singh G, Pathak H (2007) Yield and phosphorus transformations in a rice-wheat system with crop residue and phosphorus management. Soil Sci Soc Am J 71:1500–1507

Guy SO, Lauver MA (2007) No-till and conventional-till effects on spring wheat in the Palouse. Crop Manag 6:191–197

Hao MM, Hu HY, Liu Z, Dong QL, Sun K, Feng YP (2019) Shifts in microbial community and carbon sequestration in farmland soil under long-term conservation tillage and straw returning. Appl Soil Ecol 136:43–54

He J, Wang Q, Li H, Tullberg JN, McHugh AD, Bai Y, Zhang X, McLaughlin N, Gao H (2009) Soil physical properties and infiltration after long-term no-tillage and ploughing on the Chinese Loess Plateau. N Z J Crop Hortic Sci 37(3):157–166

Hendgen M, Hoppe B, Döring J, Friedel M, Kauer R, Frisch M, Dahl A, Kellner H (2018) Effects of different management regimes on microbial biodiversity in vineyard soils. Sci Rep 8:8

Hilodhari M, Das D, Baruah DC (2014) Bioenergy potential crop residue biomass in India. Renew Sust Energ Rev 32:504–512

Huang Z, Hu LC, Shi M, Li X, Dai J (2012) Changes in composition of soil organic matter after burning of straw. Acta Pedol Sinca 49:60–67

Ibewiro B, Sanginga N, Vanlauwe B, Merckx R (2000) Nitrogen contributions from decomposing cover crop residues to maize in a tropical derived savanna. Nutr Cycl Agroecosyst 57 (2):131–140

Izaurralde RC, Williams JR, Post WM, Thomson AM, McGill WB, Owens LB, Lal R (2007) Long-term modeling of soil Cerosion and sequestration at the small watershed scale. Clim Chang 80:73–90

Jain N, Bhatia A, Pathak H (2014) Emission of air pollutants from crop residue burning in India. Aerosol Air Qual Res 14(1):422–430

Jat ML (2013) Green House Gases (GHGs) emission studies in contrasting rice establishment methods under rice-wheat rotation of Indo-Gangetic plains of India. In: Annual progress report of Bayer crop science GHG project. International Maize and Wheat Improvement Centre (CIMMYT), El Batán, Texcoco, Edo. de Mexico, CP

Jat ML (2016) Advice for India's rice-wheat farmers: put aside the plow and save straw to fight pollution. https://www.cimmyt.org/advice-for-indias-rice-wheat-farmers-putaside-the-plow-and-save-straw-to-fight-pollution/

Jat ML, Gathala MK, Ladha JK, Saharawat YS, Jat AS, Sharma SK, Kumar V, Gupta RK (2009) Evaluation of precision land leveling and double zero-tillage systems in the rice-wheat rotation: water use, productivity, profitability and soil physical properties. Soil Till Res 105:112–121

Jat ML, Jat RK, Gupta R, Gopal R (2011) Conservation agriculture in cereal systems of South Asia: effect on crop productivity and carbon-based sustainability index. In: Resilient food systems for a changing world. Proceedings of the 5th world congress of conservation agriculture. Incorporating 3rd farming systems design conference, 26–29 September 2011, Brisbane Australia, pp 26–27

Jat ML, Malik RK, Saharawat YS, Gupta R, Bhag M, Paroda R (2012) Proceedings of regional dialogue on conservation agricultural in South Asia, New Delhi, India, 32 pp

Jat ML, Yadvinder-Singh HS, Singh P, Jat RK (2014) Managing cereal crop residues for sustainable crop production systems. In: Prasad R, Kumar D, Rana DS, Shivay Y, Tewatia RK (eds) Text book of plant nutrient management. Indian Society of Agronomy, Indian Agricultural Research Institute, New Delhi, pp 326–347

Jat LK, Singh YV, Meena SK, Meena SK, Parihar M, Jatav HS, Meena RK, Meena VS (2015) Does integrated nutrient management enhance agricultural productivity. J Pure Appl Microbiol 9 (2):1211–1221

Jat HS, Datta A, Choudhary M, Sharma PC, Yadav AK, Choudhary V, Gathala MK, Jat ML, McDonald A (2019) Climate Smart Agriculture practices improve soil organic carbon pools, biological properties and crop productivity in cereal-based systems of North-West India. Catena 181:104059

Jat HS, Choudhary M, Datta A, Yadav AK, Meena MD, Devi R, Gathala MK, Jat ML, McDonald A, Sharma PC (2020) Temporal changes in soil microbial properties and nutrient dynamics under climate smart agriculture practices. Soil Till Res 199:104595

Jhariya MK, Singh L (2020) Herbaceous diversity and biomass under different fire regimes in a seasonally dry forest ecosystem. Environ Dev Sustain 22:1–19. https://doi.org/10.1007/s10668-020-00892-x

Jhariya MK, Singh L (2021) Effect of fire severity on soil properties in a seasonally dry forest ecosystem of Central India. Int J Environ Sci Technol. https://doi.org/10.1007/s13762-020-03062-8

Jhariya MK, Banerjee A, Yadav DK, Raj A (2018) Leguminous trees an innovative tool for soil sustainability. In: Meena RS, Das A, Yadav GS, Lal R (eds) Legumes for soil health and sustainable management. Springer, Singapore, pp 315–345. https://doi.org/10.1007/978-981-13-0253-4_10

Jhariya MK, Banerjee A, Meena RS, Yadav DK (2019) Sustainable agriculture, forest and environmental management. Springer, Singapore, p 606. https://doi.org/10.1007/978-981-13-6830-1

Jhariya MK, Meena RS, Banerjee A (2021) Ecological intensification of natural resources for sustainable agriculture. Springer, Singapore. https://doi.org/10.1007/978-981-33-4203-3

Jiang Y, Ma N, Chen Z, Xie H (2018) Soil macrofauna assemblage composition and functional groups in no-tillage with corn stover mulch agroecosystems in a mollisol area of northeastern China. Appl Soil Ecol 128:61–70

Joseph UE, Toluwase AO, Kehinde EO, Omasan EE, Tolulope AY, George OO, Zhao C, Hongyan W (2020) Effect of biochar on soil structure and storage of soil organic carbon and nitrogen in the aggregate fractions of an Albic soil. Arch Agron Soil Sci 66(1):1–2

Kachroo D, Dixit AK, Bhat AK (2006) Decomposition of various residues and their nutrient release pattern under Alfisols of Jammu Region. J Indian Soc Soil Sci 54(3):342–344

Kajla M, Yadav VK, Chhokar RS, Sharma RK (2015) Management practices to mitigate the impact of high temperature on wheat. J Wheat Res 7(1):1–2

Kaur R, Bansal M, Sharma S, Tallapragada S (2019) Impact of in situ rice crop residue burning on agricultural soil of district Bathinda, Punjab, India. Rasayan J Chem 12:421–430

Kaushal LA, Prashar A (2020) Agricultural crop residue burning and its environmental impacts and potential causes–case of northwest India. J Environ Plan Manag 3:1–21

Khan N, Jhariya MK, Raj A, Banerjee A, Meena RS (2021a) Soil carbon stock and sequestration: implications for climate change adaptation and mitigation. In: Jhariya MK, Meena RS, Banerjee A (eds) Ecological Intensification of Natural Resources for Sustainable Agriculture. Springer, Singapore. https://doi.org/10.1007/978-981-33-4203-3_13

Khan N, Jhariya MK, Raj A, Banerjee A, Meena RS (2021b) Eco-designing for sustainability. In: Jhariya MK, Meena RS, Banerjee A (eds) Ecological intensification of natural resources for sustainable agriculture. Springer, Singapore. https://doi.org/10.1007/978-981-33-4203-3_16

Khatun R, Howlader MA, Islam MN, Alam MK, Mahmud MS, Rahman MH (2015) Exploration the causes of infectious illness and detection of antibiotic residues in warehouse poultry. Am J Food Sci Health 1(2):57–62

Kocher GS, Kaur P, Taggar MS (2017) An overview of pretreatment processes with special reference to biological pretreatment for rice straw delignification. Curr Biochem Eng 4 (3):151–163

Kumar S, Kumar P (2016) Economic impact of air pollution from agricultural residue burning on human health. In: Nautiyal et al (eds) Climate change challenge and social-economic-ecological interface-building. Springer, Cham, pp 297–313

Kumar P, Kumar S, Joshi L (2015a) The extend and management of crop residue stubbles. In: Kumar P, Kumar S, Joshi L (eds) Socioeconomic and environmental implications of agricultural residue burning: a case study of Punjab, India. Springer, New Delhi, p 144

Kumar P, Kumar S, Joshi L (2015b) Alternative uses of crop stubble. In: Kumar P, Kumar S, Joshi L (eds) Socioeconomic and environmental implications of agricultural residue burning. Springer, New Delhi, pp 69–89

Kumar S, Sharma DK, Singh DR, Biswas H, Praveen KV, Sharma V (2019a) Estimating loss of ecosystem services due to paddy straw burning in North-west India. Int J Agric Sustain 17 (2):146–157

Kumar R, Saurabh K, Kumawat N, Mishra JS, Hans H, Krishna B, Meena RS, Jha BK, Upadhyay PK, Hazra KK (2019b) Conservation agriculture: perspectives on soil and environmental management in Indo-Gangetic plains of South Asia. In: Sustainable management of soil and environment. Springer, Singapore, pp 123–168

Kumar S, Meena RS, Datta R, Verma SK, Yadav GS, Pradhan GS, Molaei A, Rahman GKMM, Mashuk HA (2020) Legumes for carbon and nitrogen cycling: an organic approach. In: Carbon and nitrogen cycling in soil, pp 337–375. https://doi.org/10.1007/978-981-13-7264-3_10

Kumar S, Meena RS, Singh RK, Munir TM, Datta R, Danish S, Singh GS, Yadav, Kumar S (2021) Soil microbial and nutrient dynamics under different sowings environment of Indian mustard (*Brassica juncea* L.) in rice based cropping system. Sci Rep 11:5289. https://doi.org/10.1038/s41598-021-84742-4

Kumari K, Prasad J, Solanki IS, Choudhary R (2014) Long-term effect of residual zinc and crop residues incorporation on soil health and crop productivity under calcareous soils of rice-wheat system in India. Int J Curr Microbiol App Sci 6(7):2401–2414

Kumawat A, Vishawakarma AK, Wanjari RH, Sharma NK, Yadav D, Kumar D, Biswas AK (2020a) Impact of levels of residue retention on soil properties under conservation agriculture in Vertisols of central India. Arch Agron Soil Sci. https://doi.org/10.1080/03650340.2020.1836345

Kumawat A, Yadav D, Samadharmam K, Rashmi I (2020b) Soil and water conservation measures for agricultural sustainability. In: Soil moisture importance. IntechOpen, London

Kushwaha CP, Tripathi SK, Singh KP (2000) Variations in soil microbial biomass and N availability due to residue and tillage management in a dryland rice agro ecosystem. Soil Till Res 56:153–166

Lal R (1976) Soil erosion on Alfisols in western Nigeria: II. Effects of mulch rates. Geoderma 16 (5):377–387

Lal R (1997) Residue management, conservation tillage and soil restoration for mitigating greenhouse effect by CO -enrichment. Soil Till Res 43:81–107

Lal R (2002) Soil carbon dynamics in cropland and rangeland. Environ Pollut 116:353–362

Lal R (2009) Use of crop residues in the production of biofuel. In: Handbook of waste management and co-product recovery in food processing. Woodhead Publishing, Cambridge, pp 455–478

Larsen E, Grossman J, Edgell J, Hoyt G (2014) Soil biological properties, soil losses and corn yield in long-term organic and conventional farming systems. Soil Till Res 139:37–45

Lehmann J, Gaunt J, Rondom M (2006) Biochar sequestration in terrestrial ecosystems–a review. Mitig Adapt Strateg Glob Chang 11:403–427

Lehmann J, Rillig MC, Thies J, Masiello CA, Hockaday WC, Crowley D (2011) Biochar effects on soil biota - A review. Soil Biol Biochem 43(9):1812–1836

Lehtinen T, Schlatter N, Baumgarten A, Bechini L, Krüger J, Grignani C, Zavattaro L, Costamagna C, Spiegel H (2014) Effect of crop residue incorporation on soil organic carbon and greenhouse gas emissions in European agricultural soils. Soil Use Manag 30(4):524–538

Logah V (2011) Chemical properties of maize residues under application of organic and inorganic nutrient sources. Int J Env Sci Dev 2:344–349

Lohan SK, Jat HS, Yadav AK, Sidhu HS, Jat ML, Choudhary M, Peter JK, Sharma PC (2018) Burning issues of paddy residue management in north-west states of India. Renew Sust Energ Rev 81:693–706

Lu X (2020) A meta-analysis of the effects of crop residue return on crop yields and water use efficiency. PLoS One 15(4):e0231740

Malhi SS, Grant CA, Johnston AM, Gill KS (2001) Nitrogen fertilization management for no-till cereal production in the Canadian Great Plains: a review. Soil Till Res 60(3–4):101–122

Mandal KG, Misra AK, Hati KM, Bandyopadhyay KK, Ghosh PK, Mohanty M (2004) Rice residue-management options and effects on soil properties and crop productivity. J Food Agric Environ 2:224–231

Mandal P, Sarkar R, Mandal A, Saud T (2014) Seasonal variation and sources of aerosol pollution in Delhi, India. Environ Chem Lett 12(4):529–534

Manjunatha K, Shirwal S, Sushilendra VP, Raghavendra V (2015) Role of Balers in agricultural crop residue management-A review. Eco Env Cons 21(3):285–293

Mechergui T, Pardos M, Jhariya MK, Banerjee A (2020) Mulching and weed management towards sustainability. In: Ecological intensification of natural resources for sustainable agriculture. Springer, Singapore. https://doi.org/10.1007/978-981-33-4203-3_8

Meena RS, Lal R (2018) Legumes for Soil Health and Sustainable Management. Springer Singapore, Singapore. pp. 541.ISBN 978–981–13-0253-4 (eBook), ISBN: 978–981–13-0252-7(Hardcover). doi:https://doi.org/10.1007/978-981-13-0253-4_10

Meena RS, Meena VS, Meena SK, Verma JP (2015) Towards the plant stress mitigate the agricultural productivity: a book review. J Clean Prod 102:552–553

Meena RS, Kumar V, Yadav GS, Mitran T (2018) Response and interaction of *Bradyrhizobium japonicum* and Arbuscular mycorrhizal fungi in the soybean rhizosphere: a review. Plant Growth Regul 84:207–223. https://doi.org/10.1007/s10725-017-0334-8

Meena RS, Lal R, Yadav GS (2020) Long term impacts of topsoil depth and amendments on soil physical and hydrological properties of an Alfisol in Central Ohio, USA. Geoderma 363:1141164. https://doi.org/10.1016/j.geoderma.2019.114164

Meena RS, Lal R, Yadav GS (2020a) Long-term impact of topsoil depth and amendments on carbon and nitrogen budgets in the surface layer of an Alfisol in Central Ohio. Catena 2020:104752. https://doi.org/10.1016/j.catena.2020.104752

Meena RS, Kumar S, Datta R, Lal R, Vijaykumar V, Brtnicky M, Sharma MP, Yadav GS, Jhariya MK, Jangir CK, Pathan SI, Dokulilova T, Pecina V, Marfo TD (2020b) Impact of agrochemicals on soil microbiota and management: a review. Land (MDPI) 9(2):34. https://doi.org/10.3390/land9020034

Mehra P, Baker J, Sojka RE, Bolan N, Desbiolles J, Kirkham MB, Ross C, Gupta R (2018) A review of tillage practices and their potential to impact the soil carbon dynamics. Adv Agron 150:185–230

Mehta CR, Sharma S, Nair R, Singh KP (2013) Impact of crop residue burning on environment and human health. Indian Farm 63(5):24–25

Milham N, Kumar P, Crean J, Singh RP (2014) Policy instruments to address air pollution issues in agriculture: implications for Happy Seeder technology adoption in India. Final Report, Australian Centre for International Agricultural Research (ACIAR), FR2014–17. https://www.aciar.gov.au/node/12171

Mishra JS, Kumar R, Ravikumar S, Kumar R, Prakash V, Rao KK, Bhatt BP (2017) Production potential of improved grain sorghum cultivars (*Sorghum bicolor*) under staggered plantings in non-traditional areas of Eastern India. Indian J Agron 62(1):74–80

MNRE (2009) Ministry of new and renewable energy resources. Government of India, New Delhi. www.mnre.gov.in/biomassrsources

MOA&FW (2019) Report on review of the scheme "promotion of agricultural mechanization for in-situ management of crop residue in States of Punjab, Haryana, Uttar Pradesh and NCT of Delhi", New Delhi, India, 4 pp

MOA&FW (2020) 3rd advance estimates of production of food grains for the year 2019–20 by Ministry of agriculture and farmers welfare, department of agriculture, cooperation and farmers welfare

Mohammadi A, Cowie A, Anh Mai TL, De La Rosa RA, Kristiansen P, Brandão M, Joseph S (2016) Biochar use for climate-change mitigation in rice cropping systems. J Clean Prod 116:61–70

Mor S, Kaur K, Khaiwal R (2016) SWOT analysis of waste management practices in Chandigarh, India and prospects for sustainable cities. J Environ Biol 37(3):327

Mulumba LN, Lal R (2008) Mulching effects on selected soil physical properties. Soil Till Res 98 (1):106–111

Mupangwa W, Twomlow S, Walker S (2012) Reduced tillage, mulching and rotational effects on maize (*Zea mays* L.), cowpea (*Vigna unguiculata* (Walp) L.) and sorghum (*Sorghum bicolor* L. (Moench)) yields under semi-arid conditions. Field Crops Res 132:139–148

Nagar R, Trivedi SK, Nagar D, Karnawat M (2020) Impact of agriculture crop residue burning on environment and soil health. Biotica Res Today 2(5 Spl):171–173

Nandan R, Singh V, Singh SS, Kumar V, Hazra KK, Nath CP, Poonia S, Malik RK, Bhattacharyya R, McDonald A (2019) Impact of conservation tillage in rice–based cropping systems on soil aggregation, carbon pools and nutrients. Geoderma 340:104–114

Nangia V, Jiantao D, Changrong Y, Xurong M, Wenqing H, Shuang L, Qin L (2010) Effects of conservation agriculture on land and water productivity in Yellow River Basin, China. Int J Agric Biol Eng 3(2):5–17

Naresh RK, Gupta RK, Rathore RS, Dwivedil A, Singh HL, Kumar V, Shukla AK, Singh V, Singh SP, Tyagil S, Singh V, Kumar V, Singh O, Mahajan NC (2017) Crop residue management and soil health with changing climate in smallholders farming: a subtropical Indian perspective. Int J Curr Microbiol App Sci 6(2):1591–1609

Nath CP, Das TK, Rana KS, Bhattacharyya R, Pathak H, Paul S, Meena MC, Singh SB (2017) Greenhouse gases emission, soil organic carbon and wheat yield as affected by tillage systems and nitrogen management practices. Arch Agron Soil Sci 63(12):1644–1660

Neina D (2019) The role of soil pH in plant nutrition and soil remediation. Appl Environ Soil Sci 2019:1–9

Pal R, Kumar R, Jalal RK, Sohane RK (2019) Assessment of Happy Seeder for Direct Sowing of Wheat without Burning of Rice Residue. Curr J Appl Sci Technol 37(1):1–4

Palansooriya KN, Wong JT, Hashimoto Y, Huang L, Rinklebe J, Chang SX, Bolan N, Wang H, Ok YS (2019) Response of microbial communities to biochar-amended soils: a critical review. Biochar 1(1):3–22

Palm C, Blanco-Canqui H, DeClerck F, Gatere L, Grace P (2014) Conservation agriculture and ecosystem services: an overview. Agric Ecosyst Environ 187:87–105

Park SI, Yang HI, Park HJ, Seo BS, Jeong YJ, Lim SS, Kwak JH, Kim HY, Yoon KS, Lee SM, Choi WJ (2020) Rice straw cover decreases soil erosion and sediment-bound C, N, and P losses but increases dissolved organic C export from upland maize fields as evidenced by δ13C. Sci Total Environ 753:142053

Pathak BS (2004) Crop residue to energy. In: Chadha KL, Swaminathan MS (eds) Environment and agriculture. Malhotra Publishing House, New Delhi, pp 854–869

Pathak S, Jain N, Bhatia A (2012) Crop residues management with conservation agriculture: potential, constraints and policy needs. Indian Agricultural Research Institute, New Delhi, 32 pp

Pimentel D (2006) Soil Erosion: a food and environmental threat. Environ Dev Sustain 8:119–137
Prasad S, Dhanya MS, Gupta N, Kumar A (2012) Biofuels from biomass: a sustainable alternative to energy and environment. Biochem Cell Arch 12(2):255–260
Prasad S, Singh A, Korres NE, Rathore D, Sevda S, Pant D (2020) Sustainable utilization of crop residues for energy generation: a life cycle assessment (LCA) perspective. Bioresour Technol 303:122964
Prasanthkumar K, Kannan SV (2018) Rice residue management for rice based cropping system in Cauvery delta zone–a review. Agric Rev 39(3):241–245
Prosdocimi M, Tarolli P, Cerdà A (2016) Mulching practices for reducing soil water erosion: a review. Earth-Sci Rev 161:191–203. https://doi.org/10.1016/j.earscirev.2016.08.006
Purakayastha TJ, Kumari S, Pathak H (2015) Characterisation, stability, and microbial effects of four biochars produced from crop residues. Geoderma 239:293–303
Rahman MA, Chikushi J, Saifizzaman M, Lauren JG (2005) Rice straw mulching and nitrogen response of no-till wheat following rice in Bangladesh. Field Crops Res 91(1):71–81
Rahman MT, Guo ZC, Zhang ZB, Zhou H, Peng XH (2018) Wetting and drying cycles improving aggregation and associated C stabilization differently after straw or biochar incorporated into a Vertisol. Soil Till Res 175:28–36
Raj A, Jhariya MK, Yadav DK, Banerjee A, Meena RS (2019a) Agroforestry: a holistic approach for agricultural sustainability. In: Jhariya MK, Banerjee A, Meena RS, Yadav DK (eds) Sustainable agriculture, forest and environmental management. Springer, Singapore, pp 101–131. https://doi.org/10.1007/978-981-13-6830-1
Raj A, Jhariya MK, Banerjee A, Yadav DK, Meena RS (2019b) Soil for sustainable environment and ecosystems management. In: Jhariya MK, Banerjee A, Meena RS, Yadav DK (eds) Sustainable agriculture, forest and environmental management. Springer, Singapore, pp 189–221. https://doi.org/10.1007/978-981-13-6830-1
Raj A, Jhariya MK, Yadav DK, Banerjee A (2020) Climate change and agroforestry systems: adaptation and mitigation strategies. Apple Academic Press, CRC Press-a Tayler and Francis Group, US & Canada, 383 pp. https://doi.org/10.1201/9780429286759
Raj A, Jhariya MK, Khan N, Banerjee A, Meena RS (2021) Ecological intensification for sustainable development. In: Ecological intensification of natural resources for sustainable agriculture. Springer, Singapore. https://doi.org/10.1007/978-981-33-4203-3_5
Ranaivoson L, Naudin K, Ripoche A, Affholder F, Rabeharisoa L, Corbeels M (2017) Agro-ecological functions of crop residues under conservation agriculture. A review. Agron Sustain Dev 37(4):26
Rao NM (2020) Air pollution effects on health. Int J Sci Dev Res 5(10):110–115
Rasoulzadeh A, Azartaj E, Asghari A, Ghavidel A (2019) Effects of plant residue management on soil properties, surface runoff, and soil loss under rainfall simulation in a semi-arid region in Iran. Arid Land Res Manag 33(2):200–211
Ravindra K, Singh T, Mor S (2019) Emissions of air pollutants from primary crop residue burning in India and their mitigation strategies for cleaner emissions. J Clean Prod 208:261–273
Rogovska N, Laird D, Cruse R, Fleming P, Parkin T, Meek D (2011) Impact of biochar on manure carbon stabilization and greenhouse gas emissions. Soil Sci Soc Am J 75(3):871–879
Roozbeh M, Rajaie M (2020) Effects of residue management and nitrogen fertilizer rates on accumulation of soil residual nitrate and wheat yield under no-tillage system in south-west of Iran. Int Soil Water Conserv Res 9(1):116–126
Saha S, Chakraborty D, Sharma AR, Tomoar RK, Bhadraray S, Sen U, Behera UK, Purakayastha TJ, Garg RN, Kalra N (2010) Effect of tillage and residue management on soil physical properties and crop productivity in maize (Zea mays)–Indian mustard (Brassica juncea) system. Indian J Agric Sci 80(8):679–685
Sahu A, Bhattacharjya S, Manna M, Patra A (2015) Crop residue management: a potential source for plant nutrients. JNKVV Res J 49(3):301–311

Sanscartier D, Dias G, Deen B, Dadfar H, McDonald I, Mac Lean HL (2014) Life cycle greenhouse gas emissions of electricity generation from corn cobs in Ontario. Can Biofuels Bioprod Biorefin 8:568–578

Saurabh K, Kanchikeri Math M, Datta SC, Thekkumpurath AS, Kumar R (2018) Nanoclay polymer composites loaded with urea and nitrification inhibitors for controlling nitrification in soil. Arch Agron Soil Sci 23:1–4

Sawlani R, Agnihotri R, Sharma C, Patra PK, Dimri AP, Ram K, Verma RL (2019) The severe Delhi SMOG of 2016: a case of delayed crop residue burning, coincident firecracker emissions, and atypical meteorology. Atmos Pollut Res 10(3):868–879

Scopel E, Findeling A, Chavez Guerra EC, Corbeels M (2005) Impact of direct sowing mulch-based cropping systems on soil carbon, soil erosion and maize yield. Agron Sustain Dev 25:425–432

Sharma UC (2016) Crop residues in India: extent, management, uses and implications. In: Pathak PK, Singh SK, Sahay CS, Sharma RK, Chaudhary M, Maity A (eds) Compendium of lecture notes of summer school. Recent approaches in crop residue management and value addition for entrepreneurship development, July 14 to August 03, 2016, ICAR-IGFRI, Jhansi, India, pp 13–21

Sharma DK, Tomar S, Chakraborty D (2017) Role of earthworm in improving soil structure and functioning. Curr Sci 113(6):1064

Sharma S, Thind HS, Sidhu HS, Jat ML, Parihar CM (2019) Effects of crop residue retention on soil carbon pools after 6 years of rice–wheat cropping system. Environ Earth Sci 78(10):296

Sheoran S, Kuma S, Kumar P, Meena RS, Rakshit S (2021) Nitrogen fixation in maize: breeding opportunities. Theor Appl Genet. https://doi.org/10.1007/s00122-021-03791-5

Sidhu BS, Beri V (2005) Experience with managing rice residue in intensive rice-wheat cropping system in Punjab. In: Abrol IP, Gupta RK, Malik RK (eds) Conservation agriculture-status and prospects. Centre for Advancement Sustain Agric (CASA), New Delhi, pp 55–63

Sidhu BS, Beri V (2008) Rice residue management: farmer's perspective. Indian J Air Pollut Control 8(1):61–67

Sidhu HS, Singh M, Singh Y, Blackwell J, Lohan SK, Humphreys E, Jat ML, Singh V, Singh S (2015) Development and evaluation of the Turbo Happy Seeder for sowing wheat into heavy rice residues in NW India. Field Crops Res 184:201–212

Singh Y, Sidhu HS (2014) Management of cereal crop residues for the sustainable rice-wheat production system in the Indo-Gangetic plains of India. Proc Indian Natl Sci Acad 80(1):95–114

Singh Y, Timsina J (2005) Crop residue management for nutrient cycling and improving and productivity in rice-based cropping systems in the tropics. Adv Agron 85:269–407

Singh Y, Khind CS, Bijay-Singh (1991) Efficient management of leguminousgreen manures in wetland rice. Adv Agron 45:135–189

Singh Y, Singh B, Meelu OP, Khind CS (2000) Long-term effects of organic manuring and crop residues on the productivity and sustainability of rice-wheat cropping system in north-west India. In: Long-term soil fertility experiments in rice-wheat cropping systems. Rice-Wheat Consortium for the Indo-Gangetic Plains, New Delhi, pp 149–162

Singh B, Yadvinder-Singh, Imas P, Xie J-C (2004) Potassium nutrition of rice-wheat cropping system. Adv Agron 81:203–259

Singh Y, Bijay-Singh, Timsina J (2005a) Crop residue management for nutrient cycling and improving soil productivity in rice-based cropping systems in the tropics. Adv Agron 85:269–407

Singh Y, Singh B, Timsina J (2005b) Crop residue management for nutrient cycling and improving soil productivity in rice-based cropping systems in the tropics. Adv Agron 85:269–407

Singh G, Jalota SK, Sidhu BS (2005c) Soil physical and hydraulic properties in a rice-wheat cropping system in India: effects of rice-straw management. Soil Use Manag 21:17–21

Singh KK, Srinivasarao C, Ali M (2008a) Phosphorous and mungbean residue incorporation improve soil fertility and crop productivity in sorghum and mungbean–lentil cropping system. J Plant Nutr 31:459–471

Singh B, Shan YH, Johnson-Beeebout SE, Singh Y, Buresh RJ (2008b) Crop residue management for lowland rice-based cropping systems in Asia. Adv Agron 98:118–199

Singh B, Shan YH, Johnson-beeebout SE, Yadvinder-Singh BRJ (2008c) Crop residue management for lowland rice-based cropping systems in Asia. Adv Agron 98:118–199

Singh Y, Gupta RK, Gurpreet-Singh, Jagmohan-Singh, Sidhu HS, Bijay-Singh (2009) Nitrogen and residue management effects on agronomic productivity and nitrogen use efficiency in rice-wheat system in Indian Punjab. Nutr Cycl Agroecosyst 84:141–154

Singh Y, Thind HS, Sidhu HS (2014) Management options for rice residues for sustainable productivity of rice-wheat cropping system. J Res Punjab Agric Univ 51(3–4):209–220

Singh Y, Singh M, Sidhu HS, Humphreys E, Thind HS, Jat ML, Blackwell J, Singh V (2015a) Nitrogen management for zero-till wheat with surface retention of rice residues in north-west India. Field Crops Res 184:183–191

Singh Y, Singh M, Sidhu HS, Humphreys L, Thind HS, Jat ML, Blackwell J, Singh V (2015b) N management for zero-till wheat with surface retention of rice residues in northwest India. Field Crops Res 184:183–191

Singh R, Kamboj NK, Gupta G, Kumar P (2017) Biased government policies contributing in crop biodiversity erosion. Marumegh 2(4):119–121

Singh S, Singh R, Mishra AK, Upadhyay S (2018) Ecological perspectives of crop residue retention under the conservation agriculture systems. Trop Ecol 59(4):589–604

Singh R, Yadav DB, Ravisankar N, Yadav A, Singh H (2020a) Crop residue management in rice–wheat cropping system for resource conservation and environmental protection in north-western India. Environ Dev Sustain 22(5):3871–3896

Singh A, Dubey SK, Pandey AK (2020b) Upscaling of crop residues management technology for doubling farm income and sustainability. In: Lead papers of resource conservation for soil security and jalshakti: farmers perspective in Bundelkhand, 3–5 February 2020, Datia, Madhya Pradesh, pp 114–117

Six J, Conant RT, Paul EA, Paustian K (2002) Stabilization mechanisms of soil organic matter: implications for C-saturation of soils. Plant Soil 241(2):155–176

Smith OH, Petersen GW, Needelman BA (2000) Environmental indicators of agroecosystems. Adv Agron 69:75–97

Soni P, Basak N, Rai A, Sundha P, Narjary B, Kumar P, Yadav G, Kumar S, Yadav R (2020) Tillage and deficit saline water irrigation with mulch in sorghum-wheat cropping system improve soil chemical and biological properties and reduce soil salinity in semi-arid Northwest India. Authorea Prepr. https://doi.org/10.22541/au.158679908.83914794

Srinivas K, Singh HP, Vanaja M, Raju AS, Sharma KL (2006) Effect of chemical composition of plant residues on nitrogen mineralization. J Indian Soc Soil Sci 54(3):300–306

Srinivasarao C, Venkateswarlu B, Lal R, Singh AK, Kundu S (2013) Sustainable management of soils of drylandecosystems of India for enhancing agronomic productivity and sequestering carbon. Adv Agron 121:253–329

Stockmann U, Adams MA, Crawford JW, Field DJ, Henakaarchchi N, Jenkins M, Minasny B, McBratney A, Courcelles VR, Singh K, Wheeler I, Abbott L, Angers DA, Baldock J, Bird M, Brookes PC, Chenu C, Jastrow JD, Lal R, Lehmann J, O'Donnell AG, Parton WJ, Whitehead D, Zimmermann M (2013) The knowns, known unknowns and unknowns of sequestration of soil organic carbon. Agric Ecosyst Environ 164:80–99

Thakuria D, Talukdar NC, Goswami C, Hazarika S, Kalita MC, Bending GD (2009) Evaluation of rice–legume– rice cropping system on grain yield, nutrient uptake, nitrogen fixation, and chemical, physical, and biological properties of soil. Bio Fertil Soils 45:237–251

Thorat TN, Agrawal KK, Kewat ML, Jha G, Silawat S (2015) Crop residue management with conservation agriculture for sustaining natural resources. JNKVV Res J 49(2):125–136

Tripathi CB, Baredar P, Tripathi L (2019) Air pollution in Delhi: biomass energy and suitable environmental policies are sustainable pathways for health safety. Curr Sci 117(7):1153

Trivedi P, Delgado-Baquerizo M, Jeffries TC, Trivedi C, Anderson IC, Lai K, McNee M, Flower K, Pal Singh B, Minkey D, Singh BK (2017) Soil aggregation and associated microbial

communities modify the impact of agricultural management on carbon content. Environ Microb 19(8):3070–3086
Turmel MS, Speratti A, Baudron F, Verhulst N, Govaerts B (2015) Crop residue management and soil health: a systems analysis. Agric Syst 134:6–16
Wang W, Dalal RC, Moody PW, Smith C (2003) Relationships of soil respiration to microbial biomass, substrate availability and clay content. Soil Biol Biochem 35(2):273–284
Wang Q, Li Y, Alva A (2010) Cropping systems to improve carbon sequestration for mitigation of climate change. J Environ Prot 1(03):207
Weng ZH, Van Zwieten L, Singh BP, Tavakkoli E, Joseph S, Macdonald LM, Rose TJ, Rose MT, Kimber SW, Morris S, Cozzolino D (2017) Biochar built soil carbon over a decade by stabilizing rhizodeposits. Nat Clim Chang 7(5):371–376
West TO, Post WM (2002) Soil organic carbon sequestration rates by tillage and crop rotation: a global data analysis. Soil Sci Soc Am J 66:1930–1946
Wilhelm WW, Johnson JMF, Karlen DL, Lightle DT (2007) Corn stover to sustain soil organic carbon further constrains biomass supply. Agron J 99:1665–1667
Woodruff NP, Siddoway FH (1973) Wind erosion control. In: Conservation tillage. Proceedings of a national conference. Soil Conservation Society of America, Ankeny, IA, pp 156–162
Woyessa YE, Bennie TP (2004) Factors affecting runoff and soil loss under simulated rainfall on a sandy Bainsvlei Amalia soil. S Afr J Plant Soil 21(4):203–208
Xiao D, Xiao SS, Ye YY, Zhang W, He XY, Wang KL (2019) Microbial biomass, metabolic functional diversity, and activity are affected differently by tillage disturbance and maize planting in a typical karst calcareous soil. J Soils Sediments 19:809–821
Yadav AS, Arora S (2018) Crop residue management in diverse agro-ecosystems for improving soil health-an overview. J Soil Water Conserv 17(4):387–392
Yadav IC, Devi NL (2018) Biomass burning, regional air quality, and climate change. Earth Syst Environ Sci. https://doi.org/10.1016/B978-0-12-409548-9.11022-X
Yadav D, Shivay YS, Vishwakarma AK, Sharma NK, Sharma VK, Kumawat A, Biswas AK (2019) Consequence of divergent crop residue and green manuring practices on soil nutrient balance: a review. J Pharmacogn Phytochem 2:903–908
Yadav D, Vishwakarma AK, Sharma NK, Biswas AK, Ojasvi PR, Kumar D, Kumawat A, Singh D (2021) Sustaining the properties of black soil of Central India through crop residue management in the conservation-agriculture based soybean-wheat system. Land Degrad Dev. https://doi.org/10.1002/(ISSN)1099-145X
Yamato M, Okimori Y, Wibowo IF, Anshori S, Ogawa M (2006) Effects of the application of charred bark of *Acacia mangium* on the yield of maize, cowpea and peanut, and soil chemical properties in South Sumatra, Indonesia. Soil Sci Plant Nutr 52:489–495
Yang X, Zheng L, Yang Q, Wang Z, Cui S, Shen Y (2018) Modelling the effects of conservation tillage on crop water productivity, soil water dynamics and evapotranspiration of a maize-winter wheat-soybean rotation system on the Loess Plateau of China using APSIM. Agric Syst 166:111–123
Zimmerman AR (2010) Abiotic and microbial oxidation of laboratory-produced black carbon (biochar). Environ Sci Technol 44(4):1295–1301

Agroecology for Agricultural Soil Management

9

N. C. Temegne, A. F. Ngome, A. P. Agendia, and E. Youmbi

Abstract

The ever-growing planet population will reach 10 billion in 2050 according to estimates. The current agricultural and food system demonstrates every day a little more its inability to feed this population adequately. More than 10.7% of the current world population suffers from chronic undernourishment. The soaring world population has resulted in multiple environmental damages: the destruction of forests, overconsumption of water reserves, extensive use of pollutants, soil degradation, etc. However, a majority (72%) of the worldwide food is cultivated and gathered by 2.5 million smallholder producers on small family farms (<1 ha). Agroecology offers concrete solutions to climate breakdown and contributes to the preservation of natural resources essential for sustainable agricultural production. The soil support for agriculture can be well managed by adopting cultivation techniques, associated with plant cover of the soil (green manures, alley or mixed cropping with agroforestry species) and vigorous biological activity, by limiting or eliminating chemical fertilizer use, prioritizing local inputs and recycling of farm by-products (manure, compost, bio-char, crop waste, household waste), maintaining inherent fertility of soil, conserving soil biodiversity, and enhancing plant nutrient availability.

Keywords

Food system · Manure · Natural resources · Nutrient availability · Soaring world population · Soil biodiversity

N. C. Temegne (✉) · A. P. Agendia · E. Youmbi
Department of Plant Biology, Faculty of Science, University of Yaounde I, Yaounde, Cameroon
e-mail: carine.temegne@facsciences-uy1.cm

A. F. Ngome
Institute of Agricultural Research for Development (IRAD), Yaounde, Cameroon

M. K. Jhariya et al. (eds.), *Sustainable Intensification for Agroecosystem Services and Management*, https://doi.org/10.1007/978-981-16-3207-5_9

Abbreviations

ADG	Aide au Développement Gembloux
AFOP	Agropastoral and Fishing Training Programme
AMF	Arbuscular Mycorrhizal Fungi
CEFRA	Centre de d'Enseignement, de Formation et de Recherche en Agroécologie
CEFREPADE	Centre Francophone de Recherche Partenariale sur l'Assainissement, les Déchets et l'Environnement
CIRAD	The French Agricultural Research Centre for International Development
ELD	Economics of Land Degradation
FAO	Food and Agricultural Organization
FNRS	National Fund for Scientific Research
GESCOD	Grand Est Solidarités et Coopérations pour le Développement
IFOAM	International Federation of Organic Agriculture Movements
IPBES	Intergovernmental Science-Policy Platform on Biodiversity and Ecosystem Services
IPM	Integrated Pest Management
IRAD	Institute of Agricultural Research for Development
ISSAEER	Institut Supérieur des Sciences Agronomiques, de L'Environnement et de l'Entrepreneuriat Rural
N	Nitrogen
NGO	Nongovernmental organization
OC	Organic carbon
OM	Organic matter
P	Phosphorus
PGPR	Plant growth-promoting rhizobacteria
SOCLA	Latin American Scientific Society of Agroecology
UN	United Nations
UNCCD	United Nations Conventions to Combat Desertification

9.1 Introduction

In 2100, the world population intended to attain 11.2 billion. However, more people (9 million) are dying of hunger each year than from AIDS, malaria, and tuberculosis altogether (UN 2018). One child (under 15 years old) dies every 5 s from hunger or related causes in 2017 (Sidhu 2020). The tremendous increase in food productivity during the past 50 years has decreased the frequency of acutely hungry people in the world. But, this agricultural system has shown many limits, among which excessive specialization and the tendency to gigantism. We can also note the explosion in the use of energy-consuming inputs and equipment and the decrease in the efficiency of the use of chemical inputs. For example, there has been stagnation in cereals (winter wheat, barley, oats, durum wheat), sunflower (*Helianthus* spp.), and vine yields in France since the late 1990s (Schauberger et al. 2018).

The substantial gains in the production of conventional farming have been also accompanied by high environmental costs/problems, which have affected the health of soils and ecosystems (FAO 2015; Kumar et al. 2020). Thus, 12 million hectares (i.e., 23 per minute) of agricultural land are damaged broadly to soil deterioration each year which represents over half (52%) of fertile soils food producers in the world (UNCCD 2015), with 78% of overall deteriorated soil localized in earthly ecosystems other than arid areas (UN 2012). Land degradation affects 1.9 billion hectares. It is costing each year between 6.3 and 10.6 trillion US dollars taken as a whole (ELD 2015). Twenty-four billion tons of fertile land is irreversibly laved or carried away (3.4 tons for each person on the earth) each year because of the erosion of the world's cultivated land (Young et al. 2015). Thus, soil deterioration has diminished the productivity of 23% of the worldwide land surface, and up to US $577 billion in annual cultivated plants of the world are at riskiness from pollinator loss (Brondizio et al. 2019; Díaz et al. 2019).

Soil degradation is a worldwide problem which is currently receiving a lot of attention (Xie et al. 2020; Khan et al. 2021a, b). However, the agribusiness model remains the model mostly taught in schools and universities and widely promoted by research centers, most producer organizations, and technical services. Smallholder farmers remain the leading providers of food (72%) but paradoxically are the first to suffer from poverty and hunger. It is therefore essential to refocus our agricultural model, particularly on peasant women to sustainably feed the populations (ADG 2016). The current challenge for agricultural policy is to combine sufficient food production for a growing community ensuring environmental restoration (FAO 2015; Banerjee et al. 2020, 2021; Raj et al. 2020). By preserving soil health, agroecology, which regenerates the functioning of ecosystems, is an effective strategy for achieving food security (FAO 2015; Jhariya et al. 2019a, b, 2021; Meena et al. 2020a, b, c).

By mastering and laboring with the interactions between land, crops, beasts, human being, and the environment in farming systems, agroecology integrates manifold dimensions of the agricultural system, enclosing ecological rehabilitation, political and social steadiness, and economical sustainability. This chapter expands the principles of agroecology, soil conservation through the management of soil composition, cultural techniques and fertilization, the constraints to adoption of agroecology, agroecology versus conventional agriculture, agroecology towards soil management and sustainability, policy and legal framework, and future roadmap of agroecology for agricultural soil management

Table 9.1 Definition of agroecology

Definition of agroecology	Reference
The implementation of ecological tenets to agriculture	Altieri (1983a)
"The application of ecological science to the study, design and management of sustainable agroecosystems"	Gliessman (1997)
An interdisciplinary process that involves a redefinition of scientific and social boundaries, which constitutes an important intellectual defiance for agricultural research	Buttel (2003)
The integrative investigation of the ecology of all food systems, including ecological, economic, and social aspects	Francis et al. (2003)
It is defined neither exclusively by scientific disciplines, social movements, nor by practices	Wezel et al. (2009)
An intrinsically transdisciplinary practice, as it binds the organization and operating of agroecosystems and fills the ditch between various disciplines as good as between theory and practice	Caporali (2011)
New agricultural template that could allegedly conciliate the economic and environmental defiance in food production	Schaller (2013)
For Pierre Rabhi, it is more than a simple agronomic option; it is linked to a deep dimension of respect for life and places the human being in his responsibility towards the living; it is both an ethics of life and agricultural practice	Lion et al. (2009), ADG (2016)
The study, application, and defense of concepts, principles, and methods aimed at the establishment of agroecosystems and sustainable food systems from the point of view productive, environmental, social, cultural, and economic	Gliessman (1997), ADG (2016)
The development of agricultural techniques to safeguard the environment and to favor the utilization of ecological theory to promote "eco-friendly" means to make food	Saj et al. (2017)
A scientific search on coming ensuring the holistic investigation of agroecosystems and agricultural commodities systems	CIDSE (2018)
One of a family of varied methods sharing a usual feature in that they involve the ecological roles of farming systems to assure long-lasting production	CIRAD (2018)
A scientific discipline, an ensemble of processes, and a societal movement	FAO (2018)
The science and method of implementing ecological notions, tenets, and acquaintance to the investigation, designing, and management of sustainable agroecosystems	IPBES (2018)
An alternative template for promoting farming systems founded on every farm being a consolidated ecosystem, in which plants and animals interact to generate suitable provisos for cultivation	Lund University (2018)
The investigation of the ecology of earthly farming systems	Nature (2018)
An interdisciplinary merger of agronomy, agriculture, scientific ecology, economics, and social sciences	Youmatter (2020)

9.2 Agroecological Concept

Agroecology is a concept defined in various ways (Table 9.1). Agroecology combines practices such as ecological farming, regenerative farming, and certain features of permaculture and thus competes with sustainable development (Youmatter 2020). With the ambition of producing knowledge and methods that make agriculture more sustainable, agroecology concentrates on the whole farming system to go beyond the scale of the plot. Indeed, it focuses on the analysis of agroecosystems and their sustainability (Stassart et al. 2012; Raj et al. 2021). Therefore, a whole of agricultural practices are aimed towards imitating nature in its field. As a science, it investigates how various constituents of the agroecosystem interact. As a whole of operations, it searches long-lasting agricultural systems optimizing and hold steady yields. As a social movement, it prosecutes multifunctional purposes for agriculture, encourages societal justness, feeds identity and culture, and reinforces the economic viability of peasant zones (FAO 2018).

Based on traditional peasant practices, agroecology links several alternatives such as organic farming, permaculture, and natural farming, without being reduced to it. These practices being drawn mainly from the traditional knowledge of the agricultural populations allow agroecology to spread quickly through communities and small family farms. The application of ecological principles to these ancestral techniques is, therefore, the basis of agroecological practices (Altieri and Nicholls 2014; Meena et al. 2020a).

Agroecology was first used by Bensin 1928, a Russian agronomist, who used it to designate "gentle" agronomic techniques in cash crops. The meaning and scope of the term have evolved. At first, it was considered as a scientific discipline linked to agricultural production, which combines ecology and agronomy. Currently, it is defined as a whole of process, a scientific discipline, and a movement (ADG 2016).

The practices are very diverse. In tropical zones, they include fertility management, soil (and water) conservation, pest control, water management, management of crops on the farm, livestock, etc. In the 1980s, agroecology appeared as an ensemble of agricultural activities. Traditional farming systems in developing countries are beginning to be recognized for their benefits in the management of natural resources. Peasants from the south are capable of combining traditional knowledge and know-how and innovation and sometimes helped by international cooperation or the scientific community. They develop and adopt specific techniques, thus gradually generating a set of agroecological practices, demonstrating that the solutions also come "from below."

As a scientific discipline, the field of study of agroecology has evolved considerably. It went from managing the plot to the ecology as well as agroecosystems management and the organization/structure of the food production system (three dimensions). In response to the green revolution, ecological movements were born in the 1960s and 1970s. It is a mixed discipline, at the crossroads of natural, social, and economic sciences which is today an alternative scientific referent. Indeed, agroecology entered into university courses in the USA in 1981 (Berkeley) and more generally from 2000, in Brazil (Santa Catarina in 2000), Belgium (University

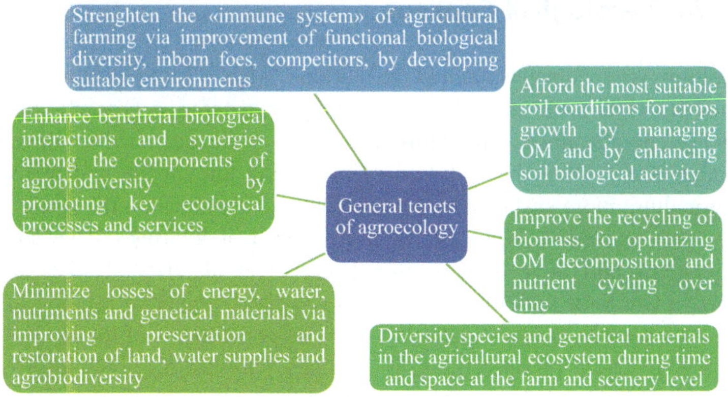

Fig. 9.1 General principles of agroecology (source: Migliorini and Wezel 2017)

Libre of Brussels in 2008), etc. Research groups like SOCLA (in 2007) in Latin America and FNRS (in 2009) in Belgium have shown their interest in this science (Altieri 1983b; ADG 2016).

The first agroecological social movements, in the south (particularly in Latin America) and the north (especially in the United States) appeared in the 1980s. It began in Mexico and Central America in the 1980s with associations (NGOs: World Neighbors, CLADES) and scientists (Bunch, Altieri) close to peasants and natives. It was presented as an alternative to industrial agriculture with high levels of chemical inputs. In Latin America, it is adopted by Via Campesina (food sovereignty, peasant agriculture, and agroecology), the network of producer actors and sympathetic organizations (Prolinnova), etc. (De Schutter 2011).

Based on the existing literature, Migliorini and Wezel (2017) summarized the principles of agroecology, as shown in Figs. 9.1 and 9.2. The success of novel farming systems depends on the application of the tenets of agroecology (Nicholls et al. 2016), making reference to the popularization of practices and serves ecology, enclosing land, water, air, and biological diversity dimensions. Thus, the straightforward implementation of a set of practices is not sufficient. Stassart et al. (2012) and Dumont et al. (2016) also append three socioeconomic tenets (Fig. 9.2) to the other tenets more linked to production and ecology.

Regarding the principles of agroecology for animal production systems, Dumont et al. (2013) complete the above tenets in Fig. 9.2. They can be summarized in two tenets: (a) adopting a management process aimed at improving animal health and (b) strengthening diversity in animal farming to enhance their resilience.

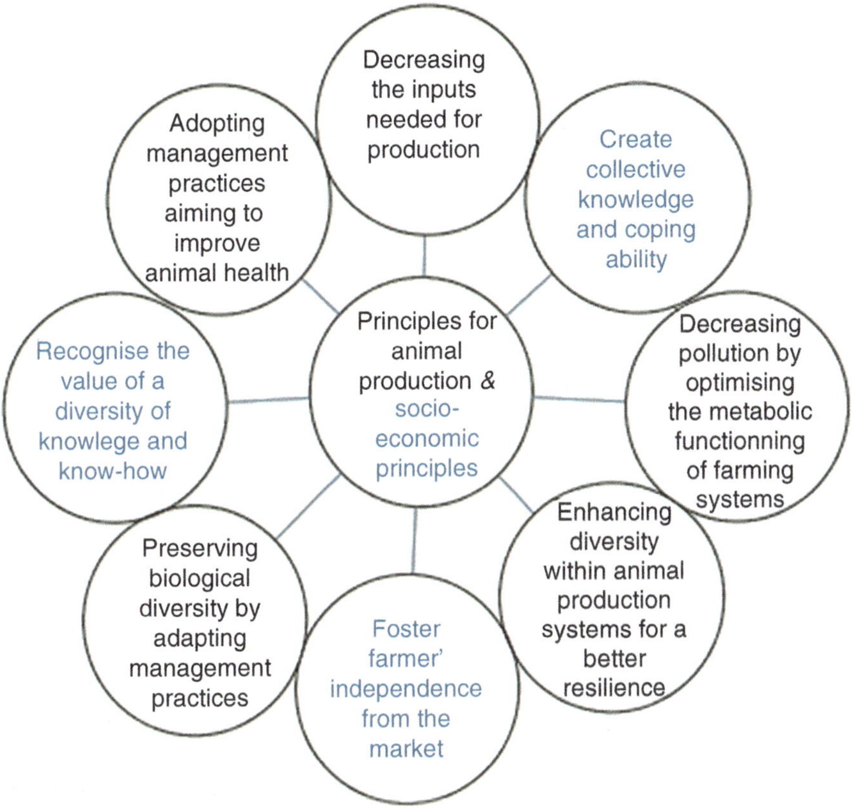

Fig. 9.2 Socioeconomic principles of agroecology and principles for animal production systems (source: Migliorini and Wezel 2017)

9.3 Agroecology and Soil Characterization

Soil is the substrate that nourishes and supports growing plants. It is composed of 25% air, 25% water, 50% solid components (by volume), and 0%, 17%, and 83% (by weight), respectively (Parker 2009). Soil is essentially a nonrenewable resource because the formation processes are prolonged while the degradation processes can be very rapid. It is a fusion between the mineral (clays) and organic (humus) with an average ratio of 16/1 by volume and 40/1 by weight. The living constitutes 0.15 to 0.2% of the soil (by weight). Soil contains 80% of the living organisms (by weight) on earth. The living organic components of the soil include the fauna (between 2.5 and 5 t ha^{-1}) consisting of earthworms, fungi, bacteria, nematodes, springtails, termites, and plants consisting mainly of roots and algae.

Soil microorganisms perform several ecosystem services and functions in the soil.

9.3.1 Soil Structuring Function

It is made through the development of stable aggregates, a macro- and a micro-porosity. Roots, mycelial hyphae, vertebrates, and invertebrates burrowing create numerous galleries in the soil resulting in high porosity of the surface layer of the soil (Huera-Lucero et al. 2020). AMFs secrete a glycoprotein (glomalin) which promotes aggregation (Vlček and Pohanka 2020). This substance cements clay particles and organic debris giving macroaggregates. Fungi and saprophytic bacteria produce exo-polysaccharides (Costa et al. 2018). Earthworms ensure the formation of the clay-humic complex. They come up every night to get litter and then they leave their droppings on the soil surface (formation of turricules). They continuously brew deep clay-rich soil with surface soil rich in humus. These earthworms consume the equivalent of their weight of soil per day (Misra et al. 2003). Termites are responsible for the creation of fecal pellets very resistant to the elements of the climate, which explains the high permeability of the oxysol from tropical rain forests. They feed on lignified plants and undecomposed wood. Their gut contains flagellated protozoa that digest cellulose and bacteria that digest lignin (Ali et al. 2019).

9.3.2 Nutrient Recycling Function

Bacteria oxidize NH_4^+ ions (*Nitrobacter*) and sulfur and ensure the chelation of trace elements (Fe, Al, etc.). The root systems of plants associated with mycorrhizae also participate in this recycling process through the networks (hyphae + roots) that they develop in the plant rhizosphere (Temegne et al. 2018, 2019; Giovannini et al. 2020).

9.3.3 Function of Decomposition, Mineralization, or Humidification of Soil Organic Matter

Many soil organisms feed on different trophic levels which abet to the intricacy of food dealings, and this conducts to effective recycling of organic matter and a net release of nutrients.

The grinding is mainly done by first- and second-order consumers. It accelerates the decomposition of residues because it blends fungi and bacteria with the residuals by increasing the zone colonizable by decomposers. The mesofauna (mites, springtail, termites, and enchytreid worms) and the macrofauna (woodlice, millipedes, beetles, ants, earthworms, snails, and slugs) abet in the grinding and recycling of organic residuals. They also drop off in the soil fecal pellets of 50–200 μm in diameter which are a prime substrate for decomposers (rich in energy and N) (Brussaard and Kooistra 2013). Earthworms also mix the upper mineral strata of the land with surface residuals (bioturbation phenomenon) and create pores and conduits allowing the passage of water and roots. Nematodes and small epigree earthworms eat the finer fractions as well as the excrement of other species.

Thysanoure, springtails, mites, myriapods, earthworms, and protoures eliminate dead roots, ensure porosity of the soil deep in the soil, and allow root respiration.

Saprophytic bacteria (first-order consumers) produce many exoenzymes (dehydrogenases, proteases, and cellulases) which allow them to degrade dead organic matter and draw their energy from a wide range of carbon products (Shalaby 2011). Numerous heterotrophic bacteria ease wrench changes of diverse nutrients other than carbon (N, P, S, K, Mg) in their cycles. They enhance and structure by generating exopolysaccharides and other metabolites that help to stick the particles together (Costa et al. 2018).

Fungi (first-order consumers) also produce many exoenzymes that break down dead organic matter. They are the primary agents for the decomposition of organic matter in exposed lands (saprophytes). They are the only organisms on earth, apart from a few bacteria in the rumen of cattle and the intestine of termites that can break down lignin from plants (the main source of humus). Fungi help to stabilize soil aggregates via their filamentous hyphae (Lehmann et al. 2020).

Actinobacteria (first consumers), less competitive than bacteria and fungi in breaking down fresh organic matter, continue to decompose the organic matter started by fungal and bacterial microflora (Matei et al. 2020). Thermophiles have a significant function in the manufacturing of compost. Between 50 and 75% of the strains secrete antibiotics which prepare ecological niches for fungi in composting (Carrasco and Preston 2020).

Browsers and shredders (first-order consumers, second-order predators) feed on bacteria, *Actinobacteria*, and soil fungi. They therefore carry out grazing. They can also consume organic matter. They feed on N-rich bacteria and reject great quantities of inorganic N and have a very important role in the recycling of mineral elements. They consist of protozoa which swallow up their prey; bacteria that enter and multiply in larger bacteria; nematodes that sweep or suck bacteria from the surface of the roots or minerals and suck the inside of the fungus with a stylus; and micro-arthropods (mites and springtails).

Micro-arthropods (second-order consumers) achieve fragmentation and restructuring physics of organic matter by chewing. This process leads to the aggregation of minerals followed by an increase of the soil surface components, which favors the bacterial activity and a more advanced decomposition of the residues (Culliney 2013).

The "grazing" nematodes (second-order consumers) are very beneficial in edaphic ecosystems. They help to control the size and structure of populations of bacteria and fungi (Ferris et al. 2004; Blanc et al. 2006). They help to speed up the recycling of nutrients.

Consumers of third-order and more (nematodes and arthropods) are the predators of different species of spiders, beetles, and ants (SWCS 2000; Menta and Remelli 2020). They can help to regulate populations of major pests.

9.3.4 Function of Facilitating the Removal of Water and Nutrients

Mycorrhizae (ecto- and endo-mychorizes) release the phosphate ions fixed by the clay-humic complex. They improve the water supply of plants through their hyphae which will draw water from the depths of the soil (Tsoata et al. 2015).

9.3.5 Atmospheric Nitrogen Fixation Function

Several soil microorganisms, free or symbiotic, have the ability to fix atmospheric N and allocate it to crops (Table 9.2).

9.3.6 Function of Protection of Plants Against the Invasion of Root Pests

Microorganisms protect plants by competing in the space in the soil. AMF creates a protective sleeve against pathogens. Beneficial nematodes compete with herbivorous nematodes. They produce the antibiotics that control *Pithium* sp. and *Pseudomonas* sp. For example, there is emission by the roots of corn in the event of insect attacks of molecules attracting entomophagous nematodes (Degenhardt et al. 2009).

The major component of soil OM is organic carbon (OC). It has a pivotal role in crop production and is the most useful single signpost of soil quality (Ngome et al. 2011a; Soil Carbon Initiative 2011). OM is a wrench element in the land, monitoring several fundamental functions (Jones et al. 2011; Kumar et al. 2020a). OC enhances the physical characteristics of land which raise the degree to which it can soak up rainfall and hold water, making it disposable for afterwards plant use, reduce leaching, and enhance microbial biomass activity and biodiversity of soil. The loss of OM in lands is caused by erosion and the raised rate of mineralization of OC in arable lands (Krasilnikov et al. 2015). Low OC level in soils leads to more crop susceptibility to disease (Altieri and Nicholls 2003; Stone et al. 2004).

Cultivation techniques have an impact on soil characteristics (Fig. 9.3). Tillage practices can be classified into three types of action: depth of fragmentation, soil turnover, and soil organic matter blending (Labreuche et al. 2007).

Tillage or plowing is a deep working operation (between 15 and 40 cm) with turning of the soil and blending of its horizons (Labreuche et al. 2007). It distributes basal dressing and amendments throughout the topsoil, controls weeds and regrowth, buries crop residues, loosens surface layers, and improves drainage (drying) of wet or drained soils. It can also be used to destroy intermediate crops (Daniel and Galardon 2008).

Pseudo-tillage or pseudo-labor is a deep working operation (between 15 and 40 cm) with the blending of horizons without turning over. The absence of inversion results in some plant debris and unburied weeds on the surface (Daniel and Galardon 2008). The presence of surface residues sharply limits erosion which provides

Table 9.2 Atmospheric nitrogen fixation function of soil microorganisms (compiled from Kitamura et al. 2011; Munk et al. 2011; Yang et al. 2016; Zeng et al. 2017; Bhowmik and Das 2018; Troost et al. 2019; Contador et al. 2020; Giraldo-Silva et al. 2020; Inomura et al. 2020; Mahmud et al. 2020; Robledo et al. 2020; Silva et al. 2020)

Types	Hosts/traits		Examples
Free	Aerobic	Phototrophs	*Cyanobacteria*: *Nostoc* spp., *Anabaena* spp., *Calothrix* spp., *Tolypothrix* spp., etc.
		Heterotrophs	*Aeschynomene* spp., *Azoarcus* spp., *Azospirillum brasilense*, *Azospirillum lipoferum*, *Azotobacter vinelandii*, *Beijerinckia indica*; *Herbaspirillum seropedicae*, *Klebsiella pneumonia*, *K. oxytoca*, *Pseudomonas putida*, etc.
	Anaerobic	Phototrophs	*Chromatium vinosum*, *Rhodobacter capsulata*, *Rhodospirillum rubrum*, etc.
		Heterotrophs	*Clostridium*, *Azotobactor*, *C. pasteurianum*, *Desulfovibrio vulgaris*, *Desulfotomaculum* spp., *Methanobacterium* spp., *Pseudomonas stutzeri*, etc.
Symbiotic	Leguminous	With root nodules	*Allorhizobium* sp., *Azorhizobium* sp., *Bradyrhizobium elkanii*, *B. japonicum*, *Ensifer meliloti*, *Mesorhizobium ciceri*, *M. lot.*, *Rhizobium etli*, *R. leguminosarum*, *R. lupine*, *R. meliloti*, *R. phaseoli*, *R. trifolii*, *R. tropici*, *Sinorhizobium fredii*, *S. meliloti*, etc.
		With stem nodules (*Sesbania*)	*Azorhizobium caulinodans*, etc.
	Cereal	Rice (*Oryza sativa* L.), sugar canes (*Saccharum* spp.)	*Azotobacter*, *Clostridia*, *Gluconacetobacter diazotrophicus*, etc.
	Others crops	Sweet potato (*Ipomoea batatas* L.), storage tubers	*Azospirillum* sp., *Bradyrhizobium* spp., etc.
	Actinorhizal symbiosis	*Casuarina* spp.	*Frankia* sp., *Parasponia* sp., etc.
	Cyanobacterial symbiosis	Azolla	*Anabaena azollae*, etc.
		Cycas	*Anabaena cycadeae*, etc.
		Lichens	*Nostoc* sp., etc.
		Mosses and liverworts	*Nostoc* sp., etc.

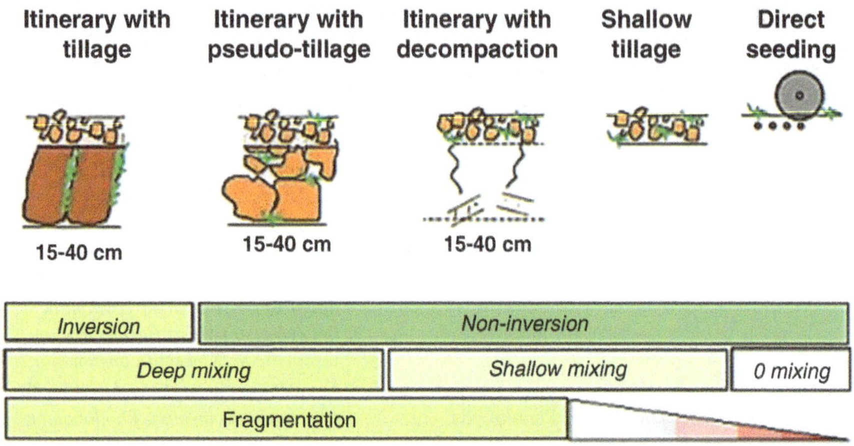

Fig. 9.3 Cultivation techniques (modified by: Daniel and Galardon 2008; Labreuche et al. 2008)

protection to the land (reducing the effect of raindrops) and the presence of more stable aggregates.

For the decompaction, the work of the soil is deep without turning, nor mixing. Like plowing, it is done at a depth of between 15 and 40 cm. This operation restructures the soil by fragmenting and lifting it. Many farmers talk about loosening or cracking, but in fact, loosening also induces soil fragmentation (Labreuche et al. 2008).

For shallow tillage, the tillage is between 0 and 8 (15) cm deep. It includes a mixture of crop residues in the volume worked but without reversal. There are several types of surface work, depending on the objectives sought: stubble cultivation, resumption of plowing, preparation of the seedbed, mechanical weeding, etc. (Labreuche et al. 2007).

Direct seeding (no tillage) is the sowing or planting of a crop without tillage. We can have three variables:

- Direct sowing without any work, i.e., no rotary hoe passing over the sowing line; the seeds are placed in the soil just after the opening disc(s); frequent cases for cereals
- Strip tillage, i.e., the passage of a rotary hoe on flat land, on a sowing strip 10 cm wide and a few centimeters deep just in front of the sowing organ; reasonably common case for weed crops
- Ridge tillage, i.e., identical to strip tillage but on hilly terrain; potato plantations and weed crops (Labreuche et al. 2007)

This minimum work results in the maintenance on the surface of almost all the crop residues and organic inputs. This technique reduces costs and time. It saves a lot of energy. The aim is to limit vertical disturbances to the ground as much as possible and to maximize the coverage by residues with minimum working technique (Daniel and Galardon 2008). But, this technique requires more technicality and observation because the most "simplified" implementations are the most demanding ones (Labreuche et al. 2014).

The characteristics of the soil have a significant influence on the effectiveness of the farming techniques adopted. In terms of soil texture, it is on clay soils that the efficacy of no-till cultivation techniques to limit erosion is most convincing (Rhoton et al. 2002). On sandy soils, their effectiveness seems lower (Quinton and Catt 2004) while on loamy soils, the results are very variable and depend mostly on other parameters such as soil cover. No-till farming techniques are useful in combating soil loss on clay soil. The effectiveness of direct seeding compared to plowing on limiting runoff is more convincing on clay soil than on loamy clay soil (77% less runoff volume with clay soil and 17% less volume runoff with loamy clay soil) (Rhoton et al. 2002). As for erosion, it is reduced to zero with direct seeding on both types of soil.

Structural stability is lower in conventional systems on all soils. However, with an augmentation of the clay content of soils, the differences between plowing and non-plowing become blurred. The less the soil is worked in-depth, the more the structural stability of the surface increases. This increase in structural stability results from the concentration of OM in the surface horizon and the increase in content in this horizon (Labreuche et al. 2007). It is the accumulation of OM on the surface by no-till farming techniques that improves the stability of aggregates. This accumulation phenomenon is directly reversible on first deep tillage or plowing (Rhoton et al. 2002). Thus, the use of occasional plowing cancels any accumulation effect on the soft soils. On clay soil, it is the high clay content which gives higher structural stability. The OM effect on the balance of aggregates is not very sensitive (Le Bissonnais and Arrouays 1997).

The stock of organic N in the soil is higher for direct seeding than for plowing whatever the horizon of the soil is considered (Mikha and Rice 2004; Wright and Hons 2005a, b).

9.4 Problems of Soil Environment

The cumulative mean loss of production during the post-World War II period caused by human-provoked land deterioration has been esteemed at 7.9% in Africa whereas it was 25% and 36.8% at Central America in accordance with ISRIC estimation (Krasilnikov et al. 2015). Each minute, 23 ha of land is lost to land degradation (12 million ha year^{-1}) (Rossi 2020). Twenty-four percent (350 lakh km^2) of the soil has deteriorated which is raising the proportion of 50–100 lakh ha year^{-1} (Vasu et al. 2020). The mean richness of indigenous species is most considered as land-based habitats have dropped by at least 20%, mainly since 1900 (Brondizio et al. 2019). So, several environmental constraints such as acidification, alkalinity, climate change, desertification, compaction, drought, erosion, nutrient deficiency, salinity, pollution, waterlogging, etc. affect the soil and reduce the area of soil available for agriculture. Most of them are caused by the intensification of food production (Altieri and

Nicholls 2015). However, to feed the growing world population, it is necessary to implement strategies to produce on these soils (Raj et al. 2019a, b).

Acidification implies the shedding of basic cations (e.g., Ca, Mg, K, Na) by leaching and their substitution with acidic compounds, primarily soluble complexes of Al and Fe, and Mn sometimes. It thus leads to aluminum, ferric, and manganese toxicity (Chérif et al. 2009; Mapiemfu-Lamaré et al. 2012; Tekeu et al. 2015). Acidification is constantly followed by a diminution in the land's ability to neutralize acid and a process of an irrevocable nature excluded during very long periods (Krasilnikov et al. 2015). An augmentation in pH and acid neutralization capacity associated with higher concentrations of basic cations, in turn, would enhance the potentialities for biological recuperation. But, given the retard in the land's reply to the diminutions in acid deposition, it will likely take several decenniums for the impacted zones to recuperate wholly (Krasilnikov et al. 2015). Soil acidity is generated by climate, acidic parent material supplying Al and Si ions, NH_4 fertilizers, OM breaking down, abduction of nutrients via harvesting of high yielding plant, and weak tampon ability from little clay and OM and Al_2SiO_5 minerals (Getachew et al. 2019).

Climate change is presumably to influence land grade and generate soil degradation by modifications in land water content (Wong et al. 2011; García-Ruiz et al. 2011; Khan et al. 2020a, b). It aggravates land deterioration, especially in low-lying coastal regions, river deltas, drylands, and permafrost regions. Climate change, land-use change, and land-use intensification have abetted to desertification and land deterioration (IPCC 2019). Across the North and the Centre of Europe, evapotranspiration raised through approximately 0.3 mm day^{-1}, which has the potential to exhaust the generally suitable land water reservoir and restrict crop growth. More recurrent and drastic droughts can conduct to a reduction in plant cover leading to the start of erosion and desertification (Jones et al. 2011). But, the precise effects of climate change on land deterioration are still unclear (Kovats et al. 2014).

Desertification is soil deterioration in arid, semiarid, and dry subhumid regions, generally recognized as drylands, arising from several elements, encompassing human actions and fluctuations of climate (UNCCD 1994; Mirzabaev et al. 2019). The range and loudness of desertification have risen in certain arid regions for the past few decades. Arid soils presently extend over about 47% worldwide and are residence of about 39% of the worldwide population (3 billion people). Desertification hotspots, as distinguished by a decrease in flora production in the space separating the 1980s and 2000s, expanded to nearly 10% of drylands influencing 620 million people in 2015 (Mirzabaev et al. 2019). According to Prince and Podwojewski (2020), desertification results in the following:

- Gulley erosion due to loss of soil cover engendered by overgrazing
- Sheet erosion exhibiting roots and slaying trees
- Forest defacement and deforestation
- Wildfire which generates biomass loss, nutrient losses via volatilization, quickened erosion, forming of water repulsive surfaces inclined to water runoff and

erosion, and rising CO_2 release and is occasionally accompanied by invasions of alien species

- Soil compaction alongside cattle paths particularly where they assemble to drink, diminished precipitation permeation, and raising runoff, which, in turn, can generate erosion
- Habitat loss that imperils indigenous species
- Dust storms and loss of topsoil engendered by bare soil in farms, particularly extensive, motorized, dryland agriculture
- Salt efflorescence generates by over-irrigation
- Unmonitored populations of savage animals that pasture and nibble helpful flora
- Bush encroachment, frequently assigned to overgrazing in dryland, modifications in fire regimes, land surrender, and CO_2 rise
- Alien species establishing
- Reduction of biological diversity engendered by habitat loss
- Overgrazing by livestock causing erosion and loss of soil C.

Human and nonhuman provoked land salinity is being an important worldwide menace to farming around. This salinization happens in watered and pluvial farming areas with the most important rates in the arid and semiarid ecosystems. Human and nonhuman induced land salinity is becoming an important worldwide menace to farming. The nonhuman-provoked land salinity are salts initially present into parent materials, mineralized floor and surface waters as well as wind-blown depots (Vargas et al. 2018). Poor irrigation and the utilization of extremely mineralized irrigation water impact approximately 3.8 million ha in Europe (Masters et al. 2005; Krasilnikov et al. 2015). Salinization has a severe effect on land functions like its capacity to proceed as a tampon and filter versus pollutants. Its involvement in the water and N cycles and its ecosystem services favor the healthiness of the environment and biological diversity (Vargas et al. 2018). Land salinization affects the agricultural production by entraining disturbances to the processes of N uptake and crop growing. The reduction of biological activity of lands is combined with the diminution of food provided by land microflora requisite for ecosystem functioning. The surrender of arable lands is linked with a high risk for land and environmental health and important ecological stress. An augmentation in land salinity further damages land ecosystem services and reduces incomes for farmers and smallholders. The loss of original vegetation and forests is the final result of the salinization of arid agricultural soils (Vargas et al. 2018). To maintain or colonize saline soils, it is recommended:

- To select and use salt-tolerant plants
- Promote the salt-tolerant pastures where livestock can help in the management and restoration of soil
- Enhance the land for growth of substitute less salt-tolerant crops
- Employ a surplus of water to rinse off salts from the land (flushing)
- Optimize the irrigation and drain management

- Make the mulching, alone or with amendments, it usually conserves yield with satisfying outcomes diminishing salinity
- Add biological land conditioners, it raises yield while diminishing salinity and can proceed better in association with other land amendments
- Reduce tillage, it lessens salinity and rises productivity
- Realize phytoremediation, it augments productivity but does not ensure positive impacts on land salinity
- Advise rotation systems
- Utilize the trees and shrubs with the ability to efficaciously proceed as bio-drains in saline (Clarke et al. 2002; Masters et al. 2005; Jhariya et al. 2018a, b; Cuevas et al. 2019).

Most hopeful for salinity is an association of amendments, conditioners, and mulching, whereas implementing rising and maintaining cover plants or rotation. Most auspicious for productivity is phytoremediation and biological conditioners whereas maintain cover crops or/and rotation (Cuevas et al. 2019).

Soil degradation and water deficiency are narrowly associated. Healthy soil has a natural ability to conserve and filter water, but this ability is lost when soil has deteriorated. Likewise, land-use modifications, like the conversion of wetlands and forests to other soil uses, disturb the water cycle and hydrological roles. Inversely, water scarcity and droughts may hasten the processes of soil deterioration (EU 2019), for example, caused by weak irrigation management and drainage and modified hydrology, leading to weaker grade lands.

9.5 Agroecology and Soil Conservation

Well-managed soils by smallholder farmers contribute to all four aspects of food security: availability, by providing the nutrients needed for plant growth (Dagnachew et al. 2020); access, by enhancing the income of family farms across more reliable crops; stability, by preserving water to allow plants to be grown almost year-round; and use, by gathering healthy and nutritious food on healthy soils (FAO 2015). Soil conservation in agroecology must promote soil protection through various techniques limiting the negative impact of harmful human intervention on the inherent structure of the soil and that of raindrops, sun, and wind. It recommends the maintenance of diversified and permanent vegetation cover, the use of mechanical or crop anti-erosion measures, and the limitation or elimination of tillage and pesticides (ADG 2016).

Soil erosion rates are higher in Asia, Africa, and South America agroecosystems (30–40 t ha^{-1} $year^{-1}$) than in the USA and Europe (17 t $ha^{-1}year^{-1}$) at the landscape level (Barrow 1991; Taddese 2001). The estimation shows that ten million ha of cropland are lost each year due to erosion (Faeth and Crosson 1994; Pimentel and Burgess 2013). The soil surfaces covered by crop biomass, the appropriate tillage, and the installation of natural anti-erosion devices (based on coconut fiber for example) contribute to fight effectively against erosion and to protect the soil.

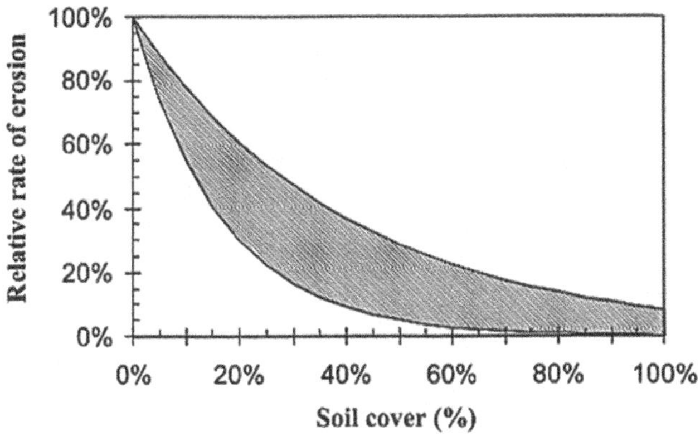

Fig. 9.4 Influence of vegetation on the relative rate of erosion (data source: ADG 2016)

High vegetation and bedding on the soil surface, whether or not linked to the use of no-till cultivation techniques, decrease the surface degradation of the soil, limit the formation and extension of a crust, and reduce the speed of diffuse runoff and water erosion (Fig. 9.4). Vegetation, therefore, helps to control runoff (Kwaad et al. 1998). Indeed, including natural and seminatural landscape components, using green manure, setting up cover crops, and relying on agroforestry are agroecological practices that contribute to soil conservation (Wezel et al. 2014; Hatt et al. 2016; Nuemsi et al. 2018). Cover crops and mulch supply nutrients to the soil. Leguminous cover crops also abet to the fixation of nitrates in the soil, the fight against weeds, the preservation of soil structure, and the conservation of humidity in the dry season or arid regions (Ngome et al. 2011b). The vegetation cover improves the porosity of the land surface. Indeed, residues kept in the upper layer of the land provide food for earthworms which rise to the surface to seize it, thus creating a natural porosity. This increased porosity makes it easier for rainwater to infiltrate, reducing runoff and erosion (Schubetzer et al. 2007). Trees, when growing among annual crops, not only change the microclimate but maintain and improve soil fertility, since their roots transport nutrients from profound land layers and make them disposable to annual plants to through their litter. This litter feeds the complex nourishing tissue of the soil. Besides, some trees enrich the soil with N and their ability to fix this element in the air (ADG 2016). For adequate soil cover, a threshold cover rate of the soil surface of 25–40% should be exceeded. In the absence of soil cover, the effectiveness of no-till farming techniques seems controversial (Kwaad et al. 1998; Heddadj et al. 2005). But, under certain conditions, no-till cultivation techniques favor the presence of plants or cover residues compared to a plowed system. In rotations with a lot of cereals, no-tillage increases the percentage of residues on 0–5 cm compared to plowing (Tebrügge and Düring 1999).

Land conservation is influenced by the type of tillage it undergoes (Fig. 9.5). Today, tillage is known as a powerful driver of the composition of microbial communities across its effect on land features (Souza et al. 2013; Degrune et al.

Fig. 9.5 Effect of cultivation techniques on soil conservation (source: Greenotec Asbl)

2017). Several authors agree that tillage, whether deep or shallow, is a threat to soils and leads to their degradation (Säle et al. 2015; Novara et al. 2019) and advocates adequate tillage (no-tillage, no turning of the soil). Tillage and secondary tillage tend to make land uniform and decrease the single microenvironments where microbial communities can live (Sengupta and Dick 2015). In no-tillage system, low soil disruption and the presence of surface residue create favorable conditions for the development of biodiversity in the soil (Daniel and Galardon 2008; Meena et al. 2020b).

But, Degrune et al. (2019) emphasize that, even if agroecological systems can favor the presence of profitable microorganisms and decrease the pressure of pathogens (Table 9.3), we cannot ultimately predict whether it will enhance agricultural productivity or other ecosystem services. There is yet sparsely proof that agricultural system favors greater microbial diversity which raises the output of agroecosystem by insuring more ecosystem roles and making it less susceptible to uttermost calamities. It emerges from the synthesis of several experimental results made by Labreuche et al. (2007) that in the absence of cover, the effectiveness of no-till farming techniques is much more controversial. Many authors agree that the

Table 9.3 Abundance and activity of earthworms according to the tillage (adapted from Tebrügge and Düring 1999; Arvalis Institut du Végétal)

Tillage practices	Number of organisms (m^{-2})	Tube volume (cm^3 m^{-2})	Biomass (g m^{-2})	Rejection of earthworms (kg m^2 $year^{-1}$)
Tillage	25	18	98	1.4
Pseudo-tillage	36	45	240	3.5
Decompaction	32	41	218	3.3
Shallow tillage	45	51	270	3.9
Direct seeding (no-tillage)	153	147	1100	11.1

use of cultivation techniques without plowing limits the formation of rills and gullies (Labreuche et al. 2007).

Unlike plowing, which tends to dilute OM in the worked horizon, no plowing concentrates OM in the surface layers. In fact, in the tests of the Arvalis Institute carried out in Boigneville from 1970 to 1998, the rate of OM was 3.6% on the surface in direct sowing whereas it was only 2% for plowing. Overall, the no-till system tends to decrease the soil porosity. Studies carried out at the Kerguéhennec experimental station, and the Boigneville station has shown a reduction in porosity of 5–10% in the unworked layers (Daniel and Galardon 2008). But, the work of Schubetzer et al. (2007) revealed that no-till cultivation techniques do not contribute to soil compaction for two main reasons: (1) this reduction in porosity remains generally limited; (2) no-till cultivation techniques favor specific mechanisms creating porosity and stabilizing the structural state which can help to reverse this trend.

According to the years, within the same experimental context, there is substantial variability in the rate of effectiveness of the same modality of cultural technique without plowing. The difference between the 2 years is 66% on average (Quinton and Catt 2004; Heddadj et al. 2005; Labreuche et al. 2007). The impact of practice can even be reversed from year to year. The test by Kwaad et al. (1998) in the Netherlands shows that for a given year, direct sowing and strip-till limit runoff on grain corn monoculture compared to plowing (−27% and −19%, respectively), while in the following year, they run more than the control mode (+15% and +50%). These differences can be linked to the variability even of the climatic years or to the crops in place at the time of the test (Quinton and Catt 2004; Rhoton et al. 2002).

The use of biopesticides in agroecology instead of the chemical pesticides commonly used in conventional agriculture contributes to the preservation of land biodiversity and therefore to land preservation. For example, the weight of earthworms is twice as high and their numbers three times higher in agroecology farming (Mader et al. 2002). Table 9.4 summarizes some biopesticides or potential biopesticides used in agroecology or organic farming in Central Africa.

Table 9.4 Some biopesticides (or potential biopesticides) used in Central Africa for sustainable agricultural production

Biopesticides	Type	Pathogen agents	Host plant	Source
Acorus calamus (L.) oil	Plant	*Prostephanus truncatus*	Corn (*Zea mays* L.)	Schmidt and Streloke (1994)
Bacillus thuringiensis	Bacteria	*Andrector ruficornis*	Potato (*Solanum tuberosum* L.)	Ambang et al. (2002)
Pseudomonas sp. (*P. fluorescens*, *P. putida*) and *Glomus deserticola*	Bacteria, fungi	*Pythium aphanidermatum*	Cowpea (*Vigna unguiculata* (L.) Walp)	Nwaga et al. (2007)
Thevetia peruviana (Pers.) K. Schum	Plant	*Cercospora arachidicola*	Groundnut (*Arachis hypogaea* L.)	Ambang et al. (2011)
Trichoderma asperellum	Fungi	*Pythium myriotylum*	Cocoyam (*Xanthosoma sagittifolium* (L.) Schott)	Mbarga et al. (2012)
Thevetia peruviana (Pers.) K. Schum	Plant	*Phytophthora megakarya*	Cocoa (*Theobroma cacao* L.)	Ngoh Dooh et al. (2014)
Streptomyces cameroonensis sp. nov.	*Actinobacteria*	*Phytophthora megakarya*	Cocoa (*Theobroma cacao* L.)	Boudjeko et al. (2017)
Trichoderma asperellum	Fungi	*Phytophthora megakarya*	Cocoa (*Theobroma cacao* L.)	Tchameni et al. (2017)
Thevetia peruviana K.	Plant	*Phytophthora infestans* and insects	Potato (*Solanum tuberosum* L.)	Dida Lontsi et al. (2019)
Trichoderma harzianum and *T. aureoviride*	Fungi	*Phytophthora colocasiae*	Taro (*Colocasia esculenta* (L.) Schott.)	Ntah et al. (2018)
Streptomyces spp. (*S. albulus*, *S. albus*, *S. gandoceansis*)	*Actinobacteria*	*Pythium myriotylum*	Cocoyam (*Xanthosoma sagittifolium* (L.) Schott)	Djuidje et al. (2019)
Eagle fern (*Pteridium aquilinum* (L.) Kuhn) and Ricin (*Ricinus communis* L.)	Plant	Fungi and insects	Lettuce (*Lactuca sativa* L.), African nightshades (*Solanum nigrum* L.), and radish (*Raphanus sativus* L.)	Mala et al. (2019)

(continued)

Table 9.4 (continued)

Biopesticides	Type	Pathogen agents	Host plant	Source
Trichoderma sp. (*T. asperellum*, *T. koningiopsis*, *T. erinaceum*, *T. gamsii*, *T. afroharzianum*, and *T. harzianum*)	Fungi	*Fusarium oxysporum*, *F. solani*, *Macrophomina phaseolina*, and *Pythium ultimum*	Common bean (*Phaseolus lunatus* L.)	Boat et al. (2020)
Dry tobacco (*Nicotiana tabacum* L.), garlic cloves (*Allium sativum* L.), onion (*Allium cepa* L.), chili fruits (*Capsicum annuum* L.), neem (*Azadirachta indica* A. Juss.) leaves and seeds, etc.	Plant	Bacteria, fungi, and insects	Okra (*Abelmoschus esculentus* (L.) Moench), lettuce (*Lactuca sativa* L.), onion (*Allium cepa* L.), eggplant (*Solanum melongena* L.), and celery (*Apium graveolens* L.).	Kacou-Amondji (2020)

9.6 Fertilization in Agroecology

The principles of land fertility management in agroecology are founded on:

- Maintaining the natural fertility of the land and the soil life through a raise in soil microbial activity to a high quantity of OM which is continuously decreasing (less than 2%)
- Minimizing external inputs by limiting considerably the use of synthetic, chemical, and harmful products to the environment, which promotes soil health (Altieri and Nicholls 2014).
- Prioritizing local inputs and the recycling of farm by-products (manure, compost, biochar, crop waste household waste) as the primary source of inputs
- Fertilization without external input is done using N-fixing species and trees. The most widely used nitrogen fixers are:
  - Symbiotic and heterotrophic bacteria like *Allorhizobium* sp., *Azorhizobium* sp., *Bradyrhizobium* sp., *Mesorhizobium* sp., *Rhizobium* sp., *Sinorhizobium* sp. (Kamtchoum et al. 2019; Mahmud et al. 2020) found in leguminous (pulses crops), *Frankia* (*Actinobacteria*) found in filao trees (*Casuarina* spp.) (Carrasco and Preston 2020)
  - Symbiotic and phototrophic bacteria (*Azolla* sp.)
  - Associative and heterotrophic bacteria (*Azospirillum* sp.) (Bhowmik and Das 2018). They can colonize many (~100) plant species
  - Nonsymbiotic and heterotrophic bacteria such as *Azotobacter* (Bhowmik and Das 2018), *Bacillus subtilis* (Efremova et al. 2020), etc.

- Nonsymbiotic and phototrophic bacteria as *Cyanobacteria* (green-blue algae)
- The best-known phosphorus solubilizers species are:
- Symbiotic fungi (mycorrhizae) as *Rhizophagus* sp., *Acaulospora* sp., *Gigaspora* sp., *Scutellospora* sp., etc. (Ngakou et al. 2012; Temegne et al. 2017, 2019; Agnolucci et al. 2019). AMF communities were influenced by the type of fertilization (Mbogne et al. 2015; Säle et al. 2015)
- Nonsymbiotic fungi like *Aspergillus* sp., *Penicillium* sp., etc.
- Nonsymbiotic and heterotrophic bacteria as *Bacillus pseudomonas* (Bhowmik and Das 2018) Fertilization with input is done by adding humus, organic/mineral elements that can be assimilated more or less quickly and microorganisms. Manure, conventional compost, earthworm humus (lombri- or vermicompost), residues from various agro-industries, shredded greenwood branches (fragmented branch wood), biochar, brush compost, fresh (green manure, tree leaves), and dry plant debris (straw in particular) are used as substantial amendments in fertilization in agroecology (Temgoua et al. 2014; Njukeng et al. 2017; Sharma et al. 2017; Billa et al. 2018). They are applied by incorporation into the top layer of the land and as a land cover with an anti-erosion and sun protection effect (but the loss of mineral elements, especially N).

The fertilization can also be carried out by the contribution of liquid manures like Supermagro, Biol, various decoctions, and purines (nettle, excrement, urine, compost, legumes, aromatic plants, ripe fruit) (Favorito et al. 2019). It is also made by adding Bokashi, natural lime or rock powders (Van Straaten 2006), growth activators, microbial inoculators, or microorganisms through all the amendments. Bokashi is an organic fertilizer based on animal fertilizer, to which straw, ash, and molasses are added. Liquid mountain microorganism and Biol is a liquid biofertilizer composed of different plants and manure (ADG 2016). The technique of Sachi also used in agroecology consists in gathering animals during a long period (e.g., 3 months), on the plot which will be cultivated to fertilize it (ADG 2016).

Figure 9.6 gives the practical indications for better use of OM. The dark green color indicates a richness in N of the soil and excellent enrichment power (type A or B). The yellowish color, on the other hand, underlines poverty in N as well as a poor enrichment quality (type C or D) of the soil. It is important to underline that the leaves with rapid decomposition have low lignin content (type A or C). The odor is also an indicator of soil quality. Indeed, an astringent smell refers to a high richness in phenols (type B or D).

9.7 Constraints to the Adoption of Agroecology

The low OM content of the soil and the imbalance of ecosystems are among the major ecological constraints of agroecology. Also, low biodiversity and the disappearance of natural enemies due to the excessive use of pesticides, aggressive

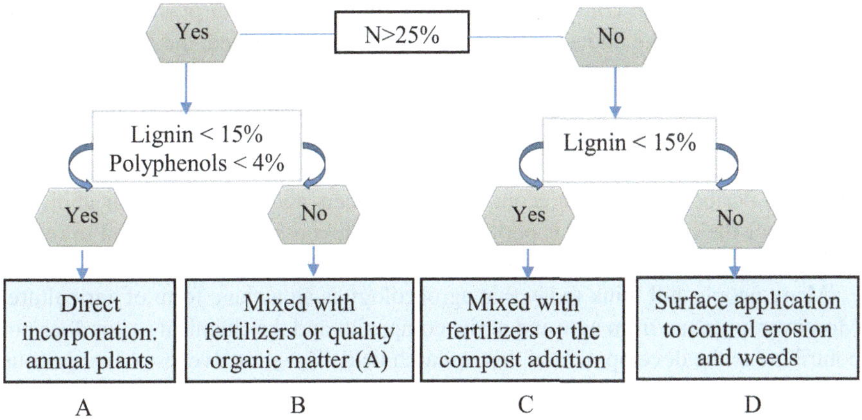

Fig. 9.6 Guide for use of organic matter (source: ADG 2016)

irrigation techniques (skate or flood irrigation), inadequate tillage practices, overgrazing, and monocultures make the soil increasingly fragile.

Adoption of agroecology is generally weak because of many technical reasons (Tittonell et al. 2012). The technical constraints of agroecology are:

- The availability of inputs (N source for humus production, seeds/plantlets for agroforestry and cover crops, water) at the local level
- The availability of equipment/tools to make and apply fertilizers (sprayer, storage of preparations (cans))
- The transport of raw materials (for compost, manure, etc., the grinder)
- The availability of labor
- The absence/insufficiency of technical knowledge
- The drop-in yield during the transition period

The scarceness of natural enemies owing to the abuse of the use of pesticides by neighboring producers who still practice conventional agriculture is also an obstacle to the adoption of agroecological practices by an ecological producer. Indeed, this producer cannot implement specific agroecological techniques since it is limited by the depletion of the ecosystem (ADG 2016).

Agroecology is a labor-intensive agriculture. The migration of young people and humans to cocoa and coffee enterprises and the mines associated with peak workloads makes the availability of labor difficult. This labor necessary for the manufacture and application of organic manure and for the control of weeds is essential only during the transition phase when a temporary fall in yields takes place (ADG 2016).

The additional cost of labor for weeding and manufacturing inputs, as well as the unattractive price (little or no differential compared to conventional), is the main economic constraint hampering the development of agroecology. The length of the transition period from conventional farming to agroecological practices is also an essential factor to take into account in raising awareness. Indeed, the drop-in yield is almost inevitable, and the duration of the transition can be extended. It varies according to the previous crop, the past practices, the state of fertility, degradation or health of the plots, the presence or absence of hedges and trees, topography, etc., with recovery being more or less rapid, but not always total. The length of the transition period can hurt the economy of producers and, therefore, their ability to provide for their families. This period is nevertheless essential for the soil to regain its balance, biodiversity, and natural fertility (ADG 2016).

Many people still think today that agroecology is an archaic form of agriculture. Moreover, pressure from agro-industrial companies and the chemical sector does not contribute to the development of this research field. The attractiveness of exogenous and higher workloads is also part of the sociocultural constraints that make difficult the adoption of agroecological practices. Also, agroecology has often been vulgarized as a whole, without appropriate tailoring to local conditions (Tittonell et al. 2012). Resistance is also psychological. The producers prefer the slight comfort of a conventional system which is not perfect but because they have mastered the workings. Many of them do not have scientific proof of the profitability of agroecological methods. Nevertheless, studies have shown that the yield was equal or even higher than that of traditional methods in the developing nations. The yield losses observed in temperate regions do not exceed 20% (ADG 2016). The lack of popularization of the results is one of the main constraints to adoption. Scientists share hardly their acquired beyond universities and research facilities. Communication with the media and decision-makers is not easy, which restricts the effect of this study (Anderson et al. 2020; DeLonge et al. 2020).

The change of political regime can be an important constraint since politics is not fixed. So, the legalization of agroecological laws and practices by the public authorities is not a guarantee of its sustainability (Murguia Gonzalez et al. 2020). Agroecology is generally considered non-priority by politicians who see it as small-scale agriculture practiced in marginal areas with few resources available for research and few trained and even fewer experienced technicians. Indeed, knowledge and practices are still very empirical, which leads to its denigration or disinterest. The absence or insufficiency of vulgarization of experience and training of farmers is also an essential constraint to the adoption of agroecology. Indeed, the subject is still poorly documented, and few scientific programs have lingered on the subject for lack of funding or interest (ADG 2016). Achieving results is dependent on substantial public funding, more specifically those that support the human aspect of the movement. They could accompany a conversion towards this movement and its associated benefits (DeLonge et al. 2020). Training is all the more complex as the agroecological solutions are local and specific to each context. This specificity is also an asset by promoting local environmental know-how and potential (ADG 2016).

9.8 Potential Solutions

Some possible solutions to address constraints to the adoption of agroecology are set out in Table 9.5.

9.9 Agroecology Versus Conventional Agriculture

Agroecology is a holistic way of farming that is less harmful to the environment as well as a natural method of food production with several economic, social, and environmental benefits (Crowder and Reganold 2015; Boeraeve et al. 2020). The primary aim of conventional agriculture is based on the use of synthesized chemicals and fertilizers to increase the productivity of a given or more plants, characteristically genetically modified in other to satisfy the ever-growing population. This technique necessitates a considerable quantity of chemicals and energy and tends to affect the natural surroundings, damages land quality, and destroys biodiversity (Savci 2012; Hooper 2016). However, to compare these two agricultural systems, several points need to be considered, i.e., production, biological diversity, land composition, erosion, water use, energy use, greenhouse gas emissions, and effect on health and environment (Table 9.6).

Conventional agriculture is carried out to fulfill the population in terms of yield since the demand in calorie- and meat-intensive regimes is estimated to double human food requests by 2050 (Mueller et al. 2012). Globally, agroecological approaches produces lower (19–41%) than conventional yield but this is dependent

Table 9.5 Potential solution for agroecological constraints (FAO 2015; ADG 2016; DeLonge et al. 2020; Murguia Gonzalez et al. 2020)

Level	Potential solutions
Ecological viewpoint	It will be needful to start the conversion with plots that are still biologically alive and to remineralize the soil by using rock powders
Economic level	Establishing a form of "labor" credit and designing and developing markets, if possible, more profitable niche markets are possible
Political level	• Increase knowledge
	• Form communal technicians
	• Reduce the distance between places of innovation (research, universities, etc.) and the places where they are applied
	• Set up a program to stake on the expertise
	• Advocating for increased capacity in agroecological research is issues to be explored
Technical level	• Facilitating access to inputs by creating farmers' enterprises or microenterprises to manufacture inputs (rock powder, seedlings and seeds, biofertilizers, phytosanitary prevention/control products, etc.)
	• Considering the human dimension of knowledge and preexisting agricultural practices
	• Paying a subsidy to the transition period by creating a conversion assistance fund

Table 9.6 Agroecology versus conventional agriculture

	Characteristics	Conventionnel agriculture	Agroecology system	References
Soil	Biodiversity	–	+	Degrune et al. (2019)
	Root length infected by AMF	–	+ (40%)	Mader et al. (2002)
	AMF spore abundance and species diversity	–	+	Oehl et al. (2004), Verbruggen et al. (2010)
	Biomass, abundance of earthworm	–	+ (1.3 to 3.2 times)	Mader et al. (2002)
	Biological activities	–	+	Peano et al. (2020)
	Nutrients	–	+	Marinari et al. (2006)
	Quality	–	+	Delate et al. (2013), Magdoff (2018)
	Water use	High quantity of water for irrigation	Organic soil retains much more water	West et al. (2014), Altieri et al. (2015), Mekonnen and Hoekstra (2016)
	Aggregate stability, respiration rates	–	+	Boeraeve et al. (2020)
	Erosion and degradation	+	–	Gomiero et al. (2011)
Production	Cropping system	Monocultures	Temporal and spatial diversification of crops	Lorenz and Lal (2014), Castellano et al. (2015), Rahman et al. (2020)
	Fertilizers	Chemical	Organic, biological	Altieri and Nicholls (2014), Mahmud et al. (2020)
	Energy to produce	+	–	Herrero et al. (2016)
	Pesticides and chemical inputs	+	–	Pfiffner and Luka (2003) Barrios et al. (2012)
	Pest abundance	+	–	Boeraeve et al. (2020)
	Cost of labor	–	+	Andriamampianina et al. (2018)
	Yield	+	– (19–41%)	Kremen and Miles (2012), Andriamampianina et al. (2018), Jouan et al. (2020)

(continued)

Table 9.6 (continued)

	Characteristics	Conventionnel agriculture	Agroecology system	References
	Prices of products	–	+ (34%)	Andriamampianina et al. (2018)
	Economic value added	–	+ (10–110%)	van der Ploeg (2020)
Environment	Pollution (water, soil, air, etc.)	+	–	Herrero et al. (2016)

on crop types and farming systems (Kremen and Miles 2012; Andriamampianina et al. 2018; Boeraeve et al. 2020). Even though common agriculture is renowned for its high returns, many environmental benefits are attached to agroecological approaches of farming (Crowder and Reganold 2015; Jouan et al. 2020). In some cases, organic agriculture has demonstrated higher yield in drought conditions and more water retention. For example, in the farming trial carried out at The Rodale Institute for 21 years (Moyer 2013), Pimentel et al. (2005) observed that in 1999, throughout the severe drought, the organic animal farming gave meaningfully higher yield (1511 kg ha^{-1}) of *Zea mays* than the conventional (1100 kg ha^{-1}) or organic legume (412 kg ha^{-1}). Besides some exceptions, agroecology generates economic value added (+10 to 110%) on farms in Europe (van der Ploeg 2020).

Agricultural health and performance are highly dependent on biodiversity. The higher the biodiversity, the more crops are naturally immune to pests and diseases without any chemical input advocated by conventional agriculture (Gomiero et al. 2011). Beyond 426 million kilograms of pesticides are being used each year with just 10% of that achieving the intended goal; this could be substantially diminished if conventional agriculture were to move to sustainable options (Sustainable Lafayette 2013). Crops in agroecological systems depend on biodiversity as it is crucial in enhancing ecological cycles. Organic farming is more abundant in nutritional elements and organisms than common farming with an increased level of biological activity (bacteria, fungi, springtails, mites, and earthworms), because of its versatility on plant rotations, diminished spreading of nutriments, and the prohibition on pesticides (Haas et al. 2001; Gomiero et al. 2011; Peano et al. 2020).

Agroecological systems are directly associated with better soil quality (Delate et al. 2013; Magdoff 2018). Sound soil ecology is observed since it promotes biodiversity, unlike monoculture, as is prescribed in conventional agriculture. Increased levels of total and OC, total N, and soluble OC are noticed in all the organic land (Wang et al. 2012). This is mainly due to the depth of the food web and quantity of biomass in the systems. The study carried out for 7 years in Italy concluded that the ecological approach exhibited meaningfully improved land nutritional and microbiological status, through an augmented level of total N, NO_3^-, and accessible P and a raised microbial biomass content and enzymatic activities (Marinari et al. 2006). Due to the global rising of agricultural production and soil becoming less disposable for plant growth, soil management is essential for the existing farms. Long-lasting techniques practiced like no-tillage system, agroforestry, and IPM help to improve the quality of the soil. Trees planted on agricultural soil aid to alleviate many of the adverse effects in agriculture, like modifying the quality of land, water, and air, preserving biological diversity, diminishing inputs by natural regulation of pests and more efficacious cycling of nutriment, and changing regional and worldwide climates (Barrios et al. 2012; Lorenz and Lal 2014).

Land erosion occurs due to nutrient loss, run-off, salinity, and drought (Issaka and Ashraf 2017). Land erosion is a menace to the growth of agriculture, particularly under uttermost climatic calamities like droughts (Gomiero et al. 2011). Agroecological agriculture improves the land composition and precludes land erosion caused by the more considerable quantity of crop material and biomass found in the land. Common agriculture, however, handles land instead of adapting to it. Lands using organic farming exhibited $<75\%$ land damage confronted to the maximum tolerance value in the area (the utmost rate of land erosion which can happen without jeopardizing sustainable plant productivity or environmental quality -11.2 t ha^{-1} $year^{-1}$). In contrast, in conventional land, the utmost tolerance value observed showed a percentage of three-time land loss (Gomiero et al. 2011). Confronted to the agroecological system, traditional plants are inefficacious at sustaining the wholeness of arable soils. Usual farming is, whereof, incapable to satisfy the requests of the increasing populations without ingurgitating an extensive quantity of soil and nonrenewable resources (Holt-Giménez et al. 2012).

Water is a renewable resource that can encounter the requirements of our present population. Water must be used efficiently because it is scarce (Mekonnen and Hoekstra 2016). Approximately 70% of water in the world is used in the agricultural sector (West et al. 2014). Cumulative demand for freshwater is pressurizing worldwide stocks. To preserve this resource, a dire renovation of methods to save water, peculiarly in agriculture, has to be developed. The richness of flora and fauna in sustainable agriculture causes organic land to characteristically hold much more water compared to that of conventional land. This augmented retention rate allows agroecological farming to generate better returns than conventional for water deficiency (Altieri et al. 2015). Nearly, 20–40% in the water holding capacity of organic farming lands when compared to conventional farming lands was recorded in heavy loess lands in a temperate climate in Switzerland. Thus, one of the main reasons for higher output in organic plants is believed to be caused by the higher water-holding capacity of the lands under ecological management (Gomiero et al. 2011).

The use of natural processes for inputs and nutrient recycling is advocated by agroecological systems to abolish the use of nonrenewable resources. The conventional system involves a significant quantity of energy to generate, prepare, and transport food (De Ponti et al. 2012). The fossil fuel-based industrial agriculture abets to greenhouse gas emissions in many ways:

- Directly by the fuel burned by farm machinery, in food processing and in transporting the mean ounce of food over a thousand miles "from farm to fork"

- Indirectly by the production of its synthetic inputs, such as N fertilizers from N and natural gas
- Finally by the breakdown of soil OM into CO_2 (through large-scale tillage and excessive synthetic inputs), which is liberated into the atmosphere as a greenhouse gas (Herrero et al. 2016)

Besides, large-scale industrial livestock farming releases massive amounts of methane (CH_4) (Eckard et al. 2010; Knapp et al. 2014). Energy effectiveness is vital to food production as it can diminish the mission of greenhouse gases and costs. About 5% of emissions of CO_2 resulting from the influence of human beings is generated by agricultural actions (Gomiero et al. 2011; Balogh 2020). The 10–12% of total worldwide emissions of greenhouse gases (5.1–6.1 Gt CO_2 eq. year^{-1} in 2005) relates from the influence of human beings, accounting for almost all the anthropogenic CH_4. One- to two-thirds of all N_2O emissions resulting from the influence of human beings is caused by agricultural actions (Gomiero et al. 2011; Balogh 2020). Therefore, agroecology can reduce this tendency than conventional agriculture. Due to land composition, conventional agriculture is ineffectual at catching C, steady production, and energy utilization to sustain the plants. Lots of machinery, pesticides, irrigation, processing, and transportation reveals that for each calorie arriving at the table, ten calories or energy has been spent. C can be stockpiled in land by the soil OM and by above the ground biomass via methods like using rotations combined to cover plants and green manures to raise soil organic material, agroforestry, and conservation-tillage agriculture (Castellano et al. 2015; Rahman et al. 2020).

Agroecology limits the usage of pesticides which is advocated in conventional agriculture. Agrochemical industries informed farmers on the profit they would make by using agrochemicals on vast scale monoculture. But, pesticides have been pointed out to have severe negative impacts on the farm farmers and consumers of the farm products (Calvert et al. 2008; Páyan-Renteria et al. 2012; Damalas and Koutroubas 2016). Also, they have negative effects on both the aquatic and terrestrial ecosystems (Sánchez-Bayo 2011; Stehle and Schulz 2015; Chagnon et al. 2015). Agroecology discourages the total eradication of pests because it will also wipe out the natural predators that are needed to keep the pests in check in a healthy ecosystem. So, agroecology tends to enrich the soil by using manure and tilled in plant residue that is using OM to maintain the biological cycle (Ge et al. 2011). The higher nutritional value such as vitamin and mineral content of crops produced from agroecological systems has been reported when compared to conventional agriculture (Rembialkowska 2007; Barański et al. 2014). Again, agroecological products have been reported to have high sugar content and have a superior structure and high metabolic integrity which makes them last longer (Bourn and Prescott 2002; Shafie and Rennie 2012; Yu et al. 2018). Agroecology can raise agricultural yield in ways that are economically, environmentally, and socially viable (Crowder and Reganold 2015).

9.10 Agroecology Towards Soil Management and Sustainability

The global level of soil degradation observed is leading to the need of managing soils in ways that maintain and improve soil resources to continue providing food, fiber, and freshwater, achieving significant inputs to energy and climate sustainability and aiding in preserving biological diversity and the whole safeguard of ecosystem goods and services (Koch et al. 2012, 2013). Soil management requires a whole method concentrated on how the soil and plants are managed, instead of an output approach that concentrates predominantly on delivering chemical solutions to nutrient and pest problems. The health and fertility of soils are essential to sustainable agriculture. If this ability is lost, then indicators like the waning in fertility, loss of species in soil biota, soil erosion, and changes in the water holding capacity can be detected (Veresoglou et al. 2015; Kay 2018). Soil health or quality is defined as the capability of the land to sustain the production and ecosystem services (Kibblewhite et al. 2008), while soil fertility is the availability of nutrients in the ground (Troeh and Thompson 2005). On the one hand, a healthy land is typified by the availability of nutrients, suitable structure, low level of salinity and toxic elements, and high resilience to harmful events (drought and flooding), resists degradation (e.g., erosion and compaction), supplies appropriate aeration and rapid water infiltration, and accepts, holds, and liberates water to crops and groundwater. On the other hand, soil richness is the balance of critical nutrients. Agroecology fosters the improvement and maintenance of physical, chemical, and biological features of the land through a set of sophisticated interrelated practices.

Primarily, the choice of plants favors the expansion of beneficial microorganisms (Hartmann et al. 2009). Microorganisms principally reside in the land rhizosphere. They quicken plant growth by various mechanisms such as boost nutrient procurement, defense versus pathogens, and modulation of phytohormone synthesis. AMF forms a significant cluster which favors plant growth, hence the sustainability of agroecosystem (Yang et al. 2014; Moreira et al. 2020). The land characteristics and land management practice applied improve their growth and efficiency in crop yield (Gianinazzi et al. 2010). The use of biofertilizers consists of applying living microorganisms to seed, crop surfaces, or land and has been reported to improve the availability of nutrients (Bhavikatti 2020). Conventionally managed agricultural lands tender to be low in AMF diversity; this has been assigned to the harmful influences of fertilization, fungicides, land cultivations, and weakness of host diversity. It has been indicated that low-input, conservation, and organic farming may improve AMF richness confronted to conventional farming (Mahmood and Rizvi 2010; Schneider et al. 2015). Some research has reported about the crop growth, raised productivity, and uptake of N and other components by inoculation with AMF (Ortas 2012; Pellegrino et al. 2011) and PGPR inoculation (Singh et al. 2011). Also, organic fertilizers like compost and manure increase the general soil richness, enhance the soil biological activity, and increase soil mineralization (Steenwerth and Belina 2008; Tao et al. 2015). The soil respiration rates, movement, and inoculum of native AMF from plots with permanent plant cover are generally higher than those from plots with shallow tillage. Maintaining permanent crop cover seems to be a better alternative than working the surface soil as a land management practice to conserve the biological fertility of the land (Turrini et al. 2017). Land content of OM and land microbial activity may impact the quantity of soil-borne pathogens and the resistance of plants to them. Some studies have revealed that organic land amendments like compost may improve the elimination of soil pathogens (Chen and Jiang 2014). OM supplies nutriments and energy to sustain various land microbial communities which rivaled with pathogens and impede their growth. Compost and various organic amendments equally have high quantities of microorganisms which can improve the diversity. Plants grown with high OM content and various active microorganism communities usually exhibit tolerance to maladies (Altieri and Nicholls 2003). Therefore, methods of agroecology like natural and little-input system may raise soil OM and improve microbial features (Ge et al. 2011).

Kirkby et al. (2014) stated that crop rotation could be used to enhance the nutrient availability of soils, thereby favoring plant growth. For instance, including legume species in the rotation permits the fixation of atmospheric N_2 and makes available a source of facilely absorbable N for the next planting season. Soil conservation and protection can be optimized by introducing cover crops which also improves the carbon content in the soil; decreased leaching, via the immobilization of N predominantly on freely drained, lighter lands; and promotes land steadiness (Dogliotti et al. 2004; Guzmán et al. 2019). Richardson et al. (2009) showed that about 40% of the assimilated microbial C occurs at root systems. Therefore, adding cover plants in the rotation is a hypothetically good idea (Wu et al. 2010; Kirkby et al. 2014). Furthermore, practicing rotation may alleviate NO_3^-leaching and enhance the effectiveness of nutrient use (Larsen 2019; Bai et al. 2020). Celette et al. (2008) reported that in temperate climates, they may also increase water infiltration over the winter period and raise water availability for the next plants. Other cropping practices like intercropping and relieve intercropping have proven to be effective in increasing soil health. For example, root exudates of some leguminous plants can enhance land P availability, solubilizing land organic P, also enhancing organic fertilization (Li et al. 2005; Darch et al. 2018). This system also enhances the land physical structure and land fertility (Darch et al. 2018). Interestingly, soil penetration and compaction resistance are weaker in these systems, and amelioration in structural steadiness is observed (Carof et al. 2007). The use of soil cover in an intercropping system reduces soil crusting and erosion (Le Bissonnais et al. 2004; Liu et al. 2017). Numerous studies have shown raised microorganism diversity, enzyme activities, and more excellent steadiness in alley cropping farming which were due to alterations in litter amount and quality and root exudates (Udawatta et al. 2008; Lacombe et al. 2009).

In agroforestry farming, nutriments are taken up and stopped from inferior land levels by tree roots and sent back to the land via falling of leaves (Rigueiro-Rodrígues et al. 2009). Thevathasan and Gordon (2004) concluded that in an agroforestry farming, fall of leaf from 6-year-old poplars ensued in mean land NO_3^- production rates in the head-to-head crop alley up to twice that confronted to lands situated 8–15 m from the tree row, and N liberate from the litter of poplar leaf was equal to 7 kg N ha^{-1} $year^{-1}$. Also, trees of red alder in the silvopastoral test farm at Henfaes near Bangor were considered to evaluate the possibility for increasing and sustaining land fertility, and the outcome revealed that the degree of N fixing was projected at 31 kg ha^{-1} $year^{-1}$ in the silvopasture treatment with densities of 400 tree stems ha^{-1} and the entire quantity of N that might hypothetically be appended to the land as an outcome of dead leaf, root, and nodule decomposition was assessed at approximately 41 kg ha^{-1} $year^{-1}$ (Teklehaimanot and Mmolotsi 2007). Roots and trunks of trees also play a role as physical barricades to diminish the flow of water on the surface and sediment (Udawatta et al. 2008).

Excepting lowland OM content, land compaction is due principally to high machinery traffic, especially tillage and intense animal treading in humid land conditions which is an important problem in modern agriculture (Hamza and Anderson 2005; Hobbs et al. 2008). The main goal of reduced or no tillage is to lessen soil disturbance and conserve OM at the top of the land surface or in the first few centimeters. Diminished CO_2 emissions, energy use and erosion, or raised land fertility and land biota activity/diversity have been mentioned as advantages for no-till or reduced tillage approaches (Gadermaier et al. 2011; Karlen et al. 2013). Mäder et al. (2012) obtained an increase in yields with reduced tillage for corn, winter, wheat, and grass-clover mixes while Berner et al. (2008) showed returns below organic conditions were 97% than the ones beneath common tillage. Also, land OC and microbial biomass were improved. Berner et al. (2008) again confronted that diminished tillage with traditional tillage in wheat and spelt plants during 3 years was capable to show an augmentation in land OM by 7.4% in the 0–10 cm land horizon. Also, they showed an up to 70% higher richness of endogeic, horizontally burrowing adult earthworms below shallow tillage, confronted to common tillage which raised land porosity, and thus enhanced water and root penetration into the land (Peigné et al. 2009).

Newer agroecological practices and approaches such as drip irrigation give an increased potential to restrict water inputs, to enhance the effectiveness of water use, and to improve satisfaction in time and space for the plant water request. It has also been found to limit the risk of soil salinization (Sun et al. 2012). Combining this irrigation technique and cover plants is beneficial and feasible by appending the cover crop rows between plants to decrease evaporation from bare land, increase land OM, decrease soil erosion, and if leguminous species are utilized increase N concentration (Lopes et al. 2011). Conclusively, protection versus wind and land erosion and surface water pollution is achieved by the integration, or reintegration, of unadulterated or seminatural landscape components like hedges and plant strips, either in or around the farm (Baudry and Jouin 2003; Wu et al. 2010). Besides, they usually ensure biological diversity preservation in soils.

9.11 Compost Manufacturing Unit, Dschang, Cameroon: A Case Study

The manufacture of inputs being one of the stages which makes agroecological systems difficult, the city of Dschang has managed to combine sanitation of the town and manufacture of compost. Indeed, Dschang is one of the most important cities in the west region of Cameroon. It is not only an agricultural production zone but a university municipality with nearly 220,000 inhabitants. Like the main Cameroonian metropolises (Yaounde and Douala), Dschang faces a significant challenge that of the management of household waste with an annual production of 40,000 tons, i.e., around 108 tons day^{-1} (CEFREPADE 2016). The city, which has only two compaction bins and two trucks for the collection of waste throughout the city, is experiencing enormous technical difficulties in removing only 20% of the deposit and bringing it back to the municipal controlled landfill. Thus, the collection rate decreased from 40% in 2007 to 10% in 2011. Furthermore, the waste produced is composed of around 80% of biodegradable materials with high humidity (65%) which makes their combustion difficult, however with a good C/N ratio ($\approx$40) favorable for composting. It is in this context that a composting project initiated in 2010 by the nongovernmental organization (NGO) ERA-Cameroon to meet a need for organic amendment and improvement of the sanitation of the city of Dschang (Temgoua et al. 2014). This project was carried out by the NGO ERA-Cameroon in partnership with the Francophone Center for Partnership Research on Sanitation, Waste and the Environment (CEFREPADE). The relay was taken in 2014 by Africompost program (2017) and Gevalor (2020) to ensure continuity.

The first composting unit in Dschang was installed in the Ngui District and the second later in the Siteu District. Ngui's unit covers an area of 3000 m^2. It is made up of a waste reception and weighing area; a sorting table; a composting area (heap fermentation and maturation area); a sieving and bagging area; an 81 m^2 drying and storage shed; and a 1000 m^2 experimental field (Temgoua et al. 2014). The working equipment consists of a sieve (12 mm), wheelbarrows, tarpaulins, buckets, forks, shovels, rakes, and machetes.

Household waste is collected using 120 kg carriers (handcrafted) and trucks in around 800 households in the city (Temgoua et al. 2014). When the waste arrives at the site (Fig. 9.7a), it undergoes a manual sorting operation (Fig. 9.7b). Then the biodegradable materials are put in heaps of 2–5 m^3 while the non-fermentable return to the landfill (Vermande et al. 2012). The technique used here is heap composting. During the process, the temperature is read daily in each pile with a metal probe thermometer. The turning is done at a frequency of once a week during the first month, then once every 2 weeks (Fig. 9.7c).

The heap is watered when the need arises, due to the presence of mold on top of the waste. During the first week of composting, the temperatures in the heaps reach 70 °C and begin to drop after 10 days. Mature compost is obtained on the platform after 90 days. Since the start of the project, compost production has continued to increase on the site through the support of Africompost and Gevalor, who ensure the continuity of the project.

Fig. 9.7 Composting process. (**a**) Waste supply, (**b**) sorting of waste and piling up, (**c**) turning heaps, (**d**) pile of compost under shelter, (**e**) compost dry sieved. Source: (**a**)–(**c**) (Ngnikam 2013); (**d**) and (**e**) (author's picture)

Some production data recorded during the project:

- From June 2013 to June 2014, 935.53 tons of waste was treated, to produce 1384 bags of compost of 50 kg each, or 69.2 tons.
- In 2016, 1750 tons of waste was processed to produce and market 136 tons of fertilizer.
- In 2018, 2,817.44 tons of waste was processed, for 402.2 tons of fertilizer. The forecast for 2019 is 6000 tons of waste managed to produce 600 tons of fertilizer. This objective was achieved through the production of the Ngui unit and the second composting unit located in the Siteu District.

To ensure the grade of the final product, at the end of the procedure, the compost is dried and sieved, and samples are taken and analyzed at the soil laboratory of IRAD in Yaounde. Thus, the total OM; total N, P, K, Ca, and Mg; as well as the heavy metals (Cd, Zn, Pb, Cu, Ni, Se) are determined. The results showed that the

Fig. 9.8 Compost marketing. (**a**) Compost weighing and packaging, (**b**) compost marketing poster in Dschang. Source: (**a**) (Ngnikam 2013; CEFREPADE), (**b**) sinotables.com

compost produced in the city of Dschang contains heavy metals (Cd, Zn, Pb, Cu) with relatively high contents but which remains below the limit values of French and Swiss standards, except for Cd, Cr, and Se. The OM rate is 20% dry matter, the total N content equal to 11 g kg^{-1}, and the C/N ratio 10.26 (Temgoua et al. 2014). Today, the center is working on optimizing the manual sorting of waste to improve the quality of the compost.

After the drying and sieving operation, the compost is weighed and packaged in 50 kg bags stored in the hangar ready for sale (Fig. 9.8). The price of a bag is set at 2000 FCFA ($1 US ~615 FCFA) and a ton at 35,000 FCFA. The period of high demand for compost was identified during the main cropping season in the locality. This leads to large volumes of compost sales from mid-January to the end of February (crop sowing period) (Vermande et al. 2012). Although the demand is sometimes higher than the supply from the Ngui unit, promotions are sometimes launched to increase farmers' awareness of the use of compost and avoid long storage periods. A plot highlighting the effect of compost use on the production of vegetables is visible next to the composting site (Fig. 9.9).

The project is not yet achieving its objectives because of many constraints encountered at several levels of the chain.

- At the collection level, the primary obstacle is technical, because of the breakdown of trucks from the municipality; there is a reduction in the volume of incoming waste and saturation of the waste disposal site.
- In terms of marketing, the delivery of compost to farmers is often limited by the availability of transport means.
- Finally, the site has no water point, and this makes the work more difficult for the workers.

In Cameroon, the composting remains in an embryonic state despite its proven advantages in waste recovery and agriculture. Several composting projects in major cities in Cameroon have failed due to investment costs and the lack of political will

Fig. 9.9 Demonstration plot with cabbage located next to the composting site (source: Scidev.net)

on the part of the competent authorities. Very tiny composting units are identified in private homes, but these do not have a real follow-up of the process. The Dschang composting platform remains one of the few that continues to operate to this day because it benefits from the support of a Partnership Agreement between the Nantes City of France and the Dschang City Council with the help of Africompost and Gevalor. Monitoring of the production chain is ensured from waste collection to the use of compost in the fields through product quality analyzes. The quantity of compost produced in Dschang remains insufficient to meet the demand of farmers. However, its quality continues to be improved by optimizing sorting to reduce the heavy metal contents. Cameroon does not have regulations on the quality of compost. Therefore, French and Swiss standards are those which are applied.

9.12 ISSAEER: A Case Study

Although agroecology is an ancestral practice in Africa, its entry into universities as a discipline was there later than in the west. ISSAEER with the support of its partners (CEFRA, AFOP, GESCOD), in its prospective, believes in agroecology as a relevant futuristic trend. It integrates agroecological practices in the training of future agropastoral entrepreneurs and the recycling and supervision of producers in the locality of Sa'a. Indeed, the institute has delimited a mini agroecological route within it for the training of students and as a demonstration plot for visitors. Figure 9.10 shows the students in the implementation of some agroecological practices. For example, the town's hilltop relief leads to the establishment of devices to combat erosion. Limiting the use of chemical pesticides resulted in a rich diversity of insects, which favor the production of good quality honey. Within its campus, it organizes workshops on agroecology and capacity building for the CEFRA team working at

Fig. 9.10 Some agroecological practices within the ISSAEER. (**a**) Establishment of crop beds by students under the supervision of experts from CEFRA and Alsace. (**b**) Arrangements of the space subject to a double slope (U-shaped ridges). (**c**) Ecological beekeeping practices—CEFRA. Source: ISSAEER

ISSAEER. The main difficulty currently facing the structure is the lack or insufficiency of funds/grants. Several of the institute's projects are seeking funding for implementation.

9.13 Policy and Legal Framework

A specialized database on various lawful structures, practices, strategies, and programs on agroecology in various nations exists at the FAO level. It is called AgroecologyLex. This database created in coordination with FAOLEX is the largest database on agriculture and renewable natural resources policies and legislation in the world. It is regularly updated. The information provided by AgroecologyLex allows users to have the full text of the document as well as a detailed summary of the content, focused mainly on the specific goal and objectives, institutional frameworks, and primary forms of support, to support transitions from conventional agriculture to agroecological approaches (FAO 2020).

Monteduro et al. (2015) stated that it was necessary to embrace a transdisciplinary oncoming to multifunctional husbandry to include the paradigm of agroecology into lawful regulations. They emphasized that this does not need an extraordinary law which aims hierarchically to integrate and unsettle current lawful areas, instead of calling for the creation of a trans-law. The trans-law gradually works to coordinate inter-legalities between the various lawful fields, while preserving their independence and by underlining their mutual historical origins.

Poyyamoli (2017) wrote that by encouraging farmers to adopt resource conservation technologies, the government has a substantial part to play. Among the areas of government intervention, he cited a few:

- Advance national policies and legal frameworks to encourage agroecological production, including the adoption of IPM. This may include adopting a national definition of agroecological production and a policy statement in support of measures to facilitate the transition to agroecological output.
- Relaunch public research in agroecology and extension programs adapted as per the requirement and situation of smallholder producers, their organization, and their connections.
- Promote convergence and collaboration between the ministers of agriculture, livestock, fisheries, environment, and forests.
- Establish a general ecological fertilization policy to support and promote all the components of ecological fertilization that the government must undergo for achieving sustainability. It should launch a green fertilization mission with sufficient financial expenditure to restore and maintain soil health.
- Public procurement of organic products should be encouraged, including the presentation of natural products at important public events.

In 2018, the FAO, in partnership with IFOAM—Organics International and the Future Policy Award of the World Future Council, worked to highlight legal and policy frameworks. These latter create environments conducive to the implementation of agroecological approaches, to help realize the plans of the 2030 Agenda for Sustainable Development and several long-lasting developing purposes (Da Silva 2018). These lawful and policy frames help protect the lives and livelihoods of smallholders and family growers and guarantee long-lasting and including systems of food production. These also perform sustainable agricultural practices that facilitate preserve and improve the natural resource base and build the ability to accommodate global warming, as well as contribute to dimming (Da Silva 2018).

9.14 Future Roadmap of Agroecology for Agricultural Soil Management

Two challenges remain to be taken up for the development of agroecology, according to De Schutter (2011). These are the increase in cultivated areas and the creation of a favorable environment for farmers. It establishes different principles capable of promoting the agroecological transition that governments should consider. These principles are nevertheless to be applied with the flexibility to be tested and reassessed according to local circumstances (environment, climate, soil condition, etc.). They must also be developed in collaboration with the beneficiaries of this development. Priority should also be given to public goods by popularizing knowledge, building storage facilities, rural infrastructure, facilitating access to resources (insurance against climate risks, education, support for farmers' organizations, and cooperatives). These investments can, in the long run, be much more sustainable than simple private goods provided to farmers/growers, when they are informed and thought out. Automating women through specific mechanisms to encourage their participation in the construction of knowledge and organizing markets (use of packaging, processing, marketing, value chains, bringing farmers together in cooperatives, etc.) to protect farmers against fluctuating prices and dumping are objectives to be achieved to ensure adoption of agroecological practices by all (De Schutter 2011).

Society must, therefore, not only be attentive to the action of agriculture on the environment but, equally, make sure to encourage it to strengthen these interactions, which means:

- The restoration of the natural agronomic functions of cultivated ecosystems
- Combating soil erosion and preserving its fertility
- Diminution in the consumption of energy, water, chemical inputs
- The use of biological interactions, ecosystem services, and potentials offered by natural resources (biodiversity, photosynthesis, etc.) while maintaining their capacity for renewal from a qualitative and quantitative point of view (Claveirole 2016)

9.15 Conclusion

Agroecology is a scientific discipline with enormous potential and the ability to lead the transition to a more inclusive, sustainable model of society based on more robust and more united social ties by relocating the economy. It embodies a credible, efficient, and human alternative while fully participating in the objectives of food sovereignty. It offers a real social transformation project that does justice to the proletariat of the countries of the south as the first food suppliers in the world through better management of agricultural soils. Agroecology improves soil fertility, biodiversity, and productivity, while reducing dependence on energy-intensive inputs. However, most agroecological techniques have, so far, a feeble integration in nowadays farming for various reasons, one being that it is described as labor-intensive. In order to satisfy the increasing request for and press on soil and water resources, it will be required to not only expand but implement eco-friendly, eco-specific, and system reposed land management techniques. Research and other support services will require to be reoriented to assist farmers better comprehend agroecology farming and perform suitable choices for land management. To nourish an increasing earth inhabitant, we need practices that supply smug feeding while preserving the environment especially the soil and that guarantee economic viability for peasants. For this reason, practices of agroecology can and should play a vital function.

References

ADG (2016) L'agroécologie : reconnecter l'homme à son écosystème. Aide au Développement Gembloux (ADG), Gembloux, p 40

Agnolucci M, Avio L, Pepe A, Turrini A, Cristani C, Bonini P, Cirino V, Colosimo F, Ruzzi M, Giovannetti M (2019) Bacteria Associated with a commercial mycorrhizal inoculum: community composition and multifunctional activity as assessed by illumina sequencing and culture-dependent tools. Front Plant Sci 9:1956. https://doi.org/10.3389/fpls.2018.01956

Ali HRK, Hemeda NF, Abdelaliem YF (2019) Symbiotic cellulolytic bacteria from the gut of the subterranean termite *Psammotermes hypostoma* Desneux and their role in cellulose digestion. AMB Express 9:111. https://doi.org/10.1186/s13568-019-0830-5

Altieri MA (1983a) Agroecology, the scientific basis of alternative agriculture. Div. of Biol. Control, U.C. Berkeley, Cleo's Duplication Services. (out of print). Berkeley

Altieri MA (1983b) The question of small farm development: who teaches whom? Agric Ecosyst Environ 9(4):401–405

Altieri MA, Nicholls CI (2003) Soil fertility and insect pests: Harmonizing soil and plant health in agroecosystems. Soil Tillage Res 72:203–211. https://doi.org/10.1016/S0167-1987(03)00089-8

Altieri MA, Nicholls CI (2014) Diffuser l'agroécologie pour la souveraineté et la résilience alimentaires. Agroécologie - enjeux et perspectives. Vol. XXI - 2014, n°3. Alternatives Sud. CETRI, Syllepse

Altieri MA, Nicholls CI (2015) Agroecology and the search for a truly sustainable agriculture, 1st edn. PNUMA, Mexico, p 291

Altieri MA, Nicholls CI, Henao A, Lana MA (2015) Agroecology and the design of climate change-resilient farming systems. Agron Sustain Dev 35(3):869–890. https://doi.org/10.1007/s13593-015-0285-2

Ambang Z, Omokolo ND, Ouzounov IS (2002) Evaluation de l'efficacité de la *Bacillus thuringiensis* contre les larves et imagos de *Andrector ruficornis* sur des plants de *Solanum tuberosum* au Cameroun. Tropicultura 20(3):113–117

Ambang Z, Ndongo B, Essono G, Ngoh JP, Kosma P, Chewachong GM, Asanga A (2011) Control of leaf spot disease caused by *Cercospora* sp on groundnut (*Arachis hypogaea*) using methanolic extracts of yellow oleander (*Thevetia peruviana*) seeds. AJCS 5(3):227–232

Anderson CR, Pimbert MP, Chappell MJ, Brem-Wilson J, Claeys P, Kiss C, Maughan C, Milgroom J, McAllister G, Moeller N, Singh J (2020) Agroecology now - connecting the dots to enable agroecology transformations. Agroecol Sustain Food Syst 44(5):561–565. https://doi.org/10.1080/21683565.2019.1709320

Andriamampianina L, Temple L, de Bon H, Malézieux E, Makowski D (2018) Multi-criteria evaluation of organic agriculture in sub-Saharan Africa using probabilistic elicitation of expert knowledge. Cah Agric 27(4):45002. https://doi.org/10.1051/cagri/2018030

Bai X, Zhang Z, Cui J, Liu Z, Chen Z, Zhou J (2020) Strategies to mitigate nitrate leaching in vegetable production in China: a meta-analysis. Environ Sci Pollut Res:1–10. https://doi.org/10.1007/s11356-020-08322-1

Balogh JM (2020) The role of agriculture in climate change: a global perspective. IJEEP 10 (2):401–408. https://doi.org/10.32479/ijeep.8859

Banerjee A, Jhariya MK, Yadav DK, Raj A (2020) Environmental and sustainable development through forestry and other resources. Apple Academic Press Inc., New York, p 400

Banerjee A, Meena RS, Jhariya MK, Yadav DK (2021) Agroecological footprints management for sustainable food system. Springer, Singapore

Barański M, Srednicka-Tober D, Volakakis N, Seal C, Sanderson R, Stewart GB, Benbrook C, Biavati B, Markellou E, Giotis C, Gromadzka-Ostrowska J, Rembiałkowska E, Skwarło-Sońta K, Tahvonen R, Janovská D, Niggli U, Nicot P, Leifert C (2014) Higher antioxidant and lower cadmium concentrations and lower incidence of pesticide residues in organically grown crops: a systematic literature review and meta-analyses. Br J Nutr 112:794–811. https://doi.org/10.1017/S0007114514001366

Barrios E, Sileshi GW, Shepherd K, Sinclair F (2012) Agroforestry and soil health: linking trees, soil biota and ecosystem services. In: Wall DH, Bardgett RD, Behan-Pelletier V, Herrick JE, Jones H, Ritz K, Six J, Strong DR, van der Putten WH (eds) Soil ecology and ecosystem services. Oxford University Press, Oxford, pp 315–330. https://doi.org/10.1093/acprof:oso/9780199575923.003.0028

Barrow CJ (1991) Land degradation. Cambridge University Press, Cambridge

Baudry J, Jouin A (2003) De la haie aux bocages. Organisation, dynamique et gestion. INRA, Paris, p 435

Bensin BM (1928) Agroecological characteristics description and classification of the local corn varieties chorotypes

Berner A, Hildermann I, Fliessbach A, Pfiffner L, Niggli U, Mäder P (2008) Crop yield and soil fertility response to reduced tillage under organic management. Soil Tillage Res 101:89–96. https://doi.org/10.1016/j.still.2008.07.012

Bhavikatti MBK (2020) Biofertilizers: a potential approach for sustainable agriculture in India. SIPN 40(63):111–118

Bhowmik SN, Das A (2018) Biofertilizers: a sustainable approach for pulse production. In: Meena RS, Das A, Yadav GS, Lal R (eds) Legumes for soil health and sustainable management. Springer, Singapore, pp 446–474. https://doi.org/10.1007/978-981-13-0253-4-14-446-474

Billa SH, Tsi EA, Ngome FA, Tata PN, Mvo DC (2018) Effect of biochar issued from crop wastes on the yield of variety 8034 cassava in the Humid Forest Agroecological Zone, Cameroon. IJHAF 2(1):13–27. https://doi.org/10.22161/ijhaf.2.1.2

Blanc C, Sy M, Djigal D, Brauman A, Normand P, Villenave C (2006) Nutrition on bacteria by bacterial-feeding nematodes and consequences on the structure of soil bacterial community. Eur J Soil Biol 42(1):S70–S78. https://doi.org/10.1016/j.ejsobi.2006.06.003

Boat MAB, Sameza ML, Iacomi B, Tchameni SN, Boyom FF (2020) Screening, identification and evaluation of *Trichoderma* spp. for biocontrol potential of common bean damping-off pathogens. Biocontrol Sci Tech 30(3):228–242. https://doi.org/10.1080/09583157.2019.1700909

Boeraeve F, Dendoncker N, Cornélis J-T, Degrune F, Dufrênea M (2020) Contribution of agroecological farming systems to the delivery of ecosystem services. J Environ Manag 260. https://doi.org/10.1016/j.jenvman.2019.109576

Boudjeko T, Tchinda RA, Zitouni M, Nana JA, Lerat S, Beaulieu C (2017) Streptomyces cameroonensis sp. nov., a geldanamycin producer that promotes Theobroma cacao growth. Microbes Environ 32(1):24–31. https://doi.org/10.1264/jsme2.ME16095

Bourn D, Prescott JA (2002) Comparison of the nutritional value, sensory qualities, and food safety of organically and conventionally produced foods. Crit Rev Food Sci Nutr 42:1–34. https://doi.org/10.1080/10408690290825439

Brondizio ES, Settele J, Díaz S, Ngo HT (2019) Global assessment report on biodiversity and ecosystem services of the Intergovernmental Science-Policy Platform on Biodiversity and Ecosystem Services. IPBES secretariat, Bonn, Germany, p XXX. https://www.ipbes.net/global-assessment-report-biodiversity-ecosystem-services

Brussaard L, Kooistra MJ (2013) Soil structure/soil biota interrelationships. Elsevier, Wageningen, p 846

Buttel FH (2003) Envisioning the future development of in USA: agroecology between extinction and multifunctionality? University of Wisconsin, Wiszconsin

Calvert G, Karnik J, Mehler L, Beckman J, Morrissey B, Sievert J, Barrett R, Lackovic M, Mabee L, Schwartz A, Mitchell Y, Moraga-McHaley S (2008) Acute pesticide poisoning among agricultural workers in the United States, 1998–2005. Am J Ind Med 5:883–898. https://doi.org/10.1002/ajim.20623

Caporali F (2011) Agroecology as a transdisciplinary science for a sustainable agriculture. In: Lichtfouse E (ed) Biodiversity, biofuels, agroforestry and conservation agriculture. Springer, Cham, pp 1–71. https://doi.org/10.1007/978-90-481-9513-8

Carof M, de Tourdonnet S, Saulas P, Le Floch D, Roger-Estrade J (2007) Undersowing wheat with different living mulches in a no-till system. I. Yield analysis. Agron Sustain Dev 27:347–356. https://doi.org/10.1051/agro:2007016

Carrasco J, Preston GM (2020) Growing edible mushrooms: a conversation between bacteria and fungi. Environ Microbiol 22(3):858–872. https://doi.org/10.1111/1462-2920.14765

Castellano MJ, Mueller KE, Olk DC, Sawyer JE, Six J (2015) Integrating plant litter quality, soil organic matter stabilization, and the carbon saturation concept. Glob Chang Biol 21 (9):3200–3209. https://doi.org/10.1111/gcb.12982

Cefrepade (2016) Développement du compostage à Dschang, pp 1–3

Celette F, Gaudin R, Gary C (2008) Spatial and temporal changes in the water regime of a Mediterranean vineyard due to the adoption of cover cropping. Eur J Agron 29:153–162. https://doi.org/10.1016/j.eja.2008.04.007

Chagnon M, Kreutzweiser D, Mitchell EAD, Morrissey CA, Noome DA, Van der Sluijs JP (2015) Risks of large-scale use of systemic insecticides to ecosystem functioning and services. Environ Sci Pollut Res 22:119–134. https://doi.org/10.1007/s11356-014-3277-x

Chen Z, Jiang X (2014) Microbiological safety of chicken litter or chicken litter-based organic fertilizers: a review. Agriculture 4(1):1–29. https://doi.org/10.3390/agriculture4010001

Chérif M, Audebert A, Fofana M, Zouzou M (2009) Evaluation of iron toxicity on lowland irrigated rice in West Africa. Tropicultura 27(2):88–92

CIDSE (2018) The principles of Agroecology – Towards just, resilient and sustainable food systems. CIDSE Belgium. https://agroecologyprinciple.atavist.com/theprinciples-ofagroecology

CIRAD (2018) Agroecology for tropical and Mediterranean farming systems. CIRAD's research position. https://www.cirad.fr/en/our-research/researchtopics/agroecology/what-s-ciraddoing

Clarke CJ, George RJ, Bell RW, Hatton TJ (2002) Dryland salinity in south-western Australia; its origins, remedies and future research directions. Aust J Soil Res 40:93–113

Claveirole C (2016) La transition agroécologique: défs et enjeux. Journal Officiel de la République Française, 114 p

Contador CA, Lo S-K, Chan SHJ, Lam H-M (2020) Metabolic analyses of nitrogen fixation in the soybean microsymbiont *Sinorhizobium fredii* using constraint-based modeling. mSystems 5(1): e00516. https://doi.org/10.1128/mSystems.00516-19

Costa O, Raaijmakers JM, Kuramae EE (2018) Microbial extracellular polymeric substances: ecological function and impact on soil aggregation. Front Microbiol 9:1636. https://doi.org/10.3389/fmicb.2018.01636

Crowder DW, Reganold JP (2015) Financial competitiveness of organic agriculture on a global scale. Proc Natl Acad Sci U S A 112(24):7611–7616. https://doi.org/10.1073/pnas.1423674112

Cuevas J, Daliakopoulos IN, del Moral F, Hueso JJ, Tsanis IK (2019) A Review of soil-improving cropping systems for soil salinization. Agronomy 9:295. https://doi.org/10.3390/agronomy9060295

Culliney TW (2013) Role of arthropods in maintaining soil fertility. Agriculture 3:629–659. https://doi.org/10.3390/agriculture3040629

Da Silva JG (2018) Celebrating the world's best policies for agroecology call for nominations. FAO/IFOAM, pp 1–5

Dagnachew M, Moges A, Kebede A, Abebe A (2020) Effects of soil and water conservation measures on soil quality indicators: the case of Geshy Subcatchment, Gojeb River Catchment, Ethiopia. Appl Environ Soil Sci 2020:1–16. https://doi.org/10.1155/2020/1868792

Damalas CA, Koutroubas SD (2016) Farmers' exposure to pesticides: toxicity types and ways of prevention. Toxics 41(1):1. https://doi.org/10.3390/toxics4010001

Daniel P, Galardon G (2008) Techniques culturale sans labour en Bretagne. Chambre d'agriculture de Bretagne/Institut Arvalis, Bretagne, p 44

Darch T, Giles CD, Blackwell MS, George TS, Brown LK, Menezes-Blackburn D, Wendler R (2018) Inter-and intra-species intercropping of barley cultivars and legume species, as affected by soil phosphorus availability. Plant Soil 427(1–2):125–138. https://doi.org/10.1007/s11104-017-3365-z

De Ponti T, Rijk B, Van Ittersum MK (2012) The crop yield gap between organic and conventional agriculture. Agric Syst 108:1–9. https://doi.org/10.1016/j.agsy.2011.12.004

De Schutter O (2011) Agroecology and the right to food. Report of the Special Rapporteur on the right to food. United Nations. www.srfood.org/images/stories/pdf/officialreports/20110308_a-hrc-16-49_agroecology_en.pdf

Degenhardt J, Hiltpold I, Köllner TG, Frey M, Gierl A, Gershenzon J, Hibbard BE, Ellersieck MR, Turlings TC (2009) Restoring a maize root signal that attracts insect-killing nematodes to control a major pest. Proc Natl Acad Sci U S A 106(32):13213–13218. https://doi.org/10.1073/pnas.0906365106

Degrune F, Theodorakopoulos N, Colinet G, Hiel M-P, Bodson B, Taminiau B, Daube G, Vandenbol M, Hartmann M (2017) Temporal dynamics of soil microbial communities below the seedbed under two contrasting tillage regimes. Front Microbiol 8:1127. https://doi.org/10.3389/fmicb.2017.01127

Degrune F, Boeraeve F, Dufrêne M, Cornélis J-T, Frey B, Hartmann M (2019) The pedological context modulates the response of soil microbial communities to agroecological management. Front Ecol Evol 7:261. https://doi.org/10.3389/fevo.2019.00261

Delate K, Cambardella C, Chase C, Johanns A, Turnbull R (2013) The long-term agroecological research (LTAR) experiment supports organic yields, soil quality, and economic performance in Iowa. Crop Manage 12:1. https://doi.org/10.1094/CM-2013-0429-02-RS

DeLonge M, Robbins T, Basche AD, Haynes-Maslow L (2020) The state of sustainable agriculture and agroecology research and impacts: a survey of U.S. scientists. J Agric Food Syst Community Dev 9(2):159–184. https://doi.org/10.5304/jafscd.2020.092.009

Díaz S, Settele J, Brondízio ES, Ngo HT, Guèze M, Agard J, Arneth A, Balvanera P, Brauman KA, Butchart SHM, Chan KMA, Garibaldi LA, Ichii K, Liu J, Subramanian SM, Midgley GF, Miloslavich P, Molnár Z, Obura D, Pfaff A, Polasky S, Purvis A, Razzaque J, Reyers B, Chowdhury R, Shin YJ, Visseren-Hamakers IJ, Willis KJ, Zayas CN (2019) Summary for policymakers of the global assessment report on biodiversity and ecosystem services of the Intergovernmental Science-Policy Platform on Biodiversity and Ecosystem Services. IPBES secretariat, Bonn, p 56. https://doi.org/10.5281/zenodo.3553579

Dida Lontsi SL, Ambang Z, Djieto CL, Chedjou Wouom BC, Heu A, Ndogho Pegalepo A (2019) Effect of aqueous extracts of *Thevetia peruviana* K. seeds on the control of late blight and pest insects of *Solanum tuberosum* L. in Cameroon. Greener J Agric Sci 9(1):14–22. https://doi.org/10.15580/GJAS.2019.1.121418176

Djuidje KPF, Ewane CA, Lerat S, Omokolo ND, Beaulieu C, Boudjeko T (2019) Evaluation of antagonistic activities against *Pythium myriotylum* and plant growth promoting traits of Streptomyces isolated from Cocoyam (*Xanthosoma sagittifolium* (L.) Schott) rhizosphere. AJCS 13 (6):920–933. https://doi.org/10.21475/ajcs.19.13.06.p1670

Dogliotti S, Rossing WAH, van Ittersum MK (2004) Systematic design and evaluation of crop rotations enhancing soil conservation, soil fertility and farm income: a case study for vegetable farms in South Uruguay. Agric Syst 80:277–302. https://doi.org/10.1016/j.agsy.2003.08.001

Dumont B, Fortun-Lamothe L, Jouven M, Thomas M, Tichit M (2013) Prospects from agroecology and industrial ecology for animal production in the 21st century. Animal 7(6):1028–1043. https://doi.org/10.1017/S1751731112002418

Dumont AM, Vanloqueren G, Stassart PM, Baret PV (2016) Clarifying the socioeconomic dimensions of agroecology: between principles and practices. Agroecol Sustain Food Syst 40 (1):24–47

Eckard RJ, Grainger C, De Klein CAM (2010) Options for the abatement of methane and nitrous oxide from ruminant production: a review. Livest Sci 130(1-3):47–56. https://doi.org/10.1016/j.livsci.2010.02.010

Efremova SY, Kulikova IV, Sukhova NV (2020) Agroecological efficiency of biomodified mineral fertilizers. E3S Web Conf 161:01115. https://doi.org/10.1051/conf/202016101115

ELD (2015) The value of land: Prosperous lands and positive rewards through sustainable land management. The Economics of Land Degradation. Available from www.eld-initiative.org

EU (2019) Desertification and drought. EU Science Hub. Accessed June 4, 2020. https://ec.europa.eu/jrc/en/research-topic/desertification-and-drought

Faeth P, Crosson P (1994) Building the case for sustainable agriculture. Environment 36(1):16–20. https://doi.org/10.1080/00139157.1994.9929138

FAO (2015) Agro ecology to reverse soil degradation and achieve food security. Food Agriculture Organization (FAO) of the United Nations, Rome, pp 1–4

FAO (2018) Family farming knowledge platform. Food Agriculture Organization (FAO), Rome

FAO (2020) Agroecology knowledge hub. Food Agriculture Organization (FAO) of the United Nations, Rome

Favorito AC, Seide EP, Kohler TR, de Carvalho WR, Pan R, de Vasconcelos ES (2019) Application of biofertilizers in cover of and its effects on organic maize. J Exp Agric Int 39(6):1–7. https://doi.org/10.9734/JEAI/2019/v39i630349

Ferris H, Venette RC, Scow K (2004) Scow soil management to enhance bacterivore and fungivore nematode populations and their nitrogen mineralisation function. Appl Soil Ecol 25(1):19–35. https://doi.org/10.1016/j.apsoil.2003.07.001

Francis C, Lieblein G, Gliessman S, Breland TA, Creamer N, Harwood R, Salomonsson L, Helenius J, Rickerl D, Salvador R, Wiedenhoeft M, Simmons S, Allen P, Altieri M, Flora C, Poincelot R (2003) Agroecology: the ecology of food systems. J Sustain Agric 22(3):99–118. https://doi.org/10.1300/J064v22n03_10

Gadermaier F, Berner A, Fliessbach A, Friedel JK, Mäder P (2011) Impact of reduced tillage on soil organic carbon and nutrient budgets under organic farming. Renew Agric Food Syst 27 (1):68–80. https://doi.org/10.1017/S1742170510000554

García-Ruiz JM, López-Moreno JI, Vicente-Serrano SM, Lasanta-Martínez T, Beguería S (2011) Mediterranean water resources in a global change scenario. Earth-Sci Rev 105(3-4):121–139. https://doi.org/10.1016/j.earscirev.2011.01.006

Ge T, Nie SA, Wu J, Shen J, Xiao HA, Tong C, Huang D, Hong Y, Iwasaki K (2011) Chemical properties, microbial biomass, and activity differ between soils of organic and conventional horticultural systems under greenhouse and open field management: a case study. J Soils Sediments 11(1):25–36. https://doi.org/10.1007/s11368-010-0293-4

Getachew A, Chilot Y, Teklu E (2019) Soil acidity management. Ethiopian Institute of Agricultural Research (EIAR), Addis Ababa, p 66

Gevalor (2020) L'unité de compostage de Dschang (Cameroun). www.gevalor.org

Gianinazzi S, Gollotte A, Binet MN, van Tuinen D, Redecker D, Wipf D (2010) Agroecology: the key role of arbuscular mycorrhizas in ecosystem services. Mycorrhiza 20(8):519–530. https://doi.org/10.1007/s00572-010-0333-3

Giovannini L, Palla M, Agnolucci M, Avio L, Sbrana C, Turrini A, Giovannetti M (2020) Arbuscular mycorrhizal fungi and associated microbiota as plant biostimulants: research strategies for the selection of the best performing inocula. Agronomy 10(1):106. https://doi.org/10.3390/agronomy10010106

Giraldo-Silva A, Fernandes VMC, Bethany J, Garcia-Pichel F (2020) Niche partitioning with temperature among heterocystous cyanobacteria (Scytonema spp., Nostoc spp., and Tolypothrix spp.) from biological soil crusts. Microorganisms 8:396. https://doi.org/10.3390/microorganisms8030396

Gliessman SR (1997) Agreocology: ecological processes in sustainable agriculture. CRC Press, Boca Raton

Gomiero T, Pimentel D, Paoletti MG (2011) Environmental impact of different agricultural management practices: conventional vs. organic agriculture. Crit Rev Plant Sci 30 (1-2):95–124. https://doi.org/10.1080/07352689.2011.554355

Guzmán G, Cabezas JM, Sánchez-Cuesta R, Lora Á, Bauer T, Strauss P, Winterd S, Zallere JG, Gómez JA (2019) A field evaluation of the impact of temporary cover crops on soil properties and vegetation communities in southern Spain vineyards. Agric Ecosyst Environ 272:135–145. https://doi.org/10.1016/j.agee.2018.11.010

Haas G, Wetterich F, Köpke U (2001) Comparing intensive, extensified and organic grassland farming in southern Germany by process life cycle assessment. Agric Ecosyst Environ 83:43–53

Hamza MA, Anderson WK (2005) Soil compaction in cropping systems: a review of the nature, causes and possible solutions. Soil Tillage Res 82:121–145. https://doi.org/10.1016/j.still.2004.08.009

Hartmann A, Schmid M, van Tuinen D, Berg G (2009) Plant-driven selection of microbes. Plant Soil 321:235–257. https://doi.org/10.1007/s11104-008-9814-y

Hatt S, Artru S, Brédart D, Lassois L, Francis F, Haubruge E, Garre S, Stassart PM, Dufrene M, Monty A, Boeraeve F (2016) Towards sustainable food systems: the concept of agroecology and how it questions current research practices. A review. Biotechnol Agron Soc Environ 20:215–224

Heddadj D, Gascuel-Odoux C, Cotinet P, Hamon Y (2005) Mode de travail du sol, ruissellement et propriétés hydrodynamiques sur un dispositif expérimental de l'Ouest de la France. Etude et gestion des sols 12(1):53–66

Herrero M, Henderson B, Havlík P, Thornton PK, Conant RT, Smith P, Wirsenius S, Hristov AN, Gerber P, Gill M, Butterbach-Bahl K, Valin H, Garnett T, Stehfest E (2016) Greenhouse gas mitigation potentials in the livestock sector. Nat Clim Chang 6(5):452–461. https://doi.org/10.1038/nclimate2925

Hobbs PR, Sayre K, Gupta R (2008) The role of conservation agriculture in sustainable agriculture. Philos Trans R Soc Lond Ser B Biol Sci 363(1491):543–555. https://doi.org/10.1098/rstb.2007.2169

Holt-Giménez E, Shattuck A, Altieri M, Herren H, Gliessman S (2012) We already grow enough food for 10 billion people and still can't end hunger. J Sustain Agric 36(6):595–598. https://doi.org/10.1080/10440046.2012.695331

Hooper D (2016) 10 things you can do to help biodiversity. https://seasonalight.com/2014/10/20/10-things-we-all-can-do-to-help-biodiversity/

Huera-Lucero T, Labrador-Moreno J, Blanco-Salas J, Ruiz-Téllez T (2020) Aframework to incorporate biological soil quality indicators into assessing the sustainability of territories in the Ecuadorian Amazon. Sustain For 12:3007. https://doi.org/10.3390/su12073007

Inomura K, Follett CL, Masuda T, Eichner M, Prášil O, Deutsch C (2020) Carbon transfer from the host diatom enables fast growth and high rate of n2 fixation by symbiotic heterocystous Cyanobacteria. Plan Theory 9:192. https://doi.org/10.3390/plants9020192

IPBES (2018) Glossary. Intergovernmental Science-Policy Platform on Biodiversity and Ecosystem Services. https://www.ipbes.net/glossary/agroecology

IPCC (2019) Summary for policymakers. In: Climate change and land: an IPCC special report on climate change, desertification, land degradation, sustainable land management, food security, and greenhouse gas fluxes in terrestrial ecosystems. IPCC, p 41

Issaka S, Ashraf MA (2017) Impact of soil erosion and degradation on water quality: a review. Geol Ecol Landsc 1(1):1–11. https://doi.org/10.1080/24749508.2017.1301053

Jhariya MK, Yadav DK, Banerjee A (2018a) Plant mediated transformation and habitat restoration: phytoremediation an eco-friendly approach. In: Gautam A, Pathak C (eds) Metallic contamination and its toxicity. Daya Publishing House, New Delhi

Jhariya MK, Banerjee A, Yadav DK, Raj A (2018b) Leguminous trees an innovative tool for soil sustainability. In: Meena RS, Das A, Yadav GS, Lal R (eds) Legumes for soil health and sustainable management. Springer, Singapore

Jhariya MK, Banerjee A, Meena RS, Yadav DK (2019a) Sustainable agriculture, forest and environmental management. Springer, Singapore

Jhariya MK, Yadav DK, Banerjee A (2019b) Agroforestry and climate change: issues and challenges. Apple Academic Press Inc., New York

Jhariya MK, Meena RS, Banerjee A (2021) Ecological intensification of natural resources for sustainable agriculture. Springer, Singapore

Jones A, Panagos P, Barcelo S, Bouraoui F, Bosco C, Dewitte O, Gardi C, Erhard M, Hervás J, Hiederer R, Jeffery S, Lükewille A, Marmo L, Montanarella L, Olazábal C, Petersen J-E, Penizek V, Strassburger T, Tóth G, Van Den Eeckhaut M, Van Liedekerke M, Verheijen F, Viestova E, Yigini Y (2011) The state of soil in Europe. Publications Office of the European Union, Luxembourg, p 71

Jouan J, Ridier A, Carof M (2020) Synergy: a regional bio-economic model analyzing farm-to-farm exchanges and legume production to enhance agricultural sustainability. Ecol Econ 175:106688. https://doi.org/10.1016/j.ecolecon.2020.106688

Kacou-Amondji H (2020) Les bio pesticides au bénéfice des maraîchers du Tchad grâce au projet HUP. FAO. http://www.fao.org/africa/news/detail-news/fr/c/426464/

Kamtchoum SM, Nchoutnji I, Temegne CN, Tonfack LB, Fofe L, Seutchueng TGT, Kemayou CM, Notche FK, Suh C, Youmbi E (2019) Comparative effect of biological fixation of nitrogen and chemical fertilizer on yield optimization of two sorghum varieties in the Western Highlands. Asian J Agric Hortic Res 4(3):1–10. https://doi.org/10.9734/ajahr/2019/v4i330023

Karlen DL, Cambardella CA, Kovar JL, Colvin TS (2013) Soil quality response to long-term tillage and crop rotation practices. Soil Tillage Res 133:54–64. https://doi.org/10.1016/j.still.2013.05.013

Kay BD (2018) Soil structure and organic carbon: a review. In: Lal R, Kimble JM, Follett RF, Stewart BA (eds) Soil processes and the carbon cycle. CRC Press, Boca Raton, pp 169–197

Khan N, Jhariya MK, Yadav DK, Banerjee A (2020a) Herbaceous dynamics and CO_2 mitigation in an urban setup- a case study from Chhattisgarh, India. Environ Sci Pollut Res 27(3):2881–2897. https://doi.org/10.1007/s11356-019-07182-8

Khan N, Jhariya MK, Yadav DK, Banerjee A (2020b) Structure, diversity and ecological function of shrub species in an urban setup of Sarguja, Chhattisgarh, India. Environ Sci Pollut Res 27(5):5418–5432. https://doi.org/10.1007/s11356-019-07172-w

Khan N, Jhariya MK, Raj A, Banerjee A, Meena RS (2021a) Soil carbon stock and sequestration: implications for climate change adaptation and mitigation. In: Jhariya MK, Meena RS, Banerjee A (eds) Ecological intensification of natural resources for sustainable agriculture. Springer, Singapore

Khan N, Jhariya MK, Raj A, Banerjee A, Meena RS (2021b) Eco-designing for sustainability. In: Jhariya MK, Meena RS, Banerjee A (eds) Ecological intensification of natural resources for sustainable agriculture. Springer, Singapore

Kibblewhite M, Ritz K, Swift M (2008) Soil health in agricultural systems. Philos Trans R Soc Lond Ser B Biol Sci 363(1492):685–701. https://doi.org/10.1098/rstb.2007.2178

Kirkby CA, Richardson AE, Wade LJ, Passioura JB, Batten GD, Blanchard C, Kirkegaard JA (2014) Nutrient availability limits carbon sequestration in arable soils. Soil Biol Biochem 68:402–409

Kitamura K, Fujita T, Akada S, Tonouchi A (2011) Methanobacterium kanagiense sp. nov., a hydrogenotrophic methanogen, isolated from rice-field soil. Int J Syst Evol Microbiol 61:1246–1252. https://doi.org/10.1099/ijs.0.026013-0

Knapp JR, Laur GL, Vadas PA, Weiss WP, Tricarico JM (2014) Invited review: Enteric methane in dairy cattle production: quantifying the opportunities and impact of reducing emissions. J Dairy Sci 97(6):3231–3261. https://doi.org/10.3168/jds.2013-7234

Koch A, McBratney A, Lal R (2012) Global soil week: put soil security on the global agenda. Nature 492:186. https://doi.org/10.1038/492186d

Koch A, McBratney A, Adams M, Field D, Hill R, Crawford J, Minasny B, Lal R, Abbott L, O'Donnell A (2013) Soil security: solving the global soil crisis. Global Pol 4(4):434–441. https://doi.org/10.1111/1758-5899.12096

Kovats RS, Valentini R, Bouwer LM, Georgopoulou E, Jacob D, Martin E, Rounsevell M, Soussana J-F (2014) Europe. In: Barros VR, Field CB, Dokken DJ, Mastrandrea MD, Mach KJ, Bilir TE, Chatterjee M, Ebi KL, Estrada YO, Genova RC, Girma B, Kissel ES, Levy AN, Mastrandrea PR, White LL (eds) Climate change 2014: impacts, adaptation, and vulnerability. Part b: regional aspects. Cambridge University Press, New York, pp 1267–1326

Krasilnikov P, Alyabina I, Arrouays D, Balyuk S, Arbestain MC, Gafurova L, Erdogan HE, Havlicek E, Konyushkova M, Kuziev R, van Liedekerke M, Medvedev V, Montanarella L, Panagos P, Ravina da Silva M, Sönmez B (2015) Status of the World's soil resources - main report. In: Regional assessment of soil changes in Europe and Eurasia. FAO/ITPS, Rome, pp 330–363

Kremen C, Miles A (2012) Ecosystem services in biologically diversified versus conventional farming systems: benefits, externalities, and trade-offs. Ecol Soc 17(4). https://doi.org/10.5751/ES-05035-170440

Kumar S, Meena RS, Jhariya MK (2020) Resources use efficiency in agriculture. Springer, Singapore, p 760

Kumar S, Meena RS, Datta R, Verma SK, Yadav GS, Pradhan GS, Molaei A, Rahman GKMM, Mashuk HA (2020a) Legumes for carbon and nitrogen cycling: an organic approach. Carbon Nitrogen Cycl Soil. https://doi.org/10.1007/978-981-13-7264-3_10.337-375

Kwaad FJPM, van der Zijp M, van Dijk PM (1998) Soil conservation and maize cropping systems on sloping loess soils in the Netherlands. Soil Tillage Res 46(1–2):13–21. https://doi.org/10.1016/S0167-1987(98)80103-7

Labreuche J, Le Souder C, Castillon P, Ouvry JF, Real B, Germon JC, De Tourdonnet S (2007). Évaluation des impacts environnementaux des techniques culturales sans labour en France. ADEME-ARVALIS, Institut du végétal-INRA-APCA-AREAS-ITB-CETIOMIFVV, p 400. http://www2.ademe.fr/servlet/getDoc

Labreuche J, Roger-Estrade J, Feix I, Viloingt T, Caboulet D, Daouze JP, Duval R, Ganteil A, Quere L, Boizart H, Jouy L, Thevenet G (2008) Les techniques culturales sans labour. Perspect Agricoles 342:38–43

Labreuche J, Roger-Estrade J, Lecomte V, Duval R, Leclech N, Longueval C, Eschenbrenner G (2014) Travail du sol et implantation des cultures. In: Colloque Faut-il travailler le sol ? du 3 avril 2014. Arvalis Institut du Végétal/INRA, Paris, pp 1–20

Lacombe S, Bradley RL, Hamel C, Beaulieu C (2009) Do tree-based intercropping systems increase the diversity and stability of soil microbial communities? Agric Ecosyst Environ 131:25–31. https://doi.org/10.1016/j.agee.2008.08.010

Larsen MR (2019) Nitrate reduction strategies: utilizing cover crops to reduce nitrate leaching, increase nitrogen retention and protect water resources. Retrieved from the University of Minnesota Digital Conservancy. http://hdl.handle.net/11299/202271

Le Bissonnais Y, Arrouays D (1997) Aggregate stability and assessment of crustability and erodability. 2. Application to humic loamy soils with various organic carbon content. Eur J Soil Sci 48(1):39–48. https://doi.org/10.1111/j.1365-2389.1997.tb00183.x

Le Bissonnais Y, Lecomte V, Cerdan O (2004) Grass strip effects on runoff and soil loss. Agronomie 24:129–136. https://doi.org/10.1051/agro:2004010

Lehmann A, Zheng W, Ryo M, Soutschek K, Roy J, Rongstock R, Maaß S, Rillig MC (2020) Fungal traits important for soil aggregation. Front Microbiol 10:2904. https://doi.org/10.3389/fmicb.2019.02904

Li W, Li L, Sun J, Guo T, Zhang F, Bao X, Peng A, Tang C (2005) Effects of intercropping and nitrogen application on nitrate present in the profile of an Orthic anthrosol in northwest China. Agric Ecosyst Environ 105:483–491. https://doi.org/10.1016/j.agee.2004.07.008

Lion R, Pascal F, Ferone G, Akalay F-Z, Breuzard S, Delapalme N, Gastambide M, Godfrain J, Harribey L, Hessel S, Josselin C, Vigier L (2009) Report for food security. Agrisud International, Paris, pp 1–12

Liu W, Luo Q, Lu H, Wu J, Duan W (2017) The effect of litter layer on controlling surface runoff and erosion in rubber plantations on tropical mountain slopes, SW China. Catena 149:167–175. https://doi.org/10.1016/j.catena.2016.09.013

Lopes CM, Santos TP, Monteiro A, Rodrígues ML, Costa JM, Chaves MM (2011) Combining cover cropping with deficit irrigation in a Mediterranean low vigor vineyard. Sci Hortic 129:603–612. https://doi.org/10.1016/j.scienta.2011.04.033

Lorenz K, Lal R (2014) Soil organic carbon sequestration in agroforestry systems. A review. Agron Sustain Dev 34(2):443–454. https://doi.org/10.1007/s13593-014-0212-y

Lund University (2018) Agroecology: a better alternative in Sub Saharan Africa. Science. www.sciencedaily.com/releases/2018/05/180504103753.htm

Mader P, Fließbach A, Dubois D, Gunst L, Fried P, Niggli U (2002) Soil fertility and biodiversity on organic farming. Science 296:1694. https://doi.org/10.1126/science.1071148

Mäder P, Berner A, Messmer M, Fliessbach A, Krauss M, Dierauer H, Clerc M, Koller M, Meier M, Schader C (2012) Reduzierte Bodenbearbeitung - deutliche Vorteile für Bodenfruchtbarkeit. Ökologie Landbau 162(2):25–27

Magdoff F (2018) Soil quality and management. In: Altieri MA (ed) Agroecology- the science of sustainable agriculture. CRC Press, Boca Raton, pp 349–364. https://doi.org/10.1201/9780429495465-18

Mahmood I, Rizvi R (2010) Mycorrhiza and organic farming. Asian J Plant Sci 9(5):241–248. https://doi.org/10.3923/ajps.2010.241.248

Mahmud K, Makaju S, Ibrahim R, Missaoui A (2020) Current progress in nitrogen fixing plants and microbiome research. Plan Theory 9:97. https://doi.org/10.3390/plants9010097

Mala C, Kekeunou S, Djoukouo N, Denis Z, Bassala J, Nukenine E (2019) Biopesticide potentialities of eagle fern (*Pteridium aquilinum*) and Ricin (*Ricinus communis*) in the protection of vegetables crops. J Exp Agric Int 35(6):1–14. https://doi.org/10.9734/jeai/2019/v35i630222

Mapiemfu-Lamaré D, Ndindeng SA, Ngome AF, Thé C, Tsoata E, Zonkeng C, Mfopou MC, Bihnchang L, Etame F (2012) Early criterion to screen maize varieties for their tolerance to aluminium toxic soil. Int J Agric For 2(4):161–165. https://doi.org/10.5923/j.ijaf.20120204.05

Marinari S, Mancinelli R, Campiglia E, Grego S (2006) Chemical and biological indicators of soil quality in organic and conventional farming systems in Central Italy. Ecol Indic 6(4):701–711. https://doi.org/10.1016/j.ecolind.2005.08.029

Masters DG, Norman HC, Barrett-Lennard EG (2005) Agricultural systems for saline soil: the potential role of livestock. Asian-Aust J Anim Sci 18(2):296–300

Matei G-M, Matei S, Mocanu V (2020) Assessing the role of soil microbial communities of natural forest ecosystem. EuroBiotechnol J 4(1):01–07. https://doi.org/10.2478/ebtj-2020-0001

Mbarga JB, Martijn Ten Hoopen G, Kuaté J, Adiobo A, Ngonkeu MEL, Ambang Z, Akoa A, Tondje PR, Begoude BAD (2012) *Trichoderma asperellum*: A potential biocontrol agent for *Pythium myriotylum*, causal agent of cocoyam (*Xanthosoma sagittifolium*) root rot disease in Cameroon. Crop Prot 36:18–22. https://doi.org/10.1016/j.cropro.2012.02.004

Mbogne JT, Temegne CN, Hougnandan P, Youmbi E, Tonfack LB, Ntsomboh-Ntsefong G (2015) Biodiversity of arbuscular mycorrhizal fungi of pumpkins (*Cucurbita* spp) under the influence of fertilizers in ferralitic soils of Cameroon and Benin. J Appl Biol Biotechnol 3(5):1–10. https://doi.org/10.7324/JABB.2015.3501

Meena RS, Lal R, Yadav GS (2020a) Long-term impact of topsoil depth and amendments on carbon and nitrogen budgets in the surface layer of an Alfisol in Central Ohio. Catena 194:104752. https://doi.org/10.1016/j.catena.2020.104752

Meena RS, Kumar S, Datta R, Lal R, Vijaykumar V, Brtnicky M, Sharma MP, Yadav GS, Jhariya MK, Jangir CK, Pathan SI, Dokulilova T, Pecina V, Marfo TD (2020b) Impact of agrochemicals on soil microbiota and management: a review. Land 9(2):34. https://doi.org/10.3390/land9020034

Meena RS, Lal R, Yadav GS (2020c) Long term impacts of topsoil depth and amendments on soil physical and hydrological properties of an Alfisol in Central Ohio, USA. Geoderma 363:1141164. https://doi.org/10.1016/j.geoderma.2019.114164

Mekonnen MM, Hoekstra AY (2016) Four billion people facing severe water scarcity. Sci Adv 2 (2):e1500323. https://doi.org/10.1126/sciadv.1500323

Menta C, Remelli S (2020) Soil health and Arthropods: from complex system to worthwhile investigation. Insects 11:54. https://doi.org/10.3390/insects11010054

Migliorini P, Wezel A (2017) Converging and diverging principles and practices of organic agriculture regulations and agroecology. A review. Agron Sustain Dev 37:63. https://doi.org/10.1007/s13593-017-0472-4

Mikha MM, Rice CW (2004) Tillage and manure effects on soil and aggregate-associated carbon and nitrogen. Soil Sci Soc Am J 68:809–816. https://doi.org/10.2136/sssaj2004.8090

Mirzabaev A, Wu J, Evans J, García-Oliva F, IAG H, Iqbal MH, Kimutai J, Knowles T, Meza F, Nedjraoui D, Tena F, Türkeş M, Vázquez RJ, Weltz M (2019) Desertification. In: Shukla PR, Skea J, Calvo Buendia E, Masson-Delmotte V, Pörtner H-O, Roberts DC, Zhai P, Slade R, Connors S, van Diemen R, Ferrat M, Haughey E, Luz S, Neogi S, Pathak M, Petzold J, Portugal Pereira J, Vyas P, Huntley E, Kissick K, Belkacemi M, Malley J (eds) Climate Change and Land: an IPCC special report on climate change, desertification, land degradation, sustainable land management, food security, and greenhouse gas fluxes in terrestrial ecosystems. IPCC, Geneva, pp 249–343

Misra RV, Roy RN, Hiraoka H (2003) On-farm composting methods. FAO, Rome

Monteduro M, Buongiorno P, Di Benedetto S, Isoni A (2015) Law and agroecology: a transdisciplinary dialogue. Springer, Berlin, p 494. https://doi.org/10.1007/978-3-662-46617-9

Moreira H, Pereir SI, Vega A, Castro PM, Marques AP (2020) Synergistic effects of arbuscular mycorrhizal fungi and plant growth-promoting bacteria benefit maize growth under increasing soil salinity. J Environ Manag 257:109982. https://doi.org/10.1016/j.jenvman.2019.109982

Moyer J (2013) Perspective on Rodale Institute's farming systems trial. Crop Manage 12(1). https://doi.org/10.1094/CM-2013-0429-03-PS

Mueller ND, Gerber JS, Johnston M, Ray DK, Ramankutty N, Foley JA (2012) Closing yield gaps through nutrient and water management. Nature 490(7419):254–257. https://doi.org/10.1038/nature11420

Munk AC, Copeland A, Lucas S, Lapidus A, Del Rio TG, Barry K, Detter JC, Hammon N, Israni S, Pitluck S, Brettin T, Bruce D, Han C, Tapia R, Gilna P, Schmutz J, Larimer F, Land M, Kyrpides NC, Mavromatis K, Richardson P, Rohde M, Göker M, Klenk HP, Zhang Y, Roberts GP, Reslewic S, Schwartz DC (2011) Complete genome sequence of *Rhodospirillum rubrum* type strain (S1). Stand Genomic Sci 4(3):293–302. https://doi.org/10.4056/sigs.1804360

Murguia Gonzalez A, Mier Y, Terán M, Cacho G, Castillo LR (2020) Policy pitfalls and the attempt to institutionalize agroecology in El Salvador 2008-2018. Agroecol Sustain Food Syst. https://doi.org/10.1080/21683565.2020.1725216

Nature (2018) Nature subjects. Nat J. https://www.nature.com/subjects/agri-ecology

Ngakou A, Ngo Nkot L, Doloum G, Adamou S (2012) Mycorrhiza-rhizobium-Vigna subterranea dual symbiosis: impact of microbial symbionts for growth and sustainable yield improvement. Int J Agric Biol 14(6):915–921

Ngnikam E (2013) Bilan du compostage des ordures ménagères dans la ville de Dschang. ERA – Cameroun. In: Atelier de capitalisation et d'échange sur la gestion décentralisée des déchets à Haïti, Port au Prince du 11 au 17 juillet 2013. www.plateforme-re-sources.org

Ngoh Dooh JP, Ambang Z, Ndongo B, Heu A, Kuate W (2014) Effect of extracts of *Thevetia peruviana* (Pers.) K. Schum on development of *Phytophthora megakarya* causal agent of black pod disease of cocoa. J Appl Biosci 77:6564–6574. https://doi.org/10.4314/jab.v77i1.11

Ngome AFE, Becker M, Mtei KM, Mussgnug F (2011a) Fertility management for maize cultivation in some soils of Western Kenya. Soil Tillage Res 117:69–75. https://doi.org/10.1016/j.still.2011.08.010

Ngome AFE, Becker M, Mtei KM (2011b) Leguminous cover crops differentially affect maize yields in three contrasting soil types of Kakamega, Western Kenya. J Agric Rural Dev Trop Subtrop 112(1):1–10

Nicholls CI, Altieri MA, Vazquez L (2016) Agroecology: principles for the conversion and redesign of farming systems. J Ecosys Ecogr S5:010. https://doi.org/10.4172/2157-7625.S5-010

Njukeng NJ, Ngome AF, Efombagn IBM, Temegne CN (2017) Response of African Nightshade (*Solanum* sp.) to cassava peel-based manure in the humid forest zone of Cameroon. Afr J Agric Res 12(22):1866–1873. https://doi.org/10.5897/AJAR2017.12315

Novara A, Stallone G, Cerdà A, Gristina L (2019) The effect of shallow tillage on soil erosion in a semi-arid vineyard. Agronomy 9:257–267. https://doi.org/10.3390/agronomy9050257

Ntah AAM, Tchameni NS, Siebatcheu EC, Ambata AHT, Sameza ML, Wansi JD (2018) Efficacy of *Trichoderma harzianum* (Edtm) and *Trichoderma aureoviride* (T4) as potential bio-control agent of taro leaf blight caused by *Phytophthora colocasiae*. IJAMBR 6:115–126

Nuemsi PPK, Tonfack LB, Taboula JM, Mir BA, Mbanga MRB, Ntsefong GN, Temegne CN, Youmbi E (2018) Cultivation systems using vegetation cover improves sustainable production and nutritional quality of new rice for Africa in the tropics. Rice Sci 25(5):286–292. https://doi.org/10.1016/j.rsci.2018.08.003

Nwaga D, Fankem H, Essono OG, Ngo Nkot L, Randrianangaly JS (2007) *Pseudomonas* and symbiotic micro-organisms as biocontrol agents against fungal disease caused by *Pythium aphanidermatum*. Afr J Biotechnol 6(3):190–197

Oehl F, Sieverding E, Mader P, Dubois D, Ineichen K, Boller T, Wiemken A (2004) Impact of long-term conventional and organic farming on the diversity of arbuscular mycorrhizal fungi. Oecologia 138:574–583. https://doi.org/10.1007/s00442-003-1458-2

Ortas I (2012) The effect of mycorrhizal fungal inoculation on plant yield, nutrient uptake and inoculation effectiveness under long-term field conditions. Field Crop Res 125:35–48. https://doi.org/10.1016/j.fcr.2011.08.005

Parker R (2009) Plant and soil Science: fundamentals and applications. Delmar Cengage Learning, Clifton Park, p 480

Páyan-Renteria R, Garibay-Chávez G, Rangel-Ascencio R, Preciado-Martínez V, Muñoz-Islas L, Beltrán-Miranda C, Mena-Munguía S, Jave-Suárez L, Feria-Velasco A, De Celis R (2012) Effect of chronic pesticide exposure in farm workers of a Mexico community. Arch Environ Occup Health 67:22–30. https://doi.org/10.1080/19338244.2011.564230

Peano C, Massaglia S, Ghisalberti C, Sottile F (2020) Pathways for the amplification of agroecology in African sustainable urban agriculture. Sustain For 12(7):2718. https://doi.org/10.3390/su12072718

Peigné J, Cannavaciuolo M, Gautronneau Y, Aveline A, Giteau JL, Cluzeau D (2009) Earthworm populations under different tillage systems in organic farming. Soil Tillage Res 104:207–214. https://doi.org/10.1016/j.still.2009.02.011

Pellegrino E, Bedini S, Avio L, Bonari E, Giovannetti M (2011) Field inoculation effectiveness of native and exotic arbuscular mycorrhizal fungi in a Mediterranean agricultural soil. Soil Biol Biochem 43:367–376. https://doi.org/10.1016/j.soilbio.2010.11.002

Pfiffner L, Luka H (2003) Effects of low-input farming systems on carabids and epigeal spiders—a paired farm approach. Basic Appl Ecol 4:117–127. https://doi.org/10.1078/1439-1791-00121

Pimentel D, Burgess M (2013) Soil erosion threatens food production. Agriculture 3:443–463. https://doi.org/10.3390/agriculture3030443

Pimentel D, Hepperly P, Hanson J, Douds D, Seidel R (2005) Environmental, energetic, and economic comparisons of organic and conventional farming systems. BioScience 55:573–582

Poyyamoli G (2017) Agroecology, ecosystems and sustainability in the tropics. Studera Press, Delhi, p 424

Prince SD, Podwojewski P (2020) Desertification: inappropriate images lead to inappropriate actions. Land Degrad Dev 31:677–682. https://doi.org/10.1002/ldr.3436

Programme Africompost (2017) Compostage des déchets au Cameroun. Fondation Goodplanet/Gevalor. https://www.goodplanet.org/fr/projet/valorisation-ordures-menageres/compostage-dechets-cameroun/

Quinton JN, Catt JA (2004) The effects of minimal tillage and contour cultivation on surface runoff, soil loss and crop yield in the long-term Woburn Erosion Reference Experiment on sandy soil at Woburn, England. Soil Use Manag 20:343–349. https://doi.org/10.1111/j.1475-2743.2004.tb00379.x

Rahman GM, Rahman MM, Alam MS, Kamal MZ, Mashuk HA, Datta R, Meena RS (2020) Biochar and organic amendments for sustainable soil carbon and soil health. In: Datta R, Meena RS, Pathan SI, Ceccherini MT (eds) Carbon and nitrogen cycling in soil. Springer, Singapore, pp 45–85. https://doi.org/10.1007/978-981-13-7264-3

Raj A, Jhariya MK, Yadav DK, Banerjee A, Meena RS (2019a) Agroforestry: a holistic approach for agricultural sustainability. In: Jhariya MK, Banerjee A, Meena RS, Yadav DK (eds) Sustainable agriculture, forest and environmental management. Springer, Singapore, p 606

Raj A, Jhariya MK, Banerjee A, Yadav DK, Meena RS (2019b) Soil for sustainable environment and ecosystems management. In: Jhariya MK, Banerjee A, Meena RS, Yadav DK (eds) Sustainable agriculture, forest and environmental management. Springer, Singapore, p 606

Raj A, Jhariya MK, Yadav DK, Banerjee A (2020) Climate change and agroforestry systems: adaptation and mitigation strategies. Apple Academic Press Inc., New York

Raj A, Jhariya MK, Khan N, Banerjee A, Meena RS (2021) Ecological intensification for sustainable development. In: Ecological intensification of natural resources for sustainable agriculture. Springer, Singapore

Rembialkowska E (2007) Quality of plant products from organic agriculture. J Sci Food Agric 87:2757–2762. https://doi.org/10.1002/jsfa.3000

Rhoton FE, Shipitalo MJ, Lindbo DL (2002) Runoff and soil loss from midwestern and southeastern US silt loam soils as affected by tillage practice and soil organic matter content. Soil Tillage Res 66:1–11. https://doi.org/10.1016/S0167-1987(02)00005-3

Richardson AE, Barea JM, McNeill AM, Prigent-Combaret C (2009) Acquisition of phosphorus and nitrogen in the rhizosphere and plant growth promotion by microorganisms. Plant Soil 321:305–339. https://doi.org/10.1007/s11104-009-9895-2

Rigueiro-Rodrígues A, McAdam J, Mosquera-Losada MR (2009) Agroforestry in Europe. Current status and future prospects. Springer, Dordrecht, 450 p. https://doi.org/10.1007/978-1-4020-8272-6

Robledo M, García-Tomsig NI, Jiménez-Zurdo JI (2020) Riboregulation in nitrogen-fixing endosymbiotic bacteria. Microorganisms 8:384. https://doi.org/10.3390/microorganisms8030384

Rossi R (2020) Desertification and agriculture. EPRS (European Parliamentary Research Service), p 12

Saj S, Torquebiau E, Hainzelin E, Pages J, Maraux F (2017) The way forward: an agroecological perspective for climate smart agriculture. Agric Ecosyst Environ 250:20–24. https://doi.org/10.1016/j.agee.2017.09.003

Säle V, Aguilera P, Laczko E, Mäder P, Berner A, Zihlmann U, van der Heijden MGA, Oehl F (2015) Impact of conservation tillage and organic farming on the diversity of arbuscular mycorrhizal fungi. Soil Biol Biochem 84:38–52. https://doi.org/10.1016/j.soilbio.2015.02.005

Sánchez-Bayo F (2011) Impacts of agricultural pesticides on terrestrial ecosystems. Ecological impacts of toxic chemicals. Bentham Science Publishers Ltd, Sharjah, pp 63–87. https://doi.org/10.2174/978160805121211110101

Savci S (2012) An agricultural pollutant: chemical fertilizer. IJESD 3(1):77–80

Schaller N (2013) Agroecology: different definitions, common principles. Center for studies and strategic foresight. French Ministry of Agriculture, Food and Forestry Ed., Paris, France. http://agreste.agriculture.gouv.fr/IMG/pdf/analyse591307anglais.pdf

Schauberger B, Ben-Ari T, Makowski D, Kato T, Kato H, Ciais P (2018) Yield trends, variability and stagnation analysis of major crops in France over more than a century. Sci Rep 8:16865. https://doi.org/10.1038/s41598-018-35351-1

Schmidt GH, Streloke M (1994) Effect of *Acorus calamus* (L.) (Araceae) oil and its main compound β-asarone on *Prostephanus truncatus* (Horn) (Coleoptera: Bostrichidae). J Stored Prod Res 30 (3):227–235. https://doi.org/10.1016/0022-474X(94)90050-R

Schneider KD, Lynch DH, Dunfield K, Khosla K, Jansa J, Voroney RP (2015) Farm system management affects community structure of arbuscular mycorrhizal fungi. Appl Soil Ecol 96:192–200. https://doi.org/10.1016/j.apsoil.2015.07.015

Schubetzer C, Bastide G, Feix I (2007) Assessment of the environmental impact of non-inversion tillage methods in France. ARVALIS - Institut du végétal, AREAS, INRA de Dijon, AgroParisTech – INRA de ParisGrignon, 58 p

Sengupta A, Dick WA (2015) Bacterial community diversity in soil under two tillage practices as determined by pyrosequencing. Microb Ecol 70:853–859. https://doi.org/10.1007/s00248-015-0609-4

Shafie FA, Rennie D (2012) Consumer perceptions towards organic food. Procedia Soc Behav Sci 49:360–367. https://doi.org/10.1016/j.sbspro.2012.07.034

Shalaby EA (2011) Prospects of effective microorganisms technology in wastes treatment in Egypt. Asian Pac J Trop Biomed 1(3):243–248. https://doi.org/10.1016/S2221-1691(11)60035-X

Sharma D, Varma VS, Yadav KD, Kalamdhad AS (2017) Evolution of chemical and biological characterization during agitated pile composting of flower waste. Int J Recycl Org Waste Agric 6(1):89–98. https://doi.org/10.1007/s40093-017-0155-9

Sidhu S (2020) A child under 15 dies every five seconds around the world – UN report. UNICEF, New York

Silva R, Filgueiras L, Santos B, Coelho M, Silva M, Estrada-Bonilla G, Vidal M, Baldani JI, Meneses C (2020) *Gluconacetobacter diazotrophicus* changes the molecular mechanisms of root development in *Oryza sativa* L. growing under water stress. Int J Mol Sci 21:333. https://doi.org/10.3390/ijms21010333

Singh JS, Pandey VC, Singh DP (2011) Efficient soil microorganisms: a new dimension for sustainable agriculture and environmental development. Agric Ecosyst Environ 140:339–353. https://doi.org/10.1016/j.agee.2011.01.017

Soil Carbon Initiative (2011) Soil carbon summit outcomes document. Soil Carbon Summit, Sidney

Souza RC, Cantão ME, Vasconcelos ATR, Nogueira MA, Hungria M (2013) Soil metagenomics reveals differences under conventional and no-tillage with crop rotation or succession. Appl Soil Ecol 72:49–61. https://doi.org/10.1016/j.apsoil.2013.05.021

Stassart PM, Baret P, J-Cl G, Hance T, Mormont M, Reheul D, Stilmant D, Vanloqueren G, Visser M (2012) L'agroécologie : trajectoire et potentiel Pour une transition vers des systèmes alimentaires durables. In: Van Dam D, Nizet J, Streith M, Stassart PM (eds) Agroécologie entre pratiques et sciences sociales. Educagri Éditions, Dijon, pp 27–51

Steenwerth K, Belina KM (2008) Cover crops and cultivation: impacts on soil N dynamics and microbiological function in a Mediterranean vineyard agroecosystem. Appl Soil Ecol 40:370–380. https://doi.org/10.1016/j.apsoil.2008.06.004

Stehle S, Schulz R (2015) Agricultural insecticides threaten surface waters at the global scale. Proc Natl Acad Sci 112(18):5750–5755. https://doi.org/10.1073/pnas.1500232112

Stone AG, Scheuerell SJ, Darby HM (2004) Suppression of soilborne diseases in field agricultural systems: organic matter management, cover cropping and other cultural practices. In: Soil organic matter in sustainable agriculture. CRC Press, Boca Raton, pp 131–177

Sun J, Kang Y, Wan S, Hu W, Jiang S, Zhang T (2012) Soil salinity management with drip irrigation and its effects on soil hydraulic properties in north China coastal saline soils. Agric Water Manag 115:10–19. https://doi.org/10.1016/j.agwat.2012.08.006

Sustainable Lafayette (2013) SustainableLafayette.Org. 13 May 2013 Web. http://www.sustainablelafayette.org/?page_id=1015

SWCS (2000) Soil biology primer, 1st edn. Soil and Water Conservation Society, Ankeny

Taddese G (2001) Land degradation a challenge to Ethiopia. Environ Manag 27:815–824. https://doi.org/10.1007/s002670010190

Tao R, Liang Y, Wakelin SA, Chu G (2015) Supplementing chemical fertilizer with an organic component increases soil biological function and quality. Appl Soil Ecol 96:42–51. https://doi.org/10.1016/j.apsoil.2015.07.009

Tchameni SN, Sameza ML, O'donovan A, Fokom R, Ngonkeu MEL, Wakam NL, Etoa FX, Nwaga D (2017) Antagonism of *Trichoderma asperellum* against *Phytophthora megakarya* and its potential to promote cacao growth and induce biochemical defence. Mycology 8(2):84–92. https://doi.org/10.1080/21501203.2017.1300199

Tebrügge R, Düring A (1999) Reducing tillage intensity — a review of results from a long-term study in Germany. Soil Tillage Res 53(1):15–28. https://doi.org/10.1016/S0167-1987(99)00073-2

Tekeu H, Ngonkeu ELM, Tandzi LN, Djocgoue PF, Bell JM, Mafouasson HA, Boyomo O, Petmi CL, Fokom R (2015) Evaluation of maize (Zea mays L.) accessions using line x tester analysis for aluminum and manganese tolerance. Int J Biol Chem Sci 9(4):2161–2173. https://doi.org/10.4314/ijbcs.v9i4.36

Teklehaimanot Z, Mmolotsi RM (2007) Contribution of red alder to soil nitrogen input in a silvopastoral system. Biol Fertil Soils 43:843–848. https://doi.org/10.1007/s00374-006-0163-9

Temegne NC, Nkou Foh TD, Taffouo VD, Ntsomboh-Ntsefong G, Youmbi E (2017) Influence of mycorrhization and phosphate fertilizer on growth of Voandzou (*Vigna subterranea* (L.) Verdc.). Int J Biol Chem Sci 11(6):2587–2593. https://doi.org/10.4314/ijbcs.v11i6.3

Temegne NC, Nkou Foh TD, Taffouo VD, Wakem G-A, Youmbi E (2018) Effect of mycorrhization and soluble phosphate on growth and phosphorus supply of Voandzou (*Vigna subterranea* (L.) Verdc.). Legum Res 41(6):879–884

Temegne NC, Gouertoumbo WF, Nkou Foh TD, Wakem G-A, Ntsomboh-Ntsefong G, Agendia AP, Taffouo VD, Youmbi E (2019) Effect of single superphosphate and arbuscular mycorrhizal fungi on growth and Bambara groundnut (*Vigna subterranea* (L.) Verdc.) yield. J Appl Sci Technol 2019:4–13

Temgoua E, Ngnikam E, Dameni H, Kouedeu Kameni GS (2014) Valorization of household waste by composting processes in Dschang town, Cameroon. Tropicultura 32(1):28–36

Thevathasan NV, Gordon AM (2004) Ecology of tree intercropping systems in the north temperate region: Experiences from Southern Ontario, Canada. Agrofor Syst 61:257–268. https://doi.org/10.1023/B:AGFO.0000029003.00933.6d

Tittonell P, Scopel E, Andrieu N, Posthumus H, Mapfum P, Corbeels M, van Halsema GE, Lahmar R, Lugandu S, Rakotoarisoa J, Mtambanengwe F, Pound B, Chikowo R, Naudin K, Triomphe B, Mkomwa S (2012) Agroecology-based aggradation-conservation agriculture (ABACO): targeting innovations to combat soil degradation and food insecurity in semi-arid Africa. Field Crop Res 132:168–174. https://doi.org/10.1016/j.fcr.2011.12.011

Troeh FR, Thompson LM (2005) Soils and soil fertility, vol 489. Wiley-Blackwell, New York, p 498

Troost K, Loeschcke A, Hilgers F, Özgür AY, Weber TM, Santiago-Schübel B, Svensson V, Hage-Hülsmann J, Habash SS, Grundler FMW, Schleker ASS, Jaeger KE, Drepper T (2019) Engineered *Rhodobacter capsulatus* as a phototrophic platform organism for the synthesis of plant sesquiterpenoids. Front Microbiol 10:1998. https://doi.org/10.3389/fmicb.2019.01998

Tsoata E, Njock SR, Youmbi E, Nwaga D (2015) Alleviation of drought effect on two mycorrhizal leguminous plants (*Tephrosia vogelii* and *Vigna subterranea*) at an early growth phase. Int J Curr Res 7(9):20424–20438

Turrini A, Caruso G, Avio L, Gennai C, Palla M, Agnolucci M, Tomei PE, Giovannetti M, Gucci R (2017) Protective green cover enhances soil respiration and native mycorrhizal potential compared with soil tillage in a high-density olive orchard in a long term study. Appl Soil Ecol 116:70–78. https://doi.org/10.1016/j.apsoil.2017.04.001

Udawatta RP, Gantzer CJ, Anderson SH, Garrett HE (2008) Agroforestry and grass buffer effects on pore characteristics measured by high-resolution X-ray computed tomography. Soil Sci Soc Am J 72(2):295–304. https://doi.org/10.2136/sssaj2007.0057

UN (2012) High-level meeting on addressing desertification, land degradation and drought in the context of sustainable development and poverty eradication. UN General Assembly A/65/861

UN (2018) World population projected to reach 9.8 billion in 2050, and 11.2 billion in 2100. https://www.un.org/development/desa/en/news/population/world-population-prospects-2017.html

UNCCD (1994) Elaboration of an international convention to combat desertification in countries experiencing serious drought and/or desertification, particularly in Africa. General Assembly, United Nations, pp 1–58

UNCCD (2015) Desertification, land degradation & drought: some global facts and figures. United Nations Conventions to Combat Desertification. Available from http://www.unccd.int/Lists/SiteDocumentLibrary/WDCD/DLDD%20Facts.pdf

Van der Ploeg JD (2020) The political economy of agroecology. J Peasant Stud. https://doi.org/10.1080/03066150.2020.1725489

Van Straaten P (2006) Farming with rocks and minerals: challenges and opportunities. An Acad Bras Cienc 78(4):731–747

Vargas R, Pankova EI, Balyuk SA, Krasilnikov PV, Khasankhanova GM (eds) (2018) Handbook for saline soil management. FAO/Lomonosov Moscow State University, Rome, p 144

Vasu D, Tiwary P, Chandran P, Singh SK (2020) Soil quality for sustainable agriculture. In: Meena RS (ed) Nutrient dynamics for sustainable crop production. Springer, Singapore, pp 41–65. https://doi.org/10.1007/978-981-13-8660-2_2

Verbruggen E, Roling WFM, Gamper HA, Kowalchuk GA, Verhoef HA, van der Heijden MGA (2010) Positive effects of organic farming on below-ground mutualists: large-scale comparison of mycorrhizal fungal communities in agricultural soils. New Phytol 186:968–979. https://doi.org/10.1111/j.1469-8137.2010.03230.x

Veresoglou SD, Halley JM, Rillig MC (2015) Extinction risk of soil biota. Nat Commun 6(1):1–10. https://doi.org/10.1038/ncomms9862

Vermande P, Bayard R, Gourdon R, Naquin P, Zahrani F, Ngnikam E, Ndame Ngangue M, Mouafo V, Tchuente R, Kayap S, Matejka G, Chambon S, Bras A, Lacour J, Emmanuel E, Morvan B, Montange D, Yonkeu S, Yiougo L, Rasoanandrasana E, Rajaomanana H, Brakez M, Baba G, Koledzi E (2012) Compostage des déchets ménagers dans les pays en développement : modalités de mise en place et de suivi d'installations décentralisées pérennes. CEFREPADE 57 p

Vlček V, Pohanka M (2020) Glomalin – an interesting protein part of the soil organic matter. Soil Water Res 15:67–74

Wang S, Li Z, Fan G (2012) Soil quality and microbes in organic and conventional farming systems. Afr J Microbiol Res 6(24):5077–5085. https://doi.org/10.5897/AJMR11.1541

West PC, Gerber JS, Engstrom PM, Mueller ND, Brauman KA, Carlson KM, Cassidy ES, Johnston M, MacDonald GK, Ray DK, Siebert S (2014) Leverage points for improving global food security and the environment. Science 345(6194):325–328. https://doi.org/10.1126/science.1246067

Wezel A, Bellon S, Doré T, Francis C, Vallod D, David C (2009) Agroecology as a science, a movement and a practice. A review. Agron Sustain Dev 29:503–515. https://doi.org/10.1051/agro/2009004

Wezel A, Casagrande M, Celette F, Vian J-F, Ferrer A, Peigné J (2014) Agroecological practices for sustainable agriculture. A review. Agron Sustain Dev 34:1–20. https://doi.org/10.1007/s13593-013-0180-7

Wong WK, Stein B, Torill E, Ingjerd H, Hege H (2011) Climate change effects on spatiotemporal patterns of hydroclimatological summer droughts in Norway. J Hydrometeorol 12 (6):1205–1220. https://doi.org/10.1175/2011JHM1357.1

Wright AL, Hons FM (2005a) Soil carbon and nitrogen storage in aggregates from different tillage and crop regimes. Soil Sci Soc Am J 69:141–147. https://doi.org/10.1002/9780470777923.ch16

Wright AL, Hons FM (2005b) Carbon and nitrogen sequestration and soil aggregation under sorghum cropping sequences. Biol Fertil Soils 41:95–100. https://doi.org/10.1007/s00374-004-0819-2

Wu J, Huang D, Teng W, Sardo V (2010) Grass hedges to reduce overland flow and soil erosion. Agron Sustain Dev 30:481–485. https://doi.org/10.1051/agro/2009037

Xie H, Zhang Y, Wu Z, Lv T (2020) A bibliometric analysis on land degradation: current status, development, and future directions. Landscape 9(28):1–37. https://doi.org/10.3390/land9010028

Yang C, Ellouze W, Navarro-Borrell A, Taheri AE, Klabi R, Dai M, Kabir Z, Hamel C (2014) Management of the arbuscular mycorrhizal symbiosis in sustainable crop production. In: Solaiman ZM, Abbott LK, Varma A (eds) Mycorrhizal fungi: use in sustainable agriculture and land restoration. Springer, Berlin, pp 89–18. https://doi.org/10.1007/978-3-662-45370-3_7

Yang G, Guo J, Li Z, Yuan Y, Zhou S (2016) Desulfotomaculum ferrireducens sp. nov., a moderately thermophilic sulfate-reducing and dissimilatory Fe(III)-reducing bacterium isolated from compost. Int J Syst Evol Microbiol 66:8

Youmatter (2020) Agro-ecology definition: History and examples. https://youmatter.world/en/definition/definitions-agro-ecology/

Young R, Orsini S, Fitzpatrick I (2015) Soil degradation: a major threat to humanity. Sustainable Food Trust, Bristol, pp 1–13

Yu X, Guo L, Jiang G, Song Y, Muminov MA (2018) Advances of organic products over conventional productions with respect to nutritional quality and food security. Acta Ecol Sin 38(1):53–60. https://doi.org/10.1016/j.chnaes.2018.01.009

Zeng L, Wooton E, Stahl DA, Walian PJ (2017) Identification and characterization of the major porin of *Desulfovibrio vulgaris* Hildenborough. J Bacteriol 199(23):e00286. https://doi.org/10.1128/JB.00286-17

Agroecology Towards Environmental Sustainability

10

Shailesh Kumar Yadav, Arnab Banerjee, Manoj Kumar Jhariya, Abhishek Raj, Nahid Khan, Ram Swaroop Meena, and Sandeep Kumar

Abstract

Agroecology refers to the process based on ecological principles to be applied in the agroecosystem for effective soil management and gain sustainable yield. The scientific application leads to a diversified agroecosystem which addresses the issue of environmental sustainability. It also focuses on various ecosystem services in the form of maintaining soil fertility, proper biogeochemical cycling, and proper nutrient exchange between crop and soil ecosystem. The process includes an integrated approach with diversified crops and animal husbandry practices all at a time. Thus, it would be successful to address the issue of food security, crisis, and help to build up climate-resilient agroecosystem. Agroecosystem is also helpful in terms of maintaining a daily livelihood, production of fuel, fodder, food for rural stakeholders, and socioeconomic well-being of

S. K. Yadav · A. Banerjee (✉)
Department of Environmental Science, University Teaching Department, Sant Gahira Guru Vishwavidyalaya, Ambikapur, India

M. K. Jhariya · N. Khan
Department of Farm Forestry, University Teaching Department, Sant Gahira Guru Vishwavidyalaya, Ambikapur, India

A. Raj
Faculty of Agriculture and Veterinary Sciences, Mewar University, Gangrar, Chittorgarh, Rafrjasthan, India

R. S. Meena
Department of Agronomy, Institute of Agricultural Sciences, Banaras Hindu University, Varanasi, Uttar Pradesh, India
e-mail: meenars@bhu.ac.in

S. Kumar
ICAR-Indian Institute of Pulses Research, Regional Station, Phanda, Bhopal, India
e-mail: sandeep.kumar5@icar.gov.in

M. K. Jhariya et al. (eds.), *Sustainable Intensification for Agroecosystem Services and Management*, https://doi.org/10.1007/978-981-16-3207-5_10

people across the globe. Thus, agroecological addresses the sustainable agriculture practice on a large scale to promote eco-friendly, self-sustaining agriculture practices. The aim of this article is to reflect an all-round aspect of agroecology along with its roadmap towards environmental sustainability.

Keywords

Agroforestry · Agroecology · Agriculture · Environmental sustainability

Abbreviations

FAO	Food and Agricultural Organization
INM	Integrated nutrient management
IPCC	Intergovernmental Panel on Climate Change
IPM	Integrated pest management
IWMP	Integrated watershed management program
N	Nitrogen
NDCs	Nationally determined contributions
OECD	Organization for Economic Co-operation and Development
P	Phosphate
SD	Sustainable development
USA	United States of America

10.1 Introduction

Agriculture accomplishes a milestone for the production of agriculture goods; besides productivity, it is concerned with the global environmental, ecological, as well as socioeconomic status of the world (Kremen and Miles 2012; Banerjee et al. 2020, 2021). Various components of the air, water, and soil are getting degraded, such as a gradual buildup of greenhouse gases concentration, soil degradation, water consumption, and the water quality (Zhang et al. 2007; Perfecto and Vandermeer 2010; Tilman et al. 2011; Hayes et al. 2011; Tscharntke et al. 2012). Modern agriculture techniques strengthen the life forms of the planet through more production. However, the principle ecological function affected by modern agriculture techniques includes an improper regulation of the climate, proper integration of the biosphere, alterations in landforms, and nutrient enrichment of water body through N (nitrogen) and P (phosphate) chemical fertilizer and causes eutrophication (Liebman and Schulte 2015; Steffen et al. 2015; West et al. 2014; Meena et al., 2018).

Agroecology refers to the integration of the agriculture and ecological system, suitable for addressing the issue of global food security and sustainability (Gliessman 2014). It ensures the uninterrupted use of ecological services and agricultural goods without altering the ecological balance (Jhariya et al. 2021). It provides a gateway for the integration of ecology, socioeconomics, and sociocultural aspects of agroecosystem (Raj et al. 2020, 2021). Subsequently, it establishes balance within the agroecosystem, stimulates increment of agricultural goods production as well as community development, and accomplished the sustainability (Velasco et al. 2019; Jhariya et al. 2019a, b; Kumar et al. 2020).

Anthropogenically promoted climate change imposes stress on human society (Khan et al. 2020a, b). On the basis of the report of the Intergovernmental Panel on Climate Change (IPCC) (Masson-Delmotte et al. 2018) in order to reduce global warming up to 1 °C, carbon neutrality should be maintained till 2060–2070. At present strongly accepted fact is that proper and sustainable land use plays a major role to minimize carbon emission, reduces the ejection of carbon, as well as introduces good substitution of fossil carbon through production of the biomass.

10.2 Core Principles of Agroecology

An eco-friendly farming system such as agroecology is characterized as homeostatic, self-sustaining, and biodiversity-rich system. Therefore, the industrial agriculture model is completely different from the principle of agroecology (Fig. 10.1). It has great resilience and self-regulation adaptability due to its heterogynous nature. If the system is diversified with flora and fauna, they offer greater resistance of the system towards climatic perturbances. Lesser utilization of synthetic inputs is another

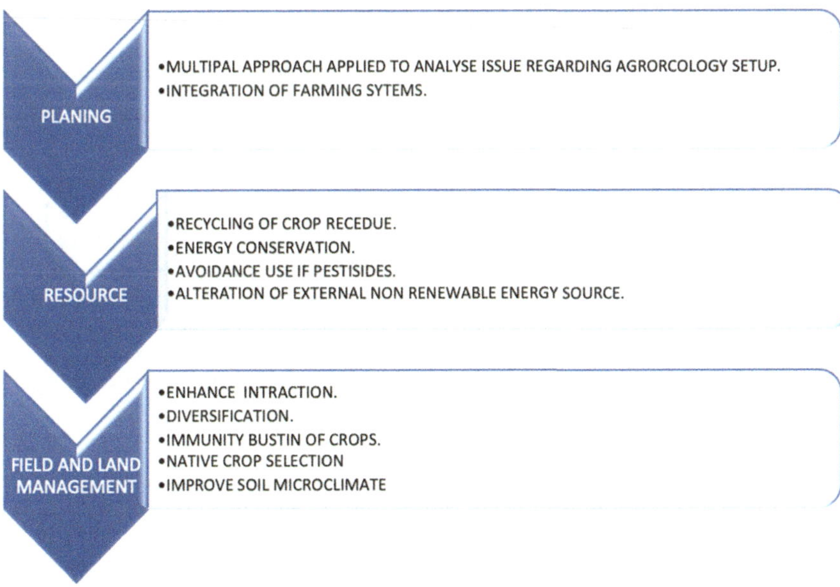

Fig. 10.1 Core principle of agroecology

characteristic of the system that helps in the recycling of nutrients. It represents a multifunction farming system capable of mitigating various climatic impacts such as climate change.

10.3 Components of Agroecology

To mitigate climate change problem, IPCC released a report with due consideration of the sustainable development (SD) goals (Roy et al. 2018; Meena and Lal 2018). The first approach was to mitigate the climate change phenomena through an agrological approach which totally depends upon traditional faming (organic farming) as well as nutrient-rich healthy food to the community of the country, along with reduction of the emission of greenhouse gases (Fig. 10.2). The second approach is

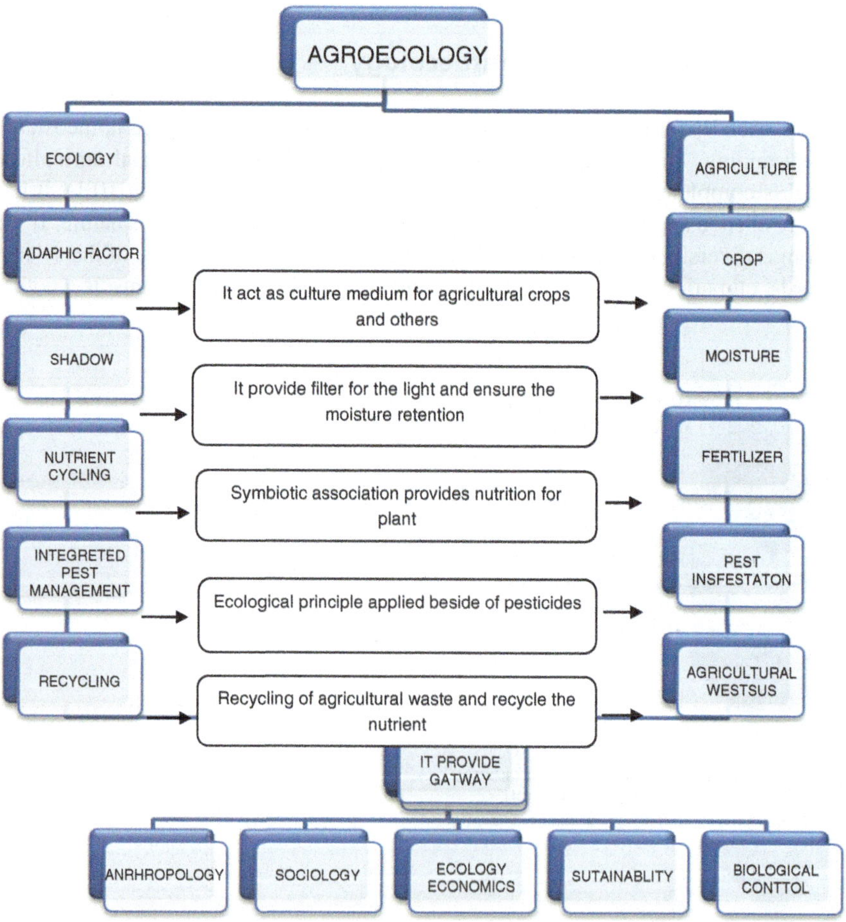

Fig. 10.2 Agroecological components and their interaction

combined with multiple objectives as well as equivalent benefits which include the regulation of biodiversity, regulate loss of nature and natural resources, and improve community health status. Change in climatic condition is a burning issue in the present existing scenario. This issue has been neglected since long time back. The third approach includes carbon sequestration/carbon removal and bioenergy/biomaterial production aiming towards maximizing the carbon balance (Roy et al. 2018).

10.4 Development of Agroecology

10.4.1 Agroecology between the 1930s and 1960s

The origin of the agroecology concept has taken place somewhere since the 1930s to 1960s. It is studied under biological science that consists of the combined study of zoology, plant physiology (crop physiology), as well as the economic value of the crop (agronomy) (Fig. 10.3). It determines the association and interaction between the flora and fauna with their environmental elements as well as agricultural crops. The agroecological concept explores the ecological importance. It also determines the harmful effects of the green revolution on ecological entity along with agro-farming systems.

10.4.2 Agroecology Between the 1970s and 1980s

At the beginning of the 1980s as well as at the end of the 1970s, agroecology has strongly influenced the social values of agronomic literature. It is also influenced by the organic farming system which ascertains the protection and management of natural resources.

10.4.3 Scientific Framework of Agroecology During 1990

During the 1990s the scientific framework of agroecology originated that relies on the integrated approach of agriculture, symbiotically associated ecological component comprising of humans, as well as existing climatic factors. Overall agroecology is a very complex system under which ecological phenomenon moves forward along with intensive agriculture activities including human activities. Agroecology denotes the dynamic interaction with their component. The principle of agroecology provides the path to achieve ecological sustainability along with optimum ecological services through various farming activities. Agroecology is the scientific framework that evaluates the interaction of integrated components such as human beings and related environmental components to achieve environmental sustainability.

In the last few years, agroecology as a scientific technology has developed more apparatus and multifunctional tool in the field of agro-economy, with the main objective to achieve a sustainable environment. Its scope is to include all the

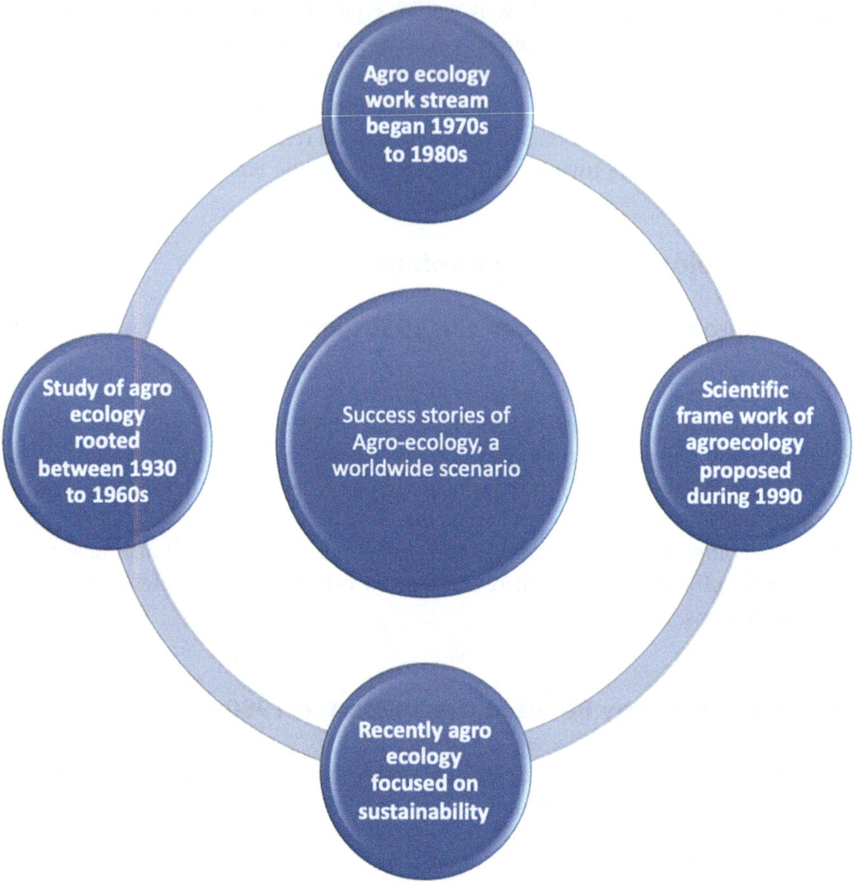

Fig. 10.3 Historical development of the agroecology

components of the food system and therefore has an interdisciplinary approach. Agroecology is a transformation in the farming community by the inclusion of ethical and religious issues in relation to the food production and consumption. On the other hand, the modern food system and farming system lead more towards unsustainability. In fact agroecological farming plays an important role to develop the food system directly through the increase and flexibility in agricultural production. Further, it encourages the reduction and recycling the food waste and re-localization of the farming system.

Agroecological processes and practices include the excess of farming, testing, and mixing of different varieties, based on the rational use of chemical fertilizer, and also develop a mathematical model to analyze and assess the irrigation system, crop protection production, and promote the farmer training. All these all practices mainly focus on decreasing the environmental impact of crop production, reducing the complex management and interaction of biological effect and production through agroecological principle (Silici 2014).

10.5 Attributes of Agroecology

There are ten agroecological elements proposed by FAO (Food and Agricultural Organization) during their regional meeting of 2015–2016. Proposed 10 agroecological elements are interlinked together and perform function freely.

10.5.1 Diversity

Agroecology has a diversified component. Apiculture, sericulture, agroforestry, horticulture, pisciculture, livestock farming, and silvipastoral production system are included under the agroecological system (Fig. 10.4). It leads to the upliftment of the socioeconomic status of the rural people and environmental services (Painkra et al. 2016).

10.5.2 Co-Creation and Sharing of Knowledge

Agroecology is concerned with the traditional knowledge for eco-friendly production of food to address the issue of food scarcity. It is the integration of the cultural, native, social, as well as ancient knowledge in a holistic manner to maintain the ecosystem balance and sustainability.

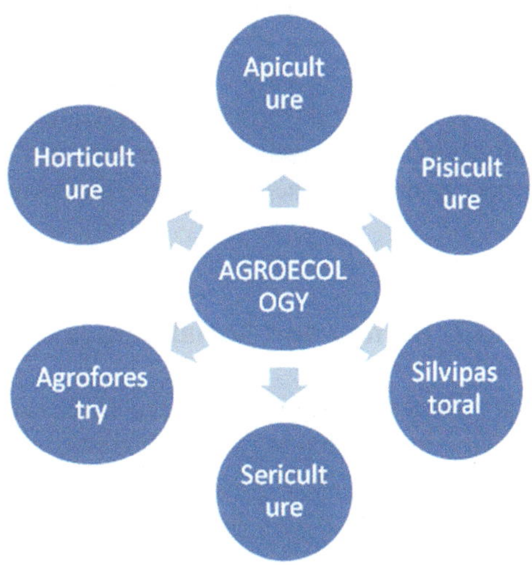

Fig. 10.4 Diversity components of agroecology

10.5.3 Synergies

For the betterment of the synergies, agroecology integrates the multiple and preferable driver elements of the farm as well as various agricultural landscapes to move towards eco-friendly production.

10.5.4 Efficiency

Agroecology ensures the maximum utilization of the resource while it reduces the exploitation of the natural resource including nonrenewable natural resource, hydrological, edaphic, as well as other resources. Agroecology promotes efficient utilization of external resources along with minimizing the cost as well as negative impact on the environment.

10.5.5 Recycling

Agroecology is the replica of the natural ecosystem that regulates the nutrient cycling, recycling of the biomass, and hydrological cycle on the arable land.

10.5.6 Resilience

Resilience refers to the restoration capability of the ecosystem which includes both socioeconomic and agroecological system. Agroecological production system has a sufficient capability to restore from natural disasters like water scarcity (drought), cloud burst condition (flood), as well as pest attack. Moreover, the enrichment of biological diversity leads to the development of the capability to resist the vulnerability of the particular species. Minimization depends upon the reduction of the input coming from the outside. It is indirectly increasing the self-sustaining potential of the producer as well as helps to reduce the economic crisis.

10.5.7 Human and Social Values

Agroecology significantly influences the socioeconomic as well as moral values like novelty, impartiality as well as honesty, and integration of the social and cultural values which leads to sustainable income. It contributes to the satisfaction of the desire and demand of the agro-produce and proper allocation of the food to consumers.

10.5.8 Culture and Food Traditions

Agroecology supports the maintenance of a balanced ecosystem which emphasizes upon traditional diet. Agroecology supports native food tradition which addresses the issue of food security.

10.5.9 Responsible Governance

Corruption less, ecoliterate, and responsible government is required which is helpful to producers to alter the production system. Unbiased utilization of arable land and natural resource merely not only contribute to social justice, and it is necessary to introduce an encouragement for long-duration contribution in sustainability.

10.5.10 Circular and Solidarity Economy

Agroecology is also reconnecting the producer and consumers through the round and form of unity and solidarity economy which gives primary priority to the local market as well as support to the agricultural development. Agroecology leads to the production of agricultural products and promotes a healthier diet for consumers.

10.6 Concept of Sustainability and Sustainable Development

SD refers to the "harvesting of natural resources to fulfill the present needs without affecting the future production and needs for the upcoming generation" (WCED 1987). Sustainability includes economic condition, social status, as well as existing environmental conditions. These types of integration deal with sustainable growth (Dyer et al. 2006; Kumar et al. 2021).

SD includes criteria, which includes environmental, social as well as economic dimension. The introduction of indicators helps to quantify various criteria related to various dimensions. The environmental dimension of agroecology includes agricultural practices towards an eco-friendly way. Agriculture practice should provide resource as well as promote proper functioning of the biogeochemical cycle. It includes disciplines which recommend organic farming without the application of chemical pesticides (Hayati et al. 2010). The economic dimension of agroecology correlates farm productivity with gross income from the crop and how much net profit is obtained through an emerged agroecological system (Hayati et al. 2010). It enhances the output by reducing dependency on external inputs. It enhances the farm income which supports to grow up living standard of farmers. Social dimension is another important dimension of agroecology. It leads to the conversion of industrial agriculture into traditional farming practices, native species conservation, and primitive knowledge-based healthy agriculture practice among the farming community. The development of agroecology leads to generate employment opportunities. It also secures the health safety issue of the engaged community (Mockshell and Kamanda 2018).

10.7 Integration Between Agroecology and Sustainable Development

Agroecology confers the platform for evaluation of the four types of agricultural dimensions like productivity, resilience, sustainability, as well as equity. The perspective of agroecology is to include the development or advanced use of innovative plant and soil science and development of high or cogent productivity with suitable development. The main aim of the agroecological food system is to provide rich biological components in the form of green and ecologically rich farming systems which lead to local and sensitive agricultural practices. Agroecology is a science that claims to the farming expert and practitioners, to operate the farming system considering the ecological and socioeconomic dimension towards organic farming movement. Therefore, sustainable agriculture comprises of two parameters like socioeconomic and agroecological aspect (Fig. 10.5). Therefore, the focus was given to the two major issues of sustainability which include the environmental integrity and economic resiliency (FAO 2013). The evaluation of the agroecological as well as socioeconomic sustainability in relation to the farming system can be done through the sustainability indicator process (Migliorini and Scaltriti 2012).

10.8 Perspectives of Agroecology

10.8.1 Social

As per agroecologist Steve Gliessman of the USA (United States of America), "agroecology includes the knowledge, education, application as well as alteration in the farming practices that lead to sustainable production of food in terms of socio-economic and ecological development. It is multidisciplinary in nature that includes various type values and knowledge that supports change in the food system. It is a cooperative effort which involves the participation of all the stockholders starting from producer and consumer" (Gliessman 2018).

Agroecology reflects social adaptation among the farming community to adopt eco-friendly farming practices. It is the science that not only detects the social change which optimizes food preservation for all community. The desired need of community belongs to the societal developmental that is not fulfilled by the modern agriculture and recent food heritage (USDA 2018). Agroecological is concerned about the inappropriate utilization of valuable and healthy food in urban and rural areas. Chemical-based agriculture and food supply series depend on various health issues. It exposes the farmer to extreme working conditions (Holt-Giménez 2017). Industrial agriculture leads to greater exposure of the community to pesticides. Pesticides come to contact with human beings through the food chain and enter

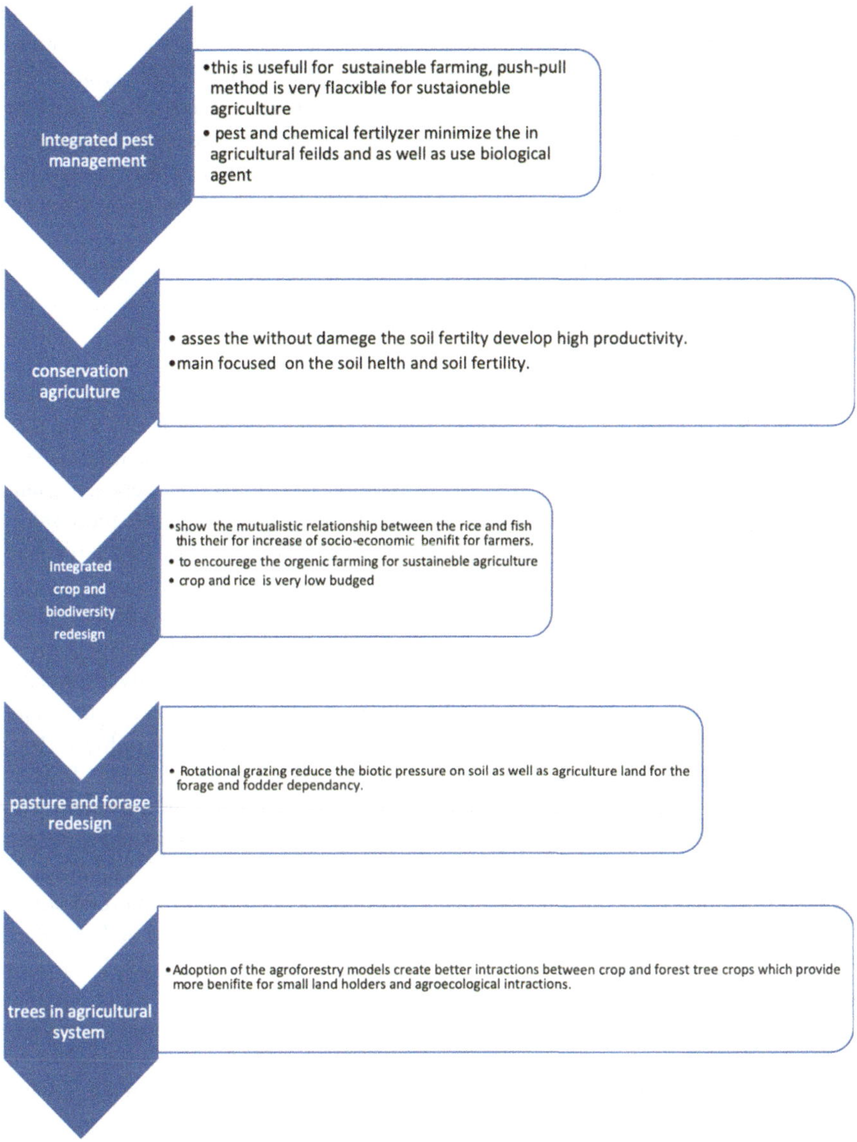

Fig. 10.5 Agroecological principles towards environmental sustainability

into the human body (Casey et al. 2015). To obtain maximum benefit agricultural industrialist pours the farmers in serious threats like unfertile land, reduction of the pollinators, and various health issues like cancer (Holt-Giménez 2017). Industrial agriculture has induced overproduction while agricultural goods, services, milk, as well as meat production and economy are attenuating since the last 15 years (USDA 2018). It has a direct impact on concerned agribusiness. Existing policies have one-way goal to promote export business. Export rely on minimum cost means one third (minimum) price of the agriculture good paid to farmers as well as continue gain over production from farmers. Agricultural-based multinational groups take benefit of trade agreements. It facilitates labor security at low cost, light working policies, as well as lesser environmental security. Policies put economic pressure on large- and small-scale farmers. Consequently, the midrib of the farm such as organic farm and small- and large-scale farm puss back towards the business trend (Howard 2009).

10.8.2 Ecological

The ecological aspect of agroecology deals with the integration and application of the ecological concept to achieve sustainable ecological services along with food security (Altieri 1995). It includes different management plans that improve the intensification of biodiversity as well as promote the co-action between plant, animal, and their nonliving entity. It privileges the soil fertility and food safety that sustains the livelihoods. A basic principle of agroecology involves the regenerative practices of a natural system.

The application of agroecological principle promotes the farmers to design mimic nature. It works towards the improvement of the edaphic factor, enhancement of the biogeochemical cycle, and restoration of degraded biological diversity, stimulates the strong interaction between living and nonliving organisms, and also regulates the optimization of the excess use of the natural resources and water (Altieri 1995; Kumar et al. 2020a; Khan et al. 2021a, b). Extension activity in the field of agroecology provides a better productive and diversified agriculture system. This productive system has better productivity as compared to traditional and fertilizer-based agriculture system (Davis et al. 2012; Sheoran et al. 2021). Moreover, it reduces the emission of greenhouse gases. It provides protection to biodiversity as well as ecological services. This system depends on less pesticide and chemical fertilizer-based practice system. Traditional forming follows the principle of agroecology farming system (Ponisio et al. 2015).

The gap between the industrial agriculture and agroecology-based agriculture could be optimized through promoting government and other financial bodies to invest research and extension, along with encouragement for the production of the seed that pertains to a maximum productivity under organic farming atmosphere, as well as promoting transfer of technology which is suitable for farmers to maintain ecological condition along with microclimate of the farm (Carlisle and Miles 2016).

Modern agriculture depends on fertilizer as well as genetically modified seeds, which leads to biodiversity loss (Dudley et al. 2017) as well as the largest contribution to water pollution (National Oceanic and Atmospheric Administration 2017). The addition of nutrients to water body causes eutrophication and increases algal bloom production (Rucinski et al. 2014) along with red tide on the sea surface. Overuse of fertilizers alters the cation exchange capacity and other basic properties of the soil (Altieri et al. 2015). It is responsible for the massive deforestation of the tropical forest (Angelsen 2010). Consequently it leads to loss of biodiversity along with the native species as well as it produces the largest greenhouse gas production (Li 2011; Raj et al. 2018a, b). Industrial agriculture caused a negative impact on biodiversity and climate. It results in climate change and various types of pollution.

10.8.3 Technical

The technical aspect of the agroecology and technology used for resource conservation has been introduced:

1. Adoption of IPM (integrated pest management) as much as possible because it plays a significant role to maintain resilience and eco-friendly diversity of biota of the agroecosystem. The use of pesticides becomes mandatory only when other substitutions have failed to control the pest.
2. Adoption of INM (integrated nutrient management) to balance the utilization of the macronutrient as well as essential organic and inorganic matter along with minimization of the erosion of soil of farmland.
3. The use of zero tillage farming that reduces the moisture loss and the soil erosibility.
4. Extension of the agroforestry system that consists of the plantation of various native forest tree species at farmland. Agriculture and forestry are symbiotically associated. Along with agriculture, it pertains to natural resource conservation.
5. In integrated farming, aquaculture plays an important role in protein production in terms of fishes and economically important hydrophytes.
6. Adoption of the scientific farming system including livestock practices such as goatery, piggery, and cow rearing along with agriculture.

10.8.4 Historical

Since 1960 the term agroecosystem has been discussed among the scientific community, but agroecology dates back in the scientific literature review for the first time in the 1930s. The concept originated through the scientific study of biological interaction for only one crop and its further application in the different-different component of the plot or farm level to the whole agroecosystem and to the wider food system. The agroecological study deals with ecological and agronomic analysis along with an interdisciplinary approach including socioeconomic and political considerations (scope of agroecological fields) for sustainable yield and productivity. From 1980 onwards agroecology gives a conceptual framework for the better performance of agroecological-related practices and processes which include both developed and developing countries and especially includes the central and South African countries. All these practices inspired the number of agroecological movements and practices which come out and consolidate in 1990. The estimation of agroecology is a multifunctional and scientific discipline process, practice, and movement which are shown in Fig. 10.1 (Wezel et al. 2009).

10.9 Agroecology-Based Application of Agroforestry

Agroforestry is the farming system which exhibited the integral approach of tree or shrub under the agriculture farmland including various agriculture crop along with livestock that give benefit to the mankind through overall interaction (interaction between tree or shrub with crop as well as livestock). Fig. 10.6 gives a typical representation of the agroforestry system including a single crop line between to tree line.

Agroforestry is a practice which involves the deliberate integration of trees or shrubs in farming landscapes involving crops or livestock in order to obtain benefits from the interactions between trees and/or shrubs and the tree and crop or livestock component (Jhariya et al. 2015; Singh and Jhariya 2016; Meena et al. 2020). As per Zomer et al. (2014), land area under agroforestry comprises of cultivation area up to 10% under tree cover along with agricultural land with more than 40% area supporting maximum population under tropical and sub-tropical conditions where poor people resides. There is a worldwide increase in the agroforestry area across various continents across the globe. For the same period, there is a large increase in the number of people living in landscapes with greater than 10% tree cover, from 746 million to over 837 million.

Agroforestry plays an important role in maintaining livelihood for rural people. Apart from maintaining livelihood, it also provides various ecological services as well as useful facilities such as facilitating multifunctional income facility along with

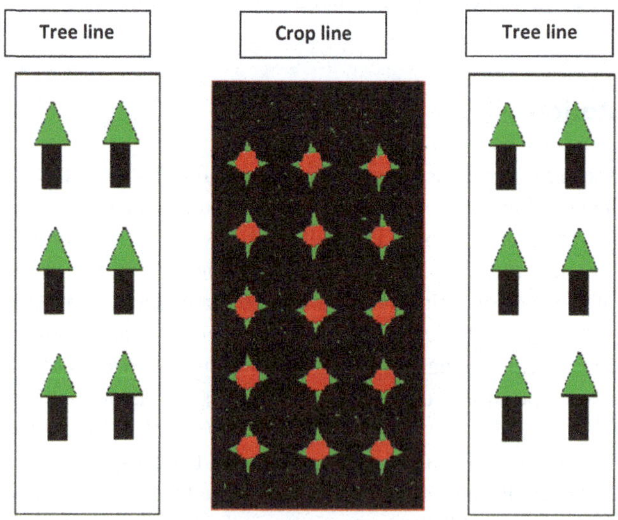

Fig. 10.6 Represent the agroforestry system

food security. Gradual increase in human population tremendous pressure occurs for need of the food and other natural resources. In order to reduce the pressure, it is necessary to replace agricultural farmland to an agroforestry farming system. It will also help to maintain the stability of the ecosystem of the entire area (Raj et al. 2019a, b).

10.10 Environmental Services of Agroecology

Agroforestry is the scientific and modern concept to ensure food security as well as to maintain ecosystem homeostasis. Agroforestry exhibits integrated approaches of both agricultural crop and forestry crops. The function of this system is to maintain the resilience of farm crop with an altered climatic condition. Agroforestry stabilizes the production of related entities by bearing various adverse conditions such as wetter as well as drier sessions. Forest tree species comprises of taproot system that ensures the availability of moisture along with sufficient nutrient at drought condition; along with taproot system, it includes adventitious root system that holds the soil; increases soil porosity, water holding capacity, and rate of infiltration; restores and improves the soil through phytoremediation and N fixation (Jhariya et al. 2018a, b). Higher plant has higher evaporation rate that evaporate the excess water within the soil and ensure the aeration of the soil. Proper aeration supports the growth of understory at farm system. Agroforestry systems are most adaptive for the dry area, and because of that, it enhances the microclimate of the particular area (Fig. 10.7).

10.10.1 Agroforestry as Livelihood Options

Dispersal of farm-based technology has become easy due to agroforestry approaches. It acts as a key element for transfer of the technology. Demographic growth stimulates demand of food, fodder, and fuel. It exerts pressure on farming land that causes degradation of soil. Balance of the soil quality with increasing demand of human beings may be facilitated through agroforestry in the current farming system. Agroforestry leads to sustainable production approaches. Demographic growth affects the landholding capacity of farmers at any country. The current scenario of farmers is that most of them are small landholder. Agroforestry diversifies the crop system which is helpful to sufficient income generation along with food for their family (Chand et al. 2011). A study of the integrated farm system based on holistic approaches applied on farm ensures more than six times more profit (Gangwar and Ravishankar 2015; Meena et al. 2020a). Moreover, the net return of the small as well as marginal landholder improves per day basis. The profit of the farmers may be increased by more than 69% due to the introduction of viable agroforestry technology. It provides facilities to farmers such as a diversified cropping system, minimizes fertilizer application, and introduces IPM (integrated pest management system), while another important aspect is that it provides green fodder

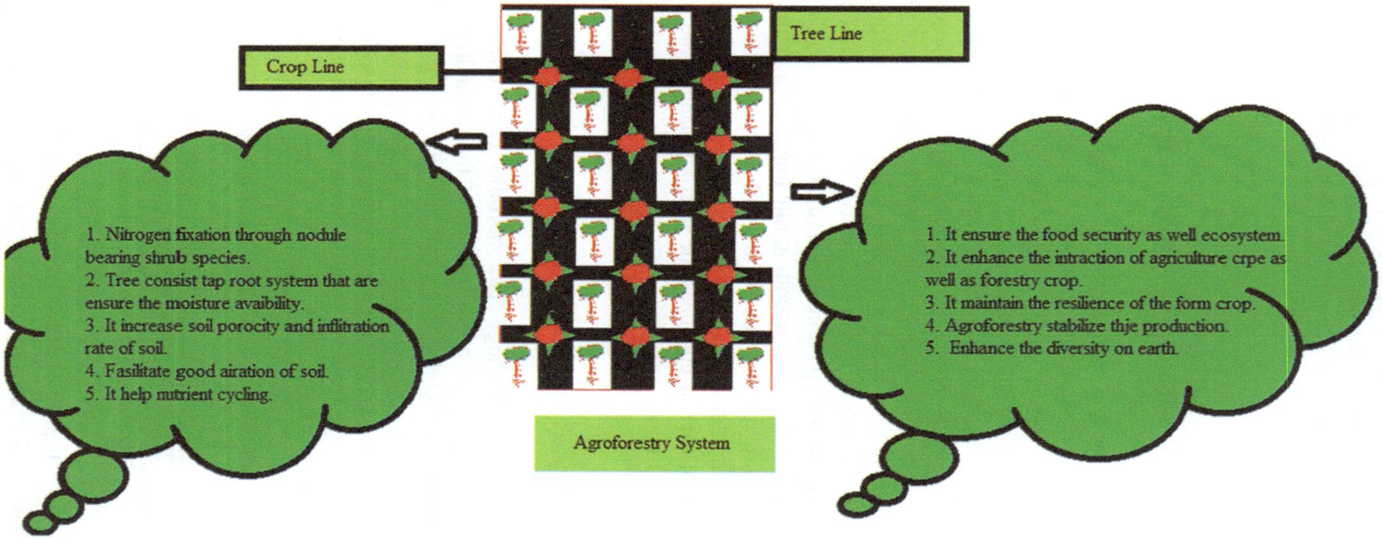

Fig. 10.7 Represented agroforestry and their environmental services

for animals throughout the year. Agroforestry is also a good representative of the agroecological farming system that ensures the availability of five F (5F) (fiber, fodder, fuel, food, as well as fertility). Its system directly stimulates sustainable production.

10.10.2 Agroforestry and Sustainable Development

Agroforestry is a SD tool under the agroecological farming system. It is the modern concepts of agriculture that exhibit integrated approaches of forestry crop along with agricultural short life cycled plant, while also including livestock farming. All components interact with each other and provide a diversified cropping system in a particular entity. It regulates smoothly microclimate of a specific area under changing climatic conditions. The agroforestry system enhances ecological services along with the socioeconomic condition of farmers. It is also a significant tool for nutrient recycling. Its main role in the ecosystem is to provide resilience to species under climatic perturbations (Mosquera-Losada et al. 2012; Nair and Garrity 2012; Catacutan et al. 2018).

Ecofriendly as well as a low-cost intervention on existing agricultural system stimulates the agroforestry as a SD option with decreasing the impact of climatic status because of industrial agriculture and intensive deforestation (FAO 2013, 2017; Meena et al. 2020b; Dinesh et al. 2017). Agroforestry contributes an important role to obtain significant sustainable goals through improving food production, an efficient way to address the issue of food security, increasing the income of the small household and reducing poverty. Further, it also contributes clean energy development. Agroforestry has an ideal approach to mitigate climate change (Waldron et al. 2017). Besides it increases tree cover along with food production. Trees provide sink of carbon because they have the ability to sequestrate atmospheric carbon dioxide. So it is an important approach to obtain NDC (nationally determined contribution) criterion (Rosenstock et al. 2018).

10.10.3 Agroforestry Policy in India

The Government of India promoted the national agroforestry rules in 2014, which has assigned the mandate of agroforestry to the ministry of agriculture and farmers, which also provided important instruction to the forestry department.

The policy states that smallholders consider that a farming effort is not like a place as a briefcase of farming activities neither as one type of cropping system which is fixed. The policy also defines agroforestry in terms of identifying the environmental profit as well as its further upgradation. The operational guideline nevertheless whose main focus is on the technical approach and implementation of such as repeated nurseries, planting matter require suitable species, capacity development and sharing cost as well as involvement of farmers.

10.11 Agroecological Practices Towards Sustainability

Addressing agroecology towards sustainability involves the introduction of efficient farm practices that leads to higher output at lower inputs. It also promotes SD approaches through the emergence of multiple farming practices based on agroecological principles. Agroecology farming practice is considered as a management tool through which farmers establish interaction with microclimate, social environment, and economic values along with their local tradition as well as culture (Silici 2014).

Some examples of farming techniques refer to biodynamic agriculture, and these parameters mostly apply to all the agroecological principles. Within the system of rice intensification and organic and conservation agriculture, some principles of agroecology were applied. A high external input is usually associated with agricultural activities for commercial crop production. On the other hand, agroecological practice for food crop production involves farmers using different-different techniques in different plots. They used to apply individual agroecological practices as per the local setup of the area. The practices include increasing soil fertility with decreasing the use of inorganic fertilizer. Agroecological practices work like a toolkit in which farmers can choose and apply different-different agricultural practices. It also depends as well as relates to the environment, socioeconomic situation, and cultural preference (Silici 2014).

10.11.1 Conservation Tillage

Zero tillage rather than moderate tillage leads to increased sustainability of soil. It helps to check soil erosion along with nutrient loss. It enhances water infiltration rate, gaseous exchange capacity, and pedogenesis process.

10.11.2 Mixing Crops in a Single Plot

Monoculture cropping system is prominent, but its ecological importance is negligible as the diversity is less so it is more vulnerable having adverse impacts. While two or more alternative crop introduced in a single culture medium, improves the nutrient utility, more resistancy to pest and diseases. Crop failure risk is less. So it spontaneously increases crop productivity along with sustainability.

10.11.3 Crop Rotation and Fallowing

This technology is used to sustain nutrients into the soil. It is necessary to adopt crop cycling with suitable alternative crop species. Alteration of crops disturbs the life cycle of pests and unwanted herbs.

10.11.4 Cover Crops and Mulching

Cover crops and mulching help to encourage sustainable practices to minimize soil degradation by soil erosion and enrichment of soil micro- and macroelements as well as ensure the viability of eco-friendly insects through adoption of IPM.

10.11.5 Crop-Livestock Integration

This is an integrated system which involves the integration of crop and livestock practices. This is a very much sustainable practice of farming of fish (aquaculture) which plays an important role for aquatic life including plants, algae, and other organisms as it helps to increase biomass.

10.11.6 Integrated Nutrient Management

It includes proper management of the nutrient cycle which is the most beneficial practice for sustainable development. It involves common uses such as fertilizer, organic compost, and cultivation of leguminous crop to promote biological N fixation with lesser use of nitrogenous fertilizer.

10.11.7 Biological Management

The pest, disease, and unusable plant through vermin management reduces the allelopathic effect of plant which affects the other plants' growth.

10.11.8 Efficient Water Harvesting System

It is very necessary to control the water consumption mostly in the dryland areas where we apply some specific techniques such as small irrigation, drip irrigation, sprinkler irrigation, rainwater harvesting, etc.

10.11.9 Manipulation of Vegetation Structure and Plant Associations

It helps to improve water use efficiency which is very important from a water sustainability perspective as well as to promote biodiversity.

10.11.10 Agroforestry

It is more useful for sustainable practice because agroforestry is a multifunction agricultural practice and it provides various multifunctional plants which help and improve the soil fertility through N fixation as well as through the macro- and micronutrients.

10.11.11 Use of Local Resources and Renewable Energy Sources, Composting, and Waste Recycling

Composting and recycling of waste help to conserve natural resource and are therefore very much effective practice for future sustainability (Silici 2014). Inclusion of new agricultural technology results in an increment of agricultural productivity as well as emergence of new era of agriculture in the Organization for Economic Co-operation and Development (OECD) countries. Sustainable farming approaches define agriculture in a way that provides continuous productivity of food, fodder, and other useful elements without eliminating their productivity. Agricultural goods produced through the agroecological principle-based approach become economically and socially viable for the next generation. Along with productivity it also contributes to environmental security for future prospective. Introduction of sustainable approaches depends upon technology at different scales of production system.

Traditional farming system and nonfertilizer-based (organic farming) farming system have locally self-regulated. There is no any standard scale for sustainable farming as well as measurement of sustainability. Sustainability vigorously depends upon farmer and technology adoption by him. Sustainability exhibited the inclusion of both technological-based as well as traditional farming system. Interaction of both farming systems displays environmental benefits, while fulfilling the demand of fodder food fibers.

10.11.12 Family Farming

This family farming program provides a small platform for the rural families (small landholders) which helps to move towards sustainable development. Family farming in terms of technological development practices helps to promote agroforestry. Proper management of watershed development and beekeeping are some of the useful technologies for the family farming in the sector of agro-industry organization to supply the agroecosystem produce to the big market (international market).

10.11.13 Organic Farming

Technologies for sustainable agriculture cover the whole spectrum of farming systems. Encouragement to the agricultural goods producer to convert their production system into an organic production system is very much required for the modern world in order to achieve sustainability. This program totally ignores the use of chemical fertilizers. It stimulates the farmer to adopt a new agriculture farming system. It is cost-efficient and promotes sustainability.

10.11.14 Agricultural Diversification

The most important agenda of the program is income generation through crop diversification. It facilitates many alternative resources for income generation. It reduces the loss through monoculture farming system. It supports integrated farming system such as dairy, apiculture, and horticulture at the same time in a particular area.

10.11.15 Medicinal Plants

This approach encourages the production of herbal medicine. It also enlightens the knowledge of the meditational plant diversity of the particular agroecological area. It is done through creating MOU (memorandum of understanding) under the various collaborating institute. Their contribution is an ultimate source of alternative income development.

10.11.16 Rainwater Collecting and Utilization for Agricultural and Forestry Purposes

Rainwater harvesting is a simple process to collect water which is stored in different-different ways such as roof of building, water storage tank, percolation tank, and recharge well, a pond, and all these storage techniques are very useful for agriculture and forestry area. Water from these sources is also used mainly for irrigation and drinking purpose. Horticulture and agroforestry are the best storage practices and rejuvenate pond for a long time use of water for multipurpose, and in this we can solve the water problem to some extent. All rainwater storage systems including farm pond and hydro-geomorphic structures and all other rainwater storage systems are necessary for scientific assessment of their potential. In farm pond we can store up to fifteen thousand liters of water through the percolation process. It is very acceptable and capable technology for storage of water. There is a scheme IWMP (integrated watershed management program) which is implemented by the department of land resource under the ministry of rural development under which water is stored and farmers can use it for irrigating their field and domestic purpose at a very low cost and also use in the lean season. The main objective of this program is to make the village people aware about rainwater harvesting, groundwater recharge, reclamation, improvement of ecology, etc. so that people can gate relief for water problems.

Today rainwater harvesting is necessary to all human beings because we have lost almost half of the water on earth. Sustainability of agroecology is maintained through avoidance of pesticides as well as chemical fertilizers. Weed is a major growth suppressor in agroecology system. It is necessary to introduce natural techniques, in which weed is converted as organic manure without the use of weedicides. In this sequence introduction of allelopathy plants which are allelopathic in nature can check the growth of other associated weeds. It produces allelopathic substances through aerial part or during decomposition. Sweet clover has an allelopathy effect along with green manure property. It contains coumarin which contributes growth suppressor of other unwanted weeds. The application of this technique assures the remediation of weed problem and enriches the soil with manure. An example of this type of system is the cultivation of fall rye allelopathy cover crop that is cultivated along with legume crop which helps to reduce weed infestation.

10.11.17 Use of Mixed-Crop Green Manures

In this model farmers allow the cultivation of cereal crop along with leguminous green manure. Cereal crop is fast-growing species, whereas the other crop is slower than cereal. Leguminous green manure crop facilitates protection against weed competition. It symbiotically remains associated with *Rhizobia* N-fixing bacteria. It provides adequate N to cereal plant. Growing cereal plant has the potential to utilize light, nutrient as well as water beside the unwanted weeds.

10.11.18 Manage the Green Manure with Care

Healthy and vigorously grown green manure has more potential to retard the growth of unwanted crops. Vigorous growth as well as healthy crop can be obtained through utilization of dominant seed. Care should be taken for paddy or other cereal crops. It must be part of crop rotation at a frequent time interval.

10.11.19 Wetland Agriculture

Wetlands are defined as areas including swampy, marshy, waterlogged area, and frequently flooded area of land and exist either naturally as well as artificially. Wetland comes under the area where a tide does not reach above 6 m. Water present in wetland areas may be freshwater as well as saltwater and brackish water (Ramsar Convention Secretariat 2013). India belongs to Asian continent having dominant complex topographic and varying climatic conditions which support diversified small and large wetland (Prasad et al. 2002). Emergence of the wetland is evaluated through an aerial survey of wetland which includes $<1\%$ as well as $>5\%$ of total geographical area. The concerned area occupies one fifth area of the existing biodiversity (Space Applications Centre 2011). Wetland extends throughout the Himalayan to Deccan plateau region.

Rwanda comes under the continent of Africa. Wetland comes under fabulous pressure due to anthropogenic and extension of agricultural activity. Population blast increases the need of food and promotes the extension of productive land. Successive increment of the human population of rural circumference like <121 persons exists per km^2 since the 1960s; during the 1990s these data denotes <262 persons per km^2, and in the twenty-first century, present data indicate <380 person per km^2 (National Statistics Institute of Rwanda 2007). Enumeration reflected that $>90\%$ of the people adopted the agriculture farming profession for their livelihood in Rwanda. Agriculture landholding decreased since the last decades before it is more than three hectares per farmer family. The current scenario of landholding has less than one hectare per household (Verdoodt and van Ranst 2006). Increased biotic potential exerts tremendous pressure on arable land as well as water resources. It decreases the productivity consequently and increases food insecurity (Bidogeza et al. 2009; Dixon and Wood 2003). To obtain maximum productivity, farmers use chemical fertilizer which increases soil acidity and alkalinity, soil erosion, and dammed water holding capacity. Sustainability of arable land decreased, and probability of food insecurity occurs (Ansoms and McKay 2010). Decreasing land productivity stimulated the Rwanda formers to expand their agricultural land for efficient productivity, while they initiate farming activities into sensitive wetland areas. The area occupied by the wetland in Rwanda has $>270,000$ hectares throughout of this 48,561 hectares (approximately 53%) area altered as arable land since 2009 (REMA 2009). In the present scenario, wetland plays a major role in farmer income as well as food security (Kanyarukiga and Ngarambe 1998).

Wetland was recognized as a significant natural recourse for the magnification of agriculture activity by the Rwanda government. It plays a significant role in policy framing to achieve the food security goal as well as increase the per capita income of farmers. Wetland management is an important tool for poverty reduction. The management of the wetland also enhances the GDP (gross domestic product) growth rate such as more than ($>0.5\%$). These were the reasons for the adoption of wetland management practices by the Rwanda government (Kanyarukiga and Ngarambe 1998). Wetland is rich in nutrients and organic matter that support diversified life forms. It is a biodiversity-rich area. It has its own endemic species that contribute a specific ecological service in the ecosystem. Alteration of endemic species causes ecosystem degradation and biodiversity loss. Wetland is also recognized as a habitat for migratory birds, and it exists as an important ecosystem which has a world-level recommendation through Ramsar Convention (REMA 2009).

The hydrological significance of wetlands includes their water quality status as health of the wetland totally depends on water quality of that small entity. Irregularity of drainage results into drought and soil erosion as well as changes in the microclimate of downstream wetland areas. Wetlands near the natural reserve or protracted area provides freshwater to the fauna of a particular reserve area. Along with water resource, it provides a habitat facility for the wild animals. Therefore, wetlands act as a buffer zone between fauna and surrounded natural vegetation.

10.11.20 Intensification of Agricultural Use of Wetlands

In various fields application of fertilizer was recommended to obtaining higher yield (Sanginga and Woomer 2009), while the objective of productivity at different localities, farmyards, as well as farmers is heterogeneous. Farming practices need to catalyze for better exchange of organic and inorganic elements by balanced and the efficient use of resources with special emphasis on soil quality differentiation under different farming systems (Okalebo et al. 2003; Vanlauwe et al. 2006). Diverse applications of organic as well as inorganic nutrients influence the higher fertility of small farmland by creating a nutrient gradient. To obtain maximum income from the field, small farmers harvested the crops along with crop residue which limits natural input in soil due to sellout into the market. Therefore, outflux of soil input produces the fertility gradient (Dugan 1990).

The knowledge of the allocation of nutrients among the different fertilizer gradients plays a functional role to enhance the productivity of agricultural crop. Sampling plots of the same crop exhibited different fertilizer responses at the same time due to different fertilizer gradients (Fofana et al. 2004). The assessment done in South Africa reveals that the average production of the grain is <1 ton per hectare on outfield while >1 tons per hectare production on infield. This leads to the recovery of the N on the outfield $<30\%$ and $> 30\%$ on the infield. Infield soil shows a dynamic soil fertility compared to that of the outfield. The presence of organic carbon intensifies the fertilizer utilization property of soil (Fofana et al. 2004). An agroecological approach such as the use of crop cycle leads to enrichment of the N into soil. For example, before taking maize crop, beans need to be cultivated. The extension of the technology and knowledge enhance the productivity and profitability of the small landholder farmers.

10.12 Conclusions

In the present times, the unprecedented growth of the human population has jeopardized the agroecosystem through intensive agriculture practices. The negative consequences of industrial agriculture production have forced the scientific community across the globe to think about low-input agriculture practices and technologies. The main objective was to develop a production unit which is eco-friendly and cost-effective. Various problems such as proper recycling of agro-wastes, altered land use, and monoculture practices have aggravated the problem of creating an unsustainable ecosystem. As a consequence of that, the soil is losing its fertility, and crop diversification is being reduced through cultivation of hybrid seeds. In this perspective agroecological principle performs the task of more production and more economic gain for the farming community keeping in mind the environmental sustainability. Innovative approaches in terms of integrated farming system in the form of mixed cropping, proper crop rotation, maintaining crop cycle, conservation agriculture, zero tillage farming, and precision agriculture do world good to maintain the environmental sustainability of the agroecosystem. Proper recycling of agro-waste could be achieved through composting. Thus, agroecological principles have immense potentiality to undermine the effects of industrial agriculture practices and lead to environmental sustainability of the agroecosystem.

10.13 Future Research and Development in Agroecology Towards Sustainability

Proper evaluation of the issues and problems related to food production systems is required along with addressing the issue of food security and crisis. This would help us formulate strategies for moving towards sustainable agriculture in the form of application of agroecological principles. Sustainability from an agroecological point of view demands a paradigm shift from industrial agriculture to traditional agriculture practices considering the environment as important. This would also help to address the mega events of climate change, global warming, etc. In present times application of technology has contributed significantly to environmental pollution and natural resource depletion. In due course of time, it would lead to social well-being as well as improves the economy of the concerned area. The major aspect of agroecology from future perspectives is proper training of the farming community to adopt the agroecological practice. It highlights the importance of traditional knowledge regarding the farming practice which helps to achieve or move towards a sustainable agriculture practice (Tripathi et al. 2015).

From future perspectives agroecological principles should have some basic properties which include the involvement of the farming community in this agricultural reform and pave the path for their own development, blending of the traditional agriculture practices with scientific knowledge and applications; techniques should be eco-friendly along with the development of sustainable development unit, a major emphasis on maintaining daily livelihood in comparison to economic profit. Further all the techniques and methods applied should be economically viable as well as improve the resource use efficiency of the locally available resources. This would reduce the farmer's dependency on the government and other people. The multidimensional approach of agroecology would help to address the issue of environmental sustainability (Tripathi et al. 2015). For future sustainability of the agroecosystem, farmers need to have access to the land to which they can manage the farming practices in an eco-friendly way. The agroecological principle should lead to equity in resource use and make good governance in terms of the market economy.

References

Altieri MA (1995) Agroecology: the science of sustainable development. Westview Press, Boulder, CO

Altieri MA, Nicholls CI, Henao A, Lana MA (2015) Agroecology and the design of climate change-resilient farming systems. Agron Sustain Dev 35(3):869–890

Angelsen A (2010) Policies for reduced deforestation and their impact on agricultural production. Proc Nat Acad Sci 107(46):19639–19644

Ansoms A, McKay A (2010) A quantitative analysis of poverty and livelihood profiles: the case of rural Rwanda. Food Policy 35(6):584–598

Banerjee A, Jhariya MK, Yadav DK, Raj A (2020) Environmental and sustainable development through forestry and other resources. Apple Academic Press Inc., CRC Press- a Tayler and Francis Group, Boca Raton, FL, p 400. https://doi.org/10.1201/9780429276026

Banerjee A, Meena RS, Jhariya MK, Yadav DK (2021) Agroecological footprints management for sustainable food system. Springer, Singapore, p 514. https://doi.org/10.1007/978-981-15-9496-0

Bidogeza J, Berentsen P, De Graaff J, Oude Lansink A (2009) A typology of farm households for the Umutara Province in Rwanda. Food Secur 1(3):321–335

Carlisle L, Miles A (2016) Closing the knowledge gap: how the USDA could tap the potential of biologically diversified farming systems. J Agric Food Syst Commun Dev 3(4):219–225

Casey JA, Kim BF, Larsen J, Price LB, Nachman KE (2015) Industrial food animal production and community health. Curr Environ Health Rep 2(3):259–271

Catacutan DC, Finlayson R, Gassner A, Perdana A, Lusiana B, Leimona B, Simelton E, Oborn I, Galudra G, Roshetko JM, Vaast P, Mulia R, Lasco R, Dewi S (2018) ASEAN guidelines for agroforestry development. ASEAN Senior Officials on Forestry 2018. www.worldagroforestry.org/region/sea/publications/detail?pubID=4392

Chand R, Lakshmi Prasanna PA, Singh A (2011) Farm size and productivity: understanding the strength of smallholders and improving their livelihoods. Econ Polit Wkly 46:26–27

Davis AS, Hill JD, Chase CA, Johann AM, Liebman M (2012) Increasing cropping system diversity balances productivity, profitability and environmental health. PLoS One 7(10):e47149

Dinesh D, Campbell B, Bonilla-Findji O, Richards M (2017) 10 best bet innovations for adaptation in agriculture: a supplement to the UNFCCC NAP technical guidelines. CGIAR Research Program on Climate Change, Agriculture and Food Security, Wageningen, The Netherlands

Dixon AB, Wood AP (2003) Wetland cultivation and hydrological management in eastern Africa: matching community and hydrological needs through sustainable wetland use. Natural Res Forum 27(2):117–129

Dudley N, Attwood SJ, Goulson D, Jarvis D, Bharucha ZP, Pretty J (2017) How should conservationists respond to pesticides as a driver of biodiversity loss in agroecosystems? Biol Conserv 209:449–453

Dugan PJ (1990) Wetland conservation: a review of current issues and action. IUCN, Gland, Switzerland

Dyer GHG, McKay MDH, Mira M (2006) From clean development to strategic sustainable development: a strategic approach to the clean development mechanism. School of Engineering, Blekinge Institute of Technology, Karlskrona, Sweden, pp 1–97

FAO (2013) In: Buttoud G, Ajayi O, Detlefsen G, Place F, Torquebiau E (eds) Advancing agroforestry on the policy agenda: a guide for decision-makers. World Agroforestry Centre, Nairobi, Kenya

FAO (2017) Module B5: production and resources. In: Corsi S, Bigi A, Borelli S, Conigliaro M, Dubois O, Luis Fernandez J, Halwart M (eds) Climate smart agriculture sourcebook. FAO, Rome

Fofana B, Breman H, Carsky RJ, Van Reuler H, Tamelokpo AF, Gnakpenou KD (2004) Using mucuna and P fertilizer to increase maize grain yield and N fertilizer use efficiency in the coastal savanna of Togo. Nutr Cycl Agroecosyst 68(3):213–222

Gangwar B, Ravishanker N (2015) Innovative approach in farming system research to adopt crop diversification. In: Dhyani SK et al (eds) Agroforestry-present status and way forward. Biotech Books, New Delhi, India, pp 487–502

Gliessman SR (2014) Agroecology: the ecology of sustainable food systems, 3rd edn. CRC/Taylor & Francis Group, Boca Raton, FL

Gliessman SR (2018) Defining agroecology. Agroecol Sustain Food Syst 42:6

Hayati D, Ranjbar Z, Karami E (2010) Measuring agricultural sustainability. In: Lichtfouse E (ed) Biodiversity, biofuels, agroforestry and conservation agriculture, sustainable agriculture reviews, vol 5. Springer, Cham. https://doi.org/10.1007/978-90-481-9513-8_2

Hayes TB, Anderson LL, Beasley VR, de Solla SR, Iguchi T, Ingraham H, Willingham E (2011) Demasculinization and feminization of male gonads by atrazine: consistent effects across vertebrate classes. J Steroid Biochem Mol Biol 27(1):64–73

Holt-Giménez E (2017) A Foodie's guide to capitalism: understanding the political economy of what we eat. NYU Press, New York

Howard PH (2009) Consolidation in the north American organic food processing sector, 1997 to 2007. Int J Sociol Agric Food 16(1):13–30

Jhariya MK, Banerjee A, Meena RS, Yadav DK (2019a) Sustainable agriculture, forest and environmental management. Springer, Singapore, p 606. https://doi.org/10.1007/978-981-13-6830-1

Jhariya MK, Banerjee A, Yadav DK, Raj A (2018b) Leguminous trees an innovative tool for soil sustainability. In: Meena RS, Das A, Yadav GS, Lal R (eds) Legumes for soil health and sustainable management. Springer, Cham, pp 315–345. https://doi.org/10.1007/978-981-13-0253-4_10

Jhariya MK, Bargali SS, Raj A (2015) Possibilities and perspectives of agroforestry in Chhattisgarh. In: Zlatic M (ed) Precious forests-precious earth. InTech, Croatia, Europe, pp 237–257. https://doi.org/10.5772/60841

Jhariya MK, Meena RS, Banerjee A (2021) Ecological intensification of natural resources towards sustainable productive system. In: Ecological intensification of natural resources for sustainable agriculture. Springer, Singapore. https://doi.org/10.1007/978-981-33-4203-3_1

Jhariya MK, Yadav DK, Banerjee A (2018a) Plant mediated transformation and habitat restoration: phytoremediation an eco-friendly approach. In: Gautam A, Pathak C (eds) Metallic contamination and its toxicity. Daya Publishing House, New Delhi, India, pp 231–247

Jhariya MK, Yadav DK, Banerjee A (2019b) Agroforestry and climate change: issues and challenges. Apple Academic Press Inc., CRC Press- a Tayler and Francis Group, Boulder, CO, p 335. https://doi.org/10.1201/9780429057274

Kanyarukiga SG, Ngarambe V (1998) Rwanda country paper: wetland characterization and classification for sustainable agricultural development. Paper presented at the proceedings of the sub-regional consultation, Harare, Zimbabwe

Khan N, Jhariya MK, Raj A, Banerjee A, Meena RS (2021a) Soil carbon stock and sequestration: implications for climate change adaptation and mitigation. In: Jhariya MK, Meena RS, Banerjee A (eds) Ecological intensification of natural resources for sustainable agriculture. Springer, Singapore. https://doi.org/10.1007/978-981-33-4203-3_13

Khan N, Jhariya MK, Raj A, Banerjee A, Meena RS (2021b) Eco-designing for sustainability. In: Jhariya MK, Meena RS, Banerjee A (eds) Ecological intensification of natural resources for sustainable agriculture. Springer, Singapore. https://doi.org/10.1007/978-981-33-4203-3_16

Khan N, Jhariya MK, Yadav DK, Banerjee A (2020a) Herbaceous dynamics and CO_2 mitigation in an urban setup- a case study from Chhattisgarh, India. Environ Sci Pollut Res 27(3):2881–2897. https://doi.org/10.1007/s11356-019-07182-8

Khan N, Jhariya MK, Yadav DK, Banerjee A (2020b) Structure, diversity and ecological function of shrub species in an urban setup of Sarguja, Chhattisgarh, India. Environ Sci Pollut Res 27 (5):5418–5432. https://doi.org/10.1007/s11356-019-07172-w

Kremen C, Miles A (2012) Ecosystem services in biologically diversified versus conventional farming systems: benefits, externalities, and trade-offs. Ecol Soc 17(4):40

Kumar S, Meena RS, Datta R, Verma SK, Yadav GS, Pradhan GS, Molaei A, Rahman GKMM, Mashuk HA (2020a) Legumes for carbon and nitrogen cycling: an organic approach. In: Carbon and nitrogen cycling in soil. Springer, Cham. https://doi.org/10.1007/978-981-13-7264-3_10. 337-375

Kumar S, Meena RS, Jhariya MK (2020) Resources use efficiency in agriculture. Springer, Singapore, p 760. https://doi.org/10.1007/978-981-15-6953-1

Kumar S, Meena RS, Singh RK, Munir TM, Datta R, Danish S, Singh GSY, Kumar S (2021) Soil microbial and nutrient dynamics under different sowings environment of Indian mustard (*Brassica juncea* L.) in rice based cropping system. Sci Rep 11:5289. https://doi.org/10.1038/s41598-021-84742-4

Li TM (2011) Centering labor in the land grab debate. J Peasant Studies 38(2):281–298

Liebman M, Schulte LA (2015) Enhancing agroecosystem performance and resilience through increased diversification of landscapes and cropping systems. Elementa 3(1):000041

Masson-Delmotte V, Zhai P, Pörtner HO, Roberts D, Skea J, Shukla PR, Pirani A, Moufouma-Okia W, Péan C, Pidcock R, Connors S, JBR M, Chen Y, Zhou X, Gomis MI, Lonnoy E, Maycock T, Tignor M, Waterfield T (2018) Global warming of 1.5°C. World Meteorological Organization, Geneva

Meena RS, Kumar S, Datta R, Lal R, Vijaykumar V, Brtnicky M, Sharma MP, Yadav GS, Jhariya MK, Jangir CK, Pathan SI, Dokulilova T, Pecina V, Marfo TD (2020b) Impact of agrochemicals on soil microbiota and management: a review. Land 9(2):34. https://doi.org/10.3390/land9020034

Meena RS, Kumar V, Yadav GS, Mitran T (2018) Response and interaction of *Bradyrhizobiumjaponicum* and Arbuscularmycorrhizal fungi in thesoybeanrhizosphere: a review. Plant Growth Regul 84:207–223. https://doi.org/10.1007/s10725-017-0334-8

Meena RS, Lal R (2018) Legumes for soil health and sustainable management. Springer, Singapore, p 541. https://doi.org/10.1007/978-981-13-0253-4_10

Meena RS, Lal R, Yadav GS (2020) Long term impacts of topsoil depthand amendments on soil physical and hydrological properties of an Alfisol in Central Ohio, USA. Geoderma 363:1141164. https://doi.org/10.1016/j.geoderma.2019.114164

Meena RS, Lal R, Yadav GS (2020a) Long-term impact of topsoil depth and amendments on carbon and nitrogen budgets in the surface layer of an Alfisol in Central Ohio. Catena 194:104752. https://doi.org/10.1016/j.catena.2020.104752

Migliorini P, Scaltriti B (2012) Evaluation of sustainability of the farms in the Agricultural Park of South Milan and their production chain. New Medit 11:53–56

Mockshell J, Kamanda J (2018) Beyond the agroecological and sustainable agricultural intensification debate: is blended sustainability the way forward? Int J Agric Sustain 16:127–149. https://doi.org/10.1080/14735903.2018.1448047

Mosquera-Losada MR, Moreno G, Pardini A, McAdam JH, Papanastasis V, Burgess PJ, Lamersdorf N, Castro M, Liagre F, Rigueiro-Rodríguez A (2012) Past, present and future of agroforestry systems in Europe. In: Nair PKR, Garrity D (eds) Agroforestry: the future of global land use. Springer, Dordrecht, Germany, pp 285–312

Nair PKR, Garrity D (2012) Agroforestry: the future of global land use. Springer, Dordrecht, Germany

National Oceanic and Atmospheric Administration (2017) Gulf of Mexico "dead zone" is the largest ever measured. https://www.noaa.gov/media-release/gulf-of-mexico-dead-zone-is-largest-evermeasured

National Statistics Institute of Rwanda (2007) Enquête intermédiaire démographique et de santé. Rapport préliminaire: NISR, Kigali, Rwanda

Okalebo JR, Palm CA, Lekasi JK, Nandwa SM, Otieno CO, Waigwa M, Ndungu KW (2003) Use of organic and inorganic resources to increase maize yields in some Kenyan infertile soils: a five year experience. In: Bationo A (ed) Managing nutrient cycles to sustain soil fertility in sub-Saharan Africa. Academy Science Publishers, Nairobi, Kenya, pp 359–372

Painkra GP, Bhagat PK, Jhariya MK, Yadav DK (2016) Beekeeping for poverty alleviation and livelihood security in Chhattisgarh, India. In: Narain S, Rawat SK (eds) Innovative technology for sustainable agriculture development. Biotech Books, New Delhi, India, pp 429–453

Perfecto I, Vandermeer J (2010) The agroecological matrix as alternative to the land-sparing/agriculture intensification model. Proc Natl Acad Sci U S A 107:5786–5791

Ponisio LC, M'Gonigle LK, Mace KC, Palomino J, de Valpine P, Kremen C (2015) Diversification practices reduce organic to conventional yield gap. Proc Royal Soc B 282(1799):20141396

Prasad SN, Ramachandra TV, Ahalya N, Sengupta T, Kumar A, Tiwari AK, Vijayan VS, Vijayan L (2002) Conservation of wetlands of India – a review. Trop Ecol 43(1):173–186

Raj A, Jhariya MK, Banerjee A, Yadav DK, Meena RS (2019b) Soil for sustainable environment and ecosystems management. In: Jhariya MK, Banerjee A, Meena RS, Yadav DK (eds) Sustainable agriculture, forest and environmental management. Springer, Singapore, pp 189–221

Raj A, Jhariya MK, Bargali SS (2018b) Climate smart agriculture and carbon sequestration. In: Pandey CB, Gaur MK, Goyal RK (eds) Climate change and agroforestry: adaptation mitigation and livelihood security. New India Publishing Agency (NIPA), New Delhi, India, pp 1–19

Raj A, Jhariya MK, Harne SS (2018a) Threats to biodiversity and conservation strategies. In: Sood KK, Mahajan V (eds) Forests, climate change and biodiversity. Kalyani Publisher, New Delhi, India, pp 304–320

Raj A, Jhariya MK, Khan N, Banerjee A, Meena RS (2021) Ecological intensification for sustainable development. In: Ecological intensification of natural resources for sustainable agriculture. Springer, Singapore. https://doi.org/10.1007/978-981-33-4203-3_5

Raj A, Jhariya MK, Yadav DK, Banerjee A (2020) Climate change and agroforestry systems: adaptation and mitigation strategies. Apple Academic Press Inc., CRC press- a Tayler and Francis Group, Boulder, CO, p 383. https://doi.org/10.1201/9780429286759

Raj A, Jhariya MK, Yadav DK, Banerjee A, Meena RS (2019a) Agroforestry: a holistic approach for agricultural sustainability. In: Jhariya MK, Banerjee A, Meena RS, Yadav DK (eds) Sustainable agriculture, forest and environmental management. Springer, Singapore, pp 101–131

Ramsar Convention Secretariat (2013) The Ramsar convention manual: a guide to the convention on wetlands (Ramsar. Iran, 1971), 6th edn. Ramsar Organisation, Ramsar, Iran

REMA (2009) Rwanda state of environment and outlook: our environment for economic development. REM Authority, Kigali, Rwanda, p 137

Rosenstock T, Wilkes A, Jallo C, Namoi N, Bulusu M, Suber M, Bernard F, Mboi D (2018) Making trees count: measurement, reporting and verification of agroforestry under the UNFCCC. CCAFS Working Paper No. 240. CGIAR Research Program on Climate Change, Agriculture and Food Security, Wageningen, The Netherlands

Roy J, Tschakert P, Waisman H, Abdul Halim S, Antwi-Agyei P, Dasgupta P, Hayward B, Kanninen M, Liverman D, Okereke C, Pinho PF, Riahi K, Suarez Rodriguez AG (2018) Sustainable development, poverty eradication and reducing inequalities. In: Masson-Delmotte-V, Zhai P, Pörtner HO, Roberts D, Skea J, Shukla PR, Pirani A, Moufouma-Okia W, Péan C, Pidcock R, Connors S, Matthews JBR, Chen Y, Zhou X, Gomis MI, Lonnoy E, Maycock T, Tignor M, Waterfield T (eds) Global warming of 1.5°C. World Meteorological Organization, Geneva, pp 445–538

Rucinski DK, DePinto JV, Scavia D, Beletsky D (2014) Modeling Lake Erie's hypoxia response to nutrient loads and physical variability. J Great Lakes Res 40:151–161

Sanginga N, Woomer P (2009) Integrated soil fertility management in Africa: principles, practices and developmental process. TSBF-CIAT, Nairobi, Kenya

Sheoran S, Kuma S, Kumar P, Meena RS, Rakshit S (2021) Nitrogen fixation in maize: breeding opportunities. Theor Appl Genet 134(5):1263–1280. https://doi.org/10.1007/s00122-021-03791-5

Silici L (2014) Agroecology what it is and what it has to offer. IIED Issue Paper IIED, London

Singh NR, Jhariya MK (2016) Agroforestry and Agrihorticulture for higher income and resource conservation. In: Narain S, Rawat SK (eds) Innovative technology for sustainable agriculture development. Biotech Books, New Delhi, India, pp 125–145

Space Applications Centre (SAC) (2011) National Wetland Atlas. SAC, Indian Space Research Organisation, Ahmedabad. Study Group on Environment, n.d. Report of the study group on environment including tourism, heritage, pollution & disastermanagement. National Capital Region Planning Board, New Delhi, India

Steffen W, Richardson K, Rockstrom J, Cornell SE, Fetzer I, Bennett EM, Sorlin S (2015) Planetary boundaries: guiding human development on a changing planet. Science 2015:1259855

Tilman D, Balzer C, Hill J, Befort BL (2011) Global food demand and the sustainable intensification of agriculture. Proc Nat Acad Sci 108(50):20260–20264

Tripathi N, Singh RK, Singh RS (2015) Agroecology and sustainability of agriculture in India: an overview. EC Agric 21:241–248

Tscharntke T, Clough Y, Wanger TC, Jackson L, Motzke I, Perfecto I, Whitbread A (2012) Global food security, biodiversity conservation and the future of agricultural intensification. Biol Conserv 151(1):53–59

USDA, Economic Reearch Service (2018) Farming and farm income. https://www.ers.usda.gov/data-products/ag-and-food-statistics-charting-the-essentials/farming-and-farm-income/

Vanlauwe B, Tittonell P, Mukalama J (2006) Within-farm soil fertility gradients affect response of maize to fertiliser application in western Kenya. Nutr Cycl Agroecosyst 76(2):171–182

Velasco CMT, Acevedo RXL, Mátyás B (2019) Comparative study of the environmental impact of models of conventional agricultural and agro-ecological agriculture in the agricultural phase of tomato cultivation. F1000Research 7:666

Verdoodt A, Van Ranst E (2006) Environmental assessment tools for multi-scale land resources information systems: a case study of Rwanda. Agric Ecosyst Environ 114(2-4):170–184

Waldron A, Garrity D, Malhi Y, Girardin C, Miller DC, Seddor N (2017) Agroforestry can enhance food security while meeting other sustainable development goals. Trop Conserv Sci 10:1–6. https://doi.org/10.1177/1940082917720667

WCED (1987) Our common future. World Commission on Environment and Development Oxford University Press, Oxford, UK

West PC, Gerber JS, Engstrom PM, Mueller ND, Brauman KA, Carlson KM, Siebert S (2014) Leverage points for improving global food security and the environment. Science 345 (6194):325–328

Wezel A, Bellon S, Thierry D, Francis C, Vallod D, David C (2009) Agroecology as a science, a movement or a practice: a review. Agron Sustain Dev 29:503–515

Zhang W, Ricketts TH, Kremen C, Carney K, Swinton SM (2007) Ecosystem services and dis-services to agriculture. Ecol Econ 64(2):253–260

Zomer RJ, Trabucco A, Coe R, Place F, van Noordwijk M, Xu JC (2014) Trees on farms: an update and reanalysis of agroforestry global extent and socio-ecological characteristics. Working paper 179. Bogor, Indonesia: world agroforestry Centre (ICRAF) Southeast Asia regional program. Doi:https://doi.org/10.5716/WP14064

Agroforestry and Its Services for Soil Management and Sustainability

11

Nahid Khan, Manoj Kumar Jhariya, Abhishek Raj, Arnab Banerjee, Ram Swaroop Meena, Surendra Singh Bargali, Shailesh Kumar Yadav, and Anita Kumawat

Abstract

Agroforestry systems (AFs) ensure greater biodiversity that intensifies ecosystem services in tangible and intangible ways. Accounting ecosystem services through well-managed agroforestry systems are other important aspects of scientific studies nowadays. AFs are an integration of trees with crops, and it also includes animal farming with the intensive land management system. In the twenty-first century, land management is one of the major challenges, and AFs have the vast potential to address and recognize these challenges as well as facilitate various services in a sustainable manner. Soil is the largest natural resource that sustains billions of life and supports a variety of flora and fauna. Agroforestry (AF) plays important role in soil health management that ensures ecological stability and environmental sustainability. In AFs interaction between aboveground and

N. Khan · M. K. Jhariya (✉)
Department of Farm Forestry, Sant Gahira Guru Vishwavidyalaya, Sarguja, Ambikapur, India

A. Raj
School of Agriculture, Lovely Professional University, Phagwara, Punjab, India

A. Banerjee · S. K. Yadav
Department of Environment Science, Sant Gahira Guru Vishwavidyalaya, Sarguja, Ambikapur, India

R. S. Meena
Department of Agronomy, Institute of Agricultural Sciences, Banaras Hindu University, Varanasi, Uttar Pradesh, India

S. S. Bargali
Department of Botany, DSB Campus, Kumaun University, Nainital, Uttarakhand, India

A. Kumawat
ICAR–Indian Institute of Soil and Water Conservation, Kota, Rajasthan, India
e-mail: anita.kumawat@icar.gov.in

M. K. Jhariya et al. (eds.), *Sustainable Intensification for Agroecosystem Services and Management*, https://doi.org/10.1007/978-981-16-3207-5_11

belowground components takes place which helps in improving the soil quality and provides shelter to many biota and soil organisms. Through AF soil management and conservation can be done and also the protection of agroecosystem at the regional and local level. The practices of sustainable soil management (SSM) make the pave for achieving the goal of sustainability. Thus, scientific AFs promise the SSM that enhances biodiversity through intensification of ecosystem services at the global scale. Soil fertility enhancement, better nutrient cycling, and higher resource use efficiency along with carbon sequestration for climate change mitigation are important services provided by AFs. AF also reduces carbon and environmental footprints by reducing greenhouse gas (GHG) emission and its sequestration and storage into both plants and soils. Thus, an effective policy and good governance are more important in achieving sustainability through adopting better scientific AFs in the tropical world. A future roadmap must be laid on adopting location-specific AF models for maintaining soil health and quality for a better sustainable world.

Keywords

Agroforestry · Carbon footprint · Ecosystem services · Soil health · Sustainability

Abbreviations

AF	Agroforestry
AFs	Agroforestry systems
C	Carbon
CSA	Climate-smart agriculture
CO_2	Carbon dioxide
N	Nitrogen
GHGs	Greenhouse gases
MPTs	Multipurpose tree species
NTFPs	Non-timber forest products
SOC	Soil organic carbon
SSM	Sustainable soil management

11.1 Introduction

Agroforestry systems (AFs) are sustainable land use farming practices that ensure ecological stability and environmental sustainability. AFs are gaining wide recognition among farmers, policy makers, and researchers due to their multifarious role in diversified products and ecosystem services (Montagnini and Nair 2004; Jhariya et al. 2015, 2019a, b, 2021a, b). In this era of climate change, agroforestry (AF) is a key component for resolving many problems around the world and became climate-smart agriculture (CSA) technique for sustainable land management and diversifying land use with different components (Raj et al. 2018). In 1977, ICRAF, the World Agroforestry Centre signified AF and focuses on the global effort for solving the land management problems which are not focused on and were often aggravate, by developments in forestry and commercial agriculture. Tree-crop integration in AFs can potentially restore soil fertility along with balancing nutrients and enriching soil organic carbon (SOC) content. These restorations would be helpful in maintaining soil physicochemical properties which in turn improve soil health and quality (Jhariya et al. 2018a, b; Meena et al. 2018; Khan et al. 2021a, b). It can enhance farm productivity, provide habitat to wildlife and manage water flow, and contribute to mitigating climate change. Integration of forest and agriculture aims at increasing and improving the restoration of biodiversity and ecosystem services (Hastings et al. 2020). Integrating crops along with trees helps in broadening restoration and also provides livelihood to the people living in that area (Khan et al. 2020a, b). Litter from trees helps in providing the organic matter to the soil, stores carbon (C) in soil, increases nitrogen (N) content in the soil surface, and also reduces water interception which helps in reducing runoff and soil erosion. Soil surface provides stratum to many micro-organisms and increases the soil quality and health (Özbolat et al. 2020).

Restoration, improvement, and sustainable management of resources help in increasing the production capacity and quality of soil. Different surveys revealed the potential of AF and its benefits which are obtained from a single land. AF diversified food, fruits, and fuelwood productions that improve livelihood standard and socioeconomic conditions of the poor farmers (Painkra et al. 2016). Different agroclimatic zones pertain to a variety of AF models that not only enhance product diversification but also intensify ecosystem services including soil health maintenance (Singh and Jhariya 2016). In this context, Handa et al. (2015) emphasized major AF models such as agrisilvicultural, agri-horticulture and agrihortisilviculture, and silvopastoral systems which were practiced in seven, six, and two agroclimatic regions of India. Further, a modern and managed AF will ensure better physicochemical properties of soil that improve soil quality and fertility as compared to the traditional agriculture systems.

AFs help in the mitigation of climate change by minimizing greenhouse gas (GHG) emissions into the atmosphere. Different AF models comprise of different tree-crop livestock's combinations that mitigate climate change and C footprint issue and their deleterious impact on the environment (Nair et al. 2009; Banerjee et al. 2021a, b, c, d). AF also includes many programs for climate change mitigation such as REDD+ (reduced emission from forest degradation and deforestation) to resolve the climate change issue (Nair and Garrity 2012). AFs include many programs and schemes under them such as watershed development, wasteland management, and rehabilitation of problematic soil and degraded lands. Many centers and researchers working in this area have worked on the study of physicochemical properties of soil. AF has the potential to work towards the conservation and restoration of land in many different ways along with increasing the productivity of the land. It highlights and focuses on land degradation reduction, soil conservation, food security, integrating natural resources, and conserving biodiversity (Dawson et al. 2012; Rathia et al. 2019; Kumar et al. 2020). Thus, AFs as a sustainable land-use system ensure soil-food-climate security through well-diversified products that intensify ecosystem services. Ecological stability and environmental sustainability are confirmed under scientific-based AF practices along with SSM (Raj et al. 2019, 2020). Therefore, the present chapter highlights the role of AF and its associated services for soil management and moves towards overall soil-environment and ecological sustainability.

11.2 Agroforestry-Mediated Ecosystem Services

AF plays a very important and vital role in the regulation of ecosystem services. It plays a multifunctional part in the ecosystem as it manages the landscape and environmental benefits and provides many economic products (Jose 2009). Tree-crop integration in AFs can improve the soil quality, reduces erosion, provides organic matter to the soil, and also enhances the biodiversity of an area. AF models help to maintain the atmospheric C content, sequestering and mitigating the C from the atmosphere into the above (tree, crop) and below ground (soil) pools (Nair et al. 2009). Trees on farmlands also help to increase the connectivity between the biological components of that landscape (Dawson et al. 2013; Meena and Lal 2018). AF contributes to increase the SOC pool and check C and environmental footprints (Zomer et al. 2016). AFs link the cultural ecosystem services, aesthetic values, and local values (Moreno et al. 2018; HLPE 2019). AF is an integrated farm management program which includes the combination of tree, crops, and livestock (Fagerholm et al. 2016). Trees are the central elements in AFs which provide a tangible form of ecosystem services as timber, fuelwood, fruits, and non-timber forest products (NTFPs) (van Noordwijk 2019; Dobie et al. 2019). AF-mediated ecosystem services in different countries of the world are depicted in Table 11.1.

AFs diversify the source of income as they provide different benefits to the land users and also fulfill their basic demand of products (e.g., food, fodder, timber, NTFPs) and also provide livelihood opportunities along with ecosystem services (Kassie 2018; FAO 2019). A sustainable AF model promises healthy ecosystem services such as soil and water regulation, soil fertility improvement, erosion control, biodiversity conservation, and C sequestration for a healthy environment (HLPE 2019). AF also diversifies faunal populations in any tropical region of the world. Diversifying birds and insects in AFs would help in enhancing biodiversity and pollination mechanisms (Dawson et al. 2013; Rathia et al. 2019).

AF helps in bridging the gap between forestry, agriculture, and animal husbandry; it allows the entire component together which helps in fulfilling the need and also regulating many ecosystem services. This smart method of cultivating crops along with forestry improves the environment and also benefits the society by maintaining the socioecological system across the world (Shin et al. 2020). AFs help in maintaining soil fertility by increasing the SOC content through C sequestration. It has different models which help the soil to rehabilitate and also give farmers more

Table 11.1 Agroforestry mediated ecosystem services in different countries of the world

Countries	Ecosystem services	References
India	Coffee-based agroforestry helps in increasing bird biodiversity	Chang et al. (2018)
China	Agroforestry practices show impact on biomass of riparian buffers	Yang et al. (2018)
Nepal	Hillside agroforestry system including maize helps in maintaining the soil health and control erosions	Tiwari et al. (2012)
Mediterranean Basin	Agroforestry systems have the greatest potential to support the diversity of social and ecological values across different agricultural landscapes	Flinzberger et al. (2020)
Sub-Saharan Africa	Agroforestry systems have a potential contribution towards climate change mitigation, increase plant productivity, resource availability to smallholders and woody perennial tree species that enhance tangible and intangible services to biodiversity	Zomer et al. (2016), Kuyah et al. (2019)
India	AF component helps in increasing the water holding capacity	Tripathi et al. (2009)
	MPTs in the agroforestry model help in improving the soil hydrophysical characteristics	Saha et al. (2007)
USA	Agroforestry and different techniques help in the reduction of runoff by 10–23% and 20% N loss reduction	Udawatta et al. (2002)
India	Cardamom-based agroforestry systems have the potential to reduce soil erosion	Mishra and Rai (2013)
	Research reveals that the agroforestry system helps in water and soil conservation by providing vegetation cover as barriers	Bundela (2007)
Synthesis of six world's case studies	Agroforestry system-linked human-natural systems contribute to social-ecological resilience	Liu et al. (2007)
Pacific Island		Ticktin et al. (2018)
World's forest landscape restoration	Integration of agriculture with forest tree helps in restoring biodiversity and ecological services and generating livelihood	Mansourian et al. (2020)
Brazil	Agroforestry is an eco-friendly and cost-effective land-use system for restoration and rehabilitation of land and environmental conservation/ enhancement at the landscape level	De Oliveira and Carvalhaes (2016)
Global perspective		Hillbrand et al. (2017)
India	Leguminous tree species help in fixing N in their leaves and through roots which help in increasing the soil fertility	Jose (2009)

production which improves the food security. AF-related ecosystem services are many and go far beyond food production. However, most of the positive effects of AF depend on sustainable management, for example, choosing suitable tree species for the purpose and local circumstances.

11.3 Function of Agroforestry

AF provides a wide range of functions directly or indirectly to the ecosystem and humankind. Productive, protective, ecological, and social functions are delivered through AFs which are depicted in Fig. 11.1 (Raj et al. 2020). It has a multifunctional role like enhancing the agricultural productivity and diversity of food and other products, timber, and NTFPs which help in generating money and provide livelihood opportunities (Lehmann et al. 2020). AF benefits in different countries of the world are depicted in Table 11.2. AF provides both productive and protective functions to the atmosphere and soil.

11.3.1 Productive Functions

AFs ensure production services by delivering various tangible products such as timber, fuelwood, other NTFPs, etc. Various products are obtained from AF models as it integrates different components such as tree, crops, and animals also. It aims at achieving the highest productivity by diversifying the land use and integrating the different components into single land. As per Wilson and Lovell (2016), AFs are recognized as sustainable agricultural practices which resolve many environmental problems and also improve the socioeconomic status of the people engaged in it. This interaction of tree and agricultural component on-field provides a wide

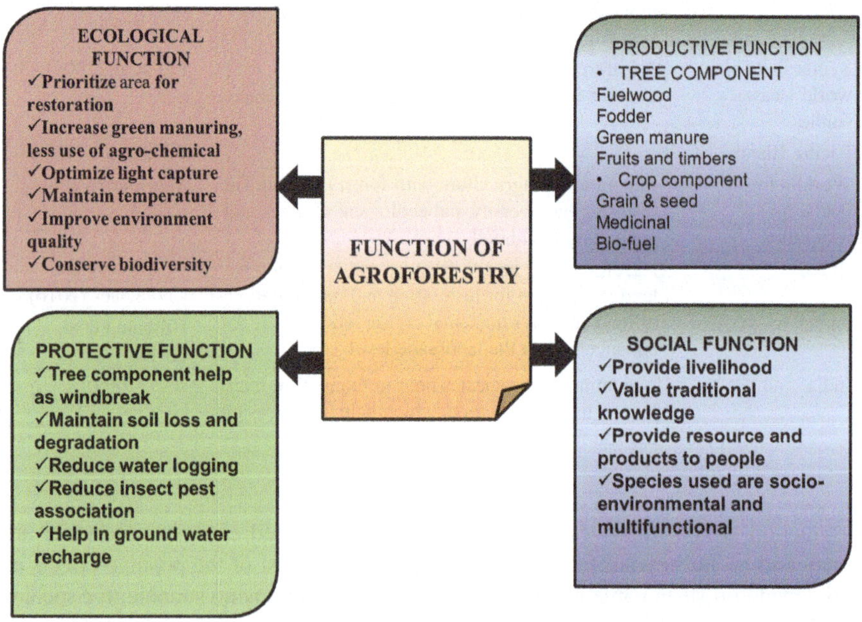

Fig. 11.1 Function of agroforestry for ecosystem sustainability (compiled: Raj et al. 2020)

Table 11.2 Agroforestry benefits in different countries of the world

Species in agroforestry systems	Benefits	Country	Reference
Melia dubia Cavanilles and *Melia azedarach*	High cation-exchange capacity (at a young stage)	West Java province	Rambey et al. (2019)
Millettia pinnata	Reported biofuel value in agroforestry system	India	Dalemans et al. (2019)
	Timber, NTFPs, and fuelwood help in conserving the environment and fulfill the requirement	Sri Lanka	De Zoysa (2001)
	Helps in food security, biodiversity conservation, climate change mitigation, and subsistence farming purposes	Asia-Pacific	Waldron et al. (2017)
Alnus nepalensis and *Gmelina arborea*	Help in N fixation Wasteland reclamation	India	Umashankar (2005)
Bombax ceiba	Slope stabilization		
Hieronyma alchorneoides	Good litter producer	Tropical region	Gonza Âlez (1996)
Stryphnodendron microstachyum	Maintain soil N value due to N_2 fixing in nature		
Vochysia guatemalensis	Good litter producer		
Vochysia ferruginea	Good litter producer		
F. albida and *A. tortilis*	Help in improving the soil fertility	Ethiopia	Desta et al. (2018)
	Provides wood energy, maintains soil fertility, and improves farmer's income along with many other tangible and intangible ecosystem services	Africa	Cheikh et al. (2014)
	Provides a diverse range of ecosystem services by connecting agriculture, forestry, and animal husbandry altogether	Asia-Pacific region	Shin et al. (2020)

range of benefits along with food, timber, and fodder, which also enhances the landscape (ICRAF 2019; Kumar et al. 2021). AF focuses on positive interactions between components to achieve maximum production by diversifying the land use. In recent time, AFs have a significant contribution to livelihood and environmental sustainability (Handa et al. 2020). AF has wider importance comprising of soil health maintenance, food security, climate change adaptation, livelihood generation, and C sequestration (Fig. 11.2) (Raj et al. 2019). AF along with rural areas provides a wide range of products to urban peoples. AFs can be applicable to small land which helps in fulfilling the small demand and provide many NTFPs and timber for marketing purposes and also plays an important role in ecosystem services.

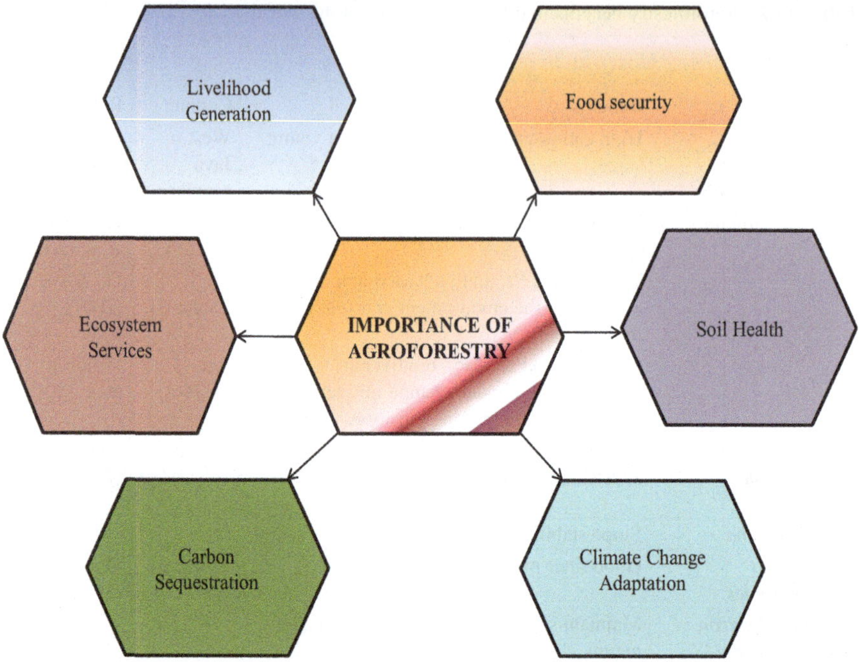

Fig. 11.2 Importance of agroforestry for the sustainable environment (compiled Raj et al. 2019)

Climate change greatly affects food production and food security at the global level. Variation in temperature and rainfall pattern had a great impact on agriculture and livestock production across the world. High temperature reduces the moisture level from soil whereas heavy rainfall increases the erosion rate of soil. AFs help in conserving and maintaining the soil moisture content through mulching and cover crops that also improve soil fertility and organic matter inputs. AFs work as a CSA which helps in reducing the impact of changing climate on soil and helps in maintaining the soil fertility which helps in increasing the production to reduce the risk of climate change on agriculture and global food security (Mosissa 2019).

11.3.2 Protective Functions

AF provides protection from high winds and heavy rainfall and conserves rainfall water by reducing runoff and soil erosion. Tree roots help in binding the soil particle and protecting it from getting eroded, and crops also reduce the interception rate on soil and also conserve the moisture content of soil. AF has a high capacity to conserve the biodiversity as it integrates trees, crops, and animals together and provides shelter to many faunas which help in conserving the biodiversity and maintain the ecosystem (Rendón-Sandoval et al. 2020). Tree-crop interaction, its composition, and variety and degree of management in AFs also influence the

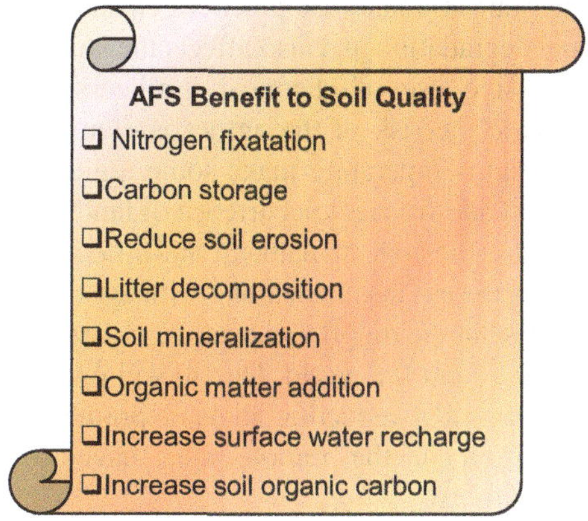

Fig. 11.3 Agroforestry benefits for soil quality (Dollinger and Jose 2018; Raj et al. 2019)

overall biodiversity and related ecosystem services (Rolim and Chiarello 2004). Indigenous species in AFs help in conserving and serve potential zones or ecological corridors for many faunas of regional areas which in turn conserve the biodiversity (Cassano et al. 2014).

Using tree species on fields which are preferred by many faunas for food and shelter can be helpful in conserving the biodiversity of many species. Many horticulture tree species provide food to many bird species and provide shelter to them which in turn result in increasing the indigenous birds and faunal species of that particular area. Kuyah et al. (2016) have reported an increase in biodiversity by 56% by increasing organisms in multi-strata as compared to the monoculture system. Integrating tree with crops improve soil organism biodiversity that plays a key role in soil fertility enhancement and C storage value. Therefore, AF has the potential to build soil fertility and maintenance of overall soil health and quality. Litter decomposition, N fixation, C storage, soil mineralization, organic matter addition, and erosion control are key benefits under AFs that form better soil health (Fig. 11.3) (Dollinger and Jose 2018; Meena et al. 2020; Raj et al. 2019). Further, AF helps in increasing the aboveground biodiversity and maintenance of the ecological system. AFs also work as shelterbelt and windbreaks to protect tree-crop livestock from the high-speed wind. These systems provide shelters to wild animals and reduce pollution by removing particulate matter in the air along with minimizing wind erosion. Farmers were made aware of the benefits and positive impacts of using windbreaks and hedgerows on farmland, and this led to increased adoption by farmers.

11.4 Constraint in Soil Health Management

Climate change and poor environmental quality affect soil health and related ecosystem services. Human-induced natural resource degradation and its unsustainable uses also lead to degradation of soil health (Jhariya and Singh 2020, 2021; Meena et al. 2020a). Intensive agriculture, overgrazing, deforestation, and other unsustainable land-use practices ensure soil degradation. Climate change has led to land degradation and also affects the soil physicochemical and biological properties. Excessive use of chemical fertilizers and unsustainable agriculture production increase the risk of the agroecosystem (Raj and Jhariya 2017). Increased rainfall and high temperature make some regions very drier and some areas due to heavy rainfall and frequent occurrence of landslides and flooding. These entire phenomena increase the risk of damage and affect the environment as well. Climate change affects the environment along with agriculture and forest production. This leads to an increase in the runoff and land degradation. Soil erosion is one of the major problems faced around the world. Due to soil erosion, the top fertile soil surface is removed that causes a reduction in land productivity. Human population growth and its expansion in the remote area have deleterious impacts on soil health and productivity.

11.5 Agroforestry for Soil Health Management

AFs help in conserving natural resources along with soil health and productivity maintenance. It improves nutrient availability for soil and soil health. Tree-crop interaction, their types, and nature and management of AFs affect the overall health and productivity of soil resources (Raj 2019; Raj and Jhariya 2020). Different soil properties such as soil structure, aggregates, and particles affect the quality and health of soil which are the important indicators for the soil health. Tree, crop, and animals all together improve the soil quality by adding organic matter to soil and help in nutrient cycling. AFs also help in improving the soil microclimate along with the improvement of the aboveground ecosystem. Thus, AF improves soil health and quality by various means which are comprehensively described below.

11.5.1 Improving Soil Organic Carbon Content

A better management of AFs enhances C storage and sequestration process. Capturing of atmospheric C and its sequestration in plant and soil are an important ecosystem service which is possible under well-managed AFs. Moreover, AF-mediated SSM practices maximize soil C sequestration that maintains SOC pools (Dollinger and Jose 2018). Therefore, SSM practices are becoming a good strategy that not only mitigates climate change but also maintains SOC pools. However, C sequestration process varies as per model to model and species to species in any agroclimatic zone of the tropical region. Decomposition and decaying of litter and other plant-based organic residues add C content into the soil. The addition of C will affect soil-inhabiting organism and diversity which influence the nutrient status and fertility of soil (Bertin et al. 2003). Thus, the type of management in AFs affects SOC. The SSM practices can enhance C content that further is modified by tree-crop interactions and nature of plant species (Banerjee et al.

2020). However, a more diverse form of C content is reported in soil of tropical AFs than in temperate region.

11.5.2 Soil Fertility Enhancement and Organic Matter Content

AFs is cultivating crops with the integration of trees and helps in maintaining as well as improving soil fertility. Soil organic matter plays a vital role in improving soil quality and helps in maintaining the soil biogeochemical cycle. In AFs, woody perennial trees help in maintaining the soil organic matter content in soil by the addition of litterfall and other debris into soil (Sikstus et al. 2020). Soil fertility is enhanced by trees that have characteristics like deep roots and having dense fine root network along with abundant mycorrhizal association with bacteria present in soil. Litter of tree leaves which decompose rapidly increase in enrichment of nutrient in soil and also maintain the soil cover and reduce the erosion rate of soil from water and wind. Integration of leguminous and N-fixing trees in AFs enhances N content and organic matter that ensures higher soil fertility. Leguminous tree and crop species help in improving the soil fertility and also increase the soil productive traits (Jhariya et al. 2018a).

AFs use marginal land, and by integrating trees, crops, and animals, altogether help in increasing soil productivity and soil fertility. The root system of trees influences the properties of soil differently. AF helps towards the reclamation of degraded land and restores soil fertility (Sileshi et al. 2020). Future demands can be fulfilled by maintaining and enhancing the fertility of soil as well as increasing tree biomass which helps in increasing the production value and increase food grain production. AFs such as intercropping and mixed cropping and legume-based cropping help in maintaining the ecological system which reduces water runoff and improves the fertility of soil and leads to sustainability of agricultural production (Raj et al. 2021).

11.5.3 Closed and Efficient Nutrient Cycling

A close type of nutrient cycling system has been observed in AFs. This nutrient cycling is quite different from the open type of nutrient cycling in which nutrient losses through leaching are observed under the sole/monocropping system. The tree component in AF increases the supply of nutrients to soil and crop by fixing N from the atmosphere into the belowground component. Trees increase the nutrient availability in subsoil surfaces by increasing the soil organic content and recycling the organic material. Deep-rooted trees help in absorbing the nutrient and minerals from the very deep soil surface and make them available for the tree. Leaf debris and crop residues also help in increasing the soil organic matter content and improve the nutrient cycling in soil which help in improving soil health (Froufe et al. 2020; Cardinael et al. 2020). Many research reported that fast-growing trees with high N-fixing capacity, such as *Calliandra calothyrsus*, *S. sesban*, and *Eucalyptus grandis*, are much more effective than the slower growing trees. AF trees have the potential to fix N which is helpful in crop yield through increasing the N content from biological N fixation and recycling N from plant residues and from manures. In AFs trees with deep root system has a high potential to intercept nutrient elements from the deep soil profile and accumulate in the subsoil and make it available for the annual crops.

11.5.4 Nitrogen-Fixing Trees (NFTs) in Agroforestry

N-fixing trees and shrubs have a great potential for conserving and restoring soil fertility and improve soil nutrient status and are very useful for soil health. Using biological N fixation and incorporating it with N-fixing tree and shrub component help in enhancing the appropriate manipulation and management of AFs. Trees with N-fixing properties and with high biomass production are preferred in AFs (Jhariya et al. 2018a). In agroecosystem trees help in N fixation and enhance the soil productivity through biological N fixation, improve nutrient cycling, and also capture nutrients from water and soil. Non-N-fixing trees also enhance the soil properties by adding organic matter to both above and belowground component as well as releasing and recycling the nutrients on land. Major N-fixing tree species in AF are *Acacia*, *Albizia*, *Casuarina*, *Erythrina*, *Gliricidia*, *Leucaena*, *Prosopis*, and *Sesbania*. Some N_2-fixing tree species for soil restoration in AFs are depicted in Table 11.3. *Sesbania sesban* trees used in AF help in increasing the water infiltration and reduce the soil bulk density. Similarly, *Acacia nilotica* is a N-fixing tree and is an integral component of AFs that also provides economically important gum for farmers (Raj and Singh 2017). These species not only provide timber and NTFPs but also improve soil fertility and health through enhancing N and C status into the soil. Roots of many leguminous tree species make a symbiotic relationship with microorganism and fix atmospheric N. Few nonleguminous species like *Casuarina* also fix N with the genus of actinomycetes, *Frankia*.

11.5.5 Multipurpose Tree Species (MPTs) in Agroforestry System

In AFs trees are grown for different purposes and provide various significant functions are termed as a multipurpose tree. This function includes both productive and protective functions. Productive functions include fuelwood, timber, fodder, food, and fibers whereas protective functions include the stain soil conservation, and water conservation, work as shelterbelt and windbreaks, preserve biodiversity, and maintain different ecosystem services (Jhariya et al. 2015). Tree performing more than one function comes under MPTs. Trees like Gliricidia sepium are grown as a live fence that produce fodder, fuelwood, and manures also. Similarly, Leucaena leucocephala are grown to fulfill the need for wood and leaf fodder. MPTs also help in stabilizing soil fertility along with providing different functions. Trees help in reducing soil erosion along the gullies and streams with the help of a strong root

Table 11.3 Nitrogen-fixing tree species for soil restoration in agroforestry systems

N_2-fixing tree species	Origin	Soil restoration	References
Acacia mangium	Australia, Indonesia, Papua New Guinea	Belongs to the genus Rhizobium, forms a symbiotic relationship with the soil bacteria, presence of root nodules for transforming the free N into an organic form	Orwa et al. (2009)
Casuarina equisetifolia	Australia, Bangladesh, Indonesia, Malaysia, New Zealand	Highly potential agroforestry species for the arid and semiarid area that also improve soil fertility	
Acacia nilotica	Ethiopia, India, Kenya, Pakistan, Saudi Arabia	A multipurpose tree having characteristics of N_2 fixation that restore the soil fertility and improve health	Jhariya et al. (2018a), Bargali and Bargali (2009)
Ailanthus excelsa	India and Sri Lanka	Fodder tree and fast-growing in nature which helps in improving soil fertility along with higher plant productivity in agroforestry system	Rajasugunasekar (2014)
Alnus nepalensis	China, India, Japan, Laos, Myanmar, Nepal	Helps in stabilizing slopes, reduces soil erosion, N fixation with genus Frankia by forming a symbiotic relationship	Sharma et al. (2007)
Eucalyptus spp.	Australia	The wood is useful and extracts of leaf in agroforestry systems used as pesticides and promote biopesticides	Rajan (1987)
Gliricidia sepium	Mexico	Topsoil erosion can be reduced by planting this tree in agroforestry systems	Coulibaly et al. (2017), Muimba-Kankolongo (2018)
Leucaena leucocephala	Colombia, Mexico, Spain, USA	Short-term nutrient enrichment by rapid decomposition of tree leaves	Muimba-Kankolongo (2018), Sinclair (2004)
Pongamia pinnata	Bangladesh, India, Myanmar, Nepal, Thailand	Biodiesel tree used in blending with nitrogenous fertilizers	Kesari and Rangan (2010)
Prosopis cineraria	Afghanistan, India, Iran, Pakistan, Sri Lanka	Help in the stabilization of sand dunes and reclaim the degraded soils	Singh and Bishnoi (2014)

system. Some tree species are helpful for stabilizing the soil erosion near streams and gullies such as Gmelina arborea, Leucaena leucocephala, Sesbania grandiflora, etc.

11.5.6 Soil Nutrient Enrichment and Erosion Control

Vegetation cover found in AFs helps in maintaining the rate of runoff and control erosion with soil nutrient enrichment. Crop and tree residues provide mulching on the soil surface that enriches soil nutrients and checks soil erosion. Trees and crops with good root systems help in increasing the soil infiltration rate and help in increasing groundwater recharge with reducing the soil loss (Atangana et al. 2014). Trees help to moderate the effect of leaching by adding base to the soil surface and reduce the acidification caused by chemical fertilizers; tree leaf litter help in adding the organic material to soil which help in maintaining the acidity of soil. Sustainable crop production has a strong relationship with soil quality, as soil health and quality affect the production capacity of land. In AF, soil conservation and nutrient enrichment are greatly affected by the type of cropping system used for the cultivation of crops. AFs can be done in land with low productivity and by integrating different components together such as crops and trees with N-fixing capacity and trees with deep taproot system that can help in nutrient cycling, improving the soil productivity and capacity of soil. Different models of AF contribute differently to soil health and productivity with the farmland. AFs provide many ecological functions and soil resilience or soil conservation along with protection of local biodiversity (Udawatta et al. 2019). The soil systems are well linked and recognized for agroecosystem sustainability and management. The management and development of agroecosystem are complex one in line with people's needs and utilization efficiency and ability to manage and balance with environmental context. The AF plays a significant role in this context and also restores the degraded landscape and improves the resilience and life support systems of the soil.

11.5.7 Soil Restoration and Wasteland Reclamation

The restoration of soil as a land management approach is an integral part of the ecological and environmental development of the degraded landscape. In this perspective, tree plantation, pasture development, and site-specific AFs seem to be promising tools towards land management and sustainability. This provides various ecosystem services as well as offers new prospects in bridging the gap between the demand and supply of commodities related to agriculture, forestry, and livestock. Besides these, the AF can enhance the social, economic, and environmental scenario within the complex socioeconomic-ecological systems of Asia-Pacific region in specific and globally in wider dimensions (Shin et al. 2020). Thus, AF restores soil health and quality by its nature and uses diversification and manure additions into the soil, through better tree-crop interaction and integration of N fixing plant, etc. These AF services help in restoring soil health and quality that ensure ecological

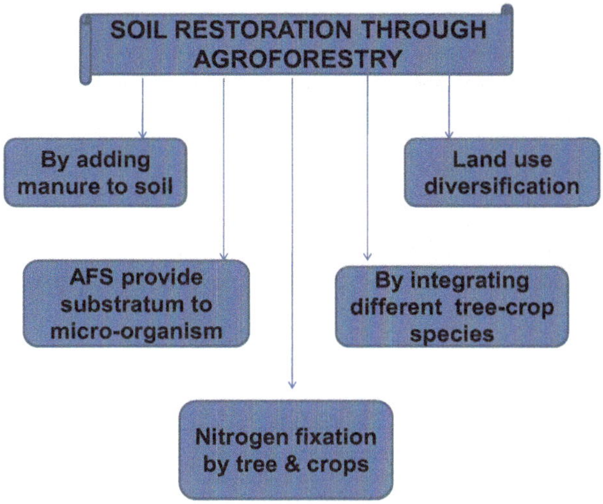

Fig. 11.4 Agroforestry systems for soil restoration (compiled: Sikstus et al. 2020; Raj et al. 2020)

stability and environmental sustainability (Fig. 11.4) (Sikstus et al. 2020; Raj et al. 2020).

11.6 Agroforestry for Climate and Food Security

Changing climate and extreme weather condition directly and indirectly affect overall AF health and productivity. Practices of intensive agriculture involve higher synthetic inputs along with heavy mechanization that not only emits GHGs but also affects soil health and fertility. Deforestation and unsustainable land-use practices lead to the emission of CO_2 into the atmosphere. Releasing GHGs into the atmosphere definitely increases the temperature which leads to global warming and climate change. Thus, an unscientific way of AF practices affects the health and productivity of soil and climate. In this context, practicing sustainable AFs ensures overall soil-food-climate security that makes sustainable environment. Therefore, tree-crop interaction, its combination, and nature of its management are key factors that affect overall soil-plant health and productivity. However, C absorption and sequestration capacity are also affected by the type of AF models and its nature of management in any agroclimatic regions of the tropical world. Varying models having varying C absorptions capacity may entirely affect the overall health and productivity of AFs. C sinks potential under AFs in the world are depicted in Table 11.4. However, humid tropics have greater potential to sequester more C as compared to other tropics due to their high rainfall characteristics. A total of 0.3 to 15.2 mega C ha^{-1} $year^{-1}$ has been reported under AFs (Nair et al. 2011). Thus, AF is a type of C farming practices that enhance C sequestration for mitigating changing climate along with fulfilling food requirement for hunger people.

Table 11.4 Carbon sink potential under agroforestry systems in the world

Country	Carbon sink potential under agroforestry systems	References
Brazil	Agroforestry systems contributed 3.2 t ha^{-1} $year^{-1}$ of biomass and 39.81 t C ha^{-1} of carbon sequestration, respectively	Viégas et al. (2019)
India	Stabilization of SOC pools is possible through mechanisms including biochemical recalcitrance and physical protection that enhances the C sequestration potential of agroforestry systems	Dhyani et al. (2020)
	C sink potential varied in between 0.29 and 15.21 Mg ha^{-1} $year^{-1}$	Dhyani et al. (2020)
	Agroforestry systems can play a major role in storing C in aboveground biomass and in soil and in belowground biomass	Murthy et al. (2013), Nair et al. (2009)
	Agroforestry system reported higher C sink potential than other treeless agriculture or pasture land-use systems under similar ecological conditions	Nair et al. (2009), Ajit et al. (2013)
	The C sequestration values under AFS in tree, crop, and soil were found to be 0.25–19.14, 0.01–0.60, and 0.003–3.98 Mg C ha^{-1} on an annual basis	Dhyani et al. (2016)
Africa	Agroforestry represents the combination of crops with trees which play an important role in C sequestration	Takimoto et al. (2008)
Tropics of the world	AFS in humid tropics may be able to sequester vegetation and soil C over 70 Mg C ha^{-1} and 25 Mg ha^{-1}, respectively	Mutuo et al. (2005)

AF provides multifarious food and non-food products that are a good source of income as well. Sustainable and ecology-oriented AFs maximize the overall plant productivity which fulfills food requirement of billions of people globally. It also provides nutritious and quality fruits by incorporating various agricultural and woody perennial horticultural trees. These woody perennial trees potentially absorb C from the atmosphere and fix into them as biomass. Thus, AF works as a good C sink along with quality food and timber productions as biomass. It provides various nutritious fruits and vegetables for people and farmers that are good source of essential nutrients. These products ensure healthy people, animals, and environment by providing protein, vitamins, and other important minerals. However, as per Roy (2011), well-managed AFs can fulfill the daily fruit (85 g) and vegetable (220 g) requirement which is reported as a balance diet for humans. Thus, AF promises food and nutritional security by generating income and employment opportunities for rural peoples (Samara 2010; Meena et al. 2020b).

11.7 Agroecosystem Sustainability Through Agroforestry

Agroecosystem is an integrated system in which productivity as well as the environmental health needs to be considered simultaneously. Principally, it involves practices of cultivating woody vegetation along with various crops. It may be in different sequences, combination, and arrangement as well as with sustainable approaches (Lundgreen and Raintree 1982; Jhariya et al. 2019c). When we discuss about the sustainability of the agroecosystem, it involves an all-round development of the soil ecosystem as well as maintaining agricultural productivity (Berhe and Retta 2015; Jhariya et al. 2021a). The fruitfulness of such an approach helps to reduce the environmental damages, natural hazards, as well as positive output from agricultural crops (FAO 2015).

AF is such an approach that is very much essential for maintaining soil health and reducing the land degradation process. The biomass contribution through AFs helps to improve the soil quality as well as the diversity of vegetation (Nair 2007; Jhariya et al. 2019b). Scientific reports reveal that the plantation of legume trees under the AFs has a significant impact towards improving soil health and productivity. Through the process of biological N fixation, more N is added into the soil under the legume crop association giving better yield and productivity (Mafongoya et al. 2004; Jhariya et al. 2018a).

Nutrient exchange is also another key ecosystem service performed by agroecosystem. It is such a process that governs the productive output of a particular crop ecosystem. AFs are a beneficial contributor in this perspective which involves an increment in the nutrient level of the soil through the nutrient pumping process. Higher availability of nutrients promotes optimum vegetative growth of the tree species under AFs. This in turn builds up the SOC pool through litter decomposition (Suprayogo et al. 2010). Some of the AFs have been reported to procure nutrients from a deeper layer of the soil and hence make it available to the crop plants for better growth and productivity.

Land degradation is the biggest issue that needs to be addressed while considering the sustainability of the agroecosystem. It was observed with gradual rise in science and technology that the misuse of natural resources is taking place at an unprecedented rate. The agroecosystem is also not out of that. It is such an important ecosystem feeding the humanity, but due to its mal-use event such as soil erosion, the sustainability of agroecosystem is challenging. To solve such problems, AF plays a key role through conservation of soil resources. It helps in maintaining soil fertility, prevents erosion, improves soil microbiota, as well as inhibits land degradation (Jhariya et al. 2015, 2019a).

11.8 Policy and Future Roadmap

Intensive agriculture, deforestation, and other unsustainable land-use practices involve soil degradation that deprives healthy ecosystem services. These practices have deleterious impacts on the environment and lead to climate change along with C and environmental footprints. Soil degradation entirely affects the overall health and productivity of the ecosystem. Poor soil affects food productions and related global challenges. In this context, practicing sustainable AFs ensure a better plant-soil health and productivity along with climate security on a long-term basis (Khan 2020a, b). Also, SSM practices restore soil fertility and maintain microbial

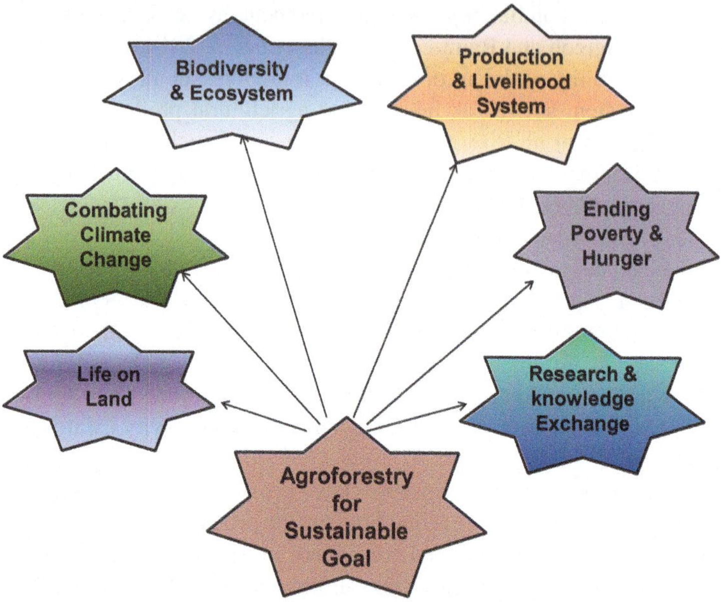

Fig. 11.5 Agroforestry system for a sustainable world (compiled: Lehmann et al. 2020; Dhyani et al. 2020)

population and rhizosphere biology along with improving soil health and quality. These sustainable land-use practices ensure ecological stability and environmental sustainability in the tropical world. In this context, a policy must be reformed to promote and make people and farmers aware of adopting AFs and their effective management to deliver fruitful results of soil-food-climate security.

Soil is the most important component for crop production and forestry and also an important component of the human environment. The success of AF production highly depends on the health and quality of the soil. The capacity of soil is to sustain and maintain the function of the living system and improve biological productivity and environment quality (Orwa et al. 2009; FAO 2015). Thus, AF makes a sustainable world by ensuring healthy biodiversity and ecosystem health. Higher plant productivity, livelihood security, climate change mitigation, poverty reduction, and soil health maintenance along with scientific research and development in AF make a sustainable world (Fig. 11.5) (Lehmann et al. 2020; Dhyani et al. 2020).

11.9 Conclusions

The soil-land-environment degradation is the biggest issue of the present time due to deforestation and faulty land-use and agricultural practices affecting the landscape and overall prosperity of human civilization and well-being. In this perspective, AF is the best solution which improves the soil system on the one hand and maintains the agricultural production on the other. Moreover, it leads to eco-environmental sustainability by reducing the agricultural inputs which deteriorate the soil, water, and environment as well as the health of the biological systems. Changing climate and related soil health problems are another burning topic of today. Climate change not only affects soil health and productivity but also influences agroecosystem services. In this context, practicing sustainable AFs will promise soil-plant-climate security which ensures to achieve the goal of sustainable development. Thus, an effective policy and good governance are needed for promoting scientific way of AF practices, and its better management improves soil health and productivity.

References

Ajit DSK, Ramnewaj HAK, Prasad R, Alam B, Rizvi RH, Gupta G, Pandey KK, Jain A (2013) Modeling analysis of potential carbon sequestration under existing agroforestry systems in three districts of Indo-Gangetic plains in India. Agrofor Syst 87:1129–1146

Atangana A, Khasa D, Chang S, Degrande A (2014) Agroforestry for soil conservation. In: Tropical agroforestry. Springer, Dordrecht, pp 203–216

Banerjee A, Jhariya MK, Yadav DK, Raj A (2020) Environmental and sustainable development through forestry and other resources. Apple Academic Press Inc., New York, p 400

Banerjee A, Meena RS, Jhariya MK, Yadav DK (2021a) Agroecological footprints management for sustainable food system. Springer, Singapore, p 514

Banerjee A, Jhariya MK, Meena RS, Yadav DK (2021b) Ecological footprints in agroecosystem- an overview. In: Banerjee A, Meena RS, Jhariya MK, Yadav DK (eds) Agroecological footprints management for sustainable food system. Springer, Singapore, pp 1–23

Banerjee A, Jhariya MK, Raj A, Yadav DK, Khan N, Meena RS (2021c) Land footprint management and policies. In: Banerjee A, Meena RS, Jhariya MK, Yadav DK (eds) Agroecological footprints management for sustainable food system. Springer, Singapore, pp 221–246

Banerjee A, Jhariya MK, Raj A, Yadav DK, Khan N, Meena RS (2021d) Energy and climate footprint towards the environmental sustainability. In: Banerjee A, Meena RS, Jhariya MK, Yadav DK (eds) Agroecological footprints management for sustainable food system. Springer, Singapore, pp 415–443

Bargali K, Bargali SS (2009) Acacia nilotica: a multipurpose leguminous plant. Nat Sci 7(4):11–19

Berhe DH, Retta AN (2015) Soil improvement by trees and crop production under tropical agroforestry systems: a review. Merit Res J Agric Sci Soil Sci 3(2):18–28

Bertin C, Yang X, Weston LA (2003) The role of root exudates and allelochemicals in the rhizosphere. Plant Soil 256(1):67–83

Bundela DS (2007) Water management in northeast India - some case studies. Karnal 2007:40–46

Cardinael R, Mao Z, Chenu C, Hinsinaer P (2020) Belowground functioning of agroforestry systems: recent advances and perspectives. Plant Soil 2020:1–13

Cassano CR, Barlow J, Pardini R (2014) Forest loss or management intensification? Identifying causes of mammal decline in cacao agroforests. Biol Conserv 169:14–22

Chang CH, Karanth KK, Robbins P (2018) Birds and beans: comparing avian richness and endemism in Arabica and robusta agroforests in India's Western Ghats. Sci Rep 8:1–9

Cheikh M, Pete S, David S, Lalisa D, Mercedes B (2014) Achieving mitigation and adaptation to climate change through sustainable agroforestry practices in Africa. Curr Opin Environ Sustain 6:8–14

Coulibaly JY, Chiputwa B, Nakelse T, Kundhlande G (2017) Adoption of agroforestry and the impact on household food security among farmers in Malawi. Agric Syst 155:52–69. https://doi.org/10.1016/j.agsy.2017.03.017

Dalemans F, Muys B, Maertens M (2019) A framework for profitability evaluation of agroforestry-based biofuel value chains: an application to Pongamia in India. GCB Bioenergy 11 (7):852–870. https://doi.org/10.1111/gcbb.12605

Dawson I, Harwood C, Jamnadass R, Beniest J (2012) Agroforestry tree domestication: a primer. World Agroforestry Centre (ICRAF), Nairobi

Dawson IK, Guariguata MR, Loo J, Weber JC, Lengkeek A, Bush D, Cornelius J, Guarino L, Kindt R, Orwa C, Russell J (2013) What is the relevance of smallholders' agroforestry systems for conserving tropical tree species and genetic diversity in circa situm, in situ and ex situ settings? A review. Biodivers Conserv 22(2):301–324. https://doi.org/10.1007/s1053-1-012-0429-5

De Oliveira RE, Carvalhaes MA (2016) Agroforestry as a tool for restoration in atlantic forest: Can we find multi-purpose species? Oecol Aust 20:425–435

De Zoysa M (2001) A review of forest policy trends in Sri Lanka. Policy Trend Report 2001. Institute for Global Environmental Strategies (IGES): Kanagawa, Japan

Desta KN, Lisanenwork N, Muktar M (2018) Physico-chemical properties of soil under the canopies of Faidherbia albida (Delile) A. Chev and Acacia tortilis (Forssk.) Hayen in park land agroforestry system in Central Rift Valley, Ethiopia. J Hortic Forestry 10(1):1–8

Dhyani SK, Ram A, Dev I (2016) Potential of agroforestry systems in carbon sequestration in India. Indian J Agric Sci 86(9):1103–1112

Dhyani SK, Ram A, Newaj R, Handa AK, Dev I (2020) Agroforestry for carbon sequestration in tropical India. Carbon Manage Trop Sub-Trop Terrestr Syst 19:313–331. https://doi.org/10.1007/978-981-13-9628-1_19

Dobie P, Zinngrebe Y, Vidal A, Gassner A, Kumar C (2019) Position on inclusion of agricultural lands as contributing to conservation of biodiversity in the post-2020 biodiversity agenda. Trees on Farms for Biodiversity programme. World Agroforestry, Nairobi

Dollinger J, Jose S (2018) Agroforestry for soil health. Agrofor Syst 92:213–219

Fagerholm N, Torralba M, Burgess PJ, Plieninger T (2016) A systematic map of ecosystem services assessments around European agroforestry. Ecol Indic 62:47–65. https://doi.org/10.1016/j.ecolind.2015.11.016

FAO (2015) Healthy soils are the basis for healthy food production. Rome, Italy

FAO (2019) The state of the world's biodiversity for food and agriculture. Bélanger J, Pilling D (eds.). FAO commission on genetic resources for food and agriculture assessments. FAO, Rome, p 572

Flinzberger L, Zinngrebe Y, Plieninger T (2020) Labelling in Mediterranean agroforestry landscapes: a Delphi study on relevant sustainability indicators. Sustain Sci. https://doi.org/10.1007/s1162-5-020-00800-2

Froufe LCM, Schwiderke DK, Castilhano AC, Cezar RM, Steenbock W, Seoane CES, Bognola IA, Vezzani FM (2020) Nutrient cycling from leaf litter in multistrata successional agroforestry systems and natural regeneration at Brazilian Atlantic Rainforest Biome. Agrofor Syst 94 (1):159–171

Gonza Âlez JE (1996) Tropical tree species for reforestation: Studies on seed storage, foliar nutrient content and wood variation. Ph.D. diss. Texas A&M Univ., College Station, TX

Handa AK, Dhyani SK, Uma (2015) Three decades of agroforestry research in India: retrospection for way forward. Agric Res J 52(3):1–10. https://doi.org/10.5958/2395-146X.2015.00028.9

Handa AK, Chavan SB, Sirohi C, Rizvi RH (2020) Importance of agroforestry systems in carbon sequestration. National Agroforestry Symposium 2020 (agroforestry system 2020) on Climate Resilient Agroforestry Systems to Augment Livestock Productivity Ensuring Environmental Biodiversity 5 & 6 March, 2020, pp 221–235

Hastings Z, Ticktin T, Botelho M, Reppun N, Kukea-Shultz K, Wong M, Melone A, Bremer L (2020) Integrating co-production and functional trait approaches for inclusive and scalable restoration solutions. Conserv Sci Pract 2(9):e250. https://doi.org/10.1111/csp2.250

Hillbrand A, Borelli S, Conigliaro M, Olivier E (2017) Agroforestry for landscape restoration. Exploring the potential of agroforestry to enhance the sustainability and resilience of degraded landscapes. FAO, Rome

HLPE (2019) Agroecological and other innovative approaches for sustainable agriculture and food systems that enhance food security and nutrition. A report by the High Level Panel of Experts on Food Security and Nutrition of the Committee on World Food Security, Rome
ICRAF (2019) About agroforestry. Available http://www.worldagroforestry.org/
Jhariya MK, Singh L (2020) Herbaceous diversity and biomass under different fire regimes in a seasonally dry forest ecosystem. Environ Dev Sustain 22:1–19. https://doi.org/10.1007/s10668-020-00892-x
Jhariya MK, Singh L (2021) Effect of fire severity on soil properties in a seasonally dry forest ecosystem of Central India. Int J Environ Sci Technol. https://doi.org/10.1007/s13762-020-03062-8
Jhariya MK, Bargali SS, Raj A (2015) Possibilities and perspectives of agroforestry in Chhattisgarh. In: Zlatic M (ed) Precious forests - precious earth. InTech, London, pp 238–257
Jhariya MK, Banerjee A, Yadav DK, Raj A (2018a) Leguminous trees an innovative tool for soil sustainability. In: Meena RS, Das A, Yadav GS, Lal R (eds) Legumes for soil sustainable management. Springer, Singapore, pp 315–345
Jhariya MK, Yadav DK, Banerjee A (2018b) Plant mediated transformation and habitat restoration: phytoremediation an eco-friendly approach. In: Gautam A, Pathak C (eds) Metallic contamination and its toxicity. Daya Publishing House, New Delhi, pp 231–247
Jhariya MK, Banerjee A, Meena RS, Yadav DK (2019a) Sustainable agriculture, forest and environmental management. Springer, Singapore, p 606
Jhariya MK, Yadav DK, Banerjee A (2019b) Agroforestry and climate change: issues and challenges. Apple Academic Press Inc., New York
Jhariya MK, Banerjee A, Meena RS, Yadav DK (2019c) Agriculture, forestry and environmental sustainability- a way forward. In: Jhariya MK, Banerjee A, Meena RS, Yadav DK (eds) Sustainable agriculture, forest and environmental management. Springer, Singapore, pp 1–29
Jhariya MK, Meena RS, Banerjee A (2021a) Ecological intensification of natural resources for sustainable agriculture. Springer, Singapore
Jhariya MK, Meena RS, Banerjee A (2021b) Ecological intensification of natural resources towards sustainable productive system. In: Ecological intensification of natural resources for sustainable agriculture. Springer, Singapore
Jose S (2009) Agroforestry for ecosystem services and environmental benefits: an overview. Agrofor Syst 76:1–10. https://doi.org/10.1007/s1045-7-009-9229-7
Kassie GW (2018) Agroforestry and farm income diversification: synergy or trade-off? The case of Ethiopia. Environ Syst Res 6:8. https://doi.org/10.1186/s4006-8-017-0085-6
Kesari V, Rangan L (2010) Development of Pongamia pinnata as an alternative biofuel crop - current status and scope of plantations in India. J Crop Sci Biotechnol 13(3):127–137
Khan N (2020a) Social and technological transformation vis-à-vis sustainability. Van Sangyan 7 (3):16–20
Khan N (2020b) Agroforestry towards sustainable agroecosystem. Van Sangyan 7(6):1–6
Khan N, Jhariya MK, Yadav DK, Banerjee A (2020a) Herbaceous dynamics and CO_2 mitigation in an urban setup- a case study from Chhattisgarh, India. Environ Sci Pollut Res 27(3):2881–2897. https://doi.org/10.1007/s11356-019-07182-8
Khan N, Jhariya MK, Yadav DK, Banerjee A (2020b) Structure, diversity and ecological function of shrub species in an urban setup of Sarguja, Chhattisgarh, India. Environ Sci Pollut Res 27 (5):5418–5432. https://doi.org/10.1007/s11356-019-07172-w
Khan N, Jhariya MK, Raj A, Banerjee A, Meena RS (2021a) Soil carbon stock and sequestration: implications for climate change adaptation and mitigation. In: Jhariya MK, Meena RS, Banerjee A (eds) Ecological intensification of natural resources for sustainable agriculture. Springer, Singapore
Khan N, Jhariya MK, Raj A, Banerjee A, Meena RS (2021b) Eco-designing for sustainability. In: Jhariya MK, Meena RS, Banerjee A (eds) Ecological intensification of natural resources for sustainable agriculture. Springer, Singapore

Kumar S, Meena RS, Datta R, Verma SK, Yadav GS, Pradhan GS, Molaei A, Rahman GKMM, Mashuk HA (2020) Legumes for carbon and nitrogen cycling: an organic approach. Carbon Nitrogen Cycl Soil. https://doi.org/10.1007/978-981-13-7264-3_10.337-375

Kumar S, Meena RS, Singh RK, Munir TM, Datta R, Danish S, Singh GS (2021) Soil microbial and nutrient dynamics under different sowings environment of Indian mustard (Brassica juncea L.) in rice based cropping system. Sci Report 11:5289. https://doi.org/10.1038/s41598-021-84742-4

Kuyah S, Öborn I, Jonsson M, Dahlin AS, Barrios E, Muthuri C, Malmer A, Nyaga J, Magaju C, Namirembe S, Nyberg Y, Sinclair FL (2016) Trees in agricultural landscapes enhance provision of ecosystem services in Sub-Saharan Africa. Int J Biodiv Sci Ecosys Serv Manage 12 (4):255–273. https://doi.org/10.1080/21513732.2016.1214178

Kuyah S, Whitney CW, Jonsson M, Sileshi GW, Öborn I, Muthuri CW, Luedeling E (2019) Agroforestry delivers a win-win solution for ecosystem services in sub-Saharan Africa- a meta-analysis. Agron Sustain Dev 39:47. https://doi.org/10.1007/s13593-019-0589-8

Lehmann LM, Smith J, Westaway S, Pisanelli A, Russo G, Borek R, Sandor M, Gliga A, Smith L, Ghaley BB (2020) Productivity and economic evaluation of agroforestry systems for sustainable production of food and non-food products. Sustain For 12(5429):1–9. https://doi.org/10.3390/su12135429

Liu J, Dietz T, Carpenter SR, Alberti M, Folke C, Moran E, Taylor WW (2007) Complexity of coupled human and natural systems. Science 317:1513–1516

Lundgreen BO, Raintree JB (1982) Sustained agroforestry. In: Nestel B (ed) Agricultural research for development: potentials and challenges in Asia. ISNAR, Hague, pp 37–49

Mafongoya P, Giller KE, Odee D, Gathumbi S, Ndufa SK, Sitompul SM (2004) Benefitingfrom N_2-fixation and managing rhizobia. In: Van Noordwijk CG, Ong CK (eds) Below-ground interactions in tropicalagroecosystems. Concepts and models with multiple plant components. CABI, Wallingford, pp 227–242

Mansourian S, Parrotta J, Balaji P, Bellwood-Howard I, Bhasme S, Bixler RP, Boedhihartono AK, Carmenta R, Jedd T, de Jong W, Lake FK, Latawiec A, Lippe M, Rai ND, Sayer J, Van Dexter K, Vira B, Visseren-Hamakers I, Wyborn C, Yang A (2020) Putting the pieces together: integration for forest landscape restoration implementation. Land Degrad Dev 31:419–429. https://doi.org/10.1002/ldr.3448

Meena RS, Lal R (2018) Legumes for soil health and sustainable management. Springer, Singapore, p 541

Meena RS, Kumar V, Yadav GS, Mitran T (2018) Response and interaction of brady rhizobiumjaponicum and arbuscular mycorrhizal fungi in the soy bean rhizosphere: a review. Plant Growth Regul 84:207–223. https://doi.org/10.1007/s10725-017-0334-8

Meena RS, Lal R, Yadav GS (2020) Long term impacts of topsoil depth and amendments on soil physical and hydrological properties of an Alfisol in Central Ohio, USA. Geoderma 363:1141164. https://doi.org/10.1016/j.geoderma.2019.114164

Meena RS, Lal R, Yadav GS (2020a) Long-term impact of topsoil depth and amendments on carbon and nitrogen budgets in the surface layer of an Alfisol in Central Ohio. Catena 194:104752. https://doi.org/10.1016/j.catena.2020.104752

Meena RS, Kumar S, Datta R, Lal R, Vijaykumar V, Brtnicky M, Sharma MP, Yadav GS, Jhariya MK, Jangir CK, Pathan SI, Dokulilova T, Pecina V, Marfo TD (2020b) Impact of agrochemicals on soil microbiota and management: a review. Landscape 9(2):34. https://doi.org/10.3390/land9020034

Mishra PK, Rai SC (2013) Use of indigenous soil and water conservation practices among farmers in Sikkim Himalaya. Indian J Tradit Knowl 12:454–464

Montagnini F, Nair PKR (2004) Carbon sequestration: an underexploited environmental benefit of agroforestry systems. Agrofor Syst 61:281–295

Moreno G, Aviron S, Berg S, Crous-Duran J, Franca A, de Jalón SG, Hartel T, Mirck J, Pantera A, Palma JHN, Paulo JA (2018) Agroforestry systems of high nature and cultural value in Europe:

provision of commercial goods and other ecosystem services. Agrofor Syst 92(4):877–891. https://doi.org/10.1007/s1045-7-017-0126-1
Mosissa D (2019) Soil and Water conservation practices and its contribution to small holder farmers livelihoods in northwest Ethiopia: a shifting syndrome from natural resources rich areas. Mod Concep Dev Agrono 3(5):000574
Muimba-Kankolongo A (2018) Food crop production by smallholder farmers in Southern Africa challenges and opportunities for improvement. Elsevier, London, p 382
Murthy IK, Gupta M, Tomar S, Munsi M, Tiwari R, Hegde GT, Ravindranath NH (2013) Carbon sequestration potential of agroforestry systems in India. J Earth Sci Clim Change 4:131
Mutuo PK, Cadisch G, Albrecht A, Palm CA, Verchot L (2005) Potential of agroforestry for carbon sequestration and mitigation of greenhouse gas emissions from soils in the tropics. Nutr Cycl Agroecosyst 71:43–54. https://doi.org/10.1007/s10705-004-5285-6
Nair PKR (2007) The coming of age of agroforestry. J Sci Food Agric 87:1613–1619. https://doi.org/10.1002/jsfa.2897
Nair PKR, Garrity D (2012) Agroforestry research and development: the way forward. Agroforestry - the future of global 515 land use. Adv Agroforestry 9:3. https://doi.org/10.1007/978-94-007-4676-3
Nair PKR, Kumar BM, Nair VD (2009) Agroforestry as a strategy for carbon sequestration. J Plant Nutr Soil Sci 172:10–23
Nair PKR, Vimala DN, Kumar BM, Showalter JM (2011) Carbon sequestration in agroforestry systems. Adv Agron 108:237–307
Orwa C, Mutua A, Kindt R, Jamnadass R, Anthony S (2009) Agroforestry tree database: a tree reference and selection guide version 4.0. International Centre for Agroforestry Research, Nairobi, Kenya
Özbolat O, Ollio I, Lloret E, Egea M, Zornoza R (2020) How agroforestry systems influence the abundance of nitrogen cycle contributing microbial genes under Mediterranean conditions? EGU General Assembly 2020. EGU2020-21485. DOI: https://doi.org/10.5194/egusphere-egu2020-21485
Painkra GP, Bhagat PK, Jhariya MK, Yadav DK (2016) Beekeeping for poverty alleviation and livelihood security in Chhattisgarh, India. In: Narain S, Rawat SK (eds) Innovative technology for sustainable agriculture development. Biotech Books, New Delhi, pp 429–453
Raj A (2019) Agroforestry and natural resource management: a linking concept. Acta Sci Microbiol 3(1):94
Raj A, Jhariya MK (2017) Sustainable agriculture with agroforestry: adoption to climate change. In: Kumar PS, Kanwat M, Meena PD, Kumar V, Alone RA (eds) Climate change and sustainable agriculture. New India Publishing Agency (NIPA), New Delhi, pp 287–294
Raj A, Jhariya MK (2020) Soil for resource management and sustainability. J Ecol Nat Resour 4 (5):000208
Raj A, Singh L (2017) Effects of girth class, injury and seasons on Ethephon induced gum exudation in Acacia nilotica in Chhattisgarh. Indian J Agrofor 19(1):36–41
Raj A, Jhariya MK, Bargali SS (2018) Climate smart agriculture and carbon sequestration. In: Pandey CB, Gaur MK, Goyal RK (eds) Climate change and agroforestry. New India Publishing Agency, New Delhi, pp 1–19
Raj A, Jhariya MK, Yadav DK, Banerjee A, Meena RS (2019) Agroforestry: a holistic approach for agricultural sustainability. In: Jhariya MK et al (eds) Sustainable agriculture, forest and environmental management. Springer, Singapore, pp 101–131
Raj A, Jhariya MK, Yadav DK, Banerjee A (2020) Climate change and agroforestry systems: adaptation and mitigation strategies. Apple Academic Press Inc., New York, p 383
Raj A, Jhariya MK, Khan N, Banerjee A, Meena RS (2021) Ecological intensification for sustainable development. In: Ecological intensification of natural resources for sustainable agriculture. Springer, Singapore
Rajan BKC (1987) Versatile eucalyptus. Diana Publishers, Bangalore, pp 164–172

Rajasugunasekar D (2014) Cultivation techniques of Ailanthus excelsa. In: Transfer of tree cultivation technologies to Krishi Vigyan Kendras (KVKs) of Tamil Nadu and Puducherry. Institute of Forest Genetics and Tree Breeding, Coimbatore, pp 19–22

Rambey R, Wijayanto N, Siregar IZ, Onrizal SA (2019) Study of agroforestry mindi planting pattern (Melia dubia cavanilles) in Selaawi Village, Garut District, West Java Province. IOP Conf Ser 374:012033

Rathia VK, Jhariya MK, Yadav DK, Banerjee A (2019) Diversity of traditional agroforestry practices of sarguja: implications for biodiversity conservation. Int J Ecol Dev 34(1):43–60

Rendón-Sandoval FJ, Casas A, Moreno-Calles AI, Torres-García I, García-Frapolli E (2020) Traditional agroforestry systems and conservation of native plant diversity of seasonally dry tropical forests. Sustain For 12:4600. https://doi.org/10.3390/su12114600

Rolim SG, Chiarello AG (2004) Slow death of Atlantic forest trees in cocoa agroforestry in southeastern Brazil. Biodivers Conserv 13:2679–2694

Roy A (2011) Requirement of vegetables and fruit. The Daily Star (A English Newspaper)

Saha R, Tomar JMS, Ghosh PK (2007) Evaluation and selection of multipurpose tree for improving soil hydro-physical behaviour under hilly eco-system of north east India. Agrofor Syst 69 (3):239–247

Samara JS (2010) Horticulture opportunities in rainfed areas. Indian J Hortic 67(1):1–7

Sharma R, Xu J, Sharma G (2007) Traditional Agro-forestry in Eastern Himalayan region: land management system supporting ecosystem services. Trop Ecol 48(2):1–12

Shin S, Soe KT, Lee H, Kim TH, Lee S, Park MS (2020) A systematic map of agroforestry research focusing on ecosystem services in the Asia-pacific region. Forests 11(368):1–23. https://doi.org/10.3390/f11040368

Sikstus G, Sumeni S, Sabodin R, Muqfi IH, Nur M, Hairiah K, Useng D, van Noordwijk M (2020) Soil organic matter, mitigation and adaptation to climate change in cocoa–based agroforestry systems. Landscape 9:323

Sileshi GW, Mafongoya PL, Nath AJ (2020) Agroforestry systems for improving nutrient recycling and soil fertility on degraded lands. In: Dagar JC, Gupta SR, Teketay D (eds) Agroforestry for degraded landscapes. Springer, Singapore. https://doi.org/10.1007/978-981-15-4136-0_8

Sinclair FL (2004) Agroforestry. Encyclopedia of Forest Sciences, pp 27–32. https://doi.org/10.1016/B0-12-145160-7/00340-9

Singh J, Bishnoi M (2014) Effect of lopping intensities on fodder and fuelwood yield of Prosopis cineraria in arid zones of Thar. Acad Arena 6(10):90–94

Singh NR, Jhariya MK (2016) Agroforestry and agrihorticulture for higher income and resource conservation. In: Narain S, Rawat SK (eds) Innovative technology for sustainable agriculture development. Biotech Books, New Delhi, pp 125–145

Suprayogo D, Hairiah K, Noordwijk M, Cadisch G (2010) Agroforestry interactions in rain fed agriculture: can hedgerow intercropping systems sustain crop yield on an ultisolin Lampung (Indonesia)? Agri 32(3):205–215

Takimoto A, Nair PK, Nair VD (2008) Carbon stock and sequestration potential of traditional and improved agroforestry systems in the West African Sahel. Agric Ecosyst Environ 125:159–166

Ticktin T, Dacks R, Quazi S, Tora M, McGuigan A, Hastings Z, Naikatini A (2018) Linkages between measures of biodiversity and community resilience in Pacific Island agroforests. Conserv Biol 32:1–11

Tiwari TP, Brook RM, Wagsta P, Sinclair FL (2012) Effects of light environment on maize in hillside agroforestry systems of Nepal. Food Secur 4:103–114

Tripathi OP, Pandey HN, Tripathi RS (2009) Litter production, decomposition and physico-chemical properties of soil in 3 developed agroforestry systems of Meghalaya, Northeast India. African J Plant Sci 3(8):160–167

Udawatta RP, Krstansky JJ, Henderson GS, Garrett HE (2002) Agroforestry practices, runoff, and nutrient loss: a paired watershed comparison. J Environ Qual 31:1214–1225

Udawatta RP, Rankoth LM, Jose S (2019) Agroforestry and biodiversity. Sustain For 11:2879

Umashankar (2005) Indigenous agroforestry tree species for conservation and rural livelihood. In: Bhatt BP, Bujarbaruah KM (eds) Agroforestry in North East India: opportunities and challenges. ICAR Research Complex for NEH Region, Umiam, pp 149–174

van Noordwijk M (ed) (2019) Sustainable development through trees on farms: agroforestry in its fifth decade. World Agroforestry (ICRAF), Bogor

Viégas LB, da Silva JMS, Pala MC, Piña-Rodrigues FCM (2019) Restoring ecological functions using agroforestry systems in riparian forests. Floresta Ambiente 26(4):1–11. https://doi.org/10.1590/2179-8087.083017

Waldron A, Garrity D, Malhi Y, Girardin C, Miller DC, Seddon N (2017) Agroforestry can enhance food security while meeting other sustainable development goals. Trop Conserv Sci 10:194008291772066

Wilson MH, Lovell ST (2016) Agroforestry-the next step in sustainable and resilient agriculture. Sustain For 8:574

Yang S, Bai J, Zhao C, Lou H, Zhang C, Guan Y, Zhang Y, Wang Z, Yu X (2018) The assessment of the changes of biomass and riparian bu_er width in the terminal reservoir under the impact of the South-to-North Water Diversion Project in China. Ecol Indic 85:932–943

Zomer RJ, Neufeldt H, Xu J, Ahrends A, Bossio D, Trabucco A, Van Noordwijk M, Wang M (2016) Global tree cover and biomass carbon on agricultural land: the contribution of agroforestry to global and national carbon budgets. Sci Rep 6:29987. https://doi.org/10.1038/srep29987

Agroecosystem Service Management and Environmental Sustainability 12

Abhishek Raj, Manoj Kumar Jhariya, Arnab Banerjee, Nahid Khan, Ram Swaroop Meena, Prabhat Ranjan Oraon, and Shailesh Kumar Yadav

Abstract

Agroecosystem means improving the agricultural ecosystem by human-induced management of trees, crops, and livestock in any land use system. Resource conservations, soil health management, minimizing environmental footprints, and climate change mitigation are key services through a healthy agroecosystem. Food demands due to burgeoning populations necessitated agricultural land expansion and intensive agricultural practices. Conversion of forest and other land use systems into agricultural land induces land degradation and leads to an increase in environmental footprints. Deforestation and other unsustainable land use practices ensure soil degradation and environmental pollutions. These unscientific and intensive agroecosystem practices lead to GHG emissions into the atmosphere causes carbon footprints. Thus, strategies for enhancing food production along with maintaining environmental health and quality are a smart choice

A. Raj
School of Agriculture, Lovely Professional University, Phagwara, Punjab, India

M. K. Jhariya (✉) · N. Khan
Department of Farm Forestry, Sant Gahira Guru Vishwavidyalaya, Sarguja, Ambikapur, Chhattisgarh, India

A. Banerjee · S. K. Yadav
Department of Environmental Science, Sant Gahira Guru Vishwavidyalaya, Sarguja, Ambikapur, Chhattisgarh, India

R. S. Meena
Department of Agronomy, Institute of Agricultural Sciences, Banaras Hindu University, Varanasi, Uttar Pradesh, India
e-mail: meenars@bhu.ac.in

P. R. Oraon
Department of Silviculture and Agroforestry, Faculty of Forestry, Birsa Agricultural University, Ranchi, Jharkhand, India

M. K. Jhariya et al. (eds.), *Sustainable Intensification for Agroecosystem Services and Management*, https://doi.org/10.1007/978-981-16-3207-5_12

of the modern day. High synthetic inputs and heavy mechanizations ensure higher production but at the cost of environmental health. Agroecosystem land expansion and practices affect other land use systems and related ecological services. These harsh and unscientific practices affect soil-food-climate security at a global scale. Thus, applying ecology-oriented sustainable agroecosystem practices ensures environmental sustainability and ecological stability. A sustainable modeling of agroecosystem will enhance biodiversity that intensifies uncountable ecosystem services. Agriculture, agroforestry, forestry, rangeland, etc. are different land use practices that build our sustainable environment. Applying eco-modeling and sustainable agroecosystem practices ensure higher production and profitability along with a healthy ecosystem. Climate-resilient agroecosystem practices and their ecological modeling enhance plant biomass productivity and soil health maintenance. These practices ensure soil fertility, higher SOC pools, healthy rhizosphere biology, and microbial populations on which entire biodiversity depends. Thus, maintaining a healthy and productive agroecosystem is the pillar of a sustainable environment that ensure a healthier world. In lieu of the above, this chapter represents the potential, perspective, and management of the agroecosystem. A principle and practices of sustainable-based agroecosystem are also discussed. A rigorous discussion is also made on climate-resilient agroecosystem practices and modeling for minimizing carbon footprint to ensure environmental sustainability at a global scale. A bit of discussion on soil-food-climate security through agroecosystem management makes this chapter more informative for policy makers worldwide.

Keywords

Agroecosystem · Biodiversity · Environmental sustainability · Ecosystem services · Footprint · Sustainable development

Abbreviations

C	Carbon
GHGs	Greenhouse gases
N	Nitrogen
NTFPs	Non-timber forest products
SOC	Soil organic carbon

12.1 Introduction

The estimated area of agroecosystem is approximately 30% of the total earth geographical land area. The agroecosystem is a type of sustainable land use practices that integrates plants and animals and blending of natural resources to make a better sustainable and viable ecosystem. This builds up the environmental sustainability and ecological stability (Banerjee et al. 2020; Raj et al. 2020; Jhariya et al. 2019a, b). The major question is "how sustainable land use system makes environmental sustainability?". Climate-smart forestry and agroecosystem land use practices work more efficiently to maximize higher productions and lesser greenhouse gas (GHG) emission along with a greater probability of environmental sustainability. The term "climate-smart forestry" is new but based on a similar concept of climate-smart agriculture and agroecosystem practices. These land use systems are based on sustainable and ecological principles that make a vibrant environment. Agroecosystem provides a variety of ecosystem services which entirely depend on sustainable land use practices and using ecology-oriented sustainable modeling (Tilman et al. 2002).

An increasing population induced a feeding requirement that directly affects land use practices. This leads to land conversion into agricultural land (agricultural land expansion) for food productions. Deforestation, illicit felling of timber, illegal forest cutting, etc. not only affect forest health and productivity but also destroy animals and people who rely upon the forest (Khan et al. 2020a, b). All these deleterious practices are used for enhancing food productivity to fulfill the global 9.8 billion people's food requirement but at the cost of environmental health. Moreover, intensive agricultural practices employ the heavy use of high synthetic fertilizers, and heavy mechanizations destroy the land and environmental health (Meena et al. 2020). This will not only affect food productions but also decline the quality of major food grains. As per one estimate, around 50% of areas are shared by intensive agricultural practices among the total arable land (Krishna 2010). An intensive agroecosystem practice affects ecosystem services including the declining of soil fertility and human-animal health (Gliessman 2015; Raj et al. 2019a, b). Agroecosystem provides multifarious and uncountable ecosystem services in both tangible and intangible ways. Timber, fuelwood, fodder (for animals), several non-timber forest products (NTFPs), etc. are tangible forms of agroecosystem services, whereas soil enrichment, fertility enhancement, soil organic carbon (C) pools, atmospheric C balance, efficient nutrient cycling, food and climate security, etc. are intangible forms of services (Kumar et al. 2020).

Emissions of GHGs due to intensive agroecosystem practices are another major problem faced by our environment. Climate change breaks a chain of environmental sustainability and ecological integrity. Extreme weather, temperature, and uncertain rainfall lead to poor crop productions and agroecosystem health. An intensive agroecosystem ensures GHG emissions and anthropogenic C footprints (Banerjee et al. 2021a, b). These unsustainable agroecosystem practices affect the overall health and productivity of soil, crops, humans, animals, and climate (Raj et al. 2018a, b). These practices not only affect global food productions but also break a sustainability chain by depriving environmental quality. In this context, sustainable agroecosystem practices ensure soil fertility and food and nutritional security along with environmental sustainability (Gaba et al. 2014; Lennon 2015). Applying eco-designing sustainable models in any land use systems makes a productive and healthy environment that leads to higher productions along with ecosystem stability.

Sustainable models in agriculture, agroforestry, forestry, and rangeland systems ensure soil-food-climate security. Climate-resilient eco-modeling promises less GHG emission and minimizes C and environmental footprint through a better sequestration process (Banerjee et al. 2021c, d). Now the question appears that "how agroecosystem modeling minimize C footprint?". Sustainable and ecological-based agroecosystem practices minimize C emission along with greater efficiency of C sequestration. These practices involve conservation agriculture, the use of cover crops, and zero-tillage and mulching practices. All these practices are based on an ecological concept that minimizes emissions and anthropogenic C footprint (Gliessman 2015; Meena et al. 2018). Thus, applying these practices not only checks C footprints but also promotes soil C pools and plant biomass (Emmerson et al. 2016; Kumar et al. 2020a).

Unsustainable land use practices lead to poor soil health and quality. Land and soil degradation are common problems which spread throughout the globe. Depriving soil quality leads to poor plant production that entirely affects overall food systems (Jhariya et al. 2018a, b). Healthy soil promotes healthy food for a healthy and sustainable ecosystem. In this context, agroecosystem management will ensure soil-food-climate productions. These practices maximize biodiversity and soil fertility, enhance SOC (soil organic carbon) pools, maximize food productions, and reacted ecosystem services (Peterson et al. 2018; Córdoba et al. 2020). Soil enrichment, efficient nutrient cycling, erosion control, C balance, water regulations, healthy rhizosphere, and microbial populations are soil-based ecosystem services mediated through sustainable agroecosystem practices. Similarly, integrating nitrogen (N)-fixing leguminous trees/crops ensures soil N availability which directly affects plant growth and development (Congreves et al. 2015). Perennial legumes and grasses/pastures hold soils, add soil C, and check soil erosion besides food and fodder productions for human-animal agroecosystem (Lal et al. 2011).

Thus, managing agroecosystem is a greater challenge today that is a prerequisite for overall ecosystem structure, functioning, and processes. The agroecosystem management ensures a healthy and sustainable environment through greater diversification, higher soil-crop productivity, and maintaining food-climate security. In lieu of the above, this chapter presents a comprehensive discussion on agroecosystem management and related ecosystem services for environmental sustainability.

12.2 Agroecosystem: Principle, Practices, and Potential

The health and productivity of agroecosystem are based on ecological principles which are rigorously discussed by policy makers, researchers, practitioners, and stakeholders. A sustainable and ecology-based agroecosystem practice produces more with less environmental hazards. Eco-designing and modeling of agroecosystem ensure soil-crop productivity with environmental sustainability. The agroecosystem practices based on ecological principles intensify the potential of ecosystem services. This represents the principle, practice, and potential of well-managed agroecosystem practices (Zhang et al. 2007). A theoretical and

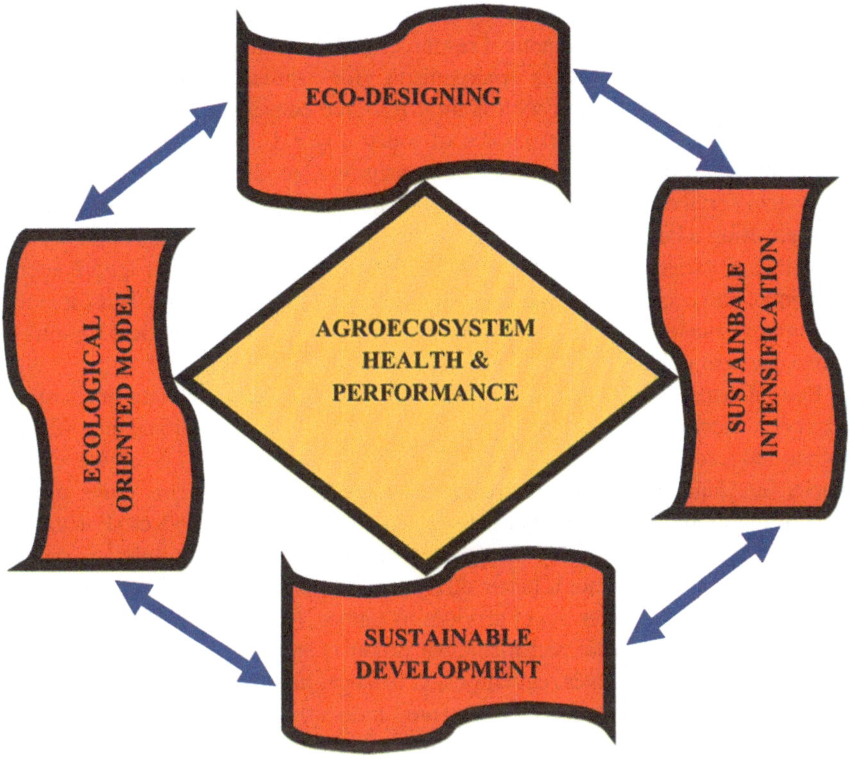

Fig. 12.1 Pillars for agroecosystem health and performance-mediated sustainable development (compiled by: Zhu et al. 2012)

hypothetical model is needed for agroecosystem management that creates a sustainable environment. The agroecosystem practices and principles depend on these theoretical models of sustainable intensifications. This model-based agroecosystem management ensures multifarious and uncountable ecosystem services. Conservation and blending of resources, high crop-soil productivity, increasing soil-water use efficiency, and providing better ecosystem services, etc. are goals, principle and significance on which agroecosystem management depends (Peterson et al. 2018; Meena and Lal 2018; Córdoba et al. 2020). A model is created to show various pillars for agroecosystem health and performance-mediated sustainable development (Fig. 12.1) (Zhu et al. 2012).

Varying location-specific models in different land use systems are based on an ecological principle that maintains ecosystem processes. The practices of organic agriculture, conservation agriculture, and integrated farming systems are based on a sustainable and ecological concept. Climate-smart agroecosystem, climate-smart forestry, sustainable soil management practices, sustainable forest management, etc. are gaining wider recognition worldwide. These all are climate-resilient land use practices that ensure soil-food-climate security along with environmental sustainability. Agroecosystem management encompasses scientific practices which employ a proper blending of natural resources and their efficient utilization. Natural resource conservation is a prerequisite for agroecosystem management. Agroecosystem management with natural resource conservation without affecting our environment is the smart choice of today's world. In this context, applying sustainable intensification in the agroecosystem is a good approach that intensifies ecosystem services rather than intensive practices (Felipe-Lucia et al. 2014). However, ensuring a better management of agroecosystem can enhance biodiversity that intensifies a variety of ecosystem services. We can't overlook and ignore social, economic, and ecological principles while managing agroecosystem. Thus, a model and strategy must be developed to ensure socioeconomic and ecological stability in agroecosystem practices (Bernués et al. 2014; Kumar et al. 2021).

Eco-intensified and scientifically managed agroecosystem delivers uncountable ecosystem services in sustainable ways. As per Ellis and Ramankutty (2008), human-induced agroecosystem-managed areas cover around 75% of the earth's glacier-free land areas. Agriculture- and pasture-based ecosystems cover around 40% of the global land areas that deliver valuable ecosystem services (Foley et al. 2005). We can't underestimate the agroecosystem services that nurture a variety of flora and fauna to maintain a better ecosystem and environment. A well-managed agroecosystem can deliver direct services as timber, fuelwood, fodder, and several NTFPs including aromatic and medicinal products, whereas soil enrichment, efficient nutrient cycling, water regulations, C balance, higher SOC pools, higher crop productivity, climate security, etc. are indirect services. Thus, managing agroecosystem will ensure all these services which are necessary for a healthy and productive ecosystem with greater environmental sustainability. Moreover, microclimate regulation, pollution reduction, pest and disease control, water purifications, erosion controls, and climate change mitigations are defined as regulatory services through well-managed agroecosystem practices. All these services not only are for ecosystem maintenance but also ensure environmental sustainability (Millenium Ecosystem Assessment, MEA 2005).

Pollination services are important among all the intangible agroecosystem services. We can't imagine the lives and population of global flora without a pollination mechanism. Animal-mediated pollinations contributed approx. 35% of the global crop productions. For instance, pollination alone regulates around 60–90% of the world plant species (Klein et al. 2007). Pollinations induce plant reproductions that provide healthier food and fruits on which humans and animals depended upon (Sundriyal and Sundriyal 2004). Wind-based pollinations control important crops like sugarcane and many cereals along with several palatable pastures/grasses. Thus, pollination mechanism controls food productions and related economy.

A well-managed agroecosystem intensifies soil-based ecosystem services. Soil sustains a variety of flora and fauna that maintain ecosystem processes. Soil C sequestration maintains SOC pools that maintain the C balance in the atmosphere and minimize climate change issues. A close type of nutrient cycling is observed in sustainable agroecosystem practices. Nutrient loss through leaching and other means are minimizing due to the presence of better tree-crop interactions in the agroecosystem. Extensive root systems of woody perennial trees can minimize nutrient loss by trapping essential nutrients. These nutrient captures by trees are utilized by plant-crop systems for proper growth and development. A well-managed agroecosystem ensures healthy soil for higher crop production. Quantity and quality of plant production ensure human-animal-climate health systems (Zhang et al. 2007). Healthy rhizosphere biology and greater microbial populations are very important for soil health and quality. Soil microbial populations also affect the organic C content that represents soil fertility. Good soil fertility is a prerequisite for a healthier and productive agroecosystem. Thus, a well-managed agroecosystem service not only maintains healthy soil but also minimizes C footprints for ensuring environmental sustainability.

Agroecosystem management also ensures disease and pest management for healthier and higher plant productivity. As per one estimate, around US $30 was lost due to animal pest attacks in various major cereal crops in the United States (Oerke 2005; Pimentel et al. 2005). However, agroecosystem management can be possible to explore predator ecology and pest behaviors (Landis et al. 2000). Thus, controlling pest attacks and disease infestation ensures a highly productive and healthier agroecosystem for a sustainable environment (Sheoran et al. 2021; Karp et al. 2013). Managing agroecosystem has the greatest potential to promote water purification, pollution reductions, air purifications, and climate regulations (Brauman et al. 2007). The agroecosystem shared 70% of the world's freshwater consumptions. Only 40% of plant productions are regulated by irrigated farming systems (UN Water (United Nations Water) 2013). Thus, a practice of agroecosystem management needs sustainable and ecology-based principles that ensure a greater potential of ecosystem and environmental sustainability.

12.3 Climate Change and Its Impact on Agroecosystem

Climate change due to C and environmental footprints induces agroecosystem health and productivity. The huge loss of health and plant productivity has been observed due to extreme weather events. Uncertain rainfall and higher temperature affect morphology, reproductive, and phenology of plants by disturbing photoperiods. Heavy fertilizer application and mechanizations deprive the land fertility and emit GHGs to the atmosphere. These uncontrolled and unscientific intensive practices lead to degradation in environmental sustainability (Alexander et al. 2015). It is a very clear concept that an intensive agroecosystem promotes higher production with the loss of environmental health and quality (Duru et al. 2015). These unorganized farming practices reduce biodiversity that negatively affects overall agroecosystem services. Similarly, Iqbal and Goheer (2008) estimated the value of GHG emissions under the practice of intensive agroecosystem. According to them, the maximum emission (4.13 Tg year^{-1}) is covered by methane followed by 0.26 Tg year^{-1} of nitrous oxide and 52.6 Tg year^{-1} of CO_2, respectively. GHG-mediated air and water pollution, soil erosion, water contaminations, heavy metal pollutions, and soil salinization are prominent risks due to intensive agroecosystem practices (Horrigan et al. 2002). Similarly, conventional and intensive agroecosystem practices lead to food and health risk due to the application of chemical and inorganic fertilizers (Hurni et al. 2015). Rangeland-mediated vegetative food and fiber productions initiated environmental footprints (Jankielsohn 2015). Climate change is also prominent in arid and semiarid climate due to increasing temperature and harsh climate that induce water shortage which directly affects microbial activities. As compared to zero tillage, tillage practices disturb soil layers and physicochemical properties that release 21% of more C into the atmosphere (Behnke et al. 2018; Meena et al. 2020).

Due to increasing population, the agricultural land expansion and deforestation have increased which resulted in tons of C release into the atmosphere. These unsustainable land use practices entirely affect atmospheric C balance, i.e., balancing between sink and source. Moreover, rising GHGs resulted in global warming and climate change phenomenon. That's why well-managed sustainable agroecosystem practices emerged as a new trend of applied science. These practices ensure soil-food-climate security by stopping higher GHG emission by minimizing unsustainable ways of agroecosystem practices.

12.4 Perspective of Agroecosystem Management

The perspective behind agroecosystem management is very clear among all researchers and policy makers. Spreading and mobilizing sustainable agroecosystem practices are of utmost importance and minimize intensive practices. To minimize higher synthetic inputs with promotions of sustainable agroecosystem practices is an important initiative. Climate-smart forestry and agroforestry, sustainable forest and soil management, and climate-resilient practices are important perspectives of agroecosystem management (Painkra et al. 2016; Singh and Jhariya 2016; Jhariya et al. 2015). Maintenance of health and productivity of soil, humans, and animals and climates are important thrust areas for agroecosystem management. The perspective of agroecosystem management is to maintain soil-food-climate security with overall environmental sustainability (Whitfield et al. 2018).

Applying ecology-based farming models in any agroecosystem ensures multifarious ecosystem services that sustain the whole biodiversity. Agroecosystem models vary as per varying topography, species types and nature, tree-crop-animal interactions, and climatic situations. As we know, agroforestry is a location-specific model that requires a variety of management for its proper functioning. Climate change mitigation through C sequestration, soil enrichment, crop productivity, human-animal health, etc. is a major perspective and goal of agroecosystem management (Bernard and Lux 2017). Moreover, a healthier and productive agroecosystem based on ecological-based management ensures environmental sustainability.

Applying chemical fertilizers disturb our food chain and ecosystem due to their residual effects on biodiversity. Moreover, these practices release various harmful gases into the atmosphere. Therefore, a sustainable agroecosystem management minimizes these negative impacts and ensures social, ecological, economic, and health security at a global scale (Gadanakis et al. 2015). Thus, enhancing agroecosystem potential in terms of better social, economic, and ecological services is possible through eco-based management practices. These are major goals and perspectives of agroecosystem management for minimizing environmental footprints to ensure sustainable development.

12.5 Sustainable Approaches of Agroecosystem Management

Around 80% of the world's population depends on agriculture for the economy (FAO 2003). In Western Africa, a mixed cropping system covers around 80% of the total agricultural system (Steiner 1982). These figures represent the importance of agroecosystem and its role in sustaining billions of lives. Intensive agroecosystem produces most but less quality and having negative impacts on the environment. This practice does not provide sustainable and higher productions for a long-term basis. Although, intensive practices release various harmful gases into the atmosphere leading to global warming and climate change issues.

Applying sustainable approaches in agroecosystem management ensures higher production and maximizing profits for a long term without disturbing environmental sustainability. Conservation agriculture, zero-tillage, the use of cover crops, and sustainable soil management practices are managing agroecosystem in efficient ways (Lemaire et al. 2014; Reese et al. 2014). These practices are a sustainable approach that ensures higher biodiversity, maximizes productions, and ensures uncountable agroecosystem services. Zero-tillage or no-tillage practices ensure lesser soil disturbance with the maintenance of organic layers. This practice is an effective sustainable approach that maintains soil structure, functions, and SOC stock in sustainable ways. Applying these sustainable soil management practices ensures a healthier and productive agroecosystem. As per Horowitz et al. (2010), a total of 35.5% of areas of total agricultural land is utilized under zero-tillage-based farming system in the United States. On the contrary, tillage practices work poorly and have negative impacts on soil health and productivity. This intensive practice destroys soil structure which results in poor soil functions and erosion in North America (Carr et al. 2012). Sustainable agroecosystem practices are a good strategy that ensures soil enrichment, greater SOC pools, and higher crop productivity and improves the socioeconomic condition of poor farmers. A sustainable approach ensures sustainable production and resource maintenance for future needs. Sustainable agroecosystem practices intensify plant-soil productivity and profitability along with the conservation of natural resources.

Conservation and management of natural resources are very essential for life maintenance. Sustainable approaches promote a proper blending of resources, i.e., space, light, essential nutrients, etc., for a healthy and productive agroecosystem. These resources are very essential for proper growth and development of plants and animals. Thus, sustainable intensification must be employed in the agroecosystem for a greater diversified productivity without depleting other natural resources.

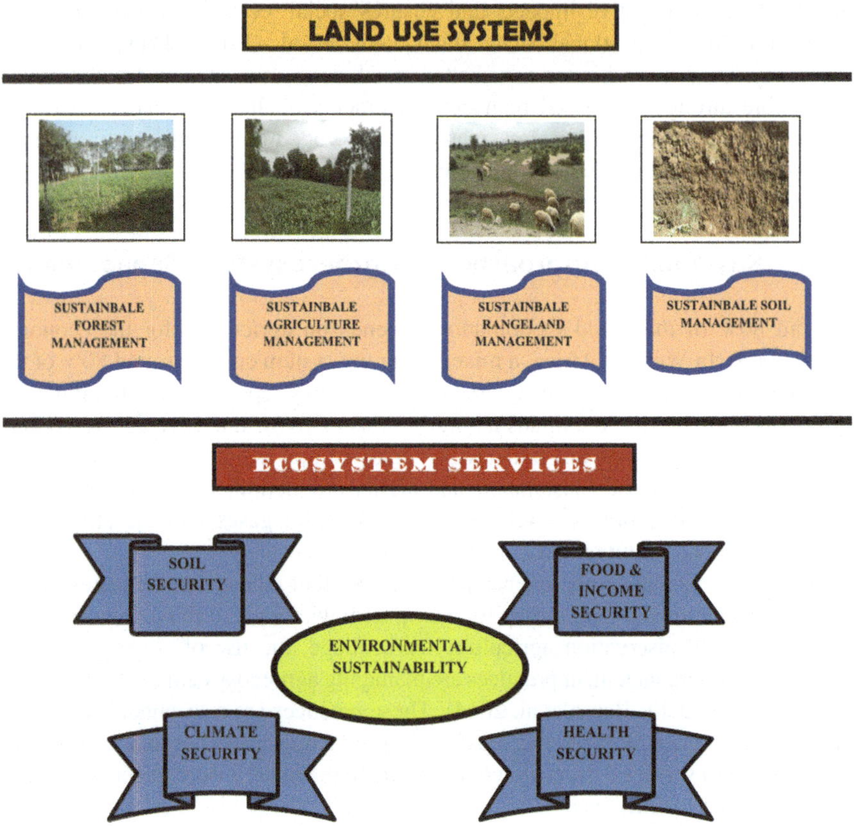

Fig. 12.2 Sustainable land use system-mediated ecosystem services for environmental sustainability (compiled by: Braimoh and Osaki 2010; Hasan et al. 2020; Xie et al. 2020)

Eco-designing and location-specific modeling are good instruments for sustainable-based agroecosystem practices. An effective design and model ensure a healthier and productive agroecosystem by minimizing C footprints for securing environmental sustainability. Ensuring higher nutrient use efficiency, soil erosion control, water purification, and other regulatory and provisioning agroecosystem services are possible through different sustainable approaches.

Sustainable land use practices, i.e., agriculture, forestry, horticulture, and rangeland, ensure food and nutritional security along with climate security through less environmental footprints. Sustainable land use system-mediated ecosystem services for environmental sustainability are depicted in Fig. 12.2 (Braimoh and Osaki 2010; Hasan et al. 2020; Xie et al. 2020). Sustainable soil management practices promise quantity and quality food productions along with greater environmental security through C sink. Sustainable and balanced application of fertilizers promotes soil-crop productivity along with soil-food-climate security in sustainable ways. Thus, applying sustainable approaches ensures better health and performance of agroecosystem with greater climate security and environmental sustainability (Ibrahim et al. 2015).

12.6 Ecological Approach for Agroecosystem Management

Applying ecological approaches in agroecosystem management are a good strategy that intensifies multifarious ecological services along with greater environmental sustainability. Ecology-based modeling and eco-designing are location specific and a prerequisite for a healthier and productive agroecosystem. Sustainable agroecosystem practices are based on an ecological concept with greater possibilities. Ecological approaches intensify agroecosystem diversity, productivity, health, and economy with greater ecological integrity. A greater ecological diversity can be possible under well-managed agroecosystem management. Soil and water conservations, erosion controls, efficient pollination mechanism, biodiversity maintenance, soil health and security, C balance, and climate change mitigations are important agro-ecological services. Environmental services through livestock-based agroecosystem and their impacts are depicted in Table 12.1.

Ecology-oriented models and agroecosystem design ensure these ecological services in sustainable ways. These practices make a pave for achieving the goal of sustainable development (Jhariya et al. 2021a, b; Meena et al. 2020a; Raj et al. 2021). Greater ecological integrity can be possible through adopting sustainable-based agroecosystem management. Applying conservation practices in agroecosystem management regulates ecological stability. Conservation agriculture

Table 12.1 Environmental services through livestock-based agroecosystem and their impacts

Livestock-based agroecosystem services	Impacts on agroecosystem	References
Higher crop diversifications and yields	Crop diversification and yield were increased when annual herbaceous crops followed pastures/forage crops which are justified by 71% of the respondents in the regions	Entz et al. (1995)
	The production is increased up to 1.0 Mg/ha under the corn-livestock/pasture system as compared to the sole corn system	Tracy and Zhang (2008)
Efficient nutrient cycling	Crop-livestock-based integrated system having close and efficient nutrient cycling as compared to the sole cropping system	Schiere et al. (2002)
Minimize plant disease	Integrating forage plants with cereal and pulse crops enhances diversification that minimizes the chance of insect pest and disease infestations. Proper rotation of forage-cereal-pulse crops was helpful to reduce soil-borne disease	Krupinsky et al. (2002)
Soil health and quality improvement	Both wheat-sheep and legume-wheat systems were significant in soil quality improvement in the regions of Australia and Alberta	Krall and Schuman (1996)

is a good driver that ensures higher plant production and soil-food-climate security with greater ecological stability (Choudhary et al. 2018). Greater soil-crop productiveness, food nutritional security, climate change mitigation, and better ecosystem health are possible through conservative agroecosystem practices. Plant diversification, minimizing soil disturbance, and improving the status of SOC are key principles of conservative agroecosystem practices (FAO 2015). Thus, soil enrichment, food security, health and economy maintenance, and water and climate regulations are indicators of healthy agroecosystem management that ensures greater ecological stability (Mutyasira et al. 2018).

12.7 Synergy Between Agroecosystem Management and Environmental Sustainability

The environment has various components such as atmosphere, lithosphere, hydrosphere, and biosphere. All these components have greater impacts on biodiversity and when combined regulate the ecosystem. Soil, plants, and animals are various elements of our environment that regulate environmental services and sustainability. Human-induced management of integrating soil, plants, and animals ensures greater diversity and performance of agroecosystem. Thus, agroecosystem management is quite necessary for resource conservation and a variety of environmental services. In this context, a question arises "how agroecosystem management performs greater environmental sustainability?". A synergy between agroecosystem management and environmental sustainability is depicted in Fig. 12.3 (Peterson et al. 2018; Córdoba et al. 2020).

The management of agroecosystem entirely reflects the soil, crop, and livestock management and its performance under human governance. These managing agroecosystem components entirely affect the health and susstainability of our environment. Of these resources, soil is an integral component of agroecosystem that regulates ecosystem processes and environmental services. The health and sustainability of our environment are linked with soil health and performance. We can't imagine the lives and environment without soil. A healthy and managed agroecosystem promotes a healthier and quality soil that ensures greater environmental sustainability. Soil provides several agroecosystem services through diversifying crops, microbes, and a variety of flora and fauna. Environmental sustainability can be possible through practicing sustainable soil-based agroecosystem management. These practices maximize productions with higher SOC pools as 1500 Gt (Lal 2004; Meena et al. 2020b; Tarnocai et al. 2009). Thus, storing SOC is higher than vegetational and atmospheric C that ensures higher environmental security. On the contrary, an unsustainable and unscientific management of agroecosystem affects SOC status by declining up to 10% in Europe (Janssens et al. 2005). Nearly, 30–50% of reduction in SOC value was observed under highly intensified agroecosystem practices in the United States (Reicosky 2003). Thus, these studies are enough to say about "how soil links with agroecosystem for better environmental sustainability?".

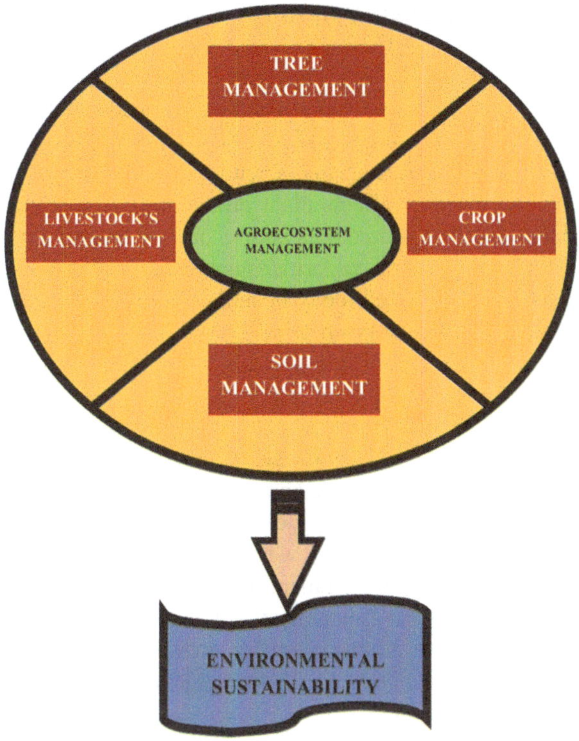

Fig. 12.3 A synergy between agroecosystem management and environmental sustainability (compiled by: Peterson et al. 2018; Córdoba et al. 2020)

Likewise, a better plant variety, disease-resistant crops, higher rotation tree species, energy plants, higher biomass producing plant species, and N-fixing tree species ensure climate security with environmental sustainability. This diversification of plant species, nature, and qualities ensures a healthier and productive agroecosystem which in turn maximizes environmental sustainability. The N-fixing legume plant maximizes the N content and regulates other essential nutrients with better SOC pools. Understanding tree-crop interaction reflects agroecosystem performance that affects the overall environmental health and sustainability. Thus, selecting tree-crop species and types and varieties along with the introduction of livestock affects agroecosystem management and services. Therefore, prior selection of species and agroecosystem models in any eco-regions, it is necessary to understand soil and environmental quality. Introducing tree species and disease-resistant plant varieties ensures a greater and diversified agroecosystem. However, woody perennial tree-crops and annual and perennial grasses/pastures hold soils and perform erosion controls. Moreover, these species provide greater climate security through C sequestration and better SOC pools. These agroecosystem services depend on tree-crop-animal interaction that regulates environmental health and sustainability.

12.8 Agroecosystem for Climate Change Mitigation: Win-a-Win Strategy

The issue of climate changes and their impacts on biodiversity have been discussed in many literatures and policy papers. Climate change makes blurred images among researchers, academicians, scientists, and farmers. Unexpected monsoon due to extreme weather affects plant productivity and related income on which farmers depend. This has created a poor economy that puts pressure on farmers to commit suicide. Thus, trends of farmer suicide are directly linked with climate change that is rigorously discussed today in India. Further, a burgeoning population necessitates food demands that promote several unsustainable land use practices. Deforestation and land conversion into agriculture, i.e., agriculture land expansion, lead to land degradation. Agricultural land expansion and its impacts on yield increment were studied in different regions of the world from 1961 to 1963 and 1989 to the 1990s (Fig. 12.4). According to the figure, declining trends of area increments were recorded from the sub-Saharan African region to East Asia under low economy countries. But the values of yield increments were rising. Thus, an inversely proportional relationship was observed between area and yield. High-income countries have less increment in areas as 2% but plant yields increased up to 92%. Around 8% of area increments reflects a 92% increment in yield. Therefore, it is clear that low economic countries having less percentage of area expansion resulting in higher yield increment. This trend is more prominent in high-economic countries (Al-Kaisi and Lal 2017).

The forest stores billion tons of C in the form of biomass and regulates C storage, flux, and ecosystem processes. Continuously felling of timber will deprive environmental quality by releasing GHG emission. Similarly, agricultural land expansion

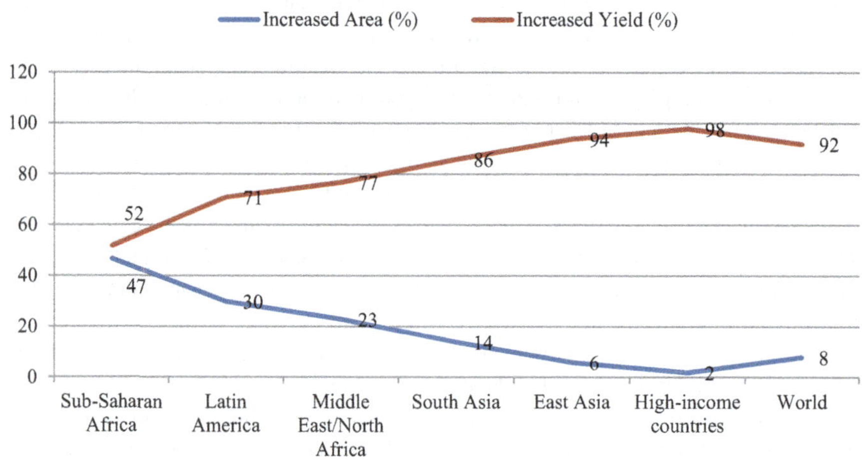

Fig. 12.4 Agricultural land expansion and its impacts on yield increment in percentage in different regions of the world during 1961 to 1963 and 1989 to the 1990s (compiled by: Al-Kaisi and Lal 2017)

affects land quality and deprives other natural resources. Intensive agroecosystem practices and heavy mechanization release C into the atmosphere. This intensified and unscientific land use practices disturb C balance in the atmosphere. Releasing excessive CO_2 increases average earth temperature causing global warming and climate change phenomenon. In this context, a climate-resilient land use practice works as a boon by stopping excessive emissions and related harmful impacts on the environment. Moreover, climate-resilient agroecosystem becomes a good strategy that signifies soil-plant productivity along with climate security (Clay and Zimmerer 2020). Therefore, the question arises that "how a managed agroecosystem minimize climate change impacts?". In this perspective, it is possible by adopting sustainable agroecosystem practices instead of intensified practices. Agroecosystem-mediated C-based environmental services to minimize C footprint are depicted in Fig. 12.5 (Platis et al. 2019).

Climate-smart agroecosystem comprises various sustainable-based agriculture practices that ensure less emission with high productivity. Adopting conservation agriculture, mulching, and use of cover crop and no-tillage practices works more efficiency (Lal 2015). These practices require less chemical input with minimum soil disturbance that enhances higher organic C with less GHG emissions. Better rice management practices in agroecosystem minimize methane emissions into the atmosphere. Similarly, less application of N fertilizers minimizes nitrous oxide and GHG emission. Climate-smart agriculture is a sustainable approach which relates to climate-resilient agroecosystem practices that are discussed in FAO, CGIAR, and World Bank platforms (Campbell et al. 2020). Maximizing crop productivity, less GHG emissions, and mitigating climate change are triple-win situations that can be possible through well-managed agroecosystem practices (FAO 2019). Thus, agroecosystem management relies on this concept based on an ecological principle for a better environment.

12.9 Constraints in Agroecosystem Management

The distribution, nature, and practices of agroecosystem vary as per varying biophysical, topography, and climatic situations. The need and potential of agroecosystem practices are based on agroecological conditions. Accordingly, management practices are required for better health and productivity of agroecosystem. Intensive practices are major hurdles and constraints in agroecosystem management. These practices destroy ecosystem health and environmental sustainability. Declining soil health, poor crop productivity, human-animal health issues, and land and environmental degradation, etc. are a major constraint under intensified agroecosystem practices. Less availability of plus tree, hybrid plants, clones, and disease-resistant varieties are also constraints that choke the agroecosystem health and productivity. Poor irrigation facility and less water supply and supply of good-quality seeds are other constraints that affect the overall management of agroecosystem. Rising temperature and extreme weather are curses for farmers due to bad impacts on crop health and production in the agroecosystem. Specially, these

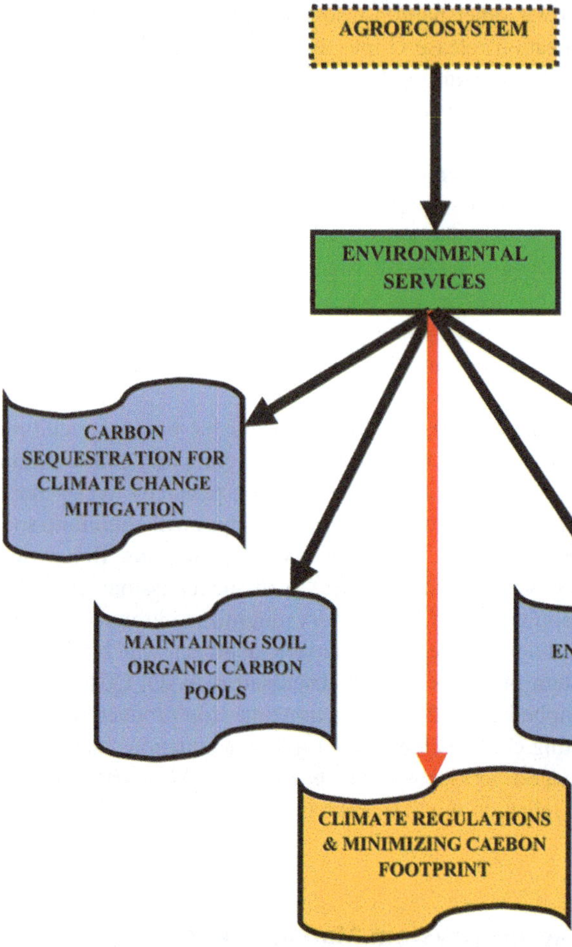

Fig. 12.5 Agroecosystem-mediated C-based environmental services to minimize C footprints (compiled by: Platis et al. 2019)

constraints are prominent in arid and dry regions of the tropical world's agroecosystem. The management is also not effective due to harsh climate and climate change issues.

Farmers' awareness is very poor in the adoption of sustainable and ecological intensification in the agroecosystem. These practices ensure a healthy and productive agroecosystem in any region. But, they adopt it poorly and practice intensive agroecosystem for bumper production. They use high synthetic inputs and heavy mechanizations in their farms. Declining soil health due to intensified practices is a major constraint and challenge of the days. Using excessive chemical fertilizers in farms not only releases GHG emissions but also deprives the quality of land/soil. In turn, poor soil delivers poor ecosystem services. Thus, we must focus and be serious about the negative impact which affects health-wealth of human beings and environmental sustainability. These practices obviously enhance production but at the cost of environmental health and sustainability. Thus, the practices of the sustainable agroecosystem, climate-resilient agroecosystem, climate-smart agriculture, etc. work more efficiently. These sustainable and ecology-based practices must be adopted and utilized for a healthy and productive environment. These constraints can be solved by the adoption of well-managed agroecosystem practices and proper awareness among farmers, researchers, stakeholders, and policy makers.

12.10 Research and Development

An authentic research is needed for the proper functioning of the agroecosystem. A large gap exists in soil-based research for better agroecosystem structure and services. Healthy soil is necessary for agroecosystem health and productivity. Research must be designed to create agroecosystem models for better soil productivity and related ecological services. Soil testing and quality are the prerequisites for the establishment and management of agroecosystem throughout the tropical world. Analysis of soil fertility and index are quite interesting indicators that determine the soil health and quality. Fertility is linked with nutrient types and its availability on which plant growth and development depend. Thus, soil testing and fertility index evaluation must be organized for a better understanding of agroecosystem management.

Climate-resilient agroecosystem should be on track for minimizing C and environmental footprints. The production functions of agroecosystem are shifted towards exploring the environmental footprints (Walbridge and Shafer 2011). Research-based eco-designed and models are promoted in agroecosystem practices to evaluate agroecosystem services (Khan et al. 2021a, b). Research must be shifted from small farms to large areas on a spatial and temporal scale (Carpenter et al. 2006; Robertson et al. 2008). In this context, USDA has initiated Long-Term Agroecosystem Research Network which comprises 18 sites for long-term studies (Karlen et al. 2014).

A research must be initiated in the context of a sustainable agroecosystem. Adopting sustainable intensification in the agroecosystem is based on ecological principles. A research must be in the ecological approach that helps in building a sustainable agroecosystem in the tropical world. Research and development should focus on soil-crop productivity enhancement for building climate-resilient agroecosystem. Further, research should be in frame of creating various eco-designed models based on location specific for agroecosystem practices. Enhancing soil fertility, C sequestration potential, maximizing crop productivity, and high plant diversifications are effectively possible through applying sustainable agroecosystem research practices (Andres and Bhullar 2016; Smith et al. 2017).

12.11 Policies and Legal Framework

An effective policy is needed for the proper management of the agroecosystem. Agroforestry performance totally depends upon the type of practices and management. Policies must be followed in favor of promoting sustainable agroecosystem practices. Practicing intensive agroecosystem leads to land degradation and affects the overall ecosystem services. Releasing GHGs and climate change are a burning issue which is possible through intensive agroecosystem practices. A legal framework is developed to check and stop these deleterious farming practices, and instead of that sustainable and ecological approaches are encouraged for better ecosystem and sustainable environment.

A policy must be framed for sustainable land use practices for better soil-crop productions. A better agroecosystem practice minimizes C footprints and promotes environmental sustainability. In this context, a policy must be developed to draw attention in various national and international organizations. For minimizing GHG emissions, climate-resilient agroecosystem practices should be promoted. Regulatory frameworks must be created in this context, and policy should be promoted in this direction (Pattey et al. 2005). Integrating leguminous and multipurpose trees in the agroecosystem will improve soil fertility, nutrient availability, and greater productivity. A policy for adopting climate-smart forestry, agroforestry, and sustainable soil, forest, and agroecosystem management should be framed in a worldwide forum. Understanding tree-crop interactions is such a good topic that covers the management practices for a healthier and productive agroecosystem. Thus, reforming policy in this direction will ensure greater agroecosystem services for a sustainable world.

12.12 Conclusions

Agroecosystem management is very necessary for ecosystem health and environmental sustainability. Deforestation for agriculture land expansion needs to be continued for satisfying global food requirements. An intensive practice in agroecosystem promotes land degradations and deprives environmental health and sustainability. These practices not only affect soil-crop productivity but also affect soil-food-climate security at a global scale. In this context, adopting sustainable agroecosystem practices enhances soil fertility and maximizes crop diversification and productivity. These practices maintain health and productivity along with greater environmental sustainability. Adopting eco-designing and sustainable modeling in agroecosystem provides uncountable ecological services. Water purification, erosion controls, pollination mechanisms, C balance, climate change mitigations, etc. are several environmental services that can be achieved through agroecosystem management. Adopting climate-smart agroecosystem and sustainable soil management practices plays a key role in greater soil-crop productivity along with environmental sustainability. A good governance and effective policy are needed to promote sustainable agroecosystem practices for better environmental quality. Higher crop productivity, soil fertility, and climate security are possible through scientific ecology-oriented research-based agroecosystem management practices. These practices not only improve productivity and profitability but also minimize C and environmental footprints and ensure sustainable development.

12.13 Future Roadmaps

Food requirement for satisfying global populations is the urgent need of today. For that, agricultural land expansion and intensification exert negative impacts on our environment. These practices ensure higher production, but degradation of soil and the environment is equally promoted. A roadmap must be created to maximize global production while ensuring environmental sustainability for a long-term basis. Conserving natural resources for satisfying present and future needs is of utmost importance. Therefore, agroecosystem management needs to be progressed in this direction to ensure a healthier and productive future environment. Similarly, a paradigm shift in agroecosystem is used to understand soil quality, plant productivity, human-animal health, and environmental quality. A roadmap should be adopted to see soil variations, and accordingly ecological and sustainable agroecosystem models are developed. Food systems should be shifted from quantity to quality productions. In this context, an effective framework should be developed for maintaining food and nutritional security for present and future generations (Gliessman 2016). Climate-resilient and pollution-free world can be created through sustainable agroecosystem practices. Involving conservation agriculture, organic agroecosystem practices, and no-tillage and mulching practices is possible under sustainable agroecosystem practices. Thus, a future roadmap must be developed to create a sustainable and pollution-free world by chemical-free agroecosystem for greater environmental sustainability.

References

Alexander P, Rounsevell MDA, Dislich C, Dodson JR, Engstrom K, Moran D (2015) Drivers for global agricultural land use change: the nexus of diet, population, yield and bioenergy. Glob Environ Chang 35:138–147. https://doi.org/10.1016/j.gloenvcha.2015.08.011

Al-Kaisi M, Lal R (2017) Conservation agriculture systems to mitigate climate variability effects on soil health. In: Al-Kaisi M, Lowery B (eds) Soil health and intensification of agroecosystems. Academic, New York, pp 79–107

Andres C, Bhullar GS (2016) Sustainable intensification of tropical agro-ecosystems: need and potentials. Front Environ Sci 4:5. https://doi.org/10.3389/fenvs.2016.00005

Banerjee A, Jhariya MK, Yadav DK, Raj A (2020) Environmental and sustainable development through forestry and other resources. Apple Academic Press Inc., New York

Banerjee A, Meena RS, Jhariya MK, Yadav DK (2021a) Agroecological footprints management for sustainable food system. Springer, Singapore

Banerjee A, Jhariya MK, Meena RS, Yadav DK (2021b) Ecological footprints in agroecosystem- an overview. In: Banerjee A, Meena RS, Jhariya MK, Yadav DK (eds) Agroecological footprints management for sustainable food system. Springer, Singapore, pp 1–23

Banerjee A, Jhariya MK, Raj A, Yadav DK, Khan N, Meena RS (2021c) Land footprint management and policies. In: Banerjee A, Meena RS, Jhariya MK, Yadav DK (eds) Agroecological footprints management for sustainable food system. Springer, Singapore, pp 221–246

Banerjee A, Jhariya MK, Raj A, Yadav DK, Khan N, Meena RS (2021d) Energy and climate footprint towards the environmental sustainability. In: Banerjee A, Meena RS, Jhariya MK, Yadav DK (eds) Agroecological footprints management for sustainable food system. Springer, Singapore, pp 415–443

Behnke GD, Zuber SM, Pittelkow CM, Nafziger ED, Villamil MB (2018) Long-term crop rotation and tillage effects on soil greenhouse gas emissions and crop production in Illinois, USA. Agric Ecosyst Environ 261:62–70

Bernard B, Lux A (2017) How to feed the world sustainably: an overview of the discourse on agroecology and sustainable intensification. Reg Environ Chang 17:1279–1290. https://doi.org/10.1007/s10113-016-1027-y

Bernués A, Rodríguez-Ortega T, Ripoll-Bosch R, Alfnes F (2014) Socio-cultural and economic valuation of ecosystem services provided by Mediterranean mountain agroecosystems. PLoS One 9(7):e102479

Braimoh A, Osaki M (2010) Land-use change and environmental sustainability. Sustain Sci 5:5–7. https://doi.org/10.1007/s11625-009-0092-2

Brauman KA, Daily GC, Duarte TK, Mooney HA (2007) The nature and value of ecosystem services: an overview highlighting hydrologic services. Ann Rev Environ Resour 32:67–98

Campbell BM, Thornton P, Zougmoré R, van Asten P, Lipper L (2020) Sustainable intensification: what is its role in climate smart agriculture? Curr Opin Environ Sustain 8:39–43

Carpenter SR, De Fries R, Dietz T, Mooney HA, Polasky S, Reid WV (2006) Millennium ecosystem assessment: research needs. Science 314:257–258

Carr PM, Mäder P, Creamer NG, Beeby JS (2012) Editorial: overview and comparison of conservation tillage practices and organic farming in Europe and North America. Renew Agric Food Syst 27:2–6

Choudhary M, Jat HS, Datta A, Yadav AK, Sapkota TB, Mondal S, Meena RP, Sharma PC, Jat ML (2018) Sustainable intensification influences soil quality, biota, and productivity in cereal-based agroecosystems. Appl Soil Ecol 126:189–198

Clay N, Zimmerer KS (2020) Who is resilient in Africa's green revolution? Sustainable intensification and climate smart agriculture in Rwanda. Land Use Policy 97:104558. https://doi.org/10.1016/j.landusepol.2020.104558

Congreves KA, Hayes A, Verhallen EA, Van Eerd LL (2015) Long-term impact of tillage and crop rotation on soil health at four temperate agroecosystems. Soil Tillage Res 152:17–28

Córdoba C, Triviño C, Toro Calderón J (2020) Agroecosystem resilience. A conceptual and methodological framework for evaluation. PLoS One 15(4):e0220349. https://doi.org/10.1371/journal.pone.0220349

Duru M, Therond O, Martin G, Martin-Clouaire R, Magne MA, Justes E, Journet EP, Aubertot JN, Savary S, Bergez JE, Sarthou JP (2015) How to implement biodiversity-based agriculture to enhance ecosystem services: a review. Agron Sustain Dev 35:1259–1281. https://doi.org/10.1007/s13593-015-0306-1

Ellis EC, Ramankutty N (2008) Putting people in the map: anthropogenic biomes of the world. Front Ecol Environ 6:439–447

Emmerson M, Morales MB, Oñate JJ, Batáry P, Berendse F, Liira J, Aavik T, Guerrero I, Bommarco R, Eggers S, Part T, Tscharntke T, Weisser W, Clement L, Bengtsson J (2016) How agricultural intensification affects biodiversity and ecosystem services. Adv Ecol Res 55:43–97

Entz MH, Bullied WJ, Katepa-Mupondwa F (1995) Rotational benefits of forage crops in Canadian prairie cropping systems. J Prod Agric 8:521–529

FAO (2003) Synthesis report on the role of agriculture on environmental externalities. Food and Agriculture Organization, Rome

FAO (2015) Conservation agriculture. Food and Agriculture Organization of the United Nations, Rome
FAO (2019) Climate-smart agriculture and the sustainable development goals: mapping interlinkages, synergies and trade-offs and guidelines for integrated implementation. FAO, Rome
Felipe-Lucia MR, Comín FA, Bennett EM (2014) Interactions among ecosystem services across land uses in a floodplain agroecosystem. Ecol Soc 19(1):20
Foley JA, DeFries R, Asner GP, Barford C, Bonan G, Carpenter SR, Chapin FS, Coe MT, Daily GC, Gibbs HK, Helkowski JH, Holloway T, Howard EA, Kucharik CJ, Monfreda C, Patz JA, Prentice IC, Ramankutty N, Snyder PK (2005) Global consequences of land use. Science 309 (5734):570–574. https://doi.org/10.1126/science.1111772
Gaba S, Bretagnolle F, Rigaud T, Philippot L (2014) Managing biotic interactions for ecological intensification of agroecosystems. Front Ecol Evol 2:29. https://doi.org/10.3389/fevo.2014.00029
Gadanakis Y, Bennett R, Park J, Areal FJ (2015) Evaluating the sustainable intensification of arable farms. J Environ Manag 150:288–298. https://doi.org/10.1016/j.jenvman.2014.10.005
Gliessman SR (2015) Agroecology: the ecology of sustainable food systems, 3rd edn. CRC Press, Boca Raton
Gliessman S (2016) Transforming food systems with agroecology. Agroecol Sustain Food Syst 40 (3):187–189. https://doi.org/10.1080/21683565.2015.1130765
Hasan S, Shi W, Zhu X (2020) Impact of land use land cover changes on ecosystem service value – a case study of Guangdong, Hong Kong, and Macao in South China. PLoS One 15(4): e0231259. https://doi.org/10.1371/journal.pone.0231259
Horowitz J, Ebel R, Ueda K (2010) No-till fasrming is a growing practice. Economic Information Bulletin No. 70. USDA Economic Research Service, Washington, DC
Horrigan L, Lawrence RS, Walker P (2002) How sustainable agriculture can address the environmental and human health harms of industrial agriculture. Environ Health Perspect 110:445–456
Hurni H, Giger M, Liniger H, Studer RM, Messerli P, Portner B, Schwilch G, Wolfgramm B, Breu T (2015) Soils, agriculture and food security: the interplay between ecosystem functioning and human well-being. Curr Opin Environ Sustain 15:25–34. https://doi.org/10.1016/j.cosust.2015.07.009
Ibrahim M, Alhameid A, Kumar S, Chintala R, Sexton P (2015) Long-term tillage and crop rotation impacts on a Northern Great Plains. Adv Crop Sci Technol 3:2
Iqbal MM, Goheer MA (2008) Greenhouse gas emissions from agro-ecosystems and their contribution to environmental change in the Indus Basin of Pakistan. Adv Atmos Sci 25:1043
Jankielsohn A (2015) The hidden cost of eating meat in South Africa: what every responsible consumer should know. J Agric Environ Ethics 28:1145–1157. https://doi.org/10.1007/s10806-015-9581-8
Janssens IA, Freibauer A, Schlamadinger B, Ceulemans R, Ciais P, Dolman AJ, Heimann M, Nabuurs GJ, Smith P, Valentini R, Schulze ED (2005) The carbon budget of terrestrial ecosystems at country scale – a European case study. Biogeosciences 2:15–26
Jhariya MK, Bargali SS, Raj A (2015) Possibilities and perspectives of agroforestry in Chhattisgarh. In: Zlatic M (ed) Precious forests-precious earth. InTech, Croatia, pp 237–257
Jhariya MK, Yadav DK, Banerjee A (2018a) Plant mediated transformation and habitat restoration: phytoremediation an eco-friendly approach. In: Gautam A, Pathak C (eds) Metallic contamination and its toxicity. Daya Publishing House, New Delhi, pp 231–247
Jhariya MK, Banerjee A, Yadav DK, Raj A (2018b) Leguminous trees an innovative tool for soil sustainability. In: Meena RS, Das A, Yadav GS, Lal R (eds) Legumes for soil health and sustainable management. Springer, New York, pp 315–345
Jhariya MK, Banerjee A, Meena RS, Yadav DK (2019a) Sustainable agriculture, forest and environmental management. Springer, Singapore
Jhariya MK, Yadav DK, Banerjee A (2019b) Agroforestry and climate change: issues and challenges. Apple Academic Press Inc., New York

Jhariya MK, Meena RS, Banerjee A (2021a) Ecological Intensification of Natural Resources for Sustainable Agriculture. Springer, Singapore

Jhariya MK, Meena RS, Banerjee A (2021b) Ecological intensification of natural resources towards sustainable productive system. In: Ecological intensification of natural resources for sustainable agriculture. Springer, Singapore

Karlen DL, Peterson GA, Westfall DG (2014) Soil and water conservation: our history and future challenges. Soil Sci Soc Am J 78:1493–1499

Karp DS, Mendenhall CD, Sandi RF, Chaumont N, Ehrlich PR, Hadly EA, Daily GC (2013) Forest bolsters bird abundance, pest control, and coffee yield. Ecol Lett 16(11):1339–1347. https://doi.org/10.1111/ele.12173

Khan N, Jhariya MK, Yadav DK, Banerjee A (2020a) Herbaceous dynamics and CO_2 mitigation in an urban setup- a case study from Chhattisgarh, India. Environ Sci Pollut Res 27(3):2881–2897. https://doi.org/10.1007/s11356-019-07182-8

Khan N, Jhariya MK, Yadav DK, Banerjee A (2020b) Structure, diversity and ecological function of shrub species in an urban setup of Sarguja, Chhattisgarh, India. Environ Sci Pollut Res 27 (5):5418–5432. https://doi.org/10.1007/s11356-019-07172-w

Khan N, Jhariya MK, Raj A, Banerjee A, Meena RS (2021a) Soil carbon stock and sequestration: implications for climate change adaptation and mitigation. In: Jhariya MK, Meena RS, Banerjee A (eds) Ecological intensification of natural resources for sustainable agriculture. Springer, Singapore

Khan N, Jhariya MK, Raj A, Banerjee A, Meena RS (2021b) Eco-designing for sustainability. In: Jhariya MK, Meena RS, Banerjee A (eds) Ecological intensification of natural resources for sustainable agriculture. Springer, Singapore

Klein AM, Vaissiere BE, Cane JH, Steffan-Dewenter I, Cunningham SA, Kremen C, Tscharntke T (2007) Importance of pollinators in changing landscapes for world crops. Proc R Soc B Biol Sci 274:303–313. https://doi.org/10.1098/rspb.2006.3721

Krall JM, Schuman GE (1996) Integrated dryland crop and livestock production systems on the Great Plains: extent and outlook. J Prod Agric 9:187–191

Krishna KR (2010) Agroecosystems of South India: nutrient dynamics, ecology and productivity. Brown Walker Press, Berkeley

Krupinsky JM, Bailey KL, McMullen MP, Gossen BD, Turkington TK (2002) Managing plant disease risk in diversified cropping systems. Agron J 94:198–209

Kumar S, Meena RS, Jhariya MK (2020) Resources use efficiency in agriculture. Springer, Singapore

Kumar S, Meena RS, Datta R, Verma SK, Yadav GS, Pradhan GS, Molaei A, Rahman GKMM, Mashuk HA (2020a) Legumes for carbon and nitrogen cycling: an organic approach. Carbon Nitrogen Cycl Soil. https://doi.org/10.1007/978-981-13-7264-3_10

Kumar S, Meena RS, Singh RK, Munir TM, Datta R, Danish S, Singh GS, Kumar S (2021) Soil microbial and nutrient dynamics under different sowings environment of Indian mustard (*Brassica juncea* L.) in rice based cropping system. Sci Rep 11:5289

Lal R (2004) Soil carbon sequestration impacts on global climate change and food security. Science 304:1623–1627

Lal R (2015) A system approach to conservation agriculture. J Soil Water Conserv 70(4):82–88

Lal R, Delgado JA, Groffman PM, Millar N, Dell C, Rotz A (2011) Management to mitigate and adapt to climate change. J Soil Water Conserv 66:276–285

Landis DA, Wratten SD, Gurr GM (2000) Habitat management to conserve natural enemies of arthropod pests in agriculture. Annu Rev Entomol 45:175–201

Lemaire G, Franzluebbers A, de Faccio Carvalho PC, Dedieu B (2014) Integrated crop–livestock systems: strategies to achieve synergy between agricultural production and environmental quality. Agric Ecosyst Environ 190:4–8

Lennon JJ (2015) Potential impacts of climate change on agriculture and food safety within the island of Ireland. Trends Food Sci Technol 44:1–10

Meena RS, Lal R (2018) Legumes for soil health and sustainable management. Springer, Singapore, p 541

Meena RS, Kumar V, Yadav GS, Mitran T (2018) Response and interaction of *Bradyrhizobiumjaponicum* and Arbuscularmycorrhizal fungi in thesoybeanrhizosphere: a review. Plant Growth Regul 84:207–223. https://doi.org/10.1007/s10725-017-0334-8

Meena RS, Lal R, Yadav GS (2020) Long term impacts of topsoil depth and amendments on soil physical and hydrological properties of an Alfisol in Central Ohio, USA. Geoderma 363:1141164. https://doi.org/10.1016/j.geoderma.2019.114164

Meena RS, Lal R, Yadav GS (2020a) Long-term impact of topsoil depth and amendments on carbon and nitrogen budgets in the surface layer of an Alfisol in Central Ohio. Catena 194:104752. https://doi.org/10.1016/j.catena.2020.104752

Meena RS, Kumar S, Datta R, Lal R, Vijaykumar V, Brtnicky M, Sharma MP, Yadav GS, Jhariya MK, Jangir CK, Pathan SI, Dokulilova T, Pecina V, Marfo TD (2020b) Impact of agrochemicals on soil microbiota and management: a review. Landscape 9(2):34. https://doi.org/10.3390/land9020034

Millenium Ecosystem Assessment (MEA) (2005) Millenium ecosystem assessment, ecosystems and human well-being. World Resources Institute, Washington, DC

Mutyasira V, Hoag D, Pendell D, Manning D (2018) Is sustainable intensification possible? Evidence from Ethiopia. Sustain For 10(11):4174

Oerke EC (2005) Crop losses to pests. J Agric Sci 144:31–43

Painkra GP, Bhagat PK, Jhariya MK, Yadav DK (2016) Beekeeping for poverty alleviation and livelihood security in Chhattisgarh, India. In: Narain S, Rawat SK (eds) Innovative technology for sustainable agriculture development. Biotech Books, New Delhi, pp 429–453

Pattey E, Trzcinski MK, Desjardins RL (2005) Quantifying the reduction of greenhouse gas emissions as a result of composting dairy and beef cattle manure. Nutr Cycl Agroecosyst 72:173–187

Peterson CA, Eviner VT, Gaudin AC (2018) Ways forward for resilience research in agroecosystems. Agric Syst 162:19–27

Pimentel D, Zuniga R, Morrison D (2005) Update on the environmental and economic costs associated with alien-invasive species in the United States. Ecol Econ 52:273–288

Platis DP, Anagnostopoulos CD, Tsaboula AD, Menexes GC, Kalburtji KL, Mamolos AP (2019) Energy analysis, and carbon and water footprint for environmentally friendly farming practices in agroecosystems and agroforestry. Sustain For 11:1664

Raj A, Jhariya MK, Harne SS (2018a) Threats to biodiversity and conservation strategies. In: Sood KK, Mahajan V (eds) Forests, climate change and biodiversity. Kalyani Publisher, New Delhi, p 381

Raj A, Jhariya MK, Bargali SS (2018b) Climate smart agriculture and carbon sequestration. In: Pandey CB, Gaur MK, Goyal RK (eds) Climate change and agroforestry: adaptation mitigation and livelihood security. New India Publishing Agency (NIPA), New Delhi

Raj A, Jhariya MK, Yadav DK, Banerjee A, Meena RS (2019a) Agroforestry: a holistic approach for agricultural sustainability. In: Jhariya MK, Banerjee A, Meena RS, Yadav DK (eds) Sustainable agriculture, forest and environmental management. Springer, Singapore

Raj A, Jhariya MK, Banerjee A, Yadav DK, Meena RS (2019b) Soil for sustainable environment and ecosystems management. In: Jhariya MK, Banerjee A, Meena RS, Yadav DK (eds) Sustainable agriculture, forest and environmental management. Springer, Singapore

Raj A, Jhariya MK, Yadav DK, Banerjee A (2020) Climate change and agroforestry systems: adaptation and mitigation strategies. Apple Academic Press Inc., New York

Raj A, Jhariya MK, Khan N, Banerjee A, Meena RS (2021) Ecological intensification for sustainable development. In: Ecological intensification of natural resources for sustainable agriculture. Springer, Singapore

Reese CL, Clay DE, Clay SA, Bich AD, Kennedy AC, Hansen SA, Moriles J (2014) Winter cover crops impact on corn production in semiarid regions. Agron J 106(4):1479–1488. https://doi.org/10.2134/agronj13.0540

Reicosky DC (2003) Tillage-induced CO_2 emissions and carbon sequestration: effect of secondary tillage and compaction. In: Garcia-Torres L, Benites J, Martinez-Vilela A, Holgado-Cabrera A (eds) Conservation agriculture. Kluwer Academic Publishers, Dordrecht

Robertson GP, Allen VG, Boody G, Boose ER, Creamer NG, Drinkwater LE, Gosz JR, Lynch L, Havlin JL, Jackson LE, Pickett STA, Pitelka L, Randall A, Reed AS, Seastedt TR, Waide RB, Wall DH (2008) Long-term agricultural research: a research, education, and extension imperative. Bioscience 58(7):640–645

Schiere JB, Ibrahim MNM, Van Keulen H (2002) The role of livestock for sustainability in mixed farming: criteria and scenario studies under varying resource allocation. Agric Ecosyst Environ 90:139–153

Sheoran S, Kuma S, Kumar P, Meena RS, Rakshit S (2021) Nitrogen fixation in maize: breeding opportunities. Theor Appl Genet. https://doi.org/10.1007/s00122-021-03791-5

Singh NR, Jhariya MK (2016) Agroforestry and agrihorticulture for higher income and resource conservation. In: Narain S, Rawat SK (eds) Innovative technology for sustainable agriculture development. Biotech Books, New Delhi, pp 125–145

Smith A, Snapp S, Chikowo R, Thorne P, Bekunda M, Glover J (2017) Measuring sustainable intensification in smallholder agroecosystems: a review. Glob Food Secur 12:127–138. https://doi.org/10.1016/j.gfs.2016.11.002

Steiner KG (1982) Intercropping in tropical smallholder agriculture with special reference to West Africa. Schriftenreihe Deutsches Gesellschaft echnische Zusammenarbeit 137:303

Sundriyal M, Sundriyal R (2004) Wild edible plants of the Sikkim Himalaya: Nutritive values of selected species. Econ Bot 55:286–299

Tarnocai C, Canadell JG, Schuur EAG, Kuhry P, Mazhitova G, Zimov S (2009) Soil organic carbon pools in the northern circumpolar permafrost region. Glob Biogeochem Cycles 23:1029–1040

Tilman D, Fargione J, Wolff B, D'Antonio C, Dobson A, Howarth R (2002) Forecasting agriculturally driven global environmental change. Science 292:281–284

Tracy BF, Zhang Y (2008) Soil compaction, corn yield response, and soil nutrient pool dynamics within an integrated crop-livestock system in Illinois. Crop Sci 48:1211–1218

UN Water (United Nations Water) (2013) Agriculture and food security. United Nations, Rome

Walbridge MR, Shafer SR (2011) A long-term agro-ecosystem research (LTAR) network for agriculture. In: Medley CN, Patterson G, Parker MJ (eds) Observing, studying, and managing change: proceedings of the fourth interagency conference in the watersheds. US Geological Survey Scientific Investigations Report # 2011-5169, pp 26–30

Whitfield S, Challinor AJ, Rees RM (2018) Frontiers in climate smart food systems: outlining the research space. Front Sustain Food Syst. https://doi.org/10.3389/fsufs.2018.00002

Xie H, Zhang Y, Choi Y, Li F (2020) A scientometrics review on land ecosystem service research. Sustain For 12:2959. https://doi.org/10.3390/su12072959

Zhang W, Ricketts TH, Kremen C, Carney K, Swinton SM (2007) Ecosystem services and dis-services to agriculture. Ecol Econ 64:253–260

Zhu W, Wang S, Caldwell CD (2012) Pathways of assessing agroecosystem health and agroecosystem management. Acta Ecol Sin 32:9–17

Ecological Intensification for Soil Management: Biochar – A Natural Solution for Soil from Agricultural Residues

13

Hana'a Burezq and Merlin K. Davidson

Abstract

The prominent concern of scientific community on sustainable agriculture merges the environmental objectives of soil management with increased food productivity to feed the ever-increasing world population. The concept of sustainable intensification has become apparent as a conspicuous outlying of this challenge. Soil is a momentous base of enriched nutrients and habitation for various microfloras. Globally, the agricultural land has been depleted, and soil quality is degraded by disproportionate addition of chemical fertilizers and other contaminants. Excessive plant and animal agricultural residues are being burnt or wasted, which can be recycled to favorable means adding benefits to sustain soil productivity. Consequently, a reformed attention is a prerequisite to preserve agricultural soil for efficient crop production by utilizing agricultural residues; biochar gives a natural solution for sustainable intensification of agricultural soil. Biochar as a soil organic amendment enriched with carbon enhances the eminence of soil and holds nutrients, thereby enhancing plant growth. In addition, it paves way for improved soil health as it affects the harmfulness, carriage, and destiny of heavy metals due to upgraded soil adsorption capacity. The improved soil properties and adsorption ability of biochar are attributed to their nutrient retention ability, high surface area, permeable nature, and ability to enhance microbial activity that leads to increased crop yield and productivity. The risk of soil compaction is minimized by biochar amendment as the stretchable asset of soil cores is decreased. Moreover, recycling agricultural residues into a precious soil nutrient makes a rural livelihood for the farming community. The productive

H. Burezq · M. K. Davidson (✉)
Desert Agriculture and Ecosystems Program, Environment and Life Sciences Research Center, Kuwait Institute for Scientific Research, Safat, Kuwait
e-mail: haborizq@kisr.edu.kw; mdavidson@kisr.edu.kw

M. K. Jhariya et al. (eds.), *Sustainable Intensification for Agroecosystem Services and Management*, https://doi.org/10.1007/978-981-16-3207-5_13

impacts of biochar amendment on crop growth and soil quality recommends biochar as a sustainable solution to withstand deficit of essential nutrients in agricultural crop productivity. This review highlights the properties of biochar and its utility in sustainable agricultural production by ecological intensification of agroecosystem services.

Keywords

Agricultural residues · Biochar · Eco-intensification · Soil management

Abbreviations

%	Percent
AC	Activated Carbon
AEC	Anion Exchange Capacity
AMF	Arbuscular Mycorrhizal Fungi
BMC	Biomass Carbon
C	Carbon
Ca	Calcium
Cd	Cadmium
CEC	Cation Exchange Capacity
CFU	Colony Forming Unit
CH_{4-}	Methane
CO_{2-}	Carbon dioxide
Cr	Chromium
EBC	European Biochar Certificate
EBF	European Biochar Foundation
GHG	Greenhouse Gas
Gt	Gigatonnes
HTT	High-Temperature Time
IBI	International Biochar Initiative
IEA	International Energy Agency
K	Potassium
Mg	Magnesium
N	Nitrogen
N_2O	Nitrous oxide
NH_4	Ammonia
NO_{3-}	Nitrate
O	Oxygen
OC	Organic Carbon
^oC	Degree Centigrade
P	Phosphorus
PAH	Polyaromatic Hydrocarbons
PO_4	Phosphate

POP	Persistent Organic Pollutants
S	Sulphur
SOM	Soil Organic Matter
t ha^{-1}	Tonnes per Hectare
UN	United Nations
WHC	Water Holding Capacity
WSC	Water-Soluble Carbon
Zn	Zinc

13.1 Introduction

Agroecosystems offer an array of ecosystem services that includes crop production for food productivity, climate regulation, soil nutrient regulation, etc. Sustainable intensification of ecosystem services can be achieved by addition or modification of biological components of ecosystem or by intensification of biological processes that supports ecosystem services for sustainable agriculture (Gaba et al. 2014; Jhariya et al. 2021a, b; Raj et al. 2021). Soil is the delicate membrane of earth, which upholds all life on earth. It is encompassed of innumerable species in a complex mixture that create a dynamic and multifaceted ecosystem. More than half of the planet's top layer of soil is vanished in the past years due to several factors. The soil quality is affected by several impacts including soil salinity, soil compaction, destruction of soil structure, and nutrient degradation (Brady and Weil 2002). Erosion effects of soil on the depletion of top fertile land lead to amplified pollution and blockage of waterways by sedimentation in streams and rivers (Carol 2001; Bear Firman 1969). Soil health and fertility continue to decline across the globe chiefly because of nutrient and carbon (C) mining, poor water management, and soil erosion; El Hanandeh 2013). Sustainability in land usage can assist to minimize harmful impact on agriculture and soil by prevention of loss in soil due to erosion and degradation leading to desertification (Khan et al. 2021a, b). Soil health is a prime concern to farming community, whose source of revenue solely hinges on agriculture that begins by laying their feet in soil (Soil Survey Staff 2014). The properties of soil, physicochemically or biologically, regulate the yield potential of many crops as well as the ecosystem (Bolan et al. 2018). Therefore, the missions are to maintain a healthy soil and their fragility from where biodiversity springs which is the prime concern for sustainability in agricultural production and ecosystem services (Meena et al. 2020; Raj et al. 2020; Khan et al. 2020a, b; Banerjee et al. 2020).

Biochar is a healthy solution to enrich soil as an amendment, which is a charcoal, prepared through pyrolysis, the biomass being heated under low-oxygen conditions, used to improve soil health (Bhattacharyya 2015). Biochar production from biological remains aptitudes to be a thought-provoking tactic in organic waste management (Kookana et al. 2011). Eco-intensification is achieved in a sustainable manner by biochar because of its ability to augment land throughput, enhance C content in soil and donate to natural intensification through manifold mechanisms and toughen flexibility of agricultural systems (Mathew and von Haden 2017).

Addition of biocharto soil promotes plant growth through recycling nutrients from organic wastes (Warnock et al. 2007). Additionally, biochar is broadly acknowledged as an effective means to sequester soil C enhancing soil fertility (Jindo et al. 2014); it could save water (moisture retention), decontaminate soils (adsorbing heavy metals/pollutants), and increase crop productivity (Burezq 2019). The role of biochar in improving soil tilth is well established; however, its role to increase soil fertility as a fertilizer is still in its infancy, but researchers have discovered as biochar is complemented with compost, animal manure, and cattle manure, it increases crop yield significantly in diverse soils and climatic conditions (Godlewska et al. 2017), as a matter of nutrient recycling (Kammann et al. 2017).

The present chapter describes the potential role of biochar towards sustainable soil management.

13.2 Soil Resource and its Problem

Soil resources are a form of environmental strength offering a series of ecosystem facilities. Various features of soil in contributing to ecosystem services include the soil formation and its functions, nutrient cycling and management, and support to soil vegetation and soil biodiversity (Stevens and Hammond 1992; Jhariya et al. 2018a, b; Raj et al. 2019a, b). Soils are a multifaceted union of minerals, water, air, organic matter, and infinite living organisms including microbes, insects, arachnids, etc. which are the decaying remains of dead living things (Allmaras et al. 1988). It forms the upper layer of land surface, capable of supporting life which is vital for life on earth. Soils play different roles in the ecosystem:

- Soils modify the atmosphere by emitting and absorbing gases (CO_2, methane, water vapor, etc.) dirt and dust (Anderson et al. 2007).
- Soils absorb and hold moisture, making it available for plant growth and other living organisms. It alters and purifies the water in terrestrial systems.
- Soil acts as living filter to clean water before it moves into an aquifer.
- Soils recycled nutrients for the living organisms in the soil to use them over and over again.
- Soils serve as substrate for growth of all kinds of plants on earth.
- Without soil agricultural crop production is not possible.
- Soils provide habitat for microbes, insects, and other animals that live in the soil that account for most of the living things on earth.
- Soils serve as an engineering source for construction of buildings, roadways, dams, and other architects. They preserve or destroy artifacts of human activities (Montanarella 2015).

Soil is an important source on earth, which provides nutrients for plant growth and anchorage for roots; therefore, it is essential for healthy crop growth and productivity (Jhariya and Singh 2020, 2021). It is a composite of organic and inorganic components that are formed by weathering of rocks constantly (Barrios 2007). Soil has several layers, with the deep and organic matter-enriched upper soil, and is trailed by a deposit with dissolved or suspended matter; underneath is the subsoil, where clay, iron, and other humic compounds accumulate after leaching from upper layers (UN 2015). Underneath is the real bedrock, which is the soil source through weathering processes (Bronick and Lal 2005). Soil is categorized based on the size of the particles it holds. The particle size of clay soil is so fine, while gravel soil is coarser. The details of soil resources, its type, and distribution around the world are imperative in global food security (Dick 2008).

13.2.1 Soil Problems

Soil is a fragile, strong, and endless natural and renewable resource produced thousand years back constantly and destroyed ultimately due to several factors (Lal 2003). Soil degradation and soil erosion are the leading causes that affect land fertility, induce nutrient wash out, and impede proper water distribution to all soil layers (Prasad and Biswas 2000). Soil degradation is a serious environmental problem gaining worldwide importance. It is the deterioration of soil quality by human misappropriation or intensification due to human actions (Lal 2003). Decline in fertility of soil is a major issue in agricultural crop production (Hartemink 2003a; Kumar et al. 2020a), as vast area of cultivable land is being affected by soil loss due to erosion and ultimately degradation with loss of nutrients and decrease in productivity (Tiziano 2016).

Soil degradation affects the functions of soil (Blum 1994) and is considered as a process happened due to misutilization of human beings that reduces the present and future capability (Baumhardt et al. 2015; Banerjee et al. 2021a, b). A chief menace to agricultural sustainability is soil degradation as it eventually deprives soil productivity (Lal 1998). Degradation of land leads to deforestation, and ultimately desertification is caused by soil erosion. Consequently, the cultivable lands with good potential of productivity are turned to desert (Graves et al. 2015). Soil erosion led to the removal of topsoil's rich in nutrients from agricultural areas, which has to be combated by addition of fertilizers, which causes pollution effects through overflow. Annual soil loss from agricultural lands accounts approximately 25 billion metric tonnes (Hobbs et al. 2008). Soil erosion is primarily caused by agents like wind and water. Soil erosion reduces soil fertility and affects crop productivity (Lal 1995). Soils themselves could sequester enough greenhouse gases (GHGs) annually equalizing 5% of annual human-made GHG emissions. Erosion degrades land, to support plants that can take in climate-warming CO_2. Land management in a much efficient manner can keep soils intact to grow more C sequestering vegetation (Ruppenthal 1995; Meena et al. 2020a; Banerjee et al. 2021c, d).

Soil holds double organic carbon level as vegetation and usually contains more than 70 billion tonnes of organic C or approximately 7% of the total global C budget. Organic matter of soil is the crucial substance related to soil, C, and nutrients. It is the amount of dead and living matter in soil including plant debris and soil microbes which is a major contributor to soil fertility (Chen et al. 2009; Eswaran et al. 1999). It acts as an elixir of plant life, by binding soil nutrients and holding them, and makes them available to plant. It upholds the structure of soil; thereby increase water infiltration, decrease evaporation, enhancing water holding potential and avoid compaction of soil. Soil houses microbes and insects and consents them to transmute plant residues and wastes to nutrients, to be used by crops (Zhao 1995). In addition, soil organic matter hastens the collapse of impurities and can bind them to particles, thereby reducing runoff risks (Pesant et al. 1987). Unfortunately, soil organic matter is being eroded and posing a serious problem for soil productivity. Soil is eroding in a much faster rate than its formation, making it unsuitable for agricultural production (Hartemink 2003b). An alternative crop establishment strategy for conservation of agriculture cropping systems is required. Smarter land management is a necessity.

13.3 Sustainable Soil Management

Sustainable soil management involves practices that create healthy soil, reduce soil erosion, and reduce need for fertilizer, pesticides, and herbicides, which include planting cover crops and adding nutrients and organic matter to soil (Wander et al. 2014). A healthy soil is attributed by high organic matter and nutrient availability, forming a nutritive base for plants. Cultivation with healthy soil reduces the nutritional inputs as nutritional requirements will be available in the soil. Healthy soils aid the roots of plants to be stronger to be tolerant to environmental stresses. The soil health can be enhanced in a supportable manner and will cause obvious advances in the upcoming years (Lorenz and Lal 2016).

Modern agriculture with enormous chemical inputs on the one hand, agricultural and soil resources being polluted by chemicals on the other hand, crop production potential is highly reduced. At this juncture, sustainable soil management aids in soil and environment protection for long-term sustainable agricultural production (Jhariya et al. 2019a, b). The sustainable management of soil resources include minimized soil erosion, augmenting soil organic matter content, reduced tillage in soil, cover cropping, crop rotation, and fostering soil nutrient balance and cycles, minimize and mitigate soil salinization, reduce soil contamination and acidification, mitigate soil compaction, preserve and enhance soil diversity, and improve soil water management.

These practices improve the soil to be healthy and ensure that it produces healthy plant canopy and profitable yield. A combination of different management tools gives crop the finest foundation of seasons to come in a proactive and sustainable way.

Biochar, generally known as agrichar, is a compound enriched with C obtained by a pyrolysis of organic biomass at 250–700 °C in limited oxygen atmosphere.

Biochar is capable of storing C for longer duration due to its nature of chemically and biologically far stable than normal C, making it tougher to convert back to CO_2. Biochar applications primarily minimize soil erosion and enhance soil organic matter content.

Minimize Soil Erosion Soil erosion caused by water and wind is recognized as a remarkable stress to soil across the globe and the eco-services provided by them. Soil erosion leads to land use changes causing depletion of topsoil and soil C (Carroll et al. 2000). Planting cover of plants to refuge soil surface and amending organic or inorganic residues is recommended to minimize erosion. Erosion of soil by water can be reduced by measures that minimizes runoffs such as raising cover crops in strips (Pimentel et al. 1995). Erosion of soil by wind can be minimized by artificial or vegetative wind breaks. Biochar amendment can minimize soil erosion according to studies, as they form macroaggregates. Biochar applied at a rate of 5% improves soil physicochemical properties and reduce soil loss (Jien and Wang 2013).

Augment Soil Organic Matter Content Soil organic matter plays a vital part in sustaining soil utilities. Soils hold the largest organic carbon source on earth and play a crucial role in climate regulation and mitigating effects of climate change (Kloot 2018). Soil houses millions of microbes and insects which play a key role in nutrient cycling by breaking down crop residue and debasing it soil nutrients (Kane 2015). Increased soil organic matter enhances soil fertility, thereby improving crop productivity (Wander et al. 1994). Enriched soil organic matter content can be attained by practices such as providing soil with dead remains of plant and animal residues and other C-rich compost (Mirsky 2017). Soil application of natural C-rich materials like biochar as amendment improves soil organic matter content and thereby fertility (Hammes and Schmidt 2009a, b).

Sustainable Soil Management by Biochar Amendment The larger surface area, pore space, high charge density, and negative surface area empowers superior aptitude to adsorb cations, to retain nutrients in soil, and to house soil beneficial microbes in plant roots (Meena et al. 2020b). The surface area of biochar is higher that aids to adsorb pollutants present in the soil. It also stabilizes biomass and soil organic matter that enhances aeration of soil, thereby increasing the activity of microorganisms and immobilizing N which in turn minimizes GHG emission. The competent role of biochar in sustainable agronomic development is immense, which needs to be more explored.

13.4 Biochar and its Role

To feed the rapidly increasing world population, enormous quantity of food crops is constantly produced, and therefore, organic residues are accumulated in large quantity annually, which should be recycled essentially (Chatterjee et al. 2017). Biochar is an optimal solution to utilize the organic waste efficiently. Biochar gained by pyrolysis of organic residues can behave as a substitute and influence soil C sequestration and thus amends the physicochemical and biological properties of soil (García et al. 2016; Zhang et al. 2017). Biochar is the produce of thermal combustion of organic biomass in the absence of oxygen at a moderate temperature of 250–700 °C (Lehmann and Joseph 2009).

The biomass is pyrolyzed to biochar, a solid component, liquid components, biofuels, oil, tar, and volatile gases Eq. (13.1). Based on the biomass used, it has a greater C content and fluctuating C to nutrient ratio (Schackley et al. 2015). Moreover, biochar is extremely porous, has larger surface area (Downie et al. 2009), and possesses dual surface charges, both positive and negative; therefore, they can absorb both positively and negatively charged particles, thereby minimizing leaching (Lehmann et al. 2006) and enhancing adsorption of nutrients (Sheng et al. 2005).

$$\text{Biomass (solid)} \xrightarrow{\text{Pyrolysis}} \text{Biochar (solid)} + \text{Biofuel (liquid)} + \text{Volatile gases} \tag{13.1}$$

13.4.1 Functions of Biochar in Soil

Biochar provides three main services when added to the soil: it is porous like sponge, soaks up water, holds and slowly releases it to soil. It attracts nutrients like magnet and behaves like a store house for exchange of nutrients to plants, and it is a substrate to provide habitat and refuge for soil microbes. The soil physical properties such as bulk density, pore space, and water retention can be altered by the permeable nature of biochar (Quin et al. 2014), whereas absorptive capacity has great promise for composting processes (Agyarko-Mintah et al. 2016) by reducing N losses, increasing C contents, and improving compost quality. The biochar increases microaggregate formation by organo-mineral interactions and resulted in measurable increases in soil C content (Schirrmann 1984). There are different ways and mechanisms of biochar functions in different type of soils and environmental conditions.

Biochar can provide nutrients to soil by mineralization behavior that release nutrients from organic matter and manures (Lal 1998). The nutrient level in biochar depends on production conditions and source of feedstock; for example, low temperature reduces emission of N_2O and N and favors increased N and P availability, and high temperature favors K availability (Lehmann et al. 2006):

- Biochar can increase physical (porosity, structure development, soil aggregation) properties and increase moisture retention and reduce leaching.
- Biochar may help to hold nutrients in sandy soils with high leaching capacity, which washes nutrients through soil media.
- Biochar favors retention of heavy metals due to its porous structure.

- Porous structure of biochar and retained nutrients provides conducive environment to microbial community and acts as a niche to be saved from predators (Qian et al. 2014).

13.4.2 Biochar and Charcoal

Charcoal and biochar are similar due to high organic carbon contents, but different in preparation and its usage. Charcoal is used as an energy source, whereas biochar is a soil amendment to enhance soil C, improve soil fertility, decontaminate soil and acts as animal and poultry feed (Thu et al. 2010). The major source for charcoal production is wood, whereas biochar can be produced from diverse organic feedstock (Lehmann and Joseph 2009). The trends of carbonization are incomplete in charcoal; in consequence charcoal has higher ash content than biochar (DeLuca et al. 2006). In biochar, six C atoms are allied together; graphite formation is likely to occur when C atoms are organized together without O or OH ions (Farrar et al. 2017). However, graphite cannot be formed from biochar as the arranged atoms of C are irregular based on the biomass used and pyrolysis temperature for biochar production.

13.4.3 Biochar Production

The carbonization of waste residues for the purpose of biochar production is known to humans since long time. Enhanced biochar technology can contribute to humankind by providing energy needs for future and also improving soil C sequestration. Biochar production involves fast pyrolysis, carbonization, and gasification (Augustínová et al. 2013).

Pyrolysis It is the thermochemical process of altering organic biomass into enriched biochar in a zero-oxygen environment. Pyrolysis is categorized (Table 13.1) as fast, intermediate, or slow, based on the parameters, i.e., temperature, residence time, heating rate, and flow rate (Gaunt and Lehmann 2008).

13.4.4 Raw Materials for Biochar

A variety of biomass contributes in biochar production. Most organic matter free from contaminants and pollutants are utilized in biochar production (Placek et al. 2016). Various types of biomass utilized for biochar production include:

- Agriculture, horticulture, and forestry byproducts: wood chips, straw, nut shells, tree bark, dried leaves, twigs, etc.
- Industrial by-products: sugarcane bagasse, paper pulp, etc.

Table 13.1 Fate of biomass between pyrolysis in products (IEA 2007)

SN	Process	Solid (biochar) (%)	Liquid (biofuel) (%)	Gases (syngases) (%)
1	Gasification High temperature (>800)	10	5(5% tar)	85
2	Fast pyrolysis Moderate temperature (500 °C) Residence time is short	12	75 (25% water)	13
3	Intermediate pyrolysis Low medium temperature Residence time is moderate hot Moderate hot vapor residence time	25	50 (50% water)	25
4	Slow pyrolysis Low moderate temperature Long residence time	35	30 (70% water)	35

- Animal by-products: chicken litter such as feathers and skin, dairy and other animal manure and skin.
- Sewage sludge.

Biochar production from agricultural and horticultural residues proposes a brilliant mode of reprocessing waste residues into valuable resources (Shinogi et al. 2003). Some of these raw material can be agricultural residues like wheat and rice straw (Naeem et al. 2017; Sun et al. 2019), cashew nut shell nuts (Abderahim et al. 2018), peanut shells (Xu et al. 2017), coconuts shells (Devenas et al. 2018), waste wood (Pituya et al. 2016), rice husks (Dunnigan et al. 2018), empty fruit bunches (Baker et al. 2015), wheat straws (Xu et al. 2017; Vaccari et al. 2011), orange wastes (Sial et al. 2019), olive pomaces (Ghouma et al. 2017), and other wastes (Mandal et al. 2004; Saleh et al. 2014; Chen and Xu 2009; El-Hanaendax). The biomass energy crops like corncobs (Demirbas 2003), tobacco stems (Pesevski et al. 2010), rice waste (Saleh et al. 2014), common reed (Komulainen et al. 2008; Kitzler et al. 2012), vine prunings (Nasser et al. 2014), tobacco stems (Pesevski et al. 2010), wood pellets (Vaughn et al. 2013), palm oil and oilseed rape (Zainal et al. 2016), date palm midribs (Nasser 2014), bioenergy residues (Asfaw and Chandraratne 2018), compost (green waste), animal manure (sheep, chicken) (Burezq 2019), sewage sludge, etc. (Sohi et al. 2010a, b) also have contributed their role in biochar production across the globe. The characteristics of biochar in different biomasses were listed in Table 13.2.

13.4.5 Methods of Biochar Application to Soil

It is well-established that biochar is commonly used as soil conditioner to advance soil health. It could be used as a stand alone amendment or integration with compost and fertilizer, which are formulated to meet specific soil amendment needs; this can reduce inputs up to 30% (Hanz 2011). The establishment of biochar system in soil

Table 13.2 Characteristics of biochar from different feedstock (compiled by: Rumi et al. 2015)

SN	Feed stock	pH	Moisture	Ash	VM	FC	C	H	N	O	Reference
1	Safflower seed cake	9.13	–	8.20	20.00	71.80	70.43	3.43	3.36	22.39	Angin (2013)
2	*Conocarpus* waste	9.67	–	5.27	–	–	76.83	2.83	0.87	14.16	Al-Wabel et al. (2013)
3	Rice straw	9.68	7.20	15.40	62.40	14.90	44.80	5.10	0.90	49.20	Pütün et al. (2004); Bakar and Titiloye (2012)
4	Pitch pine	–	–	7.90	–	–	70.70	3.40	0.60	25.50	Kim et al. 2004
5	Pine sawdust	–	5.00	0.30	77.70	16.90	50.30	6.70	0.20	42.70	DeSisto et al. (2010); Wei et al. (2006)
6	Spruce woodchips	10.9	–	31.00	–	–	74.80	0.14	0.15	4.20	Saarnio et al. (2013)
7	Corn shovers	–	2.30	58.00	12.70	28.70	33.20	1.40	0.81	8.60	Lee and Lee (2004)
8	Coconut shell	9.18	4.40	0.70	80.20	22.00	50.20	5.70	0.00	43.4	Raveendran et al. (1995); Shenbagavalli and Mahimairaja (2012a, b)
9	Peanut shell	9.50	1.90	7.80	8.10	82.20	93.61	1.99	1.05	3.35	Apaydın-Varol and Pütün (2012)
10	Pine cone	9.80	1.20	4.70	6.70	87.40	95.16	2.63	1.61	0.60	Apaydın-Varol and Pütün (2012)
11	Pea hull	8.60	–	9.30	18.10	–	81.80	2.90	2.70	3.30	Novak et al. (2009)
12	Switch grass	8.00	–	7.80	13.40	–	84.40	2.40	1.07	4.30	Novak et al. (2009)
13	Pungam cake	11.20	4.30	11.60	14.60	69.50	75.00	3.26	5.00	12.58	Chutia et al. (2014)
14	Jute dust	–	9.44	10.78	15.07	64.71	70.25	2.78	4.04	22.93	Choudhury et al. (2014)
15	Sugarcane baggase	9.30	1.30	8.57	9.17	80.97	85.59	2.82	1.11	10.48	Lee et al. (2013)
16	Coco peat	10.30	2.55	15.90	14.30	67.25	84.44	2.88	1.02	11.67	Lee et al. (2013)
17	Palm kernel shell	6.90	0.00	6.86	12.29	80.85	87.85	2.91	1.11	8.14	Lee et al. (2013)
18	Cotton seed hull	8.50	6.53	7.90	18.60	67.00	87.50	2.85	1.50	7.60	Uchimiya et al. (2011)
19	Soybean cake	–	1.50	16.80	10.10	71.60	83.95	1.48	8.32	6.25	Apaydın-Varol and Pütün (2012)
20	Sesame	–	3.40	6.80	22.00	37.80	86.64	3.10	6.93	3.09	Volli and Singh (2012)
21	Neem	–	3.70	24.50	32.00	39.80	82.34	7.89	5.76	3.57	Volli and Singh (2012)
22	Mustard	–	4.80	28.10	21.00	46.10	85.43	4.79	6.17	3.41	Volli and Singh (2012)
23	*Shorea robusta* seed	–	0.00	19.70	26.90	53.40	72.58	13.63	4.38	7.74	Singh et al. (2014)

VM, volatile matter; FC, fixed carbon (moisture, ash, volatile matter and fixed carbon in % and CHNO in wt%)

can give longer benefits than typical bioenergy alternatives (Woolf et al. 2010a). Biochar as a soil amendment gives better results when it is blended with compost that allows the biochar to absorb nutrients from the compost, and it is the combination that provides maximum benefit to the soil. There is no standard method of biochar application, it could be mixed with soil directly and subsoiled and may be blended with fertilizers and banded under the seed for fertilizer efficiency in wheat crop (Beaumont 2017). Biochar could be layered in soil. In addition, it could be mixed with compost, animal manure, and potting media and combined with fertilizers, to adsorb nutrients for slow release and control leaching. It may be banded adjacent to the root zone – rhizosphere (Graves 2017). Biochar is able to retain nutrients (carrier) on adsorbed sites due to negative charges, i.e., CEC and high surface area, leading to nutrients held for a long time and reduced requirement of fertilizer and compost for future application to crops (DeLuca et al. 2006).

The potential ways of biochar application in soil are as follows:

- Subsoil biochar to improve moisture and nutrient holding capacity.
- Low rates of biochar may be blended with fertilizers and banded under the seed to improve fertilizer use efficiency.
- The application of biochar specifically in the immediate vicinity of plants roots "rhizosphere" can increase microbial communities and nutrient availability.
- Handling of dusty biochar has health and safety risks, which can be minimized by using biochar slurry, or small biochar granules that are suitable for earthworms to ingest and mix into soils.
- Applying biochar in a layer of 1–2 inches thick and then blending it well into soil, to help developing roots and avoid removal by wind or water.

13.4.6 Advantages of Biochar Application to Soil

The biochar has the following advantages in terms of soil management (Fig. 13.1):

- Biochar is primarily recommended as soil conditioner to enrich soil fertility.
- It improves soil moisture retention and thus saves irrigation water requirement of crops.
- Enhances nutrient retention of soil and thereby reduces fertilizer application.
- Minimizes soil losses through leaching.
- Recovers soil physical health by structural development of soil.
- Long-lasting soil C stability and sequestration in soil for eras.
- Enhances soil microbial activity and other soil biota.
- Improves cation exchange capacity (CEC) and hence more availability of soil nutrients to plants.
- Mitigate emission of GHGs.
- Eliminates contaminants in soil and ultimately water ways.
- Minimizes movement of agrochemicals in soil and water ways.
- Environmentally safe as it's pure organic manure.

Fig. 13.1 Advantages of biochar in soil

13.4.7 Transformation of Biomass into Biochar

The organic biomass is transformed into biochar by the process of pyrolysis, which is the thermal degradation of biomass in high temperature resulting in the production of creation of charcoal (solid), bio-oil (liquid), and fuel gases. In the process of pyrolysis, the maximum of the mineral matter is retained, and organic compound is broken down, disentangling into volatile compounds and released as hot gas, aerosols, and residual solid component, the biochar (Fig. 13.2) (Lynch and Joseph 2010). The biochar comprises various organic matters, which mineralize at diverse rates, reliant on the soil microbes and the surrounding environment (Dari et al. 2016).

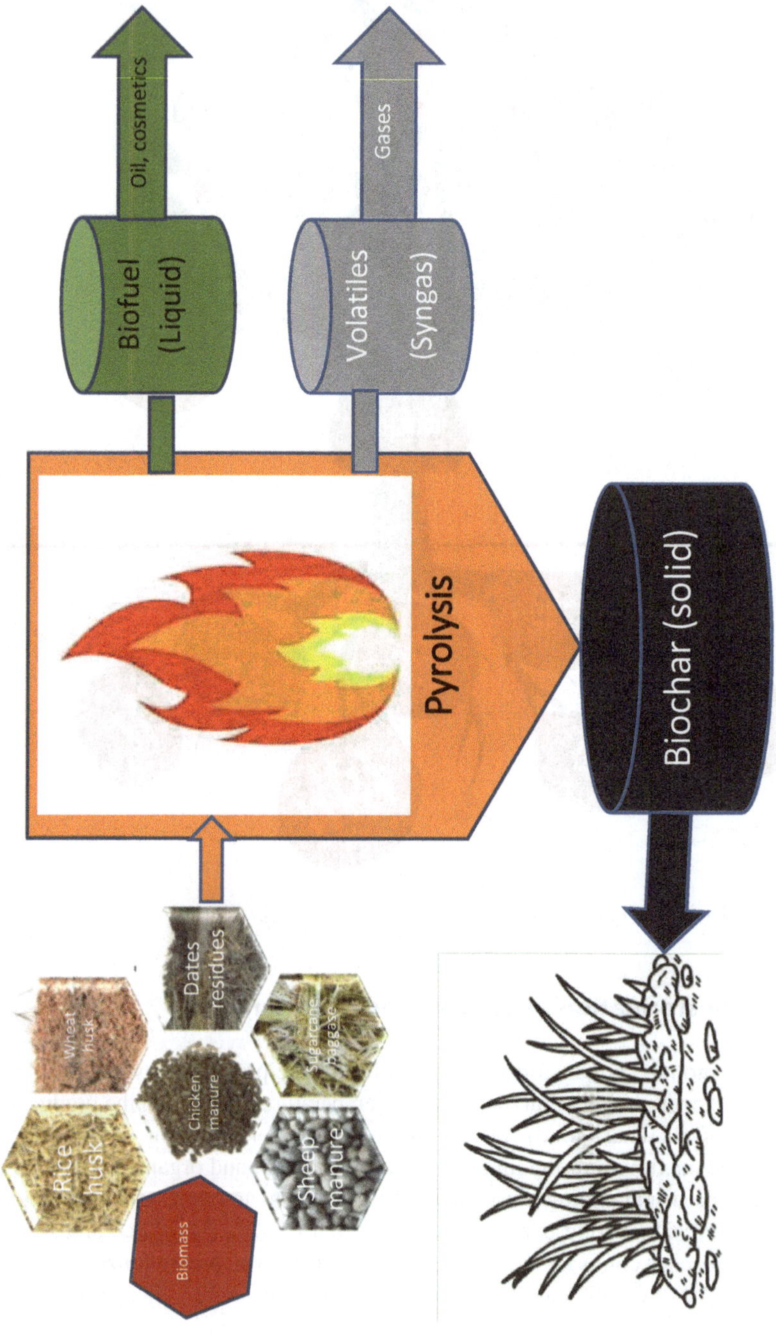

Fig. 13.2 Thermal conversion of biomass

13.4.8 Properties of Biochar

13.4.8.1 The Physical and Chemical Properties of Biochar

The physiochemical properties of biochar are dependent chiefly on the biomass utilized and the temperature at which biochar is produced. With temperature rise, the perseverance of C matrix also enhances available N and mineralizable organic compounds. C content of soil increases with high ash content in biochar. The type of pyrolyzer used, heating time, and rate are the other factors that can affect biochar properties. Plant residues, dried leaves, and organic manure biochar enhance the crop productivity response, in nonnutritive, infertile, and acidic soils due to their greater nutrient content (Wang et al. 2016). Some biochar produced at high temperature such as wood reduces the plant yield as fresh biochar adsorbs nutrients normally available for plants. However, woody biochar is also advantageous in aerating soils and can prevent runoff and leaching of chemical fertilizers. Nutrient augmentation and stimulation by pre-treating the biomass before pyrolysis resulted in plant responses in terms of increased quality and yield quantity. They also increase the abundance of beneficial microorganisms.

13.4.8.2 Physical Properties of Biochar

The physical properties of biochar subsidize its function as advice for environmental management. It greatly influences biochar functions in soils, particularly in pore size distribution, structure, and surface area. Their physical characteristics affect the soil system either directly or indirectly. Soils having distinct properties vary depending on inorganic and organic matter (Brady and Weil 2002). The existence of biochar in soil contributes to the physical nature of soil significantly, influencing structure, texture, consistency and depth through changing pore size, particle size, bulk surface area, compactness, and stuffing. Biochars influence soil properties directly through its impact on plant growth as the air and water availability is determined by physical structure of soil horizons. Biochar in soil affects the physical characteristics directly and influence the soil response to water, its combination, penetrability, swelling thickness, and its ability to hold cations and its retort to dynamics in temperature. Moreover, biological and chemical facets of soil fertility can be inferred directly from physical properties such as physical presentation of sites for chemical reactions and the provisions of protective habitats for soil microbes (Sohi et al. 2010a, b).

The physical properties of biochar are determined by processing conditions during pyrolization, and the high temperature time (HTT) and reaction residence time may be the most critical factors. As a soil amendment, biochar could effectively contribute its functions in soil, including decrease in bulk density, penetration resistance, and soil losses and increase in pH, water retention capacity, aggregation stability, retention ability of nutrients, and crop production. The physical properties of biochar influence its movement in soil ecosystem, interacting soil water, minerals, and nutrients (Schmidt and Shackley 2016). Biochar provides shelter for soil microbes from their enemies and act as a biological niche (Tomczyk et al. 2020). The density and porosity of biochar have an important role in movement of biochar in the soil environment as materials with bulk densities less than 1 g/cm^3 will float.

Bulk Density Bulk density is the mass of unit volume of an assortment of units. It depends on size, shape, and compaction of the particles assorted. The bulk density of biochar is generally 0.06–0.7 g/cm^3, which chiefly depends on the biomass used (Piash et al. 2016). The small-sized biochar particles have the ability to fill the space between the soil particles and thereby increase the bulk density

Particle Density Particle density is the mass per unit volume displaced by the top cover of the particle including internal pores. Particle density affects the ease of mobility and loss of biochar in wind and water. Grass biochar and wood biochar have a particle density of 0.25–0.3 and 0.47–0.6 g/cm^3, respectively (Pituello et al. 2014).

Pore Volume Pore volume is the volume of pores per unit weight of particles expressed as cm^3/g (Piash et al. 2016). Biochar exhibits high porosity with pores ranging from micro to macro size. Biochar with large pores aid as habitat for soil microbes, thereby improving soil quality (Thies and Rillig 2009). Biochar pores also releases vapors during pyrolysis (Lee et al. 2013).

Pore Size Distribution The pore size distribution is the relative abundance of each pore space in a representative volume of biochar. The pore size distribution is an important physical property that affects water retention potential of biochar (Kinney et al. 2012). Biochar has a wide range of pore sizes from sub-nanometer to tens of micrometers. Pores of nanometer size will not interfere in water retention, but with chemical nutrient interactions; in contrast pore of micrometer size influences on water retention ability of biochar.

Particle Size Diameter of a particle is typically measured by the particles passing through a series of different sizes. Particle size of biochar depends on biomass and its preprocessing and the production temperature and technique. Uniformly sized particles can be achieved by pelletizing and granulating (Jindo et al. 2014). Small biochar particles (0.005–1.00 mm) reduced the bulk density in sandy soil than large biochar particles (2–4 mm) and increased the maximum water holding capacity by 60%

Porosity Porosity is the degree of the percentage of barren space in a particle or bulk of particles. The porosity of biochar is hard to describe as it has a wide range of pore sizes, which vary primarily with the type of biomass used. Pore sizes are classified as macro, meso, or micro pores with relevance to physicochemical phenomenon of biochar interaction with the environment (Pituello et al. 2014).

Chemical Properties

The nature of biomass used has a great impression on the chemical properties of biochar. Biomass commonly entails minerals, cellulose, hemicellulose, lignin, and other organic compounds such as proteins, lipids, and other complex organic compounds (Liu et al. 2012a, b). The chemical properties of biochar are determined by proximate and ultimate analysis. The moisture content, volatile components (gases released while heating the biomass), solid content (solid left after evaporation of gases), and the ash (the residue after the biomass is burnt) are determined by thermal decomposition in minimal oxygen environment in proximate analysis, which differs based on the biomass utilized and the pyrolysis temperature. The elemental quantity of C, H, O, N, and S was determined in ultimate analysis. During the process of pyrolysis, the volatiles remove 50% of C but more of O and H (Dumroese et al. 2011). The organic compounds of biochar can be categorized into two components: persistent C, resistant to breakdown for decade to millennia, and mineralized C in the form of organic compounds, which are hypothetically obtainable to the plants and soil microbes. The condensates of the released gases called wood, vinegar, or smoke contain over 300 chemicals, which are valuable to plants (Smith et al. 2014).

$$\text{Biomass} \rightarrow \text{Biochar}$$

$$C_6H_{8.7}O_4 \rightarrow C_3H_{1.4}O_{0.4}$$

Persistent Carbon The persistency of C in biochar is indicated by hydrogen to C ratio. Organic C is the amount of C present after deducting C in mineral from total C content. Hydrogen content of biochar differs from 1 to 5%, with hydrogen content increasing at lower temperatures. Biochar has water-soluble components, known as labile C, that act as feed for microbes and stimulate seedling emergence from soil. Most of them are derived from volatile gases emitted due to thermal breakdown of biomass in zero-oxygen atmospheres. Those components include compounds with low molecular weight (alcohols, aldehydes, sugars, amino acids, and ketones), low molecular weight biopolymers (polysaccharides, proteins, and amino sugars), proteins, building blocks (polyphenolic/poly aromatic acids), and high molecular weight compounds such as humic acids. High-temperature biochar have water-soluble compounds chiefly with low molecular weight acids. Biochar prepared in low temperatures have higher concentration of water-soluble organics (Neal and Wagner 1983).

Ash The inorganic component of biochar, the ash, includes metals and nonmetals. Few of them have crystalline amorphous structure, and others are amorphous of 1-micron dimension. Sulphates, sulfides, carbonates, phosphates, chlorides, and oxides are amorphous crystalline compounds. Rock phosphate, sylvite, salt, struvite, calcite, dolomite, and clay include crystalline compounds. They are soluble such as sodium chloride and insoluble such as calcium sulphate (Mukherjee and Zimmerman 2013).

Heavy Metal Some biochar has certain limit of heavy metals. Biochar prepared from biomass with heavy metals like sewage effluent will hold heavy metals in them (Liu et al. 2014). Several countries have restrictions and regulations that limit the quantity of heavy metals to be accumulated in soil. For biochar certifications such as IBI or EBC, heavy metal content should be the below limits (Beesley et al. 2015).

Total and Available Nutrients Biochars prepared from different biomasses have different available nutrients. Woody biochar has low mineral content and low ash content. Biochar prepared from biomass rich in ash content has more nutrients in them, approximately 10 to 100 times, from 1 to 70 g/kg. When applied in soil, a very little of N is available for plants, though the N content of most raw materials for biochar is high. The concentration of many other elements, apart from N, is higher in biomass than that of the outcoming biochar (Ippolito et al. 2012). Phosphorus and potassium availability is generally high (36 and 54%), whereas the available N is low in biochar (less than 5%). Some of the least N available is obtained at the rate at which C is mineralized over time.

Electrical Conductivity Some of the salts of biochar will be dissolved if placed in water with the ability to conduct electricity. Electrical conductivity (EC) is a measure of total dissolved salts (TDS) and indicates the presence of nutrient availability or salts. EC is measured using EC or TDS meter, after making a mixture of biochar in water. EC is expressed normally as deciSiemen per meter or milliSiemens per meter (1dS/m = 100 mS/m). EC of soft biochar are extremely low and those of hardwood biochar are far greater (Shammi et al. 2016).

Cation Exchange Capacity (CEC) Cation exchange capacity is the potential to hold exchangeable cations such as calcium and potassium. CEC of biochar is determined by the oxygen groups present in them. Low-temperature biochar has high CEC, and high-temperature biochar has low CEC. CEC of biochar increases as it ages in soil (Pan and Richards 1989).

Anion Exchange Capacity (AEC) Anion exchange capacity is the ability of biochar to hold on anions such as phosphate through anion exchange. It retains nutrients and prevents leeching of water leading to water pollution (Nsamba et al. 2015). AEC declines on exposure of biochar to soil, and on oxidation, with a mean decrease of 54% after biochar application continually for 4 months. Some anions such as P, Mg, Ca, Fe, and Al can be retained by precipitation. Some anions such as nitrates get stuck in biochar pores and get retained.

Biological Properties of Biochar

Biochar application has shown to influence soil biological processes with some propositions for soil geochemistry (Lehmann et al. 2011). Biochar application in soil can augment soil microbial population, stimulate their activity, and alter their community in soil (Pietikainen et al. 2000). Soil microbial community in soil is affected by biochar application due to high absorption capacity and soil pH. Biochar also contains toxic carbonyl compounds and polycyclic aromatic hydrocarbons that have bactericidal and fungicidal activity (Dalvi and Ademoyero 1984). Carbon content present in the biomass, captured by photosynthesis, will reach the atmosphere eventually; the converted biochar has the potential to sequester in soil for long era, changing the biological properties of soil (Preston and Schmidt 2006). Soil amendment with biochar, which is a biomass-derived black C, determines the variations in soil microbial community and soil geochemistry (Warnock et al. 2007). Studies revealed the stimulated growth of arbuscular mycorrhizal fungi (AMF) by biochar (Rillig et al. 2010) that has been associated with enhanced crop growth (Graber et al. 2010). The labial components of biochar improve and stimulate microbial activity (Steiner et al. 2008a). Soil amendment of biochar improves soil properties by:

1. *Increased Amount and Availability of Soil Nutrients:* The increased microbial activity in soil biota due to biochar amendment increases the amount and availability of nutrients such as N, P, and metal ions (Warnock et al. 2007; Lehmann et al. 2011). In addition, enhanced soil nutrient retention and availability end in heightened plant performance and elevated tissue nutrient concentrations which allow colonization rates of microorganisms in plant roots (Jaafar et al. 2014).
2. *Favoring Soil Biota by Water Retention:* Biochar application improves soil hydration and protects soil biota against desiccation. The pores of biochar absorb moisture and make them available for soil microbes and crop growth (Liu et al. 2012a).
3. *Adsorption of Inhibitory Compounds:* Abundance of microorganisms due to biochar application increases the adsorption of compounds; otherwise, it would inhibit microbial growth (Lehmann et al. 2011). High-temperature corn stover biochar has a high adsorption capacity for catechol, which is toxic to microorganisms (Cheng et al. 2008).
4. *Refuge for Predators:* The pore sizes of biochar are larger to accommodate soil microorganisms, giving room as a refuge for predators (Warnock et al. 2007). Therefore, the pore structure within biochar offers a physical protection from soil predators and allows a better attachment for various microorganisms, which makes them less susceptible to leaching in soil and records increased microbial growth and activity in soil (Ezawa et al. 2002).

Change in Properties of Biochar

The properties of biochar depend on the concentration and configuration of organo-mineral layer that forms on the superficial of biochar, which is a function of soil, environment, and agricultural practices (Singh et al. 2010a, b). Soil type has a great role in changing the properties of biochar. The percentage of clay in soil plays a chief role in the persistence of biochar. The properties of biochar are altered by the micro as well as macroorganisms present in soil. The presence of earthworm causes fracturing of biochar and new surfaces similar to fresh biochar will be formed. Pore size influences the activity of microorganisms sheltering as refuge and altering the structure of biochar (Wang et al. 2016).

Plant Response to Biochar

Response of plants to biochar depends on the basic properties of biochar and specific soil in which it is applied. Studies reported that grass biochar (Gomez et al. 2014) increased crop yield on a single time application over time, but the yield drop as biochar ages, which was also noticed in certain studies (Ameloot et al. 2015). Biochar can uplift soil fertility and enhance plant growth and biomass yield in a very poor soil, which is demonstrated in lettuce by Dalila et al. (2017).

A positive impact of crop productivity up to 5–50 tonnes per hectare with appropriate nutrient management was reported by several researchers (Jyothi et al. 2019). The woody biochar continually applied to the soil displayed a great enhancement in soil health, fertility, and thereby crop productivity (Joseph et al. 2010).

13.4.9 Utilities of Biochar

Currently, biochar amendment to soil is being considered as a mean to sequester C while concurrently augment soil health and fertility and provide agronomic crop benefits. It influences the ecosystem by different means (Table 13.3).

13.4.9.1 Biochar as Nutrient Source

Long-term sustainability of soil fertility is fundamental for improving food security. Nutrients are required by plants for their growth either in organic or inorganic form. Inorganic fertilizers impose threat to ecosystem, while organic ones are eco-friendly (Schnug et al. 1996). Biochar is an organic nutrient source, which comprises high levels of C and manufactured by pyrolysis process (Pessenda et al. 2001) using different biomasses. Biochar is a stable state of C, is quite difficult to split, and thus will persist in soil for long years (Schmidt et al. 2002). The positive effects of biochar are due to the chemicals and nutrients present in it and its absorption capacity and ability to retain nutrients (Lehmann et al. 2011). Biochar enhances the soil organic matter content of soil, which is highly beneficial for plants (Verheijen et al. 2010). In addition, it adjusts the soil pH to neutral, thereby increasing soil CEC, which is an added advantage to soil (Verheijen et al. 2010).

Biochar is comparatively cheaper than inorganic fertilizers and is affordable for marginal scale farmers. Organic and inorganic salts like humic- and fluvic-like compounds and available N, P, and K in biochar provide it with nutrients (Schulz and Glaser 2012). Biochar produced from *Acacia saligna* biomass and saw dust at 380 °C and 450 °C were reported to contain 17.7 and 16.2% humic- and fluvic-like substances, respectively (Liu et al. 2012a, b). Fresh biochar encompasses copious nutrients and were able to release large amounts of N and P (Mukherjee and Zimmerman 2013; Zheng et al. 2012). Biochar from *Lantana camara* biomass at 300 °C were reported to contain available P, K, Na, Ca, and Mg at 0.64, 711, 1145, 5880, and 1010 mg kg^{-1}respectively.

Nutrient possession of biochar is greatly determined by the source of biomass used and pyrolytic temperature. Biochar has large surface area, high cation exchange capacity, and porous nature, which confers to the high sorptive capacity that can be exploited during nutrient recovery. Pyrolysis conducted at various temperatures has influence on biochar quality (Ippolito et al. 2015); higher pyrolysis temperature is

Table 13.3 Influence of biochar in ecosystem

SN	Influence of biochar		Reference
1	Increased crop productivity	Increased crop yield is observed within 1 or 2 years	Chan et al. (2008)
2	Provide shelter for microbes	Biochar increases soil microbial population and their activity by providing refuge for them	Jaafar et al. (2014)
3	Increase in arbuscular mycorrhizal colonization	Biochar alters soil biological properties, by interfering plant-fungi signals by detoxifying the chemicals in biochar, thereby providing refugia for fungal predators	Warnock et al. (2007)
4	Influence on N cycle	Biochar emits N_2O, depending upon the type of biochar and related microbial activities	Yanai et al. (2007)
5	Influence on soil-water holding capacity	Biochar amendment to soil, increases water retention in soil	Sohi et al. (2009)
6	Increased soil cation exchange capacity (CEC)	Biochar enhances CEC of soil. The duration required for increase in CEC and their efficiency after biochar application needs exploration	Cheng et al. (2008)
7	Impacting seed germination and early emergence of seedlings	Application of biochar enhances germination of seeds, early emergence of seedling, and their growth in several crops	Solaiman et al. (2012)
8	Dynamics in pH	The pH of biochar is influenced by the type of biomass utilized, production temperature, duration, biochar type, and application rates	Chintala et al. (2014)
9	Mobility of biochar in soil	Biochar very rarely move and get lost into the water resources and thus will stay in the soil and enhances nutritive value of soil	Sohi et al. (2009)
10	Impact of the size pf pores of biochar	The size of the biochar pores significantly alters chief soil physical properties and thereby alter the soil profile	Cheng et al. (2006)
11	Biochar decomposition as influenced by agricultural practices	Agricultural practices such as ploughing, sowing, planting, etc. cause breakdown of biochar and reduces its C storage potential	Lehmann et al. (2003)
12	Increased soil sorption capacity of pesticides and other contaminants	The toxicity, movement, and fate of pesticides such as herbicides and insecticides are altered due to biochar application as it increases the soil sorption capacity	Hiller et al. (2007)
13	Reduced GHG emissions	Biochar decrease N_2O and CH_4 emissions in crop ecosystems	Van Zwieten et al. (2010)
14	Act as animal feed	Biochar fed with diet improved growth rate of cattle and reduction of enteric methane emission	Silivong and Preston (2016)
15	Utility in aquaculture as fish feed	Biochar as feed has displayed tremendous gain in weight and supplemented feed in aquaculture	Yoo et al. (2007)

found to decrease the available N content (Koutcheiko et al. 2007) and increase available K content (Zheng et al. 2012; Koutcheiko et al. 2007). On the contrary, lower pyrolysis temperature increases nitrogen (N) and phosphorus (P) availability and increases potassium availability (Zheng et al. 2012). High pyrolysis temperature causes loss in N content which begins at about 400 °C; half of N was lost as volatiles at 750 °C (Liang et al. 2006). The total sodium and phosphorus content significantly increased from 0.12 to 0.17% with the increase of temperature from 300 to 600 °C (Zheng et al. 2012). However, available P in the biochar produced at lower temperature was much higher than the biochar's produced at high temperatures, as biochar contains less crystallized P-associated minerals in lower temperature. Moreover, the total K content increased from 3.7 to 5.02% at 300 and 600 °C, respectively, while the available K content increased with increase in pyrolysis temperature (37 and 47% at 300 and 600 °C). In addition, the ash content of biochar in poultry manure produced at 350 °C is higher than biochar produced from pine wood chip (Spokas et al. 2012). The available P in biochar made from different biomass is not proportional to the total P concentration of the biomass (Freitas et al. 2016a, b). The poultry litter biochar contains a mineral whitlockite (a sparingly soluble Ca Mg and P form) which might be used as a P fertilizer (Gurwick et al. 2013). It is concluded that the lower pyrolysis temperature increases N and P availability, whereas higher pyrolysis temperature increases the K availability from the biochar (William and Qureshi 2015).

Soil pH is a significant aspect distressing nutrient availability of biochar. The release of P and ammonia ions was pH dependent, while K and nitrates were not (Zheng et al. 2012). The calcium and magnesium concentration of biochar from corn straw was pH reliant on, exhibiting an upsurge in measure as pH lessened (Silber et al. 2010). The nutrients $PO_4{}^{3-}$ and NH_4 are pH-dependent nutrients, but K^+ and NO_3- are not. The time of application of biochar also influences on nutrient release. The nutrient release of biochar prepared from giant reed, *Arundo donax* at 300–600 °C was slow after 120 hours for ammonia and fast at 24 hours for phosphate and potassium (Zhang et al. 2013). The addition of biochar in soil increased nitrification due to phenolics sorption (Ball et al. 2010). Biochar has the capacity to adsorb different nutrients at different levels (nitrate 3.7%, ammonia 15.7%, and P 3.1%), and nutrient adsorption capacity varies based on biochar quality (Yao et al. 2012). An experiment to explore the correlation between time and concentration of water-soluble nutrients, P, ammonia, and K ions revealed that ammonia released from biochar produced from *A. donax* (giant reed) at 300–600 °C occurred within 120 hr. indicating a slow release, while phosphate and potassium ions released at 24 hr. indicating a fast release. In addition, high C mineralization and N immobilization of volatile components of biochar by soil microbes could decrease the release of nutrients (Zimmerman et al. 2011).

Biochar improves soil properties, increases nutrient cycling, and decreases leaching (Steiner et al. 2008b). It is conveyed that 15% Ca, 10% P, and 2% N in wood biochar are readily leachable with distilled water after 24 hr. There are no fixed nutrient contents in biochar due to heterogeneity of biomass; however C, H, N, and other elements (Na, Ca, Mg, and K) are common (Zhang et al. 2015). Biochar is considered as a substitute for mineral fertilizers (Glaser 2007). The macronutrients such as N, P, and K do not reflect their actual availability to plants, as they are bound in C and functional groups (Spokas et al. 2012).

Biochar is preferred over compost as:

1. Biochar is sterile and contain stable C.
2. Biochar is long lived from decades to millennia.
3. Biochar has long-term economic and environmental benefits.

The combination of biochar with fertilizers transforms the biochar into slow-release fertilizer to increase the durability of fertilizers to enhance soil fertility (Kammann et al. 2017).

13.4.9.2 Biochar as Soil Stabilizer

Biochar is the C-enriched charred organic material projected to be used as a soil stabilizer to sequester C and boost soil quality. Biochar is an effective means for clearance of organic farm residues, acting as a coherent tool for combating GHG level in a sustainable way (Sohi et al. 2010a, b). Biochar addition has a wide array of agricultural and environmental aids including waste minimization, energy production, C sequestration, and soil improvement. Biochar as a soil nutrient enhancer is a pioneering and promising factor for sustainable agriculture.

Biochar should not be taken solely as a nutrient source but also an organic soil conditioner. Biochar application to soil as a conditioner is an economically viable strategy for universal sequestration of C (Spokas et al. 2012). Biochar application in soil enhances C sequestration through soil fertility enhancement (Criscuoli et al. 2014). Biochar-enriched soils over 800 years in deserts of Brazil have a high pH and higher stable organic matter and higher productivity compared with surrounding low fertile soil (Sohi et al. 2010a, b; Lehmann and Joseph 2009). The accumulation of C in soil due to application of biochar is accredited to its persistence nature fluctuating from few years to several millennia (Lehmann et al. 2006). The evaluation of indirect biochar effect on sequestration of C may be related to lower C turnover and lower mineralization due to physical and chemical protection of other sources of organic C in soil that has commonly been addressed by short term studies (Qayyum 2012).

Application of biochar improves nutrient cycling, thereby enhancing biomass production by 20–200% (Novak et al. 2009; Major et al. 2010). Biochar has high porosity and surface area and can potentially stabilize other sources of organic C in soil through adsorption progressions, due to surface hydrophilic and phobic interactions among biochar, soil minerals, and organic compounds (Kleber et al. 2007). Biochar also subsidizes to amplified CEC in soil and can encourage organic C stabilization through organo-mineral associations (Mao et al. 2012). Moreover, biochar can promote aggregation and therefore the instantaneous steadiness of biochar particles with other sources of organic C and microaggregate (Awad et al. 2013).

13.4.9.3 Biochar in Minimizing Pollution

Biochar amendment increases the retention of major nutrients N and P in soil by minimizing leaching of nutrients due to groundwater, thereby increasing nutrient availability to crops (Lehmann et al. 2003). The nutrients that are soluble in soil are less likely to be eroded as compared to nutrients which are attached or adsorbed on the surface of soil sediments, thus saving the nutrients due to surface water flows (Lan et al. 2018). The reduction in leaching due to biochar application by its adsorption behavior is verified in a greenhouse experiment (Woolf et al. 2010a, b). Biochar from swine residues has the ability to convert the soluble inorganic P present in the biomass into the adsorbed phosphate in biochar, which leads to minimize groundwater leaching and avoid erosion (He et al. 2000).

13.4.9.4 Biochar in Waste Water Treatment

Biochar serves as an effective solution for removal of organic compounds, pesticides, metals, etc. from soil. The exceedingly permeable structure and greater surface area of biochar allows maximum surface area for the contaminants to interact with the active site of biochar. The adsorption happens by the process of C filtering in biochar. The presence of anions like hydroxyl- and carboxyl- groups in biochar functions as cation-exchanger (Lee and Lee 2004). The classical graphite structure of biochar enables the C to connect with neighboring atom or atoms from foreign molecules, which increases adsorption capacity.

13.4.9.5 Biochar in Reduction of Hazardous Material in the Environment

Biochar is efficient to adsorb contaminants present in the environment by sequestering them and altering their effects eventually, due to its resisting nature towards soil microbes as well as its binding nature of different pollutants in the environment (Xiao et al. 2011). It is reported that biochar amendment has the ability to effectually sorp pesticides in wheat and rice crop 400–2500 times than the normal soil. Biochars derived from manures from dairy farms have the capability of sorption of heavy metals like lead and other organic contaminants, which depend on the C level of soil, the properties of biochar, and interaction between soil and biochar. Zhang et al. (2010) reported that biochar derived from *Pinus radiata*, sophisticated competence for sorption and desorption of pollutant phenanthrene from the soil. Biochar is capable of adsorbing persistent organic pollutants like (POPs) as they have high affinity for naturally occurring biochar (Chen et al. 2007). Chen and Yuan (2011) revealed that application of biochar to soil contaminated with poly aromatic hydrocarbons (PAHs) can help the sorption of PAHs from the soil.

13.5 Eco-Intensification through Biochar Application

Ecological intensification encompasses better usage of ecosystem services in agricultural and horticultural ecosystems by regulating and supporting them for maximized crop yield by minimizing environmental impacts (Bommarco et al.

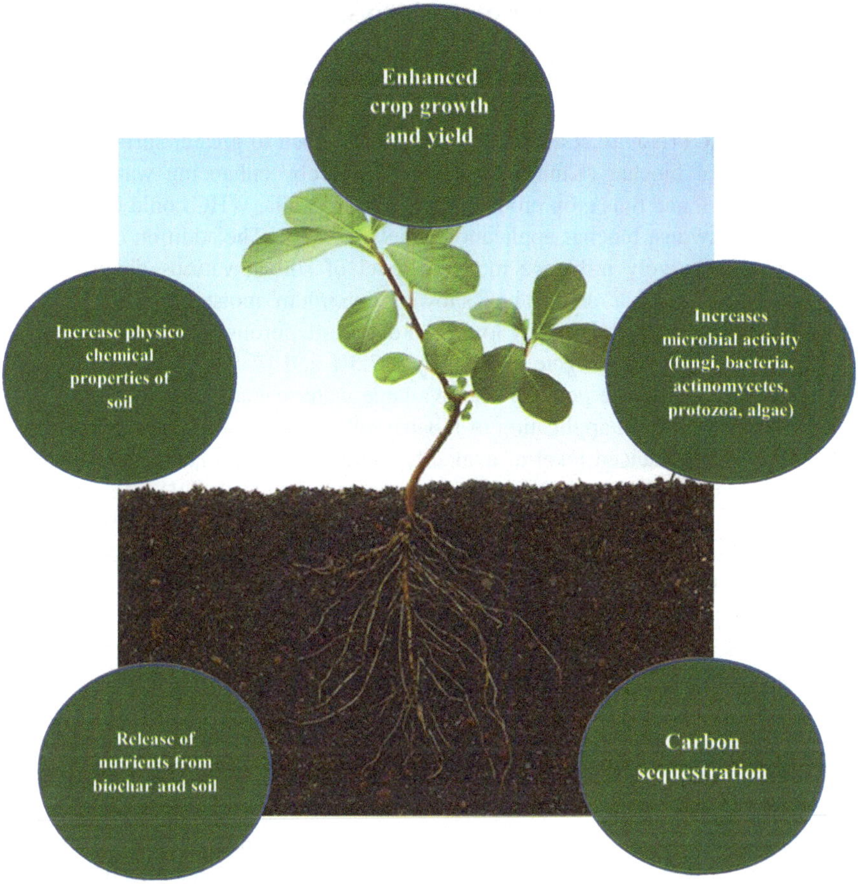

Fig. 13.3 Mechanisms of biochar in improving soil fertility

2013). Recently in agricultural sector, an increased attention on soil biochar amendment is focused on socioecological restoration including sequestering C. Biochar amendment to soil is organic, eco-friendly, and a perfect solution to utilize agricultural wastage as an effective soil fertilizer/stabilizer. The basic properties of biochar as an efficient soil conditioner are their surface area being higher with several functional groups, being porous with high nutrient possessions that aid them to act as a slow-release fertilizer (Ding et al. 2016). Biochar application has proved to improve soil fertility, promote plant growth, increase crop yield, and reduce contaminations (Fig.13.3).

13.5.1 Influence of Biochar on Soil Physical Properties

Biochar amendment influences the soil physical properties in several aspects. Addition of biochar to soil minimizes soil bulk density and improves soil water holding capacity (WHC) (Novak et al. 2009), which is ascribed to greater surface area and porous state of biochar (Kinney et al. 2012), thereby enhancing water retention potential of soil and thus crop productivity. In desert soils, WHC could be increased due to pine-sawdust biochar application (Laghari 2016). The addition of biochar is reported to positively influence moisture level of soil at various depths. A soil amendment of biochar at 5 t ha^{-1}showed maximum moisture level of 11.2% recorded at 35 cm depth in sandy loam soil due to its porous nature.

Biochar increased aggregate stability by 7–7.5% and 17–20% sandy loam coarse textured soils, where the porosity and available water capacity increased by 2–3% (Obia et al. 2018). Soil application of biochar enhances the water retention close to saturation with enhanced level of available water to plants, relative field capacity, dry bulk density, and macroporosity were noticed (Yadav et al. 2018). Soil moisture was found to increase with application of biochar. Biochar amendment of 5 t ha^{-1}followed by 20 tha^{-1} recorded average soil moisture of 24.2%. The application of biochar enhances the unit volume of soil, thereby reducing its bulk density. A minimal bulk density (1.36 g cm^3) and higher pore space (47.5%) were noticed in soil applied with 5 t ha^{-1} of biochar, while a higher bulk density (1.40 g cm^3) and low pore space of 41% were experienced in control without biochar. Dynamics of bulk density due to biochar application may lead to appreciable variation of soil water retention (Bissonnais 1996). The noticeable decline of bulk density in biochar amended soil is an indicator of increment in soil structure, aggregation, and ventilation (Atkinson et al. 2010). It is an evident fact that soil amendment of biochar enhances crop productivity by improved soil structure, augmented porosity, declined bulk density, and enhanced water retention ability, which are key factors for sustainable soil productivity that positively relates to eco-intensification (Baiamonte et al. 2015).

13.5.2 Influence of Biochar on Soil Chemical Properties

Biochar also improves the soil chemical properties like ion exchange capacity, soil aeration, organic matter content, nutrient retention, and N use efficiency (Chan et al. 2007). The use of biochar as soil amendment aids to retain plant available nutrients and decreases leaching of nutrients and agricultural chemicals (Glaser 2007; Lehmann et al. 2006; Laird et al. 2010). The general composition of biochar includes C, hydrogen, N to higher level and K, Ca, sodium, and Ma to some lesser extent (Zhang et al. 2015). There's a significant decrease in total amounts of P, Mg, Si, and N leaching using biochar, despite the simultaneous addition of swine manure (Laird et al. 2010); in contrast, total amounts of leached K and Ca increased. Baronti et al. (2014) reported an amplified retaining of total N, P, K, organic C, and nitrate in soils containing biochar as compared to control without biochar. These positive effects are mainly related to high surface area, surface charge density, and negative surface charge of biochar (Liang et al. 2006) and to biochar capacity to stimulate microbial activity (Steiner et al. 2008a).

Due to its high porosity as well as the presence of both polar and non-polar functional groups on its surface, biochar can efficiently adsorb organic molecules and nutrients (Liang et al. 2006). After its soil inclusion, char surface undergoes oxidation leading to the development of phenolic and carboxylic functional groups which represent negative pH-dependent charges (Cheng et al. 2008). Moreover, biochar increases soil pH due to its alkaline metal content (Glaser 2007). Increased pH of biochar in acidic soils increases the solubility of important elements for plants such as phosphorus, calcium, and potassium (Laird et al. 2010). This could be destructive on alkaline soils since biochar application could determine an excessive increase of soil pH (Mukherjee and Zimmerman 2013). In addition, biochar amendments enhance the soil electrical conductivity by 124.6% and cation exchange capacity by 20% (Laird et al. 2010) and reduce soil acidity by 31.9% (Keith 2016; Oguntunde et al. 2004). The chemical impact of soil amendment with biochar increases the nutrient use efficiency and paves way for more availability of nutrients to plants; thereby a curtail nutrient application and an improved ground water quality with minimized nutrient leaching and pollution (Rousk et al. 2010) positively impact on sustainable soil management which further impacts on eco-intensification of soil services.

13.5.3 Influence of Biochar in Soil Biological Properties

Biochar application to soil enhances soil biota, which plays a major role in nutrient cycling, thereby contributing to key elements for plants and crops (Pietry and Brookes 2008). Soil amendment with biochar is said to have increased the microbial load of a soil (Romaniuk et al. 2011). Biochar addition to soil increases microbial activity in soil alters structure of the microbial community and competition for available nutrients by altering the physicochemical properties of soil (Pietikainen et al. 2000; Steiner et al. 2007). Microbes, viz., bacteria and fungi, are the primary decomposers of soil organic matter constituting of 90% of soil microbial population (Turbé et al. 2010). Biochar amendment improves the activity of soil microbes and their biomass, enhances soil enzyme activity, alters soil bacteria to fungi ratio, and redefines the microbial community structure (Ahmed et al. 2016; Saxena et al. 2013). The highest population of bacteria 41–42 x 10^6 CFU, fungi 33 x 10^{-3} colonies, and 30 x 10^{-4} CFU actinomycetes was observed in biochar amended soil. The increased microbial activity in biochar-amended soil is due to a wide array of compounds on its surface instantly after pyrolysis (Pandian et al. 2016). Those compounds are metabolized by microbes as sugars and aldehydes have contributed for increase in fungal population (Rillig et al. 2010). The physical structure of biochar with its pore space provides a secure environment for increase in microbial colonies. High soil moisture in association with better aeration and aggregation of soil facilities are observed microbial growth in biochar amended soils (Milla et al.

2013). The influence of biochar on microbial activities is assorted and the below possible mechanisms are correlated:

- Biochar provides housing for soil biota in the pore spaces in their surface (Quilliam et al. 2013).
- Biochar offers enriched nutrient diet for soil microorganisms (Joseph et al. 2013).
- Biochar generates toxicity with eco-friendly persistent free radicals (Fang et al. 2014).
- Biochar alters the soil properties feasible for microbial refugia such as soil moisture content, aeration conditions, and pH (Quilliam et al. 2013).
- Biochar persuades fluctuations in soil enzyme activities that affect soil elemental cycle in relation to microbes (Lehmann et al. 2011; Yang et al. 2016).
- Biochar contains certain molecules that can signal for microbial communication. They interrupt microbial intra- and interspecific communication between microbial cells, a combination of sorption, and hydrolysis of signaling molecules (Gao et al. 2016; Masiello et al. 2013).
- Biochar augments the sorption and degradation of soil contaminants and reduces their bioavailability and toxicity to microns (Stefaniuk and Oleszczuk 2016).

Biochar amendment has increased general composition of soil biological community including beneficial microbes by 125% (Liang et al. 2010; Grossman et al. 2010). Biochar application-enhanced soil basal respiration rate by 30% following biochar application in a day or two (Steiner et al. 2008a, b). The positive role of biochar in activating soil biota is a significant factor that leads to sustainable soil management by intensifying soil eco-services.

13.5.4 Impact of Biochar on Carbon Sequestration

Soil amendment with biochar had a significant effect on soil OC content. Biochar incorporation has increased organic C content ranged from 33 to 35% (Shenbagavalli and Mahimairaja 2012a, b). The organic C buildup in soil might be due to the high C content from biochar and is attributed to low level of degradation and intractable nature of biochar in soil (Day et al. 2004). The microbial C and rhizosphere decomposition and exudates contribute to OC in soil. An experiment to determine the enhancement of OC level in soil due to biochar amendment revealed that 4.4–4.8 Kg^{-1}, while control soil contains 3.6 Kg^{-1} (Shenbagavalli and Mahimairaja 2012a, b). Biochar application in higher doses increases the WSC content by 73% and biomass carbon (BMC) content by 37%. A soil buildup of C content after biochar amendment for 3 years was reported in a study (Zimmerman et al. 2011). The negative priming effect in soil due to increased soil C in much greater extent is due to biochar application (Cheng et al. 2006). Biochar application augments the soil aggregate ability (Burrell et al. 2016).

The stable nature of biochar allows C sequestration in soil (Lehmann et al. 2006). It is estimated that biochar sequesters 5–10 Gt of C, which is equal to the present global emissions from fossil fuel use. Additionally, 40% of C to soil is contributed by biochar C (Glaser 2007). It is foreseen that the retention time of biochar is for long years and might be hundreds or thousands. Moreover, biochar can sequester soil C and contribute to GHG mitigations as it can withstand rapid microbial degradation (Lehmann et al. 2006; Thies and Rillig 2009). There is an increased C level in the atmosphere due to mass burning of fossil and decomposition of biomass. However, biochar application can aid in minimizing the emanation of CO_2 as biochar has the potential to store 50% of C from feed stock (Sohi et al. 2010a, b). Biochar is highly stable and is able to refrain emission of CO_2 and plays a vital role in controlling release of GHGs (Yanai et al. 2007). Biochar production has proved to be the best practice to sequester CO_2 from atmosphere, as C adsorbed by plants from the atmosphere stores in structure which is not released back to the environment on decomposition. The sequestered C in the biomass gets converted into a highly stable state, thereby decreasing CO_2 discharge from the soil on decomposition (Kataki et al. 2012). A study reported that the addition of 20 Kg of biochar in a kilogram of soil has the ability to reduce the emission levels of nitrogen dioxide up to 80% in grass pots and almost 50% in soybean, as biochar has the ability to adsorb and retain ammonia in soils and thus reduce the availability of N for nitrification process (Rondon et al. 2006). Reduction in GHGs due to biochar addition is influenced by the soil type, soil moisture level, and the properties of the applied biochar (Zwieten et al. 2009). Application of biochar has reduced the emission of nitrogen dioxide by 40–50% (Baiamonte et al. 2015); therefore, biochar amendment to soil has the added benefit of reduction in GHG emission (Sohi et al. 2009).

Soil amendment with biochar could be a possible strategy to reduce GHG emanation by up to 9500 million tonnes of C (Woolf et al. 2010a). Biochar can reside in soil between 11 and over 1000 years, based on the soil, ecosystem, as well as biochar properties (Singh et al. 2012). The ruminants contribute about 81% of the total GHG emission, in the form of direct gaseous excretions and through flatulence and burping by cattle emitting about 200–500 litters of methane per day. This animal-based methane is emitted through rumen microbial methanogenesis, and 90% of the GHG is produced by the cattle (Tapio et al. 2017). The addition of biochar to animal and poultry litter can reduce the GHG emission, as it can retain ammonia, and improve the composting process to reduce GHG through the process. Assuming sufficient feedstock is available to produce biochar and targeting biochar disposition on infertile land. Woolf et al. (2010b) have calculated maximum global reduction of 6.6 giga tonnes CO_2 equivalent/year from 2.27 giga tonnes biomass C. In these regards, Rondon et al. (2006) report, with the addition of biochar, 50 and 80% reduction of N_2O emissions under soybean and grass systems, respectively. Recycling agriculture residues through composting is economical and a common practice. However, during this process over half of the nitrogen (ammonia) generated through microbial decomposition is lost in gaseous state. This is not only an eminent loss of N but also pollutes the environment and emits the odor. The co-composting with biochar can reduce ammonia losses through adsorption and establishing covalent bonds between ammonia and biochar, which results in the reduction of odor and fertilizer application to crops (Okigbo 1991). The biochar when added to the feedstock compost significantly reduces N losses in the final compost product. During this trial, biochar is proved to be better in terms of retaining ammonia compared with any fresh plant material and most animal manures, but when oxidized biochar was added, the ammonia retention increased by five times.

13.5.5 Influence of Biochar on Crop Growth, Physiology, and Yield Productivity

Soil amendment of biochar has a good prospective for improved yield efficiency of agricultural crops and enhanced C sequestration (Bear 2014). Numerous researchers have substantiated the impact of biochar in augmenting the cation exchange capacity of soil nutrients (Nartemink 2003; Dume et al. 2016), stimulate activities of soil microorganisms (Aghari et al. 2016), and reduce the leaching loss of plant nutrients in soil (Debela et al. 2012). Additionally, the positive effect of biochar on root growth was highlighted by several workers (Cheraghi et al. 2009; Bonanomi et al. 2017), predominantly increased root biomass (Ding et al. 2016), and root length (Lorenz and Lal 2016; Olma et al. 2016). Biochar amendment to soil has reported to enhance crop growth parameters which ultimately result in yield attributes. Biochar shows varied responses to different crop growth and yield attributes (Chan et al. 2008). Biochar from paper mill waste amended in agricultural soil for wheat, radish, and soybean production has enhanced their biomass (Van Zwieten et al. 2010). A noteworthy reduction in dry matter content was noticed when soil is amended with biochar at a rate of 10 t ha^{-1}. Biochar amplified yield of paddy in soils with low P availability due to improved saturated conductivity of top soil, response to N and P chemicals, and xylem sap flow of a plant (Asai et al. 2009). Biochar has increased the crop yield in several crops; for instance, the yield of maize was significantly improved by 150 and 98% biochar amendment at 15 and 20 t ha^{-1}, respectively (Asai et al. 2009).

Few biochars contribute plenty of micronutrients to the soil. The pecan-shelled biochar has high level of copper (Cu), magnesium (Mg), and Zinc (Zn) than the soil (Novak et al. 2009). Biochar applied at 20 and 40 t ha^{-1}in C-deficient calcareous soils of China improved maize production by 15.8 and 17.3%, respectively (Masto et al. 2013; Zhang et al. 2010). Moreover, biochar application in nutrient poor acidic loamy soil recorded improved effect on wheat yield in the absence of mineral fertilization but recorded 20–30% more yield, when applied with high rates of mineral fertilization (Alburquerque et al. 2014). Tomato productivity was enhanced by 4.7–25.2% by biochar application (Vinh et al. 2014). The cumulative yield of maize and wheat in calcareous soil was significantly enhanced by biochar application (Liang et al. 2014). Rice husk biochar has increased root biomass, plant height, and number of leaves in lettuce (Carter et al. 2013). Biochars prepared from oak husk evaluated for 4 years consecutively on maize soybean rotation have positive sign in above ground mass and grain yield (Hottle 2013). Biochar prepared from poultry litter tested in cotton at 3000 kg ha^{-1} recorded better crop growth (Coomer et al. 2012). In contrast, biochar from maple biomass did not displayed any significant difference in root elongation of Pea (Borsari 2011), and wood chip biochar on French bean did not show any positive trend (Saxena et al. 2013).

13.5.6 Influence of Biochar on Adsorption and Release of Nutrients

Biochar, apart from retaining nutrients in a form available for plants, has empathy for both inorganic and organic compounds and may sorb toxic by-products from soils and waste waters (Decker and Corby 1971; Yu et al. 2009). Several researches have been conducted to investigate the efficiency of biochar to retain polar compounds, polar organic pesticides (Lian et al. 2011), and hydrophobic molecules such as polycyclic aromatic hydrocarbons lignin and tannin (Nair et al. 2017). The adsorptive capacity of biochar depends on its physical (its porous structure) and chemical (presence of specific surface functional groups) properties (Pignatello et al. 2006). The high surface area and presence of surface polar groups on wheat were determinant in the uptake of neutral organic contaminants, such as benzene and nitrobenzene (Chun et al. 2004; Jafri et al. 2018). In addition to organic compounds, char can also effectively bind inorganic molecules. Various studies that demonstrated biochar capacity to absorb nitrate (Mizuta et al. 2004), phosphate (Beaton et al. 1960), and metal ions (Beesley et al. 2010; Namgay et al. 2010; Uchimiya et al. 2010; Beesley and Marmiroli 2011; Fellet et al. 2011; Karami et al. 2011) found that biochar from hardwoods rapidly uptake and reduced cadmium (Cd) and Zn mobility from a polluted soil. Uchimiya et al. (2011) reported a decrease in C content of soil using broiler litter biochar as amendment. Mohan et al. (2011) used activated carbon (AC), oak bark, and oak wood char for chromium (Cr) remediation from contaminated surface water. Therefore, the highest remediation potential of biochar depends on the presence of a large amount of specific surface functional groups that serve as adsorption sites for heavy metals (Laird et al. 2010). Biochar sorptive capacity can be effectively used to mitigate diffuse pollution from agriculture and to immobilize potentially toxic organic and inorganic compounds, thereby reducing contamination from soils or wastewaters.

13.6 Mixing Biochar with Other Amendments

An increased research attention is focused on combination of biochar with compost on improved soil richness and sequestration of C. The blend of biochar and compost is widely attempted in agricultural sector for improving soil fertility and plant growth (Schultz and Glaser 2012; Prost et al. 2013). Mixing biochar with organic manures, compost, or lime before soil application has reported to increase the soil quality than application of biochar alone. Since biochar has the ability to sorb nutrients to avoid leaching, it is said to enhance the efficiency of the additional amendments added in the combination, thereby increasing the productivity (Novak et al. 2009). The synergistic effect of biochar-compost combination is achieved by the stability of biochar which promotes C sequestration, apart from retaining nutrients and enhancing fertility of soil; compost mineralizes and eventually add to the organic matter pool (Nigussie et al. 2012). The combinations of biochar and compost have contributed in waste management and nutrient recycling in oil palm industry; oil palm biochar-compost combination as fertilizer in oil palm nursery has contributed in faster establishment and growth of seedlings and soil C sequestration for long-term. A blended application of biochar and compost increased watermelon yield; displayed a positive correlation with crop productivity and soil nutrients such as available N, P, and K; and enhanced microbial diversity (Cao et al. 2017). The consequences of biochar and compost mixture depend on biochar and biomass properties (Virmani et al. 1994). The construction of oxygen containing functional groups while composting leads to increase in nutrient retention, according to some researches (Tandon 2004); therefore, biochar compost combination permits higher nutrient retention in biomass adding value to the final product by altering the physiochemical properties. The biochar compost blend seems to be a promising source of amendment and interesting alternate to inorganic fertilizer (Zheng et al. 2012).

13.7 Agricultural and Eco-Environmental Sustainability through Biochar Application

Globally, food security and environmental safety are the prime concern, emerged due to nonsustainable soil and agricultural management practices, influenced by biotic and abiotic stress factors, which resulted in declined agricultural production, excessive chemical usage, and soil contamination by pollution that impact on soil and plant health (Chukwuka and Omotayo 2009; Atkinson and Urwin 2012; Jhariya et al. 2019a, b; Kumar et al. 2020). Biochar, the C-rich organic amendment, obtained by pyrolysis of biomass is a sustainable option to challenge threats of food security and environmental safety.

Biochar application mends soil quality by retaining nutrients and skillful to muddle through GHGs to support C sequestration. More than half of biochar contains original C which is highly recalcitrant in nature, which aids in sequestration of C by locking C present in plant biomass (Lehmann and Rondon 2006). The fundamental configuration and structural arrangement of biochar are sturdily correlated with temperature, heating rate, and time maintained during biochar production (Vaccari et al. 2011). Some quantity of biofuels and volatile gases were also produced, which is utilized for energy production. Biochar amendment increases soil pH and EC which are attributed to the presence of ash residues dominated by carbonates of alkali and alkaline earth metals, some amount of silica, heavy metals, organic and inorganic N (Mankasingh et al. 2011). Biochar increases the cation exchange capacity, water holding capacity, and microbial activity owing to its high surface area and minimizes nutrient leaching by providing nutrient binding sites. Therefore, biochar amended soil demands reduced nutrient requirements with reduction in environmental pollution (Saranya et al. 2011).

Biochar application enhances soil fertility and improves soil texture and sorption of nutrients, apart from improving crop productivity (Hedley et al. 2004).

An abundant quantity of residues is generated in the form of crop, industrial, forest, weeds, animal and municipal waste, which has to be managed in a manner sustainable to environment which is essential to diminish environmental pollution. Biochar production from these residues is a twin objective for waste disposal as well as energy recovery (Walsh et al. 1999; Woolf et al. 2010a, b). The most effective method for removal of pesticides, volatile compounds, heavy metals, etc. is the use of AC in the form of biochar, where C filtering works on the principle of adsorption. The highly porous nature, larger surface area, and maximum surface area help the contaminants and impurities to interact on the active sites of biochar (Lee et al. 2004). The low thermal conductivity and water absorption ability of biochar make it suitable as a building insulating material (Schmidt et al. 2018). Biochar can be substituted for coal in energy production. The bio-oil and syngas can be promoted to bio-fuel and gasoline products (Laird et al. 2010). The major benefit of biochar amendment is that it helps to combat global warming by sequestering CO_2 from the atmosphere. It can be used for rehabilitation of destructed landforms. The application of biochar as a soil additive is an important management strategy for agricultural sustainability using degraded soils and thereby assuring food security (Asemoloye et al. 2017). Biochar has the adsorption efficiency of several persistent organic pollutants (POP's) and inorganic pollutants such as heavy metals, etc., which are basically determined by the biomass utilized by biochar production (Singh et al. 2010a, b; Mukherjee et al. 2014).

The benefits of biochar to agricultural production and environment fall on three sustainability factors, viz., utilization of sustainable biomass, sustainable production process, and sustainable usage (Elad et al. 2011). Therefore, biochar amendment alters soil health and maintains the operations of natural and managed ecosystems for sustainable agricultural productivity and eco-stability (Chintala et al. 2014; Kumar et al. 2014). Biochar thus poses several benefits to the environment, agriculture, and economy; therefore, it is highly recommended to incorporate it in agricultural ecosystems.

Sustainability is in general "all encompassing" term, a very difficult issue to measure, and the potential of biochar in sustainability is unclear (Biederman and Harpole 2013). The overall benefits and potential of biochar may have more coordinance and relevance with sustainability paradigm.

13.8 Research and Development in Biochar Application for Soil Eco-Intensification

During the past decade, awareness on biochar usage as an organic soil conditioner has increased exponentially. The changeover of locally available farm residues and biomass to biochar may have applications in crop management; thus, biochar contributes substantially to sustainable agriculture. While the benefits and opportunities of biochar usage look attractive, several bottle necks need to be addressed before widespread production and usage of biochar.

The current statistics on biochar is largely based on limited scale studies under laboratory and greenhouse conditions. Though results of few field studies were positively interpreted for biochar, a substantial and scientific awareness on large-scale field production of the supposed benefits of biochar is yet to be validated (Agegnehu et al. 2017). Biochar application has a promise for increased crop productivity, but the assumed yield benefit was controversial for different soil types, viz., nutrient poor, acidic, sodic, sandy, clayey, etc. The uniformity in literature in biochar amendment for various types of soil is to be data based (Jeffery et al. 2015). The beneficial properties of biochar largely depend on the biomass used, pyrolysis duration, temperature, and method of production; the beneficiary properties of biochar with different types of biomasses, various pyrolysis timings, and temperature should be explored. The potential role of biochar in forestry, horticulture, and other speciality crops is another area to be explored. The application of a small quantity of biochar in seedling pits aids in production of healthy seedling of these crops for landscaping and ornamental arenas, which encourages commercial and small-scale nurseries to be benefitted. The advantage of biochar with its high-water holding ability is yet to be explored in arid and semiarid regions of the world (Nair et al. 2017).

The availability of biomass as well as the economic merits, energy needs, and environmental risks if any of their large-scale production and use is yet to be explored. However, the current available statistics and indications advise the biochar could play a significant role in facing the challenges posed by threats to agroecosystem sustainability (Kanayo et al. 2020). The current indications strongly suggest continued research and development efforts in having more wider understanding on beneficiary potential as well as limitations of biochar and expanding its utility in soil and land use management (Metzger et al. 2005). In order to accomplish the target of agroecosystem sustainability, it is necessary that the beneficial role of biochar in climate change mitigation and sustainable soil management has to be strengthened and appropriately integrated.

13.9 Policy and Legal Framework

The expansion and development of biochar industry are greatly influenced by policies and regulations. Policies on biochar concern the regulations regarding biodiversity, generating impacts on biochar industry. The enlarged biomass production for biochar may threaten biodiversity, as it is essential to avoid natural ecosystems to obtain necessary biomass. Biochar may potentially improve the microbial biodiversity as they stimulate microbial growth in their porous structures (Verheijen et al. 2010). A meta-analysis of numerous researches stated that biochar has a latent consequence on plant community and their yield attributes and has variable effects on plant-associated microbes (McCormack et al. 2013). Policies on the possible effects of expansion of biochar industry and microbial biodiversity may be either stimulated or impeded (Biederman and Harpole 2013). Furthermore, biochar applied to soil can possess contaminants and pollutants such as heavy metals; therefore, regulations on soil protection and remediation impact on biochar production and the industries that supply biochar. In situations where biochar application may entail risk to the soil and ecosystem, the end user should abide with the prescribed regulations regarding soil protection. Furthermore, an environmental permit is required for production process of biochar such as storage of waste residues. Several obligations were pertained, which may hamper the progress of innovative biochar missions. A planning permission preceded by necessary public enquiry is required for each installation of biochar production unit. Moreover, policies such as employer protection, intellectual property right, etc. may hamper the development of biochar industry. However, these policies influence the development of a biochar industry that tries itself to establish itself on the market; therefore, the possible negative impacts of legislation will not affect biochar industry.

13.10 Future Perspective of Biochar Application

Global pollution of ecosystem is the chief concern affecting health and economy lately. The eco-properties of soil are extremely complex, which makes soil remediation more challenging and expensive. In this face, sustainable components like biochar and composts extant an operative and competent remediation solution. Biochar is indeed a sustainable solution that enhances soil fertility and soil C sequestration (O'Toole et al. 2016). The dynamics of biochar in diverse soil types of varied texture has to be explored. Special attention should be paid on the impression of biochar on soil basic properties such as pH, CEC, bulk density, porosity, water retention capability, SOM, and redox reactions. The moisture and nutrient retention of biochar's from various biomasses in diverse soil types should be investigated to extend the benefits of biochar globally.

Precisely, biochar along with its relative composites can achieve novel properties with addition of metal oxides, surface agents, and nanomaterial. The possibility of engineering biochar for improving specific soil deficiency is a novel approach which can enhance the biological properties of biochar. Efforts have been initiated to impregnate biochar with various inorganic fertilizers to enhance the efficiency as nutrient source.

Blending of biochar with specific added valued composts could increase its worth for soil fertility and microbial inoculation. In addition, exploring the biological activity of soil biota due to biochar application is another area to be explored. The synthesis of nanometal-assisted biochar is an innovative utility of biochar to enhance soil properties. The beneficial role of biochar application in climate-change mitigation is another area to be explored for sustainable crop productivity. Biochar is utilized as cattle feed and is reported to improve the nutrient intake efficiency, adsorb toxins and to improve animal health (Toth and Dou 2016). The European Biochar Certificate (EBC), a voluntary standard has been managing the certifying biochar quality as animal feed, which guarantees compliance with all feed limits prescribed by EU regulations and certifies sustainable, climate friendly production (Watarai and Rana 2005; Leng and Rur 2017; European Biochar Foundation (EBC), 2012). Further investigation in this area should be widened.

Significant emphasis should be focused on soil remediation, specifically on heavy metal eradication utilizing biochar-based composites. Further research should be urged on biochar-microbial interactions and their role in exclusion of heavy metals and pollutants from soil. The rate of decomposition of biochar from varied biomasses in diverse soil types is a chief area to be investigated. In addition, the potential of diverse biomasses in reduction of GHG emission is to be investigated. Available evidence on biochar research and development justifies sustained efforts in understanding more about the benefits as well as limitations of biochar and expanding its role in land management. Though, biochar amendment is not a remedy for unravelling all kind of soil-related issues, it is undoubtedly a venture that deserves attention in future soil and land management.

13.11 Future Research Area on Biochar Use as Soil Amendment

- Biochar dynamics in diverse soil types of different textures.
- Impact of biochar on soil basic properties like pH, CEC, bulk density, porosity, water retention capability, SOM, and redox reactions.
- Moisture and nutrient retention of biochar's from various biomasses in diverse soil types.
- Engineering biochar with metal oxides, surface agents, and nano-material.
- Soil remediation potential of biochar.
- Exploring the biological activity of soil biota due to biochar application.
- Application of biochar in climate change mitigation.
- Analyzing the potential of diverse biomasses in reduction of GHG emission.
- Rate of decomposition of biochar in different type of soil.

13.12 Conclusion

Feeding the growing world population is the chief concern of agriculturalists globally and the basic input for agricultural productivity is land and soil spreading over it. Extending more lands for agricultural productivity is of great need to feed the growing population globally. The best scenario is more crop productivity from limited space, provided the soil is capable of higher productivity. In olden days, farmers recognized the nutritional value of the residues and added these to the soil by several means. In modern world, the crop wastes and other residues were burnt in farms. Studies substantiate that burning can cause loss of the nutrient into the atmosphere; it can destroy the chemical and biological properties of soil too. Biochar is one of the best ways of using these residues, and it makes it more beneficial to the community. A natural material can be turned to biochar by heating under limited or without oxygen. A huge amount of the agricultural wastes is burnt to destroy residues around the world. Research on using the biochar as fertilizer or soil stabilizer to improve quality of agricultural soil is lacking. The application of biochar has been regarded as a sustainable approach in agricultural production by pyrolysis of waste materials into value-added materials in a smart way. A significant knowledge gap exists for the utilization of farm residues as economic biochar. Therefore, more research avenues are required in this area to effectively utilize farm waste in to a viable solution.

References

Abderahim A, Alfredo N, Noel D, Didier M (2018) High physical properties of biochar issued from cashew nut Shell to adsorb mycotoxins (aflatoxins and Ochratoxine a) and its effects on toxigenic Molds. Eng Technol Int J Medical Health Sci 12:472

Agegnehu G, Srivastava AK, Bird MI (2017) The role of biochar and biochar-compost in improving soil quality and crop performance: a review. Appl Soil Ecol 119:156–170. https://doi.org/10.1016/j.apsoil.2017.06.008

Aghari M, Naidu R, Xiao B, Hu Z, Mirjat MS, Hu M (2016) Recent developments in biochar as an effective tool for agricultural soil management: a review. J Sci Food Agric 96:4840–4849. https://doi.org/10.1002/jsfa.7753

Agyarko-Mintah E, Cowie A, Singh B-P, Joseph S, Van Zwieten L, Cowie A, Harden S, Smillie R (2016) Biochar increases nitrogen retention and lowers greenhouse gas emissions when added to composting poultry litter. Waste Manag 61:138–149

Ahmed A, Kurian J, Raghavan V (2016) Biochar influences on agricultural soils, crop production, and the environment: a review. Environ Rev 4:495–502

Alburquerque JA, Calero JM, Barrón V, Torrent J, del Campillo MC, Gallardo A, Villar R (2014) Effects of biochar produced from different feedstocks on soil properties and sunflower growth. J Plant Nutrition Soil Sci 177:16–25

Allmaras RR, Kraft JM, Miller DE (1988) Effect of soil compaction and incorporated crop residue on root health. Annu Rev Phytopathol 26:219–243

Al-Wabel MI, Al-Omran A, El-Naggar AH, Nadeem M, Usman ARA (2013) Pyrolysis temperature induced changes in characteristics and chemical composition of biochar produced from *conocarpus* wastes. Bioresour Technol 131:374–379

Ameloot N, Sleutel S, Das KC, Kanagaratnam J, de Neve S (2015) Biochar amendment to soils with contrasting organic matter level: effects on N mineralization and biological soil properties. GCB Bioenergy 7:135–144. https://doi.org/10.1111/gcbb.12119

Anderson PS, Reed TB, Wever PW (2007) Micro-gasification: what it is and why it works. Boiling Point 53:35–37

Angin D (2013) Effect of pyrolysis temperature and heating rate on biochar obtained from pyrolysis of safflower seed press cake. Bioresour Technol 128:593–597

Apaydın-Varol E, Pütün AE (2012) Preparation and characterization of pyrolytic chars from different biomass samples. Anal Appl Pyrolysis 98:29–36

Asai H, Samson B, Haefele S, Khamdok H, Koki K, Yoshiyuki I, Yoshio T, Shiraiwa H (2009) Biochar amendment techniques for upland rice production in northern Laos: 1. Soil physical properties, leaf SPAD and grain yield. Field Crops Res 111:81–84. https://doi.org/10.1016/j.fcr.2008.10.008

Asemoloye MD, Ahmad R, Jonathan SG (2017) Synergistic action of rhizospheric fungi with Megathyrsus maximus root speeds up hydrocarbon degradation kinetics in oil polluted soil. Chemosphere 184:1–12

Asfaw G, Chandraratne M (2018) Biochar Production from biomass Waste-Derived Material. Doi: https://doi.org/10.1016/B978-0-12-803581-8.11249-4

Atkinson CJ, Fitzgerald JD, Na H (2010) Potential mechanisms for achieving agricultural benefits from biochar application to temperate soils: a review. Plant Soil 337:1–18

Atkinson NJ, Urwin PE (2012) The interaction of plant biotic and abiotic stresses: from genes to the field. J Exp Botany 63(10):3523–3543

Augustínová J, Cvengrošová Z, Mikulec J, Vasilkovová B, Cvengroš J (2013) Upgrading of bio-oil from fast pyrolysis. 46th international conference on petroleum processing, June 7, 2013, Bratislava, Slovak Republic

Awad YM, Blagodatskaya E, Ok YS, Kuzyakov Y (2013) Effects of polyacrylamide, biopolymer and biochar on the decomposition of 14C-labelled maize residues and on their stabilization in soil aggregates. Eur J Soil Sci 64:488–499

Baiamonte G, De Pasquale C, Marsala V, Cimò G, Alonzo G, Crescimanno G, Conte P (2015) Structure alteration of a sandy-clay soil by biochar amendments. J Soils Sediments 15:816–824. https://doi.org/10.1007/s11368-014-0960-y

Bakar MSA, Titiloye JO (2012) Catalytic pyrolysis of rice husk for biooil production. Anal Appl Pyrolysis 103:362–368

Baker R, Ab R, Zahidah A, Siti J, Bahi L, Chintsong M (2015) Influence of oil palm empty fruit bunch biochar on floodwater pH and yield components of Rice cultivated on acid sulphate soil under Rice intensification practices. Plant Prod Sci 18:491–500. https://doi.org/10.1626/pps.18.491

Ball PN, MacKenzie MD, DeLuca TH, Holben WE (2010) Wildfire and charcoal enhance nitrification and ammonium-oxidizing bacteria abundance in dry montane forest soils. J Environ Qual 39:1243–1253. https://doi.org/10.2134/jeq2009.0082

Banerjee A, Meena RS, Jhariya MK, Yadav DK (2021a) Agroecological footprints management for sustainable food system. Springer Nature, Singapore, eBook ISBN: 978-981-15-9496-0, Hardcover: 978-981-15-9495-3, p 514. https://doi.org/10.1007/978-981-15-9496-0

Banerjee A, Jhariya MK, Meena RS, Yadav DK (2021b) Ecological footprints in agroecosystem-an overview. In: Banerjee A, Meena RS, Jhariya MK, Yadav DK (eds) Agroecological footprints management for sustainable food system. Springer Nature, Singapore, pp 1–23. eBook ISBN: 978-981-15-9496-0, Hardcover: 978-981-15-9495-3. https://doi.org/10.1007/978-981-15-9496-0_1

Banerjee A, Jhariya MK, Raj A, Yadav DK, Khan N, Meena RS (2021c) Land footprint management and policies. In: Banerjee A, Meena RS, Jhariya MK, Yadav DK (Eds), Agroecological footprints management for sustainable food system. Springer Nature Singapore. Pp. 221–246. eBook ISBN: 978-981-15-9496-0, Hardcover: 978-981-15-9495-3. doi:https://doi.org/10.1007/978-981-15-9496-0_7

Banerjee A, Jhariya MK, Raj A, Yadav DK, Khan N, Meena RS (2021d) Energy and climate footprint towards the environmental sustainability. In: Banerjee A, Meena RS, Jhariya MK, Yadav DK (eds) Agroecological footprints management for sustainable food system. Springer Nature, Singapore, pp 415–443. eBook ISBN: 978-981-15-9496-0, Hardcover: 978-981-15-9495-3. https://doi.org/10.1007/978-981-15-9496-0_14

Banerjee A, Jhariya MK, Yadav DK, Raj A (2020) Environmental and sustainable development through forestry and other resources. Apple Academic Press Inc., CRC Press- a Tayler and Francis Group, US & Canada, Oakville. ISBN: 9781771888110, p 400. https://doi.org/10.1201/9780429276026

Baronti S, Vaccari F, Miglietta F, Calzolari C, Lugato E, Orlandin S, Pini R, Zulian C, Genesio L (2014) Impact of biochar application on plant water relations in Vitis vinifera (L.). Eur J Agron 53:38–44. https://doi.org/10.1016/j.eja.2013.11.003

Barrios E (2007) Soil biota, ecosystem services and land productivity. Environ Econ 64:269–285

Baumhardt RL, Stewart BA, Sainju UM (2015) North America SoilDegradation. Processes, Practicesand mitigating strategies. Sustainability 7:2936–2960

Bear FE (2014) Chemistry of the soil, international edition. Soil survey staff. Keys to soil taxonomy, 12th edn. USDA-NRCS, Washington, DC

Bear Firman E (1969) Chemistry of the soil, international edition. Soil survey staff (2014) keys to soil taxonomy, 12th edn. USDA-NRCS, Washington, DC

Beaton JD, Peterson HB, Bauer N (1960) Some aspects of phosphate adsorption by charcoal. Soil Sci Soc Am Proc 24:340–346

Beaumont E (2017) FarmChar–growing the Australian biochar market. Energy farmers Australia. In: The Australia New Zealand biochar conference proceedings 10–12 august 2017, Agri futures Australia publication no 17/057, pp 19–21

Beesley L, Moreno-Jiménez E, Fellet G, Melo L, Sizmur T (2015) Biochar and heavy metals. In: Lehmann J, Joseph S (eds) Biochar for environmental management: science, technology and implementation, 2nd edn. Earthscan, London, pp 563–594. ISBN 9780415704151.http://centaur.reading.ac.uk/40801/

Beesley L, Marmiroli M (2011) The immobilization and retention of soluble arsenic, cadmium and zinc by biochar. Environ Pollut 159:474–480

Beesley L, Moreno-Jimenez E, Gomez-Eyles JL (2010) Effects of biochar and green waste compost amendments on mobility, bioavailability and toxicity of inorganic and organic contaminants in a multi-element polluted soil. Environ Pollut 158:2282–2287

Bhattacharyya T (2015) The Soil: A Natural Resource. https://www.researchgate.net/publication/305656097

Biederman L, Harpole W (2013) Biochar and its effects on plant productivity and nutrient cycling: a meta-analysis. Global Change Biol Bioenergy 5:202–214

Bissonnais Y (1996) Aggregate stability and assessment of soil crustability and erodibility: I. theory and methodology. Eur J Soil Sci 47(4):425–437. https://doi.org/10.1111/j.1365-2389.1996.tb01843.x

Blum WEH (1994) Soil resilience general approaches and definition. In: Proceedings of the 15th international Symposia on soil science, pp 233–237

Bolan N, Wijesekara H, Zahangir H (2018) Biochar application modulates soil health and fertility. In: The Australia New Zealand biochar conference proceedings, conference 14–16 august 2018. AgriFutures Australia publication no. 18/039, p 18

Bommarco R, Kleijn D, Potts SG (2013) Ecological intensification: harnessing ecosystem services for food security. Trends Ecol Evol 28:230–238

Bonanomi G, Ippolito F, Cesarano G, Nanni B, Lombardi N, Rita A (2017) Biochar as plant growth promoter: better off alone or mixed with organic amendments? Frontiers Plant Sci 8:1570. https://doi.org/10.3389/fpls.2017.01570

Borsari B (2011) A preliminary study of the effect of biochar from maple (*Acer* spp.) on root growth of selected agronomic crops.In: international symposium on growing media. Composting and Subst Anal 1013:117–122

Brady NC, Weil RR (2002) The nature and properties of soils, 13th edn. Pearson Education (Singapore), Delhi

Bronick CJ, Lal R (2005) Soil structure and management: a review. Geoderma 124:3–22

Burezq H (2019) Intensification of alfalfa (*Medicago sativa* L.) forage production using biochar, animal manure and specialty chemicals – case of Kuwait. Eur J Sci Res 154:254–268

Burrell L, Zehetner F, Rampazzo N, Wimmer B, Soja G (2016) Long-term effects of biochar on soil physical properties. Geoderma 282:96–102. https://doi.org/10.1016/j.geoderma.2016.07.019

Cao Y, Ma Y, Guo D, Wang Q, Wang G (2017) Chemical properties and microbial responses to biochar and compost amendments in the soil under continuous watermelon cropping. Plant Soil Environ 63:1–7

Carol PH (2001) Soil Erosion and Sustainable Mountain development. Mountain Res Dev 21 (1):77–83

Carroll C, Merton L, Burger P (2000) Impact of vegetative cover and slope on runoff, erosion, and water quality for field plots on a range of soil and spoil materials on central Queenslan coal mines. Aust J Soil Res 38:31328.v

Carter S, Shackley S, Sohi S, Suy TB, Haefele S (2013) The impact of biochar application on soil properties and plant growth of pot grown lettuce (*Lactuca sativa*) and cabbage (*Brassica chinensis*). Agronomy 3(2):404–418

Chan K, Van Zwieten L, Meszaros I, Downie A, Joseph S (2008) Using poultry litter biochar as soil amendments. Aust J Soil Res 46(5):437–444

Chan KY, Van Zwieten L, Meszaros I, Downie A, Joseph S (2007) Agronomic values of green waste biochar as a soil amendment. Aust J Soil Res 45:629–634

Chatterjee R, Gajjela S, Thirumdasu RK (2017) Recycling of organic wastes for sustainable soil health and crop growth. Int J Waste Resour 7:296. https://doi.org/10.4172/2252-5211.1000296

Chen B, Yuan M (2011) Enhanced sorption of polycyclic aromatic hydrocarbons by soil amended with biochar. J Soils Sediments 11:62–71

Chen H, Yao J, Wang F, Choi MMF, Bramanti E, Zaray G (2009) Study on the toxic effects of diphenol compounds on soil microbial activity by a combination of methods. JHazardMater167:846–851. Doi:https://doi.org/10.1016/j.jhazmat.2009.01.066

Chen J, Zhu D, Sun C (2007) Effect of heavy metals on the sorption of hydrophobic organic compounds to wood charcoal. Environ Sci Technol 41:2536–2541

Chen KY, Xu ZH (2009) Nutrient properties and their enhancement. In: Lehmann J, Joseph S (eds) Biochar for environmental management: science and technology. Earthscan, London, pp 67–84

Cheng CH, Lehmann J, Thies JE, Burton SD, Engelhard MH (2006) Oxidation of black carbon by biotic and abiotic processes. Org Geochem 37:1477–1488. https://doi.org/10.1016/j.orggeochem.2006.06.022

Cheng CH, Lehmann J, Engelhard MH (2008) Natural oxidation of black carbon in soils: changes in molecular form and surface change along a climosequence. Geochim Cosmochim Acta 72:1598–1610

Cheraghi SAM, Hashminejhad Y, Rahimian MH (2009) An overview of the salinity problem in Iran: assessment and monitoring technology. In: Advances in the assessment and monitoring of salinization and status of biosaline agriculture. Reports of expert consultation held in Dubai, United Arab Emirates, 26–29 Nov 2007. World Soil Resources reports no 104. FAO, Rome, p 2

Chintala R, Owen R, Kumar S, Schumacher TE, Malo D (2014) Biochar impacts on denitrification under different soil water contents. World Congress Soil Sci 6:157–167

Choudhury ND, Chutia RS, Bhaskar T, Kataki R (2014) Pyrolysis of jute dust: effect of reaction parameters and analysis of products. Material Cycles Waste Manage 16:449–459

Chukwuka KS, Omotayo OE (2009) Soil fertility restoration potentials of Tithonia green manure and water hyacinth compost on a nutrient depleted soil in South Western Nigeria using *Zea mays* L. as test crop. Res J Soil Biol 1(1):20–30

Chun Y, Sheng GY, Chiou CT, Xing BS (2004) Compositions and sorptive properties of crop residue-derived chars. Environ Sci Technol 38:4649–4655

Chutia RS, Kataki R, Bhaskar T (2014) Characterization of liquid and solid product from pyrolysis of *Pongamia glabra* deoiled cake. Bioresour Technol 165:336–342

Coomer TD, Longer DE, Oosterhuis DM, Loka DA (2012) The influence of poultry litter biochar on early season cotton growth. University of Arkansas, Arkansas

Criscuoli I, Alberti G, Baronti S (2014) Carbon sequestration and fertility after centennial time scale incorporation of charcoal into soil. PLoS One 9:e91114

Dalila T, Claudia C, Silvia B, Carla A, Francesco PV, Giuseppe L, Sara DL, Francesca F, Roberto T, Gabriella SS (2017) The effects of biochar and its combination with compost on lettuce (*Lactuca sativa* L.) growth, soil properties, and soil microbial activity and abundance. Int J Agronomy, 2017, 12.: Article ID 3158207. doi:https://doi.org/10.1155/2017/3158207

Dalvi RR, Ademoyero AA (1984) Toxic effects of aflatoxin B1 in chickens given feed contaminated with Aspergillus flavus and reduction of toxicity by activated charcoal and some chemical agents. Avian Dis 28(1):61–69

Dari B, Nair VD, Harris WG, Nair PKR, Sollenberger L, Mylavarapu R (2016) Relative influence of soil-vs. biochar properties on soil phosphorus retention. Geoderma 280:82–87. https://doi.org/10.1016/j.geoderma.2016.06.018

Day D, Evans RJ, Lee JW, Reicosky D (2004) Valuable and stable carbon co-product from fossil fuel exhaust scrubbing. American Chem Soc 49(1):352

Debela F, Thring RW, Arocena JM (2012) Immobilization of heavy metals by co-pyrolysis of contaminated soil with woody biomass. Water Air Soil Poll 223:1161–1170
Decker WJ, Corby DG (1971) Activated charcoal as a gastrointestinal decontaminant. Experiences with experimental animals and human subjects. Bull Environ Contamination Toxicol 6 (2):189–192
DeLuca TH, MacKenzie MD, Gundale MJ, Holben WE (2006) Wildfire-produced charcoal directly influences nitrogen cycling in ponderosa pine forests. Soil Sci Soc Am J 70:448–453
Demirbas A (2003) Relationships between lignin contents and fixed carbon contents of biomass samples. Energy Convers Mang 44:1481–1486
DeSisto WJ, Hill N, Beis SH, Mukkamala S, Joseph J, Baker C, Ong TH, Stemmler EA, Wheeler MC, Frederick BG, Heiningen AV (2010) Fast pyrolysis of pine sawdust in a fluidized-bed reactor. Energy Fuel 24:2642–2651
Devenas K, Neto S, Oliveira D, Gonçalves M (2018) Characterization of biochar from green coconut Shell and Orange Peel wastes. Revista Virtual de Química 10:288–294. https://doi.org/10.21577/1984-6835.20180022
Dick WA (2008) Organic-carbon, nitrogen, and phosphorus concentrations and pH in soil profiles as affected by tillage intensity. Soil Sci Soc Am J 47:102–107
Ding Y, Liu Y, Liu S, Li Z, Tan X, Huang X, Zeng G, Zhou L, Zheng B (2016) Biochar to improve soil fertility. A review. Agron Sustain Dev 36:36. https://doi.org/10.1007/s13593-016-0372-z
Downie A, Crosky A, Munroe P (2009) Physical properties of biochar. In: Lehmann J, Joseph S (eds) Biochar for environmental management: science and technology. Earthscan, London, pp 13–32
Dume B, Mosissa T, Nebiyu A (2016) Effect of biochar on soil properties and lead (Pb) availability in a military camp in south West Ethiopia. Afr J Environ Sci Tech 10(3):77–85
Dumroese RK, Heiskanen J, Englund K, Tervahauta A (2011) Pelleted biochar: chemical and physical properties show potential use as a substrate in container nurseries. Biomass Bioenergy 35:2018–2027
Dunnigan L, Ashmana J, Zhang X, Wai Kwong C (2018) Production of biochar from rice husk: particulate emissions from the combustion of raw pyrolysis volatiles. J Clean Prod 172 (20):1639–1645
El Hanandeh A (2013) Carbon abatement via treating the solid waste from the Australian olive industry in mobile pyrolysis units: LCA with uncertainty analysis. Waste Manage Res 31 (4):341–352
Elad Y, Cytryn E, Harel YM, Lew B, Graber ER (2011) The biochar effect: plant resistance to biotic stresses. Phytopathol Mediterr 50:335–349
Eswaran H, Reich P, Beinroth F (1999) Global desertification tension zones. In: Selected paper from the 10th international soil conservation organization meeting held in may 24–29, 1999, at university and the USDA-ARS-National Soil Erosion Research Laboratory, p 3
European Biochar Foundation (EBC) (2012) European biochar certificate—guidelines for a sustainable production of biochar. European Biochar Foundation
Ezawa T, Yamamoto K, Yoshida S (2002) Enhancement of the effectiveness of indigenous arbuscular mycorrhizal fungi by inorganic soil amendments. Soil Sci Plant Nutr 48:897–900. https://doi.org/10.1080/00380768.2002.10408718
Fang GD, Gao J, Liu C, Dionysiou DD, Wang Y, Zhou DM (2014) Key role of persistent free radicals in hydrogen peroxide activation by biochar: implications to organic contaminant degradation. Environ Sci Technol 48:1902–1910
Farrar MB, Wallace HM, Xu C, Nguyen TTN, Tavakkoli E, Joseph S, Bai SH (2017) Biochar based fertilizer: effects on yield, plant growth, nutrient uptake and soil fertility. Pedosphere
Fellet G, Marchiol L, Delle Vedove G, Peressotti A (2011) Application of biochar on mine tailings: effects and perspectives for land reclamation. Chemosphere 83:1262–1297
Freitas A, Nair V, Harris W, Mackowiak C (2016a) Phosphorus Release from Biochar of Various Feedstocks: Implications for Land Management

Freitas AM, Nair VD, Harris WG, Mackowiak C (2016b) "Phosphorus release from various biochar feedstocks: implications for land management," Abstract retrieved from ASA, CSSA and SSSA international annual meetings. Available online at: https://scisoc.confex.com/crops/2016am/webprogram/Paper101356.html

Gaba S, Bretagnolle F, Rigaud T, Philippot L (2014) Managing biotic interactions for ecological intensification of agroecosystems. Front Ecol Evol 2. (June 30, 2014)

Gao X, Cheng HY, Del Valle I, Liu S, Masiello CA, Silberg JJ (2016) Charcoal disrupts soil microbial communication through a combination of signal sorption and hydrolysis. Acs Omega 1:226–233

García AC, de Souza LGA, Pereira MG, Castro RN, García-Mina JM, Zonta E (2016) Structure-property-function relationship in humic substances to explain the biological activity in plants. Sci Rep 6:20798. https://doi.org/10.1038/srep20798

Gaunt JL, Lehmann J (2008) Energy balance and emissions associated with biochar sequestration and pyrolysis bioenergy production. Environ Sci Technol 42:4152–4158

Ghouma I, Jeguirim M, Guizani C, Ouederni A, Limousy L (2017) Pyrolysis of olive pomace: degradation kinetics, gaseous analysis and char characterization. Waste and Biomass Valorization. https://doi.org/10.1007/s12649-017-9919-8

Glaser B (2007) Prehistorically modified soils of Central Amazonia: a model for sustainable agriculture in the twenty-first century. Philos Trans Royal Soc B: Biol Sci 362(1478):187–196

Godlewska P, Schmidt HP, Ok YS, Oleszczuk P (2017) Biochar for composting improvement and contaminants reduction. A review. Bioresour Technol 246:193–202

Gomez JD, Denef K, Stewart CE, Zheng J, Cotrufo MF (2014) Biochar addition rate influences soil microbial abundance and activity in temperate soils. Eur J Soil Sci 65:28–39. https://doi.org/10.1111/ejss.12097

Graber ER, Harel YM, Kolton M, Cytryn E, Silber A, David DR, Tsechansky L, Borenshtein M, Elad Y (2010) Biochar impact on development and productivity of pepper and tomato grown in fertigated soilless media. Plant Soil 337:481–496

Graves D (2017) Recommended doses and location of application into soils. In The Australia New Zealand biochar conference proceedings 10–12 august 2017, Agri futures Australia publication no 17/057, pp. 29–30

Graves AR, Morris J, Deeks LK, Rickson RJ, Kibblewhite MG, Harris JA, Farewell TS, Truckle I (2015) Thetotalcostsofsoildegradationin Englandand Wales. Ecol Econ 119:399–413

Grossman JM, O'Neill BE, Tsai SM, Liang B, Neves E, Lehmann J, Thies JE (2010) Amazonian anthrosols support similar microbial communities that differ distinctly from those extant in adjacent, unmodified soils of the same mineralogy. Microb Ecol 60:192–205. https://doi.org/10.1007/s00248-010-9689-3

Gurwick NP, Moore LA, Kelly C, Elias P (2013) A systematic review of biochar research, with a focus on its stability in situ and its promise as a climate mitigation strategy. PLoS One 8:e75932

Hammes K, Schmidt MWI (2009a) Changes in biochar in soils. In: Lehmann M, Joseph S (eds) Biochar for environmental management science and technology. Earthscan, London, pp 169–182

Hammes K, Schmidt MWI (2009b) Changes in biochar in soils. In: Lehmann M, Joseph S (eds) Biochar for environmental management science and technology. Earthscan, London, pp 169–182

Hanz S (2011) Initial customer interview with Hanz Sholz. Interviewed by Mike Baranick, David McElwee and Maria Zazycki. Telephone. 21 April 2011. In: Biochar Feasibility Study presented to Dr. April Atwood, Seattle University

Hartemink AE (2003a) Land fertility decline under sisal cultivation in Tanzania. In: ISRIC technical paper 28. ISRIC, Wageningen

Hartemink AE (2003b) Soil fertility decline in the tropics with case studies on plantations/Alf. Wallingford: CABI 2003:43–48

He BJ, Zhang Y, Funk TL, Riskowski GL, Yin Y (2000) Thermochemical conversion of swine manure: an alternative process for waste treatment and renewable energy production. Trans ASAE 43:1827–1833

Hedley CB, Yule IJ, Eastwoof CR, Shephard TG, Arnold G (2004) Rapid identification of soil textural and management zones using electromagnetic induction sensing in soils. Aust J Soil Res 42:389–400

Hiller E, Fargasova A, Zemanova L, Bartal M (2007) Influence of wheat ash on the MCPA immobilization in agricultural soils. Bull Environ Contam Toxicol 79(4):478–481, ISSN 0007-4861

Hobbs PR, Sayre KD, Gupta R (2008) The role of conservation agriculture in sustainable agriculture. Philos Trans R Soc B 363:543–555

Hottle RD (2013) Quantifying the impact of biochar on plant productivity and changes to soil physical and chemical properties on a maize soybean rotation in the US. The Ohio State University, Ohio

IEA (2007) IEA bioenergy annual report 2006. International Energy Agency, Paris

Ippolito JA, Laird DA, Busscher WJ (2012) Environmental benefits of biochar. J Environ Qual 41:967–972

Ippolito JA, Spokas KA, Novak JM, Lentz RD, Cantrell KB (2015) Biochar elemental composition and factors influencing nutrient retention. In: Lehmann J, Stephen J (eds) Biochar for environmental management: science, technology and implementation, 2nd edn. Routledge; Taylor and Francis Group, London; New York, NY, pp 137–161

Jaafar NM, Clode PL, Abbott LK (2014) Microscopy observations of habitable space in biochar for colonization by fungal hyphae from soil. J Integr Agr 13:483–490

Jafri N, Wong WY, Doshi Y, Yoon LW, Cheah KH (2018) A review on production and characterization of biochar for application in direct carbon fuel cells. Process Saf Environ 118:152–166

Jeffery S, Bezemer TM, Cornelissen G, Kuyper TW, Lehmann J, Mommer L (2015) The way forward in biochar research: targeting trade-offs between the potential wins. Global Change Biol Bioenergy 7:1–13. https://doi.org/10.1111/gcbb.12132

Jhariya MK, Meena RS, Banerjee A (2021a) Ecological intensification of natural resources for sustainable agriculture. Springer Nature, Singapore. eISBN: 978-981-334-203-3, Hardcover ISBN: 978-981-334-206-6. https://doi.org/10.1007/978-981-33-4203-3

Jhariya MK, Meena RS, Banerjee A (2021b) Ecological intensification of natural resources towards sustainable productive system. In: Ecological intensification of natural resources for sustainable agriculture. Springer, Singapore. eBook ISBN 978-981-334-203-3. Hardcover ISBN: 978-981-334-202-6. https://doi.org/10.1007/978-981-33-4203-3_1

Jhariya MK, Singh L (2021) Effect of fire severity on soil properties in a seasonally dry forest ecosystem of Central India. Int J Environ Sci Technol:1–12. https://doi.org/10.1007/s13762-020-03062-8

Jhariya MK, Singh L (2020) Herbaceous diversity and biomass under different fire regimes in a seasonally dry forest ecosystem. Environ Dev Sustain 22:1–19. https://doi.org/10.1007/s10668-020-00892-x

Jhariya MK, Banerjee A, Meena RS, Yadav DK (2019a) Sustainable Agriculture, Forest and Environmental Management. Springer Nature, Singapore, p 606eISBN: 978-981-13-6830-1, Hardcover ISBN: 978-981-13-6829-5. https://doi.org/10.1007/978-981-13-6830-1

Jhariya MK, Yadav DK, Banerjee A (2019b) Agroforestry and climate change: issues and challenges. Apple Academic Press Inc., CRC Press- a Tayler and Francis Group, US & Canada. ISBN: 978-1-77188-790-8 (Hardcover), 978-0-42957-274-8 (E-book), p 335. https://doi.org/10.1201/9780429057274

Jhariya MK, Yadav DK, Banerjee A (2018a) Plant mediated transformation and habitat restoration: phytoremediation an eco-friendly approach. In: Gautam A, Pathak C (eds) Metallic contamination and its Toxicity. Daya Publishing House, A Division of Astral International Pvt. Ltd, New Delhi. ISBN: 9789351248880, pp 231–247

Jhariya MK, Banerjee A, Yadav DK, Raj A (2018b) Leguminous trees an innovative tool for soil sustainability. In: Meena RS, Das A, Yadav GS, Lal R (eds) Legumes for soil health and sustainable management. Springer, Singapore, pp 315–345. ISBN 978-981-13-0253-4 (eBook), ISBN: 978-981-13-0252-7 (Hardcover). https://doi.org/10.1007/978-981-13-0253-4_10

Jien SH, Wang CS (2013) Effects of biochar on soil properties and erosion potential in a highly weathered soil. Catena 110:225–233. https://doi.org/10.1016/j.catena.2013.06.021

Jindo K, Mizumoto H, Sawada Y, Sanchez-Monedero MA, Sonoki T (2014) Physical and chemical characterization of biochar derived from different agricultural residues. Biogeo Sciences 11:6613–6621

Joseph S, Graber ER, Chia C, Munroe P, Donne S, Thomas T, Nielsen S, Marjo C, Rutledge H, Pan GX, Li L, Taylor P, Rawal A, Hook J (2013) Shifting paradigms: development of high-efficiency biochar fertilizers based on nanostructures and soluble components. Carbon Manag 4:323–343

Joseph SD, Camps-Arbestain M, Lin Y, Munroe P, Chia CH, Hook J (2010) An investigation into the reactions of biochar in soil. Aust J Soil Res 48:501–515. https://doi.org/10.1071/SR10009

Jyothi R, Jyoti S, Pankaj S (2019) Biochar: a sustainable approach for improving plant growth and soil properties. In: Abrol V, Sharma P (eds) Biochar-an imperative amendment for soil and the environment. Intech Open, London. https://doi.org/10.5772/intechopen.82151

Kammann C, Ippolito J, Hagemann N, Borchard N, Cayuela ML, Estavillo JM, Fuertes-Mendizabal T, Jeffery S, Kern J, Novak J, Rasse D, Saarnio S, Schmidt HP, Spokas K, Wrage-Mönnig (2017) Biochar as a tool to reduce the agricultural greenhouse-gas burden-knowns, unknowns and future research needs. J Environ Engineering Landscape Manage 25 (2):114–139

Kanayo SC, Akinlolu OA, Barachel OU, Micheal DA, Adegboyega CO (2020) Biochar: a vital source for sustainable agriculture. In: Mirmajlessi SM, Radhakrishnan R (eds) Biostimulants in plant science. IntechOpen, London. https://doi.org/10.5772/intechopen.86568

Kane D (2015) Carbon sequestration potential on agricultural lands: a review of current science and available practices. In: Breakthrough strategies and solutions and National Sustainable Agriculture Coalition, p 35. http://sustainableagriculture.net.publications

Karami N, Clemente R, Moreno-Jiménez E, Lepp NW, Beesley L (2011) Efficiency of green waste compost and biochar soil amendments for reducing lead and copper mobility and uptake to ryegrass. J Hazard Mater 191:41–48

Kataki R, Das M, Chutia RS, Borah M (2012) Biochar for C sequestration and soil amelioration. In: renewable energy technology: issues and prospects, (eds G Shankar, B Das, R Blange). Excell India Publishers, New Delhi, pp 131–137

Keith G (2016) Soil acidification and the importance of liming agricultural soils with particular reference to the United Kingdom. Soil use and management. Soil Use Manag 32:390–399. https://doi.org/10.1111/sum.12270

Khan N, Jhariya MK, Raj A, Banerjee A, Meena RS (2021a) Soil carbon stock and sequestration: implications for climate change adaptation and mitigation. In: Jhariya MK, Meena RS, Banerjee A (eds) Ecological intensification of natural resources for sustainable agriculture. Springer, Singapore. eBook ISBN 978-981-334-203-3. Hardcover ISBN: 978-981-334-202-6. https://doi.org/10.1007/978-981-33-4203-3_13

Khan N, Jhariya MK, Raj A, Banerjee A, Meena RS (2021b) Eco-designing for sustainability. In: Jhariya MK, Meena RS, Banerjee A (eds) Ecological intensification of natural resources for sustainable agriculture. Springer, Singapore. eBook ISBN 978-981-334-203-3. Hardcover ISBN: 978-981-334-202-6. https://doi.org/10.1007/978-981-33-4203-3_16

Khan N, Jhariya MK, Yadav DK, Banerjee A (2020a) Herbaceous dynamics and CO_2 mitigation in an urban setup- a case study from Chhattisgarh, India. Environ Sci Pollut Res 27(3):2881–2897. https://doi.org/10.1007/s11356-019-07182-8

Khan N, Jhariya MK, Yadav DK, Banerjee A (2020b) Structure, diversity and ecological function of shrub species in an urban setup of Sarguja, Chhattisgarh, India. Environ Sci Pollut Res 27 (5):5418–5432. https://doi.org/10.1007/s11356-019-07172-w

Kim HJ, Kang BS, Kim MJ, Park YM, Kim DK, Lee JS, Lee KY (2004) Transesterification of vegetable oil to biodiesel using heterogeneous base catalyst. Catal Today 93-95:315–320

Kinney TJ, Masiello CA, Dugan B, Hockaday WC, Dean MR, Zygourakis K (2012) Hydrologic properties of biochar produced at different temperatures. Biomass Bioenergy 41:34–43. https://doi.org/10.1016/j.biombioe.2012.01.033

Kitzler H, Pfeifer C, Hofbauer H (2012) Combustion of reeds in a 3 MW district heating plant. Int J Environ Sci Dev 3:407–411

Kleber M, Sollins P, Sutton R (2007) A conceptual model of organo-mineral interactions in soils: self-assembly of organic molecular fragments into zonal structures on mineral surfaces. Biogeochemistry 85:9–24

Kloot R (2018) Using adaptive nutrient management to answer "how much fertilizer do you actually need?" NRCS webinar May 8, 2018. Sci Technol Training Library. http://www.conservationwebinars.net/listArchivedWebinars

Komulainen MP, Simi E, Hagelberg I, Ikonen S, Lyytinen S (2008) Possibilities of using the common Reed for energy generation in southern Finland. Turku University of Applied Sciences, Turku, p 21

Kookana RS, Sarmah AK, Zwieten L, van Krull E, Singh B (2011) Biochar application to soil: agronomic and environmental benefits and unintended consequences p. In: Sparks DL (ed) Advances in agronomy. Academic Press, San Diego, pp 103–143

Koutcheiko S, Monreal CM, Kodama H, McCracken T, Kotlyar L (2007) Preparation and characterization of activated carbon derived from the thermo-chemical conversion of chicken manure. Bioresour Technol 98:2459–2464

Kumar S, Meena RS, Jhariya MK (2020) Resources use efficiency in agriculture. Springer Nature, Singapore, p 760, eBook ISBN: 978-981-15-6953-1, Hardcover: 978-981-15-6952-4. https://doi.org/10.1007/978-981-15-6953-1

Kumar S, Meena RS, Datta R, Verma SK, Yadav GS, Pradhan GS, Molaei A, Rahman GKMM, Mashuk HA (2020a) Legumes for carbon and nitrogen cycling: an organic approach. In: Carbon and nitrogen cycling in soil. Springer, Singapore. https://doi.org/10.1007/978-981-13-7264-3_10.337-375

Kumar S, Nakajima T, Mbonimpa EG, Gautam S, Somireddy UR, Kadono A (2014) Long-term tillage and drainage influences on soil organic carbon dynamics, aggregate stability, and carbon yield. Soil Sci Plant Nutr 60:108–118

Laghari (2016) Recent developments in biochar as an effective tool for agricultural soil management: a review. J Sci Food Agric Hamilton 96:4840–4849

Laird DA, Fleming P, Davis DD, Horton R, Wang B, Karlen DL (2010) Impact of biochar amendments on the quality of a typical Midwestern agricultural soil. Geoderma 158:443–449. https://doi.org/10.1016/j.geoderma.2010.05.013

Lal R (1995) Erosion–crop productivity relationships for soils of Africa. Soil Sci Soc America J 59:661–667

Lal R (1998) Soil erosion impact on agronomic productivity and environment quality. Critical Rev Plant Sci 17:319–464

Lal R (2003) Axle load and tillage effects on soil degradation and rehabilitation in Western Nilgeria. Soil physical and hydrological properties. Land Degradation Rev 7:19–45

Lan T, Preston T, Leng R (2018) Feeding biochar or charcoal increased the growth rate of striped catfish (*Pangasius hypophthalmus*) and improved water quality. Livestock Res Rural Dev 28:84

Lee JY, Lee KK (2004) A short note on investigation and remediation of contaminated groundwater and soil in Korea. Eng Geol 14:123–130

Lee Y, Jung J, Park J, Hyun S, Ryu C, Gang KS (2013) Comparison of biochar properties from biomass residues produced by slow pyrolysis at 500 °C. Bioresour Technol 148:196–201

Lehmann J, Gaunt J, Rondon M (2006) Bio-char sequestration in terrestrial ecosystems–a review. Mitigation Adaptation Strategies Global Change 11(2):395419

Lehmann J, Joseph S (2009) Biochar for environmental management: An introduction. In: Lehmann J, Joseph S (eds) Biochar for environmental management: science and technology. Earthscan, London, pp 1–12

Lehmann J, Rillig MC, Thies J, Masiello CA, Hockaday WC, Crowley D (2011) Biochar effects on soil biota-a review. Soil Biol Biochem 43:1812–1836. https://doi.org/10.1016/j.soilbio.2011.04.022

Lehmann J, Rondon M (2006) Biochar soil management on highly weathered soils in the humid tropics. In: Uphoff N (ed) Biological approaches to sustainable soil systems. CRC Press, Boca Raton, pp 517–530

Lehmann J, Silva JP, Steiner C, Nehls T, Zech W (2003) Nutrient availability and leaching in an archaeological anthrosol and a ferralsol of the Central Amazon basin: fertilizer, manure and charcoal amendments. Plant Soil 249:343–357

Leng RA, Rur D (2017) Biochar in ruminant nutrition and health. In: The Australia New Zealand biochar conference proceedings 10–12 august 2017, Agri futures Australia publication no 17/057, pp 12–13

Lian F, Huang F, Chen W, Xing B, Zhu L (2011) Sorption of a polar and polar organic contaminants by waste tire rubber and its chars in single- and bi-solute systems. Environ Pollut 159:850–857

Liang B, Lehmann J, Sohi SP, Thies JE, O'Neill B, Trujillo L, Gaunt J, Solomon D, Grossman J, Neves EG, Luizão FJ (2010) Black carbon affects the cycling of non-black carbon in soil. Org Geochem 41:206–213. https://doi.org/10.1016/j.orggeochem.2009.09.007

Liang B, Lehmann J, Solomon D, Kinyangi J, Grossman J, O'Neill B, Skjemstad JO, Thies J, Luizao FJ, Petersen J, Neves EG (2006) Black carbon increases cation exchange capacity in soil. Soil Sci Soc Am J 70:1719–1730. https://doi.org/10.2136/sssaj2005.0383

Liang F, Li G, Lin Q, Zhao X (2014) Crop yield and soil properties in the first 3 years after biochar application to a calcareous soil. J Integrative Agric 13(3):525–532

Liu J, Schulz H, Brandl S, Miehtke H, Huwe B, Glaser B (2012a) Short-term effect of biochar and compost on soil fertility and water status of a dystric Cambisol in NE Germany under field conditions. J Plant Nutrition Soil Sci 175(5):1–10

Liu L, Gundersen P, Zhang T, Moa J (2012b) Effects of phosphorus addition on soil microbial biomass and community composition in three forest types in tropical China. Soil Biol Biochem 44:31

Liu T, Liu B, Zheng W (2014) Nutrients and heavy metals in biochar produced by sewage sludge pyrolysis: its application in soil amendment. Polish J Environ Stud 23(1):271–275

Lorenz K, Lal R (2016) Environmental Impact of Organic Agriculture. Adv Agronomy 139:99–152

Lynch J, Joseph S (2010) Guidelines for the development and testing of pyrolysis plants to produce biochar

Major J, Rondon M, Molina D, Riha SJ, Lehmann J (2010) Maize yield and nutrition during 4 years after biochar application to a Colombian savanna oxisol. Plant Soil 333:117–128

Mandal KG, Misra AK, Hati KM, Bandyopadhyay KK, Ghosh PK (2004) Rice residue management options and effects on soil properties and crop productivity. J Food Agric Environ 2:224–231

Mankasingh U, Choi PC, Ragnarsdottir V (2011) Biochar application in a tropical, agricultural region: a plot scale study in Tamil Nadu, India. Appl Geochem 26:218–221

Meena RS, Lal R, Yadav GS (2020) Long term impacts of topsoil depthand amendments on soil physical and hydrological properties of an Alfisol in Central Ohio, USA. Geoderma 363:1141164. https://doi.org/10.1016/j.geoderma.2019.114164

Meena RS, Lal R, Yadav GS (2020a) Long-term impact of topsoil depth and amendments on carbon and nitrogen budgets in the surface layer of an Alfisol in Central Ohio. Catena 2020194:104752. https://doi.org/10.1016/j.catena.2020.104752

Meena RS, Kumar S, Datta R, Lal R, Vijaykumar V, Brtnicky M, Sharma MP, Yadav GS, Jhariya MK, Jangir CK, Pathan SI, Dokulilova T, Pecina V, Marfo TD (2020b) Impact of agrochemicals on soil microbiota and management: a review. Land (MDPI) 9(2):34. https://doi.org/10.3390/land9020034

Mao JD, Johnson RL, Lehmann J, Olk DC, Neves EG, Thompson ML, Schmidt-Rohr K (2012) Abundant and stable char residues in soils: implications for soil fertility and carbon sequestration. Environ Sci Technol 46:9571–9576

Masiello CA, Chen Y, Gao X, Liu S, Cheng HY, Bennett MR, Rudgers JA, Wagner DS, Zygourakis K, Silberg JJ (2013) Biochar and microbial signaling: production conditions determine effects on microbial communication. Environ Sci Technol 47:11496–11503

Masto RE, Ansari MA, George J, Selvi V, Ram L (2013) Co-application of biochar and lignite fly ash on soil nutrients and biological parameters at different crop growth stages of *Zea mays*. Ecol Eng 58:314–322. https://doi.org/10.1016/j.ecoleng.2013.07.011

Mathew ED, von Haden AC (2017) Intensified agroecosystems and their effects on soil biodiversity and soil functions. In: Al-Kaisi MM, Lowery B (eds) Soil health and intensification of agroecosystems. Academic Press, London, pp 173–193., ISBN 9780128053171. https://doi.org/10.1016/B978-0-12-805317-1.00008-7

McCormack S, Ostle N, Bardgett R, Hopkins D, Vanbergen A (2013) Biochar in bioenergy cropping systems: impacts on soil faunal communities and linked ecosystem processes. Global Change Biol Bioenergy 5:81–95

Metzger MJ, Bunce RGH, Jongman RH, Mücher CA, Watkins JW (2005) A climatic stratification of the environment of Europe. Global Ecol Biogeography 14:549–563. https://doi.org/10.1111/j.1466-822X.2005.00190.x

Milla OV, Rivera EB, Huang WJ, Chien CC, Wang YM (2013) Agronomic properties and characterization of rice husk and wood biochar and their effect on the growth of water spinach in a field test. J Soil Sci Plant Nutrition 2:251–266

Mirsky SB (2017) Creating the cover crops that organic farmers need: delivering regionally adapted varieties across America. Proposal and annual report for OREI project 2015-07406. CRIS abstracts

Mizuta K, Matsumoto T, Hatate Y, Nishihara K, Nakanishi T (2004) Removal of nitrate-nitrogen from drinking water using bamboo powder charcoal. Bioresour Technol 95:255–257

Mohan D, Rajput S, Singh VK, Steele PH, Pittman CU Jr (2011) Modeling and evaluation of chromium remediation from water using low cost bio-char, a green adsorbent. J Hazard Mater 188(1–3):319–333

Montanarella L (2015) The Status of the World's Soil Resources

Mukherjee A, Lal R, Zimmerman AR (2014) Effects of biochar and other amendments on the physical properties and greenhouse gas emissions of an artificially degraded soil. Sci The Total Environ 487:26–36

Mukherjee A, Zimmerman AR (2013) Organic carbon and nutrient release from a range of laboratory-produced biochar and biochar–soil mixtures. Geoderma 193:122–130

Naeem MA, Khalid M, Aon M, Tahir GAM, Amjad M, Yand A, Akthar SS (2017) Effect of wheat and rice straw biochar produced at different temperatures on maize growth and nutrient dynamics of a calcareous soil. J Archives Agronomy Soil Sci 63:2048–2061

Nair VD, Ramachandran Nair PK, Dari B, Freitas AM, Chatterjee N, Pinheiro FM (2017) Biochar in the agroecosystem-climate-change-sustainability nexus. Front Plant Sci 8:2051

Namgay T, Singh B, Singh BP (2010) Influence of biochar application to soil on the availability of as, cd, cu, Pb, and Zn to maize (Zea mays L.). Aust J Soil Res 48:638–647

Nasser RA (2014) An evaluation of the use of midribs from common date palm cultivars grown in Saudi Arabia for energy production. BioResource 9:4343–4357

Nasser RA, Salem MZM, Al-Mefarrej HA, Abdel-Aal MA, Soliman SS (2014) Fuel characteristics of vine prunings (*Vitis vinifera* L.) as a potential source for energy production. BioResource 9:482–496

Neal JC, Wagner DF (1983) Physical and chemical properties of coal cinders as a container media component. Hort Sci 18:693–695

Nigussie A, Kissi E, Misganaw M, Ambaw G (2012) Effect of biochar application on soil properties and nutrient uptake of lettuces (*Lactuca sativa*) grown in chromium polluted soils. American-Eurasian J Agric Environ Sci 12(3):369–376

Novak JM, Lima IM, Xing B, Gaskin JW, Steiner C, Das KC (2009) Characterization of designer biochar produced at different temperatures and their effects on a loamy sand. Ann Environ Sci 3:195–206

Nsamba HK, Hale SE, Cornelissen G, Bachmann RT (2015) Sustainable Technologies for Small-Scale Biochar Production - a review. J Sustain Bioenergy Syst 5:10–31

O'Toole A, Andersson D, Gerlach A, Glaser B, Kammann CI, Kern J, Kuoppamäki K, Mumme J, Schmidt Hans-Peter Schulze M, Srocke Franziska Stenrød M, Stenström J (2016) Current and future applications for biochar. In: Shackley S, Ruysschaert G, Zwart K, Glaser B (eds) Biochar in European soils and agriculture: science and practice. Taylor & Francis Group, Abington, pp 253–280

Obia A, Mulder J, Hale SE, Nurida NL, Cornelissen G (2018) The potential of biochar in improving drainage, aeration and maize yields in heavy clay soils. PLoS One 13(5):e0196794

Oguntunde P, Fosu M, Ajayi A, Giesen N (2004) Effects of charcoal production on maize yield, chemical properties and texture of soil. Biol Fertil Soils 39:295–299. https://doi.org/10.1007/s00374-003-0707-1

Okigbo BN (1991) Development of sustainable agricultural production systems in Africa: roles. - Ibadan. International Institute of Tropical Agriculture, Ibadan, pp 26–27

Pan WP, Richards GN (1989) Influence of metal ions on volatile of pyrolysis of wood products. J Anal Appl Pyrolysis 16:117–126

Pandian K, Subramaniayan P, Gnasekaran P, Chitraputhirapillai S (2016) Effect of biochar amendment on soil physical, chemical and biological properties and groundnut yield in rainfed Alfisol of semi-arid tropics. Archives Agronomy Soil Sci 9:1293–1310

Pesant AR, Dionne JL, Genest J (1987) Soil and nutrient losses in surface runoff from conventional and no-till corn systems. Can J Soil Sci 67:835–843

Pesevski MD, Lliev BM, Zivkovic DL, Popovska VT, Srbinoska MA, Filiposki BK (2010) Possibilities for utilization of tobacco stems for production of energetic briquettes. J Agric Sci 55:45–54

Pessenda LCR, Gouveia SEM, Aravena R (2001) Radiocarbon dating of total soil organic matter and human fraction and its comparison with 14C ages of fossil charcoal. Radiocarbon 43:595–601

Piash MI, Hosaain F, Parveen Z (2016) Physico-chemical properties and nutrient content of some slow pyrolysis biochar produced from different feedstocks. Bangladesh J Sci Res 29 (2):111–122

Pietikainen J, Kiikkila O, Fritze H (2000) Charcoal as a habitat for microbes and its effect on the microbial community of the underlying humus. Oikos 89:231–242

Pietry JC, Brookes PC (2008) Relationships between soil pH and microbial properties in a UK arable soil. Soil Biol Biochem 40:1856–1861. https://doi.org/10.1016/j.soilbio.2008.03.02

Pignatello JJ, Kwon S, Lu Y (2006) Effect of natural organic substances on the surface and adsorptive properties of environmental black carbon (char): attenuation of surface activity by humic and fulvic acids. Environ Sci Technol 40:7757–7763

Pimentel D, Harvey C, Resudarmo P (1995) Environmental and economic costs of soil erosion and conservation benefits. Science 267:1117–1123

Pituello C, Francioso O, Simonetti G, Pisi A, Torreggiani A, Berti A, Morari F (2014) Characterization of chemical–physical, structural and morphological properties of biochar from biowastes produced at different temperatures. J Soils Sediments 15. https://doi.org/10.1007/s11368-014-0964-7

Pituya P, Sriburi T, Wijitkosum S (2016) Properties of biochar prepared from Acacia wood and coconut Shell for soil amendment. Environ Sci Pollut Res Int 25(2):1147–1156. https://doi.org/10.1007/s11356-017-0495-z

Placek A, Grobelak A, Kacprzak M (2016) Improving the phytoremediation of heavy metals contaminated soil by use of sewage sludge. Int J Phytoremediation 18(6):605–618

Prasad RN, Biswas PP (2000) Soil resources of India. In Singh, G. B & Sharma, B.R., eds. Fifty years of natural resource management research, pp. 13–30. New Delhi, Indian Council of Agricultural Research

Preston CM, Schmidt MWI (2006) Black (pyrogenic) carbon: a synthesis of current knowledge and uncertainties with special consideration of boreal regions. Biogeosciences 3:397–420

Prost K, Borchard N, Siemens J, Kautz T, Séquaris JM, Möller A, Amelung W (2013) Biochar affected by composting with farmyard manure. J Environ Qual 4:164–172. https://doi.org/10.2134/jeq2012.0064

Pütün AE, Apaydın E, Pütün E (2004) Rice straw as a bio-oil source via pyrolysis and steam pyrolysis. Energy 29:2171–2180

Qayyum MF (2012) Possibilities to Stabilize Organic Matter in Soil Using Various Biochar. Unpublished PhD Doctoral dissertation, Justus Liebig University Giessen, 100pp

Qian L, Chen L, Joseph S, Pan GX, Li L, Zheng J, Zhang X, Zheng J, Yu X, Wang J (2014) Biochar compound fertilizer as an option to reach high productivity but low carbon intensity in rice agriculture of China. Carbon Manage 5(2):145–158

Quilliam RS, Glanville HC, Wade SC, Jones DL (2013) Life in the 'charosphere' does biochar in agricultural soil provide a significant habitat for microorganisms? Soil Biol Biochem 65:287e293

Quin PR, Cowie AL, Flavel RJ, Mac Donald LM, Morris SG, Singh B-P, Young IM, Van Zwieten L (2014) Biochar changes soil structure and water-holding capacity - a study with x-ray μCT. Agric Ecosyst Environ 191:142–149

Raj A, Jhariya MK, Khan N, Banerjee A, Meena RS (2021) Ecological intensification for sustainable development. In: Ecological intensification of natural resources for sustainable agriculture. Springer, Singapore, eBook ISBN 978-981-334-203-3. Hardcover ISBN: 978-981-334-202-6. https://doi.org/10.1007/978-981-33-4203-3_5

Raj A, Jhariya MK, Yadav DK, Banerjee A (2020) Climate change and agroforestry systems: adaptation and mitigation strategies. Apple Academic Press Inc., CRC Press- a Tayler and Francis Group, US & Canada, Burlington. ISBN: 9781771888226, p 383. https://doi.org/10.1201/9780429286759

Raj A, Jhariya MK, Yadav DK, Banerjee A, Meena RS (2019a) Agroforestry: a holistic approach for agricultural sustainability. In: Jhariya MK, Banerjee A, Meena RS, Yadav DK (eds) Sustainable Agriculture, Forest and Environmental Management. Springer Nature, Singapore, eISBN: 978-981-13-6830-1, Hardcover ISBN: 978-981-13-6829-5. Pp. 606, pp 101–131. https://doi.org/10.1007/978-981-13-6830-1

Raj A, Jhariya MK, Banerjee A, Yadav DK, Meena RS (2019b) Soil for sustainable environment and ecosystems management. In: Jhariya MK, Banerjee A, Meena RS, Yadav DK (eds) Sustainable agriculture, Forest and environmental management. Springer Nature, Singapore, pp 189–221eISBN: 978-981-13-6830-1, Hardcover ISBN: 978-981-13-6829-5. Pp. 606. https://doi.org/10.1007/978-981-13-6830-1

Raveendran K, Ganesh A, Khilar KC (1995) Influence of mineral matter on biomass pyrolysis characteristics. Fuel 74:1812–1822

Rillig MC, Wagner M, Salem M, Antunes PM, George C, Ramke HG, Titirici MM, Antonietti M (2010) Material derived from hydro thermal carbonization: effects on plant growth and arbuscular mycorrhiza. Appl Soil Ecol 45:238–242. https://doi.org/10.1016/j.apsoil.2010.04.011

Romaniuk R, Giuffré L, Costantini A, Nannipieri P (2011) Assessment of soil microbial diversity measurements as indicators of soil functioning in organic and conventional horticulture systems. Ecol Indic 11:1345–1353. https://doi.org/10.1016/j.ecolind.2011.02.008

Rondon MA, Molina D, Hurtado M, Ramirez J, Lehmann J (2006) Enhancing the productivity of crops and grasses while reducing greenhouse gas emissions through biochar, pp 138–168

Rousk J, Baath E, Brookes PC, Lauber CL, Lozupone C, Caporaso JG, Knight R, Fierer N (2010) Soil bacterial and fungal communities across a pH gradient in an arable soil. ISME J 4:134–151. https://doi.org/10.1038/ismej.2010.58

Rumi N, Neonjyoti B, Rahul SC, Bikram B, Nirmali G, Ajitabh B, Rupam K (2015) Biochar: an overview on its production, properties and potential benefits. In: Biology, biotechnology and sustainable development. Research India Publications, New Delhi

Ruppenthal M (1995) Soil conservation in Andean cropping systems: soil Erosion and crop productivity in traditional and forage-legume based cassava cropping Systems in the South Colombian Andes. Verlag Josef Margraf, Weikersheim

Saarnio S, Heimonen K, Kettunen R (2013) Biochar addition indirectly effects N_2O emissions via soil moisture and plant N uptake. Soil Biol Biochem 58:99–106

Saleh F, Samsuddin R, Husin M (2014) Bio-fuel source from combination feed of sewage and rice waste. In: International conference on environment science engineering IPCBEE, vol 8. IACSIT Press, Singapore, pp 68–72

Saranya K, Krishnan PS, Kumutha K, French J (2011) Potential for biochar as an alternate carrier to lignite for the preparation of biofertilizers in India. Agric Environ Biotechnol 4:167–172

Saxena J, Rana G, Pandey M (2013) Impact of addition of biochar along with Bacillus sp. on growth and yield of French beans. Sci Hortic 162:351–356

Schackley S, Clare A, McCarl BA, Joseph S, Schmidt HP (2015) Economic evaluation of biochar systems: current evidence and challenges. In: Lehmann J, Joseph S (eds) Biochar for environmental management. Earthscan, London

Schirrmann U (1984) Aktivkohle und ihre Wirkung auf Bakterien und deren Toxine im Gastrointestinaltrakt. TU München, Munich

Schmidt HP, Anca-Couce A, Hagemann N, Werner C, Gerten D, Lucht W, Kammann C (2018) Pyrogenic carbon capture and storage. GCB Bioenergy 11(4):573–591

Schmidt HP, Shackley S (2016) Biochar horizon 2025. In: Shackley S, Ruysschaert G, Zwart K, Glaser B (eds) Biochar in European soils and agriculture: science. Abingdon, Taylor & Francis Group, pp 281–289

Schmidt MWI, Skjemstad JO, Jager C (2002) Carbon isotope geochemistry and nanomorphology of soil black carbon: black chernozemic soils in Central Europe originate from ancient biomass burning. Glob Biogeochem Cycles 16:1123

Schnug E, Oswald P, Haneklaus S (1996) Organic manure management and efficiency: role of organic fertilizers and their management practices. In: Rodriguez-Barrueco C (ed) Fertilizers and environment. Kluwer Academic Publishers, Dordrecht, pp 259–265

Schulz H, Glaser B (2012) Effects of biochar compared to organic and inorganic fertilizers on soil quality and plant growth in a greenhouse experiment. Soil Sci 175:410–422. https://doi.org/10.1002/jpln.201100143

Shammi M, Rahman R, Rahman MM (2016) Assessment of salinity hazard in existing water resources for irrigation and potentiality of conjunctive uses: a case report from Gopalganj District, Bangladesh. Sustain Water Resour Manage 2:369–378

Shenbagavalli S, Mahimairaja S (2012a) Characterization and effect of biochar on nitrogen and carbon dynamics in soil. Int J Adv Biol Res 2:249–255

Shenbagavalli S, Mahimairaja S (2012b) Production and characterization of biochar from different biological wastes. Plant Animal Environ Sci 2:197–201

Sheng GY, Yang YN, Huang MS, Yang K (2005) Influence of pH on pesticide sorption by soil containing wheat residue derived char. Environ Pollut 134:457–463

Shinogi Y, Yoshida H, Koizumi T, Yamaoka M, Saito T (2003) Basic characteristics of low-temperature carbon products from waste sludge. Advances Environ Res 7:661–665. https://doi.org/10.1016/S1093-0191(02)00040-0

Sial T, Lan Z, Khan M, Zhao Y, Kumbhar F, Liu J, Zhang A, Hill R, Lahori A, Mehrunisa M (2019) Evaluation of orange peel waste and its biochar on greenhouse gas emissions and soil biochemical properties within a loess soil. Waste Manag 87. https://doi.org/10.1016/j.wasman.2019.01.042

Silber A, Levkovitch I, Graber ER (2010) PH-dependent mineral release and surface properties of corn straw biochar: agronomic implications. Environ Sci Technol 44:9318–9323

Silivong P, Preston TR (2016) Supplements of water spinach (*Ipomoea aquatica*) and biochar improved feed intake, digestibility, N retention and growth performance of goas fed foliage of *Bauhinia acuminata* as the basal diet. Livestock Res Rural Dev 28:113

Singh B, Singh BP, Cowie A (2010a) Characterization and evaluation of biochars for their application as a soil amendment. Aust J Soil Res 48:516–525

Singh B, Singh BP, Cowie A L (2010b) Characterisation and evaluation of biochar for their application as a soil amendment. Soil Res 48: 516–525

Singh BP, Cowie AL, Smernik RJ (2012) Biochar carbon stability in a clayey soil as a function of feedstock and pyrolysis temperature. Environ Sci Technol 46:11770–11778

Singh VK, Soni AB, Kumar S, Singh RK (2014) Pyrolysis of Sal seed to liquid products. Bioresour Technol 151:432–435

Smith P, Bustamante M, Ahammad H, Clark H, Dong H, Elsiddig EA, Haberl H, Harper R, House J, Jafari M, Masera O, Mbow C, Ravindranath NH, Rice CW, Robledo AC, Romanovskaya A, Sperling F, Tubiello F (2014) Agriculture, forestry and other land use (AFOLU). In: Climate change 2014: mitigation of climate change. Contribution of working group III to the fifth assessment report of the intergovernmental panel on climate Change [Edenhofer O., R. Pichs-Madruga, Y. Sokona, E. Farahani, S. Kadner, K. Seyboth, A. Adler, I. Baum, S. Brunner, P. Eickemeier, B. Kriemann, J. Savolainen, S. Schlömer, C. von Stechow, T. Zwickel and J.C. Minx (eds.)]. Cambridge University Press, Cambridge

Sohi S, Krull E, Lopez-Capel E, Bol R (2010a) A review of biochar and its use and function in soil. Adv Agron 105:47–82. https://doi.org/10.1016/S0065-2113(10)05002-9

Sohi S, Lopez-Capel E, Krull E, Bol R (2009) Biochar, climate change and soil: a review to guide future research. In: CSIRO land and water science report 05/09. CSIRO, Black Mountain, pp 17–31

Sohi SP, Krull E, Lopez-Capel E, Bol R (2010b) A review of biochar and its use and function in soil. Adv Agron 105:47–82

Soil Survey Staff(2014) Kellogg soil survey laboratory methods manual. Soil survey investigations report no. 42, version 5.0. R. Burt, Soil Survey Staff (ed.). U.S. Department of Agriculture, Natural Resources Conservation Service: Washington, DC

Solaiman Z, Murphy D, Abbott L (2012) Biochars influence seed germination and early growth of seedlings. Plant Soil 353:273–287

Spokas KA, Novak JM, Venterea RT (2012) Biochar's role as an alternative N fertilizer: ammonia capture. Plant Soil 350:35–42. https://doi.org/10.1007/s11104-011-0930-8

Stefaniuk M, Oleszczuk P (2016) Addition of biochar to sewage sludge decreases freely dissolved PAHs content and toxicity of sewage sludge-amended soil. Environ Pollut 218:242e251

Steiner C, Das KC, Garcia M, Förster B, Zech W (2008a) Charcoal and smoke extract stimulate the soil microbial community in a highly weathered xanthic Ferralsol. Pedobiologia 51:359–366. https://doi.org/10.1016/j.pedobi.2007.08.002

Steiner C, Glaser B, Teixeira WG, Lehmann J, Blum WEH, Zech W (2008b) Nitrogen retention and plant uptake on a highly weathered central Amazonian Ferralsol amended with compost and charcoal. J Plant Nutr Soil Sci 171:893–899. https://doi.org/10.1002/jpln.200625199

Steiner C, Teixeira WG, Lehmann J, Nehls T, de Macedo JLV, Blum WEH, Zech W (2007) Long-term effects of manure, charcoal and mineral fertilization on crop production and fertility on a highly weathered central Amazonian upland soil. Plant Soil 291(1–2):275–290

Stevens RG, Hammond MW (1992) Soil resources: what is needed and how do we maintain these resources. American Potato J69:717–730. https://doi.org/10.1007/BF02853815

Sun H, Zhang H, Xiao H, Shi W, Muller K, Zwieten LV, Wang H (2019) Wheat straw biochar application increases ammonia volatilization from an urban compacted soil giving a short-term reduction in fertilizer nitrogen use efficiency. J Soils Sediments 19(4):1624–1631

Tandon HLS (2004) Fertilisers in Indian agriculture – from 20th to 21st century. Fertiliser Development and Consultation Organisation, New Delhi, p 239

Tapio I, Snelling TJ, Strozzi F, Wallace RJ (2017) The ruminal microbiome associated with methane emission from ruminant livestock. J Animal Sci Biotechnol 8(1):7

Thies JE, Rillig MC (2009) Characteristics of biochar: biological properties. In: Biochar for environmental management: science and technology. Routledge, Milton Park, pp 85–105

Thu M, Koshio S, Ishikawa M, Yokoyama C (2010) Effects of supplementation of dietary bamboo charcoal on growth performance and body composition of juvenile Japanese flounder, *Paralichthys olivaceus*. J World Aquacult Soc 41(S2):255–262

Tiziano G (2016) Soil degradation, land scarcity and food security: reviewing a complex challenge. Sustainability 8(3):281. https://doi.org/10.3390/su8030281

Tomczyk A, Sokołowska Z, Boguta P (2020) Biochar physicochemical properties: pyrolysis temperature and feedstock kind effects. Rev Environ Sci Biotechnol 19:191–215. https://doi.org/10.1007/s11157-020-09523-3

Toth JD, Dou Z (2016) Use and impact of biochar and charcoal in animal production systems. In: Guo M, He Z, Uchimiya SM (eds) Agricultural and environmental applications of biochar: advances and barriers. Soil Science Society of America, Inc, Madison, pp 199–224

Turbé A, De Toni A, Benito P, Lavelle P, Lavelle P, Camacho NR (2010) Soil biodiversity: functions, threats and tools for policy makers. In: Bio intelligence service, IRD, and NIOO, report for European Commission. DG Environment, Brussels, p 250

Uchimiya M, Lima IM, Klasson KT, Chang S, Wartelle LH, Rodgers JE (2010) Immobilization of heavy metal ions (CuII, CdII, NiII, and PbII) by broiler litter-derived biochar in water and soil. J Agr Food Chem 58:5538–5544

Uchimiya M, Wartelle LH, Klasson KT, Fortier CA, Lima IM (2011) Influence of pyrolysis temperature on biochar property and function as a heavy metal sorbent in soil. Agric Food Chemistry 59:2501–2510

UN (2015) Sustainable development goals. Online at http://www.un.org/sustainabledevelopment/sustainable-development-goals/

Vaccari FP, Baronti S, Lugato E, Genesio L, Castaldi S (2011) Biochar as a strategy to sequester carbon and increase yield in durum wheat. European J Agronomy 34(4):231–238

Van Zwieten L, Kimber S, Morris S (2010) Effects of biochar from slow pyrolysis of papermill waste on agronomic performance and soil fertility. Plant Soil 327:235–246. https://doi.org/10.1007/s11104-009-0050-x

Vaughn S, Kenar J, Thompson A, Peterson S (2013) Comparison of biochar derived from wood pellets and pelletized wheat straw as replacements for peat in potting substrates. Ind Crop Prod 51:437–443. https://doi.org/10.1016/j.indcrop.2013.10.010

Verheijen F, Jeffery S, Bastos AC, van der Velde M, Diafas I (2010) Biochar application to soils - a critical scientific review of effects on soil properties, processes and functions. EUR 24099 EN. Office for the Official Publications of the European Communities, Luxembourg

Vinh N, Hien N, Anh M, Lehmann J, Joseph S (2014) Biochar treatment and its effects on rice and vegetable yields in mountainous areas of northern Vietnam. Int J Agric Soil Sci 2:5–13

Virmani SM, Katyal JC, Eswaran H, Abrol I (1994) Stressed agroecosystems and sustainable agriculture. Oxford and IBH, New Delhi

Volli V, Singh RK (2012) Production of bio-oils from de-oiled cakes by thermal pyrolysis. Fuel 96:579–585

Wander MM, Traina SJ, Stinner BR, Peters SE (1994) Organic and conventional management effects on biologically active soil organic matter pools. Soil Sci Soc Am J 58:1130–1139

Wander MM, Ugarte C, Zaborski E, Phillips E (2014) Organic systems and climate change. Proposal and final report for ORG project 2010–03954. CRIS Abstracts

Wang J, Xiong Z, Kuzyakov Y (2016) Biochar stability in soil: meta-analysis of decomposition and priming effects. GCB Bioenergy 8:512–523. https://doi.org/10.1111/gcbb.12266

Warnock DD, Lehmann J, Kuyper TW, Rillig MC (2007) Mycorrhizal responses to biochar in soil-concepts and mechanisms. Plant Soil 300:9–20

Watarai S, Tana C (2005) Eliminating the carriage of Salmonella enterica serovar Enteritidis in domestic fowls by feeding activated charcoal from bark containing wood vinegar liquid (Nekka-rich). Poultry Sci 84(4):515–521

Wei L, Xu S, Zhang L, Zhang H, Liu C, Zhu H, Liu S (2006) Characteristics of fast pyrolysis of biomass in a free fall reactor. Fuel Processing Technol 87:863–871

William K, Qureshi RA (2015) Evaluation of biochar as fertilizer for the growth of some seasonal vegetables. J Bioresour Manag 2(1). https://doi.org/10.35691/JBM.5102.0011

Woolf D, Amonette JE, Street-Perrott FA, Lehmann J (2010a) Sustainable biochar to mitigate global climate change. Nature 1:1–9
Woolf D, Amonette JE, Street-Perrott FA, Lehmann J, Joseph S (2010b) Sustainable biochar to mitigate global climate change. Nat Commun 1:1–9
Xiao R, Bai J, Wang Q, Gao H, Huang L (2011) Assessment of heavy metal contamination of wetland soils from a typical aquatic-terrestrial ecotone in Haihe River basin, North China. Clean-Soil Air Water 39:612–618
Xu C, Chen HX, Xiang Q, Zhu HH, Wang S, Zhu QH, Huang DY, Zhang YZ (2017) Effect of peanut shell and wheat straw biochar on the availability of cd and Pb in a soil-rice (*Oryza sativa* L.) system. Environ Sci Pollut Res Int 25(2):1147–1156. https://doi.org/10.1007/s11356-017-0495-z
Yadav NK, Kumar V, Sharma KR (2018) Biochar and their impacts on soil properties and crop productivity: a review. J Pharmacognosy Phytochem 7(4):49–54
Yanai Y, Toyota K, Okazaki M (2007) Effects of charcoal addition on N_2O emissions from soil resulting from rewetting air-dried soil in short term laboratory experiments. Soil Sci Plant Nutr 53:181–188
Yang X, Liu J, McGrouther K, Huang H, Lu K, Guo X, He L, Lin X, Che L, Ye Z, Wang H (2016) Effect of biochar on the extractability of heavy metals (cd, cu, Pb, and Zn) and enzyme activity in soil. Environ Sci Pollut Res 23:974e984
Yao Y, Gao B, Zhang M, Inyang M, Zimmerman AR (2012) Effect of biochar amendment on sorption and leaching of nitrate, ammonium, and phosphate in a sandy soil. Chemosphere 89:1467–1471. https://doi.org/10.1016/j.chemosphere.2012.06.002
Yoo JH, Ji SC, Jeong GS (2007) Effect of dietary charcoal and wood vinegar mixture (CV82) on body composition of olive flounder *Paralichthys olivaceus*. J World Aquaculture Soc 36 (2):203–208
Yu XY, Ying GG, Kookana RS (2009) Reduced plant uptake of pesticides with biochar additions to soil. Chemosphere 76:665–671
Zainal H, Bachmann R, Loh SK (2016) Production and characterization of biochar from palm kernel Shell (PKS). In: Conference: 4th Asian conference on biomass science, at Universiti Sains Malaysia, Bayan Lepas Penang
Zhang H, Lin K, Wang H, Gan J (2010) Effect of *Pinus radiata* derived biochar on soil sorption and desorption of phenanthrene. Environ Pollut 158:2821–2825
Zhang H, Voroney R, Price G (2015) Effects of temperature and processing conditions on biochar chemical properties and their influence on soil C and N transformations. Soil Biol Biochem 83:19–28. https://doi.org/10.1016/j.soilbio.2015.01.006
Zhang P, Sun H, Yu L, Sun T (2013) Adsorption and catalytic hydrolysis of carbaryl and atrazine on pig manure-derived biochar: impact of structural properties of biochar. J Hazard Mater 244:217–224. https://doi.org/10.1016/j.jhazmat.2012.11.046
Zhang R, Zhang Y, Song L, Song X, Hanninen H, Wu J (2017) Biochar enhances nut quality of *Torreya grandis* and soil fertility under simulated nitrogen deposition. For Ecol Manag 391:321–329
Zhao QG (1995) Making new contribution to development of soil science the 21st century. Acta Pedol Sinica 32:1–13. (in Chinese with English abstract)
Zheng J, Stewart CE, Cotrufo MF (2012) Biochar and nitrogen fertilizer alters soil nitrogen dynamics and greenhouse gas fluxes from two temperate soils. J Environ Quality 41:1361–1370
Zimmerman AR, Gao B, Ahn M-Y (2011) Positive and negative carbon mineralization priming effects among a variety of biochar-amended soils. Soil Biol Biochem 43:1169–1179
Zwieten L, Singh B, Joseph S, Kimber S, Cowie A (2009) Biochar and emissions of non-CO_2 greenhouse gases from soil. In: Lehmann J, Joseph S (eds) Biochar for environmental management: science and technology. Earthscan, London, pp 227–249

Resource Conservation for Sustainable Development

14

Abhishek Raj, Manoj Kumar Jhariya, Nahid Khan, Arnab Banerjee, Poonam, Ram Swaroop Meena, and Kavita Rani

Abstract

Resources sustain the ecosystem, but its depletions are the major concern of the present times. Natural resources such as agriculture, forestry, agroforestry, soils, animals, etc. enhance the biodiversity which intensify ecosystem services in tangible and intangible ways and regulate ecosystem processes. These ecosystem services not only maintain soil-food-climate security but also make a door for achieving the goal of sustainable development. However, overexploitation, deforestation, faulty land use practices, unsustainable land management, intensive agriculture, high synthetic inputs, etc. disturb our pathway of natural ecosystem by affecting resources and its depletions. The FAO mentioned that every year around 6.5 Mha (million hectare) areas of tropical forest are converted into agricultural land due to rising populations and human needs that affects the

A. Raj
School of Agriculture, Lovely Professional University, Phagwara, Punjab, India

M. K. Jhariya (✉) · N. Khan
Department of Farm Forestry, Sant Gahira Guru Vishwavidyalaya, Ambikapur, India

A. Banerjee
Department of Environmental Science, Sant Gahira Guru Vishwavidyalaya, Ambikapur, India

Poonam
Krishi Vigyan Kendra, Navgaon (Alwar-I), Sri Karan Narendra Agriculture University (SKNAU), Jobner, Rajasthan, India

R. S. Meena
Department of Agronomy, Institute of Agricultural Sciences, Banaras Hindu University, Varanasi, Uttar Pradesh, India
e-mail: meenars@bhu.ac.in

K. Rani
Department of Microbiology, CCS Haryana Agricultural University, Hisar, Haryana, India

M. K. Jhariya et al. (eds.), *Sustainable Intensification for Agroecosystem Services and Management*, https://doi.org/10.1007/978-981-16-3207-5_14

natural resources by depriving health, quality, and quantity of other resources such as forest trees, wild animals, soil quality, etc. Soil is another important natural resource which is degraded up to 147.0 Mha in Indian land areas. Among this, water erosion, acidification, flooding, wind erosion, and salinity contributed 94, 16, 14, 9, and 6 Mha of land, whereas combination of other factors affects 7.0 Mha, respectively. This resource supports human and livestock by 18 and 15% of the global population, whereas different land use systems like agriculture, forestry, and fishery systems contribute to GDP (gross domestic product) and employment generations by 17 and 50%, respectively. Therefore, resource conservation and its management are having prime importance duse to their uncountable contributions in national and international sustainable-based development along with addressing environmental sustainability. In this context, the practices of ecology-oriented and sustainable intensification become good strategies for the conservation and management of natural resources. Contrary to intensive agriculture, the characteristics, principles, and practices of both ecological and sustainable intensification are much clear. These practices will ensure soil-food-climate security along with the maintenance of environmental sustainability and ecological stability. Thus, these practices must approach the further research and development through better methods and technology for promising resource conservation and sustainable development.

Keywords

Agroforestry · Ecology · Forestry · Natural resource · Resource conservation · Soils · Sustainable intensification

Abbreviations

C	Carbon
CO_2	Carbon dioxide
GHGs	Greenhouse gases
Mha	Million hectare
NRM	Natural resource management
NRs	Natural resources
SD	Sustainable development
SFM	Sustainable forest management
SOC	Soil organic carbon

14.1 Introduction

Nowadays, a meticulous discussion on "sustainability" draws a great attention toward the management and conservation of natural resources (NRs) which play a key role in the success of nation development. National and international agencies, policy makers, stakeholders, scientists, researchers, and academicians are involved in brainstorming discussions for three decades and having keen interest to raise this topic at the global context to achieve development on a sustainable basis without affecting our environment. The term "sustainable development" (SD) symbolizes the development in social, economic, and environmental dimensions (Ashoka et al. 2017; Jhariya et al. 2019a, b; Banerjee et al. 2020) which is linked with resource conservation and management, i.e., directly and indirectly achieved through NR conservations.

NRs play an important role in the proper functioning of the ecosystem and structure maintenance. Land resources are among them which support agriculture, forestry, livestock, pastures, and soil-inhabiting flora and fauna. Many authors have reported the extent of these resources by conducting extensive researches around the world. For example, Bhattacharyya et al. (2015) have reported that around 264.5 million hectares (Mha) of the total Indian geographical land (328.7 Mha) are supposed to be used for varying practices such as agriculture, forestry, livestock, and other biomass production system. These land use practices maintain the NRs at a certain level due to proper and efficient utilizations beyond which the resources would be degraded and depleted as represented in Fig. 14.1. For example, around 146.8 Mha of land is degraded (NBSS and LUP 2004) which has become a serious problem in the Indian context. Similarly, the annual loss of 5.30 billion tons of soil

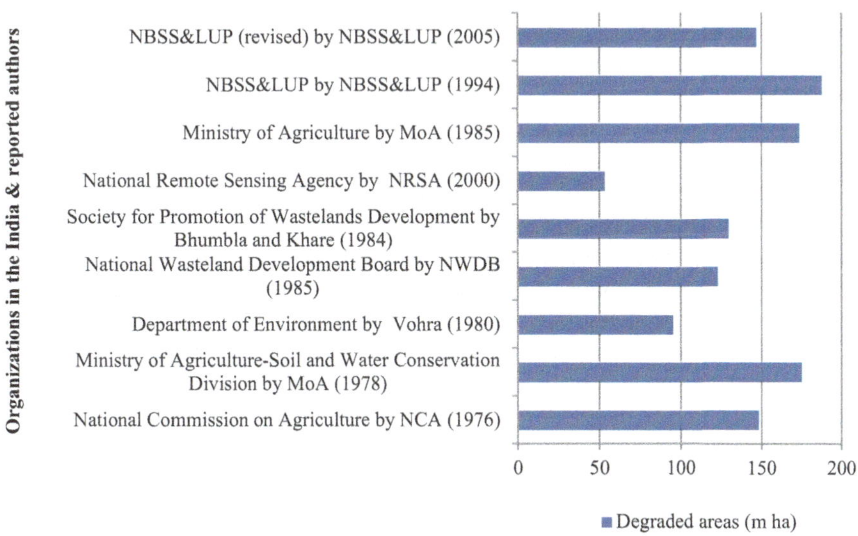

Fig. 14.1 Assessment of land degradation areas by different organizations in India

was due to erosion which was observed at the rate of 16.4 tons/ha/yr. (Dhruvanarayan and Ram 1983). Further, livestock are another important NR which is an integral part of varying land use farming systems. As per Sahay (2015), around 11.0 Mha of grasses/pastures land are utilized by 467 million populations of grazing animals which is beyond the carrying capacity, i.e., 42 heads per hectare of grazing land as compared to the optimum and sustainable number of 5 animals per hectare. Thus, overgrazing is another very important curse that resulted in poor infiltration and high runoff and soil erosions. As per Sharma (1997), overgrazing leads to 5–41- and 3–18-times higher soil erosions at the meso-scale and macro-scale levels. Thus, there is a need of the hour to emphasize on resource conservation and its management that not only ensure environmental sustainability but also build a dream of achieving the goal of SD (Bhattacharyya et al. 2015; Meena et al. 2018; Xie et al. 2020; Jhariya et al. 2021a, b; Raj et al. 2021).

NRs (forest, agriculture, agroforestry, soil, animals, etc.) are the major pools on earth involved in various ecosystem processes, and their management and conservation are very important for the maintenance of environmental suitability and ecological stability. Therefore, queries are certain in the form of (a) "How can we achieve the SD through conservation and management of these resources?" and (b) "Is there any dichotomy between development and sustainability?". A dichotomy exists between sustainability and development, i.e., while achieving development, we ignore the sustainable aspect that may affect our environment and NRs, but somehow a big concern about SD does exist, and we can't overlook it for global development without affecting and interrupting our environment. Great synergy exists between NR conservation and environmental sustainability which would be helpful in maintaining ecological integrity among biodiversity and resources such as forest, agriculture, agroforestry, soil, and animals. Conservation and management of these resources could enhance the ecosystem services by enhancing biodiversity that provides various tangible and intangible benefits to biodiversity which make healthy environment by maintaining food-climate-soil security and mitigate climate change issues through greater carbon (C) sink potential into both vegetation and soils in any agroecosystem practices (Khan et al. 2020a, b; Meena and Lal 2018).

In view of the above, this chapter comprehensively discusses about the various natural forms of resources and its contributions toward achieving the goal of global SD through the maintenance of environmental sustainability and ecological stability. Practices of resource conservation for intensifying ecosystem services, integrated issues, and proposing research and development and relevant policy behind the success of resource conservation at sustainable basis without negatively affecting our environment are also included in this chapter. This chapter will help all policy makers, planners, stakeholders, entrepreneurs, farmers, academicians, scientists, and researchers for achieving the goal of sustainability through proper practice and management of various NRs in the global context.

14.2 Natural Resources and Conservation Perspective

NR management and its conservation are having prime importance which plays a key role in regulating ecosystem processes. Forests, agriculture, agroforestry, soils, animals, etc. are important resources that are interlinked to each other and maintain the overall ecosystem structure and functions. They will enhance the biodiversity and intensify the ecosystem services (Khan et al. 2021a, b). These resources are required to be explored properly to get a better understanding of their role, importance, and conservation at global scale.

14.2.1 Forest

The term forest is gaining worldwide recognitions on several technical forums due to multifarious benefits in the context of social, cultural, economic, environment and ecology on which entire human civilization depends. However, degradation of forest is the major challenge today that not only affects overall biodiversity but also affects several ecosystem services and ecological stability (Jhariya and Singh 2020, 2021). Forest provides various tangible and intangible benefits such as timber, fuelwood, firewood, and nutritious food and fruits along with other NTFPs as direct benefits, whereas soil enrichment, biodiversity enhancement, watershed management, efficient bio-geochemical cycling, better rhizosphere biology, and climate security are delivered through intangible services. Table 14.1 shows the impact of rising carbon dioxide (CO_2) on global forest sustainability by affecting growth, biomass, and productivity.

Forest is a vital ecosystem that covers 4 billion ha lands (30% of the total land on earth) harboring many NRs, involved in ecosystem processes, and sustains many lives on our mother earth. As per world statistics, around 30.6% of world land areas are covered by forest ecosystem in which Europe and North region and South and Central America contributed majorly. It further indicates that the forests in tropics are under severe stress due to many anthropogenic, biotic, and natural factors (FAO 2016). The forests in the Asian continent are under severe threat due to various human-induced factors, deforestation, illicit felling, encroachment, implementation of developmental projects, mining, and other biotic interference. Although the rapid increase in forest covers was also observed from 2010 onward, somehow it is surprising to know that the Asian continent represents only 19.02% in forest covers which shows minimum percentage contributions than others in a global perspective.

As we know, rich biodiversity has been observed in forest ecosystem which is regulated and managed by forest to promise the resource utilization and its conservations for better environment to achieve the goal of sustainability. Forest regulates soils, animals, peoples, and other NRs that promote healthy biodiversity and intensify ecosystem services in direct (as timber, fuelwood, fodder, NTFPs, etc. which make farmer's health and income security) and indirect ways such as soil enrichments through fertility enhancement, soil microbial populations, healthy rhizosphere biology, efficient ecosystem services, higher nutrient use efficiency,

Table 14.1 Impacts of rising CO_2 on global forest sustainability by affecting growth, biomass, and productivity

Varying tree species in forests	Rising carbon dioxide level	Results	Reported authors
Forest comprised the combinations of *Alnus glutinosa* (black alder), *Betula pendula* (silver birch), and *Fagus sylvatica* (common beech) tree species	Experienced rising CO_2	The value of aboveground biomass was increased from 15.2 to 20.2 kg/m^2, i.e., overall 17 percent of increment was observed in this mixed forest	Smith et al. (2013)
Betula pendula (commonly known as silver birch)-based forest	Experienced rising CO_2	The value of LAI (leaf area index) was enhanced by 37%	
Pinus taeda (loblolly pine)- and *Liquidambar* (sweetgum)-based mixed forest stand	Experienced rising CO_2 value by 550 ppm	The value of NPP (net primary productivity) was increased by 23 ± 2 percent in these mixed stands	Norby et al. (2005)
Forest comprised the combinations of *Acer* species (maple), *Betula pendula* (silver birch), and *Populus tremuloides* (aspen) tree species	Experienced rising CO_2	The value of LAI (leaf area index) was increased in the mixed forest	Oksanen et al. (2001)
Populus nigra (commonly known as black poplar) tree species-based forest	Experienced rising CO_2	The value of NPP (net primary productivity) was increased by 225%	Gielen et al. (2001)
Conifers tree species in forest	Experienced rising CO_2	The value of biomass was increased by 130%	Saxe et al. (1998)
Pinus taeda (commonly known as loblolly pine) tree species-based forest	Experienced rising CO_2	A significant effect on seed weight was observed which was increased by 91.0%	Hussain et al. (2001)
Tree canopy mixture of *Quercus myrtifolia* (myrtle oak), *Quercus chapmanii* (chapman oak), and *Quercus geminate* (sand live oak) in oak-based natural forest stand in the region of Florida	Experienced rising CO_2	A significant effect on acorns was observed that enhanced the production potential	Stiling et al. (2004)
	Experienced rising CO_2	It was observed that the ratio of root-shoot mass was significantly enhanced	Oechel and Strain (1985)
Liriodendron tulipifera (commonly known as tulip poplar) tree species-based forest	Experienced rising CO_2	Neutral effects were experienced on root-shoot ratio	Norby et al. (1992)
Forest comprised the combinations of *Populus alba* (poplar), *Populus nigra* (cottonwood poplar), and *Populus euramericana* (Canadian poplar) tree species	Experienced rising CO_2	The value of root biomass in BGP (belowground production) was increased by 47 to 76%	Lukac et al. (2003)
	Experienced rising CO_2 value by 550 ppm	The value of fine root biomass (FRB) was increased by double	Norby et al. (2004)

(continued)

Table 14.1 (continued)

Varying tree species in forests	Rising carbon dioxide level	Results	Reported authors
Tree canopy mixture of *Quercus petraea* (sessile oak) and *Quercus robur* (common oak) in oak-based natural forest stand in the region of Northeastern France	Elevated CO_2	Basal area increment (BAI) was observed	Becker et al. (1994)
Forest of tropical regions	Experienced rising CO_2 value by 550 ppm	The value of NPP (net primary productivity) was increased by 35% in this tropical forest	Collatz et al. (1991)
Forest of temperate regions	Experienced rising CO_2 value by 370 ppm	The value of NPP (net primary productivity) was increased by 26% in this temperate forest	

watershed management, and climate change mitigation through greater potential of C sequestration along with maintenance of overall soil-food-climate security. Resource conservation is an important aspect of forest ecosystem, i.e., chances of loss of resources are minimized under complex forest systems. Forest makes good precipitation, minimizes the effects of natural calamities, regulates global water body, balances CO_2 level in the atmosphere, and maintains food and income security of poor forest fringe people whose livelihood is totally dependent upon forest resources (FAO 2006; Painkra et al. 2016).

14.2.2 Agriculture

The rising and burgeoning populations require food which exerts pressure on the cultivation and expansion of agriculture land. As per one estimate, around 40% of the total land area of the earth occupied by agricultural land practices is becoming the largest terrestrial biome globally (Zabel et al. 2014). Farmers are directly involved in cultivation practices and produce foods for satisfying billions of peoples and maintain healthy ecosystem. The rising demands of agricultural products necessitate farming practices. As per one projection, around 70–110% of demands will be expected to increase by the year 2050 due to the rising global population of approx. 9.0 billion that requires meat consumption and bio-based products (Alexandratos and Bruinsma 2012; Kumar et al. 2021; Ray et al. 2013). However, extreme weather and changing climate disturb the climatic processes such as both rising temperature and uncertain rainfall affecting the land suitability for agricultural practices. For example, desertification and soil erosion due to changing climate cause land degradation by 19–23 hectare per minute (UNCCD 2014). Similarly, urbanization leads to the shrinkage of agricultural land by 1.5 million km^2 by 2030 (Avellan et al. 2012).

Siebert et al. (2013) mentioned that the irrigated agricultural resource produces around 40% of global food through the practices on 3.1 million km^2 areas.

Agriculture stores an ample amount of resources by diversifying crops, different varieties, livestock, and soils. However, applying fertilizers and nutrients would be helpful in maintaining the health and quality of soil. In turn, healthy soil promotes diverse microbial populations and better rhizosphere biology and enhances nutrient use efficiency that facilitates the availability of essential nutrients to the plant. Plant absorbs these nutrients from soils and stores it for proper growth and development that makes diversified, quality, and nutritive food and fruit materials. Agriculture works like the two faces of a coin: One face represents positive, and the remaining other is for negative. Releasing greenhouse gases (GHGs) through faulty land use farming systems is the negative face of agriculture, whereas sequestration of atmospheric C by different agricultural practices is the positive face. As per Lal (2004), approx. 0.4–1.2 gigatons/annum of C sequestration was observed under the agricultural land use system. A better and scientific agricultural practice such as conservation and no-tillage practices would ensure resource conservation through efficient nutrient cycling and better nutrient inputs and restore nutrients through mulching practices, etc. that result in higher yield and productivity (Hansen et al. 2013; Kumar et al. 2020). Judicious use of resources such as light, space, nutrients, water, etc. and proper blending of nutrients make healthy soil and plants which is the foundation of higher productions. Thus, a synergy exists between sustainable agricultural practices and SD that can be achieved by maintaining environmental sustainability and ecological stability.

14.2.3 Agroforestry

The contributions of agroforestry systems in nation development via achieving the goal of SD are now crystal clear. Nowadays, agroforestry becomes a very popular and talkative word among policy makers, stakeholders, academicians, scientists, and farmers at national and international platforms. The practices of agroforestry systems are widely adopted and lie in various agro-climatic zones of tropics. Location specific in nature, sustainable land management practices tend to be a complex and diverse system (comprising trees, crops, and animals/pastures) and deliver various ecosystem services in both tangible (as timber, fuelwood, quality and nutritive edible fruits, fodders, and various other NTFPs including medicinal and aromatic plants) and intangible ways such as soil fertility improvement, SOC pools, better rhizosphere biology, efficient and closed nutrient cycling, and environment security for maintaining ecological stability which helps in achieving the target of resource conservation and SD. Prevalent agroforestry models comprising varying components for ecological sustainability are depicted in Table 14.2.

Agroforestry plays a key role in resource conservation and its management. In agroforestry system, a woody perennial tree gives protection/shelter to the associated crops and promotes economical outputs. Similarly, perennial tree gives protection/shelter to the associated grasses/pasture crops that in turn provide palatable nutritive

Table 14.2 Prevalent agroforestry models comprising varying components for ecological sustainability

S.N.	Prevalent agroforestry models	Admixture of varying components	Agroecological regions of the country	Reported authors
1	Taungya system	Generally comprised tree and herbaceous crop species in which crops are inter-grown in between spaces of tree species. However, these systems are practiced in three forms such as departmental Taungya, village Taungya, and leased Taungya that makes overall environment and ecological sustainability	This system is practiced in Assam, U.P., and Uttarakhand comprising both plain and hill regions	Rowbotham (1924); Ghosh and Ghosal (2019)
2	Shifting cultivation system	Ancient and primitive type of farming system in which forest is cleared and burned which is again utilized by the cultivation of varying arable crops, and again this practice is repeated. This system is also known as slash and burn methods	Mostly prevalent in the hills of N.E. states, whereas the tribal peoples of Bihar, Jharkhand, and Orissa states are also practicing this system	Sati (2019); Sahu et al. (2005); and Maithani (2005)
3	Tree used on farm boundaries and in field as scattered form	Many nitrogen-fixing leguminous MPTs are either incorporated on farm's bund or scattered in the field. Ecologically more sound providing protection to the agricultural crops from averse climatic impacts and add more soil fertility and provides additional multiple products	Prevalent in most parts of Central India having small and marginal farmers	Tiwari (2008); Jhariya et al. (2015); Raj et al. (2016)
4	Intercropping practices in varying models of AFs	Several economically and ecologically important crops, grasses/pastures, and other plants are intercropped with woody perennial trees which forms ecological integrity among various components of AFs	Whole regions of the tropics	Kumar (2010); Mthembu et al. (2018); Dollinger (2019)
5	Silvopastoral model of AFs	Important palatable pastures are integrated with woody perennial trees in this model which is an ecologically sound practice due to ameliorative potential of degraded and wasteland land throughout the tropics	This system is plasticized on most parts of dry region characterizing harsh climate and covers waste or degraded land	Singh (2008); Jose and Dollinger (2019)
6	Windbreak and shelterbelt practices in AFs	System comprised tree species on windward side which protect agricultural crops situated on leeward side that experiences less wind speed and known for ecologically protecting system based on its significance	Mostly covered dry regions experiencing harsh climate and high wind speed along with prominent soil erosion	Dagar et al. (2014); Mayrinck et al. (2019)

(continued)

Table 14.2 (continued)

S.N.	Prevalent agroforestry models	Admixture of varying components	Agroecological regions of the country	Reported authors
7	Homestead gardens (also known as kitchen gardens) in AFs	Different layers and strata are found in this system such as top layer comprised MPTs, middle layer and lower layers are composed of vegetables, spices and fruit crops which makes ecologically more diverse and complex systems that helps in producing multiple products to sustain life of many farmers	Also known as "life-supporting system" which is mostly practiced in the regions of southern India which is characterized by heavy rainfall more than 2500 mm	Kumar (2010); Galhena et al. (2013)
8	Wadi form of practices in AFs	It is also ecologically more viable due to the presence of arable crops, fruit crops and MPTs in agricultural field	Covered the regions of Rajasthan, Maharashtra, Karnataka, and Gujarat states of the Indian subcontinent	Hegde (2005)
9	Agri-silvi-apiculture under AFs	Ecologically sound and economically viable type of models comprising mustard as arable crops integrated with MPTs along with rearing practices of honey bees	Covered the regions of Rajasthan, M.P., Haryana, and Punjab states of the Indian subcontinent	Gill et al. (2010); Verma et al. (2016)
10	Plantation- and cash crop-based AFs	Combinations of varying plantation crops with different tree species	Prevalent in the humid and sub-humid regions of the tropics that comprise coffee and various fruit trees along with spices and ornamental crops	Dagar et al. (2014)
		Integration of varying plantation crops under shade trees	Mostly covered the N.E. region comprising tea plantations under important shade tree species	Selvan and Kumar (2017)
		Fast-growing tree species are integrated with various important agricultural crops	The regions of Punjab, U.P., and Haryana are characterized by this practice	Dhiman (2012); Luna (2009); Kishwan and Kumar (2013)
		Grasses/pastures and fruit tree combinations in AFs	Distributed on most parts of hill and sloppy regions	Singh (2008)
		Cocos nucifera (commonly known as coconut)-based mixed farming system	Generally covered coastal region of the Indian subcontinent	Pathak and Dagar (2000)

pastures for livestock which maintain livestock's and ecosystem health. Thus, the interactions among these tree elements in agroforestry systems are socioeconomically viable and ecologically sound that helps in enhancing better performance toward maintaining overall ecological integrity to achieve SD at the national and global context (Bugayong 2003) (Fig. 14.2). A good example, we can see in terms of nutrient conservation through efficient nutrient cycling. Leguminous nitrogen-fixing trees in an agroforestry system add some leaf litter and other residues into the soils where the microbe decomposes and decays these organic materials and releases C and other essential nutrients that can be captured by extensive root

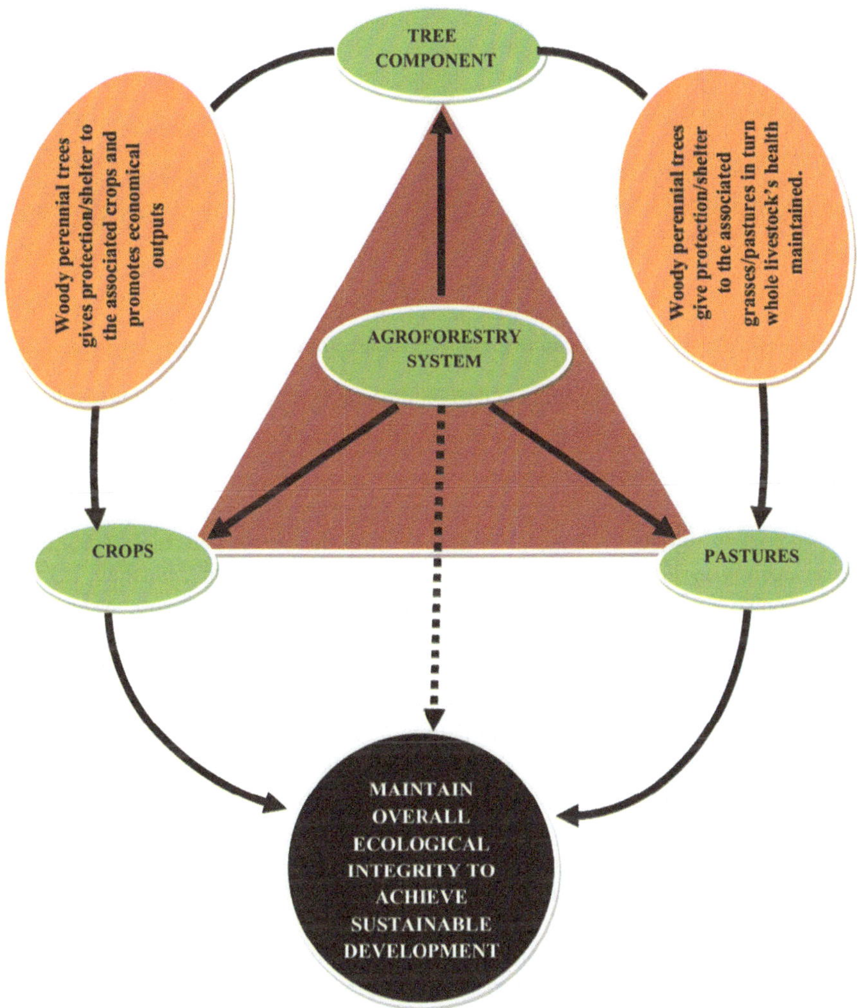

Fig. 14.2 Agroforestry performance for better ecological and sustainable development (Compiled: Bugayong 2003; Raj et al. 2020a)

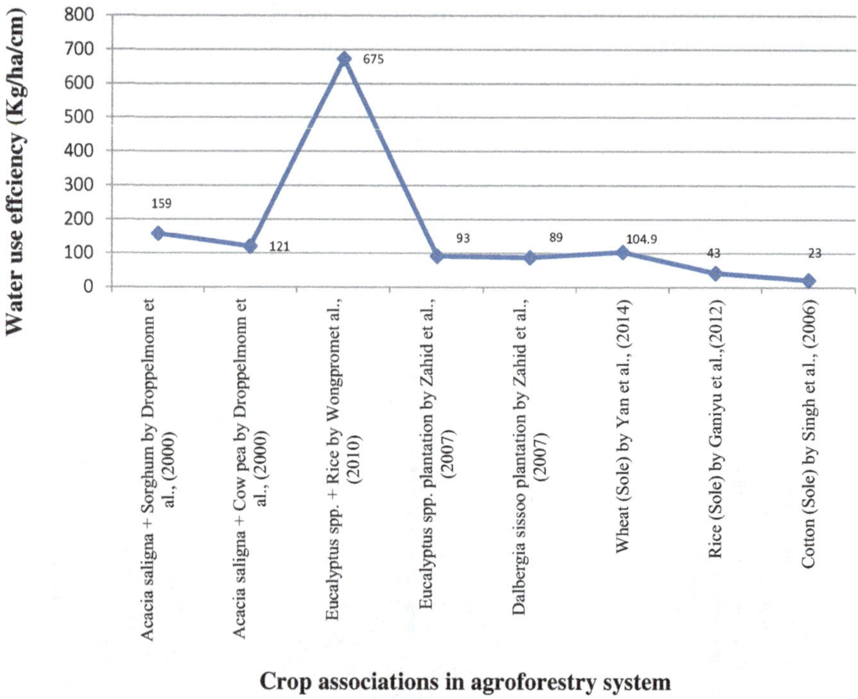

Fig. 14.3 Water resource conservation through different tree-crop associations in agroforestry practices

systems of perennial trees. The chances of nutrient loss can be minimized through a closed type of nutrient cycling, whereas in contrast these losses are highly prominent in an open type of nutrient cycling in soil cropping systems. In a nutshell, agroforestry uptake essential nutrient resources through higher extensive root systems and nutrient can be mobilized into the upper vegetation parts of tree-crop systems that helps in maintaining nutrient quality in the resulting fruits, food, vegetables, spices, etc. that make healthier people (and animals through quality and palatable fodder/pastures) and environment. This can help in achieving the SD goals through better resource conservation strategies. Soil and water conservation are another important and valuable services that were observed under well-managed agroforestry systems in varying climatic zones of the tropics (Fig. 14.3). Thus, agroforestry involves natural resource management (NRM) and promises various other uncountable services along with soil-food-climate security for better world and environmental sustainability.

Beside resource conservation technology, agroforestry is also regarded as C farming system due to capturing and storing of the atmospheric C into the vegetation and soils which makes a C balance and adds biomass which can be utilized by farmers in the form of timber and NTFPs that strengthen the farmer's socioeconomic status and livelihoods (Jhariya et al. 2018; Yadav et al. 2017).

Further tree-crop interaction determines their interim associations and their effects on resource mobilization, conservation, and management for a sustainable environment. Suitable combinations and associations of trees, crops, and animals will help in the proper and efficient utilization, conservation, and management of resources like water, nutrients, light, space, etc. for the healthy growth and development of plants which makes a better ecosystem. Of all these resources, water ranks in the prime position due to its greater importance in the management and functioning of agroforestry system. Water is involved in making ionic and nutrient solutions and also takes part in nutrient and water cycling processes (Ong et al. 2006). In this context, a question arises in the form of "How do agroforestry systems improve water availability and its use efficiency?". This question certainly has link with (a) "how can one increase the water and other nutrient use efficiency?", (b) "does agroforestry contribute in resource conservations?", and (c) "how is agroforestry involve into SD through resource conservation technology?". These all are very technical and conceptual thinking that put agroforestry in remarkable position which involvs management and conservation of water and other resources, along with their efficient and proper utilizations through higher resource use efficiency and achieve overall SD goal.

Thus, agroforestry has the greatest potential to improve barren land and resource conservation through sustainable land management practices by integrating tree, crops, and pastures/livestock simultaneously in which loss of nutrients and other resources gets minimized due to root anchorage in the soil and litter additions along with the addition of mulching and other conservation practices. This promotes SOC (soil organic carbon) pools, other essential nutrients, and microbial population and improves rhizosphere biology which frames close types of nutrient cycling, i.e., loss of leaching is minimized. Sustainable land management practices for enhancing the land fertility through nutrient cycling and water conservations along with enhancing resource use efficiency. In a nutshell, the resulted agroforestry systems could intensify the ecosystem services by enhancing biodiversity which produce multiple products as tangible and intangible ways that improve farmer's socioeconomic status. Further, it also provides food-soil-climate security through climate change mitigations that promise resource conservation, environmental sustainability, and ecological stability. In this context, Fig. 14.4 is drawn which represents the promising significance of agroforestry system in resource conservation and ecological stability (Raj et al. 2020a; Sheoran et al. 2021).

14.2.4 Animals

Integration of animals/livestock in any land use systems is more economically viable and ecologically and socially acceptable which plays a multifarious role in ecosystem processes. However, animals are interlinked with resource management and conservation. As per one estimate, livestock is represented as an important global

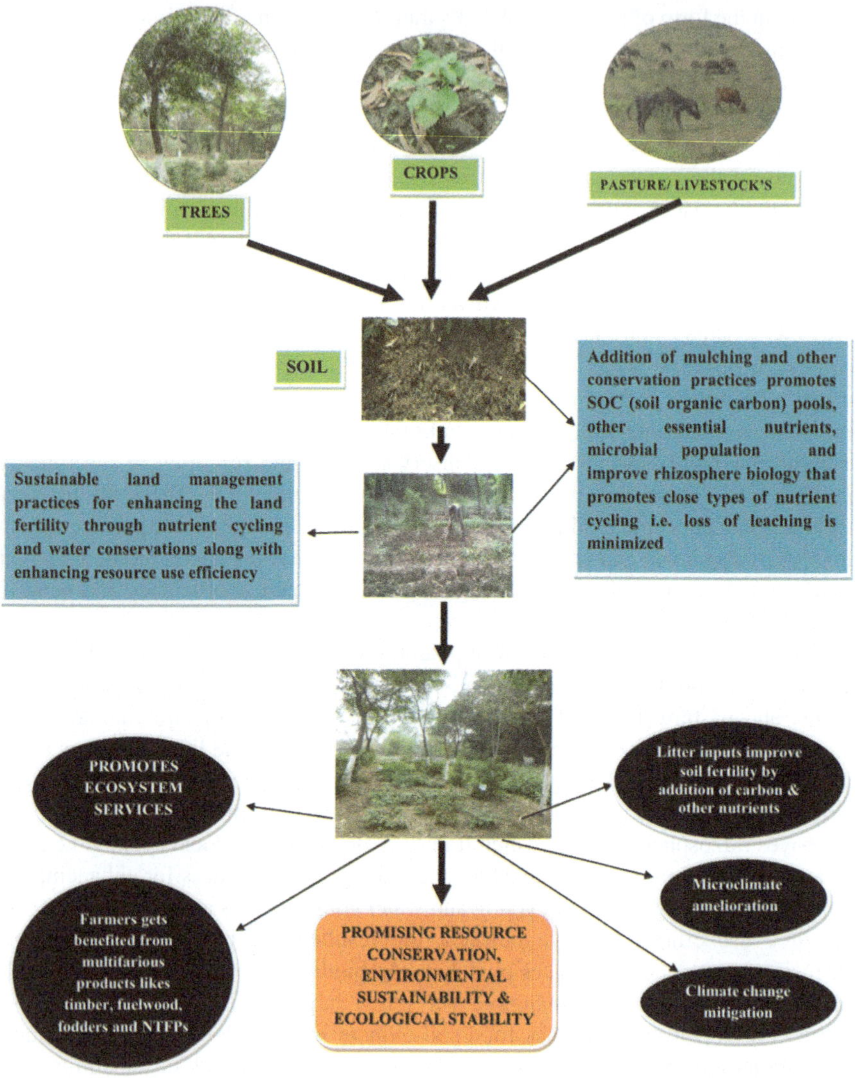

Fig. 14.4 Promising significance of agroforestry system in resource conservation and ecological stability (Compiled: Raj et al. 2020a)

asset of 1.40trillion dollar values that occupied 30% of ice-free land areas of the earth (Steinfeld et al. 2006) which provides a livelihood for about 600 million people in developing countries (Thornton et al. 2007). At present, livestock is becoming the fastest-developing sector having both positive and negative impacts on other NRs, economic growth, public health, and social security (World Bank 2009). However, an observational study used randomized control trials and empirical models to explore the relations between livestock and SD. Similarly, a positive and negative interaction exists between livestock and related SD as reflected from the other reported works (Mehrabi et al. 2020). In this context, Poore and Nemecek (2018) and Springmann et al. (2016) have studied the positive interactions between these two which is quietly focused on varying climatic impacts of unnecessary livestock's food utilizations. Approx. 26% areas of the total ice-free land surface are occupied by livestock (responsible for 13% of total GHG emissions) of which rangeland and pastures contributed 22% whereas 4% areas of cropland used for feeding purposes consumes 36% calories by plants, liable for 15% of total ground water use and contributed to water pollution about 12%, respectively (Herrero et al. 2016; Meena et al. 2020; Cassidy et al. 2013). Similarly, the populations of livestock are figured and recorded by FAOSTAT (2020), and according to this, cattle, poultry, and sheep and goat contributed 1.7, 25, and 2.2 billion populations on the earth. Integration of animals in any land use systems will intensify ecosystem services by providing tangible products in the form of meat, wool, eggs, dung cake, etc. Animals are other important NRs that build farming systems by increasing land fertility, protecting crops, and provisioning various other products. As we know, animals are integral parts of any land use farming practices which make a sustainable environment through the sustainable production of various products. Animal's excreta and dung add nutrients to the soils which improve fertility and release nutrients that are readily taken up by higher plants and provide quality food for people. Therefore, livestock's excreta add nutrient loads and improve the nutrient cycling processes that directly or indirectly affect the soil fertility status which makes healthy and quality soils (Sheldrick et al. 2002). Further, livestock's excreta works as good manures and its production and quality depends on feeding types and habit. In this context, crop residues, cereals, oil cake, varying oil seeds, etc. are the different types of livestock's concentrated feeds (de Haan et al. 1998). Animals play an important role in rangeland management for resource conservation. Rangeland areas have less fertile and degraded soils, devoid of woody perennial plants, characterized by harsh climate, and occupy dry regions where resources are minimal. These areas are utilized by animals for grazing, and the scientific management of rangeland system promotes the sustainable production of quality and palatable pastures/grasses that are feed materials of livestock and grazing animals. Thus, sustainable rangeland management practices through occupying animals would promise resource conservation along with provision of various ecosystem services that makes better ecosystem (Fig. 14.5) (Raj et al. 2020b). Therefore, integration of animals and livestock in any farming systems would be helpful in resource conservation and its efficient utilizations in proper ways. Thus, animals are directly or indirectly connected with SD.

14.2.5 Soil

Soil is the largest NRs and regarded as "soul of infinite life" that supports a variety of flora and fauna, stores essential nutrients, and harbors microbial populations that maintain rhizosphere biology. Soil promotes other resources by regulating water and nutrient cycling that not only makes nutrient-water balance but also makes it

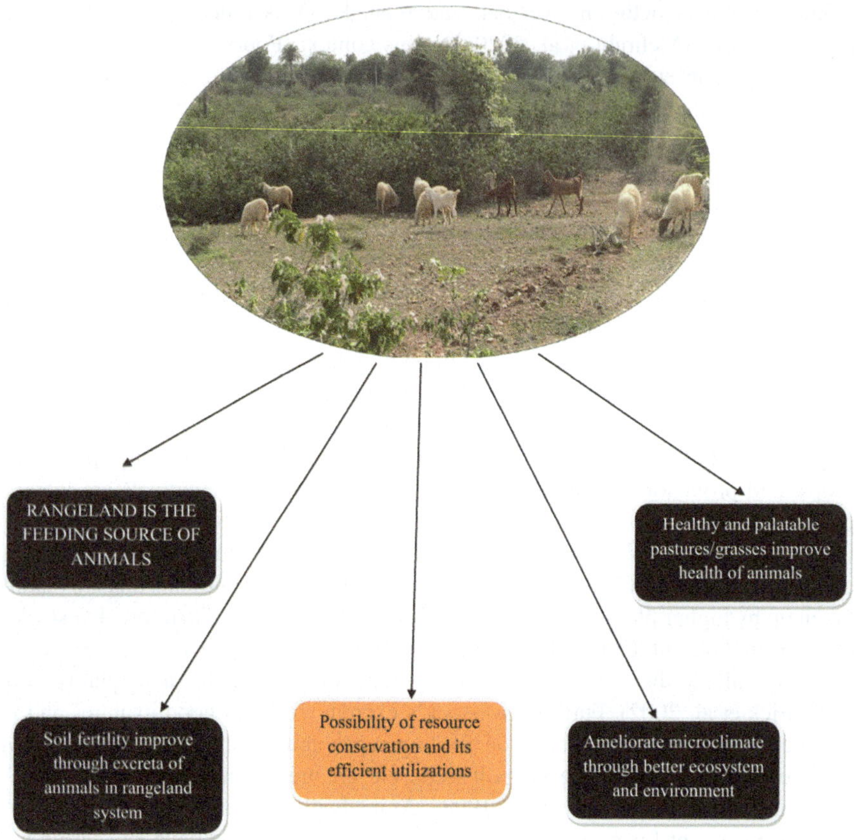

Fig. 14.5 Rangeland management for resource conservation through sustainable animal's productions (Compiled: Raj et al. 2020b)

availability to plants. Soil is one of the essential resources that stores organic C as SOC pools and essential nutrients through the decaying and decomposition of fallen leaf litter and other organic residues (Jhariya 2017a, b; Meena et al. 2020a). These nutrients can be captured by extensive root systems of higher plants which affect overall yield and productivity. The presence of organic matter and C storage in the soil depends on the rate of the decaying and decomposition of fallen plant litter and other organic residues. However, several biotic and abiotic factors play an important role in the decomposition and availability of terrestrial C pools into the soil biomes throughout the globe (Weissert et al. 2016). As per FAO and ITPs (2015), a total 1500 Pg of SOC stock was reported in the topsoil (1 meter depth) in the world, of which both peatland and wetland areas comprised the highest stock majorly in the tropical and permafrost regions (Köchy et al. 2015).

However, land degradation, unscientific land management practices, etc. are the major problems that destroy the health and quality of soils by removing the top organic layer through soil erosion (Khan et al. 2021a, b). As per FAO and ITPS (2015), around 25% of the total SOC pools were lost due to deforestation and other unsustainable land use practices. In this context, sustainable soil management practices will minimize these problems and make a sustainable environment by maintaining soil, food, and climate security (Banerjee et al. 2021a, b). Beside resource management, soil has the greatest significance for climate change mitigation through better C sequestration. Soil C sequestration is a good strategy that captures and fixes atmospheric C and maintains SOC pools that affect overall health and productivity. The extent of sequestration will depend on soil types, quality, nature, order, and suborder along with proper management practices in varying soils. If we consider the soil order, the Gelisols ranked top by storing 316 billion tons of SOC pools followed by the Inceptisols and Andisols at 190 and 20 billion tons, respectively (Eswaran et al. 2000). Similarly, tropical and boreal forest ecosystem comprises huge SOC stock as compared to other ecosystem which plays a key role in ecosystem processes and NR conservation (Pan et al. 2013). Topography, attitude, depth, and varying climatic regimes are also important parameters that affect the potential of soil C sequestration. Therefore, sustainable soil management practices will promote biodiversity which intensify ecosystem services that ensure higher yield and productivity along with resource conservation. Thus, a great nexus exists among healthy soil, resource conservation, and SD.

14.3 Sustainable Development: A Wakeup Call

The term sustainability implies the conservation and management of NRs without their further depletions and the degradation of our natural environment. Traditional knowledge systems and ecological wisdom are involved to achieve the goal of SD through resource conservation strategies without destroying the environment and sustain our natural earth ecosystem (Parotta et al. 2006; Banerjee et al. 2021c, d). The first paper on SD was published in the Brundtland Report entitled *Our Common Future* and nowadays becoming a hot topic at the global context (WCED 1987). The practices of SD depend on the principles of ensuring ecological stability and fulfilling human needs on a long-term basis without destroying environmental health. In this context, many authors have supported this principle behind the success of SD which is linked with the overall prosperity of human civilization (Holden et al. 2014).

The development of human civilization slinked with the degradation of NRs due to the unsustainable exploitation of NRs that affects overall ecosystem structure and related processes. Increasing burgeoning populations promote anthropogenic activities that result in the overuse and unsustainable utilization of NRs which affects environmental health and overall ecological stability. Resource depletion and its conservations are perfectly linked in SD. There is a great synergy between resource conservation and SD that makes a sustainable ecosphere for better environment. In this context, the question "can we achieve the goal of sustainability through resource conservation and how does a subtle change in resource management practices affect SD?" arises. This question revives in our mind which needs a good study and research to explore the better understanding of viable practices for resource conservation which helps in achieving the goal of sustainability without affecting future productions and destroying our mother nature and environment. However, SD itself represents the development at social, economic, and environmental levels (Ashoka et al. 2017). Sometimes, we forget environmental development while approaching and engaging ourselves into economic profit and development that affects our overall SD. Therefore, SD reflects a better society, sound economics, and viable environment which make a better sphere for human civilization (Reed et al. 2006). Modernization of agricultural and farming practices, deforestation activity, agricultural intensification, and overexploitation of resources exert a pressure and have negative consequences on our environment. These all are major challenges that put a pressure on policy makers and researchers to rethink over. In this context, both sustainable and ecological intensification in agriculture and other land use systems makes higher yield and productivity along with healthier environment and maintains soil-food-climate security which are prerequisite measures for ecological stability and sustainability.

14.4 Natural Resource Conservation: Rationale, Needs, and Potential

Resource depletions in agroecosystem practices are major concern today which needs effective conservation strategies and technologies for better ecosystem and ecological stability. The practices of agriculture intensification and land expansions are converting/maximizing agriculture land area with high synthetic inputs that enhance the production potential but at the cost of land degradation, resource depletions, and overall ecosystem heath through GHG emissions causing climate change. It needs practice and technology for resource conservations by promoting sustainable and ecological intensification practices in agroecosystem that helps in building environmental health and ecological integrity and achieving the goal of SD. In this context, Fig. 14.6 represents resource depletion in agriculture along with conservation strategies for SD (Kanter et al. 2018; Viccaro and Caniani 2019).

The need for NR conservation and its management is crystal clear. As we know, utilization of NRs increases due to the rising population which is expected to reach 10 billion in the year 2050. This enhances the need of NR utilization but at the cost of environment health by overexploitation of these global resources that affects the goal of sustainability. Therefore, there is a need to develop technology to ensure the proper blending and utilization of NRs such as agriculture, forestry, soils, animals, etc. which can help in establishing balance between human and NR dynamics. Thus, research and development is needed to explore the different fields of knowledge regarding sustainable agricultural and forest management that can focus on soil and climate health, wildlife management, sustainable land use change, and pollution management for maintaining environmental sustainability and ecological stability (Viccaro and Caniani 2019). However, the sustainable forest management (SFM) is a viable tool that intensifies the ecosystem services by enhancing biodiversity (Sacchelli and Bernetti 2019).

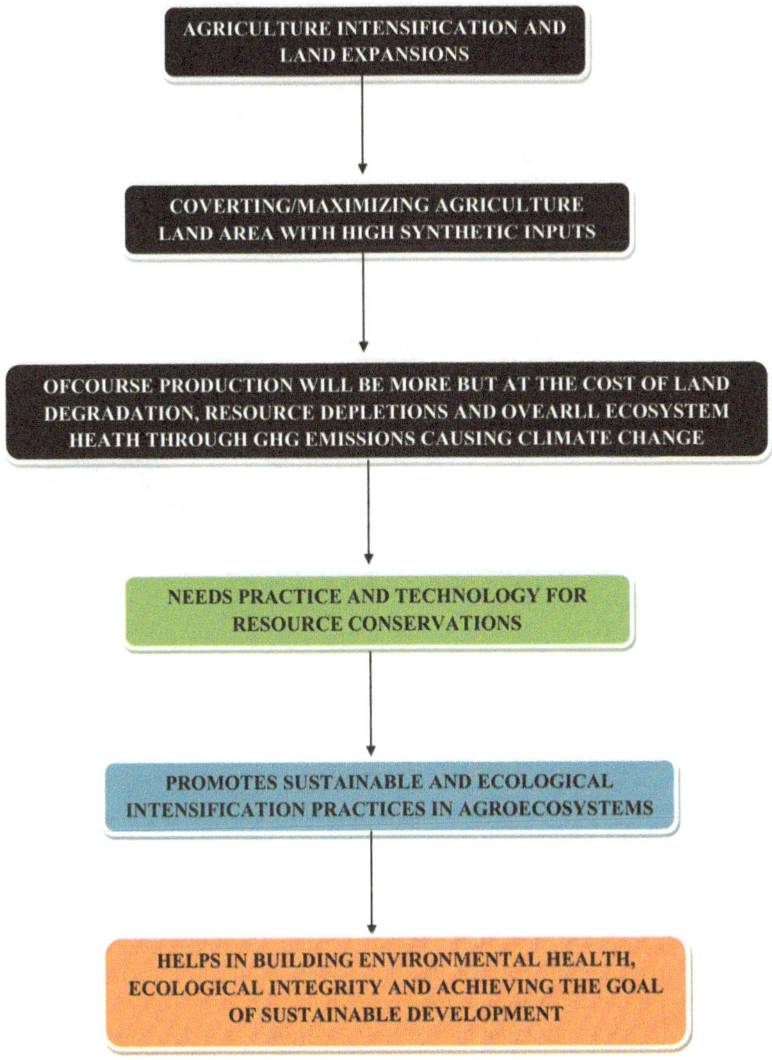

Fig. 14.6 Resource depletion in agriculture and their conservation strategies for sustainable development (Compiled: Kanter et al. 2018; Viccaro and Caniani 2019)

14.5 Sustainable Forest Management for Environmental Sustainability

Deforestation, overexploitation of forest resources, illicit felling of timbers, mining activity, etc. are the major challenges today that we need to address for better environment. In this context, the practice of SFM is a good and viable strategy

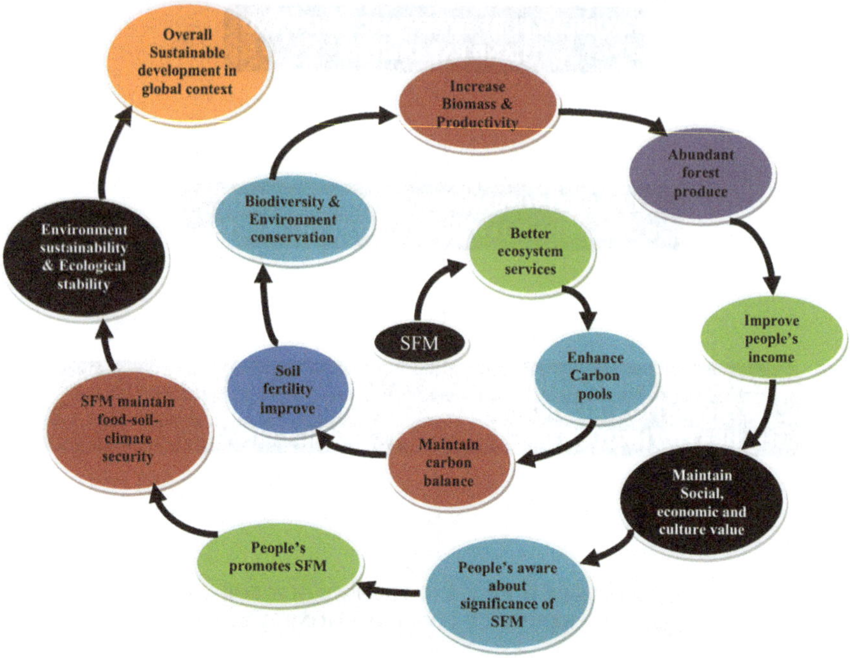

Fig. 14.7 SFM performance for better ecosystem and sustainable development (Compiled: FAO 2016; UN 2017; Pirlot et al. 2018)

that can address all these challenges and promotes biodiversity which intensifies ecosystem services and maintains overall environmental sustainability. In this context, various questions arise such as "Is SFM viable for maintaining environmental sustainability and ecological stability?" and "How does SFM build the dreams of SD at global scale?". These questions created various theories among the scientific community, and it is cent percent true that SFM is an integrated and regulatory approach that provides various forest products in sustainable ways without destroying other NRs and maintains overall environmental sustainability and ecological stability at global scale. SFM is a winner of today that assures the goal of SD. In this context, Fig. 14.7 is drawn which represents SFM performance for better ecosystem and SD (FAO 2016; UN 2017; Pirlot et al. 2018).

Nowadays, SFM is gaining wider importance due to its integrated nature of approach behind successful forest management and its protection along with resource conservation for better ecosphere (Ong and Swallow 2003). SFM is gaining wider importance not only in the Asian continent; even European countries consider it as a technique for society development and way forward to overall SD of the nation (Forest Europe 2011). This system is also initiated by global agencies such as ITTO (International Tropical Timber Organization), UN Conference on

Environment and Development, and other national and international agencies (Varma et al. 2017).

SFM is an ecological approach which represents the management, protection, and conservation of forest that promote various areas including afforestation, reforestation, improved agroforestry, and plantation activity by better silvicultural practices wherever and whenever possible. These approaches certainly affect forest health and productivity along with overall C sink capacity and SOC pools. SFM approaches increase the C sink capacity with the overall maintenance of ecological integrity, biodiversity conservation, and environmental sustainability which is enough to achieve the dream of SD without destroying future resources and its degradations. SFM jointly addresses various issues and challenges such as mitigating changing climate through better C sequestration, biodiversity conservation, and resource conservation, whereas it also solves the problems of forest fringe people by maintaining social, culture, and income security through a greater provision of tangible products on a sustainable basis (Chandel et al. 2017). Thus, SFM is one step toward achieving the sustainability of forest that is directly or indirectly linked with overall SD. Therefore, it is a systematic approach that considers a better management and practices for NR conservation and its sustainable utilization without disturbing our natural ecosystem.

14.6 Sustainable Agricultural Practices

Increasing population exerted a pressure of food production for feeding billions of people worldwide that can be possible by intensifying agricultural practices but at the cost of environmental health (Shukla and Dwivedi 2015). High synthetic inputs and additions of inorganic fertilizers can enhance the yield and bumper productions, but in due course of time, it exerts pressure on land. This, therefore, causes land degradation, loss of soil fertility, health issues, less biodiversity, and poor ecosystem services that affect the overall food-soil-climate security. As a result, the overall ecosystem gets jeopardized. In this context, the query "can sustainability be possible in agricultural intensification?" arises. Believe me, the term "sustainability" is absolutely zero, while we practice agricultural intensification that harms our ecosystem and disturbs our environmental sustainability and ecological stability. In this context, ecology-oriented farming practices comprise judicious utilization of all NRs such as water, soil, light, etc. without any synthetic inputs which helps in bumper productions on a sustainable basis. Both ecological and sustainable intensification play an important role in maintaining environmental sustainability and ecological stability that helps in achieving the goal of SD. However, sustainable and ecological intensification not only enhances the yield and productivity but also helps in the management and conservation of NRs that maintain the overall environmental sustainability. However, both Swaminathan (1995) and Thompson (1996) have emphasized the agricultural sustainability which can be ensured through efficient nutrient cycling process, greater nitrogen-fixing ability of integrated leguminous plants, better insect-pest control measures, and less use of non-renewable resources.

This, therefore, enhances farmer's wisdom and their strength in problem analyzing along with its adaptation and mitigation measures at farm level. This minimizes the unbiased use of NRs and promotes resource conservation techniques through local traditional knowledge systems. It also contributes toward raising awareness among farmers to use scientific-based farming systems that not only help in enhancing socioeconomic status but also improve the environment to achieve the SD goal.

14.7 Agroforestry Management for Ecological Sustainability

Indeed, overexploitation and depletion of NRs affect environmental sustainability and disturb the overall ecosystem (Ekosse 2009). A better management of agroforestry definitely enhances the overall productivity through ecosystem services that maintain the overall ecological sustainability. However, judicious application and blending of NRs in ecologically oriented way enhance proper resource utilization and promote yield and productivity in sustainable ways (Mukoni 2015). For example, the combination of *Citrus reticulate* (mandarin orange) + *Amomum subulatum* (cardamom plant) in agri-horticultural-based systems was a promising approach to reclaim degraded and fragile mountainous regions of Sikkim and contributed to the management and conservation of NRs in the north-eastern regions in the Indian subcontinent (Mishra and Rai 2013). Similarly, integrating nitrogen-fixing leguminous MPTs with *Triticum aestivum* (wheat) improved the soil fertility and maintains physicochemical properties in the Tarai regions of Uttarakhand (Sarvade et al. 2014b, c). However, different models of agroforestry systems have the capacity to minimize the losses of resources such as nitrogen content and soil erosions by 20 and 1–10%, respectively (Udawatta et al. 2002). Similarly, some MPTs, such as *Sesbania grandiflora* (agati), *Erythrina abyssinica*, *Acacia* spp., and *Euphorbia* spp., have a greater importance in NR conservation which is used as live fencing in the Gujarat state in India (Mathukia et al. 2016). Thus, it is quite clear that suitable tree-crop combination and blending of NRs in agroforestry systems would be helpful in both conservation and management of NRs which can be the base of sustainability.

14.8 Sustainable Soil Management for Food-Climate Security and Sustainability

Unsustainable land management, unscientific farming practices, deforestation, high synthetic inputs, intensive agricultural practices, low inputs of bio-fertilizer, etc. destroy the soil fertility and affect overall health and quality. These can be reverted through the practices of ecologically oriented sustainable soil management practices which not only maintain SOC but also improve fertility status, nutrient pools, and rhizosphere biology. In turn, soil microbial populations and other resources are enhanced that promotes yield and productivity to make a sustainable environment. Diversified flora and fauna due to sustainable soil systems would intensify ecosystem services that maintain overall soil-food-climate security. Therefore, land productivity enhancement and topsoil conservation by reducing erosions and promoting nutrient reserves and related microbial populations are the important achievements of sustainable soil management. Thus, the query as "how can we maintain these uncountable significant effects through soil resources in sustainable ways?" followed by "can we achieve these multifarious services through the practices of sustainable soil management?" as well as "can the practices of sustainable soil management help in resource conservation and SD?" appears. It is indeed cent percent right and already discussed in the previous section that sustainable soil management makes a sustainable environment which is the pillar for SD. A better soil management practice helps in maintaining soil C pools that is a good indicator of soil fertility. Similarly, soil C sequestration is effectively possible and can be enhanced by sustainable soil management practices. SOC pools and soil fertility are mostly managed than the other NRs through the regulation of water and nutrient cycling in effective ways that makes a healthy ecosystem for a better future. Similarly, healthy soil is a key for food and nutritional security which is more feasible for environmental sustainability program that can fulfill the goal of SD. The SOC stock helps to maintain the productivity of the agroecosystem managed by African farmers by compensating the fertilizer's shortage and improving soil fertility status for more productions (Lal 2004). Thus, rich SOC-based agricultural practices can enhance the overall productions by 17.60 Mt. yr.$^{-1}$ and can be mediated through sustainable management of soil (De Moraes Sá et al. 2017; Meena et al. 2020b).

In a nutshell, sustainable soil management practices are used for conserving water and other nutrients. It can be improved through additions of mulching, leaf litters, and other residues into the soils. This leads to the intensification of ecosystem services such as improving fertility through higher SOC pools and climate change mitigation, improving microbial populations and rhizosphere biology, and enhancing nutrient use efficiency and resource conservations. This, in turn, maintains ecological integrity and promotes food-climate security and its sustainability for achieving the goal of SD. In this context, Fig. 14.8 is drawn which depicts soil management practices for resource conservations, food-climate security, and SD (Raj et al. 2019).

14.9 Resource Conservation for Ecosystem Services

The conservation and management of resources are prerequisites that deliver uncountable and multifarious ecosystem services. Diverse forms of NRs surely diversify biodiversity that intensify ecosystem services. These services are delivered in both direct and indirect ways that maintains economical gains along with soil-food-climate security. We utilize timber, fuelwood, firewood, fodder (for livestock) NTFPs, medicinal plants, aromatic plants, food grains, flowers, nutritious fruits, etc. as direct benefits, whereas conservation of resources, soil enrichment through fertility enhancement, SOC pools, healthy rhizosphere biology, enhancing microbial

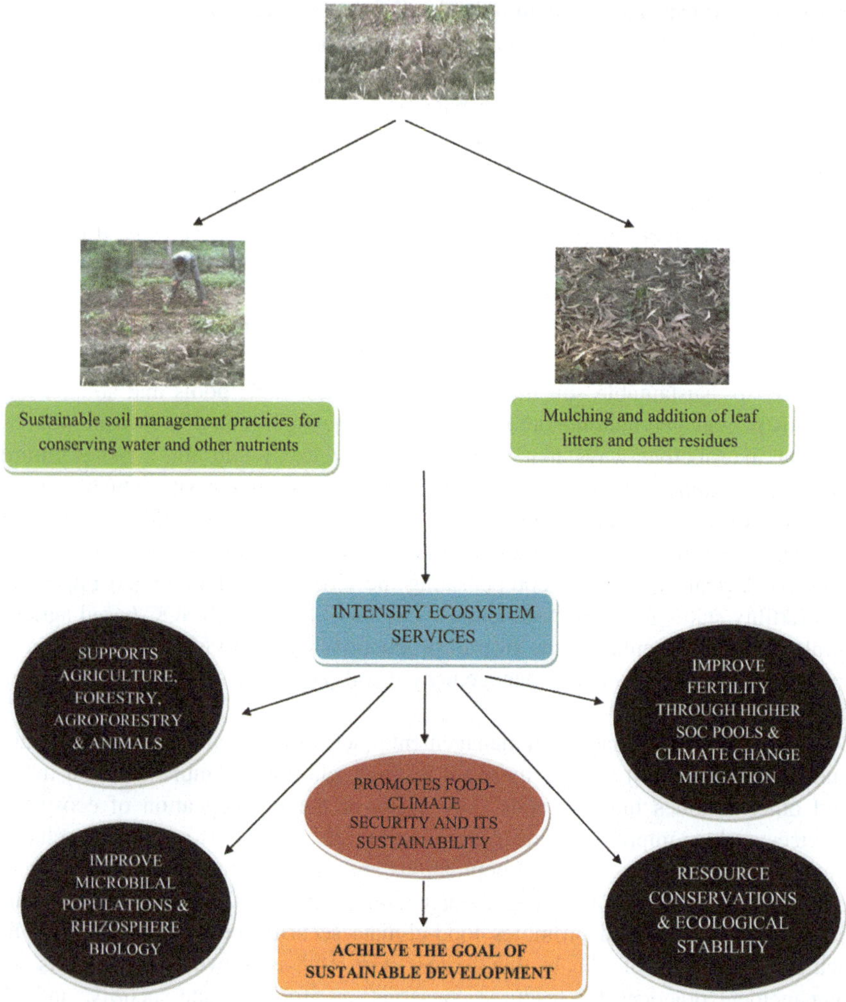

Fig. 14.8 Soil management for resource conservations, food-climate security, and sustainable development (Compiled: Raj et al. 2019)

population, better nutrient use efficiency, watershed management, efficient nutrient and water cycling, climate change mitigation through C sequestration, etc. are the indirect services. Farmer's livelihood security and socioeconomic and cultural developments are the other significant achievements that could be achieved through resource management and its conservation. Thus, conserving resources is gaining prime importance that can build our dreams of SD at global scale. Thus, ecosystem services can be intensified by promoting NR conservation by empowering community economy, enhancing productivity, and abating pollution with the overall

Table 14.3 Resource conservation-based case studies in different regions of the world

Resources	Reported case studies in regions/country	References
Water resource	Community-based water supply practices in the region of Honduras (Central America)	St Jacques (2009)
	NRM practices comprised holistic watershed management in Sukhomajri region of India	Islam and Jain (2011)
	Water resource management practices comprised Mara River water users association (MRWA) in the region of Kenya	UNDP (2012a)
	Irrigation-based agriculture sector project practiced in the region of Kenya	Islam and Jain (2011)
	The practices comprised water catchment project for water resource conservation in the region of St Lucia	St Jacques (2009)
	NRM-based locally managed marine area network (LMMAN) in the region of Fiji	UNDP (2012c)
	The practices of regional awareness of Cameroon Island (RACI) in Malaysia	St Jacques (2009)
Forest resource	The practices of joint forest management (JFM) for forest resource management throughout the regions of the Indian subcontinent	D'silva and Nagnath (2002)
	NRM-based conservation forest management (CFM) in Mexico	Bray (2003)
	A popular "monk community forest" (MCF) in the region of Cambodia	UNDP (2012b)
	Kanghua community development Center (KCDC) in China provenance	UNDP (2013)
	SFM practices in Kimana community wildlife sanctuary (KCWS) situated in the region of Kenya	Kellert et al. (2000)
Animal resource	A popular "conservancy and wildlife management (CWM)" in the region of Namibia	Jones (1999)
	NRM-based CBNRM in Okavango Delta of the Botswana	Mbaiwa (2012)
	NRM-based CBNRM in Annapurna and Makalu Barun of the region Nepal	Kellert et al. (2000)
	Impact of CAMPFIRE on local community in Zimbabwe	Mutandwa and Gadzirayi (2007)
	NRM-based conservation of animals around Lake Mburo National Park in the Uganda region of the African continent	Emerton (1999)
Fish resource	The project entitled "cooperative management of north American Pacific salmon" in Alaska	Kellert et al. (2000)
	The project entitled "cooperative management of north American Pacific salmon" in the region of Washington, DC	Kellert et al. (2000)
Wetland resource	A famous "Sepik wetlands management initiative (SWMI)" practiced in the region of Papua New Guinea	UNDP (2012d)

maintenance of environmental sustainability (Surya et al. 2020). Moreover, resource conservation-based case studies in different regions of the world are depicted in Table 14.3 which would help in enhancing ecosystem services and maintaining ecosystem stability at global scale.

14.10 Research and Development

There is an urgent need to develop approaches and management strategies that should combine both developmental efforts and conservation measures of the NRs. This would improve, maintain, and protect the natural environment and its resources for the benefit of all mankind. NRs are finite, limited, and capable of being destroyed by unsustainable use, and this can be a limiting factor on SD. Existence of some of the NRs has been there in the past, but this might not be the case in the future as it depends upon their mode of utilization. Hence, environmental education on the characteristics of NRs is required if they have to be managed in a sustainable manner so that they do not become limiting factors to SD. In this context, an integrated approach of environmental educations is a good strategic tool that creates awareness among the people for practicing the sustainable utilizations of NRs without destroying the environment. However, deficient human awareness about environmental education, bad attitude, and human greed are the major detrimental factors that affect resource management and its conservations (Matiasi 2006). Moreover, a social-ecological-based research and development would harness both environmental and ecological sustainability that promise SD at global scale (Selomane et al. 2019). A better research and development is needed to improve the practices of GHG inventory and climate modeling that would help in mitigating changing climate by reducing the emission of GHGs. Better and effective utilizations of NRs are also an important strategy that reduces the environmental problems and maintains SD.

14.11 Policy and Regulatory Frameworks

Resource deletions and its overexploitations deprive the global health and wealth. Land degradation is the major problem that affects health and quality of soil which is directly connected with the yield and productivity. Unsustainable land use practices deprive soil fertility and affect microbial populations and the overall flora and fauna diversity which play a major role in soil-food-climate security. In this context, policy must be framed to explore the better understanding of various environmental and ecofriendly practices that maintain our NRs.

Also, time-to-time scientific- and ecology-oriented practices are needed to understand tree-crop interaction, a judicious combination and utilization of NRs (such as water, soil, light, fertilizer, etc.), and identification of sustainable land use practices that would help in resource conservation. A regulatory framework must be drawn for achieving the goal of sustainability through better and efficient management practices of NRs that maintains environmental sustainability and ecological stability. Resource conservation is also connected with food and nutritional security that take part in a healthier ecosystem while mitigating the problems of hunger and malnutrition.

A policy must be established in accordance to promote SFM and sustainable soil management in regular ways. These practices would help in managing and conserving resources at present and future perspective without disturbing our environment. Policies must be reformed in accordance for achieving SD through SFM practices. Deforestation, illicit timber cutting, encroachment, declining productivity and timber quality, and poor regeneration status overall affect forest health and productivity. However, all these events are variable in nature in relation to forest issues and problems which needs effective policy, good governance, scientific research, and effective roadmap development. This is also important for implementing future strategies by adopting SFM that helps in achieving the goal of SD through biodiversity conservation, socioeconomic improvement, environmental security, and ecological stability (Fig. 14.9) (Pirlot et al. 2018). Thus, joint implementations of both technological and environmental policies are the pillar of SD that secures our ecosystem processes and environmental services (Peterson 2019).

14.12 Conclusions

Indeed, resources play an important role in ecosystem processes which sustain lives and build the dream of SD. Conservation of resources through better management practices and viable technology maintains soil-food-climate security, environmental sustainability, and overall ecological stability. Overexploitation of NRs is major greed today that must be checked for ecosystem and environmental health. In this context, the practices of SFM and sustainable soil management are becoming good strategies that not only enhance biodiversity and intensify ecosystem services but also promote the resource management and its conservation which is the pillar of SD. A policy must be framed in accordance to cause and factors of today's resource depletion and its conservation by exploring effective management practices that will build the nations by making environmental sustainability and ecological stability.

14.13 Future Perspectives

Resource depletions are a major concern today that is discussed enormously by policy makers, stakeholders, academicians, and scientists at national and international levels. Various anthropogenic activities like deforestation and intensified agricultural practices are becoming bone in neck while we are talking about sustainability. The goal of sustainability is totally dependent upon social, economic, and environmental development which is directly or indirectly connected with resource conservation (Niesenbaum 2019). Destroying resources are major hurdles behind the success of SD. In this context, an attention must be drawn to ecology-oriented sustainable practices in land use systems along with minimizing resource depletions in present and future perspective. SFM and sustainable soil management

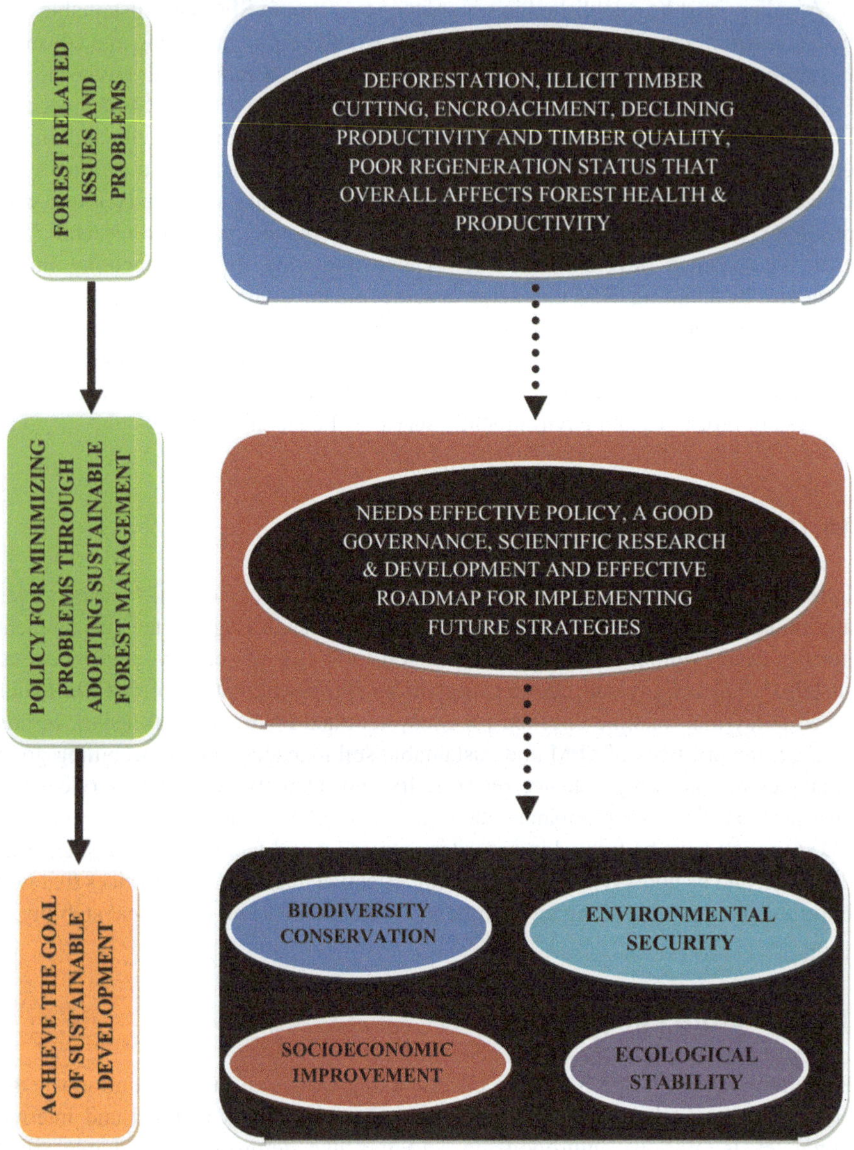

Fig. 14.9 Policies for achieving sustainable development through SFM practices (Compiled: Pirlot et al. 2018)

practices must be aimed toward future perspective that helps in maintaining sustainable uses of resources without destroying our environment. Thus, our future lies on our environmental-friendly practices that not only conserve resources but also build the path for achieving the SD goals.

References

Alexandratos N, Bruinsma J (2012) World agriculture towards 2030/2050: the 2012 revision. ESA working paper no 12–03. FAO, Rome

Ashoka P, Meena RS, Kumar S, Yadav GS, Layek J (2017) Green nanotechnology is a key for eco-friendly agriculture. J Clean Prod 142:4440–4441

Avellan T, Meier J, Mauser W (2012) Are urban areas endangering the availability of rainfed crop suitable land? Remote Sensing Letters 3:631–638

Banerjee A, Jhariya MK, Meena RS, Yadav DK (2021b) Ecological footprints in agroecosystem- an overview. In: Banerjee A, Meena RS, Jhariya MK, Yadav DK (eds) Agroecological footprints management for sustainable food system. Springer Nature, Singapore, pp 1–23. eBook ISBN: 978-981-15-9496-0, Hardcover: 978-981-15-9495-3. https://doi.org/10.1007/978-981-15-9496-0_1

Banerjee A, Jhariya MK, Raj A, Yadav DK, Khan N, Meena RS (2021c) Land footprint management and policies. In: Banerjee A, Meena RS, Jhariya MK, Yadav DK (eds) Agroecological footprints management for sustainable food system. Springer Nature, Singapore, pp 221–246. eBook ISBN: 978-981-15-9496-0, Hardcover: 978-981-15-9495-3. https://doi.org/10.1007/978-981-15-9496-0_7

Banerjee A, Jhariya MK, Raj A, Yadav DK, Khan N, Meena RS (2021d) Energy and climate footprint towards the environmental sustainability. In: Banerjee A, Meena RS, Jhariya MK, Yadav DK (eds) Agroecological footprints management for sustainable food system. Springer Nature, Singapore, pp 415–443. eBook ISBN: 978-981-15-9496-0, Hardcover: 978-981-15-9495-3. https://doi.org/10.1007/978-981-15-9496-0_14

Banerjee A, Jhariya MK, Yadav DK, Raj A (2020) Environmental and sustainable development through forestry and other resources. Apple Academic Press Inc., CRC Press- a Taylor and Francis Group, Oakville, p 400. ISBN: 9781771888110. https://doi.org/10.1201/9780429276026

Banerjee A, Meena RS, Jhariya MK, Yadav DK (2021a) Agroecological footprints management for sustainable food system. Springer Nature, Singapore, eBook ISBN: 978-981-15-9496-0, Hardcover: 978-981-15-9495-3, p 514. https://doi.org/10.1007/978-981-15-9496-0

Becker M, Nieminen TM, Geremia F (1994) Short-term variations and ¨ long-term changes in oak productivity in northeastern France. The role of climate and atmospheric CO_2. Ann des Sci For 51:477–492

Bhattacharyya R, Ghosh BN, Mishra PK, Mandal B, Rao CS, Sarkar D, Das K, Anil KS, Lalitha M, Hati KM (2015) Soil degradation in India: challenges and potential solutions. Sustainability 7:3528–3570

Bhumbla DR, Khare A (1984) Estimate of wastelands in India. Society for Promotion of Wastelands Development; Allied, New Delhi, p 18

Bray DB (2003) Mexico's community managed forests as a model for sustainable landscapes. Conserv Biol 17(3):672–677

Bugayong LA (2003) Socioeconomic and environmental benefits of agroforestry practices in a community-based Forest management site in the Philippines. In: Proceedings of the international conference on rural livelihoods forests and biodiversity, Bonn, Germany, 19–23 may 2003

Cassidy ES, West PC, Gerber JS, Foley JA (2013) Redefining agricultural yields: from tonnes to people nourished per hectare. Environ Res Lett 8:034015

Chandel AS, Yadav DK, Jhariya MK (2017) Economically and traditionally important non timber forest products of Sarguja division. Bull Environ Pharma Life Sci 6(5):32–39

Collatz GJ, Ball JT, Grivet C, Berry JA (1991) Physiological and environmental regulation of stomatal conductance, photosynthesis and transpiration: a model that includes a laminar boundary layer. Agric For Meteorology 54:107–136. https://doi.org/10.1016/0168-1923(91)90002-8

D'silva E, Nagnath B (2002) Behroonguda: a rare success story in joint forestry management. Econ Political Weekly 37(6):551–557

Dagar JC, Singh AK, Arunachalam A (2014) Agroforestry systems in India: livelihood security & ecosystem services. Advances in Agroforestry 10:400

De Haan C, Steinfeld H, Blackburn H (1998) Livestock and the environment: finding a balance. In: Study sponsored by Commission of the European Communities. Food and Agricultural Organisation of the United Nations, and the World Bank, Rome, p 115. www.fao.org/docrep/x5303e/x5303302.htm

De Moraes Sá JC, Lal R, Cerri CC, Lorenz K, Hundria M, Cesar P, Carvalho F (2017) Low-carbon agriculture in South America to mitigate global climate change and advance food security. Environ Int 98:102–112

Dhiman RC (2012) Transforming rural Uttar Pradesh through integrating tree culture on farm land: a case study of WIMCO's poplar programme. Lucknow Manage Assoc Conv J 8(1):85–98

Dhruvanarayan VVN, Ram B (1983) Estimation of soil erosion in India. J Irrig Drain Eng 109:419–434

Ekosse GE (2009) Aspects of research associated with education for sustainable development in the 21st century. In: Paper presented at the 36th international conference of the southern African Society for Education (SASE), East London, South Africa, 1–3 October. Walter Sisulu University, Mthatha

Emerton L (1999) Balancing the opportunity costs of wildlife conservation for communities around Lake Mbsuro National Park, Uganda. In: Evaluating Eden discussion paper no. 5. IIED, London

Eswaran H, Reich FP, Kimble JM, Beinroth FH, Padamnabhan E, Moncharoen P (2000) Global carbon stocks. In: Lal R, Kimble JM, Eswaran H, Stewart BA (eds) Global climate change and Pedogenic carbonates. CRC/Lewis, Boca Raton

FAO (2006) Global Forest resources assessment- Progress towards sustainable forest management. FAO forestry paper 147. Food and Agriculture Organisation of the United Nations, Rome

FAO (2016) Global forest resources assessment 2015: how are the world's forests changing? Food and Agriculture Organization of the United Nations (FAO), Rome

FAO and ITPS (2015) Status of the World's soil resources. Rome

FAOSTAT (FAO) (2020). http://www.fao.org/faostat/en/#home

Forest Europe (2011) Implementation of the Forest Europe commitments-national and Pan-European actions 2008–2011. Forest Europe, Oslo

Galhena DH, Freed R, Maredia KM (2013) Home gardens: a promising approach to enhance household food security and wellbeing. Agric Food Secur 2:8. https://doi.org/10.1186/2048-7010-2-8

Ghosh M, Ghosal S (2019) Historical geography of forestry and forest culture in sub-Himalayan West Bengal, 1757-2015. Space and Culture India 6(5):215–227. https://doi.org/10.20896/SACI.V6I5.393

Gielen B, Calfapietra C, Sabatti M, Ceulemans R (2001) Leaf area dynamics in a closed poplar plantation under free-air carbon dioxide enrichment. Tree Physiol 21:1245–1255

Gill RIS, Singh B, Kaur N, Sangha KS (2010) Agroforestry - a viable option for crop diversification in Punjab. Department of Forests and Wildlife Preservation, Punjab, p 20

Hansen J, Kharecha P, Sato M, Masson-Delmotte V, Ackerman F, Beerling DJ, Hearty PJ, Hoegh-Guldberg O, Hsu SL, Parmesan C, Rockstrom J, Rohling EJ, Sachs J, Smith P, Steffen K, Van Susteren L, Von Schuckmann K, Zachos JC (2013) Assessing "dangerous climate change": required reduction of carbon emissions to protect young people, future generations and nature (JA Añel, Ed.). PLoS One 8(12):e81648

Hegde NG (2005) Tree based Farming for Sustainable Livelihood at the workshop on Tropical Forestry and Poverty Eradication. Bangkok, Thailand. Apr. 5–6

Herrero M, Henderson B, Havlík P, Thornton PK, Conant RT, Smith P, Wirsenius S, Hristov AN, Gerber P, Gill M, Butterbach-Bahl K, Valin H, Garnett T, Stehfest E (2016) Greenhouse gas mitigation potentials in the livestock sector. Nat Clim Chang 6:452–461

Holden E, Linnerud K, Banister D (2014) Sustainable development: our common future revisited. Glob Environ Chang 26:130–139

Hussain M, Kubiske ME, Connor KF (2001) Germination of CO_2- enriched *Pinus taeda* L. seeds and seedling growth responses to CO_2 enrichment. Funct Ecol 15:344–350

Islam KMB, Jain CK (2011) Community based water resources in north East India: lessons from a global context. Allied New Delhi, 3-12

Jhariya MK (2017a) Vegetation ecology and carbon sequestration potential of shrubs in tropics of Chhattisgarh, India. Environ Monit Assess 189(10):518. https://doi.org/10.1007/s10661-017-6246-2

Jhariya MK (2017b) Influences of forest fire on forest floor and litterfall dynamics in Bhoramdeo wildlife sanctuary (C.G.), India. J Forest Environ Sci 33(4):330–341. https://doi.org/10.7747/JFES.2017.33.4.330

Jhariya MK, Banerjee A, Meena RS, Yadav DK (2019a) Sustainable agriculture, Forest and environmental management. Springer Nature, Singapore, p 606eISBN: 978-981-13-6830-1, Hardcover ISBN: 978-981-13-6829-5. https://doi.org/10.1007/978-981-13-6830-1

Jhariya MK, Bargali SS, Raj A (2015) In: Zlatic M (ed) Possibilities and perspectives of agroforestry in Chhattisgarh, precious forests – precious earth. InTech, LondonISBN:978-953-51-2175-6, pp 237–257. https://doi.org/10.5772/60841

Jhariya MK, Meena RS, Banerjee A (2021a) Ecological intensification of natural resources for sustainable Agriculture.Springer nature Singapore. eISBN: 978-981-334-203-3, hardcover ISBN: 978-981-334-206-6. https://doi.org/10.1007/978-981-33-4203-3

Jhariya MK, Meena RS, Banerjee A (2021b) Ecological intensification of natural resources towards sustainable productive system. In: Ecological intensification of natural resources for sustainable agriculture. Springer, Singapore. eBook ISBN 978-981-334-203-3. Hardcover ISBN: 978-981-334-202-6. https://doi.org/10.1007/978-981-33-4203-3_1

Jhariya MK, Singh L (2020) Herbaceous diversity and biomass under different fire regimes in a seasonally dry forest ecosystem. Environ Dev Sustain 22:1–19. https://doi.org/10.1007/s10668-020-00892-x

Jhariya MK, Yadav DK, Banerjee A (2018) Plant mediated transformation and habitat restoration: phytoremediation an eco-friendly approach. In: Gautam A, Pathak C (eds) Metallic contamination and its toxicity. Daya Publishing House, A Division of Astral International Pvt. Ltd, New Delhi, pp 231–247. ISBN: 9789351248880

Jhariya MK, Yadav DK, Banerjee A (2019b) Agroforestry and climate change: issues and challenges. Apple Academic Press Inc., CRC Press- a Taylor and Francis Group, Toronto. ISBN:978-1-77188-790-8 (Hardcover), 978-0-42957-274-8 (E-book), p 335. https://doi.org/10.1201/9780429057274

Jhariya MK, Singh L (2021) Effect of fire severity on soil properties in a seasonally dry forest ecosystem of Central India. Int J Environ Sci Technol:1–12. https://doi.org/10.1007/s13762-020-03062-8

Jones B (1999) Rights, revenues and resources: the problems and potential of conservancies as community wildlife management institutions in Namibia. In: Evaluating Eden, discussion paper no. 2. IIED, London

Jose S, Dollinger J (2019) Silvopasture: a sustainable livestock production system. Agrofor Syst 93:1–9

Kanter DR, Musumba M, Wood SL, Palm C, Antle J, Balvanera P, Dale VH, Havlik P, Kline KL, Scholes RJ, Thornton P, Tittonell P, Andelman S (2018) Evaluating agricultural trade-offs in the age of sustainable development. Agric Syst 163:73–88

Kellert SR, Mehta JN, Ebbin SA, Lichtenfeld LL (2000) Community natural resource management: promise, rhetoric and reality. Society Nat Resources 13:705–713

Khan N, Jhariya MK, Raj A, Banerjee A, Meena RS (2021a) Soil carbon stock and sequestration: implications for climate change adaptation and mitigation. In: Jhariya MK, Meena RS, Banerjee A (eds) Ecological intensification of natural resources for sustainable agriculture. Springer, Singapore. eBook ISBN 978-981-334-203-3. Hardcover ISBN: 978-981-334-202-6. https://doi.org/10.1007/978-981-33-4203-3_13

Khan N, Jhariya MK, Raj A, Banerjee A, Meena RS (2021b) Eco-designing for sustainability. In: Jhariya MK, Meena RS, Banerjee A (eds) Ecological intensification of natural resources for sustainable agriculture. Springer, Singapore. eBook ISBN 978-981-334-203-3. Hardcover ISBN: 978-981-334-202-6. https://doi.org/10.1007/978-981-33-4203-3_16

Khan N, Jhariya MK, Yadav DK, Banerjee A (2020a) Herbaceous dynamics and CO_2 mitigation in an urban setup- a case study from Chhattisgarh, India. Environ Sci Pollut Res 27(3):2881–2897. https://doi.org/10.1007/s11356-019-07182-8

Khan N, Jhariya MK, Yadav DK, Banerjee A (2020b) Structure, diversity and ecological function of shrub species in an urban setup of Sarguja, Chhattisgarh, India. Environ Sci Pollut Res 27 (5):5418–5432. https://doi.org/10.1007/s11356-019-07172-w

Kishwan J, Kumar D (2013) Future of Poplars in India. Rome, Italy: Food and Agricultural Organization of the United Nations. Available online at http://www.fao.org/forestry/6285-037bdb3 2f78ab8a552bf6b163f2ffc825.pdf

Köchy M, Hiederer R, Freibauer A (2015) Global distribution of soil organic carbon - part 1: masses and frequency distributions of SOC stocks for the tropics, permafrost regions, wetlands, and the world. Soil 1:351–365

Kumar BM (2010) Species richness and aboveground carbon stocks in the home gardens of Central Kerala, India. Agric Ecosyst Environ 140:430–440

Kumar S, Meena RS, Singh RK, Munir TM, Datta R, Danish S, Singh GS, Yadav Kumar S (2021) Soil microbial and nutrient dynamics under different sowings environment of Indian mustard (*Brassica juncea* L.) in rice based cropping system. Scientific Report 11:5289. https://doi.org/10.1038/s41598-021-84742-4

Kumar S, Meena RS, Datta R, Verma SK, Yadav GS, Pradhan GS, Molaei A, Rahman GKMM, Mashuk HA (2020) Legumes for carbon and nitrogen cycling: an organic approach. In: Carbon and nitrogen cycling in soil. Springer, Singapore, pp 337–375. https://doi.org/10.1007/978-981-13-7264-3_10

Lal R (2004) Soil carbon sequestration impacts on global climate change and food security. Science 304(5677):1623–1627

Lukac M, Calfapietra C, Godbold D (2003) Production, turnover and mycorrhizal colonization of root systems of three Populus species grown under elevated CO_2 (POPFACE). Glob Chang Biol 9:838–848

Luna RK (2009) Eucalypts in agro forestry. In: Eucalypts in India. Dehradun Publication, Dehradun

Maithani BP (2005) Shifting cultivation in north-East India policy issues and options. Published and printed by Krishna for Mittal Publications. Mohan Garden, New Delhi

Mathukia RK, Sagarka BK, Panara DM (2016) Fodder production through agroforestry: a boon for profitable dairy farming. Innov J Agric Sci 4(2):13–19

Matiasi TM (2006) Integrated environmental education and sustainable development. In: Proceedings from the International Conference on Advances in Engineering and Technology, pp 167–175. https://doi.org/10.1016/B978-008045312-5/50019-4

Mayrinck RC, Laroque CP, Amichev BY, Van Rees K (2019) Above- and below-ground carbon sequestration in shelterbelt trees in Canada: a review. Forests 10:922

Mbaiwa JE (2012) Botswana's community based natural resource management status report for 2011/2012. Kalahari Conservation Society, Gaborone

Meena RS, Kumar S, Datta R, Lal R, Vijaykumar V, Brtnicky M, Sharma MP, Yadav GS, Jhariya MK, Jangir CK, Pathan SI, Dokulilova T, Pecina V, Marfo TD (2020b) Impact of agrochemicals on soil microbiota and management: a review. Land (MDPI) 9(2):34. https://doi.org/10.3390/land9020034

Meena RS, Kumar V, Yadav GS, Mitran T (2018) Response and interaction of *Bradyrhizobium japonicum* and arbuscular mycorrhizal fungi in the soybean rhizosphere: a review. Plant Growth Regul 84:207–223. https://doi.org/10.1007/s10725-017-0334-8

Meena RS, Lal R (2018) Legumes for soil health and sustainable management. Springer, Singapore, p 541.ISBN 978-981-13-0253-4 (eBook), ISBN: 978-981-13-0252-7(Hardcover). https://doi.org/10.1007/978-981-13-0253-4_10

Meena RS, Lal R, Yadav GS (2020) Long term impacts of topsoil depth and amendments on soil physical and hydrological properties of an Alfisol in Central Ohio, USA. Geoderma 363:1141164. https://doi.org/10.1016/j.geoderma.2019.114164

Meena RS, Lal R, Yadav GS (2020a) Long-term impact of topsoil depth and amendments on carbon and nitrogen budgets in the surface layer of an Alfisol in Central Ohio. Catena 2020194:104752. https://doi.org/10.1016/j.catena.2020.104752

Mehrabi Z, Gill M, Wijk M, van Herrero M, Ramankutty N (2020) Livestock policy for sustainable development. Nature Food 1(3):160–165. https://doi.org/10.1038/s43016-020-0042-9

Mishra PK, Rai SC (2013) Use of indigenous soil and water conservation practices among farmers in Sikkim Himalaya. Indian J Tradit Knowl 12(3):454–464

MoA (1978) Indian Agriculture in Brief, 17th edn. Directorate of Economics and Statistics, Ministry of Agriculture, Department of Agriculture and Cooperation, Ministry of Agriculture and irrigation, Government of India, New Delhi

MoA (1985) Indian agriculture in brief, 20th edn. Directorate of Economics and Statistics, Ministry of Agriculture, Department of Agriculture and Cooperation, Ministry of Agriculture, Government of India, New Delhi

Mthembu BE, Everson CS, Everson TM (2018) Tree legumes-temperate grass agroforestry system effects on inorganic soil nitrogen as ecosystem services provision for smallholder farming systems in South Africa. J Crop Improv 32(2):141–155

Mukoni M (2015) Ecological sustainability: reinvigorate and emulate African principles of life or cease to exist? Int J Asian Soc Sci 5(9):514–521

Mutandwa E, Gadzirayi CT (2007) Impact of community based approaches to wildlife management: case study of the CAMPFIRE programme in Zimbabwe. Int J Sustain Dev World Ecol 14:336–344

National Bureau of Soil Survey & Land Use Planning (NBSS&LUP) (2004) Soil Map (1:1 Million Scale). NBSS&LUP, Nagpur

National Bureau of Soil Survey and Land Use Planning (NBSS&LUP) (1994) Global assessment of soil degradation (GLASOD) guidelines. NBSS&LUP, Nagpur

National Bureau of Soil Survey and Land Use Planning (NBSS&LUP) (2005) Annual report 2005, Nagpur. NBSS&LUP, Nagpur, India

NCA (1976) Report of the National Commission on agriculture. National Commission of Agriculture; Government of India, New Delhi, pp 427–472

Niesenbaum R (2019) The integration of conservation, biodiversity, and sustainability. Sustain 11:4676. https://doi.org/10.3390/su11174676

Norby RJ, DeLucia EH, Gielen B, Calfapietra C, Giardina CP, King JS, Ledford J, McCarthy HR, Moore DJP, Ceulemans R (2005) Forest response to elevated CO_2 is conserved across a broad range of productivity. Proc Natl Acad Sci U S A 102:18052–18056

Norby RJ, Gunderson CA, Wullschleger SD, O'Neill EG, McCracken MK (1992) Productivity and compensatory response of yellow popular trees in elevated CO_2. Nature 357:322–324

Norby RJ, Ledford J, Reilly CD, Miller NE, O'Neill EG (2004) Fine-root production dominates response of a deciduous forest to atmospheric CO_2 enrichment. Proc Natl Acad Sci U S A 101:9689–9693

NRSA (2000) Waste land atlas of India. Government of India Balanagar, Hyderabad

NWDB (1985) Ministry of Environment and Forests, National Wasteland Development Board Guidelines for action. Government of India, New Delhi

Oechel WC, Strain BR (1985) Native species responses to increased atmospheric carbon dioxide concentration. In: Strain BR, Cure JD (eds) Direct effects of increasing carbon dioxide on vegetation. U. S. Department of Energy, Washington, DC, pp 117–154. DOE/ER-0238

Oksanen E, Sober J, Karnosky DF (2001) Impacts of elevated CO_2 and/or O3 on leaf ultrastructure of aspen (*Populus tremuloides*) and birch (*Betula papyrifera*) in the Aspen FACE experiment. Environ Poll 115:437–446

Ong CK, Black CR, Muthuri CW (2006) Modifying forests and agroforestry for improved water productivity in the semi-arid tropics. CAB Reviews: Perspectives in Agriculture, Veterinary Science, Nutrition and Natural Resources 65:1–19

Ong CK, Swallow BM (2003) Water productivity in forestry and agroforestry. IWMI Books, Reports H032644, International Water Management Institute, pp 217–228

Painkra GP, Bhagat PK, Jhariya MK, Yadav DK (2016) Beekeeping for poverty alleviation and livelihood security in Chhattisgarh, India. In: Narain S, Rawat SK (eds) Innovative technology for sustainable agriculture development. Biotech Books, New Delhi, pp 429–453. ISBN: 978-81-7622-375-1

Pan Y, Birdsey RA, Phillips OL, Jackson RB (2013) The structure, distribution, and biomass of the World's forests. Ann Rev Ecol Evol Systematics 44:593–622

Parotta J, Agnoletti M, Johan E (2006) Cultural heritage and sustainable forest management: the role of traditional knowledge. Ministerial conference on the protection of forests in Europe. Liaison Unit, Warsaw

Pathak PS, Dagar JC (2000) Traditional agroforestry systems: an overview. In: Soni P, Srivastava AK (eds) Landmarks of botany in India. Suriya International Pub, Dehradun, pp 26–49

Peterson JM (2019) Innovation as a policy strategy for natural resource protection. Nat Res Modeling 32:e12231

Pirlot P, Delreux T, Farcy C (2018) Forests: a multi-sectoral and multi-level approach to sustainable forest management. In: Adelle C, Biedenkopf K, Torney D (eds) European Union external environmental policy. The European Union in International Affairs. Palgrave Macmillan, Cham, pp 167–187

Poore J, Nemecek T (2018) Reducing food's environmental impacts through producers and consumers. Science 992:987–992

Raj A, Jhariya MK, Khan N, Banerjee A, Meena RS (2021) Ecological intensification for sustainable development. In: Ecological intensification of natural resources for sustainable agriculture. Springer, Singapore, eBook ISBN 978-981-334-203-3. Hardcover ISBN: 978-981-334-202-6. https://doi.org/10.1007/978-981-33-4203-3_5

Raj A, Jhariya MK, Yadav DK, Banerjee A, Meena RS (2019) Soil for sustainable environment and ecosystems management. In: Jhariya MK et al (eds) Sustainable agriculture, Forest and environmental management. Springer Nature, Singapore, pp 189–221ISBN 978-981-13-6829-5. https://doi.org/10.1007/978-981-13-6830-1_6

Raj A, Jhariya MK, Yadav DK, Banerjee A, Toppo P (2020a) Agroforestry for Climate Change Mitigation, Natural Resource Management, and Livelihood Security. In: Raj A, Jhariya MK, Yadav DK, Banerjee A (Eds.), editors. Climate change and agroforestry system: adaptation and mitigation strategies. AAP: CRC Press Taylor & Francis Group, ISBN 978-177-18-8822-6. Pp 27-46

Raj A, Jhariya MK, Yadav DK, Banerjee A (2020b) Multifunctional pastures for livestock management and climate change mitigation. In: Banerjee A, Jhariya MK, Yadav DK, Raj A (eds) Environmental and sustainable development through forestry and other resources. AAP: CRC Press Taylor & Francis Group, Palm Bay, pp 75–100. ISBN 978-177-18-8811-0

Raj A, Jhariya MK, Bargali SS (2016) Bund based agroforestry using *Eucalyptus* species: a review. Curr Agric Res J 4(2):148–158

Ray DK, Mueller ND, West PC, Foley JA (2013) Yield trends are insufficient to double global crop production by 2050. PLoS One 8:e66428

Reed MS, Fraser EDG, Dougill AJ (2006) An adaptive learning process for developing and applying sustainability indicators with local communities. Ecol Econ 59:406–418

Rowbotham CJ (1924) The Taungya system in cachar division, Assam. The Indian Forester 50:7

Sacchelli S, Bernetti I (2019) Integrated management of forest ecosystem services: an optimization model based on multi-objective analysis and metaheuristic approach. Nat Res Res. https://doi.org/10.1007/s11053-018-9413-4

Sahay KB (2015) Problems of livestock population. Available online: http://www.tribuneindia.com/2000/20000411/edit.htm. Accessed 20 March 2015

Sahu SK, Nayak RK, Sarangi D (2005) Sustainable soil and land management under shifting cultivation in Orissa. Orissa Review 5:67–74

Sarvade S, Mishra HS, Kaushal R, Chaturvedi S, Tewari S (2014c) Wheat (*Triticum aestivum* L.) yield and soil properties as influenced by different Agri-silviculture systems of Terai region, northern India. Int J Bioresour Stress Manage 5(3):350–355

Sarvade S, Mishra HS, Kaushal R, Chaturvedi S, Tewari S, Jadhav TA (2014b) Performance of wheat (*Triticum aestivum* L.) crop under different spacings of trees and fertility levels. African J Agric Res 9(9):866–873

Sati VP (2019) Shifting cultivation in Mizoram, India: an empirical study of its economic implications. J Mt Sci 16:2136–2149. https://doi.org/10.1007/s11629-019-5416-9

Saxe H, Ellsworth DS, Heath J (1998) Tree and forest functioning in an enriched CO_2 atmosphere. New Phytol 139:395–436

Selomane O, Reyers B, Biggs R, Hamann M (2019) Harnessing insights from social-ecological systems research for monitoring sustainable development. Sustainability 11:1190. https://doi.org/10.3390/su11041190

Sharma KD (1997) Assessing the impact of overgrazing on soil erosion in arid regions at a range of spatial scales. Human impact on erosion and sedimentation. In: Proceedings of the international symposium of the fifth scientific assembly of the International Association of Hydrological Sciences (IAHS), Rabat, Morocco, 23 April–3 may 1997, pp 119–123

Sheldrick WF, Syers JK, Lingard J (2002) A conceptual model for conducting nutrient audits at national, regional, and global scales. Nutr Cycl Agroecosyst 62:61–72

Sheoran S, Kuma S, Kumar P, Meena RS, Rakshit S (2021) Nitrogen fixation in maize: breeding opportunities. Theor Appl Genet. https://doi.org/10.1007/s00122-021-03791-5

Shukla KH, Dwivedi NU (2015) Sustainable development in agricultural sector in India. Business Manage Rev 5(4):220–222

Siebert S, Henrich V, Frenken K, Burke J (2013) Global map of irrigation areas version 5. Rheinische Friedrich-Wilhelms-University, Bonn, Germany/Food and Agriculture Organization of the United Nations, Rome

Singh KA (2008) Resource management perspective for forage production and agroforestry system development in eastern Himalayan region: a review. Indian J Agronomy 53(4):255–266

Smith AR, Lukac M, Bambrick M, Miglietta F, Godbold DL (2013) Tree species diversity interacts with elevated CO_2 to induce a greater root system response. Glob Chang Biol 19:217–228

Springmann M, Godfray HCJ, Rayner M, Scarborough P (2016) Analysis and valuation of the health and climate change cobenefits of dietary change. Proc Natl Acad Sci U S A 113:4146–4151

St Jacques MC (2009) Community water strategies: a framework for implementation. Working document. Brace Centre for Water Resources Management McGill University, Montreal

Steinfeld H, Gerber P, Wassenaar T, Castel V, Rosales M, de Haan C (2006) Livestock's long shadow: environmental issues and options. FAO, Rome

Stiling P, Moon D, Hymus G, Drake B (2004) Differential effects of elevated CO_2 on acorn density, weight, germination, and predation among three oak species in a scrub-oak forest. Glob Chang Biol 10:228–232

Surya B, Syafri S, Sahban H, Sakti HH (2020) Natural resource conservation based on community economic empowerment: perspectives on watershed management and slum settlements in Makassar City, South Sulawesi, Indonesia. Land 9(104):1–31. https://doi.org/10.3390/land9040104

Swaminathan MS (1995) Population, environment and food security. CGIAR issues in agriculture 7. CGIAR, Washington, DC

Thompson J (1996) Sustainable agriculture and rural development: challenges for EU aid. In: EC aid and sustainable development briefing paper, no. 8. International Institute for Environment and Development, London

Thornton P, Herrero M, Freeman A, Mwai O, Rege E, Jones P, McDermott J (2007) Vulnerability, climate change and livestock— research opportunities and challenges for poverty alleviation. SAT eJournal 4(1):1–23
Tiwari SK (2008) Agro-forestry Project; Dept. of Genetics and Plant Breeding College of Agriculture, G.B. Pant University of Agriculture and Technology Pantnagar – 263145. (14–1- 2008)
Udawatta RP, Krstansky JJ, Henderson GS, Garrett HE (2002) Agroforestry practices, runoff, and nutrient loss: a paired watershed comparison. J Environ Qual 31:1214–1225
UN (2017) The UN strategies plan for forests 2017–2030. http://www.un.org/esa/forests/documents/un-strategic-plan-for-forests-2030/index.html
UNCCD (2014) Desertification - the invisible frontline. United Nations Convention to Combat Desertification, Bonn
United Nations Development Programme (2012a) Mara River water users association. Kenya. – Equator Initiative case study series, New York
United Nations Development Programme (2012b) Monks community Forest. Cambodia. – Equator Initiative Case Study Series, New York
United Nations Development Programme (2012c) Fiji locally-managed marine area network. Fiji. – Equator Initiative Case Study Series, New York, NY
United Nations Development Programme (2012d) Sepik wetlands management initiative. Papua New Guinea. – Equator Initiative Case Study Series, New York, NY
United Nations Development Programme (2013) Kanghua community development Centre. China – Equator Initiative Case Study Series, New York
Varma D, Meena RS, Kumar S (2017) Response of mungbean to fertility and lime levels under soil acidity in an alley cropping system in Vindhyan region, India. Int J Chem Struct 5(2):384–389
Verma S, Singh V, Verma DK, Giri SP (2016) Agroforestry practices and concepts in sustainable land use systems in India. Int J For Crop Improv 7(1):126–131. https://doi.org/10.15740/HAS/IJFCI/7.1/126-131
Viccaro M, Caniani D (2019) Forest, agriculture, and environmental protection as path to sustainable development. Nat Resour Res 28:1–4. https://doi.org/10.1007/s11053-019-09497-2
Vohra BBA (1980) Policy for land and water, vol 18. Department of Environment, Government of India, New Delhi, pp 64–70
WCED (1987) Our common future. World Commission on Environment and Development, Oxford University Press, Oxford
Weissert LF, Salmond JA, Turnbull JC, Schwendenmann L (2016) Temporal variability in the sources and fluxes of CO_2 in a residential area in an evergreen subtropical city. Atmos Environ 143:164–176. https://doi.org/10.1016/j.atmosenv.2016.08.044
World Bank (2009) Minding the stock: bringing public policy to bear on livestock sector development. World Bank, Agriculture and Rural Development Dept., Washington, DC. Report no. 44010-GLB
Xie H, Zhang Y, Wu Z, Lv T (2020) A bibliometric analysis on land degradation: current status, development, and future directions. Land 9:28. https://doi.org/10.3390/land9010028
Yadav GS, Babu S, Meena RS, Debnath C, Saha P, Debbaram C, Datta M (2017) Effects of godawariphosgold and single supper phosphate on groundnut (*Arachis hypogaea*) productivity, phosphorus uptake, phosphorus use efficiency and economics. Indian J Agric Sci 87 (9):1165–1169
Zabel F, Putzenlechner B, Mauser W (2014) Global agricultural land resources – a high resolution suitability evaluation and its perspectives until 2100 under climate change conditions. PLoS One 9(9):e107522. https://doi.org/10.1371/journal.pone.0107522

Anaerobic Digestate: A Sustainable Source of Bio-fertilizer

15

Sedrah Akbar, Safia Ahmed, Samiullah Khan, and Malik Badshah

Abstract

Intensive use of inorganic fertilizers has provoked risk to the well-being of humans and their surroundings including high cost, high carbon (C) footprint, giving rise to eutrophication, pollution caused by nitrate, low microbial activity in the soil, and losing the soil organic C. Moreover, bulk of the nitrogen (N) fixation is responsible for 1–2% of the world's entire energy utilization and 3–5% of the earth's natural gas expenditure. Bio-fertilizers, the substitute for synthetic fertilizers, are natural, decomposable, organic, and cost-efficient in contrast to the synthetic fertilizers and have the extensive potential for enhancing global food safety by elevating crop production and fertility of the soil. Bio-fertilizers comprise plant remnants, C-based matter, and several particular types of microorganisms. Anaerobic digestion of organic wastes yields two key products, i.e., biogas and digestate. The biogas is utilized to generate power and heat, while the digestate is valorized in sustainable farming as a bio-fertilizer and soil enhancement. Substituting synthetic fertilizers with digestate diminishes the greenhouse gas (GHG) emissions linked with fertilizer production, saves energy, and facilitates recirculation of plant minerals. Replacing 1 ton of synthetic fertilizer with digestate preserves approximately 108 tons of water and GHG emission of 4 tons CO_2-eq. Digestates have positive impacts on the physical, chemical, and biological features of the soil by introducing microbial biomass, sustaining the rhizosphere's ecology, increasing the yield of plants, supplying excessive amounts of soluble nutrients (NPK), and discharging plant growth-modulating compounds. The effect on the yield may be analogous or greater than

S. Akbar · S. Ahmed · S. Khan · M. Badshah (✉)
Sustainable Bioenergy and Biorefinery Lab, Department of Microbiology, Quaid-i-Azam University, Islamabad, Pakistan
e-mail: sahmed@qau.edu.pk; samikhan@qau.edu.pk; malikbadshah@qau.edu.pk

M. K. Jhariya et al. (eds.), *Sustainable Intensification for Agroecosystem Services and Management*, https://doi.org/10.1007/978-981-16-3207-5_15

synthetic fertilizers and animal manure. The amount of digestate produced is dependent upon the quantity and chemical composition of organic waste used for anaerobic digestion. The digestate manufactured can be both incorporated directly and additionally developed by several treatment practices leading to the production of marketable bio-fertilizers. Moreover, digestate processing minimizes storage and conveyance expenses. The current chapter deals with the constructive and versatile character of anaerobic digestates concerning the soil sustainability and plant development and its role in safeguarding the environment as sustainable and economical input for the agriculture sector.

Keywords

Bio-fertilizer · Digestate · Digestate treatment · Mineral fertilizers · Sustainability

Abbreviations

AD	Anaerobic digestion
C	Carbon
Ca	Calcium
CH_4	Methane
CO_2	Carbon dioxide
GHG	Greenhouse gas
H_2	Hydrogen
HM	Heavy metals
HRT	Hydraulic retention time
K	Potassium
Mg	Magnesium
MGRT	Minimum guaranteed retention time
N	Nitrogen
NH_3	Ammonia
NH_4^+	Ammonium
N_2O	Nitrous oxide
NO_3^-	Nitrate
P	Phosphorus
POP	Persistent organic pollutant
S	Sulfur
VFAs	Volatile fatty acids
UK	United Kingdom
US	United States

15.1 Introduction

Management of the soil fertility is a fundamental approach exploited by the cultivators encountering the decline of the soil organic matter as a repercussion of vigorous agricultural cropping techniques (Romaniuk et al. 2011; Banerjee et al. 2021; Jhariya et al. 2021a, b) and the copious usage of the mineral fertilizers. To neutralize the continuous reduction of the soil organic matter, a consistent integration of crop remains and/or organic fertilizers like farm manure and slurry can be implemented, which is also potent in enhancing the soil nutritive condition (Coutinho et al. 2006; Meena et al. 2018). These approaches are dedicated to make the agricultural structure sustainable, i.e., having the potential of being sustained at a balanced level satisfying the future necessities without depleting the natural reserves or triggering a drastic ecological destruction (Morelli 2011; Jhariya et al. 2019a, b; Kumar et al. 2020).

The cumulative global needs for food proclaim greater yields which among many other factors can be attained through the proliferated usage of fertilizers. The routine use of inorganic fertilizers has certain limitations and thus entails novel and sustainable substitutes. The key limits of concern are the globally diminishing natural resources of inorganic fertilizers, e.g., planet's P (phosphorus) supply is predicted to be entirely exhausted in 50–100 years as stated by many investigators, which would devastate the food security and crop development, and the adverse environmental effects triggered by the usage of fossil fuels for their synthesis. The boundless exploitation of synthetic fertilizers demonstrates extensive threats to nature through air, water, and soil contamination (Sujanya and Chandra 2011). As plants can't take up these perilous chemicals, they begin to concentrate in groundwater and initiate eutrophication of the water bodies. These inorganic fertilizers have negative influence on the soil by diminishing water retention capability, soil productiveness, raising salinity, inconsistency in soil minerals, and making the crops highly vulnerable to infections (Savci 2012). So as an alternative, to counteract these limitations, digestate produced through anaerobic digestion (AD) is utilized as a bio-fertilizer that is abundant in nutrients having exceptional fertilizer potential and has a substantial global capability as a sustainable substitute to the inorganic fertilizers (Al Seadi and Lukehurst 2012; Kumar et al. 2020a). Bio-fertilizers are the preparations comprising of living or dormant cells of competent microbial strains that when applied on soil aid in plant's nutrient intake by their associations in the rhizosphere (Mishra and Dadhich 2010).

Bio-fertilizers are cost-efficient and environment-friendly in nature which not merely averts destruction of the natural resources but likewise aids in liberating the plants with precipitated synthetic fertilizers and augments fertility and productiveness of soil per unit area, comparatively in a shorter period. One study has stated that exploitation of bio-fertilizers enhances crop yield approximately 10–40% by elevating the levels of proteins, important amino acids, vitamins, and nitrogen (N) fixation. The merits of utilizing bio-fertilizers comprise of being an inexpensive source of minerals, invaluable providers of macro- and micronutrients, and providers of carbon (C)-based material, discharging growth hormones, and neutralizing

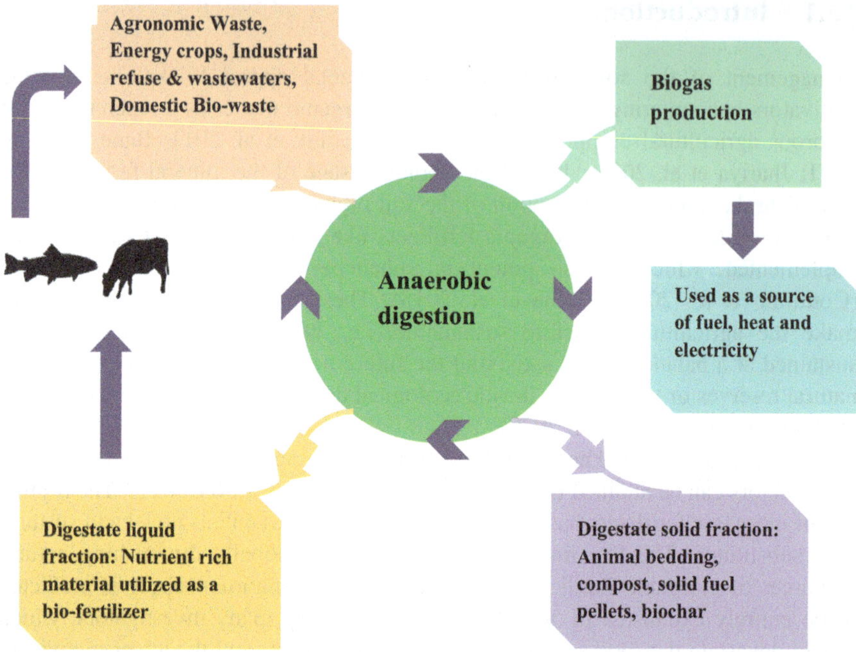

Fig. 15.1 Using biomass in circular economy to valorize anaerobic digestate (adapted from Stiles et al. 2018)

destructive influence of synthetic fertilizers (Gaur 2010). Bio-fertilizers also speed up many microbiological activities in the soil which enhances the degree of accessibility of the minerals by converting them into a form more conveniently acquired by the plants as well as transforming the nutritional minerals from an inefficacious state to a usable configuration via biological procedures. It also enhances endurance to conditions like drought and humidity stress, which highly facilitates the soil fertility making them a dynamic and powerful means for organic and sustainable agronomy (Owamah et al. 2014).

Bio-fertilizers such as AD have been acknowledged and utilized as invaluable fertilizers. It is produced along with the biogas as a coproduct in AD (Fig. 15.1; Nkoa 2014; Coelho et al. 2018), having a variable elemental constitution and substantial quantities of macro- and micro-nutrients that are vital for plant growth (Johansen et al. 2013; Nkoa 2014; Coelho et al. 2018). Anaerobic digestate is manufactured locally, having a low C footprint and anticipated manufacturing expenses, decreasing cost to agriculturalists, and diminishing greenhouse gas (GHG), and can contend with the expensive synthetic fertilizers on which the United Kingdom expends over £250 million annually. Moreover, digestates comprise a vast number of diverse microbes (Nelson et al. 2011; Vanwonterghem et al. 2014; Insam et al. 2015; Guo et al. 2015; Coelho et al. 2019). Several studies performed on the usage of digestates as a bio-fertilizer for vegetable yields specify their inflated suitability for the crop's development (Möller and Müller 2012; Meena and Lal 2018). Their usage as soil enhancer/fertilizer has shown flexible outcomes in field crops, primarily their yield, which depends on the enhancement source and the digestion procedure. Many authors have documented the usage of digestates as fertilizer for crops conveying that they are an efficient nutritional source for plants (Alburquerque et al. 2012; Möller and Müller 2012; Nkoa 2014).

The current chapter deals with the potentiality of the anaerobic digestates as a bio-fertilizer in diverse ways concerning the ecological sustainability and assessing the various aspects associated with their use as a fertilizing agent.

15.2 Basic Outline of an Anaerobic Digestion Process

AD is a technique of generating energy and income from renewable reserves while also supporting soil preservation, expansion of sustainable agronomy, improving recycling of minerals, and enhancing ecological protection (Rehl and Müller 2011; Poeschl et al. 2012). It is a biochemical procedure that takes place when easily decomposable organic C is available and oxygen is absent, analogous to the naturally occurring procedures taking place in the gut of ruminants, or in wetlands and sanitary landfills. The outcome of AD is the generation of biogas, which is a composite assortment of carbon dioxide (CO_2), methane (CH_4) (Tambone et al. 2010), and many additional gases (Ahring 2003; Achinas et al. 2017). Thus, it is a significant energy source that can be transformed into electrical or mechanical energy for public and industrial usage, while the other coproduct of AD is slurry having agricultural effects and is called digestate (Gell et al. 2011; Alburquerque et al. 2012) which is abundant in nutrients and is capable of enriching the soil. The technology of anaerobic treatment is broadly used all over the world as an economical (Govender et al. 2019) solution for the decomposable organic waste as well as for wastewater of both the domestic and commercial origin.

Digestate comprises of a blend of microbial biomass formed in the digestion course and undigested or suspended solid portion having soluble nutrients. The bulk of digestate formed is nearly similar as the feedstock size, though the mass will usually be decreased around 15%. It has all of the N, P, and K (potassium) available in the primary feedstock and thus has significance as an organic fertilizer (Nkoa 2014; Juárez et al. 2013). Reliant upon the constitution of the feedstock, around 20–95% of the organic material in the feedstock is disintegrated in the course of AD (Möller and Müller 2012).

The substrates normally utilized in AD comprise manure, waste from processed food industry, agricultural waste, energy crops, sewage slurry, and organic domestic garbage (Appels et al. 2011). The kind of feedstock utilized not merely defines the quantity of biogas generated but also the constituents and quantity of the digestate produced. The subsequent nutritional properties of the digestate are also impacted by the managing system exploited for the AD procedure (Zirkler et al. 2014; Möller and Müller 2012). Among the diverse affirmative impacts, digestate dispersion on soil may also raise the organic constituents of the soil, which aids in sustaining the soil productiveness (Masciandaro and Ceccanti 1999). AD facilitates the reprocessing of organic wastes by retaining the invaluable minerals whereas aiding to eliminate pathogens and balancing the elements that are detrimental to the environment (Logan and Visvanathan 2019).

The known biochemical route for the AD comprises four key steps, i.e., (a) hydrolysis, (b) acidogenesis, (c) acetogenesis, and (d) methanogenesis, catalyzed by microbial consortia that function sequentially throughout the procedure and are thoroughly reliant on each other (Fang et al. 2011; Kumar et al. 2021), namely, hydrolytic, acidogenic, acetogenic, and methanogenic microorganisms. Microbes involved in hydrolytic and acidogenic stages are primarily facultative anaerobes, while acetogens and methanogens are strictly anaerobes. The ultimate products formed at the end of each step act as a substrate for the subsequent step, yielding a final gaseous product containing predominantly CH_4 and CO_2 (Fig. 15.2), alongside the trace gases, i.e., hydrogen sulfide (H_2S) and hydrogen (H_2).

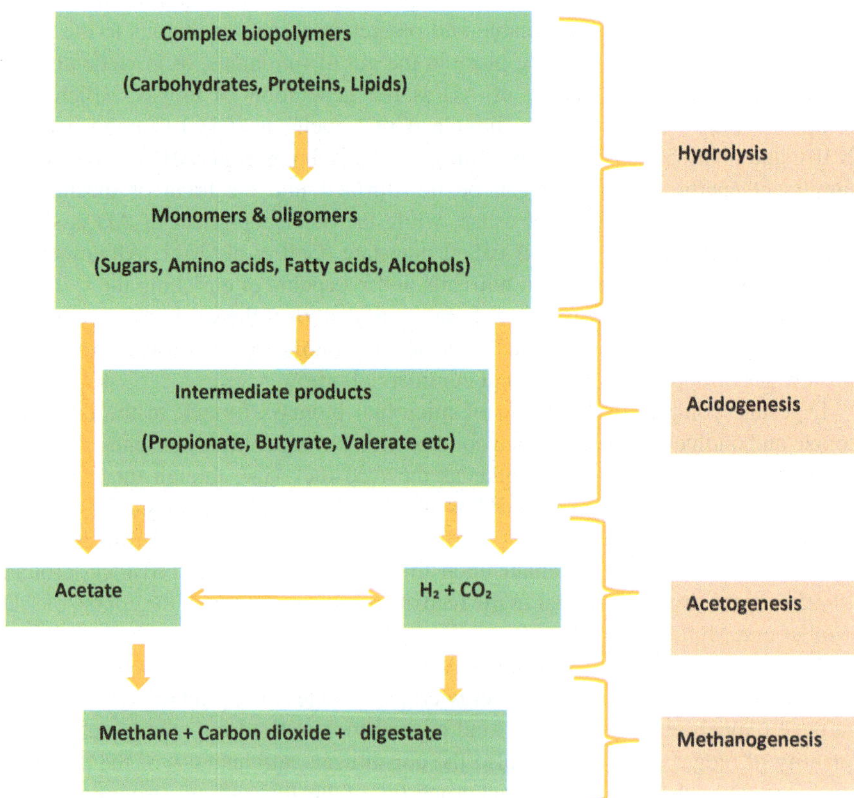

Fig. 15.2 Basic outline of different phases involved in anaerobic digestion (Compiled from Badshah 2012)

15.3 Global Scenario of Anaerobic Digestion Process

It is broadly acknowledged that the earliest AD plant which had a collection system for biogas produced was constructed in Bombay, India, to power the gas engines in 1897. In the twentieth century, it became an extensively used technique and can now be found all over the world for management of the waste streams. Since the 1990s till lately, exploration in AD has advanced at a rapid pace as a source of renewable energy that diminishes utilization of fuel wood and the associated deforestation, reduces interior air pollution, enhances soil productiveness, and most importantly overcomes the worldwide energy deficiency. In contrast to other technologies for bioenergy production, AD is feasible at several scales, i.e., small-scale anaerobic digesters producing just sufficient biogas for domestic purposes, and is highly common in countryside of developing countries, while huge centralized biogas digesters having several thousand m^3 capacities are used widespread in developed countries (Surendra et al. 2014).

15.3.1 Large-Scale Anaerobic Digesters

Large-scale anaerobic digesters which have hundreds and thousands of m^3 capacity are the most prevailing in advanced countries, as they oblige greater infrastructure and high resource funding. In majority of the situations, the resultant biogas is utilized for applications such as combined heat and power (CHP), while it is often enhanced for usage as transit fuel (Holm-Nielsen et al. 2009). Highest biogas production takes place in Europe and the United States, while other areas are also progressively establishing the technology. Biogas worldwide production in 2000 was 0.28 EJ (280,000 TJ) which rose to 1.28 EJ (1.3 million TJ) in 2014, with an average rise of 13.2% in production per annum. According to a report published by Global Market Insights, Inc., the global biogas market worth will exceed $110 billion by 2025 (IEA 2016).

Anaerobic biogas plants over the globe have expanded at the pace of 20–30% per annum with the highly proficient and well-built markets being in Europe, which being the pioneer in AD technology had over 17,400 anaerobic biogas plants of various types and volumes by 2015, extending from smaller plants on farms to bulk size co-digestion plants. EU had about 16,606 biogas plants in 2015 with overall electricity installed potential exceeding 10,100 MW (Eurostat 2017). During the previous decades, production of energy from biogas in the EU has augmented from 167 PJ to 654 PJ in the period of 2005 to 2015, respectively, with the biogas volume rising from 2.5 billion m^3 to 18 billion m^3 from year 2000 to 2015, which represents half of the worldwide biogas production. Biogas upgradation to bio-methane is carried out in 15 European countries. The majority of the plants for bio-methane production are in Germany, the United Kingdom, Sweden, and Switzerland having 185, 80, 61, and 35 plants, respectively, while in other countries beyond these, the production volumes of bio-methane are thus far marginal (Stambasky et al. 2016; Wellinger et al. 2013).

Denmark is a forerunner country in the advancement of the "centralized" or joint biogas digesters having up to 8000 m^3 massive capacity. Denmark has about 150 biogas digesters, comprising around 20 centralized plants, and is planning to boost up their capacity by approximately 50% by 2020 (Holm-Nielsen et al. 2009). The Danish production of biogas has enhanced by 40–45% in the course of 2016–2017 as compared to the production in 2015. Farm-scale AD systems have a digester volume in the range of 200–1200 m^3 and are usually fabricated in massive dairy or swine ranches (Weiland 2003), which co-digest the animal dung from 1–3 ranches with agronomic waste and other accessible organic material. Germany is the front-runner in the number of farm-scale plants, having approximately 9000 digesters, and aims to have around 10,000–12,000 digesters by 2020 (Wilkinson 2011).

In 2017, the United States had over 2100 biogas plants, comprising 250 farm-scale digesters consuming livestock dung, 38 industrial (autonomous) anaerobic digesters, and 654 biogas retrieval plants from landfills. The number of wastewater treatment plants (WWTPs) is 15,000 in the United States, in which about 1240 WWTPs are operational anaerobic plants generating biogas and are set up at large-scale amenities, processing about one to several hundred million gallons of waste-water each day (Statistics RC 2016). In 2015 and 2016, the installed electricity potential of biogas digesters was 2400 MW and 2438 MW, respectively, and produced 1030 GWh electricity. The energy potency of biogas in the United States was evaluated at 18.5 billion m^3 of biogas per annum, 7.3 billion m^3 of which is produced from manure, 8.0 billion m^3 from landfill spots, and 3.2 billion m^3 from WWTPs, which could produce approximately 41.2 TWh of power. It was assessed that nearly 13,000 biogas digesters could be constructed, of which 8241 anaerobic digesters in ranches, 1086 digesters at landfill locations, and 3681 anaerobic WWTPs. There are more than 100 biogas plants in Canada, and it has lately developed numerous schemes aiming to encourage this technology; also, Mexico's attention toward different biogas ventures and utilization of the existing biogas for energy production has been increasing (Alemán-Nava et al. 2015).

15.3.2 Small-Scale Anaerobic Digesters

Small-scale anaerobic plants are typically domestic articles that are commonly around 2–10 m^3 and are mostly found in the countryside of Asia and other developing areas, where the produced biogas through AD is utilized typically for furnaces and lamps (Surendra et al. 2014).

Asia is the front-runner of such digesters. Numerous countries in Asia, i.e., China, Pakistan, Thailand, India, Vietnam, Bangladesh, Sri Lanka, and Nepal, have massive schemes for household production of biogas. In 2014, China had an assessed 100,000 advanced biogas plants and 43 million domestic-scale digesters, producing approximately 15 billion m^3 of biogas, equivalent to 9 billion m^3 bio-methane, and giving about 324 TJ primary energy. The long-term expansion strategy for renewable energy necessitates attaining approximately 80 million domestic biogas digesters by 2020, 8000 wide-scale biogas schemes with an installed potential of 3000 MW, and biogas generation of 50 billion m^3 per annum. The biogas potency was assessed at 200–250 billion m^3 per annum (Sheoran et al. 2021; Raturi 2016; Jingming 2014). In the previous years, advanced biogas plants have been constructed, with the approximate installed electricity potential reaching 330 MW and 350 MW in 2015 and 2016, respectively (Statistics RC 2016).

The National Biogas and Manure Management Program (NBMMP) in India supports the establishment of household biogas plants to provide fuel for cooking. There were around 5 million small-scale biogas plants in 2014, in contrast to a potency of building approximately 12 million plants of biogas, which may possibly produce over 10 billion m^3 biogas per annum, i.e., around 30 million m^3 per day. The power generation capacity of these biogas plants reached 179 MW and 187 MW in 2015 and 2016, respectively (Jingming 2014). Nepal has over 330,000 domestic biogas plants supplying fuel for stoves, set up under the Biogas Support Programme which is one of the most effective biogas programs globally, that has led to a steady growth in the installed biogas plants in the preceding decade. Cambodia and Indonesia have limited biogas programs, yet they installed about 1000 biogas plants in each of the countries in 2010 (REN21 2011). In 2003, Vietnam started Biogas Programme for the animal farming industry which intended to develop industrial biogas systems, leading to 183,000 biogas plants by 2014. In 2006, Bangladesh established National Domestic Biogas and Manure Program for countryside and off-the-grid regions, which led to the installation of over 36,000 household biogas plants by 2014 and about 90,000 biogas kitchen stoves by 2015, mainly to produce cooking gas, and also there are plans to construct 100,000 small-scale biogas digesters by 2020. Installed household biogas digesters in Pakistan and Sri Lanka are around 4000 and 6000, respectively (REN21 2016).

Huge amount of wastes is accessible in Africa, but biogas generation is under developed as compared to other areas thus far. Biogas plants have been installed in various countries, i.e., Namibia, Cote d'Ivoire, Ethiopia, Burundi, Ghana, Burkina Faso, Guinea, Botswana, Lesotho, Nigeria, Kenya, Rwanda, South Africa, Uganda, Senegal, and Zimbabwe. National programs are presently executed in Tanzania, Ethiopia, Rwanda, Uganda, Cameroon, Benin, Kenya, and Burkina Faso. A Biogas Partnership Programme (ABPP) in Africa intends to develop national biogas programs in five African countries, i.e., Kenya, Uganda, Ethiopia, Burkina Faso, and Tanzania, for constructing 100,000 household digesters to ensure accessibility to energy for a half million individuals. The program has steered to the establishment of nearly 16,000 biogas plants in these five countries, i.e., 16,419 digesters in Kenya, 13,584 digesters in Ethiopia, 13,037 plants in Tanzania, 6504 plants in Uganda, and 7518 plants in Burkina Faso (Biogas 2018). The African scheme "Biogas for Better Life" intends to deliver 2 million domestic biogas plants by 2020 to replace the customary cooking fuels such as wood fuel and to offer cleaner energy to 10 million Africans for cooking. The technical biogas capacity in Africa has been assessed to allow the establishment of 18.5 million domestic biogas digesters (Austin and Morris 2012).

In Latin America, various agronomic and household biogas digesters have been installed for countryside houses, and biogas is also retrieved from numerous landfills. The Network for Biodigesters in Latin America and the Caribbean (RedBioLAC) supports the growth of small-scale biogas digesters in Costa Rica, Ecuador, Nicaragua, Bolivia, Mexico, and Peru. Bolivia is the front-runner with approximately 1000 household biogas digesters set up. Large-scale biogas digesters are constructed to utilize wastes from massive farms in Colombia, Argentina, and Honduras (Kapoor and Vijay 2013; Meena et al. 2020a). In 2015, Brazil had 127 biogas digesters consuming agronomic and industrial remains, biowaste, sewage sludge, and landfill gas, which generated around 1.6 million Nm^3/day leading to 584 billion m^3 biogas per annum, and delivering 3835 GWh of energy. The power generation capacity of these biogas plants has augmented considerably in the preceding years, reaching 196 MW and 450 MW in 2015 and 2016, respectively (Statistics RC 2016).

15.4 Microbiology of an Anaerobic Digester

In the hydrolysis step, extracellular enzymes released by hydrolytic bacteria, e.g., *Clostridium*, *Micrococcus*, *Staphylococcus*, *Bacillus*, and *Bacteroides* (St-Pierre and Wright 2014), hydrolyze the complex C-based polymers, for example, carbohydrates, proteins, and fats to sugars, amino acids, and fatty acids, respectively. In the second stage (i.e., the fermentation stage), fermentative microbes, such as *Lactobacillus*, *Escherichia*, *Bacillus*, *Desulfovibrio*, *Staphylococcus*, *Pseudomonas*, *Streptococcus*, and *Desulfobacter*, degrade the monomers and convert it into small sequence volatile fatty acids (VFAs), lactate ($C_3H_6O_3$), succinate ($C_4H_6O_4$), alcohols (C_2H_5OH), H_2, CO_2, and ammonia (NH_3). In the third stage, i.e., acetogenesis, syntrophic bacteria, for example, *Clostridium* and *Syntrophomonas wolfei*, furthermore decompose the VFAs, alcohols, and sugars transforming them into H_2 and acetate. The end stage, i.e., methanogenesis, is executed by methanogens which belong to the domain Archaea. This ending phase comprises two key categories of methanogens, i.e., hydrogenotrophic methanogens, such as *Methanosaeta* and *Methanosarcina* which generate CH_4 primarily from the H_2 gas and CO_2 (Gujer and Zehnder 1983), and acetotrophic methanogens, e.g., *Methanobacterium* and *Methanospirillum* (Sundberg et al. 2013; Zakrzewski et al. 2012) (Table 15.1) generating approximately 70% of the CH_4 primarily from the acetate. The role and association of microbial populations present in AD system are linked to the primary inoculum, kind of feedstock used, and the operational limitations (Demirel and Scherer 2008; Stronach et al. 2012). Among the operating factors, temperature has displayed a robust influence on the microbial communities, for example, the number of organisms belonging to the class *Clostridia* proliferates along the rising temperature (Sun et al. 2015; Levén et al. 2007).

The uniformity between the feedstock's quality and that of the digestate must be analyzed. The digestate comprises all of the materials not been biodegraded or decomposed during AD; hence any impurity in feedstock will persist to be in the

Table 15.1 List of some microorganisms involved in different stages of anaerobic digestion (Westerholm and Schnürer 2019)

Anaerobic digestion system			
Hydrolysis	Acidogenesis	Acetogenesis	Methanogenesis
Acetivibrio	*Acetivibrio*	*Acetobacterium*	*Methanobacterium*
Actinomyces	*Aminobacterium*	Chloroflexi	*Methanobrevibacter*
Aminobacterium	*Anaerococcus*	Cloacimonetes	*Methanocellales*
Aminomonas	*Bacillus*	*Clostridium*	*Methanococcales*
Bacteroides	*Bacteroides*	*Coprothermobacter*	*Methanocorpusculum*
Clostridium	*Butyrivibrio*	*Pelotomaculum*	*Methanoculleus*
Enterobacterium	*Caldicellulosiruptor*	*Smithella*	*Methanofollis*
Eubacterium	*Cellulomonas*	Synergistetes	*Methanogenium*
Fervidobacterium	*Clostridium*	*Syntrophaceticus*	*Methanomassiliicoccales*
Fermentimonas	*Escherichia*	*Syntrophobacter*	*Methanomethylovorans*
Fibrobacter	*Flavobacterium*	*Syntrophomonas*	*Methanomicrobium*
Fusobacteria	*Halocella*	*Syntrophospora*	*Methanopyrus*
Proteiniphilum	*Micrococcus*	*Syntrophothermus*	*Methanoregula*
Psychrobacter	*Pseudomonas*	*Syntrophus*	*Methanosaeta*
Spirochaetes	*Psychrobacter*	*Tepidanaerobacter*	*Methanosarcina*
Streptococcus	*Ruminococcus*	*Thermoacetogenium*	*Methanosphaera*
Thermotoga	*Spirochaeta*	*Thermosyntropha*	*Methanospirillum*
Trichococcus	*Thermotoga*	*Thermotoga*	*Methanothermobacter*

digestate. Feedstocks which are well developed and have excellent quality will consequently yield digestate of a good variety in contrast to the substandard quality feedstock that will yield a poor variety digestate (WRAP 2012).

15.5 Productive Applications of Anaerobic Digestion

Constructive applications of the anaerobic digesters have demonstrated that these methods deliver invaluable economic, ecological, communal, and health merits. AD methods aid in diminishing the GHG emissions, regulating water contamination, and handling the waste streams. Moreover, AD systems assist the generation of alternate and renewable energy that can enhance the quality of survival in numerous regions of the world. The additional financial stream and economic funds from the generation of biogas also build social, health, and financial welfares for the populations (Czekala 2017; Meena et al. 2020b; Lencioni et al. 2016; Global Methane Initiative 2013).

15.5.1 Financial Applications

An AD system reduces the requirement of waste carriage to a recognized waste dumping facility. Exploiting the waste in situ saves the expenses of conveyance and discarding and also makes financial advantages with the derivatives formed. The process generates a sustainable energy source that can be exploited for heating and electricity for households providing new energy forms to regions that might not have approach to the customary electrical energy resources or make income if the energy generated can be vended. Furthermore, AD scheme frequently produces fertilizers in the form of digestates that can be exploited at the spot or traded for extra revenue. In emerging countries, the biogas formed by the AD processes can also be utilized to run cooking furnaces or even as a vehicle fuel (De la Fuente et al. 2013; Czekala 2017).

15.5.2 Environmental Applications

Many of the evident merits of AD systems are ecological. By means of these systems, water and air contamination diminishes which results into progress in human health and environment. The usage of green energy is a mounting trend throughout the globe, and the biogas formed can substitute the consumption of fuel gas and oil and also depresses the GHG production from farming systems, thus refining the air quality. A peripheral advantage of the AD system is reduced agronomic runoff owing to the restraint of the manure and, consequently, reduced runoff contaminating watercourses. These runoffs percolate P, elevating eutrophication and the potentiality for water contamination in native watercourses. Moreover, exploiting manure declines the period it remains at the farm creating odor and pest problems. An additional advantage is from the usage of the digestate that is utilized as fertilizer and thus diminishes the need for inorganic fertilizers, enhancing the crop production. Numerous countries that exploit AD system have stated a drop in the usage of wood for heating which subsequently helps to reduce the problem of deforestation (Lencioni et al. 2016; Walsh et al. 2012a; Czekala 2017).

15.5.3 Social and Health Applications

Many communities that exploit AD and apprehend biogas usually don't have approach to the core power station due to remoteness or expenditures. The power generated from biogas has a remarkable social value to populations as it can alter the manner of communal interaction. AD schemes can give agriculturalists energy independence and make them self-financing. Consuming derivatives from the system, the interior air quality is amended along with boosted culinary and hygiene prospects. Also, the AD system diminishes nitrate (NO_3^-) in the groundwater and ozone, liberated into the air, refining health situation for the public (Global Methane Initiative 2013).

15.6 Fertilizer Attributes of Anaerobic Digestate

The constitution of digestates regarding plant macro- and micronutrients, C-based constituents, its living features (proportion of microorganisms) along with its fertilizer properties is governed by the source of the inflowing unprocessed organic substrate and management of the AD procedure (Risberg 2015).

15.6.1 Microbial Composition of Digestates

Digestate is a biological substance that comprises a broad variability of contrasting microbes, e.g., microbes from the antecedent AD can be detected in the digestate and are even vigorous throughout the stockpiling of the digestate (Sahlström et al. 2008; Bagge et al. 2005). Digestates from diverse AD systems consuming diverse varieties of feedstock and operative setting elucidate the microbial distinctions in profusion and variety indexes among them and are increasingly related to each other than to the unprocessed manure (Vanwonterghem et al. 2014; Satpathy et al. 2016). Major microbial populations found in digestates and animal slurries are obligate and facultative anaerobes (Gerardi 2003).

Bacterial and fungal strains segregated from the bio-fertilizer digestates typically comprise microbes such as *Pseudomonas*, *Bacteroides*, *Bacillus*, *Shigella*, *Clostridium*, *Klebsiella*, *Salmonella*, *Penicillium*, and *Aspergillus* which can be utilized in the manufacture of bio-fertilizers. The existence of these entities in the bio-fertilizer improves the productiveness of the soil. *Klebsiella* and *Clostridium* strains are free-living bacteria which carry out N fixation, whereas *Bacillus* and *Pseudomonas* perform solubilization of phosphate. *Bacillus* strains likewise perform as dissolver for trace components such as silicates and zinc besides acting as plant growth booster. *Pseudomonas* strains are also acknowledged as plant growth enhancer. *Aspergillus* and *Penicillium* fungi are similarly phosphate solubilizers (Alfa et al. 2014). The average microbial sum shows diminishing levels of coliforms in the bio-fertilizer digestate as compared to their greater amounts in the feedstock. Microbial populace has likelihood to reduce abruptly from day 1 to 7 owing to the acidic surrounding and then stay balanced throughout the bio-fertilizer production. Nonetheless, the decline in total coliforms advocates that AD may possibly eliminate pathogens of fecal source if accurately planned (Tsai et al. 2007).

Many reports have described that the metabolic heat produced by microbes in the course of bio-fertilizer production terminates pathogens (Yun et al. 2000; Crawford 1983). The resultant bio-fertilizers are not merely appropriate for usage as conditioners for soil or as fertilizers but likewise destroy soil dwelling and foliage crop pathogens (Zang et al. 1998). These bio-fertilizers accelerate the microbial numbers and their actions in the soil and upsurge the accessibility of minerals to the plants (Alburquerque et al. 2012; Johansen et al. 2013).

15.6.2 Nutrient Dynamics of Digestate

The amount of nutrients delivered to a digester by means of the feedstock is equivalent to those in the digestate. In the course of AD, biochemical fluctuations occur that amend the organic composites comprising the nutrients and augment their accessibility to vegetation (Lukehurst et al. 2010). The total C fraction of digestates fluctuates between 28 and 47% of the dry matter (Möller and Schultheiß 2015; Tambone et al. 2010). Around 80–95% of that C is organically confined with the carbonates as the lingering inorganic C (Fouda et al. 2013). The nutrient necessity can differ among diverse soils reliant upon the soil features and the agronomic operations involved. Furthermore, distinct crops have distinct needs. So the difference in mineral composition among a number of digestates may permit tailored and enhanced fertilization (Risberg 2015). As AD plants can run on several kinds of C-based raw substances and occasionally accept supplements like trace metals, the subsequent digestate exhibits a diverse nutritional constitution. The chemical makeup of digestates has high significance, as the nutrients and other constituents present in the digestates ought to assist the microbial network in the soil as well as fulfill other needs of crops in cultivable systems when utilizing digestate as fertilizer (Möller and Müller 2012).

The anaerobic discharge comprises macronutrients such as N, P, K, sulfur (S), magnesium (Mg), and calcium (Ca) and micronutrients, i.e., chlorine (Cl), iron (Fe), zinc (Zn), manganese (Mn), molybdenum (Mo), copper (Cu), and nickel (Ni) required by plants. Generally, the anaerobic digestate is copious in N, P, and K. Following the solid-liquid segregation, the fluid portion holds a greater N proportion, i.e., 70–80% of the entire ammonium (NH_4^+)-N, and lesser P fraction, i.e., 35–45%, whereas the solid portion comprises higher P fraction, i.e., 55–65% of the total P, and the residual 20–30% of the total NH_4^+-N are disseminated in the solid portion (Logan and Visvanathan 2019). Hence, the outline of two specifics arises, i.e., (1) the solid portion of digestate has a higher potentiality of soil improvement as compared to the fluid portion, while the fluid portion has higher potency as a fertilizer as compared to the solid fraction, and (2) the feedstock's quality is one of the most important elements of the qualitative significance and likely usage of the digestate.

Ammonium ratio in digestate is usually greater than the substrate provided. NH_4^+ is a plant accessible form of N that is utilized immediately by plants or promptly transformed into plant accessible NO_3^-, hence causing enhanced plant development. Substrates profuse in proteins, e.g., slaughterhouse remaining, leftover food, and manure, produce N-abundant digestates (Möller and Müller 2012). Adding NH_4^+-N to the soil can likewise generate stimulating effects, provoking microbial actions and cycling of nutrients (Gunnarsson et al. 2010).

Phosphorous and K are not expended by microbes in the system. Though, a little portion of P in the digester is transformed into orthophosphorous which is a dissolvable state of P (Topper et al. 2006). The C proportion of digestates acts as a significant stabilizer for soils which have little organic fraction and a source of energy for microbes, while the amount of C which is not decomposed will sustain the organic matter of the soil. Other macronutrients such and Ca and Mg are also present in small amounts in the digestate depending upon the type of substrate used in the process (Massé et al. 2007). Digestates also comprise bioactive hormone like composites, for example, phytohormones, which are indicator molecules in plants modulating cellular procedures, development, and protection. Vegetal hormones present in digestates are of specific importance that stimulate plant development including auxins, cytokinins, gibberellins, nucleic acids, jasmonic acid, vitamins, fulvic acid, indoleacetic acid (IAA), brassinosteroids, salicylic acid, and ethylene and escalate resistance to biotic or abiotic pressure. Numerous studies have detected the impact of digestates on hormone-directed procedures, for example, triggering the propagation and primary root development (Möller and Müller 2012).

Plants necessitate trace quantities of certain heavy metals (HM), for example, Cl, Mn, Fe, Zn, Mo, Ni, and Cu, whereas other, e.g., cadmium (Cd), chromium (Cr), mercury (Hg), and lead (Pb), are lethal for them (Banerjee et al. 2018). The core sources of this HM are livestock fodder preservatives, food managing industries, fat deposits, and municipal discharge. One study reveals that with a N burden of 150 kilogram per hectare, the burden of toxic HM (Cd, Cr, Pb) into the soil is lesser in the case of digestate application in contrast to the compost usage, whereas it was greater in some plant beneficial HM, e.g., Cu, Ni, and Zn, in contrast to the inorganic fertilizers (Pfundtner 2002). The biochemical features of a standard anaerobic digestate stated in the literature are presented in Table 15.2.

Table 15.2 Composition of anaerobic digestate

Parameters	Unit	Range	Reference
DM	% FM	1.7–11.5	Sambusiti et al. (2015), Menardo et al. (2011)
pH	–	7.5–8.1	Alburquerque et al. (2012), Sambusiti et al. (2015)
VS	% DM	62.1–77	Menardo et al. (2011), Sambusiti et al. (2015)
Ash	% DM	23–37.9	Menardo et al. (2011)
TOC	g kg^{-1} DM	273–374	Alburquerque et al. (2012)
TKN	g kg^{-1} DM	44–120	Sambusiti et al. (2015), Menardo et al. (2011)
$NH4^{+}$	g kg^{-1} DM	20–95	Alburquerque et al. (2012), Sambusiti et al. (2015)
$NH4^{+}$/TKN	%	46.2–79	Menardo et al. (2011), Alburquerque et al. (2012)
P	g kg^{-1} DM	8–42	Seppälä et al. (2013), Alburquerque et al. (2012)
K	g kg^{-1} DM	28–95	
S	g kg^{-1} DM	2.9–14.7	
Ca	g kg^{-1} DM	9–65.8	
Mg	g kg^{-1} DM	4.1– 24.6	
Na	g kg^{-1} DM	0.68–24.6	
Cl	g kg^{-1} DM	15–57	Alburquerque et al. (2012)
Fe	g kg^{-1} DM	0.46–7.9	Seppälä et al. (2013), Alburquerque et al. (2012)
Mn	g kg^{-1} DM	0.24–1.1	Alburquerque et al. (2012)
Zn	g kg^{-1} DM	0.072–2.2	Seppälä et al. (2013), Alburquerque et al. (2012)
Cu	g kg^{-1} DM	0.014–0.27	

FM fresh matter, *DM* dry matter, *VS* volatile solids, *TOC* total organic carbon, *TKN* total Kjeldahl nitrogen

15.6.3 pH of Digestate

The constancy of AD is vastly reliant upon pH. While the acidogens have high tolerance to pH values less than 6.0, the optimal pH range for methanogens is 7–8 (Raposo et al. 2012). Consequently, pH range of 6.5–7.8 is appropriate for the crucial microbial populations implicated in the procedure.

In general, digestate's pH is alkaline. Rise in pH values during AD may possibly be triggered by the generation of ammonium carbonate $(NH_4)_2CO_3$ (Georgacakis and Tsavdaris 1992), decline in the quantity of VFAs, and the elimination of CO_2 (Sommer and Husted 1995) as a consequence of the conversion of carbonate (CO_3^{2-}) and $2H^+$ to CO_2 and water. pH of digestate is similarly influenced by the accumulation of basic cations such as calcium (Ca^{2+}) and potassium ions (K^+); they heighten the digestate pH by diminishing the amount of H^+ because the solution's electric charge stability has to be neutral (Hjorth et al. 2011). Mineralization and decline of polyvalent ions such as sulfates (SO_4^{2-}) in the substrate also increase the pH. The mean slurry pH is normally around 7 (Coutteau and Sorgeloos 1992), while the digestate's average pH is around 8.3 (Table 15.3). The pH rises during the digestion process and can fluctuate, depending on the digestate's nature and the process of digestion. The final values are regardless of the initial values. The digestate's alkaline pH is a valuable feature owing to the global issues of soil acidification (Makádi et al. 2012).

15.7 Regulation of Feedstock Quality

Feedstock's quality control is the most significant approach of certifying a high-quality innocuous final product appropriate for use as a bio-fertilizer. The necessity of quality management fluctuates relying upon the kind of feedstock being utilized and the kind of threats being implicated. Contents of the biogenic wastes should be scrutinized and described prior to its delivery to the AD plants regarding (a) origin, i.e., name and location of the corporation generating the waste, the procedures through which the waste is generated, and the type of raw or processed substances being exploited; (b) in case of domestic garbage, region of gathering, whether or not source isolated and the gathering receivers, i.e., plastic baggage, paper bags, baskets, etc.; (c) chemical affirmation of contents such as macro- and micronutrients, HM, persistent organic pollutants (POPs), pH, and dry matter content; (d) explanation of the color, consistency, odor, etc.; and (e) health risks associated with managing the product or consumption as soil fertilizer or soil conditioner (Logan and Visvanathan 2019).

Beside the efficiency of AD procedure, constitution and nature of the digestate are also governed by the feedstock constitution and nature. These two are the utmost crucial aspects that strengthen the digestate's quality as a fertilizer. Hence, the key factor in managing the quality of digestate is to guarantee higher-quality feedstock. The substances utilized as feedstock shall not merely be simply decomposable but also should not be contaminated by any undesirable substances and complexes of

Table 15.3 pH of digestate obtained from various sources of feedstocks

Feedstock	Sludge from pharmacological industries	Cattle dung	Sludge from treatment plants of communal wastewater and OFMSW	Energy crops, cattle dung, and agro-industrial residues	Energy crops, cattle dung, agro-industrial residues, and OFMSW
Type of digestion process	Mesophilic (mono-digestion)	Mesophilic (mono-digestion)	Thermophilic (co-digestion)	Thermophilic (co-digestion)	Thermophilic (co-digestion)
pH of feedstock	7.0	6.9	3.5	4.8	4
pH of intermediate phase	7.5	7.2	5	7.5	8.1
pH of digestate	7.8	7.6	7.5	8.7	8.3
Reference	Gómez et al. (2007)			Pognani et al. (2009)	

chemical, physical, and biological origin. The majority of the undesirable contaminations delivered by agronomic feedstocks, excluding inactive substances, e.g., wood specks, grit, and metals, are generally disintegrated or deactivated by means of AD, hence not disturbing the appropriateness and safety of the digestate for consumption as fertilizer. Though, many of the decomposable substances exploited as feedstock in AD systems comprise substances or complexes which are precarious to living beings and their surroundings since they are capable to progress without being getting destroyed in AD plants, e.g., HM and POPs. Feedstock comprising precarious contaminants or other undesirable composites in quantities regarded to be perilous for living beings or for the surroundings must be omitted through AD when digestates are supposed to be utilized for agronomic goals. Such substances may possibly be exploited as feedstock for other practices, when the generated digestate is not utilized as fertilizer but for other commercial purposes (Al Seadi and Lukehurst 2012).

15.7.1 Feedstock Categories

Feedstock having the potential to be utilized for AD as C-based decomposable substances are derived from three chief sectors, i.e., agronomy, municipal, and commercialized or industrial sources. The variety of feedstock appropriate for AD is numerous and diverse, and billions of tonnes are generated globally, e.g., 1.2 billion tons of potent feedstock are generated within the European Union (EU) annually (WRAP 2010). The agronomy division encompasses the production of vegetation, grass fodder, sugar beet, and remnants, e.g., wheat straw and manure production by cattle, pigs, and poultry farms. The municipal zone includes domestic, restaurants, and sewage litters, whereas the commercialized and industrial zone involves waste generated from fruit and vegetal firms, manufactories, superstores, tanneries, food-handling corporations, dairy, and abattoirs (Pérez-Camacho and Curry 2018). Some of the most commonly used feedstocks and their biogas yield reported in literature are presented in Table 15.4.

15.7.2 Feedstock Characterization

A comprehensive elucidation of the feedstock delivered to a digester is a significant measure of regulating the feedstock quality. The elucidation must fulfill suitable federal protocols so as to let the system operatives evaluate its aptness as a feedstock and should comply with the prevailing practices and quality criteria for digestates intended for agronomic and horticultural usage. The system must validate concurrence of the certification with the real feedstock quality. On the basis of their features and attributes and the proposed usage of the formed digestate, a particular substance can be added or omitted from the feedstock used for AD (Wilkinson 2011).

A good feedstock composition of the macronutrients, for example, C, N, P, and S for hydrolytic and acidogenic stages is measured to be 500:15:5:3 (C: N: P: S ratio),

Table 15.4 Characteristics of various feedstocks used for anaerobic digestion (modified from Steffen et al. 1998)

Feedstock	Total solids TS (%)	Volatile solids (% of TS)	C:N ratio	Biogas yield ($m^3\ kg^{-1}$ VS)
Cow slurry	5–12	75–85	6–20	0.20–0.30
Pig slurry	3–8	70–80	3–10	0.25–0.50
Chicken slurry	10–30	70–80	3–10	0.35–0.60
Straw	70	90	90	0.35–0.45
Grass	20–25	90	12–25	0.55
Garden waste	60–70	90	100–150	0.20–0.50
Fruit waste	15–20	75	35	0.25–0.50
Whey	1–5	80–95	24.1	0.80–0.95
Wood waste	60–70	99.6	723	n.a.
Leaves	80	90	30–80	0.10–0.30
Ferment slops	1–5	80–95	4–10	0.35–0.55
Grass silage	15–25	90	10–25	0.56
Wood shavings	80	95	511	n.a.
Food remains	10	80	15	0.50–0.60

whereas for methanogenic stage, crucial proportion is ideally supposed to be 600:15:5:3 (Weiland 2001). S and P are essential in very small amounts in contrast with the C and N supplies which are required in larger fractions.

These macronutrients are vital for existence of all types of plants, animals, and microbial life. Though, animals do not utilize these nutrients proficiently, and greater amounts are defecated. Latest study results designate that 55–95% of N in livestock nutrition is expelled via urine and feces (Oenema and Tamminga 2005). Inflated amounts of P and K in livestock nutrition are likewise defecated. Hence, animal manure, slurry, and several other kinds of AD feedstocks are abundant in plant minerals, which makes the digestate a valued bio-fertilizer. Manure composition relies primarily on the gastrointestinal system of the animals, e.g., ruminants, omnivores, etc., and their nutrition. Other aspects that influence the constitution of manure comprise age, class, and gender of the animal along with the topographical and climatic situations (Risberg 2015).

Another significant factor in efficient digestion is the accessibility of micronutrients particularly Fe, Ni, and Zn in the feedstock, as various active sites of enzymes comprise metals regulating the enzymatic function. These components, when present in comparatively slight quantities, can trigger methanogenic actions (Moestedt et al. 2015). The precise quantity necessary can fluctuate for diverse categories of feedstocks (Facchin et al. 2013).

15.8 Quality Control of Digestate Used as Bio-fertilizer

The development and reprocessing of digestate as fertilizer necessitate management and regulation of the quality all through the entire AD process, i.e., from the generation of the feedstock till its final utilization. Quality management infers the usage of higher variety feedstock, pretreatment of particular feedstock varieties, regulation of the AD procedure and the procedure parameters impacting the value of digestate, digestate handling, optimum storage, and incorporation as fertilizer. Quality control has an influencing part in accomplishing the mandatory criteria for digestate quality utilized as a fertilizer and certifying the lasting sustainability and well-being of this approach (Al Seadi and Lukehurst 2012).

The inapt usage of digestate with pollutants may possibly lead to spread of diseases through the food chain, if proper and strict management are not imposed. Furthermore, regulations are obligatory to execute appropriate treatment stages and principles for safe dumping of the C-based waste, besides employing the AD technology. Principles of digestate management also support the methods used for waste management. By conforming to the guidelines, the digestate formed is considered as having discontinued being a scrap and may possibly be utilized without the necessity for being managed. Quality rules are determined to elucidate the stage whereat the management and control of waste are no longer mandatory which will offer the final consumers with the assurance that the digestate follows an accepted standard. The goal of establishing regulations for managing digestate is to safeguard human and ecological well-being by assigning criteria for the manufacture and usage of digestates in chosen applications (Makádi et al. 2012).

The majority of the protocols globally encourage usage of digestates in agronomy, forestry, or terrestrial refurbishment, which makes it the broadly executed digestate management choice. Protocols dictate digestate creators to acquire suitable accreditation comprising a declaration of compliance with the quality protocols. Likewise, protocols involve digestates to certify the eradication of pathogens and spores. The existence of biological pollutants in digestate, for example, several pathogens and spores, may lead to novel routes of transmission of pathogens and diseases among animals, humans, and their surroundings. By-products of livestock to be utilized as a feedstock necessitate particular considerations regarding the innocuous utilization of resultant digestate as a fertilizer and conditioner for soil (Al Seadi et al. 2013).

The fundamental concepts which outline the "digestate's quality" utilized as a bio-fertilizer, appropriate to substitute inorganic fertilizers in agriculture, are similar regardless of the mass and locality of the biogas system. High variety digestate suitable for utilization as fertilizer is described by crucial characteristics, e.g., proclaimed nutrient content, pH, pureness (devoid of mineral contaminations, i.e., plastic, stones, etc.), dry matter contents, consistency, sterile and innocuous for living beings, and their surroundings regarding its fraction of biotic material (pathogens) and chemical contaminants (organic/inorganic). The breakdown procedure cannot destroy all potent chemical pollutants which are delivered along the feedstock, indicating that the only means of producing greater variety digestate is to utilize feedstock not having any undesirable contaminations. Therefore, countries with established biogas field and with rules of the environment and well-being of humans and animals have established "positive lists" of feedstocks for AD which is a component of the quality control plans in these states (Nkoa 2014).

15.9 Process Parameters Associated with the Digestate Quality

Regulation of the AD process guarantees a stabilized final yield and effectual decline of pathogens through sanitization and can likewise promote the transformation of some stubborn organic contaminants. The most significant procedural factors impacting the digestate value are (1) temperature and (2) retention time, guaranteeing a balanced digestate so as to attain the corresponding efficient pathogen decrease in the course of the digestion process (Al Seadi et al. 2013).

15.9.1 Optimal Temperature for Producing a High-Quality Digestate

The interior temperature of digesters may be 45–60 °C in thermophilic digestion, 25–40 °C in mesophilic digestion, or lower than 20 °C in psychrophilic digestion, having optimal temperatures at approximately 37 °C and 55 °C for meso- and thermophilic digestion, respectively (Raposo et al. 2012). Psychrophilic temperature is infrequently exploited because of the lower reaction rate. The majority of the commercial AD plants run at the mesophilic temperatures. The warming of feedstock to a defined extent of temperature and upholding this temperature throughout a defined period offer effectual pathogen decline existing in the digestible waste (Al Seadi et al. 2013).

15.9.2 Optimal Retention Time for Producing a High-Quality Digestate

The interval of feedstock's residence in a digester (retention time), at stable procedure temperature, impacts the value of digestate. Retention time is cited as (1) hydraulic retention time (HRT) and (2) minimum guaranteed retention time (MGRT). HRT is the period in which liquid biomass feedstock persists in the digester at a constant temperature so as to make sure full stabilization of the substrate and is commonly estimated in days or dependent on the digestion of the biomass, while MGRT, commonly estimated in hours, is the minimum interval of time that any fraction of the feedstock occupies the continuous digester (Al Seadi 2001).

The HRT for mesophilic systems is between 15 and 30 days, while that for thermophilic systems is 10 to 20 days (Angelidaki et al. 1996). Though, some substances may require extensive periods to be appropriately decomposed, and so HRT might be as lengthy as 60–80 days which frequently results in improved degradability of the organic substance in the system, therefore resulting in a smaller amount of C in digestates (Bauer et al. 2009).

Incorporation of MGRT and thermo- or mesophilic temperatures in AD offers pathogen decline in manures and slurries corresponding to the EU hygiene criteria of 70 °C for 1 h and hence permitted to be exploited as feedstock. The operation must be performed in a thermophilic system or digester or in a sanitization reservoir integrated with thermo- or mesophilic temperatures (Bendixen 1999). The system operators must choose procedural temperature and retention time that are suitable for the type of feedstock to be processed and to thoroughly optimize and supervise the AD procedure and the procedural limitations.

15.10 Storage and Administration of Digestates

Storage and administration of digestate must conform to the regulations of acceptable agronomic operations and be in agreement with the state rules or legislature. The volume and timing of administration rely chiefly on soil characteristics and the propagated crops. Likewise, adequate digestate storing capability must be set up to assist digestate formation, as the land incorporation of digestates needs to be in concurrence with the season of plant development. Digestate storing restriction is greater once there is periodic restraint on crop propagation and trivial during the period of incessant plant growth all through the year (Plana and Noche 2016).

15.10.1 Storage

Digestate is produced all through the year and needs to be stockpiled till the budding season, the only suitable period for its incorporation as a fertilizer owing to the crop nutritional necessities and to evade the nutrients loss. The span of storing interval essential will be determined by the geographic region, soil nature, wintertime rainfall, crop rotation, and state protocols managing the digestate incorporation. In a moderate weather, e.g., a storing capability of 6–9 months is suggested. When digestate is stockpiled in exposed reservoirs, NH_3 and CH_4 gases are released which reduces the value of the fertilizer and escalates the discharge of potent GHGs (Möller et al. 2008). These discharges could be diminished if the digestate is concealed by a protective covering (Lukehurst et al. 2010)

Fluid portion of digestates is typically deposited in storing containers, lagoons, or flexible storing bags, whereas the solid portion of digestate is deposited in enclosed plane concrete spaces or interior of buildings. Furthermore, digestate storing amenities should be constructed with supplies and features that assure the water tightness, to evade water and soil contamination. Some other ecological actions might also be mandatory, for example, leakproof, water-resistant liners and seepage sensors to be placed at the storage amenities. Fitting of airtight storing shields diminishes the gaseous discharge of CH_4, nitrous oxide (N_2O), and NH_3 by 90%, and thus 55–100% of NH_4^+-N is maintained in the digestate (Rehl and Müller 2011). Another benefit of fitting the storage shields is that they moreover evade the buildup of raindrops, therefore thinning the digestate and expanding its volume. The covering substance of the storage amenities may be a sheath, steel, concrete, or a hovering shield of hay, plastic, or clay particles covering the fluid superficially. Essentially, appropriate storing maintains the significance and potential of digestate and averts the loss of NH_3 and CH_4 to the air, nutrient seepage, and overflow, along with the discharge of smell and aerosols (Al Seadi and Lukehurst 2012).

Temperature throughout the storing period has robust impact on the biochemical makeup of the digestate. One study has stated that in the course of 114–138 days storing of livestock digestate, the conversion of organic N to NH_4^+ was slow and trivial at less than 15 °C but amplified considerably at 20 °C. This is predominantly significant in the framework of digestate storage in hot regions where temperature may fluctuate significantly among different areas. The percentage of N loss triggered by N discharge from shielded digestate storage is reported to be 25–30% of the primary N quantity, whereas that from the exposed digestate is 60–70% (Tran et al. 2011).

15.10.2 Application over Lands

To accomplish the ecological and financial advantages from the incorporation of digestates as fertilizer, some elementary concepts of good agronomic approaches need to be achieved (Lukehurst et al. 2010; Raj et al. 2019a, b; Khan et al. 2021a, b). Digestate is essentially to be smeared in the budding season of particular crops typically from February (late winter) through spring and summertime, so as to make sure the optimal intake of the minerals by plants and to evade contamination of the groundwater by averting the nutrient leakage. It is more likely that incorporation of digestates in fall season will lead to inefficient uptake of N, except some of the crops which require N at this period of the year, e.g., oilseed rape (Plana and Noche 2016).

Digestate needs to be incorporated in fertilization strategy of the farmlands in a similar manner as inorganic fertilizers and should be smeared at precise proportions, with apparatus that safeguard leveled incorporation all through the entire fertilized region. The apparatus employed to smear digestate must reduce the surface area subjected to air and should guarantee speedy integration of the digestate into the soil and thus is primarily incorporated with the trailing hoses, with trailing shoes, or by soil injectors. These approaches of digestate integration and consideration of the meteorological conditions will diminish NH_3 volatilization (Al Seadi and Lukehurst 2012).

Rainy conditions, very high moisture, and no breeze are optimal situations for applying the digestates. Arid, sunny, and windy climate decreases the N efficacy significantly. Direct incorporation of digestate into the topsoil of grass crops offers the utmost N efficacy. Storing containers for digestates must be located in the shadow, protected from airstream. It is imperative to confirm a crusting surface on the fluid, in the storing vessel. The digestate must at all times be pushed from the lowermost of the container, to evade needless stirring which must be done merely before dissemination to ensure homogeneity of the digestate during incorporation (Fagerström et al. 2018; Al Seadi and Lukehurst 2012).

15.11 Role of Anaerobic Digestate Toward Agricultural Sustainability

Agricultural sustainability of digestates can be evaluated straight from the comparative amount of inorganic fertilizers crucial to attain the equivalent yield of crop or through comparison with the field capabilities of commonly known organic (Gong et al. 2011; Adeli et al. 2005) or commercialized mineral fertilizers. Digestates stimulate plant nourishment directly by adding plant minerals with the objective to proliferate the yield. Most of the investigations on the usage of digestates in agronomy have concentrated on its fertilizer and nutritional significance; however digestates are as well viewed an amendment because of its organic fraction and the consequent valuable impacts on the soil biochemical factors (Garg et al. 2005).

15.11.1 Impact of Digestate on Different Properties of the Soil

The long-standing influence of digestates on different characteristics of the soil is an uncharted area of research. Though, comparative to other treatment protocols, e.g., raw unprocessed manure compost or inorganic fertilizers, soils cured with digestates demonstrate maximum microbial biomass, N mineralization, and having the potency of NH_3 oxidation (Tiwari et al. 2000) which are much frequently proposed biological and chemical signs for determining the soil nature and variety (Doran and Parkin 1994; Wienhold et al. 2004). One study has published that digestates from wine industries mineralize N at a greater frequency as compared to their corresponding compost (Canali et al. 2011; Larsen et al. 2007). The other investigation, conducted by Möller et al. (2008), reflected that the soil mean N fraction in crop rotation of dinkel wheat, cereal rye, and spring wheat was 36% greater in fields modified with digestate than those treated with raw manure.

Concerning the soil physical features, many field trials have displayed diminished bulk density, amplified hydraulic conductivity, and vapor holding capability of soils when amended with digestates from agronomic waste (Garg et al. 2005). The publications regarding the instant influence of digestates on soil characteristics have frequently displayed enhancement of the soil variety modified with digestates that is verified by the expansion in microbial biomass as well as N and P proportions. These outcomes also support the previous findings of investigations showing the enhancing properties of classic anaerobic digestates and firmly designate these digestates as an efficient organic amendment resource (Crolla et al. 2013).

15.11.2 Impact of Digestates on Soil Biota

Administration of digestates may considerably alter the association and variety of microbial populations in the soil, enhancing the quality of the soil (Sullivan et al. 2006). Fluctuations in microbial groups can take place more rapidly than fluctuations in any other soil feature, due to which the analysis of microbial factors is considered a subtle sign when assessing the soil disorder and influence valuation afterward the administration of digestate (Abubaker et al. 2013). Generally, the microbial load of the soil and their metabolic actions are stirred afterward the administration of digestate, which is ascribed to intensify the C and minerals accessibility to the plants (Frąc et al. 2012). Reliant upon the dosage, digestates may comprise constituents hindering the action of NH_4^+ oxidizers in agronomic soils (Nyberg et al. 2004). The populace of bacterial as well as archaeal nitrifiers, along with the denitrifying and N-fixing communities, reacts promptly to the incorporation of NH_4^+ (Long et al. 2013). When the soil is remediated using digestate in contrast to the manure, archaeal populations in the soil are enhanced and can be tracked through some signals given out, whereas the eubacterial groups are influenced slightly (Johansen et al. 2013; Abubaker et al. 2013). Profusion of methanogenic microbes as well as acetogens shows a greater upsurge in soil biota cured with digestates as compared to the soil where raw manure has been applied. Several other studies have also established the fact that digestates in contrast to inorganic fertilizers improved, whereas unprocessed manure or other organic waste diminishes the bacterial development (Walsh et al. 2012b). The outcome, though, may fluctuate among the diverse phylogenetic communities of the soil microbes.

15.11.3 Impact of Digestates on Yield and Quality of Crops

Digestates have demonstrated to have progressive impact on harvests in comparison to without application of any soil enhancers or fertilizers. In contrast with other C-based fertilizers, for example, farm yard manure and compost, the fraction of freely accessible N for plants is greater in digestates afterward its incorporation into the soil and is predominantly vital for vegetation having small harvesting phases, e.g., spring wheat, that benefits from a swift intake of N during their primary development (Crolla et al. 2013).

The protein ratio of plants such as water lentils and tapioca leaves has been demonstrated to be greater when cured with digestates in contrast to other organic fertilizers. One publication indicates that the quality of tomato with regard to every essential and nonessential amino acid proportion and macro- and micronutrients augments in case of applying digestate in comparison with the application of synthetic fertilizer (Groot and Bogdanski 2013).

Plants developed by using digestates have growth analogous to or superior than an unfertilized control (Wentzel and Joergensen 2016). Several studies have demonstrated augmented plant development with digestate in contrast to the undigested raw materials (Möller et al. 2008; Wentzel and Joergensen 2016). Numerous investigations have revealed comparable or enhanced plant growth with digestates in contrast to the inorganic fertilizers (Haraldsen et al. 2011). Rise in plant harvest is primarily associated with the NH_4^+-N fraction of the digestates, whereas there are also other propositions that the advantages of digestates may possibly go beyond their fertilizer significance which may include impact on the soil biochemical factors or potential interaction of phytohormones and other bioactive composites found in digestates. Further study is essential to comprehend the impact of digestates on the growth and development of particular plant type. Furthermore, enhanced understanding of probable procedures for digestate impacts beyond their nutritional importance alone will assist to elevate their usage in agronomy (Pitts 2019).

In view of all these studies, it can be stated that incorporation of digestates has the potential to substantially enhance the variety of foods without being detrimental to the surroundings, which is highly essential for the sustainability of environment and healthful life.

15.12 Standards for Digestate

Digestate's quality relies on the constitution of the feedstock delivered to the digesters. To certify that quality and safety are well-maintained, the existence of undesirable substances and pollutants of biological, chemical, and physical nature in the digestate should be evaded. Digestate from agronomic and industrial feedstock usually has a superior quality and is utilized innocuously and constructively as a fertilizer. A dynamic and persistent AD procedure has a progressive influence on the quality of digestate, degrading most of the undesired complexes and contaminants to a certain degree that is delivered along with the feedstock. Particular kind of feedstock can be preprocessed by means of mechanical, chemical, or thermal practices to eliminate, decay, or deactivate undesirable contaminations. If efficient contaminant elimination cannot be assured both by preprocessing or through AD, the corresponding substance should not be utilized as a feedstock in digester, when the digestate is intended to be utilized as a fertilizer or for some other agronomic goals (Drosg et al. 2015; Al Seadi and Lukehurst 2012). The undesirable contaminations that impact the quality and safety of the digestate utilized as fertilizer are as follows:

15.12.1 Physical Contaminations

A variety of substances are regarded as physical contaminants which exist in AD feedstock, comprising of nondigestible or low-digestible substances besides the particle bulks of digestible substances. Examples of physical contamination include masses of straw in manure, animal recognition labels, baler thread, gravel, grits, rubber, metals, wood, paper, glass, etc. Such contaminations are expected to be existent in all forms of feedstock but mostly in the domestic litters, discarded food, lawn waste, hay, solid manures, and others. Organic domestic waste and food waste might also have a number of other undesirable physical contaminations comprising cutlery, plastic, packing substances, massive lawn waste, etc. The occurrence of these impurities causes aesthetic destruction of the surroundings (Logan and Visvanathan 2019).

The regulation and control of physical contamination are essentially a matter of verifying higher clarity of the feedstock which can be done both by categorization at source or by onsite segregation. As an additional safety approach, physical obstacles like strainers (Knoop et al. 2018), grit traps, or fortification screens can be fitted in the pre-storing reservoirs in AD systems. This method seems to be an ideal choice for supermarket foodstuff litter. If unit mass of the digestible substance is huge, it may be diminished by slicing, soaking, or dealing through some other methods before it gets entry into AD system (Al Seadi and Lukehurst 2012).

15.12.2 Biological Contaminations

Digestates utilized as fertilizers should present nominal threat of spreading microbes including bacteria, viruses, abdominal parasites, weeds, crop spores, and crop infections. Feedstock preference and omission of substances having higher threats of biological impurities are crucially significant methods in quality control of the digestate. Omission of particular biologically polluted feedstock relates to all categories of feedstock, comprising livestock manure or other feedstock arising from farms which have crucial threats for animal well-being. The AD procedure has a hygiene impact and is capable to deactivate majority of the pathogens existent in the mixture of feedstocks in digesters. Reliant upon the substances delivered, further hygiene processes like pasteurization or pressure decontamination can be essential and are consequently mandatory for particular substances delivered as feedstock. The stringent hygiene requires having the objective of breaking down the transmission sequence of pathogens and thus animals and plant infections (Al Seadi 2001; Lukehurst et al. 2010).

15.12.3 Chemical Contaminations

Feedstock arising from agronomic and anthropological food chains has lower chemical contamination in majority of circumstances (Govasmark et al. 2011). However, strict needs for ensuring a good quality digestate likewise suggest stringent management of the chemical substances, for example, HM and other mineral pollutants, POPs, and macroelements, i.e., N, P, and K.

HM frequently arises from anthropogenic sources in digestates (Lukehurst et al. 2010). Local wastewater emission comprises metals from metabolic litters, erosion of water pipelines, and customer goods. Industrial sewages might considerably add to metal piling. HM existent in digestate also arises from the feedstock utilized, going unaffected through AD into the digestate and ultimately into the land when utilizing digestate as fertilizer. Therefore, the majority of the countries have stringent restrictions on quantities of HM in any substance which is to be smeared over lands, while others impose restrictions on the soil fraction of such contaminants (Kupper et al. 2014).

Agronomic wastes can comprise POPs as pesticide remnants, antibiotics, or other pharmaceuticals. Industrial C-based waste, sewage slurry, domestic waste, and even food waste can comprise aromatic, halogenated, and aliphatic hydrocarbons, polychlorinated biphenyls, polycyclic aromatic hydrocarbons, etc. POPs don't decompose in the environment and are acknowledged as being lethal to the biota (Magulova 2012). Therefore, they become accumulated in the food chain due to their ecological perseverance, and thus continual subjection of humans, domesticated animals, and wildlife even to lesser quantities can possibly have destructive effects on them. It is hence particularly significant to certify higher-quality feedstock, which can be accomplished by consuming only those feedstocks which are within the acceptable limits (Lukehurst et al. 2010).

The amount and availability of macroelements for the plants should be accurately described, so as to avoid contamination from overburdening of nutrients into the soil during the reprocessing of these elements, by incorporation of digestate as a fertilizer (Al Seadi 2001).

15.13 Pros of Digestate Over Mineral Fertilizers and Unprocessed Manure

Digestate has proven to have admirable fertilizer potentiality with numerous investigations displaying analogous or greater yields attained in crops fertilized with digestate in contrast to the synthetic fertilizers or unprocessed animal manure or slurry (Nkoa 2014). The practice of digestate usage as organic fertilizer in agronomic systems appears the finest choice for its reusing capacity as it comprises substantial quantities of remnant organic C and plant minerals. Thus, land dissemination of digestate provides many advantages if incorporated through acceptable agronomic procedures and by attaining the digestate sanitation (Al Seadi and Nielsen 2002; Holm-Nielsen et al. 2009).

15.13.1 Pros of Digestate Over Mineral Fertilizers

Extensive usage of chemical or mineral fertilizers for amplified yield (Santos et al. 2012) has given rise to the degradation of soil variety, contamination by HM, and eutrophication (Owamah 2013; Zhu et al. 2012). Therefore, the significance of digestate as a bio-fertilizer is to deliver socioeconomic and environmental welfare in terms of enhancement of the soil and food variety and their safety and health of humans and animals (Johansen et al. 2013; Bakry et al. 2009) as well as it is cost-efficient and is a renewable source of plant minerals for sustainable agronomy (Grigatti et al. 2011).

The exploitation of chemical fertilizers and pesticides speeds up the acidification of the soil and also presents the threat of soiling the groundwater and atmosphere (Chun-Li et al. 2014). Furthermore, it deteriorates the roots of plants making them vulnerable to undesirable infections. Thus, efforts have been carried out concerning the manufacture of nutrient-abundant high variety bio-fertilizers (i.e., digestates) to guarantee the biosafety of vegetations. Digestates have been acknowledged as a substitute to mineral fertilizers to upsurge the productiveness of soils and crops in sustainable agricultural, which not merely maintains food safety but likewise enhances biodiversity of the soil (Raja 2013; Raj et al. 2020, 2021) that comprises a diversity of valuable bacteria and fungi aiding to promote the plant growth (Itelima et al. 2018). Utilizing digestate as a fertilizer has also been recognized to prevent the emissions occurring from synthetic fertilizers (Montes et al. 2013). Moreover, many investigations have proposed that managing the digestate in a controlled manner can repress soilborne infections (Cao et al. 2016) and assist in the deactivation of weed spores (Fernandez-Bayo et al. 2017), along with possible lessening of the nutrient percolation (Walsh et al. 2012b).

15.13.2 Pros of Digestates Over Untreated Manure

Anaerobic digestion of manure prior to its usage as fertilizer is usually assessed as progressive, as the digestate generated has greater fraction of plant minerals than the unprocessed manure, the outcome of which is greater microbial stability and more hygienic fertilizer (Al Seadi 2001). The fraction of NH_4^+ is normally greater in digestates as compared to the raw manure (Arthurson 2009) and thus has a very high significance in a fertilizer, as it is instantly accessible to the plants (Holm-Nielsen et al. 2009).

Digestate also has considerable quantities of C as compound organics from the unprocessed constituents (Franke-Whittle et al. 2014) and has enhanced capability of C sequestration in soils as compared to the unprocessed manure owing to the lesser share of simply decomposable constituents in digestate (Maucieri et al. 2017). The presence of VFAs (e.g., butyric acid, valeric acid, etc.) in the unprocessed manure can create repugnant odors, but when it is passed through AD, the sum of these compounds diminishes considerably in a way that their potentiality for creating unpleasant odors declines during storing and dissemination (Birkmose 2007).

Higher-quality digestates have negligible biological impurities such as vegetal pathogens and viable spores much lesser in comparison with the unprocessed manure and slurry (Table 15.5). Utilization of digestate as a bio-fertilizer disrupts

Table 15.5 The survival time (T_{90}) of pathogenic bacteria in digestate in comparison to the untreated raw manure (adapted from Bendixen 1999)

	Anaerobic digestion		Raw manure	
Bacteria	53 °C (h)	35 °C (h)	18–21 °C (days)	6–16 °C (days)
Salmonella typhimurium	0.7	57.6	14	41.3
Salmonella Dublin	0.6	50.4	–	–
E. coli	0.4	43.2	14	61.6
Staphylococcus aureus	0.5	21.6	6.3	49.7
Mycobacterium paratuberculosis	0.7	144	–	–
Coliform bacteria	–	74.4	14.7	65.1
Group D Streptococci	–	170.4	39.9	149.8
Streptococcus faecalis	1.0	48	–	–

T_{90}—90% decline in viable sum of a microbial populace

the sequence of spread of plant infections and weeds spores on farmlands and depresses the necessity for successive usage of herbicides or pesticides on the relevant vegetations (Lukehurst et al. 2010).

15.14 Enhancement Techniques for Digestates

Digestate management is necessary for several causes. Inapt management and dissemination of digestate may create ecological threat, either because of the seepage of NO_3^- into the soil or water or because of the eventual gaseous loss of NH_3, CH_4, and N_2O. Moreover, digestate has greater water fraction, thus making it challenging to be handled, transported, and spread in the fields (Bauer et al. 2009). Additional advancement and conditioning have a significant part in the quality control of digestates by allowing reduction of water fraction, their nutritional management, appropriate storing, transport, cost of application and thus improving the overall quality of digestates. Setting up a digestate handling equipment owes to the unfeasibility of digestate dissemination nearby the biogas reactors and the afterward long distance carriage for its removal (Plana and Noche 2016). Hence, the important objectives of managing digestates are to upsurge the significance of the digestate, to make novel marketplaces for digestate goods, to diminish reliance on on-site terrestrial incorporation, and possibly to decrease the operational cost of the amenities (Logan and Visvanathan 2019).

Currently numerous treatment approaches are being exploited in the biogas field, and the choice of the appropriate method greatly depends on the digestate features, locality, indigenous conditions, energy demands, final consumer, and the investment charges. Processing of the digestate is not a regular or distinct procedure but relies on the needs of a particular biogas system, native conditions, and appropriate protocols and can comprise of a solitary procedure or a blend of numerous methods (Plana and Noche 2016) which can differ significantly from manufacturer to manufacturer and occasionally also fluctuate from the data presented. The advancement technologies described in this chapter are simply envisioned to deliver a primary valuation and an outline of the several choices. For appropriate management of the digestate, several enhancement procedures can be implemented at three crucial steps, i.e., predigestion, in-vessel treatment, and post-digestion (Logan and Visvanathan 2019) (Fig. 15.3).

15.15 Predigestion Enhancement Practices

AD has a sterilizing impact on the processed feedstock and hence the digestate's quality. Though the majority of the usual pathogens and viruses are destroyed in the course of mesophilic and thermophilic breakdown (Qi et al. 2018), additional sanitation can be prerequisite for some particular type of feedstock, before being introduced into the digesters, which helps avoiding the adulteration of the whole feedstock and saves the additional expenses of pasteurizing the complete digester

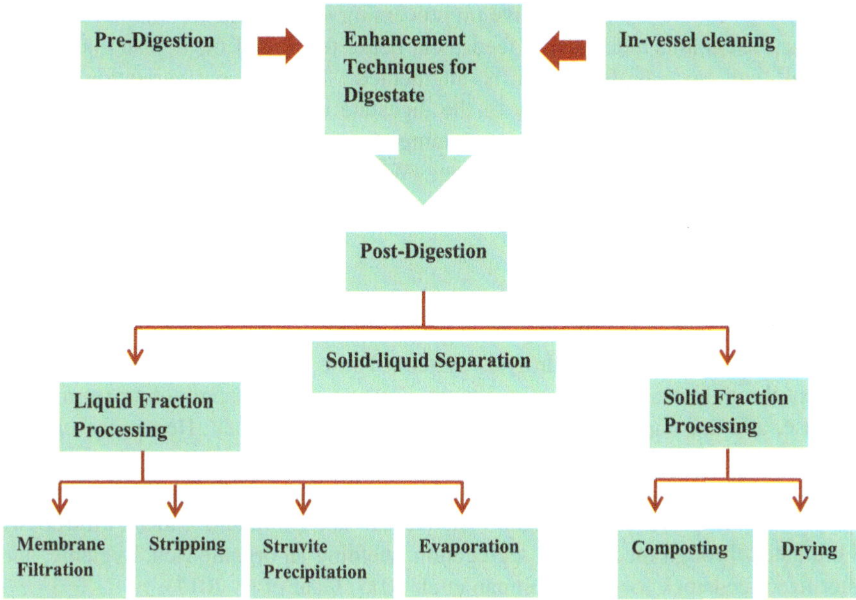

Fig. 15.3 Summary of different enhancement methods for digestate (adapted from Logan and Visvanathan 2019)

mass. For some particular type of feedstock, pretreatment is carried out at the production location of the feedstock, thus diminishing any potential biological threat related with the transportation of unsterilized matter (Al Seadi and Lukehurst 2012).

Feedstock can be pretreated with a number of methods so as to heighten the potential of AD by raising the accessibility of easily decomposable organic matter to the microbes involved in AD (Bruni et al. 2010). These kinds of treatments are generally employed at the site of AD plants, most commonly for substances having higher lignocellulosic and hemicellulose content (Hjorth et al. 2011). Some pretreatment techniques used for the enhancement of digestate are as follows:

15.15.1 Thermal Hydrolysis

Thermal hydrolysis is a practice involving vapor pretreatment of the feedstock through elevated pressure and temperature. The feedstocks are heated up and pressurized through steam in a reaction chamber prior to being promptly depressurized, leading to the disintegration of cellular organization in the biomass. Hence the organic material is delivered to the system in a disintegrated form, which results in highly efficient digestion with improved biogas formation and enhanced digestate quality (Zhang et al. 2018).

To make sure the procedure is effectual both thermally and economically, the process necessitates dewatered feedstock having 15–16% dry solids. As a result, dewatering scheme is an imperative preprocessing step. While the thermal hydrolysis practice exploits a dewatered feedstock, better feeding of the digester is accomplished, and so gas formation is augmented. Through pasteurization of the hydrolyzed digestate, the quality of the digestate is enhanced, and thus it can be easily dewatered to attain products having greater dry solids fraction, allowing secure storage, management, and carriage (WRAP 2012).

15.15.2 Autoclave Systems

Digester feedstocks are pretreated through autoclave system (a pressurized chamber) analogous to the thermal hydrolysis system. These autoclaves treat the feedstock within, through steam at a continuous temperature and pressure aiding to sanitize, cleanse, and decompose the C-based material in the feedstock. Hence, the organic material is delivered to the system in a disintegrated form, which results in highly efficient digestion with improved biogas formation and enhanced digestate quality. Afterward the processing, inorganic substances and pollutants can be effortlessly eliminated through mechanical segregation, yielding an organic-rich, hygienic, and sterilized feedstock for AD (Holtman et al. 2017; Gaur et al. 2017).

15.15.3 Enzymatic Liquefaction

In this method, enzymes are employed to dissolve and additionally decompose the cellular organization of the feedstock, which are previously thermally pretreated and "exposed" for enzymes. The method involves three phases to degrade and segregate the organic material from within the feedstock preceding the digestion. Phase 1 includes non-pressurized thermic treatment employing either boiling water or vapor, which helps "opening" the feedstock making it available to enzymes. In the second step, enzymes are employed to dissolve and degrade the cellular organization of the feedstock. In the third step of the processing, the developed feedstock is being assimilated. After assimilation, the constituent parts are segregated in a way that C-abundant fluid for terrestrial incorporation can be effortlessly isolated from inorganic matter and other physical impurities (WRAP 2012; Gunes et al. 2019).

15.16 In-vessel Cleaning

Feedstock may have contaminations, e.g., plastic, timber, fibers, gravel or sand, metallic remains, and solid fruit remnants. While the digestion procedure itself includes substantial blending and stirring, the digestion container will perform as a depository for all feedstock. Dense substances tend to settle down, whereas lightweight substances move up the container and get trapped in a scum and froth film (Logan and Visvanathan 2019). In-vessel cleaning methods can be exploited to eliminate pollutants from the digesters, both refining digestate quality and averting the accumulation of inerts. Gravel and dense solid matter aggregating at the bottom-most of digesters is guided by a rotary scraper technique to the brink of the digester where they are eliminated and segregated from digestate. The segregated digestate is reverted back to the breakdown procedure. The isolated gravel and solids may be exploited as a cumulative modification for possible terrestrial remediation. Floating substances, e.g., plastics and tatters, can similarly be eliminated by means of a rotational skimmer. Substances are shoved toward the brink of the reactor where they are destroyed and isolated from the digestate. The isolated digestate is reverted back to the breakdown process, whereas the segregated solids are moreover reprocessed or discarded to landfills (Slorach et al. 2019; Logan and Visvanathan 2019).

15.17 Posttreatment of Digestates

Digestates can be utilized as fertilizer with no additional treatments once they are removed from the digester and subsequent chilling. As digestates commonly have little dry matter proportion, its storage, transportation, and incorporation are costly, making the processing and bulk mitigation of the digestate an appealing choice. Digestates can be treated through a number of diverse technologies. Treating digestates can have a number of objectives, liable on native requirements. If the objective is to improve the demand and quality of the digestate and to yield a consistent bio-fertilizer, this is termed as conditioning of digestate. If the objective is to eliminate minerals and C-based materials from the processed effluent, digestate may be treated by methods analogous to the treatment systems for wastewater. In the majority of circumstances, it is obligatory to perform both conditioning and treatment so as to initiate a sustainable digestate development (Drosg et al. 2015). From a practical perspective, refining the digestates can be partial, typically aiming to mitigate the size of the digestate, or can be complete, i.e., digestate's segregation into solid constituents, concentrates of inorganic minerals, and sterilized water (Al Seadi et al. 2013).

15.17.1 Partial Treatment of Digestates

Partial treatment or processing of digestates is a cost-effective, low-energy demanding, and an efficient conditioning technology (Dumitru 2014). It involves solid-liquid segregation procedure and is typically the first stage in treatment of digestate, splitting it up into a condensed P-abundant solid part and N-abundant liquid portion, aiming to dewater the digestate, and permits handling the minerals separately. Therefore, solid-liquid segregation offers lesser carriage charges, because of the lesser water fraction, along with simpler storing environments (Fuchs and Drosg 2013). A range of solid-liquid segregation methods are available including decanter centrifuges, sieve belt filter, flocculation or precipitation, screens and filters, screw press separators, and floatation. The solid portion can then be smeared directly as fertilizer on agronomic lands or may be allowed to dry or even composted for intermediary storage and better transportation. The solid part, without being treated further, can also be vended as a P-abundant fertilizer. The liquid portion, comprising N and K, can be incorporated as fluid fertilizer or blended with feedstock having higher solid proportion and reintroduced to the digesters (Al Seadi et al. 2013).

15.17.2 Complete Treatment of Digestates

Complete treatment or processing of digestate incorporates diverse number of techniques and strategies at different levels (Braun et al. 2008). Membrane procedures, for example, nano-filtration and ultrafiltration, succeeded by reverse osmosis are exploited for retrieval of nutrients. The fluid digestate on the other hand can also be refined through aerobic treatment. Other probability for condensing the digestate is evaporation, exhausting the extra heat from biogas system. Stripping and precipitation of struvite are also exploited to diminish the N ratio of the digestate (Plana and Noche 2016; Marti et al. 2008).

15.17.3 Screw Press Separators

The screw press is a frequently employed method most commonly exploited in moderate to wide-ranging biogas systems with digestates having greater fiber fraction. In this method, a screw pressurizes fibers against the cylinder-shaped mesh which allows drainage of the liquid portion. As the width of the screw expands, compression rises with the spread of fibers in the sieve, and eventually, the fiber portion leaves through the separator's exit, where automatically the resistance is calibrated. This technology is established, vigorous, and simple. In contrast to the decanter separators, screw press centrifuges cannot segregate slurry portion of the digestate. If digestate primarily comprises of the fiber portion, the quantity of solid fraction that will cumulate is reliant upon the digestate's dry matter fraction. The benefits of this method over the decanter separators are lower financial expenses and lower energy intake (Fuchs and Drosg 2010; Bolzonella et al. 2018).

15.17.4 Decanter Centrifuges

Decanter centrifuges are exploited to isolate any colloids or other tiny particles from the digestate. Furthermore, these centrifuges can also be utilized to segregate bulk of the P confined within the solid part in digestates. Decanter centrifuges and screw press centrifuge have achieved acceptance, particularly among agriculturalists exporting their surplus of nutrients to other regions (Fechter and Kraume 2016). Decanter centrifuges involve a rapidly spinning screw conveyor positioned in a slowly spinning jacket cylinder. The digestate arrives into the centrifuge through a pivotal channel and is supplied to the mid of the centrifuge. Through centrifugal force, the particles are segregated. The segregation activity is influenced by the particles mass and form, variance in density between the particles and liquid, and the liquid viscosity. Particles which are segregated cumulate on the cylinder walls and are conveyed and additionally compacted by a bolt. At one side, the solid portion exits the decanter through the ending outlet, while the clarified fluid exits the decanter through another side (Drosg et al. 2015).

15.17.5 Belt Filter Presses

Belt filters exploited for digestate treatment are of two types, i.e., belt filter press and vacuum belt filter. It comprises of a sealed sphere with a fabric belt twisted around the drums. Digestate is incorporated constantly at the starting point of the belt sieve. The initial pre-dewatering takes place through gravity. In the subsequent phase, materials are forced between two filter belts. Consequently, fluctuating automated power is employed in order to desiccate the filtered cake additionally and then lastly eliminated from the belt filter by a powered appliance. The belts' filters are then mist washed and are then re-exploited for filtration. Vacuum belt filters are another choice in which digestate is administered onto a mesh belt and vacuity is incorporated underneath, whereas water is drawn via the filter and the mesh cake resides on the belt. The level of segregation can be calibrated through contact pressure, flowing speed, and state of the belts along with the quantity of deflection rollers. For digestate treatment with belt filters, the introduction of precipitating or flocculating mediators is crucial so as to increase the separation efficacy. Different aspects that impact the separation efficacy are different features of the digestate, quantity and varieties of precipitating/flocculating mediators administered, and mesh sizes of the sieve. Benefits of this technique over the screw press are greater separation competence while having lesser energy requirements as compared to a decanter separator (Fechter and Kraume 2016; Kavitha et al. 2019).

15.17.6 Flotation

The concept of flotation is that "the stimulating potential of suspended solids is amplified by the adherence of tiny air bubbles to the particles suspended and thus causing them to float." Air-supplemented fluid is introduced into the flotation reservoir along with the introduction of flocculants. Flotation slurry is generated superficially on the sink which is then skimmed off. The benefit of flotation is providing an uncontaminated, nearly particle-free fluid portion, which, for example, can be simply utilized in membrane filtration. The flotation slurry has a greater water ratio and may need to be additionally condensed (Herbes et al. 2020).

Generally, flotation apparatus inhabits 30–50% lesser area than typical sedimentation apparatus as the lifting energy is considerably greater than the sedimentation energy. Two distinct flotation methods are present, i.e., flotation through decompression or via gassing. In the first method, air-saturated pressurized water is introduced into the flotation compartment. The abrupt decrease in pressure causes the generation of microbubbles. In the second method, air is introduced through distinct outlets directly which generates smaller air bubbles. For an effective flotation procedure, introduction of flotation agents analogous to precipitating or flocculating agents is required, creating a cluster of the suspended solids confined in the digestate. The introduction of flocculating agents may be required in many circumstances, for example, for separating a significantly lower dry matter fraction of the digestate or when a particle-free fluid portion is needed to be further processed (Herbes et al. 2020; Monfet et al. 2018; Drosg et al. 2015).

15.17.7 Precipitating Agents/Flocculants

Precipitating agents as well as flocculants are commonly introduced to digestate so as to enhance the separation efficacy of, for example, suspended solids or P by altering their physical state in basically any solid-liquid segregation method. The usage of chemicals for improvement of segregation is a comparatively novel method for digestate treatment. The chemicals frequently employed for elimination of P are aluminum sulfate, ferric chloride, ferric sulfate, and lime. For additional accumulation of clotted particles, the introduction of polymers may be required in certain conditions (Al Seadi et al. 2013).

Generally, though P is condensed in the solid portion of any solid-liquid segregation procedure, the separation efficacy can be amplified considerably through introduction of precipitating or flocculating agents. The concept of flocculation is that tiny suspended solids/particles in digestates have most commonly negative charge and so persist in the suspension. Here the precipitating and flocculating agents play their role. Ions carrying positive charge agglomerate around the particles, which causes the development of bigger particles. As a result, bigger particles made by flocculation can be removed more simply. Introduction of the precipitating or flocculating agents can be done both independently in the mixing containers preceding the solid-liquid segregation or in-line, showing that they are introduced straight into the transference channels, where blending methods are incorporated to deliver adequate instability (Meixner et al. 2015).

15.18 Further Treatment of the Fluid Portion

Afterward solid-liquid segregation, the fluid portion preserves some nutrients as well as suspended solids. Although a considerable quantity of the liquid portion may be employed in agronomy as soil enhancer or fertilizer, additional processing of the digestate is carried out to utilize the by-product. For example, rise in amount of nutrients can yield a high-grade fertilizer. However, increasing the demand for fluid digestates beyond its agronomic use is significant to produce more opportunities (Rigby and Smith 2011). A number of technologies can be employed in treatment of the fluid portion, for example, precipitation, membrane technology, struvite precipitation, evaporation, stripping, and biological treatments (Drosg et al. 2015).

15.18.1 Membrane Filtration

Membrane filtration technology is an alternate process for the treatment of fluid digestate. It is a physical segregation method in which the fluid digestate to be purified flows through a membrane or a molecular filter that performs as a physical blockade retaining the impurities, however permitting water to pass through. Hence, this process offers two products, i.e., nutrients and refined process water (Klink et al. 2007). Depending on the particular membrane choice, the porous membrane segregates pollutants at a molecular level from the digestate, yielding permeate which is potentially appropriate to be directly released to waterways and a congregate having the potential to be incorporated as a fertilizer. In membrane technology, a diverse number of membranes with diminishing pore sizes are typically employed. Examples include micro-filtration, ultrafiltration, nano-filtration, and reverse osmosis (Waeger et al. 2010).

Microfiltration has the ability to segregate particles having diameter of up to 0.1 μm, whereas ultrafiltration has the capability to retain dissolvable macromolecules and other large particles having diameters less than 0.01 micrometer (μm). Through nano-filtration and reverse osmosis, uncontaminated water having solubilized salts can be as well segregated and have the ability to retain tiny particles and ions (Logan and Visvanathan 2019).

15.18.2 Ammonia Stripping

Stripping is employed to eliminate or retrieve N from the liquid digestate in the form of NH_3. The instability of NH_3 in a water solution may be boosted up by elevating the temperature and pH (Guštin and Marinšek-Logar 2011). Thus, in digestate treatment, surplus heat is utilized for heating up the digestate, whereas pH can be raised by degassing, to eliminate CO_2, or by adding alkali (Drosg et al. 2015). The input substance is spread in a column, and gas is employed in the countercurrents in the form of air or steam. Air stripping is less energy demanding as compared to steam stripping but necessitates more lye. Digestate in gas stripping is provided heat in order to move into a stripping column which is packed with filling matter to expand the surface area offered for NH_3 mass transfer to the stripping gas column from the fluid digestate. CO_2 is eliminated as a result of pretreatment, thus lowering the buffer capacity. Subsequently, NH_3 is retrieved from the gaseous phase by means of a sulfuric acid (H_2SO_4) scrubber, thus producing an invaluable commercial class ammonium sulfate $(NH_4)_2SO_4$ fertilizer. The gas scrubbed can be used again in the stripping column (Jamaludin et al. 2018).

Vapor stripping requires elevated temperatures, in which NH_3 along with vapor can be directly condensed to generate NH_3 water having concentration of up to

25–35%. Effective solid-fluid segregation and a higher-level cleaning energy and maintenance may be required. A huge benefit of NH_3 stripping is the retrieval of a consistent and unpolluted N fertilizer. Furthermore, such fluid fertilizers can be utilized in digestate treatment to supplement other portions of the digestates with a standardized N concentration so as to proliferate its market value (De la Rubia et al. 2010).

15.18.3 Struvite Precipitation

Struvite is a term normally employed for the chemical composite, i.e., Mg ammonium phosphate having the molar ratio of 1.3:1:0.9, which can be utilized as a mineral fertilizer (Vidlarova et al. 2017). As NH_3 is almost all the time present in high quantities in digestate, it is supplemented with Mg oxide (MgO) and phosphoric acid (H_3PO_4). Moreover, pH is moderately raised to 8.5–9.0. The resultant struvite is retrieved as a solid substance, suitable for usage either as a fertilizer or a feedstock for the manufacture of fertilizers, as N, P, and Mg are invaluable plant nutrients. The addition of chemicals can be carried out either in the first stage with succeeding segregation through centrifugation, or supplementation of chemical and precipitation of the struvite crystals take place in the same container. P has limited worldwide resources, and as a result, recovery of struvite is expected to become highly significant in the coming era (Driver et al. 1999). This method does not usually eliminate all of the NH_4 from the digestate, due to the presence of scarce amount of P in the digestate (Munir et al. 2017; Jia et al. 2017).

15.18.4 Evaporation

Digestate evaporation can be a remarkable option if adequate surplus heat is accessible at the biogas installation to concentrate the digestate or increase the dry matter portion. Evaporation is usually employed for liquid or whole digestate processing and intends to preserve the minerals and a percentage of the moistness trapped in the digestate, thus producing a nutrient-abundant concentrate, while the condensate is retrieved. In such a method, specifically the fibers are separated to diminish the potential blockage of the evaporators or heat exchangers. Higher temperatures will result in the discharge of NH_3, which is controlled by diminishing the pH of digestate, usually with acid administration, e.g., sulfuric acid, preceding the evaporation. This method permits the digestate liquid to be transformed into a concentrated fertilizer (Bai et al. 2017; Tampio et al. 2016).

15.19 Further Treatment of the Solid Portion

When digestate is dewatered, partly stabilized solid portion may be directly exploited as a soil conditioner or a bio-fertilizer. As the solid part still comprises some decomposable material, microbial actions may still be vigorous, and release of odor can also take place. To diminish the ecological influence and to develop a marketable and stabilized bio-fertilizer, additional treatments, for example, composting and drying, are suggested for steadiness of the organic substance (Logan and Visvanathan 2019).

15.19.1 Composting

In composting, microorganisms decompose and convert C-based matter to compost in aerobic environments, causing the transformation of NH_3 to NO_3^- which is a highly stable and extremely mobile N source for plants (Botheju 2010). Compost is a model bio-fertilizer as it gradually liberates nutrients and displays good results as a soil enhancer. Composting can elevate N, P, Mg, and Fe, along with the total organic C, pH, and moistness of the soil. Temperature within the composting procedure can increase up to 70 °C or even more because of the strength of microbial activities; therefore pasteurization can be accomplished. Yet the capability to accomplish pasteurization will be reliant upon the composting procedure and related procedural regulations (Tambone et al. 2010; Bustamante et al. 2013; Zeng et al. 2016). Nevertheless, as the solid portion of digestate is moist and partly decomposed previously, supplementation of bulking substance, for example, woodchips, is required for a steady composting procedure to happen, assisting the air to move into the pile of compost. Moreover, the bulking material has progressive impacts, e.g., rise in concentration of minerals, reducing the electrical conductivity, reduction of N in the course of composting, and minimization of HM proportions in the ultimate product (Kharrazi et al. 2014).

15.19.2 Drying

The objective of drying is to vaporize the water in digestate making it dry and stabilized and also reducing the total mass. It furthermore elevates the nutrients' concentration whereas decreasing the moisture content and amount of N, making the storage and conveyance effortless. Unprocessed digestate displays greater discharge of NH_3 as compared to dewatered digestate (Maurer and Müller 2012). In several circumstances, electrical energy is generated at the biogas installations, employing the surplus heat for drying. It must be considered that the emissions from digestate dryers comprise dust, NH_3, and other unstable materials, and therefore exhaust gas scrubbing methods should be employed to diminish the discharges, e.g., acid scrubbers. On the other hand, NH_3 discharges can be evaded by acidification of the digestate prior to drying.

Solar drying is also incorporated for decreasing the humidity of solid portion of digestate or generating a concentrate of fluid digestate (Knoop et al. 2018). Furthermore, this method preserves energy; however large space is requisite. The desiccated solid portion may be pelletized for enhanced marketability and is made accessible as a bio-fertilizer (Keotiamchanh 2018).

15.20 Future Prospective of Digestates

In addition to the fertilizer potential or amendment features of digestate, currently there are also several other means for its utilization. These novel approaches are highly resourceful producing the likelihood of appropriate exploitation of digestates with diverse qualities. A recent potential exploitation of digestate is its usage as solid fuel when dried. Pellets made from digestate have enhanced mechanical strength. The caloric significance of these pellets is analogous to the caloric significance of the wood. Additives' addition moreover enhances the value of digestate pellets. One more remarkable option of digestate exploitation is the incorporation of digestate effluent as a substitute of freshwater and minerals for the production of bioethanol. One study has published 18% improved bioethanol generation with digestate effluents in contrast to freshwater exploitation (Gao and Li 2011). Synthesis and retrieval of VFAs from digestates are getting significant responsiveness because of their greater potency as a renewable C source, beside their extensive use in medicine, food, chemical production, bioplastics, and bio-H_2 production (Atasoy et al. 2018).

Higher pH range of digestate presents a solution to the worldwide acidification issues of the soil. Moreover, when digestate is utilized for remediating lands polluted with HM, its basicity can raise the pH of the soil, thus improving the immobility of HM (Peng and Pivato 2019). Sometimes, when the digestate quality is not appropriate for utilization as a fertilizer, for example, containing elevated levels of chemical impurities, and when such exploitation is banned by national legislature, their utilization for energy goals, e.g., co-combustion for energy production, might be an ultimate choice. Pyrolysis is moreover an evolving preference for management of digestate that transforms the C-based material into char, bio-oil, or synthetic gas in O_2-free conditions. Char has the potency to be exploited as a soil amendment or an energy source, whereas synthetic gas and bio-oils have higher caloric estimates and could be utilized as types of renewable energy. One other possibility for digestate is their usage as a covering substance for the sanitary landfills. Nevertheless, digestate requires dewatering and stabilization before dumping to make sure that it meets the strict criteria established for landfills (Monlau et al. 2015). The digestate's incorporation in reed beds as well as microbial fuel cells is also achieving marketability.

15.21 Conclusion

Reliance on the chemical fertilizers is not merely perilous for human intake but can likewise disrupt the environmental stability and have adverse influence on the human body when the food grown with these inorganic fertilizers are consumed.

Hence, the concern is now shifting toward the usage of bio-fertilizers. Digestate, a bio-fertilizer from biogas installations produced through good practices, is a top-quality product, appropriate and innocuous for usage as fertilizer in farming. Exploitation of digestates as bio-fertilizer reuses the minerals and organic materials and saves expenditures to the agronomists. The selection and stringent quality regulation of the resources exploited as feedstock for AD is the first and most significant stage of managing the digestate quality ensuring environmental and financial profits from exploitation of digestate as a bio-fertilizer. The incorporation of bio-fertilizers comprising useful microbes stimulates the crop throughput to a greater degree. The substantial decline of weed seeds and animal and plant pathogens by means of AD treatment lowers their spread and thus also increases well-being of animals. This provides digestate with a substantial benefit over the unprocessed feedstock. Its usage as bio-fertilizer adds to the conservation of the natural reservoirs of fossil P, a greatly valued but quickly diminishing reserve on our planet. This chapter specifies that substituting mineral fertilizers with digestates has the potentiality to sustain or increase crops yield and simultaneously decrease the potency for nutrients loss to the surroundings. This may eventually diminish agronomic reliance on mineral fertilizers and the energy and financial expenditures linked with their usage. Hence, AD should also be acknowledged as a pollution reduction technology besides being a source of renewable energy and hence plays a crucial part in sustainability and output of the soil and shields the environment being an ecofriendly and economical input for the agronomists.

References

Abubaker J, Cederlund H, Arthurson V, Pell M (2013) Bacterial community structure and microbial activity in different soils amended with biogas residues and cattle slurry. Appl Soil Ecol 72:171–180

Achinas S, Achinas V, Euverink GJW (2017) A technological overview of biogas production from biowaste. Engineering 3(3):299–307

Adeli A, Varco JJ, Sistani KR, Rowe DE (2005) Effects of swine lagoon effluent relative to commercial fertilizer applications on warm-season forage nutritive value. Agron J 97 (2):408–417

Ahring BK (2003) Perspectives for anaerobic digestion. In: Biomethanation. Springer, Berlin, pp 1–30

Al Seadi T (2001) Good practice in quality management of AD residues from biogas production. Report made for the International Energy Agency, Task

Al Seadi T, Lukehurst C (2012) Quality management of digestate from biogas plants used as fertiliser. IEA Bioenergy 37:40

Al Seadi T, Nielsen BH (2002) Manure based biogas System-Danish experience. In: Organic Diversion Symp., Ontario vol. I, pp. 1–15

Al Seadi T, Drosg B, Fuchs W, Rutz D, Janssen R (2013) Biogas digestate quality and utilization. In: The biogas handbook. Woodhead Publishing, Sawston, pp 267–301

Alburquerque JA, De la Fuente C, Campoy M, Carrasco L, Nájera I, Baixauli C, Bernal MP (2012) Agricultural use of digestate for horticultural crop production and improvement of soil properties. Eur J Agron 43:119–128

Alemán-Nava GS, Meneses-Jácome A, Cárdenas-Chávez DL, Díaz-Chavez R, Scarlat N, Dallemand JF, Parra R (2015) Bioenergy in Mexico: status and perspective. Biofuels Bioprod Biorefin 9(1):8–20

Alfa MI, Adie DB, Igboro SB, Oranusi US, Dahunsi SO, Akali DM (2014) Assessment of bio-fertiliser quality and health implications of anaerobic digestion effluent of cow dung and chicken droppings. Renew Energy 63:681–686

Angelidaki I, Hendriksen HV, Mathrani IM, Schmidt JE, Sørensen AH, Ahring BK (1996) The biogas process. Lecture notes for: Energy from biomass (6362)

Appels L, Lauwers J, Degrève J, Helsen L, Lievens B, Willems K, Dewil R (2011) Anaerobic digestion in global bio-energy production: potential and research challenges. Renew Sust Energ Rev 15(9):4295–4301

Arthurson V (2009) Closing the global energy and nutrient cycles through application of biogas residue to agricultural land–potential benefits and drawback. Energies 2(2):226–242

Atasoy M, Owusu-Agyeman I, Plaza E, Cetecioglu Z (2018) Bio-based volatile fatty acid production and recovery from waste streams: current status and future challenges. Bioresour Technol 268:773–786

Austin G, Morris G (2012) Biogas production in Africa. In: Bioenergy for sustainable development in Africa. Springer, Dordrecht, pp 103–115

Badshah M (2012) Evaluation of process parameters and treatments of different raw materials for biogas production. Lund University, Lund

Bagge E, Sahlström L, Albihn A (2005) The effect of hygienic treatment on the microbial flora of biowaste at biogas plants. Water Res 39(20):4879–4886

Bai X, Li Z, Ni J, Yun Y, Zhou X (2017) Nutrient recovery of liquid digestate from anaerobic digestion of human excrement by evaporation. Appl Eng Agric 33(2):227–234

Bakry MA, Soliman YR, Moussa SA (2009) Importance of micronutrients, organic manure and biofertilizer for improving maize yield and its components grown in desert sandy soil. Res J Agric Biol Sci 5(1):16–23

Banerjee A, Jhariya MK, Yadav DK, Raj A (2018) Micro-remediation of metals: a new frontier in bioremediation. In: Hussain C (ed) Handbook of environmental materials management. Springer, Cham

Banerjee A, Meena RS, Jhariya MK, Yadav DK (2021) Agroecological footprints management for sustainable food system. Springer, Singapore, p 514

Bauer A, Mayr H, Hopfner-Sixt K, Amon T (2009) Detailed monitoring of two biogas plants and mechanical solid–liquid separation of fermentation residues. J Biotechnol 142(1):56–63

Bendixen HJ (1999) Hygienic safety–results of scientific investigations in Denmark (sanitation requirements in Danish biogas plants). In: IEA bioenergy workshop: hygienic and environmental aspects of AD: Legislation and experiences in Europe

Biogas A (2018) Africa biogas partnership programme. http://www.africabiogas.org

Birkmose T (2007) Digested manure is a valuable fertilizer. In: European biogas workshop: the future of biogas in Europe, Vol. 3, pp 89–94

Bolzonella D, Fatone F, Gottardo M, Frison N (2018) Nutrients recovery from anaerobic digestate of agro-waste: techno-economic assessment of full scale applications. J Environ Manag 216:111–119

Botheju PDC (2010) Effects of free oxygen on anaerobic digestion (Doctoral dissertation, Doctoral thesis at Norwegian University of Science and Technology, Trondheim, Norway, vol 188, pp 245–255

Braun R, Weiland P, Wellinger A (2008) Biogas from energy crop digestion. In: IEA bioenergy task (Vol. 37, pp 1–20)

Bruni E, Jensen AP, Angelidaki I (2010) Comparative study of mechanical, hydrothermal, chemical and enzymatic treatments of digested biofibers to improve biogas production. Bioresour Technol 101(22):8713–8717

Bustamante MA, Restrepo AP, Alburquerque JA, Pérez-Murcia MD, Paredes C, Moral R, Bernal MP (2013) Recycling of anaerobic digestates by composting: effect of the bulking agent used. J Clean Prod 47:61–69

Canali S, Di Bartolomeo E, Tittarelli F, Montemurro F, Verrastro V, Ferri D (2011) Comparison of different laboratory incubation procedures to evaluate nitrogen mineralization in soils amended with aerobic and anaerobic stabilized organic materials. J Food Agric Environ 9:540–546

Cao Y, Wang J, Wu H, Yan S, Guo D, Wang G, Ma Y (2016) Soil chemical and microbial responses to biogas slurry amendment and its effect on Fusarium wilt suppression. Appl Soil Ecol 107:116–123

Chun-Li W, Shiuan-Yuh C, Chiu-Chung Y (2014) Present situation and future perspective of bio-fertiliser for environmentally friendly agriculture. Annual Reports, pp 1–5

Coelho JJ, Prieto ML, Dowling S, Hennessy A, Casey I, Woodcock T, Kennedy N (2018) Physical-chemical traits, phytotoxicity and pathogen detection in liquid anaerobic digestates. Waste Manag 78:8–15

Coelho JJ, Prieto ML, Hennessy A, Casey I, Woodcock T, Kennedy N (2019) Determination of microbial numbers in anaerobically digested bio-fertilisers. Environ Technol 2019:1–11. https://doi.org/10.1080/09593330.2019.1645214

Coutinho J, Dalla Costa L, Borin M, Battilani A, Dolezal F, Bizik J, Plauborg FL (2006) One single value for maximum N application from organic residues: Is it technically and environmentally sound? In: ESA Congress

Coutteau P, Sorgeloos P (1992) The use of algal substitutes and the requirement for live algae in the hatchery and nursery rearing of bivalve molluscs: an international survey. J Shellfish Res 11:467–467

Crawford JH (1983) Composting of agricultural wastes-a review. Process Biochem 18:14–18

Crolla A, Kinsley C, Pattey E (2013) Land application of digestate. In: The biogas handbook. Woodhead Publishing, Cambridge, pp 302–325

Czekala W (2017) Processing digestate from agricultural biogas plant to obtain financial and environmental benefits. In: HAICTA, pp 188–195

De la Fuente C, Alburquerque JA, Clemente R, Bernal MP (2013) Soil C and N mineralisation and agricultural value of the products of an anaerobic digestion system. Biol Fertil Soils 49 (3):313–322

De la Rubia MÁ, Walker M, Heaven S, Banks CJ, Borja R (2010) Preliminary trials of in situ ammonia stripping from source segregated domestic food waste digestate using biogas: effect of temperature and flow rate. Bioresour Technol 101(24):9486–9492

Demirel B, Scherer P (2008) The roles of acetotrophic and hydrogenotrophic methanogens during anaerobic conversion of biomass to methane: a review. Rev Environ Sci 7(2):173–190

Doran JW, Parkin TB (1994) Defining and assessing soil quality. Sustain Environ 35:1–21

Driver J, Lijmbach D, Steen I (1999) Why recover phosphorus for recycling, and how? Environ Technol 20(7):651–662

Drosg B, Fuchs W, Al Seadi T, Madsen M, Linke B (2015) Nutrient recovery by biogas digestate processing. IEA Bioenergy, Dublin, p 711

Dumitru M (2014) Studies concerning the utilisation of digestate in biogas plants. Sci Pap Ser 14:1

Eurostat SA (2017) Your key to European statistics

Facchin V, Cavinato C, Fatone F, Pavan P, Cecchi F, Bolzonella D (2013) Effect of trace element supplementation on the mesophilic anaerobic digestion of foodwaste in batch trials: the influence of inoculum origin. Biochem Eng J 70:71–77

Fagerström A, Al Seadi T, Rasi S, Briseid T (2018) The role of anaerobic digestion and biogas in the circular economy. IEA Bioenergy

Fang C, Boe K, Angelidaki I (2011) Anaerobic co-digestion of by-products from sugar production with cow manure. Water Res 45(11):3473–3480

Fechter M, Kraume M (2016) Digestate treatment techniques. CzasopismoTechniczn

Fernandez-Bayo JD, Achmon Y, Harrold DR, McCurry DG, Hernandez K, Dahlquist-Willard RM, Simmons CW (2017) Assessment of two solid anaerobic digestate soil amendments for effects on soil quality and biosolarization efficacy. J Agric Food Chem 65(17):3434–3442

Fouda S, von Tucher S, Lichti F, Schmidhalter U (2013) Nitrogen availability of various biogas residues applied to ryegrass. J Plant Nutr Soil Sci 176(4):572–584

Frąc M, Oszust K, Lipiec J (2012) Community level physiological profiles (CLPP), characterization and microbial activity of soil amended with dairy sewage sludge. Sensors 12(3):3253–3268

Franke-Whittle IH, Walter A, Ebner C, Insam H (2014) Investigation into the effect of high concentrations of volatile fatty acids in anaerobic digestion on methanogenic communities. Waste Manag 34(11):2080–2089

Fuchs W, Drosg B (2010) Technologiebewertung von Gärrestbehandlungs-und Verwertungskonzepten. na

Fuchs W, Drosg B (2013) Assessment of the state of the art of technologies for the processing of digestate residue from anaerobic digesters. Water Sci Technol 67(9):1984–1993

Gao T, Li X (2011) Using thermophilic anaerobic digestate effluent to replace freshwater for bioethanol production. Bioresour Technol 102(2):2126–2129

Garg RN, Pathak H, Das DK, Tomar RK (2005) Use of fly ash and biogas slurry for improving wheat yield and physical properties of soil. Environ Monit Assess 107(1-3):1–9

Gaur V (2010) Biofertilizer–necessity for sustainability. J Adv Dev 1:7–8

Gaur RZ, Khan AA, Suthar S (2017) Effect of thermal pre-treatment on co-digestion of duckweed (Lemna gibba) and waste activated sludge on biogas production. Chemosphere 174:754–763

Gell K, van Groenigen J, Cayuela ML (2011) Residues of bioenergy production chains as soil amendments: immediate and temporal phytotoxicity. J Hazard Mater 186(2-3):2017–2025

Georgacakis D, Tsavdaris A (1992) Combined anaerobic digestion of animal slurries and food industry liquid by-products. REUR Technical Series (FAO)

Gerardi MH (2003) The microbiology of anaerobic digesters. Wiley, London

Global Methane Initiative (2013) Successful applications of anaerobic digestion from across the world

Gómez X, Cuetos MJ, García AI, Morán A (2007) An evaluation of stability by thermogravimetric analysis of digestate obtained from different biowastes. J Hazard Mater 149(1):97–105

Gong W, Yan X, Wang J, Hu T, Gong Y (2011) Long-term applications of chemical and organic fertilizers on plant-available nitrogen pools and nitrogen management index. Biol Fertil Soils 47 (7):767

Govasmark E, Stäb J, Holen B, Hoornstra D, Nesbakk T, Salkinoja-Salonen M (2011) Chemical and microbiological hazards associated with recycling of anaerobic digested residue intended for agricultural use. Waste Manag 31(12):2577–2583

Govender I, Thopil GA, Inglesi-Lotz R (2019) Financial and economic appraisal of a biogas to electricity project. J Clean Prod 214:154–165

Grigatti M, Di Girolamo G, Chincarini R, Ciavatta C, Barbanti L (2011) Potential nitrogen mineralization, plant utilization efficiency and soil CO_2 emissions following the addition of anaerobic digested slurries. Biomass Bioenergy 35(11):4619–4629

Groot LD, Bogdanski A (2013) Bioslurry-brown gold? A review of scientific literature on the co-product of biogas production. Food and Agriculture Organization of the United Nations (FAO)

Gujer W, Zehnder AJ (1983) Conversion processes in anaerobic digestion. Water Sci Technol 15 (8-9):127–167

Gunes B, Stokes J, Davis P, Connolly C, Lawler J (2019) Pre-treatments to enhance biogas yield and quality from anaerobic digestion of whiskey distillery and brewery wastes: a review. Renew Sust Energ Rev 113:109281

Gunnarsson A, Bengtsson F, Caspersen S (2010) Use efficiency of nitrogen from biodigested plant material by ryegrass. J Plant Nutr Soil Sci 173(1):113–119

Guo J, Peng Y, Ni BJ, Han X, Fan L, Yuan Z (2015) Dissecting microbial community structure and methane-producing pathways of a full-scale anaerobic reactor digesting activated sludge from wastewater treatment by metagenomic sequencing. Microb Cell Factories 14(1):33

Guštin S, Marinšek-Logar R (2011) Effect of pH, temperature and air flow rate on the continuous ammonia stripping of the anaerobic digestion effluent. Process Saf Environ Prot 89(1):61–66

Haraldsen TK, Andersen U, Krogstad T, Sørheim R (2011) Liquid digestate from anaerobic treatment of source-separated household waste as fertilizer to barley. Waste Manag Res 29 (12):1271–1276

Herbes C, Roth U, Wulf S, Dahlin J (2020) Economic assessment of different biogas digestate processing technologies: a scenario-based analysis. J Clean Prod 255:120282

Hjorth M, Christensen KV, Christensen ML, Sommer SG (2011) Solid–liquid separation of animal slurry in theory and practice. In: Sustainable agriculture, vol 2. Springer, Dordrecht, pp 953–986
Holm-Nielsen JB, Al Seadi T, Oleskowicz-Popiel P (2009) The future of anaerobic digestion and biogas utilization. Bioresour Technol 100(22):5478–5484
Holtman KM, Bozzi DV, Franqui-Villanueva D, Offeman RD, Orts WJ (2017) Pilot scale high solids anaerobic digestion of steam autoclaved municipal solid waste (MSW) pulp. Renew Energy 113:257–265
IEA (2016) World energy outlook 2016, Executive Summary. International Energy Agency, pp 1–8
Insam H, Gómez-Brandón M, Ascher J (2015) Manure-based biogas fermentation residues–friend or foe of soil fertility? Soil Biol Biochem 84:1–14
Itelima JU, Bang WJ, Onyimba IA, Oj E (2018) A review: bio-fertiliser; a key player in enhancing soil fertility and crop productivity. J Microbiol Biotechnol Rep 2:22–28
Jamaludin Z, Rollings-Scattergood S, Lutes K, Vaneeckhaute C (2018) Evaluation of sustainable scrubbing agents for ammonia recovery from anaerobic digestate. Bioresour Technol 270:596–602
Jhariya MK, Banerjee A, Meena RS, Yadav DK (2019a) Sustainable agriculture, forest and environmental management. Springer, Singapore, p 606
Jhariya MK, Yadav DK, Banerjee A (2019b) Agroforestry and climate change: issues and challenges. Apple Academic Press Inc., New York, p 335
Jhariya MK, Meena RS, Banerjee A (2021a) Ecological intensification of natural resources for sustainable agriculture. Springer, Singapore
Jhariya MK, Meena RS, Banerjee A (2021b) Ecological intensification of natural resources towards sustainable productive system. In: Ecological intensification of natural resources for sustainable agriculture. Springer, Singapore
Jia G, Zhang H, Krampe J, Muster T, Gao B, Zhu N, Jin B (2017) Applying a chemical equilibrium model for optimizing struvite precipitation for ammonium recovery from anaerobic digester effluent. J Clean Prod 147:297–305
Jingming L (2014) The future of biogas in China. China biogas society Berlin Germany. Retrieved from www.abfz.de
Johansen A, Carter MS, Jensen ES, Hauggard-Nielsen H, Ambus P (2013) Effects of digestate from anaerobically digested cattle slurry and plant materials on soil microbial community and emission of CO_2 and N_2O. Appl Soil Ecol 63:36–44
Juárez MFD, Waldhuber S, Knapp A, Partl C, Gómez-Brandón M, Insam H (2013) Wood ash effects on chemical and microbiological properties of digestate-and manure-amended soils. Biol Fertil Soils 49(5):575–585
Kapoor R, Vijay VD (2013) Evaluation of existing low cost gas bottling systems for vehicles use adaption in developing economies. Public deliverable EU FP7 VALORGAS project (grant agreement no. 241334)
Kavitha S, Devi TP, Kannah RY, Kaliappan S, Banu JR (2019) Post-treatment methods for organic solid wastes. https://doi.org/10.2166/9781780409740_0323
Keotiamchanh S (2018) Food waste anaerobic digestate treatment and management strategies. Master's Thesis. Environmental Engineering and Management, Asian Institute of Technology, Thailand
Khan N, Jhariya MK, Raj A, Banerjee A, Meena RS (2021a) Soil carbon stock and sequestration: implications for climate change adaptation and mitigation. In: Jhariya MK, Meena RS, Banerjee A (eds) Ecological intensification of natural resources for sustainable agriculture. Springer, Singapore
Khan N, Jhariya MK, Raj A, Banerjee A, Meena RS (2021b) Eco-designing for sustainability. In: Jhariya MK, Meena RS, Banerjee A (eds) Ecological intensification of natural resources for sustainable agriculture. Springer, Singapore
Kharrazi SM, Younesi H, Abedini-Torghabeh J (2014) Microbial biodegradation of waste materials for nutrients enrichment and heavy metals removal: an integrated composting-vermicomposting process. Int Biodeterior Biodegrad 92:41–48
Klink G, Salewski C, Bolduan P (2007) VomGärrestzumNährstoffkonzentrat (From digestate to nutrient concentrate). Verfahrenstechnik 10:46–47

Knoop C, Dornack C, Raab T (2018) Effect of drying, composting and subsequent impurity removal by sieving on the properties of digestates from municipal organic waste. Waste Manag 72:168–177

Kumar S, Meena RS, Jhariya MK (2020) Resources use efficiency in agriculture. Springer, Singapore, p 760

Kumar S, Meena RS, Datta R, Verma SK, Yadav GS, Pradhan GS, Molaei A, Rahman GKMM, Mashuk HA (2020a) Legumes for carbon and nitrogen cycling: an organic approach. Carbon Nitrogen Cycl Soil. https://doi.org/10.1007/978-981-13-7264-3_10

Kumar S, Meena RS, Singh RK, Munir TM, Datta R, Danish S, Singh GS, Kumar S (2021) Soil microbial and nutrient dynamics under different sowings environment of Indian mustard (Brassica juncea L.) in rice based cropping system. Sci Rep 11:5289. https://doi.org/10.1038/s41598-021-84742-4

Kupper T, Bürge D, Bachmann HJ, Güsewell S, Mayer J (2014) Heavy metals in source-separated compost and digestates. Waste Manag 34(5):867–874

Larsen T, Luxhøi J, Magid J, Jensen LS, Krogh PH (2007) Properties of anaerobically digested and composted municipal solid waste assessed by linking soil mesofauna dynamics and nitrogen modelling. Biol Fertil Soils 44(1):59–68

Lencioni G, Imperiale D, Cavirani N, Marmiroli N, Marmiroli M (2016) Environmental application and phytotoxicity of anaerobic digestate from pig farming by in vitro and in vivo trials. Int J Environ Sci Technol 13(11):2549–2560

Levén L, Eriksson AR, Schnürer A (2007) Effect of process temperature on bacterial and archaeal communities in two methanogenic bioreactors treating organic household waste. FEMS Microbiol Ecol 59(3):683–693

Logan M, Visvanathan C (2019) Management strategies for anaerobic digestate of organic fraction of municipal solid waste: current status and future prospects. Waste Manag Res 37(1):27–39

Long A, Heitman J, Tobias C, Philips R, Song B (2013) Co-occurring anammox, denitrification, and codenitrification in agricultural soils. Appl Environ Microbiol 79(1):168–176

Lukehurst CT, Frost P, Al Seadi T (2010) Utilisation of digestate from biogas plants as bio-fertiliser. IEA Bioenergy, pp 1–36

Magulova K (2012) Stockholm convention on persistent organic pollutants: triggering, streamlining and catalyzing global scientific exchange. Atmos Pollut Res 3(4):366–368

Makádi M, Tomócsik A, Orosz V (2012) Digestate: a new nutrient source–review. Biogas 2012:295–310

Marti N, Bouzas A, Seco A, Ferrer J (2008) Struvite precipitation assessment in anaerobic digestion processes. Chem Eng J 141(1-3):67–74

Masciandaro G, Ceccanti B (1999) Assessing soil quality in different agro-ecosystems through biochemical and chemico-structural properties of humic substances. Soil Tillage Res 51 (1-2):129–137

Massé DI, Croteau F, Masse L (2007) The fate of crop nutrients during digestion of swine manure in psychrophilic anaerobic sequencing batch reactors. Bioresour Technol 98(15):2819–2823

Maucieri C, Barbera AC, Vymazal J, Borin M (2017) A review on the main affecting factors of greenhouse gases emission in constructed wetlands. Agric For Meteorol 236:175–193

Maurer C, Müller J (2012) Ammonia (NH3) emissions during drying of untreated and dewatered biogas digestate in a hybrid waste-heat/solar dryer. Eng Life Sci 12(3):321–326

Meena RS, Lal R (2018) Legumes for soil health and sustainable management. Springer, Singapore, p 541

Meena RS, Kumar V, Yadav GS, Mitran T (2018) Response and interaction of Bradyrhizo biumjaponicum and arbuscular mycorrhizal fungi in the soybean rhizosphere: a review. Plant Growth Regul 84:207–223. https://doi.org/10.1007/s10725-017-0334-8

Meena RS, Lal R, Yadav GS (2020a) Long term impacts of topsoil depth and amendments on soil physical and hydrological properties of an Alfisol in Central Ohio, USA. Geoderma 363:1141164. https://doi.org/10.1016/j.geoderma.2019.114164

Meena RS, Lal R, Yadav GS (2020b) Long-term impact of topsoil depth and amendments on carbon and nitrogen budgets in the surface layer of an Alfisol in Central Ohio. Catena 194:104752. https://doi.org/10.1016/j.catena.2020.104752

Meixner K, Fuchs W, Valkova T, Svardal K, Loderer C, Neureiter M, Drosg B (2015) Effect of precipitating agents on centrifugation and ultrafiltration performance of thin stillage digestate. Sep Purif Technol 145:154–160

Menardo S, Gioelli F, Balsari P (2011) The methane yield of digestate: effect of organic loading rate, hydraulic retention time, and plant feeding. Bioresour Technol 102(3):2348–2351

Mishra BK, Dadhich SK (2010) Methodology of nitrogen bio-fertiliser production. J Adv Dev Res 1(1):3–6

Moestedt J, Malmborg J, Nordell E (2015) Determination of methane and carbon dioxide formation rate constants for semi-continuously fed anaerobic digesters. Energies 8(1):645–655

Möller K, Müller T (2012) Effects of anaerobic digestion on digestate nutrient availability and crop growth: a review. Eng Life Sci 12(3):242–257

Möller K, Schultheiß U (2015) Chemical characterization of commercial organic fertilizers. Arch Agron Soil Sci 61(7):989–1012

Möller K, Stinner W, Leithold G (2008) Growth, composition, biological N2 fixation and nutrient uptake of a leguminous cover crop mixture and the effect of their removal on field nitrogen balances and nitrate leaching risk. Nutr Cycl Agroecosyst 82(3):233

Monfet E, Aubry G, Ramirez AA (2018) Nutrient removal and recovery from digestate: a review of the technology. Biofuels 9(2):247–262

Monlau F, Sambusiti C, Ficara E, Aboulkas A, Barakat A, Carrere H (2015) New opportunities for agricultural digestate valorization: current situation and perspectives. Energy Environ Sci 8 (9):2600–2621

Montes F, Meinen R, Dell C, Rotz A, Hristov AN, Oh J, Dijkstra J (2013) Special topics—mitigation of methane and nitrous oxide emissions from animal operations: II. A review of manure management mitigation options. J Anim Sci 91(11):5070–5094

Morelli J (2011) Environmental sustainability: a definition for environmental professionals. J Environ Sust 1(1):2

Munir MT, Li B, Boiarkina I, Baroutian S, Yu W, Young BR (2017) Phosphate recovery from hydrothermally treated sewage sludge using struvite precipitation. Bioresour Technol 239:171–179

Nelson MC, Morrison M, Yu Z (2011) A meta-analysis of the microbial diversity observed in anaerobic digesters. Bioresour Technol 102(4):3730–3739

Nkoa R (2014) Agricultural benefits and environmental risks of soil fertilization with anaerobic digestates: a review. Agron Sustain Dev 34(2):473–492

Nyberg K, Sundh I, Johansson M, Schnürer A (2004) Presence of potential ammonia oxidation (PAO) inhibiting substances in anaerobic digestion residues. Appl Soil Ecol 26(2):107–112

Oenema O, Tamminga S (2005) Nitrogen in global animal production and management options for improving nitrogen use efficiency. Sci China Ser C Life Sci 48(2):871–887

Owamah HI (2013) Heavy metals determination and assessment in a petroleum impacted river in the Niger Delta Region of Nigeria. Albanian J Agric Sci 12(1):129

Owamah HI, Dahunsi SO, Oranusi US, Alfa MI (2014) Fertilizer and sanitary quality of digestate bio-fertiliser from the co-digestion of food waste and human excreta. Waste Manag 34 (4):747–752

Peng W, Pivato A (2019) Sustainable management of digestate from the organic fraction of municipal solid waste and food waste under the concepts of back to earth alternatives and circular economy. Waste Biomass Valorizat 10(2):465–481

Pérez-Camacho MN, Curry R (2018) Regional assessment of bioeconomy options using the anaerobic biorefinery concept. In: Proceedings of the Institution of Civil Engineers-Waste and Resource Management (Vol. 171, No. 4). Thomas Telford Ltd., pp 104–113

Pfundtner E (2002) Limits and merits of sludge utilisation–land application. In: Conference proceedings of impacts of waste management. Legislation on Biogas Technology. Tulln, pp 1–10

Pitts J (2019) Digestates from food waste and lignocellulosic materials: effects on plant growth (Doctoral dissertation, University of Hawaii at Manoa)

Plana PV, Noche B (2016) A review of the current digestate distribution models: storage and transport. WIT Trans Ecol Environ 202:345–357

Poeschl M, Ward S, Owende P (2012) Environmental impacts of biogas deployment–Part II: life cycle assessment of multiple production and utilization pathways. J Clean Prod 24:184–201

Pognani M, D'Imporzano G, Scaglia B, Adani F (2009) Substituting energy crops with organic fraction of municipal solid waste for biogas production at farm level: a full-scale plant study. Process Biochem 44(8):817–821

Qi G, Pan Z, Sugawa Y, Andriamanohiarisoamanana FJ, Yamashiro T, Iwasaki M, Umetsu K (2018) Comparative fertilizer properties of digestates from mesophilic and thermophilic anaerobic digestion of dairy manure: focusing on plant growth promoting bacteria (PGPB) and environmental risk. J Mater Cycl Waste Manage 20(3):1448–1457

Raj A, Jhariya MK, Yadav DK, Banerjee A, Meena RS (2019a) Agroforestry: a holistic approach for agricultural sustainability. In: Jhariya MK, Banerjee A, Meena RS, Yadav DK (eds) Sustainable agriculture, forest and environmental management. Springer, Singapore, p 606

Raj A, Jhariya MK, Banerjee A, Yadav DK, Meena RS (2019b) Soil for sustainable environment and ecosystems management. In: Jhariya MK, Banerjee A, Meena RS, Yadav DK (eds) Sustainable agriculture, forest and environmental management. Springer, Singapore, p 606

Raj A, Jhariya MK, Yadav DK, Banerjee A (2020) Climate change and agroforestry systems: adaptation and mitigation strategies. Apple Academic Press Inc., New York, p 383

Raj A, Jhariya MK, Khan N, Banerjee A, Meena RS (2021) Ecological intensification for sustainable development. In: Ecological intensification of natural resources for sustainable agriculture. Springer, Singapore

Raja N (2013) Biopesticides and bio-fertilisers: ecofriendly sources for sustainable agriculture. J Biofertil Biopestici 4:e112

Raposo F, De la Rubia MA, Fernández-Cegrí V, Borja R (2012) Anaerobic digestion of solid organic substrates in batch mode: an overview relating to methane yields and experimental procedures. Renew Sust Energ Rev 16(1):861–877

Raturi AK (2016) Renewables 2016 Global status report

Rehl T, Müller J (2011) Life cycle assessment of biogas digestate processing technologies. Resour Conserv Recycl 56(1):92–104

REN21 (2011) Renewables 2011 global status report. REN21 Secretariat

REN21 (2016) Global status report. Paris, 2016

Rigby H, Smith SR (2011) New markets for digestate from anaerobic digestion. WRAP Report: ISS001-001

Risberg K (2015) Quality and function of anaerobic digestion residues (Vol. 2015, No. 40)

Romaniuk R, Giuffré L, Costantini A, Nannipieri P (2011) Assessment of soil microbial diversity measurements as indicators of soil functioning in organic and conventional horticulture systems. Ecol Indic 11(5):1345–1353

Sahlström L, Bagge E, Emmoth E, Holmqvist A, Danielsson-Tham ML, Albihn A (2008) A laboratory study of survival of selected microorganisms after heat treatment of biowaste used in biogas plants. Bioresour Technol 99(16):7859–7865

Sambusiti C, Monlau F, Ficara E, Musatti A, Rollini M, Barakat A, Malpei F (2015) Comparison of various post-treatments for recovering methane from agricultural digestate. Fuel Process Technol 137:359–365

Santos VB, Araújo AS, Leite LF, Nunes LA, Melo WJ (2012) Soil microbial biomass and organic matter fractions during transition from conventional to organic farming systems. Geoderma 170:227–231

Satpathy P, Steinigeweg S, Cypionka H, Engelen B (2016) Different substrates and starter inocula govern microbial community structures in biogas reactors. Environ Technol 37(11):1441–1450

Savci S (2012) An agricultural pollutant: chemical fertilizer. Int J Environ Sci Dev 3(1):73

Seppälä M, Pyykkönen V, Väisänen A, Rintala J (2013) Biomethane production from maize and liquid cow manure–effect of share of maize, post-methanation potential and digestate characteristics. Fuel 107:209–216

Sheoran S, Kuma S, Kumar P, Meena RS, Rakshit S (2021) Nitrogen fixation in maize: breeding opportunities. Theor Appl Genet. https://doi.org/10.1007/s00122-021-03791-5

Slorach PC, Jeswani HK, Cuéllar-Franca R, Azapagic A (2019) Environmental and economic implications of recovering resources from food waste in a circular economy. Sci Total Environ 693:133516

Sommer SG, Husted S (1995) A simple model of pH in slurry. J Agric Sci 124(3):447–453

Stambasky J, Pflüger S, Deremince B, Scheidl S, de la Vega N, Conton M (2016) Annual statistical report of the European biogas association. European Biogas Association, Brussels

Statistics RC (2016) International renewable energy agency (IRENA)

Steffen R, Szolar O, Braun R (1998) Feedstocks for anaerobic digestion. Institute of Agrobiotechnology Tulin, University of Agricultural Sciences, Vienna

Stiles WA, Styles D, Chapman SP, Esteves S, Bywater A, Melville L, Chaloner T (2018) Using microalgae in the circular economy to valorise anaerobic digestate: challenges and opportunities. Bioresour Technol 267:732–742

St-Pierre B, Wright ADG (2014) Comparative metagenomic analysis of bacterial populations in three full-scale mesophilic anaerobic manure digesters. Appl Microbiol Biotechnol 98 (6):2709–2717

Stronach SM, Rudd T, Lester JN (2012) Anaerobic digestion processes in industrial wastewater treatment, vol 2. Springer, New York

Sujanya S, Chandra S (2011) Effect of part replacement of chemical fertilizers with organic and bio-organic agents in ground nut, Arachis hypogea. J Algal Biomass Util 2(4):38–40

Sullivan TS, Stromberger ME, Paschke MW, Ippolito JA (2006) Long-term impacts of infrequent biosolids applications on chemical and microbial properties of a semi-arid rangeland soil. Biol Fertil Soils 42(3):258–266

Sun L, Pope PB, Eijsink VG, Schnürer A (2015) Characterization of microbial community structure during continuous anaerobic digestion of straw and cow manure. Microb Biotechnol 8 (5):815–827

Sundberg C, Al-Soud WA, Larsson M, Alm E, Yekta SS, Svensson BH, Karlsson A (2013) 454 pyrosequencing analyses of bacterial and archaeal richness in 21 full-scale biogas digesters. FEMS Microbiol Ecol 85(3):612–626

Surendra KC, Takara D, Hashimoto AG, Khanal SK (2014) Biogas as a sustainable energy source for developing countries: opportunities and challenges. Renew Sust Energ Rev 31:846–859

Tambone F, Scaglia B, D'Imporzano G, Schievano A, Orzi V, Salati S, Adani F (2010) Assessing amendment and fertilizing properties of digestates from anaerobic digestion through a comparative study with digested sludge and compost. Chemosphere 81(5):577–583

Tampio E, Marttinen S, Rintala J (2016) Liquid fertilizer products from anaerobic digestion of food waste: mass, nutrient and energy balance of four digestate liquid treatment systems. J Clean Prod 125:22–32

Tiwari VN, Tiwari KN, Upadhyay RM (2000) Effect of crop residues and biogas slurry incorporation in wheat on yield and soil fertility. J Indian Soc Soil Sci 48(3):515–520

Topper PA, Graves RE, Richard T (2006) The fate of nutrients and pathogens during anaerobic digestion of dairy manure. Pennsylvania State, Department of Agricultural and Biological Engineering G, 71

Tran MT, Vu TKV, Sommer SG, Jensen LS (2011) Nitrogen turnover and loss during storage of slurry and composting of solid manure under typical Vietnamese farming conditions. J Agric Sci 149(3):285–296

Tsai SH, Liu CP, Yang SS (2007) Microbial conversion of food wastes for bio-fertiliser production with thermophilic lipolytic microbes. Renew Energy 32(6):904–915

Vanwonterghem I, Jensen PD, Ho DP, Batstone DJ, Tyson GW (2014) Linking microbial community structure, interactions and function in anaerobic digesters using new molecular techniques. Curr Opin Biotechnol 27:55–64

Vidlarova P, Heviankova S, Kyncl M (2017) Contribution to the study of ammonia removal from digestate by struvite precipitation. IOP Conf Ser 92(1):012072

Waeger F, Delhaye T, Fuchs W (2010) The use of ceramic microfiltration and ultrafiltration membranes for particle removal from anaerobic digester effluents. Separat Purif Tech 73 (2):271–278

Walsh JJ, Jones DL, Edwards-Jones G, Williams AP (2012a) Replacing inorganic fertilizer with anaerobic digestate may maintain agricultural productivity at less environmental cost. J Plant Nutr Soil Sci 175(6):840–845

Walsh JJ, Rousk J, Edwards-Jones G, Jones DL, Williams AP (2012b) Fungal and bacterial growth following the application of slurry and anaerobic digestate of livestock manure to temperate pasture soils. Biol Fertil Soils 48(8):889–897

Weiland P (2001) VDI-Berichte Nr. 1620, Grundlagen der Methangärung-Biologie und Substrate (Basics of methane fermentation-Biology and substrates). Biogas als regenerative Energie-Stand und Perspektiven, Hannover, Germany, pp 19–20

Weiland P (2003) Production and energetic use of biogas from energy crops and wastes in Germany. Appl Biochem Biotechnol 109(1-3):263–274

Wellinger A, Murphy JD, Baxter D (eds) (2013) The biogas handbook: science, production and applications. Elsevier, London

Wentzel S, Joergensen RG (2016) Effects of biogas and raw slurries on grass growth and soil microbial indices. J Plant Nutr Soil Sci 179(2):215–222

Westerholm M, Schnürer A (2019) Microbial responses to different operating practices for biogas production systems [working title]

Wienhold BJ, Andrews SS, Karlen DL (2004) Soil quality: a review of the science and experiences in the USA. Environ Geochem Health 26(2):89–95

Wilkinson KG (2011) A comparison of the drivers influencing adoption of on-farm anaerobic digestion in Germany and Australia. Biomass Bioenergy 35(5):1613–1622

WRAP (2010) Final report: waste arisings in the supply of food and drink to households in the UK, March, 2010

WRAP (2012) Enhancement and treatment of digestates from anaerobic digestion, desk top study on digestate enhancement and treatment. Pell Frischmann Consultants Ltd, London

Yun YS, Park JI, Suh MS, Park JM (2000) Treatment of food wastes using slurry-phase decomposition. Bioresour Technol 73(1):21–27

Zakrzewski M, Goesmann A, Jaenicke S, Jünemann S, Eikmeyer F, Szczepanowski R, Schlüter A (2012) Profiling of the metabolically active community from a production-scale biogas plant by means of high-throughput metatranscriptome sequencing. J Biotechnol 158(4):248–258

Zang W, Han DY, Dick WA, Davis KR, Hoitink HAJ (1998) Compost and compost water extract-induced systemic acquired resistance in cucumber and arabidopsis. Phytopathology 88:450e5

Zeng Y, De Guardia A, Dabert P (2016) Improving composting as a post-treatment of anaerobic digestate. Bioresour Technol 201:293–303

Zhang J, Li N, Dai X, Tao W, Jenkinson IR, Li Z (2018) Enhanced dewater ability of sludge during anaerobic digestion with thermal hydrolysis pretreatment: new insights through structure evolution. Water Res 131:177–185

Zhu HN, Yuan XZ, Zeng GM, Jiang M, Liang J, Zhang C, Jiang HW (2012) Ecological risk assessment of heavy metals in sediments of Xiawan Port based on modified potential ecological risk index. Trans Nonferrous Metals Soc China 22(6):1470–1477

Zirkler D, Peters A, Kaupenjohann M (2014) Elemental composition of biogas residues: Variability and alteration during anaerobic digestion. Biomass Bioenergy 67:89–98

Soil Fertility Status and Sugarcane Growth Performance in the Mangrove Ecosystem of Nigeria

16

J. E. K. Aroh

Abstract

Mangrove forests are globally distributed, including Nigeria with the third largest mangrove area in the world. Mangroves perform diverse ecological and productive functions wherever they exist. They are considered to be the most efficient ecosystem in carbon synthesis on Earth. But the nature of their habitat with features of both terrestrial and aquatic environments imposes on them certain challenges that are management-oriented, such as reclamation as practiced in some countries. The agricultural use of mangrove swamps requires thorough understanding of their peculiarities and potential in productive systems, as well as adaptability of crops that can be grown, all of which result from research data in specific localities across the globe. Physical and chemical properties of three soil units or sites in their natural setting in the mangrove (*Rhizophora* spp.) forest area of Ogonokom-Abua, Rivers State, Nigeria, were determined and then evaluated relative to benchmark fertility criteria. This was to provide an insight into their expected impact on plant growth and, hence, their potential for sugarcane cultivation. The data from laboratory soil analyses showed varying degrees of limitations, including the generally coarse or sandy (63–75% sand) textures with predisposition to poor soil-air-water relations and plant nutrient deficiencies in intensive agricultural production. Soil unit or site I in uplifted or reclaimed platform that is no longer subject to flooding by saline tidewater appears to have moderate prospect in agricultural production. Nutrient status, though low coupled with severe acidity (pH 4.8) and salinity (20.0dSm^{-1}), soil unit/site I can be improved upon with appropriate management practices for sustainable sugarcane production in the study area and similar environments in the Niger Delta. Soils of units II and III which are prone to recurrent flooding can only be cultivated to

J. E. K. Aroh (✉)
Niger Delta University, Wilberforce Island, Bayelsa State, Nigeria

M. K. Jhariya et al. (eds.), *Sustainable Intensification for Agroecosystem Services and Management*, https://doi.org/10.1007/978-981-16-3207-5_16

naturally adaptive crop species that will also withstand or tolerate extreme levels of salt stress or salinity (21.0–36.5dSm^{-1}), sodium and aluminium toxicities (11.59–12.75 ESP and 20.73–38.10% Al-sat., respectively) and, particularly, waterlogging. Differences in soil properties were substantially reflected in growth performance of sugarcane crop in the study area. In soil unit or site I, the plants produced higher cumulative average number of tillers, nodes and green leaves than the overall average and achieved mean stalk length, stalk girth, leaf length, leaf width and leaf sheath length that exceeded the 3-monthly average values at 3, 6, 9 and 12 months after transplanting (MAT). Growth indices of the plants in soil unit III were the least, while those in unit II were midway between units I and III. Furthermore, soil unit I was significantly different from unit II in leaf width and leaf sheath length as well as from unit III in stalk length, stalk girth, number of nodes, internode length, number of green leaves, leaf length, leaf width and leaf sheath length. Soil unit II was also significantly different from unit III in stalk girth, number of nodes and green leaves, leaf length and leaf sheath length. Thus, there is no gainsaying about the progressive deterioration in soil conditions with the attendant reduction in growth of the crop from sites I to III. Except for soil unit I that is fairly fertile and can be cultivated to sugarcane crop but at great costs in managing soil acidity, salinity and nutrient deficiency challenges, especially potassium, soil units II and III are only marginally fertile and should better be preserved to perform other vital ecological roles rather than their instant use for agricultural production. However, in the event that soil site II is reclaimed and prevented from recurrent flooding by saline tidewater in the area, together with unit I, and given proper management, these sites would provide extensive land area in the mangrove belt for mechanized sugarcane cultivation at commercial scales to boost the sugar industry in Nigeria. An integrated multidisciplinary policy framework, pragmatic and collaborative implementation strategies and suitable management practices for sugarcane cultivation are all that are needed to assure sustainable socio-economic human development in the region.

Keywords

Acidity · Fertility criteria · Flooding · Mangrove · Sugarcane · Sodic-saline · Toxicity

Abbreviations

Al-sat.	Aluminium saturation
BIOPAMP	The Biodiversity and Protected Areas Management Programme
BS	Base saturation
C:N	Carbon/nitrogen ratio
CEC	Cation exchange capacity
DMRT	Duncan's multiple range test
EA	Exchangeable acidity

EC	Electrical conductivity
ECEC	Effective cation exchange capacity
EDTA	Ethylenediaminetetraacetic acid
EIA	Environmental impact assessment
ERC-m	Electric resistance conductivity-meter
ERL	Expected reserve life
ESP	Exchangeable sodium percentage
FAO	Food and Agriculture Organization of the United Nations
FDALR	Federal Department of Agricultural Land Resources
GhG	Greenhouse gas
GPS	Global positioning system
IITA	International Institute of Tropical Agriculture
IMFM	Integrated mangrove forest management
IPCC	Intergovernmental Panel on Climate Change
LGAs	Local government areas
MAP	Months after planting
MAT	Months after transplanting
NCRI	National Cereals Research Institute
NEEDS	National Economic Empowerment and Development Strategy
NSDC	National Sugar Development Council
OC	Organic carbon
P-ROW	Pipeline right-of-way
ROS	Reactive oxygen species
SCF/B	Standard cubic feet per barrel
SFM	Sustainable forest management
USAID	United States Agency for International Development
VOCs	Volatile organic compounds
WCED	World Commission on Environment and Development

16.1 Introduction

Mangrove forests constitute a climax vegetation type or biome found mostly in intertidal tropical and subtropical climatic zones around the globe occurring between 25°N and 25°S (Kathiresan and Bingham 2001; Pariona 2017). They are found in 118 countries and territories and cover roughly 1.7×105 km^2 along shorelines in river deltas. In aggregated estimates of the country data where mangroves exist based on continents, Valiela et al. (2001) reported that, by far, the largest proportions of mangroves occur in Asia and the Americas. Countries with the largest areas of mangroves are Indonesia having 4.25×10^4 km^2 followed by Brazil with 1.34×10^4 km^2, Nigeria 1.05×10^4 km^2 and Australia 1.00×10^4 km^2. Clearly, mangrove areas in Nigeria are the largest in Africa and third in the world. In Nigeria, mangrove forests spread across some 5400–6000 km^2 in the Niger Delta region where they are most abundant.

As wetlands, mangrove swamps perform enormous ecological and productive functions. They stabilize and protect coastlines, enrich coastal waters with sediments, yield commercial forest products and support marine fisheries. Inasmuch as the ecological functions of mangroves are vital in nature, their productive capacity is no less important, but they have not been fully exploited to complement the environmental roles for the benefit of mankind. Mangroves and similar wetlands are considered to be the most productive ecosystems in the world, including their use for the cultivation of lowland rice (*Oryza sativa*)—a major food crop of global importance (Isirimah et al. 1989; Reddy et al. 2000). Also, mangrove swamps offer great prospect for massive forest trees' establishment (Aluko et al. 2001).

In spite of the extensiveness of mangrove forest soils in Rivers State occurring in 11 out of 23 local government areas (LGAs) of the state (Ayolagha and Onuegbu 2002), they are uncultivated and have remained untapped due to a multiplicity of natural constraints (Eshett 1993). Upland areas of Rivers State covering just about one-third of the total landmass are densely populated and keenly contested for different forms of land use including farming, housing and industrial and infrastructural development. Farm fallow periods in the state have been reduced from a range of 4–6 (Hartoungh 1966) to 2–3 years (Allison-Oguru et al. 2002) in, for example, the Ogoni, Abua, Ikwerre and Etche areas of the state resulting in extensive soil fertility impoverishment and land degradation leading to low soil productivity (Pathak 2008; Meena et al. 2020b). With a population density of 261 persons per square kilometre in Rivers State (Akpogomeh and Atemie 2002), pressure on land has necessitated the need to explore the agricultural potential of mangrove forest soils in the state.

But the agricultural productivity of the soils would depend upon their fertility status and whether or not the crop(s) to be cultivated would withstand or tolerate the perceived soil constraints in the environment (Kumar et al. 2020b; Banerjee et al. 2020, 2021; Khan et al. 2021a, b). This brings to fore the necessity to understand the peculiarities of the soils through the generation of scientific data on ambient soil conditions and research trials on growth performance of select crops for cultivation (Jhariya et al. 2019a, b, 2021a, b). Although sugarcane is, strictly speaking, not a staple food crop, it is fast becoming an important agricultural crop and foreign exchange earner all over the world, especially in India, Thailand and Cuba. The selection of sugarcane varieties for resistance or tolerance to flooding, drought, diseases, pests, salinity and other natural phenomena has been and is still a great challenge to scientists in the major sugarcane-producing nations of the world as the demand for sugar and several industrial by-products of the crop continues to escalate on a global scale (Gomathi and Thandapani 2004; Hemaprabha et al. 2004). Moreover, in Nigeria, more than 95% of the refined sugar currently consumed in the country is imported (NCRI 2007). This precarious situation spells doom for the nation's economy while denying its citizens the various industrial raw materials from the sugarcane crop that are needed for human development.

At present, there is scarcely any information, whatsoever, about the cultivation of sugarcane in the mangrove swamps of Rivers State or elsewhere in Nigeria. Therefore, the cultivation of sugarcane in mangrove swamps requires scientific investigation, and the choice of sugarcane was deliberate. The objective of the present study was to evaluate inherent soil properties in the mangrove ecosystem of Rivers State and relate these to growth performance of sugarcane in the area. It is thought that the results from the study will provide the basis for sugarcane cultivation in mangrove swamps in the region and boost the sugar industry leading to diversification of the economy and increase in foreign exchange earnings of the nation—Nigeria.

16.2 Mangrove Soil Habitat

As prelude to the discussion in the current work, the term "mangroves" has three interrelated connotations: (1) a group of woody perennial plants typified by the red and white mangrove trees such as *Rhizophora*, *Laguncularia* or *Avicennia* species occurring as a forest formation or structure; (2) an ecological setting, habitat or system—i.e. an ecosystem—in which mangrove trees dominate the vegetation; and (3) all biological organisms both plants and animals including microflora and microfauna as well as nonliving entities such as soil, water, etc. existing in a mangrove ecosystem. The term, therefore, refers to all of the component parts in the definition as being important members of the mangrove community. Mangrove forests, in themselves, constitute a biome defined as an association of flora and fauna coexisting in an area and having similar climatic and edaphic requirements. Biomes usually spread over a wide landmass and are, physically, limited or confined by natural barriers including mountains, rivers, depressions and soil or water conditions such as saline or brackish water environment in the case of mangroves. Sometimes referred to as provinces, biochures or regions, biomes have a distribution somewhat parallel to the lines of latitude although the borders are often undulating but generally influenced and defined by climatic variables, particularly temperature, rainfall and humidity.

The Niger Delta consists of a number of distinct ecological zones or biomes such as coastal ridges, barrier islands, freshwater swamp forests, lowland rainforests and saltwater or saline swamp forests housing the mangrove vegetation (NDES 1997; Nyananyo 2002; Kinako 2008). The mangrove biome or belt is of utmost importance in the Niger Delta because of the six valuable mangrove species that have been identified in the zone (Nyananyo 1999). These are *Rhizophora racemosa*, *R. harisonii*, *R. mangle*, *Laguncularia racemosa*, *Acrostichum aureum* and *Avicennia germinans* or *A. africana*. The structures formed by these species, together with other associated plant species, animals and microorganisms, have produced an active mangrove forest ecosystem with huge resources beneficial to humans.

The geology of the Niger Delta that has the most extensive areas of mangrove forest in Nigeria has been described variously by some authors and many of these with concordance in stratigraphic structure. The region is filled with tertiary deposits, which may be up to 12,000 m in thickness. These depositional materials typical of deltaic environments—marine, mixed and continental—can be observed as the

Benin, Agbada and Akata Formations (Okonny 2002). The mangrove belt, itself, is found in the tidal flats or basins that are filled with three deltaic sediments of Quaternary to Recent (Pleistocene, Holocene) geologic age and are found to depths of 25–50 cm from the surface (Ayolagha and Onuegbu 2002).

These more recent sediments constitute the parent materials upon which the soils have eventually developed. These are (1) soils of recently deposited "soft mud", (2) soils of "peaty clay" material and (3) soils of "saline sands". On these sediments which were originally covered by open tidewater, mud bands develop. The creation of mud bands above low-level tidewater encourages further deposition of mud, and this, in turn, provides a suitable substratum for more mangroves to establish. At the inner curves of the open-ended large river channels where the current velocity of tidewater is least, point bars of "soft mud" are formed which carry dense stands of tall mangroves—*Rhizophora racemosa*—that grow to heights of 25–40 meters. The "soft mud" soils are submerged by tidewater at high tide but exposed at low tide, and the tidal range is from 1 to 2 m depth.

Upon the death of the tall mangrove trees (*Rhizophora racemosa*), the woody fibres of the roots remain in the soil without appreciable decomposition due to poor aeration in waterlogged condition. After two or three generations of *Rhizophora racemosa*, the accumulated organic fibres are converted into peaty clay, a material that is extremely high in organic matter, is quite firm and can be cut into blocks for the construction of platforms or dykes and to reclaim swamps. Soils of "peaty clay" known locally as "Chicoco" or "Chikoko" can no longer support the vigorous growth of *R. racemosa*, and subsequent generations are the stunted mangrove species such as *Rhizophora mangle* and *R. harisonii* that are about 1.5–2.5 m tall. Other mangrove species anchored in "peaty clay" include *Avicennia africana, Laguncularia racemosa* and the fern plant *Acrostichum aureum* as well as sedges and a grass plant – *Paspalum* sp.

The "saline sands" are transitional soils between the mangrove forest swamp proper and the adjoining beach ridge zone. They are derived from three sources: (1) "beach ridges", (2) "Sombreiro-Warri Deltaic Plain" remnants near Degema and Buguma, and (3) "coastal plain sands" remnants near Port Harcourt and the Ogoni lands (Ayolagha and Onuegbu 2002), more so the "Ogoni Sands" (Aroh 2003) to which they are closely related in both physical and chemical properties. In all cases, the "saline sands" arise from the erosion of sandy deposits originally above high tidewater level. They are dominantly sands (~75%) and contain lower amounts of organic matter and nutrients than the "soft mud" and "peaty clay" (Ayolagha and Onuegbu 2002; Meena et al. 2020c). The "saline sands" support a rather sparse growth of short *Rhizophora mangle*, sometimes with *R. harisonii*, *Avicennia africana* and *Laguncularia racemosa*. These mangrove species sometimes cause accumulation of peat. There is usually a thick ground cover of *Paspalum* sp. along with sedges in areas where salinity is relatively mild.

The various mangrove species are, therefore, instrumental and strongly related to the different soils found in the mangrove belt. All the three types of soil, especially the soft mud, contain high amounts of sulphides (pyrite—FeS_2) and polysulphides produced by the reduction of sulphates contained in saline tidewater, specifically in flooded conditions in soil. Seawater contains large amounts of sulphates and chlorides (Singer and Munns 1996; Keren 2000). Upon drainage as during ebb (low) tide, or when the land has prograded and is then raised above high tidewater level through sediment deposition, the abundant mangrove rootlets or fibres provide an excellent organic substrate for the bacteria *Thiobacillus ferrooxidans* which causes the oxidation of sulphates to sulphuric acid (Bloomfield and Zahari 1982; Dent 1986; Isirimah and Ojanuga 1987; Ayolagha and Onuegbu 2002). This results in increased soil acidity, i.e. a sharp drop in pH from about 7 (neutral) in flooded condition to an extremely acid (pH 2 or less) condition in soil (Okonny et al. 1999; Keren 2000). For this reason, soils of recently deposited "soft mud" in the mangrove belt and all such wetland soils are regarded as "potential acid sulphate soils" or "cat clays" (Isirimah et al. 1989; Ritsema et al. 2000) to indicate their high tendency to generate acid ions (H^+, Al^{3+}) or components upon drainage.

The floor of the mangrove forest is home to innumerable small fauna and is, ultimately, fundamental to the entire food chain up to human. Many invertebrates in this habitat live in burrows, and these include polychaetes, crabs, certain shrimps, clams and eel-like fish (Nyananyo 2002). Other members of the mangrove community include phytoplankton, zooplankton and mudskippers. Birds and fishes as well as shrimps feed on algae, decomposing leaves and their associated microfauna. Some common animals in the mangrove ecosystem include periwinkles (*Tympanotonus fuscatus* and *T. radula*) and the mangrove whelks (*Thais callifea* and *Semifuscus gasar*) (Powell 1985, 1987). At the creek edges, stilt roots of mangrove trees are covered with oysters and barnacles. Richness of stilt roots' fauna increases close to the sea and includes colourful sponges, sea anemones, soft corals and hydroids (Nyananyo 2002). In fact, fresh water ecology does not support as much biodiversity as salt marshes, and this explains why mangroves are usually accorded the highest sensitivity rating in relation to conservation efforts (Murday et al. 1988).

The mangrove belt provides breeding grounds and nurseries for juveniles of many marine fish species which later move into other coastal and marine habitats (Chinda 2002), thus increasing fish population and diversity in the region. Through the build-up of detritus from leaf litter, the periodic release of larvae from a tremendous variety of species and the abundance of their invertebrate life, mangrove habitats contribute much of the organic matter that makes its way into nearby waters that are otherwise nutrient-poor, thereby making them productive for fisheries. In addition, mangrove forest ecosystems provide essential ecological functions including habitat for wildlife, groundwater recharge, shoreline stabilization, flood control and water quality improvement through biogeochemical transformations (Reddy et al. 2000; Raj et al. 2019a). They also protect low-lying coastal lands from saltwater intrusion, storm surges and sea waves (Nyananyo 2002). This is more so with their unique ecological roles of shoreline protection and as shelterbelt to inland areas inter alia (Reddy et al. 2000).

The three groups of *Rhizophora* (Rhizophoraceae) in the mangrove swamps are colonizing as well as stabilizing species. These plants achieve these feats with the aid of their stilt roots that trap debris carried down the course of the various river systems—Niger, Benue, Forcados and Nun—as they empty into the Atlantic Ocean via some 20 or more estuaries (Nyananyo 2002). In particular, the *Rhizophora* spp. roots, in trapping debris brought back by tidewater, coupled with their organic litter effectively raise the mangrove basin above low and high water levels, and this results in the reclamation of land from the sea as it progrades seaward (Bell-Gam 2002). Gradually, in a vegetation succession from *Rhizophora racemosa* through *R. mangle* to a mixed transitional vegetation type, the forest changes to a true lowland rainforest in a landward direction. Nyananyo (2002), therefore, described mangroves as keystone components of the intertidal environment in which they make disproportionately large contributions to the forest community structure, composition and processes as well as the formation of the soils associated with them. Whereas the mangrove vegetation contributes to the types of soil found in the tidal swamps, the soils themselves determine species diversity and abundance of the vegetation.

Moreover, salt marshes such as mangrove swamps and other wetlands are considered to be the most productive ecosystem in carbon synthesis on Earth. Compared to upland ecosystems, rates of photosynthesis are higher in wetlands than in terrestrial ecosystems, while rates of decomposition are typically lower in the former due to anaerobic soil condition, and so organic matter tends to amass (Mausbach and Richardson 1994; Meena et al. 2020a). The estimated mean primary productivity of a wetland ecosystem is approximately 1300 g C m^{-2} $year^{-1}$ (Houghton and Skole 1990), which is higher than most terrestrial tropical and subtropical ecosystems. Schlesinger (1997) estimated net accumulation of C in some peatland ecosystems to be in the range of 11 to 105 g C m^{-2} $year^{-1}$. The production and accumulation of organic matter in wetland soils serve crucial environmental roles. Wetland soils function as global C sinks through C-sequestration, while, under drained conditions, they act as a veritable source of C to the rest of the biosphere (Reddy et al. 2000; Kumar et al. 2020a). The same is true for nitrogen (N), phosphorus (P) and sulphur (S). Thus, wetland soils including mangroves play crucial roles in the global C, N, P and S cycles, which are fundamental to all biological life forms on Earth.

The agricultural potential of mangrove soils has not been fully exploited in Rivers State in the Niger Delta region. The "soft mud" supporting the tall mangroves (*Rhizophora racemosa*) is too fluid to provide firm anchorage for plant roots and lacks oxygen for plant respiration in waterlogged condition, especially crop plants. The "peaty clay" materials seem to be the potentially productive soils including swamp rice production (Isirimah et al. 1989). The extensive area covered by the "peaty clay" (~ 90% of the mangrove forest zone) is an obvious advantage for massive rice cultivation, especially with rain water in the wet season when salinity level is lowered through leaching. The "saline sands" that constitute soils of the current study provide some of the most productive paddy rice soils in the mangrove forest swamps, but the fertility status is too low that yield levels are unlikely to justify the cost of reclamation (Ayolagha and Onuegbu 2002). According to these authors, they have serious management problems in terms of water control, drainage and reclamation needs. Aroh et al. (2019) identified waterlogging and the attendant deprivation of free oxygen in soil as the most daunting constraints to plant growth in the mangrove ecosystem in Rivers State.

16.3 Sugarcane Cultivation in Mangrove Ecosystem

As wetlands, mangrove swamps have been widely reported for the cultivation of paddy or lowland rice – a major food crop of global importance (Isirimah et al. 1989; Reddy et al. 2000). Also, mangrove swamp soils offer great prospect for massive forest trees establishment (Aluko et al. 2001). There have been several reports of aquaculture land use in Indonesia's mangrove swamps. In addition, mangrove swamp reclamation for sugarcane cultivation has been extensively practiced in Indonesia and the Philippines (FAO 1994).

However, in Nigeria where mangrove forests extend over vast areas in the Niger Delta region, the soils are uncultivated due to a diversity of natural constraints, such as flooding and salinity issues. Studies reported by Zhao and Yang-Rui (2015) showed that some genotypes and/or cultivars of sugarcane exhibit tolerance to certain stress conditions, such as water deficit and low temperature, as well as show superlative radiation and nutrient use efficiency. It could be possible that sugarcane would withstand some of the perceived constraints associated with the mangrove ecosystem in Rivers State, Nigeria.

The present study is the first attempt in a series of efforts aimed at cultivating sugarcane in the mangrove ecosystem of Rivers State and, indeed, the Niger Delta region. The intention is to determine inherent soil conditions in the study area and relate these to growth indices of sugarcane. It is thought that, as preliminary as the investigation may seem, results from the work will provide insight into ambient soil properties in the study area and the direction of management requirements that would lead to commercial sugarcane farming in the area, which will boost the sugar industry as well as improve household, local and national revenue earnings in the country.

16.4 Ogonokom-Abua, Rivers State, Nigeria: A Case Study

In agriculture, climatic variables and ecological peculiarities such as temperature, sunshine hours, day length, precipitation, drought, floods and salinity issues among others constitute natural factors that dictate, to a large extent, the types of crop to be grown in any locality. For example, salinity and flood events represent natural disturbances that influence the distribution of food and forage crops on the globe due to the detrimental effects they provoke on most land plants (Bailey-Serres and Voesenek 2008; Colmer and Voesenek 2009; Jhariya et al. 2019c). So the adaptability of crops to the natural environment will continue to be a critical consideration in agriculture. In the present case study, the cultivation of sugarcane in the mangrove swamps of Rivers State, Nigeria, is one such effort to experiment on the adaptability of the crop to climatic elements and unmodified soil conditions in the environment being investigated.

The study area, Ogonokom-Abua in Abua-Odual Local Government Area, is situated in the mangrove forest belt around the Degema "Hulk" axis of Rivers State, Nigeria. The State is located in the southernmost part of Southeastern Nigeria, approximately between latitudes 4°15' and 6°47'N and longitudes 5°22' and 7°37'E in the Niger Delta region (Fig. 16.1). The study site has global positioning system (GPS) coordinates as 4.764966 N and 6.746107E (Fig. 16.2).

The state has a typical humid tropical (monsoon or equatorial) climate with two distinct seasons in the year as wet or rainy and dry spells (Oyegun and Ologunorisa 2002). Mean annual rainfall is over 4000 mm in the coastal towns of Opobo and Bonny decreasing inland to 3000 mm in the mid-Delta and to slightly less than 2400 mm in the northern parts of the state (Valiela et al. 2001). Temperatures are uniformly high throughout the year, with mean monthly minimum and maximum temperatures of 23 °C and 33 °C, respectively. The highest temperatures occur from December to May, with a brief spell of cold, dry, dust-laden "Harmattan" from late December to mid-January. The state also has high relative humidity, especially in the rainy season, varying from 60 to 90% with the highest and lowest values in the months of July and January, respectively. With heavy clouds during the rainy season and the Harmattan haze early in the dry season around December to January, the average daily bright sunshine hours available to crop plants is in the neighbourhood of 4.2 in the state capital, Port Harcourt. The deduced day length for the Niger Delta region varies from 11.33 to 12.19 hours; the longest days occur between April and August.

The test crop in the study, sugarcane (*Saccharum* spp. hybrids), is a climate-specific tropical giant grass plant widely grown in both Northern and Southern hemispheres in over 120 countries worldwide (Sanghera et al. 2019), preferably between 40°N and 30°S. The most decisive factors in sugarcane growth are temperature, sunlight, soil moisture and soil nutrients. Sugarcane can be grown on a variety of soil types, ranging from light sandy to heavy clayey, but the ideal type is well-drained, well-structured loam or loamy clay which is sufficiently air permeable for root establishment and penetration and rich in available nutrients. Sugarcane thrives well in a water-soaked ecology and requires a lot of soil moisture during its growth. The minimum annual precipitation required to achieve maximum yields in cane under rainfed agriculture is from 1500 to 1600 mm, of which as much as 30% and 70% should occur in the dry and rainy seasons, respectively. Although sugarcane requires a lot of soil moisture, it cannot tolerate prolonged wetness in its root zone or flooding for more than 2 weeks.

Being a C_4 photosynthetic plant, sugarcane is known to utilize, relatively, the greater amount of solar radiation and ranks high in dry matter production and yield (Sanghera et al. 2019). Temperature is the primary factor which drives shoot emergence, leaf appearance and stalk elongation of the crop (Inman-Bamber 1994). The sugarcane plant generally requires a long warm growing season with high incident radiation and a dry, cool, but sunny weather during ripening and harvesting. Ideal temperature requirement of sugarcane ranges from 20 to 26 °C.

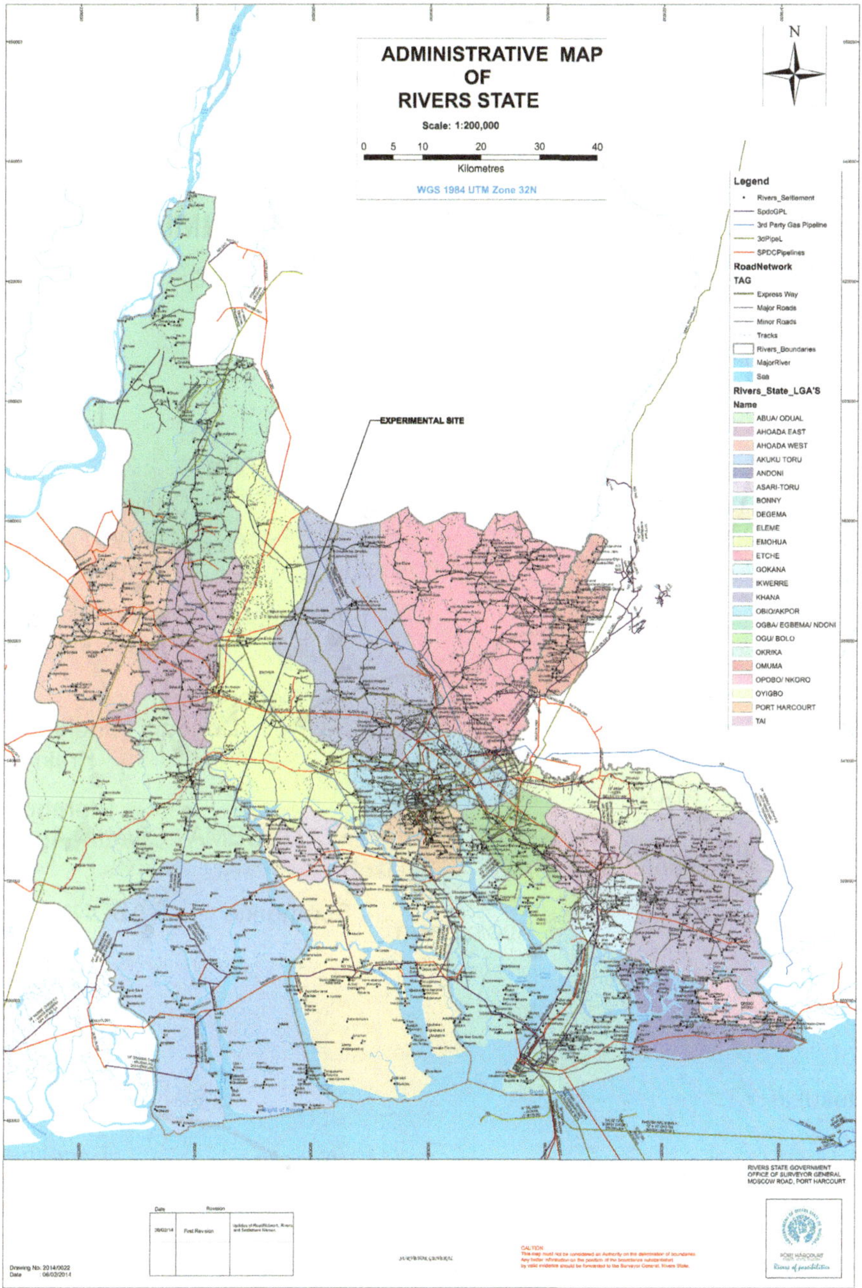

Fig. 16.1 Location map of the study area in Rivers State, Nigeria

Fig. 16.2 Location map showing experimental site in Ogonokom-Abua, Rivers State, Nigeria

Photosynthetic activities of the plant are attenuated, and growth retardation occurs if temperatures drop to between 15 and 18 °C during the vegetative phase. It is tender and susceptible to cold conditions in which tops are killed by temperatures a little below freezing point. Low temperatures below 10 °C delay germination of cane setts although low night temperatures between 7 and 10 °C and low day temperatures of 18–22 °C during the maturing period aid the ripening process and contribute to sucrose accumulation. Sugarcane is a photophylous, short-day plant, and since flowering in the crop is photosensitive, the total amount of sunshine hours could be very crucial; for example, a night period of 11.15–11.25 h is necessary for floral induction to occur, which is undesirable in commercial sugarcane farming for sugar yield. Evidently, climatic factors are most decisive for sugarcane cultivation in order to achieve maximum cane yield for the sugar industry.

Edaphic factors are also essential in sugarcane cultivation, and these are related to soil properties such as depth, type or texture, moisture and nutrient contents and pH among others. By and large, the experimental site in the present study satisfies the climatic requirements of sugarcane. Therefore, the growth response of the crop would be expected to reflect the fertility or productive capacity of the soils in the experiment. It is anticipated that sugarcane can be grown in mangrove soils and results of the study will lead to massive cultivation of the crop in the mangrove forest belt of Rivers State and similar environments in the Niger Delta region.

Fig. 16.3 The tidal flats/basins showing mangrove vegetation (*Rhizophora mangle*) in Ogonokom-Abua, Rivers State, Nigeria

Fig. 16.4 Mixed vegetation of mangrove trees, ferns and grasses at the experimental site in the tidal flats/basins in Ogonokom-Abua, Rivers State, Nigeria

16.5 Materials and Methods

16.5.1 Study Site Description and Field Operations

The vegetation types at the study site are shown in Figs. 16.3 and 16.4 comprising, essentially, the short red mangrove tree species—*Rhizophora mangle* – and a mixture of grasses and fern plants. The plants are rooted in soils of the "saline sands" in the mangrove belt of Rivers State. They are semi-diurnally or twice daily flooded by tidewater at roughly 6-hourly intervals, and the tidal range varies from <15 cm to 1.5 m depth (Afolabi 1998).

The experimental site was qualitatively delineated into three units or sites representing experimental blocks or replicates based on differences in water regime, viz.:

Block/Replicate I > Soil Site No. I (non-flooded, i.e. previously flooded but now raised far above the flood level of tidewater or 0% flood incidence).

Block/Replicate II > Site No. II (partially flooded and drained daily, i.e. only on the lower portions or one-half of the land area all through the year or 50% flood incidence).

Block/Replicate III > Site No. III (completely flooded and drained daily during every high tide all through the year or 100% flood incidence).

The plot size was as follows:

Plot size per block = 6 meters (length) × 5 meters (width) = 30 m^2 according to specifications by Busari (2004) and Amosun (2001).

Planting materials were obtained from the National Cereals Research Institute (NCRI), Badeggi, Niger State, that has the national mandate for sugarcane research in Nigeria. The plants were spaced 1.00 m along rows and 1.25 m between rows as described by Amosun (2000) with 25 cm end-lap on each row. Two rows were provided to accommodate four sugarcane setts in each experimental unit. The blocks were separated by 3 m wide alleys. All planting materials were raised under a shaded nursery in the vicinity of the experimental site. Nursery bags measuring 40 × 35 cm were filled with 5 kg of topsoil. These were devoid of any form of basal fertilizer or lime application, whatsoever, in order to determine the natural fertility level of the soils in the mangrove ecosystem being investigated. The nursery bags were watered with 750 ml of rain water twice weekly. Sprouted sugarcane plants were transplanted to the field 30 days after planting (30 DAP) into nursery bags, and these got established in the field 45 days after planting (45 DAP).

The experiment spanned 12 calendar months from time of transplanting to the field up to maturation of the "plant crop". Plant sampling was done at 3, 6, 9 and 12 months after transplanting (MAT). In an experimental unit, sampling or observation was made on three (3) plants at random, and the average value was recorded for each growth parameter of interest that was determined. Measurements or observations were made on the following growth parameters: (i) number of tillers; (ii) cane (stem or stalk) height/length; (iii) stem girth; (iv) number of nodes; (v) internode length; (vi) number of green leaves; (vii) leaf length; (viii) leaf width; and (ix) length of leaf sheath. Leaf measurements were made on 3rd to 6th top visible dewlap (TVD) leaves; these have been identified as the three sensitive index tissues for sugarcane nutrient content (NCRI 1992) and are, therefore, dependable growth indicators. Soil sampling was done to 30 cm depth from the surface at the time of transplanting sprouted setts (i.e. 30 DAP), and the soil sample was air-dried and passed through 2 mm mesh sieve for laboratory analyses.

16.5.2 Laboratory Soil Analyses

Soil texture was determined by the hydrometer method as described by Nelson and Summers (1996); soil pH was by electrometric method with glass electrode pH-meter in suspension using distilled water in 1:1 ratio as described by Thomas (1996); organic carbon was by the Walkley and Black (1934) dichromate wet oxidation method as described by IITA (1979); total nitrogen was by the semimicro Kjeldahl digestion and steam distillation method as described by IITA (1979); carbon/nitrogen ratio was obtained from results of organic carbon and total nitrogen; and available phosphorus was by the Bray 1-P method as described by IITA (1979).

Exchangeable cations were leached with neutral 1 N ammonium acetate ($NH_4C_2H_3O_2$) buffered at pH 7.0 followed by Kjeldahl digestion of adsorbed ammonium ions for determination of exchangeable basses, namely, Ca, Mg, K and Na as described by IITA (1979). Calcium and magnesium were determined by EDTA titration method (IITA 1979), while sodium and potassium were done by flame spectrophotometric method (IITA 1979). Exchangeable acidity was determined by the titration method using 1 N potassium chloride (KCl) as described by IITA (1979). Cation exchange capacity (CEC) was obtained by summation of exchangeable Ca, Mg, K and Na in soil samples as described by IITA (1979).

Exchangeable sodium percentage (ESP) was obtained from exchangeable Na content expressed as percentage of total exchangeable cations. Effective cation exchange capacity (ECEC) was obtained by summation of exchangeable cations Ca^{2+}, Mg^{2+}, K^+ and Na^+ and exchangeable acidity $Al^{3+} + H^+$ in the samples as described by IITA (1979).

Aluminium saturation was obtained as the percentage of ECEC occupied by exchangeable Al in the sample. Base saturation was obtained as the portion of the ECEC occupied by exchangeable bases (CEC) in percentage. Electrical conductivity was determined in soil extract for pH determination using the electric resistance conductivity-meter (ERC-m) measured at 25 °C as described by Rhoades (1982).

16.6 Results and Discussion

16.6.1 Soil Physical and Chemical Properties

Data on soil physical and chemical properties are shown in Tables 16.1 and 16.2, respectively. Soil properties in the study were evaluated with regard to criteria for soil fertility classes prescribed for agricultural production in Nigeria by the Federal Department of Agricultural Land Resources (FDALR 1990). Results of the evaluation are presented in Table 16.3 ranging from extremely low to extremely high. In soil, a property described as high or low does not necessarily translate to positive or negative impact, as the case may be, on plant growth. A soil property or characteristic entails a range of values that contribute positively to productivity, hence desirable, and a range that contributes negatively, i.e. undesirable (Singer and Ewing 2000). Some of the properties in which high values are usually undesirable include soil reaction or pH in which both low and high levels are detrimental to plant growth, organic carbon, C:N ratio, exchangeable acidity, ECEC, Al-saturation, ESP or sodicity and EC or salinity, among others. These soil attributes need to be taken in their proper contexts as explained below.

Laboratory analysis of particle size distribution or texture (Table 16.1) shows that all the soils had sandy loam textures indicating extremely high content of sand size particles and very low clay, while silt fractions were moderate. They were earlier described as remnants of the "coastal plain sands" (Ayolagha and Onuegbu 2002). They are closely related to, indeed, typified by the "Ogoni Sands" in Rivers

Table 16.1 Particle size distribution or texture of the soils in the study area

	Sand	Silt	Clay	
Soil unit (Site)	%			Textural Class name
I	75	17	8	Sandy loam
II	67	26	7	Sandy loam
III	63	30	7	Sandy loam

Table 16.2 Average values of soil chemical characteristics in the study area

		P	Total N	Org C		Ca	Mg	K	Na	CEC	EA	ECEC	EA/ ECEC	Al-Sat	ESP	BS	EC
Soil unit/ site	pH (H_2O)	mg. kg^{-1}	%		C:N ratio	$Cmol_c.kg^{-1}$							%				dSm^{-1}
I	4.8	17.46	0.12	1.46	12.27	3.75	1.98	0.13	0.13	5.98	1.10	7.08	15.54	12.95	2.16	83.49	20.0
II	4.7	17.48	0.18	2.57	14.28	7.10	3.90	0.18	1.46	12.64	4.33	16.96	25.53	20.73	11.59	76.57	21.0
III	4.5	15.71	0.22	3.53	16.41	7.33	4.11	0.15	1.61	13.18	12.70	25.88	49.07	38.10	12.75	58.78	36.5

Table 16.3 Status of the soils according to FDALR (1990) fertility criteria

S/No	Soil property/characteristic	Soil unit I	Soil unit II	Soil unit III
		Fertility status/potential		
1	Particle size distribution (Texture)	Low(−)	Low(−)	Low(−)
2	Soil reaction or acidity/ alkalinity (pH)	Very low (−)	Very low (−)	Very low (−)
3	Available phosphorus (Bray 1-P) ($mg.Kg^{-1}$)	Moderate (+)	Moderate (+)	Moderate (+)
4	Organic carbon (OC) (%)	Moderate (+)	Very high (−)	Very high (−)
5	Total nitrogen (TN) (%)	Moderate (+)	Moderate (+)	High(+)
6	Carbon/nitrogen (C:N) ratio	Low(+)	Low(+)	Moderate (−)
7	Exchangeable calcium ($Cmol_c.Kg^{-1}$)	Low(−)	Moderate (+)	Moderate (+)
8	Exchangeable magnesium ($Cmol_c.Kg^{-1}$)	Moderate (+)	High(+)	High(+)
9	Exchangeable potassium ($Cmol_c.Kg^{-1}$)	Very low (−)	Very low (−)	Very low (−)
10	Exchangeable sodium ($Cmol_c.Kg^{-1}$)	Low(+)	High(−)	High(−)
11	Cation exchange capacity (CEC) ($Cmol_c.Kg^{-1}$)	Very low (−)	Moderate (+)	Moderate (+)
12	Exchangeable acidity (EA) ($Cmol_c.Kg^{-1}$)	Very low (+)	Moderate (−)	Very high (−)
13	Effective cation exch. capacity (ECEC) ($Cmol_c.Kg^{-1}$)	Very low (+)	Low(+)	Moderate (−)
14	Exch. Acidity/effective cation exch. capacity (%)	Moderate (+)	High(−)	Extremely high(−)
15	Percentage aluminium saturation (Al-sat.) (%)	Fairly high (+)	Very high (−)	Extremely high(−)
16	Exchangeable sodium percentage (ESP) (%)	Low(+)	High(−)	Very high (−)
17	Percentage base saturation (BS) (%)	Very high (+)	High(+)	Moderate (+)
18	Electrical conductivity (EC or salinity) ($dS.m^{-1}$)	Extremely high(−)	Extremely high(−)	Extremely high(−)
19	Available soil depth or depth to groundwater (cm)	High(+)	Very low (−)	Extremely low(−)
20	Freedom from flooding (frequency and duration)	Very high (+)	Low(−)	Very low (−)
Total number of positive and negative impacts		(14 + ve), (6-ve)	(8 + ve), (12-ve)	(6 + ve), (14-ve)

KEY: (+) = Positive impact and (−) = Negative impact on plant growth

State that are generally low in nutrient content (Aroh 2003). The relatively insufficient proportions of clay size fractions suggest that the soils will have weak soil structures and poor soil-water-air relations. They are not only liable to excessive drainage and poor water storage but, more importantly, have low levels of inorganic colloidal materials and, thus, are likely to lack the essential plant nutrients for growth and development of cultivated plants.

Based on texture, the soils in the present study may be generally adjudged low in terms of fertility status and barely suitable for agricultural production (Table 16.3). In spite of their common unfitness, soil unit III with the least average sand content of 63% and highest silt of 30% can be considered to have a vista of optimism in agriculture, though meagre, on account of soil texture alone and nothing else that contributes to productivity.

Data on soil chemical characteristics (Table 16.2) indicate that soil pH was generally in the acid region and ranged from 4.4 to 4.9 with average values of 4.8, 4.7 and 4.5 in soil units I, II and III, respectively. These types of soil are described as being strongly to extremely acid (FDALR 1990) and very low in fertility status (Table 16.3). High acidity or low pH in soil constitutes a major constraint to plant growth through its influence on the availability of certain nutrient elements required for development. The majority of cultivated crop plants are known to thrive within pH 5.5 to 8.5, preferably best in the range 6.5 to 7.0 (Tisdale and Nelson 1975). Acidity itself at pH 5.5 or lower inhibits the growth of sensitive plant species, though it has little effect on insensitive species even at pH as low as 4 (Singer and Munns 1996).

Acid mineral soils are frequently high in Al, manganese and iron, which are toxic in excessive amounts, especially Al (Rahman et al. 2018). This pH effect is often compounded and often overshadowed by Al toxicity, Mn toxicity, Ca deficiency and Mo deficiency. These soil acidity-related factors tend to always occur together, and they also interact enhancing one another's effect to cause more damage to growing plants. Such complimentary negative effects include the reduced availability of phosphorus in acid soils high in iron and aluminium, as well as decreased levels of manganese in soils high in organic matter which also have high pH, and a decline in the availability of molybdenum with decreases in soil pH or high acidity (Tisdale and Nelson 1975; Singer and Munns 1996). Soil pH <5 in the soils, thus, suggests the likelihood of Al and Mn toxicities, Ca and Mo deficiencies as well as reduced availability of phosphorus, while pH <5.5 additionally indicates the possibility of Zn, K and S deficiencies (Tisdale and Nelson 1975). This may have been responsible for the sub-optimal levels of P, Ca, and K in soils of the study area.

In particular, coastal soils and sediments contain reduced sulphur such as organic-S, Fe-sulphide and a range of metastable polysulphides collectively known as "acid-volatile sulphides" or simply "sulphidic materials" such as greigite (Fe_3S_4) and pyrite (FeS_2) (Ritsema et al. 2000). Their relative proportions depend upon the environment of accumulation, but FeS_2 is dominant in marine and brackish water environments where the actual concentration of sulphides is at peak levels such as the case under investigation. High inputs of organic matter into wetland soils, along with oxic surface and anoxic sub-surface zones, potentially allow sulphur to play a critical role in the biogeochemistry of wetlands, including mangroves.

As coastal wetland soils are naturally high in sulphidic materials, when these are exposed to atmospheric oxygen as in soil unit I and during low or ebb tide in units II and III, sulphuric acid is produced in oxidation reactions, and the pH of the soil is drastically decreased, i.e, the acidity is severely increased to the detriment of growing plants. Such soils are regarded as "potential acid sulphate soils" or "cat clays" (Isirimah and Ojanuga 1987; Okonny et al. 1999; Ritsema et al. 2000) to indicate their high tendency to generate acid components upon drainage. From the definition of sulphidic materials by Dent (1986) and Ritsema et al. (2000), units II and III qualify as "potential acid sulphate" soils that remain neutral in waterlogged condition and will generate acid components upon drainage, while unit I is described as "ripe acid sulphate" soil that has passed the acid generation stage but still has high acidity.

Notwithstanding the widespread limitation due to high acidity, soil unit I with pH 4.8 affords a relatively higher prospect in agriculture as less liming material will ordinarily be required to reduce the acidity level to some manageable value above pH 5.0. Greater amounts of lime will be needed in soil unit II with pH 4.7 and much more in unit III with pH 4.5. The foregoing statements are deliberate and in gross disregard to the recurrent waterlogging to which soil unit III is naturally prone. In such saturated condition, soil unit III will remain neutral at pH 7 or so (Singer and Munns 1996; Reddy et al. 2000). From the results, the high acidity levels in soil units II and III may have been due to air-drying for laboratory analysis, which is likely the case if they are drained for cultivation, thus requiring large amounts of lime for effective remedy.

Available P ranged from 14.79 to 18.08 $mg.kg^{-1}$, being lowest in soil unit III. Average values were 17.46, 17.48 and 15.71 $mg.kg^{-1}$ in soil units I, II and III, respectively (Table 16.2). P is largely immobilized in waterlogged conditions (Reddy et al. 2000) as made manifest in soil unit III. Soils with available P values below 10 $mg.kg^{-1}$ are marginally suitable for crop production (FAO 1976). The three soil units are, thus, generally rated as moderate with available P values of 7.1–20.0 $mg.kg^{-1}$ (Table 16.3) and will technically qualify as reasonably suitable for agricultural production (FDALR 1990) at least to varying degrees, especially soil unit III with the least content of available P. Even so, as one of the macronutrients (NPK) needed by plants in large amounts, the soils will require normal P-fertilizer application for sustainable sugarcane production.

Organic carbon (OC) content ranged from 1.31 to 3.72% with average values of 1.46, 2.57 and 3.53% in soil units I, II and III, respectively (Table 16.2). They are rated as moderate (1.10–1.5%) in soil unit I and very high (>2.0%) in units II and III (Table 16.3). After two or three generations of the tall mangrove trees (*Rhizophora racemosa*) anchored in semisolid or unconsolidated "soft mud" soil, the accumulated organic fibres are converted into hard "peaty clay" soil that anchors the short mangrove trees (*Rhizophora mangle*) as exemplified by soil units II and III in the present study. The higher values of OC in soil units II and III, thus, reflect a greater store of organic materials, apparently, due to saturation or near saturation state of these soil units in which aerobic decomposition is usually impeded (Singer and Munns 1996). Being restricted to fermentation processes as their energy source in the absence of free oxygen, anaerobic microbes decompose organic matter only partially so that organic matter stockpiles in the soil as seen in soil units II and III.

Besides, rates of photosynthesis in wetlands are higher than in terrestrial ecosystems, while rates of decomposition are typically lower due to anaerobic soil condition; hence, organic matter tends to accumukate (Mausbach and Richardson 1994). Organic litter in wetlands is a major support structure for soil microbes, being a source of energy and nutrients as well as containing high numbers (10^9–10^{10} g^{-1} dry matter) of bacteria and fungi (Kjoller and Struwe 1980; Kjoller et al. 1985). Thus, organic litter and microbial biomass account for a significant amount of OC in wetland substrate (DeBusk 1996). Its decomposition or mineralization to release mineral N and other nutrients is critical to plant growth in such soils, with N being the core nutrient element required for vegetative growth in plants.

Total nitrogen (TN) ranged from 0.14 to 0.22% with mean values of 0.12, 0.18 and 0.22% in soil units I, II and III, respectively (Table 16.2). These values are rated as moderate with 0.101–0.200% in soil units I and II and high with 0.201–0.300% in soil unit III (Table 16.3). In waterlogged condition as in soil unit III, mineralized N remains reduced as ammonium ion (NH_4^-), which is highly mobile in anaerobic soil because of its high concentration in solution. The high water content that allows N to diffuse rapidly in soil solution makes it to become a major exchangeable cation for plant uptake (Singer and Munns 1996). Due to low N requirements of anaerobic microorganisms, wetland soils usually accumulate NH_4^-N which supports most of the N requirements of wetland plants (Reddy et al. 2000). As a major nutrient element central to plant growth and development, cultivated plants will be reasonably supplied with N in soils of the present study, particularly so in soil units II and III.

The C:N ratio ranged from 11.43 to 16.91 with mean values of 12.27, 14.28 and 16.0 in soil units I, II and III, respectively (Table 16.2). The soils are considered as low with C:N ratios of 10.0–14.0 in soil units I and II and moderate with 15.0–19.0 in unit III (Table 16.3). The ideal C:N ratio of stable soil organic matter such as humus, and in the uppermost organic layer of the soil, is placed around 10 or 12 (Tisdale and Nelson 1975). It can even range between 10 and 16 with the higher ratios in undecomposed fragments of plant litter and organic materials in peat deposits where decomposition is hindered by anaerobic conditions (Baldock and Nelson 2000). Carbon:nitrogen ratio above 25, which is the upper limit for effective mineralization, is indicative of poor organic decay resulting in net immobilization of N and minimal N turnover in soil. As such, notwithstanding the higher OC and TN contents in soil units II and III in the preceding discussion, soil unit I indicates greater N yield on a weight-for-weight basis than units II and III. However, the soils in the study generally have good outlook for sufficiency in N supply to growing plants in agriculture, which is more in soil unit I and decreasing through unit II to unit III although, when drained to facilitate aerobic organic decay, the reverse will likely be the case.

Exchangeable cations varied greatly in the soils wherein exchangeable calcium ranged from 3.10 to 8.95 $Cmol_c.kg^{-1}$ with mean values of 3.75, 7.10 and 7.33 $Cmol_c.kg^{-1}$ in soil units I, II and III, respectively (Table 16.2). These are, accordingly, rated as low in soil unit I and moderate in units II and III (Table 16.3). The lower limit for fertile soils in Nigeria is placed at 4 $Cmol_c.kg^{-1}$ calcium (FDALR 1990). As such, soil units II and III present better conditions favourable to crop production than unit I that had less than adequate calcium content.

Exchangeable magnesium ranged from 1.70 to 5.16 with average values of 1.98, 3.90 and 4.11 $Cmol_c.kg^{-1}$ in soil units I, II and III, respectively (Table 16.2). The soils are rated as moderate in soil unit I and high in units II and III (Table 16.3). All the soil units had more than 0.5 $Cmol_c.kg^{-1}$ magnesium regarded as critical for crops to be grown (FDALR 1990). Magnesium, being a vital component of chlorophyll, is involved in P and carbohydrate metabolism, the synthesis of oils, as well as the activation of a number of plant enzyme systems (Tisdale and Nelson 1975). It is, thus, considered as adequate for healthy plant growth such as sugarcane in soils of the present investigation, especially soil units II and III.

Exchangeable K ranged from 0.09 to 0.20 with mean values of 0.13, 0.18 and 0.15 $Cmol_c.kg^{-1}$ in soil units I, II and III, respectively (Table 16.2). They are generally rated as very low with <0.2 $Cmol_c.kg^{-1}$ in soil units I, II and III (Table 16.3). When K levels are below the minimum 0.2 $Cmol_c.kg^{-1}$ necessary for crop production (FDALR 1990), they indicate deficiency levels. Plant requirements for K are usually quite high (Singer and Munns 1996). This is because the role of K in plant nutrition is tremendous as it is involved in most catalytic reactions including carbohydrate formation and breakdown, translocation of starch, N metabolism and synthesis of proteins. It is also involved in neutralization of physiologically important organic acids, activation of various enzymes, promotion of the growth of meristematic tissues and adjustment of stomatal movement and water relations in plants (Tisdale and Nelson 1975). Therefore, the overall effects of K deficiency on plant growth are, of course, the result of the accompanying physiological disturbances within the plant system, which are reflected in several ways culminating in gross reduction in crop growth, development and yield.

Exchangeable sodium ranged from 0.12 to 1.62 with average values of 0.13, 1.46 and 1.61 $Cmol_c.kg^{-1}$ in soil units I, II and III, respectively (Table 16.2). The soils are rated as low with 0.1–0.3 $Cmol_c.kg^{-1}$ in soil unit I and high with 0.8–2.0 $Cmol_c.kg^{-1}$ in units II and III (Table 16.3). Unlike Ca, Mg and K as major nutrient elements, the requirement of plants for Na does not usually match them quantity-for-quantity. At high concentrations in soil, sodium is toxic to certain plants, especially if calcium levels are low to just moderate as in the present soils. However, the overriding adverse effects of high sodium are on the soil's physical properties relating to water movement and retention. Soils that have high amounts of exchangeable sodium readily lose their structure and become dispersed and impermeable (Singer and Munns 1996). Furthermore, sodium is known to be the chief culprit by combining with chloride ions that are in abundant supply in coastal marshlands as the ones in the present study to cause severe salinity problems that reduce the chances of survival in sensitive plants. Therefore, the rather high exchangeable sodium contents of 1.46 and 1.61 $Cmol_c.kg^{-1}$ in soil units II and III, respectively, portend a perilous environment for plant growth. Thus, soil unit I with average sodium content of 0.13 $Cmol_c.kg^{-1}$ and considered low by FDALR (1990) rating is considered more conducive to sugarcane growth than soil units II and III.

Cation exchange capacity (CEC) of the soils ranged between 5.05 and 15.90 $Cmol_c.kg^{-1}$ with average values of 5.98, 12.64 and 13.18 $Cmol_c.kg^{-1}$ in soil units I, II and III, respectively (Table 2). The soils in the present study, according to FDALR (1990) standards, are rated as very low with <6.0 $Cmol_c.kg^{-1}$ in soil unit I and just fairly moderate with 12.1–25.0 $Cmol_c.kg^{-1}$ in units II and III in their CEC levels (Table 16.3). As mineral soils with CEC values below the required minimum of 16 $Cmol_c.kg^{-1}$ for cultivation, their agricultural sustainability will be highly compromised, especially soil unit I that will require large doses of K-fertilizer application to sustain agricultural productivity. Soil units II and III, given proper management, stand better chances of sustaining agricultural production on account of their higher CEC values per se in the absence of any other limitation to plant growth and development.

Exchangeable acidity (EA) ranged from 1.05 to 22.83 $Cmol_c.kg^{-1}$ with average values of 1.10, 4.33 and 12.70 $Cmol_c.kg^{-1}$ in soil units I, II and III, respectively (Table 16.2), and was rated as very low with <2.0 $Cmol_c.kg^{-1}$ in soil unit I, moderate with 4.1–6.0 $Cmol_c.kg^{-1}$ in unit II and very high with >8.0 $Cmol_c.kg^{-1}$ in unit III (Table 16.3). The higher EA values in soil units II and, especially, III indicate an abundance of H^+ and Al^{3+} ions in the exchange complex and a tendency towards further acidification in the soils. Given the not-too-good outlook in CEC levels by being just moderate in soil units II and III, and being also extremely acidic with pH 4.5–4.7, these two soils represent great risks to agricultural production including sugarcane cultivation. Heavy and repeated lime application is necessary to reduce soil acidity in these soils if they are drained for cultivation, particularly in soil unit III wherein yields in sugarcane production may not justify the cost of reclamation works and remediation measures.

ECEC ranged from 6.20 to 33.29 with average values of 7.08, 16.96 and 25.88 $Cmol_c.kg^{-1}$ in soil units I, II and III, respectively (Table 16.2). The soils are rated as very low with <10. 0 $Cmol_c.kg^{-1}$ in unit I, low with 10.0–20.0 $Cmol_c.kg^{-1}$ in unit II and moderate with 20.1–40.0 $Cmol_c.kg^{-1}$ in unit III (Table 16.3). The results are true reflections of the high CEC and EA values in soil units II and III but which do not necessarily translate to, or reflect availability of plant nutrients in these soils. In particular, the contribution of EA to total ECEC ranged between 15.19 and 68.58% in the soils, with average values of 15.54, 25.53 and 49.07% described as moderate, high and extremely high in soil units I, II and III, respectively (Table 16.3). The results show that soil unit I has a relatively small portion of the ECEC of ~16% that is occupied by acid-inducing ions (H^+ + Al^{3+}) and which corresponds to the higher pH value of pH 4.8 discussed earlier.

Conversely, soil units II and III have as much as 26 and 49%, respectively, of the exchangeable cations contributed by acid components (H^+ + Al^{3+}) that will not add to but rather inhibit the nutrient-supplying capacity of these soils. The corresponding pH values in these soils were reported as 4.7 and 4.5 in units II and III, respectively. In view of the lower EA value despite its lesser CEC level reported earlier, soil unit I can be considered more suitable for agricultural production including sugarcane than unit II and much less unit III.

Aluminium saturation (Al-sat) ranged from 10.72 to 62.93 with mean values of 12.95, 20.73 and 38.10% in soil units I, II and III, respectively (Table 16.2). The soils are rated as moderate to fairly high in soil unit I with 10.0–14.0%, very high in unit II with 15.0–20.0% and extremely high in unit III with >20.0% (Table 16.3). Al-sat levels above 10% are regarded as high and suggest the possibility of immobilization of certain nutrient elements in soil such as Ca and Mo (Singer and Munns 1996) which leads to nutrient imbalance and reduced plant growth. More than 15% of the exchange capacity occupied by Al in the soil spells doom for many crop species. Moreover, values ≥15% Al-sat indicates Al toxicity in which the element Al becomes directly harmful to plants. Beyond 20%, Al-sat makes soil sterile, in which case exchangeable H^+ itself remains minor even in acid soils (Singer and Munns 1996). The results show that soil unit I with the least average Al-sat level of 12.95% provides a more tolerable soil environment for sugarcane growth than is the case in unit II with 20.73% Al-sat. Soil unit III is considered as inhospitable to sugarcane and most other cultivated plants due to the extremely high Al-sat level of 38.10% in the soil exchange complex.

Exchangeable sodium percentage (ESP) ranged from 1.74 to 15.49 with mean values of 2.16, 11.59 and 12.75% in soil units I, II and III, respectively (Table 16.2). The soils are rated as low with <10.0% in unit I, high with 10.0–12.0% in unit II and very high with 12.1–15.0% in unit III (Table 16.3). Soils that contain Na as a significant proportion of their total exchangeable cations are termed sodic and have ESP of more than 10% or so (Levy 2000), which may be up to 15% (Singer and Munns 1996). Therefore, while soil units II and III are sodic, unit I is not, which is a positive attribute for healthy crop growth. In this respect, once more, soil unit I provides a more conducive growth medium for the test canes, whereas units II and III are uncongenial, especially the latter.

Percentage base saturation (%BS) ranged from 31.42 to 86.13 with average values of 83.49, 76.57 and 58.78% in soil units I, II and III, respectively (Table 16.2). They are rated as very high with >80.0% in soil unit I, high with 60.1–80.0% in unit II and moderate with 40.0–60.0% in unit III (Table 16.3). The fertility implications of these results point to the relatively higher agricultural potential of soil unit I followed rather feebly by unit II and, lastly, unit III. Located in the Degema "Hulk" axis of Rivers State, these saline sands were described generally as poor in nutrient contents (Ayolagha and Onuegbu 2002), but as shown in the present study, soil units I and II indicate some potential in cane cultivation or agriculture, generally, on the basis of their high base saturation status, especially soil unit I.

Electrical conductivity (EC) or salinity level ranged from 21.0 to 38.0 $dS.m^{-1}$ with average values of 20.0, 21.0 and 36.5 $dS.m^{-1}$ in soil units I, II and III, respectively (Table 16.2). The soils are rated as extremely high with >4 $dS.m^{-1}$ in soil units I, II and III (Table 16.3). In soil, electrical conductivity of 4 $dS.m^{-1}$ represents the upper limit for salinity tolerance in most plants. Thus, a soil or water sample is termed saline if the electrical conductivity is ≥4 $dS.m^{-1}$ (Singer and

Munns 1996; Keren 2000). Threshold salt concentrations for classifying crop tolerance to salinity were reported by Maas (1986, 1990) as (i) sensitive with <1.5 dS.m^{-1} for bean (*Phaseolus vulgaris*), carrot (*Daucus carota*) and lettuce (*Lactuca sativa*); (ii) moderately sensitive with 1.5–3.0 dS.m^{-1} for corn (*Zea mays*), alfalfa (*Medicago sativa*), broccoli (*Brassica oleracea*) and potato (*Solanum tuberosum*); (iii) moderately tolerant with 3.0–6.0 dS.m^{-1} for soybean (*Glycine max*), wheat (*Triticum aestivum*) and squash (*Cucurbita maxima*); and (iv) tolerant with 6.0–10.0 dS.m^{-1} for barley (*Hordeum vulgare*), guinea corn (*Sorghum bicolor*) and, particularly, sugar beet (*Beta vulgaris*) as high as 10.0 dS.m^{-1} and up to 350.0 mg.L^{-1} dissolved chloride (Cl^-) salts in soil solution.

Having regard to the above reference figures, it can be assumed that none of the above-mentioned crops will likely survive in an environment with EC or salinity levels of 20.0, 21.5 and 36.5 dS.m^{-1} in soil units I, II and III, respectively, which will be extremely unbearable for most cultivated crop plants. The results indicate that all the soil units in the study exhibit overkill levels of dissolved salts or salinity in soil solution and this state of condition amounts to a hazardous environment for plant growth, particularly in soil unit III. It was found in the course of the present investigation as reported earlier that soil units II and III were sodic with ESP values of 11.59 and 12.75%, respectively. As sodic soils, in addition to being saline, they qualify as sodic-saline thereby compounding two devastating growth conditions and the situation being worse in soil unit III that has the greatest sodium and salt contents.

Soil depth and waterlogging or oxygen availability are critical factors in the use of soil for agricultural production. These are described as "land qualities" and, at the same time, "soil characteristics" (FAO 1997) that are germane to the fitness of soil for cultivation; hence, they were considered in the present work. Available soil depths in the study are rated as high with >60.0 cm in soil unit I, very low with 15.0–30.0 cm in unit II and extremely low with <15.0 cm in unit III (Table 16.3). Freedom from flooding incidence or frequency is described as non-flooded or very high in soil unit I, partially flooded and drained daily or low in unit II and totally flooded and drained daily or very low in unit III (Table 16.3). It needs restating that soil unit II is intermediate between units I and III where it represents the current land-water interface with alternating 6-hourly oxidation-reduction reactions in soil at the experimental site. One-half of the land area in soil unit II is subject to flooding, while the other half is free from flooding.

In general, the results indicate that soil unit or site I located in uplifted or naturally reclaimed parts of the mangrove ecosystem in the study area has relatively less devastating growth conditions and described earlier by Aroh et al. (2019) in a related study as fairly fertile for the cultivation of the test crop, sugarcane. Soil units/sites II and III in the study were described as marginally fertile and are uncongenial to plant growth largely due to extreme levels of salinity and, particularly, waterlogging in soil following frequent flooding by tidewater. Until these are reclaimed, naturally or artificially, they can only be cultivated to naturally adaptive crops that would withstand these constraints to plant growth in the mangrove ecosystem under investigation.

16.6.2 Sugarcane Growth Indices on Mangrove Soils

Sugarcane has essentially five growth phases, viz. germination, tillering or formative, grand growth, maturity and ripening (Sanghera et al. 2019). Bull (2000) lists them as germination, early growth, maturation, flowering and ratooning, while others consider these as germination or sprouting, tillering, internode elongation, maturation or sucrose accumulation and flowering stages. All of the phases overlap in sugarcane and contribute to both cane and sugar yield (Bull 2000). Data on average growth indices of the test sugarcane crop in the present study are shown in Table 16.4.

16.6.2.1 Tiller Formation or Production

The sugarcane plant has a number of tillers, stems or stalks bunched together at the point of planting to form a "stool". The stool consists of 5–20 stalks at different stages of growth, and they are usually erect, semi-erect, spreading or tangled at maturity running up to a height of 3–10 meters (Bull 2000; Khan et al. 2017). In the present study, tiller production ranged from 1.63 to 2.63 with cumulative average values of 2.30, 2.07 and 1.93 per stool in soil units I, II and III, respectively (Table 16.4). In general, tiller formation declined from soil unit I through II to III, which indicates better growth conditions in soil unit I. The relative advantages of soil unit I over units II and III were its lower values or toxic levels of sodium and aluminium, freedom from flooding and better soil depth for plant root development. It also recorded higher amounts of available phosphorus, total nitrogen, lower exchangeable acidity and salinity levels as shown in Table 16.2, which must have combined to provide more conducive growth conditions than soil units II and, especially, III.

The tillering phase in sugarcane is succeeded by a period of rapid cell division resulting in extensive growth. Hemaprabha et al. (2004) referred to the tillering and rapid stalk elongation phases in sugarcane as "formative" and "grand growth" stages, respectively. The formative phase is identified as the critical water demand period (Gascho and Shih 1983), and the grand growth phase of rapid internode elongation starts from 3 to 6 months after planting (MAP) in early maturing varieties (Thomas and Rozeff 1989) and from about 5 or 6 months in late maturing varieties that lasts till the eighth or ninth month (Bull 2000). Thus, the most active uptake of nutrients, especially N and K, as well as water takes place within the first 6 months during tillering and early internode elongation to cause height increase in sugarcane (Sanghera et al. 2019).

The general outlook is one in which the average number of tillers increased from 3 to 12 MAT even as increases over time as well as within the soil units were inconsistent or irregular. Early formed tillers give rise to thicker and heavier stalks, while late-formed tillers either die or remain short or simply immature (Sanghera et al. 2019) which might have accounted for the unpredictable pattern in tiller production recorded in the study. Sanghera et al. (2019), reported that, of the total number of tillers produced in a sugarcane stool, only 40–50% survives by 150 days to form millable cane. Despite considerable differences in soil properties recorded in

Table 16.4 Field data on sugarcane growth parameters in Ogonokom-Abua, Rivers State, Nigeria

Sugarcane growth parameter	Soil site (Soil unit)	3-monthly average 3 MAT	6 MAT	9 MAT	12 MAT	**Cumulative site total**	**Cumulative site mean**
Number of tillers/stool	I	2.00	2.31	2.25	2.63	9.19	**2.30**
	II	1.69	1.69	2.25	2.63	8.26	**2.07**
	III	1.63	2.13	1.81	2.13	7.70	**1.93**
	Quarterly total	**5.32**	**6.13**	**6.31**	**7.39**	**Grand total 25.15**	**Grand mean 2.10**
	Quarterly mean	**1.77**	**2.04**	**2.10**	**2.46**		
Single stem length/stool (cm)	I	39.58	54.48	68.73	83.18	245.97	**61.49**
	II	37.64	46.61	49.80	62.98	197.03	**49.26**
	III	44.17	58.35	47.56	52.74	202.82	**50.71**
	Quarterly total	**121.39**	**159.44**	**166.09**	**198.90**	**Grand total 645.82**	**Grand mean 53.82**
	Quarterly mean	**40.46**	**53.15**	**55.36**	**66.30**		
Single stem girth/stool (cm)	I	5.08	5.76	5.80	5.59	22.23	**5.56**
	II	4.81	4.94	4.33	4.43	18.51	**4.63**
	III	5.11	4.88	2.83	2.91	15.73	**3.93**
	Quarterly total	**15.00**	**15.58**	**12.96**	**12.93**	**Grand total 56.47**	**Grand mean 4.71**
	Quarterly mean	**5.00**	**5.19**	**4.32**	**4.31**		
Number of nodes/stem	I	2.19	6.31	11.81	14.81	35.12	**8.78**
	II	2.56	5.75	9.69	11.94	29.94	**7.49**
	III	3.69	7.50	6.94	7.69	25.82	**6.46**
	Quarterly total	**8.44**	**19.56**	**28.44**	**34.44**	**Grand total 90.88**	**Grand mean 7.57**
	Quarterly mean	**2.81**	**6.52**	**9.48**	**11.48**		
Single internode length/ stem (cm)	I	2.33	3.64	3.93	3.00	12.90	**3.23**
	II	3.48	3.49	3.23	2.79	12.99	**3.25**
	III	3.76	4.49	2.91	2.86	14.02	**3.51**

(continued)

Table 16.4 (continued)

Sugarcane growth parameter	Soil site (Soil unit)	3-monthly average 3 MAT	6 MAT	9 MAT	12 MAT	Cumulative site total	Cumulative site mean
	Quarterly total	**9.57**	**11.62**	**10.07**	**8.65**	**Grand total 39.91**	**Grand mean 3.33**
	Quarterly mean	**3.19**	**3.87**	**3.36**	**2.88**		
Number of green leaves/stem	I	7.06	7.19	6.38	5.69	26.32	**6.58**
	II	6.38	6.38	5.00	5.13	22.89	**5.72**
	III	6.06	6.06	3.63	3.00	18.75	**4.69**
	Quarterly total	**19.50**	**19.63**	**15.01**	**13.82**	**Grand total 67.96**	**Grand mean 5.66**
	Quarterly mean	**6.50**	**6.54**	**5.00**	**4.61**		
Single leaf length/stem (cm)	I	106.62	115.21	117.29	114.40	453.52	**113.38**
	II	89.06	96.95	100.31	96.00	382.32	**95.58**
	III	89.29	87.11	64.01	59.49	299.90	**74.98**
	Quarterly total	**284.97**	**299.27**	**281.61**	**269.89**	**Grand total 1135.74**	**Grand mean 94.65**
	Quarterly mean	**94.99**	**99.76**	**93.87**	**89.96**		
Single leaf width/stem (cm)	I	2.30	2.58	2.96	2.74	10.58	**2.65**
	II	2.58	2.03	2.09	2.04	8.74	**2.19**
	III	2.00	2.09	1.34	1.39	6.82	**1.71**
	Quarterly total	**6.88**	**6.70**	**6.39**	**6.17**	**Grand total 26.14**	**Grand mean 2.18**
	Quarterly mean	**2.29**	**2.23**	**2.13**	**2.06**		
Length of single leaf sheath/stem (cm)	I	21.64	25.04	23.74	22.74	93.16	**23.29**
	II	15.81	22.66	20.74	20.08	79.29	**19.82**
	III	24.54	22.82	13.72	13.13	74.21	**18.55**
	Quarterly total	**61.99**	**70.52**	**58.20**	**55.95**	**Grand total 246.66**	**Grand mean 20.56**
	Quarterly mean	**20.67**	**23.51**	**19.40**	**18.65**		

Note: MAT = Months after transplanting

Table 16.5 Quarterly average tiller production and plant height in test canes from three soil sites

Soil unit (site)	Tiller production				Plant height			
	3 MAT	6 MAT	9 MAT	12 MAT	3 MAT	6 MAT	9 MAT	12 MAT
I	2.00a	2.31a	2.25a	2.63a	39.58a	54.48ab	68.73a	83.18a
II	1.69a	1.69a	2.25a	2.63a	37.64a	46.61b	49.80b	62.98ab
III	1.63a	2.13a	1.81a	2.13a	44.17a	58.35a	47.56b	52.74b
	NS	NS	NS	NS	NS			

Means with the same letter(s) are not significantly different according to Duncan's multiple range test (DMRT) at $p \leq 0.05$; MAT = Months after transplanting.

the study, critical observation of the results shows that within each of the 3, 6, 9 and 12 MAT quarterly intervals, there were no statistically significant differences between soil units I, II and III in the production of tillers as shown in Table 16.5.

16.6.2.2 Plant Height or Stem Length

Plant height ranged from 37.64 to 83.18 cm with mean cumulative values of 61.49, 49.26 and 50.71 cm per stool in soil sites I, II and III, respectively (Table 16.4). Similar to tiller production, soil unit I recorded the greatest plant heights, but it was difficult to distinguish cane heights between sites II and III. However, no single soil unit recorded superlative plant height consistently from 3 to 12 MAT. Whereas plant height generally increased with time from 3 to 12 MAT in soil units I and II, this was not the case with cane plants in soil unit III which recorded decreased cane heights from 6 to 12 MAT. The initial excellent gain in plant height experienced in soil unit III increasing from 44.17 to 58.35 cm at 3 and 6 MAT must have been facilitated by an abundance of soil moisture. However, it may seem that the debilitating effect of salts in soil unit III of 36.5 $dS.m^{-1}$ became overbearing from 6 to 9 MAT growth period of the canes resulting in the decrease in plant height which continued up to 12 MAT.

Being a typical glycophyte, sugarcane exhibits stunted growth or no growth at all under salinity stress whereby its yield declines to 50% or more as compared to its actual potential (Wahid et al. 1997; Akhtar and Rasul 2003; Wiedenfeld 2008). High salt environment adversely affects plant growth due to alterations in water relations, ionic and metabolic perturbations, generation of reactive oxygen species (ROS) and tissue damage (Patade et al. 2012). Others are dislocation of enzymes involved in sugar metabolism (Gomathi and Thandapani 2005), and the plant responds with an altered expression of stress-responsive genes that may ameliorate the detrimental effects of salinity with varying degrees of success. The foregoing assertion is substantiated by the number of experimental plots with sugarcane plants that dried up after 6 months of transplanting in soil unit III that was saturated with lethally saline tidewater in contrast to canes on platforms free from flooding as shown in Figs. 16.5, 16.6, 16.7, 16.8, 16.9 and 16.10.

In general, there was a steady increase in average plant height at 3, 6, 9 and 12 MAT time periods only in soil units I and II. In particular, the steep increase in

Fig. 16.5 Sugarcane growth performances in ridges (right and left) and flat beds (middle) in soil site I (non-flooded)

Fig. 16.6 Sugarcane growth performances in ridges (right and left) and flat beds (middle) in soil site III at low (ebb) tide

plant height in soil unit I suggests quite favourable growth conditions than in unit II that showed a similar but less radical growth. The distinction is more obvious between soil units I and III. It will be recalled that soil unit III was prone to complete flooding by tidewater twice daily and was, thus, characterized by reduction reactions that were unfavourable to plant growth.

Fig. 16.7 Sugarcane plants in soil unit/site III at high (full) tide without plants in the middle plot (flat bed)

Fig. 16.8 Contrast between sugarcane growths in ridges (middle) and flat beds without plants (left and right) in soil unit/site III at high (full) tide

Flood events represent a natural disturbance affecting crop and forage production worldwide due to the detrimental effects they provoke on most land plants (Bailey-Serres and Voesenek 2008; Colmer and Voesenek 2009). In waterlogged soil, oxygen (O_2) becomes depleted if microbes and plant roots consume it in respiration faster than it is replaced from the atmosphere. Consumption of O_2 is accelerated by warmth, sufficient moisture, readily decomposable organic matter – all of which

Fig. 16.9 Panoramic view of the experimental plot at high (full) tide with soil unit/site III in the foreground

Fig. 16.10 Panoramic view of experimental plot at low (ebb) tide with soil unit/site III in the foreground

were applicable at the study site – and anything else that promotes respiration (Singer and Munns 1996).

The direct effect of O_2 depletion in flooded soil is the reduction of plant C fixation or rate of photosynthesis (O'Shea 2000). In the short term, photosynthesis can drop as a result of a restriction of CO_2 uptake due to stomata closure following leaf dehydration and turgor loss in guard cells (Else et al. 1996). If flooding continues in time, a decrease in the photosynthetic capacity of mesophyll cells leads to a further reduction of photosynthesis. In flood-sensitive species like tomato (*Solanum lycopersicum*), pigeon pea (*Pisum sativum*), sunflower (*Helianthus annuus*) and tobacco (*Nicotiana tabacum*), a few hours after the soil becomes flooded, water uptake by roots is reduced (Bradford and Hsiao 1982; Jackson and Drew 1984). The reduction of water absorption under flooding is a consequence of a reduction of the root hydraulic conductivity (Islam and Mac Donald 2004) that seems to be associated with the acidification of the cell cytoplasm and the gating of aquaporins (Tournaire-Roux et al. 2003). So the reduction of water uptake by flood-sensitive species under water excess in soil is paradoxically similar to the response of leaf wilting as observed under drought situation (Else et al. 1996) which was the case in the test canes that dried up and died on flat beds in soil unit III as earlier shown in Figs. 16.7 and 16.8.

Besides, the rate of transpiration follows the same trend as in photosynthesis. Transpiration rates by sugarcane plants are drastically curtailed under flooded conditions as cane leaves assume tightly curled positions similar to those under drought conditions, indicating reduced supply of moisture within the cane tissues (Humbert 1968). In flood-sensitive species unable to tolerate flooding, plants die without recovery even when the water recedes. In such flood-sensitive species, even short-term flooding can be devastating enough to reduce yield significantly. In sugarcane, growth rates are drastically reduced indicating that photosynthates are not stored during this period of the cane plant's struggle for survival, thus stimulating the plants to approach "induced" or "forced" maturity.

Results from the present study show that average plant heights per stool in soil sites I, II and III were significantly different at 6, 9 and 12 MAT in which soil unit III was plainly different from soil unit II at 6 MAT, soil unit I different from soil units II and III at 9 MAT and soil unit I different from soil unit III at 12 MAT (Table 16.5). The outstanding performance of the canes in soil unit I is clearly demonstrated, especially from 9 to 12 MAT.

16.6.2.3 Stalk Girth

Cane girth ranged from 2.83 to 5.80 cm with cumulative average values of 5.56, 4.63 and 3.93 cm per stool in soil units I, II and III, respectively (Table 16.4). As it was with tiller production and cane height, stalk girth was highest in soil unit I followed distantly by unit II and lastly unit III. On the whole, only soil unit I maintained a fairly consistent increase in stalk girth from 3 up to 9 MAT and just dropped during the maturation period from 9 to 12 MAT. According to Sanghera et al. (2019), ripening and maturation phase in a 12-month crop lasts for about 3 months starting from 270 to 360 days, i.e. 9–12 months of growth. The reduction in stalk growth rates, either in length or diameter, is followed by an increase in the rate of sucrose accumulation (Mamet and Galwey 1999). Sugar synthesis and rapid accumulation of sugar take place during this phase, while vegetative growth is reduced, and as ripening advances, simple sugars, the monosaccharides or fructose and glucose,

Table 16.6 Quarterly average stalk girth and number of nodes in test canes from three soil sites

	Stalk girth (cm)				Number of nodes			
Soil unit (site)	3 MAT	6 MAT	9 MAT	12 MAT	3 MAT	6 MAT	9 MAT	12 MAT
I	5.08a	5.76a	5.80a	5.59a	2.19b	6.31ab	11.81a	14.81a
II	4.81a	4.94a	4.33b	4.43a	2.56ab	5.75b	9.69ab	11.94a
III	5.11a	4.88a	2.83c	2.91b	3.69a	7.50a	6.94b	7.69b
	NS	NS						

Means with the same letter(s) are not significantly different according to Duncan's multiple range test (DMRT) at $p \leq 0.05$; MAT = Months after transplanting

are converted into cane sugar or disaccharide sucrose. So, the drop in stalk girth from 9 to 12 MAT in soil unit I was not unusual.

However, the generally unimpressive performance in stalk girth expansion rates may not be unconnected with adversities in growth conditions at the study site. Sugarcane commonly responds to various stresses differently depending upon the stress severity and the developmental stage. For instance, different varieties of sugarcane possess different sensitivities to the stresses that alter the rates of photosynthesis, structural growth and sucrose accumulation (O'Shea 2000). Among the various stresses, high temperature and water stresses in sugarcane are of key importance and may cause drastic impact during germination, early growth phases as well as flowering and maturation stages (Tao et al. 2006; Sanghera and Kumar 2018). Sanghera et al. (2019) reported strong relationship between sugarcane height and girth in response to environmental stresses.

The results depict a generally poor performance in girth expansion in the plants even in the light of the scant increase in soil unit I. The generally low pH levels in the soils could have impacted negatively on the plants. Sugarcane thrives best around pH 6.5 although it can tolerate a considerable range from pH 6.0 to 7.7 at the most. Moreover, the sandy nature of the soils may have contributed to poor growth in the crop. Sugarcane requires fertile light clay soil, i.e. clay loam or alluvial soil for its growth which was less than ideal at the experimental site. It is, therefore, obvious that stalk girth was adversely impacted in the canes generally but more so in soil units II and III in the present experiment. Statistical analysis shows that at 9 MAT, soil units I, II and III were significantly different from each other. At 12 MAT, both soil units I and II were significantly different from unit III (Table 16.6).

16.6.2.4 Node Production or Number of Nodes

Average number of nodes produced by the plants ranged from 2.19 to 14.8 with cumulative mean values of 8.78, 7.49 and 6.46 nodes per stalk in soil units I, II and III, respectively (Table 16.4). Soil unit I recorded the highest cumulative average number of nodes that was much more than the overall average of 7.57 nodes per stem, while sites II and III had numbers that fell below this value, which was worse in site III. As mentioned earlier, in soil unit III that was subject to recurrent flooding by saline tidewater, the initial tolerance of the plants to soil constraints up to 6 MAT

Fig. 16.11 Sugarcane plants in soil site I at 6 months after planting (6 MAT)

collapsed as they grew older. The sensitivity of crops to soil salinity, for example, changes from one stage of growth to the next (Keren 2000). Most plants are relatively salt tolerant during germination but become more sensitive during emergence, early growth and, possibly, much later in their development (Keren 2000) as evidenced in the present study.

Node production in each of soil units I and II during the period 9–12 MAT virtually doubled that in unit III. The results are yet another demonstration of better growth conditions that prevailed in soil unit I over those in unit II, which itself had more complementary soil conditions than those in unit III with extreme growth-limiting conditions. The situation was different from what obtained in soil units I and II that continually produced additional nodes to bring about substantial height increases in the plants as time progressed.

In spite of the poor performance of the plants in soil unit III, node production generally increased over time in all the soil units with quarterly average values of 2.81, 6.52, 9.48 and 11.48 nodes per stem at 3, 6, 9 and 12 MAT, respectively, except for a decrease in unit III at 9 MAT. However, the increases in the number of nodes produced were not quite compatible with gradual increase of time or age of the canes as seen in Figs. 16.11 and 16.12. Under favourable conditions, cane stalks grow rapidly at the rate of almost 4–5 nodes per month (Sanghera et al. 2019). Bull (2000) reported that there can be up to 20 or 30 nodes on a single sugarcane stem at maturity.

Quarterly average values of 2.81, 6.52, 9.48 and 11.48 nodes per stem at 3, 6, 9 and 12 MAT, respectively, and cumulative site average of 8.78, 7.49 and 6.46 nodes per stem in soil units I, II and III, respectively, indicate less than satisfactory performance in the environment. The all-time high figure of 14.84 nodes per stalk

Fig. 16.12 Sugarcane plants in soil unit/site I at 9 months after planting (9 MAT)

produced in soil unit I at 12 MAT was, itself, far less than the minimum 20 nodes per stalk as reported by Bull (2000).

Results on node production show that there were significant differences between soil units I, II and III during the period 3–12 MAT (Table 16.6). The number of nodes in soil unit I was significantly different from that in unit III at 3, 9 and 12 MAT. Soil unit II was also significantly different from soil unit III in node production at 6 and 12 MAT. On the other hand, nodes from soil units I and II were not significantly different throughout the period of cane growth from 3 to 12 MAT. It means that node production in soil unit II was fairly comparable to that in unit I even as the performance was much better in the latter. Without doubt, the canes in soil unit I performed overall best in the number of nodes produced, followed by those in unit II, while unit III recorded the least numbers.

16.6.2.5 Internode Length

Internode length ranged from 2.33 to 4.49 cm with cumulative average values of 3.23, 3.25 and 3.51 cm in soil units I, II and III, respectively (Table 16.4). In this instance, internode lengths in soil unit III outstripped those in units I and II, which was made possible by the matchless performance at 3 and 6 MAT. However, internode length at the periodic intervals was quite irregular such that there was no clear-cut pattern among the soil units. Nevertheless, internode lengths considered over the 12-month period across the three soil units show an overall increase from 3 to 6 MAT and a decline from 9 to 12 MAT, except in soil unit I that they increased up to 9 MAT.

In particular, the performance of the plants in internode length increase beyond 6 MAT up to 9 MAT was unimpressive. The grand growth stage in sugarcane lasts from 3 to 9 months to give way to maturation from 9 to 12 months when ripening occurs in a 12-month crop. The grand growth phase is the period of rapid internode

Table 16.7 Quarterly average internode length and number of green leaves in test canes from three soil sites

Soil unit (site)	Internode length (cm)				Number of green leaves			
	3 MAT	6 MAT	9 MAT	12 MAT	3 MAT	6 MAT	9 MAT	12 MAT
I	2.33b	3.64a	3.93a	3.00a	7.06a	7.19a	6.38a	5.69a
II	3.48ab	3.49a	3.23a	2.79a	6.38a	6.38a	5.00ab	5.13a
III	3.76a	4.49a	2.91a	2.86a	6.06a	6.06a	3.63b	3.00b
		NS	NS	NS	NS	NS		

Means with the with same letter(s) are not significantly different according to Duncan's multiple range test (DMRT) at $p \leq 0.05$

elongation when the basic yield of each cane stem is formed, with daily length increase of up to 1–2 cm and monthly increase may be as high as 50–60 cm (Sanghera et al. 2019). The drop in internode length soon after 6 MAT may have been caused by the severity of growth inhibitors in the environment such as repeated soil saturation or flooding resulting in O_2 depletion in soil unit III and parts of soil unit II. Other inhibitors were the unbearable Na and Al toxicities and salt effects on cane tissues in soil units II and III, as well as nutrient deficiencies and/or imbalance coupled with the scorching action of acid soil on the canes in soil unit I.

Across the three soil units and within the quarterly intervals, there were significant differences in internode length between soil units I and III only at 3 MAT (Table 16.7). At all other periods, there were no significant differences between soil units I, II and III as internode length in the canes was grossly impacted by growth-restricting soil conditions at the three sites, particularly from 6 to 12 MAT reflecting varying degrees of severity.

16.6.2.6 Number of Green Leaves

Average number of green leaves produced by the plants ranged from 3.00 to 7.19 in which cumulative mean values were 6.58, 5.72 and 4.69 per stalk in soil units I, II and III, respectively (Table 16.4), thus indicating a decline from units I to III. The same pattern of reduced number of green leaves from soil units I to III can be observed within each of the 3, 6, 9 and 12 MAT time periods. However, here, while canes from soil unit I clearly produced green leaves in excess of the 3-monthly averages of 6.50, 6.54, 5.00 and 4.61 at 3, 6, 9 and 12 MAT, respectively, those from soil unit II did the same only at 12 MAT, and those from soil unit III never got to achieve this feat all through the duration of the experiment. According to Bull (2000), there can be between 5 and 15 green leaves on a single stem. In the light of obvious drawbacks in growth conditions identified in the present study, the implication is that, in spite of its peculiar constraints relating to high soil acidity, soil unit I provided relatively milder limitations to the plants than did unit II, which itself was more tolerable than unit III.

Though leaf production appeared to be constant from 3 to 6 MAT in soil units II and III, the general fall in numbers from 9 to 12 MAT was substantial enough to influence the pattern of growth. In particular, green leaf production was grossly hampered in soil unit III, down from an average of 6.06 green leaves per stalk at 3 MAT to just 3.0 at 12 MAT. Statistical analysis reveals that canes from soil units I and III recorded significant differences in green leaf production at 9 MAT, while the plants in unit III differed from those in units I and II at 12 MAT (Table 16.7).

In essence, cane plants in soil unit III seemed to have had much more unfavourable soil conditions than those from units I and II, among which was flooding that was accentuated by high salt content of 36.5 $dS.m^{-1}$. The lack of free oxygen to plant roots and possible presence of toxic gases such as methane in the flooded soil of unit III were all detrimental to plant growth (Singer and Munns 1996; Else et al. 1996; Islam and Mac Donald 2004), coupled with the reduction of C fixation in photosynthesis (Humbert 1968). In flood-sensitive species, closure of stomata with or without leaf dehydration and reduction of transpiration can occur within just a few hours of flooding. The resultant effect of flooding is, therefore, gross reduction in growth and vigour of growing plants as demonstrated here by scanty green leaves in the canes which also reflected the poor health status of the canes and, according to Bradford and Hsiao (1982) and Else et al. (1996), death in worse scenarios in soil unit III as shown earlier in Figs. 16.7 and 16.8.

16.6.2.7 Leaf Length

Average leaf length ranged from 59.49 to 117.29 cm with cumulative mean values of 113.38, 95.58 and 74.98 cm per stalk in soil units I, II and III, respectively (Table 16.4), decreasing from soil units I to III. Both of soil units I and II recorded leaf lengths that were more than the overall average of 94.65 cm per stalk, while unit III fell below this figure. It is quite remarkable that only in soil unit III did leaf length fail to get higher than the initial figure of 89.29 cm per stalk at 3 MAT, and this is an expression of severe constraints to plant growth in the unit. Bull (2000) reported leaf lengths varying between 90 and 200 cm in mature leaves and that leaf width is strongly correlated with stem diameter and leaves become both longer and wider as the stem expands, until a stable leaf size is attained.

In soil unit III, the poor performance by the canes in leaf length can only be attributed to the overwhelming effects of flooding and salinity stress, in particular, and generally toxic levels of Al and Na mentioned earlier. Related studies have shown that growth limitations in combination have additive effects of individual stresses. Quite commonly, a combination of stresses aggravates the effect of an individual stress, and the same mechanisms involved in response to a single stress are raised under the combined stress condition (Sanghera et al. 2019). Statistical analysis shows that cane leaf lengths in soil unit I were significantly different from those in unit III all through from 3 to 12 MAT. The same was observed between leaf lengths in soil units I and II but only at 3 and 6 MAT, while those from soil unit II were significantly different from those in soil unit III at 9 and 12 MAT (Table 16.8).

16.6.2.8 Leaf Width

Single leaf width ranged from 1.34 to 2.96 cm with cumulative mean values of 2.65, 2.19 and 1.71 cm per stalk in soil units I, II and III, respectively (Table 16.4). Soil unit I recorded cane leaf widths that surpassed the 3-monthly mean values of 2.29,

Table 16.8 Quarterly average leaf length and leaf width in test canes from three soil sites

Soil unit (Site)	Leaf length (cm)				Leaf width (cm)			
	3 MAT	6 MAT	9 MAT	12 MAT	3 MAT	6 MAT	9 MAT	12 MAT
I	106.62a	115.21a	117.29a	114.40a	2.30a	2.58a	2.96a	2.74a
II	89.06b	96.95b	100.31a	96.00a	2.58a	2.03b	2.09b	2.04ab
III	89.29b	87.11b	64.01b	59.49b	2.00a	2.09ab	1.34b	1.39b
					NS			

Means with the with same letter(s) are not significantly different according to Duncan's multiple range test (DMRT) at $p \leq 0.05$

2.23, 2.13 and 2.06 cm at 3, 6, 9 and 12 MAT, respectively. Leaf width in soil unit II exceeded the quarterly average value of 2.29 but only at 3 MAT. Soil unit III remained the third best all through the 3 to 12 MAT time periods, except at 6 MAT, and this conforms to the earlier assertion that the site had more than just a few growth limitations. In actual scores, the canes in soil unit I performed best, followed closely by those in unit II, while in unit III, leaf width expansion suffered ruinously due to adverse soil conditions from 6 to 12 MAT.

Indeed, leaf width plummeted in soil unit III from 2.00 cm at 3 MAT to a dismal 1.39 cm per stalk at 12 MAT. There is, therefore, no denying the progressive deterioration in soil conditions with attendant reduction in growth performance of sugarcane from soil units I to III. Except in soil unit I in which leaf width increased from 3 to 9 MAT, there were decreases in cane leaf widths at different times in units II and III. Comparatively, cane leaf widths in soil unit I were significantly different from those in unit II at 6 and 9 MAT and from those in unit III at 9 and 12 MAT (Table 16.8). In this instance, soil unit II was more like unit III as there were no significant differences between the two sites in cane leaf widths over time.

16.6.2.9 Leaf Sheath Length

Average length of single leaf sheath produced by the canes ranged from 13.13 to 25.04 cm with cumulative average lengths of 23.29, 19.82 and 18.55 cm per stalk in soil units I, II and III, respectively (Table 16.4). Leaf sheath length decreased from soil units I to III. The drop in leaf sheath length from soil units I to III corresponds with the decline in number of green leaves, leaf length and leaf width discussed earlier. Over time and across the three soil sites, quarterly average leaf sheath length generally decreased from 20.67 to 18.65 cm per stalk from 3 to 12 MAT, notwithstanding the superior performance in unit III at 3 MAT.

The early advantage of greater leaf sheath length produced in soil unit III can be likened to those of stalk length, number of nodes and internode length. All these were attributable to the apparent sufficiency of available soil moisture and were overturned by the compounding effects of excess water and high salts or salinity in the soil. Sugarcane responds to various stresses differently, depending upon the stress severity and the developmental stage (Sanghera et al. 2019). Clearly, soil

Table 16.9 Quarterly average leaf sheath length in test canes from three soil sites

Soil unit (site)	Leaf sheath length (cm)			
	3MAT	6 MAT	9MAT	12MAT
I	21.64a	25.04a	23.74a	22.74a
II	15.81a	22.66a	20.74a	20.08a
III	24.54a	22.82a	13.72b	13.13b
	NS	NS		

Means with the same letter(s) are not significantly different according to Duncan's multiple range test (DMRT) at $p \leq 0.05$.

conditions in unit III negatively impacted on leaf sheath length rather severely as the plants aged from 3 up to 12 MAT.

Critical observation of the data shows that there were significant differences in leaf sheath length between soil units I and II on the one hand and unit III on the other at 9 and 12 MAT (Table 16.9). The results indicate the codominance of soil units I and II over unit III with respect to leaf sheath length in the canes, even as actual leaf sheath lengths were greater in soil unit I.

To sum up, growth parameters of the test crop, sugarcane, were copiously influenced by soil conditions at the study site. In soil unit or site I, the plants produced the higher cumulative average number of tillers, nodes and green leaves than the overall average and achieved mean stalk length, stalk girth, leaf length, leaf width and leaf sheath length that exceeded the 3-monthly average values at 3, 6, 9 and 12 months after transplanting (MAT). Growth indices of the plants in soil unit III were the least, while those in unit II were midway between units I and III. Furthermore, soil unit I was significantly different from unit II in leaf width and leaf sheath length as well as from unit III in stalk length, stalk girth, number of nodes, internode length, number of green leaves, leaf length, leaf width and leaf sheath length. Soil unit II was also significantly different from unit III in stalk girth, number of nodes and green leaves, leaf length and leaf sheath length. Thus, there is no doubt about the progressive deterioration in soil conditions with the attendant reduction in growth of the crop from sites I to III. The results are in agreement with the earlier assertion in Sect. 16.6.1 that soil conditions worsened from sites I to III and that sites II and III could only be cultivated to crops that are naturally tolerant to the major constraints identified in the study.

16.7 Research and Development for Sugarcane Cultivation on Mangrove Soils

The mangrove ecosystem is a highly complex, dynamic and fragile environment that requires cautious considerations in the management and utilization of its resources. There is, therefore, the need to establish a common understanding of the importance and direction of research interventions for their utilization and the capacity of the mangrove ecosystem to withstand the burden of exploitation and self-regeneration.

The planet Earth is made up of two broad components – the human and other environmental resources comprising the physical, chemical and biological properties of the natural environment (Raj et al. 2020, 2021). While the physical, chemical and biological properties of the natural environment have the capacity to support a wide range of human needs and functions, their ability to continuously provide this support depends largely on the knowledge acquired by human beings about their nature and characteristics and how to plan and manage them properly (Adeniyi 1999; Raj et al. 2019a, b). The human history has shown that economic progress in any country has generally followed the advancement and application of science and technology (S&T) research that is controlled by and responsive to the needs and natural endowments of the country.

Thus, research in all its ramifications is the cornerstone of a nation's social and economic progress, that is, the proven capacity of that society to translate scientific knowledge through its judicious and determined transformation of environmental resources, into economic productivity and prosperity (Abiodun 1999). S&T have their footing firmly grounded in research, and altogether, they utilize natural resources to foster economic affluence and stability in a nation, which manifests in wholesome human development on a long-term basis, i.e. sustainable development. As Ravetz (1971) noted "...the prosperity and economic independence of a firm or nation does not rest so much in its existing factories as in its research and development (R&D) laboratories". Thus, R&D are twin terms representing two sides of the same coin and cannot be separated one from the other if the coin is to have economic value or relevance towards sustaining humanity.

For individual human beings, human societies or nations, development means improvement in living standards through economic and social advancement, which enables the individuals or people of a nation to realize their full potential, build self-reliance and self-confidence and lead lives of dignity and fulfilment over a long period of time as in perpetuity. According to Chinsman (1999), it is a value word used to describe the process of economic and social transformation, and development becomes real only if it is enduring, i.e. sustainable in the long-run.

The sustainable development concept focuses around the two issues of economic growth and environmental protection. The need for environmental protection becomes very fundamental in that natural resources are not unlimited or infinite (Khan et al. 2020a, b). Creation as a whole is finite, and so are the resources that nature can provide for our needs; thus, their sensible exploitation becomes imperative. This is borne out of the realization that human survival depends largely on the health of the natural resources and the environment in which they exist. As Ndukwe (2000) succinctly put it, "... no development can take place without an environment that provides the natural resources for utilization".

In the present context, sustainable development means the utilization or exploitation of the mangrove forest environment in such a way that entails maximum use of the resources therein and, at the same time, ensuring environmental stability in the area. Therefore, R&D efforts should be hinged on better understanding of the mangrove ecosystem and its utility to provide products and services for humankind in the face of imminent perturbations as a result of human use, i.e. anthropogenic causes. The following discussion of R&D shall focus on the mangrove ecosystem, sugarcane cultivation and the necessity for ecosystem stability in the region.

16.7.1 Research on the Mangrove Ecosystem as a Reservoir of Renewable Natural Resources

A mangrove forest represents a pool of renewable natural resources. A "renewable natural resource" is defined here as one that has inherent capacity to naturally increase or regenerate itself with usage over a reasonably short period of time or assisted to do so through human management intervention. With the countless number of natural resources that abound in mangroves and the complex, dynamic and fragile nature of the ecosystem, the foremost prerequisite for their continuing exploitation by mankind is the collection of relevant information about the existence, characteristics and quantum of the various renewable natural resources that are available for transformation into goods and services to serve human needs.

It is needless to catalogue renewable natural resources stored-up in mangroves or specifics of research objectives needed, but as Abiodun (1999) stated, "...of higher importance is what one does with that information, i.e., process it, analyze it, digest it, and apply the results obtained thereof, as may be appropriate, to arrive at an informed decision" and he concluded by saying "that is what is called research and its application in decision-making". In this respect, investigative research on the mangrove ecosystem needs to be comprehensive enough to the extent that exploitation of any of the resources should be well guided by its ability to regenerate itself and based on decisions driven by real needs for human development and not for luxury due to abundance of the resource(s) or ease of access. The sanctity of the mangrove ecosystem as a living entity with limited ability to sustain itself over a reasonable period of time should be respected, or else, it will degenerate, literally collapse and die, i.e. become extinct and may never be restored fully in our lifetime.

16.7.2 Research on Sugarcane Cultivation in Mangrove Ecosystem

The history and literature pertaining to sugarcane cultivation on mangrove soils in Nigeria is simply empty, without any record whatsoever. The existing four sugar factories and the National Cereals Research Institute (NCRI) that is charged with the mandate for sugarcane research in Nigeria are all located in the northern parts of the country and none in the south, let alone the mangrove belt. As narrated in an NCRI report, besides operating within the North Central Zone of the country comprising Nasarawa, Niger, Kwara, Kogi, Plateau, Taraba and Benue States as well as the Federal Capital Territory in Abuja, the Institute conducts research in nine (9) outstations located in different parts of the country (NCRI 1998). The locations (and states) include Ibadan (Oyo), Amakama-Olokoro (Abia), Uyo (Akwa-Ibom), Ubo-Ukuku (Akwa-Ibom), Warri (Delta), Bacita (Kwara), Mokwa (Niger), Birnin-Kebbi (Kebbi), Numan (Adamawa) and Yandev (Benue).

The corporate mission of NCRI is the provision of efficient research support to local cane growers as well as the Nigerian sugar industry (NCRI 1998). Research project areas, according to NCRI reports (NCRI 2001, 2007), include:

- Sugarcane breeding and varietal improvement.
- Sugarcane agronomy.
- Plant nutrition and soils.
- Crop protection.
- Crop processing economics and utilization.
- Sugarcane seed production and related matters.

As part of its activities, NCRI also conducts on-farm adaptive multi-locational trials in order to ensure that results are applicable to different ecologies in the country, particularly in the north-central zone. But suffice it to state that the reach is awfully narrow as many states of the Federation are inadvertently excluded. This is more so as, according to NCRI report, over 60 potential sugarcane estate sites have been identified all across the country (NCRI 1998). Yet, production is dismal in relation to demand as there are only few government-owned sugarcane estates with large expanses of unutilized land areas and relatively low factory-installed capacities in Nigeria.

The production outlook over the years shows that whereas industrial cane output for the sugar industry continued to decline, chewing cane produced mostly by local farmers increased astronomically. This is understandable since industrial cane with slim, hard rind and less succulent and juicy tissue is unattractive to local farmers that service the needs of the chewing populace. Because of the consistently high chewing cane output in the country, the Sugarcane Research Programme of NCRI is now focusing some of its attention on how more of the chewing cane produced could be diverted to sugar manufacture (NCRI 1995).

Also, in spite of the fact that sugarcane can be cultivated all across Nigeria, its processing into local products is limited to only a part of the north and has not attracted wide patronage outside the traditional areas of northern Nigeria. An obvious reason for the apathy is that such technologies were developed without proper understanding of the time-tested traditional processing methods and due consideration for socio-economic needs of target end-users.

From the foregoing narration, the following sugarcane research interventions may be advanced generally:

1. Intensification of the existing project lines as listed earlier.
2. Expansion of outstations for multi-locational trials from the current 9 to all states of the Federation, their number being 36, and the Federal Capital Territory in Abuja.
3. Aggressive breeding and selection of chewing cane cultivars for adoption and processing in the sugar industry.

4. Research related to new technologies for sugar processing into local products should be driven by the capabilities and socio-economic background of target end-users.
5. For sugarcane cultivation on mangrove soils, in particular, the following areas of research priorities are necessary:
   (a). Breeding for resistance or tolerance to frequent waterlogging, salt-stress and high acidity in soil.
   (b). Soil fertility and plant nutrition, including toxicity issues related to sodium, aluminium and, possibly, manganese.
   (c). Pests and diseases.
   (d). Land preparation methods and farm management practices.
   (e). Irrigation water management on sites free from flooding, the soils being extremely sandy in texture.

16.7.3 Research on Mangroves for Ecosystem and Productive Functions Towards Sustainability

In view of the obvious changes likely to be introduced by vegetation clearance, sugarcane cultivation and farm management activities in the mangrove ecosystem, certain factors require attention. Reclamation of mangroves for sugarcane cultivation has been widely reported in the literature. Reclamation, itself, is an exogenous process with some attendant alterations to ecosystem integrity and functions, depending on the magnitude and scale of reclamation to be undertaken. Inasmuch as reclamation efforts have yielded some positive results in some countries of the world, peculiarities of the local environment and the capacity of the soils to withstand such physicochemical and biological disturbances are all research questions to be answered.

From the case study reported in the present work, it was shown that sites prone to frequent flooding by saline tidewater were unfit for sugarcane cultivation due to adverse effects in waterlogged soil condition. Naturally reclaimed portions in the landscape arising from progradation or accretion were shown to be better suited to the crop *albeit* they would necessitate huge management inputs to ameliorate the effects of high acidity, salt stress, nutrient imbalance and/or deficiencies. Since climatic factors in the study area were earlier alluded to as being adequate for sugarcane cultivation, edaphic, i.e. soil conditions become fundamental to the use of the mangrove ecosystem for agricultural production. These issues can only be resolved from relevant research findings. Indeed, it is here suggested that further experiments be conducted in the study area with ameliorative measures to alleviate the identified negative effects of extreme acidity, salinity and nutrient imbalance or deficiencies at reclaimed sites in order to make them productive in sustainable agriculture.

With regard to the recommendation to introduce sugarcane into uplifted or reclaimed portions of the mangrove environment, the salient questions would be the impact on the ecosystem and interrelationships with other resources to bring about multiple land use that assures sustainability both for the numerous natural resources available and for socio-economic human development. The right agroforestry practices that would be compatible with economic realities, increase the productive capacity of the soils and enhance environmental health need to be in place, and this can only be provided through dedicated research (Jhariya et al. 2015; Singh and Jhariya 2016). This is the right way to go with respect to exploitation and protection of mangrove ecosystems in the present study area.

16.8 Threats to Mangrove Ecosystems in Nigeria

16.8.1 Brief Global Overview

Globally, mangroves are one of the most threatened tropical and subtropical biomes in which about 35% of habitats have already been destroyed (Ramirez et al. 2016). The rates of disturbance and variety of the human-induced influences on mangrove ecosystems have been steadily increasing, such that large proportions of the world's mangroves are threatened with extinction (Kusmana 2015). The main causes of mangrove destruction as currently happening in the world can be broadly distinguished as (1) over-exploitation by the traditional users for fuel wood, charcoal, pole and uncontrolled forest concession airs; (2) destructive actions resulting from activities such as conversion to agriculture, mining or mineral extraction and aquaculture among others; and (3) pollution and natural disasters (Kusmana 2015).

Major exploitation and conversions have resulted in the destruction of more than 50% of Indonesia's mangrove areas, and the same applies to India, the Philippines and Vietnam. For example, increasing population growth and economic development in Indonesia have resulted in the destruction, even disappearance, of many mangrove areas through land conversion to fish ponds, industrial estates, transportation and recreation infrastructure, resettlement, tin mining, agricultural activities and other land uses. Mangroves are being cut down faster than rainforests in the Americas as well. At the regional level, Asia suffered the largest net loss, more than 1.9 million ha since 1980, mainly due to changes in land use from 1980 to 1990. North and Central America and Africa also contributed significantly to the decrease in mangrove areas at the global level, with losses of about 690,000 and 510,000 ha, respectively, over the last 25 years. However, FAO (2007) reported that although mangroves still face major threats, the rate of loss has recently been decreasing from some 187,000 ha lost annually in the 1980s (−1.04% per year) to 102,000 ha annually (−0.66% per year) during the 2000–2005 period. The figures suggest that during the past 25 years, about 3.6 million hectares have been lost corresponding to some 20% of the global mangrove area.

Overall, more than one-third of the world's mangroves have already vanished over the past 60 years (Hamilton and Casey 2016). Despite conservation measures being deployed in many nations of the world such as Australia, mangroves continue to be lost at a global rate of about 0.2% per year. Only Bangladesh and

Guinea-Bissau, out of the top 15 countries for mangrove area, experienced no net loss from 2000 to 2012 (Hamilton and Casey 2016). On the global picture, deforestation rate of 3.5% in Nigeria was the highest in the world (Byer 1999), and that included the mangrove forests. With escalation in population figures in the last two decades, the current situation in the country is much worse.

Losses in mangrove areas across the world have not been without grave environmental, economic and social costs. Besides the quantum and variety of commercial forest products derivable from mangrove forests, they provide habitats for numerous organisms including plants and animals that contribute to human livelihood, health and well-being. Their ecological role as buffer zones for terrestrial and marine ecosystems cannot be overemphasized. Their destruction implies a denial of the countless number of products and services they render to humankind. For instance, destruction of mangrove forests occurring globally has led to changes such as increased sea level, thus affecting terrestrial areas causing large-scale flooding and submergence. In the opinion of Valiela et al. (2001), accretion rates in mangrove forests may be large enough to compensate for the present-day rise in sea level. This, however, may not compensate for the massive losses and destruction to existing land features and habitations that took ages or decades to form in coastal areas.

But, more importantly, the global decline in mangroves due to anthropogenic activities raises concerns about the fate of the large organic carbon (C) deposits stored up within the soils (Lovelock et al. 2015). Several studies have suggested that degradation and removal of vegetated coastal habitats have the potential to disturb soil-C down to depths of 1 m, thus leading to its re-mineralization to carbon dioxide (CO_2) gas (Pendleton et al. 2012; Fourqurean 2012). Because mangrove soil-C deposits take thousands of years to form, once disrupted, they cannot be regained over meaningful human timescales by just restoring the forest. As a result, the re-mineralization of mangrove soil-C may add significantly to the component of anthropogenic greenhouse gas (GhG) emissions designated as "arising from land use change" and still unaccounted for in global C inventories (IPCC 2007). This, in the main, will affect the global C-cycle, the consequences of which are plainly monumental.

16.8.2 The Nigerian Experience

The mangrove ecosystem in Nigeria is bedevilled with a multitude of challenges that are either natural or human-induced. Natural threats are those related to coastal erosion in some areas and siltation in others, in addition to the cyclic rise and fall in sea level. Anthropogenic or man-induced causes range from localized threats from the natives inhabiting the area to the extraordinary from external forces. By and large, anthropogenic causes are those that pose the most formidable threats to the existence of the mangrove ecosystem, itself, as well as the very essence of life and human well-being in the Niger Delta as a whole.

Mangrove forests and forests in general, as well as open waters such as rivers, are lawfully the exclusive preserve of the Federal Government vide the Petroleum Act of

1969 and the Land Use Act No. 6 that came into force in 1978. But, except for designated conservation areas such as forest and game reserves that are under the management of state governments (Onuoha 1999; Ndukwe 2000), citizens have unfettered access to forest resources in the country, including the mangrove forest ecosystem. Because of the laissez-faire position of both federal and state governments in the administration of forest resources and subliminal sumptuous attitude of citizens, forests in the country have been subjected to indiscriminate and wasteful exploitation resulting in uncontrolled deforestation and degradation of forestlands. The same applies to water resources, and some aquatic organisms have already gone extinct. The non-inclusion of mangroves as a special forest type in the National Forest Policy has left mangrove forests the worse off, and today, mangrove resources also face imminent extinction in Nigeria.

Threats to the mangrove forests from external sources are mainly those associated with the oil industry in the country. By the nature of petroleum-hydrocarbon formations, these are products of geophysical and biochemical processes on marine organisms of yesteryears buried deep down in layers of sedimentary rocks. The Niger Delta is a region of sedimentation, with the mangrove belt as a zone of recent and continuing deposition. Thus, the mangrove belt and adjoining lowland areas constitute the epicentre or hub of the oil industry in Nigeria. Threats to environmental resources arising from the oil industry would, invariably, impact negatively on the mangrove ecosystem more than elsewhere in Nigeria.

Although it is possible that other sedimentary basins in Nigeria may contain substantial quantities of petroleum reserves, the Niger Delta has, so far, yielded all of Nigeria's oil and gas resources from wells both in land and offshore platforms. Oil and gas revenue from the Niger Delta region accounts for about 90% of Nigeria's foreign exchange earnings and over 80% of the national income (Alamieyeseigha 2003), making Nigeria the ninth oil-producing nation in the world (Ajakaiye and Akande 1998; Kusmana et al. 2008). In addition, the Niger Delta has been identified as a geological gas province in association with crude oil reserves. Expectedly, gas/oil ratios (GOR) of between 800 and 1000 standard cubic feet per barrel (SCF/B) are common with crude oil produced in the region, while the expected reserve life (ERL) of natural gas in Nigeria is currently placed at 250 years (Kaladumo 1996).

Inasmuch as any treatise on threats from the oil industry in Nigeria is more than or, at least, comparable to the subject of a standard textbook considering the economic and political dimensions, a brief contextual outlook on their occurrence and environmental implications may be attempted at each stage of the operations in the industry as reported by several authors. The petroleum industry is segmented into four major sectors, namely, prospecting or exploration, production or exploitation, refining and marketing, while transportation cuts across the various operations. Threats to the biota begin with the search for oil and gas in seismic surveys using small explosive charges and cutting traverses or tracts across mangrove forests, other vegetation types and farmlands. Their impact can be judged from the fact that in one single year, there could be as much as 10,000 km of seismic lines cut through mangrove vegetation and farmlands in the region (Kaladumo 1996; Imevbore 2001; Bassey 2001).

During exploitation as in drilling, the spread of drilling mud mixed with chemicals and oil, the establishment of rigs and the associated exploitation activities all contribute to disturb the mangrove environment. The installation of oil rigs and platforms is, itself, another area of concern considering the multiplicity of such structures in accordance with the multiple locations of oil deposits in Nigeria. In addition, crude oil is pretreated prior to being transported from the rig to the refinery, and the extent of pretreatment depends on the type of crude oil. In Nigeria, this includes removal of gas and brine. The salinity of formation water, i.e. water associated with the crude in Nigerian wells, can be as high as 15,000–60,000 ppm, even though the actual volume is generally small, about 1–5% (Kaladumo 1996). Even in the sea, this high concentration of salt will harm living organisms. It is important to note that while the high initial salinity may not have acute effects on aquatic life forms, the total salt load pumped into the river is quite another matter altogether, especially in the mangrove ecosystem that is naturally saline.

Still on exploitation, serious blowouts have occurred in the industry wherein the well explodes and the oil either burns itself out or is forced out of the rig. In one type of blowout which is the more common and known as "gas cutting", the rig hits a very high-pressure gas zone, and because the gas pressure is so great, it uproots the rig top (Kaladumo 1996). Although this is often prevented successfully by continuous mud-logging surveys, it is important to realize that the risks are high in the sort of multiple well formations that exist in Nigeria.

Furthermore, crude oil wells in Nigeria contain oil along with the associated gas and water amidst shale. Thus, a froth of oil with gas is what comes out to the surface, creating the need to be separated from the gas and water. In Nigeria and most other petroleum-producing nations where insufficient investment has been made in infrastructure to utilize the natural gas associated with crude oil, flaring is employed to dispose of the gas which is routed to an elevated vertical chimney or horizontal stack and burnt off at the tip in what is generally known as "gas flaring" in the oil industry (Ubani and Onyejekwe 2013). The quantum of gas flared annually in the Niger Delta is enormous (Ayoola 2011; Ajugwo 2013). A 1995 World Bank Report had it that Nigeria flared more gas than any other country in the world (World Bank 1995).

At the present production rate of two million barrels of crude daily, an average of 1.8 billion SCF of associated gas is produced daily. The ratio of gas flared to gas utilized is currently put at 80:20, thus rendering a copious 1.44 billion SCF of gas flared away daily into the Niger Delta environment, especially the mangrove ecosystem (Kaladumo 1996; Davidson 2000; Onuoma et al. 2015). This is equal to an annual atmospheric load of 525.6 billion SCF of gaseous pollutants. One can only imagine what has happened in the last 60 years or more (1957–2020) of gas flaring in Nigeria.

The gas contains, essentially, methane, some volatile organic compounds (VOCs) as well as sulphur dioxide and other sulphur compounds which are all detrimental to human health and the environment. The impact on surface water and groundwater sources is no less mind-boggling as reported by Nwankwo and Ogagarue (2011) as well as human habitation according to Ekpoh and Obia (2010). A typical flare in the Port Harcourt region has the ground devoid of vegetation around it from 30 to 40 m radius (Alakpodia 2000; Efe 2003). The area is not colonized by any vegetation and remains bare without further clearing operations.

The area fringing the "no life" zone has low bird and insect populations. Higher vegetation was either suppressed or damaged over some considerable distance after the composition of the ground herbs had become normal or nearly so following stoppage of the flare. Also, air, soil and leaf temperatures were found to have increased, while relative humidity of the air decreased within about 110 m of six flare sites in Port Harcourt (Deekor 2002; Efe 2003; Gogoi and Baruoh 2012). Physiological effects observed on plants showed shortening of apical internodes, some leaf distortion causing crinkling and curling or "pebbling" and reduction in leaf size as well as in chlorophyll content (Imevbore 2001; Ogidiolu 2003).

In general, the effects of gas flaring on the environment may be summarized as follows:

- Thermal impact.
- Chemical impact.
- Light or visual impact causing night-daylight illusion or aberration.
- Sound or noise pollution, especially to wildlife and humans.
- Air pollution.
- Water pollution and acid rain.
- Biodiversity loss.

Details about the impacts of gas flaring on land or soil, air, water and both terrestrial and marine organisms are too numerous to be listed and are as diverse as their sources. Of greatest concern, globally, are the issues of global warming-cum-climate change, greenhouse gasses and ozone layer depletion resulting from gas flaring (Kaladumo 1996; Ismail and Umukoro 2012). There are, however, some recent efforts at harnessing the natural gas resources of the region in various utilization programmes. But these belated attempts at minimizing gas flaring have been an outcome of the global awareness and outcry about the environment rather than some deliberate legal and policy measures by the government. Deadline for stoppage of gas flaring has been shifted time and time again due to pressure from the oil companies operating in Nigeria. The issue of gas flaring has been highly politicized (AGOC 2003) for some trumped-up rationalization and claims of greater national economic interests. Notwithstanding the plentitude of public outrage and global condemnation, Nigeria lacks the political will to enforce regulations to eliminate gas flaring in the country. And so, despite legislative provisions for gas re-injection, Nigeria continues to be infamous for contributing 25% of all the gas flared in the world (World Bank 1995). And all environmental components, particularly human well-being such as economic, social and demographic structures, are the worse for it.

Equally worrisome and overwhelming is the threat of oil spillage arising from transporting operations in the petroleum industry in Nigeria. First, the transportation and distribution systems involved in the production of crude oil are relatively complex with cross-country pipelines extending over several thousands of kilometres, and these are either buried in the ground or placed on the land surface. The pipes are laid across farmlands, waterways and fishing grounds, while some pipes cross communities and living quarters in the Niger Delta.

In the mangrove belt, besides extensive vegetation clearance, excavations are made through the forest in order to bury the pipes conveying crude oil or refined petroleum products. Excavation of soil and riverbed "dredge spoil" and consequent deposition in the immediate vicinity are major sources of pollution and degradation to the mangrove ecosystem. These excavations are about 60 m wide in forestlands as in the mangroves, but in areas where they pass through agricultural lands, the size is reduced to 15 m (Kaladumo 1996). In this process, valuable agricultural lands or mangrove forests may be lost in pipeline "right-of-way" (P-ROW) or "wayleaves".

These are pipes from the rigs or oil wells to the flow stations and from the flow stations to the refineries or oil terminals for foreign trade, i.e. for export. It is important to state that, except for the Kaduna Refinery located in the northern part of the country, all three other refineries and six oil terminals are located in the Niger Delta region. For example, crude oil for export is transported in a criss-cross of pipelines to Bonny and Forcados terminals covering about 6000 km, mostly through mangrove vegetation which is the basin of oil deposits in Nigeria.

In this process of transportation, crude oil is spilled into the environment through accidental discharge, improper handling, burst pipes due to corrosion or equipment failure and, occasionally, sabotage or vandalism by disgruntled elements in the region. In all cases, a deluge of oily substances are poured out into the environment, and these all find their way to the mangrove ecosystem as the final sink even from spills on higher lands in the region. It was estimated that, in the oil-producing states of Nigeria, an average of one oil spill occurred every week (Alamieyeseigha 2003). The total number of annual spills has gradually increased since 1957, from approximately 250 spills per year to 500 spills per year (Kusmana et al. 2010). Accordingly, the Nigerian Federal Ministry of Environment estimated 13 million barrels of oil spilled during extraction processes.

Meanwhile, the location of many drilling rigs and platforms and the necessity of supplying the workers with food and other accessories including power have caused considerably increased water transportation system with a number of vessels loading and discharging materials. One would be amazed at the widespread oil pollution from refined oils encountered along the creeks and estuaries linked to the mangrove ecosystem arising from the transportation system alone. This is beside the amount of oil discharged into the sea from tankers in what is referred-to as "ballasting" in which the tanker is cleared of the oil dregs from offloaded stuff, and these find their way through waves to shore in the mangroves.

The impacts of oil from the petroleum industry, particularly crude, on environmental components in the mangrove ecosystem are simply countless. Suffice it to say that these hydrocarbon products have a tendency to adhere to surfaces and to build-up as deposits in mangrove swamps causing restricted gaseous exchange for fishes and other living organisms as recorded by Nwankwo and Ogagarue (2011). The residue left in tanker compartments after discharge is simply crude oil with enlarged wax content. Beaches that are used for repeated disposition of lumps of this heavy viscous oil or tar become blackened that, in chronic cases, causes the death of shoreline organisms (Kaladumo 1996). Oil spill on coastal vegetation showed clusters of dying mangroves in proximity to where oil had washed ashore (Kusmana et al. 2010). According to these workers, petroleum has toxicological impacts on mangrove trees and causes direct physical damage.

It was shown in Australia that mullets in polluted waters acquired a kerosene-like tint on exposure to the product. Similarly, oil-contaminated eggs of lobsters were known not to hatch, while mullet breeding grounds showed that areas of oil activity were devoid of mullet fry (Imevbore 2001). Furthermore, it was recently observed that the deforestation of mangrove swamp forests for pipelines is a frequent cause of death to the edible crabs (*Callinectes gladiator* and *C. latimanus*) in the area (*op cit.*). Possible causes include the high temperatures (>45 °C) over exposed sediments, deoxygenation arising from micro-stratification of the shallow surface waters as well as poisoning from hydrogen sulphide and/or methane gas. Again, the destruction of mangrove forests, especially the "soft mud" soil sites occupied by *Rhizophora racemosa* in the present study area, removes the substrate for the edible oyster (*Crassostrea gasser*) (*op cit.*). On average, studies showed that mangrove fatality rates were highest during the first 6 months of a spill; thus, fresh crude oil spills had greater toxicity than "weathered" oil (Kusmana et al. 2010).

In general, the myriad threats from anthropogenic or industrial sources to the mangrove ecosystem in Rivers State, as in the entire Niger Delta region, represent tendencies of barefaced annihilation of natural resources in the region. Sadly enough, the implementation of environmental laws including the mandatory environmental impact assessment (EIA) provisions and their violations are treated with half-hearted sanctions as far as the oil industry is concerned in Nigeria. Considering the numerous forest products obtainable from mangroves, including fisheries and the invaluable ecological services rendered by them, it behoves the government at all levels of administration in Nigeria to swiftly develop policies, legal instruments, programmes and strategies to protect and sustainably manage the mangrove forest ecosystem in the country in the interest of economic revitalization and environmental stewardship as well as safety.

As suggested by Carroll and Turpin (2002), a very careful examination should be made by the environmental protection agency and ecologists before any area of mangroves is allowed to be altered in any way or allocated for any form of development, including the activities of the oil industry. They also recommended that extensive mangrove areas should be reserved in which the habitat will be protected whereby total protection is given to areas of special interest by establishing mangrove national parks or similar restricted areas. It is simply unthinkable that an entire ecological zone with such rarity of organisms and services as the mangrove belt could be left to the whims and caprices of multinational oil companies to exploit relentlessly only to be abandoned as wasteland as is the case today with the first ever oil well discovered in 1956 at Oloibiri in Bayelsa State in the Niger Delta that is worse than a shadow of itself – despoiled and derelict with no life forms in existence and the community without the slightest social amenity.

16.9 Policy and Legal Framework Related to Sugarcane Cultivation on Mangrove Soils in Nigeria

16.9.1 National Agricultural Policy and Governance Issues

Right from independence in 1960, Nigeria operated its agricultural programmes that were largely subsumed in its economic development policies basically targeted at earning foreign exchange through export trade. Although commercial sugarcane cultivation for the local market actually started in the late 1950s, agricultural produce for export were mainly cotton (*Gossypium hirsutum*, *G. barbadense*), groundnuts (*Arachis hypogaea*), benniseed (*Sesamum indicum*), rubber (*Hevea brasiliensis*), cocoa (*Theobroma cacao*) and palm oil and palm kernel from the oil palm tree (*Elaeis guineensis*). The first ever agricultural policy document for Nigeria was launched in 1988 aimed at redressing the underdevelopment of the agricultural subsector in spite of huge investments, streamlining policies in all tiers of government and ensuring policy stability in agriculture up to the year 2000 AD. The implementation of the policy was bedevilled by several constraints but more of poor execution and governance issues resulting in unrealized policy objectives and goals such that Nigeria could hardly feed its citizens.

In some frantic efforts, the Operation Feed the Nation programme was operated from 1976 to 1979 and another, the Green Revolution from 1980 to 1983. The recent National Economic Empowerment and Development Strategy (NEEDS) document of 2000 and the new agricultural policy launched in 2001 both are focused on reducing the pervasive poverty level currently estimated at 70%, ensuring national food security, attainment of self-sufficiency in basic food items, enhancement of employment opportunities and achieving a high growth rate in the economy through agricultural and rural development programmes. Specific objectives of the new agricultural policy are to ensure:

- Self-sufficiency in basic food production and attainment of national food security.
- Production of agricultural raw materials for industries.
- Production and processing for export.
- Application of modern processing technologies.
- Generation of gainful employment.
- Enhancement of the quality of life of rural dwellers.
- Protection of agricultural land from degradation and environmental pollution for sustainable agricultural production.

Granted that there have been several disruptions in governance over the years, the various approaches seem to have made little or no significant impact on agricultural productivity as evidenced by obvious and persistent national food and industrial raw material deficits in the country. The experiences of large increases in national food import bills, rise in domestic food prices, decrease in agricultural exports and escalating unemployment statistics over the years are all indications of unmistakable crisis situation in the country. According to Ajakaiye and Akande (1998), unless emphasis is placed on proactive policy-oriented research which should be the basis for macroeconomic and other policies in agriculture, the sector may find it difficult to play its proper roles of providing food to households and agricultural raw materials to industries, as well as improving the welfare of the rural population through increased employment opportunities and income in Nigeria. In addition, policy researchers are generally unaware of scientific advances in agriculture tucked away in research institutes and, hence, do not take them into account in policy analysis. Consequently, the necessary policy environments that will promote the rapid adoption of scientific advances are not consciously articulated for consideration by policymakers in the country.

Even with the wide-ranging new agricultural policy objectives, Nigeria has never exported sugar to earn foreign exchange in its history to date. This is largely because it still has not been able to satisfy local demand for the product despite efforts to better the situation. Ravaged by these distress features, the economies of both the federal and state governments are in dire need of urgent consciousness for national survival by way of reappraisal of policy objectives, basic strategies and implementation measures in the agricultural subsector, including the sugar industry. It is hoped that, with reorientation in policy framework, a world view of the demand and contribution of the sugar industry to national economy will help to generate enough awareness and political will on the part of various tiers of government to explore ways of keying into the ever-expanding global sugar market.

16.9.2 Policy on Mangrove Forests

The land resources of Nigeria include the mangrove forests that occupy extensive areas in the Niger Delta region, particularly in Rivers State spanning some 5000–6000 km^2 of land area. Perhaps by design or default, few people would argue that agriculture is the primary use of land for food production and, in most regions of the world, the best land is reserved for that purpose. This would be a right approach as the limited acreage of good agricultural land in Rivers State, for instance, makes it a very precious asset which should be conserved in the best possible way. As such, in most places, afforestation and reforestation are confined to marginal lands that are deemed to have low potential for crop farming.

And so, whereas Forestry policies are developed for other vegetation types in Nigeria, the mangroves are regarded as wasteland when viewed from the uncontrolled nature of degradation that takes place occasioned by deforestation and, for that matter, pollution from the oil industry. Deforestation has been propelled by demand for bushmeat, timber, fuel wood and other forest products. In the face of indiscriminate exploitation, fragmentation and degradation, no policy has been enacted in Nigeria with the sole purpose to conserve the endangered mangrove forests. Unintentionally affecting mangrove conservation in 1989, the National Policy on Environment was developed in Nigeria (USAID 2008). This policy resulted in 14.2% of Nigeria's land mass qualifying as protected area, securing 988 nationally protected reserves and 12 protected areas under international conventions of the Biodiversity and Protected Areas Management Programme (BIOPAMP) (Willoughby 2015). Despite federally declaring these 1000 sites protected, under-regulation and mismanagement have resulted in anthropogenic exploitation of reserve resources (USAID 2008).

According to the Nigerian Department of National Parks, the reserves collectively represent a variety of Nigeria's most essential ecological zones; among these signified zones are tropical and wetland forests (USAID 2008). Wetlands, generally, include the mangroves. The World Bank proposed the Andoni Game Reserve covering 124 km^2 in the mangrove-freshwater swamps in Rivers State (World Bank 1995). Yet, there is no single national park or protected area that is legally constituted and administered in the mangrove belt of Nigeria. At least three forest reserves can be created in the state, viz.: (i) tropical rainforest for the management of its bountiful resources, especially some high-quality timber species; (ii) freshwater swamp forest also for timber and other forest products including raffia palm; and (iii) saline or brackish water forest for the management and conservation of mangroves including wildlife. Considering the limitation on available land in Rivers State, it has been suggested that forest reserves as small as 5–10 ha could be created in the mangrove belt as it would be unrealistic to expect that only areas of several square kilometres should be considered for reservation (Kinako 2008), if for no other reason, at least, to protect the mangrove ecosystems and their abundant natural resources in the region.

Worse still, no deliberate policy or programme has been developed for crop cultivation in the mangrove belt of Nigeria till date, let alone sugarcane despite the yawning gap between national output of sugar and demand for the product in the country. Modern trends in forestry revolve around incorporation of non-timber land use into scientific forestry practices. In the face of fierce competition with other forms of land use for scarce land space in Rivers State, one cannot but agree with Beale (1970) that the thrust of forestry must be directed towards new involvements. A new and more dynamic forest policy that provides for multiple land use in mangrove forests is necessary now, more than ever before, in Nigeria.

16.9.3 Sugarcane Programme Implementation and Constraints

Sugarcane has not been given the desired attention in Nigeria's economic drive despite the fact that the crop can be grown in commercial quantities in all states of the Federation. The result of this neglect leading to the collapse of the few sugar manufacturing industries remains the reason for the massive importation of sugar to meet local demand in Nigeria. As far back as 1961, the Nigerian Sugar Company at Bacita was incorporated and then commissioned in 1964. Some 12,500 ha tract of land was acquired for the company, about half of which is currently under cultivation, with an installed annual capacity of 40,000 tonnes. While Bacita was being developed, Nigeria's demand for sugar had increased tremendously to 350,000 tonnes annually. This led to the establishment of the Savannah Sugar Company at Numan, which was incorporated in 1973 with an installed capacity of 65,000 tonnes per annum. Current production of cane is from just 5000 ha out of a total area of 22,000 ha originally acquired for cultivation.

Again, the Lafiagi Sugar Company was established in 1975–1976 and was halted due to a number of reasons, including the proposed hydroelectric power project at Lokoja that never took off. Sugarcane cultivation at the site, however, continued on about 200 ha land, while harvests were processed at Bacita, producing roughly 2000 tonnes of sugar annually. Furthermore, in an attempt to satisfy local demand for sugar, an agreement for the Sunti Sugar Company was signed in 1979 with a factory capacity of 100,000 tonnes of sugar per annum and 7500 tonnes of fodder yeast. Although the company has virtually taken off, it has continued to produce sugarcane from just about 350 ha land in relation to 15,000 ha originally acquired which it sells to Bacita.

While still not being able to meet its local demand for sugar, Nigeria issued a gazette in 2007 which, among other things, directed the establishment of the Biofuel Commission to oversee the successful operation of the biofuel programme in the country. The programme anticipated the use of 10% ethanol fuel for use by the year 2010. One of the major raw materials for ethanol production is sugarcane. There are other raw material sources such as grains, potatoes, etc. Of all these sources, sugarcane made the most sense as the base for the programme, since using any of the other sources would mean competing with the staple food needs of the citizens. For this programme to succeed, the Federal Government of Nigeria directed the Ministry of Agriculture and other relevant agencies to embark on massive farming and other necessary activities that would provide the requisite enabling environment. The sheer volume of sugarcane crop necessary to achieve the desired 10% blend of petrol and other petroleum products by the year 2010, coupled with the sugar needs of the country, made sugarcane a crop for the future. But, sadly enough, the programme has still not materialized in the country till date.

The combined sugar production from the existing sugar mills at Bacita and Numan, as well as the more recent sugar processing mini-factory at Lafiagi, has hovered between 10,000 and 50,000 tonnes in the last three decades. Thus, domestic production has been able to satisfy only about 3–5% of Nigeria's total demand. Indeed, in more recent times from 1996 to 2020, domestic production has plummeted with virtually all locally consumed sugar being imported into the country. Nigeria has been the loser for the neglect of the sugar industry because sugarcane farming and sugar processing are definite stimulants to tackle the unemployment scourge in the country. There is a large market for locally produced sugar which, if available, will redress the imbalance and mitigate import dependency to meet domestic demand of the product.

Fortunately, there seems to be an increasing official awareness at state government levels to introduce sugarcane on their list of crops for economic rejuvenation. The necessity of evolving more dynamic strategies involving judicious planning based on well-articulated policy initiatives designed deliberately to raise the level of agricultural production in the country seems to be unfurling as a new direction in the federal and state bureaucracies. Admittedly, through the efforts of both the National Cereals Research Institute (NCRI) at Badeggi and the National Sugar Development

Council (NSDC) in the nation's capital city of Abuja, states such as Jigawa, Bauchi, Kano and Katsina now devote large expanses of land to industrial cane production with a view to establish sugar processing mini-plants. These efforts are, however, still in their gestation stages and do not substantially contribute to the overall sugar production statistics in the country just yet.

16.9.4 Proposed Policy Thrust for Sugarcane Cultivation on Mangrove Soils in Nigeria

A new national agricultural policy direction is here proposed for Nigeria that includes and focuses on the export potential of sugarcane through the utilization of research advances, aggressive expansion in total hectares under cultivation across the country, increased processing facilities, marketing and, above all, private-sector participation down to community and household levels. Also, state and local governments should follow suit by formulating meaningful and workable guidelines in implementing the national policy adapted to their peculiar environments and resources. This is more so because the federal government tends to concentrate on the articulation of the national development strategy for the agricultural and other subsectors of the economy, while state and local governments implement such strategies (Ajakaiye and Akande 1998; Ndukwe 2000). Since agriculture is on the concurrent legislative list, state governments make their own policies to align and complement federal government strategies. It is on this basis that the following discussion centres on Rivers State as may be applicable to other states of the Federation in Nigeria.

A policy for the development of the sugar industry in Rivers State needs to be in place with the following objectives:

- Production of sugar in sufficient quantities as its contribution to national efforts to meet local demand for the product.
- Development of the sugar industry to facilitate industrialization in the sugar downstream sector.
- Promotion of processing technologies for the sugar industry.
- Encouragement of research and training in sugarcane cultivation.
- Providing entrepreneurship and employment generation in the sugar industry.
- Ensuring environmental health and standards in the sugar industry.
- "Additionally", strategies for achieving the policy objectives in the sugar industry in Rivers State shall include:
- The inclusion of sugarcane on the list of crops grown in the state and development of programmes for increased production, with special reference to the State Ministry of Agriculture and the Agricultural Development Programme (ADP).
- Conduct feasibility studies/surveys of locations in the state suitable for sugarcane cultivation.
- Sensitize farmers in the state to take to sugarcane farming.

- Ease of access to land and provide security of land rights or tenure.
- Establishment of "seed" multiplication centres and "seed banks" in the state for the availability of planting materials to farmers, preferably in collaboration with the National Cereals Research Institute (NCRI) at Badeggi, Niger State, Nigeria.
- Formal take-over from NCRI and replication of the brown sugar processing mini-plant at Ibaa in Emohua Local Government Area of Rivers State.
- Provision of incentives to sugarcane farmers, such as soft loans, grants, subsidies, tax relief or holiday, etc.
- Institution of a buy-back mechanism for sugarcane from farmers and raw sugar from processing plants for refining.
- Provision of training and skills development in the sugar industry through postgraduate scholarships, training workshops and in-service courses.
- Set standards for the quality of processed sugar and evaluate and monitor environmental safety in the industry.

With these policy objectives and strategies, and the quantum of sugarcane crop necessary to achieve the desired 10% blend of petrol and other petroleum products earlier proposed, coupled with the ever-increasing sugar needs of the country, it can be asserted that sugarcane remains a crop for the future in Nigeria.

16.10 Management Options of Mangrove Ecosystem Towards Agricultural Sustainability

The concept of sustainable development was enunciated, propagated and popularized in the late 1980s by the report of the World Commission on Environment and Development (WCED). The document – *Our Common Future* – which has become known as the Brundtland report stipulates "...economic development which meets the needs of the present generation without compromising the ability of future generations to meet their own needs using the same resources" (WCED 1987). The report set the global agenda for discussions and decisions about the relationship between economic growth and environmental resources. That environmental protection should constitute an integral part of the development process and cannot be considered in isolation. In doing so, however, the idea is not to protect the environment as a museum or in its pristine condition but to see it as a complex system that can be geared to productive requirements in such a way that our current utilization of it for development does not compromise its potential to meet the needs of future generations. This calls for a harmonious relationship between economic development processes and ecological or environmental protection issues.

The management of resources associated with mangroves can best be categorized to achieve (i) ecological and (ii) productive objectives, where productive use includes agriculture, in this case, sugarcane cultivation for economic development. The ecological roles of mangroves have been widely recognized to include (1) coastal protection, habitat for wildlife, nursery grounds for coastal fisheries, ecotourism, rapid cycling of nutrients, a protective barrier and reducing coastal erosion, storm surges and strong winds; (2) spawning, feeding and nursery area for many economically important marine organisms especially fishes; and (3) habitat for wildlife such as birds, primates, amphibians, reptiles and mammals among several others (Kusmana 2015). According to this author, the ecological value of mangroves lies in the fact that they constitute an interface unique ecosystem between marine and terrestrial environments characterized by high productivity and rapid cycling of nutrients that contribute a major share of the energy requirements of offshore ecosystems. For these and many more reasons, it may be argued that the ecological functions of mangroves far outweigh their agricultural value.

Still, the productive potential of mangrove ecosystems cannot be overemphasized, including being the most productive ecosystem on Earth in their capacity for carbon synthesis. Such exceptional potential needs to be explored maximally for sustainable human development. Moreover, the concept of sustainable development referred-to earlier prescribes that exploitation of natural resources should be on sustained-yield basis, thus implying that the annual harvest from a forest, for example, should balance the annual growth rate or renewal of the stock in the estate. In this context, sustained-yield management principle dictates that the annual harvest of mangrove trees for timber or whatever should not exceed the annual rate of regrowth or regeneration in the forest estate. Indeed, Melo et al. (2020) stated that management of mangrove forests is an important aspect in the effort to conserve the environment in coastal areas.

This involves reforestation and/or afforestation efforts as well as other silvicultural practices to maintain productivity and ecosystem stability. This applies to every other resource from mangrove forests in the country, including fisheries and similar exploitative activities. For that matter, agricultural land use in the mangrove ecosystem as being recommended in the present work deserves more than a casual consideration in order to achieve sustainability in both economic development indices and environmental health, in this case, productivity in sugarcane cultivation as well as environmental stability. It is against this background that the management and utilization of the mangrove ecosystem are discussed below.

At present, there is no crop farming on mangrove soils in the Niger Delta region. The "soft mud" soils may be described as too fluid to provide firm foothold for crop plants, while the "peaty clay" soils are markedly high in raw organic matter, and the "saline sands" that constitute the soils in the current study area are coarse-textured, poor in nutrient contents, etc. Besides, these soil types are prone to recurrent flooding by saline tidewater, thereby provoking unfavourable growth conditions or constraints for most cultivated crop plants. However, results from the current experiment, which is novel in Nigeria's mangroves, have shown that sugarcane crop can be cultivated and would grow to maturation stages in some parts of the mangrove landscape free from flooding represented by soil site/unit I. Those soils subject to waterlogging such as sites/units II and III in the study can only be cultivated to crops that would tolerate or withstand uncongenial growth conditions arising from flooding and salt stress or high salinity.

It is likely that if flooding was curtailed in soil unit II, sugarcane growth performance could have been as good as that in unit I since the results were quite comparable in a number of growth parameters determined in the experiment. This implies reclamation works on soil unit II whereby the soils would be exposed to atmospheric oxidation processes to facilitate decomposition of the huge amount of organic matter and nutrient mineralization to aid plant growth at such site. With the advantage of its extensive coverage, occupying about 90% of the mangrove forest belt in Rivers State, large areas of farmland can be provided for the cultivation of sugarcane in the area. Together with the naturally reclaimed soils of unit I constituting the fringe lands bordering the lowland rainforest, soil units I and II would afford substantial land area for mechanized sugarcane farming at commercial scales to boost the sugar industry in Nigeria.

The issue of soil management then comes to question. It was earlier reported that, besides flooding, the soils had high acidity, i.e. low pH, salinity and nutrient imbalance and/or deficiencies, particularly potassium. It stands to reason that appropriate farm management practices would involve lime and fertilizer applications. Coincidentally, lime application serves dual purposes on agricultural lands – to control or remedy high acidity and salinity in soil. It is common knowledge that calcium-containing lime material provides relatively more efficient and cheaper remedies to these soil conditions, preferably gypsum ($CaSO_4.2H_2O$). Additionally, since soil textures are essentially sandy with tendencies for poor water retention, supplemental water source in the form of small-scale irrigation scheme would be inevitable for both salt leaching and provision of water for increased soil productivity in order to achieve a sustainable agricultural enterprise in the study area.

Sustainability in agriculture similarly applies to forest management practices. Sustainable forest management (SFM), in this case mangroves, is the process of managing forests to achieve clearly specified objectives of management with regard to the production of a continuous flow of desired forest products and services without undue reduction of its inherent values and future productivity and without unnecessary adverse effects on the physical and social environment. SFM was described as the stewardship and use of forests and forestlands in ways and at rates that maintain their biodiversity, productivity, regeneration capacity, vitality and potential to fulfil, at present and in the future, relevant ecological, economic and social functions at local, national and global levels and that do not cause damage to other ecosystems (Jhariya et al. 2018, 2019a). Under the SFM option, certain specific management programmes have been advanced and practiced as reported by some workers, such as tourist destination (Spalding and Parrett 2019), seafood production in China (Szuwalski et al. 2020), seaweed farming in Indonesia and carbon sequestration business in Southeast Asia and the Philippines (Suratman 2008; Gevana 2010).

The SFM option is practiced to provide wood and non-wood forest products, ecotourism, agroforestry, silvoaquaculture or agrosilvopastoral system. In this system, mangroves with various valuable resources and environmental services must be managed on a sustainable basis (Kinako 2008). Multiple-use management of the forest estate on sustained-yield basis is the way to go in maximizing the use of forests in Nigeria. Viewed this way, all mangrove functions relating to ecological, economic and social benefits must be treated equitably such that a compromise is reached in mangroves exploitation programmes which allow sustainable yield and reasonable semblance to an undisturbed or non-harvested forest ecosystem (Chong 2006).

According to Kusmana (2015), preservation or maintaining a completely undisturbed or unexploited state may be a desirable management option for certain localities or for parts of some extensive mangrove vegetation in which such areas serve as refuge for fauna and flora as well as biological resources for restoring portions where management policies have failed or accidents have occurred. In other words, preservation of some portions of a mangrove area as buffer zones may be advantageous to an overall sustained use management plan. Mangrove swamps adjacent to developing areas need to be reserved for their value to fisheries, maintaining equilibrium in the food chain, biodiversity conservation, stability of the natural environment and aesthetic reasons. As suggested by Aroh et al. (2020), and in the present work, certain portions of the mangrove ecosystem in the study area need to be preserved to perform vital ecological buffering roles rather than for agricultural use.

Indeed, the multifarious functions performed by mangroves coupled with the sensitive, fragile and dynamic nature of the ecosystem demand that management of the resources should be based on both "preservation" and "conservation" principles. Preservation refers to total protection of a natural resource without consumptive or exploitative use. Such management option applies to resources that are either endangered or threatened and, thus, prevented from going extinct. Conservation, on the other hand, implies sensible consumption or exploitation of a natural resource in a manner that is not only equal to its regenerative capacity but, more importantly, that enhances the regeneration process for optimal and continuous use. The realization and optimization of resource renewal provide the basis for differentiating good from bad management practices. As such, appropriate management tools geared towards achieving both productive and ecological objectives, maximally, are essential to efficient utilization of mangrove ecosystems.

Mangroves are renewable natural resources that are amenable to programmed management and utilization of biological species and ecosystems in satisfying human needs through such activities as agriculture, forestry, fisheries, animal husbandry, relaxation, etc. Specifically, agroforestry entails the art and science of utilization of biodiversity in wild and/or managed systems of plants and animals in order to meet human needs for food, shelter, clothing and drugs as well as satisfy man's cultural and spiritual aspirations. Inasmuch as agriculture's claim to land is unchallenged in most cases, a balanced integration of agriculture and forestry can be achieved with each contributing to the economic viability of the other and to the maintenance of a healthy environment. Referring to a former FAO Director, Pathak (2008) reported thus: "Forestry alone, by virtue of its biological characteristics, is able to produce while conserving and to conserve while producing".

The combination of agriculture, i.e. sugarcane cultivation along with the exploitation of other forest resources including wood and non-wood products, and the performance of certain critical ecological roles by mangroves remains the harbinger of sustainable development in the region. In this dynamic mix of interests and imperatives, ingenuity in policy formulation, implementation strategies, as well as laws and regulatory mechanisms will continue to be the hallmarks of socioeconomic prosperity. Management skills, schemes and models as well as programme monitoring and evaluation will then be veritable tools to drive the processes for achieving socio-economic development in the mangrove region of Rivers State and, indeed, Nigeria.

Kusmana (2015) proposed a sustainable forest management option known as the integrated mangrove forest management (IMFM) system that involves five dimensions of scientific knowledge, technical considerations, coordination, consultation and ecological interrelationship. The IMFM option is designed for wood and non-wood forest products, ecotourism, agroforestry systems including silvoaquaculture, agrosilvopastoralism, etc. such that various resources are managed on sustained-yield basis, while mangrove functions related to ecologic, economic and social benefits are treated equitably. It is a fusion of interdisciplinary expertise geared towards a common development agenda.

Thus, the integrated management option is the product of far-sighted forest management policies, and it involves long-term planning. An integrated mangrove forest management system for the study area would need to adopt the following processes among others:

1. Inventories of natural resources in mangrove ecosystems.
2. Formulation of specific feasibility reports for sustainability.
3. Manpower development in various disciplines.
4. Convergence of interrelated policy objectives.
5. Programme implementation, monitoring and evaluation.
6. Laws and law enforcement.
7. Consultation with and involvement of local communities as major stakeholders and end-users of mangrove resources.

In particular, mangrove forest management practices should be based on the needs and aspirations of all stakeholders starting from the processes of planning until implementation as well as monitoring and evaluation. Harmonized collaboration among stakeholders such as government, private sector and communities should be implemented in order to manage the mangrove ecosystems in Nigeria. In terms of government which involves several sectors, there should be good coordination among them concerning duties, authority and responsibilities either for horizontal or vertical sector integration to achieve set goals. Also, community participation in programmes of mangrove forest management is pivotal to achieve sustainable development and to the avoidance of conflict and crisis in the Niger Delta region of Nigeria as reported by Fubara (1983), Ibaba (2001), Bassey (2001) and Pettiward (2018).

The local communities in the vicinity of the mangrove swamps should be sensitized and encouraged to embark on deliberate mangrove re-establishment programmes. This group of people benefits directly from mangroves through fishing activities, harvesting of timber and poles for construction, fuel wood and use of plant parts for medicinal purposes. They, therefore, should be brought to the awareness that the loss of a very small mangrove area must be compensated for by a planting programme in their respective localities. Achieving these goals will require public awareness campaigns, intensive research, community training and coastal education on environmental sustainability (Adegbehin and Nwaigbo 1992). Altogether, mangrove forest resources and their cautious utilization in the Niger Delta region of Nigeria will not only assure environmental stability but also sustainable socio-economic human development in the country.

16.11 Conclusions

The mangrove vegetation type is globally distributed and found in peculiar habitats along coastal fringes in tropical and subtropical regions. The uniqueness of their environment as being extremely inhospitable predisposes them to ingenious management methodologies and protocols. They are, thus, regarded as marginal or unproductive lands and subjected to neglect, massive destruction and conversion to several unsustainable land uses in many countries where they exist. Yet, mangroves are recognized as the most productive ecosystem on Earth in terms of carbon synthesis or photosynthetic activity in comparison with terrestrial ecosystems. The onus, therefore, lies in human's ability to unlock the huge potential in mangroves to achieve sustainable socio-economic human development.

In the present study, sugarcane crop performance has been determined vis-a-vis soil characteristics in terms of inherent fertility status of mangrove soils in their natural state without any human intervention through supportive or ameliorative agricultural management practice such as fertilizer and manure application, liming, etc. Growth indicators of the crop were substantially influenced by soil conditions in which the plants in soil unit I performed best and those in unit III were the least, while the results in unit II were midway between units I and III. The crucial deductions from the study are that, except for soil unit I in uplifted platforms free from flooding and considered as fairly fertile that may be cultivated to sugarcane crop but at great costs in managing soil acidity, salinity and nutrient deficiencies, particularly potassium, units II and III are only marginally fertile and should better be preserved to perform other vital environmental buffering roles rather than their instant use for agricultural production in the study area. If, however, soil unit II is reclaimed and prevented from frequent flooding by saline tidewater in the area, together with unit I, these two sites would provide extensive land area for mechanized sugarcane cultivation at commercial scales to boost the sugar industry in Nigeria.

With the possibility of cultivating sugarcane in the mangrove ecosystem side by side with relevant agroforestry practices and coastal administration in the study area, appropriate policy objectives, implementation strategies, programmes and management tools need to be fashioned and executed by government in collaboration with the private sector and, in particular, communities in the immediate neighbourhood of development actions. The best approach would be integrated, broad-spectrum, (i.e. interdisciplinary) management instruments that aim to provide and guarantee improved socio-economic well-being of the citizens and increased national income as well as environmental stability or health within the mangrove ecosystem in the Niger Delta region.

16.12 Future Perspective

In Nigeria, the sugarcane crop is grown in all the agro-ecological zones of the country, except the mangrove ecology which is neglected by government at all levels of administration and, thus, subjected to uncontrolled exploitation and pollution. The mangrove forest belt occupies expansive areas in the Niger Delta region awaiting systematic utilization to contribute to sustainable development in the country. The present study has shown that sugarcane crop can be cultivated and would grow to maturation stages in certain portions of the mangrove ecosystem. As such, the vast mangrove forest belt offers great prospect for the cultivation of sugarcane.

In view of the huge shortfall in local demand for sugar in Nigeria and the sheer volume of the product required to make up for the deficit coupled with the immense expanse of arable land across the country including the mangrove forest belt, large-scale cultivation of sugarcane is, undoubtedly, the way to go in order to create jobs for the teeming population and grow the economy for sustainable development. With relevant policy framework in view of an earlier plan for petrol blend in automobiles, sugarcane remains a choice crop for the future in Nigeria.

Acknowledgements The author is immensely grateful to the director and management staff of the National Cereals Research Institute (NCRI), Badeggi, Niger State, especially the Head and Field Officers of the Sugarcane Research Programme and the librarian for providing sugarcane planting materials and research reports for the study. My special appreciation goes to Dr. Akaamaa C. Wada of NCRI for his invaluable encouragement and some assistance in the literature search and link to the publishers. I thank the laboratory staff of the Niger Delta University, Wilberforce Island, Bayelsa State, for soil and tidewater analyses in the study.

References

Abiodun AA (1999) Research: the foremost prerequisite for sustainable development in Nigeria. Ch. 2. In: Adeniyi PO (ed) Research capacity building for sustainable development in Nigeria. Unilag Consult, University of Lagos, Lagos, pp 13–28

Adegbehin JO, Nwaigbo LC (1992) Mangrove resources in Nigeria: use and management. Nature Resource 26(2):13–20

Adeniyi PO (1999) Research and sustainable development. Ch. 1. In: Adeniyi PO (ed) Research capacity building for sustainable development in Nigeria. Unilag Consult, University of Lagos, Lagos, pp 3–12

Afolabi D (1998) The Nigerian ecosystem. In: Third regional workshop of the Gulf of Guinea large marine ecosystem (GOGLME). Nigerian Institute of Oceanography and Marine Research, Lagos

AGOC (2003) Gas flaring, oil companies and politics in Nigeria. Alexander's Gas and Oil Connections (AGOC) Publications 6(20):15–16

Ajakaiye DO, Akande SO (1998) National Agricultural Research Programme in Nigeria: the efficacy of policy studies. Ch. 1. Nigerian Institute of Social and Economic Research (NISER), Ibadan, 1–4

Ajugwo AO (2013) Negative effects of gas flaring: the Nigerian experience. J Envt Poll Human Health 1(1):6–8

Akhtar SA, Rasul E (2003) Emergence, growth and nutrient composition of sugarcane sprouts under NaCl salinity. Biol Plants 46:113–117

Akpogomeh OS, Atemie JD (2002) Population profile. In: Alagoa EJ, Derefaka AA (eds) The land and people of Rivers state: eastern Niger Delta. Onyoma Research Publications, Port Harcourt, pp 121–147

Alakpodia IJ (2000) Soil characteristics under gas flaring in the Niger Delta, southern Nigeria. *Geo-studies forum.* Int J Environ Policy 1:1–10

Alamieyeseigha DSP (2003) The Environmental Challenges of Developing the Niger Delta. Paper presented by the Executive Governor of Bayelsa State, Nigeria on the occasion of the 1st *Newsafrica Global Lecture Series*, London, November. 25, 2003. 32p

Allison-Oguru EA, Berepubo NA, Kalio AE (2002) Agriculture. In: Alagoa EJ, Derefaka AA (eds) The land and people of Rivers state: eastern Niger Delta. Onyoma Res Publications, pp 361–386

Aluko AP, Owunobi JJ, Fagbenro JA (2001) Physico-chemical properties and potentials of some wetland soils of the mangrove ecosystem for sustainable forest trees establishment. Management of Wetland Soils for sustainable agriculture and environment. In: Proceedings of the 27th Annual Conference of the Soil Science Society of Nigeria, Calabar, Nov 5–9 2001. *Soil Science Society of Nigeria* pp 237–242

Amosun A (2000) Determination of the optimum fertilizer rates for chewing cane production. In: Sugarcane research Programme report. *National Cereals Research Institute* (NCRI), Badeggi, Niger State, Nigeria. Annual review meeting 2–6 April, 2000, pp 21–26

Amosun A (2001) Determination of the optimum fertilizer rates for chewing cane production. In: Sugarcane research Programme report. *National Cereals Research Institute* (NCRI), Badeggi, Niger state, Nigeria. Annual review meeting 4–5 June, 2001, pp 26–28

Aroh JEK (2003) Classification of the "Ogoni Sands" in the Niger Delta in Nigeria: (I) taxonomic. In: Ojeniyi SO, Ano AO, Asawalam DO, Chukwu GO (eds) Land degradation, agricultural productivity and rural poverty: environmental implications. Proceedings of the 28th Annual Conference of the Soil Science Society of Nigeria, Umudike Nov. 4–7, 2003. *Soil Science Society of Nigeria,* pp 353–361

Aroh JEK, Mbakwe R, Wada AC (2019) Soil quality evaluation and classification system: criteria for determining fertility classes for agricultural production. SCHOLEGE Int J Multidis Allied Stud 6(12):137–155. URL:link.thescholedge.org/1133

Aroh JEK, Mbakwe R, Wada AC (2020) Effect of soil site on sugarcane (*Saccharum officinarum* L.) growth over time in the mangrove (*Rhizophora* spp.) Forest ecosystem of Ogonokom-Abua, Rivers state, Nigeria. SCHOLEGE Int J Multidis Allied Stud 7(2):45–73

Ayolagha GA, Onuegbu BA (2002) The soils of Rivers state. In: Alagoa EJ, Derefaka AA (eds) The land and people of Rivers state: eastern Niger Delta. Onyoma Research Publications, Port Harcourt, pp 19–42

Ayoola TJ (2011) Gas flaring and its implications for environmental accounting in Nigeria. J Sustain Dev 4(5):244–250

Bailey-Serres J, Voesenek LACJ (2008) Flooding stress: acclimations and genetic diversity. Annu Rev Plant Biol 59:313–339

Baldock JA, Nelson PN (2000) Soil organic matter. In: Sumner ME (ed) Handbook of soil science. CRC Press LLC, Boca Raton, Sect. B Ch 2 p 30

Banerjee A, Jhariya MK, Yadav DK, Raj A (2020) Environmental and sustainable development through forestry and other resources. Apple Academic Press Inc., CRC Press- a Tayler and Francis Group, US & Canada, Oakville, p 400. ISBN: 9781771888110. https://doi.org/10.1201/9780429276026

Banerjee A, Meena RS, Jhariya MK, Yadav DK (2021) Agroecological footprints management for sustainable food system. Springer Nature, Singapore, eBook ISBN: 978-981-15-9496-0, Hard-cover: 978-981-15-9495-3, p 514. https://doi.org/10.1007/978-981-15-9496-0

Bassey N (2001) Oil and gas exploration and exploitation in the Niger Delta: the social and demographic challenges. In: Paper presented at the second *Nigerian Academy of Science*

Forum on the oil and gas industry in Nigeria in the 21st century held at Port Harcourt, Nigeria, January 19, 2001

Beale JA (1970) Forestry in the environmental decades. J of Forestry 68(9):639–641

Bell-Gam W (2002) Relief and drainage. In: Alagoa EJ, Derefaka AA (eds) The land and people of Rivers state: eastern Niger Delta. Onyoma Research Publications, Rivers State, pp 43–52

Bloomfield C, Zahari AB (1982) Acid sulphate soils. Outlook on Agriculture II(2):48–54

Bradford KJ, Hsiao TC (1982) Stomatal behaviour and water relations of waterlogged tomato plants. Plant Physiol 70:1508–1513

Bull T (2000) The sugarcane plant. In: Manual of cane growing. Ch. 04. Elsevier Scientific Publishing Co., Amsterdam, pp 71–82

Busari AD (2004) Sugarcane and sugar industry in Nigeria: the bitter and sweet lessons. Spectrum Books Limited, Ibadan, p 279

Byer TA (1999) Issues of sustainable management of the Nigerian environment. Keynote address by resident representative, World Bank, Nigeria. Conservation of Nigerian resources and the threatened environment. In: Proceedings of the 26th annual Conference of the forestry Association of Nigeria held in Maiduguri, Bornu state Nigeria. *Forestry Association of Nigeria* (FAN) 19th-23 April, 1999.

Carroll B, Turpin T (2002) Environmental impact assessment handbook. A practical guide for planners, developers and communities. Thomas Telford Publishing, London, p 175

Chinda AC (2002) Marine and riverine resources. In: Alagoa EJ, Derefaka AA (eds) The land and people of Rivers state: eastern Niger Delta. Onyoma Research Publications, Rivers State, pp 83–117

Chinsman B (1999) Strategies for promoting and strengthening north-south and south-south cooperation in research capacity-building for National Sustainable Development. Ch. 30. In: Adeniyi PO (ed) Research capacity building for sustainable development in Nigeria. Unilag Consult, University of Lagos, Lagos, pp 329–343

Chong VC (2006) Sustainable utilization and management of mangrove ecosystems of Malaysia. Aquat Ecosyst Health Manage 9(2):249–260

Colmer TD, Voesenek LACJ (2009) Flooding tolerance: suites of plants traits in variable environments. Functional Plant Biol 36:665–681

Davidson JT (2000) Oil and gas production in Nigeria: problems and challenges. Nigerian Institute of Social and Economic Research (NISER), Ibadan, pp 40–43

DeBusk WF (1996) Organic matter turnover along a nutrient gradient in the Everglades. Ph.D. dissertation. *University of Florida, Gainesville, FL.* (Unpublished)

Deekor TN (2002) The impact of gas flaring on wetland soils of the Niger Delta: a case study of Ebocha and Obigbo. Unpublished M.Sc. Thesis. University of Port Harcourt

Dent DL (1986) Acid sulfate soils: a baseline for research and development. ILRI Publ. 39. International Institute of Land Reclamation and Improvement, Wageningen

Efe SI (2003) Effect of gas flaring on temperature and adjacent vegetation in the Niger Delta environment. Int J Environ 1(1):91–101

Ekpoh IJ, Obia AE (2010) The role of gas flaring in the rapid corrosion of zinc roofs in the Niger Delta region of Nigeria. Environmentalist 30(4):347–352

Else MA, Tiekstra AE, Croker SJ, Davies WJ, Jackson MB (1996) Stomatal closure in flooded tomato plants involves abscisic acid and a chemically unidentified anti-transpirant in xylem sap. Plant Physiol 112:239–247

Eshett ET (1993) The wetland soils of Nigeria: properties, classification and traditional land-use practices. Studies on land-water interactions, Wetlands and Ecotones, pp 227–244

FAO (1976) A framework for land evaluation. FAO, Rome. Food and Agriculture Organization (FAO) Bulletin 32

FAO (1994) Mangrove forest management guidelines. In: Forest resources development branch, Forest resources division FAG forestry department FAO forestry paper 117

FAO (1997) Land quality indicators and their use in sustainable agriculture and rural development. Land and Water Bulletin5.Food and Agriculture Organization (FAO), Rome

FAO (2007) State of the World's forests 2007. Food and Agriculture Organization of the United Nations, FAO Rome

FDALR (1990) The Reconnaissance Soil Survey of Nigeria (1 : 650,000). Soil Report: Vols. 1, 2, 3, 4. Rating for Soil Data Interpretation. Federal Department of Agricultural Land Resources. Federal Ministry of Agriculture, Nigeria. Pp 372–374

Fourqurean JW (2012) Sea grass ecosystems as a globally significant carbon stock. Nat Geosci 5:505–509

Fubara DMJ (1983) Physical conquest and development of the Niger Delta. Commemoration lecture. 1st Convocation and Installation of Chancellor, Rivers State University of Science and Technology, Port Harcourt. (Unpublished)

Gascho GJ, Shih SF (1983) In: Teare ID, Peet MM (eds) Sugarcane: crop: water relations. John Wiley & Sons, New York, pp 445–479

Gevana D (2010) Carbon sequestration potential of mangrove Forest in the Philippines. In: Conference: XXIII IUFRO world congress 2010 project: Blue Carbon in the Philippines

Gogoi N, Baruoh KK (2012) Effects of natural gas flare on growth, flowering and yield of Rice. http://www.envfor.nic.in/paryaabs/v19n12/plant.html

Gomathi R, Thandapani PV (2004) Sugar metabolism and carbon partitioning of sugarcane genotypes under salinity stress condition. Sugar Tech 6:151–158. https://doi.org/10.1007/BF02942716

Gomathi R, Thandapani PV (2005) Salt stress in relation to nutrient accumulation and quality of sugarcane genotypes. Sugar Tech 1:39–47

Hamilton S, Casey D (2016) Creation of high spatiotemporal resolution global database of continuous mangrove forest cover for the 21st century: a big-data fusion approach. Glob Ecol Biogeogr 25:729–738

Hartoungh JCC (1966) Report on the agricultural development of the Niger Delta special area: agro-sociology. Niger Delta Development Board (NDDB), Port Harcourt, pp 7.1–7.17

Hemaprabha G, Nagarajan GR, Alarmelu S (2004) Response of sugarcane genotypes to water deficit stress. Sugar Tech 6:165–168

Houghton RA, Skole DL (1990) Carbon. In: Turner BL, Clark WC, Kates RW, Richards JF, Matthews JT, Meyer WB (eds) The earth as transformed by human action. Cambridge University Press, Cambridge, pp 393–408

Humbert RP (1968) The growing of sugarcane Ch. VI. In: Drainage, Revised edn. Elsevier Publishing Company, Amsterdam, p 779

Ibaba IS (2001) Understanding the Niger Delta crisis: the oil industry and the Niger Delta. Jivac Publishing Company, Port Harcourt, pp 22–43

IITA (1979) Selected methods of soil and plant analysis. Manual series no. 1, 3rd edn. International Institute of Tropical Agriculture (IITA), Ibadan, 70p

Imevbore AMA (2001) The impact of oil production on the biota of the Niger Delta. In: Paper presented at the second *Nigerian Academy of Science Forum* on the oil and gas industry in Nigeria in the 21st century held at Port Harcourt, Nigeria. January 19, 2001

Inman-Bamber NG (1994) Temperature and seasonal effects on canopy development and light interception of sugarcane. Field Crop Res 36:41–51

IPCC (2007) AR4 climate change 2007: the physical sciences basis. Intergovernmental panel on climate change. In: Solomon S, Qin D, Manning M, Chen Z, Marquis M, Averyt KB, Tignor M, Miller HL (eds) Contribution of working group I to the fourth assessment report of the intergovernmental panel on climate change. Cambridge Univ. Press, New York, p 996

Isirimah NO, Ojanuga AG (1987) Potential acid sulphate peats of the Niger Delta wetland. Soil Survey and Land Evaluation 7(3):147–156

Isirimah NO, Zuofa K, Ayolagha G (1989) Transferable technologies for improved management of acid sulphate soils for increased rice production. In: Babalola O (ed) Proc. 17th annual Conference of the Soil Science Society of Nigeria held at Nsukka, Nigeria. Soil Science Society of Nigeria (SSSN), Nsukka, pp 78–87

Islam MA, Mac Donald SE (2004) Ecophysiological adaptation of Black spruce (*Picea mariana*) and tamarack (*Larix laricina*) seedlings to flooding. Trees 18:35–42

Ismail OS, Umukoro GE (2012) Global impact of gas flaring. Energy Power Eng 4:290–302

Jackson MB, Drew M (1984) Effects of flooding on growth and metabolism of herbaceous plants. In: Kozloski TT (ed) Flooding and plant growth. Academic Press Inc., Orlando, pp 47–128

Jhariya MK, Banerjee A, Meena RS, Yadav DK (2019b) Sustainable agriculture, Forest and environmental management. Springer Nature Singapore Pte Ltd., Singapore, eISBN: 978-981-13-6830-1, Hardcover ISBN: 978-981-13-6829-5. . Pp. 606. https://doi.org/10.1007/978-981-13-6830-1

Jhariya MK, Banerjee A, Yadav DK, Raj A (2018) Leguminous trees an innovative tool for soil sustainability. In: Meena RS, Das A, Yadav GS, Lal R (eds) Legumes for soil health and sustainable management. Springer, Singapore, ISBN 978-981-13-0253-4 (eBook), ISBN: 978-981-13-0252-7 (Hardcover), pp 315–345. https://doi.org/10.1007/978-981-13-0253-4_10

Jhariya MK, Bargali SS, Raj A (2015) Possibilities and perspectives of agroforestry in Chhattisgarh. In: Zlatic M (ed) Precious forests-precious earth. InTech, London, pp 237–257. https://doi.org/10.5772/60841ISBN: 978-953-51-2175-6. 286 pages

Jhariya MK, Meena RS, Banerjee A (2021a) Ecological Intensification of Natural Resources for Sustainable Agriculture. Springer Nature, Singapore. eISBN: 978-981-334-203-3, Hardcover ISBN: 978-981-334-206-6. https://doi.org/10.1007/978-981-33-4203-3

Jhariya MK, Meena RS, Banerjee A (2021b) Ecological intensification of natural resources towards sustainable productive system. In: Ecological Intensification of Natural Resources for Sustainable Agriculture. Springer, Singapore. eBook ISBN 978-981-334-203-3. Hardcover ISBN: 978-981-334-202-6. https://doi.org/10.1007/978-981-33-4203-3_1

Jhariya MK, Yadav DK, Banerjee A (2019c) Agroforestry and Climate Change: Issues and Challenges. Apple Academic Press Inc., CRC Press- a Taylor and Francis Group, US & Canada, Oakville. ISBN: 978-1-77188-790-8 (Hardcover), 978-0-42957-274-8 (E-book), p 335. https://doi.org/10.1201/9780429057274

Jhariya MK, Yadav DK, Banerjee A, Raj A, Meena RS (2019a) Sustainable Forestry under Changing Climate. In: Jhariya MK, Banerjee A, Meena RS, Yadav DK (eds) Sustainable agriculture, forest and environmental management. Springer Nature, Singapore, pp 285–326. ISBN: 978-981-13-6830-1, Hardcover ISBN: 978-981-13-6829-5. Pp 606

Kaladumo COK (1996) The implications of gas flaring on the Niger Delta environment. In: Paper presented at the 8th biennial international seminar on the petroleum industry and the Nigerian environment held at the Hotel Presidential, Port Harcourt, Rivers state, Nigeria. 18–21 Nov. 1996. Rivers State Petroleum Pollution Bureau Office. Government House, Port Harcourt

Kathiresan L, Bingham BL (2001) Biology of mangroves and mangrove ecosystems. Adv Marine Biol 40:81–251

Keren R (2000) Salinity. In: Sumner ME (ed) Handbook of soil science. CRC Press LLC, Boca Raton, Sect. G Ch 1 pp 1–25

Khan AQ, Tadesse KA, Robe BL (2017) A study on morphological characters of introduced sugarcane varieties (*Saccharum* spp., hybrid) in Ethiopia. Int J Plant Breeding and Genetics 11:1–12. https://doi.org/10.3923/ijpbg.2017.1.12

Khan N, Jhariya MK, Raj A, Banerjee A, Meena RS (2021a) Soil carbon stock and sequestration: implications for climate change adaptation and mitigation. In: Jhariya MK, Meena RS, Banerjee A (eds) Ecological intensification of natural resources for sustainable agriculture. Springer, Singapore. eBook ISBN 978-981-334-203-3. Hardcover ISBN: 978-981-334-202-6. https://doi.org/10.1007/978-981-33-4203-3_13

Khan N, Jhariya MK, Raj A, Banerjee A, Meena RS (2021b) Eco-designing for sustainability. In: Jhariya MK, Meena RS, Banerjee A (eds) Ecological intensification of natural resources for sustainable agriculture. Springer, Singapore. eBook ISBN 978-981-334-203-3. Hardcover ISBN: 978-981-334-202-6. https://doi.org/10.1007/978-981-33-4203-3_16

Khan N, Jhariya MK, Yadav DK, Banerjee A (2020a) Herbaceous dynamics and CO_2 mitigation in an urban setup- a case study from Chhattisgarh, India. Environ Sci Pollut Res 27(3):2881–2897. https://doi.org/10.1007/s11356-019-07182-8

Khan N, Jhariya MK, Yadav DK, Banerjee A (2020b) Structure, diversity and ecological function of shrub species in an urban setup of Sarguja, Chhattisgarh, India. Environ Sci Pollut Res 27 (5):5418–5432. https://doi.org/10.1007/s11356-019-07172-w

Kinako PDS (2008) Harmonizing forestry use objective in formulating Forest development Programme. In: Isirimah NO, Monsi A, Kinako PDS (eds) Strategies for sustainable agricultural development in Nigeria: focus on Niger Delta states, Revised edn. Davidstones Publishers Ltd., Port Harcourt, pp 66–76

Kjoller A, Struwe S (1980) Micro-fungi of decomposing red alder leaves and their substrate utilization. Soil Biol Biochem 12:425–431

Kjoller A, Struwe S, Vestberg K (1985) Bacterial dynamics during decomposition of alder litter. Soil Biol Biochem 17:463–468

Kumar S, Meena RS, Datta R, Verma SK, Yadav GS, Pradhan GS, Molaei A, Rahman GKMM, Mashuk HA (2020a) Legumes for carbon and nitrogen cycling: an organic approach. Carbon and Nitrogen Cycling in Soil:337–375. https://doi.org/10.1007/978-981-13-7264-3_10

Kumar S, Meena RS, Jhariya MK (2020b) Resources use efficiency in agriculture. Springer Nature, Singapore, eBook ISBN: 978-981-15-6953-1, Hardcover: 978-981-15-6952-4, p 760. https://doi.org/10.1007/978-981-15-6953-1

Kusmana C (2015) Integrated sustainable mangrove Forest management. JPSL 5(1):1–6

Kusmana C, Istomo C, Wibowo SW, Budi IZ, Siregar T, Tiryana SS (2008) Manual of mangrove Silviculture in Indonesia. International Cooperation Agency (KOICA), Jakarta

Kusmana C, Istomo C, Wibowo SW, Budi IZ, Siregar T, Tiryana SS (2010) Konservasi dan Pengelolaan mangrove. In: Presented on: Konferensi dan Pameran Nasional "Penyelamatan Hutan Pantai dan Peningkatan Kesejahteraan Masyarakat Pesisir", 23 November 2010

Levy GJ (2000) Sodicity. In: Sumner ME (ed) Handbook of soil science. CRC Press LLC, Boca Raton, Sect. G Ch 2 pp 27–63

Lovelock CE, Cahoon DR, Friess DA, Guntenspergen GR, Krauss KW, Reen DA, Rogers K, Saunders ML, Sidik F, Swales A, Sainlilan N, Thuyen le X, Triet T (2015) The vulnerability of indo-Pacific mangrove forests to sea-level rise. Nature 526:559–563

Maas EV (1986) Salt tolerance of plants. Appl Agric Res 1:12–26

Maas EV (1990) Crop salt tolerance. In: Tanji KK (ed) Agricultural salinity assessment and management. American Society of Civil Engineers, New York, pp 262–304

Mamet LD, Galwey NW (1999) A relationship between stalk elongation and earliness of ripening in sugarcane. Exp Agric 35:283–291

Mausbach MJ, Richardson JL (1994) Biogeochemical processes in hydric soil formation. Current topics in wetland biogeochemistry. *Wetland Biogeochemistry Institute*. Louisiana State University 1:68–127

Meena RS, Kumar S, Datta R, Lal R, Vijaykumar V, Brtnicky M, Sharma MP, Yadav GS, Jhariya MK, Jangir CK, Pathan SI, Dokulilova T, Pecina V, Marfo TD (2020a) Impact of agrochemicals on soil microbiota and management: a review. Land (MDPI) 9(2):34. https://doi.org/10.3390/land9020034

Meena RS, Lal R, Yadav GS (2020b) Long term impacts of topsoil depth and amendments on soil physical and hydrological properties of an Alfisol in Central Ohio, USA. Geoderma 363:1141164. https://doi.org/10.1016/j.geoderma.2019.114164

Meena RS, Lal R, Yadav GS (2020c) Long-term impact of topsoil depth and amendments on carbon and nitrogen budgets in the surface layer of an Alfisol in Central Ohio. Catena 2020194:104752. https://doi.org/10.1016/j.catena.2020.104752

Melo RH, Kusmana CE, Nurrochmat DR (2020) Short communication: mangrove forest management based on multi-dimension scaling (RAP-Mforest) in Kwandang sub-district, North Gorontalo District, Indonesia. Biodiversitas 21:1352–1357

Murday M, Murday JR, Sexton WJ, Powell CB (1988) Atlas of environmental sensitivity to spilled oil for operational areas of NNPC/GOCON OML 49. Chevron Petroleum Incorporated, San Ramos River

NCRI (1992) Sugar cane research Programme report for 1991 cropping year. In: In-house review meeting March 1992. National Cereals Research Institute, Badeggi, Niger State, Nigeria

NCRI (1995) Sugarcane research Programme report. In: L.D. Busari (Ed.). *National Cereals Research Institute* (NCRI), Badeggi, Niger state, Nigeria. Annual review meeting, 2-5 May, 1995. 55p

NCRI (1998) Sugarcane research programme report. In: Busari LD (ed) Annual research review meeting 9–12 February, 1998. National Cereals Research Institute (NCRI), Badeggi, Niger State, Nigeria, p 50

NCRI (2001) Sugarcane research Programme report. In: Busari LD (ed) Annual review meeting 4–5 June, 2001. National Cereals Research Institute (NCRI), Badeggi, Niger State, Nigeria 58p

NCRI (2007) Sugarcane research division report. Sugarcane research Programme. In: Wada AC (ed) Annual research, planning and extension review meeting 10–11 May, 2007. National Cereals Research Institute (NCRI), Badeggi, Niger State, Nigeria, 17p

NDES (1997) Niger Delta environmental survey (NDES). Final report, phase I: environmental and socio-economic conditions. NDES, Lagos

Ndukwe OU (2000) Elements of Nigerian environmental Laws.Ch 1. In: Environment, pollution and sustainable development. University of Calabar Press, Calabar, pp 1–17. ISBN: 978-007-057-5

Nelson OM, Summers LE (1996) Total carbon, organic carbon and organic matter. In: Spark DL (ed) Methods of soil analysis, part 3. Chemical methods, vol 5. SSA Book series, Madison, pp 961–1010

Nwankwo JT, Ogagarue D (2011) Effects of gas flaring on surface and ground waters in Delta state, Nigeria. J Geology and Mining Res 3(5):131–136

Nyananyo BL (1999) Vegetation. In: Alagoa EJ (ed) The land and people of Bayelsa state: Central Niger Delta. Onyoma Research Publications, Port Harcourt, pp 44–57

Nyananyo BL (2002) Forest resources. In: Alagoa EJ, Derefaka AA (eds) The land and people of Rivers state: eastern Niger Delta. Onyoma Research Publications, Port Harcourt, pp 63–81

O'Shea M (2000) Sugar accumulation.Ch 04. In: Manual of cane growing. Elsevier Scientific Publishing Co., Amsterdam, pp 82–88

Ogidiolu A (2003) Effects of gas flaring on soil and vegetation characteristics in oil producing region of the Niger Delta, Nigeria. Int J Ecol Environ Dynamics 1:47–53

Okonny IP (2002) Geology. In: Alagoa EJ, Derefaka AA (eds) The land and people of Rivers state: eastern Niger Delta. Onyoma Research Publications, Port Harcourt, Rivers State, Nigeria, pp 7–18

Okonny IP, Ayolagha G, Dickson AA (1999) Soils of Bayelsa state. In: Alagoa EJ (ed) The land and people of Bayelsa state: Central Niger Delta. Onyoma Research Publications, Port Harcourt, pp 9–29

Onuoha CM (1999) Environmental impact of Forest depletion and suggested sustainable management strategies in Nigeria. In: Proceedings of the 26th annual Conference of the forestry Association of Nigeria held in Maiduguri, Bornu state Nigeria. *Forestry Association of Nigeria* (FAN) 19th-23 April, 1999, pp 50–61

Onuoma OF, Alex OL, Olawole OO, Adelanmi A (2015) The environmental implication of gas flaring in Sapele Community of Delta State, Nigeria. Int J Oil, Gas and Coal Engineering 3 (5):56–61

Oyegun CU, Ologunorisa YE (2002) In: Alagoa EJ, Derefaka AA (eds) Weather and climate. The land and people of Rivers state: eastern Niger Delta. Onyoma Research Publications, Port Harcourt, pp 53–62

Pariona A (2017) What Is A Mangrove Habitat? Environment. World Atlas, Aug. 1, 2017. worldatlas.com/articles/what-is-a-mangrove-habitat.html

Patade VY, Bhargava S, Suprasanna P (2012) Transcript expression profiling of stress responsive genes in response to short-term salt or PEG stress in sugarcane leaves. Molecular Biol Reports 39:3311–3318. https://doi.org/10.1007/s11033-011-1100-z

Pathak MK (2008) Harmonizing forestry use objective in formulating Forest development Programme. In: Strategies for sustainable agricultural development in Nigeria: focus on Niger Delta states. Revised edn. N.O. Isirimah, A. Monsi, and P.D.S. Kinako (Eds.). Davidstones Publishers Ltd., Port Harcourt. Pp. 66–76

Pendleton L, Donato DC, Murray BC, Grooks S, Jenkins WA, Sifleet S, Craft C, Fourqurean JW, Kauffman JB (2012) Estimating global 'blue carbon' emissions from conversion and degradation of vegetated coastal ecosystems. PLoS One 7:e43542

Pettiward J (2018) The Mangrove Action Project. Regenerating Mangroves. https://www.synchronicityearth.org/the-mangrove-action-project

Powell CB (1985) The decapods Crustacea of the Niger Delta. The mangrove ecosystem of the Niger Delta. In: Wilcox B, Powell CB (eds) Proceedings of a workshop in Port Harcourt. University of Port Harcourt Press, Port Harcourt, pp 13–18

Powell CB (1987) Mollusks of the Niger Delta. Impacts of pollution on biological resources within the Niger Delta basin. In: Imevbvbore AMA (ed) Final report on the environmental pollution monitoring of the Niger Delta basin of Nigeria, vol 6. Niger Delta Basin Development Authority (NDBDA), Port Harcourt, pp 388–410

Rahman MA, Lee SH, Ji HC, Kabir AH, Jones CS, Lee KW (2018) Importance of mineral nutrition for mitigating Aluminium toxicity in plants on acidic soils: current status and opportunities. Int J Mol Sci 19:30–73

Raj A, Jhariya MK, Banerjee A, Yadav DK, Meena RS (2019a) Soil for sustainable environment and ecosystems management. In: Jhariya MK, Banerjee A, Meena RS, Yadav DK (eds) Sustainable agriculture, Forest and environmental management. Springer Nature, Singapore, eISBN: 978-981-13-6830-1, Hardcover ISBN: 978-981-13-6829-5. Pp. 606, pp 189–221. https://doi.org/10.1007/978-981-13-6830-1

Raj A, Jhariya MK, Yadav DK, Banerjee A, Meena RS (2019b) Agroforestry: a holistic approach for agricultural sustainability. In: Jhariya MK, Banerjee A, Meena RS, Yadav DK (eds) Sustainable agriculture, Forest and environmental management. Springer Nature, Singapore, eISBN: 978-981-13-6830-1, Hardcover ISBN: 978-981-13-6829-5. Pp 606., pp 101–131. https://doi.org/10.1007/978-981-13-6830-1

Raj A, Jhariya MK, Khan N, Banerjee A, Meena RS (2021) Ecological intensification for sustainable development. In: Ecological intensification of natural resources for sustainable agriculture. Springer, Singapore, eBook ISBN 978-981-334-203-3. Hardcover ISBN: 978-981-334-202-6. https://doi.org/10.1007/978-981-33-4203-3_5

Raj A, Jhariya MK, Yadav DK, Banerjee A (2020) Climate change and agroforestry systems: adaptation and mitigation strategies. Apple Academic Press Inc., CRC Press- a Taylor and Francis Group, US & Canada, Oakville. ISBN: 9781771888226, p 383. https://doi.org/10.1201/9780429286759

Ramirez M, Hughes TR, Acosta JR, Lochhead J (2016) Large-Scale Sugarcane Production in El Salvador. *VOICES—on the border*—33pp

Ravetz JR (1971) Scientific knowledge and its social problems. Clarendon Press, Oxford, 449p

Reddy KR, D'Angelo EM, Harris WG (2000) Biogeochemistry of wetlands. In: Sumner ME (ed) Handbook of soil science. CRC Press LLC, Boca Raton, Sect G Ch 4 pp 89–119

Rhoades JD (1982) Soluble salts. In: Page AL, Miller RH, Keeney DR, Baker DE, Ellis R Jr, Rhoades JD (eds) Methods of soil analysis, part 2. Chemical and mineralogical properties, 2nd edn. Soil Science Society of America, Madison WI, pp 561–573. Agronomy No. 9

Ritsema CJ, van Mensvoort MEF, Dent DL, Tan Y, van den Bosch H, van Wijk ALM (2000) Acid sulphate soils. In: Sumner ME (ed) Handbook of soil science. CRC Press, Boca Raton, Sect. G Ch 5 pp 121–154

Sanghera GS, Kumar A (2018) Recent perspectives towards enhancing drought tolerance in sugarcane. J Pl Sci Res 34(1):23–34

Sanghera GS, Malhotra PK, Singh H, Bhatt R (2019) Harnessing plant biotechnology and physiology to stimulate agricultural growth. In: Climate change impact in sugarcane agriculture and mitigation strategies. Springer, New York, pp 99–114

Schlesinger WH (1997) Biogeochemistry- an analysis of global change, 2nd edn. Academic Press, Orlando, FL

Singer MJ, Ewing S (2000) Soil quality. In: Sumner ME (ed) Handbook of soil science. CRC Press, Boca Raton, Sect. G Ch 11 pp 271–298

Singer MJ, Munns DN (1996) Soils: an introduction, 3rd edn. Prentice-Hall Inc, Upper Saddle River, NJ

Singh NR, Jhariya MK (2016) Agroforestry and Agrihorticulture for higher income and resource conservation. In: Narain S, Rawat SK (eds) Innovative Technology for Sustainable Agriculture Development. Biotech Books, New Delhi, pp 125–145. ISBN: 978-81-7622-375-1

Spalding M, Parrett CL (2019) Global patterns in mangrove recreation and tourism. Mar Policy 110:103540

Suratman MN (2008) Carbon sequestration potential of mangroves in Southeast Asia. In: Managing Forest ecosystems: the challenge of climate change. Springer, Cham, pp 297–314. https://doi.org/10.1007/978-1-4020-8343-3_17

Szuwalski C, Jin X, Shan X, Clavelle T (2020) Marine seafood production via intense exploitation and cultivation in China: costs, benefits, and risks. PLoS One 15(1):e0227106. https://doi.org/10.1371/journal.pone.0227106

Tao F, Yokozawa M, Xu Y, Hayashi Y, Zhang Z (2006) Climate changes and trends in phenology and yields of field crops in China, 1981–2000. Agric For Meteorology 138:82–92

Thomas GW (1996) Soil pH and soil acidity. In: Bigham JM (ed) Methods of soil analysis, part 3. Chemical methods. Soil Science Society of America and American Society of Agronomy, Madison, Wisconsin, pp 475–490

Thomas JR, Rozeff N (1989) Nutritional status survey of sugarcane in Texas. J Am Soc Sugarcane Tech 9:81–90

Tisdale SL, Nelson WL (1975) Soil fertility and fertilizers, 3rd edn. Collier Macmillan Publishers, London, 407 p

Tournaire-Roux C, Sutka M, Javot H, Gout E, Gerbeau P, Luu DT, Richard RB, Maurel C (2003) Cytosolic pH regulates root water transport during anoxic stress through gating of aquaporins. Nature 425:393–397

Ubani EC, Onyejekwe IM (2013) Environmental impact analysis of gas flaring in the Niger Delta region of Nigeria. American J Indust Res 4(2):246–252

USAID (2008) Nigeria Biodiversity and Tropical Forest Assessment: Maximizing Agricultural Revenue in Key Enterprises for Targeted Sites (Markets). 2008. Publication no. 620-C-00-05-00077-00. *Chemonics International Inc.*, June 2008. Web. 9 Mar. 2017. http://pdf.usaid.gov/pdf_docs/PNADN536

Valiela I, Bowen JL, York JK (2001) Mangrove forests: one of the World's threatened major tropical environments. BioSci 51(10):807–815

Wahid A, Rao A-ur-R, Rasulh E (1997) Identification of salt tolerance traits in sugarcane lines. Field Crop Res 54(1):9–17. https://doi.org/10.1016/S0378-4290(97)00038-5

Walkley A, Black CA (1934) Determination of soil organic matter and proposed modification of the chromic acid titration method. Soil Sci 37:29–38

Wiedenfeld B (2008) Effects of irrigation water salinity and electrostatic water treatment for sugarcane production. Agric Water Manag 95:85–88

Willoughby S (2015) Biodiversity and protected areas management Programme (BIOPAMA). In: Missouri botanical garden open Conference systems, TDWG 2015 ANNUAL CONFERENCE. Building: Windsor hotel room: oak room

World Bank (1995) Defining an environmental development strategy for the Niger Delta, vol 1. The World Bank, Washington, D.C.. May 30, 1995

World Commission on Environment and Development (WCED) (1987) Our common future. Oxford Univ. Press, Oxford and New York

Zhao D, Yang-Rui L (2015) Climate change and sugarcane production: potential impact and mitigation strategies. Int J Agronomy. Article ID:547386:10 pages. https://doi.org/10.1155/2015/547386

17 Managing Natural Resources Through Ecological Intensification in Oil-Rich Niger Delta

Aroloye O. Numbere

Abstract

The Niger Delta region is a coastal area and rich in natural resources. It is the largest producer of crude oil in Africa and has the largest mangrove forest in the Atlantic region. Oil and gas exploration activities had impoverished the land and the water, which in turn had eliminated most fishing and farming activities in the area. Severely hit from the oil economy is the agro-industry, which had suffered huge losses in terms of land pollution, land despoliation, and land fragmentation. These situations have affected the ecology of the region. Ecological intensification is as an aspect of human-based ecosystem management, a situation where nature heals itself from detrimental activities and continues to render ecosystem services. It highlights the need to use less fertilizers and more organic manure to reduce the toxicity in the environment. It also introduces a new farming technique called "agro-mangrove forestry" as a way of salvaging abandoned forest and decommissioned oil fields as a way of sustainable management of forest resources to benefit present and future generations. Presently less than 5% of mangrove forest in the region is used for farming. Therefore, the deliberate protection of mangrove forest will help to boost its ecosystem services and thus enhance local economy.

Keywords

Agriculture · Crude oil · Ecological intensification · Niger Delta · Pollution

A. O. Numbere (✉)
Department of Animal and Environmental Biology, University of Port Harcourt, Port Harcourt, Nigeria

M. K. Jhariya et al. (eds.), *Sustainable Intensification for Agroecosystem Services and Management*, https://doi.org/10.1007/978-981-16-3207-5_17

Abbreviations

DRMT	Deforestation Research Monitoring Team
GBRMPA	Great Buffer Reef Marine Park Authority
HIPPO	Habitat loss, invasive species, population growth, and overexploitation
IPM	Integrated pest management
MPA	Mangrove protected area
NPK	Nitrogen-phosphorus-potassium
OML	Oil mining license

17.1 Introduction

Ecological intensification involves actively managing farmland to increase the intensity of the ecological processes that support production, such as biotic, pest regulation, nutrient cycling, and pollination. It means making smart use of nature's functions and services, at field and landscape scales, to enhance agricultural productivity and reduce reliance on agrochemicals and the need for further land use conversion (Bommarco et al. 2013; Tittonell 2014; Jhariya et al. 2021a, b). The practice of ecological intensification is not common in the Niger Delta though it is deliberately practiced in developed countries to reduce the use of chemicals, which have a negative effect on the environment. However, there is sustainable agricultural practice in some areas in Nigeria, where farmers practice agroforestry (Oyewole and Sennuga 2020), which promotes ecological intensification. Ecological intensification is common in smallholder agriculture in Africa (Tittonell and Giller 2013; Folberth et al. 2014). The natural resources of the Niger Delta are made up of the following among others: crude oil, fishery, local craft, and wildlife.

Crude Oil Oil and gas exploration in the Niger Delta impacts agroforestry and agricultural practice (Ejiba et al. 2016) in the sense that there is a migration of people from agrarian to industrial economy (Leo 2012). Proliferation of the oil industry has taken over agricultural land, which leads to the loss of ecosystem services of the forest. This has taken away the practice of agro-silvicultural system, a system where we have crops and woody perennials growing side by side with each other. In many rural communities, local farmers plant their crops along fertile mangrove swamps within the rows of mangrove or rainforest (e.g., pineapple farm in Buguma and Plantain farm in Okrika, all in the Niger Delta region). These farms don't use fertilizers but rely on manure from animal farms or poultry (Kathiresan 2020). The mangrove swamp being a biodiversity hotspot is rich in nutrients needed for the growth of the crops and the trees. This provides food supplements for the people who consume and also sell the harvested farm products. Pollination activities go on because the crops attract a diversity of insects around the forests such as bees, wasp, flies, and butterflies, which pollinate the flowers in the vicinity (Numbere

2020a). The destruction of the mangroves to create a way for building industrial complexes, offices, and living apartments takes away the function of the trees. Communities that are situated along the coastal embark on fish farming for subsistence and to make money. The forest is also rich in wildlife animals such as antelopes, monkeys, and bush rats. These are a source of livelihood to the local people who capture and sell them to buyers (Elum et al. 2016).

Fishery The Niger Delta is a fishing hub due to its coastal nature. Fishing is a major occupation of the people of this area and a source of income by natives (Onyena and Sam 2020). There are numerous aquatic resources which include fish, periwinkles, crabs, prawns, and mussels. Aquaculture is practiced in this region because of the availability of numerous water bodies. The natural coastal terrain is ideal for constructing fish ponds and rearing different varieties of fish. This reduces production cost because there is a readily available source of water. Then there is also easy drainage for the waste water from the ponds to be evacuated back to the river after treatment. Fingerlings can also be collected from the river during high tide, thus reducing the cost of buying from artificial ponds; however, this depends on the species of fish reared.

Local Craft Raw materials from mangroves are used to manufacture local crafts such as basket, hat, mat, and medicinal products. These products are sold to earn some income by natives. Raffia palm tree grows in mangrove forest and is the source of the basket, mat, and straw hat, while leaves of a mangrove *Acrostichum aureum* are believed to have antiviral and antibacterial properties and have been used to produce drugs in other parts of the world (Gajula et al. 2020). Stems used for scaffold during construction are also derived from wetland. The stem of red mangroves (*Rhizophora* species) is used to produce firewood for cooking.

Wildlife Plants and animals that are not tamed and found roaming freely in the forest. The Niger Delta is rich in biodiversity (Ringim et al. 2016) and ranges from invertebrates to vertebrates. Some vertebrate organisms found here include monkeys, apes, birds, snakes, crocodiles, baboons, and lions. The bird species found include crow, stork, cattle egret, eagle, vultures, and kingfisher. There are many land and water insect species that are yet to be identified and classified. There are dragonflies, grasshoppers, crickets, beetles, butterflies, etc. Some of these insects are used as food in the Niger Delta (Ebenebe et al. 2020). There are also different species of crabs found in the mangrove forest such as West African mangrove crabs (*Goniopsis pelii*) and fiddler crab (*Uca tangeri*), blue crabs, etc. There are some small tree crabs found in the mangrove forest that are yet to be classified. They don't grow large and remain small all their life with unique features. The problem of this rich supply of biodiversity is that their population continues to decline rapidly over the years due to anthropogenic activities of oil and gas exploration, sand mining, dredging, and infrastructural development projects that are executed within and around the mangrove forest area. Many coastal areas had been converted to terrestrial areas to accommodate industrial projects. A classical example is the loss of many lepidopteran families because of air pollution which had pushed them to a state of extinction (Extinctions 2013). The number of lepidopterists in the region is few, so there are little or no data on the population of butterflies.

17.2 Oil Economy in the Niger Delta

The people of the Niger Delta mainly engage in fishing and farming and have an agrarian economy with income derived from the sale of farm products (Babatunde 2020). Fishing is a traditional occupation that supplies their daily protein needs. An agrarian society dwells more on producing and maintaining farmlands, and food production is the main focus of this type of economy. The origin of this kind of society dates back to the time of the ancient hunter-gatherer community where crude implements such as stones and spears were deployed for hunting of wild animals. Some of these animals are placed in cages, domesticated, and raised for meat production. The gathering aspect is the crop farming that supplies farm products; here seeds with long or short life cycle are planted to provide food. Mixed farming is also done where animals are kept in cages, while crops are planted and are supplied with manures from the poultry and animal farms. The soil of the region is loamy and rich in humus and nutrients which make the area to have limited use of chemical fertilizers.

Characteristics of an agrarian society as practiced in the Niger Delta include:

Surplus Food Supply: The Niger Delta is rich in agricultural products such as cassava that is used to produce a local staple food called "garri" (Iyagba and Anyanwu 2012; Meena et al. 2020). We also have crops like maize, vegetables, fruits, and cowpea. These products are harvested and sold in the market to make some income; alternatively, these farm products are consumed by both humans and their livestock. The upland area of the Niger Delta has a wide landscape that is fertile and suitable for agricultural activities. Similarly, the coastal areas have a lot of water bodies that grade from highly saline, called saltwater, to lowly saline known as freshwater, whereas the mixture of salt and freshwater gives the estuary. These areas have thousands of fish species and seafood (periwinkle, mussel, crabs, and prawn) that are consumed or sold in the market to earn some income. Because of the presence of fertile land and rich fishery resources, food scarcity is minimal with low starvation level.

1. More Social Organization: The communities in the Niger Delta are organized into societies and run by traditional rulers and kings since the colonial era (Chizea and Osumah 2015). Traditional chiefs and paramount rulers oversee the day-to-day affair of the community, while the family heads run the various compounds that make up the villages. In most communities, headship is hereditary and runs through the line of a particular family, while in other communities, headship is based on age, where the oldest person becomes the ruler of that community or compound.

2. Less Technological Advancement: Method of farming is by manual means of using hoes and cutlasses to make ridges and mounds. Cutlass is used for cutting the grasses before they are burned. The level of technological advancement here is low because of the lack of government intervention to assist members of the community to acquire modern equipment. In addition, lack of funds has also prevented them from procuring modern equipment on their own (Kergna et al. 2020). In the past, low-scale farming can feed a family and generate some income. Currently, population increase has necessitated the need to acquire modern farming tools to facilitate large-scale farming. This machinery includes tractors, bulldozers, caterpillars, planters, mowers, and harvesters and is used in felling trees, clearing bushes, planting, and harvesting crops. Mechanized agriculture enables large-scale farming to occur, which leads to increase in food production and income generation for local farmers. For instance, mechanized agriculture improved yam and cassava productions (Nkakini et al. 2006). Increase in farming activities increases commercial activities that require raw materials from farm and forest products. In most communities, people come together to establish cooperative unions to pull their resources together to embark on large-scale agriculture. There are also different market days from Monday to Friday in many communities where people come to display their goods. This increases the profit of the people because buyers and sellers come from near and far communities to buy goods and render services. This sudden increase in wealth had resulted in the financial empowerment of the people to embark on more adventurous projects such as the construction of modern houses in many rural communities.
3. Depletion of Soil Fertility: Overcropping is the consequence of limited land space due to land loss from urbanization and soil degradation from industrial pollution. A major problem encountered by many farmers is poor soil fertility, which has increased the use of chemical fertilizers. However, the problem of chemical fertilizers is that it flows into the water bodies (Ibrahim et al. 2020) leading to increase in harmful chemicals in the bodies of sea organisms. It also contaminates the water and prevents it from being used for domestic purpose such as washing and bathing. Overuse of chemicals can increase the toxicity of soil leading to the death of some soil microbes and bioaccumulation of harmful chemicals within the food chain. An alternative way of improving soil fertility is the use of organic manure; this improves soil fertility, texture, and soil structure (Khan et al. 2021a, b). The soil of the Niger Delta is loamy and dark brown and well watered from the high rainfall in the region.

An oil economy operates in a community where oil revenues drive the major aspect of the economy (Nwajiaku-Dahou 2012) such as Nigeria that earns its major foreign exchange from the sale of crude oil (Fig.17.1). Nigeria is the largest producer of crude oil in Africa, and the Niger Delta area in particular is where all the oil resources originate (Numbere 2018). The Niger Delta has refineries and oil prospecting firms scattered all over the region. The presence of numerous industries and oil fields takes up land that would have been used for agriculture. The forests are

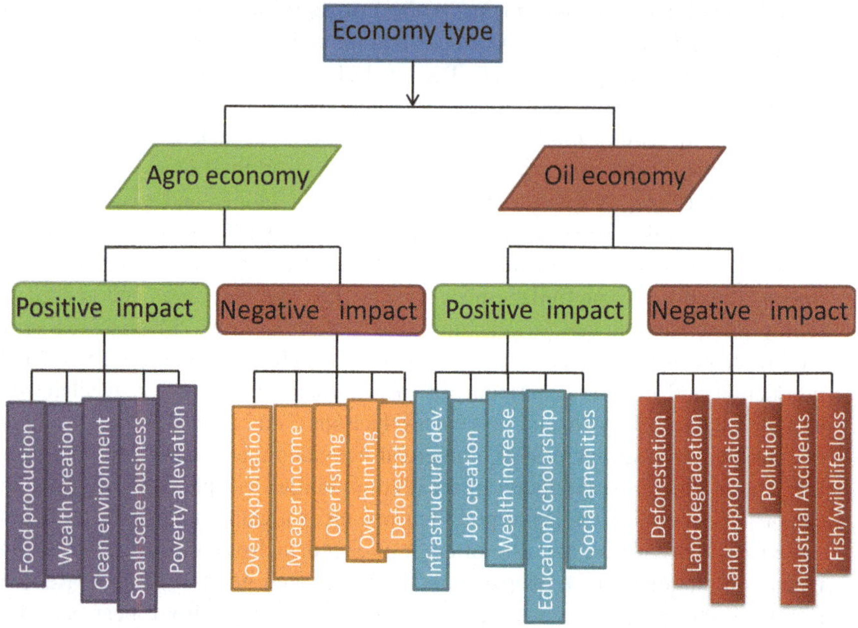

Fig. 17.1 Comparison of economy types practiced in the Niger Delta, Nigeria

cleared to make right of way passage for crude oil and gas pipelines. Oil economy is a society that dwells on crude oil resources as its source of wealth. Under this system, oil industries are given license to explore and drill crude oil and gas at offshore and onshore oil fields. Oil mining license (OML) is given to both public and private investors to explore for crude oil. The wealth derived from working in the oil industry had made a lot of rural farmers to migrate into cities to search for white-collar and industrial jobs. This had changed the wealth dynamics in many communities. A lot of people abandon their farms to work in oil industries sited in their localities because they feel they can earn more money by working in industries than what they will get from farming. In addition, manual farming goes with strenuous activities during land preparation, weeding, and harvesting. Apart from the wealth that comes from industries, many people would have continued their farming activities, but the problem is that most farmlands are already depleted by high chemical toxicity from constant crude oil spills and overuse of fertilizers and overcropping. Similarly, many fertile lands had been impoverished by pollution from numerous oil spills (Onwuka 2005).

17.2.1 Movement from Agrarian to Oil Economy and its Impact on Ecological Processes

Establishment of oil companies in various oil-producing communities has resulted in the conversion of agrarian-based economy to industry-based economy with the following results:

1. Increase in wealth creation.
2. Loss of agricultural land.
3. Depletion of soil fertility and degradation of soil structure.
4. Employment of youths into oil companies.
5. Safety and health risk in consuming crops from polluted land.
6. Decline in fish catch due to water pollution and general changes in water quality.
7. Poverty.
8. Proliferation of local militia groups.

Ecological processes are the interactions between plants, animals, and the non-living components of the environment like climate and land (e.g., rocks). These processes are affected by oil pollution (Holdway 2002; Barron et al. 2020). Ecological processes are crucial for maintaining healthy ecosystems and supporting the long-term persistence of biodiversity (Jhariya et al. 2019a, b; Meena et al. 2020a). Holistically, ecological processes produce organic matter through litter fall and death of animals; furthermore, it leads to the transfer of carbon and nutrients, accelerates soil formation, and enables organisms to reproduce and interact via food chains (Jhariya and Singh 2020, 2021; Kumar et al. 2020a). Four fundamental ecological processes of ecosystems are (1) water cycle, (2) nutrient or biogeochemical cycle, (3) energy flow, and (4) community dynamics, i.e., it shows how the composition and structure of an ecosystem change after a disturbance or succession.

The movement of agro-based economy to an industry-based economy has severe impact on ecological processes as follows:

1. Increase in Wealth Creation: Proliferation of industries in rural settings has changed the wealth dynamics of the area by creating well-paying jobs for members of the host community. Already there is a law in existence that mandates companies to employ low-level job force exclusively from members of the host community such as clerks, cleaners, housekeeping staff, and labor hand. These jobs are aimed at empowering unskilled members of host communities so that they can take care of their families and start small-scale businesses that will sustain the family. They can also use the money realized to buy farm products to restart their farming activities. Mass employment of community members by industries reduces overcropping, thereby saving the land from degradation and loss in fertility.
2. Loss of Agricultural Land: The establishment of industries in rural areas takes up land that would have been used for agriculture. This situation can cause the drop in food production leading to food scarcity. It also leads to the loss of livelihood

systems of farmers. Reduction of agricultural produce also lead to the reduction in raw materials for industrial productions of other goods. Scarcity of agriculture products leads to the high cost of available products.

3. Soil Depletion: Soil depletion is caused by numerous factors, one of which is incessant pollution caused by oil spills from pipelines conveying crude oil and other petrochemical products on onshore and offshore routes (e.g., Barron et al. 2020). Toxic waste products are deposited in bushes and on fallow land in a clandestine manner by some oil companies. Deforestation also takes away trees, which consolidate and protect the soil against erosion. The trees act as windbreaks against rainstorms that blow away topsoil and crops in farmland (Raj et al. 2019a, b, 2020). Gully erosion leads to soil fragmentation, which washes away crops and separates farmland into smaller fragments, thereby making it a risky venture for farmers to continue farming. Erosion also flushes away soil nutrients down the drains and into rivers (Ikehi et al. 2014). Leaching also occurs where soil nutrients from topsoil are washed down the soil profile beyond the reach of plant roots. Seismic activities by oil industries also contribute to the depletion of the soil when vibrations of dynamite explosion disintegrate the soil structure leading to soil liquefaction. Soil liquefaction leads to landslides, which covers and buries large farmland (Numbere 2018).
4. Drop in Fishing Activity: Proliferation of industries around coastal areas leads to increase in aquatic pollution from oil spills and industrial wastes. Construction of crude oil platforms on top of water bodies and the placement of pipelines underneath the sea create a lot of damages on the seabed and all aspects of the water column. Water pollution leads to contamination of aquatic organisms and death of fishes. An example is the death of thousands of fishes along the Bonny River in the Niger Delta region. It was reported that the fishes were killed as a result of contamination from toxic waste dumped into the water body, which increased the metal load (cadmium, lead, mercury, etc.) to hazardous levels. Continuous death of aquatic organisms will eventually lead to a decline in fish population and thus fishing business due to lack of harvest by fisher folks.
5. Poverty: Most community members don't have college degrees, which put them at a disadvantaged position to secure industrial or white-collar jobs. They therefore engage in agricultural activities to earn a living. Destruction of farmland takes away their livelihood system and prevents them from providing for their families. Poverty breeds other destructive societal behavior such as theft, violence, cultism,and militancy because an idle mind is a devil's workshop; thus, jobless youth go around fomenting trouble which most of the time lead to inter- or intra-communal conflicts. These conflicts destabilize the communal life and prevent agricultural activities from going on when locals abandon their farms and flee to neighboring towns so as not to be caught in the crisis. After some years, the abandoned farms would be overtaken by weeds and bushes.
6. Increase in Anti-social Vices: Loss of livelihood system pushes many into crimes to survive the excruciating hardship. Youths that were formerly farmers now engage in petty theft and join militant groups to terrorize oil companies operating in the area. They also engage in kidnapping for ransom. Most of the farmers and

fisher folks are not well educated and they lack the requisite qualification to apply for industrial jobs, Therefore, they resort to violence, cultism, oil bunkering, robbery, arson, and murder. These activities lead to an entropic state which further eliminates any form of livelihood opportunities left.

17.2.2 Impact of Artisanal Refinery on the Environment

The presence of oil industries and their infrastructure such as pipelines, oil well heads, and rigs in rural communities had led to the setting up of illegal artisanal refineries, which is the siphoning of crude oil from punctured pipelines and producing fuel through boiling of the crude in large drums within the mangrove forest. The destruction of pipelines to siphon crude oil leads to massive oil spills into the wetland environment. This has resulted in fire leading to the death of many persons in the Niger Delta region. These fires burn violently and destroy anything along its path as a result of the leakage of gas. Establishment of numerous artisanal refineries in the region is one of the greatest environmental disasters of the twenty-first century because of the colossal and collateral damages done to the environment. Burning of crude to eliminate sulfur content had resulted in atmospheric pollution leading to suffocating soot in the skies of the region. The deposition of dry and wet soot on any exposed items leading to black soot everywhere even on mangrove leaves (Fig. 17.2). The soot also mixes with rainwater and falls as acid rain, which increases the acid level in soil and water. High acidity in soil leads to the death of crops and loss of livelihood system for farmers. Similarly, high acid content in water changes the water quality leading to changes in the physiology of aquatic organisms which affect their reproductive function leading to population decline.

The life of those engaged in oil theft is also at risk because they are directly exposed to the smoke that comes from the burning crude. The river is also polluted to the extent that no aquatic life can survive in them because of the toxic environment. Artisanal refinery is dangerous to the environment because it destroys every aspect of the environment, i.e., land, air, and water, and damages the organism cohesion in the ecosystem through the destabilization of food chains.

17.2.3 Increase in Militancy

Unemployed youths form criminal gangs that terrorize their communities by engaging in kidnapping of oil company workers and demanding for ransom from their families. They also engage in violent protests where they attack oil companies and vandalize their property because of dissatisfaction with employment of their members. They also block the gates of the company till they are compensated. Sometimes the protest leads to clashes with law enforcement agencies resulting in injuries and death. Some examples are the military invasions of some Niger Delta communities, e.g., Odi, Umuechem, and Ogoni, as a result of the people's agitation for fair distribution of the oil wealth coming from their communities. During these

Fig. 17.2 Soot from burning crude oil settles on mangrove leaves at Okrika Jetty, Niger Delta

invasions, whole communities are burned down and people maimed and killed by security forces. During these crises, all economic activities are lost, and people become refugees in neighboring communities.

17.3 Human Versus Environmental Ecology

Human Ecology: Human beings are at the center of everything in the environment. This is because they can make or mar the natural equilibrium that exists in the environment through their actions and inactions. Human beings are the major problems of biodiversity globally, and thus human ecology is an interdisciplinary and trans-disciplinary study of the relationship between humans and their natural, social, and built environment (Khan et al. 2020a, b; Banerjee et al. 2020, 2021; Meena et al. 2020b). Humans occupy land upon which they build their homes. They practice agriculture on land and tap crude oil and gas beneath the earth. The earth also provides other resources such as iron, aluminum, cobalt, diamond, gold, etc.

Environmental ecology is the branch of biology which studies the interactions among organisms and their environment. It includes interaction of organisms with each other and with the abiotic components of their environment. Ecological systems do not have intrinsic homeostatic mechanisms. An important change occurs at the level of the individual. Thus, the fitness of the individual depends on their homeostatic processes. Natural selection does not operate at levels above the individual. But they are the collective result of natural selection operating independently on all the interacting species. The lack of homeostasis in ecological systems is one reason that non-equilibrium processes are important in ecology (Mori 2011). An example of non-equilibrium is disturbance such as population dynamics and plant population structure and organization (Illius and O'Connor 1999). The implication of this belief is that human impacts can be ignored or accepted as a natural process. Rather some ecologists suggest that we distinguish between disturbance and degradation, the latter showing human disruption of ecological processes. Another important principle is the effect of species diversity in the function of ecological systems, and this diversity-complexity provides ecosystem services (Jhariya et al. 2019c, d). The finite nature of resources shows that ecological mechanisms are governed by resource limitation. For example, population growth changes when the number of individuals exceeds the carrying capacity (k). Competition is inherently related to limiting resources. Furthermore, ecology is a science of interaction because any change in one aspect of the ecosystem affects another part of the system. Nothing happens in isolation, for example, there is a synergistic effect on many environmental issues such as climate change and acid rain. Regional impacts expand and interact with other ecosystems as well.

17.3.1 Processes of Human Ecology

These include concentration, centralization, segregation, invasion, and succession.

1. Concentration: It is a process where a given area becomes concentrated due to its fertility and rich resources. For example, oasis is an area where water is found that benefits people.
2. Centralization: This has to do with the gathering of people in places called focal points which are resource rich and beneficial to the economic interest of the people. For instance, nodal cities with huge natural and human resources are available and with high infrastructural development.
3. Segregation: This is when units cluster together based on homogeneity of ideas or goals. Homogeneous ideals such as customs and educational and occupational goals all constitute segregation. The Niger Delta is oil rich and a host of some refineries and has a concentration of people who are seeking for industrial jobs.
4. Invasion: It is the migration of people into areas with high economic benefit that is rich in resources. The Niger Delta region is rich in crude oil and gas and hosts numerous oil prospecting industries. This makes it a choice area for many job seekers. For instance, Bonny town is the host to the liquefied natural gas (LNG) company and has people from all walks of life residing in the town.

Fig. 17.3 Building of living apartments right in the mangrove swamp at Eagle Island, Niger Delta

5. Succession: Ecologically speaking succession is a change in the species composition of communities over time. It is the result of a variety of abiotic (physical and chemical) and biotic agents of change. In human ecology, succession is when a particular land use is replaced with another land use. For instance, the building of a house on sand-filled mangrove forest completely replaces the land use (Fig. 17.3).

The concept of succession establishes equilibrium in land use and distribution after invasion. There is natural and human-induced succession (Jin et al. 2019). Natural succession is a process of environmental change after disturbance that occurs without the intervention of humans, whereas human-induced (artificial) change is the deliberate replacement of a given environment to different type. For instance, after an incident of forest pollution or wildfire, there comes a period of regeneration of species which gradually takes over the disturbed environment. However, for the artificial succession, a mangrove forest can be cut and replaced with a sand fill. Similarly, a river can be dredged and converted to terrestrial area, and a wetland can be scooped away and converted to grassland. The implication of all these human-induced changes is the migration and disappearance of native species.

17.3.2 Hydrocarbon Pollution Ecology in the Niger Delta

Hydrocarbons consist solely of atoms of carbon and hydrogen. Fossil fuels and the many petroleum products consist of hydrocarbons. Smallest hydrocarbons exist in a gaseous state, whereas larger hydrocarbons are liquids, while those containing over 20 carbon atoms are normally solids. Some hydrocarbons from petroleum are known to cause health hazards to wildlife and people, e.g., polycyclic aromatic hydrocarbons (PAH). Pollution can be chemical, physical, or biological material that can affect water, air, and soil negatively. Pollution can come from point source, which means they come from a particular location, e.g., gas flare, smoke from smoke stack, and effluents from sewer pipes, or it can be non-point source, which regards pollution as cumulative and coming from multiple inputs over large areas, e.g., farms, city streets, and residential quarters. Water pollution can take the following forms: toxic chemical pollution, physical pollution by sediments, thermal pollution, nutrient pollution, and pollution caused by disease-causing organisms. Hydrocarbon pollution is caused by toxic chemicals from crude oil and poison organisms in the environment. This poison enters the food chain to contaminate humans. Pollution is always known to have negative consequences on the environment, but based on new research findings, pollution can have some positive effects on the ecosystem as shown by studies on productivity, decomposition, and herbivory. Pollution is the entry into the environment of substances that are toxic and inimical to health and safety of living organisms. It is the alteration of the well-being of the environment via the addition of dangerous substances that destabilize environmental stability. Therefore, hydrocarbon pollution is an addition into the environment of hydrocarbons that are harmful to living organisms. Hydrocarbon is a chemical compound that is made up of hydrogen and carbon joined together by bonds. This compound is a component of crude oil which are derived from oil wells and refined to produce fossil fuels and other petrochemical products. Crude oil is useful to human as a major source of fossil fuel-generated energy that powers automobiles and engines in industries. However, the negative impact of hydrocarbon pollution in the environment outweighs their usefulness because of their long-lasting effect on the environment. Hydrocarbon pollution is among the four factors of environmental degradation; the rest are habitat loss and fragmentation, invasive species, population growth, and overexploitation (HIPO).

In the Niger Delta, hydrocarbon pollution comes mainly from oil and gas exploration which occurs almost in every community in the region. The land is made up of sedimentary rock, which is rich in crude oil due to organic deposits (i.e., mangrove plants and animal matter) that have decomposed to form peat thousands of years ago. The peat is what forms the crude oil that is presently being explored to produce petroleum. Most often oil wells are found in coastal areas occupied by mangrove forest, which puts the mangroves at the mercy of exploratory activities. Hydrocarbon pollution in the mangrove forest is a common denominator of the environment, which places the mangrove in a state of constant degradation. Globally, mangrove forest and hydrocarbon pollution have a negative relationship. This is because numerous studies have shown that crude oil kills mangrove and causes mutation (Naidoo 2016). This negative relationship has existed in the Niger Delta since the discovery of the first crude oil well in Oloibiri community in June 1956. There are no organisms found in the Niger Delta without a trace of pollutants due to the widespread pollution of the environment. All the parts of mangroves have some amount of hydrocarbon content such as the leaves, seeds, stem, and roots (Numbere 2019b). This occurs when the roots inside the swamp absorb chemicals and transfer the contaminants to plant parts. Research has shown that no soil sample collected around the coast is devoid of some amount of pollutants. This is as a result of long years of deliberate or accidental spills on the land and in the water, which has led to the movement of pollutants into the mangrove and the bodies of coastal organisms via food chain vertically and horizontally. Years of research by the author have shown that several biota of the mangrove forest have been found to contain some amount of hydrocarbons and heavy metals (Table 17.1).

17.3.2.1 Positive Interaction Between Mangroves and Hydrocarbon Pollution Observed in the Mangrove Forest of the Niger Delta

Some studies have shown that mangrove forests do not only have a negative relationship with hydrocarbon pollution but also have positive interactions. This aspect of mangrove pollution is new because not many studies have found the usefulness of hydrocarbon pollution in the growth of mangrove forest because of the devastation it causes, but three studies reflect that there are some positive relationships between mangrove forest and hydrocarbon pollution. These studies include:

1. Productivity Study: The outcome of this study revealed that mangrove forests growing in highly polluted sites have a higher productivity than their counterparts growing in lowly polluted sites. This is based on a 3-year study that found out that hydrocarbon pollution accelerates litter fall through massive defoliation leading to higher calculated productivity and productivity/biomass ratio (Numbere and Camilo 2018).

$$\frac{\text{Dry weight of litter gDW}}{0.15211\text{m}^2} \tag{17.1}$$

Similarly, higher structural characteristics (tree height, diameter at breast height, and aboveground biomass) were found in highly polluted sites than lowly polluted sites. The highly polluted site is a mangrove forest close to a major refinery with high-level oiling activities resulting in constant oil spills from a concatenation of pipelines criss-crossing the terrain. The mean THC of the soil in this location is 2500 mg/kg. The lowly polluted site has one oil wellhead facility and no pipelines. The oiling activities are minimal and the mean soil THC is 300 mg/kg. Both sites have similar geography but are miles apart.

2. Decomposition Study: This study was aimed at determining the rate of decomposition of mangrove leaves collected from trees in highly and less polluted sites. Fresh leaves were collected and weighed to be 20 g each and sealed in a litter bag and placed on the ground of both forests for 455 days to decompose into humus. The decomposition rate constant was calculated (Eq. 17.2) from the weight of the oven-dried leaves.

Table 17.1 Studies on THC and heavy metal concentration of biota of mangrove ecosystem in the Niger Delta

Species	Taxa	Location	Heavy metals mg/kg	THC	Author
West African red mangrove crabs (*G. pelii*)	Crustacean	Okrika	Zn: 9.9–993.4 Cd: 2.6–14.27 Pb: 11.9–151.9	4.0 ± 0.5–39.5 ± 2.9	Numbere (2019a)
Red mangroves (*Rhizophora racemosa*)	Plant	Buguma	–	Seed: 25.07 ± 0.77 Leaf: 25.38 ± 0.46 Stem: 2.31 ± 0.08 Root: 0.47 ± 0.03	
White mangroves (*Avicennia germinans*)	Plant	Buguma	–	Seed: 0.66 ± 0.02 Leaf: 5.90 ± 0.24 Stem: 1.08 ± 0.02 Root: 1.17 ± 0.02	
Black mangroves (*Laguncularia racemosa*)	Plant	Buguma	–	Seed: 21.64 ± 0.77 Leaf: 20.22 ± 0.93 Stem: 1.08 ± 0.02 Root: 1.85 ± 0.49	
Fiddler crab (*Uca tangeri*)	Crustacean	Eagle Island	Zn: 8.0–83.57 Cd: 1.45–10.06 Pb: 0.001–44.5	0.02 ± 0.00–0.39 ± 0.32	
Sardine (*Sardinella maderensis*)	Fish	Bonny	Zn: 258.0 ± 40.0 Cd: 0.5 ± 0.2 Pb: 0.7 ± 0.1	–	Obeka and Numbere (2020)
Mullet (*Liza falcipinnis*)	Fish	Bonny	Zn: 109.3 ± 1.6 Cd: 0.04 ± 0.03 Pb: 0.9 ± 0.2	–	
Tilapia (*Sarotherodon melanotheron*)	Fish	Bonny	Zn: 175.1 ± 36.3 Cd: 0.03 ± 0.02 Pb: 7518.3 ± 7516.3	–	
Periwinkle (*Tympanotonus fuscatus*)	Crustacean	Okrika		0.25 ± 0.002	Numbere (2020a)

$$Y = Yo\, e^{(-kt)} \tag{17.2}$$

The result indicates that leaves placed on the ground of the highly polluted site had a lower rate of decomposition (0.000658) than leaves at the lowly polluted site (0.0000175). The outcome of this study revealed that hydrocarbon pollution does not prevent decomposition and nutrient redistribution processes as often reported in the literature. This study rather showed that hydrocarbon pollution slowed the decomposition of mangrove leaves (Numbere and Camilo 2017). This also means microbial actions still go on in highly polluted soils, which made the mangroves to have robust growth and good tree structural characteristics in the presence of massive pollution.

3. Herbivory Study: This study like the other two studies was carried out in highly and less polluted sites. The aim of the study was to find out which forest will have higher leaf herbivory in an exclusion experiment that was conducted for 2 years (Numbere and Camilo 2019). In this experiment, leaf area eaten (LA_{eaten}) was calculated by subtracting leaf area after herbivory (LA_{after}) from original leaf area before herbivory (LA_{before}) Eq. (17.3). A second experiment (i.e., cafeteria experiment) was carried out to determine the feeding preference of 20 West African red mangrove crabs (*Goniopsis pelii*) on leaves collected from highly and lowly polluted sites. In the first experiment, a total of 453 leaves were sampled between 6 months and 2 years, and results showed that there was significant difference ($P < 0.05$) in herbivory of leaves between highly and lowly polluted sites. The result further revealed that there was more leaf herbivory in highly polluted (4521.69 cm^2) sites than less polluted (2769.83 cm^2) sites. Similarly, the results of the second experiment revealed that more leaves from the highly polluted plot were consumed by crabs, which were the most dominant herbivore in the mangrove forest. The outcome of both experiments revealed that leaves from highly polluted sites were more palatable than leaves from less polluted sites. The outcome of this study showed that mangroves growing in highly polluted environment prevent pollutants from entering its system via its root. Another reason is that the mangrove converts pollutants within its system into less harmful form that makes it palatable for crab herbivores. This supports the hypothesis that mangroves have a positive interaction with hydrocarbon pollution that is why they are well adapted to the polluted environment. However, more studies are needed to verify this claim in other regions of the world that have mangrove growing in hydrocarbon-polluted environment.

$$LA_{eaten} = LA_{before} - LA_{after} \tag{17.3}$$

17.4 Ecosystem-Based Resource Management

Ecosystem-based resource management is a new way of managing reserves to benefit biodiversity and people (Raj et al. 2018a, b). It is a strategy for protecting or restoring the function, structure, and species composition of an ecosystem while providing for its sustainable socioeconomic use (Jhariya et al. 2018a, b).

The principles of ecosystem management (Cunningham and Saigo 2001) include the following:

1. Ecosystems are dynamic and change during succession, maintaining biodiversity and essential ecosystem processes.
2. Ecosystems are subject to predictable disturbances, e.g., forest fires, so management needs to be flexible, which is called adaptive management (i.e., it means that every program is regarded as experimental and subject to change as new information becomes available).
3. Humans are integral part of all ecosystems. Therefore, it is necessary to maintain meaningful stakeholder and public involvement and facilitate collective decision-making. It should also consider human needs and promote sustainable economic development and communities. This should be based on conscious experimentation and research.
4. Ecosystem requires constant monitoring of populations. Here we need to collect data to know whether our management system is working or not or whether the population is increasing or decreasing.
5. Managing across whole landscapes, watershed, or regions over an ecological time scale.
6. Utilizing cooperative institutional arrangement.

It is also an effort to oversee resource harvesting processes that will have less effect on the ecosystems and ecological processes that provide the resources. It aims to protect certain forested areas, restore ecologically important habitats, and consider landscape patterns. It also considers the functional integrity and succession patterns of the forest. Ecosystem-based resource management has its root in sustainable development, which considers present use as well as future benefits of resource (Slocombe 1993; Reid and Rout 2020). This management system looks at the environment holistically rather than as a separate disjointed unit. This is because the entire environment is intertwined around humans. The soil, water, and atmosphere are all linked and depend on one another. Thus, the ecosystem will not function effectively if these aspects of the environment are not safeguarded to ensure the survival of living organisms.

Ecosystem management goals (Cunningham and Saigo 2001) are:

1. Maintain viable populations of native species in situ.
2. Represent within protected areas all native ecosystem types across their natural range of variation.

3. Protect essential ecological processes such as nutrient cycles, succession, hydrologic processes, etc.
4. Manage over long enough time periods to sustain the evolutionary potential of species and ecosystems.
5. Accommodate human use and occupancy within these constraints.

17.4.1 Establishment of Protected Areas

Protected areas are a management strategy to safeguard forest and its resources from wanton and destructive exploitation. This is because humans are the main culprit of biodiversity loss and population declines leading to species extinctions (Venter et al. 2014). Protected areas are thus established to reduce anthropogenic impact by restructuring access. Three ways of rendering protection to forests are:

1. By Law: Here the state or federal government passes a bill or edict through the House of Assembly to protect public lands from destruction. This pattern of protection is ideal for mangrove forest in the Niger Delta region because of reckless oil and gas exploitation and construction activities which had resulted in the loss of 5% of its mangrove forest (Wang et al. 2016). Mangrove protected areas (MPA) can be established especially in public forest and overexploited areas. Access into such areas should be restricted completely to avoid unsustainable environmental practices. These public forests are often found on the boundaries of cities with easy access by encroachers. Since the forests are not protected, they are often used as public latrines and conveniences as well as a dumping ground for waste products by residents and traders (Numbere 2019c). Therefore, in the MPA, only controlled access should be allowed to ensure that harvesting of trees for firewood and other forest resources are monitored. Another key issue that destroys many mangrove forest and coastal areas in the Niger Delta is that they are sold out by community agents to private and public investors. There are instances where rivers close to shorelines are apportioned to community members and later sold out due to lack of land. The water bodies are then filled with "chikoko" or hardened mangrove soil used to fill up the river before building on it. When this happens, it constricts the water channel and changes the river course leading to seasonal flooding during peak tides. As more water bodies are sold out with time, a small creek eventually becomes totally converted to a terrestrial area that is occupied by buildings. To end this practice, the authorities can intervene to pass an edict banning the sale of mangrove forest, swamps, and rivers. And such areas should be declared a mangrove protected area. The government can compensate the indigenes by providing low-cost housing for them. Furthermore, the government can also buy up such forests, swamps, and rivers from the original owners and make those areas protected areas. Furthermore, two categories of protected areas can be established depending on the state of the species or resources present whether they are threatened or endangered or at the verge of extinction. They are:

(a) Strict Protection: This area includes nature reserves, wilderness, national parks, and monuments. In this type of protection, no resource extraction, no oil and gas exploration, and no developmental project are executed because the main aim is to conserve all the biodiversity present for the sake of preserving them for future generations (de Lima et al. 2020; Wang et al. 2020). However, activities such as scientific, research work, training, and educational activities are allowed to study and understand the nature of the mangrove ecosystem and how best to monitor and preserve them. During these events, the population dynamics of species is studied, and new species are found especially those that are rare and endangered.

(b) Limited Protection: These areas include managed wildlife sanctuaries and resource conservation areas, national forest, etc. In this type of protection, some resource extraction such as hunting, fishing, logging, and grazing are allowed to enable people to have the opportunity to see nature at its best and to appreciate it. The problem of this type of protection method is that there is a high human impact on biodiversity, which may lead to the decline of some species. The goal of this method of forest management is to allow for the multiple use of natural resources aimed at appeasing competing human interest, which if not allowed will lead to clandestine exploitation of resources that may jeopardize the aims of conservation and worsen the condition of the forest.

17.4.2 Establishment of Restoration Sites

Restoration is the art and science of bringing back a destroyed environment to its original state. Ecological restoration has to do with active manipulation of nature to re-create conditions that existed before anthropogenic disturbance. The best form of environmental management is the prevention of any destructive activities that impede the smooth working of the ecosystem. But when this effort fails, the next available option is to restore the devastated environment. For instance, mangrove forest polluted as a result of oil spill from punctured crude oil pipelines can be restored (Das 2017). This can be done by re-establishing the original environment or something similar to it that would perform ecosystem functions. To establish restoration sites, it is expedient to understand the science of restoration ecology.

Restoration ecology is intended to repair or reconstruct ecosystems damaged by humans or natural forces. It intends to accelerate the recovery of impoverished land (Dobson et al. 1997). It is also a form of land management (Hobbs and Harris 2001). It is a process of intentionally altering a site to establish a defined indigenous (natural) historic ecosystem. Restoration ecology started 50 to 60 years ago in the Midwestern USA where prairie is converted to cropland being restored. Two methods of carrying out restoration are as follows:

1. Passive Method: It is the process of allowing nature to heal itself. It is to stop degradation and allow succession to take place. This method is predictable and attains its climax and involves natural recovery through succession. Here the land is bought from indigenous people and allowed for succession to take place (Villa et al. 2019; Barros et al. 2020).
2. Active Method: This is required for areas that have been severely impacted by human activities:
   (a) Replacement: Instead of going back to the original which is impossible, it involves taking it to a different direction to create a replacement. It is a way of establishing a new habitat type because we can't establish a historic ecosystem, for example, planting grass in a former mine field.
   (b) Rehabilitation: In this method, we are trying to restore the original ecosystem, but it can't be fully restored because most of the original species have all gone extinct. It is an attempt to rebuild elements of structure or function in the ecological system without necessarily achieving complete restoration to its original condition. It aims to reverse deterioration of a resource even if it cannot be restored completely. Example is the rehabilitation of prairie along highways to mimic the natural ecosystem in Illinois and Missouri, USA, by the Illinois and the Missouri Department of Transport, respectively. The new habitat can carry out environmental and ecological functions such as air purification via CO_2 absorption, soil stabilization, and other ecosystem services beneficial to humans and the environment.
   (c) Remediation: It is a process of cleaning chemical contaminants from a polluted area by physical or biological methods as a first step toward protecting human and ecosystem health (e.g., Dhaliwal et al. 2020). Incineration and bioremediation are methods of cleaning crude oil-contaminated soils (Okoh et al. 2020). The use of living organisms such as water hyacinth and ground mangrove plant parts is effective in absorbing heavy metals and other toxins from polluted water (Abdullah et al. 2020). Microorganisms (bacteria and fungi) are also used for remediation purposes and are derived in nature (e.g., mushroom) or produced in the laboratory. They are used to destroy many dangerous chlorinated compounds. Addition of fertilizers to soil to encourage plant and microbial growth can also be used to clean surface pollutants.
   (d) Reclamation: This is used to describe chemicals or physical manipulations carried out in severely degraded sites, such as open-pit mines or large-scale construction.
   (e) Re-creation: Attempts to construct a new biological community on a site that is extremely disturbed that there is almost nothing left to restore (e.g., Dondajewska et al. 2020). The new system may be modeled on what we think was there before human disturbance or it may be something that never existed on that site but we think fits current conditions. Often private developers and the government are expected to mitigate damage caused in one area by re-creating a comparable biological community in another place.

(f) Restoration: It is the attempt to fully restore the original ecosystem. An example is the Rio Grande in the Colorado river ecosystems (Gerlak et al. 2013). It is a major river system that has a lot of snow that builds up on the mountain, which led to a natural disturbance. This is because when it melts it leads to a major spring flood. So a series of dams were placed in the river to help regulate water flow by taking the spring flood out of the ecosystem. In the 1990s, people noticed some unintended negative consequences, so the government tried to restore major floods. From the foregoing, it is thus necessary to establish several restoration sites to counter the quickly accelerating human degradation activities of oil and gas exploration and exploitation leading to hydrocarbon pollution and deforestation and urbanization leading to habitat loss and landscape fragmentation. The establishment of restoration sites is a deliberate management strategy to create safe zones for plants and animals and the reconversion of abandoned and decommissioned project sites to their original habitats.

Activities needed for restoration in the Niger Delta include de-silting of drains, canalization of blocked waterways, reconstruction of drained wetlands, and the replanting of degraded mangrove forests. These things are to be done so as to reverse human actions and to restore the system. Some areas would be difficult to recover because of the establishment of permanent structures; therefore, artificial restoration sites can be created near city limits or places that have infrastructural development that cannot be moved away to a different environment. Restoration sites can be created in numerous abandoned oil fields scattered across the Niger Delta area. Mangrove forest can be regenerated via natural or artificial means of seedling propagation and recruitment. This is because deliberate establishment of restoration sites will help the environment to heal faster as compared to when it is allowed to recuperate naturally. An example of an artificial ecosystem is the turning of an abandoned dump near an existing treatment plant in Arcata, California, USA, into marsh and wildlife sanctuary, Humboldt Bay in 1974 by city residents and faculty from Humboldt State University, which has turned into a model for other communities (Higley 1989).

17.5 Human-Based Resource Management: Win-Win Ecology

Humans are the major problems of biodiversity; therefore, to effectively manage natural resources, humans are to be kept out as much as possible from overexploiting the natural environment to help reduce their negative impact (Jones et al. 2018). People destroy forests through logging, mining, and introduction of exotic species. In reality, humans cannot be separated from nature and have a huge effect on ecological patterns and processes, which in turn affect them. To solve this problem, reserves can be created to give maximum advantage to plants and animals (Niemeyer et al. 2020). This can be done by creating zones that depict different levels of resource exploitation. The reserve can be divided into three areas, namely:

1. Core Zone: No resource extraction is allowed.
2. Buffer Zone: Some level of use allowed for local people (e.g., resource extraction and tourism), which will engender a win-win ecology. This is when the environment and species win.
3. Transition Zone: It will have higher level of resource use, i.e., commercial building and ecotourism. For instance, people are allowed to go to the buffer zone to hunt or farm in the Amazon forest (Paiva et al. 2020). This is a practical example of how to create a gradient of human action and disturbance. This aims at assuring the people that a forest reserve is a good thing that needs to be protected. It allows plants and animals to adapt to the gradient and creates an opportunity for source-sink dynamics where source can maintain the sink population. This seems to be a more successful approach in the management of humans and animals. This kind of management is proposed for the mangroves of the Niger Delta. An example is the Great Barrier Reef ecosystem management off the east coast of Australia called the Great Buffer Reef Marine Park Authority (GBRMPA). Here oil and gas exploration is banned and preservation zones are established. Only scientists are allowed to visit and also breeding sites are threatened and endangered species are established.

17.5.1 Win-Win Ecology

Win-win ecology is a deviation from the previous way of considering the environment as a single entity. It is rather a form of ecology where the environment wins and human wins. For example, the loss of farmland birds as a result of the loss of fallow land can be stopped by applying land management strategies (Tarjuelo et al. 2020). In the previous way, conservation of plants and animals is paramount without consideration of human aspect. Over the years, it was observed that absence of humans in the planning of forest conservation produced poor results. This is because when humans become the custodians of the forests, there will be low exploitation and increased protection against incessant exploitation. This is because humans are the main destroyers of biodiversity; thus, incorporating them into the protection of the environment is more result oriented. Win-win ecology is likened to limited protection where humans who live around the forest are charged to protect it, but allowed to farm, hunt, and cut trees to satisfy their subsistence and financial needs. The people act like guardians of the forest to prevent intruders from plundering the resource of the forest. Humans are at the center of all forest management plans and can make or mar its long-term goals of sustainability.

In rural communities, indigenes have no other source of livelihood apart from fishing, farming, and hunting. The forest and rivers provide the only source of natural resources from which they survive. Therefore, taking it away without an alternative means of survival puts them on a collision course with the authorities. In win-win ecology, buffer zones are created to enable indigenes to tap some resources from their environment in a sustainable manner.

17.6 Creating Ecological Equilibrium through Controlled Disturbances

Previously the definition of disturbance is any phenomenon that disrupts the natural order of things in the environment. But current observations have shown that some disturbances don't end up in negative consequences, but can have beneficial effects. For instance, disturbance has a role in shaping coastal environment (Schroeder 2000) and coral reef (Nyström et al. 2000). In nature, there is a breakdown and buildup process which ensures a form of pseudo-stability or equilibrium. Both natural and anthropogenic activities change the natural dynamics of the ecosystem. In an ecosystem, the building up process can be growth and reproduction of plants and animals for living things and formation of volcanic hills and mountains from volcanic eruptions and outpouring magma from under the earth surface from tectonic movement for inanimate objects, whereas the breakdown processes involve the leveling of hills and mountains and the death and destruction of organisms through natural or anthropogenic forces.

In the Niger Delta region, the major forces of degradation of its mangrove forest are urbanization, invasive species, and pollution from industrial activities. Studies have shown that physical cutting of mangrove trees to create cities gives no chance for the survival of the mangroves. Similarly, invasive palms were brought into the mangrove forest by humans, but for over half a century, the mangroves have been struggling to adapt to their presence without much success. But studies have shown that the mangroves are better adapted to pollution than the palms. This is revealed by some studies which indicate that mangroves have robust growth in highly polluted sites (Numbere and Camilo 2017, 2018, 2019). Based on these findings, the complete removal of pollutants from the swamps of the mangrove forest will spell doom for the mangroves in the long run. Fifty years of oil and gas exploration activities in the Niger Delta, resulting to spillages of millions of barrels of crude oil into the rivers and swamps, had made the mangrove forest to be stronger and resistant to pollution. Based on these unique situations, it is suggested that the remediation of degraded forest could involve the application of some low-level pollutants as a form of mangrove management in the Niger Delta. Similarly, polluted soils can be collected from the field and applied to nursery beds to aid seedling regeneration and adaptation, especially if the parent trees are situated in similar environment. This is because soils and seeds derived from the same area will have faster acclimation in terms of growth and development than soils and seedlings brought from two different environments.

17.7 Ecological Intensification in the Niger Delta

Ecological intensification do occur in the Niger Delta because of its long-term practice of organic farming, where manures from plant and animal matter are utilized with limited use of agrochemicals. Waste from poultry farms is often collected and used to boost soil nutrients. Leaves and plant parts are also gathered and used as mulch on seed beds and ridges to supply needed soil nutrients. This facilitates the natural growth of farm crops such as plantain, maize, cassava, yams, and pumpkin plant, which are staple food in the area. Inorganic fertilizers are not commonly used by local farmers except in government-owned large farms to help facilitate growth. The advantage of using organic manure is because it reduces soil chemical toxicity and encourages the proliferation of microbes, which facilitate decomposition and nutrient cycling. Mulching is a cultural practice embarked on by farmers in the Niger Delta (Umeh and Nwachukwu 2019). After the harvesting season, farmers gather leaves and plant materials and bury them in shallow holes to make compost. Some of these materials are also burned to form charcoal which is applied to the soil to boost its fertility. However, during seasons of poor yield and low productivity as a result of overcropping, NPK fertilizer is usually added to the soil. Similarly, pesticides, herbicides, and insecticides are added to the soil to destroy pests, weeds, and insects, respectively. Two types of pesticides are inorganic and natural organic pesticides. The inorganic pesticides include compounds of arsenic, copper, lead, and mercury. They are highly toxic and non-destructive and remain in the soil permanently. On the other hand, natural organic pesticide or botanicals are derived from plants, e.g., nicotine, rotenone, pyrethrum, turpentine, phenols, and other natural hydrocarbons are effective pesticides, but the synthetic forms (pentachlorophenol) are more toxic than natural forms. Pesticides are bad because they affect non-target species and facilitate resistance and resurgence of pests. They also create new pests and are persistent and mobile in the environment.

Already the Niger Delta environment is highly polluted, and the addition of pesticide would be one too many because of the severe impact they have on human health when they migrate into the food chain. Short-term effects of these pesticides can cause acute poisoning or illnesses caused by high-dose and accidental exposures, while long-term effects cause more serious conditions such as cancer, birth defects, immunological problems, Parkinson's disease, and other chronic degenerative diseases (e.g., Garrigou et al. 2020). Nevertheless, ecological intensification provides alternative ways to farming rather than using chemicals. This is because chemical-free methods are the best. This includes behavioral changes that will reduce the use of chemicals and the promotion of cultural methods such as physical hand picking of pests from farms or the use of natural enemies to fight the pest. Here ecological principles of the food chain can be applied to eliminate insect pests, for instance, use of biological controls such as wasps, lady birds, praying mantises, lizards, ducks, and chickens can be introduced to the farm to eat pests. Crop rotation can also be adopted to prevent pest population from building up in the farm. This involves the planting of both monocotyledonous (maize) and dicotyledonous (cowpea) crops together. The cowpea is a legume and provides nitrogen through its root nodules; thus, pests that attack maize may not attack cowpea. For example, soybean or corn rotation is effective and economical against white-fringed weevils. Then mechanical cultivation can turn over the soil to suppress weed and soil insects. Furthermore, burning of crop residues and replanting with a cover crop can suppress both weeds and insect pests. Some insects are beneficial to the farm crops by providing ecosystem services such as crop pollination, which facilitates fruiting and seed production (Painkra et al. 2016).

Integrated pest management (IPM) promotes ecological intensification. It is an ecologically based pest control strategy that involves the coordination of several control strategies that takes into consideration the advantages and disadvantages of pest prevention methods that will have less impact on the environment (e.g., Oke et al. 2020). It takes into consideration the population threshold of pests that will have limited effect on the environment. It also reduces the dependence on chemicals and relies on the use of ecological principles to control farm pests. The practice of IPM involves monitoring, use of natural enemies, habitat modification, and limited use of pesticides. Monitoring the factors that lead to the proliferation of pests will enhance their specific control without applying unnecessary treatment that will negatively impact the ecosystem. For instance, insects or rodents can be observed to determine what improves their reproduction and to cut off such lifelines to reduce their population. Trap can be used to capture and eliminate them. Trap crops can be planted ahead of the main crops to attract insect pests that are later sprayed with pesticide and eliminated completely, thus making the farm to be free from their attack.

Natural enemies such as predators, parasites, and diseases can be used to eliminate farm pests (Horrocks et al. 2020). For instance, lady bird beetles can be released in the greenhouse to control other pests. Bacteria-based pesticides are used on vegetables and crops. The non-use of pesticides saves natural enemies from performing their positive ecological functions.

Habitat modification is a habit that keeps pests away from their basic needs such as food, shelter, and water. This involves the use of pest-resistant varieties and crop rotation. The use of low-impact pesticides is encouraged to reduce their adverse environmental impact. It involves the use of less hazardous targeted pesticides. IPM avoids broad-spectrum ecologically disruptive products. IPM relies on preventive practices that encourage growth and diversity of beneficial organisms and enhances plant defenses and vigor.

17.7.1 Introduction of Agro-Mangrove Forestry in the Niger Delta

Agroforestry is a collective name for a land use system in which forest and shrubs are grown with agri-crops or pasture and livestock in both spatial arrangement and a time sequence and in which there is both an economic and ecological interaction between the trees and non-tree components of the system (Amadu et al. 2020; Muchane et al. 2020). It is the combination of agriculture and forestry (Fig. 17.4).

The distinguishing factor between agriculture and forestry is the integration of multipurpose wooden species and the utilization of the entire soil profile resource (Singh and Jhariya 2016).

In agroforestry, there are three basic elements of management:

1. Agri-crops.
2. Multipurpose woody species.
3. Farm animals.

The integration of woody species with farm crops eliminates the spatial separation between farm and forest. Both practices are combined in time and space to ensure little competition for nutrients, water, and light but rather complement one another in their different requirements so as to achieve an elevated output per unit area. Woody perennial component is significant for the system to be designated as agroforestry (Fig. 17.4). Agro-mangrove forestry is beneficial to humans by producing food crops and animal products (Fig. 17.5).

Five types of agroforestry can be identified as follows:

1. Agrosilvicultural System: In this system, crops and woody perennials grow together and include improved shifting cultivation (Singh et al. 2017). It includes alley cropping and a mixture of plantation crops. This method is also called *taungya* system where food crops and trees grow together. It is also known as shelter beds and mainly practiced in the savanna where the aim is to have trees that will protect the farm belt from winds. This system is used to fight desert encroachment (Fig. 17.4).
2. Silvopastoral System: It is a combination of pastureland, livestock, and woody perennial cropping (Gomes et al. 2020). It is also called fodder bank. Here trees are grown around farmland as shrubs or pasture, and they integrate products of farmland and woody species.
3. Agrosilvopastoral System: It is a combination of woody species, agricultural crops, and livestock. This is an integrated production of foodstuff, livestock, and woody perennials that is commonly found around home farming layout.

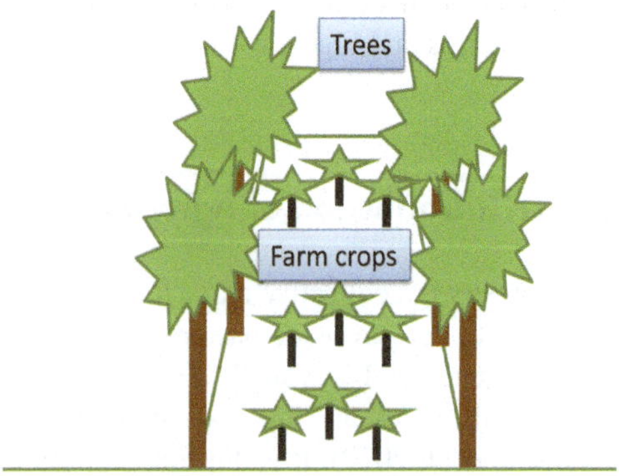

Fig. 17.4 Conceptual diagram of agroforestry practice where trees grow side by side with farms crops

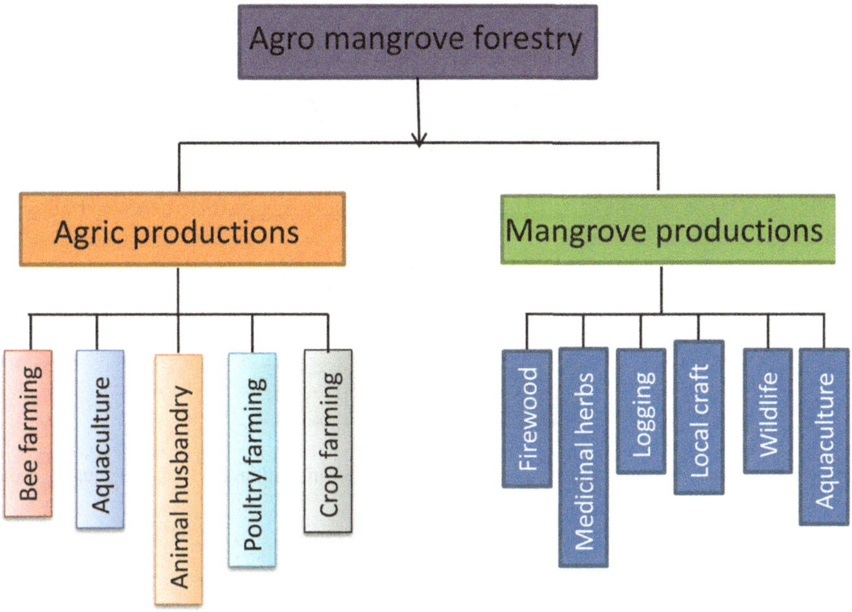

Fig. 17.5 Comparison of agriculture and mangrove forest productions in the Niger Delta

4. Entomoforestry: This is a combination of insects and trees (e.g., Shin et al. 2020). Under this system, we have (1) sericulture, production of silk and mulberry trees, and (2) apiculture, bee farming.
5. Aqua Forestry: It involves fish farming in a forest. For example, aqua-cultural practice in a mangrove forest is common in the Niger Delta area. It is a viable system practiced in other parts of the world such as Vietnam and SE Asia (Fig. 17.6).

The pond is enriched from the residue arising from the farm. There is no external input, and an important feature of this type of agroforestry is that it utilizes the entire soil profile.

Agro-Mangrove Forestry The purpose is to conserve the environment and ecosystem services (e.g., Jalloh et al. 2012). Mangrove forest is a sanctuary for biodiversity from invertebrates to vertebrates as well as from microorganisms to macroorganisms. However, because of the high salinity of its soil, most crops will not grow on mangrove soil, except there is desalination, the removal of salt from water by distillation, freezing, or ultrafiltration. But freshwater mangrove forest has low salinity that can tolerate the growth of other aquatic non-halophytic plants. This less saline environment can be used for farming activities. Similarly, brackish water environments in an estuary have low salinity levels and therefore can be used for planting agricultural crops and for raising freshwater (cartilaginous) fishes. An

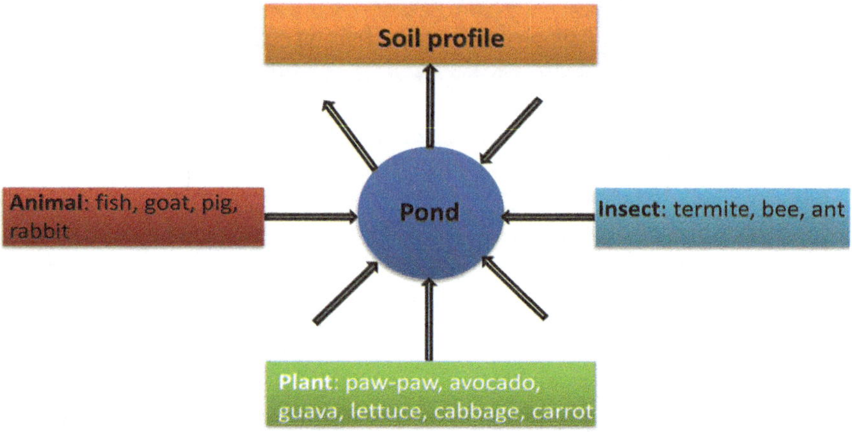

Fig. 17.6 Agro-mangrove forestry, a combination of organisms cultured on soil and aquatic media in mangrove forest

estuary is a place where a river empties into the sea. The propagation of crops in mangrove forest is a novel branch of agriculture that is already ongoing in some locations in the Niger Delta at small-scale level. There are limited studies on agro-mangrove forestry in the region. However, there are some examples of the planting of edible crops in mangrove forest found by the author. These include the growing of pineapples, coconuts, and date palms on cleared mangrove forest near a bridge connecting Tema in the Kalabari region (Fig. 17.7a) and plantain farm near a brackish river at Abam Ama, Okrika, all in the Niger Delta (Fig. 17.7b).

During a field reconnaissance visit to the plantain farms, it was found out that the plantains were growing very well. The salinity of the river was 1.59 parts per thousand. This shows that it is an estuary because it is not too salty for the plantains. The physico-chemistry of the plantain farm is shown in Table 17.2. Plantains usually grow in upland less saline and alkaline soils rich in manure. Since urbanization has taken up most farmland, some farmers had resorted to farming in mangrove forest. Coastal mangrove agriculture provides a new opportunity for farming activities in the region. This is because there are still vast amount of untouched mangrove forests that can be utilized for agroforestry. Aquaculture, animal husbandry, and poultry farms are also practiced in mangrove forest in some rural areas. Aquaculture, which is the art of growing seafood (fish, prawns, oysters, etc.) for subsistence or economic purposes, is the major farm activity that is practiced in coastal areas of the Niger Delta.

This chapter thus proposes the increased practice of agro-mangrove forestry in the Niger Delta in order to produce more agricultural crop for the ever-increasing population of people migrating into the region to search for white-collar jobs.

Fig. 17.7 Agro-mangrove forestry practice in the Niger Delta: (**a**) pineapple farm in a mangrove forest by the riverside of Tema Kalabari area and (**b**) plantain farm by the riverside in Abam Ama, Okrika

17.7.2 Control of Deforestation to Enhance Ecological Intensification

Deforestation is a process by which an area is deprived of existing natural vegetation and resources and can be brought about by a systematic felling of trees in place of industrial estate. It is also the willful removal of trees from the ground surface, which is a major destroyer of mangrove forest in the Niger Delta. Control of deforestation is a form of sustainable development that involves multiple-level private and public decision-making (Furumo and Lambin 2020). Forests are cut down in the Niger Delta region to construct and set up buildings, industrial estates, roads, and infrastructures. Another key reason why mangroves are cut is because of their use in producing firewood, which provides cooking energy for members of the local communities. Forest products are also sold to earn some income by rural dwellers.

Table 17.2 Physico-chemistry of sample area at Abam Ama Okrika, Niger Delta, Nigeria

Sample	Sample ID	Salinity (ppt)	Temp (°C)	Coordinates
Water	Ok1	1.59	25.3	N0453.318′; E006 53, 903’
Water	Ok2	1.59	25.2	N04 45.603; E007 04. 429’

The removal of trees has a negative effect on the environment because of the loss of its ecosystem services. Therefore, reduction or elimination of deforestation will enhance ecological intensification by allowing the forest to carry out its natural functions. Trees absorb carbon dioxide which helps reduce global warming, climate change, and sea level rise and the melting of the polar caps. The presence of forest also helps to stabilize the soil and landscape against flooding and erosion.

Causes of deforestation in the Niger Delta are as follows:

1. Death and Decay: Trees are living things and die due to root condition, lack of nutrients, unavailability of light (photosynthesis), and diseases. These deficiencies may predispose the trees to attacks by fungi, which may result in their death and elimination.
2. Wind: This may reduce the forest population or completely wipe it out when trees are blown down by high-velocity winds.
3. Human: They are major architect of forest destruction directly or indirectly. Anthropogenic activities that destroy forests include clearance of trees for livestock farming, shifting cultivation, infrastructural development, and selection and elimination of some species.
4. Mining and road construction.
5. Oil and gas exploration and exploitation.
6. Bush fires.

Deforestation can be controlled through the following ways:

1. Forest education: Here forestry program requires trained persons with skills and education, which will enable them to enlighten the public on the importance of forest and the negative consequences of deforestation.
2. Forest regulation: Promulgation of necessary laws to protect the forest and prevent deforestation will help to save the forest from imminent destruction. It allows forest resources to be sustainably exploited to provide necessary goods and services while still under protection. Forest regulation is thus aimed at determining the amount, frequency, and time of felling of forest trees for timber or firewood and other wood resources.
3. Forest guards: They are uniformed officials of the forestry commission that are charged with the responsibility of policing the inside and outside of the forest estate to prevent deforestation and other acts of economic sabotage.
4. Rangers: They supervise the forest guards and other uniform workers within the forest zone. Among other functions, they educate the people about the advantage of afforestation and the disadvantage of deforestation. They also check and give

out permits to timber exploiters, saw millers, and firewood producers before they cut any tree in the forest.

17.7.3 Promotion of Ecological Intensification in the Niger Delta

Ecological intensification can be promoted in the Niger Delta through the reduction in the use of chemicals in farming. Already the Niger Delta environment is filled with toxic materials from various sources of pollution due to the proliferation of oil and gas prospecting industries, which flare gas and spill crude oil into the environment. Ecology-based environmental management allows nature to take its course on remediation with the reduction of new anthropogenic impacts, and agro-biodiversity can accelerate ecological processes in plant production (Wan et al. 2020). This is because natural equilibrium involves processes that help to heal the environment from adverse human impact. This process can be promoted by allowing succession to take its course in bringing back a destroyed environment to its original state.

In the Niger Delta, mangrove stem is used to produce firewood, which is the primary source of cooking energy (Numbere 2020b). Cutting of trees for firewood has led to over 50 years of deforestation without any form of re-planting. Therefore, to restore the mangroves, a major tree planting program needs to be instituted to replace lost trees. This is because less value is placed on replanting of mangrove trees as compared to ornamental trees in the area. Mangrove propagules are usually planted after remediation of a polluted site by environmental agencies contracted to do the job. Ministries of environment in the region are mainly interested in monitoring sites where oil spills have occurred rather than carrying out remediation and tree planting. Field observations had shown that the clearing of mangrove forest for building and the cutting of trees for firewood had damaged mangrove forest more than pollution in the Niger Delta. The recommendation is that more mangrove propagules should be planted after deforestation. This will enable the forest to recuperate after some years, which will promote ecological activities.

17.8 Research and Development toward Ecological Intensification in the Niger Delta

The study and practice of ecological intensification is not too common in ecology curriculum in schools across the Niger Delta. However, increased awareness through training and education is necessary to bring this topic to the front burner. This will involve the introduction of ecological intensification as part of the course outline for biological and agricultural students in tertiary institutions. Similarly, practical classes can be scheduled to enable students to gain field experiences. Research topics can be crafted around this topic especially in agro-mangrove forestry. Some possible areas of research and development in this field may include the following topics:

1. Soil quality analysis.
2. Seedling growth.
3. Tree health monitoring.
4. Air pollution monitoring.
5. Water quality analysis.
6. Food chain dynamics.
7. Landscape fragmentation studies.
8. Impacts of industrial activities.
9. Agro-mangrove forestry studies.
10. Irrigation and water conservation.
11. Pest resurgence.
12. Insect pollination.

In the Niger Delta, oil and gas exploration activities influence agriculture and forestry in many ways. This is because farming activities in the region take place in the forest where numerous oil wells are situated. The exploration for crude oil results in pollution, degradation, and contamination, which has a ripple effect on the vegetation cover. These events negatively impact the soil quality which has necessitated the use of chemicals such as fertilizers by farmers to boost the soil quality as a way of stopping land degradation. There is a need to regularly monitor oiling activities around farmlands in the region because it is a major threat to agriculture. It is recommended that a research unit be set up to constantly collect field data which will be used to study the impact of exploration on agroforestry. In the same vein, recommendations are to be made to remediate destroyed and abandoned land so as to revert it to farming activities.

A major problem of industrialization is deforestation, which has often been mentioned as one of the biggest threats of mangrove forest globally (e.g., Barlow et al. 2016). To stop this activity and promote ecological intensification, it is recommended that a deforestation research and monitoring team (DRMT) be set up by institutions to give monthly updates on the number of trees lost monthly and the amount of land devastated by industrial activities. They will also study the patterns and extent of forest decline over the years using satellite and field data. This research team is to collaborate with local and international experts to promote alternative ways to stop or reduce the rate of deforestation in the Niger Delta region. This is because there is currently no real-time data on deforestation apart from some studies that had used GIS and Landsat data to calculate the extent of deforestation in the region (James et al. 2007; Wang et al. 2016). Most of these studies are old and don't reflect the current situation on ground. But a real-time study by research teams stationed close to the area can always provide current data, which would be used to make the best prediction of forest and farmland losses. Why these studies are significant is because environmental health is correlated with forest health (Jennings and Gaither 2015). Similarly, food scarcity is also correlated to soil conditions. Destroying farmland and eliminating forest will rather make humans the carbon sink in place of plants. This is because there will be an exchange of roles of ecosystem services between plants and humans in terms of air purification. Clean air boosts other ecological activities such as pollination by insect pollinators. In the Niger Delta, the lepidopteran families are at the brink of extinction as a result of over 50 years of sustained oil and gas exploration. The increased atmospheric pollution has killed off large populations of butterflies. In addition, loss of trees has also led to the loss of their habitat, which has made a large population of them to migrate into the cities from the forest zones. Several years ago, thousands of monarch butterflies were spotted flying into urban areas. Further, it seemed to be an annual breeding migration, but today it is believed that those daily migrations were due to the loss of their natural habitat as a result of deforestation and air pollution. These migrations were the first warning sign that were missed by researchers in the area. Now there are no more migrations and no more butterflies around the forest and urban areas. This shows the level of environmental degradation, which had necessitated the need for more effort to be put in stimulating interest in this field of ecological intensification. Pollination boosts agricultural production when insects pollinate flowers of food crops. In some places where insects are scarce, additional cost is incurred when bees are hired by farmers to pollinate their crops.

Urban ecology is also another branch of study that brings in ecological practice into cities to help reduce anthropogenic impact on the environment (Verma et al. 2020). This is a form of compensation for the loss of trees and vegetation cover as a result of the execution of urban development projects. Trees are planted around the cities and farming activities are done within city limits. Presence of trees around cities helps to purify the air and provide aesthetic sites for visitors and urban dwellers. There should also be the expansion of organic farms around the cities to boost food supply.

17.9 Conclusions and Recommendation

The Niger Delta operates an oil economy due to its huge crude oil resources and the large numbers of oil industries that operate in the region. The area is dotted with both offshore and onshore oil wells. But the hard truth is that crude oil is a nonrenewable resource that once exhausted cannot be recovered in a lifetime because it takes thousands and millions of years to form. This fact has made the call for the use of renewable resources such as solar, biomass, and geothermal energies more urgent than ever. However, the aspect that this chapter is concerned about is the damage done to the ecosystem by the activities of these oil companies. This includes the pollution of air, water, and land and the resultant effect on human health. Since the mainstay of the nation's economy is oil, there would be no reduction in exploration for more oil wells in the nearest future. This is because of the huge loss in foreign currency that may accrue from the stoppage in the sale of crude oil. More drilling and expansion of oil fields is done with increased vigor to meet up with both local and international crude oil demands. The problem is that no proper environmental impact assessment is usually done to study the trend of land conversion and land degradation as a result of oiling activities. Many farmlands have been lost to pollution leading to loss of the livelihood system of the rural dwellers. The establishment of oil industries in many oil-producing communities has made a lot of farmers abandon their farms in search of industry-based jobs. The migration of farmers and fisher folks into urban areas for white-collar jobs has also created a lot of fallow lands that are left to degrade. However, agro-based economy is the solution to hunger and poverty in Africa. This is because small-scale farmers can form partnerships among themselves to buy more land for agriculture so as to increase productions.

There should also be soil quality improvement through the administration of organic manure to soil rather than the use of chemical fertilizers to forestall negative environmental consequences. A key aspect of the ecosystem is the human factor. This is because humans can make or mar the environment. Thus, deliberate effort should be made to enable people in rural and urban communities to act as custodians of the environment. This kind of resource management is a win-win ecology, where the environment wins and human wins. Deforestation and pollution are two major disturbances of the ecosystem in the Niger Delta. Both need to be stopped to help the mangrove forest and other vegetation to recuperate. Afforestation should be encouraged to re-populate deforested areas. Waste land should also be remediated and used for agriculture to boost food production. The recovery of waste land will increase the amount of land space available for agricultural activities. Similarly, polluted water bodies should be cleaned to enable their ecosystem functions to run smoothly so as to accelerate the proliferation of phyto- and zooplankton, which are the food materials needed for the proliferation of fishes and other aquatic organisms. This will increase the fish population and increase harvest of fisher folks. Good farmland and cleaner rivers will strengthen ecological intensification. Cities too can be made greener through aggressive tree planting campaigns and establishment of crop farms, organic gardens, and pollination sites for insects and butterflies to breed. All these eco-based activities can be incorporated into town planning. Lastly, clean energy systems such as solar and biomass and geothermal should be explored in order to reduce pressure on the use of fossil fuels.

17.10 Future Perspectives and Sustainable Management

Sustainable development involves environmental protection, economic well-being, and social equity (e.g., Verma et al. 2020). The Niger Delta is an area that is rich in crude oil and other natural resources. But oil and gas exploration has led to the despoliation of the land and water making it unfit for farming and fishing activities, respectively. Land, air, and water pollution have led to the loss of livelihood opportunities. This has made people look for other well-paying jobs in the cities. The conversion of farmland and mangrove forest to construction sites is rampant in the rural areas. Young farmers abandon their farms and go to cities to look for white-collar and industrial jobs. Deforestation is the major decimator of mangrove forest as well as associated species. The loss of the forest makes the soil bare for erosion and flooding, which washes away large quantities of soil into the river. The use of chemicals in farms should be done moderately to prevent bioaccumulation within the food chain and eventually to human through food and water consumed (Table 17.2).

Future perspective of ecological intensification is the creation of an opportunity for nature to take its fill course and heal itself. This can be encouraged by promoting less use of agrochemicals and intensified use of manure and cultural means of eliminating farm pests. Industrial pollution and emission of liquid and solid waste into the environment should be stopped, while already polluted sites should be remediated. This chapter proposes the reconversion of abandoned polluted land into agricultural farms, which will benefit the society. The practice of sustainable development is important in order to maintain ecological equilibrium, which will help prevent reckless exploitation of natural resources, and to allow future generations to benefit from the common heritage of a fairly good environment (Kumar et al. 2020; Raj et al. 2021). This involves the deliberate protection of the environment because of its direct and indirect value to humanity. Some of these services include soil and land detoxification through microbial activity, air purification through carbon sequestration by plants, and pharmacological properties by medicinal plants (e.g., *Acrostichum aureum*). All aspects of the environment are unique because they play different roles. Thus, protection of nature ensures the economic well-being of humans. Agroforestry is a new tool that can convert many abandoned forests into a productive venture for members of the community. Mangrove forest, in particular, has great potential because it is the largest forest in Africa. Therefore, it is the submission of this chapter that polluted mangrove sites be remediated, rehabilitated, and converted to farms after conducting thorough soil conditioning processes (desalination, introduction of organic manure, etc.) and detoxification. All human beings both present and future have the right to a better life on a clean earth. It is therefore the duty of humans to join hands to protect it.

References

Abdullah SRS, Al-Baldawi IA, Almansoory AF, Purwanti IF, Al-Sbani NH, Sharuddin SSN (2020) Plant-assisted remediation of hydrocarbons in water and soil: application, mechanisms, challenges and opportunities. Chemosphere 247:125932

Amadu FO, Miller DC, McNamara PE (2020) Agroforestry as a pathway to agricultural yield impacts in climate-smart agriculture investments: evidence from southern Malawi. Ecol Econ 167:106443

Babatunde AO (2020) Local perspectives on food security in Nigeria's Niger Delta. The Extractive Industries and Societies. https://doi.org/10.1016/j.exis.2020.07.011

Banerjee A, Jhariya MK, Yadav DK, Raj A (2020) Environmental and sustainable development through forestry and other resources. Apple Academic Press Inc., CRC Press- a Taylor and Francis Group, US & Canada, Toronto, p 400ISBN: 9781771888110. https://doi.org/10.1201/9780429276026

Banerjee A, Meena RS, Jhariya MK, Yadav DK (2021) Agroecological footprints management for sustainable food system. Springer Nature, Singapore, p 514, eBook ISBN: 978-981-15-9496-0, Hardcover: 978-981-15-9495-3. https://doi.org/10.1007/978-981-15-9496-0

Barlow J, Lennox GD, Ferreira J, Berenguer E, Lees AC, Mac Nally R, Parry L (2016) Anthropogenic disturbance in tropical forests can double biodiversity loss from deforestation. Nature 535 (7610):144–147

Barron MG, Vivian DN, Heintz RA, Yim UH (2020) Long-term ecological impacts from oil spills: comparison of Exxon Valdez, Hebei Spirit, and Deepwater horizon. Environ Sci and Tech 54 (11):6456–6467

Barros TC, Elias F, Romano LL, Ferreira J (2020) Natural recovery of plant species diversity in secondary forests in eastern Amazonia: contributions to passive forest restoration. Brazilian J Bot 43(1):165–175

Bommarco R, Kleijn D, Potts SG (2013) Ecological intensification: harnessing ecosystem services for food security. Trends in Ecol and Evo 28(4):230–238

Chizea B, Osumah O (2015) Two sides of a coin: traditional rulership and the mitigation of nonstate security threats in Nigeria. African Security 8(2):75–95

Cunningham WP, Saigo BW (2001) Environmental science: a global concern, 6th edn. Mc Graw-Hill, New York, pp 1–646

Das S (2017) Ecological restoration and livelihood: contribution of planted mangroves as nursery and habitat for artisanal and commercial fishery. World Dev 94:492–502

Dhaliwal SS, Singh J, Taneja PK, Mandal A (2020) Remediation techniques for removal of heavy metals from the soil contaminated through different sources: a review. Environ Sci and Pollut Res 27(2):1319–1333

Dobson AP, Bradshaw AD, Baker AÁ (1997) Hopes for the future: restoration ecology and conservation biology. Science 277(5325):515–522

Dondajewska R, Gołdyn R, Kowalczewska-Madura K, Kozak A, Romanowicz-Brzozowska W, Rosińska J et al (2020) Hypertrophic lakes and the results of their restoration in Western Poland. In: Polish River basins and lakes–part II. Springer, Cham, pp 373–399

Ebenebe CI, Ibitoye OS, Amobi IM, Okpoko VO (2020) African edible insect consumption market. In: African edible insects as alternative source of food, oil, protein and bioactive components. Springer, Cham, pp 19–51

Ejiba IV, Onya SC, Adams OK (2016) Impact of oil pollution on livelihood: evidence from the Niger Delta region of Nigeria. J Sci Res and Rep 12:1–12

Elum ZA, Mopipi K, Henri-Ukoha A (2016) Oil exploitation and its socioeconomic effects on the Niger Delta region of Nigeria. Environ Sci Pollut Res 23(13):12880–11288

Extinctions BP (2013) And the contribution of deforestation to climate change. Mar Policy 40:194–204. Source: terra-i (2013) www. terra-i. org/terra-i

Folberth C, Yang H, Gaiser T, Liu J, Wang X, Williams J, Schulin R (2014) Effects of ecological and conventional agricultural intensification practices on maize yields in sub-Saharan Africa under potential climate change. Environ Res Lett 9(4):044004

Furumo PR, Lambin EF (2020) Scaling up zero-deforestation initiatives through public-private partnerships: a look inside post-conflict Colombia. Glob Environ Chang 62:102055

Gajula H, Kumar V, Vijendra PD, Rajashekar J, Sannabommaji T, Basappa G (2020) Secondary metabolites from mangrove plants and their biological activities. In: Biotechnological utilization of mangrove resources. Academic Press, London, pp 117–134

Garrigou A, Laurent C, Berthet A, Colosio C, Jas N, Daubas-Letourneux V, Lebailly P (2020) Critical review of the role of PPE in the prevention of risks related to agricultural pesticide use. Safety Sci 123:104527

Gerlak AK, Zamora-Arroyo F, Kahler HP (2013) A delta in repair: restoration, binational cooperation, and the future of the Colorado River Delta. Environment: Sci Poll Sustain Dev 55(3):29–40

Gomes FJ, Pedreira BC, Santos PM, Bosi C, Lulu J, Pedreira CG (2020) Microclimate effects on canopy characteristics of shaded palisadegrass pastures in a silvopastoral system in the Amazon biome of Central Brazil. Euro J of Agro 115:126029

Higley JM (1989) Waterbird and raptor use of the Arcata Marsh and Wildlife Sanctuary, Humboldt County, California, 1984–1986 (Doctoral dissertation, Humboldt State University)

Hobbs RJ, Harris JA (2001) Restoration ecology: repairing the earth's ecosystems in the new millennium. Restor Ecol 9(2):239–246

Holdway A (2002) The acute and chronic effects of wastes associated with offshore oil and gas production on temperate and tropical marine ecological processes. Mar Pollut Bull 44 (3):185–203

Horrocks KJ, Ward D, Suckling DM (2020) Can natural enemies of current insect pests provide biotic resistance to future pests? Agri For Entomol 22(1):20–29

Ibrahim M, Ndatsu A, Yisa KM (2020) Effects of flood on crop farmers in riverine area in Niger state Nigeria. ATBU J Sci Tech Edu 8(2):311–319

Ikehi ME, Onu FM, Ifeanyieze FO, Paradang PS (2014) Farming families and climate change issues in Niger Delta region of Nigeria: extent of impact and adaptation strategies. Agri Sci 5(12):1140

Illius AW, O'Connor TG (1999) On the relevance of nonequilibrium concepts to arid and semiarid grazing systems. Ecol Appl 9(3):798–813

Iyagba AG, Anyanwu SO (2012) Problems and prospects of cassava production in Rivers state, Nigeria: a case study of Oyigbo local government area. Agri Biol J North Ame 3(7):296–301

Jalloh A, Roy-Macauley H, Sereme P (2012) Major agro-ecosystems of west and Central Africa: brief description, species richness, management, environmental limitations and concerns. Agri Eco Environ 157:5–16

James GK, Adegoke JO, Saba E, Nwilo P, Akinyede J (2007) Satellite-based assessment of the extent and changes in the mangrove ecosystem of the Niger Delta. Mar Geod 30:249–267. https://doi.org/10.1080/01490410701438224

Jennings V, Gaither CJ (2015) Approaching environmental health disparities and green spaces: an ecosystem services perspective. Int J Environ Res Pub Health 12(2):1952–1968

Jhariya MK, Yadav DK, Banerjee A (2018a) Plant mediated transformation and habitat restoration: phytoremediation an eco-friendly approach. In: Gautam A, Pathak C (eds) Metallic contamination and its toxicity. Daya Publishing House, A Division of Astral International Pvt. Ltd, New Delhi, pp 231–247. ISBN: 9789351248880

Jhariya MK, Banerjee A, Yadav DK, Raj A (2018b) Leguminous trees an innovative tool for soil sustainability. In: Meena RS, Das A, Yadav GS, Lal R (eds) Legumes for soil health and sustainable management. Springer, Singapore, pp 315–345. ISBN 978-981-13-0253-4 (eBook), ISBN: 978-981-13-0252-7 (Hardcover). https://doi.org/10.1007/978-981-13-0253-4_10

Jhariya MK, Banerjee A, Meena RS, Yadav DK (2019a) Sustainable agriculture, Forest and environmental management. Springer Nature Singapore Pte Ltd., Singapore, p 606. https://doi.org/10.1007/978-981-13-6830-1. eISBN: 978-981-13-6830-1, Hardcover ISBN: 978-981-13-6829-5

Jhariya MK, Yadav DK, Banerjee A (2019b) Agroforestry and climate change: issues and challenges. Apple Academic Press Inc., CRC Press- a Taylor and Francis Group, US & Canada, Toronto. ISBN: 978-1-77188-790-8 (Hardcover), 978-0-42957-274-8 (E-book), p 335. https://doi.org/10.1201/9780429057274

Jhariya MK, Yadav DK, Banerjee A, Raj A, Meena RS (2019d) Sustainable forestry under changing climate. In: Jhariya MK, Banerjee A, Meena RS, Yadav DK (eds) Sustainable agriculture, Forest and environmental management. Springer Nature, Singapore, pp 285–326. eISBN: 978-981-13-6830-1, Hardcover ISBN: 978-981-13-6829-5. Pp 606

Jhariya MK, Banerjee A, Meena RS, Yadav DK (2019c) Agriculture, forestry and environmental sustainability- a way forward. In: Jhariya MK, Banerjee A, Meena RS, Yadav DK (eds) Sustainable agriculture, Forest and environmental management. Springer Nature, Singapore, pp 1–29. eISBN: 978-981-13-6830-1, Hardcover ISBN: 978-981-13-6829-5. Pp 606

Jhariya MK, Meena RS, Banerjee A (2021a) Ecological intensification of natural resources for sustainable agriculture. Springer Nature, Singapore. eISBN: 978-981-334-203-3, Hardcover ISBN: 978-981-334-206-6. https://doi.org/10.1007/978-981-33-4203-3

Jhariya MK, Meena RS, Banerjee A (2021b) Ecological intensification of natural resources towards sustainable productive system. In: Ecological intensification of natural resources for sustainable agriculture. Springer, Singapore. eBook ISBN 978-981-334-203-3. Hardcover ISBN: 978-981-334-202-6. https://doi.org/10.1007/978-981-33-4203-3_1

Jhariya MK, Singh L (2020) Herbaceous diversity and biomass under different fire regimes in a seasonally dry forest ecosystem. Environ Dev Sustain 22:1–19. https://doi.org/10.1007/s10668-020-00892-x

Jhariya MK, Singh L (2021) Effect of fire severity on soil properties in a seasonally dry forest ecosystem of Central India. Int J Environ Sci Technol. https://doi.org/10.1007/s13762-020-03062-8

Jin G, Chen K, Wang P, Guo B, Dong Y, Yang J (2019) Trade-offs in land-use competition and sustainable land development in the North China plain. Technol Forecast Soc Chang 141:36–46

Jones KR, Venter O, Fuller RA, Allan JR, Maxwell SL, Negret PJ, Watson JE (2018) One-third of global protected land is under intense human pressure. Science 360(6390):788–791

Kathiresan K (2020) Bioprospecting potential of mangrove resources. In: Biotechnological utilization of mangrove resources. Academic Press, London, pp 225–241

Kergna AO, Nientao A, Ongoiba B, Sidibe M (2020) Mechanization and skill development for productivity growth, employment and value addition. Insights from Mali (p. 92). FARA Research Report 5 (18)

Khan N, Jhariya MK, Raj A, Banerjee A, Meena RS (2021a) Soil carbon stock and sequestration: implications for climate change adaptation and mitigation. In: Jhariya MK, Meena RS, Banerjee A (eds) Ecological intensification of natural resources for sustainable agriculture. Springer, Singapore. eBook ISBN 978-981-334-203-3. Hardcover ISBN: 978-981-334-202-6. https://doi.org/10.1007/978-981-33-4203-3_13

Khan N, Jhariya MK, Raj A, Banerjee A, Meena RS (2021b) Eco-designing for sustainability. In: Jhariya MK, Meena RS, Banerjee A (eds) Ecological intensification of natural resources for sustainable agriculture. Springer, Singapore. eBook ISBN 978-981-334-203-3. Hardcover ISBN: 978-981-334-202-6. https://doi.org/10.1007/978-981-33-4203-3_16

Khan N, Jhariya MK, Yadav DK, Banerjee A (2020a) Herbaceous dynamics and CO_2 mitigation in an urban setup- a case study from Chhattisgarh, India. Environ Sci Pollut Res 27(3):2881–2897. https://doi.org/10.1007/s11356-019-07182-8

Khan N, Jhariya MK, Yadav DK, Banerjee A (2020b) Structure, diversity and ecological function of shrub species in an urban setup of Sarguja, Chhattisgarh, India. Environ Sci Pollut Res 27 (5):5418–5432. https://doi.org/10.1007/s11356-019-07172-w

Kumar S, Meena RS, Datta R, Verma SK, Yadav GS, Pradhan GS, Molaei A, Rahman GKMM, Mashuk HA (2020a) Legumes for carbon and nitrogen cycling: an organic approach. In: Carbon and nitrogen cycling in soil. Springer, Singapore. https://doi.org/10.1007/978-981-13-7264-3_10.337-375

Kumar S, Meena RS, Jhariya MK (2020) Resources use efficiency in agriculture. Springer Nature, Singapore, p 760eBook ISBN: 978-981-15-6953-1, Hardcover: 978-981-15-6952-4. https://doi.org/10.1007/978-981-15-6953-1

Leo M (2012) Vision 20: 2020 initiative and the Nigerian oil and gas industry: an empirical analysis of the environmental dimension. In: Nigeria annual international conference and exhibition. Society of petroleum engineers

de Lima DO, Crouzeilles R, Vieira MV (2020) Integrating strict protection and sustainable use areas to preserve the Brazilian Pampa biome through conservation planning. Land Use Policy 99:104836

Meena RS, Kumar S, Datta R, Lal R, Vijaykumar V, Brtnicky M, Sharma MP, Yadav GS, Jhariya MK, Jangir CK, Pathan SI, Dokulilova T, Pecina V, Marfo TD (2020b) Impact of agrochemicals on soil microbiota and management: a review. Land (MDPI) 9(2):34. https://doi.org/10.3390/land9020034

Meena RS, Lal R, Yadav GS (2020) Long term impacts of topsoil depth and amendments on soil physical and hydrological properties of an Alfisol in Central Ohio, USA. Geoderma 363:1141164. https://doi.org/10.1016/j.geoderma.2019.114164

Meena RS, Lal R, Yadav GS (2020a) Long-term impact of topsoil depth and amendments on carbon and nitrogen budgets in the surface layer of an Alfisol in Central Ohio. Catena 2020194:104752. https://doi.org/10.1016/j.catena.2020.104752

Mori AS (2011) Ecosystem management based on natural disturbances: hierarchical context and non-equilibrium paradigm. J Appl Ecol 48(2):280–292

Muchane MN, Sileshi GW, Gripenberg S, Jonsson M, Pumariño L, Barrios E (2020) Agroforestry boosts soil health in the humid and sub-humid tropics: a meta-analysis. Agri Econ Environ 295:106899

Naidoo G (2016) Mangrove propagule size and oil contamination effects: does size matter? Mar Pollut Bull 110(1):362–370

Niemeyer J, Barros FS, Silva DS, Crouzeilles R, Vale MM (2020) Planning forest restoration within private land holdings with conservation co-benefits at the landscape scale. Sci Total Environ 717:135262

Nkakini SO, Ayotamuno MJ, Ogaji SOT, Probert SD (2006) Farm mechanization leading to more effective energy-utilizations for cassava and yam cultivations in Rivers state, Nigeria. Appl Energy 83(12):1317–1325

Numbere AO (2018) The impact of oil and gas exploration: invasive nypa palm species and urbanization on mangroves in the Niger River Delta, Nigeria. In: Threats to mangrove forests. Springer, Cham, pp 247–266

Numbere AO (2019a) Bioaccumulation of total hydrocarbon and heavy metals in body parts of the west African red mangrove crab (*Goniopsis pelii*) in the Niger Delta, Nigeria. Int Lett Nat Sci 75:1–12

Numbere AO (2019b) Bioaccumulation of Total hydrocarbon content by three mangrove species (*Rhizophora, Laguncularia, Avicennia*) in the Niger Delta, Nigeria. J Pet Environ Biotechnol 10 (1):1–6

Numbere AO (2019c) Municipal solid waste disposal in mangrove forest: environmental implication and management strategies in the Niger Delta, Nigeria. In: Municipal solid waste management. IntechOpen, London

Numbere AO (2020a) Analysis of total hydrocarbon and heavy metal accumulation in sediment, water and associated organisms of mangrove ecosystem in the Niger Delta. Indian JSci Tech 13 (26):2678–2688

Numbere AO (2020b) Utilization of the mangrove Forest for sustainable renewable energy production. Progress Petro Sci 3(3):324–329

Numbere AO, Camilo GR (2017) Mangrove leaf litter decomposition under mangrove forest stands with different levels of pollution in the Niger River Delta, Nigeria. Afr J Ecol 55(2):162–167

Numbere AO, Camilo GR (2018) Structural characteristics, above-ground biomass and productivity of mangrove forest situated in areas with different levels of pollution in the Niger Delta, Nigeria. Afr J Ecol 56(4):917–927

Numbere AO, Camilo GR (2019) Mangrove leaf herbivory along a hydrocarbon pollution gradient in a mangrove Forest (*Rhizophora racemosa*) in the Niger River Delta, Nigeria. J Pet Environ Biotechnol 10(2):1–8

Nwajiaku-Dahou K (2012) The political economy of oil and 'rebellion' in Nigeria's Niger Delta. Rev of Afr Pol Eco 39(132):295–313

Nyström M, Folke C, Moberg F (2000) Coral reef disturbance and resilience in a human-dominated environment. Trends Ecol Evol 15(10):413–417

Obeka C, Numbere AO (2020) Heavy metal concentration and public health risk in consuming *Sardinella maderensis* (sardine), *Sarotherodon melanotheron* (Tilapia), and *Liza falicipinis* (mullet) harvested from Bonny River, Nigeria. J Ocean Mar Sci 11(1):1–10

Oke OA, Kolawole RO, Akinsola OA, Olajide-Taiwo LO, Oladigbolu AA, Ajayi EO (2020) The first effort at adopting integrated Pest management IPM to contain the infestation of the *Tuta absoluta* in Nigeria. In: Sustainable Management of Invasive Pests in Africa. Springer, Cham, pp 57–64

Okoh E, Yelebe ZR, Oruabena B, Nelson ES, Indiamaowei OP (2020) Clean-up of crude oil-contaminated soils: bioremediation option. Int J Environ Sci Tech 17(2):1185–1198

Onwuka EC (2005) Oil extraction, environmental degradation and poverty in the Niger Delta region of Nigeria: a viewpoint. Int J Environ Stud 62(6):655–662

Onyena AP, Sam K (2020) A review of the threat of oil exploitation to mangrove ecosystem: Insights from Niger Delta, Nigeria. Glob Ecol Conserve 22:e00961

Oyewole SO, Sennuga SO (2020) Factors influencing sustainable agricultural practices among smallholder farmers in Ogun state of Nigeria. Asian J Adv Agric Res 14:17–24

Painkra GP, Bhagat PK, Jhariya MK, Yadav DK (2016) Beekeeping for poverty alleviation and livelihood security in Chhattisgarh, India. In: Narain S, Rawat SK (eds) Innovative Technology for Sustainable Agriculture Development. Biotech Books, New Delhi, pp 429–453. ISBN: 978-81-7622-375-1

Paiva PFPR, Ruivo MDLP, da Silva Junior OM, Maciel MDNM, Braga TGM, de Andrade MMN et al (2020) Deforestation in protect areas in the Amazon: a threat to biodiversity. Biodivers Conserv 29(1):19–38

Raj A, Jhariya MK, Banerjee A, Yadav DK, Meena RS (2019b) Soil for sustainable environment and ecosystems management. In: Jhariya MK, Banerjee A, Meena RS, Yadav DK (eds) Sustainable agriculture, Forest and environmental management. Springer Nature, Singapore, pp 189–221. eISBN: 978-981-13-6830-1, Hardcover ISBN: 978-981-13-6829-5. Pp. 606. https://doi.org/10.1007/978-981-13-6830-1

Raj A, Jhariya MK, Bargali SS (2018b) Climate smart agriculture and carbon sequestration. In: Pandey CB, Gaur MK, Goyal RK (eds) Climate change and agroforestry: adaptation mitigation and livelihood security. New India Publishing Agency (NIPA), New Delhi, pp 1–19. ISBN: 9789-386546067

Raj A, Jhariya MK, Harne SS (2018a) Threats to biodiversity and conservation strategies. In: Sood KK, Mahajan V (eds) Forests, climate change and biodiversity. Kalyani Publisher, New Delhi, p 381. Pp. 304-320

Raj A, Jhariya MK, Khan N, Banerjee A, Meena RS (2021) Ecological intensification for sustainable development. In: Ecological intensification of natural resources for sustainable agriculture. Springer, Singapore, eBook ISBN 978-981-334-203-3. Hardcover ISBN: 978-981-334-202-6. https://doi.org/10.1007/978-981-33-4203-3_5

Raj A, Jhariya MK, Yadav DK, Banerjee A (2020) Climate change and agroforestry systems: adaptation and mitigation strategies. Apple Academic Press Inc., CRC Press- a Taylor and Francis Group, US & Canada, Burlington, p 383. https://doi.org/10.1201/9780429286759. ISBN: 9781771888226

Raj A, Jhariya MK, Yadav DK, Banerjee A, Meena RS (2019a) Agroforestry: a holistic approach for agricultural sustainability. In: Jhariya MK, Banerjee A, Meena RS, Yadav DK (eds) Sustainable agriculture, Forest and environmental management. Springer Nature, Singapore, eISBN: 978-981-13-6830-1, Hardcover ISBN: 978-981-13-6829-5. Pp. 606, pp 101–131. https://doi.org/10.1007/978-981-13-6830-1.

Reid J, Rout M (2020) The implementation of ecosystem-based management in New Zealand–a Māori perspective. Mar Policy 117:103889

Ringim AS, Sulaiman IM, Lyakurwa JV (2016) Implementation of integrated coastal zone management approach in the Niger Delta, Nigeria: a review. Int Res J Environ Sci Stud 1(3):43–55

Schroeder WW (2000). Disturbances: their role in shaping coastal environments. In: Rodriguez GR, Brebbia CA, Perez-Martell E (eds) Environmental coastal regions III. WIT Press, Southampton. pp 431–440

Shin S, Soe KT, Lee H, Kim TH, Lee S, Park MS (2020) A systematic map of agroforestry research focusing on ecosystem services in the Asia-pacific region. Forests 11(4):368

Singh NR, Arunachalam A, Dobriyal MJ, Bhusaraand JB, Gunaga RP (2017) Crop biomass and yield patterns of dominant agro forestry systems of Navsari district, Gujarat, India. Indian J Agrofor 19(2):72–78

Singh NR, Jhariya MK (2016) Agroforestry and Agrihorticulture for higher income and resource conservation. In: Narain S, Rawat SK (eds) Innovative Technology for Sustainable Agriculture Development. Biotech Books, New Delhi, pp 125–145. ISBN: 978-81-7622-375-1

Slocombe DS (1993) Implementing ecosystem-based management. Bio Sci 43(9):612–622

Tarjuelo R, Margalida A, Mougeot F (2020) Changing the fallow paradigm: a win–win strategy for the post-2020 common agricultural policy to halt farmland bird declines. J Appl Ecol 57 (3):642–649

Tittonell P (2014) Ecological intensification of agriculture—sustainable by nature. Curr Opin Environ Sust 8:53–61

Tittonell P, Giller KE (2013) When yield gaps are poverty traps: the paradigm of ecological intensification in African smallholder agriculture. Field Crops Res 143:76–90

Umeh OJ, Nwachukwu I (2019) Behavioural approaches of rural women farmers to mitigation and adaptation measures of climate change in Abia state, Nigeria. In: University initiatives in climate change mitigation and adaptation. Springer, Cham, pp 111–129

Venter O, Fuller RA, Segan DB, Carwardine J, Brooks T, Butchart SH, Possingham HP (2014) Targeting global protected area expansion for imperiled biodiversity. PLoS Biol 12(6): e1001891

Verma P, Singh R, Singh P, Raghubanshi AS (2020) Urban ecology–current state of research and concepts. In: Urban ecology. Elsevier, Amsterdam, pp 3–16

Villa PM, Martins SV, Rodrigues AC, Safar NVH, Bonilla MAC, Ali A (2019) Testing species abundance distribution models in tropical forest successions: implications for fine-scale passive restoration. Ecol Engineering 135:28–35

Wan NF, Su H, Cavalieri A, Brack B, Wang JY, Weiner J, Jiang JX (2020) Multispecies co-culture promotes ecological intensification of vegetable production. J Clean Prod 257:120851

Wang P, Numbere AO, Camilo GR (2016) Long term changes in mangrove landscape of the Niger River Delta, Nigeria. Am J Environ Sci 12(3):248–259

Wang W, Fu H, Lee SY, Fan H, Wang M (2020) Can strict protection stop the decline of mangrove ecosystems in China? From rapid destruction to rampant degradation. Forests 11(1):55

Climatic Risks on Fruit Quality, Health, and Livelihoods: A Nigerian Case of Rural Women in Fruit Farming Business

18

Angela Oyilieze Akanwa and Uchechukwu B. Okoli

Abstract

In recent times, the production of fruits and vegetable products across the globe has gained attention due to its added nutritional value to human health, food security, and livelihood. However, climate volatility has impacted the agricultural sector in Africa with increasing destructive effects of temperature rise, droughts, and floods that promote the growth of pests and diseases. This has placed the fruit's/crop's health, food security, and rural livelihoods at great jeopardy. Since majority of the rural women farmers occupy about 40–50% of smallholder farms in sub-Saharan Africa making fruit farming their major occupation and source of livelihood, hence, their farm products have become primary targets of climate crisis. This present chapter examined the climatic risks on fruit quality, health, and livelihoods using three Nigerian communities as a case study. It investigated rural small-scale women fruit farmers/sellers and the common fruits sold in local markets in Ideato North Local Government in Imo State, Nigeria. A mixed method approach was adopted using qualitative and quantitative methods followed by laboratory analysis to determine and quantify the concentration levels of heavy metals and pesticides in sampled fruits (pineapple, orange, and banana) in each community (Uruala, Akokwa, and Obodoukwu). Findings from the study showed that (1) about 114 women were largely involved in fruit farming and sales in the study area; (2) fruit farming business is a viable source of livelihood for these rural women with which they support their families financially; (3) climate change has influenced warmer temperatures during dry seasons that has increased heat, pest, and diseases; (4) post-harvest/storage facilities are inefficient thereby threatening the quality of fruits sold at the local markets and

A. O. Akanwa (✉) · U. B. Okoli
Department of Environmental Management, Faculty of Environmental Sciences, Chukwuemeka Odumegwu Ojukwu University, Uli, Anambra, Nigeria

M. K. Jhariya et al. (eds.), *Sustainable Intensification for Agroecosystem Services and Management*, https://doi.org/10.1007/978-981-16-3207-5_18

placing rural women livelihoods at risks; and (5) the analyzed sample fruits confirmed that 7 out of 15 toxic chemicals were detected in fruits and pesticides (aldrin, endosulfan, DDD, and HCB) and heavy metals (lead, nickel, and cadmium) were above WHO/FAO standard, hence placing community health at risk.

This chapter provided the rationale for agribusiness organizations to authenticate the efforts of female rural fruit farmers/sellers and their economic contribution to both their families and rural development. Also, there is a need to create a support system for community-based research projects to invest in their business so as to increase the market value while combating the numerous challenges these women face such as climate risks, poor post-harvest/storage facilities, low technology, poor education, lack of improved species, lands, and time. Finally, there is a need to boost rural fruit farming for maximal food production, nutrition, and health stability. It is expedient to note that this study does not only address local small-scale women fruit farmers/sellers in Nigeria. However, its findings are generalizable to sub-Saharan African and Asian economies where large volumes of women livelihoods spring from diverse agricultural activities and particularly their silent contributions to fruit agribusiness which promotes human health and well-being.

Keywords

Climate change · Fruit quality · Heavy metals · Health risk · Livelihoods · Pesticides · Women fruit farmers/sellers

Abbreviations

CBPR	Community-based participatory research
CCSP	Climate Change Science Program
CO_2	Carbon dioxide
FAO	Forest and Agriculture Organization
FMEAN	Federal Ministry of Environment Abuja, Nigeria
GDP	Gross domestic product
GHG	Greenhouse gases
IPCC	Intergovernmental Panel on Climate Change
MMT	Million metric tons
NEEDS	National Environmental, Economic, and Development Study
NIMET	Nigerian Meteorological Agency
NOAA	US National Oceanic and Atmospheric Administration
NPCC	National Policy on Climate Change
PACJA	Pan African Climate Justice Alliance
UNSCN	United Nations System Standing Committee on Nutrition
WHO	World Health Organization

18.1 Introduction

Africa is spread over a large regional space with diverse natural resources, human population, fast-growing cities, and more importantly vast arable lands that support agricultural production. In Nigeria, agriculture was heavily practiced as the major foreign exchange (cash crops and citrus fruits) long before the nation started crude oil exploration from 1958 at Oloibiri, Bayelsa State. Over the years, agriculture has employed about 70% of the Nigerian population much more than any sector and made an additional 40% to the Nigerian gross domestic product (GDP) (NEEDS 2010).

In this present study, the term "fruit quality" denotes the attention given to the presence of toxicants in fruits and their possible risks on human health when consumed. The health safety of fresh fruits is greatly impacted by climatic conditions during production, especially the handling procedures during harvest, post-harvest, and storage, hence determining the health status of the fresh fruit (Poonam and Jhariya 2017; Elliot 2019). Farming fruits/fruit tree is an ancient trademark that has existed side by side with farming other root crops and since humans and animals have become dependent on fruit consumption (Elliot 2019). Common fresh fruits produced in Nigeria include mango (*Mangifera indica*), pineapple (*Ananas comosus*), banana (*Musa* sp.), Orange (*Citrus sinensis*), and guava (*Psidium guajava*), among others. However, aside from its medicinal, health, nutritional, and environmental benefits, they are also employed in the manufacture of cosmetics, processed foods, and other household needs.

These benefits have triggered its demand and consumption levels globally. In recent times, it has gained improved technological and scientific approaches in the Global North regions that have made it a competitive venture in agribusiness. In 2017, China was the highest producer of fruits/vegetables at 554.29 million metric tons (MMT). This was followed by India which produced 127.14 MMT, while Nigeria occupied the sixth position and the major producer in Africa with about 16.41 MMT of vegetable crops (Shahbandeh 2020). However, critical views have emerged over the years with studies questioning the health risks associated with fruit quality and its consumption safety (D'mello 2003).

According to the Fourth National Climate Assessment (2017), climate change has been responsible for global rising temperatures that have resulted in extreme heat, drought, wildfires, floods, and other weather disasters. More importantly, rising temperatures escalate the growth and spread of microorganisms causing diseases that increasingly compromise the quality of food/fruit products, and hence, food security and livelihoods are at risk in agrarian rural areas in Nigeria.

It is unfortunate that developing countries like Nigeria are grossly affected by global warming being a heavily dominant agrarian and natural resource-extract region (Akanwa and Joe-Ikechebelu 2019). Basically, African agricultural sector is mostly at the subsistence level with small-scale farms occupying a larger portion of the sector. This ranges between 15 m and 30 m rural farmers that produce over 90% of the nation's food crops worth almost a hundred billion dollars though these farmers at their small-scale levels own farms barely above one hectare of land for their households (Mgbenka et al. 2016). The situation is even more challenging in Nigeria since females make up over 70% of the rural farmers, in addition to the fact that over 60% of the indigenous households in Nigeria are catered for with finance sourced from sales of farm products by these female farmers (Igbine 2018).

In addition, changing climate has influenced erratic farming seasons making it difficult for farmers to keep up with growing and harvest seasons (IPCC Land Use Report 2019; Nigerian Meteorological Agency (NIMET) 2018; Kumar et al. 2020). This aids deterioration of food crops/fruits during harvest and storage periods with rapid infestation of pest, diseases, and invasion of toxic substances (Sobukola et al. 2010). Hence, a hotter weather in tropical climate escalates increased carbon dioxide (CO_2) levels, thereby affecting food production levels and its edible condition while promoting health risks due to its impaired quality (USGCRP 2014; Khan et al. 2020a, b, 2021a, b).

Together, fruit and vegetable quality can become affected by toxicants and heavy metals promoting health risks. This has become a worldwide health concern since fruits can be locally grown or traded between countries. In response to the increasing health risks from consumed fruits and vegetables, there is a clarion call for research studies to assess the quality of locally farmed fruit and vegetables (Slavin and Lloyd 2012), since contaminated fruits contain toxic substances which are considered to be a worldwide threat being responsible for environmental contamination due to their high toxicity and persistency in the environment (Tonkinwise 2015). Notably, the increasing dangers of a global vulnerable climate have aided the invasion of toxic atmospheric substances, thus the planetary health and well-being based on extreme concentration levels of these substances.

It is noted that heavy metals can have divergent reactions at low and high temperatures. At lower temperatures, they can act as micronutrients since they are at lower concentrations levels; unfortunately, they become toxic under increased concentration levels when consumed in foods. Toxic substances such as heavy metals can be present in fruits' body parts and be consumed since it serves as variable sources of nutritional value to human. However, their compromised quality may harbor accumulated poisonous substances at high concentration levels as a consequence of environmental and hygiene factors during harvest and storage (Antisari et al. 2015). Consequently, the increased bio-toxic implications of these metals depend on a wide range of concentration and oxidation levels of heavy metals along with their inherent makeup, derivations, and means of accumulation in food especially in regions where climate vulnerability and changes remain persistent (Antisari et al. 2015; Kumar et al. 2020a; Banerjee et al. 2018).

Clearly, climate change makes farming risky for rural and impoverished farmers who are domiciled in local communities particularly rural women farmers who play vital roles to ensure food security (Altieri Miguel and Koohafkan 2008). Unfortunately, rural women involved in the farming and sale of fruits and vegetables are mostly affected by the damage of changing climate on food/fruits as this threatens the quality (Brown et al. 2015; Poonam and Jhariya 2019; Banerjee et al. 2020, 2021) and prices and can impact community health and livelihoods and endanger food security as well (USGCRP 2014; Sobukola et al. 2010).

In addition, climate change, food quality, health risks, poverty, and livelihood problems have been identified and increasingly characterized as "bad problems" (Catrien et al. 2019; Meena et al. 2020b). This study suggested a CBRA (community-based research approach) as a conceptual framework to address these prevalent interlinked wicked issues. ACBRA is an advanced intervention approach which is necessary to firmly address "bad" issues since they are progressive, interlinked, and intractable showing resistance to solutions (Tonkinwise 2015). Global warming is a bad problem that has been declared a global emergency, and it cuts across socioeconomic and ecological complexities affecting the core fabric of the society with prolonged and more damaging consequences (Churchman 1967; Jhariya et al. 2019a, b; Raj et al. 2020, 2021).

From the foregoing, there is a need for policy regulation, adaptive governance (Gunderson and Light 2006), enforcement, and rural intervention toward poor and vulnerable Nigerian women fruit farmers/sellers who are faced with myriad limitations in agriculture. Hence, this study investigated the climatic risks on fruit quality, health, and livelihoods in rural small-scale women fruit farmers/sellers in Ideato North Local Government in Imo State, Nigeria.

18.2 Global Climatic Risks with Reference to Nigeria

Scientific evidences and documented proofs are conclusive that global warming is changing the world's climates. This is accompanied by persistent damaging consequences driven by extreme weather events that are aggravating serious environmental and developmental issues such as food insecurity, threatened livelihoods, and conflicts among others in the Global South region (Akanwa and Joe-Ikechebelu 2019).

IPCC (2007) revealed that anthropogenic actions particularly the mining of crude oil, industrialization, and deforestation have triggered global warming. Globally, the last century has shown that human activities have altered mean temperature to about 0.6 °C, sea level by 15–20 cm, and high rainfall over land by 2%; consequently, the twenty-first century will definitely bring more fatal occurrences due to human actions that change the climate (IPCC 2007). PACJA (Pan African Climate Justice Alliance) (2009) informed that before the industrial expansion period, human actions have initially increased the temperature by 0.8 °C and if global resilient measures are not taken, the earth's average temperature would hit 2 °C by 2060.

The levels of changes experienced are consequent upon the level of greenhouse gases (GHG) released since IPCC (2007) report also projected that the world's mean temperature will increase from 1.1 to 5.4 °C by 2100. Unfortunately, this will result in increases and changes in overall rainfall patterns and timing. Obviously, from all indications, it is expected globally that there would be quick levels in our changing climate in the twenty-first century compared to that which has been experienced in the last 10 millennia (CCSP Final Report 2008).

Nigeria has a rich ecological diversity visible in its fast-growing population of >200.96 million inhabitants, a geographical area of 923,768 km^{-2}, 36 autonomous states, >500 different ethnic groups, and varied mineral resources (Nigerian Population Review 2019; Hughes et al. 1997). This reflects the nation's highly variable climatic conditions, which makes it an epitome of sub-Saharan conditions.

The African continent is experiencing varied temperature increase and observed climatic changes. Sylla et al. (2016) estimated that persistent and unpredictable climate changes ranging from 1.5 to 6.5 °C will affect rainfall intensity in West Africa. These climatic changes will escalate diverse impacts in West Africa, especially Nigeria. These will promote negative impacts combined with their vulnerable geographical location, low-income and institutional capacities, and its livelihood that is hugely sourced from natural resources. The levels of climatic impact are influenced by industrial productivity levels surrounding the use of fossil fuels, GHG emissions, and mean global temperature (NEEDS 2010).

The study by PACJA (Pan African Climate Justice Alliance) (2009) estimated that economically it would cost Africa by 2040 an average global temperature of 1.5 °C and averagely a total expense of 1.7% of the continent's GDP. Unfortunately, the predominant negative impact of global warming in sub-Saharan Africa is its agriculture sector which is the continent's major economic activity. The agriculture systems have become vulnerable to unpredictable, high-intensity, and erratic rainfall patterns associated with severe temperature changes (Ching et al. 2011). Hence, food production and sustenance of marginalized and vulnerable groups have become threatened (Williams et al. 2017).

Historically, changing climate in Nigeria was observed with scientific evidence of global warming with rise in temperature levels since 1901. The average temperature for over a century was examined from 1901 to 2005 which revealed that temperature for over a century ranged hit 1.1 °C and an average of 26.6 °C (Odjugo 2010a, b). This explains an increase in the world average temperature of 0.74 °C since 1860. Obviously, this happened to be the initial period that determines the beginning of scientific temperature measurement and record in Nigeria (IPCC 2007). The elongated temperature rise indicated a rise in temperature in the country examined from about 26.2 °C in the year 1951 to about 27.6 °C over the years. Further, the Nigerian Meteorological Agency (NIMET) (2018) authenticated the fact that Nigeria has experienced changing climate with confirmed records for 79 years showing elongated temperature rises and climate change.

The average temperature rise in Nigeria by 1.4–1.9 °C has recorded varying effects in different zones in north and southern Nigeria. The low and high precipitation spread has made certain areas highly desertified and others increasingly flooded. Northern cities such as Sokoto, Kano, Nguru, and Maiduguri located in semi-arid regions have been vulnerable to drought, high temperatures, and desertification. However, these northern cities are experiencing increasing temperatures and decreasing rainfall pattern (Odjugo 2010a, b). Also, southeast and southwest zones have become highly flooded due to increasing rainfall and rising temperatures (Akanwa and Ezeomedo 2018).

The climatic risks have further aggravated the agricultural production, livestock, and water systems in these Nigerian zones. Also, most Nigerian cities located along River banks are vulnerable to global warming with observable changes in climate where excessive flooding and overflow of rivers. In addition, there were predictions that have been made that global warming and the resultant effects of a changed climate will cause huge losses in Nigeria's GDP to about 30% in 2050 with projected cost of US$100–460 billion dollars (DFID 2009). This calls for immediate sustainable adaptation and mitigation strategies that should be able to eradicate an estimated loss of 2–11% of the nation's GDP in 2020.

18.3 Nigeria Contribution to Climate Change, Response, and Resolution

In Nigeria, the economic sector is heavily reliant on fossil fuel and agricultural activities, and these contribute immensely to GHG emissions that culminate in climate change problems. Both sectors (fossil fuel and agriculture) are major emitters of carbon. This makes Nigeria a minimal contributor to global carbon emissions and therefore a target of climate change consequences (Akanwa and Joe-Ikechebelu 2019). NEEDS (2010) indicated that Nigeria contributes to the release of GHG minimally though the culminated GHG emissions (in CO_2 is equivalent) for three major GHG were CO_2, CH_4, and N_2O. This covered five major sectors which were energy, industry, agriculture, land use change and forestry—LUCF—and waste. This was totaled to about 330,946 Gg CO_2e in 2000.

The data determined by Obioh (2003) was employed as a starting point for comparison in the estimation of GHG emission for the year 2000 in Nigeria. Further, the IPCC (2007) procedures were additionally applied in these estimates as evident in Table 18.1. Here, the emitted GHG were unfairly spread across the three gases whereby the total CO_2 amounted to 214,523 Gg (65%) of the national GHG emissions. Also, methane (CH_4) was 109,319 Gg CO_2e (33%), and nitrous oxide (N_2O) totaled 7104 Gg (CO_2e) (2%). For the energy sector, the total CO_2 emission was 108,000 Gg CO_2 in 1995, and this was then predicted to increase to about 186,000 Gg CO_2 in 2020 and 232,000 Gg CO_2 in 2030 in the baseline scenario and 359,000 Gg CO_2 in 2050, at a mean yearly growth level of 2.2%. The total increase from baseline till 2030 is 887,000 Gg CO_2 (NEEDS 2010).

Also, with further emphasis on the energy sector, Adebulugbe (2003) projected GHG emissions from 1995 to 2030, and the IPCC default CO_2 emission factors were adopted. The projection showed that CO_2 emission trends used both the baseline and GHG abatement scenarios in the analysis of total CO_2 emissions. The total CO_2

Table 18.1 Summary of GHG emission in Nigeria (in Gg) (adapted: NEEDS 2010)

Sectors	CO_2 emission	CH_4	N_2O	CO_2e
Energy	115,038	50,508	2960	168,506
Industry	2101	–	–	2,101
Agriculture	–	57,730	2664	60,394
LUCF	97,384	184	–	97,568
Waste	–	897	1480	2377
Total	214,523	109,319	7104	330,946

amounted to about 108,000 Gg CO_2 in 1995 and is expected to rise to 232,000 Gg CO_2 by year 2030 in the baseline scenario, at an average annual growth rate of 2.2%. In terms of energy supply, total primary energy consumption of Nigeria was 1270 PJ in 1995, projected to 1360 PJ in year 2000, 1718 PJ in 2010, 2800 PJ in 2020, and 3140 PJ in 2030 in the baseline scenario (NEEDS 2010).

18.4 Women in Fruit Farming and Livelihoods

Globally, the agricultural sector has made inestimable contributions to the world economy where over 70% of the income generated and livelihoods of impoverished people domiciled in local communities are sustained through the farming of root crops, fruits, vegetable, and livestock (Guidelines on Metals and Alloys Council of Europe (GMACE) 2001). Women are increasingly playing vital roles in agricultural production globally.

There are research studies with future trends that agricultural sustainability will be fostered by a new generation of women farmers. These women will make even more significant addition in the agricultural sector (Bock 2004; McNally 2001; Trauger 2004). Globally, according to United Nations (2009), it has been determined that in major farming activities, the first five prominent workers were females. However, in recent times, the number of women actively involved in farm production and processing has increased tremendously over the years, with given estimates that about 40% of small-scaled farming and post-harvest activities are mostly carried out by women around the globe (Food and Agriculture Organization of the United Nations (FAO) 2011; Jarosz 2011; Gordon et al. 2002).

The activities of women in agricultural activities vary across continents such as 20% in Latin America, while there are over 50% of women farmers spread across Africa and Asia who have made vital contributions necessary for agricultural sustainability (Doss and SOFA Team 2011). In America, the census data in 2012 revealed that females active in agriculture peaked 969,672 which covers 30% of total participants in the agricultural sector and also a rising of 14.4% from 2002 to 2013 over the years and still increasing (Census of Agriculture 2012).

In Africa, Uganda precisely, about 40% of the working population in small-scale farms was women (Uganda Bureau of Statistics). Similarly, in Kenya, women who participate in agriculture cover about 30% of the 10,000 agribusiness operators which is overtly high in Kenya when compared to other African countries (World Bank 2016). Also, in Ethiopia, women farmers who are active members of the farmer cooperatives are about 20% in addition to the fact that women make up a major proportion of the total number of farmers in Ethiopia (Woldu and Tadesse 2015).

Additionally, there are conclusive results indicating that through agriculture women have made sustained significant socioeconomic contributions to their family well-being though without being given global recognition (Brandth 1995; Fink 1992).

Accordingly, the World Bank (2013) estimated that agribusiness has contributed immensely to African economy where over 300 billion dollars was generated yearly.

Plate 18.1 Showing African women fruit farmers working on farms and harvesting fruits such as African cherry, pineapple, watermelon, pears, oranges, paw-paw, and vegetables among others

Africa has proved its ability in fruit farming of indigenous fruit species grown from home gardens and sold at local markets (Akinnifesi et al. 2008). Some of the common fruits consumed in Africa were shown in Plate 18.1. WHO (2002) estimated that in sub-Saharan Africa, an individual mean of 36 g and 90 g were consumed daily in both Eastern and Western Africa, respectively (WHO 2002). Shockingly, these daily consumption rates are inadequate and do not measure to the maximum nutrient value of 200 g or 73 kg per person required by FAO (WHO 2003; Agricultural Research Council (ARC) 2013). This deficiency in low nutrient intake could be responsible for malnutrition experienced by vulnerable groups especially mothers and their young which make up about 30% of inhabitants in sub-Saharan Africa (UNSCN 2010).

Agriculture and forestry are major contributors of food, income supply, and livelihoods for a large population of over 70% Nigerians. This is why Nigeria placed half of its 71 m hectares of arable farmlands to the production of fruits and root crops, thereby generating over 40% and 45% to its national economy and GDP, respectively (NEEDS 2010; National Bureau of Statistics 2017). In Uganda, agricultural and fisheries sector represented 71.9% of Uganda's GDP (Central Intelligence Agency 2020) and 48% of its employment in 2017 (World Bank 2020).

Fruit and vegetable farming attract higher incomes and labor per hectare of land cultivated. Hence, small-scaled female farmers maximize land space to ensure efficient food and fresh fruit farming and supply (Williams et al. 2017). Fruits are rich sources of nutrients and vitamins used to complement other classes of foods that are low in nutritional content (Kehlenbeck et al. 2013). African indigenous fruits such as baobab (*Adansonia* spp.), safou (*Dacryodes edulis*), black plum (*Syzygium cumini*), and tamarind (*Tamarindus* spp.) are vital supplements that offer diverse high and low nutritional contents such as vitamins among others (Stadlmayr et al. 2013).

Fruit farming has been practiced alongside with root crops in farms/gardens or as wild fruits on small-scale levels to ensure nutrition, healthy living, and income generation/livelihoods in sub-Saharan Africa (Meena et al. 2020; Jamnadass et al. 2011). Women were majorly involved in fruit production, processing, storage, and selling at the local markets (Schreckenberg et al. 2006).

Unfortunately, small-scale women farmers are besieged by challenges that limit their contribution to the agricultural sector especially in rural communities. These challenges include shortage of farmlands and labor; inadequate farming equipment; lack of expertise, loans, training, and markets; limited access to land ownership; sexual and domestic violence; and emotional abuse and family/household care infringes on their time (Giovarelli et al. 2013; Doss et al. 2019).

Again, women farmers have lower economic returns than men from sale of their food crops, while men have opportunities to sale their products at larger markets where they obtain higher financial outputs from their sales (Coles and Jonathan 2011). This leaves women farmers with limited participation in the crop farming mostly for their family satisfaction or sale at the rural markets.

Together, socioeconomic, finance, developmental, and gender constraints are stated as huge issues that limit women production and livelihood in fruit/crop farming in Africa and Nigeria. However, the globe is intrigued by the spider web problems of climate change, where West Africa and particularly Nigeria are presently faced with climate emergencies based on observed weather vagaries (Mora et al. 2013). Increased temperatures and erratic rainfall patterns over the past century have subjected the weather conditions to high levels of vulnerability and unpredictability with high risks that threaten human health and security of crop plants, forests, livestock, and livelihoods (Akanwa and Ezeomedo 2018; NEEDS 2010).

18.5 Women Fruit Farmers and Fruit Farming Problems in Ideato North LGA in Imo State, Nigeria

Ideato North LGA, Imo State, is located in the southeast of Nigeria with a population estimate of 215,100 inhabitants (National Population Commission 2006). It has an area of 190 km^2 and a density of 1132 km^2. It is bounded by Abia State on the east, by the River Niger and Delta State on the west, by Anambra State to the north, and by Rivers State to the south (Fig. 18.1).

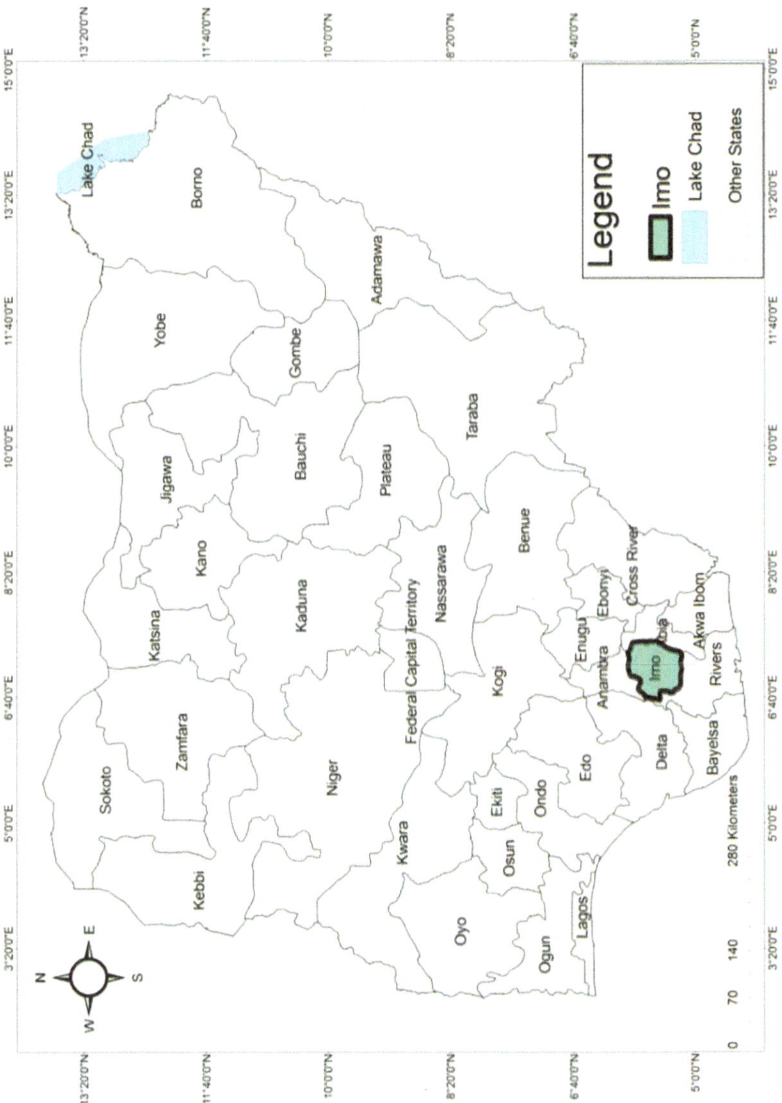

Fig. 18.1 Map of Nigeria showing Imo State

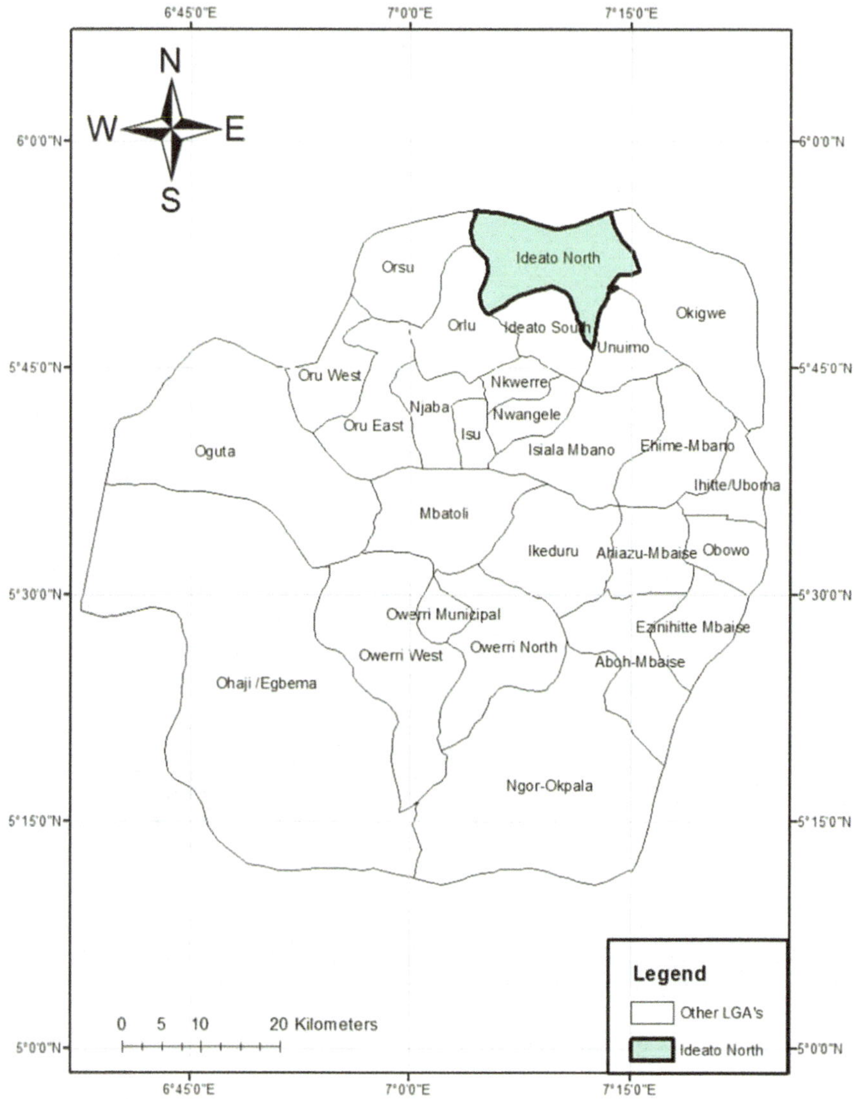

Fig. 18.2 Map of Imo State showing the study area—Ideato North LGA

Ideato North Local Government has its headquarters at Urualla which happens to be a major local government area in Imo State Nigeria (Fig. 18.1). Its geographical coordinates are latitude 5°45′ N and 6°0′ N and longitude 7°15′ E and 7°0′ E. This is about 50 km from Owerri—the Imo State capital. The area consists of towns and villages as shown in Fig. 18.2 which includes Isiokpo, Osina, Akpulu, Akokwa, Ezeamazi, Eluama, Ezihe, and Obodoukwu.

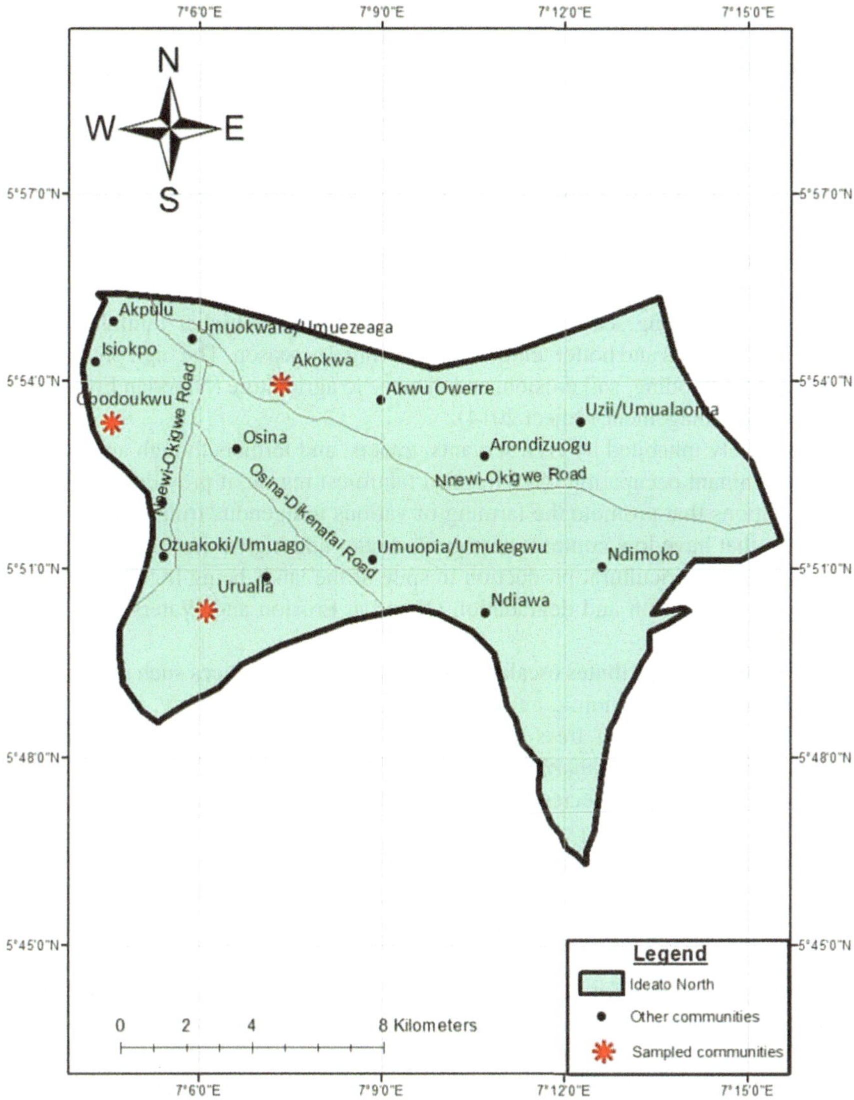

Fig. 18.3 Map of the study area showing the locations of studied communities

The study covered three communities in Ideato North LGA (Fig. 18.3), namely, Akokwa, Obodoukwu, and Urualla. The dominant geological formation is the Eocene-thick Imo Shale formation of Paleocene age and overlain by the lignite-clay seams of the Oligocene-Miocene Ogwashi-Asaba formation. The soils are derived from coastal plain sands of sedimentary rocks of about 5480 m thick and are highly weathered and dominated by sandy soils with little percentages of clay, loam, and silt (Okoroafor et al. 2018; Meena et al. 2020a).

The wet season begins in April and lasts until October with annual rainfall varying from 1500 mm to 2200 mm (60–80 inches) (Climate Data 2012). An average annual temperature above 20 °C (68.0 F) creates an annual relative humidity of 75%. The humidity reaches 90% in the wet season. The dry season experiences 2 months of Harmattan from late December to late February. The hottest months are between January and March (Climate Data 2012). However, there have been remarkable weather changes and vulnerability with high levels of uncertainty in the patterns, intensity, and duration of seasons as accurately predicted by IPCC (2007). The growing seasons have slightly changed with higher rainfall intensity during wet seasons and hotter temperatures during dry season. The high precipitation has triggered flooding, soil erosion, and damage to agriculture (Nigerian Erosion and Watershed Management Project 2014).

It is largely inhabited by civil servants, traders, and farmers though agriculture is the predominant occupation. As a tropical rainforest region, it provides the ecological conditions that promote the farming of various indigenous fruits and root crops and soils that have low contents of organic matter and high water storage capacity. This promotes agricultural production in spite of the lands being highly susceptible to accelerated erosion and degradation (Nigerian Erosion and Watershed Management Project 2014).

Together, these attributes escalate several farms/crop practices such as homestead garden, farms, and economic trees on crop lands, improved fallow, and multistory crop. The common fruit trees include bananas, mango, pear (*Pyrus communis*), guava, avocado (*Persea americana*), coconut (*Cocos nucifera*), and melon (*Cucumis melo*). These fruit tree species are either grown as farms within family compounds and farms in forest areas or cultivated as semi-wild and protected farms. Some of the local names of the indigenous trees are udara (African cherry), oroma (orange), ube (local pear), unere (banana), unere ezi (paw-paw), and ugiri (bush mango).

These fruit trees are mostly intercropped on small farm lands with other food crops usually within living premises. They serve as sources of food, nutrients, vitamins, and fodder for livestock, vegetable, and medicine (Oboho and Anyia 1992). These farms are mainly managed by family labor—mostly women. A large percent of women (70%) are active in food production, while 80% are farmers, and 100% of women are involved in agribusiness involving 60–90% sale food products (Fresco 1998). Women are involved in cultivating, weeding, harvesting, storage, and sale of fruit products. The fruit products from home farms or gardens are primarily grown for local consumption and sold at the local markets.

The subsistence nature of these fruit and vegetable farms revealed the absence of land space, poor access to hired labor, lack of improved seeds, poor equipment and technology, low training, finance, and few local markets (Grady 2016; Jiggins et al. 1996; Feed the Future 2015). It has been revealed that women ownership to lands provides the incentive to invest in their land (O'Sullivan 2017). These limitations can lead to the decline in agricultural productions that will affect rural development (Kofi Annan Foundation 2015; Mbabazi et al. 2015; Fan et al. 2013).

The agribusiness in Africa has generated over $300 billion yearly, and if small-scaled farmers are supplied with their basic needs, their income will rise to about $1 trillion annually by 2030 (World Bank 2013). Unfortunately, increased impacts of climate change on rural fruit farming business have placed fruit products on high risks as destructive effects of erratic weather patterns which affects fruit quality and sales particularly for rural impoverished farmers located in vulnerable environments (Altieri Miguel and Koohafkan 2008).

Climate crisis affects levels of agricultural productivity particularly fruits and may hinder its ability to combat food and nutrition and insecurity (Guidelines on Metals and Alloys Council of Europe (GMACE) 2001). In addition, fruit loss and damage occur at all levels during and after the harvest periods of fresh fruits especially in the procedures of harvest, storage, and transportation to their local market for sales. These processes combined with the tender nature of fruits make it susceptible to physical damages/changes such as but not limited to coloration, abscission, and odor that can reduce quality and facilitate disease spread. Notably, the poor physical infrastructures and sanitary conditions at local markets expose fruits to flies, dust, and emissions from vehicles. These are all avenues for fruit contamination and poisoning. Consequently, the fruit quality is depreciated, prices are reduced, and women livelihoods are endangered (Duku et al. 2019; Bhardwaj 2012).

It is known that toxic substances like heavy metals and other chemicals such as pesticides of organochlorine and organophosphate origin are potential health risk to the consumers. Food consumers can be affected through food contamination as this provides the buildup of these toxic substances in vital human organs over elongated periods (Chibowski 2000; Appenroth 2010). Toxic substance accumulation in plants depends on though is not limited to plant species, genetics, and other varied factors or conditions that affect the fruits even while it is sold in the market or/and consumed.

Despite these health and livelihood risks, there is a rising demand for fruits and root crops which has prompted the United Nations (UN) to demand for greater farming activities to augment global food shortages and malnutrition in the next three decades (United Nations 2009). Hence, the central thrust of this study is to investigate the climatic risks on fruit quality, health, and livelihoods of rural small-scale women fruit farmers/sellers in Ideato North Local Government in Imo State, Nigeria. The objectives of the study include:

1. To assess the level of female involvement in fruit farming business in the study area
2. To determine if fruit farming business is a major source of livelihood
3. To ascertain the impact of a changing climate on fruit farming business
4. To identify post-harvest and storage methods of fruit products in the area
5. To assess the environmental conditions of the markets where the sampled fruits are sold

6. To examine the chemical contents/toxic substances contained in the sampled fruits sold in the selected markets in the study area
7. To promote sustainable and best practices on fruit agribusiness

18.6 Materials and Method

18.6.1 Data Gathering

This study employed a mixed method approach as an advanced methodological research which is necessary to firmly address various issues (Peters 2017). Wicked problems are complex, involving multiple possible causes. They are linked with social and institutional uncertainties that exist side by side with limited knowledge about their nature and solutions (Rittel and Webber 1973). However, there is a need to utilize an approach that will examine the internal dynamics and bring multiple solutions that will firmly address its negative consequences (Peters 2017; Torfing et al. 2012).

The study also requires multiple and robust data that would provide concise and in-depth information from community contribution in order to inform policy and further research that would serve as generalized findings as well. Primary data sources include questionnaires, in-depth interviews, observation, photographs, GPS, and Google Earth aerial photograph. Secondary sources of data include journals, newsletters, and reports on WHO, FAO, UN, and the World Bank and documents on global warming in the country such as NIMET among others.

18.6.2 Sampling Technique

Probability and non-probability sampling methods were used in sample selection. The sampling methods employed were multi-stage, random, and purposive sampling methods. These were applied since the researcher played varied roles as an investigator and a facilitator with proper identification and selection of key informants during the participatory research (Guest and Macqueen 2007; Merriam 2009; Patton 2002, 2008; Takhar and Tipping 2008). It reflected people's perception, values, experiential knowledge, and differing paradigms to avoid bias which is critical in decision-making locally and generally on climate risks and women in fruit business.

Three communities (Akokwa, Urualla, and Obodoukwu) were purposively studied based on a large distribution of fruit farmers. A total 150 fruit farmers were randomly picked from the list of fruit farmers in each community compiled for this study to avoid bias. This was made up of 50 fruit farmers, from each of the 3 communities. Information extracted includes the socioeconomic variables and farming conditions among others using structured questionnaires. Focus group discussion (FGD) was also employed to interact with ten (10) women fruit sellers from each local market adding up to 30 women fruit sellers. Unstructured questionnaires were used to interview the women on the implications of global warming on fruit quality, sales, and livelihood.

Additionally, key informant interview (KII) was used to select ten key informants in order to solicit their responses. These included elders, processors, transporters, community leaders, youths, and other women who were also interviewed. Moreover, direct field observations and photographs were documented during market surveys. The data collected was processed using the necessary simple measures of data distribution and presentation.

18.6.3 Sample Collection

Common fruits consumed in the study area were selected as samples to be studied. The selected fruits were (banana, orange andpineapple) totaling nine fruits sample three from each local market in the study areas. Fruit samples were sourced from their local farms and purchased from the local markets as well. The collected samples of fruits were in their best quality without any physical observable threats.

18.6.4 Sample Preparation

The analyzed fruits were fresh and purchased from their local fruit markets. It was properly washed with deionized water and cleansed with blotting paper for removal of all sorts of dirt. The samples then prepared and shred into small pieces; a homogenized and accurate amount was weighed as required for the analysis.

18.6.5 Determination of Heavy Metals

The heavy metals analyzed include lead (Pb), iron (Fe), cadmium (Cd), zinc (Zn), arsenic (Ar), copper (Cu), and nickel (Ni). The analysis was conducted using Agilent FS240AA atomic absorption spectrometer following the method of APHA (American Public Health Association 1995).

18.6.6 Determination of Pesticides

Organochlorine and organophosphate pesticide content that was tested in the fruit samples includes DDD, aldrin, lindane, HCB, endosulfan, DDE, t-nanochlor, and dichlorvos. The samples were prepared for GC analysis PCB using the Soxhlet extraction method (AOAC 1990) (Plate 18.2).

Plate 18.2 Google Earth Satellite Image of study areas showing the studied markets (Source: Google Earth Image 2020)

18.7 Results and Discussion

The data related to socioeconomic conditions, fruit farming, and climate risks were gathered from questionnaires covering the three communities, namely, Akokwa, Urualla and Obodoukwu. Also, in-depth interviews, FGD, observation, photographs, and laboratory analysis were presented and discussed.

Table 18.2 summarized the collected data for 50 fruit farmers/marketers in Akokwa community. From the results, 10% of the sampled respondents were less than 20 years, 30% were between 20 and 30 years, 42% were between 31 and 40 years, 12% were between 41 and 50 years, and 6% were within the age of 50 years and above. This result showed that most of them were matured and could provide necessary information regarding the study since they fell within the age group of 20–40 years. The gender of the respondents revealed that 84% of the fruit farmers/marketers were women, while men were 16%. This showed that there are more women involved in fruit farming/marketing business in the study area than men.

Further, the educational levels of the farmers/marketers showed that 60% had their first school leaving certificate, 38% had WAEC certificate, 2% were NCE holders, and none had a graduate degree. This exposed the fact that though there was no farmer with a graduate degree, they were literate and collaborated in this participatory research. Also, 10% of the respondents were not married, 80% were married, 8% have lost their spouses, and 2% have lost their marriage. This showed that majority of the respondents were married and fruit farming/marketing was their main source of livelihood. Further, household sizes of the respondents showed that

Table 18.2 Summary of fruit farmers/marketers data in Akokwa community

Variables	Frequency (50)	Percentage (%)
Age		
Less than 20 years	05	10
20–30 years	15	30
31–40 years	21	42
41–50 years	06	12
50 years and above	03	06
Fruit farmers/marketers		
Male	08	16
Female	42	84
Education level		
FSLC	30	60
WAEC/WASC	19	38
NCE/OND	01	02
BSC/HND	00	00
MSC/MBA	00	00
Marital status		
Single	05	10
Married	40	80
Widowed	04	08
Divorced	01	02
Separated	00	00
Household size		
1–4	18	36
5–9	22	44
10–14	10	20
Years in fruit business		
1–2	21	42
2–5	15	30
5–10	08	16
10 years and above	06	12
Income level (monthly)		
₦1,000–10000	05	10
₦11,000–₦20,000	08	16
₦21,000–₦30,000	16	24
₦31000–₦40,000	20	40
₦41,000–50,000	02	04
₦50,000 and above	06	12
Source of fruit products		
Farmers	02	16
Farm and sell	22	36
Buy from farmers	18	44
Buy from fruit sellers	08	04
Storage life of fruits		

(continued)

Table 18.2 (continued)

Variables	Frequency (50)	Percentage (%)
1–5 days	28	56
1 week	10	20
2 weeks and above	12	24
Method of preservation		
Open drying	05	10
Kept in cool place	45	90
Place of fruit storage		
Outside the house	10	20
Within closed areas	35	70
Open and airy area	05	10
Market condition		
Sell with open table/trays	08	16
Sell in local baskets	18	36
See with umbrella	24	48
Sell in shops	00	00
Demand level of fruits		
High	32	64
Low	04	08
Very high	14	28
Very low	00	00
Farm size		
0.5–1 hectares	25	50
1.0–2 hectares	15	30
1.0–5 hectares	10	20
Climate impacts		
Pre-harvest period	18	36
Harvest period	08	16
Post-harvest period	24	48
Climate risks on environmental factors		
Heat/high temperature	21	42
Cold/high rainfall	14	28
Humidity	07	14
Winds	08	16
Climate risks on fruit quality		
Dryness (orange)	20	40
Softness (banana)	18	36
Taste (pineapple)	12	24

36% had 1–4 members, 44% had 5–9 members, and 20% had 10–14 members. Results showed that the fruit farmers had large families ranging between 1 and 9 members. Obviously, fruit framing business in the study area could benefit farmers financially while supporting their large families.

The study further indicated that 42% of respondents have been in fruit farming/ selling for a duration of 1–2 years, 30% for 2–5 years, 16% for 5–10 years, and 12% for 10 years and above. This indicated that the majority of the fruit farmers/marketers have joined the business within the space of 1–5 years probably to improve their financial capacity. Others have been in the business for over 15 years giving them the experience to provide relevant information for the study.

The range of monthly income of the respondents indicated that 10% of the respondents earn between ₦1000 and 10,000, 16% earn between ₦11,000 and 20,000, 24% earn between ₦21,000 and 30,000, 40% earn between ₦31,000 and 40,000, and 4% earn between ₦41,000 and 50,000, while 6% earn ₦50,000 and above monthly. This showed that over 70% of the respondents earn between ₦1000 and 30,000. Hence, it generates income. The sources of the fruit products were determined whereby respondents indicated that 36% of the respondents farm and sell fruits, 44% buy fruits from farmers, and 4% buy from other fruit sellers, while only 16% have their own farm. This showed that 44% of the respondents are fully involved in fruit agribusiness.

Moreover, the storage life of fruit products after harvest showed that 56% last for 1–5 days, 20% can be stored for 1 week, and 24% is stored for 2 weeks and above. Also, 10% indicated that fruit products are preserved by open drying, while 90% are preserved in a cool place. Further, 20% indicated that the fruit products are stored outside their buildings, 70% store within closed areas, and 10% are kept in open area. The market conditions showed that 16% of the respondents sell their fruit products in open tables and trays, 36% in local baskets, and 48% with umbrellas and none sell in built shops. The demand for fruit products showed that 64% are in high demand, 28% in very high demand, and 8% in low demand. This showed that fruit products are in high demand in the study area.

Further, the farm sizes for fruit farming were determined by the respondents where 50% of the farms were between 0.5 and 1 hectares, 30% fell between 1 and 2 hectares, and 20% occupied 2–5 hectares of lands. This identified that available fruit farms were mainly within 1 hectare of land. Also, respondents indicated that climate risks affect fruit farming/marketing to the point that 48% damages were felt during post-harvest or storage, 36% during pre-harvest, and 16% during harvest period. The changing climate also affects the fruit farms with high temperatures at 42%, followed by cold/high rainfall at 28%, humidity 14%, and winds 16%. The respondents also identified that fruit quality is affected by climate change where 40% of oranges were affected, followed by banana 36% and pineapple 24%. This showed that in Akokwa climate risks were felt more on orange and banana fruits.

Table 18.3 summarized the derived information of 50 fruit farmers/marketers in Urualla community. The age distribution of the fruit farmers indicated that 6% were <20 years, 12% were between 20 and 30 years, 44% were between 31 and 40 years, 16% were between 41 and 50 years, and 20% were within 50 years and above. This explained that a large number of the fruit farmers were notably adults engaged in fruit farming business. The 76% of the fruit farmers/marketers were women, while 24%were men. This result further buttressed the undeniable fact that more women were participants in fruit farming/marketing in the study area than men.

Table 18.3 Summary of fruit farmers/marketers data in Urualla community

Variables	Frequency (50)	Percentage (%)
Age		
Less than 20 years	03	06
20–30 years	06	12
31–40 years	21	44
41–50 years	08	16
50 years and above	12	20
Fruit farmers/marketers		
Male	12	24
Female	38	76
Education level		
FSLC	32	64
WAEC/WASC	18	36
NCE/OND	00	00
BSC/HND	00	00
MSC/MBA	00	00
Marital status		
Single	08	16
Married	38	76
Widowed	04	08
Divorced	00	00
Separated	00	00
Household size		
1–4	12	24
5–9	17	34
10–14	21	42
Years in fruit business		
1–2	25	50
2–5	12	24
5–10	08	18
10 years and above	05	08
Income level (monthly)		
₦1000–10000	10	20
₦11,000–₦20,000	12	24
₦21,000–₦30,000	27	54
₦31000–₦40,000	01	02
₦41,000–50,000	00	00
₦50,000 and above	00	00
Source of fruit products		
Farmers	14	28
Farm and sell	25	50
Buy from farmers	08	16
Buy from fruit sellers	03	06
Storage life of fruits		

(continued)

Table 18.3 (continued)

Variables	Frequency (50)	Percentage (%)
1–5 days	10	20
1 week	22	44
2 weeks and above	18	36
Method of preservation		
Open drying	12	24
Kept in cool place	38	76
Place of fruit storage		
Outside the house	14	28
Within closed areas	30	60
Open and airy area	06	12
Market condition		
Sell with open table/trays	12	24
Sell in local baskets	23	46
See with umbrella	15	30
Sell in shops	00	00
Demand level of fruits		
High	30	60
Low	08	16
Very high	12	24
Very low	00	00
Farm size		
0.5–1 hectares	12	24
1–2 hectares	18	36
2–5 Hectares	20	40
Climate impacts		
Pre-harvest period	15	30
Harvest period	10	20
Post-harvest period	25	50
Climate risks/environmental factors		
Heat/high temperature	22	44
Cold/high rainfall	16	32
Humidity	09	18
Winds	03	06
Climate risks on fruit quality		
Dryness (orange)	17	34
Softness (banana)	20	40
Taste (pineapple)	13	26

Further, the educational levels of the respondent showed that 64% had their first school leaving certificate and 36% had WAEC certificate. Also, 16% of the respondents were single, 76% married, and 4% widowed. This showed a high portion of married people. The household sizes of the respondents showed that

24% had 1–4 members, 34% had 5–9 members, and 42% had 10–14 members. This revealed that the respondents had large families that were supported by fruit farming and marketing. Also, 50% of respondents have been in fruit farming/selling for a duration of 1–2 years, 24% for 2–5 years, 18% for 5–10 years, and 8% for 10 years and above. This indicated that fruit farmers/marketing business is lucrative with high participants.

The range of monthly income of fruit farmers showed that 20% of them earn between ₦1000 and 10,000, 24% earn between ₦11,000 and 20,000, and 54% earn between ₦21,000 and 30,000, while 2% earn ₦31,000–40,000 monthly. This showed that over 70% of the respondents earn between ₦1000 and 30,000. The sources of the fruit products were determined whereby respondents indicated that 28% of the respondents were mainly farmers, 50% both farm and sell fruits, 16% buy fruits from farmers, and 6% buy from other fruit sellers.

Further, the storage life of fruit products after harvest showed that 20% last for 1–5 days, 44% can be stored for 1 week, and 36% are stored for 2 weeks and above. Also, 24% indicated that fruit products are preserved by open drying, while 76% are preserved in a cool place. Nearly 28% respondent indicated that the fruit products are stored outside their buildings, 60% store within closed areas, and 12% are kept in open area.

The market conditions showed that 24% of the respondents sell their fruit products in open tables and trays, 46% in local baskets, and 30% with umbrellas and none sell in built shops. The demand for fruit products showed that 60% are in high demand, 16% in very high demand, and 24% in low demand. Further, the farm sizes for fruit farming were determined by the respondents where 24% of the farms were between 0.5 and 1 hectares, 36% fell between 1 and 2 hectares, and 40% occupied 2–5 hectares of lands. This identified that fruit farms were mainly within (2–5) hectares of land.

Also, respondents indicated that climate risks affect fruit farming/marketing to the point that 50% damages were felt during post-harvest or storage, 30% during pre-harvest, and 20% during harvest period. The changing climate also affects the fruit farms with high temperatures at 44% followed by cold/high rainfall at 32%, humidity 18%, and winds 6%. The respondents also identified that fruit quality is affected by climate change where 34% of oranges were affected, followed by banana 40% and pineapple 26%. This showed that in Urualla community climate risks were felt more on banana fruits.

Table 18.4 summarized the analyzed information of 50 fruit farmers/marketers in Obodoukwu community. Analyses showed that 4% of the sampled fruit farmers were less than 20 years, 16% were between 20 and 30 years, 28% covered 31–40 years, 34% fell within 41–50 years, and 18% were 50 years and above. This result identified that a large number of fruit farmers were within the age distribution of 31–50 years. Also, there were more women fruit farmers/marketers (68%) than men (32%) in the study area. Again, women fruit farmers/marketers in Obodoukwu just like in other two studied locations (Akokwa and Urualla) have outnumbered the men population.

Table 18.4 Summary of fruit farmers/marketers' data in Obodoukwu community

Variables	Frequency (50)	Percentage (%)
Age		
Less than 20 years	02	04
20–30 years	08	16
31–40 years	14	28
41–50 years	17	34
50 years and above	09	18
Fruit farmers/marketers		
Male	16	32
Female	34	68
Education level		
FSLC	42	84
WAEC/WASC	08	16
NCE/OND	00	00
BSC/HND	00	00
MSC/MBA	00	00
Marital status		
Single	04	08
Married	40	80
Widowed	06	12
Divorced	00	00
Separated	00	00
Household size		
1–4	04	08
5–9	15	30
10–14	31	62
Years in fruit business		
1–2	09	18
2–5	12	24
5–10	07	14
10 years and above	22	56
Income level		
₦1000–10000	13	26
₦11,000–20,000	35	70
₦21,000–30,000	02	04
₦31000–40,000	00	00
₦41,000–50,000	00	00
₦50,000 and above	00	00
Source of fruit products		
Farmers	20	40
Farm and sell	15	30
Buy from farmers	10	20
Buy from fruit sellers	05	10
Storage life of fruits		

(continued)

Table 18.4 (continued)

Variables	Frequency (50)	Percentage (%)
1–5 days	15	30
1 week	23	46
2 weeks and above	12	24
Method of preservation		
Open drying	20	40
Kept in cool place	30	60
Place of fruit storage		
Outside the house	15	30
Within closed areas	25	50
Open and airy area	10	20
Market condition		
Sell with open table/trays	25	50
Sell in local baskets	14	28
See with umbrella	11	22
Sell in shops	00	00
Demand level of fruits		
High	25	50
Low	17	34
Very high	08	16
Very low	00	00
Farm size		
0.5–1 hectares	04	08
1.0–2 hectares	11	22
1.0–5 hectares	35	70
Climate impacts		
Pre-harvest period	12	24
Harvest period	18	36
Post-harvest period	20	40
Climate risks/environmental factors		
Heat/high temperature	26	52
Cold/high rainfall	18	36
Humidity	06	12
Winds	00	00
Climate risks on fruit quality		
Dryness (orange)	15	30
Softness (banana)	20	40
Taste (pineapple)	15	30

Further, the educational levels of the respondent showed that 84% had their first school leaving certificate and 16% had WAEC certificate. Also, 8% of the respondents were single, 80% married, and 12% widowed. This showed that married people were higher. The household sizes of the respondents showed that 8% had 1–4 members, 30% had 5–9 members, and 62% had 10–14 members. This revealed that the respondents had large families that were supported by fruit farming and marketing. Also, 18% of respondents have been in fruit farming/selling for a duration of 1–2 years, 24% for 2–5 years, 14% for 5–10 years, and 56% for 10 years and above. This indicated that fruit farming business has gained stability with 56% respondents for over 10 years.

The range of monthly income of the respondents indicated that 26% of the respondents earn between ₦1,000 and 10,000, 70% earn between ₦11,000 and 20,000, and 4% earn between ₦21,000 and 30,000 monthly. The sources of the fruit products were determined whereby respondents indicated that 40% of the respondents were mainly farmers, 30% both farm and sell fruits, and 20% buy fruits from farmers, and 10% buy from other fruit sellers.

Further, the storage life of fruit products after harvest showed that 30% last for 1–5days, 46% can be stored for 1 week, and 24% is stored for 2 weeks and above. Also, 40% indicated that fruit products are preserved by open drying, while 60% are preserved in a cool place; 30% indicated that the fruit products are stored outside their buildings, 50% are stored within closed areas, and 20% are kept in open area.

The market conditions showed that 50% of the respondents sell their fruit products in open tables and trays, 28% in local baskets, and 22% with umbrellas and none sell in built shops. The demand for fruit products showed that 50% are in high demand, 34% in very high demand, and 16% in low demand. Further, the farm sizes for fruit farming were determined by the respondents where 8% of the farms were between 0.5 and 1 hectares, 22% fell between 1 and 2 hectares, and 70% occupied 2–5 hectares of lands. This identified that the sizes of fruit farms were mainly within 2–5 hectares of land.

Also, respondents indicated that climate risks affect fruit farming/marketing to the point that 24% damages were felt during post-harvest or storage, 36% during pre-harvest, and 40% during harvest period. The changing climate also affects the fruit farms with high temperatures at 52%, followed by cold/high rainfall at 36% and humidity 12%. The respondents also identified that fruit quality is affected by climate change where 30% of oranges were affected, followed by banana 40% and pineapple 30%. This showed that in Obodoukwu community climate risks were felt more on banana fruits.

18.8 Changing Climate and Implications on Fruit Farms, Their Quality, and Health Risks

A background description of the studied towns in Ideato North LGAs showed that they were known predominately for agricultural production. In Akokwa town, there is an old saying among indigenes that “Akokwa na oru” meaning “Akokwa and farmwork.” This is seen in the diverse agricultural production of vast crops such as yams, cocoyam, cassava, fruit trees, and vegetables. Also, the community is involved in animal farming of goats, pigs, chickens, and sheep. The farm products are usually sold at the popular daily Orie market. Akokwa lies on the border between

Ideato LGA in Imo State and Aguata LGA in Anambra State. Banana fruits are plenty in occurrence from August to November and oranges by December.

Urualla literally means "profit derived from land." This referred to its ancient ancestors who were basically crop farmers and hunters. It has four major villages where 50% of the populations are farmers and artisans (Nigerian Erosion and Watershed Management Project 2014). It is the administrative headquarters for Ideato North LGA, and it is next to Akokwa town. They grow fruits such as banana, mango, pineapple, avocado, and oranges and other crops such as maize, legumes, and cassava among others. These products are sold in the Afor-Urualla market.

Obodoukwu town has a population of about 10,000 people. It is comprised of nine villages that were known for livestock farming, and palm tree forms 60% of the forest reserves. Here, fruit trees that are grown include palm oil, palm fruits, palm kernel, kola nut, and breadfruit among others. These formed the major cash products produced in large quantities and sold at the largest local Eke market in Obodoukwu. However, all the agricultural production in these towns is still at the subsistence level. In addition, the agricultural products are mostly meager coupled with poor management and storage facilities. The increasing temperatures triggered by global warming compromise the quality of these fruits (Akanwa and Joe-Ikechebelu 2019). Unfortunately, higher temperatures cause rapid deterioration of fruit quality since the higher rates of respiration caused by higher temperatures lead to a higher use of sugars by the plants. As a result, the sugar level in the harvested fruit is rapidly decreased affecting the market value, and hence, the livelihoods of the women smallholder farmers are greatly threatened (Hatfield and Prueger 2015; Food and Agriculture Organization of the United Nations (FAO) 2011; Simpson 2017).

18.9 Women Involvement in Fruit Agribusiness and Livelihood in Ideato North LGA

Further, findings revealed that a large population of fruit farmers/marketers involved in fruit agribusiness in these three communities (Akokwa, Urualla, and Obodoukwu) was women with a mean of 36, while the men were 12. Also, majority of these women were married and between the ages of 31 and 40 in Akokwa and Urualla, but in Obodoukwu, the women were much older 41–50. These women had large families, given that Akokwa had 5–9, Urualla had 10–14, and Obodoukwu had 10–14 households.

It is noted that in Ideato LGA, a greater population of the women were involved in fruit farming business and outnumbered the men about three times over. Majority of them were married and had a minimum of 5–9 and a maximum of 10–14 families. The large population of women is a proof that women are indispensable in agricultural production especially in fruit agribusiness. The interview with informants and discussion with farmers and market women collaborated that women were majorly involved as farm workers and harvesters; they also store and sell the fruit products from their farms. It was also noted that majority of the women were married and had large households. This could be traceable to the culture of early marriage in the

African-Nigerian setting for women. In addition to the fact that majority of these women attended only primary and secondary schools without a tertiary/university degree. Invariably, the culture and minimal exposure to higher educational pursuits can promote the birth of many children; thus, large households become inevitable.

More so, results stated that a large proportion of the women had been in fruit farming business between 1 and 2 years in Akokwa (42%) and in Urualla (50%), but in Obodoukwu (56%), more people have been active for 10 years and above. Accordingly, about 40% of women in Akokwa earned incomes within the ranges of ₦31,000–40,000, while 54% of women in Urualla earned between ₦21,000 and 30,000 monthly. In Obodoukwu, 70% of the women earned between ₦11,000 and 20,000. It was noted that majority of the women farmers had attained highest results in their West African Examination Council (WAEC) in Akokwa and Urualla (38%) and (36%), respectively, while Obodoukwu was lowest in WAEC (16%) and highest in FSLC (84%).

Findings showed that majority of the women were younger, earned more money, and seemed more educated in Akokwa and Urualla towns, but in Obodoukwu, the women were older, earned lesser, and were likely to be less educated. However, the differences in the results between Akokwa, Urualla, and Obodoukwu could be deduced from the fact that Obodoukwu had higher features of a rural setting. The higher rurality index of Obodoukwu could explain its distinctive results why the women had larger families, were less educated, had lesser levels of income generated, and were even older in age from 41 to 50 years (Table 18.4). But, Akokwa town lies at the border between Imo and Anambra State, while Urualla is the administrative headquarters, and together, these advantages attracted younger and more educated women into fruit farming business.

Also, it was noted that in Akokwa and Urualla, there were higher numbers of young women who were attracted to the business within 1–2 years. This can also be deduced from its income levels too where the women in Akokwa and Urualla could earn as high as ($102.56 US) and ($76.92 US) monthly from fruit agribusiness, respectively. But in Obodoukwu, it held a different scenario where the largest population of women had been established in fruit business for over 10 years. This also revealed that fruit farming and marketing generates income in Akokwa and Urualla towns, so that a large population of women are drawn into fruit business between 1 and 2 years. Also, it is a dependable source of income and livelihood where rural women within their mid-life (41–50) years could earn $51.28 US monthly.

The high levels of demand for fruit products in Akokwa, Urualla, and Obodoukwu communities indicated that 64%, 60%, and 50% were demanded, respectively. This makes the business attractive, labor-intensive, and on regular demand throughout the year. This will demand a lot of hard work on the farms, post-harvest processes, and marketing. Notably, these women are married with many children to feed and various responsibilities at the same time, and they need to maintain the high demand levels in order to remain in the business of fruit farming.

Further, findings showed that majority of the women in Akokwa community 44% buy fruit products from farmers, 36% farm and sell fruits, 4% buy from fruit sellers, and 16% are mainly farmers. In Urualla, 50% of women farm and sell fruits, 16% buy from farmers, 6% buy from fruit sellers, and 28% are farmers, whereas in Obodoukwu majority (40%) of the women are farmers, 30% farm and sell, 20% buy from farmers, and 10% buy from fruit sellers. The farm sizes also revealed that in Akokwa 50% of the farmers use 0.5–1 hectares but in Urualla 40% of the farm sizes are between 2 and 5 hectares. In Obodoukwu, 70% of the farm sizes were between 2 and 5 hectares. It was observed that in the communities, majority of the fruit trees were grown within their houses, others at nearby farms, and some at forest farms far from their houses as seen in Plates 18.3 and 18.4.

This showed that a large proportion of the women in Akokwa, Urualla, and Obodoukwu farm and sell fruit products (36%, 50%, and 30%), respectively. However, Obodoukwu has the highest number of farmers, and Akokwa has the highest number of women who buy fruit from farmers. Urualla has the highest number of women who farm and sell fruits. This confirmed that fruit business is the main source of income generation and livelihood for 114 women in Ideato LGA.

Notably, the female farmers and other women informants interviewed indicated that many parts of Akokwa town are not fertile for agricultural production except with the application of fertilizers. This could be responsible for the high number of women farmers who buy fruits and by observation their farms were located within their houses (50%) of the farms were about 0.5–1 hectares. It added to the fact that it is densely populated compared to other towns in Ideato North LGA and this has affected the availability of lands for agricultural production.

Several informants also pointed out that fruits were brought to Akokwa from other rural villages in Ideato North LGA such as Umualoma, Arondizuogu, and Umuoji. Moreover, Obodoukwu has a typical rural background, and majority of the farms by observation were located outside their homes providing lands for vast agricultural production. This explained why it had the largest concentration of farmers compared to other studied areas, and 70% of the farm sizes were between 2 and 5 hectares. Urualla town had the highest number of women fruit farmers and marketers, and 40% of the farm sizes were between 2 and 5 hectares. Its rural and administration position provides Urualla the advantages of large parcels of land for farming and fruit business.

18.10 Implications of a Changing Climate on Fruit Farming and Livelihood in Ideato LGA

Further, it was confirmed that climate change known as "Ntuhari Igwe" in the local dialect has affected fruit farming business where 48%, 50%, and 40% of the impact were felt during post-harvest period in Akokwa, Urualla, and Obodoukwu, respectively.

Plate 18.3 Showing indigenous fruit trees growing inside community living premises; examples include banana, coconut, and mango trees in the study area

Plate 18.4 Showing indigenous fruit trees growing outside community buildings, others in farms combined with root crops and far away from their homes and in forest areas as wild fruit trees. Examples are mostly oranges and mango in the study area

Findings from the respondents in Akokwa (48%), Urualla (50%), and Obodoukwu (40%) showed that the highest climatic risk on fruit farming were high temperatures. Women farmers indicated that they have detected changes particularly hotter dry seasons and colder wet seasons. This fluctuation in temperature exacerbates extreme heat which affects the fruits especially the post-harvest periods. Apparently, the high temperatures affect mostly banana fruits in Urualla and Obodoukwu communities, while oranges were affected in Akokwa community. During the FGD with the market women, they further informed that:

> Ntuhari igwe which refers to climate change in their local dialect has increased high temperatures and this has deteriorated the fruit quality mostly during harvest and storage periods affecting the flavor and taste of the fruits.

It is noted that the exposure of fruits to high temperatures just before and after harvest periods lowers the quality of the fruit during storage or before it is sold in the market. This is because fruits' tissues continue to live even after harvest, and hence their rate of chemical decomposition is faster under increased temperatures threatening their quality. The women farmers informed that due to the effect of high temperature, the fruits before and after harvest were monitored to avoid the outbreak of pest and diseases before complete maturation/ripening.

Consequently, increased temperatures hasten the premature weakening of fruit parts and, hence, escalate the growth of microorganisms and diseases that destroy fruit quality. These inflicted damages by microorganisms on fruits are mainly in the physical loss of edible matter either partially or totally destroying the quality of the fruit. Unfortunately, when fruit quality is partially or completely damaged by high temperatures, income levels and livelihoods are eventually affected. Most of the women fruit sellers complained during interviews that temperature effects have affected their sales hugely in the last couple of years, and one of the women informed said:

> Extreme temperature has inflicted serious damage on the fruit products during storage so much that the fruits become too soft or dried up especially the bananas and oranges while the pineapples change taste. We are left with no other option than to sell at lower prices and this affects our income levels. Most of us are forced to sell to other fruit sellers in order not to lose completely. The remaining fruit products are either thrown away or eaten by my household.

Unfortunately, the time, labor, risks, limitations, and capital expended in growing these fruits to maturity are not commeasurable to the income generated when sold most times. Further, the fruit products were mostly preserved and stored in cool and closed areas. This was done to preserve the quality of the fruits from temperature effects as a result of increased heat/hot weather. However, the storage lives of the fruit products (banana, orange, and pineapple) showed that majority of them were kept within a week before being sold to enable ripening and to escape damages from the farms especially from insects, diseases, pest, and water loss.

According to Singh (2009), bananas have a limited life span after its harvest which ranges between 4 and 10 days when mature green and 2 and 4 days when ripe. Both green and ripe bananas are easily affected by cold, and it deteriorates at temperature less than 13 °C. Also, during harvest, oranges (citrus) can easily be punctured, and this gives access to blue and green mold disease. Oranges can be stored about 3 weeks under the right atmosphere. Heat/high temperature causes moisture to deplete fast before and after harvesting oranges. Also, pineapples are stored at about 7–13 °C (45–55 °F), and relative humidity is 85–95%. It is sensitive; therefore, it should be stored for 2–5 weeks (Srivastava and Kumar 2002).

The storage and preservation methods were to minimize damage. But during storage, fruit decay still takes place since excessive heat sponsors quick infection of microorganisms such as bacteria and fungi. Also, the women farmers informed that during harvest which is carried out manually, the fruits can sustain fractures aiding damages. For example, banana is harvested with cutlass, orange with handheld poles, and pineapple with knives, and they are transferred with buckets, bags, pans, and baskets. This process has a high propensity to cause injuries providing access for organisms to penetrate through their natural openings, injuries, or intact skin (Ramyan and Mohd 2018). Oftentimes, the infection may be active during the

Plate 18.5 Showing local market in Akokwa community where fruit products are sold in open areas and umbrellas. The fruit products were displayed in open trays and baskets and stored in wooden cupboards to aid ripening. The sanitary conditions of the market are poor

growth periods in the farm though it may not be visible until it is harvested and stored or ripens under high temperatures.

The study also revealed the local market conditions where these fruit products are sold. The wooden box where the fruits were stored to ripen was in poor conditions (Plate 18.5). The market floors were bare and unclean. The market conditions showed absence of sanitary practices with high vulnerability for contamination based on the fact that 48% of the women sell fruit products with umbrellas in Akokwa local market. In Urualla community, 46% sold in local baskets, while 50% sell fruits with open trays and tables in Obodoukwu local market as shown in Plates 18.5, 18.6, and 18.7.

Plate 18.6 Showing local market in Obodoukwu community where fruit products are sold under umbrellas and on the floor and makeshift buildings. The fruit products were displayed in unpleasant environmental conditions

18.11 Effect of Fruit Quality and Its Associated Health Risks

The prevailing consequences of climatic risks on fruit farming, quality, and livelihoods gathered from quantitative and qualitative data were also authenticated with laboratory analysis and interpreted using tables and charts. The results of toxic substances of fruit samples from local markets in Ideato North were grouped into two categories, namely, pesticides and heavy metals.

The data obtained from laboratory analysis of pesticide residues from fruit samples and their coordinates are shown on Table 18.5. Findings showed that the parameters are observed to be highest in aldrin concentrations in pineapple sold from

Plate 18.7 Showing the local market in Urualla community where fruit products were sold in open areas, on the floor, on tables, and under makeshift buildings and umbrellas. The environmental conditions of the market were poor as well

Orie Akokwa market and endosulfan concentrations in oranges sold from Urualla market and pineapple sold from Orie Akokwa market. HCB concentration exceeded limits in pineapple in Orie Akokwa market and orange and banana sold in Orie Akokwa market. Also, DDD concentration was exceeded in pineapple and banana sold from of an Urualla market.

The concentrations of aldrin, HCB, DDD, and endosulfan pesticide residues in these fruits were relatively above the WHO/FAO standard for maximum pesticide residue limit. Table 18.6 showed the mean value of pesticide residue in selected fruit samples sold in major markets in Ideato North (Fig. 18.4).

Table 18.5 Pesticide residues in fruit samples from local markets in the study area

Fruit samples	Pineapple			Orange			Banana			
Fruit sample location / Parameter	Sample A (Orie (Akokwa)	Sample B (Eke Obodoukwu)	Sample C (A for Urualla)	Sample A (Orie Akokwa)	Sample B (Eke Obodoukwu)	Sample C (A for Urualla)	Sample A (Orie Akokwa)	Sample B (Eke Obodoukwu)	Sample C (A for Urualla)	WHO/ FAO MRL
DDD	0.000	0.000	0.000	0.000	0.0018	0.000	0.000	0.000	0.000	0.400
Aldrin	0.3270	0.0008	0.0000	0.000	0.0150	0.0000	0.000	0.0150	0.0000	0.100
Lindane	0.0000	0.3182	0.2962	0.0000	0.3040	0.0000	0.2881	0.2464	0.2881	0.400
HCB	1.7150	0.0012	0.0000	1.9547	0.0000	0.0000	1.9201	0.0000	0.0000	0.100
Endosulfan	0.4709	0.0092	0.0000	0.0000	0.0057	1.8671	0.0000	0.0000	0.0000	0.200
DDE	0.3631	0.0000	0.3522	0.000	0.1731	0.000	0.2747	0.1796	0.2747	0.500
t-Nanochlor	0.000	0.1191	0.000	0.000	0.1189	0.000	0.000	0.0317	0.000	0.500
Dichlorvos	0.0124	0.0085	0.0124	0.0124	0.0085	0.0124	0.0124	0.0085	0.0124	0.300
GPS locations X	5°54′6.642″N	5°52′10.6932″N	5°54′42.9948″N	5°54′6.642″N	5°52′9.6348″N	5°54′44.2188″N	5°54′5.526″N	5°52′11.9532″N	5°52′11.9532″N	
GPS locations Y	7°4′45.0228″E	7°4′43.1688″E	7°6′38.3112″E	7°4′45.0228″E	7°4′43.0068″E	7°6′38.5812″E	7°4′42.924″E	7°4′42.6612″E	7°4′42.6612″E	

Table 18.6 The mean level of pesticide residue in some selected fruits sold in local markets in the study area

Component	Pineapple	Orange	Banana	FAO standard
Lindane	0.3072	0.2801	0.2379	0.4
p-p′-DDE	0.3577	0.2621	0.2875	0.5
Aldrin	0.1639	0.015	0.015	0.1
t-Nanochlor	0.1652	0.2428	0.1213	0.5
Endosulfan	0.2401	0.9364	0	0.2
HCB	0.8581	1.9547	1.9201	0.1
P′p′-DDD	0.4526	0.2271	0.4509	0.4
Dichlorvos	0.0173	0.0172	0.0173	0.3

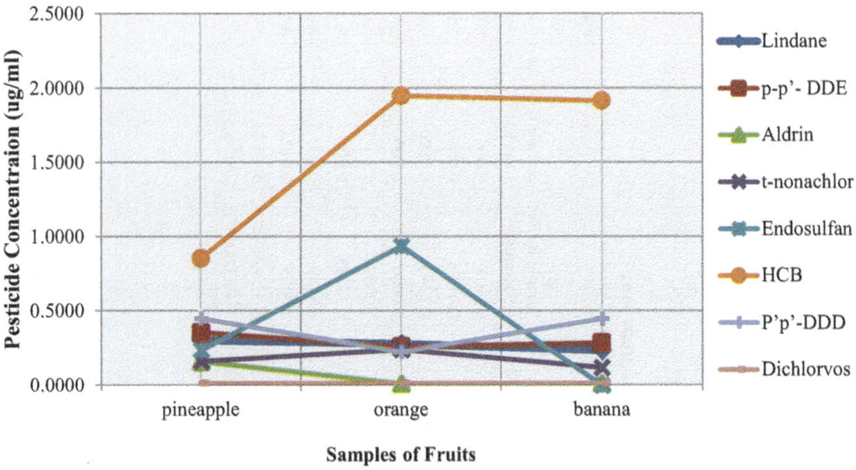

Fig. 18.4 Variation of pesticide in fruits

18.11.1 Heavy Metal Parameters

The information determined from the laboratory analysis of heavy metal residues from fruit samples and their coordinates are shown on Table 18.7. Findings revealed that the heavy metal parameters were observed to be highest in cadmium concentration of pineapple sold in Orie Akokwa and Afor Urualla, in orange sold in Orie Akokwa and Afor Urualla, and in banana sold in Orie Akokwa and Afor Urualla. Lead concentration exceeds the acceptable limits of pineapple sold in Eke Obodoukwu and Orie Akokwa, also in oranges sold in Eke Obodoukwu and Orie Akokwa, and in banana sold in Eke Obodoukwu, Orie Akokwa, and Afor Urualla.

Nickel concentration exceeds limits of pineapple sold in Eke Obodoukwu and Orie Akokwa and oranges sold in Eke Obodoukwu market. Cadmium, lead, and nickel concentrations exceeded the acceptable limits of WHO/FAO standard in the fruit samples listed above in Ideato North. Table 18.8 showed the mean value of sampled fruits in Ideato North LGA (Fig. 18.5).

Table 18.7 Heavy metal residues in fruit samples from local markets in the study area

Fruit samples		Pineapple			Orange			Banana			
Fruit sample location / Parameter		Sample A (Orie) (Akokwa)	Sample B (Eke Obodoukwu)	Sample C (A for Urualla)	Sample A (Orie Akokwa)	Sample B (Eke Obodoukwu	Sample C (A for Urualla)	Sample A (Orie Akokwa)	Sample B (Eke Obodoukwu)	Sample C (A for Urualla)	WHO/ FAO MRL
Iron		0.000	0.134	0.602	0.516	0.684	0.745	0.663	0.665	0.770	3.9
Cadmium		0.3270	0.087	0.846	0.812	0.287	0.781	0.918	0.260	0.657	0.3
Copper		0.0000	0.227	0.787	0.595	0.344	0.716	0.619	0.326	0.994	2.63
Zinc		1.7150	0.984	2.000	2.185	1.421	1.945	1.530	1.058	2.417	3.7
Lead		0.4709	0.995	0.110	0.662	0.964	0.133	0.845	0.723	0.274	0.24
Nickel		0.3631	0.470	0.00	0.000	0.813	0.00	0.000	0.00	0.018	0.08
Arsenic		0.000	0.056	0.00	0.006	0.090	0.003	0.098	0.00	0.00	0.25
GPS locations	X	5°54′6.642″N	5°52′10.6932″N	5°54′42.9948″N	5°54′8.244″N	5°52′9.6348″N	5°54′44.2188″N	5°54′5.526″N	5°52′11.9532″N	5°52′11.9532″N	
	Y	7°4′45.0228″E	7°4′43.1688″E	7°6′38.3112″E	7°4′41.9988″E	7°4′43.0068″E	7°6′38.5812″E	7°4′42.924″E	7°4′42.6612″E	7°4′42.6612″E	

Table 18.8 The mean of heavy metals in some selected fruits sold in three major markets in the study area

Parameters	Pineapple	Orange	Banana
Iron	0.3897	0.6483	0.6993
Cadmium	0.5203	0.6267	0.6117
Copper	0.5167	0.5517	0.6463
Zinc	1.5087	1.8503	1.6683
Lead	0.5123	0.5863	0.6140
Nickel	0.3880	0.2710	0.0060
Arsenic	0.0260	0.0070	0.0000

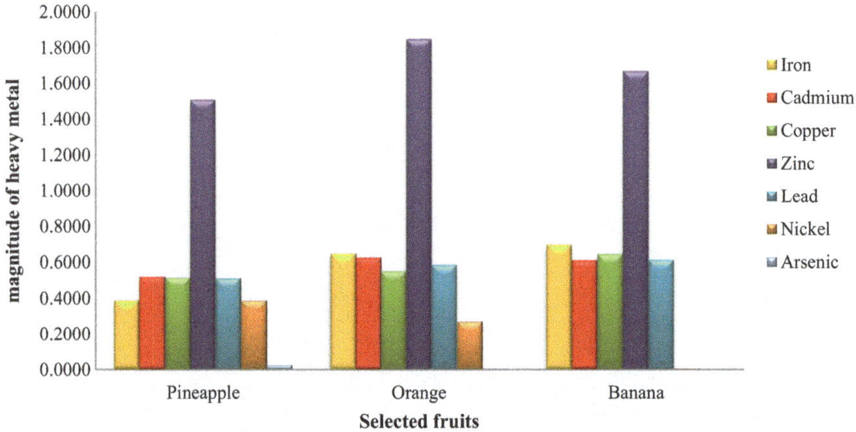

Fig. 18.5 Variation of heavy metals in selected fruits in the study area

Findings for pesticides showed that aldrin, endosulfan, HCB, p′p′, and DDD concentrations exceeded the acceptable standards of WHO/FAO in the sampled pineapples, oranges, and bananas in Akokwa, Obodoukwu, and Urualla towns. Also, the heavy metal results confirmed Pb, Ni, and Cd concentrations were above the acceptable limits of WHO/FAO in the sampled pineapples, oranges, and banana in the study area.

The results from the analysis confirmed that the fruits contain pesticides and heavy metals with slightly 50% of the samples analyzed containing both pesticide and heavy metal residue above WHO/FAO standard level. There are serious health concerns since 7 out of 15 tested parameters were above the maximum permissible residue limits of WHO/FAO in fruits. The results obtained in this study were also authenticated by other research works which confirmed that sample fruits contained high levels of pesticides and heavy metals (Ihesinachi and Eresiya 2014; Jallow et al. 2017).

Aldrin is an organochlorine pesticide that is extremely toxic to the point of terminating human life and survival and has a high ability to remain in the atmosphere over extended periods of time (Weaver et al. 2012). This is responsible for high aldrin concentrations in the environment, and it easily impacts plants. Also, it can be exposed to animals that feed on contaminated plants or animals that live in contaminated water. The consumed levels of aldrin in humans have been given to cause convulsions and other symptoms of fever such as headache and pains in the muscle areas among others (Savage et al. 1988). The detection of aldrin is an indication that the fruits have health risks.

Notably, the menace caused by endosulfan is of great concern whereby the WHO placed it in the category of "Averagely Toxic" based on a rat LD50 of 80 mg/kg. Endosulfan has the capacity to persist in the atmosphere for elongated periods enabling it to contaminate edible substances like plants and animals eaten by human through the process of bio-accumulation. It targets the human nervous system and compromises the organs to malfunction (Briz et al. 2011). Studies have shown that HCB residues bio-accumulate in the blood of pregnant mothers from where it is passed on through the placenta and affects the newborn's development (Li et al. 2014).

Again, DDD has been grouped in the class of possible human carcinogen since if it is contained in the fruit, it can result to cancer which is harmful to human health. DDD pesticide residue is a metabolite of DDT pesticide residue (USEPA 2007).

Cadmium causes toxic damage to human health by deteriorating the kidney function and causes hypercalciuria via renal injury and dysfunction of the heart, thereby damaging the genes responsible for development and reproduction. Also, the kidney can be affected when it is exposed to the toxic effect of cadmium. When food substances have been exposed to toxic heavy metals, they can result in acute gastrointestinal health risks with symptoms of diarrhea (Guidelines on Metals and Alloys Council of Europe (GMACE) 2001).

Lead has been associated with metal poisoning in both children and adults that damages the nervous systems and the gastrointestinal tract; hence, it has been grouped as a classic disease (Markowitz 2000). When food products are exposed to large quantities of lead, it could be responsible for varied birth and growth defects, it added to major problems affecting mental and physical development and emotional stability even to cessation of human life (Martin and Griswold 2009).

Nickel can cause adverse effect on human health when consumed in food, such as kidney diseases and cancer of the respiratory tract (Oller 2003). The outcome of the questionnaire distributed showed that a large number of the women sell fruits under poor market conditions and this gives rise to contamination. Also, majority of these fruits are bought from neighboring villages and sold at the local market. Hence, the health safety of these fruits sold in the study area becomes questionable since these villages are subject to similar climate changes.

High temperatures attributed to a changing climate affects these fruits either through diseases, pests spread, or poor storage conditions become severely affected either by, but not limited to physiological disorders, abscission, off-flavors, color, softening, and discoloration (Duku et al., 2019; Bhardwaj 2012). Unfortunately, females involved in rural agricultural production do not have access to scientific and technological means of fruit protection.

It is expedient to protect the quality of these fruits, so that losses through decay can be minimized while protecting human health. Further, fresh fruits, vegetables, and root crops can absorb large water content even to about 95%. However, the fruit quality after the harvest periods is further reduced since their large water content and food stored are gradually lost depending on the storage conditions and temperature which determine the evaporation rate (Arias and Toledo 2000). Unfortunately, the loss of water and food stored in the fruits and vegetables cause it to decompose rapidly making it unfit for human consumption.

In most small-scale farms and garden, importance is usually not given to fruits and vegetables after harvest and during storage (Rahman et al. 2009). These women farmers find post-harvest operations rather cumbersome, and hence, attaining good-quality storage of fruits and vegetables though vital seems almost an impossible task for them (Gordon et al. 2002), since rural women farmers lack refrigerators and other improved strategies to store fruit products before sale at the local markets to consumers (IPCC Land Use Report 2019).

When the storage periods of fruits do not meet up with the maximum temperature and humidity requirements, damages become inevitable. This grossly affects women farmers' inputs such as time, money, and labor in addition to other silent socioeconomic and gender-related limitations. Unfortunately, they encounter losses, hardship, food insecurity, and livelihood risks as well. All these uprising issues are sponsored by climate change which triggers quick growth of pests, diseases, and thus rapid decomposition in fruits.

Notably, when humans are exposed to toxic contents in fruits, it may be harmless at first dose, depending on the particular toxic agent present. But repeated and accumulated doses over time may affect physiological functioning of the body or health disorders. These health and livelihood risks necessitate the need for improved storage and production of high-quality fruits, which is grossly lacking in rural women farmers in developing countries like Nigeria (Stathers et al. 2013).

Appropriate precautions should be taken at the time of transportation and regular monitoring of toxic substances such as heavy metals and pesticides in fruits. This is to disallow the accumulation of these toxic substances linked to food products consumed by humans in any way in order to avoid consumption of contaminated food stuffs. There should be increase in awareness of dangers of toxic substance contamination in food intake to consumers and fruit sellers. Regular monitoring team by the Ministry of Environment and the Ministry of Agriculture is advised to checkmate the market conditions of the study area to make sure that fruits are sold in a hygienic manner free from health hazards.

18.12 Community Adaptation and Mitigation of Climatic Risks

Clearly in Nigeria, climate change has brought about the insurgence of mainly erratic rainfall patterns and extreme high temperatures (Nelson et al. 2010). This has affected agricultural production especially fruit and vegetable farming (Okeleye et al. 2016) and rural farmers who happen to be largely women dependent on the sale of these fruits in their local markets for survival (Williams et al. 2017; Datta 2013).

Notably, excessive heat damages fruits and affects the physical qualities, thereby causing internal damages and physiological disorders in pre-harvest seasons; thereby, their yield and quality are greatly reduced as confirmed by this present study on the communities in Ideato North LGA (Spaldon et al. 2015; Deuter 2014). However, most rural areas have improvised adaptable measures toward climate change impacts on fruit farming and delivery to consumers. Small-scale farmers have reverted to employ intervention measures during cultivation, growth, and post-harvest seasons for better practices to preserve fruit quality (Datta 2013).

Other climate-resilient solutions to rural fruit farming and livelihoods in Africa include the introduction of initiatives in rural areas. Fruits for Livelihoods (2019), an operational example of an initiative practiced in Embu County, Kenya, is an initiative of Mount Kenya Environmental Conservation (MKEC). The project encouraged and equipped poor farmers to grow three major fruits (mango, avocado, and macadamia nut) from a low-capacity level using scientific approaches since these fruits have high health values, so farmers can generate more income from sales at their local markets.

Similarly, another example is the ICRAF initiatives operational in numerous communities in Cameroon where over diverse indigenous fruit trees are farmed, for example, safou (*Dacryodes edulis*) and bush mango (*Irvingia gabonensis*) (Kehlenbeck et al. 2013).

Also, crop intensification or diversification is practiced by farmers in Africa. This refers to the integration between fruit trees and coffee farms near their dwelling places so the fruit trees can protect the coffee with its shade (World Cocoa Foundation 2015). Further, participatory domestication is taught to local farmers whereby farmers combine and grow healthy fruits, while it aids in propagating trees from cuttings or grafts (Degrande 2006; Leakey 2014). Also, farmers are trained on skills on managing their farms from a business perspective which promotes agribusiness and benefits all concerned farmers.

The introduction of resistant fruit species or re-inventing the right climate, for example, in Kaduna, Nigeria, homegrown apples were produced under unfavorable weather conditions where mean annual temperatures were 25.2 °C and in dry seasons as high as 40 °C. Whereas the acceptable global climate condition for growing apples is 21–24 °C, it can even tolerate winter temperatures as low as −40 °C (Premium Times 2019). Growing apple fruits under controlled environment is possible and can reduce the amount spent on importation of apples in Nigeria.

According to the world's top export list, the international purchases of imported apples cost a total $8.2 billion in 2018. Nigeria ranked the 47th largest importer of apples in the world costing about $29.7 million. In 2014, it was recorded that Nigeria imported 51,200 metric tons of apples which increased to 57,500 in 2015 but decreased quickly to 35,400 in 2016 (United States Department of Agriculture 2020). However, after the decrease in 2016, suddenly the importation rates rose a little to 35,900 metric tons in 2017 and 40,000 in 2018. Arguably, the consistent increase in the importation rates of apples is an indication of a high demand for the fruit and, more so, a dire need for improved technology for growing apples in Nigeria to reduce importation costs.

18.13 Practices Toward Eco-intensification of Fruit Farming in Nigeria for Agricultural Sustainability

According to the World Bank (2018), more than half of Nigerians cannot live within the global poverty limit which is $2 per day. In addition, Nigeria has a population of about 200.96 million people, and out of this population, about 23.1% are in dire need of jobs and a source of livelihood. The present administration has set targets on how to provide employment to the Nigerian populace through agriculture. Two targets were placed in motion; The first was to employ agriculture as a means to ensure that Nigeria food importation cost covering about 6 billion dollars yearly is drastically reduced. The second was to grow food production levels beyond local consumption to exportation levels, thereby reducing sole dependency of the Nigerian economic sector on crude oil which has been the main source of exportation and foreign exchange (Wheatly 2017).

This is expedient since the population of employable youths in Nigeria is fast rising at over 3.5m annually and a heightened unemployment rate above 60% is a glaring threat to national development. There is a need to introduce fruit farming as a viable means of job allocation beyond small-scale levels.

These smallholders' farms can be organized into cooperatives where subsistence farmers can grow to commercial levels. Babban Gona introduced his operation located in Kaduna State, Northern Nigeria, where uneducated farmers are taught to employ new strategies especially interviews and tests specially prepared for them. Then they are further taught on managerial skills and techniques to become experts in conflict resolution and equally provided with aids to create small cooperatives that can become large commercial agriculture.

Each farmer present is given a loan worth $1000 (USD) which covers a year's worth of farm facilities such as seeds and fertilizer. It helps to complete a planting season for cereals such as maize and rice. An experienced supervisor observes this process with keen interests using digitalized mobile apps or accessories which can store and extract information such as chlorophyll levels in crops. Also, the farmer is provided with services during harvest to manage the products such as storage facilities and marketing strategies that will yield greater income from sales of crops to about 2.3 times the national average.

New memberships of 8000 farmers were added to the project in 2016, and in 2017, the membership rose to over 20,000 farmers who farmed on 14,000 hectares. The operator is expectant that new memberships may double the previous numbers and projections are aimed at 1 million farmers by 2025. It is also projected that 1 billion (USD) will be required to carry out this project that would provide employment to over 5 million people as farmers (Wheatly 2017).

18.14 Research and Developmental Activities

Findings from studies showed that climate change will have huge impacts in sub-Saharan Africa specially Nigeria. It is also confirmed that climate change will affect various sectors in Nigeria especially water resources, agriculture (crop production and livestock), forest and forestry, coastal areas, energy, human health, settlements, tourism, population, industry, transport, and gender issues. Notably, it is expedient that adaptation strategies should be set in motion to combat its consequences (Oladipo 2008).

The most vulnerable groups are farmers, fishermen, elderly, children, and the poor. It is critical that adaptation, mitigation, and resilience strategies should be incorporated in policy planning toward the agricultural sector (Raj et al. 2019a, b; Jhariya et al. 2021a, b). This is particularly pertinent in Nigeria because of its dependency in agricultural production coupled with the increasing population growth. It is noted that agricultural processes have huge potentials for global warming due to deforestation. Obviously, in Nigeria, the economic and social benefits are pursued while neglecting the associated environmental-health consequences.

There is a need for community-based participatory approach where these women's contributions and challenges in fruit agribusiness can be voiced. This will aid in the implementation of expedient innovations that will transform their subsistence-based approach in fruit business to agro-based commercial and industrialized approach. Finally, planetary health concerns should be taken into consideration regarding pre-harvest, harvest, post-harvest, and sanitary conditions that would minimize the growth of microorganisms and disease growth and, hence, the quality of fruits transported to the market to be sold.

18.15 Policy and Legal Framework

Nigeria has taken various strides toward resilience in global warming and its attendant climate change problems. This was through several policies and strategic interventions that would be employed can provide adaptive as well as mitigative measures. The First National Communication (FNC) was given in November, 2003 (Federal Government of Nigeria 2003), followed by a Second National Communication (SNC) in December 2009, and a National Adaptation Strategy and Action Plan (NASPA) was concluded in 2011.

In 2015, a National Policy on Climate Change was developed as a response strategy to address the situation. Notably, Nigeria has taken strides toward the reduction of gas flaring by 8%, and further improvements are still ongoing where gas is used to generate electricity for new power stations (ICEED 2015; Olumide 2018). NPCC is working smartly toward a resilient climate change in Nigeria. Its sectorial adaptation and mitigation program covers energy, agriculture, water, coastal areas, forestry, land use, transport, health, culture, tourism, population, human settlement, and ICT. The documented details of the action plan and implementation framework are elaborated in the national policy (NPCC 2015).

18.16 Conclusion

Findings from this study showed that climate change has hugely affected fruit farming business negatively in the studied agrarian communities in Ideato North LGA. This effect is particularly on small-scale and poor women farmers whose livelihoods are dependent on fruit business. These women farmers play important roles in providing essential vitamins, minerals, dietary fiber, and, hence, food security to the rural population.

In addition, the women were faced with diverse constraints such as unfavorable climate, lack of technological innovations, poor sanitary requirements, sub-standard market facilities, lack of infrastructures and inadequate labor/farm lands, and lower prices of fruit products among others. These constraints particularly climate change have affected the farming periods especially post-harvest season which affected the quality of the fruit products, livelihood of these women, and, hence, associated community health risks. Findings showed that these women were married with large families and they support them with sales from fruit products.

Further, 7 out of 15 analyzed toxic substances revealed that these fruits sold at the local markets in Ideato LGA were contaminated with pesticides aldrin, DDD, HCB, and endosulfan and heavy metals lead, cadmium, and nickel all above the recommended standards of WHO/FAO. There is a need for policy regulation, adaptive governance, enforcement, and rural intervention toward poor and vulnerable Nigerian women fruit farmers/sellers. This will require community mobilization using the innovative alternative participatory research approach at all levels of the government. There is need for enviornmental education on health risks, provision of facilities especially sustainable post-harvest, storage strategies, sanitary market spaces in rural areas, youth entrepreneurship, and climate-smart agriculture solutions.

18.17 Future Perspectives

Sub-Saharan Africa, particularly Nigeria, needs to tremendously increase its present levels of agricultural productivity to meet the teeming population demands and check nutrition insecurity. Nigeria must attain these prospects while pursuing resilient intervention tools to minimize climate change crisis. An inclusive, sci-tech-enabled strategy can be instrumental in providing improved livelihoods to Africa's 250 million smallholder farmers. This will provide greater opportunities for women employment and income generation in agribusiness and at the same time curb the consequences of climate change.

In addition, Food and Agriculture Organization of the United Nations (FAO) (2011) reported that women represent between 60 and 79% of Nigeria's poor rural working group. Despite the fact that women have large community groups in farming, they are oftentimes not provided with their own lands unlike their male counterpart. Women should be given access to farmlands to promote agriculture. Also, rural women lack education, and as a result, it hinders their ability to apply technological advancement and ICT in agriculture. Consequently, their chances of getting educated are greatly reduced since about 60% of the poorest women in Nigeria never attended school and 94% of them are illiterate (Egbo 1997). Also, there is a need to provide both physical (roads and electricity) and human infrastructure (extensions, financial agents, agro-dealers, and agent networks), human networks, and human interaction in order to improve market efficiency, transparency, aggregation, and integration, to deliver inputs to farmers, and to deliver farm products to market.

References

Adebulugbe A (2003) Inventory of greenhouse gases for selected sectors in Nigeria. Technical report submitted to the Federal Ministry of Environment

Agricultural Research Council (ARC) (2013) Production guidelines for summer vegetables. Vegetables and Ornamental Plants Institute, Pretoria

Akanwa AO, Ezeomedo IC (2018) Changing climate and the effect of gully erosion on Akpo community farmers. J Ecol Nat Res 2(6):2–12

Akanwa AO, Joe-Ikechebelu N (2019) The developing world's contribution to global warming and the resulting consequences of climate change in these regions: a Nigerian case. IntechOpen, London. https://doi.org/10.5772/intechopen.85052

Akinnifesi FK, Sileshi G, Ajayi OC, Chirwa PW, Kwesiga F, Harawa R (2008) Contributions of agroforestry research and development to livelihood of smallholder farmers in Southern Africa. Fruit medicinal, fuelwood and fodder tree systems. Agric J 3(1):103–121

Altieri Miguel A, Koohafkan P (2008) Enduring farms: climate change, smallholders, and traditional farming communities. Third World Network, Penang

American Public Health Association (1995) 3111B, Direct air-acetylene flame method, standard methods for the examination of metals, 20th ed. APHA, AWWA, WEF

Antisari LV, Orsini L, Marchetti G, Vianello G (2015) Heavy metals accumulation in vegetables grown in urban gardens. Agron Sustain Dev 35:1139–1147

AOAC (1990) Association of analytical chemistry. Methods Chem Anal:2217–2280

Appenroth KJ (2010) Definition of heavy metals and their role in biological systems chapter 2, Dornburgerstr 159, 07743, Jena, Germany, 2010

Arias CJ, Toledo J (2000) Manual de manejo poscosecha de frutas tropicales". FAO, TCP/PER/6713, Lima, Peru, 2000

Banerjee A, Jhariya MK, Yadav DK, Raj A (2018) Micro-remediation of metals: a new frontier in bioremediation. In: Hussain C (ed) Handbook of environmental materials management. Springer, Cham

Banerjee A, Jhariya MK, Yadav DK, Raj A (2020) Environmental and sustainable development through forestry and other resources. Apple Academic Press Inc., New York

Banerjee A, Meena RS, Jhariya MK, Yadav DK (2021) Agroecological footprints management for sustainable food system. Springer, Singapore

Bhardwaj ML (2012) Challenges and opportunities of vegetable cultivation under changing climate scenario. A training manual on vegetable production under changing climate scenario, pp 13–18

Bock B (2004) Fitting in and multi-tasking: Dutch farm women's strategies in rural entrepreneurship. Sociol Rural 44(3):245–260

Brandth B (1995) Masculinity in transition: gender images in tractor advertisements. J Rural Stud 11(2):123–133

Briz V, Molina-Molina JM, Sanchez-Redondo S, Rodriguez-Farre E, Sunol C (2011) Differential estrogenic effects of the persistent dieldrin, endosulfan and lindane in primary neuronal cultures. Toxicol Sci 120(2):413–427

Brown ME, Antle, JM, Backlund, P, CarrER, Easterling WE, Walsh MK, Ammann C, Attavanich W, Barrett CB, Bellemare MF, Dancheck V, Funk C, Grace K, Ingram JSI, Jiang H, Maletta H, Mata T, Murray A, NgugiM, Ojima D, O'Neill B, Tebaldi C (2015) Climate change, global food security, and the U.S. Food System, 146

Catrien JAM, Art T, Dewulf, Biesbroek R (2019) A critical assessment of the wicked problem concept: relevance and usefulness for policy science and practice. Polic Soc 38(2):167–179

CCSP Final Report (2008) The effects of climate change on agriculture, land resources, water resources, and biodiversity in the United States. U.S. climate change science program synthesis and assessment product 4.3 May 2008

Census of Agriculture (2012) US summary and state data census report 2012 for agriculture. US

Central Intelligence Agency (2020) Africa: Uganda. CIA gov. Retrieved 29 January 2020

Chibowski S (2000) Studies of radioactive contamination and heavy metals contents in vegetables and fruit from Lublin, Poland. Pol J Environ Stud 9(4):249–253

Ching L, Edwards S, El-Hage SN (2011) Climate change and food systems resilience in sub-saharan Africa. Food and Agriculture Organization of the United Nations (FAO), Rome

Churchman CW (1967) Wicked problems. Manag Sci 13(4):141–142

Climate Data (2012) Owerri, Imo State weather averages from 1982-2012

Coles C, Jonathan M (2011) Gender and agricultural value chains, a review of current knowledge and practice and their policy implications. ESA Working Paper No. 11-05. http://www.fao.org/3/a-am310e.pdf

D'mello JPF (2003) Food safety: contaminants and toxins. Scottish Agricultural College, Edinburgh, p 480

Datta S (2013) Impact of climate change in Indian horticulture—a review. Int J Environ Sci Technol 2(4):661–671

Degrande A (2006) Farmers' fruit tree-growing strategies in the humid forest zone of Cameroon and Nigeria. Agroforestry System

Deuter P (2014) Defining the impacts of climate change on horticulture in Australia. Reports Commissioned by the Garnaut Climate Change Review

DFID (2009) Impact of climate change on Nigeria's economy. Federal Government of Nigeria, 2003: First National Communication to the UNFCC

Doss C and SOFA Team (2011) The role of women in agriculture. ESA Working Paper No. 11-02. Rome, Italy: Agricultural Development Economics Division, Food and Agriculture Organization of the United Nations. http://www.fao.org/docrep/013/am307e/am307e00.pdf

Doss C, Kieran C, Kilic T (2019) Measuring ownership, control and use of assets, feminist economics, pp 1–25

Duku CA, de Groot M, Budding-Polo, Demissie T (2019) Sesame Uganda: climate change risks and opportunities. CRAFT

EgboB (1997) Female literacy and life choices in rural Nigeria. In: Edward V, Corson D (eds) Encyclopedia of language and education, vol 2. Springer, New York, pp 215–223

Elliot NM (2019) Defining and describing fruit quality of Mango, Its development and protection in fruit production. Acta Sci Agric 3(9):38–41

Fan S, Brzeska J, Keyzer M, Halsema A (2013) From subsistence to profit, transforming smallholder farms. International Food Policy Research Institute, Washington, DC

Federal Government of Nigeria (2003) First national communication to the UNFCC

Feed the Future (2015) Mechanization and access to markets empower women farmers in Bangladesh. Field story in feed the future newsletter. https://www.feedthefuture.gov/article/mechanization-and-access-markets-empower-women-farmers-bangladesh

Fink D (1992) Agrarian women: wives and mothers in rural Nebraska, 1880-1940. The University of South Carolina Press, Columbia

Food and Agriculture Organization of the United Nations (FAO) (2011) The state of food and agriculture 2010–2011: Women in agriculture: Closing the gender gap for development. FAO, Rome, Italy

Fourth National Climate Assessment (2017) Summary findings. Volume II: Impacts, risks, and adaptation in the United States. U.S. Global Change Research Program

Fresco LO (1998) Higher agricultural education: An opportunity in rural development for Women. Sustainable development department, Food and Agricultural Organization (FAO), for the United Nations, p 4

Fruits for Livelihoods (2019) An initiative of Mount Kenya Environmental Conservation (MKEC) in Embu County, Kenya. Funded by Guernsey Overseas Aid. 20 Million Trees for Kenya's. Oxford

Giovarelli R, Wamalwa B, Leslie H (2013) Land tenure, property rights, and gender. USAID Issue Brief. USAID. https://www.land-links.org/issue-brief/land-tenure-property-rights-and-gender

Gordon AA, Swetman, Albright K (2002) Women in post-harvest operations: Reducing the drudgery." Issues Paper – 6. Kent, England: Department for International Development. http://r4d.dfid.gov.uk/PDF/Outputs/CropPostHarvest/Issuepaper6.pdf

Grady B (2016) Kellogg's new push to help smallholder farming women. GreenBiz article. https://www.greenbiz.com/article/kelloggs-new-push-help-women-smallholder-farmers

Guest G, Macqueen KM (2007) Re-evaluating guidelines in qualitative research. In: Guest G, Macqueen KM (eds) Handbook for team-based qualitative research. Altamira Press, New York, pp 205–226

Guidelines on Metals and Alloys Council of Europe (GMACE) (2001) Download from www.foodcontentmaterials.com/materials/coe%20metals%20policy%20statement.pdf

Gunderson LH, Light SS (2006) Adaptive management and adaptive governance in the everglades ecosystem. Policy Sci 39(4):323–334

Hatfield JL, Prueger JH (2015) Temperature extremes: effect on plant growth and development. Weather Clim Extremes 10:4–10

Hughes JB, Daily GC, Ehrlich PR (1997) Population diversity: its extent and extinction. Science 278:689–692

ICEED (2015) Nigeria adopts climate change policy document

Igbine, L (2018) Nigerian centenary story: challenges and contributions of women farmers to growth and development. AgroNigeria. https://agronigeria.ng/2018/03/02/nigerian-centenary-story-challenges-contributions-women-farmers-growth-development/

Ihesinachi K, Eresiya D (2014) Evaluation of heavy metals in orange, pineapple, avocado pear and paw paw from a farm in Kaani, Bori, Rivers State. Int Res J Publ Environ Health 1(4):87–94

IPCC Land Use Report (2019) IPPC Special report on climate change and land. Summary for policymakers approved draft. https://www.Ipcc.ch

Jallow MFA, Awadh DG, Albaho MS, Devi VY, Ahmad N (2017) Monitoring of pesticide residue in commonly used fruits and vegetables in Kuwait. Int J Environ Res Public Health 14(8):833

Jamnadass RH, Dawson IK, Franzel S, Leakey RRB, Mithöfer D, Akinnifesi FK, Tchoundjeu Z (2011) CIFOR, forests, food security. Int For Rev 13(3):338–354

Jarosz L (2011) Nourishing women: toward a feminist political ecology of community supported agriculture in the United States. J Feminist Geogr Place Cult 3:307–326

Jhariya MK, Banerjee A, Meena RS, Yadav DK (2019a) Sustainable agriculture, forest and environmental management. Springer, Singapore

Jhariya MK, Yadav DK, Banerjee A (2019b) Agroforestry and climate change: issues and challenges. Apple Academic Press Inc., New York

Jhariya MK, Meena RS, Banerjee A (2021a) Ecological intensification of natural resources for sustainable agriculture. Springer, Singapore

Jhariya MK, Meena RS, Banerjee A (2021b) Ecological intensification of natural resources towards sustainable productive system. In: Ecological intensification of natural resources for sustainable agriculture. Springer, Singapore

Jiggins J, Samanta RK, Olawoye JE (1996) Improving women farmers. Access to extension services. In: Improving agricultural extension: a reference manual. FAO, Rome

Kehlenbeck K, Asaah E, Jamnadass R (2013) Diversity of indigenous fruit trees and their contribution to nutrition and livelihoods in sub-Saharan Africa: examples from Kenya and Cameroon. In: Fanzo J et al (eds) Diversifying food and diets: using agricultural biodiversity to improve nutrition and health. Routledge, New York

Khan N, Jhariya MK, Yadav DK, Banerjee A (2020a) Herbaceous dynamics and CO_2 mitigation in an urban setup- a case study from Chhattisgarh, India. Environ Sci Pollut Res 27(3):2881–2897. https://doi.org/10.1007/s11356-019-07182-8

Khan N, Jhariya MK, Yadav DK, Banerjee A (2020b) Structure, diversity and ecological function of shrub species in an urban setup of Sarguja, Chhattisgarh, India. Environ Sci Pollut Res 27 (5):5418–5432. https://doi.org/10.1007/s11356-019-07172-w

Khan N, Jhariya MK, Raj A, Banerjee A, Meena RS (2021a) Soil carbon stock and sequestration: implications for climate change adaptation and mitigation. In: Jhariya MK, Meena RS, Banerjee A (eds) Ecological intensification of natural resources for sustainable agriculture. Springer, Singapore

Khan N, Jhariya MK, Raj A, Banerjee A, Meena RS (2021b) Eco-designing for sustainability. In: Jhariya MK, Meena RS, Banerjee A (eds) Ecological intensification of natural resources for sustainable agriculture. Springer, Singapore

Kofi Annan Foundation (2015) Combatting hunger–scaling-up investment in African agriculture. Switzerland, Geneva

Kumar S, Meena RS, Jhariya MK (2020) Resources use efficiency in agriculture. Springer, Singapore

Kumar S, Meena RS, Datta R, Verma SK, Yadav GS, Pradhan GS, Molaei A, Rahman GKMM, Mashuk HA (2020a) Legumes for carbon and nitrogen cycling: an organic approach. Carbon Nitrogen Cycl Soil. https://doi.org/10.1007/978-981-13-7264-3_10.337-375

Leakey RRB (2014) Agroforestry: participatory domestication of trees. In: Van Alfen NK (ed) Encyclopedia of agriculture and food systems. Elsevier, Philadelphia

Li C, Cheng Y, Tang Q (2014) The Association between prenatal exposure to organochlorine pesticides and thyroid hormone levels in newborns in Yancheng, China. Environ Res 129:47–50

Markowitz M (2000) Lead poisoning. Pediatr Rev 21(10):327–335

Martin S, Griswold W (2009) Human health effects of heavy metals. Environ Sci Technol 15:1–6

Mbabazi J, El-hadj M, Bah M, Verdier-Chouchane A (2015) Africa Competitiveness Report 2015, Chapter 2–transforming Africa's Economies. African Development Bank. http://www3.weforum.org/docs/WEF_ACR_2015/Africa_Competitiveness_Report_2015.pdf

McNally S (2001) Farm diversification in England and Wales – what can we learn from the farm business survey? J Rural Stud 17(2):247–257

Meena RS, Lal R, Yadav GS (2020) Long term impacts of topsoil depth and amendments on soil physical and hydrological properties of an Alfisol in Central Ohio, USA. Geoderma 363:1141164. https://doi.org/10.1016/j.geoderma.2019.114164

Meena RS, Lal R, Yadav GS (2020a) Long-term impact of topsoil depth and amendments on carbon and nitrogen budgets in the surface layer of an Alfisol in Central Ohio. Catena 194:104752. https://doi.org/10.1016/j.catena.2020.104752

Meena RS, Kumar S, Datta R, Lal R, Vijaykumar V, Brtnicky M, Sharma MP, Yadav GS, Jhariya MK, Jangir CK, Pathan SI, Dokulilova T, Pecina V, Marfo TD (2020b) Impact of agrochemicals on soil microbiota and management: a review. Land (MDPI) 9(2):34. https://doi.org/10.3390/land9020034

Merriam SB (2009) Qualitative research: a guide to design and implementation. Jossey-Bass, San Francisco

Mgbenka RN, Mbah EN, Ezeano CI (2016) A review of smallholder farming in Nigeria: need for transformation. Agric Eng Res J 5(2):19–26

Mora C, Frazier AG, Longman RJ, Dacks RS, Walton MM, Tong EJ, Sanchez JJ, Kaiser LR, Stender YO, Anderson JM, Ambrosino CM, Fernandez-Silvia I, Giuseffi LM, Giambelluca TW

(2013) The projected timing of climate departure from recent variability. Nature 502 (7470):183–187. https://doi.org/10.1038/nature12540
National Bureau of Statistics (2017) Nigeria gross domestic product report. Agriculture and Forestry Data Report (Q3,2017), November. Federal Government of Nigeria (FGN). http://wwwnigeriastat.gov.ng/nada/index.php/catalog
National Population Commission (2006) Population distribution by sex, state, LGAs and senatorial district: 2006 Census Priority Tables Abuja, Nigeria
NEEDS (2010) National Environmental, Economic and Development Study for Climate Change in Nigeria (Final Draft) Federal Ministry of Environment Abuja, Nigeria (special climate change unit) September 2010
Nelson V, Morton JF, Chancellor T, Burt P, Pound B (2010) Climate change, agriculture and Fairtrade: Identifying the challenges and opportunities. NRI working paper series
Nigerian Erosion and Watershed Management Project (2014) Final Report Environmental and Social Management Plan (ESMP) for the Umueshi Gully Erosion Site, Ideato South, Imo State, Nigeria prepared for Imo State Nigeria Erosion and Watershed Management Project (IMONEWMAP)
Nigerian Meteorological Agency (NIMET) (2018) Seasonal rainfall predictions. In: A report by NIMET; 2018. p 47
Nigerian Population Review (2019) World population prospects. http://worldpopulationreview.com
NPCC (2015) National policy on climate change in Nigeria. Federal Ministry of Environment. Department of Climate Change, FCT, Abuja, Nigeria
O'Sullivan M (2017) Gender and property rights in Sub-Saharan African. A Review of constraints and effective intervention. Policy research working Paper 8250. World Bank, Washington, DC
Obioh IB (2003) Trends in greenhouse gas emissions in Nigeria: 1988 – 2000 (Technical Report) Ibadan, Nigeria: Nigeria Environmental Study Action Team
Oboho EG, Anyia OO (1992) Agroforestry practice in semi-arid zone of Nigeria. In: Proceedings of the 22nd annual conference of the forestry. Association of Nigeria, Kano
Odjugo AO (2010a) General overview of climate change impacts in Nigeria. J Hum Ecol 29 (1):47–55
Odjugo PAO (2010b) Regional evidence of climate change in Nigeria. J Geogr Reg Plan 3 (6):142–150
Okeleye SO, Olorunfemi FB, Sogbedji JM, Aziadekey M (2016) Impact assessment of flood disaster on livelihoods of farmers in selected farming communities in Oke-Ogun region of Oyo state, Nigeria. Int J Sci Eng Res 7(8):2067–2083
Okoroafor OO, Akinbile CO, Adeyemo AJ (2018) Determination of soils erodibility factor (k) for sites in Imo State, Nigeria. Resour Environ 8(1):6–13
Oladipo E (2008) Climate change and sustainable livelihoods – the greening options. Paper presented at the First National Summit on the Environment, Abuja Nigeria
Oller (2003) Toxicological profile for nickel. Agency for toxic substances. http://www.atsdr.cdc.gov
Olumide I (2018) Can Nigeria Climate Change Policy Influence Adaptation Strategy. Nigeria News Brief and Action Alert. Nigeria Climate scorecard
PACJA (Pan African Climate Justice Alliance) (2009) The economic cost of climate change in Africa. http://www.preventionweb.net
Patton MQ (2002) Qualitative research and evaluation methods, 3rd edn. Sage, Thousand Oaks
Patton MQ (2008) Utilization- focused evaluation, 4th edn. Sage, Thousand Oaks
Peters BG (2017) What is so wicked about wicked problems? A conceptual analysis and a research program. Polic Soc 36(3):385–396
Poonam, Jhariya MK (2017) Knowledge, adoption and managerial perspectives of banana growing communities in Durg (C.G.), India. J Hum Ecol 59(1):10–19
Poonam, Jhariya MK (2019) Socio- psychological-techno and economical profile of banana growing community and its impact on adoption. Int J Chem Stud 7:31–35

Premium Times (2019) Special report: How Nigerian farmers grow apples locally amidst 'harsh' weather. Accessed on Wednesday, March 11, 2020

Rahman M, Sulfikar, Akhter S, Rahman MA, Sharmin E (2009) Capacity strengthening of rural women in carrying out post-harvest activities of vegetables and fruits towards food security. Department of Agricultural Extension Education, Bangladesh Agricultural University, Mymensingh

Raj A, Jhariya MK, Yadav DK, Banerjee A, Meena RS (2019a) Agroforestry: a holistic approach for agricultural sustainability. In: Jhariya MK, Banerjee A, Meena RS, Yadav DK (eds) Sustainable agriculture, forest and environmental management. Springer, Singapore

Raj A, Jhariya MK, Banerjee A, Yadav DK, Meena RS (2019b) Soil for sustainable environment and ecosystems management. In: Jhariya MK, Banerjee A, Meena RS, Yadav DK (eds) Sustainable agriculture, forest and environmental management. Springer, Singapore

Raj A, Jhariya MK, Yadav DK, Banerjee A (2020) Climate change and agroforestry systems: adaptation and mitigation strategies. Apple Academic Press Inc., New York

Raj A, Jhariya MK, Khan N, Banerjee A, Meena RS (2021) Ecological intensification for sustainable development. In: Ecological intensification of natural resources for sustainable agriculture. Springer, Singapore

Ramyan, Mohd TA (2018) Factors affecting fruits, vegetable and its quality. J Med Plants Stud 6 (6):16–18

Rittel HW, Webber MM (1973) Dilemmas in a general theory of planning. Policy Sci 4:155–169

Savage EP, Keefe TJ, Mounce LM, Heaton RK, Lewis JA, Bucar PJ (1988) Chronic neurological sequelae of acute poisoning. Arch Environ Health 43:38–45

Schreckenberg K, Awono A, Deegrande A, Mbosso C, Ndoye O, Tchoundjeu Z (2006) Domesticating indigenous fruit trees as a contribution to poverty reduction. For Trees Livelihood 16:35–51

Shahbandeh M (2020) Leading global producers of fresh vegetables worldwide in 2017. Statista. http://www.statista.com

Simpson BM (2017) Preparing smallholders farm families to adapt to climate change. Pocket guide 2: managing crop resources. Catholic Relief Services, Baltimore

Singh IS (2009) Harvest handling and processing of fruits and vegetables. Westvillie Publishing House, New Delhi, pp 25–22

Slavin JL, Lloyd B (2012) Health benefits of fruits and vegetables. Adv Nutr 3(4):506–516

Sobukola OP, Adeniran OM, Odedairo AA, Kajihausa OE (2010) Heavy metals levels of some fruits and leafy vegetables from selected markets in Lagos, Nigeria. Afr J Food Sci 4 (2):389–393

Spaldon S, Samnotra RK, Chopra S (2015) Climate resilient technologies to meet the challenges in vegetable production. Int Res Curr Acad Rev 3(2):28–47

Srivastava RP, Kumar S (2002) Fruits and vegetable preservation principles and practice. International Book Distribution Company, Charbagoh, pp 936–939

Stadlmayr B, Charrondiere UR, Eisenwagen S, Jamnadass R, Kehlenbeck K (2013) Nutrient composition of selected indigenous fruits from sub-Saharan Africa. J Sci Food Agric 9 (11):2627–2636

Stathers T, Richard L, Brighton M, Mvumi (2013) Post-harvest agriculture in a changing climate. In: Rural 21: Focus. http://www.rural21.com/uploads/media/rural2013_01-S12-14.pdf

Sylla MB, Nikiema PM, Gibba P, Kebe I, Klutse NAB (2016) Climate change over West Africa: recent trend future projections. In: Yaro JA, Hesselberg J (eds) Adaptation to climate change and variability in rural West Africa. Springer, Cham

Takhar J, Tipping J (2008) Using focus groups for strategic planning in a SME Unit. J Cont Edu Health Profes 28:113–114

Tonkinwise C (2015) Design for Transitions - from and to what? Academia.edu. Retrieved 9 November 2017

Torfing J, Peters BG, Pierre J, Sorensen E (2012) Interactive governance. Advancing the paradigm. Oxford University Press, Oxford

Trauger A (2004) Because they can do the work: women farmers and sustainable agriculture. Gend Place Cult 11(2):289–307

United Nations (2009) Food production must double by 2050 to meet demand from world's growing population, innovative strategies needed to combat hunger, experts tell second committee. press release. The United Nations, New York

United States Department of Agriculture (2020) Nigeria fresh apples imports by year. https://www.indexmundi.com/agriculture/?country=ng&commodity=apples&graph=imports

UNSCN (2010) The sixth report on the world nutrition situation: progress in nutrition, New York, and United Nations Standing Committee on Nutrition. http://www.unscn.org/files/Publications/RWNS6/html/index.

USEPA (2007) Pesticides in water, soil, sediment, biosolids and tissue by HRGC/HRMS. USEPA Office of Science and Technology, Washington DC, p 20460

USGCRP (2014) Ch. 7: food safety, nutrition, and distribution. In: The impacts of climate change on human health in the united states: a scientific assessment. U.S. Global Change Research Program, Washington, DC, pp 189–216

Weaver TB, Ghadiri H, Hulugalle NR, Harden S (2012) Organochlorine pesticides in soil under irrigated cotton farming systems in vertisols of Namoi Valley, North-western New South Wales, Australia. Chemosphere 88:336–343

Wheatly J (2017) Nigeria's 'great farm' model bears fruit in time of high insecurity. Financial Times, November, 7, 2017. https://www.ft.com/content/a32b0c58-a75a-11e7-ab66-21cc87a2edde

WHO (2002) The world health report 2002: Reducing Risks, Promoting Healthy Life. http://www.who.int/whr/2002/en/

WHO (2003) Diet, nutrition and the prevention of chronic diseases: report of a joint WHO/FAO expert consultation, Geneva, 28 January – 1 February 2002, Technical Report Series 916, WHO/FAO. http://whqlibdoc.who.int/trs/who_trs_916.pdf

Williams PA, Crespo O, Essegbey GO (2017) Economic implications of a changing climate on smallholder pineapple production in Ghana. J Econ Sustain Dev 8:34–43

Woldu T, Tadesse F (2015) Women's participation in agricultural cooperatives in Ethiopia. International Association of Agricultural Economists Conference, August 9-14, 2015, Milan, Italy

World Bank (2013) Africa's food markets could create one trillion dollar opportunity by 2030. Press release. Washington: World Bank. http://www.worldbank.org/en/news/press-release/2013/03/04/africas-food-markets-could-create-one-trillion-dollar-opportunity-2030

World Bank (2016) Women, agriculture and work in Africa full report. http://www.worldbank.org

World Bank (2018) Poverty headcount ratio at national poverty lines-Nigeria. World Bank Global Working Group. http://data.worldbank.org

World Bank (2020) Employment to population ratio. National estimate. Uganda Data

World Cocoa Foundation (2015) Cocoa market update. World Cocoa Foundation, Washington, DC

Allelopathic Effect of *Taraxacum officinale* L. on Germination and Physiology of Wheat

19

Syeda Naila, Zahoor ul Haq, Abdulllah, and Abdul Salam

Abstract

Allelopathic plants affect other plants in their vicinity by releasing chemicals in many ways. The main factors that drag this phenomenon are allelochemicals, which after release effect the plant positively or negatively. The present study was carried out to find the allelopathic effect of dandelion (*Taraxacum officinale* L.) on wheat (*Triticum aestivum* L. var. Janbaz). Fresh dandelion was taken from the wheat fields, and their different parts were shade dried for experimental purposes. All the parts of dandelion were grounded into powder, filtered for the extract, and applied to wheat seeds germinated on twofold filter papers in petri dishes. Five replicates were taken for each treatment. After a few days of germination, it was found that dandelion extract had a prominent effect on wheat germination. The present study will provide a baseline for future researchers toward analyzing the allelopathic potential of different weeds on wheat varieties. It will guide the future researches regarding the amount of extract and biomass that should be allowed in wheat fields.

S. Naila
Department of Botany, University of Peshawar, Peshawar, Pakistan

Z. u. Haq
Department of Botany, Shaheed Benazir Bhutto University, Wari Campus, Pakistan

Abdulllah
Department of Plant Sciences, Quaid-i-Azam University, Islamabad, Pakistan

A. Salam (✉)
Department of Plant Sciences, Quaid-i-Azam University, Islamabad, Pakistan

Zhejiang Key Lab of Crop Germplasm, Department of Agronomy, College of Agriculture and Biotechnology, Zhejiang University, Hangzhou, China

M. K. Jhariya et al. (eds.), *Sustainable Intensification for Agroecosystem Services and Management*, https://doi.org/10.1007/978-981-16-3207-5_19

Keywords

Allelopathy · Allelochemicals · Growth · Sustainability · Wheat · Weed

Abbreviations

ANOVA	Analysis of variance
CE	Common era
CV	Coefficient of variance
Df	Degree of freedom
F	F test
LSD	Least significance difference
MS	Mean square
P	Probability value
SS	Sum of square

19.1 Allelopathy: An Introductory Remark

Allelopathy word is derivation of *allelon*, which means "each other," and "pathos," which means "to suffer" (Gross 1999). Plants with allelopathic activities affect other flora in their vicinity through emitting chemicals by several methods. These plants possess chemicals in their leaves and fruits that are leached out to soil when the leaves fall off. These chemicals are then taken up by other plants with soil water. The chemicals can also be released through roots and transfer to soil and taken up by other plants.

Allelopathy can also be referred to the beneficial or harmful impacts of one plant on the other by producing chemical constituents and emitting into the surrounding (Cheema et al. 2004; Hadi et al. 2013). Any biological process through which an organism releases biochemicals is known as allelopathy. Organisms existing in the surrounding can be affected by these biochemicals with factors like the germination, survival, growth, and reproduction (Banerjee et al. 2020, 2021; Meena et al. 2020). Biochemicals are termed as allelochemicals. These allelochemicals affect the target organisms and community beneficially, e.g., for the management of agricultural activity like crop protection, control of weed, or crop restoration (positive allelopathy), or harmfully, e.g., autotoxicity, soil sickness, or biological infestation (negative allelopathy). The allelochemicals are one kind of secondary metabolites not necessary for the allelopathic individual's metabolism (i.e., development, reproduction, and growth). Defense against herbivory of the plant is the major importance of allelochemicals with negative allelopathic effect (Stamp 2003).

Wheat is grown on 215 million hectares area yearly, which makes it a more widely grown staple food crop than any other. Wheat is an economical food for

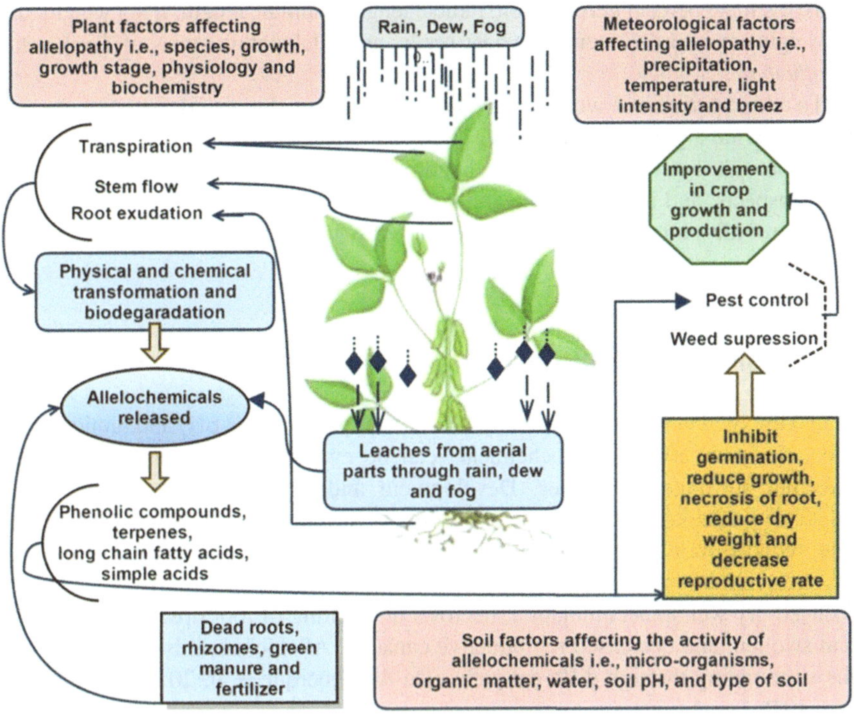

Fig. 19.1 Overall process of allelopathy and factors affecting allelopathy (Compiled: Shah et al. 2016)

majority of population of the world. About 2.5 billion people prefer to eat wheat in 89 countries, and its products are rich in carbohydrates (https://wheat.org/wheat-in-the-world). Weed infestation remains the biggest loss and a chronic problem in least yield production at Pakistan (Khaliq et al. 2013). About 48% wheat yield losses are due to weeds (Khan and Haq 2002). However, the loss due to weeds depends on the type and density of a particular type of weed, its emergence time, and its life cycle (Fahad et al. 2015).

Many factors affect the process of allelopathy like plant taxon, soil, and environment. Precipitation, season, temperature, air, and light intensity are the environmental factors affecting allelopathic procedure. Soil factors comprise microorganisms, pH, water, type of soil, and organic matter which affect allelopathy. Growth stage, plant species, biochemistry, and physiology of plant are also responsible for process of allelopathy (Rice 2012). Overall process has been shown in Fig. 19.1. The main objectives of the chapter are as follows:

- To evaluate the allelopathic potential of *Taraxacum officinale* in relation to its various parts, i.e., leaves, flower, and root on germination and physiology of wheat.

- To evaluate percent germination, radical length, plumule length, fresh weight, dry weight, moisture content of wheat seeds, and allelopathic stress of *Taraxacum officinale*.
- To add experimental work on weeds and its allelopathic nature on crops.
- To provide a guideline for further allelopathic studies.
- To provide outcomes that help in handling *Taraxacum officinale*, its physiological aspects, and its importance about wheat.
- To provide statistical data on production versus reduction of wheat.

19.2 Allelochemicals and Historical Background of Allelopathy

In making of allelochemicals, abiotic factors (temperature and pH) and biotic factors (plants) are involved. Allelochemicals can be evaporated by stomata in leaves, affecting surrounding species. Development and growth of plants are largely affected by allelochemicals. Some of the effects are retarded or inhibited germination rate; reduced root or shoot and radical or coleoptiles extension; curling of the root axis; discoloration, increased number of seminal roots, lack of root hairs, and reduced dry weight accumulation; necrosis or swelling of root tips; seeds darkened and swollen; and decreased reproductive capacity. Allelochemicals are set free into the surrounding through different paths (De Albuquerque et al. 2011), as shown in Fig. 19.2.

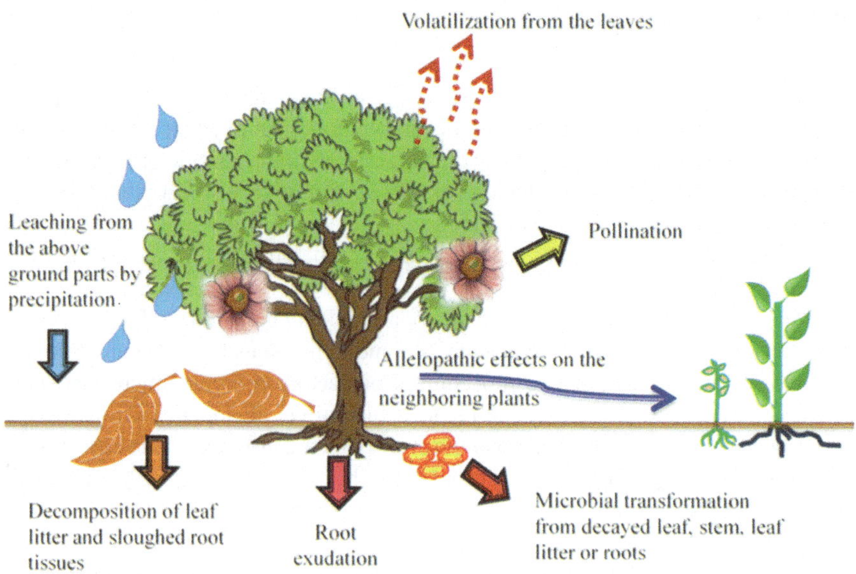

Fig. 19.2 Release of allelochemicals from plant through certain ways (Compiled: Manoel et al. 2011)

1. Release of volatile compounds from upper parts.
2. Washing of above-ground parts, that is, flower, stem, and leaves, with dew, fog, and rain.
3. Decay of plant residues.
4. Exudation of roots.

Allelochemicals comprise of several chemical families and based on chemical similarity are divided into subsequent 14 groups (Rice 1974): straight-chain alcohols, aliphatic aldehydes, water-soluble organic acids, and ketones; polyacetylenes and fatty acids; simple unsaturated lactones; benzoquinone and complex quinones; cinnamic acid; simple benzoic acid, phenols, and its derivatives; tannins; coumarin; steroids; cyanohydrins and alkaloids; peptides and protein; sulfide; and glucosinolates and nucleosides. Plant hormones (ethylene salicylic acid and gibberellic acid) are also assumed as allelochemicals. In recent years, the advance technology has made it possible to separate and identify very small concentration of allelochemicals. Figure 19.3, 19.4, and 19.5 represents some plant allelochemicals structures (Cheng and Cheng 2015).

The Austrian professor, Hans Molisch, coined the term allelopathy in 1937 for the first time. The famous book where the term is mentioned is *Der Einflusseiner Pflanze auf die andere- Allelopathie*. The term was used by him to explain biochemical interactions through which growth of neighboring plants are affected (Reigosa et al. 2006). A scientific work has been published in the journal of *Science* in 1971 by Whittaker and Feeny, who reported that the study of allelochemicals is chemical interactions between organisms (Willis 2007; Whittaker and Feeny 1971).

Further, a broad definition was presented by Rice (1984). In his research, he elaborated the term allelopathy to add all direct harmful or beneficial effect of a plant on the surrounding organisms (Rice 1984). Other researcher used the term in broader way to include the chemical interaction among organisms, in the next 10 years. In 1996 the International Allelopathy Society (IAS) defined allelopathy as the process which occurs in algae, plants, fungi, bacteria, and secondary metabolites that are produced which affect the development and growth of biological systems and agriculture (Roger 2006). The same term allelopathy was also used by zoologists for chemical interaction between invertebrates, i.e., sponges and corals (Willis 2007). The effects of one plant species on another species were noticed by common folks long before the usage of allelopathy term. The inhibitory effects of *Amaranthus* on *Medicago sativa* were observed by Theophrastus. *Shennong Ben Cao Jing* is a book on agriculture and medicinal plants presented in China in the first century CE, in which the author listed 267 plants with pesticidal abilities, along with allelopathic effects (Chou 2006).

In agriculture, a problem called soil sickness caused by crop plant exudates was proposed by the Swiss botanist, De Candolle, in 1832. According to the ecologist, the effects of competition and allelopathy could not be differentiated. Great efforts were made by some researchers in 1970, while in the 1990s other researchers stated that the effects were usually mutually dependent and could not easily be differentiated (Willis 2007). In the *Journal of Chemical Ecology*, two papers

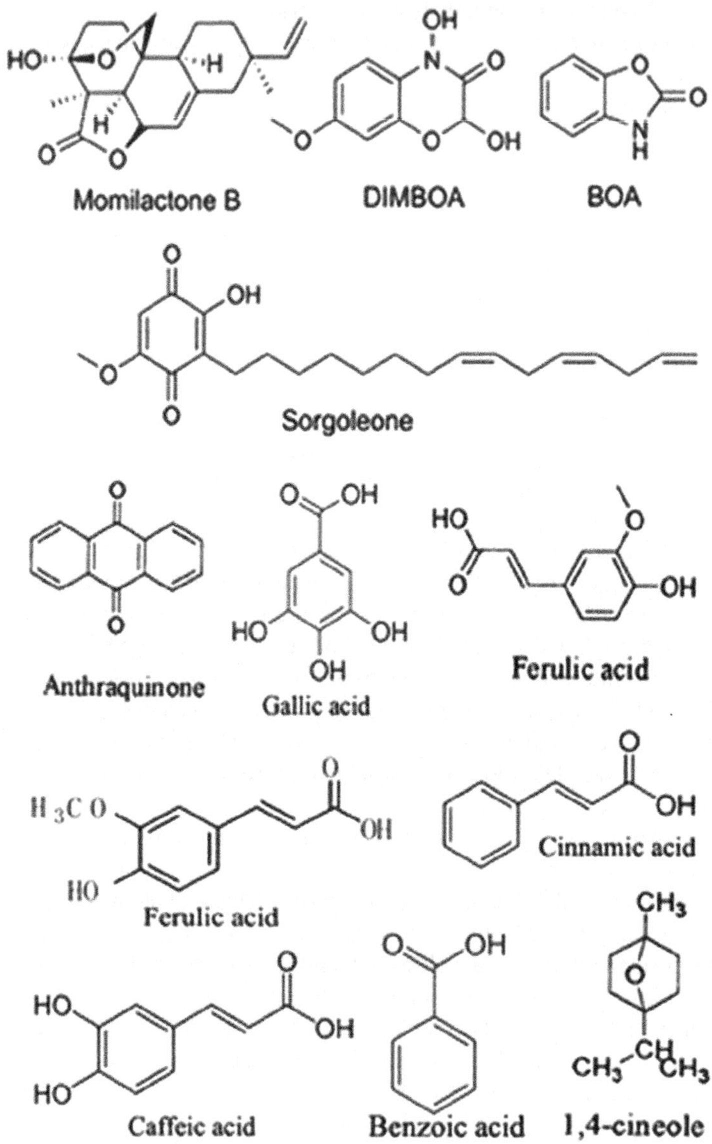

Fig. 19.3 Structure of some plant allelochemicals

presented by Liu and Lovett in 1993 reflected information regarding barley plants (*Hordeum vulgare* L.) and introduced a process for direct examination of allelochemicals and stated that allelopathic effects are not included in other competitive effects (Liu and Lovett 1993a, b). They used barley crop and introduced a

Fig. 19.4 View of wheat crop in Khyber Pakhtunkhwa, Pakistan (Photo: Mr. Afnan)

Fig. 19.5 A view of *Taraxacum officinale*

process to examine the allelochemicals directly. In Table 19.1, research work about allelopathic plants and their effect has been listed.

19.3 Allelopathic Impact and Agricultural Productivity

The significance of allelopathy in agroecosystems has attracted global attention in recent times. Allelopathy comprises of negative and positive impacts of one plant on another across the environment; however, majority of analyses appear to emphasize on its harmful effects. In controlled and natural ecosystems, it performs a main function. Allelopathic analyses were restricted to the impact of one plant on another in initial period, lacking information about the chemicals accountable for such

Table 19.1 Review of allelopathic plants and their effects from published research articles

Allelopathic plant	Impact	Reference
Juglans nigra interplanted with *Zea mays*	*Zea mays* has reduced yield because of production of juglone, an allelopathic compound from *J. nigra*	Jose et al. (1995)
Leucaena interplanted with *Triticum aestivum*, *Curcuma longa*, *Z. mays*, and *Oryza sativa*	*Triticum aestivum* and *Curcuma longa* had reduced yield, but *Z. mays* and *O. sativa* had increased yield	Dagar and Tewari (2017)
Acer rubrum, *Quercus michauxii*, *Magnolia virginiana*, and *Juniperus virginiana*	Wood extracts reduced *Lactuca sativa* seeds more than *Juglans nigra* extracts	Rathinasabapathi et al. (2005)
Eucalyptus and *Azadirachta indica*	If *T. aestivum* was grown within 5 m, a spatial allelopathic relationship exists	Saha et al. (2018)
Vitex agnus-castus, *Acer negundo*	Leachates restrict the propagation of *Digitaria eriantha* but enhance the propagation of *Andropogon*	Croom (1837)
Mangifera indica	*Cyperus rotundus* tubers completely inhibited by dried *Mangifera indica* leaf powder	El-Rokiek et al. (2010)
Ailanthus altissima	Ailanthone has shown nonselective post-emergence herbicidal activity like paraquat and glyphosate	Bhowmik (2003)
Secale cereale, *Festuca*, and *Triticum aestivum*	Weeds prohibition	Rehman et al. (2006)
Brassica oleracea var. *italica*	*Brassica oleracea* var. *italica* restrict other cruciferous crops that follow	Kalavrouziotis et al. (2008)
Echinochloa colona	*Oryza sativa* crop inhibition	Cutler and Cutler (1999)
Raphanus sativus var. *oleiformis*	Cover crop remnant reduction of weeds	Croom (1837)
Helianthus tuberosus	Restrain the growth of weeds in its vicinity	Cutler and Cutler (1999)
Helianthus annuus and *Fagopyrum esculentum*	Cover crop remnants inhibited weed pressure in *Vicia faba* crop	Richards (2001)
Medicago rigidula	*Triticum aestivum* growth inhibition and autotoxicity in *Medicago rigidula*	Krall and Legg (2012)
Crotalaria juncea	*Amaranthus hybridus* and *Lactuca sativa* growth inhibition of different green vegetables	Croom (1837)
Trianthema portulacastrum	*Amaranthus viridis* growth promoted	Al Sherif and Gharieb (2011)
Rhazya stricta	*Zea mays* growth inhibited	Musharaf et al. (2011)
Xanthium strumarium	*Vigna radiata* inhibited	Musharaf et al. (2011)
Alliaria petiolata	Inhibition of arbuscular mycorrhizal fungi colonizing on *Acer saccharum*	Evans (2006)

(continued)

Table 19.1 (continued)

Allelopathic plant	Impact	Reference
Jatropha curcas	*Nicotiana tabacum* and *Zea mays* inhibited by extracts of roots and leaves	Ma et al. (2011)
Cichorium intybus	*Amaranthus retroflexus* and *Echinochloa crus-galli* inhibited	Mansour et al. (2014)
Vincetoxicum nigrum	Invasive in northeastern United States and Canada; suppressed various weeds	DiTommaso et al. (2005)
Tephrosia vogelii	Three narrow leaf weed species and *Zea mays* growth inhibition	Wang et al. (2011)
Euphorbia	*Cicer arietinum* inhibited	Tanveer et al. (2010)
Digitaria sanguinalis	*Zea mays* and *Helianthus annuus* inhibited. *Triticosecale* not inhibited when dry *Digitaria sanguinalis* remnant was mixed into soil	Cutler and Cutler (1999)
Acacia dealbata	In Northwest Spain, native understory plants inhibited	Lorenzo et al. (2011)
Ageratina adenophora	Plants in non-native areas inhibited by volatiles. Plants in native areas showed no inhibitory growth	Cutler and Cutler (1999)
Parthenium hysterophorus	Cereal crops effected by aqueous extracts negatively	Rashid et al. (2008)
Tectona grandis	*Echinochloa colona* and Cyperaceae members inhibited by leaf extracts. *Oryza sativa* not affected	Flamini (2012)
Polypogon monspeliensis	*Triticum aestivum* inhibited by leaf extracts and mulch	Shah et al. (2018)

impacts. Allelopathy became a multidisciplinary subject by dynamic participation of scientists belonging to several fields. Crop allelopathy can be beneficial to lessen severe issues in the recent agricultural productions like environmental pollution, human health concerns, unsafe products, soil sickness, reduction in crop diversity, and decrease of crop productivity (Khanh et al. 2005; Kumar et al. 2020a; Meena et al. 2020a). The term allelopathy is considered only in terms of negative interactions, since its initiation in 1937. But of late, it is being realized that, if suitably controlled, this phenomenon can be utilized for enriching the crop productivity. The results (Figs. 19.6, 19.7, 19.8, 19.9, 19.10, 19.11 and 19.12) showing the enhancement of crop production due to allelopathic interactions are rising. This exploitation can be attained by controlling of toxic agricultural weeds, management of pests, control of crop diseases, and maintenance of nitrogen in croplands.

Plants produce allelochemicals which own a vast variety of biological actions proceeding to a diversified interaction with plants and microbes. These interactions lead to recognition of new molecules prepared to attack plant in respect to protection of plant. Recently, the practical application of developing pest-resistant cultivars by

Fig. 19.6 Control treatment

Fig. 19.7 Root 5 g

applying analyses to catch whether allelochemicals and resistance is significant statistically. Recent findings have revealed that allelochemicals frequently convey plant resistance to pathogens, insects, second by their liberation into the surrounding, regulate growth, development, and distribution of plants as well as weeds (Wink 1987). For herbicidal purpose, several plants showed inhibitory effect against weeds. Rizvi et al. (1980) stated that *Coffea arabica* seed extract showed strong inhibition. The 1,3,7-T, an alkaloid, was extracted which was enhancing the weeds' inhibition (Rizvi et al. 1981). On the other side, many plants produce allelochemical before or after a pathogenic attack which reduce the plants tolerance to diseases (Rice 1984). If

Fig. 19.8 Root 10 g

Fig. 19.9 Leaves 5 g

allelochemicals can trigger disease resistance in plants, exploitation of their fungitoxic activities needs to be explored. Yet, this part of allelopathy, as related with others, which has persisted was chiefly ignored. Consequently, fungitoxicity of chemicals, concerned in other features of allelopathy, was tested against several plant pathogenic fungi. The antifungal activity of 1,3,7-T was assessed in vitro against *Helminthosporium maydis*, a parasitic pathogen of maize where it was proved to have effective impact on the species (Rizvi et al. 1988).

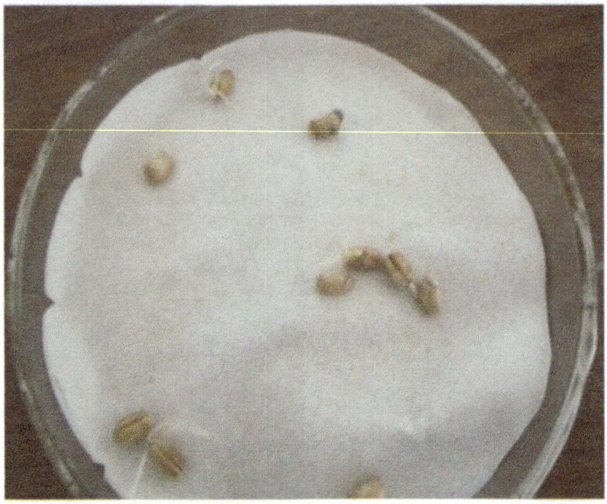

Fig. 19.10 Leaves10 g

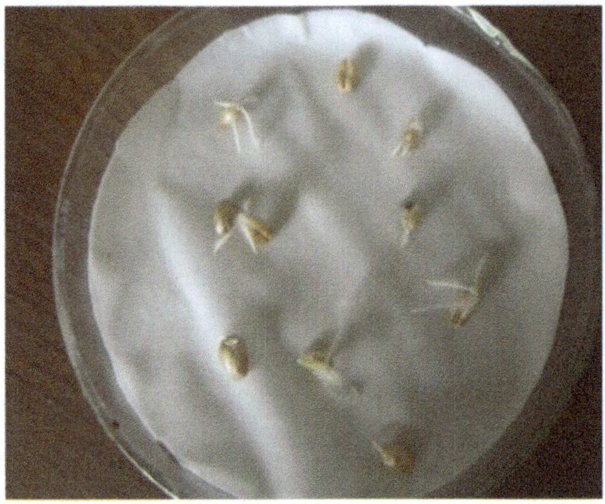

Fig. 19.11 Flower 5 g

19.4 Wheat Productivity and Distribution in Pakistan

Wheat belongs to family Poaceae and native to the Southwest Asia and Mediterranean region. A distichous spike type of inflorescence is present. Spikelets are solitary at the nodes of the tough or fragile rachis which are laterally compressed. 2–6(−9) flowers are present in inflorescence, and the upper 1 or 2 florets are usually sterile. Asymmetrical, subequal, chartaceous, or rarely membranous glumes occur. It is grown in the world for human consumption. Aside from human food, chief parts of it are utilized for livestock forage and biofuel. Wheat is high in carbohydrates, protein, and vitamins and an important source of calories. Wheat flour is used to make chapatti, bread, noodles, biscuit, and pasta. Flour quality depends upon starch

Fig. 19.12 Flower 10 g

and protein (Meena et al. 2020b; Ahmad and Hassan 2015). Based upon solubility, wheat grain protein is classified as monomeric (gliadins, globulins, albumins) and polymeric (glutenins) (Hurkman et al. 2009). Rheological properties, elasticity, and viscosity of flour is determined by balance of two proteins, i.e., monomeric and polymeric (Labuschagne et al. 2009).

Being an agricultural country, Pakistan grows wheat as the major crop cultivated on an area of 8.0339 million hectares with grain production of 19.183 million tons and average grain yield of 2388 kg/ha during 2002–2003 (Anon 2005). With 651 million tons production in 2010, wheat became the third most produced cereal after *Z. mays* (844 million tons) and *O. sativa* (672 million tons). Out of total production in the world, Pakistan is contributing about 3.72% and stands as the eighth largest wheat producer. Weeds perform an extensive function for low yield in Pakistan, among other causes (Khan et al. 2018). Forty-five weed species belonging to 16 Angiospermae families were recorded from Pakistan (Qureshi and Bhatti 2001; Siddiqui et al. 2010). It is estimated by scientists that 17–25% production losses are due to allelopathic effect (Shah et al. 2006).

Various prominent findings by different researchers revealed the facts in these aspects. Naseem et al. (2009) documented the extract of sunflower and its effect on wheat productivity. The allelopathic response of rice wheat crop by *Phalaris minor* was recorded by Om et al. (2002). Tanveer et al. (2010) applied *Euphorbia helioscopia* L. extract on wheat, *Cicer arietinum* L., and *Lens culinaris* Medic. Shahid et al. (2006) enlisted the effect of different weeds extract on wheat. The *Prosopis juliflora* extract was applied on wheat, and different parameters, i.e., leaf, shoot, and roots, were investigated by Siddiqui et al. (2009).

19.5 Ecology and Distribution of Dandelion (*Taraxacum officinale*) in Pakistan

Dandelion is an herbaceous, flowering perennial plant from Asteraceae or Compositae family. It is an inhabitant of temperate regions in the world, found in lawns, shores of water ways, on roadsides, disturbed banks, and moist areas. Dandelion is known for its weedicide nature but is utilized in food items and as a medicinal plant also (tonic, diuretic, and slightly aperient). It is applicable for treatment of gallstones, pile, kidney, and liver mixtures. It can be eaten as green salad and vegetable. From its flower, dandelion wine is obtained (Cavieres et al. 2008).

This plant has attractive yellow-colored flower heads that become spherical balls of silver tufted fruits which spread by wind named "blow balls" and "clocks" (in both British and American English). Blooming period is from April to September. The dandelion is commonly colonized in disturbed areas and propagates through windblown seeds, and seed germinates from the seed bank. For several years, seeds exist viable in seed banks and can germinate up to 9 years. The production rate of seed is 54–172 per head and exceed than 5000 per plant in a year. A crowd stand of dandelions can produce 97,000,000 seeds per hectare in a year. When released, seeds could be propagated by the wind near to hundred meters from their source. Germination of seeds are not dependent on cold temperatures before germinating, but seeds well germinates when the bulk of soil is up to 2.5 cm (0.98 in) (Fig. 19.5).

Some previous research works showed the allelopathic potential of dandelion on other plants. Jankowska et al. (2012) from their study verified the extract from the leaves of *Lolium westerwoldicum* on wheat. Marian et al. (2017) investigated that diluted extracts of *Cirsium vulgare* and dandelion sharply repressed the growth and germination of corn and beans. In this chapter allelopathic activity of *T. officinale* was investigated by applying on crop of wheat variety Janbaz.

19.6 Allelopathic Importance of Dandelion

Allelopathy plays an important role in agroecosystems. It has many applications in the appropriate crop's cultivation, weeds, insects, and disease control. According to Khanh et al. (2005), allelopathy of crops may be useful in the mitigation of various problems related to crops such as unsafe products, crop diversity depletion, reductions in crop production, soil sickness, environmental pollution, and most importantly human health. Dandelion is one among such plant species which has allelochemicals. Jankowska et al. (2012) used dandelion extract against *Lolium westerwoldicum.* The extract inhibited the germination capacity and energy. It inhibits leaf blade, leaf sheath, and roots growth. Therefore, dandelion allelochemicals could be used for various purposes in agriculture.

19.7 Allelopathic Effect of Dandelion on Wheat: A Case Study from Pakistan

Dandelion was collected in spring from wheat fields (Peshawar). For air drying, the collected leaves, roots, and inflorescence were placed apart in shade. Powders of the parts were obtained by crushing separately. Activity of plant parts was evaluated by soaking 5 and 10 g of every grinded leaves, roots, and inflorescence in 100 ml distilled water for 7 days, respectively, at 25 °C. Extracts were obtained by filtering the material. The seeds of wheat (variety Janbaz) were germinated in petri plates (10 seeds/plate) having twofold of filter paper moisturised with the plants extracts. For control data, distilled water was utilized. Five replicates per treatment were set. Data was recorded for percentage germination, growth of radical and plumule (mm), fresh weight, dry weight, and moisture content. The data was statistically analyzed using MSTATC program.

19.8 Impact of Dandelion and Allelochemicals on Wheat in Pakistan

The research experiments were performed in the Ecophysiology Laboratory, University of Peshawar, to study the allelopathic impact of dandelion on wheat. Results indicated the existence of allelochemicals in the roots, leaves, and inflorescence of dandelion which reduced and repressed the germination, moisture, and seedling growth of wheat (Table 19.2). Data revealed that the highest percentage of germination was acquired in control test and the lowest was noted in root 10 g test. The study of Hadi et al. (2013) is in line with our results, indicating the extract of roots of *Desmostachya bipinnata* inhibits the growth of wheat. As compared to 5 g, the entire 10 g extract parts inhibited the seed germination (Fig.19.13). By increasing material of plant concentration (from 5 g to 10 g), the allelopathic impact also enhances. Least significant differences (LSD) value for interaction between plant parts and concentration showed that the mean values are not significantly different from one another (Tables 19.3 and 19.4).

Table 19.2 Allelopathic impact of plant dandelion parts extracts on growth and germination parameters of wheat seeds

Treatment	GP	PL (mm)	RL (mm)	FW (gm)	DW (gm)	MC (%)
Control	94	26.82	42.114	1.248	0.506	146.64
Leaves 5 g	64	4.419	8.928	0.788	0.498	58.23
Leaves 10 g	54	1.12	3.66	0.498	0.456	9.21
Flower 5 g	72	5.29	13.977	0.676	0.474	42.61
Flower 10 g	66	3.03	7.687	0.774	0.5	54.8
Root 5 g	46	3.6825	6.415	0.714	0.478	49.37
Root 10 g	40	2.54	4.742	0.584	0.436	33.94

Note: *GP* germination percentage, *PL* plumule radical length, *FW* fresh weight, *RL* root length, *DW* dry weight, *MC* moisture contents

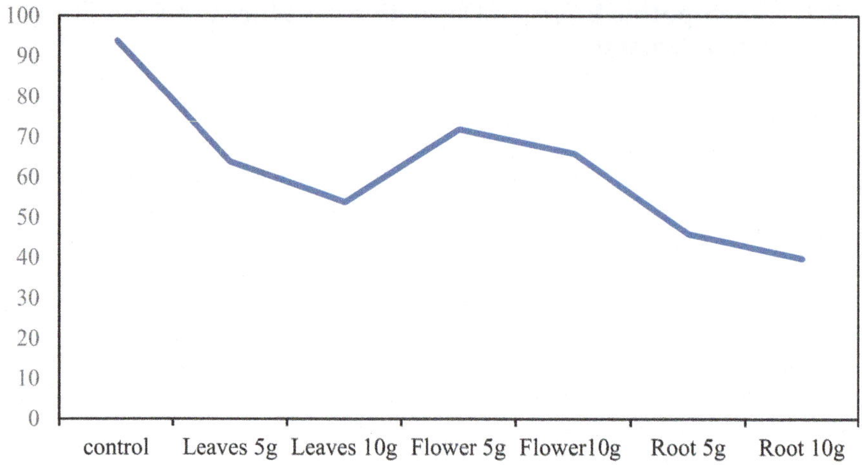

Fig. 19.13 Wheat germination percentage

Table 19.3 ANOVA for percent germination

Source	DF	SS	MS	F	P
Plant parts	2	3071.1	1535.56	8.92	0.0007
Concentration	2	12537.8	6268.89	36.40	0.0000
Interaction	4	2222.2	555.56	3.23	0.0232
Error	36	6200.0	172.22		
Total	44	24031.1			

Grand mean-72.444, CV-18.12

Table 19.4 LSD (percent germination) all-pairwise comparisons test of interaction for plant parts and concentration

Plant parts	Concentration	Mean	Homogenous groups
Flower	Control	96.000	A
Root	Control	96.000	A
Leaves	Control	94.000	A
Flower	5 g	88.000	A
Flower	10 g	66.000	B
Leaves	5 g	64.000	BC
Leaves	10 g	54.000	BCD
Root	10 g	48.000	CD
Root	5 g	46.000	D

Alpha–0.05, Standard error for comparison (SE)–8.2999, Critical T value–2.028, Critical value for comparison (CV)–16.833
There are four groups (A, B, C, D) in which the mean value is not significantly different from one another

Control treatment showed the highest plumule length, in which 10 g leaves extract depicted lowest plumule length. The mean values of interaction do not significantly vary from one another (Tables 19.5 and 19.6, Fig. 19.13). Hadi et al.

Table 19.5 ANOVA for plumule length

Source	DF	SS	MS	F	P
Parts	2	16.52	8.26	0.23	0.7972
Concentration	2	3977.41	1988.70	54.91	0.0000
Interaction	4	133.49	33.37	0.92	0.4622
Error	36	1303.78	36.22		
Total	44	5431.20			

Grand mean–9.9311, CV–60.60

Table 19.6 LSD (plumule length) all-pairwise comparisons test of interaction for plant parts and concentration

Plant parts	Concentration	Mean	Homogenous groups
Leaves	Control	26.820	A
Root	Control	23.040	A
Flower	Control	19.640	A
Flower	5 g	5.460	B
Leaves	5 g	4.400	B
Root	5 g	3.360	B
Flower	10 g	3.020	B
Root	10 g	2.540	B
Leaves	10 g	1.100	B

Alpha–0.05, Standard error for comparison (SE)–3.8061, Critical T value–2.028; Critical value for comparison (CV)–7.7191
There are two groups (A and B). In both the groups, the mean value is not significantly different from one another

Table 19.7 ANOVA for radical length

Source	DF	SS	MS	F	P
Parts	2	189.8	94.92	2.46	0.0994
Concentration	2	9775.4	4887.70	126.84	0.0000
Interaction	4	194.5	48.63	1.26	0.3028
Error	36	1387.3	38.54		
Total	44	11547.0			

Grand mean–16.518, CV–37.58

(2013) matched our results that allelopathic effect increases with increased plant material. The maximum and minimum radical length was reported in control and leaves of 10 g tests, respectively. LSD value showed that mean values are nonsignificant for interaction (Tables 19.7 and 19.8, Figs. 19.14 and 19.15). In allelopathic studies, several researchers reported the inhibition of radical growth in different seeds, i.e., Hamayun et al. (2005), Khan et al. (2006), and Anjum and Bajwa (2007), which were also comparable with the present findings. As compared to 5 g concentrations, 10 g concentrations inhibited plumule length. Tanveer et al. (2010) and Samreen et al. (2009) also reported that high allopathic effect is the cause of more plant material.

Table 19.8 LSD (radical length) all-pairwise comparisons test of interaction for plant parts and concentration

Plant parts	Concentration	Mean	Homogenous groups
Leaves	Control	42.060	A
Flower	Control	37.320	AB
Root	Control	32.280	B
Flower	5 g	11.160	C
Leaves	5 g	8.900	CD
Flower	10 g	6.140	CD
Root	5 g	4.740	CD
Root	10 g	3.860	CD
Leaves	10 g	2.200	D

Alpha–0.05, Standard error for comparison (SE)–3.9261, Critical T value–2.028, Critical value for comparison (CV)–7.9625

There are four groups (A, B, C, D) in which the means are not significantly different from one another

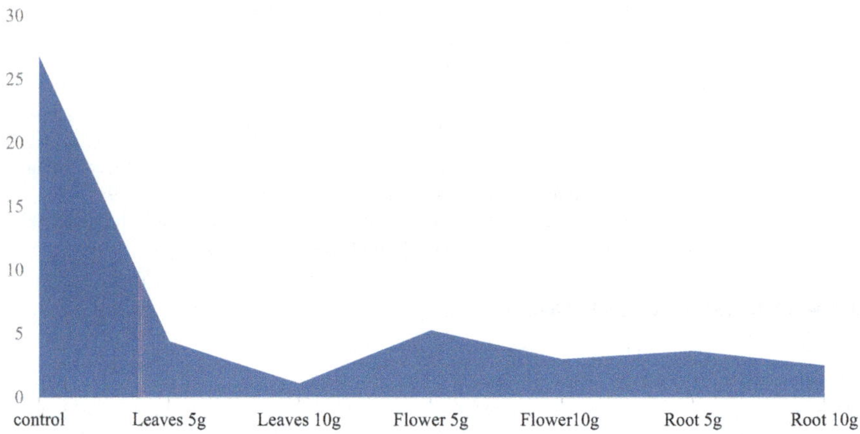

Fig. 19.14 Plumule length of wheat

The highest fresh weight of seed was shown in control treatment (1.248), whereas the lowest was shown by leaves 10 g treatment (0.498)1. LSD was also analyzed for interaction between plant parts and concentration (Tables 19.9 and 19.10; Fig. 19.16).

The highest and lowest dry weight were given by control and root 10 g treatment (0.506 and 0.436). The mean values for interaction showed nonsignificant difference (Tables 19.11 and 19.12; Fig. 19.17).

The percent moisture content of seeds was calculated. Control showed high percentage (146.64), while leaves 10 g showed lowest percentage (9.21). Mean value showed nonsignificant difference for interaction (Tables 19.13 and 19.14; Fig. 19.18).

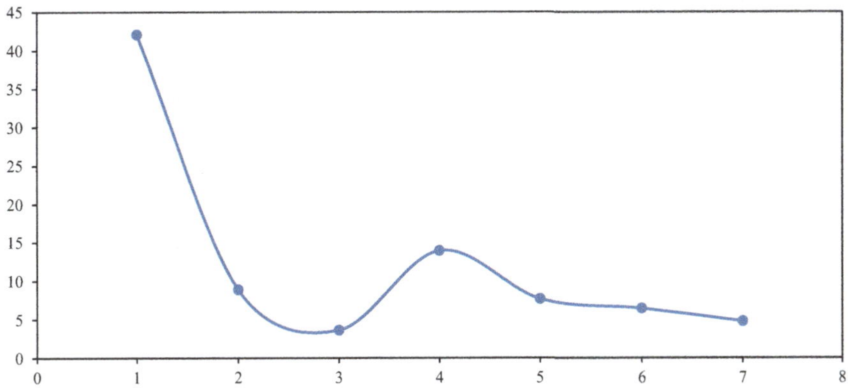

Fig. 19.15 Radical length of wheat

Table 19.9 ANOVA for fresh weight of seeds

Source	DF	SS	MS	F	P
Parts	2	0.03656	0.01828	0.50	0.6081
Concentration	2	2.13606	1.06803	29.46	0.0000
Interaction	4	0.22401	0.05600	1.54	0.2101
Error	36	1.30508	0.03625		
Total	44	3.70171			

Grand mean–0.8444, CV–22.55

Table 19.10 LSD (fresh weight of seed) all-pairwise comparisons test of interaction for plant parts and concentration

Plant parts	Concentration	Mean	Homogenous groups
Leaves	Control	1.2480	A
Root	Control	1.1720	A
Flower	Control	1.0300	AB
Leaves	5 g	0.7880	BC
Flower	10 g	0.7740	C
Root	5 g	0.7140	C
Flower	5 g	0.6760	C
Leaves	10 g	0.6180	C
Root	10 g	0.5800	C

Alpha–0.05, Standard error for comparison (SE)–0.1204, Critical T value–2.028, Critical value for comparison (CV)–0.2442
There are three groups (A, B,C) in which the means are not significantly different from one another

Fig. 19.16 Fresh weight of wheat seeds

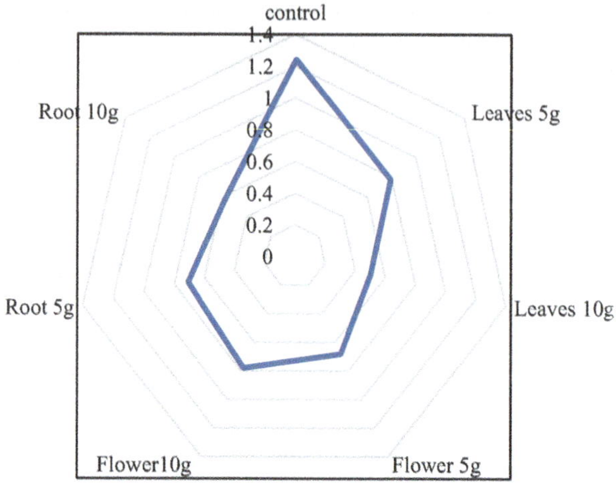

Table 19.11 ANOVA for dry weight of seeds

Source	DF	SS	MS	F	P
Parts	2	0.00608	0.00304	0.49	0.6168
Concentration	2	0.00750	0.00375	0.60	0.5523
Interaction	4	0.05614	0.01404	2.26	0.0818
Error	36	0.22364	0.00621		
Total	44	0.29336			

Grand mean–0.4809, CV–16.39

Table 19.12 LSD (dry weight of seed) all-pairwise comparisons test of interaction for plant parts and concentration

Plant parts	Concentration	Mean	Homogenous groups
Root	Control	0.5600	A
Leaves	Control	0.5060	AB
Flower	10 g	0.5000	AB
Leaves	5 g	0.4980	AB
Root	5 g	0.4780	AB
Flower	5 g	0.4740	AB
Leaves	10 g	0.4560	B
Root	10 g	0.4360	B
Flower	Control	0.4200	B

Alpha–0.05, Standard error for comparison (SE)–0.0498, Critical T value–2.028, Critical value for comparison (CV)–0.1011
There are two groups (A and B) in which the means are not significantly different from one another

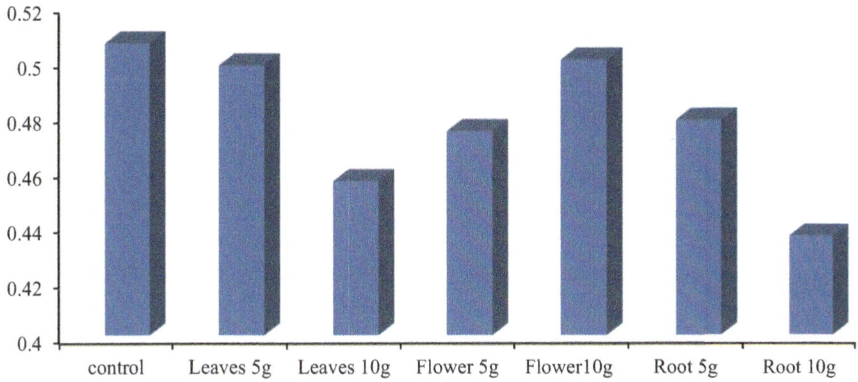

Fig. 19.17 Dry weight of wheat seeds

Table 19.13 ANOVA for moisture content of seeds

Source	DF	SS	MS	F	P
Parts	2	40	20.2	0.02	0.9786
Concentration	2	90,420	45209.8	48.41	0.0000
Interaction	4	1492	372.9	0.40	0.8078
Error	36	33,621	0.40		
Total	44	125,573			

Grand mean–81.635, CV–37.43

Table 19.14 LSD (moisture content) all-pairwise comparisons test of interaction for plant parts and concentration

Plant parts	Concentration	Mean	Homogenous groups
Leaves	Control	145.22	A
Flower	Control	144.80	A
Root	Control	144.75	A
Leaves	5 g	60.65	B
Flower	10 g	57.63	B
Root	5 g	52.36	B
Flower	5 g	46.45	B
Root	10 g	45.28	B
Leaves	10 g	37.58	B

Alpha–0.05, Standard error for comparison (SE)–19.328, Critical T value–2.028, Critical value for comparison (CV)–39.199

There are two groups (A and B) in which the means are not significantly different from one another

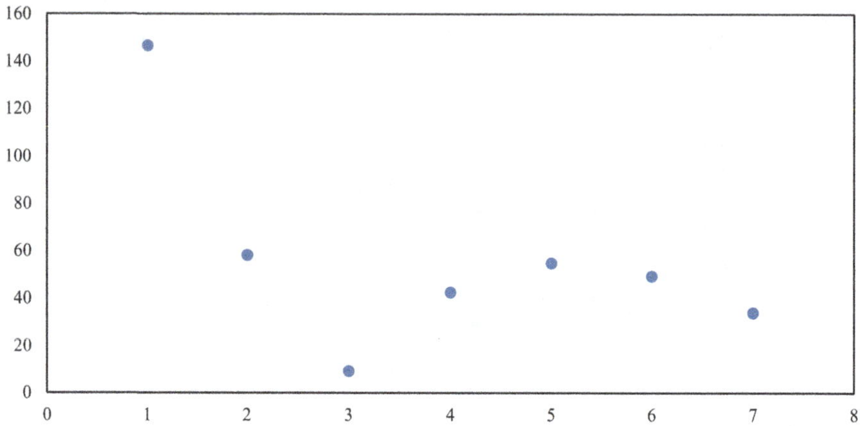

Fig. 19.18 Percent moisture content of wheat seeds

19.9 Experimental Outputs toward Agricultural Sustainability

The experimental outputs can be used in a certain way toward agricultural sustainability. Different scientists have established a distinctive scheme of utilizing the plant portions, roots and leaves, of allelopathic species to manufacture a mixture of agrochemicals to substitute traditional insecticides, fungicides, or herbicides, resulting in neglecting the remaining impacts of agrochemicals and dropping the environmental deterioration. The allelopathic genes from dandelion can be inserted into wheat which has the allelopathic ability to defeat its competitive weeds in the field by applying advanced biotechnology techniques. Hence, allelopathy has performed chief function in sustainable agriculture (Chou 2010).

In the production of biological pesticides and herbicides, allelochemicals extracted from allelopathic crops can benefit. In the formation of sustainable agriculture, a system cultivated with allelopathic crops performs central function. In upcoming agricultural making, the insertion of allelopathic characters from elevation with influential allelopathic ability to the focus crops will increase the efficiency of crop allelopathy (Khanh et al. 2005) (Fig. 19.19).

19.10 Management Aspects

Allelopathy can be utilized for better weed management either through directly using natural allelopathic interaction or as natural herbicides. Putnam and Duke (1978) explored for the first time the possibility of using allopathic plants to suppress weed growth such as agricultural crops, including the development of weed-suppressive crops and use as intercrops, or cover crops for suppression of weed (Putnam and Duke 1978). The introduction of using allelopathy into crops through breeding

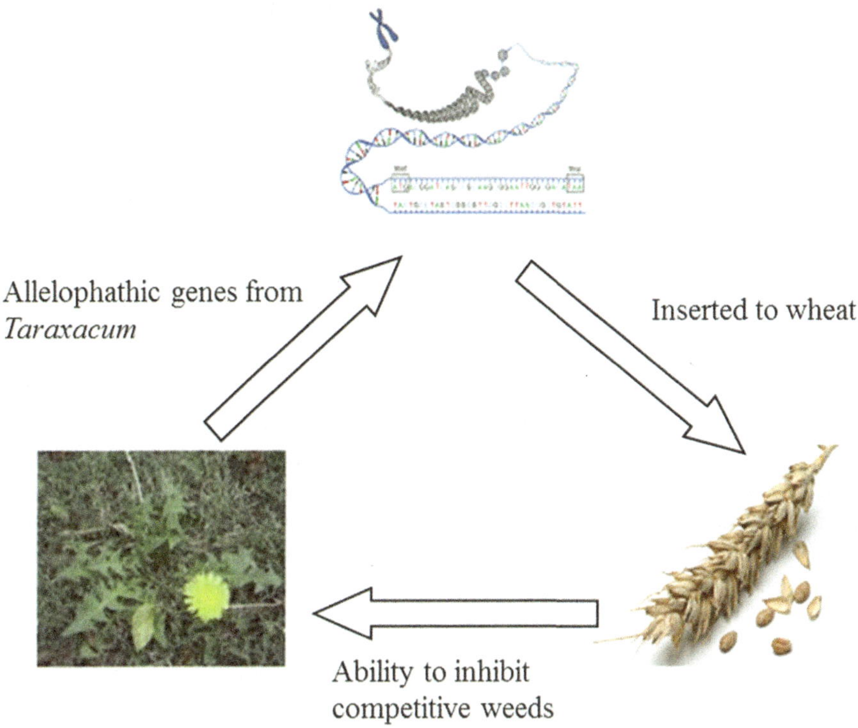

Fig. 19.19 The allelopathic genes from *T. officinale* can be inserted into wheat which has the allelopathic ability to defeat its competitive weeds in the field by applying advanced biotechnology techniques

and/or genetic manipulation could develop alternative weed management systems (Weston and Duke 2003). Still, the development of allelopathic crops by engineering natural products has many problems (detoxification, multigenetic trait, and autotoxicity). Allopathic cover crop as weed management has received greater attention in recent years because of its cost-effectiveness and environmental friendliness. Cover crops conquest the weeds, which could be accredited to physical, chemical, and biological issues (Brust et al. 2014). Surface mulch, live or desiccated, reduces the quantity and changes the value of light for photosynthesis and disturbs the germination and growth of weed species (Cheema et al. 2009; Mechergui et al. 2021). In general, mulch adjusts soil moisture, improves infiltration, and reduces evaporative water loss which leads toward soil sustainability (Jhariya et al. 2021a, b; Khan et al. 2021a, b). Allelochemicals are easily released from cover crop species in the wet soil profile (Jabran et al. 2015). For complete weed management, cover crop cultivation must be collectively performed with other cultural practices for weed control (Abdin et al. 2000). A limitation of every allelopathic plant is that it controls or defeats a very narrow group of weeds. Therefore, good agronomic practices, such as commercial crop rotation, cover crop, and sanitation practices for integrated weed management would also be necessary (Teasdale and Mohler 2000; Jhariya et al. 2019a, b; Kumar et al. 2020). The recognized properties of cover crops are to benefit the environment and sustainable agriculture (Raj et al. 2019a, b, 2020, 2021). However, there are many disadvantages and possible problems with the use of cover crops, for example, cost of establishment, nitrate leaching (if vegetables), dropping the temperatures of the soil in spring, which is squeezed out of the soil moisture in the spring, and the unknown effects of the release of natural phytotoxins in an environment that can increase insects and diseases (Lawley et al. 2012). Allelopathic plants as mulch or assimilation into soil to manage weeds in crop production can be sustained (Ashfaq et al. 2014). Allelopathic substances present in the crop can reduce the need for weed management, mostly the use of herbicides. Allelochemicals directly deals with the physiological functions of the "receiver plant" as seed germination, root shoot growth, and symbiotic efficiency or indirectly deals through improver or synergistic impact with its pathological infection, insects' injury, and environmental stress. Many crop plants with their allelopathic potential could be used as cover, smother, and green manure crop for controlling weeds by performing desired operations in the cultural practices and cropping systems. These can be properly switched or intercropped with other crops to control the desired weeds (including parasitic ones) (Dahiya et al. 2017) (Table 19.15).

19.11 Future Perspective

The projections of work on several areas referred in this chapter have been deliberated in a piece section. It can be expressed that the results reflected possible ways of determining the ability of allelochemicals for crop improvement. Allelopathy is a composite procedure. Up to now, numerous allelochemicals have been recognized. Significant additional study is needed in the area of allelochemical research, because of the different sensitivities of different receptors to the same allelochemicals and various allelopsthic activities of different allelochemicals. Meagerly is recognized about the transference and biodegradation of allelochemicals in soil, the formation of practical methods of utilizing allelochemicals, the population genetics of allelopathic species, the variety of the community of soil microbes that is preserved in their existence, the speedy modification of weeds to neglect them, or the character of signal transduction against herbivore defense. These parameters should be emphasized on upcoming explorations. Significant research exhibited that allelopathy has sound application potential in agricultural manufacture. A number of allelopathic crops have been utilized in agricultural production up till now, but the uses are restricted in regional areas and small scale. In current years, the structure and mode of action of numerous allelochemicals have been intensely exposed, and this has put an effective basis for projects where allelochemicals are utilized to attain the elementary structures or templates for producing new synthetic herbicides.

Table 19.15 Review of crop plants and their allelopathic effects on weeds

Crop species	Weed species	Allelochemical source	References
Brassica juncea	*Amaranthus retroflexus* *Polygonum aviculare*	Extract	Ercoli et al. (2005)
Helianthus tuberosus	*Amaranthus retroflexus* *Digitaria sanguinalis* *Solanum nigrum* *Chenopodium album*	Unknown	Vidotto et al. (2008)
Hordeum vulgare	*Brassica kaber* *Descurainia sophia* *Stellaria media* *Thlaspi arvense* *Bromus tectorum*	Extract of roots	Bertholdsson (2004)
Nicotiana tabacum	*Chamomilla recutita*	Roots extract	Ashihara et al. (2008)
Allium cepa	*Kochia scoparia*	Mature plants	Djurdjevic et al. (2004)
Avena sativa	*Bromus tectorum* *Thlaspi arvense* *Chenopodium album*	Plants extracts	Reeleder et al. (2004)
Oryza sativa	*Heteranthera limosa* *Bacopa rotundifolia* *Cyperus difformis* *Echinochloa crus-galli*	Rice straw, aqueousextracts	ChuiHua et al. (2004); Jung et al. (2004); Lee et al. (2004); Song et al. (2004)
Triticum aestivum	*Amaranthus albus* *Avena fatua* *Ipomea hederacea* *Chenopodium album*	Plant phenolics, wheat straw and residues, root Exudates	Belz and Hurle (2004); Krogh et al. (2006); Mathiassen et al. (2006)
Vicia villosa	*Ipomoea lacunose* 13 weed species	Seed extracts	Ercoli et al. (2005); Hill et al. (2007)
Fagopyrum esculentum	*Chenopodium album*	Aqueous extracts, growing plants, and germinating seeds	Xuan and Tsuzuki (2004)

19.12 Conclusion

The percentage of seedling germination, plumule and radical growth, dry and fresh weight, and moisture content in wheat reduced in all the treatments (roots and leaves extract especially). Similarly, the factors are affected by increasing plant parts material (from 5 g to 10 g). From these findings, it is suggested that the inhibitory effect increases with increase of plant material. These results reflect similarity with other researchers. This study shows that dandelion has allelochemicals (water leachable) and strong allelopathic effect specifically against the wheat variety Janbaz. Extracts from roots showed high allelopathic effect followed by leaves and inflorescence. This report is also supported by the work of researchers in the field of toxic potential of aqueous extracts. From this study, it is concluded that some toxic allelochemicals may be present in tested plant parts by which the growth and development of test species can be inhibited. In future studies on allelopathy, these allelochemicals need to be identified and analyzed.

References

Abdin OA, Zhou XM, Cloutier D, Coulman DC, Faris MA, Smith DL (2000) Cover crops and interrow tillage for weed control in short season maize (*Z zea mays*). Eur J Agron 12(2):93–102

Ahmad M, Hassan F (2015) Response of spring wheat (*Triticum aestivum* L.) quality traits and yield to sowing date. PLoS One 10(4):e0126097

Al Sherif EA, Gharieb HR (2011) Allelochemical effect of *Trianthema portulacastrum* L. on *Amaranthus viridis* L. supports the ecological importance of allelopathy. Afr J Agric Res 6 (32):6690–6697

Anjum T, Bajwa R (2007) The effect of sunflower leaf extracts on *Chenopodium album* in wheat fields in Pakistan. Crop Prot 26:1390–1394

Anon (2005) FAOSTAT database for agriculture. Food and Agricultural Organization. http://faostat.Fao.org/faostat/collection?subset=agriculture

Ashfaq M, Ali A, Haider MS, Ali M, Ullah A, Mubashar U, Mubashar H, Saleem I, Sajjad M (2014) Allelopathic association between weeds extract and rice (*Oryza sativa* L.) seedlings. J Pure Appl Microbiol 8:573–580

Ashihara H, Sano H, Crozier A (2008) Caffeine and related purine alkaloids: biosynthesis, catabolism, function and genetic engineering. Phytochemistry 69(4):841–856

Banerjee A, Jhariya MK, Yadav DK, Raj A (2020) Environmental and sustainable development through forestry and other resources. Apple Academic Press Inc., CRC Press- a Tayler and Francis Group, Oakville, p 400. ISBN: 9781771888110. https://doi.org/10.1201/9780429276026

Banerjee A, Meena RS, Jhariya MK, Yadav DK (2021) Agroecological footprints management for sustainable food system. Springer Nature, Singapore, p 514, eBook ISBN: 978-981-15-9496-0, Hardcover: 978-981-15-9495-3. https://doi.org/10.1007/978-981-15-9496-0

Belz RG, Hurle K (2004) A novel laboratory screening bioassay for crop seedling allelopathy. J Chem Ecol 30(1):175–198

Bertholdsson NO (2004) Variation in allelopathic activity over 100 years of barley selection and breeding. Weed Res 44(2):78–86

Bhowmik PC (2003) Challenges and opportunities in implementing allelopathy for natural weed management. Crop Prot 22(4):661–671

Brust J, Weber J, Gerhards R (2014) Do cover crop mixtures have the same ability to suppress weeds as competitive monoculture cover crops? Julius-Kühn-Archiv 443:422–430

Cavieres LA, Quiroz CL, Molina-Montenegro MA (2008) Facilitation of the non-native *Taraxacum officinale* by native nurse cushion species in the high Andes of Central Chile: are there differences between nurses. Funct Ecol 22(1):148–156

Cheema ZA, Khaliq A, Saeed S (2004) Weed control in maize (*Zea mays* L.) through sorghum allelopathy. J Sustain Agric 23:73–86

Cheema ZA, Mushtaq MN, Farooq M, Hussain A, Din IU (2009) Purple nutsedge management with allelopathic sorghum. Allelopathy J 23:305–312

Cheng F, Cheng Z (2015) Research Progress on the use of plant allelopathy in agriculture and the physiological and ecological mechanisms of allelopathy. Front Plant Sci 6:1020

Chou CH (2006) Introduction to allelopathy. Allelopathy 1:1–9

ChuiHua K, XiaoHua X, Bin Z, Fei H, ChaoXian Z, MaoXin Z (2004) Two compounds from allelopathic rice accession and their inhibitory activity on weeds and fungal pathogens. Phytochemistry 65(8):1123–1128

Croom HB (1837) A catalogue of plants, native or naturalized, in the vicinity of New Bern, North Carolina: with remarks and synonyms. GP Scott and Company, Printers, New York

Cutler HG, Cutler SJ (1999) Allelopathy of velvet bean: determination and identification of L-DOPA as a candidate of allelopathic substances. In: Biologically active natural products. CRC Press, Boca Raton, pp 49–64

Dagar JC, Tewari VP (2017) Evolution of agroforestry as a modern science. In: Agroforestry. Springer, Singapore, pp 13–90

Dahiya S, Kumar S, Khedwal RS, Jakhar SR (2017) Allelopathy for sustainable weed management. J Pharmacognosy Phytochem 6:832–837

de Albuquerque MB, dos Santos RC, Lima LM, de Albuquerque Melo Filho P, Nogueira RJMC, Da Câmara CAG, de Rezende RA (2011) Allelopathy, an alternative tool to improve cropping systems-a review. Agron Sustain Dev 31(2):379–395

DiTommaso A, Lawlor FM, Darbyshire SJ (2005) The biology of invasive alien plants in Canada. 2. *Cynanchumrossicum* (Kleopow) Borhidi [= *Vincetoxicumrossicum* (Kleopow) Barbar.] and *Cynanchumlouiseae* (L.) Kartesz& Gandhi [= *Vincetoxicumnigrum* (L.) Moench]. Canadian J Plant Sci 85(1):243–263

Djurdjevic L, Dinic A, Pavlovic P, Mitrovic M, Karadzic B, Tesevic V (2004) Allelopathic potential of *Allium ursinum* L. Biochem Syst Ecol 32(6):533–544

El-Rokiek KG, El-Masry RR, Messiha NK, Ahmed SA (2010) The allelopathic effect of mango leaves on the growth and propagative capacity of purple nutsedge (*Cyperus rotundus* L.). J Am Sci 6(9):151–159

Ercoli L, Masoni A, Pampana S (2005) Weed suppression by winter cover crops. Allelopathy J 16 (2):273–278

Evans JA (2006) Impacts of herbivores and plant communities on establishment and spread of *Alliariapetiolata* (garlic mustard) in Michigan. Michigan State University, Michigan

Fahad S, Hussain S, Chauhan BS, Saud S, Wu C, Hassan S, Tanveer M, Jan A, Huang J (2015) Weed growth and crop yield loss in wheat as influenced by row spacing and weed emergence times. Crop Prot 71:101–108

Flamini G (2012) Natural herbicides as a safer and more environmentally friendly approach to weed control: a review of the literature since 2000. Stud Nat Prod Chem 38:353–396

Gross E (1999) Allelopathy in benthic and littoral areas case studies on allelochemicals from benthic cyanobacteria and submerged macrophytes. In: Inderjit KM, Dakshini M, Foy CL (eds) Principles and practices in plant ecology Allelochemical interactions, 179Ð199. CRC Press, Boca Raton

Hadi F, Ali G, Rashid A (2013) Allelopathic potential of *Desmostachya bipinnata* (L.) P. Beauv. On wheat varieties (Ghaznavi and Tatara). Scholarly J Agric Sci 3(8):313–316

Hamayun M, Hussain F, Afzal S, Ahmad N (2005) Allelopathic effect of *Cyperus rotundus* and *Echinochloa crus-gaili* on seed germination, and plumule and radical growth in maize (*Zea mays* L.). Pak J Weed Sci Res 11:81–84

Hill EC, Ngouajio M, Nair MG (2007) Allelopathic potential of hairy vetch (*Vicia villosa*) and cowpea (*Vigna unguiculata*) methanol and ethyl acetate extracts on weeds and vegetables. Weed Tech 21(2):437–444

Hurkman WJ, Vensel WH, Tanaka CK, Whitehand L, Altenbach SB (2009) Effect of high temperature on albumin and globulin accumulation in the endosperm proteome of the developing wheat grain. J Cereal Sci 49:12–23

Jabran K, Mahajan G, Sardana V, Chauhan BS (2015) Allelopathy for weed control in agricultural systems. Crop Prot 72:57–65

Jankowska J, Ciepiela GA, Sosnowski J, Kolczarek R, Jankowski K (2012) The allelopathic effect of *Taraxacumofficinale* FG Wigg on the seed's germination and initial growth of *Loliumwesterwoldicum* R. Br. Acta Agrobot 62:207–212

Jhariya MK, Banerjee A, Meena RS, Yadav DK (2019a) Sustainable agriculture, forest and environmental management. Springer Nature, Singapore, p 606eISBN: 978-981-13-6830-1, Hardcover ISBN: 978-981-13-6829-5. https://doi.org/10.1007/978-981-13-6830-1

Jhariya MK, Meena RS, Banerjee A (2021a) Ecological intensification of natural resources for sustainable agriculture. Springer Nature, Singapore. eISBN: 978-981-334-203-3, Hardcover ISBN: 978-981-334-206-6. https://doi.org/10.1007/978-981-33-4203-3

Jhariya MK, Meena RS, Banerjee A (2021b) Ecological intensification of natural resources towards sustainable productive system. In: Ecological intensification of natural resources for sustainable agriculture. Springer, Singapore. eBook ISBN 978-981-334-203-3. Hardcover ISBN: 978-981-334-202-6. https://doi.org/10.1007/978-981-33-4203-3_1

Jhariya MK, Yadav DK, Banerjee A (2019b) Agroforestry and climate change: issues and challenges. Apple Academic Press Inc., CRC Press- a Tayler and Francis Group, Oakville, p 335. ISBN: 978-1-77188-790-8 (Hardcover), 978-0-42957-274-8 (E-book). https://doi.org/10.1201/9780429057274

Jose S, Gillespie AR, Siefert JE (1995) The micro environmental and physiological basis for temporal reductions in crop production in an Indiana alley cropping system. In: Ehrenreich JH, Ehrenreich DL, Lee HW (eds) Proceedings of the 4th north American agroforestry conference: growing a sustainable future. University of Idaho, Moscow, pp 54–56

Jung WS, Kim KH, Ahn JK, Hahn SJ, Chung IM (2004) Allelopathic potential of rice (*Oryza sativa* L.) residues against *Echinochloa crusgalli*. Crop Prot 23(3):211–218

Kalavrouziotis IK, Robolas P, Koukoulakis PH, Papadopoulos AH (2008) Effects of municipal reclaimed wastewater on the macro-and micro-elements status of soil and of *Brassica oleracea* var. Italica, and *B.oleracea* var. Gemmifera. Agric Water Mgt 95(4):419–426

Khaliq A, Gondal MR, Matloob A, Ullah E, Hussain S, Murtaza G (2013) Chemical weed control in wheat under different rice residue management options. Pak J Weed Sci Res 19(1):1–14

Khan M, Haq N (2002) Wheat crop yield loss assessment due to weeds. Sarhad J Agric 18:443–448

Khan MA, Iqbal Z, Hussain M, Rahman I (2018) Allelopathic effect of leaves on germination of wheat (*Triticum aestivum* L.). Bangladesh J Bot 47(1):151–154

Khan MA, Nawab K, Din S, Hussain N, Gul B (2006) Allelopathic proclivities of tree leaf extracts on seed germination and growth of wheat and wild oats. Pak J Weed Sci Res 12(4):265–269

Khan N, Jhariya MK, Raj A, Banerjee A, Meena RS (2021a) Soil carbon stock and sequestration: implications for climate change adaptation and mitigation. In: Jhariya MK, Meena RS, Banerjee A (eds) Ecological intensification of natural resources for sustainable agriculture. Springer, Singapore. https://doi.org/10.1007/978-981-33-4203-3_13. eBook ISBN 978-981-334-203-3. Hardcover ISBN: 978-981-334-202-6

Khan N, Jhariya MK, Raj A, Banerjee A, Meena RS (2021b) Eco-designing for sustainability. In: Jhariya MK, Meena RS, Banerjee A (eds) Ecological intensification of natural resources for sustainable agriculture. Springer, Singapore. https://doi.org/10.1007/978-981-33-4203-3_16. eBook ISBN 978-981-334-203-3. Hardcover ISBN: 978-981-334-202-6

Khanh TD, Chung MI, Xuan TD, Tawata S (2005) The exploitation of crop allelopathy in sustainable agricultural production. J Agron Crop Sci 191(3:172–184

Krall JM, Legg DE (2012) Interference of Tifton burclover residues with growth of burclover and wheat seedlings. Agron J 104(4):982–990

Krogh SS, Mensz SJM, Nielsen ST, Mortensen AG, Christophersen C, Fomsgaard IS (2006) Fate of benzoxazinone allelochemicals in soil after incorporation of wheat and rye sprouts. J Agr Food Chem 54(4):1064–1074

Kumar S, Meena RS, Jhariya MK (2020) Resources use efficiency in agriculture. Springer Nature, Singapore, p 760. https://doi.org/10.1007/978-981-15-6953-1. eBook ISBN: 978-981-15-6953-1, Hardcover: 978-981-15-6952-4

Kumar S, Meena RS, Datta R, Verma SK, Yadav GS, Pradhan GS, Molaei A, Rahman GKMM, Mashuk HA (2020a) Legumes for carbon and nitrogen cycling: an organic approach. In: Carbon and nitrogen cycling in soil. Springer, Singapore, pp 337–375. https://doi.org/10.1007/978-981-13-7264-3_10

Labuschagne MT, Elago O, Koen E (2009) Influence of extreme temperatures during grain filling on protein fractions, and its relationship to some quality characteristics in bread, biscuit, and durum wheat. Cereal Chem J 86:61–66

Lawley YE, Teasdale JR, Weil RR (2012) The mechanism for weed suppression by a forage radish cover crop. Agron J 104(2):205–214. https://doi.org/10.2134/agronj2011.0128

Lee SB, Ku YC, Kim KH, Hahn SJ, Chung IM (2004) Allelopathic potential of rice germplasm against barnyardgrass. Allelopathy J 13(1):17–28

Liu D, Lovett J (1993a) Biologically active secondary metabolites of barley I. developing techniques and assessing allelopathy in barley. J Chem Ecol 19:2217–2230

Liu D, Lovett J (1993b) Biologically active secondary metabolites of barley II. Phytotoxicity of barley allelochemicals. J Chem Ecol 19:2231–2244

Lorenzo P, Palomera-Pérez A, Reigosa MJ, González L (2011) Allelopathic interference of invasive *Acacia dealbata* link on the physiological parameters of native understory species. Plant Ecol 212(3):403–412

Ma Y, Chun J, Wang S, Chen F (2011) Allelopathic potential of *Jatropha curcas*. Afr J Biotech 10 (56):11932–11942

Manoel BA, Roseane CS, Liziane ML, Pericles M, Rejane N, Claudio C (2011) Allelopathy, an alternative tool to improve cropping systems- a review. Agron Sust Dev 31(2):379–395

Mansour SA, Khafagi OMA, Mossa ATH, El-Sanhoury RM (2014) Allelopathic effects of some botanical extracts, compared to the herbicide atrazine, against germination of selected weeds. Egyptian Acad J Biol Sci H Botany 5(1):21–38

Marian M, Voşgan Z, Roşca OM, Mihalescu L (2017) Allelopathy relationship between plants and their use in organic farming. In: IOP Conf Series: Materials Sci Engineering 200(1):012039

Mathiassen SK, Kudsk P, Mogensen BB (2006) Herbicidal effects of soil-incorporated wheat. J Agr Food Chem 54(4):1058–1063

Mechergui T, Pardos M, Jhariya MK, Banerjee A (2021) Mulching and Weed Management towards sustainability. In: Ecological intensification of natural resources for sustainable agriculture. Springer, Singapore. https://doi.org/10.1007/978-981-33-4203-3_8. eBook ISBN 978-981-334-203-3. Hardcover ISBN: 978-981-334-202-6

Meena RS, Kumar S, Datta R, Lal R, Vijaykumar V, Brtnicky M, Sharma MP, Yadav GS, Jhariya MK, Jangir CK, Pathan SI, Dokulilova T, Pecina V, Marfo TD (2020b) Impact of agrochemicals on soil microbiota and management: a review. Land (MDPI) 9(2):34. https://doi.org/10.3390/land9020034

Meena RS, Lal R, Yadav GS (2020) Long term impacts of topsoil depth and amendments on soil physical and hydrological properties of an Alfisol in Central Ohio, USA. Geoderma 363:1141164. https://doi.org/10.1016/j.geoderma.2019.114164

Meena RS, Lal R, Yadav GS (2020a) Long-term impact of topsoil depth and amendments on carbon and nitrogen budgets in the surface layer of an Alfisol in Central Ohio. Catena 2020194:104752. https://doi.org/10.1016/j.catena.2020.104752

Musharaf K, Farrukh H, Shahana M (2011) Allelopathic effects of Rhazyastrictadecne on seed germination and seedling growth of maize. African J Agric Res 6(30):6391–6396

Naseem M, Aslam M, Ansar M, Azhar M (2009) Allelopathic effects of sunflower water extract on weed control and wheat productivity. Pak J Weed Sci Res 15(1):107–116
Om H, Dhiman SD, Kumar S, Kumar H (2002) Allelopathic response of *Phalaris minor* to crop and weed plants in rice–wheat system. Crop Prot 21(9):699–705
Putnam AR, Duke WB (1978) Allelopathy in agroecosystems. Annu Rev Phytopathol 16 (1):431–451
Qureshi R, Bhatti GR (2001) Determination of weed communities in wheat field of district Sukhur. Pak J Bot 33:109–115
Raj A, Jhariya MK, Banerjee A, Yadav DK, Meena RS (2019b) Soil for sustainable environment and ecosystems management. In: Jhariya MK, Banerjee A, Meena RS, Yadav DK (eds) Sustainable Agriculture, Forest and Environmental Management. Springer Nature, Singapore, pp 189–221. eISBN: 978-981-13-6830-1, Hardcover ISBN: 978-981-13-6829-5
Raj A, Jhariya MK, Khan N, Banerjee A, Meena RS (2021) Ecological intensification for sustainable development. In: Ecological intensification of natural resources for sustainable agriculture. Springer, Singapore. https://doi.org/10.1007/978-981-33-4203-3_5. eBook ISBN 978-981-334-203-3. Hardcover ISBN: 978-981-334-202-6
Raj A, Jhariya MK, Yadav DK, Banerjee A (2020) Climate change and agroforestry systems: adaptation and mitigation strategies. Apple Academic Press Inc., CRC Press- a Tayler and Francis Group: Burlington. 383, ISBN: 9781771888226. https://doi.org/10.1201/9780429286759
Raj A, Jhariya MK, Yadav DK, Banerjee A, Meena RS (2019a) Agroforestry: a holistic approach for agricultural sustainability. In: Jhariya MK, Banerjee A, Meena RS, Yadav DK (eds) Sustainable agriculture, Forest and environmental management. Springer Nature, Singapore, pp 101–131. eISBN: 978-981-13-6830-1, Hardcover ISBN: 978-981-13-6829-5
Rashid H, Khan MA, Amin A, Nawab K, Hussain N, Bhowmik PK (2008) Effect of *Parthenium hysterophorus* L. root extracts on seed germination and growth of maize and barley. Am J Plant Sci Biotech 2(2):51–55
Rathinasabapathi B, Ferguson J, Gal M (2005) Evaluation of allelopathic potential of wood chips for weed suppression in horticultural production systems. Hort Science 40(3):711–713
Reeleder RD, Capell BB, Roy RC, Grohs R, Zilkey B (2004) Suppressive effect of bark mulch on weeds and fungal diseases in ginseng (*Panaxquinquefolius* L.). Allelopathy J 13(2):211–232
Rehman M, Hansen JL, Zemetra RS (2006) Hybrids and Amphiploids of *Aegilops cylindrica* with *Triticum aestivum* L.; production morphology and fertility. Pak J Biol Sci 9:1563–1566
Reigosa MJ, Pedrol N, Gonzalez L (2006) Allelopathy: a physiological process with ecological implications. Springer, Dordrecht
Rice EL (1974) Allelopathy. Academic Press, New York
Rice EL (1984) Allelopathy, 2nd edn. Academic Press, New York, p 1422
Rice EL (2012) Allelopathy. Academic press, New York
Richards AJ (2001) Does low biodiversity resulting from modern agricultural practice affect crop pollination and yield? Annals Botany 88(2):165–172
Rizvi SJH, Mukerji D, Mathur SN (1980) A new report on a possible source of natural herbicide. Ind J Exp Biol 18(7):777–778
Rizvi SJH, Mukerji D, Mathur SN (1981) 1,3,7-Trimethylxanthine, a new natural herbicide. Agril Biol Chem 54:1255–1256
Rizvi SJH, Singh VK, Rizvi V, Waller GR (1988) Geraniol, an allelochemical of possible use in integrated pest management. Plant Prot 3:112–114
Saha D, Marble SC, Pearson BJ (2018) Allelopathic effects of common landscape and nursery mulch materials on weed control. Front Plant Sci 9:733
Samreen U, Hussain F, Sher Z (2009) Allelopathic potential of *Calotropis procera* (AIT.). Pak J Plant Sci 15(1):7–14
Shah AN, Iqbal J, Ullah A, Yang G, Yousaf M, Fahad S, Khan A (2016) Allelopathic potential of oil seed crops in production of crops: a review. Environ Sci Pollut Res 23(15):14854–14867

Shah N, Ahmad I, Imran M, Gul H, Babar MJ, Shah WA, Anwar S (2018) Allelopathic effect of *Callistemon lanceolatus* DC. Against two cultivars of *Triticum aestivum* L. Pure Appl Biol 7 (2):783–790

Shah SR, Ijaz AK, Shah SA (2006) Study of medicinal plants among weeds of wheat and maize in Peshawar region. Pak J Weed Sci Res 12:191–197

Shahid M, Ahmad B, Khattak RA, Hassan G, Khan H (2006) Response of wheat and its weeds to different allelopathic plant water extract. Pak J Weed Sci Res 12(1–2):61–68

Siddiqui I, Bajwa R, Huma Z, Javaid A (2010) Effect of six problematic weeds on growth and yield of wheat. Pak J Bot 42(4):2461–2471

Siddiqui S, Bhardwaj S, Khan SS, Meghvanshi MK (2009) Allelopathic effect of different concentration of water extract of *Prosopsis juliflora* leaf on seed germination and radicle length of wheat (*Triticum aestivum* Var-Lok-1). Am Eur J Sci Res 4(2):81–84

Song HK, Ahn JK, Ahmad A, Hahn SJ, Kim SH, Chung IM (2004) Identification of allelochemicals in rice root exudates at various phenological phases and their influence on barnyardgrass. Allelopathy J 13(2):173–188

Stamp N (2003) Out of the quagmire of plant defense hypotheses. Q Rev Biol 78(1):23–55

Tanveer A, Rehman A, Javaid MM, Abbas RN, Sibtain M, Ahmad AUH, Aziz A (2010) Allelopathic potential of *Euphorbia helioscopia* L. against wheat (*Triticum aestivum* L.), chickpea (*Cicer arietinum* L.) and lentil (*Lens culinaris* medic.). Turkish J Agric For 34(1):75–81

Teasdale JR, Mohler CL (2000) The quantitative relationship between weed emergence and the physical properties of mulches. Weed Sci 48(3):385–392

Vidotto F, Tesio F, Ferrero A (2008) Allelopathic effects of *Heliathus tuberosus* L. germination and seedling growth of several crops and weeds. Biol Agr Hort 26:55–68

Wang R, Yang X, Song Y, Zhang M, Hu L, Su Y, Zeng R (2011) Allelopathic potential of *Tephrosiavogelii* Hook. F.: laboratory and field evaluation. Allelopathy J 28(1):53–62

Weston LA, Duke SO (2003) Weed and crop allelopathy. Crit Rev Plant Sci 22(3–4):367–389

Whittaker RH, Feeny PP (1971) Allelochemics: chemical interaction between species. Science 171:757–770

Willis RJ (2007) The history of allelopathy. Springer Science & Business Media, Dordrecht

Wink M (1987) Chemical ecology of quinolizidine alkaloids. In: Waller GR (ed) Allelochemicals: role in agriculture and forestry, ACS Symp. Ser. 330, Amer. Chern. Soc., Washington, DC524–33

Xuan TD, Tsuzuki E (2004) Allelopathic plants: buckwheat (*Fagopyrum* spp.). Allelopathy J 13 (2):137–148

Watershed Sustainability for Agricultural Intensification

20

Surendra Kumar Chandniha, Mitthan Lal Kansal, and Diwakar Naidu

Abstract

In this study, a framework for assessing water sustainability in a watershed is suggested and applied in Piperiya watershed of Chhattisgarh, India. The framework involves the various issues and responses related to watershed hydrology (H), environment (E), life (L), and policy (P). Further, in hydrology, quantity and quality aspects are considered. A methodology has been proposed for assessing the quantity-based water sustainability index using the concepts of dependable flow and the performance parameters such as reliability, resilience, and vulnerability. The proposed methodology introduces the concept of index of optimism of the decision-maker while estimating the performance parameters. For assessing the water quality sustainability index, the weightage of quality parameters is assessed through the analytical hierarchy process (AHP), and thereafter, the quality index is assessed using the concept of fuzzy logics. A new rule base (involving five parameters and five states) is suggested for aggregation of quality parameters. The sustainability aspects related to E, L, and P are assessed using the concept of pressure, state, and response. This has been assessed using the historical secondary data. After assessing the sustainability at H, E, L, and P levels, a new fuzzy rule base (involving four parameters and five states) is

S. K. Chandniha (✉)
Soil and Water Engineering, BRSM Collage of Agricultural Engineering and Technology and Research Station, IGKV, Mungeli, Chhattisgarh, India

M. L. Kansal
Department of Water Resources Development and Management, Indian Institute of Technology, Roorkee, India

D. Naidu
Computer Science, BRSM Collage of Agricultural Engineering and Technology and Research Station, IGKV, Mungeli, Chhattisgarh, India

M. K. Jhariya et al. (eds.), *Sustainable Intensification for Agroecosystem Services and Management*, https://doi.org/10.1007/978-981-16-3207-5_20

suggested to aggregate the HELP parameters and to assess the overall water sustainability index for the watershed. The proposed methodology is explained through the case study of Piperiya watershed.

Keywords

Environment · Hydrology · Life · Policy · Reliability · Resilience · Sustainability · Vulnerability · Watershed

Abbreviations

AHP	Analytical hierarchy process
BOD	Biochemical oxygen demand
CCMEWQI	Canadian Council of Ministers of the Environment Water Quality Index
COD	Chemical oxygen demand
DO	Dissolved oxygen
E	Environment
H	Hydrology
HDI	Human development index
HQI	Hydrological Quantity Index
HWQI	Hydrological Water Quality Index
IWRM	Integrated Water Resources Management
L	Life
P	Policy
SIUWM	Sustainability index for integrated urban water management
TSS	Total suspended solids
WSI	Water Sustainability Index
WWI	Water Wealth Index

20.1 Introduction

According to Bruntland Commission report, sustainable development is the development which meets the needs of present without compromising the ability of future generations to meet their own needs (Brundtland 1987).Sustainable development relates three major sectors – economic, environmental, and social (Jhariya et al. 2019a, b, 2021a, b; Meena and Lal 2018). In order to have sustainable development, it should be based on equitable, bearable, and viable considerations as shown in Fig. 20.1.

In order to achieve continuous sustainable development in a watershed, it is desired that natural resources such as water are assessed and utilized efficiently (Wagener et al. 2010; Meena et al. 2018). Since the water requirements and

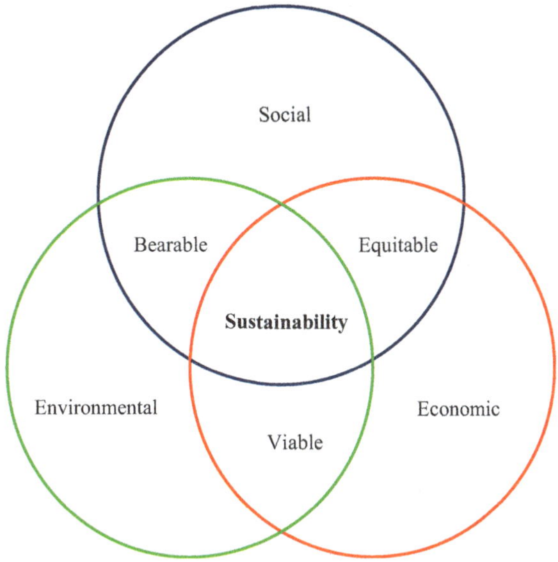

Fig. 20.1 Scope of sustainable development

availability vary in space and time, it is desired to manage the water resources so as to satisfy the demand on sustainable basis (Clark et al. 2010; Clark et al. 2014; Kumar et al. 2021; Parmar and Keshari 2012; Kumar et al. 2020). It is important to note that it is not only the quantity of water that is important but the quality of water should also be good. In many developing and developed cities, nonpayment for water services undermines the financial health of water utilities; in poorer areas, it also hinders the water utilities ability to serve poor households (Aguilar-Benitez and Saphores 2008; Aguilar-Benitez and Saphores 2010). Further, in order to achieve sustainability, it is necessary to consider social, economic, and environment aspects of water resources. However, it is difficult to bring all these indicators on a single platform. Therefore, attempts have been made by various researchers (Pitt and Clark 2008) to quantify the water sustainability in a watershed in the form of a Watershed Sustainability Index (WSI) which integrates the hydrology (H), environment (E), life (L), and policy (P) (HELP) aspects under three parameters: *pressure*, *state*, *and response*. *Pressure* addresses the human activities imposed on the watershed, *state* assesses the quality of the watershed in the base year of study as well as the quality and quantity of natural resources, while *response* examines the society's level of desire to address ecological problems in the watershed (Chaves and Alipaz 2007; Sheoran et al. 2021; Catano et al. 2009). In order to integrate the HELP issues as well as the existing pressures and policy responses in one quantitative, dynamic, and aggregated indicator, attempts have been made by various researchers (e.g., Raskin et al. 1996; Loucks and Gladwell 1999; Salameh 2000; Lawrence et al. 2002; Sullivan 2002; Meena et al. 2020; Chaves and Alipaz 2007). This philosophy is further studied, and the index framework is modified by incorporating the newer scientific knowledge such as fuzzy mathematics. The proposed framework is applied in Piperiya watershed of Chhattisgarh state in India to assess its water sustainability.

Under the proposed framework, HELP is considered to be the major indicators. In order to assign the weightage of these indicators in the overall WSI, expert opinion, literature review, and analytic hierarchy process (AHP) approach have been adopted. Further, hydrological indicator is subdivided into two components, i.e., quantity and quality, and the sustainability has been assessed for quantity and quality separately. Environmental indicator based on land use/land cover changes in the watershed level is incorporated. Life indicator, associated with human development index (HDI) in terms of income, is considered. Policy indicator is based on HDI during the study period. All the indicators deal with pressure, state, and response parameters. Finally, fuzzy-based approach has been suggested for estimation of overall WSI.

20.2 Watershed Sustainability Index and Its Application

Watershed Sustainability Index has considered mainly four indices and further sub-indicators which are enlisted in details below:

20.2.1 Hydrology

H1—Water quantity parameters.

Reliabsility-, Resilience-, and Vulnerability-Based WSI.

F1—Annual 75% dependable flow basis study.

F2—Pre-monsoon 75% dependable flow basis study.

F3—Monsoon 75% dependable flow basis study.

F4—Post-monsoon 75% dependable flow basis study.

F5—Winter 75% dependable flow basis study.

H2—Water quality parameters.

Q1—pH.

Q2—Total suspended solids (TSS).

Q3—Dissolved oxygen (DO).

Q4—5-day biochemical oxygen demand (5-day BOD).

Q5—Chemical oxygen demand (COD).

20.2.2 Environment

P3—Pressure (averaged growth rate per year of agriculture area and urban population in the watershed).

S3—State (growth rate per year of natural vegetation during the study period).

R3—Response (growth rate per year of forest area during the study period).

20.2.3 Life

P4—Pressure (growth rate of per capita income during the study period in terms of HDI_{income}).

S4—State (human development index weighted by district area involved in watershed).

R4—Response (on the basis of evolution of watershed HDI_{income}).

20.2.4 Policy

P5—Pressure (growth rate of per year $HDI_{education}$ during the study period).

S5—State (watershed institutional capacity is considered in terms of IWRM).

R5—Response (on the basis of evolution of IWRM expenditure in the study period).

The abovementioned parameters are utilized for estimation of WSI using hydrology, environment, life, and policy indicators. A new water sustainability index framework has been suggested for estimation of WSI which can be calculated by single numerical value with the help of MATLAB-based program. A detailed methodology has been illustrated in Fig.20.2.

20.3 Research Framework of Watershed Sustainability Index

Various researchers have quantified the water sustainability in terms of WSI. Some of the important works carried out in this area are summarized in Table 20.1.

Out of all the abovementioned indices, the most popular index particularly at watershed level is as suggested by Chaves and Alipaz (2007). Various researchers (Catano et al. 2009; Garriga and Agustí 2010; Meena et al. 2020a; Brown 2012; Kumar et al. 2020a; Cortés et al. 2012; Firdaus et al. 2014) have used this as basis for assessing the water sustainability at watershed level. Further, most of the researchers have considered the equal weights for all parameters/indicators. These indicators have further sub-indicators with equal weights under each of the indicator. However, it varies from case to case basis, on the basis of expert's opinion, and with the risk-taking attitude of the decision-maker(s) which are subjective in nature. Fuzzy mathematics is a branch of mathematics which can be used to quantify the qualitative parameters. This study also adopts the same structure as suggested by Chaves and Alipaz (2007) but with modification at sub-indicator levels as well with new fuzzy-based proposed method of aggregation for assessing WSI. The structure of WSI for a watershed is presented in Fig. 20.2.

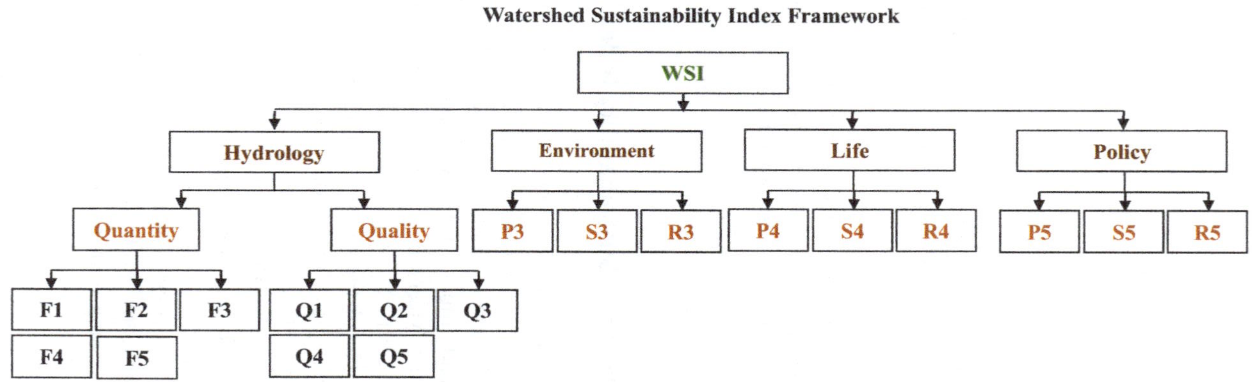

where,

WSI- Watershed Sustainability Index

Reliability, Resilience & Vulnerability Based WSI
F1- Annual 75% dependable flow basis study
F2- Pre-monsoon 75% dependable flow basis study
F3- Monsoon 75% dependable flow basis study
F4- Post-monsoon 75% dependable flow basis study
F5- Winter 75% dependable flow basis study

Fuzzy Rule Based WQI
Q1- pH
Q2- Total Suspended Solids (TSS)
Q3- Dissolved Oxygen (DO)
Q4- 5-days Biochemical Oxygen Demand (5-days BOD)
Q5- Chemical Oxygen Demand (COD)

Fuzzy Rule Based Environmental Index)
P3- Pressure (Averaged growth rate per year of agriculture area and urban population in the watershed)
S3- State (Growth rate per year of natural vegetation during the study period)
R3- Response (Growth rate per year of forest area during the study period)

Fuzzy Rule Based Life Index
P4- Pressure (Growth rate of percapita income during the study period in terms of HDI_{income})
S4- State (Human development index weighted by district area involved in watershed)
R4- Response (On the basis of evolution of watershed HDI_{income})

Fuzzy Rule Based Policy Index
P5- Pressure (Growth rate of per year $HDI_{education}$ during the study period)
S5- State (Watershed institutional capacity is considered in terms of IWRM)
R5- Response (On the basis of evolution of IWRM expenditure in the study period)

Fig. 20.2 Flow chart of watershed sustainability framework

Table 20.1 Summary of various water sustainability indices reported during last few years

Reference	Description	Remark(s) (if any)
Lawrence et al. (2002)	An international water poverty index (IWPI) is suggested that links household welfare with water availability and indicates the degree to which water scarcity impact the human being. Main components are *resources (R), access (A), capacity (C), use (U), and environment (E).* Relative position of a country is accessed through following equation: $\mathrm{IWPI} = \frac{x_i - x_{\min}}{x_{\max} - x_{\min}}$ Where, x_i, x_{max}, and x_{min} are the values of various sub-indicators/indicators for the ith country, highest and the lowest value among various countries	Various countries are listed with relative positions for each indicator. However, reliability and variability in space and time need further study
Sullivan and Jeremy (2003)	Water poverty index is established with subcomponents as access to water, water quantity, quality and variability, water uses, capacity for water management, and environmental aspects. They considered these parameters in terms of *resources (R), access (A), capacity (C), use (U), and environment (E).* The proposed index is $\mathrm{WPI} = \frac{\sum_{i=1}^{n} w_i X_i}{\sum_{i=1}^{n} w_i}$	The index considered social, financial, human, physical, and natural capitals of the country to assess the RACUE of that country. Has the same limitations as that of Lawrence et al. (2002)
Sullivan and Meigh (2005)	Introduced the climate variability index in terms of *resources (R), access (A), capacity (C), use (U), environment (E), and geospatial components (G)* $\mathrm{CVI} = \frac{r_r R + r_a A + r_c C + r_u U + r_e E + r_g G}{r_r + r_a + r_c + r_u + r_e + r_g}$ Where R, A, C, U, E, and G are the resource, access, capacity, use, environment, and geospatial components	This has been assessed for various countries in 2000 and predicted for 2030
Sullivan et al. (2006)	Water wealth index (WWI) comprises the following components: *Food security, health, productivity, institutional and human capacity, and environment* $\mathrm{WWI} = \frac{w_f F(\mathrm{NBE}, I) + w_h H(\mathrm{NBE}, I) + w_p P(\mathrm{NBE}, I) + w_e E(\mathrm{NBE}, I)}{w_f + w_h + w_p + w_e}$ NBE: Natural baseline endowment F: Food availability and vulnerability of supply related to water H: Human health issues related to water supply and sanitation P: Productivity, based on economic value and employment I: Institutional capacity including water-related infrastructure E: Ecosystem integrity, measure of disruption due to anthropogenic activity	–
Hernández (2007)	Water sustainability index for Canary Islands is built upon eight major components, namely, *natural resources, infrastructure, water quality,*	–

(continued)

Table 20.1 (continued)

Reference	Description	Remark(s) (if any)
	water quantity, efficiency, technology + research, education + social, and water economics. These are further subdivided into sub-indicators. Pressure-state-response approach is utilized for index calculation	
Chaves and Alipaz (2007)	Watershed sustainability index (WSI) developed using pressure-state-response function. Indicators used are *hydrology, environment, life, and policy issues* $\mathrm{WSI} = \frac{(H+E+L+P)}{4}$	Equal weights are given to all the indicators
Gine and Perez et al. (2008)	Aggregated index to assess water and poverty linkage to produce a holistic tool for policy making is suggested. Water poverty is expressed as a function of *availability of water resources (R), access to water (A), effective water management (C), multiple use of water (U), and water for ecological services (E).* Score between 0 and 1 (0.25, 0.50, 0.75, 1.0) is assigned to each of the parameter of every indicator $\mathrm{WPI} = \frac{(R+A+C+U+E)}{5}$	Equal weights are assigned to each indicator
Ali (2009)	Arab water sustainability index is proposed as a policy tool in terms of *water crowding, dependency, scarcity, and environmental sustainability.* Principal components analysis is used, which describes whole data set affording data reduction with minimum loss of original information	Different weightage for each indicator is suggested
Juwana et al. (2009)	West Java water sustainability index (WJWSI) is developed to outline major water issues in West Java, Indonesia. Components of WJWSI include *water resources, water provision, capacity, and human health* which are further subdivided into sub-indicators	–
De Carvalho et al. (2009)	Sustainability index for integrated urban water management (SIUWM) is developed which consists of 5 components which disaggregate into 20 sub-indicators and ultimately into 64 variables $SI_i = \frac{\sum_{i=1}^{n} Wxi\,X_i}{\sum Wxi} \times 100$ Where SI = overall sustainability index score for a particular urban area; X_i = variable; and W_{xi} = attribute weight	A framework is developed for overall sustainability index for urban area. Weights are considered on a scale of 0 to 5. Further, sensitivity analysis is also carried out
Boyacioglu (2010)	Water quality index is introduced, and samples are analyzed for pH, total dissolved solids, chlorides, nitrate-nitrogen, dissolved oxygen, biochemical oxygen demand, sulfates, and boron. The results are taken monthly over 2 years for five monitoring sites	Drinking, aquatic, and agricultural water uses are considered. The study modifies the Canadian Council of Ministers of the

(continued)

Table 20.1 (continued)

Reference	Description	Remark(s) (if any)
	$WQI = \frac{(F_1^2+F_2^2+F_3^2)}{1.732}$ Where F1: The number of variables whose objectives Are not met (scope) F2: The frequency by which the objectives are Not met (frequency) F3: The amount by which the objectives are not Met (amplitude)	environment water quality index (CCMEWQI)
Soodand William (2011)	Watershed sustainability index (WSI) is built based *on social, environment, and biodiversity indicators.* Watershed sustainability is then calculated using concepts of reliability, resilience, and vulnerability $S_{\text{watershed}} = S_{\text{Rel.}} \times S_{\text{Res.}} \times (1 - \text{Relative } S_{\text{Vul.}})$ Where $S_{\text{watershed}}$ = watershed sustainability; S_{Rel} = reliability; S_{Res} = resilience; $S_{\text{Vul.}}$ = vulnerability	–

20.4 Indicators for Water Sustainability Framework

Brief description of the indicators and sub-indicators useful for the assessment of WSI is discussed as follows.

20.4.1 Hydrology

One of the most important parameters for water sustainability in a watershed is its hydrological parameters. This consists of quantity and quality of water available in the watershed. For quantity point of view, sustainability may be considered on the basis of dependability of flow. Most of the water resources projects are designed on the basis of 75% dependable flow in a river. Therefore, in this study 75% dependable flow is considered as a threshold value. If the flow is above the threshold value, it means the flow is surplus, and if the flow is below the threshold value, it means deficit condition. Keeping this heuristic in mind, reliability, resilience, and vulnerability of the flow time series are calculated and used for estimation of water sustainability in a watershed as suggested by Sood and William (2011). However, while estimating the vulnerability of a particular flow value, the index of optimism of the decision-maker is incorporated rather than simply considering the minimum, average, or maximum value.

The second phase of hydrology consideration is the *water quality*. In the present study, as the water quality data was limited, only five water quality parameters, i.e., pH, total suspended solid (TSS), dissolved oxygen demand (DO), 5-day biochemical oxygen demand (5-day BOD), and chemical oxygen demand (COD), are considered for estimation of water quality index (WQI). Analytical hierarchy process (AHP) has been suggested for assigning the weights of each sub-indicators. While using the AHP approach, relative weights are considered on the basis of expert opinion and literature review. Further, fuzzy membership function (triangular as well as trapezoidal) is suggested for each water quality parameters.

20.4.2 Environment

The environmental parameters of a watershed are linked with its land use change in terms of environment pressure index (EPI). The EPI is assessed on the basis of 16-year time period. EPI has been determined on the basis of growth rate in agricultural area and urban population in the watershed. For the calculation of EPI, averaged growth rate is considered. Further, different membership functions have been suggested for classification of the scores. In the state sub-indicator, the growth rate per year of natural vegetation has been considered during the period of study. In response parameter, growth rate per year of forest area is considered during the period of study.

20.4.3 Life

Life indicator is assessed in the form of human development index (HDI_{income}) in terms of income as well as education. The pressure, state, and response sub-indicator has been considered for estimation of life indicator. For pressure, growth rate per year HDI_{income} has been considered during the period of study. In state, HDI weightage by area is considered in the watershed level. In the response, evolution in watershed HDI is considered in the period of study. Further, fuzzy rule base is proposed using the three parameters (pressure, state, and response) with five states such as poor, average, good, very good, and outstanding for estimation of final score of "life" parameter score.

20.4.4 Policy

In policy indicator, also pressure, state, and response sub-indicator has been considered for estimation of life indicator. For pressure, growth rate per year $HDI_{education}$ has been considered during the period of study. In state, watershed institutional capacity is considered in terms of IWRM by different departments, i.e., department of water resources, irrigation department, department of soil and water conservation, etc. In response, evolution of IWRM expenditure in the watershed is considered in the study period duration. Again, fuzzy rule base has been suggested for estimation of policy indicator score. Further, overall WSI has been calculated using the combination of HELP indicator with fuzzy rule base.

20.5 Quantitative Assessment for Indicator of Watershed Sustainability Framework

20.5.1 Hydrological Quantity Index (HQI)

Score of hydrological indicators (in terms of quantity) is assessed on the basis of time series of the discharge available at the outlet of the watershed. In order to consider the water sustainability in a watershed, a threshold value of 75% dependable flow is considered. This means that if the flow available is equal to or more than the 75% dependable flow, then the watershed is sustainable from water quantity point of view. For sustainability assessment, five-time frames are considered, i.e., annual and four seasons – pre-monsoon, monsoon, post-monsoon, and winter. The performance is assessed on the basis of three parameters, i.e., reliability, resilience, and vulnerability. The threshold value is taken as 75% dependable flow. Thereafter, Hydrological Quantity Index (HQI) is assessed using the following equation:

$$\text{HQI} = \sqrt[3]{\text{Reliability} \times \text{Resilience} \times (1 - \text{Relative Vulnerability})} \tag{20.1}$$

20.5.2 Hydrological Water Quality Index (HWQI)

Two procedures are suggested for estimation of HWQI. Steps of the first procedure are as follows:

1. Identify the water quality parameters that affect the HWQI.
2. Estimate weights of these parameters (AHP is suggested in this study Ref. Appendix VI).
3. Prepare the membership function for each of the quality parameter (triangular/trapezoidal membership functions are used in this study) (Table 20.1).
4. Fuzzy score of each parameter in a sample was studied, and a weighted score of the sample is estimated (using weight of each parameter calculated in step 2). This is termed as the HWQI for that sample.
5. In order to covert this value in qualitative term, a five-point scale is adopted to defuzzify the fuzzy score (output) as per the details given in Table 20.2.

In the second procedure, the following steps are implemented:

1. Identify the water quality parameters that affect the HWQI.
2. Prepare the membership function for each of the quality parameter (triangular/trapezoidal membership functions are used in this study).
3. Information related to membership function of each parameter (Table 20.1) and the related overall water quality as mentioned in Table 20.2 are incorporated. Table 20.2 is derived from the mutually exclusive states of 126 as shown in

Table 20.2 Membership functions and their ranges for each water quality parameter

pH	*Ranges*	*Function*
Poor	0–0–5–5.6	Trapezoidal
Average	5.5–5.75–6.1	Triangular
Good	6–6.25–6.5	Triangular
Very good	6.4–6.7–7	Triangular
Outstanding	6.9–7–7.1	Triangular
Very good	7–7.5–8.1	Triangular
Good	8–8.25–8.5	Triangular
Average	8.4–9–9.5	Triangular
Poor	9.4–10–14–14	Trapezoidal
TSS	*Ranges*	*Function*
Outstanding	0–0–15–21	Trapezoidal
V. Good	20–25–31	Triangular
Good	30–45–61	Triangular
Average	60–80–101	Triangular
Poor	100–130–150–150	Trapezoidal
DO	*Ranges*	*Function*
Outstanding	7–8–10–10	Trapezoidal
V. Good	6–6.5–7.1	Triangular
Good	4–5–6.1	Triangular
Average	3–3.5–4.1	Triangular
Poor	0–0–1.5–3.1	Trapezoidal
BOD	*Ranges*	*Function*
Outstanding	0–0–0.5–1.1	Trapezoidal
V. Good	1–2–3.1	Triangular
Good	3–3.5–4.1	Triangular
Average	4–5–6.1	Triangular
Poor	6–8–10–10	Trapezoidal
COD	*Ranges*	*Function*
Outstanding	0–0–2.5–6	Trapezoidal
V. Good	5–7.5–11	Triangular
Good	10–15–21	Triangular
Average	20–25–31	Triangular
Poor	30–40–50–50	Trapezoidal
Output	*Range*	*Function*
Poor	0–0–0.30–0.4	Trapezoidal
Average	0.30–0.45–0.60	Triangular
Good	0.5–0.65–0.8	Triangular
V. Good	0.7–0.8–0.9	Triangular
Outstanding	0.8–0.9–1–1	Trapezoidal

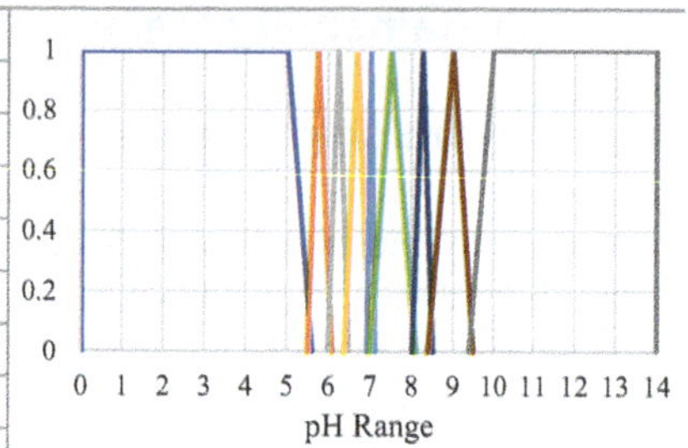

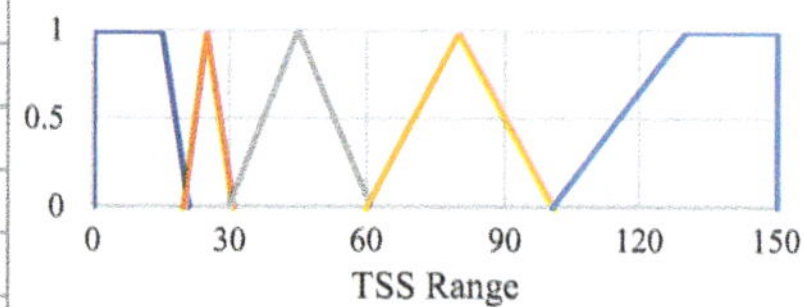

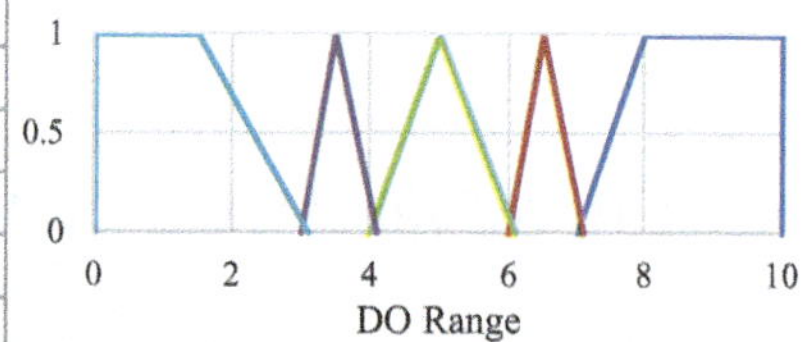

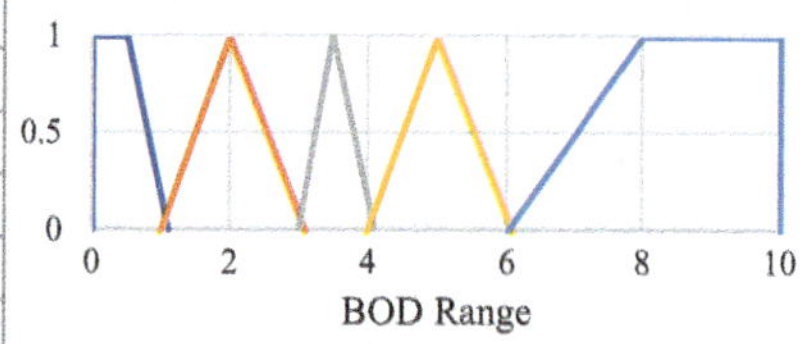

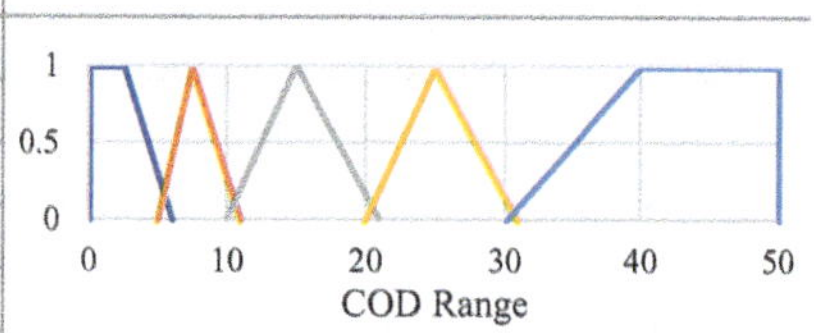

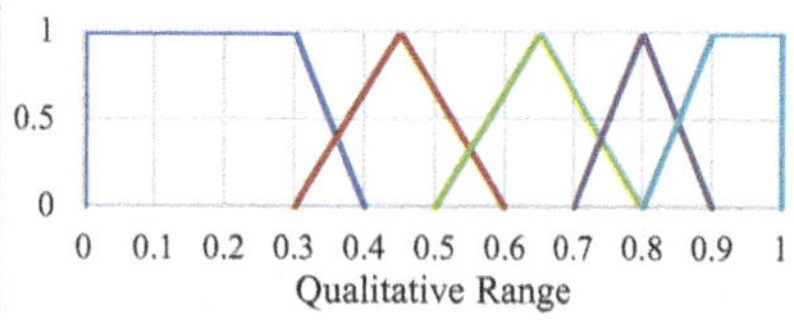

Appendix VI. The fuzzy model uses Mamdani's approach of fuzzy interface system.

4. The individual sample values were used as input, and defuzzified score of overall HWQI was studied. This numerical value can be converted in to qualitative term (output) as mentioned in Table 20.2.

20.5.2.1 Identified Water Quality Parameters

In order to estimate the HWQI, the first step is to identify the river water quality parameters which affect the human health. The various parameters that affect the human health are pH, TSS, DO, BOD, COD, nitrate, turbidity, fecal coliforms, etc. In this case study, from water quality point of view, the secondary water quality data of five parameters (pH, TSS, DO, BOD, and COD) are considered.

20.5.2.2 Membership Functions

Membership functions are prepared by its possible ranges which are suggested by Chaves and Alipaz (2007). It can be modified through fuzzy triangular and trapezoidal membership functions shown in Table 20.2.

After assessing the individual pH, TSS, DO, BOD, and COD water quality indicators, these are aggregated to assess the HWQI. For aggregation, a 5-point scale (Outstanding, Very Good, Good, Average, and Poor) is considered as output. It may be noticed that for the 5 indicators, there can be $5^5 = 3125$ possible states of these parameters. Out of these 3125 possible states, there are only 126 ($\frac{(n+r-1)!}{(n-1)!r!}$, where r is number of parameters and n is the number of categories) mutually exclusive independent states. Keeping these 126 mutually exclusive states (Appendix – VI) in mind, following rule base (Table 20.2) is suggested for assessing the HWQI. If more parameter values are available, one can consider the same and modify the procedure accordingly. Further, a fuzzy rule base is suggested for estimation of HWQI as shown in Table 20.3. A similar methodology was also suggested by Joshi and Kansal (2015).

Table 20.3 Proposed fuzzy rule base using five sub-indicators with five categories

Rule base for HWQI				
Poor	Average	Good	Very Good	Outstanding
If one or more components **"poor"** (irrespective of other components)	If one or more components **"average"** and no component is **"poor"**	1. Three or more components **"good"** and not more than one component **"outstanding"** 2. Two components **"very good,"** no component **"outstanding"**	One or two components **"outstanding,"** no more than two are **"good"** No component is **"poor"** and **"average"**	Three or more components **"outstanding"** No component is **"poor"** and **"average"**
		No component is **"poor"** and **"average"**		

20.5.2.3 Estimation of Defuzzified (Crisp) Score

In order to estimate defuzzified value of each parameter, fuzzy logic interface has been used in MATLAB-R2013a software. After completion of fuzzy model, all input parameters associated with defuzzified value, and it is variable with input values. Further, AHP weights are used as factor for estimation of final score of HWQI.

20.6 Quantitative Assessment of Environmental Indicator

The environmental indicator is studied under three parameters, i.e., pressure, state, and response. Chaves and Alipaz (2007) have suggested an average value of the three as the environmental indicator. **Environmental pressure** is quantified using the average of percentage variation of watershed agricultural area (from 1990 to 2006) and percentage variation of urban population in the watershed (1990–2006, interpolated). Here, population variation is estimated on the basis of 1991, 2001, and 2011 census data. The environmental pressure is fuzzified as shown in Table 20.3. **Environmental state** parameter is quantified on the basis of percentage change in natural vegetation (forest and shrubs) in the watershed during the period 1990–2006. On the basis of percentage change in natural area, it is assigned an environmental state indicator value as suggested by Chaves and Alipaz (2007) and fuzzified accordingly as shown in Table 20.4. **Environmental response** parameter is quantified by taking average of the watershed conservation area (%) and the best management practices (BMPs) (area %). Again, in this study the values are fuzzified as shown in Table 20.3.

Further, in order to quantify the crisp value (defuzzified score) of environmental indicator, a fuzzy rule base is proposed for three parameters (pressure, state, and response) with five categories (Poor, Average, Good, Very Good, and Outstanding) as shown in Table 20.5. Table 20.5 is derived from the mutually exclusive states of 35 as shown in Appendix I.

20.7 Quantitative Assessment of Life Indicator

Life indicator reflects the watershed as human life quality. It is represented in terms of HDI. HDI combines the three dimensions, i.e., life expectancy, expected years of schooling, and decent standard of living. UNDP suggests the following method for calculating the HDI:

$$\text{Life Expectancy Index (LEI)} = \frac{\text{LE-20}}{\text{85-20}}$$

$$\text{Education Index (EI)} = \frac{\text{MYSI} + \text{EYSI}}{2}$$

Table 20.4 Membership functions and their ranges for environment indicator

Pressure	*Range (%)*	*Function*	
Very low	0–0–1–2	Trapezoidal	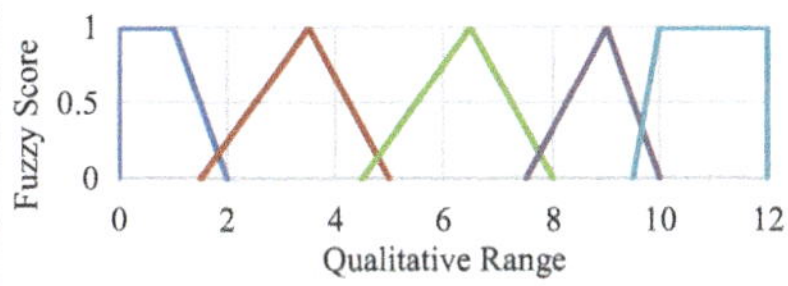
Low	1.5–3.5–5	Triangular	
Medium	4.5–6.5–8	Triangular	
High	7.5–9–10	Triangular	
Very high	9.5–10–12–12	Trapezoidal	
State	*Range (%)*	*Function*	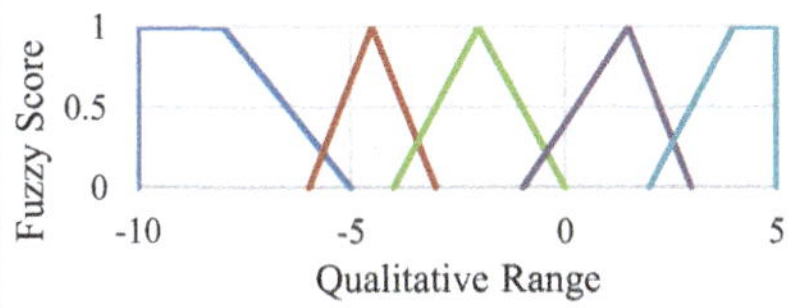
Poor	−10–(−10)–(−8)–(−5)	Trapezoidal	
Average	−6–4.5–3	Triangular	
Good	−4–(−2)–0	Triangular	
V. Good	−1–1.5–3	Triangular	
Outstanding	2–4–5–5	Trapezoidal	
Response	*Range (%)*	*Function*	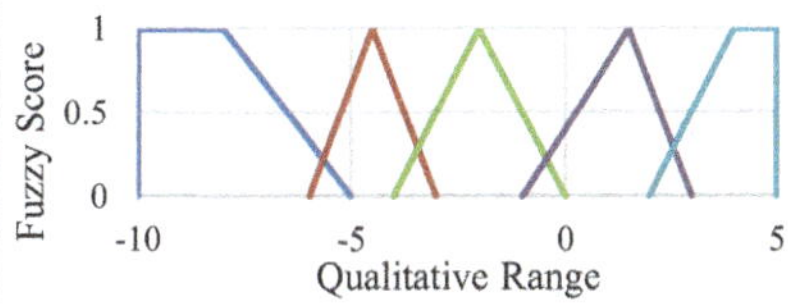
Poor	−10–(−10)–(−8)–(−5)	Trapezoidal	
Average	−6–(−4.5)–(−3)	Triangular	
Good	−4–(−2)–0	Triangular	
V. Good	−1–1.5–3	Triangular	
Outstanding	2–4–5–5	Trapezoidal	
Output	*Range*	*Function*	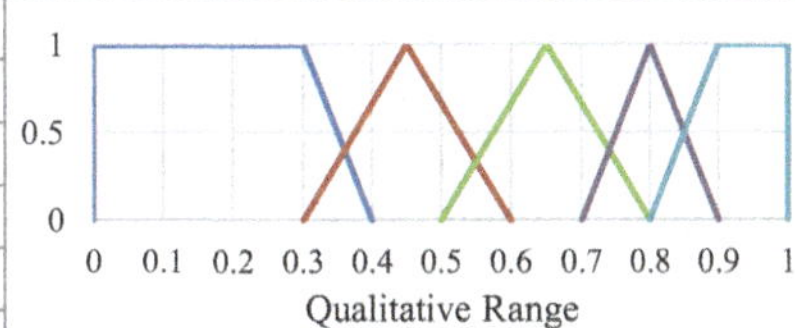
Poor	0–0–0.30–0.4	Trapezoidal	
Average	0.30–0.45–0.60	Triangular	
Good	0.5–0.65–0.8	Triangular	
V. Good	0.7–0.8–0.9	Triangular	
Outstanding	0.8–0.9–1–1	Trapezoidal	

Table 20.5 Proposed fuzzy rule base using three sub-indicators with five categories

Rule base for environmental/life/policy indicator				
Poor	Average	Good	Very Good	Outstanding
One or more components **"poor"**	One or more components **"average"**	1. Two or more components **"good"**	Two or more components **"very good"** Or All three components different Or No component is **"very good,"** and not more than one is **"good"**	Two or more components are **"outstanding"**
Irrespective of other components	No component is **"poor"**	No component is **"poor"** and **"average"**	No component is **"poor"** and **"average"**	No component is **"poor"** and **"average"**

$$\text{Mean Year of Schooling Index (MYSI)} = \frac{\text{MYS}}{15}$$

$$\text{Expected Year of Schoolind Index (EYSI)} = \frac{\text{EYS}}{18}$$

$$\text{Income Index (II)} = \frac{\ln\ (\text{GNIpc}) - \ln\ (100)}{\ln\ (75000) - \ln\ (100)}$$

LE: life expectancy at birth.

where.

MYS: mean year of schooling.

(year that a person 25 years of age or older has spent in schools)

EYS: expected years of schooling.

(year that a 5-year-old child will spend in schools throughout his life)

GNIpc: Gross national income at purchasing power parity per capita.

Finally, the HDI is considered as the geometric mean of the three normalized indices. This can be mathematically represented as.

$\text{HDI} = \sqrt[3]{\text{LEI} \cdot \text{EI} \cdot \text{II}}$

Similar to environmental indicator, it is also subdivided into three parameters, i.e., pressure, state, and response. **Pressure** parameter is quantified on the basis of variation in watershed's HDI income subindex from 1990 to 2006. **State** parameter is estimated as the watershed HDI weighted by the district area involved in the watershed. **Response** parameter is estimated by evolution in the watershed HDI in the watershed. The various ranges of life pressure, response, and states are fuzzified as shown in Table 20.6.

Further, quantification of crisp value (defuzzified score) of life indicator, the same fuzzy rule base has been proposed using three sub-indicators (pressure, state, and response) and five categories (Poor, Average, Good, Very Good, and Outstanding) as shown in Table 20.5.

20.8 Quantitative Assessment of Policy Indicator

Similar to life indicator, policy indicator is also subdivided into three parameters, i.e., pressure, state, and response. **Policy pressure** parameter is quantified on the basis of variation in watershed's $\text{HDI}_{\text{education}}$ from 1990 to 2006. **Policy state parameter** is estimated on the basis of watershed institutional capacity in Integrated Water Resources Management (IWRM). **Policy response** is assumed by estimating the evolution of the expenditures in water resources management in the watershed. It reflects the response by stakeholders, decision-, and policy makers for water resources problems. If the higher expenditure is to be involved in the watershed, that means watershed will meet its water-related problems under IWRM plans. The various ranges of life pressure, response, and states are fuzzified as shown in Table 20.7.

Table 20.6 Membership functions and their ranges for life indicator

Pressure	*Range (%)*	*Function*	
Very low	0–0–1–2	Trapezoidal	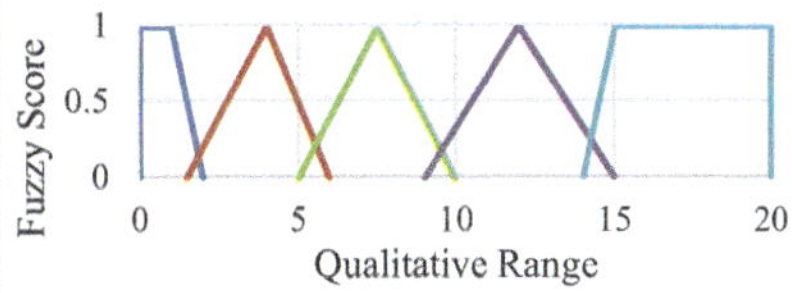
Low	1.5–4–6	Triangular	
Medium	5–7.5–10	Triangular	
High	9–12–15	Triangular	
Very high	14–15–20–20	Trapezoidal	
State	*Range (%)*	*Function*	
Poor	0–0–0.1–0.2	Trapezoidal	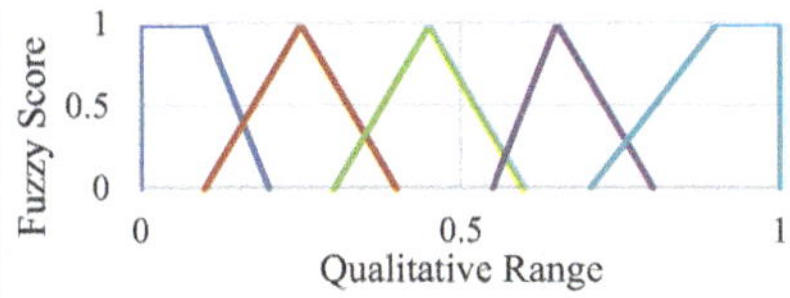
Average	0.1–0.25–0.4	Triangular	
Good	0.3–0.45–0.6	Triangular	
V. Good	0.55–0.65–0.8	Triangular	
Outstanding	0.7–0.9–1–1	Trapezoidal	
Response	*Range (%)*	*Function*	
Poor	0–0–1–2	Trapezoidal	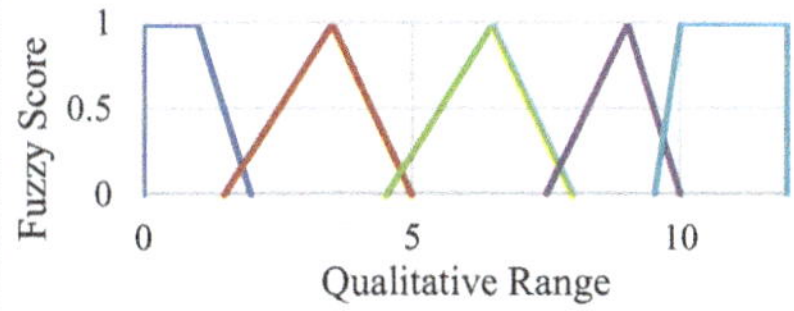
Average	1.5–3.5–5	Triangular	
Good	4.5–6.5–8	Triangular	
V. Good	7.5–9–10	Triangular	
Outstanding	9.5–10–12–12	Trapezoidal	
Output	*Range*	*Function*	
Poor	0–0–0.30–0.4	Trapezoidal	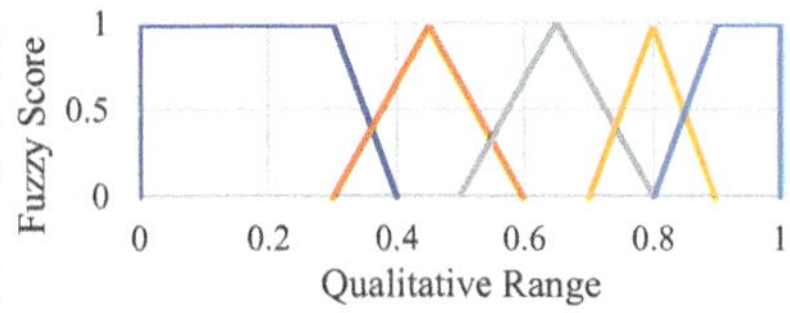
Average	0.30–0.45–0.60	Triangular	
Good	0.5–0.65–0.8	Triangular	
V. Good	0.7–0.8–0.9	Triangular	
Outstanding	0.8–0.9–1–1	Trapezoidal	

20.9 Quantitative Assessment of Overall WSI

After assessing the individual H, E, L, and P indicators, these are aggregated to assess the overall WSI. For aggregation, again a 5-point scale (Outstanding, Very Good, Good, Average, and Poor) is considered. It may be noticed that for the 4 indicators, there can be $4^5 = 625$ possible states of these parameters. Out of these 625 possible states, there are only 70 ($\frac{(n+r-1)!}{(n-1)!r!}$, where r is number of parameters and n is the number of categories) mutually exclusive independent states. Keeping these 70 mutually exclusive states in mind, following rule base (Table 20.8) is suggested for assessing the WSI.

Table 20.7 Membership functions and their ranges for policy indicator

Pressure	*Range (%)*	*Function*
Very low	0–0–1–2	Trapezoidal
Low	1.5–3.5–5	Triangular
Medium	4.5–6.5–8	Triangular
High	7.5–9–10	Triangular
Very high	9–10–12–12	Trapezoidal
State	*Range (%)*	*Function*
Poor	0–0–0.1–0.2	Trapezoidal
Average	0.1–0.25–0.4	Triangular
Good	0.3–0.45–0.6	Triangular
V. Good	0.5–0.65–0.8	Triangular
Outstanding	0.7–0.9–1–1	Trapezoidal
Response	*Range (%)*	*Function*
Poor	0–0–1–2	Trapezoidal
Average	1.5–3.5–5	Triangular
Good	4.5–3.5–5	Triangular
V. Good	7.5–9–10	Triangular
Outstanding	9–11–12–12	Trapezoidal
Output	*Range*	*Function*
Poor	0–0–0.30–0.4	Trapezoidal
Average	0.30–0.45–0.60	Triangular
Good	0.5–0.65–0.8	Triangular
V. Good	0.7–0.8–0.9	Triangular
Outstanding	0.8–0.9–1–1	Trapezoidal

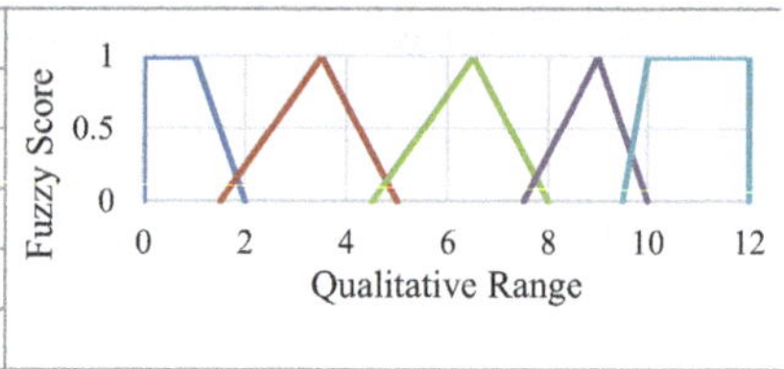

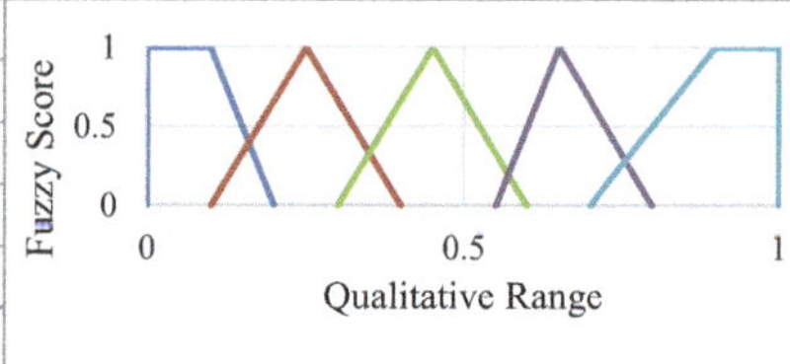

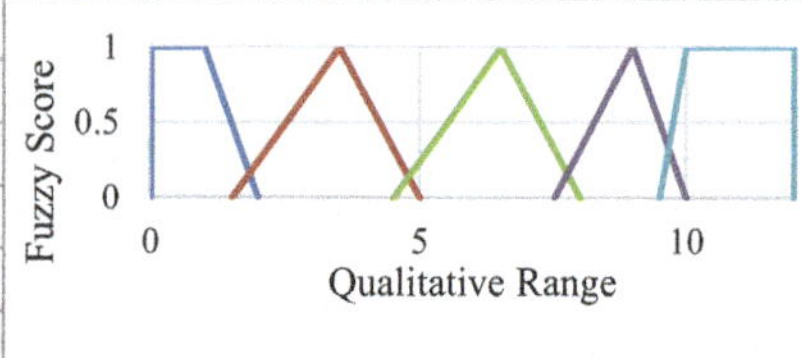

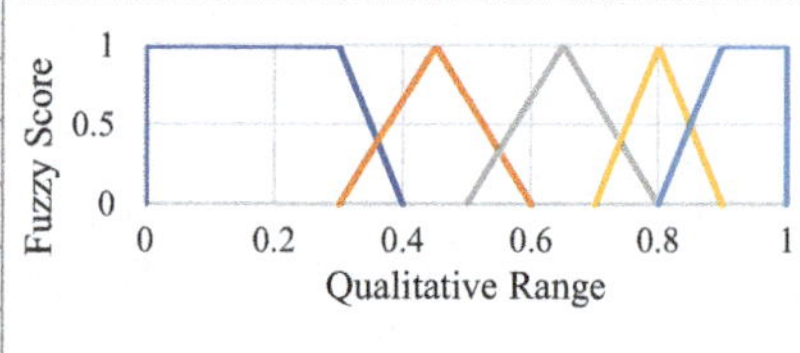

20.10 Case Study Related to Sustainable Watershed Framework

20.10.1 Quantitative Assessment of Hydrological Indicator

20.10.1.1 Hydrological Quantity Index (HQI)

For the estimation of quantity (sub-indicator) of hydrology, long-term flow data of Piperiya watershed are considered. Annual and seasonal flows are estimated for the case study during 1990–2006. Further, 50%, 75%, and 90% dependable flows are calculated. However, 75% dependable flow is considered as baseline (threshold) for water sustainability. For assessment of HQI in Piperiya watershed, the concepts of reliability, resilience, vulnerability, and relative vulnerability are utilized. The concepts of reliability, resilience, and vulnerability are explained by Kansal et al. (2015). Here it may be noted that vulnerability is quantified on basis of extent as well as duration. Further, three criteria of decision-making have been considered under the index of optimism (neutral, pessimistic, and optimistic). On this basis, the decision-maker (DM) can select the HQI, i.e., if the DM is optimistic in nature, he will consider the minimum value of relative vulnerability; if the DM is neutral in

Table 20.8 Proposed fuzzy rule base for overall assessment of WSI using HELP

Rule base for overall WSI				
Poor	Average	Good	Very Good	Outstanding
One or more components **"poor"**	One or more components **"average"**	1. Three or more components **"good"** and not more than one component **"outstanding" OR** 2. Two components **"very good,"** no component **"outstanding"**	1. One or less components **"outstanding,"** not more than two are **"good" OR** 2. Two or more components **"very good,"** not more than one component **"good" OR** 3. One or less components **"very good,"** at least two components **"outstanding" OR** 4. Two components **"very good,"** no components **"good"**	Three or more components **"outstanding"**
Irrespective of other components	No component is **"poor"**	No component is **"poor"** and **"average"**	No component is **"poor"** and **"average"**	No component is **"poor"** and **"average"**

nature, he will consider the average value of relative vulnerability; and if the DM is pessimistic in nature, he will consider the maximum value of the relative vulnerability for estimation of HQI. Average annual and seasonal flows (pre-monsoon, monsoon, post-monsoon, and winter) and the lines of 50%, 75%, and 90% dependability are shown in Fig. 20.3. Further, the extent from threshold at 75% dependability line is also indicated in Fig. 20.3.

Using the flow values as shown in Fig. 20.3, the various performance parameters like reliability, resilience, vulnerability, and relative vulnerability are estimated on annual and seasonal basis. The summary of these parameters is shown in Table 20.9.

In this study, neutral approach has been adopted, the value of HQI is about **0.67** which indicates the **Good** condition of quantity under hydrology indicator.

20.10.1.2 Hydrological Water Quality Index (HWQI)

In order to estimate the HWQI, the first step is to identify the river water quality parameters which affect the human health. The various parameters that affect the human health are pH, TSS, DO, BOD, COD, nitrate, turbidity, fecal coliforms, etc. In this case study, from water quality point of view, the secondary water quality data of five parameters (pH, TSS, DO, BOD, and COD) are considered. River water quality data is taken from Chhattisgarh Environment Conservation Board (CECB), Report 2011, which has given the range of values after collecting daily water quality

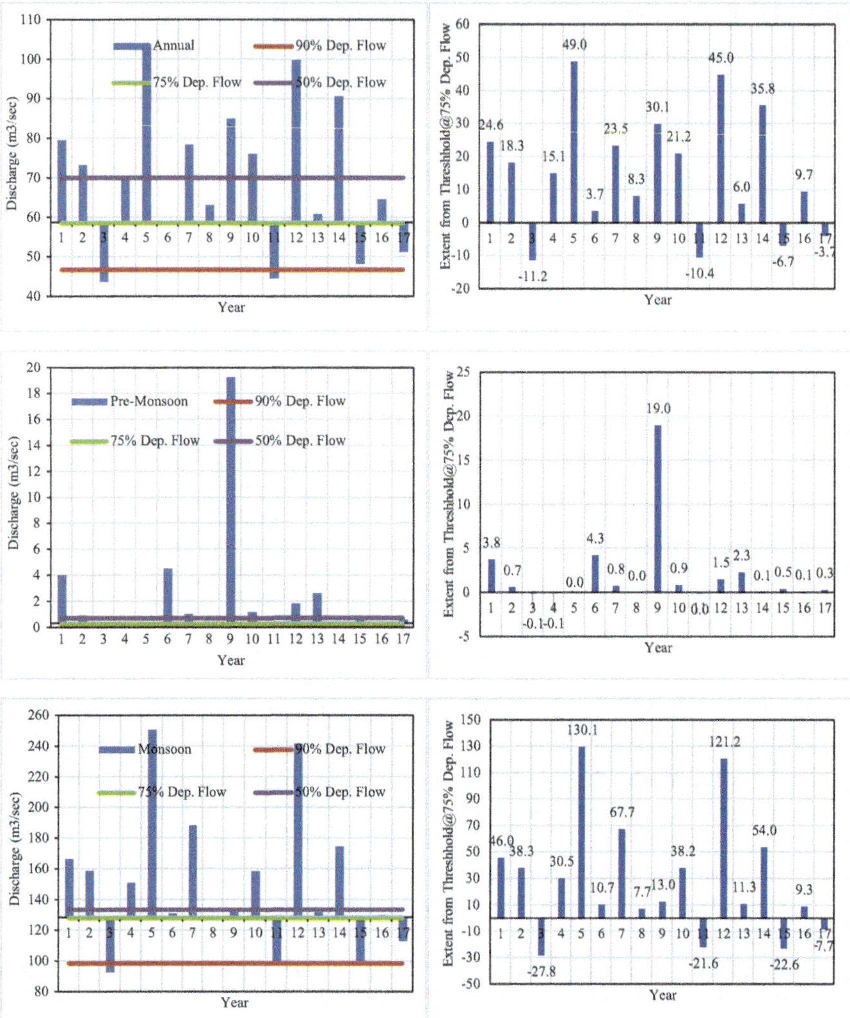

Fig. 20.3 Observed discharge and its extant from threshold (75% dependable flow) in annual and pre-monsoon, monsoon, post-monsoon, and winter season during 1990–2006

data from 2008 to 2010. The site location is near Hasdeo barrage and near Urga village. The HQI is assessed by both the methods, i.e., by (1) **AHP fuzzy** and (2) **total fuzzy rule base**.

AHP Fuzzy-Based Approach

Weightage Estimation

Relative weights of each parameter are decided by expert opinion and literature review. Further, AHP methodology has been adopted in this case study for relative weights estimation. Further, consolidated weights are estimated for each of the water quality parameter as shown in Tables 20.10, 20.11, 20.12, 20.13, 20.14.

The final weights for pH, TSS, DO, BOD, and COD are 0.16, 0.22, 0.18, 0.23, and 0.21, respectively, which are used as factors for each of the water quality parameters. Further summation of all the resultant scores may be considered as HQI. Results are summarized in Tables 20.15, 20.16, 20.17, 20.18, 20.19, and 20.20.

As per AHP fuzzy-based approach, the value of HQI (near Hasdeo barrage) varies from 0.579 (Good) to 0.739 (Very Good), whereas HQI near Urga village varied from 0.468 (Average) to 0.708 (Very Good).

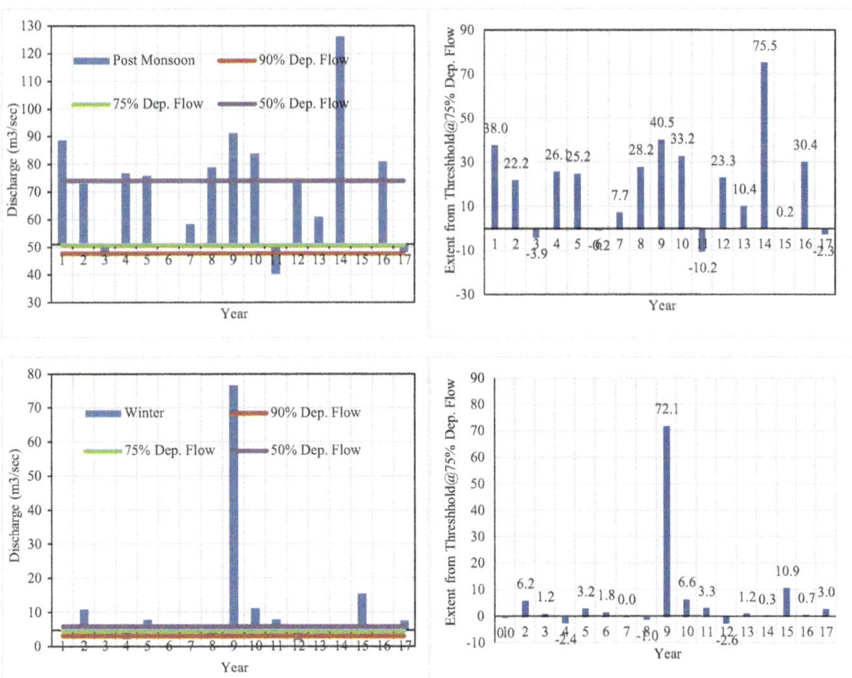

Fig. 20.3 (continued)

Table 20.9 Different parameters and its results as per 75% dependable flow

Particular	Annual	Pre-monsoon	Monsoon	Post-monsoon	Winter
Reliability	0.76	0.76	0.76	0.76	0.76
Resilience	0.75	0.50	0.75	0.75	1.00
Vulnerability extent	8.02	0.05	19.92	4.16	1.52
Vulnerability duration	1.00	1.33	1.00	1.00	1.00
Rel. Vulnerability extent	0.71	0.60	0.72	0.41	0.58
Rel. Vulnerability duration	0.25	0.33	0.25	0.25	0.25
Rel. Vul. (avg.)	0.48	0.46	0.48	0.33	0.41
HQI (neutral approach)	0.67	0.58	0.67	0.73	0.77
HQI (pessimistic approach)	0.54	0.53	0.54	0.70	0.68
HQI (optimistic approach)	0.75	0.63	0.75	0.75	0.83

Table 20.10 AHP comparison matrix for weight determination

Parameters	TSS	pH	DO	BOD	COD
TSS	1.00	4/5	4/5	2/3	2/3
pH	5/4	1.00	3/2	1	1
DO	5/4	2/3	1.00	4/5	8/9
BOD	3/2	1	5/4	1.00	6/5
COD	3/2	1	9/8	5/6	1.00
SUM	6.36	4.47	5.66	4.33	4.80

Table 20.11 Final weights for each parameter

TSS	pH	DO	BOD	COD	AVG.
0.157	0.179	0.141	0.162	0.146	0.157
0.197	0.224	0.265	0.231	0.208	0.225
0.197	0.149	0.177	0.185	0.188	0.179
0.225	0.224	0.221	0.231	0.250	0.230
0.225	0.224	0.196	0.192	0.208	0.209
1.000	1.000	1.000	1.000	1.000	1.000

Table 20.12 Predefined values of lambda as per sample size

No of Parameters (n)	3	4	5	6	7	8
Lambda	Par-1	Par-2	Par-3	Par-4	Par-5	SUM
	0.9982	1.0046	1.0129	0.9968	1.0037	5.02

Table 20.13 Consistency statistics for case study

Consistency	Index
CI (Consistency Index)	0.004
RI (Random Consistency Index)	1.188
CR (Consistency Ratio)	**0.003**
Status	Consistent

Table 20.14 Criteria for consistency

Consistency ratio < 0.1	Consistence
Consistency ratio > 0.1	Non-consistence

Table 20.15 Water quality (minimum values) at near Hasdeo barrage during 2008–2010

Quality variables	Weights	Observed values (minimum)	Fuzzy score	Final minimum score
pH (pH unit)	0.16	6.9	0.8	0.126
Suspended solids (mg/lit)	0.22	42	0.65	0.146
Dissolved oxygen (mg/lit)	0.18	6	0.s65	0.116
BOD (mg/lit)	0.23	1.2	0.8	0.184
COD (mg/lit)	0.21	8	0.8	0.167
Total	1.00	Final score		0.739

Table 20.16 Water quality (maximum values) at near Hasdeo barrage during 2008–2010

Quality variables	Weights	Observed values (maximum)	Fuzzy score	Final maximum score
pH (pH unit)	0.16	7.5	0.8	0.126
Suspended solids (mg/lit)	0.22	89	0.45	0.101
Dissolved oxygen (mg/lit)	0.18	7.6	0.917	0.164
BOD (mg/lit)	0.23	3.3	0.65	0.150
COD (mg/lit)	0.21	38	0.184	0.038
Total	1.00	Final score		0.579

Fuzzy-Based Approach

In the all fuzzy rule-based approach, no specific weights are required. In the case study, five inputs and one output are used in fuzzy interface. Fuzzy rule base is proposed for five sub-indicators and five categories which are discussed in Table 20.2. Further, results are summarized using the proposed methodology and are shown in Tables 20.21, 20.22, 20.23, 20.24, 20.25, and 20.26.

Table 20.17 Water quality (average values) at near Hasdeo barrage during 2008–2010

Quality variables	Weights	Observed values (average)	Fuzzy score	Final average score
pH (pH unit)	0.16	7.2	0.8	0.126
Suspended solids (mg/lit)	0.22	65.5	0.45	0.101
Dissolved oxygen (mg/lit)	0.18	6.8	0.8	0.143
BOD (mg/lit)	0.23	2.25	0.8	0.184
COD (mg/lit)	0.21	23	0.45	0.094
Total	1.00	Final score		0.648

Table 20.18 Water quality (minimum values) at near Urga village during 2008–2010

Quality variables	Weights	Observed values (minimum)	Fuzzy score	Final minimum score
pH (pH unit)	0.16	6.8	0.8	0.126
Suspended solids (mg/lit)	0.22	52	0.65	0.146
Dissolved oxygen (mg/lit)	0.18	5.3	0.65	0.116
BOD (mg/lit)	0.23	1.8	0.8	0.184
COD (mg/lit)	0.21	16	0.65	0.136
Total	1.00	Final score		0.708

Table 20.19 Water quality (maximum values) at near Urga village during 2008–2010

Quality variables	Weights	Observed values (maximum)	Fuzzy score	Final maximum score
pH (pH unit)	0.16	7.4	0.8	0.126
Suspended solids (mg/lit)	0.22	133	0.174	0.039
Dissolved oxygen (mg/lit)	0.18	7.4	0.912	0.163
BOD (mg/lit)	0.23	4.1	0.45	0.104
COD (mg/lit)	0.21	48	0.174	0.036
Total	1.00	Final score		0.468

Using the second all fuzzy-based approach, the HQI (near Hasdeo barrage) is found to vary from 0.26 (Poor) to 0.76 (Very good), whereas HQI (near Urga village) is found to vary from 0.21 (Poor) to 0.67 (Good).

Therefore, the overall score of HWQI in average condition is about (0.63 + 0.335)/2 = 0.485, which indicates the "Average" condition of water quality in hydrology indicator.

Further, the overall hydrology index is about (0.67 + 0.485)/2 = 0.58, which indicates the **Average** condition.

Table 20.20 Water quality (average values) at near Urga village during 2008–2010

Quality variables	Weights	Observed values (average)	Fuzzy score	Final average score
pH (pH unit)	0.16	7.1	0.911	0.143
Suspended solids (mg/lit)	0.22	92.5	0.45	0.101
Dissolved oxygen (mg/lit)	0.18	6.35	0.8	0.143
BOD (mg/lit)	0.23	2.95	0.8	0.184
COD (mg/lit)	0.21	32	0.194	0.041
Total	1.00	Final score		0.612

Table 20.21 Water quality (minimum values) at near Hasdeo barrage during 2008–2010

Quality variables	Observed values	Fuzzy output	Category
pH (pH unit)	6.9	0.80	Very good
Suspended solids (mg/lit)	42	0.65	Good
Dissolved oxygen (mg/lit)	6	0.65	Very good
BOD (mg/lit)	1.2	0.80	Very good
COD (mg/lit)	8	0.80	Very good
Total	Overall defuzzified score	0.76	Very good

Table 20.22 Water quality (maximum values) at near Hasdeo barrage during 2008–2010

Quality variables	Observed values	Fuzzy output	Category
pH (pH unit)	7.5	0.80	Very good
Suspended solids (mg/lit)	89	0.45	Avg.
Dissolved oxygen (mg/lit)	7.6	0.92	Outstanding
BOD (mg/lit)	3.3	0.65	Good
COD (mg/lit)	38	0.18	Poor
Total	Overall defuzzified score	0.26	Poor

Table 20.23 Water quality (average values) at near Hasdeo barrage during 2008–2010

Quality variables	Observed values	Fuzzy output	Category
pH (pH unit)	7.2	0.80	Very good
Suspended solids (mg/lit)	65.5	0.45	Avg.
Dissolved oxygen (mg/lit)	6.8	0.80	Very good
BOD (mg/lit)	2.25	0.80	Very good
COD (mg/lit)	23	0.45	Avg.
Total	Overall defuzzified score	0.43	Avg.

Table 20.24 Water quality (minimum values) at near Urga village during 2008–2010

Quality variables	Observed values	Fuzzy output	Category
pH (pH unit)	6.8	0.80	Very good
Suspended solids (mg/lit)	52	0.65	Good
Dissolved oxygen (mg/lit)	5.3	0.65	Good
BOD (mg/lit)	1.8	0.80	Very good
COD (mg/lit)	16	0.65	Good
Total	Overall defuzzified score	0.67	Good

Table 20.25 Water quality (minimum values) at near Urga village during 2008–2010

Quality variables	Observed values	Fuzzy output	Category
pH (pH unit)	7.4	0.80	Very good
Suspended solids (mg/lit)	133	0.17	Poor
Dissolved oxygen (mg/lit)	7.4	0.91	Outstanding
BOD (mg/lit)	4.1	0.45	Good
COD (mg/lit)	48	0.17	Poor
Total	Overall defuzzified score	0.21	Poor

Table 20.26 Water quality (average values) at near Urga village during 2008–2010

Quality variables	Observed values	Fuzzy output	Category
pH (pH unit)	7.1	0.91	Outstanding
Suspended solids (mg/lit)	92.5	0.45	Avg.
Dissolved oxygen (mg/lit)	6.35	0.80	Very good
BOD (mg/lit)	2.95	0.80	Good
COD (mg/lit)	32	0.19	Poor
Total	Overall defuzzified score	0.24	Poor

20.10.2 Quantitative Assessment of Environment Indicator

Pressure

In the environmental pressure sub-indicator, two major issues at watershed level are considered, i.e., agriculture area and population growth per year. Land use/land cover pattern of Piperiya watershed is shown in Table 20.27.

Agricultural Area

Percentage growth rate of agriculture area per year during 1990–2006 is about 3.96%.

Percentage growth rate of urban population (Table 20.28) per year during 1990–2006 is about 9.03%.

Therefore, averaged growth rate of environment pressure index (EPI) = (3.96% + 9.03%)/2.

EPI = 6.49% which indicate **Medium** level of pressure in the environment.

Table 20.27 Land use/land cover changes during three decades (1990–2011)

S No	LULC	1990 Area (km^2)	1990 Area (%)	2001 Area (km^2)	2001 Area (%)	2006 Area (km^2)	2006 Area (%)	2011 Area (km^2)	2011 Area (%)
1	Settlement	26	1	34	1	40	2	45	2
2	Agriculture	496	21	755	31	924	38	1093	45
3	Forest	1209	50	1055	44	936	39	818	34
4	Shrubs	527	22	384	16	320	13	255	11
5	Barren land	132	5	156	6	163	7	171	7
6	Water	25	1	30	1	32	1	33	1
	Total	2415	100	2415	100	2415	100	2415	2415

Table 20.28 Urban population of Piperiya watershed

Year	Urban population in Piperiya watershed (area weightage)
1991	319,020
2001	1,153,546
2006	1,271,759
2011	1,350,568

State

In the state sub-indicator, growth rate of natural vegetation (forest and shrubs) was considered during 1990–2006. The total natural vegetation area in 1990 is about 1735 km^2, and in 2006 it is estimated as 1256km^2. This means that the growth rate of natural vegetation during the study period is about −2.0%, which represent the low impact and hence the **Good** condition of state sub-indicator of environment.

Response

In the response sub-indicator, growth rate of forest area has been considered during 1990–2006. The total forest area in 1990 is about 1209 km^2, and 2006 it is estimated as 936km^2. This means that the growth rate of forest area during the study period is about −1.59% which represents the low adverse impact and hence the **Good** condition of response sub-indicator of environment. The defuzzified value of environment indicator is about 0.65, which indicates the **Good** condition of environment indicator as shown in Fig.20.4.

20.10.3 Quantitative Assessment of Life Indicator

Pressure

Chhattisgarh state came into existence in year 2000. Development and life changed drastically from 2001 onwards. HDI_{income} has increased from 0.278 in 2000–2001 to 0.471 in 2005–2006. This means that it recorded a growth rate of about 11% during the period. In 2005, HDI_{income} in the watershed is about 0.386 which is about 82% of

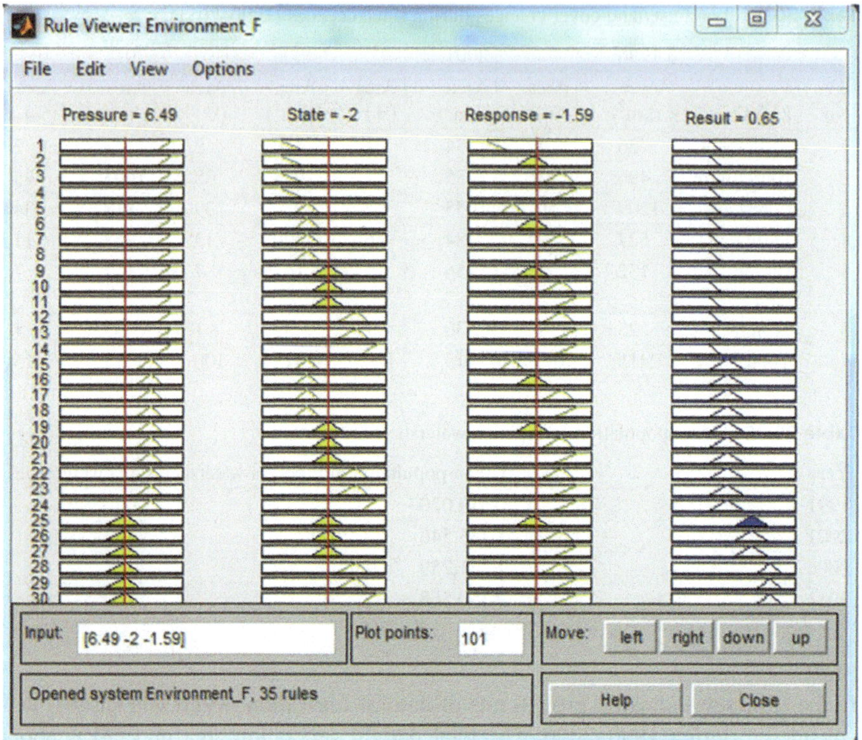

Fig. 20.4 Fuzzy interface for estimation of environment indicator score

Table 20.29 Calculation of area weighted watershed HDI (Source: Chhattisgarh Human Development Report 2005)

Districts in involved in WS	Area (km^2)	Area weightage	HDI	Weighted HDI
Bilaspur	386	0.160	0.449	0.072
Korba	342	0.142	0.625	0.089
Koriya	1585	0.656	0.391	0.257
Anuppur/Shahdol*	102	0.042	0.525	0.022
Area weighted HDI in watershed				0.439

the state HDI_{income}. In the absence of available data for the watershed, it is assumed that the HDI_{income} in watershed is about 9.1% (0.82*11 = 9.1). This represents **Medium** pressure in the life indicator.

State

State represents the watershed HDI. Therefore, in order to assess the state sub-indicator of Life in the watershed, area weighted HDI is required. The district-wise HDI values in the watershed are shown in Table 20.29.

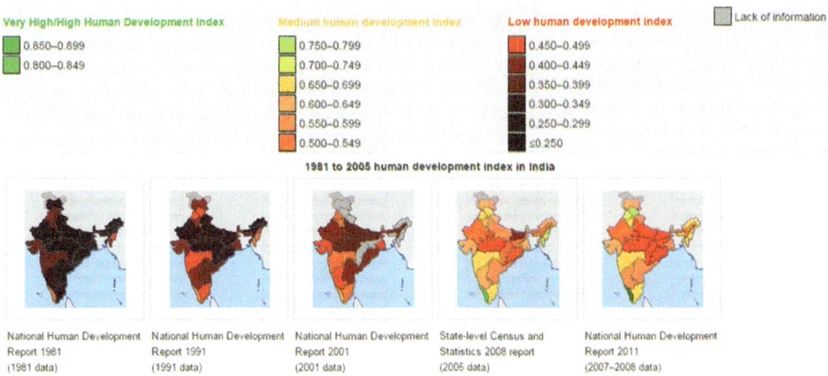

Fig. 20.5 Spatial map of HDI for Indian regions during the period of 1981–2008 (https://en.wikipedia.org/wiki/List_of_Indian_states_and_territories_by_Human_Development_Index#cite_note)

Area weighted HDI in the watershed comes out to be 0.439, which indicate the **Good** condition of state sub-indicator of life.

Response

Response sub-indicator is estimated by evolution in watershed HDI during the study period (1990–2006). In this case, since the quantitative values are not available and only the spatial maps (Fig. 20.5) are available for the whole country, the estimation of evolution is estimated on the basis of this map.

The HDI value of watershed in 1990 is about 0.150, and in 2005 it is about 0.439. The growth rate per year in HDI in the watershed was estimated as 6.94% which indicate the **Good** condition of response sub-indicator of life. The defuzzified value of life parameter is about 0.633, which indicates the **Good** condition of life indicator as shown in Fig. 20.6.

20.10.4 Quantitative Assessment of Policy Indicator

Pressure

Pressure parameter is quantified on the basis of variation in watershed $HDI_{education}$ during the study period. In 2006, $HDI_{education}$ for the watershed has been estimated as 0.70. In 1990, it is assumed that the $HDI_{education}$ in the watershed was almost the same as that of the country as a whole, which is about 0.33. Therefore, $HDI_{education}$ index is assumed to increase from 0.33 to 0.70. This means a growth rate of 4.8% annually during the period 1990–2006, which indicates **Medium** level of pressure to the policy indicator.

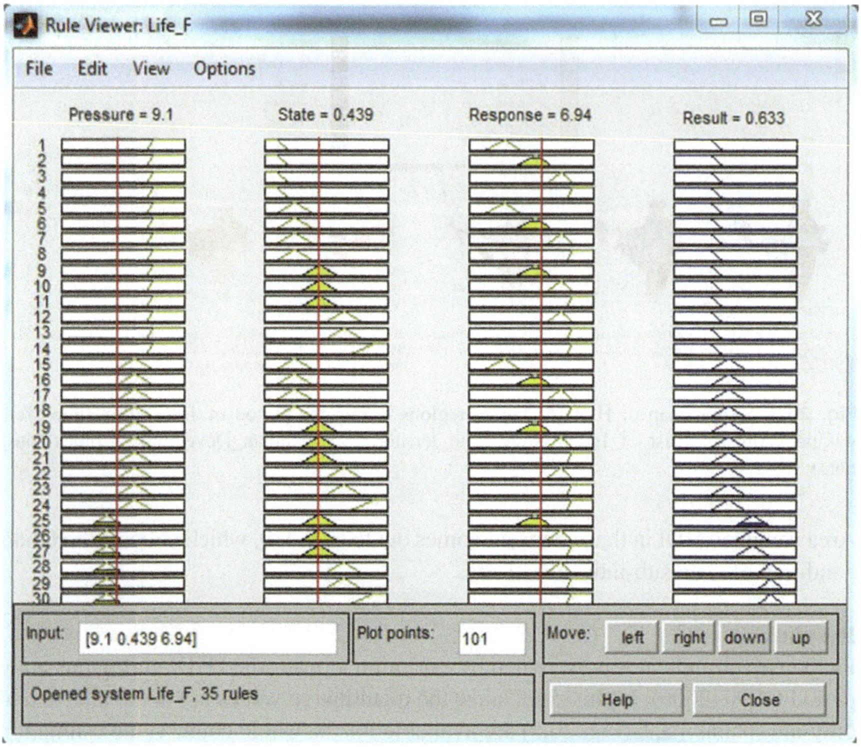

Fig. 20.6 Fuzzy interface for estimation of life indicator score

State

After becoming the full-fledged stat in 200, the awareness has increased about the integrated water resource management (IWRM) in the state. The government has established new departments like department of water resources, irrigation department, department of soil and water conservation, etc., and has improved the capacity in terms of human resource in the area of IWRM. Therefore, from the policy point of view, it is considered as **Good** (0.5) state of policy.

Response

Policy response is estimated by estimating the evolution of the expenditures in the area of water resources management in the watershed. In the year 2001–2002, the expenditure on water sector was about 284 crore rupees, which increased to 436 (present worth of 641 crore at 8% annual) for the whole state. This means that it recorded a growth rate of 9% in the state. Assuming the same growth rate in the watershed, i.e., 9% in the watershed, it can be considered as **'Very Good'** level of response. Further, awareness among stakeholders is high, and the decision- and policy makers are serious to deal with the water resources problems in the area.

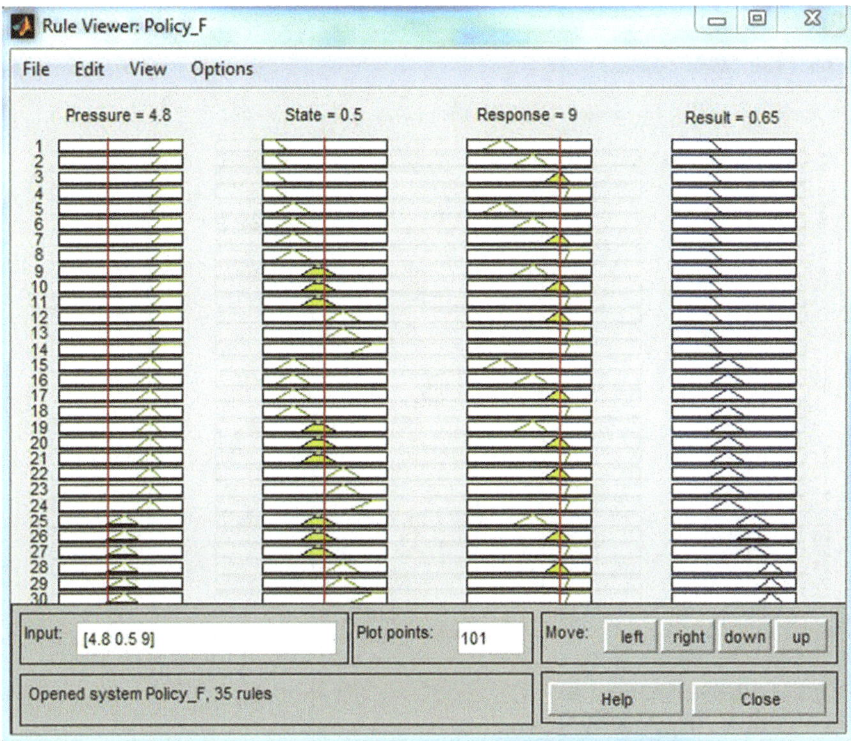

Fig. 20.7 Fuzzy interface for estimation of policy indicator score

Therefore, it is considered as very good. The defuzzified value of policy indicator is about 0.65 which indicates **Good** condition of policy indicator as shown in Fig. 20.7.

20.11 Overall WSI

In order to carry out the overall WSI, all the defuzzified scores have been utilized in fuzzy interface. The individual scores of hydrology, environment, life, and policy are about 0.58, 0.65, 0.633, and 0.65, respectively. The final WSI score for the watershed is estimated to be **0.65** which indicate the **Good** condition of watershed sustainability as shown in Fig. 20.8.

20.12 Conclusion

In this chapter, a framework is proposed for estimation of WSI in a watershed. This index is based on four indicators – hydrology (H), environment (E), life (L), and policy (P). Further, in hydrology, quantity and quality aspects are considered. A

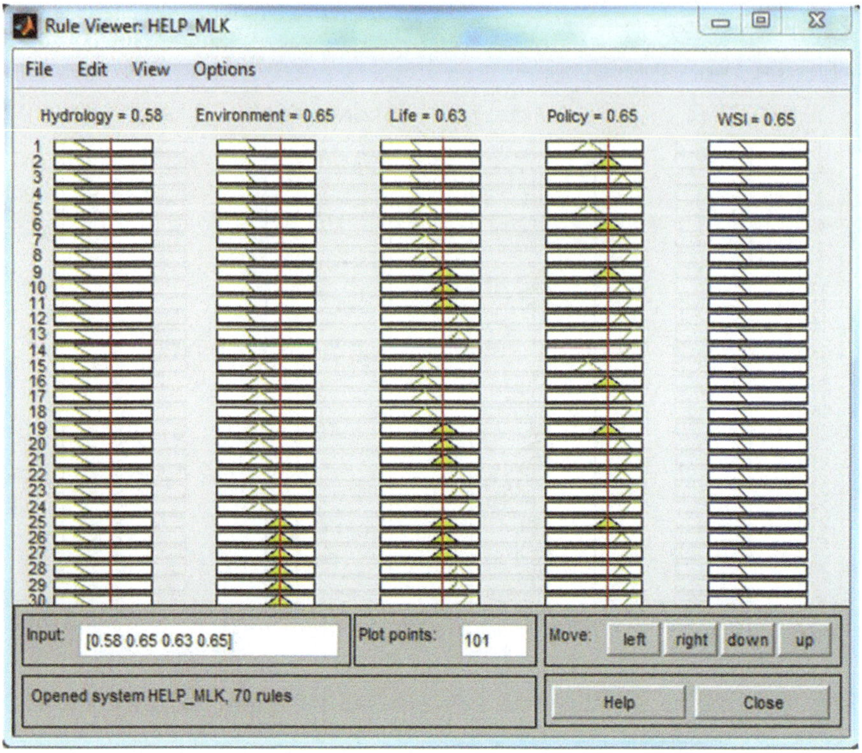

Fig. 20.8 Fuzzy interface for estimation of Life indicator score

methodology has been proposed for assessing the quantity-based water sustainability index using the concepts of dependable flow and the performance parameters such as reliability, resilience, and vulnerability. The proposed methodology introduces the concept of index of optimism of the decision-maker while estimating the performance parameters. For assessing the water quality sustainability index, the weightage of quality parameters is assessed through the analytical hierarchy process (AHP) and thereafter, the quality index is assessed using the concept of fuzzy logics. A new rule base (involving five parameters and five states) is suggested for aggregation of quality parameters. The sustainability aspects related to E, L, and P are assessed using the concept of pressure, state, and response. This has been assessed using the historical secondary data. After assessing the sustainability at H, E, L, and P levels, a new fuzzy rule base (involving four parameters and five states) is suggested to aggregate the HELP parameters and to assess the overall water sustainability index for the watershed. The proposed methodology is explained through the case study of Piperiya watershed.

The proposed framework is applied on the Piperiya watershed of Chhattisgarh state in India. Quantification of hydrology indicator in the watershed shows a level of 58%, environment a level of 65%, life of 63.3%, and policy of 65% with an overall defuzzified score of 0.65. This shows that the watershed sustainability is of the order of 65% which can be considered as **Good**.

20.13 Future Perspective of Watershed Sustainability

Water crisis is a major issue nowadays across the globe. The world scenario reflects that the major portion of the earth would be deficient in terms of freshwater availability. So, conservation approach for proper management of water resource is the need of the hour. Watershed development is a step forward toward sustainable management of freshwater resource. Developing watershed in a particular area is governed by the local topography, relief feature, as well as various environmental attributes (Banerjee et al. 2021a, b, c, d; Meena et al. 2020b). Watershed is a structure that provides various benefits to the human civilization and therefore encompasses the concept of hydrology, environment, life, and policy. Future research perspectives should be aimed toward considering the aforesaid components in order to achieve sustainability in water conservation. In this regard, modeling approach needs to be upgraded properly to develop different water sustainability indices along with the framework. Use of diverse indicators is the key for successful modeling of watershed development. Some technology-oriented indicators need to be framed to maintain the balance of ecology and environment of a particular area (Raj et al. 2021; Khan et al. 2021a, b). Using sustainability index approach in water resource management is very much essential for future perspective in order to promote agricultural intensification. Further, research toward developing new water sustainability indices, new modeling approach, would be required in the upcoming century for sustainable management of water resource.

References

Aguilar-Benitez I, Saphores JD (2010) Public accountability and performance of two border water utilities. Water Policy 12(2):203

Aguilar-Benitez I, Saphores JD (2008) An analysis of nonpayment of water services in Nuevo Laredo, Mexico. Water Resources Res 44(8)

Ali HM (2009) Development of Arab water sustainability index using principal component analysis. In: The 13th international water technology conference, IWTC13, Hurghada, Egypt, pp 1563–1579

Banerjee A, Jhariya MK, Meena RS, Yadav DK (2021b) Ecological footprints in agroecosystem- an overview. In: Banerjee A, Meena RS, Jhariya MK, Yadav DK (eds) Agroecological footprints management for sustainable food system. Springer Nature, Singapore, pp 1–23. https://doi.org/10.1007/978-981-15-9496-0_1. eBook ISBN: 978-981-15-9496-0, Hardcover: 97-981-15-9495-3

Banerjee A, Jhariya MK, Raj A, Yadav DK, Khan N, Meena RS (2021c) Land footprint management and policies. In: Banerjee A, Meena RS, Jhariya MK, Yadav DK (eds) Agroecological footprints management for sustainable food system. Springer Nature, Singapore, pp 221–246. https://doi.org/10.1007/978-981-15-9496-0_7. eBook ISBN: 978-981-15-9496-0, Hardcover: 978-981-15-9495-3

Banerjee A, Jhariya MK, Raj A, Yadav DK, Khan N,Meena RS (2021d) Energy and climate footprint towards the environmental sustainability. In: Banerjee A, Meena RS, Jhariya MK, Yadav DK, Agroecological footprints management for sustainable food system. Springer Nature Singapore. Pp. 415–443. https://doi.org/10.1007/978-981-15-9496-0_14eBook ISBN: 978-981-15-9496-0, Hardcover: 978-981-15-9495-3

Banerjee A, Meena RS, Jhariya MK, Yadav DK (2021a) Agroecological footprints management for sustainable food system. Springer Nature, Singapore, p 514eBook ISBN: 978-981-15-9496-0, Hardcover: 978-981-15-9495-3. https://doi.org/10.1007/978-981-15-9496-0

Boyacioglu H (2010) Utilization of the water quality index method as a classification tool. Environ Monit Assess 167(1–4):115–124

Brown C (2012) Introduction to water, poverty, and ecology: a vision for sustainability. In: Integrating ecology and poverty reduction. Springer, New York, pp 109–111

Brundtland GH (1987) Report of the World Commission on Environment and Development: Our Common Future. United Nation (UN)

Catano N, Marchand M, Staley S, Wang Y (2009) Development and Validation of the Watershed Sustainability Index (WSI) for the Watershed of the Reventazón River. Comcure Report, pp 1–54

Chaves HM, Alipaz S (2007) An integrated indicator based on basin hydrology, environment, life, and policy: the watershed sustainability index. Water Resources Manage 21(5):883–895

Clark SE, Baker KH, Treese DP, Mikula JB, Siu CY, Burkhardt CS (2010) Sustainable Stormwater management: infiltration vs. surface treatment strategies. Water Environment Research Foundation, London

Clark SE, Lenhart JH, de Ridder SA, Barrett ME, Malina JF Jr, Hunt S, Jarrett AR (2014) Performance of BMPs. In: Bridges. Dial Books, New York

Cortés AE, Oyarzún R, Kretschmer N, Chaves H, Soto G, Soto M, Maturana H (2012) Application of the watershed sustainability index to the Elqui river basin, north-Central Chile. ObrasyProyectos 12:57–69

De Carvalho SCP, Carden KJ, Armitage NP (2009) Application of a sustainability index for integrated urban water management in southern African cities: case study comparison-Maputo and Hermanus. Water SA 35(2):144–151

Firdaus R, Nobukazu N, Idris A (2014) Sustainability assessment of humid tropical watershed: a case of BatangMerao watershed, Indonesia. Procedia Environ Sci 20:722–731

Garriga RG, Agustí PF (2010) Improved method to calculate a water poverty index at local scale. J Environ Engineering 136(11):1287–1298

Hernández M (2007) A water sustainability index for the Canary Islands. In: Water resources management in water stressed environments and islands: the challenge of Ecoadaption

Jhariya MK, Banerjee A, Meena RS, Yadav DK (2019a) Sustainable agriculture, Forest and environmental management. Springer Nature, Singapore, p 606. eISBN: 978-981-13-6830-1, Hardcover ISBN: 978-981-13-6829-5. https://doi.org/10.1007/978-981-13-6830-1

Jhariya MK, Meena RS, Banerjee A (2021a) Ecological intensification of natural resources for sustainable agriculture. Springer Nature, Singapore. eISBN: 978-981-334-203-3, Hardcover ISBN: 978-981-334-206-6. https://doi.org/10.1007/978-981-33-4203-3

Jhariya MK, Meena RS, Banerjee A (2021b) Ecological intensification of natural resources towards sustainable productive system. In: Ecological intensification of natural resources for sustainable agriculture. Springer, Singapore. eBook ISBN 978-981-334-203-3. Hardcover ISBN: 978-981-334-202-6. https://doi.org/10.1007/978-981-33-4203-3_1

Jhariya MK, Yadav DK, Banerjee A (2019b) Agroforestry and climate change: issues and challenges. Apple Academic Press Inc., CRC Press- a Tayler and Francis Group, US & Canada, Toronto, p 335. ISBN:978-1-77188-790-8 (Hardcover), 978-0-42957-274-8 (E-book). https://doi.org/10.1201/9780429057274

Joshi H, Kansal ML (2015) Sustainability assessment of rural water supply schemes in Uttarakhand. M. Tech Thesis, Department of Water Resources Development and Management, IIT Roorkee

Juwana I, Perera BJC, Muttil N (2009) Conceptual framework for the development of West Java water sustainability index. In: 18th world IMACS congress and MODSIM09 international congress on modelling and simulation, pp 792–978

Kansal ML, Chandniha SK, Tyagi A (2015) Distance based water sustainability assessment using SPI for the state of Chhattisgarh in India. American Society of Civil Engineers, Reston

Khan N, Jhariya MK, Raj A, Banerjee A, Meena RS (2021a) Soil carbon stock and sequestration: implications for climate change adaptation and mitigation. In: Jhariya MK, Meena RS, Banerjee A (eds) Ecological intensification of natural resources for sustainable agriculture. Springer, Singapore. eBook ISBN 978-981-334-203-3. Hardcover ISBN: 978-981-334-202-6. https://doi.org/10.1007/978-981-33-4203-3_13

Khan N, Jhariya MK, Raj A, Banerjee A, Meena RS (2021b) Eco-designing for sustainability. In: ecological intensification of natural resources for sustainable agriculture. In: Jhariya MK, Meena RS, Banerjee A (eds) . Springer, Singapore. eBook ISBN 978-981-334-203-3. Hardcover ISBN: 978-981-334-202-6. https://doi.org/10.1007/978-981-33-4203-3_16

Kumar S, Meena RS, Jhariya MK (2020) Resources use efficiency in agriculture. Springer Nature, Singapore, p 760. eBook ISBN: 978-981-15-6953-1, Hardcover: 978-981-15-6952-4. https://doi.org/10.1007/978-981-15-6953-1

Kumar S, Meena RS, Singh RK, Munir TM, Datta R, Danish S, Singh GS, Yadav Kumar S (2021) Soil microbial and nutrient dynamics under different sowings environment of Indian mustard (*Brassica juncea* L.) in rice based cropping system. Scientific Report 11:5289. https://doi.org/10.1038/s41598-021-84742-4

Kumar S, Meena RS, Datta R, Verma SK, Yadav GS, Pradhan GS, Molaei A, Rahman GKMM, Mashuk HA (2020a) Legumes for carbon and nitrogen cycling: an organic approach. In: Carbon and nitrogen cycling in soil. Springer, Singapore, pp 337–375. https://doi.org/10.1007/978-981-13-7264-3_10

Lawrence P, Meigh J, Sullivan C (2002) The water poverty index: an international comparison. In: Keele economics research papers. www.keele.ac.uk/depts/ec/web/wpapers/kerp0219.pdf

Loucks DP, Gladwell JS (1999) Sustainability criteria for water resource systems. Cambridge University Press, Cambridge

Meena RS, Kumar S, Datta R, Lal R, Vijaykumar V, Brtnicky M, Sharma MP, Yadav GS, Jhariya MK, Jangir CK, Pathan SI, Dokulilova T, Pecina V, Marfo TD (2020b) Impact of agrochemicals on soil microbiota and management: a review. Land (MDPI) 9(2):34. https://doi.org/10.3390/land9020034

Meena RS, Kumar V, Yadav GS, Mitran T (2018) Response and interaction of *Bradyrhizobium japonicum* and arbuscular mycorrhizal fungi in the soy bean rhizosphere: a review. Plant Growth Regul 84:207–223. https://doi.org/10.1007/s10725-017-0334-8

Meena RS, Lal R (2018) Legumes for soil health and sustainable management. Springer Singapore, Singapore, p 541.ISBN 978-981-13-0253-4 (eBook), ISBN: 978-981-13-0252-7(Hardcover). https://doi.org/10.1007/978-981-13-0253-4_10

Meena RS, Lal R, Yadav GS (2020) Long term impacts of topsoil depth and amendments on soil physical and hydrological properties of an Alfisol in Central Ohio, USA. Geoderma 363:1141164. https://doi.org/10.1016/j.geoderma.2019.114164

Meena RS, Lal R, Yadav GS (2020a) Long-term impact of topsoil depth and amendments on carbon and nitrogen budgets in the surface layer of an Alfisol in Central Ohio. Catena 2020194:104752. https://doi.org/10.1016/j.catena.2020.104752

Parmar DL, Keshari AK (2012) Sensitivity analysis of water quality for Delhi stretch of the river Yamuna, India. Environ Monit Assess 184(3):1487–1508

Pitt R, Clark SE (2008) Integrated storm-water management for watershed sustainability. J Irrigation Drainage Engineering 134(5):548–555

Raj A, Jhariya MK, Khan N, Banerjee A, Meena RS (2021) Ecological intensification for sustainable development. In: Ecological intensification of natural resources for sustainable agriculture. Springer, Singapore.,eBook ISBN 978-981-334-203-3. Hardcover ISBN: 978-981-334-202-6. https://doi.org/10.1007/978-981-33-4203-3_5

Raskin PD, Hansen E, Margolis RM (1996) Water and sustainability. In: Proc natural resources forum. Wiley Online Library, New York, pp 1–15

Salameh E (2000) Redefining the water poverty index. Water Int 25(3):469–473

Sheoran S, Kuma S, Kumar P, Meena RS, Rakshit S (2021) Nitrogen fixation in maize: breeding opportunities. Theor Appl Genet. https://doi.org/10.1007/s00122-021-03791-5

Sood A, William FR (2011) Developing a framework to measure watershed sustainability by using hydrological/water quality model. J Water Resource Protection 2011(3):788–804

Sullivan C (2002) Calculating a water poverty index. World Dev 30(7):1195–1210

Sullivan C, Jeremy M (2003) Considering the water poverty index in the context of poverty alleviation. Water Policy 5(5–6):513–528

Sullivan C, Meigh J (2005) Targeting attention on local vulnerabilities using an integrated index approach: the example of the climate vulnerability index. Water Sci Technol 51(5):69–78

Sullivan CA, Vörösmarty CJ, Craswell E, Bunn S, Cline S, Heidecke C, Meigh JR (2006) Mapping the links between water, poverty and food security. In: The limits to growth: a report for the Club of Rome's project on the predicament of mankind by Donella H. meadows and others, 2nd edn. Universe Books, New York. 1974

Wagener T, Sivapalan M, Troch PA, McGlynn BL, Harman CJ, GuptaHV KP, Rao PSC, Basu NB, Wilson JS (2010) The future of hydrology: an evolving science for a changing world. Water Resources Res 46(5)

Impact of Climate Change on Insects and their Sustainable Management

21

I. Merlin Kamala and I. Isaac Devanand

Abstract

Global warming and its threatening impression on global yield of agricultural and horticultural crops and food security have engrossed the scientific attraction across the continent. Insects are arthropods, with greater adaptive mechanisms for survival in diverse habitats. The climatic variations due to global warming influence the insect diversity by disturbing their ecosystem. Being poikilothermic, insects are greatly affected by the alterations in abiotic factors with heavy impact owed by elevated temperature. Insect experiences higher fecundity rate and increased life cycles with rapid growth rate causing outbreak which severely affects agricultural production due to climatic variations. Globally 40% of food production is minimized by pests, and the reforecasting pest population is essential to ensure global food security. The pest management strategies should focus on reducing crop losses induced by pests by enhancing services of ecosystem and the flexibility of crop ecosystem in the face of climate change. This review highlights the impact of climatic factors on behavior of insects and possible tactics to mitigate climate-induced changes in insects for their effective management.

Keywords

Climate change · Ecosystem · Environment · Global warming · Insects · Management

I. M. Kamala (✉)
Department of Agricultural Entomology, Tamil Nadu Agricultural University, Coimbatore, Tamil Nadu, India

I. I. Devanand
Department of Agricultural Extension, Annamalai University, Chidambaram, Tamil Nadu, India

M. K. Jhariya et al. (eds.), *Sustainable Intensification for Agroecosystem Services and Management*, https://doi.org/10.1007/978-981-16-3207-5_21

Abbreviations

C:N	Carbon/Nitrogen ratio
CO_2	Carbon dioxide
IPCC	Intergovernmental Panel on Climate Change
IPM	Integrated Pest Management
N	Nitrogen

21.1 Introduction

Global Warming or climate change is the chief environmental theme to date owing to its intricacy, ambiguity, randomness, and variable effects over place and time. It is the change in the global weather worldwide over time as a result of usual inconsistency or due to human activities as defined by the Intergovernmental Panel on Climate Change (IPCC 2014). Recent trends in Earth's weather report that an average of 4 °C increase in temperature will be attained by the year 2050 (Paris Contribution Map 2016). The ever-changing weather patterns threaten food production and climate change impacts globally (Jhariya et al. 2019a, b, 2021; Meena et al., 2020; Raj et al. 2020, 2021). Without drastic immediate action, it will be challenging and exorbitant to adapt its impact in future (Banerjee et al. 2020).

Climate change triggers universal swings in temperature, carbon dioxide (CO_2) levels, and precipitation patterns resulting into inerratic and risky climatic patterns (Kumar et al. 2020a; Khan et al. 2020a, b). It is apparent that weather change has momentous influence on agricultural production worldwide (Lobell and Field 2007) and endures its impacts in forthcoming years (Challinor et al. 2014; Beddington et al. 2012). Global warming influences the pest incidence and distribution directly or indirectly across the globe affecting the crop yields (Lamichhane et al. 2015; Juroszek and von Tiedemann 2011; Macfadyen et al. 2018). Pests are organisms (insects, plants, microbes, animals) capable of reducing the eminence and measure of the produces and their by-products. The most diverse group of organisms, the insects, comprises millions and millions of designated species representing the most population of living organisms (Chapman 2006).

They are excellent indicators of climate change driven by various factors in an ecosystem. The mounting problems of changing weather globally might have grave penalties on the abundance and diversity of arthropods, influencing crop production as well as food security. Insects have retorted to warming in all potential anticipated means from phenological variations to dynamics in distribution by undergoing evolutionary changes (Menendez 2007). Pest management is, therefore, an integral part of maintaining viability in these systems. Insect reactions to the change in ecosystems are decisive for considering the response of agroecosystems to climate change (Kumar et al. 2020). Predictions of the number and distribution of insect pests will help to figure out pest management approaches to lighten the adversities of climate change on crop production.

Change in insect behavior due to weather variations influences agricultural productivity in several ways. As climate change progresses, crop damages due to pests will compound and interact with plant stress and severely affect the plants. Increased pest population and their activity will stress crop plants and accelerate the peril of yield loss. Insects, being poikilothermic animals, are overdelicate to change in temperature. The unpredictable change in climate could result in outbreaks, new pest emergence or resistant biotypes migration, biodiversity variations, species extinction, and change in host shift and invite insect-borne diseases. These impacts would positively uplift insect damage and down lift crop yields, increase the cost on crop protection, and thereby affect the economy.

Climate change affects the physiology and incidence of insect pests in wide array of crops (Painkra et al. 2016). The world food production is at loss by 40% due to pests; the reduction of crop pest incidence is gaining severe attention for assured global food security (Heeb et al. 2019). Novel invasive pests are entering several new provinces of the world as a result of climate change without proper monitoring and management tactics; they have the potential to become key pests (Jessica et al. 2008). Climate change has its own impact and amplified issues on insect transmitted diseases also. Global warming severely impacts on crop protection from insects and thereby food security, exclusively in developing countries where sustainable food production is an urgent needs (Sharma 2014). Consequently, dynamics in population of pests occur in different agroecosystems and ecological zones. Moreover, species extinction is accelerated due to climate change, and the rate of species extinction is projected because of global warming (Sinervo et al. 2010; Urban 2015). Therefore, to mitigate biodiversity losses, environmental policies and management practices in amending climate change impacts should be modulated (Brooke 2008).

Consequently, it is the necessity to have a rigorous view on the probable adverse influence of global warming in pest management and toward invent apposite action to ease the global warming effects of food security (Sharma and Prabhakar 2014).

The climatic change impacts on pests may include (Table 21.1):

- Variations in abundance and diversity of crop insect pests.
- Fluctuations in geographical spread of crop insect pests.
- Deviations in synchrony between insect pests and their host crops.
- Changes in host plant resistance.
- Changes in insect biotypes.
- Changes in tritrophic interactions.
- Fluctuations in the profusion and for aging motion of natural enemies.
- Rapid population growth and number of generations of insects.
- Increased overwintering insects.
- Impact on extinction of species.
- Amplified peril of invasive pests.
- Augmented menace of insect-transmitted diseases.

Table 21.1 Global warming influence on insects (Modified from Bhagat et al. 2015)

Weather parameter	Pests	Influence on host plant	Influence on insect	References
Elevated temperature	*Spodoptera litura*	Declined nutritional quality of foliage	Insect development declined with increased temperature till 30°C but increased after 35°C	Srinivasa Rao et al. (2015)
Reduced temperature	*Rhaphalosiphum maidis*	Heavy crop yield loss	Development of immature stages and longevity of adults were declining till 25 °C. Highest fecundity response at 20 °C and reduced to 50% as temperature increased till 30 °C	
Elevated CO_2	Peanut aphid, *Aphis craccivora*	Heavy crop yield loss	Increased fecundity and decreased developmental time and longevity at elevated CO_2 of 500 ppm than ambient conditions	Fand et al. (2012)
	Foliage feeding lepidopterans	Reduction in nitrogen level in plant tissues, enhanced level of carbon-based defenses like tannins, declined level of nitrogen-based defenses such as alkaloids	Enhanced foliage feeding to gain nitrogen for body metabolism leads to slower growth development and increased life stage development	
	Gypsy moth, *Lymantria dispar*	Minimal leaf water content, enhanced soluble sugar level, and low nutritional quality of plant foliage	Increased larval feeding, reduced larval weight gain, and prolonged development	
Drought and water stress due to high temperature	Midge, *Stenodiplosis sorghicola* Spotted stem borer, *Chilo partellus*	Resistance breakdown against target pests in sorghum host plant leading to heavy yield loss		
	Cotton bollworm, *Helicoverpa armigera*	Increased susceptibility of cotton to bollworms		

(continued)

Table 21.1 (continued)

Weather parameter	Pests	Influence on host plant	Influence on insect	References
Erratic weather patterns	Sugarcane woolly aphid, *Ceratovacuna lanigera*	Reduced cane recovery and yield loss		Joshi and Viraktamath (2004)
	Rice plant hopper, *Nilaparvata lugens*	Crop failure		IARI News (2008)
	Mealy bug, *Phenacoccus solenopsis*	Heavy yield loss of 40–50%		Dhawan et al. (2007)
	Papaya mealy bug, *Paracoccus marginatus*	Heavy yield loss		Tanwar et al. (2010)

- Introduction of alternative host plants.
- Reduced efficacy of crop protection technologies.

Incessant and intense vigilance of insect population and behavior primarily in recognizably delicate sections may provide few first signals of a biotic retort to climate change. The adverse effects of climate change will diverge across provinces, crops, and pest species. A vast figure of replicas and etiquettes has been intended to quantify the climate change impacts on environment and food security (Figs. 21.1 and 21.2). Comprehensive knowledge on adaptive mechanisms of insect species to withstand changing weather pattern is utmost needed for scheduling research and development energies in IPM strategies for the future (Sharma 2010).

21.2 Weather Parameters that Influence on Insect Behavior

21.2.1 Rising Temperature

Temperature is documented as the leading abiotic factor directly affecting herbivorous insects. Global warming causing rise in temperature could influence insect population which infests crops in several composite ways. Rise in temperature can possibly disturb insect behavior, survival, development, geographic range, and population growth. Insects, being poikilothermic, have their body temperature roughly the same as that of environment, and hence the life stages of insects are strongly dependent on temperature for survival (Edward et al. 2004). As the global temperature rises, there are higher chances for the insects to complete their life cycles in faster rate. The speedier the completion of life cycle of an insect, the greater will be the abundance of pests. Insect pests such as whiteflies, beet armyworms, cabbage

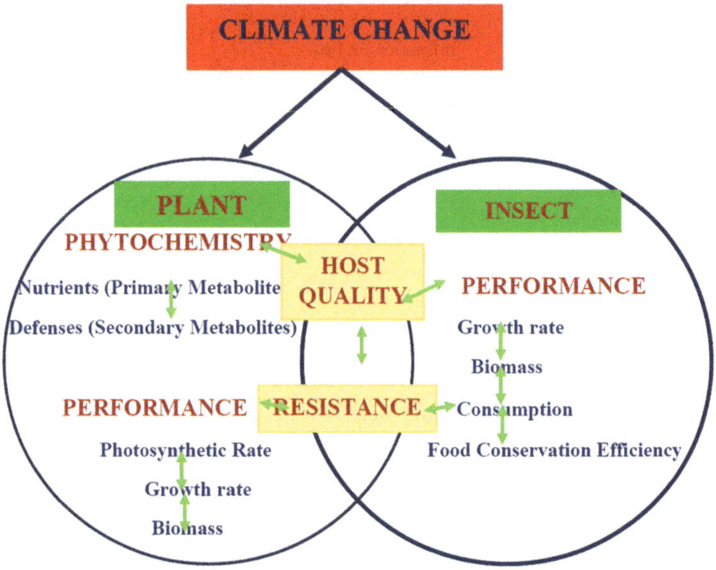

Fig. 21.1 Conceptual framework illustrating global warming impacts on crop and insect pests

loopers, and leaf miners develop quickly when the average temperature is about 85–90 °F and develop much slower under cool, winter conditions. Majority of the existing insect's species will have influence on at least few degrees of changes in temperature and thereby imparting multiple effects in different life stages. The consequences of increase in temperature on insects could directly impact on insect's composition and performance or indirectly as interceded by crop host, herbivorous pests, and their natural enemies (Thomson et al. 2010). Rise in temperature has led to augmented North relocation of few insects, amplified growth and rate of oviposition, probability of insect outbreaks, and introduction of foreign species and destruction of insects (Sable and Rana 2016). Increased rise in temperature impacts insects chiefly by enlarging the geographical area of distribution, amplified period of overwintering, higher growth rate in population, increased number of generations per year extending the evolving period of life stages, changes in synchronization between crop and pest, changes in interspecific interactions, augmented peril of incursions by pest refugees and insect borne diseases, new introduction of alternative hosts, etc. (Bale and Hayward 2010). In addition, the diverse impacts constitute phenological changes in insects, community composition, and distribution in ecosystem which leads to species extinction (Walther et al. 2002). Moreover, higher temperature decreases the effectiveness of microbes, viz., fungi and bacteria, in biological control of insects, which is a remarkable disadvantage in ecofriendly pest management (Sable and Rana 2016).

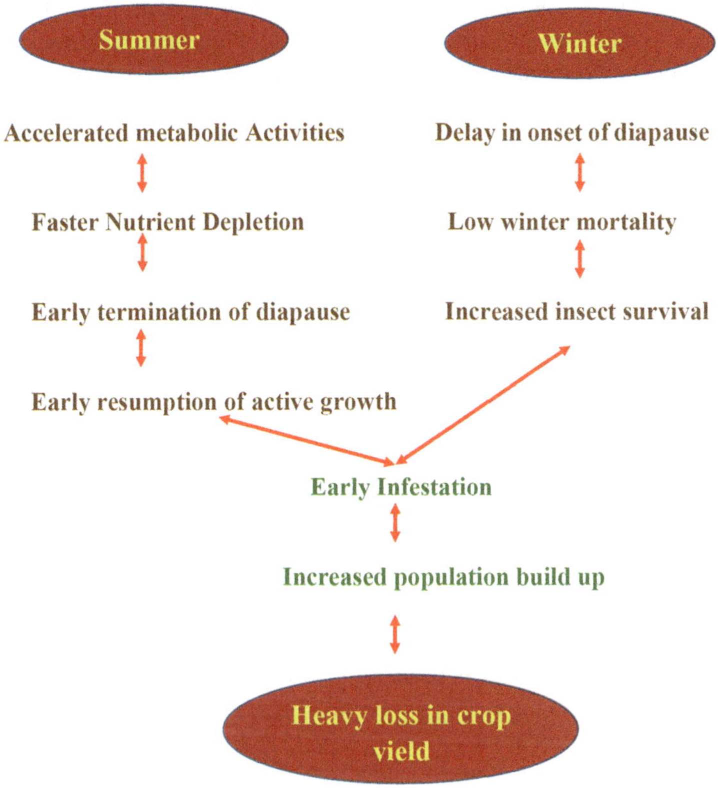

Fig. 21.2 Season-wise influence of global warming on insect survival

21.2.2 Increased CO_2 Level

Increasing concentrations of CO_2 on plants are a major influence of climate change which influences insects ultimately. High atmospheric CO_2 accelerates photosynthetic rate and increases the proportion of carbohydrates relative to N in plant leaves. Increased CO_2 content in the atmosphere enhances the C:N ratio of plants as plant tissues accumulate nonstructural C in plant tissues, which results in 15–20% reduction in N content in plant tissues, by which nutritional quality is limited for insects (Coviella and Trumble 1999). Hence, insects are forced to accelerate their food to compensate the food intake (Meena et al. 2020a; Holton et al. 2003). The predicted outcome from insects due to the rise in C:N ratio is compensatory feeding by adapting to new alternative hosts, increased accretion, plant defensive chemicals, and pest competition for survival. However, the insects respond variably to increased CO_2 level among species. The larvae of insects need N for their growth and establish their tissues; nonetheless, adults can reproduce and survive on a high-carbohydrate diet and would migrate to high CO_2 host and feed on more voraciously, causing enormous damage to host plants (Litvak et al. 2002). Amplified C:N ratio in plant tissues ratios will slow the insect development with prolonged duration of life stages, susceptible to attack by parasitoids. A major pest of soybean, the Japanese beetle (*Popillia japonica*) extended its life span by 8.25%, and their females laid twice the quantity of eggs when fed on soybean plants raised under elevated CO_2, due to amplified sugar levels of glucose, fructose, and sucrose in plant foliage (OíNeill et al. 2008). Aphids that intake sap under elevated CO_2-conditioned plants have a high rate of fecundity due to higher carbohydrate level (Chen et al. 2005). An increased food intake by lepidopterans and multiplication ability by sucking insects occur due to elevated CO_2, while a decreased development potential of insects, their response to pheromones, and parasitism potential of parasitoids were observed (Das et al. 2011; Boullis et al. 2016). The pine sawfly (*Neodiprion lecontei*) reared on plants under elevated CO_2 concentration pronounced increased N utilization potential (Williams et al. 2000).

21.2.3 Low Temperature

Low temperature processes, viz., cooling and freezing, severely disturb the insect's physiology and behavior. It can cause variation in the chemical constituents of insect physiology. They maintain the body fluids below the melting point and cause dehydration of cells. An increased metabolism, slowing down movement, feeding, and reproduction of insects are noticed due to low temperature (Mullen and Arbogast 1979). A minimal dynamic in temperature also impacts the pre-ovipositional, ovipositional, and survival of insects. The rice pest, brown plant hopper (*Nilaparvata lugens*), performed higher survival rate between 25 and 35 °C but reduced at 40 °C. Moreover, the egg-laying rate of females was advanced at 35–40 °C and lesser at 25–30 °C. A boost in the population of the rice pests, green leaf hopper, *Nephottetix cinctipes* and Striped rice stem borer, *Chilo suppressalis* was noticed due to low temperature (Narayanasamy et al. 2015). A drastic increase in population of rice earhead bug (*Leptocorisa acuta*) is noticed due to 3 °C decrease in temperature (Reji and Chander 2008). The pine beetle (*Dendroctonus ponderosae*) establishes well and disperses in warm winter as compared to deep cold winter (Bleikar et al. 2017).

21.2.4 Humidity

Temperature and humidity go hand in hand to influence the physiology and behavior of insects directly as well as indirectly. The direct influence may be by preventing and motivating insect activity, their dispersal, growth rate, and survival in adverse climatic conditions. Indirect influence includes variations in plant development, plant phenology, host nutritional quality, and predators, parasitoids, and microbial activity (Porter et al. 1991). Relative humidity (RH) can influence the insect physiology and thereby their progress. The embryo formation, egg development, and egg hatching were affected by low relative humidity that can prevent structural deformation as the cuticle is softened (Guarneri et al. 2002). Various insects respond to humid conditions in different ways. Humidity, apart from influencing the insect behavior, also disturbs the physiological mechanism. It causes deformities in insects, and larval mortality will also increase. The impact of RH on the egg hatchability of the bamboo borer (*Dinoderus minutus* Fabricius) revealed that low egg hatchability at 20% and 85% RH levels and shrinkage of eggs were noticed due to moisture loss leading to embryo and chorion shrinkage which prevents larval release (Norhisham et al. 2013). The variations in atmospheric humidity stress the insects, and they produce immune responses. Humidity can cause mortality, fecundity, oviposition rate, sex ratio, and mutagenic effects to certain extent. The cuticular hydrocarbons in integument and the sub-elytra chamber extend a vigorous part in conservation of water to survive drought conditions.

21.2.5 Precipitation/Drought

Distribution and frequency of rainfall impact pest incidence directly due to alteration in the humidity level. The predictions on climate change advise that the rainfall frequency would weaken, while its intensity increases due to global warming, which would lead to high precipitation, drought, etc. The receipt of average rainfall will decline in some regions, leading to rise in occurrence of droughts. The tiny insects such as thrips, aphids, whiteflies, leafhoppers, mites, etc., will be washed away in rains, and therefore the crop escapes the incidence of these deadly sucking insects (Pathak et al. 2012). The infestation of tiny sucking aphids in most crops including wheat is suppressed by rainfall or sprinkler irrigation that mimics rainfall (Daebeler and Hinz 1977; Chander 1998). The correlation of *Helicoverpa armigera* outbreak and rainfall from the period of the 1940s to 1960s showed that November rainfall favored higher infestation of *H. armigera* (Hub.) damage (Lever 1969). The larval population of *Agriotes* is higher with severe rainfall; in contrast, the species *Lecanopsis formicarium* belonging to family Coccoidea, was not affected due to rainfall (Karuppaiah and Sujayanad 2012). Crops such as groundnut, cotton, chillies, and coriander before, after, and along with tobacco lead to higher incidence of *S. litura* under heavy rainfall (Chari et al. 1993). The population of oriental army-worm (*Mythimna separata* Walker) increased as extended drought period follows heavy rainfall.

21.2.6 Gases

Insects pronounced varied retorts toward diverse gases level. Increased or decreased concentration of gases such as O_2 and CO_2 might cause hypoxia or hyperoxia kind of situations, which is noticed in insects like *S. americana*, *T. moliteor*, *C. vomitoria*, *C. trichopterus*, *D. melanogaster*, *Manduca* species, etc. The insects, in return, produce compensatory changes by opening their spiracles for ventilation when exposed to hypoxia or hyperopia. In severe cases, enhanced oxidative stress or reduced survival rate was recorded (Harrison et al. 2006). The extremes of temperatures (15–35 °C) and concentration of oxygen (10%, 21%, and 40%) diverge the development rate of tsetse fly (*Glossina pallidipes*) puparia (Ciss et al. 2013).

21.3 Effect of Global Warming on Insect Behavior

21.3.1 Effect on Insect Biology

Temperature is an important abiotic factor altering insect biology, by making them more active, expanding their population, and causing severe economic loss. Predictions suggest that a rise of 2 °C temperature might result in few supplementary life stages in a term (Yamamura and Kiritani 1998). Elevated temperatures could fasten the process of egg development in insects with additional generation per year (Scott et al. 2010). The life span of aphid (*Aphis gossypii*) Glover extends to 20–22 days at 10–25 °C, but at 30 °C it will take only 6–9 days to complete their life cycle. An increased fecundity, speedy egg development, and increase in weight were experienced in the cricket (*Gryllus texensis*) exposed to enhanced temperature for 6 days. Moreover, an enhanced activity of phenol oxidase and lysozyme-like enzymes was noticed, which can lead to emergence of disease-resistant crickets. The commencement and termination of diapause with subsequent variations in voltinism were observed and ultimately lead to disruption in developmental cycles due to increased temperature (Ayres and Lombardero 2000). Globally, climate change is expected to increase the surface temperature of the upper layer of soil by 1.6–3.4 °C by 2100, affecting soil insects severely. An increased number of eggs (324.3 ± 112.3 eggs/female) were laid by brachypterous females of rice brown plant hopper (*Nilaparvata lugens* Stal.) on rice plants treated with enhanced CO_2 (570 ± 25 ppm) than 380 ppm ambient CO_2 (231.7 ± 31.8 eggs) (Prasannakumar et al. 2012).

21.3.2 Effect on Insect Population Dynamics

Climate change highly impacts the magnitude of insect population, their growth and development, incidence and distribution, and their outbreak under favorable conditions (Juroszek & von Tiedemann, 2012). A rapid increase in population is experienced in insects dwelling in colder region as they undergo lower winter mortality, leading to higher development rates and frequent reproduction. It also causes increased insect populations due to warmer winter temperatures (Harrington et al. 2001). Elevation in temperature regimes has negative correlation with delicate natural enemy population. About 10 °C rise in temperature declines winter mortality of *Nezara viridula* by 15%. The predator of brown plant hopper (*Cyrtorhinus lividipennis*) has negative effect at 40 °C with the BPH (*N. lugens*) 17 times more tolerant than the predator. Climate change has been reported to increase the population of 15 species of polyphagous heteropteran bugs, of which 12 species cause major outbreaks (Tomokuni et al. 1993).

The warm weather caused population outbreak of the pyralid rice pest, *Chilo suppressalis*, which widened its distribution due to increased fecundity (Kiritani and Morimoto 2004; Kiritani 2006). The light-trap observations of rice pests, viz., rice stem borer (*C. suppressalis*), green rice leafhopper (*Nephotettix cincticeps*), and small brown plant hopper (*Laodelphax striatellus*), reflect a declining trend due to climate change (Yamamura et al. 2006). In sugarcane belts of India, the outbreak of the sugarcane woolly aphid (*Ceratovacuna lanigera* Zehntner) was recorded due to changes in climatic variables in the region (Joshi and Viraktamath 2004; Srikanth 2007). Population dynamics is an important population ecology criterion that focuses on factors that upset the population densities of insects. The development, incidence, and dispersal of insects were vastly affected due to seasonal variations. Dynamics in temperature regimes alters the life stage development, molting stages, and survival, which subsequently impacts on population size and density of insects. The genetic composition of insects is also affected due to seasonal variations, which impact on their level of host plant exploitation (Bale and Hayward 2010). The unpredictable changes in weather may upset ecological balance in the insect pest and their existing natural enemy population in any ecosystem (Rao et al. 2009; IPCC (Intergovernmental Panel on Climate Change) 2007a).

21.3.3 Effect on Insect Diversity

Insects encompass the largest group of the animal kingdom and render major ecosystem services (Kremen et al. 1993; Kannan and James 2009). The health status of an ecosystem is indicated by the insect diversity in a habitat, as they are excellent pointers of alterations in an ecosystem (Gregory et al. 2009) and show an imperative part in food chains. The potential changes in weather vastly influence the relative abundance of insect pests, and the specific species that cannot withstand the changes may vanish in due course (Thomas and Blanford 2003). The Western Ghats of India is the center that attracts diverse rare, endemic, exotic colorful butterflies of the world (Hampson 1908; Anand and Pereira 2008). The diminution of vegetative cover due to climatic variations as well as anthropogenic actions has minimized butterfly population, which is under threat at present time (Costanza et al. 1997; Sachs 2008; Meena et al. 2020b; Sidhu and Mehta 2008). The diversity of ecosystems vulnerable to climate change needs to be studied to decrease the losses due to insect pests (Newton et al. 2009). Speciation usually occurs between 100 and 10,000,000 years, providing 10–10,000 new species annually. Most of the existing insect species on Earth have become extinct. The extinction rates of insects are 100–1000 times greater due to increasing alterations in weather reporting about 45–275 species vanishing per day (Sharma 2010).

21.3.4 Effect on Insect Phenology

Insect phenology displays a varied response to global warming with widespread shifts toward earlier seasonal activity. Seasonal variations have led to ecological swing in period, with change in species and their phenology. Shift in phenology is the easiest form to monitor climate change in insects and most documented for varied series of organisms (Root et al. 2003). Insects pass through their larval stages in a quicker rate and reach adult stage early due to rise in temperature. The phenological responses of insects to climate change are more complex, with many species having prolonged activity and others delayed activity periods. An advance in larval and adult advent timings and an increase in flight duration length are usual responses of insect phenology (Menéndez 2007). The insect order Lepidoptera represents a better example for phenological changes. In UK, 26–35 species have advanced their initial appearance due to changes in butterfly phenology (Roy and Sparks 2000). Early emergence and onset of many migrants of aphids was reported from the UK (Harrington et al. 2001). The climate change effect on butterflies, beetles, fly, etc., showed that all the tested species displayed fluctuations in their initial appearance over years due to increased spring temperature (Gordo and Sanz 2011). Insect phenology deviations can be extensively studied through long-term experiments with diverse sowing dates and monitoring the pest show up through light trap, suction trap, or pheromone trap. The Rothamsted Insect Survey reflected that the spring flights of potato aphid (*Myzus persicae* Sulcer) began 2 weeks prior for every 1°C rise. Diverse phenological changes in insects such as being overactive, earlier migration and reproduction, early emergence from hibernation, and increased growth rate were evidenced in several insect-recording schemes by long-term monitoring (Roy and Sparks 2000). Moreover, a vast number of insect species being active throughout the year with more annual number of generations are witnessed due to rise in temperature. The pest onset on a crop ecosystem is also altered due to climate change according to a long-term data analysis on phenology (Pathak et al. 2012). Global warming can upsurge or decline harmonization among insects, their host plant, and associated natural enemies (Jessica et al. 2008).

21.3.5 Effect on Overwintering Survival

The insects have limited ability to adapt with surrounding environmental temperature changes as they are poikilothermic. To withstand these changes, they have adapted several behavioral avoidance and physiological adaptations like diapause to survive thermal stress (Bale and Hayward 2010). Diapause is a resting period with development events initiated and postponed by environmental aspects such as temperature, rainfall, humidity, and photoperiod. Diapause, being an adaptive feature, regulates insect life cycle, when they face difficulty in their survivals. Diapause occurs as aestivation and hibernation to survive in both high and low temperatures correspondingly (Chapman 1998). Global warming is predicted to occur greatest at high latitudes (IPCC (Intergovernmental Panel on Climate Change) 2007b; IMD

2010). Consequently, substantial variations were experienced in insects that undergo winter diapause due to global warming (Bale and Hayward 2010).

21.3.6 Effect on Insect Coloration

Insect coloration is an adaption by insects to maintain the heat. It plays a vital role to absorb solar energy, to fuel their flying mechanism, as well as to amend the body temperature. Principally, darker shades have the ability to absorb heat and paler shades to avoid or restrict heating. Dark-colored insects dwelling in cooler region have the potentiality to captivate more solar energy than pale-shaded insects, to intensify their body temperature. In contrast, insects living in arid region have to defend themselves to avoid overheating. Light-colored insects can reflect light to guard their body being overheated and sustain active for longer duration, and, therefore, they are likely to dwell in tropical region. A study at Europe with 366 butterfly species and 107 dragonfly species clearly depicts the pale-shaded insects dominating the southern part of Europe with warm weather, and dark-shaded insect's rule over the Northern region with cold weather (Dirk 2014). A change in color of ladybird beetles from the Netherland coast from black to red was noticed due to climate change. Red imitates more energy; therefore, the ladybirds stay cooler (Sara 2011).

21.3.7 Effect of Parasitism

The survival rates of a parasitoid hang on the vulnerability of the host insect. Elevated temperatures have higher probability of favoring the host than its parasitoid. Higher rate of parasitism of *Spodoptera littoralis* by the parasitoid *Microplitis rufiventris* was recorded at lower temperature, 20 °C (68 °F), than at higher temperature, 27 °C (80.6 °F) (Thomas and Blanford 2003). Likewise, higher parasitism of the spruce budworm (*Choristoneura fumiferana*) was noticed at lower temperatures (Harrington et al. 2001).

21.3.8 Effect of Insect-Vectored Plant Diseases

Climate change causes speedy multiplication of insect vectors which may lead to more occurrences of insect-transmitted diseases (Petzoldt and Seaman 2007). A quick development of disease pathogens in insect shippers was reported due to rise in temperature. The development and spread of malaria pathogen require temperature above 16 °C (60.8°F), and progress of the pathogen that causes malaria (*Plasmodium falciparum* protozoa) will be doubled at 5 °C (9 °F) increase in temperature (Sharma et al. 2010). The early colonization of virus-bearing aphids which are the chief vectors of potato viruses in Europe, which causes higher occurrence of viral diseases of potato, was advocated by higher temperature (Robert et al. 2000). Deadly human diseases transmitted by mosquitoes are chiefly influenced by seasonal variations, as the eggs are laid in standing water, viz., puddles, ponds, lakes, pools, etc. The intensity of the stagnant water alters with weather parameters such as temperature, humidity, rainfall, wet weather, etc., characteristically attracting a huge number of mosquitoes. As a female mosquito acquires infection by drawing blood from an infected host, it transmits to different hosts by another bite (Irfan 2011).

21.3.9 Effect on Immune Genetic Responses

The stress due to the surrounding environmental change alters the insect immune system by upsurging stress hormone in some insects. Insects utilize their malphigian tubules functioning as to avoid atmospheric stress due to climate change. A stronger action of insect immune parameters along with phenoxides and anti-bacteria was experienced due to temperatures at higher rates than lower rates (Catalan et al. 2012). Extreme temperature, along with body weight, affects the development of ovary and reproductive flexibility in butterfly (*Pararge aegeria*) (Fresquet and Lazzari 2011). Deviation from upper and lower temperature regimes causes damage to the nervous and endocrine system of insects, influencing elicitation of different heat shock proteins, thereby causing developmental and behavioral changes (Chambers and Schneider 2012; Adamo and Lovett 2011; Overgaard and Sorenson 2008). The metabolic rate of puparia of tsetse fly varies under variable temperatures and oxygen concentration (Neven 2000).

21.3.10 Effect on Crop-Pest Interaction

Climate change has extensive impact on plant-insect communications, and the physiology of both insect pests and plants is tampered. A reduction in larval weight, prolonged feeding duration, and extended development period are shown by gypsy moth attacking the red maple (*Acer rubrum* L.) and sugar maple (*A. saccharum* L.) due to altered metabolism. The severe outbreaks of the insects observed in new alternate hosts are due to their high vulnerability to hosts or inability of their natural enemies to spot the larvae in a new host environment (Williams et al. 2000). Temperature and related abiotic factors, viz., humidity and photoperiod, have been found to profoundly affect plant life stages such as root and stem elongation, flowering, fruiting, and seeding (Cleland et al. 2007). Climate change causes intensification in temperature which will eventually accelerate the life cycles of many species of plants (Willis et al. 2008; Fitter and Fitter 2002; Parmesan and Yohe 2003), which is positively correlated with behavioral patterns with the associated insect pests. These changes in plant life stages and eventually in insect life stages due to global warming can significantly aggravate the adverse environmental and commercial costs (Timoney 2003; Millenium Ecosystem Assessment 2005).

21.3.11 Effect of Climate Change on Food Security

The biggest challenge ahead for agricultural scientists is to enhance the present food productivity rate to at least twofold to meet the demands of the increasing world population with dwindling natural resources (Deka et al. 2008; Raj et al. 2019a, b). The exasperating pest-related issues under changing climatic regimes are predicted to intensify, the yield losses impacting severe threat in food production relying more on agriculture. Climate change is likely to affect the pollination of flowers on insects by distrusting the plant-pollinator synchronization (Kudo et al. 2004), thereby reducing world food production to one-third, increasing the risk of world food production (Klein et al. 2007). Reduced pollination by insects will cause key inference on global food and nutritional security (FAO 2008). The rural livelihoods that are dependent on pollination for their survival for food production will be directly affected (IPCC (Intergovernmental Panel on Climate Change) 2007b; Chahal et al. 2008).

21.4 Effect of Climate Change on Chemical Ecology and Tritrophic Interactions

The global change of climate has been found to exert positive and negative response on the tritrophic interactions between crop, insect pest, and their natural scavengers through physiological fluctuation associated with suitability of the host and their nutritional status (Roth and Lindroth 1995; Hare 1992; Coviella and Trumble 1999; Guitierrez et al. 2008). An herbivorous insect depend on both host plant and environmental conditions to complete its development. Seasonal variations diverge nutrient level of plants, which impacts insect growth, population, and the access of prey for natural enemies (Selvaraj et al. 2013; Walther et al. 2002; Sharma 2016; Dhillon and Sharma 2007; Boullis et al. 2015). Plants grown in elevated CO_2 and temperature levels with minimal precipitation have the altered nutritional quality which influences the capability of the natural enemies feeding on these hosts (Moore and Allard 2008).

Global warming favor crops that are susceptible to pests or nonresistant cultivars with greater insect pest infestation (Lobell and Gourdji 2012; De Lucia et al. 2012). The production of plant secondary metabolites and other plant defensive traits will be affected and conducive for insects. Unpredicted climatic conditions minimize defense potential of plants due to stress accumulated in the absence of adaptation (Coley and Markham 1998; Niziolek et al. 2012; Zavala et al. 2008). The amendments in chemical status of host plants cause pest outbreak in plants grown in temperate regimes (Harrington et al. 2001). The nutritional status of plants grown under enhanced CO_2 regimes cause extended larval duration, weak life stages, and rapid mortality (Coviella and Trumble 1999; Sharma et al. 2016). Elevated CO_2 level may accumulate N, thereby minimizing N-based defenses such as alkaloids and increasing C accumulation with increased carbon-based defenses such as tannins (Sharma et al. 2016). Enhanced food ingestion by herbivores up to 40% was observed as the N level of foliar is very low due to elevated CO_2 level (Guitierrez et al. 2008). Similarly, explosion of herbivores occurs due to unusual drought spell which minimizes the protein profile of the plant (Chen et al. 2005). An upsurge in carbohydrate level is evidenced in plants grown in elevated CO_2 conditions attributed to increased photosynthesis and recorded heavier foliage damage (Sharma et al. 2016). The plant defense system is profusely altered causing a drop in its immunity level toward insect pests due to climate change (Dhaliwal et al. 2004, 2010), which is notably witnessed by early incidence of *H. armigera* infestation in cotton and pulses under Indian conditions (Sharma 2010). Under elevated CO_2 conditions, the signaling ability of plant defensive pathways mediated by jasmonic acid (JA) in soybean plants is disrupted and thus becomes vulnerable to pest attack (Zavala et al. 2008). The maize plants grown in elevated CO_2 conditions were susceptible to insects pests such as Western corn rootworm (*Diabrotica virgifera* Leconte) and the Japanese beetle (*Popillia japonica*) as the defensive cysteine protein inhibitor (Cysp PIS) production was minimized. In addition, the emission of herbivore-induced plant volatiles (HIPVs) (Arneth et al. 2010; Gouinguene' and Tulings 2002) becomes the key factor that attracts natural enemies to a crop ecosystem. It is greatly reduced under higher temperature and elevated CO_2 regimes (Bruce and Picket 2011). Temperature will confuse the volatile profile, and hence olfactory discernment of the natural enemies for host location will be affected (Helmig et al. 2007; Yuan et al. 2009; Thomson et al. 2010). Furthermore, the resistance of chickpea plants to herbivores is influenced by higher CO_2 level as it alters the defensive oxalic and malic acid levels in plants (Selvaraj et al. 2013; Sharma et al. 2016). To withstand new atmospheric condition established due to adverse altered climatic conditions, introduction of new crops and cultivators is a major strategy for adaption (Moore and Allard 2008; Benedict 2003). This will definitely alter plant and insect growth rates, positively or negatively, which in turn will influence the natural enemies. A greater generation turnover will result into greater injury to plants.

21.5 Impact of Climate Change on Beneficial Insects

21.5.1 Impact of Climate Change on Natural Enemies

Climate change can have assorted effects on natural enemies of insect pests. Natural enemies are beneficial insects that keep the insect pests under check naturally. They include predators that kill and consume the insect pests and parasitoids which are organisms that survive in host expense and later destroy the host. Climate change impacts abruptly on the interaction between pest and natural enemies ensuing both positive and negative impacts in pest status. Ecological imbalances and disruptions happen when the trophic levels respond contrarily to weather conditions (Hance et al. 2007). Diverse climatic conditions respond differently for parasitoids, predators, and parasites. Enumerating global warming effects on the efficacy of natural enemies will be a major concern in future pest management agendas.

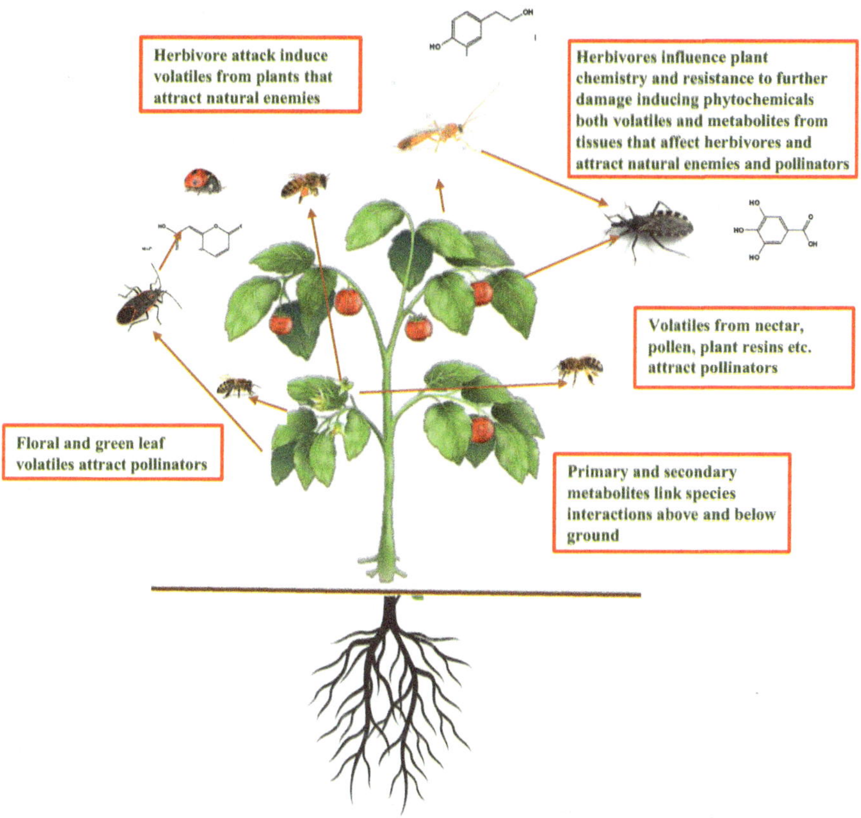

Fig. 21.3 Global change effects on plant-insect interactions: the role of phytochemistry

The knowledge and understanding of insect pests and natural enemies reciprocating to erratic climatic changes is essential for biocontrol programs. Global warming rises or minimizes the status of natural enemies/pest species and alters their relationship (Fig. 21.3). Temperature and CO_2 variations impacting the herbivore quality will directly impact the fitness of natural enemies (Dhaliwal et al. 2010; Selvaraj et al. 2013). A minimal parasitism or predation of pests by natural enemies is noticed due to enhanced plant growth which gives room for the pests to hide from natural enemies (DeLucia et al. 2012; Coviella and Trumble 1999).

Temperature has a distinct impact by either encouraging or discouraging predator activity. Insect pests and their natural enemies may retort differently to changes in temperatures. The vulnerability of aphids to natural enemies may increase due to climate change (Awmack et al. 1997). The reproduction of pea aphid (*Acyrthosiphon pisum*) surpasses the proportion at which ladybird beetle (*Coccinella septempunctata*) can consume it, when the temperature is below 11 °C (51.8 °F). Above this temperature the situation is reversed. Temperatures up to 25 °C will augment the management of aphids by coccinellids (Freier and Triltsch 1996).

Hence, low temperature has positive activity on natural enemies. Increase in temperatures might have an adverse consequence on fragile natural enemies such as small predators and hymenopteran parasitoids. The notorious rice pest, brown plant hopper is 17 times more lenient than its predator, *Cyrtorhinus lividipennis* at 40 °C, while predacious spider, *Pardosa pseudoannulata*, was more tolerant at 40 °C (Kambrekar et al. 2015). Higher temperatures reduce the efficacy of natural enemies of spruce budworm (*Chloristoneura fumiferana*) (Harrington et al. 2001).

Herbivorous insects may expand their range of survival due to global warming. As a consequence, they may migrate into new locations where their natural enemies are not present. The pink bollworm of cotton (*Pectinophora gossypiella* Saunders) expands its array on cotton into regions that are not amiable at present, and their parasitoid, *Trichogramma toideabactrae*, may not follow them to new locations (Guiterrez et al. 2008). Monophagous parasitoids have extreme effects as they will have difficulty in adapting to a new host (Hance et al. 2007). Moreover, host-specific and specialist hymenopteran parasitoids are more delicate to variations in the exact time of their host emergence or development rates and stages, except tachinid flies that feed on several hosts and are less vulnerable to asynchrony with their host convinced by host change. The fecundity rate and sex ratio of ichneumonid larval parasitoid of *H. armigera, Campoletis chlorideae* is influenced by temperature (Dhillon and Sharma 2009). If the emergence of host occurs before parasitoids, parasitism could be reduced. The insect pests as host insects for natural enemies undergo susceptible life stages more rapidly at higher temperature reducing the parasitism openings for parasitoids.

The influence of elevated CO_2 on natural enemies is principally indirect and intervened by the alteration in nutritional quality of hosts that feed on plants. The change in natural enemy fitness, development, mortality, and abundance could be the effects which differ between parasitoids and predators. With regard to parasitoids, the specialists that are host specific are likely to be more adversely affected than generalists that survive on a variety of host insects. The population of aphid parasitoid, *Diaeretiella rapae* (M'Intosh) reduced by 50% with short-lived adults in elevated CO_2 condition (Klaiber et al. 2013). Augmented larval and pupa mortality of parasitoid, *Cotesia melanoscela* Ratz., was observed under elevated CO_2 condition (Roth and Lindroth 1995). Researches are limited to explore on tritrophic system under elevated CO_2 conditions under controlled conditions with elevated CO_2 to explore the tritrophic system effects.

The cotton bollworm larvae (*H. armigera*) fed on pea plants (*Pisum sativum*) grown on elevated CO_2 of 700 ppm were smaller sized, and the predatory activity of the omnivorous bug (*Oechalia schellenbergii*) was more on these larvae due to their small size and thus easily available. Under elevated CO_2 levels, increased rate of predation was observed due to higher prey vulnerability (Coll and Hughes 2008). In contrast, the parasitism rates of *Aphidius matricariae*, on aphids, remain unchanged in elevated CO_2 though there is copiousness of peach potato aphid (*Myzus persicae*) with an increase in CO_2 and temperature (Bezemer et al. 1998).

Drought and heavy precipitation heavily influence the natural enemy activity. The populations of armyworm (*M. separata*) increased during extended periods of drought, and natural enemy activity declined due to the adverse effects of drought (Sharma et al. 2002). The immune response of mealy bugs was improved in water stress conditions due to drought, thereby reducing mealy bug parasitism (Calatayud et al. 2002).

Quantifying the impacts of climate change on the effectiveness of natural enemies will be a key anxiety for the pest management programs in the future. The rising variation in climate decreases the potential of natural enemies in checking the pests and thereby increases the susceptibility of host plants to herbivores. However, activity of certain natural enemies will be increased under climate change conditions. The duration of development of the aphid parasitoid (*Lysiphlebia japonica* Ashmead) and lepidopteran parasitoid (*Cotesia plutellae* Kurd.) has declined under elevated CO_2 levels (Feng et al. 2011; Vuorinen et al. 2004). The pentatomid bug (*Oechalia schellenbergii* Guerin) preys on *H. armigera* easily due to its reduced size when fed with plants grown in elevated CO_2.

The generalist predators are efficient in managing crop pests under varied climatic conditions than their specialist counterparts. A higher rate of parasitism was noticed in braconid parasitoid in parasitizing *Aphidius picipes* (Nees) under elevated CO_2. Higher consumption of prey by the coccinellid predator, *Leis axyridis*, feeding on *Aphis gossypii* Glover was noticed under conditions of elevated CO_2 levels (Chen et al. 2005). Controversially, no substantial impression of prominent CO_2 levels due to consumption of coccinellid predator (*Harmonia axyridis* Pallas) on aphid pest (*Sitobion avenae* Fab) was noticed by Chen et al. (2007).

Parasitoid emergence from eggs is perilous due to higher temperature in response to global warming, and the prior arrival of parasitoids than the host would result in a marked or extinction of parasitoid population due to unavailability of the host (Grabenweger et al. 2007). The availability of host at the susceptible stage will mismatch with natural enemy availability due to its early emergence. This disproportion on arrival pattern of hosts or parasitoids as a result of climatic variations will drastically influence the host plant development (Thomson et al. 2010). A weak synchrony between host and parasitoid was documented in earlier studies (Chen et al. 2007). A decreased level of parasitism was observed due to early advent of dormanting parasitoids of horse chestnut leaf miners (*Cameraria ohridella*) under field conditions (Grabenweger et al. 2007).

21.5.2 Impact of Climate Change on Pollinators

Insects perform a crucial role in providing various ecosystem services, of which the most important one is pollination. Pollinators are keystone species as they care about humankind by increased food production, security, and maintain biodiversity (Rathee and Dalal 2017). Honeybees and other pollinators are a vital part of agroecosystems as these tiny creatures decide the seed set of various crops by their pollination services. Insects are excellent pollinators for 267 economically important crops (Sidhu and Mehta 2008; Murugan 2006). Honeybees are the chief contributors of pollination services to approximately 73% of the world's cultivated crops; the others being flies to an extent of 19%, wasps by 5%, beetles by 15%, birds by 4% butterflies, and moths by 4% (Abrol 2009). The pollinators receive natural resources from flower such as nectar, pollen, or both and in turn pollinate the flowers. The natural terrestrial ecosystems as well as human-made ecosystems were benefited by this mutualism that evolved centuries before. Insect pollination (entomophily) contributes to one-third of the world human food (Klein et al. 2007). The economic value of these services by insects was predicted to be about $153 billion per year.

Global warming, an evolving universal subject, with the capability to distress achievement of agricultural ecosystems has its influence on insect pollinators at countless capacities and competence to afford pollination service (Costanza et al. 1997). Pollination is one of the 15 chief bionet work amenities under risk at present from escalating pressures exercised by increasing population, exhausting natural resources, and changing global climate (Millennium Ecosystem Assessment Report 2005). Previous research on pollinators emphasizes the steady decline of the abundance in population, topographical range, and pollinating events of chief pollinators such as honeybees, butterflies and moths, wasps, etc., due to changing climate (FAO 2009). The imperative events of plant life cycle such as flowering, pollination, fruiting, etc., have been found to be affected intensively by the climatic regimes such as temperature, availability of moisture due to precipitation, etc. (Cleland et al. 2007). The life cycle of pollinators is synchronized with phenological measures of the host plant to coincide with pollination. Disruption in routine climate is anticipated to upset the plant-pollinator synchronization by changing the phenological actions ultimately affecting the pollination level (Kudo et al. 2004; Ricketts et al. 2008). The quality and quantity of pollination services rendered by insects have manifold consequences on ecosystem stability, species diversity, food security, and pliability to variation in climate (FAO 2009).

Honeybee is the chief pollinator of countless wild plants and crops but faces solemn perils due to climate change, which is one of the chief grounds for the *colony collapse disorder* which is a phenomenon that threatens honeybee colonies characterized by sudden colony death with reduction in healthy bees inside a hive. Appropriate awareness on bee behavior to climate change setting is a key factor to overcome this issue. A study was attempted by remotely monitoring the hive weight during the flowering period of 2016 with usual weather as well as 2017 with severe drought and high temperatures. The study revealed the reduction in flowering for 3 weeks in 2017 compared to 2016, with a weight gain of 7.67 and 18.92 kg, respectively, severely impacting on bee population as well as pollen and nectar reserves causing food stress for bees (Flores et al. 2019). The C:N ratio of plant tissues is expected to be modified, feasibly leading to changes in nectar composition due to the elevated CO_2. Furthermore, a threat of change in plant community structure, particularly in the extents of C_3 and C_4 plants in a territory, was noticed due to eminent levels of CO_2 in the atmosphere (Bazzaz 1990). But direct consequence of augmented CO_2 concentrations of atmosphere on honeybees is not yet recorded. Hegland et al. (2009) studied the magnitudes of temperature-based changes in pollinator and plant relations and found that both pollinator activity and flowering of plant timings were severely affected by temperature. The differences in response to changed temperatures by plants and insect visitation create temporal and spatial disparities, exerting severe consequences for the species involved. Alterations in weather parameters mismatch with insect visitations, thereby affecting plants by reduced pollen deposition and fruit set due to less availability of food to pollinators. The visual responses of both plants and pollinators to rising temperature peninsula indicated a great incompatibility between the occurrence of the mutualistic associates (Gordo and Sanz 2011). The pollinators *Apis mellifera* and *Pieris rapae* advanced their pollinating period affected by temperature resulting in sequential disparity with their host plants (Hegland et al. 2009). In Japan, advancement in flowering during spring occurs for honeybees, whereas queen emergence of bumblebees was not affected by temperatures and created a gap in pollinator activity. Consequently, higher/lower temperature responses and discrepancies in pollinator-host collaborations may increase among species and regions (Hegland et al. 2009).

The bee species present in Sahara Desert (*Apis mellifera sahariensis*) is well adapted to extreme arid condition, where they require abundant supply of water to raise their larvae at brood temperatures of 34 °C and 35 °C. In warmer environment, the moisture of flowers dries, and the honeybees die due to starvation of water. The weather predictions for the future state that some regions of the world might experience more severe dryness, which affects the survival of bee forage plants and thereby the honeybee (Conte and Navajas 2008). The possible effects of forthcoming variations in weather on bee responding to drift in various abiotic factors will significantly alter the pollination services and thereby world food security (Corbet et al. 1993).

21.6 Effect of Climate Change on Integrated Pest Management Strategies

21.6.1 Host Plant Resistance Breakdown

Host plant resistance works with mechanisms like antixenosis, antibiosis, and tolerance, i.e., an ecofriendly pest management aspect, wherein minimal pest damage can be achieved (Dhaliwal and Dilawari 1993). Host plant resistance is expression that is significantly influenced by abiotic features such as temperature, carbon dioxide levels, moisture, air pollution, etc., which impact on the interaction between host plants, pests, and natural enemies (Sharma et al. 2010). Plants become susceptible to pest attack as their defense system is weakened under abiotic stress (Rhoades 1985). In India, a severe yield loss is experienced in sorghum due to midge (*Stenodiplosis sorghicola*) and stem borer (*Chilo partellus*) (Sharma et al. 2005). The chemical profile of few crops is altered due to insect damage as tissues are not as much of appropriate for development and survival of insect pests (Sharma 2002).

21.6.2 Transgenic Crops

Insect-resistant transgenics expressing *Bacillus thuringiensis* (Bt) insecticidal protein is a recent advancement in integrated pest management (Kranti et al. 2005). The level of Bt toxin protein was reduced in the transgenic plants during periods with abiotic stresses like high temperature, elevated CO_2levels, drought, precipitation, etc., leading to reduction in resistance level of plants (Dong and Li 2007). Bt cottons were seriously damaged by bollworm (*Heliothis virescens*) in the USA due to high temperature (Kaiser 1996). The damage to Bt cotton by *H. armigera* and *H. punctigera* in Australia is due to condensed production of Bt toxins (Hilder and Boulter 1999). The Bt toxin Cry1 Ac level reduces as plant gets older, leading to susceptibility of the crop to bollworm in advanced stages of crop growth (Kranti et al. 2005). The reasons for insect management in transgenic crops are due to development of resistance to toxin by insects, insufficient production of toxin, and environmental consequences on transgenic expression (Sharma and Ortiz 2000). Therefore, the effect of global warming on transgenic plants requires urgent attention.

21.6.3 Biopesticides and Synthetic Insecticides

Climate change causes increased variability in insect damage. High temperatures make the environment drier and low temperature colder with increase in quantity and intensity of rainfall that is responsible for the region's wetness. Natural biopesticides derived from plants such as entomo-pathogenic fungi, bacteria, virus, nematodes, as well as synthetic chemical pesticides are extremely delicate to the environmental regimes. Sensitive issues in pest management such as environmental pollution, health hazards to living things, pest resistance, and resurgence are outputs of inappropriate handling of synthetic pesticides. Temperature is a major issue that affects pesticide toxicity either positively or negatively, thereby affecting efficacy. Diflubenzuron caused rapid mortality of insects at higher temperatures of 35 °C (Amarasekare and Edelson 2004). The relation between temperature and pesticide efficacy chiefly hinges on pesticide mode of action, method of pesticide application, quantity of pesticide ingested, or in contact and the target pest species (Johnson 1990).

21.6.4 Biological Control

The relation between natural enemy and pests changes considerably due to erratic climatic conditions leading to upsurge or downsurge in the level of individual pest species. The diurnal pattern of different group of insects and their interspecific interaction will be altered due to global warming, which modifies the success of natural enemies in pest management (Young 1982; Hill and Dymock 1989). Enumerating the influence of climate change on the efficacy of natural enemies in pest management should be focused on upcoming pest management scenarios. Oriental army worm (*M. separata*) populations increase during extended period of drought followed by heavy rainfall as the decline in natural enemy activity due to drought and rainfall (Sharma et al. 2002). Aphid management by coccinellids will be enhanced for temperature up to 25 °C (Freier and Triltsch 1996). Abiotic stresses not only affect natural enemy development but also egg-laying behavior and sex ratio (Dhillon and Sharma 2007). The positive and negative interface between insect pests and natural enemies should be discovered in detail to device appropriate methods in integrated pest management.

21.7 Strategies to Mitigate the Effects of Climate Change in Pest Management

Diversity and shifts in species abundance as outcome of drastic climatic change in this era have the potential to degrade the effectiveness of insect pest management strategies. Therefore, sharpening the prevailing monitoring agendas and developing new effective programs to detect probable changes in population ecology, pest distribution, pest migration, damage assessment, and yield loss were urgently needed (Sharma 2014). The fluctuation in pest survival approaches requires wider and robust studies to develop innovative IPM choices or disseminate existing ones to new for practical applicability (Dhaliwal et al. 2010). The sensitive issues on pest management, viz., 3Rs pest resistance, resurgence, and residues, were the consequences of use of synthetic insecticides (Dhaliwal et al. 2004). The indiscriminate usage of synthetic pesticides has imparted environmental pollution globally as well as caught the attention toward "Go Green" strategies for resources management (Khan et al. 2021a, b; Banerjee et al. 2021a, b, c, d). In this aspect, numerous bio-products have been developed and used as environmentally safe products. However, the nonchemical strategies of pest management are highly sensitive to the atmosphere. Rise in temperature and UV radiation and a decline in RH may render the ineffectiveness of the biological strategies in pest management (Zvereva and Kozlov 2006; Niziolek et al. 2012).

The various components of IPM such as host plant resistance, bio-products, natural enemies, nano-suspensions, and cultural agronomic practices offer a viable option to manage pests, but the comparative effectiveness of most of these components is severely affected by changing weather globally. Biological control, involving predators and parasitoids, considered as an effective component of IPM programs, with no harm to the environment is harshly exaggerated by variations in climate, shattering the affiliated pest and natural enemies. More or less all the IPM strategies that include cultural practices, biocontrol aspects, host plant resistance as well as synthetic chemicals are extremely sensitive to the environment.

Consequently, developing suitable approaches for pest management to be effective in global warming scenario in future is under priority. For sustainability in agriculture and to alleviate the influence of climate change on agricultural production, assessing the impact of global warming on crop production and developing climate smart crops to withstand drastic climatic conditions are required. Additionally, advanced/delayed cropping methods and cropping systems need to be explored to reduce the risk of climate change on crop production. Climate and crop-associated models have to be developed for land use measures, soil productivity, and familiarizing pesticides to any crop ecosystem (Jhariya et al. 2018a, b). The alarming point due to global warming is the shift of insect pests to new territories, where absence of their natural enemies will lead to pest outbreak. The chief task ahead in this regard is to develop effective prediction models that would pave the way for successful insect pest management. Developing models to anticipate the variations in topographical distributions and dynamics in population of insect pests and tactics to be amended to minimize crop losses is utmost needed. The awareness of climate, crop, and insect pests is imperative to develop warning systems to be cautious to avert outbreak of pests and to evade monetary losses. The pest premonition systems based on weather changes are verdict support tools that aid growers to assess the threat of pest outbreak under erratic climate regimes. With pest forewarning systems and possible prediction models, it is possible to develop climate smart plants that are reformed to these extreme climatic circumstances for sustainable agriculture and food security.

Thus, there is a necessity to envisage and plot the inclinations of deviations to geographical distribution and their impact in progress, occurrence, and dynamics in population of insect species, comprehend the impact of variation in climate on cropping patterns and the stimulus on the richness of insect pests, recognize the bases of resistance and engineering the resistant genes in commercial cultivars, investigate climate change impacts on transgenic crops, evaluate the effectiveness of diverse pest management strategies under dissimilar environmental settings with possible prediction models, and develop appropriate approaches to alleviate climate change implications.

21.8 Sustainable Management of Insect Population under Changing Climate

The heavy pest pressure and its impact on crop losses have extended beyond farm to national and international food security instigating ecological, fiscal, and communal sustainability. It is a difficult task to understand and deal with the problems of abiotic stresses and crop pest management under changing climate. Therefore, immediate action is needed on geographical levels as climate change regimes on pests are threatening food security. Adaption to the climate change is a strategy to overcome the risk and reduce the potential hazard imposed by global warming consequences (Howden et al. 2007).

Climate smart agriculture (CSA) as promoted by FAO (2009) is an innovative tactic to reorient the entire agricultural system to warrant food security in the dynamics of weather parameters. It is an approach aiming to diminish pest-induced crop losses, augment ecosystem services, curtail intensity of greenhouse gas emission per unit of food produced, and strengthen the resilience of agricultural system in relation to climate change (Chakraborty and Newton 2011). In this strategy, IPM is adapted as a sound approach to adopt the changing climate (Lamichhane et al. 2015; Juroszek and von Tiedemann 2011). It suggests recommendations in multiple stakeholder levels, to guide producers, extension agents, geographical scales, researchers, and policy makers in developing strategies against induced global changes of crop pests (Macfadyen et al. 2018).

Climate smart pest management (CSPM) is an innovative strategy that comprehends a set of interdisciplinary approaches to adapt to changing environmental conditions with supports from extension and research, which influences the distribution of crop, pests, and natural enemies. It is an effective concept of pest management minimizing the threats of existing and new crop pests in agricultural production, thereby increasing the resilience of livelihood of farmers and food security in changing climate. It improves overall greenhouse balance and reduces pest-related yield losses (Cock et al. 2016). CSPM leads farmers with information to proactively act on pest management aspects in a cost-effective manner, viz., crop diversification, cultural and physical practices, enriching natural habitats, etc., will uplift the health of a farm and reduce susceptibility to pests (Hoffmann et al. 2008). Additionally, pest monitoring and pest risk for casting will be done due to climate change, and farmers will be educated to act immediately on cost-effective environmentally safe pest management tactics (Agrhymet 2013).

21.9 Policy and Legal Framework

Specific laws to promote holistic form of agricultural production in organic way and IPM need to be implemented for pest management in changing climate scenario. The necessity to implement nonconventional agricultural strategies that involve IPM as a major component under a specific policy linked to climate change should be given prior attention. A recognized enunciation of IPM principles and operative tactics to pest management should be unified with climate change policy and its implementing rules and regulations or as an alteration to the national law. The prevailing law should be swatted to consolidate integrated and low toxic pest management in the context of adaptations to variation in climate and its mitigation. For sustainability in agriculture through effective pest management, IPM strategies should be allied with subjects that provide a solution to farmers' concerns such as erratic climate changes that affect crop productivity tremendously, aiming at low production costs by providing proper subsidies for fertilizers, bio-pesticides, and better access to market. A proper articulation of policies that are in line with climate change predictions requires utmost attention.

21.10 Future Perspectives

The future area of investigation for contesting the insect pest issues with changing weather parameters can be considered into the following categories:

1. *Breeding Climate-Resilient Varieties*: The influence of climate change and other environmental conditions can be minimized by breeding novel varieties with augmented resistance to abiotic anxieties. In view of the early or delay in season, the crop production can be delayed or postponed for better productivity. Therefore, breeding varieties adaptable to changing weather scenario to overcome pest and disease incidence and pliable for all other stresses is of utmost importance in agricultural research.
2. *Alteration in Crop Sowing Dates*: The alteration in sowing dates of crops caused by global weather change will alter host-plant synchrony. Various host-plant interactions under diverse sowing period, viz., early late or nominal conditions, are required to be explored to endorse ideal sowing dates for reduced pest pressure and increased yield.
3. *Rescheduling of Crop Calendars*: Most of the scheduled IPM practices will be noneffective for managing pest under changing weather condition. Therefore, the crop calendar should be rescheduled rendering to fluctuating crop ecosystem. The growers should be advised to shift the insect pest management tactics correlating the anticipated deviations in pest incidence and the degree of economic losses in view of varying weather regimes.
4. *GIS-Based Risk Mapping of Crop Pests*: Geographic Information System (GIS) is an empowering technology that aids entomologist, agronomist, and meteorologist to correlate pest incidence with biographical and geographical pest management programs. GIS helps to study on the effect of climate change on incidence, development, and dynamics of insect pests and aids to predict and map trends of potential variations in topographical distribution of agroecological hotspots of pest outbreak and forthcoming zones of pest peril.
5. *Screening of Pesticides with a Novel Mode of Action*: Several investigations by eminent researchers have reported the inducement of Salicylic acid-related plant defenses, which augment plant health and vigor by tolerating adverse abiotic stresses due to the application of neonicotinoid pesticides as a management tactic for sucking pests. This provides an awareness to propose researches in enhancing host plant resistance to stress tolerance in plants by pesticide application. Investigations on diverse pesticides for their role in upsurging abiotic stresses in plants need to be focused in a wider angle. More such compounds have to be recognized for future pest management.

21.11 Conclusion

The global challenge faced by humanity in the near future will be inevitable to double the food production to feed the flourishing upsurge in world populace by utilizing minimal land area, water, and nutrients along with the capability to withstand drastic change in climate. The erratic changes in weather will have grave penalties on diversity and copiousness of arthropods and the degree of crop loss due to insect pests which impact severely in agricultural crop productivity. The negative impacts of fluctuating weather conditions leading to warmer temperatures, rise in

CO_2 levels, change in precipitation level, and amplified frequency of droughts will have a devastating consequence on pest abundance, leading to advent of invasive novel pests. Pest outbreaks might occur in frequent, predominantly during prolonged drought followed by heavy rainfall. Insect multiplication increases with rise in CO_2 and temperature. Insect pests depend on plants for their living and respond differently to different atmospheric climate changes according to the change in their plant hosts. Hence, due to climate change, communications between the insect pests and their host plants will be altered resulting in susceptibility of some resistant cultivars to insect pests that may display vulnerable response due to climate change. Due to global warming, there is an upsurge in the number of insect pest population, outbreak of insects, and intensification in the number of insect generations which would increase the damage in plants, decrease the crop yields, increase the cost on crop protection, and thereby affect the economy. Further, climate change impacts heavily on the natural enemies, viz., predators and parasitoids along with their insect hosts, resulting in a complex dynamic situation, which would upset the equilibrium between insect pests and their natural enemies leading to more frequent outbreaks. The antagonistic effects of alterations in climate regimes on the efficacy of natural enemies are an area to be dealt more passionately which will be a great concern in future pest management tactics. In this regard, the correlation between crop production cost and the crop outcomes will be altered in response to shift in climate paradigms, which will severely impact the economic threshold levels. A wider variability in climatic parameters will directly influence the pest incidence, population, and their multiplication. Forecast on variations in geographical distribution and dynamics in insect pest population will greatly aid to acclimatize pest management strategies to mitigate the adverse effects of climate change on crop productivity. Therefore, there is a need to have a strenuous look at the expected effects of global warming on insects, mitigate their damage to crops, and formulate apt actions to alleviate the impact of climate change on food security.

References

Abrol DP (2009) Plant-pollinator interactions in the context of climate change - an endangered mutualism. J Palynol 45:1–25

Adamo SA, Lovett MME (2011) Some like it hot: the effects of climate change on reproduction, immune function and disease resistance in the cricket *Gryllustexensis*. J Exp Biol 214:1997–2004

Agrhymet (2013) Centre regional agrhymet—Nos produits. http://www.agrhymet.ne/produit.html. Accessed 16 May 2018

Amarasekare KG, Edelson JV (2004) Effect of temperature on efficacy of insecticides to differential grasshopper (Orthoptera: Acrididae). J Econ Entomol 97:1595–1602

Anand T, Pereira GN (2008) Butterflies-the flying jewels of the Western Ghats. http://www.daijiworld.com/chan/exclusive_arch.asp?ex_id=1003

Arneth A, Sitch S, Bondeau A, Butterbach-Bahl K, Foster P (2010) From biota to chemistry and climate: towards a comprehensive description of trace gas exchange between the biosphere and atmosphere. Biogeosciences 7:121–149

Awmack CS, Woodcock CM, Harrington R (1997) Climate change may increase vulnerability of aphids to natural enemies. Eco Entomol 22:366–368

Ayres MP, Lombardero MJ (2000) Assessing the consequences of global change for forest disturbance from herbivores and pathogens. Sci Tot Env 262:263–286

Bale JS, Hayward SAL (2010) Insect overwintering in a changing climate. J Exp Biol 213:980–994

Banerjee A, Jhariya MK, Meena RS, Yadav DK (2021b) Ecological footprints in agroecosystem-an overview. In: Banerjee A, Meena RS, Jhariya MK, Yadav DK (eds) Agroecological footprints management for sustainable food system. Springer, Singapore, pp 1–23. https://doi.org/10.1007/978-981-15-9496-0_1

Banerjee A, Jhariya MK, Raj A, Yadav DK, Khan N, Meena RS (2021c) Land footprint management and policies. In: Banerjee A, Meena RS, Jhariya MK, Yadav DK (eds) Agroecological footprints management for sustainable food system. Springer, Singapore, pp 221–246. https://doi.org/10.1007/978-981-15-9496-0_7

Banerjee A, Jhariya MK, Raj A, Yadav DK, Khan N, Meena RS (2021d) Energy and climate footprint towards the environmental sustainability. In: Banerjee A, Meena RS, Jhariya MK, Yadav DK (eds) Agroecological footprints management for sustainable food system. Springer, Singapore, pp 415–443. https://doi.org/10.1007/978-981-15-9496-0_14

Banerjee A, Jhariya MK, Yadav DK, Raj A (2020) Environmental and sustainable development through forestry and other resources. Apple Academic Press Inc., CRC Press- a Tayler and Francis Group, Boulder, CO, p 400. https://doi.org/10.1201/9780429276026

Banerjee A, Meena RS, Jhariya MK, Yadav DK (2021a) Agroecological footprints management for sustainable food system. Springer, Singapore, p 514. https://doi.org/10.1007/978-981-15-9496-0

Bazzaz FA (1990) The response of natural ecosystems to the rising global CO_2 levels. Ann Rev Ecol System 21:167–196

Beddington J, Asaduzzaman M, Clark M, Fernández A, Guillou M, Jahn M, Erda L, Mamo T, Van Bo N, Nobre CA, Scholes R, Sharma R, Wakhungu J (2012) Achieving food security in the face of climate change: final report from the commission on sustainable agriculture and climate change. CGIAR research program on climate change. Agriculture and Food Security (CCAFS), Copenhagen, Denmark

Benedict JH (2003) Strategies for controlling insect, mite and nematode pests. In: Chrispeels MJ, Sadava DE (eds) Plants, genes and crop bio-technology. Jones and Bartlet Publishers, Sudbury, MA, pp 414–442

Bezemer TM, Jones TH, Knight KJ (1998) Long-term effects of elevated CO_2 and temperature on populations of the peach potato aphid *Myzuspersicae* and its parasitoid *Aphidiusmatricariae*. Oecologia 116(1–2):128–135. https://doi.org/10.1007/s004420050571

Bhagat S, Nivedita C, Chiranthan C (2015) Climate change: pest incidence in agricultural crops. In: Chattopadhyay C, Prasad D (eds) Dynamics of crop protection and climate change. Studera Press publications, New Delhi, India

Bleikar K, Smith GD, Humble L (2017) Cold tolerance of mountain pine beetle (Coleoptera: Curculionidae) eggs from historic and expanded ranges. Environ Entomol 46(5):1165–1170. https://doi.org/10.1093/ee/nvx127

Boullis A, Francis F, Verheggen FJ (2015) Climate change and tritrophic interactions: will modifications to greenhouse gas emissions increase the vulnerability of herbivorous insects to natural enemies? Environ Entomol 44(2):277–286

Boullis AC, Detrain F, Francis FJ, Verheggen J (2016) Will climate change affect insect pheromonal communication? Curr Opin Insect Sci 17:87–91

Brooke C (2008) Conservation and adaptation to climate change. Conserv Biol 22(6):1471–1476

Bruce TJA, Picket JA (2011) Perception of volatile blends by herbivorous insects-finding the right mix. Phytochemistry 72:1605–1611

Calatayud PA, Polanía MA, Seligmann CD, Bellotti AC (2002) Influence of water stressed cassava on *Phenacoccusherreni*and three associated parasitoids. Entomol Exp App 102:163–175

Catalán TP, Wozniak A, Niemeyer HM, Kalergis AM, Bozinovic F (2012) Interplay between thermal and immune ecology: effect of environmental temperature on insect immune response and energetic costs after an immune challenge. J Insect Physiol 58(3):310–317
Chahal SK, Bains GS, Dhaliwal LK (2008) Climate change: mitigation and adaptation. Proceedings of international conference on climate change, biodiversity and food security in the south Asian region, 3–4 November, 2008, Punjab State Council for Science and Technology, Chandigarh and United Nations Educational, Scientific and Cultural Organization, New Delhi, India, 12
Chakraborty S, Newton AC (2011) Climate change, plant diseases and food security: an overview. Plant Pathol 60:2–14. https://doi.org/10.1111/j.1365-3059.2010.02411.x
Challinor AJ, Watson J, Lobell DB, Howden SM, Smith DR, Chhetri N (2014) A meta-analysis of crop yield under climate change and adaptation. Nat Clim Chang 4:287–291. https://doi.org/10.1038/nclimate2153
Chambers MC, Schneider DS (2012) Pioneering immunology: insect style. Curr Opin Immunol 24 (1):10–14
Chander S (1998) Infestation of root and foliage/earhead aphids on wheat in relation to predators. Indian J Agric Sci 68:754–755
Chapman AD (2006) Numbers of living species in Australia and the World. Australian Biological Resources Study, Canberra, Australia
Chapman RF (1998) The insects-structure and function, 4th edn. Cambridge University Press, Cambridge, UK, p 788
Chari MS, Raghupathirao G, Ramaprasad G, Sreedhar U (1993) Pest epidemics I tobacco: a recent trend. Lead papers of the National Seminar on changing scenario in pest and pest management in India. Plant Protection Association of India, Hyderabad, India, pp 96–102
Chen FJ, Ge F, Parajulee MN (2005) Impact of elevated CO_2 on tri-trophic interaction of *Gossypium hirsutum*, *Aphis gossypii*, and *Leis axyridis*. Environ Entomol 34:37–46
Chen FJ, Wu G, Parajulee MN, Ge F (2007) Impact of elevated CO_2 on the third trophic level: a predator *Harmonia axyridis*(Pallas) and a parasitoid *Aphidiuspicipes* (Nees). Biocontrol Sci Tech 17:313–324
Ciss M, Parisey N, Dedryver CA, Pierre JS (2013) Understanding flying insect dispersion: multiscale analyses of fragmented landscapes. Ecol Inform 14:59–63
Cleland EE, Chuine I, Mooney HA, Schwartz MD (2007) Shifting plant phenology in response to global change. Trends Ecol Evol 22(7):357–365
Cock MJW, Tang R, Liu Z, Wan H, McGillivray LA, Thomas SE, Cameron KH, Zhang F (2016) The main agricultural pests and diseases of China and implications for the use of remote sensing for their management. CAB Rev 11:14. https://doi.org/10.1079/pavsnnr201611014
Coley PD, Markham A (1998) Possible effects of climate change on plant/herbivore interactions in moist tropical forests. Clim Chang 39:455–472
Coll M, Hughes L (2008) Effects of elevated CO_2 on an insect omnivore: a test for nutritional effects mediated by host plants and prey. Agri Eco Environ 123:271–279
Conte YL, Navajas M (2008) Climate change: impact on honey bee populations and diseases. Rev Sci Tech Off Int Epiz 27:499–510
Corbet SA, Fussell M, Ake R, Fraser A, Gunson C, Savage A, Smith K (1993) Temperature and the pollinating activity of social bees. Ecol Entomol 18:17–30
Costanza R, d'Arge R, de Groot R, Farber S, Grasso M, Hannon B, Limburg K, Naeem S, O'Neill RV, Paruelo J, Raskin GR, Sutton P, van der Belt M (1997) The value of the world's ecosystem services and natural capital. Nature 387:253–260
Coviella CE, Trumble JT (1999) Effects of elevated atmospheric carbon dioxide on insect-plant interactions. Conserv Biol 13:700–712
Daebeler F, Hinz B (1977) Compensation of aphid injury to sugarbeet by means of supplementary rain. Arch Phytopathol Pflanzenschutz 13(3):199–205
Das DK, Singh J, Vennila S (2011) Emerging crop Pest scenario under the impact of climate change. J Agril Phy 11:13–20

Deka S, Byjesh K, Kumar U, Choudhary R (2008) Climate change and impacts on crop pests- a critique. ISPRS archives XXXVIII-8/W3 workshop proceedings: impact of climate change on agriculture, pp 147–149

DeLucia EH, Nabity PD, Zavala JA, Berenbaum MR (2012) Climate change: resetting plant-insect interactions. Plant Physiol 160:1677–1685

Dhaliwal GS, Arora R, Dhawan AK (2004) Crop losses due to insect pests in Indian agriculture: an update. Indian J Ecol 31:1–7

Dhaliwal GS, Dilawari VK (1993) Advances in host resistance to insects. Kalyani Publishers, New Delhi, India

Dhaliwal GS, Jindal V, Dhawan AK (2010) Insect pest problems and crop losses: changing trends. Indian J Ecol 37:1–7

Dhawan AK, Singh K, Saini S, Mohindru B, Kaur A, Singh G, Singh S (2007) Incidence and damage potential of mealybug, *Phenacoccus solenopsis* Tinsley, on cotton in Punjab. Indian J Ecol 34:110–116

Dhillon MK, Sharma HC (2007) Effect of storage temperature and duration on viability of eggs of *Helicoverpa armigera* (Lepidoptera: Noctuidae). Bull Entomol Res 97:55–59

Dhillon MK, Sharma HC (2009) Temperature influences the performance and effectiveness of field and laboratory strains of the ichneumonid parasitoid, *Campoletis chlorideae*. Biol Control 54:743–750

Dirk Z (2014) Global warming favours light-coloured insects in Europe. Nat Commun 5:3874. https://doi.org/10.1038/ncomms4874

Dong HZ, Li WJ (2007) Variability of endotoxin expression in Bttransgeniccotton. J Agron Crop Sci 193:21–29

Edward BM, Michelle NT, Caroline SA, Richard LL (2004) Divergent pheromone mediated insect behaviour under global atmospheric change. Glob Chang Biol 10:1820–1824

Fand BB, Kamble AL, Kumar M (2012) Will climate change pose serious threat to crop Pest management- a critical review. Int J Scientific Res Pub 2:1–14

FAO (2009) Global action on pollination services for sustainable agriculture. Food and Agriculture Organization of the United Nations, Rome, Italy

FAO, Food and Agriculture Organization of the United Nations (2008) Rapid assessment of pollinators. FAO, Rome, Italy

Feng L, Gao F, Ge F (2011) Effects of elevated CO_2 and plant genotype on interactions among cotton, aphids and parasitoids. Insect Sci 18:451–461

Fitter AH, Fitter RS (2002) Rapid changes in flowering time in British plants. Science 296:1689–1691

Flores JM, Sergio GL, Gamez V, Marai IR, Manuel AO, Franscisco JQ (2019) Effect of the climate change on honey bee colonies in a temperate Mediterranean zone assessed through remote hive weight monitoring system in conjunction with exhaustive colonies assessment. Sci Total Environ 653(25):1111–1119

Freier B, Triltsch H (1996) Climate chamber experiments and computer simulations on the influence of increasing temperature on wheat-aphid-predator interactions. Aspects Appl Biol 45:293–298

Fresquet N, Lazzari CR (2011) Response to heat in *Rhodniusprolixus*: the role of the thermal background. J Insect Physiol 57(10):1446–1449

Gordo O, Sanz JJ (2011) Phenology and climate change: a long-term study in a Mediterranean locality. Oecologia 146(3):484–495

Gouinguené SP, Turlings TCJ (2002) The effects of abiotic factors on induced volatile emissions in corn plants. Plant Physiol 129:1296–1307

Grabenweger G, Hopp H, Jackel B, Balder H, Koch T, Schmolling S (2007) Impact of poor host-parasitoid synchronisation on the parasitism of *Camerariaohridella* (Lepidoptera: Gracillariidae). Eur J Entomol 104:153–158

Gregory PJ, Johnson SN, Newton AC, Ingram JSI (2009) Integrating pests and pathogens into the climate change/food security debate. J Exp Bot 60:2827–2838

Guarneri AA, Lazzari C, Diotaiuti L, Lorenzo MG (2002) The effect of relative humidity on the behaviour and development of *Triatoma brasiliensis*. Phys Entomol 27:142–147. https://doi.org/10.1046/j.1365-3032.2002.00279
Guitierrez AP, Ponti L, d'Oultremont T, Ellis CK (2008) Climate change effects on poikilotherm tritrophic interactions. Clim Chang 87:S167–S192
Hampson GF (1908) The moths of India-supplementary papers to the volume in the Fauna of British India. Ser.III, Pt.X. J Bombay Nat Hist Soc 18:257–271
Hance T, van Baaren J, Vernon P, Boivin G (2007) Impact of extreme temperatures on parasitoids in a climate change perspective. Annu Rev Entomol 52:107–126
Hare JD (1992) Effects of plant variation on herbivore-natural enemy interactions. In: Fritz RS, Simms EL (eds) Plant resistance to herbivores and pathogens: ecology, evolution and genetics. University of Chicago Press, Chicago, IL, pp 278–298
Harrington R, Fleming RA, Woiwod IP (2001) Climate change impacts on insect management and conservation in temperate regions: can they be predicted? Agri Forest Entomol 3:233–240
Harrison J, Frazier MR, Henry JR, Kaiser A, Klok CJ, Rascón B (2006) Responses of terrestrial insects to hypoxia or hyperoxia. Res Physiol Neurobiol 154(1–2):4–17
Heeb L, Jenner E, Mathew JWC (2019) Climate-smart pest management: building resilience of farms and landscapes to changing pest threats. J Pest Sci 92:951–996. https://doi.org/10.1007/s10340-019-01083-y
Hegland SJ, Nielsen A, Lazaro A, Bjerknes A, Totland O (2009) How does climate warming affect plant-pollinator interactions? Ecol Lett 12:184–195
Helmig D, Ortega J, Duhl T (2007) Sesquiterpene emissions from pine trees identifications, emission rates and flux estimates for the contiguous United States. Environ Sci Technol 41:1545–1553
Hilder VA, Boulter D (1999) Genetic engineering of crop plants for insect resistance - a critical review. Crop Protec 18:177–191
Hill MG, Dymock JJ (1989) Impact of climate change: agricultural/horticultural systems. DSIR Entomology Division Submission to the New Zealand Climate Change Program. Department of Scientific and Industrial Research, Auckland, New Zealand, p 16
Hoffmann AA, Weeks AR, Nash MA, Mangano GP, Umina PA (2008) The changing status of invertebrate pests and the future of pest management in the Australian grains industry. Aust J Exp Agric 48:1481–1493
Holton MK, Lindroth RL, Nordheim EV (2003) Foliar quality influences tree-herbivore-parasitoid interactions: effects of elevated CO_2, O_3 and plant genotype. Oecologia 137:233–244
Howden SM, Soussana JF, Tubiello FN, Chhetri N, Dunlop M, Meinke H (2007) Adapting agriculture to climate change. Proc Acad Sci USA 104:19691–19696. https://doi.org/10.1073/pnas.0701890104
IARI (2008) Brown plant hopper outbreak in Rice. News 24
IMD (2010) Annual climate summary. India meteorological department, Pune. Government of India, Ministry of Earth Sciences, pp. 27
IPCC (2014) In: Pachauri RK, Meyer LA (eds) Climate change: synthesis report. Contribution of working groups I, II and III to the fifth assessment report of the intergovernmental panel on climate change. IPCC, Geneva, Switzerland, p 151
IPCC (Intergovernmental Panel on Climate Change) (2007a) Climate change 2007: the physical science basis. IPCC, Geneva, p 996
IPCC (Intergovernmental Panel on Climate Change) (2007b) In: Parry ML, Canziani OF, Palutikof JP, van der Linden PJ, Hanson CE (eds) Climate change-impacts, adaptation and vulnerability. Cambridge University Press, Cambridge, UK, p 976
Irfan U (2011) Climate change may make insect-borne diseases harder to control reprinted from climatewire with permission. Environment & Energy Publishing LLC, Washington, DC
Jessica J, Hellmann JEB, Britta GB, Jeffrey SD (2008) Five potential consequences of climate change for invasive species. Conserv Biol 22(3):534–543

Jhariya MK, Banerjee A, Meena RS, Yadav DK (2019a) Sustainable agriculture, forest and environmental management. Springer, Singapore, p 606. https://doi.org/10.1007/978-981-13-6830-1

Jhariya MK, Banerjee A, Yadav DK, Raj A (2018b) Leguminous trees an innovative tool for soil sustainability. In: Meena RS, Das A, Yadav GS, Lal R (eds) Legumes for soil health and sustainable management. Springer, Cham, pp 315–345. https://doi.org/10.1007/978-981-13-0253-4_10

Jhariya MK, Meena RS, Banerjee A (2021) Ecological intensification of natural resources towards sustainable productive system. In: Ecological intensification of natural resources for sustainable agriculture. Springer, Singapore. https://doi.org/10.1007/978-981-33-4203-3_1

Jhariya MK, Yadav DK, Banerjee A (2018a) Plant mediated transformation and habitat restoration: phytoremediation an eco-friendly approach. In: Gautam A, Pathak C (eds) Metallic contamination and its toxicity. Daya Publishing House, New Delhi, India, pp 231–247

Jhariya MK, Yadav DK, Banerjee A (2019b) Agroforestry and climate change: issues and challenges. Apple Academic Press Inc., CRC press- a Tayler and Francis Group, Boulder, CO, p 335. https://doi.org/10.1201/9780429057274

Johnson DL (1990) Influence of temperature on toxicity of two pyrethroids to grasshoppers (Orthoptera: Acrididae). J Econ Entomol 83:366–373

Joshi S, Viraktamath CA (2004) The sugarcane woolly aphid, *Ceratovacuna lanigera* Zehntner (Hemiptera: Aphididae): its biology, Pest status and control. Curr Sci 87:307–316

Juroszek P, von Tiedemann A (2011) Potential strategies and future requirements for plant disease management under a changing climate. Plant Pathol 60:100–112

Juroszek P, von Tiedemann A (2012) Plant diseases, insect pests and weeds in a changing global climate: a review of approaches, challenges, research gaps, key studies and concepts. J Agric Sci 151(2):163–188. https://doi.org/10.1017/S0021859612000500

Kaiser J (1996) Pests overwhelm Bt cotton crop. Nature 273:423

Kambrekar DN, Guledgudda SS, Anand K, Mohan K (2015) Impact of climate change on insect pests and their natural enemies. Karnataka J Agric Sci Spl Issue 28(5):814–816

Kannan R, James DA (2009) Effects of climate change on global diversity: a review of key literature. Trop Ecol 50:31–39

Karuppaiah V, Sujayanad GK (2012) Impact of climate change on population dynamics of insect pests. World J Agric Sci 8(3):240–246

Khan N, Jhariya MK, Raj A, Banerjee A, Meena RS (2021a) Soil carbon stock and sequestration: implications for climate change adaptation and mitigation. In: Jhariya MK, Meena RS, Banerjee A (eds) Ecological intensification of natural resources for sustainable agriculture. Springer, Singapore. https://doi.org/10.1007/978-981-33-4203-3_13

Khan N, Jhariya MK, Raj A, Banerjee A, Meena RS (2021b) Eco-designing for sustainability. In: Jhariya MK, Meena RS, Banerjee A (eds) Ecological intensification of natural resources for sustainable agriculture. Springer, Singapore. https://doi.org/10.1007/978-981-33-4203-3_16

Khan N, Jhariya MK, Yadav DK, Banerjee A (2020a) Herbaceous dynamics and CO_2 mitigation in an urban setup- a case study from Chhattisgarh, India. Environ Sci Pollut Res 27(3):2881–2897. https://doi.org/10.1007/s11356-019-07182-8

Khan N, Jhariya MK, Yadav DK, Banerjee A (2020b) Structure, diversity and ecological function of shrub species in an urban setup of Sarguja, Chhattisgarh, India. Environ Sci Pollut Res 27 (5):5418–5432. https://doi.org/10.1007/s11356-019-07172-w

Kiritani K (2006) Predicting impacts of global warming on population dynamics and distribution of arthropods in Japan. Popln Ecol 48:5–12

Kiritani K, Morimoto N (2004) Invasive insect and nematode pests from North America. Global Environ Res Assoc Res Initiatives Environ Stud 2004:75–88

Klaiber J, Najar-Rodriguez AJ, Dialer E, Dorn S (2013) Elevated carbon dioxide impairs the performance of a specialized parasitoid of an aphid host feeding on *Brassica* plants. Biol Control 66:49–55

Klein AM, Vaissiere B, Cane JH, Steffan-Dewenter I, Cunningham SA, Kremen C, Tscharntke T (2007) Importance of pollinators in changing landscapes for world crops. Proc Roy Soc B 274:303–313

Kranti KR, Naidu S, Dhawad CS, Tatwawadi A, Mate K, Patil E, Bharose AA, Behere GT, Wadaskar RM, Kranti S (2005) Temporal and intraplant variation in Cry1Ac expression of Bt cotton and its influence on the development of cotton bollworm, *Helicoverpa armigera* (Hubner), (Noctuidae, Lepidoptera). Curr Sci 89:291–298

Kremen C, Colwell RK, Erwin TL, Murphy DD, Noss RF, Sanjayan MA (1993) Terrestrial arthropod assemblages: their use in conservation planning. Conserv Biol 7:796–808

Kudo G, Nishikawa Y, Kasagi T, Kosuge S (2004) Does seed production of spring ephemerals decrease when spring comes early? Ecol Res 19:255–259

Kumar S, Meena RS, Datta R, Verma SK, Yadav GS, Pradhan GS, Molaei A, Rahman GKMM, Mashuk HA (2020a) Legumes for carbon and nitrogen cycling: an organic approach. In: Carbon and nitrogen cycling in soil. Springer, Cham, pp 337–375. https://doi.org/10.1007/978-981-13-7264-3_10.337-375

Kumar S, Meena RS, Jhariya MK (2020) Resources use efficiency in agriculture. Springer, Singapore, p 760. https://doi.org/10.1007/978-981-15-6953-1

Lamichhane JR, Barzman M, Booij K, Boonekamp P, Desneux N, Huber L, Kudsk P, Langrell SRH, Ratnadass A, Ricci P, Sarah J-L, Messéan A (2015) Robust cropping systems to tackle pests under climate change. A review. Agron Sustain Dev 35:443–459. https://doi.org/10.1007/s13593-014-0275-9

Lever RJW (1969) Do armyworm follow the rain? Wild Crops 21:351–352

Litvak ME, Constable JVH, Monson RK (2002) Supply and demand processes as controls over needle monoterpene synthesis and concentration in Douglas fir [*Pseudotsuga menziesii* (Mirb.) Franco]. Oecologia 132:382–391

Lobell DB, Field CB (2007) Global scale climate–crop yield relationships and the impacts of recent warming. Environ Res Lett 2:014002. https://doi.org/10.1088/1748-9326/2/1/014002

Lobell DB, Gourdji SM (2012) The influence of climate change on global crop productivity. Plant Physiol 160:1686–1697

Macfadyen S, McDonald G, Hill MP (2018) From species distributions to climate change adaptation: knowledge gaps in managing invertebrate pests in broad-acre grain crops. Agric Ecosyst Environ 253:208–219. https://doi.org/10.1016/j.agee.2016.08.029

Meena RS, Kumar S, Datta R, Lal R, Vijaykumar V, Brtnicky M, Sharma MP, Yadav GS, Jhariya MK, Jangir CK, Pathan SI, Dokulilova T, Pecina V, Marfo TD (2020b) Impact of agrochemicals on soil microbiota and management: a review. Land 9(2):34. https://doi.org/10.3390/land9020034

Meena RS, Lal R, Yadav GS (2020) Long term impacts of topsoil depth and amendments on soil physical and hydrological properties of an Alfisol in Central Ohio, USA. Geoderma 363:1141164. https://doi.org/10.1016/j.geoderma.2019.114164

Meena RS, Lal R, Yadav GS (2020a) Long-term impact of topsoil depth and amendments on carbon and nitrogen budgets in the surface layer of an Alfisol in Central Ohio. Catena 194:104752. https://doi.org/10.1016/j.catena.2020.104752

Menendez R (2007) How are insects responding to global warming? Tijdschriftvoor Entomologie 150:355–365

Millennium Ecosystem Assessment (2005) Ecosystems and human Well-being: synthesis. Island Press, Washington, DC

Moore BA, Allard GB (2008) Climate change impacts on forest health. Forest Health & Biosecurity Working Papers, forestry department food and agriculture Organization of the United Nations. Working paper FBS/9E. FAO, Rome, Italy, p 35

Mullen MA, Arbogast RT (1979) Time temperature-mortality relationships for various stored-product insect eggs and chilling times for selected commodities. J Econ Entomol 72(4):476–478

Murugan K (2006) Bio-diversity of insects. Curr Sci 91:1602–1603

Narayanasamy M, Kennedy JS, Geetha Lakshmi V (2015) Effect of temperature on life history parameters of brown plant hopper (*Nilaparvatalugens* Stal). Afr J Agric Res 10(38):3678–3685. https://doi.org/10.5897/AJAR2015.10094
Neven LG (2000) Physiological responses of insects to heat. Postharvest Biol Technol 21 (1):103–111
Newton AC, Begg G, Swanston JS (2009) Deployment of diversity for enhanced crop function. Ann Appl Biol 154:309–322
Niziolek OK, Berenbaum MR, DeLucia EH (2012) Impact of elevated CO_2 and temperature on Japanese beetle herbivory. Insect Sci 20(4):513–523
Norhisham AR, Abood F, Rita M, Hakeem KR (2013) Effect of humidity on egg hatchability and reproductive biology of the bamboo borer (*Dinoderus minutus* Fabricius). Springerplus 2 (9):1–6. https://doi.org/10.1186/2193-1801-2-9
OfNeill ZAR, DeLucia EH, Berenbaum MR (2008) Longevity and fecundity of Japanese beetle (*Popillia japonica*) on foliage grown under elevated carbon dioxide. Plant-Insect Interact 37 (2):601–607
Overgaard J, Sorenson JG (2008) Rapid thermal adaptation during field temperature variations in *Drosophila melanogaster*. Cryobiology 56:159–162
Painkra GP, Bhagat PK, Jhariya MK, Yadav DK (2016) Beekeeping for poverty alleviation and livelihood security in Chhattisgarh, India. In: Narain S, Rawat SK (eds) Innovative technology for sustainable agriculture development. Biotech Books, New Delhi, India, pp 429–453
Paris Contributions Map @World Resources (2016) CAIT Paris Contributions Map. Retrieved 2016-11-14
Parmesan C, Yohe G (2003) A globally coherent fingerprint of climate change impacts across natural systems. Nature 421:37–42
Pathak H, Aggarwal PK, Singh SD (2012) Climate change impact, adaptation and mitigation in agriculture: methodology for assessment and application. Division of Environmental Sciences, Indian Agricultural Research Institute, New Delhi, India, pp 111–130
Petzoldt C, Seaman A (2007) Climate change effects on insects and pathogens. Fact sheet. http://www.climateandfarming.org/clrcc.php
Porter JH, Parry ML, Carter TR (1991) The potential effects of climate change on agricultural insect pests. Agric Forest Meteorol 57:221–240
Prasannakumar NR, Subhash C, Madan P (2012) Assessment of impact of climate change with reference to elevated CO_2 on rice brown plant hopper, *Nilaparvata lugens* (Stal.) and crop yield. Curr Sci 103:1201–1205
Raj A, Jhariya MK, Banerjee A, Yadav DK, Meena RS (2019b) Soil for sustainable environment and ecosystems management. In: Jhariya MK, Banerjee A, Meena RS, Yadav DK (eds) Sustainable agriculture, forest and environmental management. Springer, Singapore, pp 189–221
Raj A, Jhariya MK, Khan N, Banerjee A, Meena RS (2021) Ecological intensification for sustainable development. In: Ecological intensification of natural resources for sustainable agriculture. Springer, Singapore. https://doi.org/10.1007/978-981-33-4203-3_5
Raj A, Jhariya MK, Yadav DK, Banerjee A (2020) Climate change and agroforestry systems: adaptation and mitigation strategies. Apple Academic Press Inc., CRC press- a Tayler and Francis Group, Boulder, CO, p 383. https://doi.org/10.1201/9780429286759
Raj A, Jhariya MK, Yadav DK, Banerjee A, Meena RS (2019a) Agroforestry: a holistic approach for agricultural sustainability. In: Jhariya MK, Banerjee A, Meena RS, Yadav DK (eds) Sustainable agriculture, forest and environmental management. Springer, Singapore, pp 101–131
Rao GGSN, Rao AVMS, Rao VUM (2009) In: Aggarwal PK (ed) Trends in rainfall and temperature in rainfed India in previous century. In: global climate change and Indian agriculture case studies from ICAR network project. ICAR Publication, New Delhi, pp 71–73
Rathee M, Dalal PK (2017) Impact of climate change on pollinators, pollination and food security. National Seminar on Agrometeorology for sustainable development with special emphasis on

agrometeorological practices for climate resilient farming and food security (AGMET-2017) at: CCSHAU, Hisar

Reji G, Chander S (2008) A degree-day simulation model for the population dynamics of the rice bug, *Leptocorisa acuta* (Thunb.). J Appl Entomol 132:646–653

Rhoades DF (1985) Offensive-defensive interactions between herbivores and plants: their relevance in herbivore population dynamics and ecological theory. American Nat 125:205–238

Ricketts TH, Regetz J, Steffan-Dewenter I, Cunningham SA, Kremen C, Bogdanski A, Gemmill-Herren B, Greenleaf SS, Klein AM, Mayfield MM, Morandin LA, Ochieng A, Viana BF (2008) Landscape effects on crop pollination services: are there general patterns? Ecol Lett 11 (5):499–495

Robert Y, Woodford JAT, Ducray-Bourdin DG (2000) Some epidemiological approaches to the control of aphid-borne virus diseases in seed potato crops in northern Europe. Virus Res 71:33–47

Root TL, Price JT, Hall KR, Schneider SH, Rosenzweig C, Pounds JA (2003) Fingerprints of global warming on wild animals and plants. Nature 421:57–60

Roth SK, Lindroth RL (1995) Elevated atmospheric CO_2: effects on phytochemistry, insect performance and insect parasitoid interactions. Glob Change Biol 1:173–182

Roy DB, Sparks TH (2000) Phenology of British butterflies and climate change. Glob Chang Biol 6:407–416

Sable MG, Rana DK (2016) Impact of global warming on insect behavior - a review. Agric Rev 37 (1):81–84

Sachs J (2008) Common wealth: economics for a crowded planet. Allen Lane, London

Sara R (2011) Ladybugs changed color in response to climate change. https://www.sciencemag.org/news/2011/11/ladybugs-changed-color-response-climate-change. Retrieved on 10.11.2019

Scott NJ, Gregory PJ, Mcnicol JW, Oodally Y, Zhang X, Murray PJ (2010) Effects of soil conditions and drought on egg hatching and larval survival of the clover root weevil (Sitonalepidus). Appl S Ecol 44:75–79

Selvaraj S, Ganeshamoorthi P, Pandiaraj T (2013) Potential impacts of recent climate change on biological control agents in agro-ecosystem: a review. Int J Biodivers Conserv 5(12):845–852

Sharma CH (2014) Climate change effects on insects: implications for crop protection and food security. J Crop Improv 28(2):229–259

Sharma CH, Prabhakar CS (2014) Impact of climate change on Pest management and food security. In: Integrated pest management. Elsevier, Amsterdam, pp 23–36

Sharma HC (2002) Host plant resistance to insects: principles and practices. In: Sarath Babu B, Varaprasad KS, Anitha K, Prasada Rao RDVJ, Chandurkar PS (eds) Resources management in plant protection, Plant protection Association of India, vol 1. Rajendranagar, Hyderabad, India, pp 37–63

Sharma HC (2010) Effect of climate change on IPM in grain legumes. In: Fifth International Food Legumes Research Conference (IFLRC V), and the Seventh European Conference on Grain Legumes (AEP VII), 26ñ30 April 2010, Anatalaya, Turkey

Sharma HC (2016) Climate change vis-a-vis pest management. Proceedings in conference on national priorities in plant health management, February 4–5, 2016, Tirupati. pp 17–25

Sharma HC, Dhillon MK, Kibuka J, Mukuru SZ (2005) Plant defense responses to sorghum spotted stem borer, *Chilopartellus* under irrigated and drought conditions. Int Sorghum Millets Newsletter 46:49–52

Sharma HC, Ortiz R (2000) Transgenics, pest management, and the environment. Curr Sci 79:421–437

Sharma HC, Srivastava CP, Durairaj C, Gowda CLL (2010) Pest management in grain legumes and climate change. In: Yadav SS, Neil DL, Redden R, Patil SA (eds) Climate change and management of cool season grain legume crops. Springer, Dordrecht, The Netherlands, pp 115–140

Sharma HC, Sullivan DJ, Bhatnagar VS (2002) Population dynamics of the oriental armyworm, *Mythimna separatea* (Walker) (Lepidoptera: Noctuidae) in south-Central India. Crop Prot 21:721–732

Sharma HC, War AR, Pathania M, Sharma SP, Akbar SMD, Munghate RS (2016) Elevated CO_2 influences host plant defense response in chickpea against *Helicoverpa armigera*. Arthropod Plant Interact. https://doi.org/10.1007/s11829-016-9422-3

Sidhu AK, Mehta HS (2008) Role of butterflies in the natural ecosystem with special reference to high altitude (Pangi Valley, Himachal Pradesh). In: proceedings of international conference on climate change, biodiversity and food security in the south Asian region, 3–4 November, 2008. Punjab State Council for Science and Technology, Chandigarh and United Nations Educational, Scientific and Cultural Organization, New Delhi, pp. 36

Sinervo B, Mendez-De-La-Cruz F, Miles D, Heulin B, Bastiaans E, Villagrán-Sant Cruz M, Lara Resendiz R, Martínez-Méndez N, Calderón-Espinosa M, Meza-Lázaro R (2010) Erosion of lizard diversity by climate change and altered thermal niches. Science 328(5980):894–899

Srikanth J (2007) World and Indian scenario of sugarcane woolly aphid. In: Mukunthan N, Srikanth J, Singaravelu B, Rajula Shanthy T, Thiagarajan R, Puthira Prathap D (eds) Woolly aphid management in sugarcane, vol 154. Extension Publication, Sugarcane Breeding Institute, Coimbatore, India, pp 1–12

Srinivasa Rao M, Swathi P, Rama Rao CA, Rao KV, Raju BMK, Srinivas K, Manimanjari D, Maheshwari M (2015) Model and scenario variations in predicted number of generations of Spodoptera litura fab. On Peanut during future climate change scenario. PLoS One 10(2): e0116762. https://doi.org/10.1371/journal.pone.0116762

Tanwar RK, Jeyakumar P, Vennila S (2010) Papaya mealybug and its management strategies. New Delhi: National Centre for integrated Pest management, technical bulletin 22

Thomas MB, Blanford S (2003) Thermal biology in insect parasite interactions. Trends Ecol Evol 18:344–350

Thomson LJ, Macfadyen S, Hoffmann AA (2010) Predicting the effects of climate change on natural enemies of agricultural pests. Biol Control 52:296–306

Timoney KP (2003) The changing disturbance regime of the boreal forest of the Canadian prairie provinces. Forest Chronicle 79:502–516

Tomokuni M, Yasunaga T, Takai M, Yamashita I, Kawamura M, Kawasawa T (1993) A field guide to Japanese bugs: terrestrial heteropterans (in Japanese). Zen-Nokyo, Tokyo

Urban M (2015) Accelerating extinction risk from climate change. Science 348(6234):571–573

Vuorinen T, Nerg AM, Ibrahim AM, Reddy GV, Holopainen JK (2004) Emission of *Plutellaxylostella*induced compounds from cabbages grown at elevated CO2 and orientation behavior of the natural enemies. Plant Physiol 135:1984–1992

Walther GR, Post E, Convey P, Menzel A, Parmesan C, Beebee TJC, Fromentin JM, Hoegh-Guldberg O, Bairlein F (2002) Ecological responses to recent climate change. Nature 416:389–395

Williams R, Norby RJ, Lioncln DE (2000) Effects of elevated CO_2 and temperature grown red and sugar maple on gypsy moth performance. Glob Chang Biol 6:685–695

Willis CG, Ruhfel B, Primack RB, Miller Rushing AJ, Davis CC (2008) Phylogenetic patterns of species loss in Thoreaufs woods are driven by climate change. Proc Natl Acad Sci U S A 105:17029–17033

Yamamura K, Kiritani K (1998) A simple method to estimate the potential increase in the number of generations under global warming in temperate zones. Appl Entomol Zool 33:289–298

Yamamura K, Yokozawa M, Nishimori M, Ueda Y, Yokosuka T (2006) How to analyze long-term insect population dynamics under climate change: 50-year data of three insect pests in paddy fields. Popl Ecol 48:31

Young AM (1982) Population biology of tropical insects. Plenum Press, New York, p 511
Yuan JS, Himanen SJ, Holopainen JK, Chen F, Stewart CN (2009) Smelling global climate change: mitigation of function for plant volatile organic compounds. Trends Ecol Evol 24(6):323–331
Zavala JA, Casteel CL, Delucia EH, Berenbaum MR (2008) Anthropogenic increase in carbon dioxide compromises plant defence against invasive insects. Proc Natl Acad Sci U S A 105:5129–5133
Zvereva EL, Kozlov MV (2006) Consequences of simultaneous elevation of carbon dioxide and temperature for plant-herbivore interactions: a metaanalysis. Glob Change Biol 12:27–35

Ecointensification in Agriculture Under Changing Climate

22

Papiya Nandi Palit

Abstract

Climate change has a significant impact over the productivity of agroecosystem on a global basis. There is urgent need of more production to feed the ever-growing human population. In order to maintain agroecosystem and environmental sustainability, ecological intensification seems to serve the purpose to some extent. Various approaches such as climate smart agricultural practices, eco-friendly farming technologies along with latest technology in the form of precision farming, system of rice intensification, and biofertilizer technology are gaining more importance day by day. Such approaches are helping to improve the productivity along with environmental security. To mitigate the effect of changing climate on agriculture sector, approaches at socioeconomic level combined with traditional knowledge in terms of conventional technologies need to be adopted. Research and developmental activities from future perspective need to be implemented towards mitigation and management of the negative impacts of changing climate over the agroecosystem. Therefore, ecointensification approaches stand to be the most fruitful and suitable under the context of agricultural sustainability.

Keywords

Climate smart agriculture · Ecointensification · Food security · Sustainability

P. N. Palit (✉)
Department of Geography, Barjora College, Bankura, West Bengal, India

M. K. Jhariya et al. (eds.), *Sustainable Intensification for Agroecosystem Services and Management*, https://doi.org/10.1007/978-981-16-3207-5_22

Abbreviations

C	Carbon
CH_4	Methane
CO_2	Carbon dioxide
CSA	Climate smart agriculture
DNRA	Dissimilarity nitrate reduction to ammonium
EI	Ecological intensification
GDP	Gross domestic product
GHG	Greenhouse gas
N	Nitrogen
N_2O	Nitrous oxide
SI	Sustainable intensification

22.1 Ecointensification: An Overview

Ecological intensification (EI) is described as utilizing natural processes to reduce human contravention with nature and products such as fertilizers and pesticides while sustaining or enhancing production of food per unit area (Bommarco et al. 2013; Kremen 2020; Jhariya et al. 2021). In the edges of field, plantation of flower-rich habitats can assist the natural foes of crop pests, lowering crop damage, and utilization of pesticide. Legumes as intercropping or cover-cropping to populate fields with bacteria which fix nitrogen (N) to improve soil fertility and enhanced crop yield (Garibaldi et al. 2019; Jhariya et al. 2018a, b; Kremen 2020). These methodologies are chief and mandatory for the biophysical component of landscapes of agriculture to obtain joint goals of popularizing biodiversity together with human well-being, besides transformative social, economic, and political changes (Garibaldi et al. 2019; Kumar et al. 2020; Kremen 2020). Environmental sustainability can be achieved by EI (1) by removing or reducing obstructive environmental externalities from farming (negative impact that runs off the farm, negative impact on other individuals, polluting water with pesticides and fertilizers that minimize fish catch); (2) by promoting natural processes that regenerate the ecosystem services on which farmers depend (Kremen and Miles 2012; Zhang et al. 2007; Kremen 2020); and (3) providing ecosystem services, such as good water quality and biodiversity conservation, which gives benefits to the society at a large scale (Meena et al. 2020; Zhang et al. 2007; Kremen 2020).

In plywood industries, eucalyptus (*E. tereticornis*) possesses huge economic value due to its major use in various sectors. Eucalyptus is one of the highest adapted species in India for plantations because of the high returns, supportive government policies, and assured market. *Eucalyptus tereticornis* is highly accepted by farmers due to its capacity to survive in marginal and affected soils and resistance to termites and pests. In Haryana, India, several plantation drives and farms (agroforestry) participate to bring out around 45% of entire outside forest region under eucalyptus plantation. Kumar et al. (2019) endorse *E. tereticornis* plantation as a potential candidate for carbon (C) mitigation and sustainable production. Biomass accumulation was observed highest in the form of fuel wood and other parts that involve leaves, twigs, bole, and roots, and the total approximate dry biomass obtained was 225 mg/ha using allometric model (nonlinear).

22.1.1 Sustainable Intensification: Its Role in Climate Smart Agriculture

The EI basically requires knowledge-oriented optimal management of ecological functions of nature together with biodiversity to enhance performance in the system of agriculture and efficiency of it which will ultimately lead to improvement of farmers' earnings and hence livelihood too (Raj et al. 2021). Actually it is aimed for feeding everyone globally. This review is aimed for revealing various aspects of EI for sustainability, its applications in favor of farmers, and to mitigate global food crises and many allied areas. The biodiversity is an asset of nature which if utilized properly and sustainably can enrich us with various products of use. Unfortunately in this diplomatic world, we generally take nature and its assets for granted, and we go for harming forest and other patches for building houses. Ironically we are the most dependent on biodiversity like starting from morning tea to rice in dinner, everything is the modified natural product, and hence we must conserve biodiversity for maintaining ecological balance and achieving more and more profit in terms of food and money from the ideas and implications of EI.

By 2050, the global population in context of hungry people may reach 2 billion people. To combat these, climate smart agriculture (CSA) is in focus and a reliable approach which provides mitigation challenges and adaptation facing humanity (FAO 2013a; Campbell et al. 2014). Patterns of food consumption are changing as the average population in the world is getting richer, and a shifting of diet was observed which is consuming more food along with more meat. Food systems are crucial in global warming and are significant in 19–29% of global emissions, a major amount contributed from agricultural production processes (CH_4 and N_2O) along with indirect change in land cover governed by agriculture. Rise in competition for the water, land, and other needs is gradually increasing day by day in food production process. Several challenges are caused by climate change in agriculture especially in developing countries (Vermeulen et al. 2012). There are three goals in CSA: (1) supporting elevation of income, development, and food security by elevating agricultural productivity, (2) enhanced adaptive capacity at several levels and, (3) minimizing emission of greenhouse gas along with elevating C sinks. All the objectives are highly dedicated towards productivity along with adaptive capacity of smallholder farming systems (low input) in the developed countries. Important part of CSA is detecting potential synergies along with trade-offs between the goals (Neufeldt et al. 2013).

22.1.2 Climate Change Adaptation

Developing the adaptive capacity through CSA is one of the three objectives of the service providers to farmers themselves along with some institutions that are capable of effective response for the long lasting climate change together with potential to manage the risk of climate variability (Raj et al. 2018a, b; Meena et al. 2020a). Among the variation actions to build up adaptive capacity, crucial part is ecosystem service buildup in agriculture system. This promotes ecosystem resilience through plant, water, and soil nutrient management, improved irrigation and water storage facilities, varieties of crops those are stress tolerant (heat, salinity, drought, flood), farm enterprises diversification together with developing capacity of institution for supporting collective action, local adaptation planning should be undertaken together that disseminate knowledge (Bennett et al. 2014; Kumar et al. 2020a; Campbell et al. 2014). Climate change can influence negatively over the food sector in rural along with urban populations by minimizing agricultural incomes and disrupting markets along with elevated environmental risk (Vermeulen 2014; Khan et al. 2020a, b).

22.1.3 Climate Change Mitigation

Active participation of the food system in the form of food production unit has led to global warming and is responsible for around 19–29% of emissions globally. The huge amount of emission takes place from the agriculture sector along with alteration of land use (Vermeulen et al. 2012). Providing the requirement that elevates production within several countries which come under developing category, greenhouse gas (GHG) emissions from agriculture are most forecasted to increase, largely because of continuous increment of livestock production, fertilizer use, as well as change of land cover (Bennett et al. 2014). Nevertheless, the sustainable intensification (SI) initiative, targeting the advancing capabilities of production, is vital to fulfill the CSA objective: attaining lower N_2O (nitrous oxide) as well as CH_4 (methane) release per unit as regards output. In agricultural land existence, SI has the huge capability to act as mitigation source for the changes in land cover, precisely forests which are C rich as well as wetlands (Wollenberg et al. 2011). Yield at low scale during less-intensive farming may create local environmental well-being. This plan of action may be needed where land has been cleared into other land uses to settle accounts with lower yields locally, moving forward towards higher environmental influences. Globally, entire crop yields of cereal as well as oil crops production have elevated by 135% from 1961 to 2005, while the cropland region shows increment about 27% (Burney et al. 2010). Nonetheless, elevated efficiencies because of intensification can build up incentives for the farming community (Ewers et al. 2009; Rudel et al. 2009). Combination of intensification can thus be implemented along with policies and price incentives to strengthen its land-sparing influence (Angelsen and Kaimowitz 2001). Previous initiatives for the protection of forests indicate that monitoring the interface between forest farm relies on unification of measures: land tenure-related institutions, forest governance, land zoning, as well as enforcement of boundaries of forest (Wollenberg et al. 2011; Robinson et al. 2014). Forest protection along with good governance may be the suitable option for attractive agricultural livelihood to motivate farmers to avoid encroachment and forest degradation (Sassen et al. 2013). For attaining the mitigation objective of CSA, there is urgent need to target more than simple goal of intensifying agriculture. Reality gets recognized in both the SI and CSA concepts. Either land sharing or the land sparing, SI as well as CSA may involve the analysis of the trade-off which would include understanding of more fruitful objective as well as policy exploring the mechanisms of market which elevates initiatives in sharing or sparing (Garnett et al. 2013).

22.1.4 Case Study on Climate Smart Agriculture

22.1.4.1 Banana-Coffee Intercropping

According to Laderach et al. (2011), *Arabica* coffee cultivation may be influenced by climate change at higher altitudes because of low temperature. Jaramillo et al. (2011) illustrate that rise in temperatures not only stays specific to crop physiological effect but also elevates pest along with disease pressure. Around the 1950s globally, acceleration in the coffee production took place, and several public authorities promoted high input monocropping systems for small land holding farmers. Nonetheless, studies in East Africa illustrates that plot revenue may increase through banana intercropping by more than 50% (Van den Berg et al. 2017) under unfertilized as well as fertilized conditions. By nature coffee is understory shade-tolerant species. According to Jassogne et al. (2013), apart from shade, bananas also provide protection to the coffee plant from various diseases such as coffee leaf rust. Besides that, mitigation gets contributed by banana intercropping which incorporates up to 15–30 t C/ha in soil environment.

22.1.4.2 Livestock Systems Intensification

Mitigation as well as adaptation gets influenced by the systems in livestock production of SI. These systems are highly variable in terms of production efficiency, spatially as well as productivity. Havlik et al. (2014) reported worldwide economic modelling research. As per their report transitions are governed by economic incentives having outcomes from transitions in demand as well as prices of relative factor between now and 2030. Further, such transitions would also influence the ruminant animal along with their relative distribution between rangeland and other agricultural unit in the same agroecology zone. Under the intensification process, there will be more output in the form of meat and milk gains. Such changes would minimize release of 736 Mt. CO_2 equiv. y^{1} (near about 10% of all emissions in the agricultural sector), through avoidance along with land use conversion up to 162 Mha.

22.1.4.3 Agroforestry in Livestock Diet Intensification

Intensification in the domesticated animals is achieved through optimum nutrition by using *Leucaena leucocephala* as a feed material for the animals. Milk yield per day can be triple on addition of minute quantity of leaf material of *Leucaena* which leads to increase in weight, hence elevating economic gain, as well as minimizing the production of CH_4 (Thornton and Herrero 2010). According to Albrecht and Kandji (2003), C sequestration can be elevated by agroforestry trees. Adapting this option has high potentiality towards substantial mitigation as intensified diets would considerably minimize the counting of ruminants required for coping with demand in future for milk as well as meat.

22.1.4.4 Model Structure in the Sahel

Engineering stone bunds together with contours are an efficient path to water harvest together with minimization of erosions by runoff. Although, when combined with several other techniques for management of land such as Zai pits (compost filled shallow bowls filled or manure in that crops planted), double yield of millet or sorghum can be achieved compared with lands (unimproved) may reach more than 1 t per ha (Landolt 2011; Bayala et al. 2012). Developed land management often results into increment of tree cover together with advanced soil fertility along with ground water levels. This permits farmers to use small pots to grow vegetables on near wells, thus elevating both their income and diet diversity. So, the utilization of stone bunds could thus result in nutritional benefits, supporting farmers to fight with changing weather condition (wetter or drier climate adaptation). More manure used results in soil fertility and rise in vegetal cover to increase the mitigation process. Consequently, such approaches lead to SI. However, some examples illustrate that in CSA, SI is a cornerstone which raises the resource utilization as well as promote mitigation along with reduced emission and high productive output (Thornton and Lipper 2014). For promotion of SI, various approaches of CSA, livestock maintenance, along with associated local climate can facilitate SI. Thus, sustainability should be a key component in the intensification process under CSA approach. On the long run basis, SI and CSA would help in sustainable food production and gaining food security and well-being of the society (Garnett et al. 2013).

22.1.5 Models of Ecological Intensification

Intensification practice has sustainable and ecological perspective. The various attributes have been precisely represented in Table 22.1.

22.2 Climate Change: An Outline

Catastrophic changes in environment negatively influence the natural systems and productions in agriculture area along with human health (Arunanondchai et al. 2018; Banerjee et al. 2020; Raj et al. 2020). Elevation in population across the globe

Table 22.1 Characteristic features of agricultural SI along with EI (Compiled: Xie et al. 2019)

Type of Intensification	Characteristics	References
Agricultural sustainable intensification	Without negatively influence land and environment, elevate production	Baulcombe et al. (2009), Pretty and Bharucha (2014)
	For minimizing the environmental damage, by building up the natural, human, and social assets along with utilizing best technologies available together	Pretty (2008)
	Maintaining nutrient balance in soil, developing the return of labor along with land	Ruerd and Lee (2000)
	Output along with input of production system of livestock to control just to enhance productivity during environmental integrity together with maintaining system	Gibon et al. (1999)
	During protecting the natural resources, increase production	Pretty (1997)
Ecological intensification	Resource together with the efficiency of investment get improved	Tittonell and Giller (2013); CIRAD (2008)
	A system which makes use of ecological services along with processes	CIRAD (2008); Tittonell (2014)
	In the process of reducing external inputs, production gaps get reduce; based on the local agricultural system knowledge	Clay (2018)
	Support production system for positive influence on grain production side by side reducing the negative influence on environment	Cassman (1999)
	Elevate production of food during minimizing external inputs along with decreasing negative influence on the environment, together with utilizing ecological process, ecosystem services	Wezel et al. (2015)

correlates with increase in demand for food, owing to solicitude about the global environment stability. Productivity of agricultural field has higher influence from the perspective of soil fertility, pollution, and availability of water (Noya et al. 2018). However, with instantaneous changes in conditions of environment, the barbarous impacts on the productivity of plants are moving towards greater intensification which would suffer direct and indirect consequences. Ongoing practice of fossil fuel consumption together with deforestation results in the elevation of proportion of CO_2 in atmosphere that has jumped up to 400μmol^{-1}. Emission of CO_2 leads to gradual increment in the average global temperatures along with GHG's negative influence (Vaughan et al. 2018). The negative impacts of climate change together with environmental variation on the whole agroecosystem would be severe (FAO; UNICEF; WFP; WHO 2017).

22.3 Climatic Factors and their Adverse Effect on Agriculture

The influence of variations in environment along with climate change is on the whole forecasted on the quantity of the stress spells, as well as their influence on daily life, along with damages in agricultural crops (UNICEF and WHO 2017). Due to catastrophic conditions in environment, in developing countries, the yields in agricultural sector predominantly suffer. Hence, high temperature along with abundance of CO_2 storage forced scientists to come up with new plauns of action to persevere with less forecasted challenges. Climate smart crop cultivators are highly required to manage these restrictions and assure food security. Yields along with growth of plants are chiefly influenced by abiotic stresses. Most of the time, plant undergoes several stresses under natural climate conditions (Meena et al. 2020b). However, the UV-B, gas release, light intensities, and flooding, along with chemical and physical factors, trigger more stresses under abiotic factors.

22.3.1 Average Temperature Increase

As temperate climate, upraised temperature can influence insect population expansion. Upraised temperature may have impact on insect survival, geographic range, and development along with population size. It could also have impact on insect physiology. During such circumstances, some insects opt several years for the completion of life cycle (cicadas, arctic moths). On the other hand, few insects advance at a quick rate with certain range of temperature based upon degree days (diamond back moth, cabbage maggot, Colorado potato beetle, onion maggot, aphides, European corn borer). As a result, crop damage expands. Migratory pests may migrate earlier. However, natural enemy-host relationship may influence the outcome into minimized parasitism. Upraised temperature may influence change in insect gender ratios of thrips. Insect population expansion will occur because of minimal insect's mortality during winter as it leads to warmer winter. Excess temperature geographically may lead to shifting of cropping pattern leading to alteration in the pest population at higher altitudes. Fossil records revealed an increment in the feeding intensity level as well as insect species diversity with elevation in temperature. Upraised temperature can lead to reduce insect community (aphids) in few crops that are unable to grow under elevated temperature. Such repeated circumstances might be conducive for elevated task of natural enemies of that pest, further negatively influencing its population.

22.3.2 Changes in Precipitation Patterns

During the twenty-first century, food security together with climate change is the matter of great concern. Approximately, around 815 million people are negatively influenced by malnutrition and improper development, and thus it is a challenge to promote sustainable events to acquire the goal in universal scale of minimizing or removing hunger till 2030. Catastrophic weather seems to impact food security along with agricultural yields. Around the world, decline in major crop production was found to be correlated with rise in temperature. End of the present century would record decline in crop production with gradual increase of average temperature from 2.6 °C to 4 °C. Indication of minimization of productivity in crops, the potential danger to security of food takes a speedy level up along with the growing world's population.

Around 2050, there would be a rise in population up to 9 billion along with 85% increase in requirement of food. Schemes for present-day cropping, climatic impacts get worsened with low-scale variation along with leveled up input concentration along with unstable productivity because of changing crop environment. Dhankher and Foyer (2018) described that arise in frequency of heavy rainfalls as well as drought, temperature up-down, salinity, and insect pest attacks are anticipated to sabotage productivity of crops which leads to starvation threats. This variation in rainfall crop adaptability also suffered along with variations in temperature. Campbell et al. (2014) suggest that in present day and the major target is to minimize the pressure of food security. Variations in climate majorly impact the plant physiology. Stress upon plants gets elevated by environmental extremes as well as climate variability (Thornton and Lipper 2014). Insects physically get dislodged because of rain drops from their hosts such as thrips, leafhoppers, cut worms, plant hoppers, etc., meanwhile drowning others towards death, e.g., rice stem borers, mealy bugs, pupae of fruit fly, *Helicoverpa*, and *Spodoptera*. For stem borers as well as termites, flooding could act as a control measure.

22.3.3 Rising Atmospheric Concentration of CO_2

CO_2 possesses both negative and positive influences. Physiological influence in positive ways expected from CO_2 is through acceleration of photosynthesis. C_3 crops get highly influenced such as wheat and rice when compared to C_4 plants such as grasses and maize. Variations in proportions of CO_2 will directly influence through variations in temperature and radiation along with the precipitation. Nonetheless, changes can be brought by indirect influences in soil moisture as well as infestation by diseases together with pests due to acceleration in temperature and relative humidity. These indirectly influence through elevation of temperature that will minimize crop timing, elevate the rates of crop respiration, minimize efficiency of fertilizer utilization and evapotranspiration, as well as accelerate pest infestation. A common consensus with yields of main season (Kharif) crop is enhancement due to excessive CO_2 levels. Nonetheless, forecast of higher yield gets reduced for the Rabi crops under elevated temperatures. Excessive proportions of CO_2 in the atmosphere have indirect influence on populations of insect. Under higher proportion of CO_2, soybean crop had 57% increment in insect damage (Japanese beetle,

Mexican bean beetle, leafhopper, root worm, etc.) than initial phase. This triggers elevation in the amount of simple sugars in the leaves which stimulates more feeding by insects. Elevation in C/N ratio in tissues of the plant due to elevated CO_2 level might slow the development of insect along with accelerated life stages of insect pests which may become vulnerable to be damaged by parasitoids. In the present scenario with the degree of GHG emissions, several of the vital pests' population would expand at the end of the twenty-first century.

Large leaf area as well as duration, branching, length of root as well as stem, leaf thickness, and tillering as well as dry weight are popular impacts of elevated CO_2 on most plants. Several suggestions from scientists revealed that rising CO_2 level would enlarge canopy size along with density resulting into greater biomass of huge nutritional quality. Foliar disease was promoted when combined with humidity and increased canopy growth leading to diseases such as leaf spots, rusts, and blights along with powdery mildews. In the nutrient cycle, the litter plant decomposition is a vital factor as well as part of the pathogen's saprophytic survival. Litter's shoot up C: N ratio has a significant impact on growth of plants that is escalated under CO_2. Increment in plant biomass and decomposition of litter at slow rate along with leveled up winter temperature could elevate survival of pathogen on over residues of wintering crops as well as enhance the quantity of available initial inoculums for the infection of subsequent crops. Due to impacts of escalated CO_2 level, two vital trends have come out for interaction in host-pathogen relationship in selected fungal patho-systems. The first one is the delay in the initial establishment of pathogen due to modifications in susceptibility of host and/or aggressiveness of the pathogens. The second vital observation was escalation in the fecundity in presence of escalated CO_2 level.

22.3.4 Pollution Levels

22.3.4.1 Soil

Pool of soil C is just over 5 times the pool of atmosphere along with 6.5 times of pool of biotic components (Lal 2014). Globally, from land perspective, it appears to be third in position which is at the edge of high risk of degradation accompanying half of global population (Glover and Reganold 2010). Soil under cultivation has been reported to lose 25% and 75% organic C across the globe (Lal 2014). Further, still few systems of agriculture are efficient to perform C capture along with its sequestration. Demand for soil nutrient from crop production is elevated due to agricultural intensification. To cope with these demands, use of synthetic fertilizers is correlated along with the higher energy consumption, environment, and human health (Jones et al. 2013). Jones et al. (2013) mentioned that the fertilization of phosphorus (P), N, and potassium (K) reload few nutrients eliminated by producers intensively, and several of the species regained mineral nutrients inadequately, with gloomy insinuation for soil health along with security towards nutrition. Drop in health of soil is a worldwide challenge under climate change and insecurities for food along with loss of environmental quality (McBratney et al. 2014). For agricultural systems, soil is a boon as it participates as a worldwide C sink. Conditions of soil decline during soil loss due to erosion along with loss of C, nutrients, and organic matter in soil (Khan et al. 2021a, b), although farming seems to be a chief source for global loss of soil along with loss in production up to 0.3% (Montgomery 2007; den Biggelaar et al. 2003).

22.3.4.2 Water

Wastewater is a serious concern nowadays for the entire globe. Generation of wastewater from various industries is polluting various surface water bodies at an unprecedented rate. In this connection the use of bioremediation technology is doing the world good in order to remove the contamination. Modified bioreactors and genetic engineering are used in Japan for industrial purposes along with other wastewater treatment. Various enzymatic activities are utilized, and the enzymes which are employed are phosphate esterases, aminopeptidases, and catalase. Lactases are a fungal enzyme used to degrade contaminants in effluent of paper and pulp industry as these enzymes dechlorinate chlorinated phenolic compounds under unfavorable conditions such as high temperature, organic solvents, low pH, etc. Lactase is an extracellular fungal extracted from *Botrytis cinerea* or *Trametes versicolor* (Berry et al. 2014). Several microorganisms are utilized in many places for removing metals from the wastewater of industries. Some microbes release extracellular compounds (polymer complex) and store metals such as copper, nickel, iron, cadmium, and uranium. The stored metals can be extracted from biomass by treating with HCl. A schematic diagram showing microbes that accumulate metals in industrial wastewaters is presented in Fig. 22.1.

Pollutants present in water affect many organisms. Water pollution is categorized as nonpoint source and point source pollution such as agricultural waste, community wastewater, and industrial waste that contribute to water pollution. Diarrhea is one of the major reasons of poor health, worldwide, along with mortality cases due to use of contaminated water. Introduction of excess cadmium in the water bodies is extremely harmful for the consumption of humans as it caused itai-itai (ouch-ouch) disease. Cadmium around 350–3500 mg is lethal dose for humans by oral route, although 3 mg dose of cadmium for adults possesses not so significant effects (Krajnc 1987; Chattopadhyay et al. 2020a). Cadmium chronic exposure through oral route negatively influences the resorption function of the proximal tubules, and early stage symptoms are rise in concentration of protein in low molecular weight in excretion (urine), explained as tubular proteinuria (Krajnc 1987; Chattopadhyay et al. 2020a).

22.3.4.3 Air

Pollution originates at a place, and then it migrates and ends up at a very different place. Pollution in air comes in several ways which can be categorized into two groups, natural along with anthropogenic air pollution. Air pollution by anthropogenic activities comes from motor vehicles, fossil fuel (burning), and industries; on other hand, air pollution naturally is triggered by sand dust, bushfires, and volcanic eruption. Usually carbon monoxide is generated from incomplete combustion

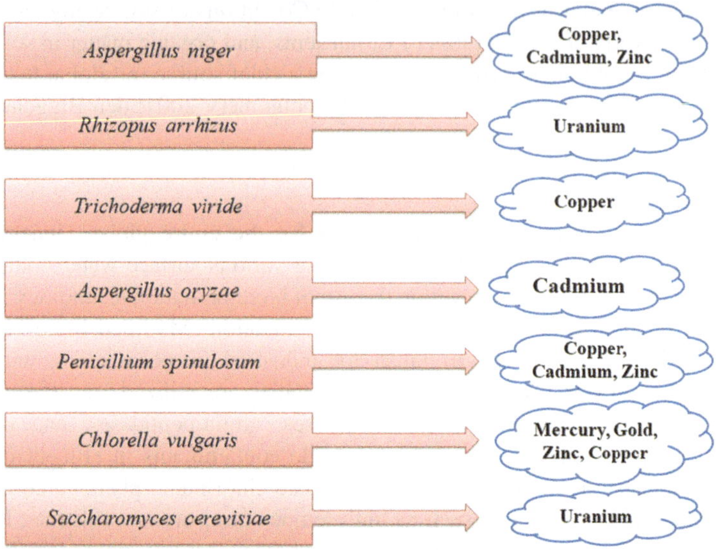

Fig. 22.1 Schematic diagram showing microbes that accumulate metals in industrial wastewaters (Compiled: Chattopadhyay et al. 2020a)

together with N oxides. Fossil fuel combustion is majorly responsible for air pollution which emits sulfur dioxide (Pope et al. 2004; Chattopadhyay et al. 2020b).

22.4 Climate Change Influence in Agriculture

Agriculture happens to be the main pillar of socioeconomic development of India. It has the potentiality to raise the gross domestic product (GDP) and employment opportunities. However, under the prevalence of changing climate, the GDP and employment opportunities show a declining trend up to 40% in 1983 to 24% till 2001. In terms of employment opportunities, the reduction is 63–57% during the same time period. Such events reduce the productive output as well as land properties for the rural farmers. This leads to lowering of agricultural efficiency and other socioeconomic problems. This therefore increases the vulnerability of the earth's ecosystem towards changing climate. From Indian perspective 550 gm/capita appears to be the food consumption rate in comparison to China which has already reached up to 980 gm/capita. The per capita food consumption is even higher in the USA (2850 gm). The problem is further bigger where animal need-based consumption level is 210 million tons (Mt) which is the present production rate. This therefore necessitates the practices of CSA to make the agroecosystem compatible with the sustainable development.

22.5 Agriculture and Food Supply

Following a restricted illustration of the calories per capita, worldwide agriculture, in the present, produces sufficient for the entire global population to flourish (FAO, WFP, IFAD 2012). Adding on this, the globe is keeping on to the processing of "triple burden" of (a) undernutrition (calories consumption along with protein insufficient amount), (b) malnutrition (insufficient amount of consumption of several other vital nutrients), and (c) overnutrition (consuming calories immoderately). However, this challenge hermitically entangles with several equivalent recalcitrant poverty challenges and insecurities in energy together with breached planetary boundaries as described by Rockstrom et al. (2009). According to Maxwell et al. (2012), crises of protracted food are fitting to norm. Respective shortage might indulge furthermore than crop failure or elevating prices. In Somalia during 2011, the incident of famine has brushed all the pillars (four) of food security, where production shock was there along with crisis in malnutrition and access shock (Maxwell 2012). A strike in food price, during 2007–2008, just as an example, was created by a mixture of elevation of prices of oil accompanied by market regulation along with speculative activity. Even, there is forecast to expand of protracted shortages due to various causes along with lack of comprehensible and potential recovery (Foresight 2011).

22.6 Agriculture and Food Security

Now, overnutrition has a negative influence on health of over a billion people globally. Somewhere in some point, precisely in two generations, world war afterwards, that put forward food rationing persists as a legacy since 1954, for the most part, degree of obesity during mid 1980s in the UK showed an increasing trend from 6 to 24% and for USA it was 3 to 35% (CMO 2013). In rapid developing countries, many wealthier groups of people are further on related transition towards overweight together with obese populations as explained by several papers (Pretty 2013; Samson and Pretty 2006; Lang and Rayner 2012; Foresight 2011). Mexico is categorized as the largest soft drinks consumer that has given rise to the degree of obesity (Carolan 2013). Undernutrition persists in the countries (developing), notwithstanding economic gain (Frayne et al. 2014), and at the same time, overnutrition along with associated concerns in public health is highly recorded (Peer et al. 2014). Diabetes along with hypertension in Indian regions becomes common knowledge as chief human health concerns (Shetty 2012). Everywhere on Africa, notwithstanding uninterrupted incidence of diabetes (Peer et al. 2014), undernutrition explained by Frayne et al. (2014) along with obesity cases is accelerating (Muthuri et al. 2014).

Growth in the agriculture sector reduces the poverty along with environmental degradation. Agriculture-based countries need proper economic development in terms of economic growth in relation to agricultural output. Without such objectives they would be suffering from the issue of hunger and poverty (de Janvry and Sadoulet 2010; Christiansen et al. 2011). Conflict management through proper planning in the agroecosystem production would address the issue on world basis (Allouche 2011).

22.6.1 Approaches to Combat Climate Change

Combating climate change, two methodologies are involved such as cultural methodologies and conventional methodologies which are discussed below:

22.6.1.1 Cultural Methodologies

Several researchers have reported that screening of the plan of actions performed by farmers to cope with climatic variations for plant adaptation. Plenty useful initiatives are practiced by farmers, involving the abiotic factors such as altered plantation along with harvest duration, abundance of crops which have short life cycles, techniques in irrigation, and crop rotation along with schemes in variation in cropping. Along with adaptability of crop under climatic stress circumstances, such initiatives appeared to be extremely fruitful (Duku et al. 2018; Marcinkowski and Pinieuski 2018; Teixeira et al. 2018; Deligios et al. 2019). Advancing in sowing duration and implication of drought-resistant cultivars along with the cultivation of new crops are few vital plans to minimize the danger of climatic variations as well as provide better adaptability to crop plants for satisfying food safety together with security (Ali and Erenstein 2017). Some other plant adaptation initiatives include different modes of crop management which have the capacity to upgrade development of crop under several environmental stresses. The pickings of sowing period and duration, planting density as well as practices for optimum irrigation are vital techniques to handle weather stresses (Battisti et al. 2018). To support plants for attaining better adaptability along with minimizing negative impacts of global warming, biofertilizers play a crucial role. It furnishes substantial energy to plants which is fruitful to sustain the fertility of soil while increasing productivity. Therefore, the importance of fertilizer in nurturing the world is undeniable (Henderson et al. 2018).

22.6.1.2 Conventional Techniques

In the field of crop betterment and development under various stresses, plant breeding displays dynamic techniques. It gives a path to potentially guarantee for food security as well as safety under unfavorable weather variations that help plants to escape from numerous stresses through a vital phase of plant growth by advancing stress-resistant cultivars (Blum 2018). For polymorphism, assortment, inbreeding, assessment as well as recombination towards attaining plant perfection, genetic divergence analysis is utilized or preferred, to define accomplished inbreeding, among chief aspects. Genetic divergence analysis is often seen as a very crucial method for the advancement of new cultivars based on genetic distance and similarities (Raza et al. 2018, 2019). For the researches in genetic studies, landraces are a remarkable source. For example, in data bank, a wheat landrace was kept which comprises variance in broader genetic as well as vital basis for stress resistance, as it consists adjustable cultivars to stresses in the diverse environmental condition (Lopes et al. 2015).

22.7 Adaptation of Agriculture to Climate Change

Organic wastes of municipal, agricultural, and industrial origin undergo anaerobic fermentation and produce biogas. Biogas is combination of several gases, and fermentation of organic matters often generates digestate and biogas. Biogas is mixture of hydrogen, hydrogen sulfide, dioxide, N, ammonia, and CH_4. The amount of CH_4 ranges from 50% to 75% which depends on the type of raw material (Jerzak et al. 2016; Kuznia et al. 2019). Agricultural digestate further can be used as manure. Elimination of nondigestible materials is obtained by utilizing some physical barriers such as stone traps, screens, and sieves. Later these can be dried and converted into pellet form to be utilized in industrial applications, for example, co-combustion for CHP generation. To enhance quality of biogas, municipal waste can be mixed along with agricultural waste (Kuznia et al. 2019). Several modes contribute in enhancement of productivity of crops which entirely is an outcome of EI. Knowledge base and practical initiative are the agroecological intensification, to requite and saturate the needs of marginal farmers so as to enhance production by utilizing more productive tools along with techniques for environmental sustainability (Jhariya et al. 2019a, b). It is biological mechanisms encircled approach to pests and disease suppression along with enhanced total crop photosynthesis for better yield. The soil nutrient cycles need proper management for more productive and healthy crops, substantially with these profitable endophytes (Cote et al. 2008).

In recent agricultural processes, endophytes maintain a balance among growing demand and production in agriculture. To remove negative impacts along with enhanced crop yield and assurance of elevation in agricultural productivity, extensification and intensification augment EI. Microorganisms that reside in the plant roots participate in enhancement of crop production and EI. Current challenge for the researchers will be to design some systems which can optimize profitable plant-endophyte bacterial relationships. The goal should be to harbor the reservoir of desired endophytic bacterial populations which are potential candidate to restore soil sources along with stabilization and optimum levels (Dheeman et al. 2017).

22.7.1 Indian Examples

India as a developing country possesses around 700 million rural population directly depending on sectors which are climate sensitive (agriculture and forests together with fisheries) and few natural resources such as coastal zones, water, biodiversity, mangroves, and grasslands for their subsistence as well as livelihoods. Furthermore, the adaptive capability of farmers of dry land, nomadic shepherds, forest dwellers as well as fisherfolk is very low (Ravindranath and Sathaye 2002; Sathaye et al. 2006). Climate change is likely to have influence on all the natural ecosystems together with socioeconomic systems as reported by the National Communications Report of India to the UNFCCC (Ministry of Environment and Forests 2004; Sathaye et al. 2006).

These are the impacts of climate change on the biodiversity of India:

- Adverse effect on agriculture, health, forestry, and infrastructure.
- Temperature rise by 3–4 °C towards the end of the twenty-first century.
- Reduction in wheat and rice yields.
- Rainfall patterns and quantities in periods of drought in some regions, more rainfall in Central India, and reduced rain in the northeast, leading to changes in forestry and vegetation. More intense rain spells in the Ganga, Krishna, and Godavari.
- Reduced number of rainy days in the western parts of the Gangetic basin.
- 70% of vegetation vulnerable to change
- Adverse impact on wildlife and other biological species.

22.7.2 Replacing Rice Cultivation by Other Crops

In Madagascar, experiments on farm initiated the system of rice intensification (SRI), whereby present norms were radically amended for paddy rice such as minimized plant population, upgraded soil with organic matter, minimized application of water as well as prior transplantation of plants (young). The adoption of these four general principles has shown to lead the way to considerable increase in yields with less water inputs both internally and externally. In the 1980s since its inception, SRI principles get well adapted as well as implemented to a variety of further crops, incorporating teff, wheat, sugarcane, finger millet, and numerous pulses along with turmeric. These approaches highlight the changes in resource use as well as crop planting application combined with design. To meet the preferences of farmers as well as local context, adaptation is appreciated. For that reason, participatory models regarding development as well as dissemination becomes crucial for popularization of SRI, outcomes as adaptations which enable farmers to increase yields, minimize resource use along with buffer against challenges in paths whatever is convenient for them (Krupnik et al. 2012). SRI remains little bit controversial largely because yield elevation remains only partially illustrated. There are some fundamental queries in relation to basic agronomy of SRI as well as to claims of higher yields (Dobermann 2004; Sheehy et al. 2004, 2005). Although in some parts of India, experimental weighing up between SRI and conventional recommended practice showed that SRI has 42% higher yields. This is an outcome of plant physiological processes which result into incorporation of longer panicles adding more grains per panicle basis with higher rate of grain filling. Further, it also ensures deep rooted system with proper distribution and larger leaves (Thakur et al. 2010). In another study of SRI in India, the comparison was made between SRI and conventional cultivation across 13 rice-growing states which reflected 12% and 54% higher yields in the former, along with water use efficiency (Palanisami et al. 2013). SRI practice seems to be highly flexible in combination with agroecologically based management initiative. Elevation in yields in Mali was reported by Styger et al. (2011) without addition of expensive inputs. Through adaptation of SRI principle in Africa, works have been done under local condition to improve the potentiality of SRI towards increasing yield and food security of the region. Results reveal that 66% more yield was recorded from the SRI plots than control plots. SRI plots utilized fewer seeds substantially per hectare with 30% lesser use of inorganic fertilizer and 10% lesser water use. Further, a narrative review by Turmel et al. (2011) that comprises 72 studies related to comparison of SRI along with conventional practice reported better performance of SRI than conventional practice.

Direct seeding is the other system in the field of rice cultivation that save water as well as labor, whereby rice is sown and sprouted into field directly rather than raising along with transplanting seedlings, although, since long time, direct seeding has been practiced primarily to rejuvenate or reciprocate water scarcity in Asia (Rao et al. 2007). In Sri Lanka, 95% of entire rice is grown under seeding either wet or dry condition under the direct seeding process (Weerakoon et al. 2011). Weed infestations were relatively low, and the yield varied under direct seeded systems. The requirement for the development as well as dissemination of advanced cultivars as suggested by reviews is very much essential on case to case basis (incorporating herbicide-resistant as well as early maturing varieties). Such approaches develop nutrient management as well as the provision of high-quality herbicides (Farooq et al. 2011).

22.7.2.1 Crop Variety Improvements

Advancement in terms of improved variety of crops having enhanced yield as well as pest resistance, agricultural intensification shows the pathway of sustainability. Yield advancement in chief staple food crops of agriculture has been reported for cassava (36%), wheat (208%), potato (78%), paddy rice (109%), and maize (157%) during 1960–2000 (Pingali and Raney 2005). These crops were keys to minimize malnutrition of protein energy (especially undernourishment) within developing world (Gomez et al. 2013) along with more output and reduction of food prices. The chief advancement in technology came from international network of public sector bodies, i.e., CGIAR. Advanced germplasm from dominant source precisely for rice, wheat, as well as maize was developed, although, in the 1990s 36% of all varietal releases were to be found on CGIAR crosses, and 26% of all modern varieties had few contents of CGIAR (Evenson and Gollin 2003). Basically, global well-being from conventional plant breeding in the mid to late twentieth century was the outcome of the international spread of germplasm. Strategic decision by countries regarding the investment to improve the plant traits under plant breeding enabled many countries to overcome spillover impacts of international investment in the improvement of the crops (Pingali and Raney 2005).

22.8 Management and Policy Implication for Ecointensification in Agriculture Under Changing Climate

Adopting specific policies for screening of agricultural systems that come up with public and private goods is very much essential. To rise up the output, proper policies have given good results as observed in the Asian Green Revolutions. Still, intensification may take in trade-offs between provisioning ecosystem services (food production) along with regulating as well as supporting assistance (Firbank et al. 2011). Major question is: can it also address challenges such as better natural capital and nutritional security along with social-ecological resilience? Policymakers at global scale targeted broad objectives towards sustainability. FAO has mentioned about such issues which require upgrading nutrition along with advance nutritional product by emphasizing research and development. Such policies should be inclusive for smallholders, concentrating various crop species under integrated system (FAO 2013a). Subsequently, consciousness generation has been attempted on global basis (FAO 2013b) as well as "save and grow" models (FAO 2011) which would enhance the natural assets, nurturing resilience along with yield.

In agriculture sector policymakers have found some difficulties in the issues of pesticides application and its contamination with the drinking water management. Therefore, amendment in the existing regulatory framework is very much essential to combat such issues. Practicing control over new crop varieties, farsightedness for realizing potential well-being for more food crops. An "authoritative assessment" by Fedoroff (2010) regarding genetically modified crop safety has been reported which clearly states about the toxicity and other associated impacts on agroecosystem, although they hypothesized about minimization of the complications of the regulatory process that is needed to be done. Encouragement for positive practices has been harder. SI practices as well as the concept offer several possibilities to cope with changing climate along with broad challenges of agricultural production and sustainability (Banerjee et al. 2021a, b). Norse (2012) put the spotlight on the need for generating an evidence base to keep up decision-making on low C agriculture as well as mentioned about the requirement to create consciousness of existing initiatives that can call for higher eco-environment-ecological gain on long-term basis (Banerjee et al. 2021c, d).

22.9 Research and Development in Ecointensification in Agriculture Under Changing Climate

Various research approaches have been aimed towards ecointensification practices in agriculture under changing climatic condition. Approaches include soil management, water management, and maintenance of overall agroecosystem. In the terrestrial ecosystem, N cycling for conserving N, the dissimilarity nitrate reduction to ammonium (DNRA) approach has been adopted in soil. Commonly, in highly reduced environments such as rumen of ruminant animal, sludge, and sediments, DNRA has been studied to observe the catalytic process by obligatory and facultative fermentative bacteria (Tiedje et al. 1983; Bonin 1996; Pandey et al. 2020). Diverse group of anaerobic chemolithoautotrophic and heterotrophic bacteria carries out the two-step process for the DNRA (Dalsgaard and Bak 1994; Silver et al. 2001; Brunet and Garcia-Gil 1996; Pandey et al. 2020). There are several studies reported on DNRA and its methodologies. Yang et al. (2015) studied DNRA in the natural salt marsh ecosystem using 15 N isotope analysis and found that microbial processing of NO^{-3} acts as filters in salt marshes for mitigating N input upslope. In tropical forest, USA, Silver et al. (2001) studied DNRA using 15 N tracing technique and found that rates of the combination of N_2O and N_2 are three times lesser than DNRA, and fluxes from nitrification and denitrification account for 75% turnover of the nitrate pool. Zhang et al. (2013) examined DNRA in subtropical zones in China using 15 N tracing technique, and results reveal that NO-3 immobilization capacity enhanced as elevation in C:N ratio and soil organic C content and minimized the risk of runoff from soil and N leaching. Pandey et al. (2018) studied DNRA in rice field in Australia using 15 N isotope analysis and quantitative PCR and found that in paddy soil, DNRA has a crucial role for N retention, as it considers up to 55% of the entire NO-3 reduction through N application. It is clear and accepted now that DNRA is a potential candidate to reduce N_2O emissions, along with protecting NO-3 from leaching in surface water and ground water and soil enrichment with NH_4 N available to primary procedures and heterotrophic microorganisms. The entire two-step process is actually happening due to nitrate, ammonifiers, which participate in reduction of NO-3 to NH_4. NO-3 gets reduced to NO-2 in the first step, and following that NO-2 gets reduced to NH_4 in the second step (Neubauer and Gotz 1996; Pandey et al. 2020). The DNRA is generally known as fermentative DNRA as it involves two steps which include respiratory (electron transfer chain) and fermentative (substrate level phosphorylation) process all together. Both the steps of NO-3 conversion take place in periplasm as Nrf and Nap are periplasmic in nature (membrane bound), and conservation of energy takes place through ETC, thus known as respiratory DNRA. Metagenomics analysis reveals the role of mixed population in creating redox environment where components of mixed population are denitrifying and DNRA bacteria community (Fig. 22.2). Among those bacteria, some take part in NO-3 dissimilation, and the remaining give substrates in fermentation such as hydrogen, acetate, formate, etc. (Kraft et al. 2014; Van de Berg et al. 2017; Pandey et al. 2020).

Large amount of runoff (wastewater) is generated in steel industries when cold rolling is used to come up with special features together with hardness and thickness reduced on steel. Such effluents contain heavy metals which are toxic towards environment and thus treated through preliminary treatment accompanied by lime treatment inside high-density sludge reaction tanks. After treatment, sludges can employ in wastelands to enhance soil fertility together with preventing soil erosion. These treated sludges are rich in nutrients which are favorable to microorganisms in soil and supports biological activity which leads to vegetation growth together with organic matter buildup (Mishra et al. 2020).

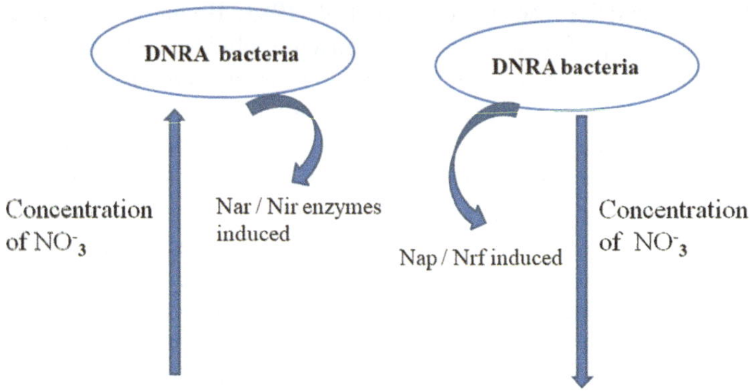

Fig. 22.2 Mechanism of DNRA bacteria

22.10 Future Perspectives/Directives of Ecointensification in Agriculture Under Changing Climate

Trends of advancement of world agricultural production on the basis of the utilization of information technology, global positioning as well as geographic information systems, automated systems in management along with control of technological process, and robotics together with integrated precision farming are the latest developments (Shpaar et al. 2009; Izmaylov and Lobachevsky 2010; Izmaylov et al. 2017; Lachuga et al. 2012). The service of unmanned aerial vehicles for monitoring as well as remote sensing in the agricultural fields for crops becomes a routine operation along with which provides data to build digital maps of fields as well as the formation of map tasks for the differentiated application of plant protection fertilizers along with products (Mikhaylenko 2015).

Technological assistance to farmers is required in different phases of crop growth. Socially, politically, and economically, farmers are the worst sufferer. Several challenges in the agricultural sectors have been detected which demands proper approach to mitigate such challenges. With respect to several crop details, knowledge base regarding information acquisition, flow, numerous inputs such as market availability, weather prophecy, and geospatial data is required. The monitoring stage involves modules such as remainder, during various stages keeping track of plant growth, irrigation planner, problem detector, crop profit calculators as well as calamity check. Evapotranspiration method is used to calculate the water requirement of a plant per day with devised algorithm's assistance.

22.11 Conclusion

EI is an umbrella term, and it is a path for sustainability by adapting some recent trends. Preservation and maintenance of ecosystems within limited land should be the primary objective. Agricultural wastes are not waste anymore, instead they are being used to produce biogas, and the remaining digested products are utilized in agriculture as manures. Sustainability is required to maintain the ecosystem, match the increasing demand, and get better yield. Biopesticides are prepared for the protection of plants against pests via an eco-friendly approach. Pests damage crops in huge amounts and have a negative impact on crop yield. Treated industrial sludge is being used to restore waste lands. Adding such materials in soil sometimes prevents soil erosion. EI provides various opportunities for enhancing crop productivity to protection. Soil is one of the key components which can directly affect the crop health and yield. The utilization of cover crops and functional biodiversity is also essential for uplifting the profit margin and to ensure better economic growth of the pillars of our country (farmers). Proper management and policy making are needed to cover the limitations, and to overcome innovation is the key step. Agriculture is the art of farmers which needs to be enhanced to get the real picture of sustainable development.

References

Albrecht A, Kandji ST (2003) Carbon sequestration in tropical agroforestry systems. Agric Ecosyst Environ 99(1–3):15–27. https://doi.org/10.1016/S0167-8809(03)00138-5

Ali A, Erenstein O (2017) Assessing farmer use of climate change adaptation practices and impacts on food security and poverty in Pakistan. Clim Risk Manag 16:183–194

Allouche J (2011) The sustainability and resilience of global water and food systems: political analysis of the interplay between security, resource scarcity, political systems and global trade. Food Policy 36:S3–S8

Angelsen A, Kaimowitz D (eds) (2001) Agricultural technologies and tropical deforestation. CABi, Wallingford, UK

Arunanondchai P, Fei C, Fisher A, McCarlBA WW, Yang Y (2018) How does climate change affect agriculture. In: The Routledge handbook of agricultural economics, vol 191. Routledge

Banerjee A, Jhariya MK, Meena RS, Yadav DK (2021b) Ecological footprints in agroecosystem-an overview. In: Banerjee A, Meena RS, Jhariya MK, Yadav DK (eds) Agroecological footprints management for sustainable food system. Springer, Singapore, pp 1–23. https://doi.org/10.1007/978-981-15-9496-0_1

Banerjee A, Jhariya MK, Raj A, Yadav DK, Khan N, Meena RS (2021c) Land footprint management and policies. In: Banerjee A, Meena RS, Jhariya MK, Yadav DK (eds) Agroecological footprints management for sustainable food system. Springer, Singapore, pp 221–246. https://doi.org/10.1007/978-981-15-9496-0_7

Banerjee A, Jhariya MK, Raj A, Yadav DK, Khan N, Meena RS (2021d) Energy and climate footprint towards the environmental sustainability. In: Banerjee A, Meena RS, Jhariya MK, Yadav DK (eds) Agroecological footprints management for sustainable food system. Springer, Singapore, pp 415–443. https://doi.org/10.1007/978-981-15-9496-0_14

Banerjee A, Jhariya MK, Yadav DK, Raj A (2020) Environmental and sustainable development through forestry and other resources. Apple Academic Press Inc., CRC press- a Tayler and Francis Group, Boulder, CO, p 400. https://doi.org/10.1201/9780429276026

Banerjee A, Meena RS, Jhariya MK, Yadav DK (2021a) Agroecological footprints management for sustainable food system. Springer, Singapore, p 514. https://doi.org/10.1007/978-981-15-9496-0

Battisti R, Sentelhas PC, Parker PS, Nendel C, Gil MDS, Farias JR, Basso CJ (2018) Assessment of crop-management strategies to improve soybean resilience to climate change in southern Brazil. Crop Pasture Sci 69(2):154–162

Baulcombe D, Crute I, Davies B, Dunwell J, Gale M, Jones J, Pretty J, Sutherland W, Toulmin C (2009) Reaping the benefits: science and the sustainable intensification of global agriculture. The Royal Society, RS policy document 11/09, pp 1–69

Bayala J, Sileshi GW, Coe R, Kalinganire A, Tchoundjeu Z, Sinclair F, Garrity D (2012) Cereal yield response to conservation agriculture practices in drylands of West Africa: a quantitative synthesis. J Arid Environ 78:13–25. https://doi.org/10.1016/j.jaridenv.2011.10.011

Bennett EM, Carpenter SR, Gordon LJ, Ramakutty N, Balvanera P, Campbell BM, Spierenburg M (2014) Resilient thinking for a more sustainable agriculture. Solutions J 5(5):65–75

Berry TD, Filley TR, Blanchette RA (2014) Oxidative enzymatic response of white-rot fungi to single-walled carbon nanotubes. Environ Poll 193:197–204. https://doi.org/10.1016/j.envpol.2014.06.013

Blum A (2018) Plant breeding for stress environments. CRC Press, Boca Raton, FL

Bommarco R, Kleijn D, Potts SG (2013) Ecological intensification: harnessing ecosystem services for food security. Trends Ecol Evol 28(4):230–238. https://doi.org/10.1016/j.tree.2012.10.012

Bonin P (1996) Anaerobic nitrate reduction to ammonium in two strains isolated from coastal marine sediment: a dissimilatory pathway. FEMS Microbiol Ecol 19(1):27–38

Brunet RC, Garcia-Gil LJ (1996) Sulfide-induced dissimilatory nitrate reduction to ammonia in anaerobic freshwater sediments. FEMS Microbiol Ecol 21:131–138

Burney JA, Davis SJ, Lobell DB (2010) Greenhouse gas mitigation by agricultural intensification. Proc Nat Acad Sci 107(26):12052–12057

Campbell BM, Thornton P, Zougmoré R, Van Asten P, Lipper L (2014) Sustainable intensification: what is its role in climate smart agriculture? Curr Opin Env Sustain 8:39–43. https://doi.org/10.1016/j.cosust.2014.07.002

Carolan MS (2013) Reclaiming food security. Routledge, Abingdon, UK

Cassman KG (1999) Ecological intensification of cereal production systems: yield potential, soil quality, and precision agriculture. Proc Nat Acad Sci 96(11):5952–5959. https://doi.org/10.1073/pnas.96.11.5952

Centre de Coopération Internationaleen Recherche Agronomique Pour le Développement (CIRAD) (2008) CIRAD Strateg vision. Montpellier, France, CIRAD, p 2008

Chattopadhyay P, Chattopadhyay P, Palit D (2020a) Effect of environmental pollution on health and its prevention: an overview. In: Environmental and sustainable development through forestry and other resources, vol 229. Routledge, Abingdon, UK. https://doi.org/10.1201/9780429276026

Chattopadhyay P, Chattopadhyay P, Palit D (2020b) Innovative approach to sustainable development and biodiversity conservation. In: Environmental and sustainable development through forestry and other resources, vol 267. Routledge, Abingdon, UK. https://doi.org/10.1201/9780429276026

Christiansen L, Demery L, Kuhl J (2011) The (evolving) role of agriculture in poverty reduction – an empirical perspective. J Dev Econ 96:239–254

Clay N (2018) Seeking justice in Green revolutions: synergies and trade-offs between large-scale and smallholder agricultural intensification in Rwanda. Geoforum 97:352–362. https://doi.org/10.1016/j.geoforum.2018.09.021

CMO (Chief Medical Officer) (2013) Chief medical Officer's annual report 2012: our children deserve better: prevention pays. UK Government, London

Côte F, Tomekpé K, Staver C, Depigny S, Lescot T, Markham R (2008) Agro-ecological intensification in Banana and plantain (Musa spp.): an approach to develop more sustainable cropping systems for both smallholder farmers and large-scale commercial producers. In:IV International

Symposium on Banana: International Conference on Banana and Plantain in Africa: Harnessing International 879 (pp. 457–463). Doi:https://doi.org/10.17660/ActaHortic.2010.879.50
Dalsgaard T, Bak F (1994) Nitrate reduction in a sulfate-reducing bacterium, *Desulfovibriodesulfuricans*, isolated from rice paddy soil: sulfide inhibition, kinetics, and regulation. Appl Environ Microbiol 60:291–297
de Janvry A, Sadoulet E (2010) Agricultural growth and poverty reduction: additional evidence. World Bank Res Obs 25:1–20
Deligios PA, Chergia AP, Sanna G, Solinas S, Todde G, Narvarte L, Ledda L (2019) Climate change adaptation and water saving by innovative irrigation management applied on open field globe artichoke. Sci Total Environ 649:461–472
den Biggelaar C, Lal R, Wiebe K, Eswaran H, Breneman V, Reich P (2003) The global impact of soil erosion on productivity. II: effects on crop yields and production over time. Adv Agron 81:49–95
Dhankher OP, Foyer CH (2018) Climate resilient crops for improving global food security and safety
Dheeman S, Maheshwari DK, Baliyan N (2017) Bacterial endophytes for ecological intensification of agriculture. In: Endophytes: biology and biotechnology. Springer, Cham, pp 193–231. https://doi.org/10.1007/978-3-319-66541-2_9
Dobermann AA (2004) A critical assessment of the system of rice intensification (SRI). Agric Syst 79:261–281
Duku C, Zwart SJ, Hein L (2018) Impacts of climate change on cropping patterns in a tropical, sub-humid watershed. PLoS One 13:e0192642
Evenson RE, Gollin D (2003) Assessing the impact of the green revolution: 1960–1980. Science 300:758–762
Ewers RM, Scharlemann JP, Balmford A, Green RE (2009) Do increases in agricultural yield spare land for nature? Global Change Biol 15(7):1716–1726
FAO (2011) Save and grow: a policymaker's guide to the sustainable intensification of smallholder crop production. FAO, Rome, Italy
FAO (2013a) The state of food and agriculture. FAO, Rome, Italy
FAO (2013b) Climate-smart agriculture sourcebook. FAO, Rome, Italy
FAO; UNICEF; WFP; WHO (2017) The state of food security and nutrition in the world: building resilience for peace and food security (2018) food and agriculture Organization of the United Nations (FAO). FAO, Rome, Italy
FAO, WFP, IFAD (2012) The state of food insecurity in the world. FAO, Rome, Italy
Farooq M, Siddique KHM, Rehman H, Aziz T, Lee D, Wahid A (2011) Rice direct seeding: experiences, challenges and opportunities. Soil Tillage Res 111:87–98
Fedoroff NV (2010) Radically rethinking agriculture for the 21st century. Science 327:833–834
Firbank L, Bradbury RB, McCracken DI, Stoate C (2011) Delivering multiple ecosystem services from enclosed farmland in the UK. Agric Ecosyst Environ 166:65–75
Foresight (2011) The future of global food and farming. Final project report. Government Office for Science London, London
Frayne B, Crush J, McLachlan M (2014) Urbanization, nutrition and development in southern African cities. Food Security 6:101–112
Garibaldi LA, Pérez-Méndez N, Garratt MP, Gemmill-Herren B, Miguez FE, Dicks LV (2019) Policies for ecological intensification of crop production. Trends Ecol Evol 34(4):282–286. https://doi.org/10.1016/j.tree.2019.01.003
Garnett T, Appleby MC, Balmford A, Bateman IJ, Benton TG, Bloomer P, Burlingame B, Dawkins M, Dolan L, Fraser D, Herrero M, Hoffmann I, Smith P, Thornton PK, Toulmin C, Vermeulen SJ, Godfray HCJ (2013) Sustainable intensification in agriculture: premises and policies. Science 341(6141):33–34
Gibon A, Sibbald AR, Flamant JC, Lhoste P, Revilla R, Rubino R, Sørensen JT (1999) Livestock farming systems research in Europe and its potential contribution for managing towards

sustainability in livestock farming. Livestock Prod Sci 61(2–3):121–137. https://doi.org/10.1016/S0301-6226(99)00062-7

Glover JD, Reganold JP (2010) Perennial grains: food security for the future. Issues Sci Technol 26:41–47

Gomez MI, Barrett CB, Raney T, Pinstrup-Andersen P, Meerman J, Croppenstedt A, Carisma B, Thompson B (2013) Post-green revolution food systems and the triple burden of malnutrition. Food Policy 42:129–138

Havlík P, Valin H, Herrero M, Obersteiner M, Schmid E, Rufino MC, Mosnier A, Thornton PK, Bottcher H, Conant RT, Frank S, Fritz S, Fuss S, Kraxner F, Notenbaert A (2014) Climate change mitigation through livestock system transitions. Proc Nat Acad Sci 111(10):3709–3714. https://doi.org/10.1073/pnas.1308044111

Henderson B, Cacho O, Thornton P, van Wijk M, Herrero M (2018) The economic potential of residue management and fertilizer use to address climate change impacts on mixed smallholder farmers in Burkina Faso. Agric Syst 167:195–205

Izmaylov AY, Lobachevskiy YP (2010) Management and information support by innovative technological processes in plant industry. Avtomatizatsiyaiinformatsionnoeobespechenieproizvodstvennykhprotsessov v sel'skomkhozyaystve: Sbornikdokladov Mezhdunarodnoynauchno-tekhnicheskoykonferentsii. Vol. 1. Moscow: VIM: 47–58. (In Russian)

Izmaylov AY, Smirnov IG, Kolesnikova VA, Marchenko LA (2017) Substantiation of parameters of unmanned aerial vehicles for pesticides and fertilizers application in precision farming system. Mechanization Agric Conserv Resour 63(5):168–170

Jaramillo J, Muchugu E, Vega FE, Davis A, Borgemeister C, Chabi-Olaye A (2011) Some like it hot: the influence and implications of climate change on coffee berry borer (*Hypothenemushampei*) and coffee production in East Africa. PloS One 6(9):e24528. https://doi.org/10.1371/journal.pone.0024528

Jassogne L, van Asten PJ, Wanyama I, Baret PV (2013) Perceptions and outlook on intercropping coffee with banana as an opportunity for smallholder coffee farmers in Uganda. Int J Agric Sustain 11(2):144–158. https://doi.org/10.1080/14735903.2012.714576

Jerzak W, Kuźnia M, Szajding A (2016) Experimental studies and the chemical kinetics modelling of oxidation of hydrogen sulfide contained in biogas. Procedia Eng 157:222–229. https://doi.org/10.1016/j.proeng.2016.08.360

Jhariya MK, Banerjee A, Meena RS, Yadav DK (2019a) Sustainable agriculture, Forest and environmental management. Springer, Singapore, p 606. https://doi.org/10.1007/978-981-13-6830-1

Jhariya MK, Banerjee A, Yadav DK, Raj A (2018b) Leguminous trees an innovative tool for soil sustainability. In: Meena RS, Das A, Yadav GS, Lal R (eds) Legumes for soil health and sustainable management. Springer, Cham, pp 315–345. https://doi.org/10.1007/978-981-13-0253-4_10

Jhariya MK, Meena RS, Banerjee A (2021) Ecological intensification of natural resources towards sustainable productive system. In: Ecological intensification of natural resources for sustainable agriculture. Springer, Singapore, pp 1–28. https://doi.org/10.1007/978-981-33-4203-3_1

Jhariya MK, Yadav DK, Banerjee A (2018a) Plant mediated transformation and habitat restoration: phytoremediation an eco-friendly approach. In: Gautam A, Pathak C (eds) Metallic contamination and its toxicity. Daya Publishing House, New Delhi, pp 231–247

Jhariya MK, Yadav DK, Banerjee A (2019b) Agroforestry and climate change: issues and challenges. Apple Academic Press Inc., CRC press- a Tayler and Francis Group, Boulder, CO, p 335. https://doi.org/10.1201/9780429057274

Jones DL, Cross P, Withers PJA, DeLuca TH, Robinson DA, Quilliam RS, Harris IM, Chadwick DR, Edwards-Jones G (2013) Nutrient-stripping: the global disparity between food security and soil nutrient stocks. J Appl Ecol 50:851–862

Khan N, Jhariya MK, Raj A, Banerjee A, Meena RS (2021a) Soil carbon stock and sequestration: implications for climate change adaptation and mitigation. In: Jhariya MK, Meena RS, Banerjee

A (eds) Ecological intensification of natural resources for sustainable agriculture. Springer, Singapore, pp 461–489. https://doi.org/10.1007/978-981-33-4203-3_13

Khan N, Jhariya MK, Raj A, Banerjee A, Meena RS (2021b) Eco-designing for sustainability. In: Jhariya MK, Meena RS, Banerjee A (eds) Ecological intensification of natural resources for sustainable agriculture. Springer, Singapore, pp 565–595. https://doi.org/10.1007/978-981-33-4203-3_16

Khan N, Jhariya MK, Yadav DK, Banerjee A (2020a) Herbaceous dynamics and CO_2 mitigation in an urban setup- a case study from Chhattisgarh, India. Environ Sci Pollut Res 27(3):2881–2897. https://doi.org/10.1007/s11356-019-07182-8

Khan N, Jhariya MK, Yadav DK, Banerjee A (2020b) Structure, diversity and ecological function of shrub species in an urban setup of Sarguja, Chhattisgarh, India. Environ Sci Pollut Res 27 (5):5418–5432. https://doi.org/10.1007/s11356-019-07172-w

Kraft B, Tegetmeyer HE, Sharma R, Klotz MG, Ferdelman TG, Hettich RL, Geelhoed JS, Strous M (2014) The environmental controls that govern the end product of bacterial nitrate respiration. Science 345(6197):676–679

Krajnc EI (1987) Integrated criteria document cadmium: effects: appendix to report Nr. 758476004. RIVM

Kremen C (2020) Ecological intensification and diversification approaches to maintain biodiversity, ecosystem services and food production in a changing world. Emerg Topics Life Sci 4 (2):229–240. https://doi.org/10.1042/ETLS20190205

Kremen C, Miles A (2012) Ecosystem services in biologically diversified versus conventional farming systems: benefits, externalities, and trade-offs. Ecol Soc 17(4):40

Krupnik TJ, Shennan C, Settle W, Demont M, Ndiaye AB, Rodenburg J (2012) Improving irrigated rice production in the Senegal River valley through experiential learning and innovation. Agric Syst 109:101–112

Kumar P, Mishra AK, Kumar M, Chaudhari SK, Singh R, Singh K, Rai P, Sharma DK (2019) Biomass production and carbon sequestration of Eucalyptus tereticornis plantation in reclaimed sodic soils of north-west India. Indian J Agric Sci 89(7):1091–1095

Kumar S, Meena RS, Datta R, Verma SK, Yadav GS, Pradhan GS, Molaei A, Rahman GKMM, Mashuk HA (2020a) Legumes for carbon and nitrogen cycling: an organic approach. In: Carbon and nitrogen cycling in soil. Springer, Singapore. https://doi.org/10.1007/978-981-13-7264-3_ 10.337-375

Kumar S, Meena RS, Jhariya MK (2020) Resources use efficiency in agriculture. Springer, Singapore, p 760. https://doi.org/10.1007/978-981-15-6953-1

Kuźnia M, Magiera A, Jerzak W, Pielichowska K, Sikora J (2019) Biogas production from agricultural and municipal waste. In: E3S web of conferences, vol 108. EDP Sciences, Les Ulis, France, p 02010. https://doi.org/10.1051/e3sconf/201910802010

Lachuga YF, Gorbachev IV, Ezhevskiy AA, Izmaylov AY, Elizarov VP, Artyushin AA, Lobachevskiy YP, Kryazhkov VM, Antyshev NM, Babchenko VD, Beylis VM, Golubkovich AV, Grishin AP, Evtyushenkov NE, Zhalnin EV, Zhuk AF, Kolesnikova VA, Levina NS, Lichman GI, Marchenko NM (2012) i dr. Sistema mashinitekhnologiydlyakompleksnoymekhanizatsiiiavtomatizatsiisel'skokhozyaystvennogopr-oizvodstvana period do 2020 goda [System of machinery and technologies for complex mechanization and automation of agricultural production for the period until 2020.]. Vol. 2. Zhivotnovodstvo. Moscow: VIM, 2012: 212. (In Russian)

Laderach P, Lundy M, Jarvis A, Ramirez J, Portilla EP, Schepp K, Eitzinger A (2011) Predicted impact of climate change on coffee supply chains. In: The economic, social and political elements of climate change. Springer, Berlin, pp 703–723. https://doi.org/10.1007/978-3-642-14776-0_42

Lal R (2014) Abating climate change and feeding the world through soil carbon sequestration. In: Dent D (ed) Soil as world heritage. Springer, Berlin, pp 443–457

Landolt M (2011) Stone lines against desertification. Rural 21:36–37

Lang T, Rayner G (2012) Ecological public health: the 21st century's big idea? BMJ 345:e5466

Lopes MS, El-Basyoni I, Baenziger PS, Singh S, Royo C, Ozbek K, Aktas H, Ozer E, Ozdemir F, Manickavelu A, Ban T, Vikram P (2015) Exploiting genetic diversity from landraces in wheat breeding for adaptation to climate change. J Exper Bot 66(12):3477–3486

Marcinkowski P, Piniewski M (2018) Effect of climate change on sowing and harvest dates of spring barley and maize in Poland. Int Agrophys 32:265–271

Maxwell D (2012) Food security and its implications for political stability: a humanitarian perspective. Paper prepared for workshop on'food security and its implications for global stability' Cornell University: Ithaca, NY, 18–19 June 2012

Maxwell D, Russo L, Alinovi L (2012) Constraints to addressing food insecurity in protracted crises. Proc Natl Acad Sci U S A 109:12321–12325

McBratney A, Field DJ, Koch A (2014) The dimensions of soil security. Geoderma 213:203–213

Meena RS, Kumar S, Datta R, Lal R, Vijaykumar V, Brtnicky M, Sharma MP, Yadav GS, Jhariya MK, Jangir CK, Pathan SI, Dokulilova T, Pecina V, Marfo TD (2020b) Impact of agrochemicals on soil microbiota and management: a review. Land (MDPI) 9(2):34. https://doi.org/10.3390/land9020034

Meena RS, Lal R, Yadav GS (2020) Long term impacts of topsoil depth and amendments on soil physical and hydrological properties of an Alfisol in Central Ohio, USA. Geoderma 363:1141164. https://doi.org/10.1016/j.geoderma.2019.114164

Meena RS, Lal R, Yadav GS (2020a) Long-term impact of topsoil depth and amendments on carbon and nitrogen budgets in the surface layer of an Alfisol in Central Ohio. Catena 2020194:104752. https://doi.org/10.1016/j.catena.2020.104752

Mikhaylenko IM (2015) Unmanned small aircraft in agriculture. Agrofizika 2:16–24

Ministry of Environment and Forests (2004) India's initial national communications to the United Nations framework convention on climate change. Ministry of Environment and Forests, New Delhi

Mishra AK, Banerjee P, Mishra S, Chattopadhyay P, Gupta R (2020) Suitability of cold rolling mill-effluent sludge as a soil amendment for reclamation of degraded lands. In: Hussain C (ed) Handbook of environmental materials management. Springer, Cham. https://doi.org/10.1007/978-3-319-58538-3_186-1

Montgomery DR (2007) Soil erosion and agricultural sustainability. Proc Nat Acad Sci USA 104:13268–13272

Muthuri SK, Francis CE, Wachira LM, Leblanc AG, Sampson M, Onywera VO, Tremblay MS (2014) Evidence of an overweight/obesity transition among school-aged children and youth in sub-Saharan: a systematic review. PLoS One 9:e92846

Neubauer H, Götz F (1996) Physiology and interaction of nitrate and nitrite reduction in Staphylococcus carnosus. J Bacteriol 178(7):2005–2009. https://doi.org/10.1128/jb.178.7.2005-2009.1996

Neufeldt H, Jahn M, Campbell BM, Beddington JR, DeClerck F, De Pinto A, Gulledge J, Hellin J, Herrero M, Jarvis A, LeZaks D, Meinke H, Rosenstock T, Scholes M, Scholes R, Vermeulen S, Wollenberg E, Zougmoré R (2013) Beyond climate-smart agriculture: toward safe operating spaces for global food systems. Agric Food Security 2(1):1–6. https://doi.org/10.1186/2048-7010-2-12

Norse D (2012) Low carbon agriculture: objectives and policy pathways. Environ Dev 1:25–39

Noya I, González-García S, Bacenetti J, Fiala M, Moreira MT (2018) Environmental impacts of the cultivation-phase associated with agricultural crops for feed production. J Clean Prod 172:3721–3733

Palanisami K, Karunakaran KR, Amarasinghe U, Ranganathan CR (2013) Doing different things or doing it differently: rice intensification practices in 13 states of India. Econ Polit Wkly XLVIII 48:51–67

Pandey A, Suter H, He JZ, Hu HW, Chen D (2018) Nitrogen addition decreases dis-similatory nitrate reduction to ammonium in rice paddies. Appl Environ Microbiol 84(17):e00870–e00818. https://doi.org/10.1128/AEM.00870-18

Pandey CB, Kumar U, Kaviraj M, Minick KJ, Mishra AK, Singh JS (2020) DNRA: a short-circuit in biological N-cycling to conserve nitrogen in terrestrial ecosystems. Sci Total Environ 2020:139710. https://doi.org/10.1016/j.scitotenv.2020.139710

Peer N, Kengne A, Motala AA, Mbanya JC (2014) Diabetes in the Africa region: an update. Diabetes Res Clin Pract 103:197–205

Pingali P, Raney T (2005) From the green revolution to the gene revolution. How will the poor fare? ESA working paper no. 05–09. November 2005. Agricultural and Economics Division, FAO. ftp://ftp.fao.org/docrep/fao/008/af276e/af276e00.pdf

Pope CA III, Burnett RT, Thurston GD, Thun MJ, Calle EE, Krewski D, Godleski JJ (2004) Cardiovascular mortality and long-term exposure to particulate air pollution: epidemiological evidence of general pathophysiological pathways of disease. Circulation 109(1):71–77. https://doi.org/10.1161/01.CIR.0000108927.80044.7F

Pretty J (2008) Agricultural sustainability: concepts, principles and evidence. Philos Trans Royal Soc B: Biol Sci 363(1491):447–465. https://doi.org/10.1098/rstb.2007.2163

Pretty J (2013) The consumption of a finite planet: Well-being, convergence, divergence and the nascent green economy. Environ Res Econ 55:475–499

Pretty J, Bharucha ZP (2014) Sustainable intensification in agricultural systems. Ann Botany 114 (8):1571–1596. https://doi.org/10.1093/aob/mcu205

Pretty JN (1997) The sustainable intensification of agriculture. Nat Res Forum 21(4):247–256. https://doi.org/10.1111/j.1477-8947.1997.tb00699.x

Raj A, Jhariya MK, Bargali SS (2018b) Climate smart agriculture and carbon sequestration. In: Pandey CB, Gaur MK, Goyal RK (eds) Climate change and agroforestry: adaptation mitigation and livelihood security. New India Publishing Agency (NIPA), New Delhi, India, pp 1–19

Raj A, Jhariya MK, Harne SS (2018a) Threats to biodiversity and conservation strategies. In: Sood KK, Mahajan V (eds) Forests, climate change and biodiversity. Kalyani Publisher, New Delhi, India, pp 304–320

Raj A, Jhariya MK, Khan N, Banerjee A, Meena RS (2021) Ecological intensification for sustainable development. In: Ecological intensification of natural resources for sustainable agriculture. Springer, Singapore, pp 137–170. https://doi.org/10.1007/978-981-33-4203-3_5

Raj A, Jhariya MK, Yadav DK, Banerjee A (2020) Climate change and agroforestry systems: adaptation and mitigation strategies. Apple Academic Press Inc., CRC press- a Tayler and Francis Group, Boulder, CO, p 383. https://doi.org/10.1201/9780429286759

Rao AN, Johnson DE, Sivaprasad B, Ladha JK, Mortimer AM (2007) Weed management in direct-seeded rice. Adv Agronomy 93:153–255

Ravindranath NH, Sathaye JA (2002) Climate change: vulnerability, impacts and adaptation. In: Climate change and developing countries. Springer, Dordrecht, pp 63–95

Raza A, Mehmood SS, Ashraf F, Khan RSA (2019) Genetic diversity analysis of Brassica species using PCR-based SSR markers. Gesunde Pflanzen 71(1):1–7

Raza A, Shaukat H, Ali Q, Habib M (2018) Assessment of RAPD markers to analyse the genetic diversity among sunflower (*Helianthus annuus* L.) genotypes. Turkish J Agric-Food Sci Technol 6(1):107–111

Robinson BE, Holland MB, Naughton-Treves L (2014) Does secure land tenure save forests? A meta-analysis of the relationship between land tenure and tropical deforestation. Glob Environ Chang 29:281–293. https://doi.org/10.1016/j.gloenvcha.2013.05.012

Rockstrom J, Steffen W, Noone K, Persson A, Chapin FS III, Lambin EF, Lenton TM, Scheffer M, Folke C, Schelnhuber HJ, Nykvist B, de Wit CA, Hughes T, van der Leeuw S, Rodhe H, Sorlin S, Snyder PK, Costanza R, Svedin U, Falkenmark M, Karlberg L, Corell RW, Fabry VJ, Hansen J, Walker B, Liverman D, Richardson K, Crutzen P, Foley JA (2009) A safe operating space for humanity. Nature 461:472–475

Rudel TK, Schneider L, Uriarte M, Turner BL II, DeFries R, Lawrence D, Geoghegan J, Hecht S, Ickowitz A, Lambin EF, Birkenholtz T, Baptista S, Grau R (2009) Agricultural intensification and changes in cultivated areas, 1970–2005. Proc Nat Acad Sci 106(49):20675–20680

Ruerd R, Lee DR (2000) Combining internal and external inputs for sustainable intensification (no. 567-2016-39014). ageconsearch.umn.edu
Samson C, Pretty J (2006) Environmental and health benefits of hunting lifestyles and diets for the Innu of Labrador. Food Policy 31:528–553
Sassen M, Sheil D, Giller KE, ter Braak CJ (2013) Complex contexts and dynamic drivers: understanding four decades of forest loss and recovery in an east African protected area. Biol Conserv 159:257–268. https://doi.org/10.1016/j.biocon.2012.12.003
Sathaye J, Shukla PR, Ravindranath NH (2006) Climate change, sustainable development and India: global and national concerns. Curr Sci 90(3):314–325
Sheehy JE, Peng S, Dobermann A, Mitchell PL (2004) Fantastic yields in the system of rice intensification: fact or fallacy? Field Crops Res 88:1–8
Sheehy JE, Sinclair TR, Cassman KG (2005) Curiosities, nonsense, nonscience and SRI. Field Crops Res 91:355–356
Shetty P (2012) Public health: India's diabetes time bomb. Nature 485:S14–S16
Shpaar D, Zakharenko AV, Yakushev VP, et al. (2009) Precision agriculture. St. Petersburg, Pushkin: 397. (In Russian)
Silver WL, Herman DJ, Firestone MK (2001) Dissimilatory nitrate reduction to ammonium in upland tropical forest soils. Ecology 82(9):2410–2416. https://doi.org/10.1890/0012-9658(2001)082[2410:DNRTAI]2.0.CO;2
Styger E, Aboubacrine G, Attaher MA, Uphoff N (2011) The system of rice intensification as a sustainable agricultural innovation: introducing, adapting and scaling up a system of rice intensification practices in the Timbuktu region of Mali. Int J Agric Sustain 9:67–75
Teixeira EI, de Ruiter J, Ausseil AG, Daigneault A, Johnstone P, Holmes A, Tait A, Ewert F (2018) Adapting crop rotations to climate change in regional impact modelling assessments. Sci Total Environ 616:785–795
Thakur AK, Uphoff N, Antony E (2010) An assessment of physiological effects of system of rice intensification (SRI) practices compared with recommended rice cultivation practices in India. Experim Agric 46:77–98
Thornton PK, Herrero M (2010) Potential for reduced methane and carbon dioxide emissions from livestock and pasture management in the tropics. Proc Nat Acad Sci 107(46):19667–19672. https://doi.org/10.1073/pnas.0912890107
Thornton PK, Lipper L (2014) How does climate change alter agricultural strategies to support food security? vol 1340. International Food Policy Research Institute (IFPRI)
Tiedje JM, Sexstone AJ, Myrold DD, Robinson JA (1983) Denitrification: ecological niches, competition and survival. Antonie Van Leeuwenhoek 48(6):569–583
Tittonell P (2014) Ecological intensification of agriculture—sustainable by nature. Curr Opin Environ Sustain 8:53–61. https://doi.org/10.1016/j.cosust.2014.08.006
Tittonell P, Giller KE (2013) When yield gaps are poverty traps: the paradigm of ecological intensification in African smallholder agriculture. Field Crops Res 143:76–90. https://doi.org/10.1016/j.fcr.2012.10.007
Turmel M, Turner B, Whalen JK (2011) Soil fertility and the yield response to the system of Rice intensification. Renewable Agric Food Syst 26:185–192
Unicef, World Health Organization (2017) The state of food security and nutrition in the world 2017: Building resilience for peace and food security
Van den Berg EM, Elisário MP, Kuenen JG, Kleerebezem R, van Loosdrecht M (2017) Fermentative bacteria influence the competition between denitrifiers and DNRA bacteria. Front Microbiol 8:1684. https://doi.org/10.3389/fmicb.2017.01684
Vaughan MM, Block A, Christensen SA, Allen LH, Schmelz EA (2018) The effects of climate change associated abiotic stresses on maize phytochemical defenses. Phytochem Rev 17(1):37–49

Vermeulen SJ (2014) Climate change, food security and small-scale producers: analysis of findings of the fifth assessment report (AR5) of the intergovernmental panel on climate change (IPCC)
Vermeulen SJ, Campbell BM, Ingram JS (2012) Climate change and food systems. Ann Rev Environ Res 37:195–222. https://doi.org/10.1146/annurev-environ-020411-130608
Weerakoon WMW, Mutunayake MMP, Bandara C, Rao AN, Bhandari DC, Ladha JK (2011) Direct-seeded rice culture in Sri Lanka: lessons from farmers. Field Crops Res 121:53–63
Wezel A, Soboksa G, McClelland S, Delespesse F, Boissau A (2015) The blurred boundaries of ecological, sustainable, and agroecological intensification: a review. Agron Sustain Dev 35 (4):1283–1295. https://doi.org/10.1007/s13593-015-0333-y
Wollenberg E, Campbell BM, Nihart A, Holmgren P, Seymour F, Sibanda L, von Braun J (2011) Actions needed to halt deforestation and promote climate-smart agriculture. In: Peetermans J (ed) Greenhouse gas market report 2011 Asia and beyond: the roadmap to Global Carbon & Energy Markets. Genève, Switzerland, International Emissions Trading Association, pp 95–100
Xie H, Huang Y, Chen Q, Zhang Y, Wu Q (2019) Prospects for agricultural sustainable intensification: a review of research. Land 8(11):157. https://doi.org/10.3390/land8110157
Yang WH, Traut BH, Silver WL (2015) Microbially mediated nitrogen retention and loss in a salt marsh soil. Ecosphere 6(1):1–15. https://doi.org/10.1890/ES14-00179.1
Zhang J, Zhu T, Meng T, Zhang Y, Yang J, Yang W, Müller C, Cai Z (2013) Agricul-tural land use affects nitrate production and conservation in humid subtropicalsoils in China. Soil Biol Biochem 62:107–114. https://doi.org/10.1016/j.soilbio.2013.03.006
Zhang W, Ricketts TH, Kremen C, Carney K, Swinton SM (2007) Ecosystem services and dis-services to agriculture. Ecol Econ 64(2):253–260. https://doi.org/10.1016/j.ecolecon.2007.02.024

Management of Agroecosystem for Food Security: An Overview

23

Pallavi Chattopadhyay, Pinaki Chattopadhyay, and Debnath Palit

Abstract

With the escalating world population along with urbanization and changing consumption patterns, the global need for food is projected to elevate. Econometric models predict that global cereal demand will increase by 1.3% annually through 2015; cereal yields must increase by 1% annually to meet this demand. This scenario projects 50 Mha (million hectares) rise in cereal production area. Climate smart agriculture (CSA) is a new and trending approach towards improving livelihood and food security. CSA is often defined as a combination of practices and technologies. Practicing CSA leads to reduction of greenhouse gas (GHG) emission. CSA helps farmers to meet the global food demand by coping changing climatic conditions. CSA and sustainable intensification are complementary with each other. Under the scenario of changing climate, there is increase in competition for energy, water, labor, and land for food production. Many agricultural practices contribute in formation of GHG (anthropogenic). CSA possesses three objectives: enhancing agricultural productivity, developing capacity to adapt at multiple levels, and eliminating emission of GHG along with encouraging carbon sinks. Sustainable intensification involves improving and maintaining soil biodiversity and monitoring and balancing the biogeochemical cycles. Negative influence of the ongoing agricultural activities involves acidification, erosion, soil structure loss, soil organic matter reduction, gradual buildup of toxic elements in soil, biodiversity loss, and land utilization for nonagricultural

P. Chattopadhyay
Amity Institute of Biotechnology, Amity University Uttar Pradesh, Lucknow, Uttar Pradesh, India

P. Chattopadhyay
Department of Botany, West Bengal State University, Barasat, West Bengal, India

D. Palit (✉)
Department of Botany, Durgapur Government College, Durgapur, West Bengal, India

M. K. Jhariya et al. (eds.), *Sustainable Intensification for Agroecosystem Services and Management*, https://doi.org/10.1007/978-981-16-3207-5_23

purposes. These influences further the enhancement of soil quality from the agronomic side. Physical and chemical properties of soil are governed by glomalin, and economical and ecological importance in this aspect is the actual outcome of mycorrhiza. Glomalin is recalcitrant, difficult to dissolve in water, and heat resistant and forms soil aggregates. This therefore promotes the productivity of the soil and helps to cope with the food security issues.

Keywords

Agroecosystem management · Climate smart agriculture · Food security · Glomalin · Soil health

Abbreviations

C	Carbon
CO_2	Carbon dioxide
CSA	Climate smart agriculture
GHG	Greenhouse gas
Mha	Million hectares
N	Nitrogen
WUE	Water use efficiency

23.1 Introduction

In agroecosystem some critical factors in relation to agricultural production along with ecosystem services include human, water, soil, and climate management (Banerjee et al. 2020). The biggest challenge in the agricultural sector is to search strategies which could support productivity, sustainability, and integrity of the ecosystem (Meena et al. 2018; Blesh and Drinkwater 2013; Jhariya et al. 2019a, b, 2021; Kumar et al. 2020). Variability in climate leads to risk and uncertainty, challenging the regulations of water and soil resource managements in the agriculture sector (Meena and Lal 2018; Khan et al. 2021a, b). The event of soil degradation is becoming evident due to rise in extreme precipitation which are the major concerns for both the irrigated and rainfed agriculture. Reports from 2014 North America Regional aspects reveal that carbon dioxide (CO_2) amount and temperature elevation in cities, vulnerable climate extreme, infestations of pests, frequency of wildfire increase, change in land use, and drought in some areas along with pollution have created stress over North American ecosystem in the form of water scarcity by anthropogenic practices. Major crops yield get disturbed, and lower outcome was observed because of uncertainties in climatic elements (Romero-Lankao et al. 2014; Morton 2014). Uncertainty in climatic changes leads to uneven water distribution (excess and crises) in many regions which can negatively influence crops and their yield. Fluctuations in temperature pattern alter the amount of precipitation and distribution along with length of seasons for cultivation (Johnson et al. 2010; Morton 2014).

Social-ecological benefits from agroecosystem are gained in the form of essential goods (fuel, fiber, and food) along with services such as biodiversity management, climate regulation, conservation of soil and water, and carbon (C) sequestration (Liu et al. 2017; Raj et al. 2019a, b, 2020).

To fulfill the demand for bioenergy and food along with cropland, high-intensity usage and reclamation (Song and Pijanowski 2014; Song and Deng 2017) have promoted the rise in consumption of water (Lang et al. 2018; Liu et al. 2017) and natural ecosystem degradation (Lang and Song 2018; Khan et al. 2020a, b). Development in water use efficiency (WUE) is necessary for global perspective along with addressing regional ecological security, food, and water (Deng et al. 2015; Kumar et al. 2021). Evaluating and monitoring of the WUE are required on a long-term and large-scale basis to attain agroecosystem improvement in sustainable way. The ongoing changes in climate lead to uneven spread of water resources globally (Lang et al. 2017) which has aggravated the problem of water and food shortages (Smajgl et al. 2016).

Agroecosystem WUE is explained as the ratio of net ecosystem production (NEP), consumption of water (actual consumption, evapotranspiration), crop C gain (net primary production), and gross primary production (GPP) to NEP (Han et al. 2013; Lu et al. 2016; Wagle et al. 2016). Developing WUE will automatically give the outcome as improvement in crop yield along with minimization of water consumption (i.e., famously said, "more crop per drop"), crucial for adaption for rising demand in food and scarcity of water in rainfed and irrigated agroecosystems (Marris 2008; Monaghan et al. 2013).

Liu and Song (2020) observed that WUE differs with respect to irrigation, adjustments in crop structure, and variabilities in climate. In order to achieve WUE developing management practices in a sustainable way, water-efficient practices in the sectors of irrigation system, field management, planting structure, and administrative water allocation need to be reoriented considering the economic advantages along with environmental burdens. All strategies of WUE must be integrated in a system of knowledge exchange to allow farmers to modify their water usage. Crop breeding also participates in improving WUE by stimulating biophysical vigor to enhance the harvest index along with rise in resistance against drought (Molden et al. 2010). Zak et al. (2000) studied and concluded that CO_2 enrichment has the ability to significantly impact N (nitrogen) cycle and mineralization along with dry wet of microorganisms. Species of plant which is utilized as cover crops mostly need labor inputs with regard to initiation of processing to ease protection towards woody species out of aerial legumes. For physiological needs, during summer days, competitions are observed between woody crops and cover crops (Lehmann et al. 2000; Sheoran et al. 2021). Regarding operation of biological N fixation process, N is originated from ecosystem as it involves decomposition of biomass with regard to root turnover and litter. Forecast of impact on climate change on the ecosystem describes lowering in quality of litter and its rate of decomposition supporting higher mobility of nutrients to litter from plant parts having high ratio of C and N (Kumar et al. 2020a; Strain and Bazzaz 1983; Jhariya et al. 2018).

Biological N fixation is key process for maintaining soil health and fertility through various forms of sustainable approaches. Further, in agroecosystem, incorporation of leguminous species could be worthy to promote this process and help in N enrichment of the soil (Dhakal et al. 2016; Stevens 2001; Bargali 2016). The present chapter deals with various challenges of agroecosystem along with sustainable management practices to address the issue of food security.

23.2 Agroecosystem and Its Challenges

Agroecology is explained as implementation of ecological science to scrutinize, manage, and design sustainable agroecosystem (Raj et al. 2021). Due to introduction of *Green Revolution* in India, the country focused on technological innovation in agriculture (Swaminathan 2007a; Gangwar et al. 2019). Agroecology is the principle that addresses the ecological approach in agricultural system which involves biodiversity, cycling of nutrient, energy flow, and population regulation of livestock along with dynamic equilibrium. The natural balance of agroecosystem is altered due to improper application of fertilizers and hazardous and non-eco-friendly farm operations. Zaks and Kucharik (2011) suggest that this amount of disruption is brought through socioeconomic pressures. Further, they mentioned that new technologies in the form of precision farming along with optimum inputs in farm parameters are necessary for sustainability of the agroecosystem.

Agriculture has widespread impact over food security, biodiversity, economy, gross happiness index, ecology, climate change, environmental conservation, gross domestic product, etc. (Food and Agriculture Organization 2000; Gangwar et al. 2019; Banerjee et al. 2021a, b, c, d). The Indian agriculture sector is targeting towards the direction of several agroecological techniques with an objective of sustainable agriculture and rural development.

During the 1970s, Green Revolution mainly concentrated on higher yields by utilizing scientific knowledge together with progressive farming techniques (Swaminathan 2007a; Gangwar et al. 2019). With time interval, the society became technology driven, and thus Indian farmers are revealing interest towards climate-smart organic farming along with eco-friendly approaches that reduce environmental burden (Swaminathan 2007b). Farmers in some places of India such as Tamil Nadu, Punjab, Maharashtra, Uttar Pradesh, Gujarat, and Uttarakhand are following environment-friendly practices to promote evergreen revolution (Gangwar and Tyagi 2016). Walker and Schulze (2008) studied that climate change has negative influence on agroecosystem around the maize belt of South Africa.

Agriculture is primary stage of any country for ensuring livelihood and food security. Climate change contributed by greenhouse gas (GHG) emissions usually takes place from industrial activities as well as stiff and shift in lifestyle. Irrigation requirement in semiarid and arid regions is measured to increase by 10% with per 1 °C elevation in temperature. This, in turn, leads to elevated sea level which can bring vulnerable situations in livelihoods of all coastal committees along with fishermen (NMSA 2010; Yadav et al. 2018).

23.3 Food Security Along with its Challenges

Hunger in any society resulted in reduction of working efficiency, lesser resistance towards disease, and elevation in mortality rate, anemic patients, etc. Modern agriculture practice contributes towards uneven development in several regions and is featured by exploitation of natural resources in tribal areas with low productivity and yield leading to low per capita crop output along with decline in asset values from common property resources. Hunger has relations with chronic food insecurity. Public policy is required ensuring adequate production and its distribution to all the segments of the society with nutritious foods. Elevation in production cost, storage, and transport food grains and regional imbalances had inclined production base towards high-cost from low-cost economy (Saxena 2018; Meena et al. 2020).

Future estimations reveal that elevation in demand for agricultural products would be around 50% till 2030 along with increase in global population (Bruinsma 2003; Wheeler and von Braun 2013). Climate change has several influences on global food security. As per Food and Agricultural Organization (FAO) of the United Nations, the issue of food security can be discussed as (1) proper quality and quantity of available food along with domestic production supplies, (2) to obtain a nutritious diet for every individual who has reach to adequate and appropriate food items, (3) to achieve a nutritional well-being state by proper utilization of clean water, healthcare, nutritious adequate diet, and proper sanitation, as well as (4) stability, household, food security, and accessibility (FAO 1996; Wheeler and von Braun 2013).

23.4 Food Security and its Problem

Generally, climate change results in elevation of temperature and leads to change in rainfall pattern. Changing climate is imposing elevation in sea level and rise in land surface temperature. Flood risk increases on agricultural lands in and around coastal regions due to rise in sea level. Food security is multilayered as it contains social, biophysical, and economic factors. Climate change may influence decline in crop yields in South Asia and Africa by 8% till 2050. In South Asia, it is predicted to change by −11% (sorghum) and − 16% (maize), whereas for Africa it will change by −10% (millet), −17% (wheat), −15% (sorghum), and − 5% (maize) (Knox et al. 2012; Wheeler and von Braun 2013).

23.5 Sustainable Approaches Towards Addressing Agroecosystem Management and Food Security

Soil microbiology is an essential part of sustainable agroecosystem. Microbes in agroecology play a vital role such as plant nutrition, soil formation, and suppression of plant pathogens and pests. Modern industrialized agriculture system is putting pressure on the agroecosystem in the form of natural resource depletion, increase in intensity of agriculture pollution, as well as loss of germplasm diversity. Conventional farming practices such as spray of pesticide, monocropping, intensive tillage, and fertilizer spray are harmful for soil microbiota. The microbial flora of soil may differ depending upon the biotic and abiotic conditions such as pH, temperature, soil texture, structure, and moisture content (Campbell et al. 1999; Yadav et al. 2018). The biotic flora in soil may involve both prokaryotic (archaea, eubacteria, and actinomycetes) and eukaryotic organisms (algae, yeast, fungi, and protozoa) along with organic material in soil ecosystem that governs the status of soil microbiota (Shannon et al. 2002; Meena et al. 2020a; Yadav et al. 2018). Introduction of optimum organic inputs alters microbial population, biological nutrient transformation, and soil food web (Nannipieri et al. 1990; Vanhala and Ahtiainen 1994; Stockdale et al. 2002; Yadav et al. 2018).

23.5.1 Glomalin in Agriculture

Glomalin is a sugar protein compound (glycoprotein) that contributes in soil formation. Glomalin in soil is directly proportional to soil quality, i.e., the more the glomalin in soil, the better the soil quality. Glomalin content in soil increases depending upon the degree of association between arbuscular mycorrhizal fungi and plant. The fungi possess hairlike filaments known as hyphae that help in enhancement of plant root reach. Several researchers have reported that arbuscular mycorrhizal fungi are capable to perform biodegradation of pollutants in soil. Wu et al. (2011) reported that in pot culture of maize (*Zea mays*), enhancement of PAH uptake by roots was observed and rises in translocation to its shoots while having an association with *F. mosseae*. Arbuscular mycorrhizal fungi association of *F. mosseae* and *Lolium perenne* (plant) has reported biodegradation of phenanthrene (Corgie et al. 2006). Benefits of glomalin have been described in Fig. 23.1. In in vitro hairy root culture method, Verdin et al. (2006) observed the dissolution of anthracene and its accumulation inside fungal species (*R. irregularis*) having association with *Cichorium intybus*. In in vitro hairy root culture, the storage of soil contaminants such as phenanthrene, dibenzothiophene, and anthracene was reported by Aranda et al. (2013), in arbuscular mycorrhizal fungal structure (*R. custos*) having an association with *Daucus carota*. In pot culture method, Liu et al. (2004) observed the benzo[α]-pyrene biodegradation by arbuscular mycorrhizal fungi association among *F. mosseae* and *Lolium perenne*. A hypothesis was proposed by Young et al. (2012) that massing of glomalin-related soil protein (GPRS) and/or ergosterol in soil has positive correlation with hydrophobicity of soil. A biomolecule in

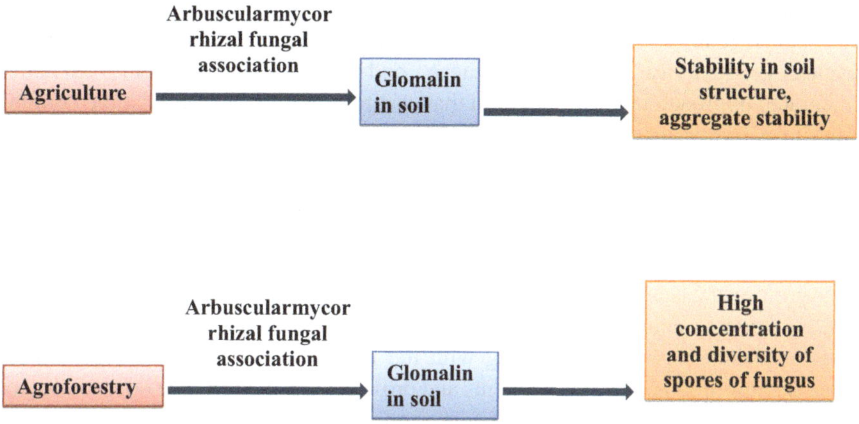

Fig. 23.1 Benefits of glomalin in agroecosystem

filamentous fungi, ergosterol decreases permeability in huge variety of biological membranes. In land management treatments, fungus rules the water repellency.

In Uttarakhand, Nautiyal et al. (2019) studied the glomalin content in various depths (0–30 cm) of soil with silty loam texture having cropping system of *Eleusine coracana*, *Amaranthus* spp., *Glycine max*, *Malus domestica*, *Prunus persica*, *P. domestica*, etc. In Poland, Anna et al. (2017) studied the glomalin content in soil at depth (0–15 cm) in brown podzolic soil type having different cropping system. Yang et al. (2017) evaluate the impacts of heavy metal existence on growth of arbuscular mycorrhizal association, aggregate distribution, aggregate stability, soil properties, and correlation at several depths of soil (0–40 cm). Results indicate that during ecological restoration, removal of heavy metal from soil and systematic management practices are required to guarantee the maximum benefits from relationship of arbuscular mycorrhizal fungi and plant which can be assisted by following inception of indigenous plant associated with arbuscular mycorrhizal fungi. Presence of lead (Pb) exhibits toxicity for arbuscular mycorrhizal fungal growth and has a negative influence on soil properties such as inhibition of mycorrhizal colonization, GRSP, soil organic matter, and hyphal length density together with soil organic C. Arbuscular mycorrhizal fungi play role in soil organic C and GRSP together with soil organic matter storage which rules the aggregate formation together with particle size distribution in soil contaminated with heavy metals.

23.5.2 Agroforestry Towards Sustainable Agroecosystem

An agroecosystem is a unit of agricultural productions which involves both biotic and abiotic factors to work in harmony to feed the human population. To attain a sustainable agroecosystem, incorporation of ecological principles will be beneficial along with less reliability on nonrenewable sources that would help proper energy and nutrient cycling as well as optimum production of agricultural food products. Low nutrient loss from system might be achieved by continuing closed nutrient cycles. Productivity, resilience, balance, and stability can be achieved under integrated approach in agroecosystem that ensures continuation of dynamic equilibrium to achieve sustainability. Sustainability from agroecosystem perspective has been defined as optimum utilization of resources giving sustainable yield to benefit the present and future generation. In order to develop sustainable agroecosystem, organic farming acts as a promising tool (Bagyaraj 2014; Meena et al. 2020b; Yadav et al. 2018).

Agroforestry is an agricultural initiative mainly on the basis of diversification of the components in agroecosystem production (trees and shrubs, woody perennials, livestock, and/or plus crops) along with the intensification of the agroecological relationships among these components and provides a transformative opportunity for enhancing the sustainability in organic farming (Painkra et al. 2016; Singh and Jhariya 2016; Rathia et al. 2019). Specialized agri-food system is unable to assure even distribution of added value in supply chain (HLPE 2019; Rosati et al. 2020), and consumers often perceive system as incapable of expressing typicality and quality (IPES-Food 2016). Specialized agriculture has often been identified as source of several problems, from loss of biodiversity to soil erosion and compaction which is an outcome of agroecosystem simplification (Bastian 1999; Dupraz et al. 2005; Nair et al. 2008, 2011; Rosati et al. 2020).

Loss of biodiversity results in disappearance of the traditional species and germplasm, breed of animals, and crops varieties leading to destabilization of agroecosystems and elevation in the requirement for external inputs (Tsiafouli et al. 2015; Rosati et al. 2020) along with pest control (Stamps and Linit 1998; Rosati et al. 2020). Soil organic matter reduction occurs due to the absence of manure and livestock, removal of tree, and rise in frequency together with depth of tilling which results in less soil fertility, water retention capacity, and permeability. These conditions led to soil erosion and drought, enhance the requirement for external inputs, and cause water pollution and emission of C in atmosphere (Caon and Vargas 2017; Rosati et al. 2020).

Five phases of transitions heading towards sustainable food system were identified and explained by Gliessman (2014). Among those five, the first three phases would operate at the agroecosystem level and examine the (1) increase in input use efficiency, (2) practicing agriculture with agroecological alternatives and substitute conventional input, and (3) redesigning of the agroecosystem and enhancing the spatial along with temporal diversification of its entire components which support and promote ecological relations among the components. The remaining phases among the five govern at the food system level through (4) redesigning a setup for direct contact between producers and consumers and (5) on a basis of equity, participation, justice and locality, and construction of a global food system. The initial two phases can be explained as "incremental," whereas the other remaining three phases can be put into "transformative" nature.

23.5.3 Climate Smart Agriculture

Climate smart agriculture (CSA) is a new and trending approach towards improving livelihood and food security (Raj et al. 2018a, b). CSA is often defined as a combination of practices and technologies. Practicing CSA contributes in reducing GHG emission. CSA helps farmers to meet the global food demand by withstanding the changing climatic conditions (Partey et al. 2018). CSA and sustainable intensification are complementary with each other. There is a gradual increase in competition for energy, water, labor, and land for food production. Change in climate pattern affects the agriculture system. Many agricultural practices contribute in GHG emission. CSA possesses three objectives which includes enhancing agricultural productivity, developing capacity to adapt at multiple levels, and eliminating emission of GHG through C sink approach. Under the system of CSA, sustainable intensification is an approach which involves improving and maintaining soil biodiversity, monitoring and balancing the biogeochemical cycles, as well as maintaining the agricultural productivity (Campbell et al. 2014). CSA possesses three pillars described in Fig. 23.2.

Some countries had applied CSA concept such as Peru for genetic diversity, Tanzania for agroforestry, and China for sustainable grazing approaches (Rosenstock et al. 2016).

In different parts of sub-Saharan Africa and South Asia, it is estimated that around 2.4 billion people are going to live in developing countries by 2050. Agriculture would be the sole source of income and important economic sector in these regions (Wheeler and von Braun 2013; Lipper et al. 2014). The agriculture sector may involve CSA pathways by following right practices, investments, and policies which lead to reduction in poverty and food insecurity in a short span of time while coping with climate change leading to food security for long span of time

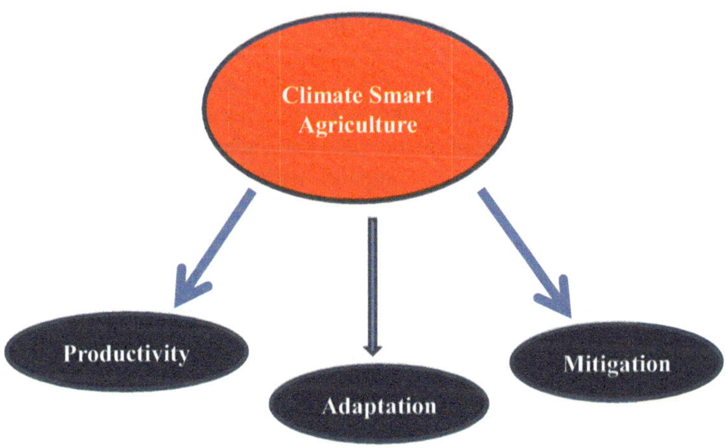

Fig. 23.2 Pillars of climate smart agriculture (Modified: Totin et al. 2018)

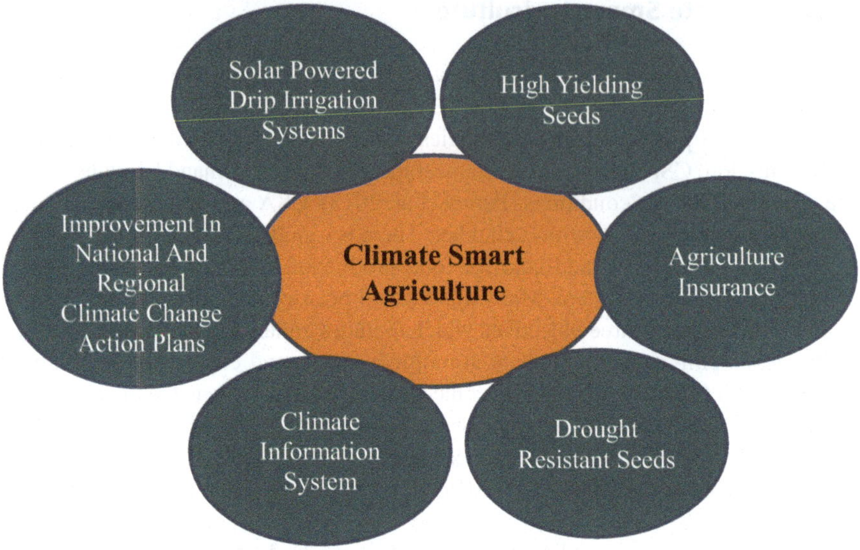

Fig. 23.3 Positive outcomes of climate smart agriculture

(Lipper et al. 2014). Positive outcomes of CSA have been mentioned in Fig. 23.3. In CSA, the diverse climate funding (financing) sources are the Green Climate Fund (GCF), the adaptation fund, and the Global Environment Fund (GEF) (Financing Climate Smart Agriculture 2013; Lipper et al. 2014).

Climate change has negative impact on agricultural sector due to climatic extreme such as drought, flood, high temperature, heavy rainfall, etc. Uneven distribution of climate change leads to water scarcity in several regions (Porter et al. 2014). Mainly the marginalized ethnic groups, poor producers, and the landless are at high risk due to food insecurity caused by climate change, making it a potential threat as food inaccessibility and decrease in agricultural income and yield (Lipper et al. 2014). The ultimate goal of CSA is to endorse the efforts from all levels from local to global to use agricultural system in sustainable way to achieve proper nutrition and food security for people through potential mitigation and adaptations. Spotting the cost of strategies for low-emission growth compared with the conventional path of high-emission growth can assist to join agricultural development attempts that create mitigation co-benefits to the climate finance sources.

CSA possesses three main goals such as to (1) enhance income, development, and food security and increase agriculture productivity in sustainable way; (2) from farm to national levels, build resilience together with adaptation towards climate change; and 3) modify opportunities to decrease emission of GHG from agriculture in comparison to past trends (FAO 2013; Lipper et al. 2014). CSA highlights the agricultural system to endorse mitigation, productivity, and adaptation by using ecosystem services such as livestock, agroforestry, aquaculture, integrated crop, water management, landscape approaches, improved pest, improved grassland management, nutrient management, forestry management, decrease in tillage and diverse varieties and breeds, restoration of degrading lands, N fertilizer use introducing trees in agricultural system, and manure management involving the use of anaerobic biodigesters (Lipper et al. 2014).

23.5.4 IOT (Internet of Things) and Data Science in Agroecosystem Management and Food Security

IoT frameworks are utilized to get an easy view and operation along with interactive database. Information created by sensors get transmitted to the platform (IoT-based M2M platform) or server, over network, accessible from remote location for later proceedings and monitoring. Client machine is employed to access transmitted data and process and notify the user-based filtered information (Arias et al. 2014; Baranwal et al. 2016). Smart agricultural device can be formed to achieve large amount of database through integration of agricultural information cloud along with wireless communication and/or RFID, an information sensing technique of IoT. Utilizing vitalization technology, hardware resources get integrated into cluster, distribution of resources, and equilibrium of loads to boost up competence of resource devices and delegate control (Ke 2013; Baranwal et al. 2016). IoT is able to provide efficient security system for private fields together with farm productions, leading to improved security and monitoring of postharvest and preharvest grain using communication and information technology based on smart agriculture. Parent et al. (2020) studied the management of nutrient in lowbush blueberry by optimized yields modeling by Markov chain algorithm obeying Gaussian processes from dosages of fertilizer, tissue test, and soil test. It is trained to fixed meteorological features from history or earlier data, promising rising yield through median factor of 1.5 machine learning along with analysis from compositional data and also Markov chains permits.

Baranwal et al. (2016) have engineered a device based on IoT and by using Python scripts to analyze the received information and transmit to the user. The respective device can be monitored and controlled from remote areas and can be applied in cold stores and grain stores for security purpose and in agricultural fields. The device can serve in rodent identification and threats to crops along with transmitting real-time notification based on data (information) analysis and required action taken without intervention of humans. Agricultural modernization can be achieved by combining latest technologies such as wireless sensor networks and Internet of Things with traditional methodology. These products can be employed for water irrigation, water quality monitoring, intelligent greenhouses, and scientific pest, disease, and soil monitoring. Researchers are now focused on various IoT-based security devices development although little amount of work is done in agricultural sector.

RPi Libraries and Python scripts control the devices, and application of Python programming language helps in implementation of smart security system. A script is developed in Python accompanying API written in uURL, which is utilized as collected data which is required for further processing and transmits to Thing Worx IoT platform server. PTC LLC provides an IoT-based platform, Thing Worx, to contribute in machine to machine services along with application based on IoT. cURL is basically a computer software project scripted by C language, which gives library and command line tool for the transferring employing its library "libcurl" which hold up common range of procedures involving HTTPS, IMAP, HTTP, SMTP, FTP, POP3, FTPS, and TELNET. A website link will be shared to the user after processing of data, in company with information and timestamp, and based on the calculated distance by ultrasonic ranging device, activation of repeller will occur with a specific frequency in the range 30 kHz to 65 kHz, which is unpleasant to rodents.

The technological protocol of unmanned aerial vehicles (UAVs) implementation in the system of precise farming involves sequential interrelated operations: monitoring along with soundings of crops (putting into service, light UAVs equipped in the company of multispectral cameras), followed by obtaining, processing, and then transmitting information for management of crops and differentiated implementation of pesticides and fertilizers on the report of a specified treatment program (Izmaylov et al. 2017). Employing UAVs for remote sensing along with monitoring of agricultural crops becomes an accustomed protocol and provides data to develop digital maps of fields and genesis of map tasks for application of fertilizers and plant protection products (Izmaylov et al. 2017). Some information on IoT solution assistances in the agriculture field is given in Table 23.1.

23.5.5 IOT in Sustainable Agroecosystem Management

Gangwar et al. (2019) proposed a cost-effective and smart infrastructure that is employed to improve the management together with monitoring of agroecological resources. Resource management in agroecological system is described as interconnected system of components targeting to achieve, disseminate, and format operational information in league with precision farming. Precision farming is already popular in industrialized countries and is followed by India, Brazil, and China along with some African countries. ZigBee is beneficial when integrating information from several acquisition sensor nodes along with computational processing and storage of heterogenous data. The model illustrates that multi-sink WSN architecture of several nodes shares the alike medium together with distinct packet rates together with packet measurements. Multi-sink approach helps in conveying the equilibrium of energy and traffic load in between entire nodes of sensors to lengthen the entire lifetime of the respective network. Theoretical model aided 21 sensors of agriculture already interlinked in grid topology procuring an entire data of soil and crop together with climate for the entire field scale level. This multidimensional and multiparametric duration sequence investigates the power efficiency, topology, and data integration for the WSN networks and architecture for resource management in agroecological system. Generally, farmers require information regarding soil nutrients, weather, crop health, and soil moisture for

Table 23.1 Some IoT solution assistance in agriculture (Modified: Elijah et al. 2018)

IoT Solution	KAA	Cropx	Phytech	Farmlogs	OnFarm	EZfarm	MbeguChoice	Semios
Assistances	Come up with monitoring of remote crop, status on produce, together with livestock feeding, resource mapping, warehousing, together with smart logistics, livestock tracking, and foresight analytics for livestock together with crop	Come up with software services adaptive irrigation which increase crop yield and energy price together with water-saving assistance	Come up with plant IoT platform for recommendation, direct sensing, and plant status together with data analytics	Software for farm management for crop health imagery together with activity recording (automatic)	A tool for farm management to collect data from several varieties of sources and analyze and display	Management of water focused on IBM project and plant health together with soil monitoring	An application which assists farmers to reach better seeds and seed varieties together with drought tolerance from multiple supplies	Network coverage focused, disease, pest, and frost together with orchard irrigation
Comment	IoT cloud platform (open) which communicate large cases of IoT usage together with rise in IoT product development	Covers annual subscription and price of sensors for cloud service that permits monitoring of soil anytime and anywhere	Come up with decision support services and enhance yield together with optimized irrigation	Come up with subscription at three levels which are free, essential, and premium	Free, enterprise, and standard subscription at three levels and functionalities at each level	Nairobi, Kenya, trial project	Kenya farmers focused	Come up with event notification of real-time monitoring services

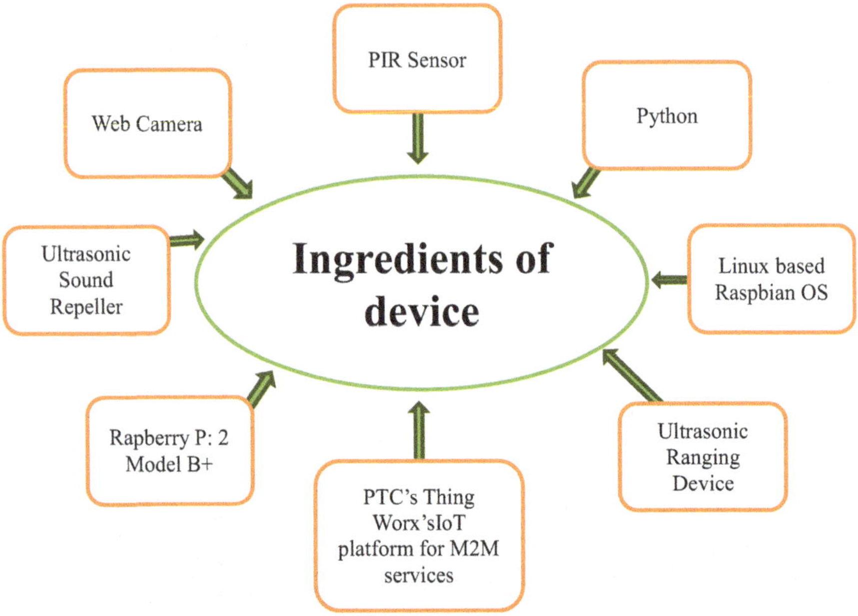

Fig. 23.4 IoT systems for food security (Modified: Baranwal et al. 2016)

managing crops efficiently. Thus, agroecological sensor networks have sensors based on some parameters. The recipe of device proposed by Baranwal et al. (2016) for the architecture of IoT based food security and agricultural field has been described in Fig. 23.4.

To study and observe variations in WUE (water use efficiency) in a large-scale spatiotemporal, ET (example MOD16A3) and remote sensing (RS)-derived NPP (example MOD16A3) products are employed here (Xia et al. 2015). For detecting crop distribution, technology based on Moderate Resolution Imaging Spectroradiometer (MODIS), Thematic Mapper (TM), Huanjing (HJ) charge coupled device (CCD), operational land imager (OLI) sensors on landset, and enhanced thematic mapper plus (ETM+) are utilized such as enhanced vegetation index (EVI), land surface water index (LSWI), multitemporal or time series vegetation indices, normalized difference vegetation index (NDVI), etc. (Dong et al. 2016; Arvor et al. 2011; Hao et al. 2014). Sensors used in agroecosystem management are described in Fig. 23.5.

23.6 Agroecosystem Sustainability and Intensification

Due to agricultural pollution, nutrient loss, and soil erosion, sustainability of soil is in huge demand and need of the hour (Zentner et al. 2004; Jhariya et al. 2018). Soil is itself a mixture of many components. In agricultural region, equilibrium is required

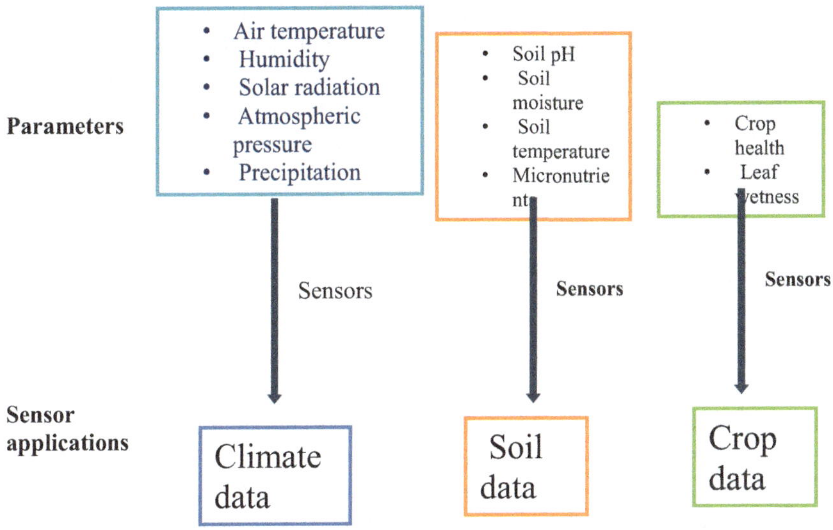

Fig. 23.5 Sensors in agroecosystem management (Modified: Jhariya et al. 2018)

in between production and elevation in demand, which are maintained by endophytes. Reducing negative influence and higher crop yield along with assurance of rise in productivity in agriculture, extensification along with intensification augment ecological intensification. Plant roots harbor microorganisms which contribute in development in production of crop along with ecological intensification.

23.7 Legal and Policy Framework Towards Agroecosystem Management and Food Security

Proper execution and development of proven technologies can be uplifted worldwide through SSNC, TAC, and SSC. The C amount in soil of agroecosystem and its involvement in climate change adaptation along with mitigation play crucial role towards food security which helps in gaining better recognition from international policy perspective. Initiatives such as CSA together with C farming are highly efficient for upgrading C stock in soil and correlated ESs in agroecosystems, although it is crucial that land managers and farmers are appreciated with rewards for practices of soil C enrichment as these are linked with additional costs. Activities to enrich soil with C in agroecosystems are well recognized, and their execution is supported in relation to land under the banner of soil SDGs, UNCCD's LDN, and INDC framework which kept the matter forward in front of UNFCCC together with 4p1000. Stimulus may involve deposit for ESs such as subsidies, certifications, and taxes. Although further research is required in policy measures, effectivity regarding rise in SOC stock on farms is possible mainly by stimulating huge amount of smallholders (Lorenze and Lal 2018).

Agricultural GHG emission from pasture and cropland is also taken by Intergovernmental Panel on Climate Change (IPCC) for emissions of nitrous oxide (N_2O) contributed by agricultural soil (management of residues in crop, fertilizer along with application of manure), CO_2 emission contributed by energy use (machinery use, production of fertilizers, etc.), and paddy cultivation contributing to methane emission (Porter et al. 2014). Commonly, sequestration in land-based systems economically approximates 2.5% in the climate mitigation dollars (Buchner et al. 2015). CSA earlier is brought forward in 2010 by FAO and intended to yield mitigation benefits and adaptation while enhancing productivity (FAO 2013). Although, management to improve land, restoration and conservation activities that enhance C accumulation as well as maintenance of global forest, agricultural land and wetland as natural climate solution (Griscom et al. 2017). Apart from these, climate-resilient agriculture (CRA) circumscribes the resilient activities and adaptation incorporation within agriculture, which enhances the ability of agroecosystem to give response with respect to different disturbances in relation with climate by assuring quick recovery and resisting damage (Rao et al. 2016).

Climate-resilient activities, strategy, and crop diversification face several challenges in the Corn Belt of the USA (Roesch-McNally et al. 2018). Disturbances include flood, pest outbreak, drought, cold/hot waves, and rainfall pattern in an erratic way with so many threats fueled along with changing climate. Improved management and judicious involvement for resilience of natural resources, genetic resources, water, soil, and land are achieved through adoption of best management practices. UNFCCC, Convention on Biological Diversity (CBD), and UNCCD are the three Rio Conventions which are crucial in arranging the governance of the international LDN together with the successful implementation context as stated by Akhtar-Schuster et al. (2017). The main target of LDN is developing land resource base such as land resources-associated nature capital stock and services in regard to ecosystem (Orr et al. 2017). Further, LDN is utilized as an indicator for land management to achieve sustainability (Kust et al. 2017). Land degradation decreases C sequestration potential along with lowering in multiple good and services offered by ecosystem. Collaboration of CBD, UNFCCC, and UNCCD actions is utilized as a step forward to LDN. C stock in soil is reduced, while agricultural land degradation would diminish the part of agroecosystem in soil for climate change mitigation together with food security.

In agriculture, treatment involving the C stock in soil of agroecosystems evolved as passing the IPCC lifetime, as detected by examined evidences together with initial one published around 1990 as studied by Porter et al. (2017). Although, Porter et al. (2017) advised, influence of adaptation and climate change on total green-house-gas grants, especially with respect to soil C amount, measured crucial research points with regards to food-security after publication of recent enquired report 2013–2014. Creation and adaptation of excellent practices for an integrated approach of climate mitigation along with productive agenda by agribusiness, ministries of agricultural sector, and financial institutions (World Bank Supports and contribute in CSA) is the need of the hour (Dickie et al. 2014). Changes in supply chain are needed when there is rise in demand for local food. Management of demand such as dietary changes and reduction in food wastes should be interlinked with optimization of agricultural products (Scherer and Verburg 2017). Few success in the field of implementation of CSA is present in Peru (genetic diversity), Tanzania (agroforestry), and also in China (sustainable grazing) (FAO 2013). Smallholder CSA pilot project, a part in the Mitigation of Climate Change in Agriculture (MICCA) Programme, established a crop-livestock and tree farming system (integrated) in Kenya.

23.8 Conclusion

To meet global food demands, food security is the basic requirement which can be achieved through agroecosystem management. Economic output of agriculture is the major source of income in agricultural-dependent countries. Agroecosystem can be managed using IoT, glomalin, and agroforestry system practices for food security. IoT can assist human beings in agriculture to maintain the quality and increase in yield. Biosensors are immensely used in collecting data of the climate, soil, and crop by considering some parameters. These data can help human or farmer to take action quickly according to the requirements.

23.9 Future Perspective of Agroecosystem Management and Food Security

At the present context, agroecosystem management and food security are the biggest challenges of the century. In the scientific world, one needs to develop innovativeness for sustainable management of agroecosystem. Sustainability in agroecosystem demands the management of soil and water resource, maintenance of crop diversity, input management, and coping with climate change. Future research and development should be aimed for better utilization of the various eco-friendly approaches under diverse conditions. More exploration is required in the field of agroforestry, potential application of glomalin on large-scale basis, and developing and adoption of CSA practices along with greater involvement IoT interface to improve the productivity of the agroecosystem.

Incorporation of legumes for multipurpose use leads to socioeconomic development and increase in agro-productivity of farming communities in various agroecological zones in the world which needs to be focused in development and research activities. Agroecosystem performance lowers without proper crop rotation as well as with irregularity in biotic and abiotic components. Non-judicious practices of crop rotation often move towards lowering in crop yield and productivity with gradual fall in the quality of soil with moderate infiltration of several biotic components (Dumanski et al. 1998; Jhariya and Yadav 2017), although mixed cropping can be a good alternative of monoculture as it provides huge biomass turnover and huge productivity during annual rotation system (Drinkwater et al. 1998; Jhariya et al. 2018). In order to achieve sustainability in food production system, eco-friendly approaches need to be stimulated at the grassroot level for its effective implementation. By adopting suitable intensification practices, there can be increase in yield without damaging the environment. This however required modernized approaches with latest tools and techniques which will bring sustainability in the agroecosystem for the upcoming century.

References

Akhtar-Schuster M, Stringer LC, Erlewein A, Matternicht G, Minelli S, Safriel U, Sommer S (2017) Unpacking the concept of land degradation neutrality and addressing its operation through the Rio conventions. J Environ Manag 195:4–15. https://doi.org/10.1016/j.jenvman.2016.09.044

Anna G, Karolina G, Jarosław G, Jerzy K (2017) Effect of different agricultural management practices on soil biological parameters including glomalin fraction. Plant Soil Environ 63(7):300–306. https://doi.org/10.17221/207/2017-PSE

Aranda E, Scervino JM, Goday P, Reina R, Ocampo JA, Wittich RM, Garcia-Romera I (2013) Role of arbuscular mycorrhizal fungus Rhizophagus custos in the dissipation of PAHs under root – organ culture conditions. Environ Poll 181:182–189

Arias JFC, Barajas YJC, Lopez JLR (2014) Wireless sensor system according to the concept of internet of things. Int J Adv Comput Sci Inform Technol 3(3):327–343

Arvor D, Jonathan M, Meirelles MSP, Dubreuil V, Durieux L (2011) Classification of MODIS EVI time series for crop mapping in the state of Mato Grosso, Brazil. Int J Remote Sensing 32 (22):7847–7871

Bagyaraj (2014) Microorganisms in sustainable agriculture. Proc Indian Nat Sci Aca 80(2):357

Banerjee A, Jhariya MK, Meena RS, Yadav DK (2021b) Ecological footprints in agroecosystem-an overview. In: Banerjee A, Meena RS, Jhariya MK, Yadav DK (eds) Agroecological footprints management for sustainable food system. Springer, Singapore, pp 1–23. https://doi.org/10.1007/978-981-15-9496-0_1

Banerjee A, Jhariya MK, Raj A, Yadav DK, Khan N, Meena RS (2021c) Land footprint management and policies. In: Banerjee A, Meena RS, Jhariya MK, Yadav DK (eds) Agroecological footprints management for sustainable food system. Springer, Singapore, pp 221–246. https://doi.org/10.1007/978-981-15-9496-0_7

Banerjee A, Jhariya MK, Raj A, Yadav DK, Khan N, Meena RS (2021d) Energy and climate footprint towards the environmental sustainability. In: Banerjee A, Meena RS, Jhariya MK, Yadav DK (eds) Agroecological footprints management for sustainable food system. Springer, Singapore, pp 415–443. https://doi.org/10.1007/978-981-15-9496-0_14

Banerjee A, Jhariya MK, Yadav DK, Raj A (2020) Environmental and sustainable development through forestry and other resources. Apple Academic Press Inc., CRC Press- a Tayler, Boulder, CO

Banerjee A, Meena RS, Jhariya MK, Yadav DK (2021a) Agroecological footprints management for sustainable food system. Springer, Singapore, p 514. https://doi.org/10.1007/978-981-15-9496-0

Baranwal T, Nitika R, Pateriya PK (2016) Development of IoT based smart security and monitoring devices for agriculture. 6th international conference- cloud system and big data engineering (confluence), IEEE

Bargali K (2016) Leguminous plants of Kumaun Himalaya: diversity, distribution, threats and management. In: Chand R, Leimgruber W (eds) Globalization and marginalization in mountain regions, perspectives on geographical marginality, vol 1, pp 199–204. https://doi.org/10.1007/978-3-319-32649-8-15

Bastian O (1999) Notwendigkeit and Mittelokologischer Planung. In: Bastian O, Schreiber K-F (eds) Analysis und okologischebewertung der Landschaft. Spektrum Akademischer Verlag GmbH, Berlin, pp 13–25

Blesh J, Drinkwater LE (2013) The impact of nitrogen source and crop rotation on nitrogen mass balances in the Mississippi River basin. Ecol Appl 23(5):1017–1035

Bruinsma (2003) World Agriulture: towards 2015/2030: an FAO perspective. Earthscan, London

Buchner BK, Trabacchi C, Mazza F, Abramskiehn D, Wang D (2015) Global landscape of climate finance 2015. https://climatepolicyinitiative.org/publication/global-landscape-of-climate-finance-2015/

Campbell BM, Thorton P, Zougmore R, Asten P, Lipper L (2014) Sustainable intensification: what is its role in climate smart agriculture. Curr Opin Environ Sustain 8:39–43

Campbell CA, Lafond GP, Biederbeck VO, Wen G, Schoenau J, Hahn D (1999) Seasonal trends in soil biochemical attributes: effects of crop management on a black chernozem. Can J Soil Sci 79:85–97

Caon L, Vargas R (2017) Threats to soils: global trends and perspectives. A contribution from the intergovernmental technical panel on soils. Global soil partnership food and agriculture Organization of the United Nations. G. Pierzynski and Brajendra (eds), global land outlook working paper

Corgie SC, Fons F, Beguiristain T, Leyval C (2006) Biodegradation of phenanthrene, spatial distribution of bacteria pollutions and dioxygenase expression in the mycorrhizosphere of Lolium perenne inoculated with Glomus mosseae. Mycorrhiza 16:207–212. https://doi.org/10.1007/s00572-006-0049-6

Deng X, Singh RB, Liu J, Güneralp B (2015) Water use efficiency and integrated water resource management for river basin. PCE 89:1–2

Dhakal Y, Meena RS, Kumar S (2016) Effect of INM on nodulation, yield, quality and available nutrient status in soil after harvest of green gram. Leg Res 39(4):590–594

Dickie A, Streck C, Roe S, Zurek M, Haupt F, Dolginow A (2014) Strategies for mitigating climate change in agriculture: abridged report. Climate focus and California environmental associates, prepared with the support of the climate and land use Allinace. pp. 1–85. Report and supplementary materials available at: www.agriculturalmitigation.org

Dong J, Xiao X, Menarguez MA, Zhang G, Qin Y, Thau D, Moore B III (2016) Mapping paddy rice planting area in northeastern Asia with Landsat 8 images, phenology-based algorithm and Google earth engine. Remote Sens Environ 185:142–154

Drinkwater LE, Wagoner P, Sarrantonlo M (1998) Legume-based cropping systems have reduced carbon and nitrogen losses. Nature 396(6708):262–265

Dumanski J, Desjardins RL, Tarnocai C, Moreal C, Gregorich EG, Kirkwood V, Campbell CA (1998) Possibilities for future carbon sequestration in Canadian agriculture in relation to land use changes. J Clim Res 40(1):81–103

Dupraz C, Burgess P, Gavaland A, Graves A, Herzog F, Incoll L, Liagre F (2005) Synthesis of the SAFE (Silvoarable agroforestry for Europe) project

Elijah O, Rahman TA, Orikumhi I, Leow CY, Hindia MN (2018) An overview of internet of things (IoT) and data analytics in agriculture: benefits and challenges. IEEE Int Things J 5 (5):3758–3773

FAO (1996) Rome declaration on world food security and world food summit plan of action, World Food Summit, Rome 13–17 November

FAO (2013) Climate-smart agriculture sourcebook. FAO, Rome, Italy

Financing Climate Smart Agriculture (2013) Climate smart agriculture (Climate- smart agriculture sourcebook module 14, FAO, 2013). FAO, Rome, Italy, pp 375–406

Food and Agriculture Organization (2000) The socio – economic impact of agricultural modernization, food and agriculture organization of the United. Nations, Rome, Italy

Gangwar DS, Tyagi S (2016) Challenges and opportunities for sensor and actuator networks in Indian agriculture", In: 8th international conference on computational intelligence and communication networks, 23–25 Dec, 2016, Tehri, Uttarakhand, India, pp. 38–42

Gangwar DS, Tyagi S, Soni SK (2019) A conceptual framework of agroecological resource management system for climate-smart agriculture. Int J Environ Sci Technol 16(8):4123–4132

Gliessman (2014) Agroecology: the ecology of sustainable food systems, 3rd edn. CRC Press, Taylor Francis Group, Boca Ratan, FL

Griscom BW, Adams J, Ellis PW, Houghton RA, Lomax G, Miteva DA, Schlesinger WH, Shoch D, Siikamäki JV, Smith P, Woodbury P, Zganjar C, Blackman A, Campari J, Conant RT, Delgado C, Elias P, Gopalakrishna T, Hamsik MR, Herrero M, Kiesecker J, Landis E, Laestadius L, Leavitt SM, Minnemeyer S, Polasky S, Potapov P, Putz FE, Sanderman J, Silvius M, Wollenberg E, Fargione J (2017) Natural climate solutions. Proc Natl Acad Sci U S A 114:11645–11650

Han Q, Luo G, Li C, Ye H, Chen Y (2013) Modeling grassland net primary productivity and water-use efficiency along an elevational gradient of the northern Tianshan Mountains. J Arid Land 5 (3):354–365

Hao P, Wang L, Niu Z, Aablikim A, Huang N, Xu S, Chen F (2014) The potential of time series merged from Landsat-5 TM and HJ-1 CCD for crop classification: a case study for bole and Manas counties in Xinjiang, China. Remote Sens 6(8):7610–7631

HLPE (2019) Report on agroecological and other innovative approaches for sustainable agriculture and food systems that enhance food security and nutrition. Extract from the report: summary and recommendations. International assessment of agricultural knowledge, science and technology for development agriculture at a crossroads: synthesis report (island press, 2009)

IPES-Food (2016) From uniformity to diversity: a paradigm shift from industrial agriculture to diversified agroecological system. International Panel of experts on sustainable food system

Izmaylov AY, Smirnov IG, Kolesnikova VA, Marchenko LA (2017) Substantiation of parameters of unmanned aerial vehicles for pesticides and fertilizers application in precision farming system. Mechanization Agric Conserv Res 63(5):168–170

Jhariya MK, Banerjee A, Meena RS, Yadav DK (2019a) Sustainable agriculture, forest and environmental management. Springer, Singapore, p 606. https://doi.org/10.1007/978-981-13-6830-1

Jhariya MK, Banerjee A, Yadav DK, Raj A (2018) Leguminous trees an innovative tool for soil sustainability. In: Legumes for soil health and sustainable management. Springer, Singapore, pp 315–345

Jhariya MK, Meena RS, Banerjee A (2021) Ecological intensification of natural resources towards sustainable productive system. In: Ecological intensification of natural resources for sustainable agriculture. Springer, Singapore. https://doi.org/10.1007/978-981-33-4203-3_1

Jhariya MK, Yadav DK (2017) Invasive alien species: challenges, threats and management. In: Rawat SK, Narain S (eds) Agriculture technology for sustaining rural growth. Biotech Books, New Delhi, pp 263–285

Jhariya MK, Yadav DK, Banerjee A (2019b) Agroforestry and climate change: issues and challenges. Apple Academic Press Inc., CRC press- a Tayler and Francis Group, Boulder, CO, p 335. https://doi.org/10.1201/9780429057274

Johnson WC, Werner B, Guntenspergen GR, Voldseth RA, Millett B, Naugle DE, Tulbure M, Carroll RWH, Tracy J, Olawsky C (2010) Prairie wetland complexes as landscape functional units in a changing climate. Bioscience 60:128–140

Ke T (2013) Smart agriculture based on cloud computing and IoT. J Convergence Infor Tech 8 (2):313–324

Khan N, Jhariya MK, Raj A, Banerjee A, Meena RS (2021a) Soil carbon stock and sequestration: implications for climate change adaptation and mitigation. In: Jhariya MK, Meena RS, Banerjee A (eds) Ecological intensification of natural resources for sustainable agriculture. Springer, Singapore. https://doi.org/10.1007/978-981-33-4203-3_13

Khan N, Jhariya MK, Raj A, Banerjee A, Meena RS (2021b) Eco-designing for sustainability. In: Jhariya MK, Meena RS, Banerjee A (eds) Ecological intensification of natural resources for sustainable agriculture. Springer, Singapore. https://doi.org/10.1007/978-981-33-4203-3_16

Khan N, Jhariya MK, Yadav DK, Banerjee A (2020a) Herbaceous dynamics and CO_2 mitigation in an urban setup- a case study from Chhattisgarh, India. Environ Sci Pollut Res 27(3):2881–2897. https://doi.org/10.1007/s11356-019-07182-8

Khan N, Jhariya MK, Yadav DK, Banerjee A (2020b) Structure, diversity and ecological function of shrub species in an urban setup of Sarguja, Chhattisgarh, India. Environ Sci Pollut Res 27 (5):5418–5432. https://doi.org/10.1007/s11356-019-07172-w

Knox J, Hess T, Daccache A, Wheeler T (2012) Climate change impacts on crop productivity in Africa and South Asia. Environ Res Lett 7(3):034032

Kumar S, Meena RS, Datta R, Verma SK, Yadav GS, Pradhan GS, Molaei A, Rahman GKMM, Mashuk HA (2020a) Legumes for carbon and nitrogen cycling: an organic approach. In: Carbon and nitrogen cycling in soil. Springer, Singapore. https://doi.org/10.1007/978-981-13-7264-3_10.337-375

Kumar S, Meena RS, Jhariya MK (2020) Resources use efficiency in agriculture. Springer, Singapore, p 760. https://doi.org/10.1007/978-981-15-6953-1

Kumar S, Meena RS, Singh RK, Munir TM, Datta R, Danish S, Singh GSY, Kumar S (2021) Soil microbial and nutrient dynamics under different sowings environment of Indian mustard (*Brassica juncea* L.) in rice based cropping system. Sci Rep 11:5289. https://doi.org/10.1038/s41598-021-84742-4

Kust G, Andreeva O, Cowie AL (2017) Land degradation neutrality: concept development, practical applications and assessment. J Environ Manag 195:16–24

Lang Y, Song W (2018) Trade-off analysis of ecosystem services in a mountainous karst area, China. Water 10(3):300

Lang Y, Song W, Deng X (2018) Projected land use changes impacts on water yields in the karst mountain areas of China. Phys Chem Earth Parts A/B/C 104:66–75

Lang Y, Song W, Zhang Y (2017) Responses of the water-yield ecosystem service to climate and land use change in Sancha River basin, China. Phys Chem Earth, Parts A/B/C 101:102–111

Lehmann J, da Silva Jr JP, Trujillo L, Uguen K (2000) Legume cover crops and nutrient cycling in tropical fruit tree production. Acta Hortic 531:65–72

Lipper L, Thornton P, Campbell BM, Baedeker T, Braimoh A, Bwalya M, Caron P, Cattaneo A, Garrity D, Henry K, Hottle R (2014) Climate-smart agriculture for food security. Nat Clim Chang 4(12):1068–1072

Liu SL, Luo YM, Cao ZH, Wu LH, Ding KQ, Christie P (2004) Degradation of benzo [a] pyrene in soil with arbuscular mycorrhizal alfalfa. Environ Geochem Health 26(2):285–293. https://doi.org/10.1023/B:EGAH.0000039592.80489.e5

Liu Y, Song W (2020) Modelling crop yield, water consumption, and water use efficiency for sustainable agroecosystem management. J Clean Prod 253:119940

Liu Y, Song W, Deng X (2017) Spatiotemporal patterns of crop irrigation water requirements in the Heihe River basin, China. Water 9(8):616

Lorenz K, Lal R (2018) Importance of soils of agroecosystems for climate change policy. In: Carbon sequestration in agricultural ecosystems. Springer, Cham, pp 357–386

Lu Y, Zhang X, Chen S, Shao L, Sun H (2016) Changes in water use efficiency and water footprint in grain production over the past 35 years: a case study in the North China plain. J Clean Prod 116:71–79

Marris E (2008) More crop per drop: farmers' yields in the developing world are often limited by unreliable rains. Improving their harvests will require plant breeders, agronomists and geneticists to pull together--but can these experts work out their differences? Nature 452 (7185):273–278

Meena RS, Kumar S, Datta R, Lal R, Vijaykumar V, Brtnicky M, Sharma MP, Yadav GS, Jhariya MK, Jangir CK, Pathan SI, Dokulilova T, Pecina V, Marfo TD (2020b) Impact of agrochemicals on soil microbiota and management: a review. Land (MDPI) 9(2):34. https://doi.org/10.3390/land9020034

Meena RS, Kumar V, Yadav GS, Mitran T (2018) Response and interaction of *Bradyrhizobium japonicum* and arbuscular mycorrhizal fungi in the soybean rhizosphere: a review. Plant Growth Regul 84:207–223. https://doi.org/10.1007/s10725-017-0334-8

Meena RS, Lal R (2018) Legumes for soil health and sustainable management. Springer, Singapore, p 541. https://doi.org/10.1007/978-981-13-0253-4_10

Meena RS, Lal R, Yadav GS (2020) Long term impacts of topsoil depth and amendments on soil physical and hydrological properties of an Alfisol in Central Ohio, USA. Geoderma 363:1141164. https://doi.org/10.1016/j.geoderma.2019.114164

Meena RS, Lal R, Yadav GS (2020a) Long-term impact of topsoil depth and amendments on carbon and nitrogen budgets in the surface layer of an Alfisol in Central Ohio. Catena 194:104752. https://doi.org/10.1016/j.catena.2020.104752

Molden D, Oweis T, Steduto P, Bindraban P, Hanjra MA, Kijne J (2010) Improving agricultural water productivity: between optimism and caution. Agric Water Manag 97(4):528–535

Monaghan JM, Daccache A, Vickers LH, Hess TM, Weatherhead EK, Grove IG, Knox JW (2013) More 'crop per drop': constraints and opportunities for precision irrigation in European agriculture. J Sci Food Agric 93(5):977–980

Morton LW (2014) The science of variable climate and agroecosystem management. J Soil Water Conserv 69(6):207A–212A

Nair PKR (2011) Agroforestry systems and environmental quality: introduction. J Environ Qual 40 (3):784–790. https://doi.org/10.2134/jeq2011.0076

Nair PKR, Gordon AM, Mosquera-Losada M-R (2008) Agroforestry. In: Jorgensen SE, Fath BD (eds) Ecological engineering. Encyclopedia of ecology, vol 1. Elsevier, Oxford, UK, pp 101–110

Nannipieri PAOLO, Grego S, Ceccanti B, Bollag J, Stotzky G (1990) Ecological significance of the biological activity in soil. In: Soil biochemistry, vol 6. Routledge, Abingdon, UK

National Mission for Sustainable Agriculture (2010) Department of Agriculture and Cooperation. Ministry of Agriculture, New Delhi

Nautiyal P, Rajput R, Pandey D, Arunachalam K, Arunachalam A (2019) Role of glomalin in soil carbon storage and its variation across land uses in temperate Himalayan regime. Biocatal Agric Biotech 21:101311. https://doi.org/10.1016/j.bcab.2019.101311

Orr BJ, Cowie AL, Castillo Sanchez VM, Chasek P, Crossman ND, Erlewein A, Louwagie G, Maron M, Metternicht GI, Minelli S, Tengberg AE, Walter S, Welton S (2017) Scientific conceptual framework for land degradation neutrality. A report of the science-policy interface. United Nations convention to combat desertification (UNCCD), Bonn, Germany, pp 1–128

Painkra GP, Bhagat PK, Jhariya MK, Yadav DK (2016) Beekeeping for poverty alleviation and livelihood security in Chhattisgarh, India. In: Narain S, Rawat SK (eds) Innovative technology for sustainable agriculture development. Biotech Books, New Delhi, India, pp 429–453

Parent SÉ, Lafond J, Paré MC, Parent LE, Ziadi N (2020) Conditioning machine learning models to adjust lowbush blueberry crop management to the local agroecosystem. Plan Theory 9(10):1401

Partey ST, Zougmoré RB, Ouédraogo M, Campbell BM (2018) Developing climate-smart agriculture to face climate variability in West Africa: challenges and lessons learnt. J Clean Prod 187:285–295

Porter JR, Howden M, Smith P (2017) Considering agriculture in IPCC assessments. Nat Clim Chang 7:680–683

Porter JR, Xie L, Challinor AJ, Cochrane K, Howden SM, Iqbal MM, Lobell DB, Travasso MI (2014) Food security and food production systems. In: IPCC. 2014. Climate change 2014: impacts, adaptation, and vulnerability. Part a: global and sectoral aspects. Contribution of working group II to the fifth assessment report of the intergovernmental panel on climate change. Cambridge University Press, Cambridge, UK, pp 485–533

Raj A, Jhariya MK, Banerjee A, Yadav DK, Meena RS (2019b) Soil for sustainable environment and ecosystems management. In: Jhariya MK, Banerjee A, Meena RS, Yadav DK (eds) Sustainable agriculture, forest and environmental management. Springer, Singapore, pp 189–221

Raj A, Jhariya MK, Bargali SS (2018b) Climate smart agriculture and carbon sequestration. In: Pandey CB, Gaur MK, Goyal RK (eds) Climate change and agroforestry: adaptation mitigation and livelihood security. New India Publishing Agency (NIPA), New Delhi, India, pp 1–19

Raj A, Jhariya MK, Harne SS (2018a) Threats to biodiversity and conservation strategies. In: Sood KK, Mahajan V (eds) Forests, climate change and biodiversity. Kalyani Publisher, New Delhi, India, pp 304–320

Raj A, Jhariya MK, Khan N, Banerjee A, Meena RS (2021) Ecological intensification for sustainable development. In: ecological intensification of natural resources for sustainable agriculture. Springer, Singapore, eBook ISBN 978-981-334-203-3. Hardcover ISBN: 978-981-334-202-6. doi:https://doi.org/10.1007/978-981-33-4203-3_5

Raj A, Jhariya MK, Yadav DK, Banerjee A (2020) Climate change and agroforestry systems: adaptation and mitigation strategies. Apple academic press Inc., CRC press- a Tayler and Francis Group, Boulder, CO, p 383. https://doi.org/10.1201/9780429286759

Raj A, Jhariya MK, Yadav DK, Banerjee A, Meena RS (2019a) Agroforestry: a holistic approach for agricultural sustainability. In: Jhariya MK, Banerjee A, Meena RS, Yadav DK (eds) Sustainable agriculture, forest and environmental management. Springer, Singapore, pp 101–131

Rao CS, Gopinath KA, Prasad JVNS, Prasannakumar Singh AK (2016) Climate resilient villages for sustainable food security in tropical India: concept, process, technologies, institutions, and impacts. Adv Agron 140:101–214

Rathia VK, Jhariya MK, Yadav DK, Banerjee A (2019) Diversity of traditional agroforestry practices of Sarguja: implications for biodiversity conservation. Int J Ecol Dev 34(1):43–60

Roesch-McNally GE, Arbuckle JG, Tyndall JC (2018) Barriers to implementing climate resilient agricultural strategies: the case of crop diversification in the U.S. Corn Belt. Glob Environ Change 48:206–215

Romero-Lankao P, Smith JB, Davidson D, Diffenbaugh N, Kinney P, Kirshen P, Kovacs P, Villers Ruiz L (2014) North America chapter 26 in climate change 2014: impacts, adaptation, and vulnerability volume II: regional aspects IPCC work group II, IPCC 5th assessment report

Rosati A, Borek R, Canali S (2020) Agroforestry and organic agriculture. Agrofor Syst 936:1–17

Rosenstock TS, Lamanna C, Chesterman S, Bell P, Arslan A, Richards M, Rioux J, Akinleye AO, Champalle C, Cheng Z, Corner-Dolloff C, Dohn J, English W, Eyrich AS, Girvetz EH, Kerr A, Lizarazo M, Madalinska A, McFatridge S, Morris KS, Namoi N, Poultouchidou N, Ravina da Silva M, Rayess S, Ström H, Tully KL, Zhou W (2016) The scientific basis of climate-smart agriculture: a systematic review protocol. CCAFS working paper no. 138. Copenhagen, Denmark: CGIAR research program on climate change, agriculture and food security (CCAFS). https://hdl.handle.net/10568/70967

Saxena NC (2018) Hunger, under-nutrition and food security in India. In: Poverty, chronic poverty and poverty dynamics. Springer, Singapore, pp 55–92. https://doi.org/10.1007/978-981-13-0677-8_4

Scherer L, Verburg PH (2017) Mapping and linking supply- and demand-side measures in climate-smart agriculture. A review. Agron Sustain Dev 37:66. https://doi.org/10.1007/s13593-017-0475-1

Shannon D, Sen AM, Johnson DB (2002) A comparative study of the microbiology of soils managed under organic and conventional regimes. Soil Use Manag 18:274–283

Sheoran S, Kuma S, Kumar P, Meena RS, Rakshit S (2021) Nitrogen fixation in maize: breeding opportunities. Theor Appl Genet 134(5):1263–1280. https://doi.org/10.1007/s00122-021-03791-5

Singh NR, Jhariya MK (2016) Agroforestry and Agrihorticulture for higher income and resource conservation. In: Narain S, Rawat SK (eds) Innovative technology for sustainable agriculture development. Biotech Books, New Delhi, India, pp 125–145

Smajgl A, Ward J, Pluschke L (2016) The water–food–energy nexus–Realising a new paradigm. J Hydrol 533:533–540

Song W, Deng X (2017) Land-use/land-cover change and ecosystem service provision in China. Sci Total Environ 576:705–719

Song W, Pijanowski BC (2014) The effects of China's cultivated land balance program on potential land productivity at a national scale. Appl Geogr 46:158–170

Stamps and Linit (1998) Plant diversity and arthropod communities: implications for temperate agroforestry. Agrofor Syst 39:73–89

Stevens PF (2001) Angiosperm phylogeny, website version 9, June 2008. Assessed on March 15, 2012

Stockdale EA, Shepherd MA, Fortune S, Cuttle SP (2002) Soil fertility in organic farming systems–fundamentally different? Soil Use Manag 18:301–308

Strain BR, Bazzaz FA (1983) Terrestrial plant communities. In: Lemon ER (ed) CO_2 and plants. Westview, Boulder, CO, pp 177–222

Swaminathan MS, (2007a) The crises of Indian agriculture, national policy for farmers, Government of India, Assessed May 20, 2017

Swaminathan MS (2007b) National policy for farmers, Department of Agriculture and Cooperation. Ministry of Agriculture, Government of India

Totin E, Segnon AC, Schut M, Affognon H, Zougmoré RB, Rosenstock T, Thornton PK (2018) Institutional perspectives of climate-smart agriculture: a systematic literature review. Sustainability 10(6):1990

Tsiafouli MA, Thébault E, Sgardelis SP, De Ruiter PC, Van Der Putten WH, Birkhofer K, Bjornlund L (2015) Intensive agriculture reduces soil biodiversity across Europe. Glob Chang Biol 21(2):973–985

Vanhala PT, Ahtiainen JH (1994) Soil respiration, ATP content and photobacterium toxicity test as indicators of metal pollution in soil. Environ Toxicol Water Qual 9:115–121

Verdin A, Sahraoui ALH, Fontaine J, Grandmougin-Ferjani A, Durand R (2006) Effects of anthracene on development of an arbuscular mycorrhizal fungus and contribution of the symbiotic association to pollutant dissipation. Mycorrhiza 16(6):397–405. https://doi.org/10.1007/s00572-006-0055-8

Wagle P, Gowda PH, Xiao X, Anup KC (2016) Parameterizing ecosystem light use efficiency and water use efficiency to estimate maize gross primary production and evapotranspiration using MODIS EVI. Agric Forest Meteorol 222:87–97

Walker NJ, Schulze RE (2008) Climate change impacts on agro- ecosystem sustainability across three climate regions in the maize belt of South Africa. Agric Ecosyst Environ 124 (1-2):114–124. https://doi.org/10.1016/j.agee.2007.09.001

Wheeler T, von Braun J (2013) Climate change impacts on global food security. Science 341 (6145):508–513

Wu FY, Yu XZ, Wu SC, Lin XG, Wong MH (2011) Phenanthrene and pyrene uptake by arbuscular mycorrhizal maize and their dissipation in soil. J Hazardous Mater 187(1-3):341–347. https://doi.org/10.1016/j.jhazmat.2011.01.024

Xia L, Wang F, Mu X, Jin K, Sun W, Gao P, Zhao G (2015) Water use efficiency of net primary production in global terrestrial ecosystems. J Earth System Sci 124(5):921–931

Yadav SK, Soni R, Rajput AS (2018) Role of microbes in organic farming for sustainable agro-ecosystem. In: Microorganisms for green revolution. Springer, Singapore, pp 241–252

Yang Y, He C, Huang L, Ban Y, Tang M (2017) The effects of arbuscular mycorrhizal fungi on glomalin-related soil protein distribution, aggregate stability and their relationships with soil properties at different soil depths in lead-zinc contaminated area. PloS One 12(8):e0182264. https://doi.org/10.1371/journal.pone.0182264

Young IM, Feeney DS, O'Donnell AG, Goulding KW (2012) Fungi in century old managed soils could hold key to the development of soil water repellency. Soil Biol Biochem 45:125–127. https://doi.org/10.1016/j.soilbio.2011.10.007

Zak DR, Pregitzer KS, Curtis PS, Vogel CS, Holmes WE, Lussenhop J (2000) Atmospheric CO_2, soil-N availability, and allocation of biomass and nitrogen by Populus tremuloides. Ecol Appl 10:34–46

Zaks DP, Kucharik CJ (2011) Data and monitoring needs for a more ecological agriculture. Environ Res Lett 6(1):014017

Zentner RP, Campbell CA, Biederebeck VO, Selles F, Lemke R, Jefferson PG, Gan Y (2004) Long term assessment of management of an annual legume green manure crop for fallow replacement in the brown soil zone. Can J Plant Sci 83:475–482

The manufacturer's authorised representative in the EU is Springer Nature Customer Service Centre GmbH, Europaplatz 3, 69115 Heidelberg, Germany. If you have any concerns regarding our products, please contact ProductSafety@springernature.com

Printed and bound by CPI Group (UK) Ltd, Croydon, CR0 4YY
30/09/2026
02225330-0019